Informal and formal classification of Embryonta (land plants)

Informal group names	Division	Class	Common names/(examples)	Chapter where covered
bryophytes	**Bryophyta**	**Hepaticae** **Anthocerotae** **Musci**	liverworts hornworts (*Anthoceros*) mosses	27
vascular plants	**Tracheophyta**			
lower vascular plants		**Rhyniopsida** **Zosterophyllopsida** **Trimerophytopsida** **Progymnospermopsida** **Psilopsida** **Lycopsida** **Sphenopsida** **Filicopsida**	earliest, rootless vascular plants (*Rhynia, Hornea*) pre-lycopods (*Asteroxylon*) pre-megaphyllous plants (*Psilophyton*) spore-bearing pre-gymnosperms (*Archaeopteris*) whisk ferns (*Psilotum*) club mosses (*Lycopodium*) horsetails (*Equisetum*) ferns	28 28 29
gymnosperms		**Coniferopsida** **Cycadopsida** **Gnetopsdida**	conifers and related forms cycads and cycad-like plants (*Gnetum, Ephedra*)	30
flowering plants		**Angiospermopsida** *Monocotyledonae*[†] *Dicotyledonae*[†]	angiosperms monocots dicots	3, 8–17 31, 32

*Extinct early vascular plant groups, known mainly from fossils of Devonian age; all disappeared before the end of the Paleozoic Era.

[†]Subclasses

Botany

Botany

Peter M. Ray
STANFORD UNIVERSITY

Taylor A. Steeves
UNIVERSITY OF SASKATCHEWAN

Sara A. Fultz
STANFORD UNIVERSITY

With the Assistance of

Mark Jacobs
SWARTHMORE COLLEGE

SAUNDERS COLLEGE PUBLISHING

Philadelphia   New York   Chicago
San Francisco   Montreal   Toronto
London   Sydney   Tokyo   Mexico City
Rio de Janeiro   Madrid

Address orders to:
383 Madison Avenue
New York, NY 10017

Address editorial correspondence to:
West Washington Square
Philadelphia, PA 19105

Text Typeface: 10/11 Melior
Compositor: Waldman Graphics, Inc.
Color Separator: Lithoprep Corporation
Acquisitions Editor: Mike Brown
Developmental Editor: Amy Satran
Project Editor: Carol Field
Copy Editor: Ruth Low
Managing Editor & Art Director: Richard L. Moore
Design Assistant: Virginia A. Bollard
Text Design: Phoenix Studio, Inc.
Cover Design: Richard L. Moore
Text Artwork: Alan D. Singer, J & R Technical Services, Inc.,
    and Tom Mallon
Page Layout: Carol Bleistine
Black and White Photo Conversions: Peter M. Deierlein, P.M.D. Photography
Production Manager: Tim Frelick
Assistant Production Manager: Maureen Read

Cover Credit: "Thistles and poppies in Maine" © 1982 by Susan Kuklin

Library of Congress Cataloging in Publication Data

Ray, Peter M. (Peter Martin)
    Botany.

    "Developed from the . . . fifth edition of Botany by Carl L. Wilson, Walter E. Loomis, and Taylor A. Steeves"—
        Bibliography: p.
        Includes index.
        1. Botany. I. Steeves, Taylor A., 1926–
II. Fultz, Sara A. III. Title.
QK47.R317 1982   581   81-53072
ISBN 0-03-089942-7

BOTANY   ISBN 0-03-089942-7

345   071   987654321

CBS COLLEGE PUBLISHING
Saunders College Publishing
Holt, Rinehart and Winston
The Dryden Press

PREFACE

Botany, or plant biology, is both one of the oldest and currently one of the most exciting and rapidly developing of the sciences. Long recognized as important to medical, agricultural, and environmental issues by the people directly involved, plant biology is now acquiring much wider public prominence. This is due to mushrooming corporate and governmental interest in exploiting new techniques such as plant cell culture and genetic engineering for plant improvement to increase agricultural and forest productivity, a goal of great importance to national economies and vital future food and energy supplies.

This textbook introduces the student both to traditional or classical botanical knowledge and to some of the genetic and molecular aspects currently being publicized because of potential practical applications. A brief survey of these exciting new areas is included in the first chapter; their fuller explanation in later chapters is presented as a part of the more general knowledge about plant function and structure needed to understand and evaluate the potential applications. We give, however, equal attention to the ecological and evolutionary or systematic aspects of botany, which will actually be of more direct use and value to the average person than the currently more publicized molecular biology and physiology.

This book has been developed from the well-known and extensively used fifth edition of *Botany* by Carl L. Wilson, Walter E. Loomis, and Taylor A. Steeves. While the basic plan of organization and philosophy of presentation is similar to that of the Wilson/Loomis/Steeves text, the authors of this new book have almost entirely rewritten the material and have extensively added to as well as subtracted from the topics and points covered and illustrated, reflecting our view of changing emphasis within the field. For example, the sections on cell and higher plant structure, function, and development have been extensively modified to reflect current interests and points of

view about energetics, transport processes, metabolic and developmental controls, and so forth. In these chapters and in the chapters on genetics and prokaryotes, current approaches to plant improvement and genetic engineering are also explained. To the treatment of ecology a new chapter on physiological and genetic aspects of ecological adaptation has been added, and coverage of ecosystem dynamics and population ecology has been expanded in the last two chapters of the book, including discussions of agricultural as well as natural ecosystems. New material on molecular evolution and the origin and early evolution of life has been added in Chapter 21, which introduces the survey of major plant groups, and a classification scheme influenced by the endosymbiotic theory of plastid origins and by recent ultrastructural and molecular evidence about relationships is presented. The two chapters on fungi in the Wilson/Loomis/Steeves text have been condensed into one, but new chapters on plant symbioses and on plant diseases, currently important areas in which fungi figure prominently, have been added. Material on extinct early land plant groups and on their modern survivors has been condensed into one chapter, but with addition of recent insights into vascular plant evolution due to new fossil discoveries and interpretations. To conclude the systematic survey, a new chapter on economically important species and families of angiosperms, the object of considerable current interest, has been added.

The breadth of information encompassed in appreciating plants as organisms and their role in nature and agriculture often makes it difficult for a student reading a general textbook to "see the forest for the trees." One problem is the chemical and biochemical background needed to understand current concepts of how plants function. We have provided a new appendix covering the botanically important elements of chemistry, and in Chapters 5 and 6 we give a very elementary introduction to basic aspects of plant biochemistry. Another problem even for those aspects of botany not so dependent on technical background is the sheer mass of terms and points of information surrounding hundreds of thousands of kinds of organisms, with innumerable variations in form and structure, thousands of uses, and numerous roles in the ecosystem. In common with the fifth edition of the Wilson, Loomis and Steeves text, our philosophy in choosing what to present and how to present it has been to give, for each topic or group of organisms, first a general description of principles and features which integrates structure and function, uses a relatively minimal amount of chemistry, and introduces only the botanical terminology we feel is required to define the basic concepts clearly. This material we call the "first-level" presentation. In many cases we follow it with "second level" material, in smaller type, comprising more detailed or specialized descriptive information, or chemically more complex facts and more sophisticated concepts, pointing toward current views of the mechanisms of particular functions, often pertinent to currently prominent applications. We also mention important uncertainties and gaps in our information and understanding, so the reader will be aware how numerous and important the opportunities are for future research.

Depending on their interests and the topics they wish to emphasize, we expect that some teachers will consider the second-level material as essential reading, while others will feel it can be skipped in an introductory exposure to the topic, remaining available as additional optional background for students who wish to explore it. The first-level presentation depends only upon preceding first-level material (and the concepts in the chemistry appendix) as background, so it can be read and understood throughout the book even if second-level sections have been skipped. The chapter summaries generally cover only first-level material (if any points from the second level are included, they appear in parentheses).

The time available in botany and biology courses today usually does not allow all of the topics in this book to be covered even at our "first level" of presentation; entire chapters or blocks of chapters will often have to be omitted. The book is so written and organized that various chapters can be omitted without leaving the reader unaware of basic information needed for comprehending other aspects of botany. For example, since elementary plant and cell physiological functions are introduced in Chapters 1 and 4, for a course emphasizing systematic or ecological aspects of botany one could omit the mainly biochemical Chapters 5 through 7, plus the physiological and anatomical Parts III and IV, and concentrate on Parts V through IX or VI through IX. On the other hand, since Chapter 2 gives a brief survey of the plant and fungal kingdoms, for courses empha-

sizing cellular or physiological aspects of botany it is possible to omit the systematic survey (Parts VII and VIII) and still have the student aware of the major plant groups whose members are considered in Parts II through VI in connection with various plant functions. Furthermore, Chapter 21 presents a general introduction to the evolutionary history of plants that will give the student some idea of this aspect of botany even if Parts VII and VIII are not covered. Similarly, Chapter 3 gives very elementary background on plant development, affording some general understanding of how plants grow and reproduce even if the specific chapters on development (Part IV) or reproduction (Part V) cannot be covered. In courses with only lmited time for ecology one can choose between covering ecological adaptation (Chapter 19), ecosystem dynamics (Chapter 33), or plant communities (Chapter 34).

The basic introductions to plant and cell form, functions, development, and diversity in Chapters 1 through 4 will also make it easier for teachers to take up topics in a different order than they appear in this book. For example, using the elementary background in Chapters 1 through 4 or this plus the cell physiological material in Chapters 5 through 7, either the systematic (Chapters 22 through 32), the ecological (Chapters 19, 33, and 34), or the genetic and evolutionary (Chapters 18 through 21) aspects of botany could be covered before (or without) taking up the relatively extensive information and concepts about seed plant structure and function (Chapters 8 through 17).

Reading lists for each chapter given at the end of the book have been carefully selected to provide additional information on specific topics noted with the references. Rather than include a separate glossary, we have provided an index/glossary, which includes brief definitions of many terms plus a citation of the page number on which the term is introduced and more completely defined.

We thank Carl Wilson for valuable critical comments on all chapters of this book during various stages of preparation. For comments on specific chapters we are indebted to many colleagues including Olle E. Björkman, Richard W. Holm, Harold A. Mooney, Naomi Franklin, Margaret W. Steeves, and Robert T. Wilce, as well as a number of other reviewers named in the acknowledgments that follow. Errors and omissions remain, of course, our responsibility and we shall be very grateful to readers who point them out to us. We are much indebted to our many colleagues who generously provided materials for illustrations, as listed in the figure acknowledgments at the end of the book.

Peter M. Ray
Taylor A. Steeves
Sara A. Fultz

Acknowledgments

The authors and publisher would like to express their appreciation for the opinions, advice, criticism, and support of a number of botanists who have assisted in the development of the book.

Kenneth Barker
NORTH CAROLINA STATE UNIVERSITY (RALEIGH)

Patricia Bonamo
SUNY-BINGHAMTON

Earl Camp
TEXAS TECH UNIVERSITY

Leeds Carluccio
CENTRAL CONNECTICUT STATE COLLEGE

Iris Charvat
UNIVERSITY OF MINNESOTA

Robert Costello
UNIVERSITY OF WISCONSIN (MILWAUKEE)

Theodore Delevoryas
UNIVERSITY OF TEXAS

Robert Embree
UNIVERSITY OF IOWA

Douglas Fratianne
THE OHIO STATE UNIVERSITY

William Ghiorse
CORNELL UNIVERSITY

Margaret Goodhue
COMMUNITY COLLEGE OF DENVER, NORTH CAMPUS

Larry R. Griffing
UNIVERSITY OF SASKATCHEWAN

A. Tyrone Harrison
UNIVERSITY OF NEBRASKA

Alan Heilman
UNIVERSITY OF TENNESSEE

Richard Holm
STANFORD UNIVERSITY

Peter Jankay
CALIFORNIA STATE POLYTECHNIC

Donald Kaplan
UNIVERSITY OF CALIFORNIA (BERKELEY)

Robert Keck
INDIANA UNIVERSITY-PURDUE UNIVERSITY (INDIANAPOLIS)

Carl Keener
THE PENNSYLVANIA STATE UNIVERSITY

Meredith Lieux
LOUISIANA STATE UNIVERSITY

Gustave Mehlquist
UNIVERSITY OF CONNECTICUT

I. K. Mills
MONTANA STATE UNIVERSITY

Harold Mooney
STANFORD UNIVERSITY

Glen Moore
BRIGHAM YOUNG UNIVERSITY (RETIRED)

David Newman
MIAMI UNIVERSITY

Elray Nixon
STEPHEN F. AUSTIN STATE UNIVERSITY

Glenn Patterson
UNIVERSITY OF MARYLAND

Javier Penalosa
BOSTON UNIVERSITY

Richard Pippen
WESTERN MICHIGAN UNIVERSITY

Robert Platt
THE OHIO STATE UNIVERSITY

Mark Schedlbauer
TEXAS A&M UNIVERSITY

Howard Schneiderman
MONSANTO COMPANY (FORMERLY OF UNIVERSITY OF CALIFORNIA, IRVINE)

Gerald Van Dyke
NORTH CAROLINA STATE UNIVERSITY (RALEIGH)

Michael Walsh
UTAH STATE UNIVERSITY

Thomas Watson
UNIVERSITY OF MONTANA

Peter Webster
UNIVERSITY OF MASSACHUSETTS

Robert Wilce
UNIVERSITY OF MASSACHUSETTS

Albert Will
BROWARD COMMUNITY COLLEGE

John Youree
FLORIDA JUNIOR COLLEGE AT JACKSONVILLE

Contents Overview

CONTENTS

ix

6 EVOLUTION AND ADAPTATION

7 NON-VASCULAR PLANTS

8 VASCULAR PLANTS

9 ECOSYSTEMS AND PLANT COMMUNITIES

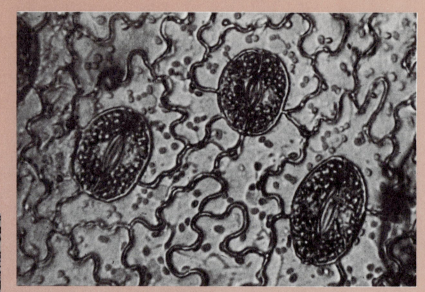

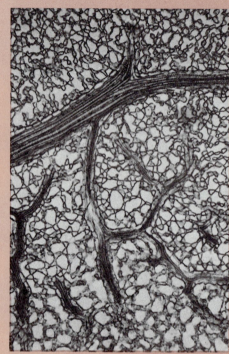

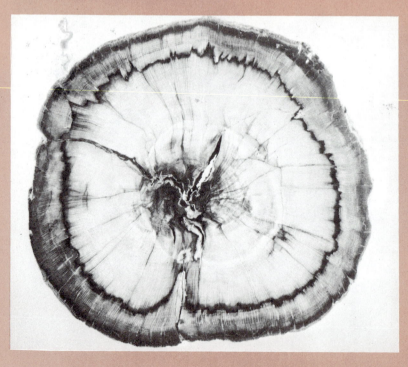

Botany

Chapter 1

BASIC PLANT FUNCTIONS: DISCOVERY AND UTILIZATION

Plants power the biological and human world. Directly or indirectly, they feed the human race and all other animals on earth. They provide most of our energy supply—not only currently produced fuels such as wood, but also "fossil fuels" such as coal and oil, which stem from the growth of plants in past ages. Plants also offer one of our principal hopes for a renewable energy resource to replace fossil fuels as their supply runs out. And plants provide many other basic commodities, such as lumber; cotton and other fibers; beverages such as coffee, tea, cocoa, wine, and beer; and the raw material for various plastics. It is important, therefore, to understand plants and how they provide so many valuable resources. Botany, or plant science, embodies our knowledge of these matters, acquired by the scientific methods of observation and experimentation.

Botanical knowledge underlies our efforts to improve plants and their management, in endeavoring to assure an adequate supply of food, fiber, and forest products for the earth's ever-growing population. Botanical knowledge is also an integral element of ecology, the study of the environment, on which efforts to assure livable surroundings for ourselves and future generations are based.

An elementary understanding of botany is significant and useful to many people not involved occupationally with plant products or environmental impact work. For example, botanical principles underlie the science of horticulture, or gardening. Botanical knowledge helps home gardeners to understand better what they are doing and, hence, to do it better. Basic botanical knowledge also stimulates a greater appreciation, enjoyment, and understanding of nature, both in the wild (such as in national parks) and in domestic parks and gardens. Today the desire to experience nature and the out-of-doors is growing at a phe-

nomenal pace, as more and more people discover how it enriches their lives aesthetically and deepens and satisfies their spirits. The huge sums of money spent nationally each year to acquire and protect parks and natural areas, and to visit them, demonstrate the importance attached to the nature experience, to which botany contributes so much.

METHODS AND SUBFIELDS OF BOTANY

Like scientific knowledge in general, botanical information and concepts are obtained by methods of careful, systematic observation and measurement, combined with intentional experimentation where this is possible. Those aspects of botany that are based primarily upon observation are often called **descriptive** subfields. They seek, first, to ascertain what is "out there," what we have to deal with, and, second, to classify and interpret this diversity of nature to make it comprehensible and meaningful. Plant **taxonomy** (from Greek *taxis*, arrangement, and *nomos*, law), or **systematic botany,** aims at answering the following kinds of questions: What different kinds of plants exist? How can they be recognized and distinguished from one another? What name should be used for each? How did the diversity of plants arise? Plant **morphology** (from Greek *morphe*, form, and *logos*, reason) and plant **anatomy** explore, respectively, the external form and the internal structure of plants, how these features arise during growth, and how they may have originated and become modified during the historical development or evolution of plants. Plant **cytology** is the investigation and interpretation of the structure of cells, the fundamental biological units of which almost all organisms are composed.

Many of the structural features that are important to our understanding of plants, for example, most cells, are not visible to the unaided eye and can be observed only with the aid of the

Figure 1-A. A field of sugarcane on the Island of Hawaii. Sugarcane is the world's most efficient energy-converting crop; as much as 20% of the photosynthetically active radiant energy striking the plants is converted into the energy stored in carbohydrates.

3

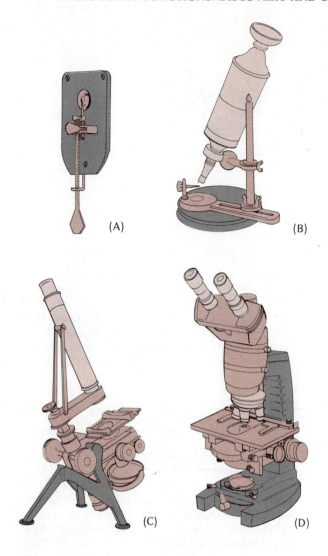

Figure 1-1. Microscopes: a bit of their evolution. (A) Single-lens microscope (ca. 1670) of Anton van Leeuwenhoek, the Dutchman who discovered microbes, consisted of a small highly curved lens in a metal block, about 5 cm (2 in) long, really just a very powerful magnifying glass. The object to be observed is carried on the pin. (B) Compound microscope of Robert Hooke (ca. 1665), who first saw and described plant cells, possessed two lenses at opposite ends of a tube, plus a device for illuminating the observed specimen from a lamp (not shown). Grew and Malpighi used similar microscopes. (C) Nineteenth century research microscope used much-improved lenses corrected for distortion and color aberrations, and possessed a lens system (the *condenser*) beneath the specimen to illuminate it better. (D) The two eyepieces of a modern binocular research microscope make it much easier to study the specimen carefully; lenses are further improved and objective (lower) lenses with different magnifications can be interchanged by rotating the nosepiece. Special optical systems such as "phase contrast" can be employed to bring delicate cell structures into view that are nearly invisible with plain light. These features as well as the modern electron microscope are described in Chapter 4.

microscope. Its invention in the seventeenth century and improvement especially in the nineteenth century (Figure 1-1) were crucial to the growth of our concepts of what plants are and how they develop. Since 1950 the electron microscope, which reveals structures much smaller than can be seen with the light microscope, has similarly expanded our understanding of the internal structure of cells and has become the basic observational tool of plant cytology.

Inquiry into the life processes or functions of plants is made possible by experiments, intentional disturbances that an investigator imposes upon the system he or she is studying. Experimentation is the means by which scientists can establish whether different factors or phenomena are causally connected. The subfields of botany whose concepts rely primarily upon experimentation are commonly called the **experimental** branches. Plant **physiology** concerns the internal functions of plants, such as how they capture and transform energy and how they grow. Plant physiology relies extensively upon ideas and techniques of measurement taken from the physical sciences, since many of the life processes of plants are of a chemical or physical nature not directly perceptible to our senses, and can be adequately comprehended only with the aid of chemical and physical concepts. Plant **genetics** concerns the processes of inheritance in plants. Genetics rests first upon numerical or statistical methods of interpreting the results of inheritance experiments, but chemical methods and concepts are playing an ever-increasing role in this field too.

The experimental subfields of botany also depend heavily upon knowledge and techniques from the descriptive subfields, especially concerning the cellular structure of plants and the substructure of individual cells. The descriptive subfields use information on physiological functions, extensively employ concepts and techniques of genetics, and utilize other kinds of experiments where appropriate and feasible, in attacking the problems with which they deal. Thus the distinction between descriptive and experimental aspects of botany is not hard and fast but is really a matter of degree, the so-called experimental subfields simply relying more heavily upon experimental evidence, by the nature of their material, than the descriptive subfields do. An intermediate subfield is plant **ecology,** which deals with the relation of plants to their environment. Plant ecology uses experiments to determine how plants are affected by and respond to environmental factors, but relies mainly upon observations to understand the makeup of vegetation, upon which it is usually impractical to experiment at large.

The Concept of Scientific Experimentation

The basic idea of a scientific experiment is to

Figure 1-2. Example of a controlled experiment: response of dwarf pea plants to the plant growth hormone, gibberellin (GA). All the plants received an alcohol solution containing the indicated amount of GA, placed onto the first leaf 7 days prior to the photograph. The control plants are on the left. They received a solution containing no GA. The second group of plants from the left received 0.0001 μg (10^{-10} g) of GA, too small an amount to stimulate growth significantly. The dose of GA was increased in each successive group up to 0.1 μg per plant at the far right. The results indicate that 0.0005 μg of GA is the minimal amount for significant stimulation, and maximum stimulation of growth is obtained with 0.05 μg per plant. [A microgram (μg) = 10^{-6} g, i.e., one millionth of a gram.]

set up a situation in which the effect of a single factor or treatment on the subject or process of interest can be positively determined. This is accomplished by holding, insofar as possible, all other factors and conditions constant, while varying just the particular factor or treatment whose effect is to be determined, so that if any differences appear between the treated and the untreated ("control") subjects or specimens, these differences can reasonably be attributed to the treatment (Figure 1-2). This simplification of nature into one cause-and-effect relation at a time, although artificial, is a very powerful aid in analyzing the complexity of the real world and deducing how it works.

An experimental science progresses by using experiments to test a **hypothesis,** or possible explanation of some phenomenon or causal relationship. From a hypothesis we can deduce certain predictions that should be true if the hypothesis is true. For example, if we propose that one process, A, is necessary for the operation of a second process, B, we can predict that any factor or treatment that affects A should influence B in the same manner. We can test this by designing experiments to determine the effect on B of any factors or treatments known to influence A. If the results of these experiments agree with the prediction, this supports the hypothesis and encourages us to try to extend or refine it by introducing additional ideas. These will lead to additional predictions that can be tested. If an extensive set of predictions proves true and can be verified by

independent experimenters, the hypothesis will come to be regarded as probable and will eventually become an established concept. If some of the predictions are disproved by experimental tests, we must modify the hypothesis to account for these results or replace it with a different hypothesis that can account for them and proceed from that one. In this way, step by step, an evergrowing body of concepts and principles about a subject is built up.

Not all scientific concepts, however, develop so prosaically. Some important discoveries are made serendipitously, that is, while looking for something else. For example, the type of plant growth hormone whose effect is shown in Figure 1-2 was originally discovered by Japanese scientists studying a disease of rice. One symptom of this disease is abnormally tall growth of infected plants. In trying to explain this, the scientists found that the disease organism produces a growth-stimulating substance. This discovery led eventually to the recognition of a major class of plant hormones, the gibberellins.

The evolution of scientific thinking can be illustrated to advantage by the history of how the processes by which plants feed themselves (and us) were discovered. The conceptions about plant nutrition that were held in ancient times represent what reasonable people even today would probably think without benefit of the results of scientific experimentation. The emerging concepts form the very core of our understanding of plants and their role in nature and human affairs.

DISCOVERY OF THE PLANT MODE OF NUTRITION

Aristotle and Theophrastus

Although people must have known and used plants for as long as the human race has existed, the first serious studies of the nature of plants that we know of were made by the Greek philosopher Aristotle (384–322 B.C.) and his student Theophrastus of Eresus (ca. 371–285 B.C.) in Athens (Figure 1-3). Aristotle's teachings probably had more influence upon the secular thinking of western Europeans than any other ideas until post-Renaissance times. Since plants spring from the soil, it seemed evident to Aristotle that their substance must come from the soil. "As many flavors as there are in the flesh of different fruits," he wrote, "so many, it is plain, prevail also in the earth." He regarded the soil as equivalent to a vast stomach that prepares and supplies the food of plants. This view later became known as the **humus theory** of plant nutrition (humus refers to organic matter in the soil). The beneficial effect of applying manures to the soil, which had been discovered even before classical times, fit in with the humus theory because it seemed obvious that manure supplies food for the growth of plants.

Whereas Aristotle's contributions to botany are scattered through works on other subjects, Theophrastus concentrated on the investigation of plants. Toward the end of his life (ca. 308–305 B.C.) the knowledge of plants that he had accumulated and taught was written down in two treatises called "Enquiry into Plants" and "Causes of Plants," which have come down to us. Since these are the first known works devoted systematically to plants, Theophrastus is properly regarded as the "father of botany." By careful observations he acquired an extensive knowledge of the morphology or form of plants, for example, the shapes and arrangements of leaves, and how they differ

among different kinds of plants. Theophrastus did not advance ideas about the nature of plant growth that differed in any important way from Aristotle's. From the modern point of view Theophrastus' most interesting contribution was to point out how different kinds of plants grow in, and are characteristic of, different environments or habitats such as mountains, streamsides, plains, and marshes. Theophrastus thus appears to have been the first ecologist.

Early Experiments to Determine the Origin of Plant Material

Although the Aristotelian humus theory of plant nutrition was widely believed until well into the nineteenth century, some contrary ideas arose even in classical times. In a philosophical novel or romance called the *Recognitions of Clement*, which was widely circulated in the Middle Ages but had been composed early in the Christian era by a now unknown author(s) and was translated from Greek into Latin shortly after 400 A.D., there appears a passage advocating that water rather than earth is the essential substance required for plant growth. It asserts that if one were to put a known weight—for example, 100 pounds—of soil into a tub, then grow repeated crops of grain and other plants in that soil until 100 pounds of produce had been accumulated, the soil would still remain in the tub, implying that the plant material is derived not from the soil itself but from the water that one adds to it. Cardinal Nicolas of Cusa (1401–1464) presented the same argument in a widely read book that he wrote in 1450. The test described in these writings was apparently only a "thought experiment" whose probable results seemed obvious to the authors. The first person known to have actually performed this type of experiment was the Belgian physician and early chemist Johann Baptista Van Helmont (1578–1644; Figure 1-4). His results, il-

Figure 1-3.   (A) Aristotle (384–322 B.C.) and (B) Theophrastus (ca. 371–285 B.C.)

Figure 1-4. J.B. Van Helmont (1578–1644).

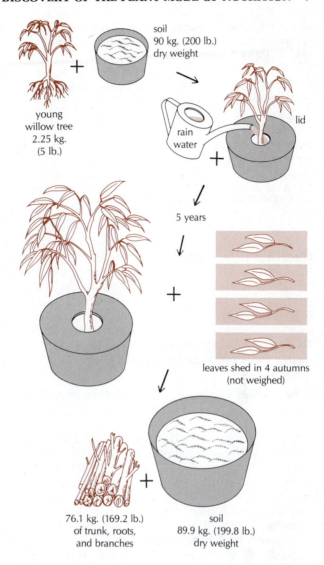

soil
90 kg. (200 lb.)
dry weight

young
willow tree
2.25 kg.
(5 lb.)

rain
water

lid

5 years

leaves shed in 4 autumns
(not weighed)

76.1 kg. (169.2 lb.)
of trunk, roots,
and branches

soil
89.9 kg. (199.8 lb.)
dry weight

Figure 1-5. Van Helmont's willow experiment (ca. 1600). During the 5-year experiment only rain water (the purest water available) was added to the tub of soil, which was kept covered with a lid to prevent dust and dirt from falling into it. At the end of 5 years the willow had increased in weight by 164 lb 3 oz (not counting its crops of leaves that were shed each autumn, which were not weighed), whereas the soil when dried and reweighed was found to have lost but 2 oz in dry weight.

lustrated in Figure 1-5, show indeed that most of the weight of a plant is not derived from the soil.

Because Van Helmont had taken care to keep out dust and had added to the tub nothing but the purest water available, he concluded that the substance of his willow tree had arisen "out of water only." Before Van Helmont's work was published (posthumously, in 1648), the English scientist Robert Boyle (1627–1691) had also become interested in the origin of the substance of plants. He had his gardener perform a number of experiments comparable to Van Helmont's but using squash plants. They yielded similar results, which after learning of Van Helmont's data, Boyle published in 1661 as confirmation of Van Helmont's conclusion that plant material arises from water. Although both Boyle and Van Helmont had experimented upon air and other gases as substances (Van Helmont indeed coined the name gas for them, and Boyle discovered one of the basic laws of gas behavior), in designing and interpreting their plant experiments they both failed to consider the possibility that their plants might take material from the air. This oversight, which turns out to have been a serious one, illustrates how conventional thinking—in this case the Aristotelian concept that the roots are the nutritive organs of a plant—can limit the intellectual vision even of able scientists.

During the same period other experimenters, using different approaches, were being led to conclusions opposed to that of Van Helmont and Boyle. A German chemist, Johann Rudolf Glauber, in 1656 reported obtaining from manure the chemical called nitre (saltpeter or potassium nitrate; origin of the modern term nitrogen). When applied to the soil, nitre stimulated plant growth as well as or even better than manure. Glauber concluded that nitre must be the "principle" of vegetation. An Englishman, John Woodward, in 1699 published the results of experiments in which he had grown spearmint plants with their roots in different kinds of water (Table 1-1). Although all his plants had free access to water and consumed comparable amounts of it, their growth varied greatly, being poorest in the purest water

and best in water intentionally contaminated with "earthy" material. Woodward therefore concluded that "earth, rather than water, is the matter that constitutes vegetables." He correctly reasoned from the data in Table 1-1 that plants evaporate into the air most of the water that they consume, but his conclusion about the source of plant material was not warranted because he did not measure how much "earth"-like material the plants actually absorbed, as Van Helmont and Boyle did. What Woodward's results really

TABLE 1-1 Growth and Water Consumption by Spearmint Plants with Roots in Water from Different Sources (John Woodward, 1699)[a]

Water Source	Growth of Plants (g)	Water Consumed (g)
Rain	1.1	194
Thames River	1.7	161
Hyde Park conduit	2.3	216
Same + leaf mold	5.6	295

[a]Woodward's data have been reduced to a common basis of 1.82 g initial weight for plants in each treatment. The growth period was 11 weeks. Phil. Trans. Roy. Soc. 21, 382 (1699).

showed was that something from the earth other than water was important to plant growth. In view of Glauber's results Woodward might have guessed this something to be nitre, but he failed to make the connection. It would take more than a century, and considerable improvement in the crude chemical concepts then available, for his results to be properly explained.

First Studies of Structures and Processes within Plants

The Italian Marcello Malpighi (1628–1694) and the Englishman Nehemiah Grew (1641–1721) (Figure 1-6) made, in the 1670s, the first careful microscopic studies of the internal structure or anatomy of plants. The anatomical features that they discovered (Figure 1-7) were a revelation to scientists of the times, but did not really show how plants feed themselves. Mainly on intuitive grounds, however, Malpighi advanced the idea that the leaves of a plant are its nutritive organs. This was a break with Aristotelian thinking, which regarded roots, as we have seen, as the nutritive organs of plants, leaves merely providing

protection and shade for other more delicate organs such as flowers and fruits. Malpighi described one experimental result in support of his proposal, namely, that if the first green leaves of a squash seedling are cut off, the seedling will not grow, even though its roots still have free access to water and to whatever other nourishment the soil provides. Malpighi thought that the sap that flows from the soil via the roots to the leaves becomes "elaborated" by the leaves into food material that the plant uses for its growth.

The first person to study extensively the processes in plants by experimental methods was the English scientist, Stephen Hales (1677–1761), a small-town parish priest (Figure 1-8). In his book *Vegetable Staticks* (staticks meant what we

Figure 1-6. (A) Marcello Malpighi (1628–1694); (B) Nehemiah Grew (1641–1721).

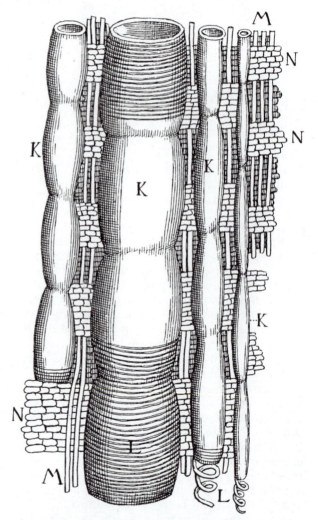

Figure 1-7. Malpighi's drawing of cells in the wood of an oak stem. The vertically running tubular structures, called vessels, we now know conduct water (Malpighi thought they conducted air). The thin vertical structures are fibers, helping to support the stem. The horizontal rows of small cells, looking like bricks, are called rays. From Malpighi's *Anatome Plantarum, Pars Altera* (1679).

Figure 1-8. Stephen Hales (1677–1761).

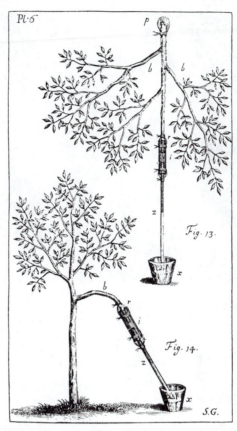

Figure 1-9. Apparatus used by Stephen Hales to detect pressure changes in branches and trunks of trees and vines during sap flow. The glass tube and connection contain water. The cup contains mercury, which becomes drawn up (point marked *z*) indicating that a suction develops in the branches when water is evaporating from their leaves (explained further in Chapter 10). From Hale's *Vegetable Staticks* (1727).

now call physics), published in 1727, Hales described many ingenious measurements he had devised for investigating the flow of sap in roots and stems of plants (Figure 1-9) and the evaporation of water from their leaves. Hales also wondered where the substance of plants orginates. He decomposed plant materials such as wood and seeds by heating them severely in a closed container and observed that they broke down to release what appeared to be "air." At that time no one knew that air is not a single substance but a mixture of several gases. Hales reasoned that if plant material can be broken down into air, "plants probably draw through their leaves some part of their nourishment from the air." He attempted to show that a plant absorbs air, using an apparatus similar to that shown in Figure 1-12, but this was not successful because (as Hales could not realize) the plant gives off another gas as rapidly as it absorbs the component of air that it uses in its nutrition.

Discovery of Oxygen Production

Later in the same century another Englishman (who subsequently emigrated to Pennsylvania), Joseph Priestley (1733–1804), recognized the existence in air of a component (later named oxygen) necessary for combustion and for the respiration of animals. If a candle were burned in a vessel of air until the flame went out, a mouse or bird placed in that air would quickly die of suffocation. In 1772 Priestley reported to the Royal Society in London that a sprig of mint would *not* die when placed in air that had been spent by burning a candle in it. To the contrary, in such air the plant would grow and the air would then, to his astonishment, again support a candle flame or the life of an animal. In terms of the chemical concepts of his day, Priestley inferred that a plant can "purify" the air of harmful material (at that time widely called "phlogiston") that combustion or the respiration of an animal releases.

Priestley's conclusion that the earth's vegetation constantly restores the air that human and animal respiration, and combustion, has rendered unfit for breathing excited immediate attention. But other scientists obtained results contradicting Priestley's, and a dispute began. This was settled by Jan Ingenhousz (1730–1799; Figure 1-10), a successful Dutch physician who had come to London and become interested in Priestley's findings. In his book *Experiments on Vegetables*, published in 1779, Ingenhousz established that plants purify the air only in sunlight. He showed that only the green parts of plants, especially the leaves, have this capacity, nongreen organs such as flowers and roots "injuring" the air just like animals or combustion. In 1779 Ingenhousz still explained his findings in terms of the concept of phlogiston. But by 1784 the great French scientist Antoine Lavoisier, father of modern chemistry, had discarded the notion of phlogiston and arrived at an understanding of combustion and respiration as

Figure 1-10 (left). Jan Ingenhousz (1730–1799).

Figure 1-11 (right). Théodore de Saussure (1767–1845).

processes that consume oxygen from the air. It then became evident that the "pure air" that plants release is oxygen.

Photosynthetic Plant Nutrition

A Swiss scientist, Jean Senebier, concluded in 1782 that plants produce oxygen only when carbon dioxide (then called "fixed air") is available to them. He reasoned that the plant must gain substance in this process. By 1796 Ingenhousz, in his "Essay on the Food of Plants and the Renovation of Soils," could argue that the basic food-making activity of plants is a conversion of carbon from carbon dioxide into plant material, with the liberation of oxygen.

A complete elementary picture of plant nutrition was first achieved, however, in 1804 by another Genevan Swiss, Theodore de Saussure (1767–1845; Figure 1-11), in his book *Chemical Researches on Vegetation*. By careful quantitative measurements made with the kind of apparatus illustrated in Figure 1-12, de Saussure showed conclusively that green plants produce oxygen only when they consume carbon dioxide, that this carbon dioxide comes from the air, and that its uptake adds carbon to the plant. He thus confirmed Hales' conjecture that leaves draw nourishment from the air. De Saussure determined that plants in the light gain about twice as much dry matter as the weight of the carbon that they take from the air (carbon dioxide absorbed minus oxygen released). Since the only other source for an increase in weight in de Saussure's apparatus was water taken up by the roots, de Saussure concluded that plants convert water, along with carbon dioxide from the air, into dry matter in their food-making process, a partial return to Van Helmont's much earlier conclusion. Because of the requirement of light for this process, it eventually became called **photosynthesis** (Greek *photos*, of light).

Mineral Nutrition

De Saussure also investigated the inorganic or "mineral" material found in plants, which remains as ash after plant material is burned. He showed that it comes from the soil in which the plant has grown, and concluded that this material, although minor in amount, is just as necessary for plant growth as carbon dioxide and water are. This explained the results of Woodward (Table 1-1) and the minor, but in fact important, loss of soil weight in Van Helmont's experiment (Figure 1-5). It explained Glauber's findings with saltpeter or nitre: de Saussure found that his plants did not take nitrogen from the air (even though by then air was known to contain 79% nitrogen) but got this element from the soil. The need for nitrogen and other chemical elements from the soil has become known as the **mineral nutrition** of plants. It explains soil fertility and the beneficial effects on plant growth of fertilizers such as manures, which had been fundamentally misunderstood in the humus theory of plant nutrition. The humus theory persisted, however, in popular belief and in many scientific writings for several decades after de Saussure's book appeared. Not until about 1850 did the requirement of plants for mineral nutrients become generally accepted and lead to the rational development of modern commercial fertilizers for agricultural use.

Light as Energy

The final concept needed for an elementary understanding of the economy of plants and their contribution to nature was to appreciate the role of light as an energy source for photosynthesis. This was recognized in 1845 by the German biologist Robert Mayer, who wrote, "plants absorb one form of energy, light, and put forth another, chemical [energy]." He saw that this transformation of energy fuels the energy needs of humans and the entire living world. By this time the green pigment that occurs in leaves and other green organs of plants had been named **chlorophyll.** Before long it would be shown that chlorophyll plays the essential role of absorbing light energy for photosynthesis.

PLANT SCIENCE TODAY AND IN THE FUTURE

Current botanical research strives to add further to our knowledge of plants and their activities, both to augment our understanding of the world and to provide information of potential use in improving our plant resources and their management. Plant taxonomists, for example, continue to search for new kinds of plants, especially in the tropics where many still await discovery. Geneticists work to combine, by crossing or hybridization, the hereditary features of different cultivated or wild plants to produce new, improved varieties for use in agriculture and gardening.

In recent decades, spectacular increases in agricultural production have been achieved by this approach, especially with wheat and rice. This has been done by combining hereditary factors, or genes, for resistance to important plant diseases with other genes for a partially stunted or "semidwarf" habit of growth (Figure 1-13), which permits the crop to be planted more densely and fertilized more heavily than the traditional varieties can be, increasing the yield of grain per acre (Table 1-2). This movement toward "high-yielding" varieties, known as the **Green Revolution,** is benefiting agriculture worldwide and has played a major part in staving off, at least for the moment, the catastrophic famines to which continued rapid world population growth threatens to lead.

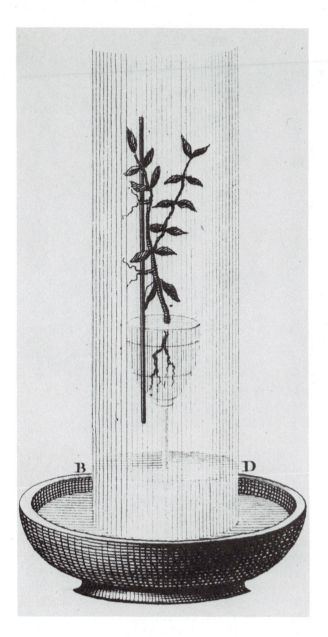

Figure 1-12. De Saussure's apparatus for measuring gas exchanges of a plant. The plant, supported by a stick, has its roots in a wineglass of water and is enclosed in a large inverted jar dipping into a pan of mercury covered with a thin layer of water (BD). Changes in the mercury level, properly corrected for changes in barometric pressure, allowed changes in the total amount of air to be measured; analyses of oxygen and carbon dioxide content performed on samples of the air before and after a period of contact with the plant permitted de Saussure to calculate the amount of carbon dioxide it took in and oyxgen it gave off. In order to observe appreciable oxygen production, de Saussure had to add extra carbon dioxide to the air at the start of the experiment (the natural carbon dioxide content of air is so low that a plant can produce only a tiny amount of oxygen when enclosed in a small volume of unmodified air). From de Saussure's *Recherches Chimiques sur la Végétation* (1804).

TABLE 1–2   Average wheat yields in The Punjab (India), 1962–70.[a] **These data illustrate how rapidly grain yields have increased by introducing new high-yielding semidwarf varieties and along with them the fertilization and irrigation practices needed for realizing their yield potential.**

Year	% of Area under Semidwarf Varieties	Average Yield kg/ha[b]
1962–66 (mean)	0	1290
1966–67	4	1520
1967–68	35	1860
1968–69	58	2180
1969–70	71	2240

[a]From D. S. Athwal, 1971, Quart. Rev. Biol. 46, 1–34.
[b] 1 kilogram per hectare equals 0.9 lb./acre

Figure 1-13. Comparison of semidwarf (center rows) and standard (to either side) wheat varieties. Farmers were able to see for themselves how much more the new varieties produced, and quickly adopted the new seeds.

Pest Control

Control of pests, including weeds, is one of the most important preoccupations of agriculture and forestry, for an estimated 35% of the potential world agricultural production is lost in the field to the combined influences of pests and diseases. In recent decades the principal emphasis in pest control has, of course, been on chemical poisons, or pesticides, the use of which has led to now well-recognized, serious ecological problems. Advances in botanical knowledge and thinking have opened up prospects for alternative control systems and strategies.

We now recognize that plants are actually engaged in their own game of chemical warfare against diseases and insect pests. Plants produce special substances (often called "secondary products") whose odors repel insects or whose tastes discourage feeding by insects, as well as products that are outright toxic to insects or to disease-causing organisms. Because these "botanicals" are natural products, their use for crop protection should be safer, environmentally, than the use of synthetic chemical poisons.

In the field of weed control enormous strides have been made, starting with the discovery by plant physiologists of plant hormones and the finding that synthetic compounds related to plant hormones can act as weed killers or herbicides. Some of these exhibit a degree of selectivity that permits them to be applied in the field without harming the crop itself. Selective herbicides not only are used for agricultural weed control but have made possible the introduction of zero-tillage agriculture (Figure 1-14), greatly reducing soil erosion and saving much fuel energy formerly needed for cultivation. Some agriculturalists expect this to become the predominant farming technique in the United States by the year 2000.

Another important class of herbicides comprises compounds that act as specific poisons of photosynthesis. These compounds, and the synthetic plant hormones, are relatively nontoxic to animals and humans, which are not photosynthetic and whose hormone systems differ from those of plants.

One possible future strategy for weed control without synthetic herbicides is the use of plant diseases that specifically attack weeds (biological control). Another possibility is to exploit the phenomenon called allelopathy, or chemical warfare between plants, wherein a given plant produces secondary products that are inhibitory or toxic to other plants that compete with it.

Control of Development

To understand exactly how a complex organism such as a plant develops from a simple egg or other reproductive cell remains one of the most

Figure 1-14.   Selective weed-killer action. A cornfield that would otherwise show severe emergence of witchweed remains free of this parasitic pest due to a subsurface application of the herbicide trifluralin four weeks previously. With the help of such agents, crops may now be planted and grown without plowing the soil at all, a farming technique called "zero-tillage agriculture."

important goals of developmental botany, as also of animal developmental studies. We know that hormones are involved since they can profoundly influence development. Hormone effects on plant development have already been put to a number of important practical uses. These include improving the quantity of fruit yielded by fruit crops by increasing, through hormone treatment, the number of fruits that begin development and decreasing the number that drop before completing their growth.

The size of some fruits, such as grapes, is now routinely increased by hormone treatments. Hormonal control of flowering is also possible in some crops, such as pineapple, where it is of great help in enabling growers to bring all the plants in a field into fruit at once so the entire crop can be harvested at one time (Figure 1-15). Knowledge of hormone effects on ripening is important in the postharvest storage of produce for transport in good condition to distant markets and for extending the season during which fresh fruits and vegetables are available to consumers. The better we come to understand how plants regulate their developmental phenomena, the better we can expect to control these processes for beneficial purposes.

Enhancing the Autotrophic Efficiency of Plants

Exciting ideas for the future improvement of crop productivity are coming from new discoveries about plant nutrition. It used to be thought that photosynthesis was a basically similar process in all higher plants and algae. Recently it has been discovered that certain plants such as corn, sorghum, and sugarcane (Figure 1-A) possess a special kind of photosynthesis, called C4 photosynthesis, which has a substantially higher productive capacity than ordinary photosynthesis. If C4 photosynthesis could be introduced into other major food crops such as wheat, rice, and potatoes, their productivity might be markedly increased.

Water is a seriously limiting resource in many agricultural areas. In recent years plant physiologists and ecologists have made huge strides in understanding how plants control **transpiration,** or the evaporation of water from their leaves, the principal cause of agricultural water consumption. This new knowledge, which includes the discovery that a hormonal system can regulate water loss, offers attractive possibilities for de-

Figure 1-15. Pineapples being harvested from plants sprayed with a plant hormone to induce all the plants to flower at the same time. Without this treatment, individual plants flower at widely different times, so the fruits must be harvested one by one as they appear. Hormone treatment enables the entire crop to be harvested at once, a much more economical operation.

creasing the enormous water cost of agricultural production.

Another important opportunity for improving plant efficiency is through mineral nutrition. One of the principal means by which productivity in developed countries has been increased in recent decades is by increasing the rate of fertilization. However, at the high rates of fertilization now practiced, over half of the nutrients applied to the soil are not used by crops but eventually wash away and create environmentally damaging pollution. Current fertilization has also become prodigiously expensive in terms of energy, which is needed in large amounts to produce and distribute fertilizers.

Better understanding of the nutrient uptake mechanisms of roots is suggesting ways in which fertilizer uptake could be increased so that less fertilizer would need to be used. For example, associations, called **mycorrhizas,** between roots and

soil microorganisms are now recognized as important aids to nutrient uptake by wild plants growing in soils with marginal nutrient supplies. It is becoming clear that crop plants can also develop mycorrhizal associations. It may be possible to improve the efficiency of nutrient uptake by supplying crop roots with superior strains of mycorrhizal organisms, so that more of the fertilizer applied to the soil would be absorbed by the crop. This technique might also enable some presently unused nutrient-poor soils to be brought into agricultural production, and help in reclaiming for useful purposes wastelands such as strip-mined areas, which often have such a nutrient-impoverished soil that most plants can hardly grow.

Nitrogen is the mineral nutrient that most often limits agricultural and natural production. Since late in the nineteenth century it has been known that plants of the legume group, such as beans, peas, alfalfa, and clover, can obtain their nitrogen from the air, by means of special bacteria that live in their roots, conducting the process of **nitrogen fixation.** In recent decades a variety of wild plants unrelated to the legumes have been found to harbor nitrogen-fixing microorganisms (Figure 1-16), and major breakthroughs have been made in understanding the mechanism of nitrogen fixation and its genetic specification. These advances have helped foster the idea of transferring the nitrogen fixation capability (or the capacity to associate with appropriate microorganisms) to important grain crops, such as wheat and rice, and of improving the feeble nitrogen-fixing associations that seem to occur in some crops, such as corn.

Unconventional Gene Transfers, or "Genetic Engineering"

The idea of transferring genes for desired physiological characteristics between unrelated plants, as mentioned several times above, would have semed pure fantasy as recently as a decade ago, as then the only known means of gene exchange in higher organisms was sexual hybridization, which is possible only between closely related organisms. Advances in genetic techniques that can be utilized with bacteria, however, have provided ideas for unconventional methods of gene transfer between plant cells. For example, certain viruses can transfer genes from one type of bacterial cell to an unrelated type, and there is now evidence that this kind of gene transfer can occur in plants.

Improved knowledge of the hormonal controls of plant growth opens up the prospect of another unconventional method. By suitable treatment with plant hormones, cells from many plant tissues can be grown and multiplied in the test tube (Figure 1-17), in an artificial culture medium

Figure 1-16. Plants of *Myrica gale,* a nonlegume that can harbor nitrogen-fixing microbes in its roots. These 1-year-old plants were grown in a culture solution lacking combined nitrogen. The plants on the left, whose roots contain the special microorganism, are growing healthily while those on the right, which lack the microorganism, are severely stunted due to nitrogen deficiency.

containing sugar, mineral nutrients, the necessary hormones, and often certain vitamins. By manipulating the hormone and nutrient supply in the medium, it is now possible to get buds or plant embryos to form from cell cultures and, from these, to regenerate entire plants. Techniques have also been developed for inducing cultured cells to fuse with one another. Fusion of cells from different plants, called somatic hybridization (somatic refers to nonreproductive, or vegetative, cells), can be carried out between cells of completely unrelated organisms (Figure 1-18). Hybrid cells between certain plants have been induced to divide and multiply, and in at least a few cases plants have been successfully regenerated from the hybrid cells (Figure 1-19). By this technique it may become possible to transfer valuable genes between unrelated plants. Enthusiasts foresee being able to construct future crop plants that are physiologically made to order, by such methods of genetic engineering. Practical implementation of these ideas may be far in the future, but it is exciting that we now have the concepts and at least some of the techniques needed to begin work on them.

OTHER USES OF PLANT CELL AND TISSUE CULTURES

By cell multiplication in cultures, millions of cells can be produced from a few original cells, in a relatively short time. If complete plants can be regenerated from these numerous cells, the possibility arises of multiplying a plant that possesses a unique, desirable combination of genes much more rapidly and extensively than is possible by ordinary seed production or other traditional means of propagation. This procedure, called **cloning,** will probably be of great advantage in making new genetic varieties (produced by either conventional or unconventional breeding methods) more rapidly available for agricultural use and also in plant breeding itself, by enabling plants with desired genes to be rapidly multiplied for use in further breeding steps.

Cell and tissue cultures may also enable geneticists to search more effectively for useful variant genes, or **mutations.** Any particular kind of mutation is usually very rare, one in a million individuals being a typical frequency, which makes it difficult or impossible to search for it by growing ordinary populations of plants. However, among the many millions of cells that can be cultivated in a single cell culture, even very rare mutations should occur. The problem is to find a way of picking out cells that possess the desired kind of mutation. This is not simple, because of the nature of hereditary systems of plant cells, but plant physiologists have found ways to obtain what are called **haploid** cell cultures, in which any mutation that occurs can express its effect on the behavior of the cells. We can then select for cells that carry a desired mutation, by altering the culture medium in such a way that the mutant cells can survive while the rest of the cells cannot. For example, to select for mutants that tolerate abnormal amounts of salt, we might add a high concentration of salt to the culture medium, and simply let grow whatever cells are able to grow in this medium. We would hope then to regenerate, from these cells, plants bearing genes for salt tolerance which can be used to develop genetic varieties that can grow in salt-ridden soils presently unusable for agriculture.

Plants and the Energy Supply

Since plants are the earth's natural solar energy converters, and are far less expensive than any man-made solar energy device, can we turn

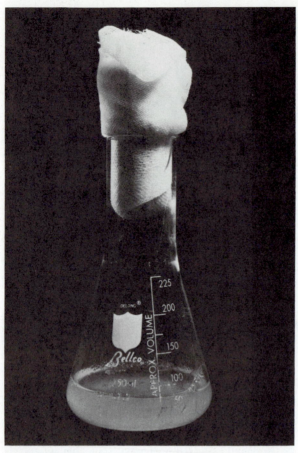
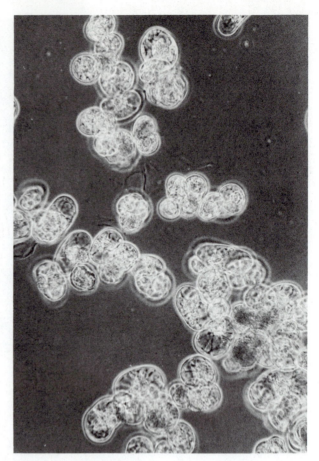

Figure 1-17.   Cell suspension culture derived from soybean. (A) Culture flask containing milky suspension of cells in liquid culture medium. Flask is shaken mechanically during growth of the culture to keep cells suspended. (B) Cells from the suspension culture, 200 ×. Nucleus and cytoplasmic strands (Chapter 4) can be seen within many of the cells.

to them for help with the energy crisis that is developing as fossil fuels become depleted? The seemingly simplest way is to use currently produced plant material ("biomass") as fuel; this is what many people in the less developed countries are and have been doing since time immemorial. However, the current consumption of fossil fuels by industrialized nations such as the United States actually exceeds by a large margin the entire photosynthetic production of their land area (and about half of this output is already being harvested for agricultural and forest products), so there is no realistic possibility of supplanting fossil fuels, at the present level of consumption, with biomass. Nevertheless, the utilization of some presently unused biomass and the production of some additional biomass specifically for energy, for example, fuelwood tree plantations, could help significantly with the energy problem, especially if public and industrial energy consumption can be reduced. The most important potential

source of biomass for energy seems to be the substantial amount of waste from the lumbering and wood-processing industries, which is currently harvested, but unused, biomass.

The difficulties and probable low efficiency and high cost of collecting and directly utilizing diffusely distributed crude biomass for energy on a modern industrial scale promote interest in more specialized possible techniques for deriving energy from plant photosynthesis, which a few examples will illustrate. One approach that is already a practical reality is to let yeasts ferment to alcohol the sugar from plant photosynthesis, then use this alcohol (after concentrating it by distillation) as a liquid fuel. This is being done in Brazil, using sugar from sugarcane, to obtain a partial substitute for petroleum; "gasohol" (gasoline with some alcohol mixed in) is also now being used in the United States. Yeasts, which are nonphotosynthetic microscopic plants (Chapter 2), convert the energy of sugar into that of alcohol with a

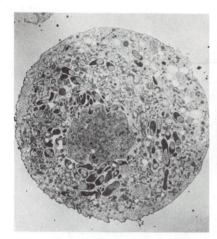
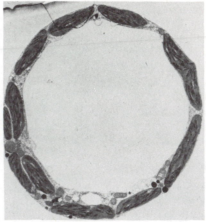
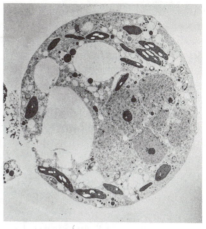

Figure 1-18. Somatic hybridization between a carrot cell (left) (4000 ×) and a pea cell (center) (6000 ×) yields a hybrid cell (right) (2500 ×) with characteristics representing a mixture of the two fusion parents. The cell walls of the carrot and pea cells were removed prior to protoplast fusion.

Figure 1-19. Young plants of a cross between the potato and the tomato, obtained by protoplast fusion followed by division and growth of hybrid protoplasts. The plants are not very vigorous, as might be expected from their bizarre genetic makeup, but some of them have produced flowers.

rather high efficiency. An even better conversion of photosynthetic products into alcohol might be achieved if other microorganisms could be introduced into the process to break down the woody material (cellulose) of plant biomass into sugar, so that the yeasts could convert this material into alcohol also. This kind of technique is visualized as a possible way of converting into useful fuel presently unused industrial wastes such as those from lumbering and sawmill operations.

Certain plants manufacture secondary products of an oil-like nature which contain relatively high proportions of substances similar to those found in petroleum (Figure 1-20). These might be pressed out and prepared for use as fuels much more simply and efficiently than the processing required for the direct use of crude biomass as fuel or for alcoholic fermentation and distillation. Since some of these plants are adapted to arid conditions, one proposal is to plant them on desert land that is currently not used for agriculture, then harvest fuel from this land. The trouble is that without water, even plants that are adapted to dry conditions grow only slowly, so the harvest and the amount of energy converted might be small in relation to the expense of planting and taking care of the crop. However, in more humid regions this kind of crop might contribute helpfully to the energy supply, if areas could be found that are not needed for the production of other commodities.

Research on the mechanism of photosynthesis to be considered in Chapter 7 shows that this process in essence uses solar energy to take hydrogen from water, then employs this hydrogen to convert carbon dioxide into sugar. In certain circumstances plant cells can be made to release their photosynthetic hydrogen as hydrogen gas, which is very combustible and could be used as a substitute for "natural" gas. This too has its complications and is not yet practical, but the general idea is receiving considerable attention and research effort at the present time.

The various lines of current research noted here are explained more fully, in the context of their respective areas of botanical knowledge, in subsequent chapters.

Figure 1-20. An oil-yielding shrub, *Euphorbia lathyrus*. This plant, as well as others, is suggested for cultivation for direct production of oil useful for fuel, to substitute for petroleum fuels now used.

SUMMARY

(1) Botany is important to our understanding of the world, our utilization of its resources, and our enjoyment of life, because (a) plants are the ultimate source of all human food and of our principal renewable energy resources, (b) plant photosynthesis supports the biological communities of nature, and (c) plants comprise an aesthetically important part of both natural and man-modified landscapes.

(2) Descriptive subfields of botany use scientific methods of observation to describe and understand the different kinds of plants, their structure, and their development. Microscopes are among the principal research tools of these fields. Experimental subfields of botany use controlled experiments to deduce how plants function. Many of these experiments and their interpretation utilize ideas and techniques from chemistry and physics.

(3) The discovery of how green plants feed themselves was the key to our understanding of plants. Essential milestones of this discovery were as follows. (a) Van Helmont (prior to 1644) and Boyle (1661): plant substance does not come mainly from soil. (b) Malpighi (1674): leaves produce a plant's food. (c) Woodward (1699): plant substance does not arise just from water, but needs "earth." (d) Hales (1727): plant substance comes partly from air. (e) Priestley (1772): plants give off oxygen. (f) Ingenhousz (1779): oxygen production requires sunlight. (g) Senebier (1782): oxygen production requires carbon dioxide. (h) de Saussure (1804): plants incorporate carbon from carbon dioxide, plus hydrogen and oxygen

from water, in making their food materials, a process now called photosynthesis. (i) Mayer (1845): photosynthesis converts sunlight energy into chemical energy. (j) de Saussure (1804): plants need minor but essential amounts of mineral nutrients such as nitrogen from the soil, accounting for soil fertility.

(4) Current botanical research of special importance includes plant genetics or the study and manipulation of hereditary characteristics; relationships between plants and their pests and diseases; plant development and its genetic and hormonal control, including the growth of plants and plant cells in artificial cultures and the possibility of gene transfer by nonsexual means; environmental relations of plants, for example, their water and mineral nutrient supply and their tolerance of adverse conditions; and photosynthetic energy conversion and its improvement. These areas of research afford notable current or future applications to important practical problems, some of which are briefly discussed.

Chapter 2

NATURE, KINDS, AND NAMES OF PLANTS

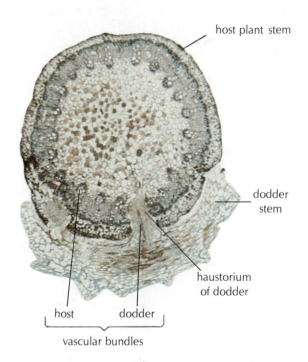

host plant stem

dodder stem

haustorium of dodder

host   dodder

vascular bundles

Figure 2-A. (Left) Illustrating nutritional extremes within the plant kingdom, snake-like orange shoots of dodder (*Cuscuta*), a nonphotosynthetic, heterotrophic flowering plant, entwine and parasitize a green photosynthetic plant. The dodder stems, bearing only rudimentary scale-like nongreen leaves visible here and there in the picture, penetrate the host plant by means of specialized branches called *haustoria,* illustrated in the microscopic view of a cross section of a host stem, shown above. The conspicuous oval cell clusters arranged in the form of a ring within the stem are its *vascular bundles* or strands of water- and food-conducting tissue, composed of cells similar to those illustrated in Figure 1-7. These vascular strands run lengthwise in the stem and are thus cut crosswise in this cross section. Note that the dodder's haustoria connect to these vascular bundles, tapping the food produced and transported within the host plant. Dark strands of vascular tissue are also visible within the main stem and haustoria of the dodder, since it, like its host, is a vascular plant (a distinction explained later in this chapter).

In everyday experience we distinguish plants as living organisms that do not move about and do not eat, a distinction going back at least to Aristotle (died 322 B.C.). Among the organisms answering this description and conventionally regarded as plants are many that do not possess the photosynthetic mode of nutrition described in Chapter 1. To understand the full diversity of plant life and its role in nature, we need first to consider the various nutritional strategies found among living organisms.

MAJOR KINDS OF BIOLOGICAL NUTRITION

Autotrophic Nutrition

The green plant mode of nutrition, in which the plant makes its own organic carbon compounds or foodstuffs from inorganic materials such as carbon dioxide, water, and mineral elements from the soil, and thus does not require preformed organic matter, is called **autotrophic nutrition** (Greek *autos*, self, and *trophē*, nourishment). As noted in Chapter 1, green plants use light as the energy source needed to produce foodstuffs, so their kind of autotrophy is called **photoautotrophy.**

The autotrophic organisms of a biological community or population of organisms are called its **primary producers,** the producers of its food and energy supplies. Besides photoautotrophy, we recognize a second kind of autotrophic nutrition, called **chemoautotrophy,** by which certain bacteria grow. These organisms obtain energy for food production from inorganic chemicals in their environment. "Iron bacteria," for example, use oxygen from the air to oxidize soluble iron compounds occurring in natural waters (Figure 2-1), a process that releases energy.

Since inorganic sources of energy are not abundant in most places on the earth's surface, chemoautotrophy is of very minor significance in the overall

Figure 2-1. Ferric deposits caused by iron bacteria along the Lower Falls of the Grand Canyon of the Yellowstone River.

economy of nature compared with photoautotrophy. However, at certain locations on the ocean bottom, where there is no light for photosynthesis, spring waters emerge containing hydrogen sulfide, which certain chemoautotrophic bacteria can use as an energy source. These primary producers support a diverse population of animals living off the organic matter formed by the bacteria.

Heterotrophic Nutrition

Organisms that live by utilizing preformed organic matter (food) are **heterotrophic** (Greek *heteros*, other, and *trophē*, nourishment). In ecological parlance, heterotrophs are **consumers.** Being nonphotosynthetic, they lack chlorophyll and are usually not green (if green, they possess some green pigment other than chlorophyll). Animals are heterotrophs: their typical method of obtaining food by eating it and then breaking down or digesting it in internal digestive organs is called **holozoic** nutrition. A number of kinds of organisms traditionally regarded as plants are heterotrophic, for example, molds.

While the minimum definition of a heterotroph is an organism that requires from outside itself a supply of food such as sugar, as an energy source and as raw material for making its own cell constituents, many heterotrophs also require in their diet certain *specific* organic compounds such as vitamins. These compounds are not energy sources but are needed for particular cell functions, and cannot be synthesized by the organism from its energy food. An organism requiring a specific organic nutritional supplement is called an **auxotroph** (Greek *auxo,* increase), whereas a heterotroph requiring only a general energy and growth food

such as sugar, plus mineral nutrients, is a **prototroph** (Greek *protos,* first or elementary). Some bacteria and molds are prototrophs, but many heterotrophic plants and virtually all animals are auxotrophs.

Heterotrophic organs of higher plants, such as roots, are at least in some cases auxotrophic, because to grow them in isolation (for example, in test-tube culture) one must supply not only sugar and mineral nutrients but also certain vitamins, which evidently are furnished normally by the green, photosynthetic part of the plant. Surprisingly, some entire autotrophic plants, although self-sufficient for energy and bulk food material, require specific organic nutrients from outside and must therefore be termed auxotrophs: most notably, many single-celled algae (see below) require one or more vitamins, especially vitamin B_{12}, from the water in which they grow.

Some kinds of bacteria use externally obtained organic compounds to make cell constituents but not as an energy source, or vice versa. These kinds of nutrition and the applicable vocabulary are considered in Chapter 22.

The majority of heterotrophic plants are **saprotrophic,** (saprophytic) meaning that they grow on and decay dead organic matter such as the bodies of dead animals or plants (Greek *sapros,* rotten). Saprotrophs do not "eat"; in contrast to animals, they break down (digest) insoluble potential food materials such as wood and protein outside their bodies and then absorb the soluble digestion products (such as sugars) into their cells. Primitive single-celled animals or protozoa, such as the ameba, in contrast, engulf or ingest insoluble food particles into their cells and digest this food material intracellularly, a mode of nutrition called **phagotrophic** (Greek *phagein*, to eat).

Symbiosis

Symbiosis denotes a situation in which different organisms live together in a close nutritional relationship. A **parasite** is a heterotrophic organism that obtains its food from another living organism called its host. Parasitism is one kind of symbiosis, where only the parasite derives benefit, while the host suffers a depletion of its resources, if not outright disease. Many and diverse symbiotic associations exist, however, in which both partners benefit nutritionally from the relationship; this is called **mutualism.** Many mutualistic associations have developed between an autotrophic green plant and a heterotrophic organism, where the green plant supplies the heterotroph with organic matter (food), while the heterotroph furnishes some other benefit to the green plant. An important example is the association between the roots of legume crops, such as beans and peas, and certain bacteria called *Rhizobium,* which furnish nitrogen to the plant from the air and, in return, obtain food from the legume plant.

THE PRINCIPAL KINDS OF PLANTS

Most of the nutritional strategies just explained occur among organisms that are traditionally regarded as plants. We shall now look briefly at the principal kinds of plants, to see in what forms different kinds of nutrition are found and to acquire a basic background about plant diversity that will be needed in subsequent chapters.

Thallophytes

Many different kinds of plants can be categorized as thallophytes. These are simple or relatively simple plants whose vegetative body (called a **thallus**) lacks supporting and conducting tissues and is usually not differentiated into distinct organs such as roots, leaves, and stems. Some thallophytes are single-celled and others are multicellular, ranging in size from microscopic to quite large forms such as the larger seaweeds. Many thallophytes reproduce and disperse themselves by means of single-celled, microscopic reproductive units called **spores** (example shown in Figure 2-5). Thallophytes include the following major groups.

(1) **Bacteria.** These are small, single cells or chains of cells (Figure 2-2) whose structure is very simple compared with that of cells of most of the following groups. They reproduce by the simple division of each cell into two (binary fission); some of them can also form specialized spores. Most bacteria are heterotrophic—either saprotrophic or parasitic—but a few are photoautotrophic and a few others, as previously noted, are chemoautotrophic, a mode of nutrition known only among bacteria. Saprotrophic bacteria occur everywhere, as decay organisms, while parasitic bacteria cause many diseases of animals and humans and some diseases of higher plants. Although bacteria were traditionally grouped with plants because they do not ingest food like animals do, they differ fundamentally from most green plants in cell structure and are now classified in a separate kingdom from plants.

(2) **Algae.** These are chlorophyll-bearing, photoautotrophic, mostly aquatic thallophytes. **Blue-green algae** (see Figure 2-3A), which are common in fresh waters, are microscopic and have small cells similar in structure to those of bacteria. They are now often called **cyanobacteria** ("blue bacteria") and grouped with the true bacteria mentioned above. The other kinds of algae, which include the most important primary producers of fresh waters and the open ocean, have cells that are more similar in structure to those of higher plants. They range from single cells and chains (filaments) of cells (Figure 2-3B) to large seaweeds (Figure 2-4), which are important primary producers in coastal seawaters.

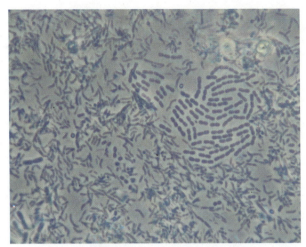

Figure 2-2. A mixed bacterial culture from the surface film of a stagnant pool, single cells and chains (250 ×).

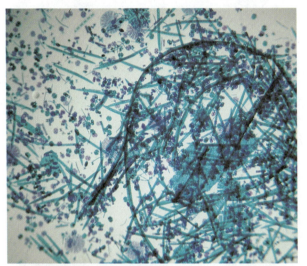

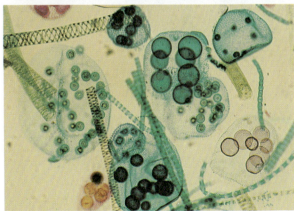

Figure 2-3. Examples of photoautotrophic microorganisms traditionally called algae. (A, top) Mixed population of single-celled and filamentous organisms from a fresh-water pond, mostly blue-greens (cyanobacteria) (20 ×). (B, bottom) Some longer filamentous green algae (*Spirogyra*) and motile green algal colonies (*Volvox*) (40 ×). The red and purple colors in some of the organisms result from staining in preparation for photography.

(3) **Fungi.** These are heterotrophic, nongreen thallophytes, primarily saprotrophs and parasites. Most fungi grow in the form of a delicate filamentous plant body called a **mycelium,** illustrated by the growth of molds (Figure 2-5). Some fungi are aquatic (e.g., water molds); many are terrestrial, but usually grow in relatively damp situations because their mycelium is not protected against desiccation. Molds and mushrooms (Figure 2-6), along with saprotrophic bacteria, in nature serve importantly as **decomposers** which aid in the decay of biological wastes and dead organisms, returning the carbon of these wastes to the air as carbon dioxide and releasing into the environment the mineral nutrients that they contain, in inorganic forms which photosynthetic plants can

Figure 2-4.   Macroscopic algae (seaweeds) growing on a rock at the seashore, between low and high tide lines; the rock is about 0.6 m (2 ft) high. Clumps of olive-green branching shoots at right and in left center are "rockweeds" (brown algae); the dark red leaf-like fronds (red algae) at the center and far left include plants of the thin membranous *Porphyra* ("nori") commonly used in Japanese food. Other brown and red algae can also be seen. When the tide is in, all these fronds float in the water, absorbing light for photosynthesis.

Figure 2-5.   Branching tubular vegetative body, or *mycelium,* of a mold, bearing spores (single-celled reproductive structures) at the tips of many of its branches. (× 250).

reuse. Another group of saprotrophic fungi is the yeasts, which ferment beer and wine and make bread rise. Parasitic fungi include rusts, mildews, and many other diseases of higher plants.

(4) **Slime molds.** These are peculiar organisms which, although associated by name with fungi, differ fundamentally. Instead of mycelial growth, slime molds develop a semifluid body that moves about by a process of flow, feeding phagotrophically upon bacteria. Slime molds occur commonly on damp, decaying vegetable matter such as wood and fallen leaves, where they reproduce by forming spores (Figure 2-7).

(5) **Lichens.** These are symbiotic, probably mutualistic associations of fungi with algae, creating a fully autotrophic, compound organism with its own distinctive morphology (Figure 2-8) and capable of a very spartan life. Lichens occur nearly everywhere on land, commonly forming crust- and leaf-like growths on bare rocks, on walls and roofs, and on bark, or diminutive shrub-like growths on tree branches and, especially in arctic and subarctic regions, on open ground.

Bryophytes

Bryophytes are primitive photoautotrophic land plants that lack roots and true conducting or supporting tissues. They reproduce by spores.

Figure 2-6.   Mushrooms such as these are the spore-producing structures of the species of fungi from which they grow.

Figure 2-7. Slime molds, showing the semi-fluid body (*plasmodium*; A) and spore cases (*sporangia*; B) of two different species.

Figure 2-8. A mixed population of foliose (leaf-like) and crustose (crust-like) lichens growing on a dead spruce in the Alaskan interior.

Figure 2-9. A moss plant, Mt. Wellington, Hobart, Tasmania. The capsules at the ends of the thin stalks enclose special tissues that produce spores.

The most familiar bryophytes are the **mosses,** which have a short leafy stem (Figure 2-9) and are common both in wet places such as bogs and springs and in intermittently moist shady locations such as on bark and protected rocks or soil.

Although thallophytes and bryophytes do not have a conducting tissue system (vascular system) comparable to that of vascular plants (see below), some of the larger marine algae or kelps do develop a specialized system of cells for transporting photosynthetic products from the leafy, photosynthetic parts of the plant to its organ of attachment to the ocean bottom many meters away. Their plant body is noticeably differentiated into functionally specialized organs but is nevertheless commonly called a thallus, although this usage conflicts with the basic definition of a thallus given above. Some of the larger mosses develop cell systems plainly specialized for water conduction, although less so than the water-conducting cells of vascular plants. The distinction between vascular and nonvascular plants is not completely sharp as regards the possession of conducting cells, but there is no real overlap between the thallophyte or bryophyte groups and the true vascular plants considered below, because their respective patterns of growth and reproduction are quite different.

Vascular Plants

Vascular plants possess a conducting (vascular) system for transporting water and photosynthetic products. Vascular plant bodies are differentiated into stems, leaves, and (usually) roots and usually possess strengthening tissues that aid in mechanical support. These and other features fit them for life on dry land; most land plants belong to this group, often loosely referred to as the "higher plants." Most vascular plants are photoautotrophic, but a few are saprophytic or parasitic (Figure 2-A).

(1) **Lower vascular plants.** These comprise several groups of seedless, spore-reproducing

plants that made up most of the earth's land vegetation before and during the "coal ages" but are relatively unimportant today. The most abundant modern group of lower vascular plants is the **ferns** (Figure 2-10).

(2) **Seed plants.** These, the most important vascular plants, reproduce and disperse themselves by seeds, which are multicellular structures that contain an embryonic plant. **Gymnosperms** bear their seeds "naked," that is, not enclosed within a fruit. This group includes the cone-bearing, needle-leaved trees (conifers; Figure 2-11). **Angiosperms** or **flowering plants** produce their seeds enclosed within a fruit, which arises from a flower (Figure 2-12). Flowering plants are by far the most abundant and diversified land plants. They fall into two great subgroups, called **monocots** (short for monocotyledons) and **dicots** (dicotyledons) which, among other differences, have, respectively, one (mono-) and two (di-) first leaves or cotyledons within the seed. Monocots include the grasses (among which are the cereal grains such as wheat, rice, and corn), rushes and bulrushes, lilies, palms, bananas, and orchids. Dicots include most broad-leaved plants including fruit trees, most vegetable plants, and broad-leaved trees. The differences between monocots and dicots are explained more completely in subsequent chapters.

Figure 2-11. A group of pine trees: sugar pine (*Pinus lambertiana*) (left) and ponderosa pine (*P. ponderosa*) (right and foreground).

Figure 2-10. A tree fern, New Zealand.

Figure 2-12. An orange tree in flower and fruit.

PLANTS VERSUS ANIMALS

The preceding summary reveals various groups of organisms that blur or break down the traditional distinction between plants and animals. Many single-celled and some multicellular algae, for example, are motile like animals, despite having an autotrophic mode of nutrition closely similar to that of higher plants. Fungi, on the other hand, are heterotrophic like animals, although they are nonmotile. Another distinction commonly made between animals and plants, which also dates from the teaching of Aristotle and Theophrastus, is that plants lack senses and feelings. There is indeed no evidence that plants possess feelings, which is no wonder since plants lack a nervous system, the seat of the feelings and sensations of animals. But plant physiological research shows that plants do sense many features of their environment and respond to them in a variety of ways, including movements and developmental changes such as flowering, which are important in ecological adaptation. Primitive animals such as the protozoa, on the other hand, also lack a nervous system and may be compared, in their senses and environmental responses, more with plants than with higher animals.

The following features are shared most consistently by the groups of organisms, discussed in the preceding section, that are traditionally regarded as plants. (1) Plants usually have either autotrophic or saprotrophic, not phagotrophic or holozoic, nutrition. (Those plants that are parasitic plainly resemble saprotrophic or autotrophic forms.) (2) Plant cells are normally surrounded by a jacket of skeletal or structural material called a **cell wall,** which animal cells lack.

Even these characteristics, however, are not completely universal among ''plant'' groups. Some of the motile single-celled algae can engulf food particles phagotrophically, and some do not develop a true cell wall. The slime molds are not only motile but phagotrophic, as noted, and their vegetative (feeding) stage does not possess cell walls. But they reproduce by spores like many fungi do. The kinds of organisms just mentioned may be included within the animal kingdom by zoologists even though they have been regarded as plants by many botanists. Thus at the level of lower organisms, no hard and fast distinction can be maintained between the traditional plant and animal kingdoms.

The Evolution of Plant Groups

The concept that made understandable the biological diversity that exists in nature and the difficulties encountered in trying to classify it in simple or arbitrary ways, such as distinguishing between plants and animals, is the principle of **organic evolution.** Until the middle of the nineteenth century most people believed, in accord with traditional religious teachings, that all the different kinds of organisms on earth had been individually and separately created by a divine act, and had reproduced themselves without change ever since. However, botanists and zoologists had begun to perceive extensive and basic similarities between the organisms that could be regarded as falling within a common class or group, similarities that did not seem understandable if each kind of organism had originated independently of all others. In 1859 the English naturalist Charles Darwin proposed that these similarities are the result of different kinds of organisms having descended from common ancestors by a process of gradual change in their hereditary characteristics. This process came to be known as organic evolution.

Most systematic botanists today consider that the organisms that have traditionally been called plants and included in the field of botany actually represent several distinct lines of evolution, or kingdoms. Bacteria and blue-green algae comprise one kingdom (called the Monera), and fungi are regarded as another kingdom (Mycota). The algae are sometimes placed into a third kingdom, called Protista, which also includes the unicellular animals or protozoa. This reflects a belief that the more plant-like and animal-like unicells, as well as the types with intermediate characteristics, are all part of a primitive evolutionary group from which probably arose the fungi, the multicellular animals, and the higher plants (including both bryophytes and vascular plants), as separate evolutionary lines (Figure 2-13). This view of biological diversity overcomes the problems of the traditional distinction between plants and animals, but is by no means free of difficulties when plant groups and their resemblances are examined more closely, as we shall do from Chapter 21 onward. A more detailed classification will then become necessary.

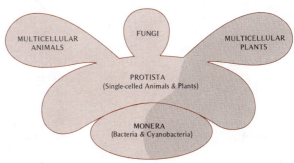

Figure 2-13. The five kingdom classification scheme. The darker shading represents photoautotrophic organisms.

THE NAMES OF PLANTS

Common Names

Every language contains vernacular or common names for plants about which its speakers have a need to talk. Common names usually differ between languages; for example, what in English is called a *bean* is in French *haricot*, in German *Bohne*, and in Spanish *frijol*. Furthermore, even in one language different common names may exist for the same plant; for example, green peppers are called mangoes in parts of the American midwest. This example also illustrates that one and the same common name can be used for entirely different plants, for the name mango more widely refers to a tropical fruit, entirely unrelated to the green pepper. A common name that creates international confusion is *corn*, which to English- and German-speaking Europeans has always meant cereal grains in general, especially wheat and rye, but in America has come to mean what European settlers originally called "Indian corn," or *maize* (from the Spanish common name, *maíz*). Because of their ambiguities, in science common names must be employed with caution. For example, the unambiguous name maize is preferred, in scientific writing, to the ambiguous name corn.

Scientific Names

Many kinds of plants lack a genuine vernacular name, because they were not noticed during the development of our language but only after the beginning of scientific study and exploration. As early as the sixteenth century, botanists began to realize that to deal with such plants as well as to circumvent the multiplicity and frequent ambiguity of the common names of more familiar plants, they needed to invent standardized scientific names that could be recognized by scholars of all countries and would be unique for each kind of plant. Because the scholarly language of the sixteenth to eighteenth centuries was Latin, from the outset scientific names were written in Latin form. Indeed many of them were taken originally from the classical Latin or Greek names for the kind of plant in question.

As plants were studied more carefully botanists began to recognize distinguishable individual kinds, called **species,*** which fall into groups of obviously similar species. Such a group became known as a **genus** (pronounced "jean'-us"; plural, *genera*, pronounced "jeh'-neh-ra"). Examples of species and genera familiar from everyday experience are the different kinds (species) of oaks and the different species of pines (Figure 2-11), oak and pine each being a common name for what botanists identify as a genus. The species belonging to any one genus were recognized by Darwin in 1859 as representing a set of evolutionarily closely related organisms, that is, organisms with a not very distant common ancestor. However, as early as 1596 the concept of genus and species had been inaugurated by the Swiss botanist Caspar Bauhin (1560–1624; Figure 2-14) as the basis for naming plants. His work is often regarded as the beginning of the field of plant taxonomy as a science.

The Swedish botanist Carl Linnaeus (1707–1778; Figure 2-15) introduced, just before 1750, the system of binomial or two-word scientific names that became accepted as the modern scientific convention. Each species is designated by a Latin or Latinized genus name, capitalized, followed by a specific epithet that is normally not capitalized. For example, *Quercus alba* denotes the oak whose American common name is white oak, *Quercus* being the genus name for oaks and *alba* being its specific epithet, which means white in Latin. A specific epithet may be a descriptive adjective such as *alba*, *lutea* (yellow), *pilosa* (hairy), and so forth. It may refer instead to a geographic region in which the species occurs, for example, the southeastern live oak, *Quercus virginiana*, occurs in Virginia. Or a species may be named in honor of a person, usually a deserving botanist or botanical explorer, for example, *Quercus kelloggii*, the California black oak, after Albert Kellogg (1813–1887), a pioneer California botanist. Some specific names are derived from vernacular plant names, for example, *Quercus robur*, the European white oak which the Romans called *robur*.

Figure 2-14. Caspar Bauhin (1560–1624).

---

*Note that the word *species* is both singular and plural: one species, several species. The deceptively similar, seemingly singular form *specie* actually means coins.

Figure 2-15. A youthful Carl Linnaeus (1707–1778), no doubt tired from a full day of plant classification.

Genus name and specific epithet parallel somewhat the relationship between surname and given name, respectively, of a person. The surname denotes the group (human family); the given name denotes the specific individual within that group. Like a human given name, a specific epithet by itself does not identify anything; it must be used along with a genus name, since a particular specific epithet can be employed for species belonging to different genera. For example, besides *Quercus alba* there are *Salix alba*, the white willow, and *Morus alba*, the white mulberry. However, once a genus name has been mentioned, species belonging to it may be designated by their specific epithets following the initial of the genus name, for example, *Q. virginiana* or *Q. alba*, as long as this does not create confusion between different genera whose names begin with the same letter.

In contrast to specific epithets and also to human surnames, every genus name in botany must be unique. Thus the species *Quercus robur* and *Quercus utahensis* must (if these names are correct) belong to the same genus and should be evolutionarily related, whereas John Jones and Tim Jones need not belong to the same family or indeed be related at all. Because of its uniqueness, a genus name can be used by itself, to denote the entire group (e.g., *Quercus*, all the oaks) or, more loosely, an undesignated species of oak. The expression *Quercus sp.* (sp., abbreviation for species) means some particular species of oak, whose name the author either does not know or does not consider important, in the given circumstances,

to designate. Genus names may be derived in any of the ways mentioned for species names, but must be in the form of a Latin noun.

The scientific names in use today for plants go back, by international agreement among botanists, only to 1753, the date of publication of Linnaeus' treatise called "Species Plantarum" (species of plants; Figure 2-16), in which he listed with binomial names all the plant species then known. A system of logical rules to govern the naming of plants, the International Code of Botanical Nomenclature, was later developed by international conventions of botanists called International Botanical Congresses, which have been held at intervals since 1864. These rules specify how a name for a newly discovered plant must be published in order to be valid, and they mandate which name must be used when, as often happens, more than one scientific name is applied to a given species or the same name is applied to different species or different genera. To help botanists check on how particular scientific names should be employed, the complete formal scientific name includes, after the specific epithet, the name (or an abbreviation of the name) of the author who coined

alba.   10. QUERCUS foliis oblique pinnatifidis: finubus angulisque obtufis.
  Quercus foliis fuperne latioribus oppofite finuatis, finubus angulisque obtufis. *Gron. virg.* 117..
  Quercus alba virginiana. *Catesb. car.* 1. *p.* 21.*t.*21.*f.*2.
  Habitat in Virginia. ♂

Efculus.   11. QUERCUS foliis pinnato-finuatis lævibus, fructibus feffilibus. *Roy. lugdb.* 80.
  Quercus parva f. Fagus græcorum & Efculus. *Bauh. pin.* 420.
  Habitat in Europa auftrali. ♂

Robur.   12. QUERCUS foliis deciduis oblongis fuperne latioribus: finubus acutioribus; angulis obtufis. *Hort. cliff.* 448. *Fl. fuec.* 784. *Mat. med.* 426. *Hall. h.* ⸗ 1.152. *Roy. lugdb.* 80. *Dalib. parif.* 293. *Gmel. fiv.* 150.
  Quercus cum longo pediculo. *Bauh. pin.* 420.
  Quercus. *Fuchf. hift.* 229. *Læf. pruff.* 211. *t.* 69.
  Habitat in Europa. ♃

Ægilops.   13. QUERCUS foliis ovato-oblongis glabris ferrato-repandis.
  Quercus calyce echinato, glande majore. *Bauh. pin.* 420.
  Cerri glans Ægilops afpris. *Bauh. hift.* 1. *p.* 77. *fructus.*

Ha-

Figure 2-16. Part of a page from Linnaeus' *Species Plantarum* (1753). Here he lists four species of oak (*Quercus*), numbered 10–13, using short multi-word Latin descriptions that were at that time considered the proper formal scientific names of species. In the margin to the left of each formal name he added, in italics, what he at first called a "trivial" single name intended as an informal shorthand for the specific epithet. The genus name plus his "trivial" epithet subsequently became accepted as the standard scientific name (e.g., *Quercus alba*, no. 10 in the list). Below each numbered formal name Linnaeus lists other (multi-word) names that had been used by previous botanists for the same species; for example, names followed by *Bauh. pin.* were used by Caspar Bauhin in his *Pinax Theatri Botanici* (1623). After the synonyms is a statement of the species' geographical occurrence ("habitat"), so far as known to Linnaeus.

that particular scientific name; for example, the American white oak is formally *Quercus alba* L., where L. stands for Linnaeus, who introduced this name in his "Species Plantarum" (1753).

Subspecific Names

Many species contain recognizable hereditary variations, some of which may be named. Forms of plant species that differ fairly substantially from one another and usually occur in different geographic areas are called **varieties** (technically, Latin *varietas*). A variety possesses a varietal epithet that follows the species' specific epithet but is preceded by the abbreviation "var.," for example, *Prunus angustifolia* var. *watsoni*, the sand plum of the southern plains states, which is a variety of the chickasaw plum (*P. angustifolia*) of the southeast.

Most cultivated plants have given rise, under domestication, to different genetic strains or stocks which, in common parlance, are called horticultural varieties, usually shortened merely to "variety." To eliminate the confusion between references to this kind of variant and to the geographical races or botanical varieties of wild plants defined above, horticultural varieties are technically called **cultivars.** Cultivar names are in vernacular languages such as English, not Latinized as are the names of botanical varieties. The cultivar name, which can consist of more than one word, is preceded by the abbreviation "cv." and is not in italics or underlined as genus names and specific and varietal epithets conventionally are. For example, the "Kentucky Wonder" string bean is *Phaseolus vulgaris* cv. Kentucky Wonder.

Group Names above the Genus Level

The genus–species method of naming, introduced by Bauhin, began the system of plant classification elaborated by subsequent systematic botanists, who gathered genera into larger and larger groupings reflecting more and more general kinds of resemblances and, it was later presumed (after Darwin), more and more distant common ancestry. Genera are grouped into botanical **families,** the names of which conventionally end in -aceae. For example, in the Fagaceae, the beech family, fall oaks (*Quercus*), as well as beeches (*Fagus*), chestnuts (*Castanea*), and so forth. Families are grouped into **orders** with names ending in -ales, for example, Fagales for the order that includes the Fagaceae. Above the order are the **class** (e.g., angiosperms, or flowering plants) and, above that, the **division** or **phylum** (e.g., vascular plants, called Tracheophyta). Categories intermediate between these levels sometimes must be used; for example, dicots, in which the order Fa-

gales falls, and monocots are **subclasses** under angiosperms.

Many of the currently accepted botanical families of north temperate-zone plants were already recognized, at least approximately, by Linnaeus, because of relatively definite resemblances between their genera. The major units of classification, however, have been changed radically since the time of Linnaeus, as factual knowledge and understanding of plants, especially lower plants, have increased. Changes in primary classification continue to be made today as different views about evolutionary relationships of plant groups develop from new knowledge. Most of the groups named in the preceding survey of plant diversity were used as major units of formal plant classification in the earlier part of this century. While some of these units, such as bryophytes and angiosperms, continue to be formally recognized, the majority of them, although still commonly used as informal names, have been split into several divisions in formal classifications, or given up altogether, to express better the apparent complexity of plant evolutionary history that we will examine from Chapter 21 onward.

Plant Identification

Since the resemblances among the genera in a given botanical family are usually relatively obvious, at least among vascular plants, families are the most useful higher categories for the purposes of identifying the name of an unfamiliar vascular plant. A botanical identification manual or **flora** is organized by grouping, under families, all the genera that occur in a given geographic region. The first task in identifying an unknown plant is to determine the family to which it belongs, then the genus within that family, and finally the species within that genus. Breaking the problem of identification into these successive levels or hierarchies affords a definite, analytical procedure for accomplishing what might otherwise be a bewildering task.

The botanist who prepares a manual normally devises and provides, as an aid to identification, an **analytical key** (commonly called just a "key") to the families covered, then under each family a key to the genera in it, and under each genus a key to the species in it. As illustrated by the key to species of birch given in Table 2-1, a key analyzes in a binary-choice fashion the characteristics by which one can distinguish the different named groups within a given unit of classification, in this case the different species of *Betula*. Keys lead one to focus successively on each characteristic useful in narrowing down the choice of possible families, genera, and (in turn) species to which a plant might belong and, thus, enable its identity to be determined expeditiously.

TABLE 2-1 Key to Birch (*Betula*) Tree Species in North America[a]

I. Leaves with mostly 9–12 pairs of veins (5–9 in *B. nigra*); cones[b] oval, held upward; seed[b] bears a wing narrower than, to at most as wide as, the seed itself
    A. Bark dark, furrowed, not separating into layers; cone scales[b] lacking hairs   *B. lenta* (sweet birch)
    B. Bark yellowish, silvery, or reddish, separating into papery layers; cone scales hairy
        1. Bark yellowish or silvery; leaves 8–11 cm long, with 9–12 pairs of veins and doubly toothed edges   *B. lutea* (yellow birch)
        2. Bark light reddish-brown; leaves 4–8 cm long, with 5–9 pairs of veins, often more or less cut into lobes   *B. nigra* (red or river birch)
II. Leaves with fewer than 9 pairs of veins; cones oblong to elongated, horizontal or hanging downward; wing broader than the seed that bears it
    A. Bark red-brown to chestnut brown   *B. occidentalis* (water birch)
    B. Bark white
        1. Bark chalky, not separating into papery layers; tip of leaf long-pointed   *B. populifolia* (gray birch)
        2. Bark peeling off in papery layers; leaf with a sharp but not long-pointed tip
            a. Young twigs and leaves somewhat hairy; middle lobe of cone scale longer than the lobes on either side   *B. papyrifera* (paper birch)
            b. Leaves and twigs without hairs; middle lobe of cone scale not longer than lateral lobes   *B. pendula* (European white or weeping birch)

[a]Adapted from R. J. Preston, *North American Trees*, and C. S. Sargent, *Manual of the Trees of North America*. To simplify the key, shrubby and subshrubby species of *Betula*, and two tree species of rather restricted distribution, are omitted. The key includes the commonly planted European species, *B. pendula*.
[b]The "cones" of birches are actually fruit-containing catkins or specialized branches, their "scales" are modified leaves called bracts, and their "seeds" are actually winged fruits. These points are ignored in the key for simplicity.

SUMMARY

(1) Heterotrophic organisms use preformed organic matter (food) obtained directly or indirectly from autotrophic organisms (primary producers), principally green plants. Saprotrophs are plant-like heterotrophs that obtain food from dead plant and animal bodies and biological wastes, by breaking these materials down outside the saprotroph's body and absorbing the breakdown products. Parasites obtain food from a living host. Symbiotic organisms associate together in a close nutritional relationship, for example, a green photosynthetic plant with a heterotrophic organism fed by the green plant.

(2) Organisms traditionally regarded as plants include bacteria; algae (primitive photosynthetic, mostly aquatic, green plants); fungi (primitive heterotrophic nongreen plants); lichens (symbiotic associations of fungi with algae); bryophytes (mosses and related nonvascular land plants); and vascular plants (land plants with a conducting or vascular system) including ferns, gymnosperms (conifers, etc.), and angiosperms (flowering plants). Features common to most organisms of these groups are autotrophic or saprotrophic nutrition and the presence of a cell wall. Many primitive plants are motile; among lower organisms no sharp dividing line can be drawn between plant-like and animal-like forms. Bacteria, algae, fungi, and land plants (bryophytes plus vascular plants) are now usually considered separate kingdoms or distinct major lines of evolutionary descent.

(3) For scientific accuracy each plant species is designated by a binomial scientific name consisting of a genus name followed by a specific epithet, in Latin form. A genus is a group of similar species, probably derived from a common ancestral species through evolution. Subspecific variations that are named include the variety, a naturally occurring variant form usually encountered in a limited geographic region, and the cultivar or horticultural variety, a genetic variant developed and maintained under human cultivation.

(4) In formal classification, according to the degree of resemblance and the inferred closeness of common evolutionary ancestry, genera are grouped successively into families, orders, classes, divisions or phyla, and kingdoms, often with subdivisions between these levels. The number and makeup, especially of the higher categories, are subject to considerable change as concepts of evolutionary relationships change.

Chapter 3

THE SEED PLANT PLAN AND ITS DEVELOPMENT

Even a casual observer of plant life must be impressed by its variety and diversity—grass and pine trees, celery and potatoes, palms and cacti. Visiting a flower show or botanical garden or perusing a nursery catalog will reveal hundreds of kinds of plants, each distinctive in color, form, growth habit, winter hardiness, and other characteristics. Despite their diversity, all these seed plants carry on the same basic activities and share a similar construction, which is modified for different functions in different plans. This chapter examines the general forms and modes of development of the seed plant body, some of its modifications that are significant in survival and reproduction, and the origin of this plant body during seed germination.

THE PLANT PLAN

The plant body comprises fundamentally an axis, consisting of the **root** plus the **shoot** (Figure 3-1). The shoot in turn is made up of **stem** and **leaves.** The root anchors the plant and absorbs water and mineral nutrients from the soil. The leaves serve for photosynthesis (food production), while the stem serves primarily for support and conduction. Specialized parts of the shoot, such as flowers, function in reproduction.

A young stem or twig of a flowering plant is marked by the presence of **nodes,** the points on a stem where a leaf or leaves are attached. The intervals between the nodes are called **internodes.** The oldest leaves are at the base of the shoot; the youngest, near the tip. A bud usually occurs at the base of each leaf, in the angle between the leaf and stem, called the **leaf axil.**

For a given species the leaves are usually arranged on the stem in a constant manner (Figure 3-2). If only one leaf is attached at a node, the leaves are **alternate.** If two leaves occur at the same node, on opposing sides of the twig, the leaves are **opposite.** In some plants more than two leaves occur at the same node, a **whorled** arrangement.

Although the root and the shoot share many common structural features, the root bears no appendages comparable to leaves and, consequently, has no nodes or internodes.

The root, stem, and leaf, along with the parts of the flower, are called organs, by analogy with the organs of specialized function in the animal body, such as the heart, stomach, and lungs. Organs are composed of tissues, and these, in turn, of cells, considered in later chapters.

Woody and Herbaceous Plants

Trees and shrubs, whose stems live for a number of years and increase in diameter each year by the addition of woody tissue, are called **woody plants.** Because the tissues of a woody stem become relatively strong and rigid, the plant can support a considerable bulk of branches, twigs, and leaves. In contrast, plants with a relatively short-lived aboveground stem that is comparatively thin, soft, and nonwoody are called **herbs,** or **herbaceous plants.** Plants whose stems are not self-supporting, but either trail on the ground or climb by attaching to other plants or supports, are **vines.** Among vines both woody and herbaceous forms occur. Grapevines, for example, have woody stems, whereas the common morning glory, gourds, pole beans, and scarlet runner beans are herbaceous vines.

Types and Durations of Life Cycles

Annuals are plants whose entire life cycles, from germination to seed production, take place within one growing season. Among the annuals are many of the world's most useful plants, such

Figure 3-A. Expanding leaf bud (left) and mixed bud (right) of lilac. The mixed bud contains both embryonic flowers (small spherical structures) and embryonic leaves.

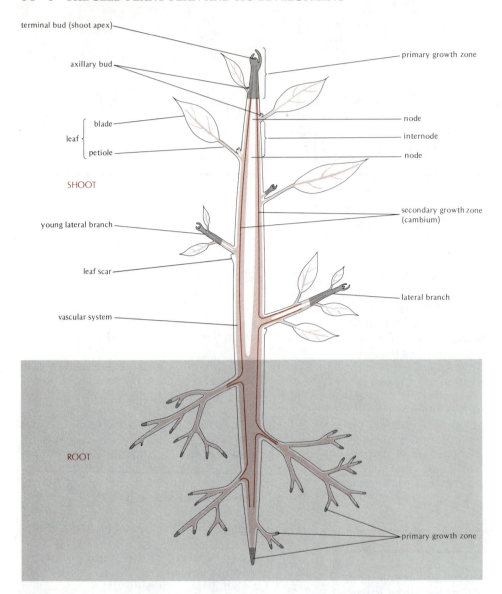

terminal bud (shoot apex)

axillary bud

leaf {
  blade
  petiole
}

SHOOT

young lateral branch

leaf scar

vascular system

ROOT

primary growth zone

node

internode

node

secondary growth zone (cambium)

lateral branch

primary growth zone

Figure 3-1. Principal organs and growth zones of a young woody plant. Stem and roots, shown as if cut in half lengthwise, are exaggerated in diameter for clarity, especially in the regions where secondary growth has occurred. Growth zones are indicated in grey, the vascular system in brown. The root system of an actual plant would be much more extensively branched at this stage of shoot development.

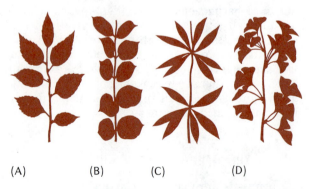

(A)   (B)   (C)   (D)

Figure 3-2. Leaf arrangements on the shoot. (A) Alternate. (B) Opposite. (C) Whorled. (D) Leaves on lateral short shoots, called "spurs" in horticultural parlance.

as the cereal grains, peas, beans and soybeans, buckwheat, flax, jute, and tobacco.

Biennials are plants that normally do not bloom until the growing season after that in which the seed germinates. In the first season the shoot of many biennials takes the form of a cluster of leaves borne at ground level by a very short stem, a form called a **rosette** (Figure 3-3A). In its second year the plant sends up an elongated stem bearing flowers and seeds (Figure 3-3B, C). Many garden food plants—beet, celery, cabbage, carrot, and turnip—are rosette biennials. Because they are harvested for food during the first growing season, however, the average gardener seldom sees them flower and mature.

Perennials live from year to year and are either woody (trees, shrubs, and woody vines) or herbaceous. The aboveground shoots of herbaceous perennials die at the end of each growing season, but the plants persist by means of underground stems or roots, from which new aerial shoots grow up each season. Herbaceous perennial food plants include potatoes, yams, and asparagus; ornamental herbaceous perennials include peony, delphinium, primula, dahlia, and lupine.

The distinction between annuals and perennials is sharper in temperate regions than in the tropics and subtropics. In temperate climates, for example, the castor bean grows as an annual, dying with the autumn frosts; in the tropics it can grow for years and develop considerable woody tissue.

Many plants experience distinct **vegetative** and **sexually reproductive** growth phases. During vegetative growth, food materials produced by photosynthesis are used to increase the size of the vegetative organs (shoot and root) or are stored for future use. In the sexually reproductive phase, photosynthetic products and/or previously stored food reserves are expended to produce flowers, fruits, and seeds. Some plants, including many annuals such as corn and garden beans, pass through a period of vigorous vegetative growth, then into a brief reproductive phase, and subsequently die. Biennials pass their first year in the vegetative phase and complete the reproductive phase in their second year.

Some perennials, such as bamboos and "century plants" (Figure 3-19), grow vegetatively for many years before suddenly flowering, fruiting, and dying. Such plants, as well as annuals and biennials, which produce but a single crop of seeds in the individual's life, are called **monocarpic.** In contrast, most shrubs, trees, and herbaceous perennials pass through a strictly vegetative or **juvenile** phase of limited duration (as much as 20 years in some trees), then continue vegetative growth along with repeated cycles of reproduction for the rest of their lives, a behavior termed **polycarpic.**

Woody Plant Forms

Large woody plants with a single erect stem are **trees;** smaller woody plants, usually with several stems emerging from the ground, are **shrubs.** Some plants that grow as shrubs in one region or location may become trees in another location or area where the climate is milder. Many woody plants, however, grow only as shrubs or only as trees wherever they are able to survive.

Trees vary in **habit**—that is, in general appearance and form of branching (Figure 3-4). Differences in the extent of lateral branch development and in the vigor of growth of a main vertical shoot or **leader** are responsible for this variation.

Trees such as ash, maple, and poplar, which lose their leaves at the end of each growing season, are **deciduous. Evergreen** trees and shrubs, in contrast, shed each year's crop of leaves gradually only after 1 or more years and hence are never barren of foliage. Although people living in

Figure 3-3. (A) Herbaceous rosette shoot. (B) Beginning to bolt. (C) Bolted and in flower. This progression occurs during the first year in many annuals. Biennials usually grow as rosettes (A) their first year, then bolt in the second. Many herbaceous perennials pass through this cycle each year, starting from an underground bud.

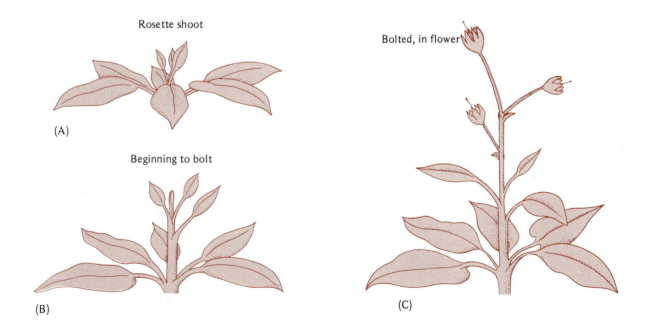

Rosette shoot

(A)

Beginning to bolt

(B)

Bolted, in flower

(C)

(A) (B) (C)

Figure 3-4. Growth habit in trees. (A) Excurrent growth of larch (*Larix laricina*). (B) Deliquescent growth of white oak (*Quercus alba*). (C) Columnar growth in the coconut palm (*Cocos nucifera*).

northern regions often use "evergreens" to mean needle-leaved trees and shrubs (conifers), many tropical and subtropical broad-leaved trees and shrubs are evergreen. For example, the species of *Magnolia* native to the southern United States, such as *Magnolia grandiflora*, are broad-leaved evergreens, whereas the hardy *Magnolia* species planted as ornamental trees or shrubs in colder regions are deciduous. On the other hand, the larch is a deciduous needle-bearing tree, in contrast to the familiar pines, firs, and hemlocks with evergreen needles. When in leaf, deciduous plants bear leaves only on the portion of each branch that was produced during the current growing season, called the **twig** (Figure 3-5A).

DEVELOPMENT OF THE SHOOT

The mode of growth of shoots is basically similar in plants as different as corn and an oak tree. It can be approached by examining a young branch, or twig, of a deciduous tree.

Growth of Woody Shoots

A twig (Figure 3-5A) bears a **terminal bud** at its tip and smaller **lateral,** or **axillary, buds** in each leaf axil. The buds of most woody plants are covered by a number of overlapping **bud scales.** These serve to protect the delicate tissues within the bud from mechanical injury and desiccation, especially during the winter. (There is no basis for the popular belief that bud scales protect

against winter cold.) When leaves fall in autumn, a **leaf scar** remains where each leaf was attached (Figure 3-5B). Under winter conditions, the axillary buds of the twig are found directly above these leaf scars.

In a terminal bud cut in half lengthwise (Figure 3-6), one can observe a number of embryonic leaves attached to a very telescoped embryonic stem tipped by a dome-shaped mass of tissue, the **shoot apex,** or **apical meristem.** The shoot apex is composed of cells capable of active division, termed meristematic cells. They initiate the tissues of the stem and also give rise to new embryonic leaves. In spring, when growth begins, the embryonic structures within a terminal bud enlarge (Figure 3-7) into a new twig similar to the one at the tip of which this bud was located. Axillary buds develop similarly into lateral twigs. In some plants all the leaves that unfold in the spring already exist within the bud in embryonic form. In others, additional embryonic leaves and internodes initiated by the shoot apex also enlarge as growth proceeds.

Shoot elongation in most temperate-zone woody plants, particularly older trees, is limited to the first weeks of the growing season, after which further activity of the shoot apex produces the structures within another terminal bud, while lateral buds develop in the leaf axils. During the next growing season, these buds grow into twigs, similarly producing lateral and terminal buds. In this manner lengthwise growth and branching of the shoot continue year by year to a potentially unlimited extent.

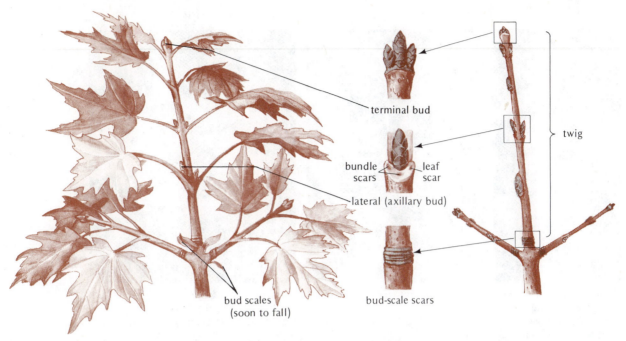

Figure 3-5. A maple twig and its parts. (Left) Summer condition. (Right) Winter condition.

As buds expand in spring their bud scales, which first spread apart (Figure 3-7), eventually drop off, leaving a ringlike set of small scars called **bud-scale scars** (Figure 3-5). Because each terminal bud leaves a set of bud-scale scars behind when it expands, one can determine the age of any part of a small branch by counting the number of sets of bud-scale scars between that part and the tip of the branch (Figure 3-8).

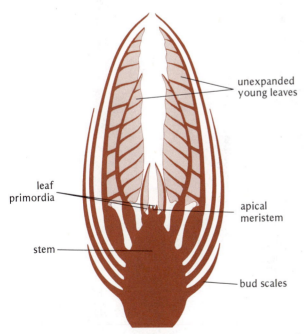

Figure 3-6. Longitudinal section through a terminal bud of buckeye (*Aesculus*).

Some trees, notably many oaks and young beech trees, depart from this growth pattern. Many of the resting terminal buds formed after the spring flush of growth break dormancy in midsummer, putting out a second flush often more elongated than the spring growth. These shoots (Figure 3-9) are traditionally called lammas shoots because they appear around Lammas, a now little-known holiday (August 1). Even less seasonally timed growth occurs in tropical trees, which often put out several growth flushes a year, at irregular intervals, alternating with periods of bud dormancy. Trees and shrubs generally will respond to defoliation or pruning back by putting out new growth from buds that would otherwise remain dormant until the next season. By this means the tree restores the proper balance between leaves and nonphotosynthetic organs that is necessary for adequate nourishment of the latter.

Short and Long Shoots

Some woody plants, such as apples, quinces, and barberries, produce two distinct types of vegetative shoots called long shoots and short shoots. Long shoots are elongated twigs like those already described, generally developing from the terminal bud of the branch. Short shoots develop from axillary buds and consist of leaves (often plus flowers) clustered on a stem that hardly elongates at all (Figure 3-23A), very similar to a rosette shoot in many herbaceous plants (Figure 3-3A). If the long shoot(s) on a given branch is removed, for example, by pruning, the terminal buds of one or more short shoots may grow out to form long shoots. This dual shoot system affords a simpler and more compact growth form than if all axillary buds develop into long shoots when they begin to grow. Understanding the two types of shoots is of practical importance when pruning apples and certain other fruit trees in which flowers are produced only by short shoots: if these are cut off or are all forced to become long shoots, the tree fails to bear.

(A)   **(B)**   **(C)**   **(D)**

Figure 3-7. Bud enlargement and shoot expansion in early spring (Norway maple, *Acer platanoides*). The changes shown take some 2–3 weeks, depending upon the warmth of the weather. The enlarged bud scales still attached at the base of each new twig in (D) will soon be shed.

Figure 3-8. Seasonal changes in an elm twig, showing expansion of flower and leaf buds. (A) Winter condition. (B) Early spring, flower buds expanded, leaf buds beginning to swell. (C) Summer condition, shoots fully expanded. (D) Autumn condition, after leaf fall, dormant winter buds fully developed. The unexpanded axillary leaf buds remaining in (C) enable replacement shoots to develop if the expanded shoots are damaged or destroyed.

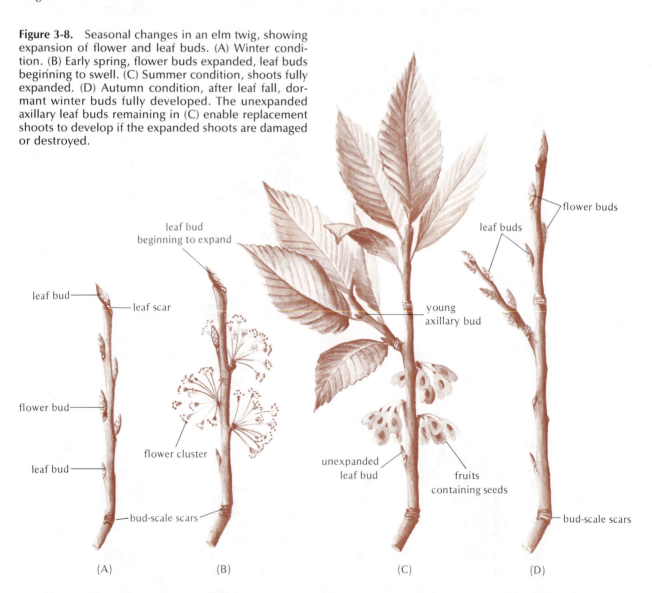

(A)   (B)   (C)   (D)

Figure 3-9. Lammas shoots on a young tree of Valley oak (*Quercus lobata*), photographed on August 1. The dark foliage is borne by shoots that expanded in spring, while the still-expanding summer (lammas) shoots have a whitish appearance at this stage.

Growth of Herbaceous Shoots

Herbaceous shoots grow similarly to woody shoots except that they form no massive terminal bud, new leaves and internodes initiated at the shoot apex expanding without delay throughout the growing season. The internodes of rosette shoots (Figure 3-3A) do not elongate, so successive leaves remain clustered together. Rosette plants include biennials, mentioned above, as well as many annuals (e.g., lettuce and radish) and herbaceous perennials (e.g., primroses). Axillary buds of herbaceous shoots may develop into lateral shoots during the season in which they are produced; however, in many cases they expand only if the main shoot tip is damaged or removed.

Flowering

Besides vegetative shoots, flowers are also initiated within terminal or lateral buds and sooner or later expand like vegetative organs do. Many woody and herbaceous perennial plants that flower in the spring initiate their flowers during the preceding growing season or during the winter, within winter buds. **Mixed buds,** as found in apple and lilac, develop (Figure 3-A) into leafy

Figure 3-10. Trunks of the Bigtree or Sierra Redwood, *Sequoiadendron giganteum,* an impressive example of the results of secondary growth. Continued production of wood and bark over the tree's life span (sometimes exceeding 2000 years) leads to the massive trunks seen in the picture. The thick fibrous bark protects the trunk against fires; both bark and wood contain copious reddish tannin-like materials (hence the name redwood), that confer resistance to invasion by decay fungi. These features may be responsible for the Sierra Redwood's unusual longevity.

shoots bearing flowers. **Flower buds,** for example, in peach, cherry, red maple, and elm (Figure 3-8), are specialized to produce only embryonic flowers. Buds that develop into strictly vegetative shoots are termed **leaf buds** (Figure 3-8).

Depending on the kind of plant, herbaceous shoots may initiate flowers or a flowering shoot (called an **inflorescence**) either in lateral or terminal buds or in buds on underground organs discussed later. Annuals often initiate a flower or inflorescence from the main shoot apex, terminating its vegetative growth and leading eventually to the death of the plant.

Secondary Growth

The stems of many plants also grow in diameter—particularly those of trees, which continue to increase in girth throughout life, some attaining impressive size (Figure 3-10). This lateral growth is called **secondary growth** to distinguish it from **primary growth,** or growth in length, considered above, which is localized at the tips of shoots (and also roots). Secondary growth takes place to a limited extent in many herbaceous stems, even in annuals. In woody plants secondary growth begins in the twig shortly after its expansion from the bud, that is, during its first season, but the consequences of secondary growth usually become apparent externally only in subsequent years as the branch or trunk thickens and develops a corky bark.

Most monocots, such as grasses, lilies, and palms, do not show secondary growth. This is why the trunk of a palm tree (Figure 3-4C) is no wider at the bottom than at the top, even though the base of the trunk is much older. In plants having secondary growth the branches of the shoot system become progressively thicker toward the base (Figure 3-4A, B) because secondary growth has been going on longer there.

Secondary growth is due mainly to the activity of a thin meristematic cell layer within the stem called the **vascular cambium,** which is located at the boundary between wood and bark, and adds new layers of conducting and supporting tissue to the stem. (Chapter 13).

Significance of the Plant Mode of Growth

The potentially unlimited growth of the shoot (and similarly of the root) in both extent and number of parts is one of the most striking contrasts between vascular plants and vertebrate animals. An animal undergoes a period of development as an embryo, during which its basic organs and tissues form. Subsequently, after birth or hatching, growth continues as an enlargement and elaboration of the structures formed in the embryo, and usually to a strictly limited ultimate size. A plant, too, begins its life with a period of early embryonic development, for example, during the formation of the seed within which an embryo seed plant develops. This simple embryo (Figure 3-34), in contrast to that of an animal, possesses almost none of the organs or tissues of the mature plant; these features instead arise progressively, after germination, by the continued embryonic activity of the shoot and root apices. Development continues indefinitely, subject to limitations imposed by the environment, and to internal limitations in the case of annuals, biennials, and herbaceous perennials. This mode of development provides constant renewal and replacement of the functional tissues and organs of the plant, allowing for the unlimited longevity of perennial plants as compared with animals. The capacity for continued growth also substitutes, to some extent, for a plant's lack of mobility. For example, although a plant cannot roam in search of light, water, and mineral nutrients, its ever-expanding root system can find and exploit untapped regions of the soil by growing into them, and the shoot can grow toward favorable light conditions.

Figure 3-11. Types of simple leaves. (A) Unlobed, pinnately veined (apple). (B) Pinnately lobed (black oak). (C) Palmately veined and palmately lobed (maple). (D) Parallel-veined (grass; the leaf sheath, or basal part of the leaf encircling the stem, is also characteristic of grass leaves).

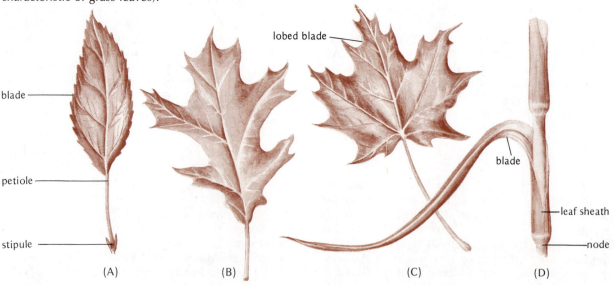

(A) (B) (C) (D)

THE LEAF

The growth of leaves, in contrast to that of stems, is usually strictly limited, or **determinate.** A young leaf soon grows to mature size and stops growing, usually functioning for only one or a few seasons and then falling away.

Leaf Structure

The typical leaf has two parts: a thin, flat, expanded photosynthetic portion, the **blade;** and (if present) a stalk, or **petiole** (Figure 3-11). A pair of leaflike or scalelike structures, called **stipules,** sometimes occurs at the base of the petiole and is considered a part of the leaf. Stipules are usually small, but in some plants such as the garden pea (Figure 3-20B), they are large and supplement the leaf as photosynthetic organs. Some stipules drop as the leaves expand; others persist throughout the growing season.

Veins appear in the blade as lines or ridges. Veins are composed of conducting (vascular) and supporting tissue. The venation, or arrangement of veins, distinguishes the two great subdivisions of flowering plants. In monocotyledons, the chief veins are parallel or nearly so. This arrangement is termed **parallel venation** (Figure 3-11D). The veins form a netlike pattern in leaves of dicotyledons, such as maple, geranium, apple, oak, and sunflower, which have **net venation.** In most net-veined leaves a single, strong vein, the midrib, runs the length of the leaf, with the main lateral veins arranged in a featherlike pattern, called **pin-**nate venation (Figure 3-11A; Latin *pinna,* feather). Leaf blades having several equally strong veins from the base, diverging like the fingers from the palm of the hand, are termed **palmately veined** (Figure 3-11C).

Simple and Compound Leaves

The blade of a **simple leaf** (Figure 3-11) is undivided, although the margin may be indented in various ways. In a **compound leaf** (Figure 3-12) the blade is divided into parts, termed leaflets. Palmately compound leaves have leaflets attached at one point, the tip of the petiole, as in horse chestnut, red clover, and Virginia creeper (Figure 3-12A). Pinnately compound leaves have leaflets arranged along a central axis, like a feather, as in ash (Figure 3-12B), potato, tree of heaven (*Ailanthus*), and walnut. To distinguish a leaflet of a compound leaf from a simple leaf, one can determine the location of axillary buds or lateral branches that have grown out from them. Axillary buds occur at the base of the leaf, whether it is simple or compound, and not out on the leaf itself, such as where leaflets of a pinnately compound leaf are attached to the leaf axis (the **rachis,** a prolongation of the petiole).

THE ROOT

A plant's original or primary root system arises from the lower end (called the **radicle;** Figure 3-34) of the embryo during germination of the seed. Near its tip the root has an apical meristem in which cell division provides new cells for growth of the young root. A tough cell layer called

Figure 3-12. Compound leaves. (A) Palmately compound leaf of Virginia creeper (*Parthenocissus quinquefolia*). (B) Pinnately compound leaf of ash (*Fraxinus americana*).

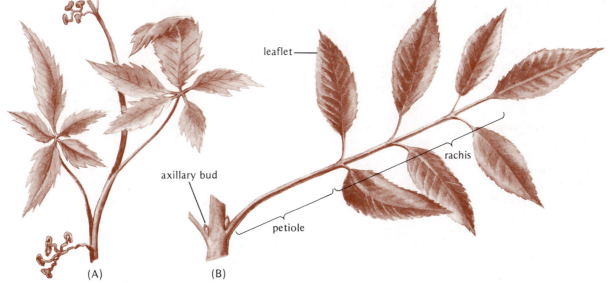

leaflet

rachis

axillary bud

petiole

(A)   (B)

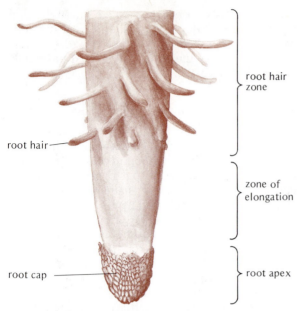

Figure 3-13. Magnified external view of a root tip, showing major zones. Root hairs arise as outgrowths of surface cells of the root, and serve to improve water and nutrient uptake by the root (Chapter 11).

the **root cap** (Figure 3-13) covers the apical meristem of a root, protecting it from mechanical damage as the root forces its way through the soil. Lateral or branch roots arise and grow out in the region behind the root tip.

If the primary root continues to grow and becomes the plant's chief root system, it is called a **tap root** (Figure 3-14A), as is found in many herbaceous plants such as dandelion, carrot, and alfalfa. A number of forest trees, such as oak, hickory, and the conifers, also have tap roots, at least when young. In other species, including many herbaceous perennials and especially grasses, the growth of the primary root is soon equaled or surpassed by the growth of its branches, or of adventitious roots (see below), resulting in a **fibrous root system** (Figure 3-14B).

ADVENTITIOUS ORGANS

The basic plant development described above is modified in many species by the formation of **adventitious organs,** structures produced in an unusual position or at an unusual time in the plant's development. Adventitious buds are shoots that arise on stems at locations other than shoot tips or leaf axils, that is, from neither terminal nor axillary buds. Adventitious buds can also appear on roots. Adventitious roots may arise on stems as a result of wounding, or where a stem touches the soil, or on the cut ends of shoots placed in soil (cuttings).

Aerial Roots

Many kinds of plants produce adventitious roots on their stems above ground level, serving the plants in various ways. Many tropical plants produce thickened **aerial roots** that absorb and probably store dew and rainwater. In a Malayan orchid found on tree bark, the stem is only 2–4 cm (an inch or so) long, the leaves are reduced to brownish scales, and the roots are green and constitute the only photosynthetic organs of the plant. Many vines form adventitious roots that anchor or attach the climbing stems, for example, poison ivy, English ivy (Figure 3-15), and the trumpet vine (*Campsis radicans*). Some plants form on their stems adventitious **prop roots,** which contribute support in addition to that provided by the roots originating underground. Corn produces adventitious roots from several nodes above the soil (Figure 3-16), comprising a sizable

(A)   (B)

Figure 3-14. Types of root systems. (A) Taproot system (dandelion). (B) Fibrous root system (grass).

Figure 3-15. Aerial adventitious roots of English ivy (*Hedera helix*).

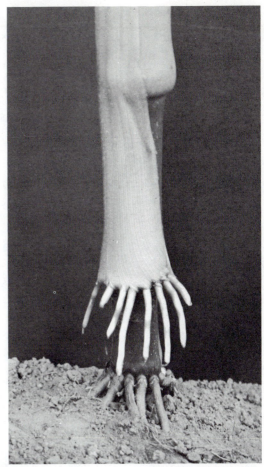

Figure 3-16. Prop roots of corn, arising adventitiously at basal nodes of the stem. The upper set will eventually reach the ground and help hold the increasingly bulky stalk erect.

proportion of its total root system at maturity. Conspicuously long and large prop roots are formed by various tropical trees such as certain palms, mangroves, and the banyan tree (Figure 3-17). The mangroves, which grow along tropical coasts and tidal river banks, produce stilt-like prop roots from both the main trunk and the branches. These roots may form almost impenetrable thickets.

Adventitious Shoots

Canada thistle (Figure 3-18), bindweed, milkweed, and many other perennial weeds multiply by forming adventitious buds on their roots. Aerial stems also frequently arise from the horizontal roots of woody plants such as wild plums, Osage orange, poplars, sumac, elderberry, and lilac, leading to thickets or dense groves of the shrub or tree in question.

Another important role of adventitious bud formation is in enabling an injured plant to **regenerate**—that is, restore, or partly restore—lost parts. For example, a portion of a dandelion tap root left in the soil following attempts to remove the plant by hand digging forms adventitious buds

that shortly produce new leaves above the soil. Many broad-leaved trees will form adventitious buds in the bark of the trunk or branches below a point of breakage or fire injury and, after cutting, will produce adventitious sprouts from the stump. This can be important in the regeneration of forest in cut-over and burned areas.

PLANT ORGAN MODIFICATIONS

In many plants, the roots, stems, or leaves have become modified into special organs that contribute to the plants' ability to survive and to compete successfully by functioning in ways unusual for roots, stems, or leaves. Modifications for food or water storage, or for overwintering or against other adverse environmental conditions, are common examples. In other plants, organ modifications provide a means of vegetative reproduction—the multiplication of a plant from vegetative organs as opposed to seeds or other sexually produced reproductive structures.

Figure 3-17. Prop roots of the banyan tree (*Ficus bengalensis*) of India.

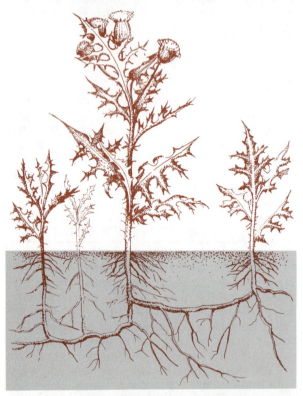

Figure 3-18. Adventitious (or "sucker") shoots arise from adventitious buds on a horizontal root (Canada thistle, *Cirsium arvense*).

Figure 3-19. Century plant (*Agave americana*, photographed in Texas) with a rosette of large succulent water-storing leaves. *Agave* grows vegetatively (plant on the left) for a decade or so but not as long as a century, contrary to the popular myth expressed in the plant's common name. It then bolts, flowers, sheds its seeds and dies (plant on the right in photo), exemplifying a monocarpic perennial growth cycle.

Figure 3-21. Branchlets of "ground pine" (*Lycopodium complanatum*) showing photosynthetic scale leaves.

Leaf Modifications

Thickened and **fleshy leaves** in which water can be stored are common, especially in plants of arid and semiarid regions. Such plants, cultivated in rock gardens and greenhouses as ornamentals, are called **succulents.** (Succulents also include plants that store water in a modified stem rather than in leaves; see below.) Among the leaf succulents are the century plant (Figure 3-19), ice plant, jade plant, and hens-and-chickens, often grown as houseplants.

Tendrils are organs that hold up a vine by coiling around a support. The leaf petiole, the entire leaf, leaflets of a compound leaf, or even stipules may be modified as tendrils (Figure 3-20). Leaf parts transformed into tendrils become long and slender, lacking a photosynthetic blade, and they respond to contact with a foreign object by coiling. Some tendrils are not modified leaves but modified branches (Figure 3-20D).

Scale leaves are much-reduced leaves lacking an expanded blade (Figure 3-21). Some conifers such as juniper and cypress have green, photosynthetic scale leaves, interpreted as a modification reducing the amount of leaf surface for evaporation. Pines, spruces, firs, and most other conifers of course have needle-like photosynthetic leaves, a less extreme modification. Similar modifications are encountered among flowering plants adapted to dry environments such as deserts (Chapter 20).

Scale leaves on other plants are often nonphotosynthetic and serve a protective function. Examples are bud scales (Figure 3-6) and scale leaves that occur at the nodes of underground shoots such as asparagus spears (Figure 3-22A), or rhizomes and tubers. The small unspecialized scale leaves of many underground shoots may be regarded as merely a suppression of leaf development to avoid the production of expanded leaves, which are useless underground and, by encumbering the shoot, would hold back its growth through the soil.

Figure 3-20. Tendrils of various types. (A) Entire leaf (yellow vetchling, *Lathyrus aphaca*): the stipules are the photosynthetic organs of this plant. (B) Distal leaflets of a pinnately compound leaf are modified as tendrils (pea). (C) Leaf petioles act as tendrils (*Clematis*). (D) Branch system serving as a tendril (grapevine, *Vitis vinifera*). Figure 3-12(A) shows still another type of tendril, a modified adventitious root.

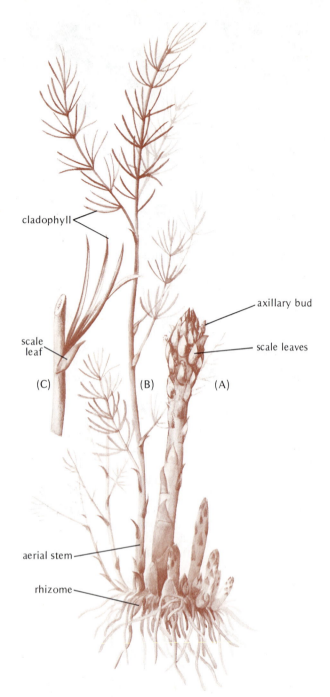

cladophyll

axillary bud

scale
leaf

scale leaves

(C)   (B)   (A)

aerial stem

rhizome

Figure 3-22. Asparagus shoots at different stages of development. (A) "Spears" emerging from below ground, bearing protective scale leaves. (B) Portion of fully developed branch, bearing reduced scale leaves and photosynthetic cladophylls (leaf-like green branches which are doubly compound). (C) Much magnified view of part of a branchlet with several of the needle-like cladophylls in the axil of a scale leaf.

Bulb scales are fleshy scale leaves that form part of the compact type of underground shoot known as a bulb, such as the onion. Bulb scales are storage organs, accumulating sugar or starch for the future use of the plant.

Spines, thorns, and prickles discourage animals and humans from molesting plants possess-

ing them. **Spines** are modified leaves. Spines of barberry (Figure 3-23A), gooseberry, and cacti are modified entire leaves. Leaf parts, such as leaflets or stipules, can be modified into spines, as in the black locust and in some species of *Acacia*, a woody legume. The spines of acacias of Central America not only protect the plants against foraging by herbivorous animals, but also provide living quarters for certain ants which protect the plants from insect pests (Figure 3-23B).

The thorny or spine-like structures of the rose and blackberry, technically termed **prickles,** are merely outgrowths of superficial stem tissues. Although their function is the same, prickles are morphologically quite different from spines or from thorns, which are modified branches (see below).

Aboveground Stem Modifications

Thorns (Figure 3-23C) are sharp-pointed defense structures that arise from a leaf axil or at a branch tip, often bear scale leaves at nodes, and therefore are modified branches.

Although most young stems are green and photosynthesize to some extent, the stems of some plants are specially modified to replace the leaves as the plants' photosynthetic organs. **Cladophylls** (Greek *klados*, shoot; *phyllon*, leaf) or cladodes are green branches that develop a leaf-like form. Asparagus branches (Figure 3-22) illustrate cladophylls developing in the axils of scale leaves on a shoot that lacks photosynthetic leaves. Some asparagus cladophylls are needle-like rather than flattened like ordinary leaves. Asparagus branch systems with their cladophylls superficially resemble highly divided compound fern-like leaves and are used decoratively under the name "asparagus fern."

Photosynthesis is largely or entirely restricted to the stems of various succulents having reduced leaves or none at all, such as cacti and similar plants (Figure 3-24). Their fleshy, water-storing stems enable **stem succulents** to live for lengthy periods without an external water source. The giant saguaro (*Carnegiea gigantea*), a large, tree-like cactus of the deserts of Arizona and northern Mexico (Figure 3-25), reportedly stores enough water in its stem to last for several years.

Underground Stem Modifications

A **rhizome** (from Greek *rhiza*, root) (Figures 3-22, 3-26) is a stem that grows underground, usually horizontally, in a more or less elongated manner (in contrast to the very compact form of tubers, corms, and bulbs, discussed below). Rhizomes are often somewhat thickened and fleshy for the storage of food reserves, and they usually produce adventitious roots at intervals along their

(A) (B) axillary bud (C)

Figure 3-23. Spines and thorns. (A) Spine of Japanese barberry (*Berberis thunbergii*), a modified leaf (actually, the midrib of the original leaf) in the axil of which a short shoot develops. (B) Spines of *Acacia nicoyensis* (from Costa Rica) which are modified stipules inhabited by ants; the holes drilled by the ants lead into their nests within each thorn. Note the *Acacia* leaf, which is *doubly compound*: the leaflets emanating from the rachis are themselves pinnately compound. (C) Thorns of hawthorn (*Crategus*), which are short lateral shoots that develop from axillary buds of the previous season's growth.

Figure 3-24. Various cacti, examples of stem succulents.

Figure 3-25. The saguaro cactus (*Carnegiea gigantea*), one of the largest stem succulents, in its native habitat (Arizona).

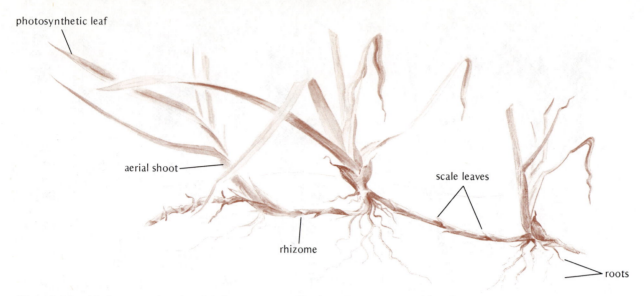

Figure 3-26. Underground and aerial shoot system of sedge (*Carex*), a grass-like plant. The rhizome bears scale leaves, and gives rise to leafy aerial shoots, each of which after one or two seasons dies and is replaced by new ones that have meanwhile developed from further extensions of the rhizome.

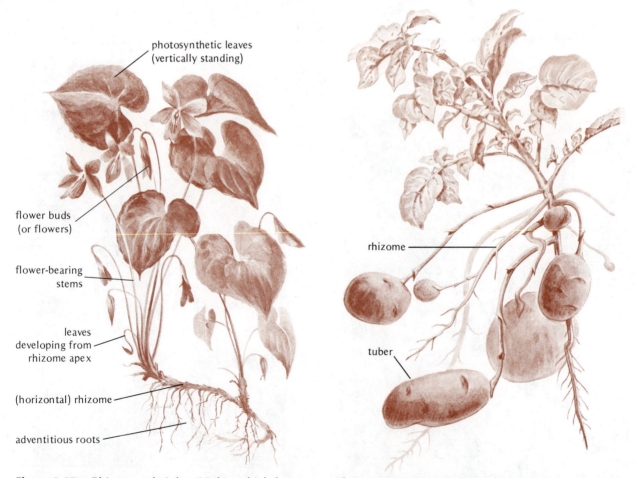

Figure 3-27. Rhizome of violet (*Viola*), which bears photosynthetic leaves directly, and gives rise to flowering stems.

Figure 3-28. Potato plant, showing rhizomes arising from axillary buds near the base of the stem, and developing tubers at their tips.

lengths. Rhizomes can be recognized as stems rather than roots because they possess nodes (bearing leaves or scale leaves) and internodes, whereas roots do not.

The rhizome comprises the permanent axis of many herbaceous perennial plants, whose aerial shoots are only short-lived as noted earlier. It is the rhizome that overwinters, allowing the plant to survive and grow year after year. Such plants commonly multiply by branching of their rhizomes, so the rhizome is also a means of vegetative reproduction. Some plants with more permanent aerial shoots develop very elongated rhizomes, sometimes called **stolons**, serving specifically for vegetative reproduction, as considered further in Chapter 15.

The type of rhizome found in irises, water lilies, violets (Figure 3-27), and many orchids bears the plants' photosynthetic leaves; such plants may form aerial shoots only to produce flowers. Many ferns grow by similar leaf-bearing rhizomes, which produce no aerial shoots at all (ferns being nonflowering plants as noted in Chapter 2). Rhizomes of other herbaceous perennials bear, instead, scale leaves at their nodes and give rise every year to leafy aerial shoots from terminal or axillary buds borne by the rhizome (Figure 3-26).

A **tuber** is a short, much-enlarged portion of stem, containing stored food reserves. The most familiar example is the potato tuber, which arises underground as an enlargement of the tip of a rhizome or stolon (Figure 3-28). The surface of the tuber is marked at one end by a scar, which was the point of attachment to the stolon, and by scattered "eyes." Each eye is a node bearing a minute scale leaf, in the axil of which are several (usually three) small buds. The portions of the tuber between consecutive eyes correspond to its internodes. An aboveground stem tuber is kohlrabi

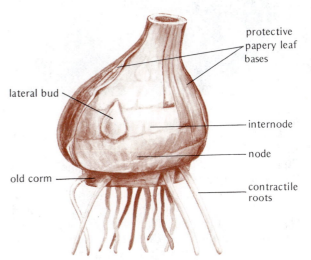

Figure 3-30. Corm of *Gladiolus.*

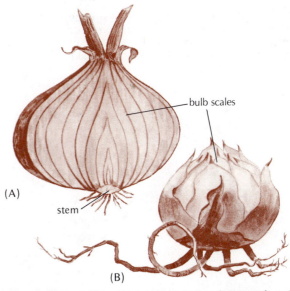

Figure 3-31. Bulbs. (A) Layered bulb of onion (longitudinal section). (B) Scaly bulb of lily. The bulb scales of onion are actually the succulent *bases* of leaves, while in lily they are succulent scale leaves.

Figure 3-29. Kohlrabi plant, showing the aboveground tuberous stem that is used as a green vegetable.

Figure 3-32. Root system of sweet potato (*Ipomea batatas*) showing storage roots used as a staple vegetable.

(Figure 3-29), one of the many edible members of the cabbage group of vegetables.

A **corm** is the underground tuber-like base of the vertical stem of plants such as crocus and gladiolus (Figure 3-30). This enlarged food-storing structure possesses nodes and internodes, but does not bear fleshy scale leaves like a bulb (see below), with which people frequently confuse it.

The corm serves for overwintering and lasts but a single growing season. In gladiolus, for example (Figure 3-30), as growth begins in the spring, one or more buds on the corm develop into shoots bearing leaves and flowers. The food used in early shoot growth comes from the corm, which shrivels as growth proceeds. After the leaves mature one or more new corms form by lateral enlargement of the basal part of the new stem, just above the old, exhausted corm. The large corms of the taro and dasheen plants are extensively used as a starchy food in parts of the tropics.

A **bulb** is a very short underground stem bearing thickened, fleshy bulb scales, which are food-storing modified leaves as discussed earlier. Onion and hyacinth bulbs are said to be layered (Figure 3-31A), because the scales wrap around the bulb and form a series of rings visible when the bulbs are cut across. Most lilies have scaly bulbs, in which the scales do not encircle the bulb, but are small, fleshy, and rather loosely attached (Figure 3-31B).

Most bulbs multiply vegetatively by forming smaller bulbs (bulblets) in the axils of the outer scales. When mature, these are known as offsets and can be removed and planted to grow into flowering-sized bulbs.

Root Modifications

The roots of biennial and perennial plants frequently accumulate foods and, especially in cultivated varieties, become much enlarged. In the turnip, parsnip, carrot, and sugar beet, the tap root becomes fleshy by secondary growth. Other important food plants that have edible **storage roots** are the yam (*Dioscorea*) and the sweet potato

Figure 3-33. Aerating roots of an Indonesian mangrove, *Sonneratia*. The tree grows along seacoasts in the intertidal zone, rooted in water-logged, oxygen-depleted mud (A). Specialized lateral roots, growing vertically upward into the air, possess surface openings near the tip (B), allowing uptake of oxygen for the root system below.

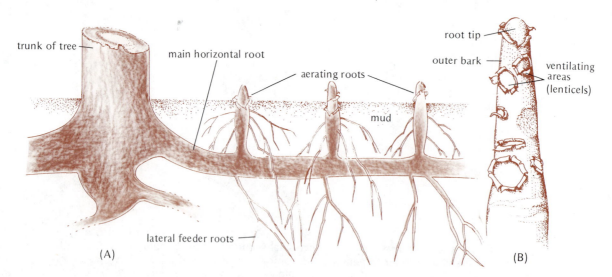

trunk of tree

main horizontal root

aerating roots

mud

lateral feeder roots

(A)

root tip

outer bark

ventilating areas (lenticels)

(B)

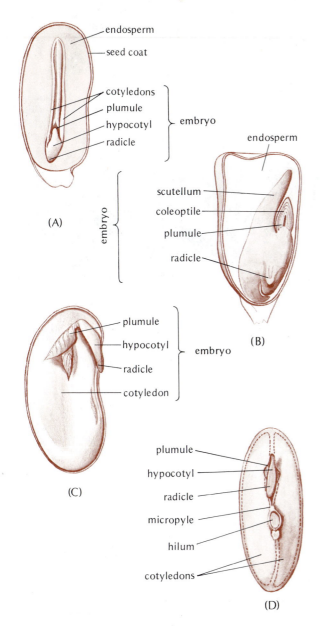

Figure 3-34. Example of seed structure. (A) Castor bean, a dicot possessing endosperm. (B) Corn, a monocot with endosperm. (C, D) Common bean (*Phaseolus vulgaris*), a dicot with no endosperm, its food reserves stored instead in the embryo's thick fleshy cotyledons (C, internal side view with one of the cotyledons removed; D, external view, edge on, showing by dotted lines the location of cotyledons and embryo axis within seed.)

[*Ipomoea batatas* (Figure 3-32), unrelated to the white, or Irish, potato, the edible part of which is a modified stem as noted above].

Other kinds of modified roots mentioned above are the thickened and sometimes photosynthetic roots of epiphytes (plants that grow on the trunks or branches of trees, especially in the tropics) and the adventitious roots of vines, specialized for attaching the stem to supports. **Aerating roots** grow up from buried or submerged roots of various swamp plants such as some of the mangroves (Figure 3-33) and serve for the uptake of oxygen from the air by the root system.

Contractile roots are remarkable roots produced by bulb and corm plants, which help keep these modified stems in position underground. Each year upward growth in the formation of a new corm or bulb tends to bring it closer to the soil surface (Figure 3-30). However, adventitious contractile roots grow down from the base of the new corm or bulb. Once each root is anchored in the soil below, its upper part (near the bulb or corm) begins to contract lengthwise, pulling the bulb or corm gradually downward toward its former position. Contractions as great as 50% (i.e., to one-half its initial length) have been recorded for some contractile roots, which expand substantially in width in the peculiar kind of growth process that causes the lengthwise contraction. The primary root of various seedlings, such as carrot, also contracts to some extent shortly after germination, a behavior that helps position the seedling shoot advantageously for its further development.

THE SEED

The diverse structures that a mature seed plant may possess all trace back, developmentally, to the seed. To understand plant growth more fully, we therefore need to examine the seed and its germination. As previously noted, the seed contains a relatively rudimentary embryo plant (Figure 3-34). The embryo consists of a short root or **radicle,** connected via a short length of stem or **hypocotyl** to a shoot tip at which are located one or more "seed leaves" or **cotyledons** plus a shoot apex or minute bud called the **plumule.** The plumule is located between the cotyledons if there are more than one (2 in dicots, up to about 10 in gymnosperms, only 1 in monocots).

Since the embryo is nonphotosynthetic, the seed must contain food reserves such as starch, protein, and fat for the embryo to use in its early growth or **germination.** These reserves, which make up most of the food value of important human food sources such as cereal grains and bean seeds, are stored (depending upon the seed) either in the cotyledons or in a special storage tissue called the **endosperm,** in which the embryo is embedded. In the former case, of which beans or peas are an example, the cotyledons are usually thick and fleshy, and occupy most of the seed's volume (Figure 3-34C, D). On the other hand, if a storage endosperm is present, it may make up much of the seed's volume and the cotyledon(s) will usually be relatively small or thin (Figure 3-34A), as with a cereal grain such as wheat or corn (Figure 3-34B). Hard, protective **seed coats** enclose the embryo, plus endosperm if present, to complete the seed.

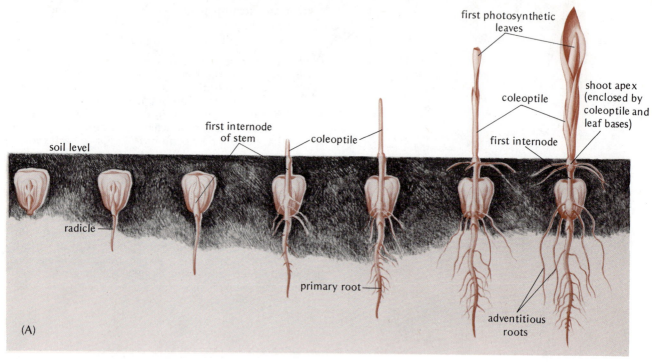

Figure 3-35. Hypogeous seed germination. (A) Corn. (B) Pea.

Figure 3-36. Epigeous germination in squash (*Cucurbita*). Note that the squash seedling develops at the base of the hypocotyl a peculiar spur (probably a modified root) that holds the seed coats, helping the hypocotyl hook pull the cotyledons and shoot tip out of the seed coats leaving the latter below ground. Most epigeously germinating seeds do not have this feature.

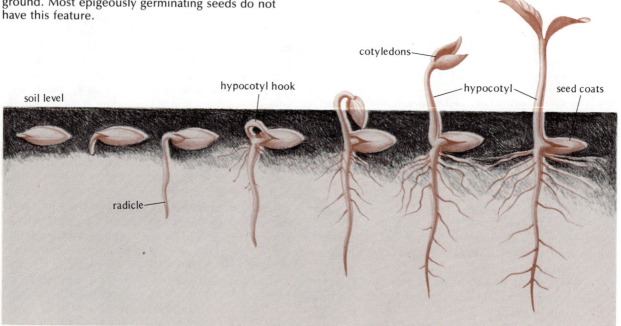

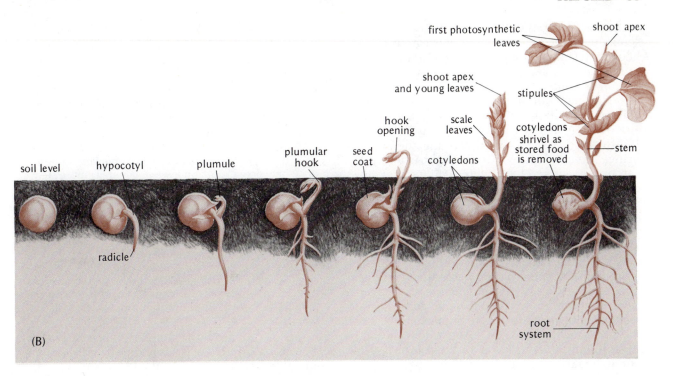

first photosynthetic leaves
shoot apex
shoot apex and young leaves
stipules
hook opening
scale leaves
cotyledons shrivel as stored food is removed
stem
plumule
plumular hook
seed coat
cotyledons
soil level
hypocotyl
radicle
root system

(B)

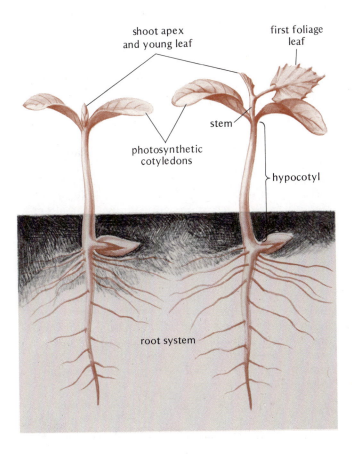

shoot apex and young leaf
first foliage leaf
stem
photosynthetic cotyledons
hypocotyl
root system

Germination

After a seed is exposed to a moist environment and a favorable temperature, the first visible sign of germination is elongation of the radicle or root, pushing its way out of the seed through the seed coats. The young root quickly grows downward and begins to form branch roots, anchoring itself in the soil. The shoot then begins to emerge from the seed and grow upward toward the soil surface in different ways depending upon the kind of plant.

In seeds such as corn and pea, which undergo **hypogeous** germination (Greek *hypo*, under, and *geo*-, earth), the young shoot or plumule above the cotyledons simply grows out and upward from within the seed coats, leaving the cotyledons where they were (Figure 3-35). Beans, squash, and many other seeds undergo **epigeous** germination (Greek *epi*, over, and *geo*-, earth), wherein the cotyledons are brought aboveground. Usually the hypocotyl first elongates, doubling up as it does so and emerging from within the seed coats as an elbow, the **hypocotyl hook** (Figure 3-36). This hook elongates further and literally elbows its way upward through the ground, pulling the cotyledons and often the enclosing seed coats after it. In this case the first structure to appear at the surface is the bend of the hypocotyl hook, which soon after pulls the cotyledons aboveground. The hook then straightens out, and the plumule grows forth from between cotyledons (or beside the cotyledon in a monocot) (Figure 3-36).

The behavior in epigeous germination enables the seedling to bring its bulky cotyledons and delicate shoot tip aboveground without having to push them forward through the soil. It brings the cotyledons into the light where they can turn green and begin photosynthesis, as they do in most plants showing epigeous germination (the common bean, *Phaseolus vulgaris*, is an exception; its cotyledons just shrivel and drop once emptied of their stored food materials). Seeds having hypogeous germination use their cotyledons only as storage organs or for absorbing stored material from the endosperm; the seedling begins to photosynthesize only when green leaves expand from the plumule. Hypogeously germinating seeds do, however, have adaptations to cope with the mechanical problem of raising a delicate shoot tip through the soil. Many of them form a **plumular hook** (Figure 3-35B) which functions like the hypocotyl hook of an epigeously germinating seed but involves the true stem rather than the hypocotyl and pulls the young foliage leaves and shoot apex, but not the cotyledons, upward. Grass seeds such as corn (Figure 3-35A), on the other hand, form no hook-like organ but possess a hol-low, pointed, modified leaf called the **coleoptile** (Greek *koleos*, sheath) which encloses the rest of the plumule and functions by elongating to penetrate through the soil spearwise. Soon after it emerges into the light, the coleoptile stops growing, its tip splits open, and the first foliage leaves of the grass grow upward into the air through the passageway that the coleoptile has created.

The hypocotyl is anatomically somewhat intermediate between a root and a stem, generally more stem-like than root-like. However, in plants such as carrot and radish that form by secondary growth a thickened storage "root" below a rosette shoot, the thickening involves the hypocotyl of the seedling as well as, usually, the upper part of the young tap root proper. Secondary growth changes the internal structures of roots and stems to resemble one another, so that by the time a radish or carrot has developed, it is hardly possible to discern which part of it was derived from the hypocotyl and which from the root proper.

At germination the general shape and size of a seedling do not reveal whether it will become a mighty tree or a delicate flower, a rank weed or a fruitful food plant. All of this depends on its postgermination growth and development, covered earlier in this chapter, and, in the last analysis, rests on the plant's hereditary makeup or genes and how they specify its development. These topics are considered in later chapters.

SUMMARY

(1) The shoot and root increase in length at their tips by primary growth, involving cell division and organ initiation at an apical meristem; dicot and gymnosperm stems and roots increase in diameter due to cell division in a cambium (secondary growth).

(2) The stem consists of nodes, bearing leaves, and internodes. Leaves are alternate, opposite, whorled, or in a rosette arrangement and may be simple or compound, consisting of blade (or leaflets), petiole, and (if present) stipules.

(3) Plants are herbaceous or woody, depending upon the extent and woodiness of their secondary growth. Depending on stature, plants are herbs, shrubs, trees, and vines; shrubs and trees are woody plants, while vines can be either herbaceous or woody.

(4) Annual plants live one growing season; biennials, two; and perennials, more than two and often indefinitely. Annuals and biennials produce a single crop of seeds at the end of their life spans, as do monocarpic perennials. Polycarpic perennials produce seeds repeatedly, year after year, following a nonreproductive juvenile phase. Herbaceous perennials normally die down to underground organs every winter. Woody perennials are either evergreen or deciduous.

(5) Shoots grow from buds, condensed embryonic shoots that may give rise to either vegetative shoots, flowers, or both (mixed buds). Lateral branches normally arise from axillary buds. Many of the organs for a woody shoot's seasonal growth flush are intiated within winter buds during the preceding season.

(6) Roots do not possess nodes and internodes. The root apical meristem is covered by a root cap. According to the root's pattern of branching, a tap root or fibrous root system is distinguished.

(7) Adventitious roots may arise anywhere on the shoot, but are commonly formed by underground stems. Aerial adventitious roots arise on the shoot for anchorage of climbing plants and as prop roots. Adventitious buds arise on roots and on older parts of the stem for shoot regeneration after injury or loss and for vegetative reproduction.

(8) Leaves are modified in various plants as scale leaves (including bud scales), fleshy water-storing structures (succulent leaves), tendrils, and spines.

(9) Stems are modified in various plants as thorns, photosynthetic and water-storing structures (succulent stems), and tendrils; and as underground rhizomes, tubers, corms, and bulbs for food storage and overwintering of perennial herbs, and also for vegetative reproduction.

(10) Roots are modified for food storage and for other specialized functions such as aeration of a submerged root system.

(11) The seed is enclosed by seed coats and contains an embryo consisting of root (radicle), seed leaf or leaves (cotyledons), shoot tip (plumule), and a section of stem (hypocotyl) between the cotyledon(s) and the root. In addition, the seed often contains a food storage tissue (endosperm); if not, then its food is stored in its cotyledons.

(12) Seed germination is either epigeous or hypogeous, according to whether or not the cotyledons are brought aboveground. Most seedlings pull their shoot tip, young leaves, and cotyledons (if epigeous) through the soil to the surface by developing a hook-shaped hypocotyl or plumule. In epigeous seedlings the cotyledons often become the plants' first photosynthetic organs, whereas in hypogeous seedlings they serve only for food storage and/or absorption from the endosperm.

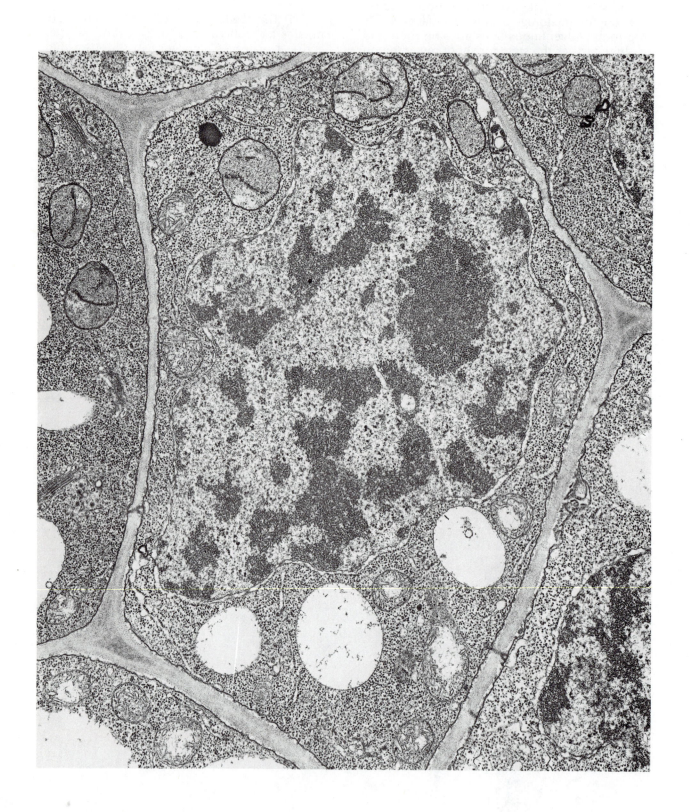

Chapter 4

ORGANIZATION AND FUNCTION OF THE CELL

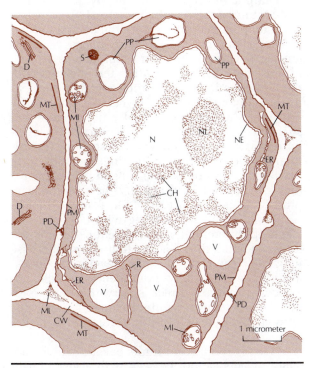

Figure 4-A and B. Cell from region of cell division in embryonic pea stem (14,500 ×). CH, chromatin; CW, cell wall; D, dictyosome (Golgi body); ER, endoplasmic reticulum; MI, mitochondrion; ML, middle lamella; MT, microtubule; N, nucleus; NE, nuclear envelope; NL, nucleolus; PD, plasmodesma; PM, plasma membrane; PP, proplastid; R, ribosomes; S, spherosome (lipid body or liposome); V, vacuole.

The cell is the fundamental organizational unit of plant and animal bodies, not only because distinguishable cells compose almost all tissues and organs but also because the cell is the smallest living entity that can autonomously maintain itself, grow, and reproduce. The cell is therefore the unit element of biological survival, growth and reproduction. These generalizations comprise the **cell theory,** which reduces to a common denominator the diverse and seemingly disparate structures and functions of animals, plants, and microbes. This universality implies that the elementary principles underlying and explaining life can be discovered by seeking to understand the functions of cells.

Cells were first seen, and named, by the English scientist Robert Hooke in 1664 while examining thin slices of common cork (the outer bark of the cork oak tree; Figure 4-1) and of various green vegetables under his microscope (Figure 1-1). The cell consists of living material or **protoplasm** enclosed within a **membrane,** a very thin boundary restricting the exchange of materials between the cell and its surroundings and enabling the cell to maintain a composition different from that of its environment. As noted in Chapter 2, plant cells are usually sheathed by a jacket of solid material, called the **cell wall,** located outside the cell membrane.

Protoplasm is composed primarily of proteins, dissolved or dispersed in water. Proteins are very large molecules, thousands of times larger than water (H_2O) molecules. Protoplasmic proteins are responsible for the biochemical reactions or **metabolism** of cells, on which the basic cellular functions of maintenance, growth, and division depend.

Within the protoplasm of many cells is a region called the **nucleus** (plural, nuclei) containing deoxyribonucleic acid (DNA), hereditary material that codes for or specifies a cell's components. Not all cells contain a discernible nucleus, however.

Prokaryotic cells lack a membrane-bounded nucleus (Figure 4-2). Their DNA is localized in a relatively ill-defined region called a nucleoid or

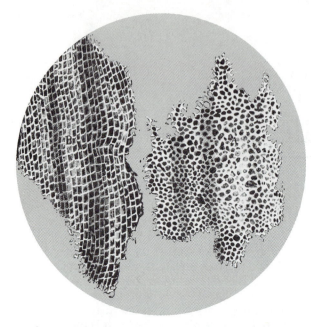

Figure 4-1.   Robert Hooke's drawings of cork cells (from his book, *Micrographia,* 1664).

nuclear region. Bacteria and blue-green algae, or cyanobacteria, are prokaryotic cells.

Eukaryotic cells contain a well-defined nucleus (Figure 4-3), delimited by its own bounding membrane, the **nuclear envelope** (Figure 4-A). Eukaryotic cells also possess a number of other intracellular membranous structures (Figure 4-A) called membranous organelles, which are largely or entirely lacking in prokaryotic cells. Plants, animals, and most microorganisms other than bacteria are eukaryotic. Eukaryotic cells differ from prokaryotic cells in their manner of division during cell reproduction and in some basic features of their metabolic machinery.

Cellular structure is not quite universal among living organisms. In certain algae and in many fungi, for example, the protoplasm, containing numerous nuclei, extends uninterruptedly throughout the plant body without being divided into separate cells. This condition is called coe-

nocytic. The body of most slime molds is simply a mass of protoplasm called a plasmodium containing many nuclei. In these exceptions to the basic cellular organization, the protoplasm contains organelles and nuclei closely resembling those of ordinary eukaryotic cells. These organisms are therefore probably evolutionary specializations in which the usual division into cellular units has been abandoned.

OBSERVING CELL STRUCTURE

The development of the cell theory depended on the invention and continual improvement of the **compound microscope,** the now familiar laboratory microscope. Utilizing at least two separate lens components (Figure 4-4), the compound microscope can produce a clear image at much greater magnifications than a simple lens or magnifying glass. Fully exploiting the advantages of the compound microscope necessitated developing techniques of sectioning, that is, cutting tissue into thin slices about one cell thick (Figure 4-5) so that details of interest are not obscured by overlying masses of cells. To avoid disrupting the cellular structure during sectioning, methods were developed for (a) fixation, chemical treatments that kill the cells and solidify their protoplasm, preserving its structure in a form as nearly lifelike as possible; and (b) staining with dyes that are taken up differentially by different cellular structures, enabling the observer to discern otherwise invisible features. Recently developed optical techniques, especially interference-contrast microscopy, permit direct observation of protoplasmic structures within isolated living cells, without fixation or staining (example in Figure 4-3).

There are inherent limits beyond which very small objects cannot be seen with a visible-light microscope, no matter how great its power of magnification. Many important cellular structures are too small to be discerned by light microscopy; they were recognized only with the development of the **electron microscope,** which affords a much higher resolution than the light microscope.

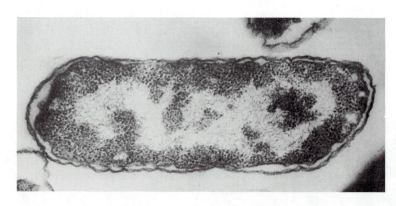

Figure 4-2.   Electron micrograph of section through a bacterial cell (*Hemophilus influenzae*). Note the cell wall, the plasma membrane, and the centrally located, irregularly shaped nuclear region composed of DNA fibrils. The dark granules outside the nuclear body are ribosomes (35,300 ×).

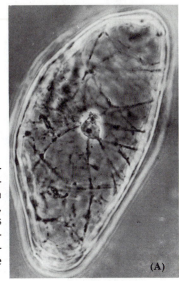

Figure 4-3. Phase-contrast light micrograph of a suspension cultured carrot cell, with a central nucleus containing a nucleolus, and strands of cytoplasm traversing the vacuole.

Electron Microscopy

The electron microscope (Figure 4-6) uses an electron beam focused by magnetic "lenses." The electron beam can penetrate only very thin objects, much thinner than a cell. Hence fixed cells must be cut into ultrathin sections by a delicate technique, then stained with heavy metals (usually osmium, lead, or uranium) that are taken up by membranes and other cell structures, causing them to appear dark in the resulting picture (Figure 4-A, for example). This method, by which most of the ultrastructure (structure finer than can be resolved with the light microscope) of cells has been revealed, is called **transmission electron microscopy** because the electron beam is transmitted through the specimen. The fixation and staining technique is particularly critical in this method and, for many plant materials, is difficult to carry out successfully.

An alternative method is **scanning electron microscopy,** in which electrons are reflected from the surface of the specimen when the electron beam is passed over it. This method does not require sectioning and is particularly good for obtaining three-dimensional views of the shapes and organization of cells in tissues (for example, Figures 13-13, 13-15).

The Sizes of Cells and Their Substructures

Robert Hooke estimated the diameter of cork cells to be about 0.001 in., or 0.04 mm. Hence a cubic inch of cork must contain about 1 billion cells, emphasizing the astronomical numbers of cells that comprise the bodies of larger plants and animals. To express cell dimensions in convenient units, scientists use the micrometer (or micron; abbreviation, μm), which is one-thousandth

of a millimeter. Mature plant cells are commonly 25–50 μm wide and may often be substantially longer (100–200 μm). Most animal cells are considerably smaller (10 μm or less), as are the actively growing and dividing cells in the growing points of shoots and roots. Most bacterial cells are substantially smaller still, about 1 μm. Figure 4-7 relates these sizes to those of objects visible with the naked eye.

The nucleus and the largest organelles of plant cells, the plastids, are easily visible with the light microscope (Figure 4-3). They range from about 3 to 10 μm in diameter and are therefore larger than many bacterial cells. Most cell organelles other than plastids are smaller than 0.5 μm and are therefore at or below the limit of resolution (about 0.3 μm) of the light microscope.

In examining the finer substructure of the cell by electron microscopy, the nanometer (nm), which is one-thousandth of a micrometer, becomes a more convenient unit of measure. Cellular membranes, for example, are about 7 nm thick.

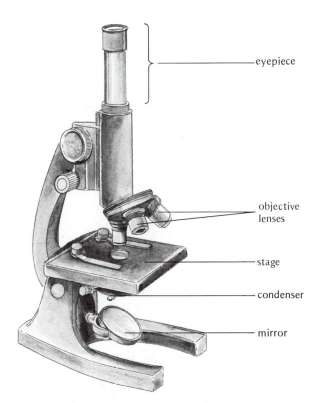

Figure 4-4. A plain monocular compound microscope. Light is reflected by the mirror up through the condenser. The lens system of the condenser focuses the illumination on the specimen (located on the stage). The objective lens creates a magnified image of the specimen within the tubular body of the microscope. The eyepiece, or ocular lens, magnifies this image in turn and brings it to focus at the observer's eye. A binocular research microscope is illustrated in Figure 1-2.

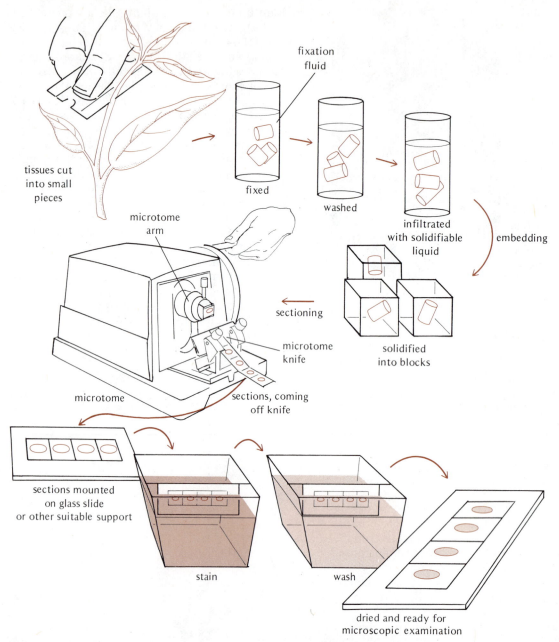

Figure 4-5.   Protocol of fixation and sectioning for microscopy. The microtome's arm, carrying the block of embedded tissue, moves up and down while slowly advancing, thus causing the knife to slice very thin sections from the face of the block. The nature of the fixation fluids, embedding media, microtome and knife, support for sections, and stains used for light and electron microscopy differ (glass slides, for example, cannot be used for electron microscopy, sections being supported instead on small metal grids), but the basic procedure is similar. The fixation, washing, embedding, and staining often involve several steps rather than just one each as shown here.

At the other end of the size scale, some plant cells are large enough to be visible with the naked eye, as illustrated without magnification in Figure 4-7. Very long cells such as root hairs and fiber cells can be perceived but not resolved clearly by eye because their diameters are too small (about 10–20 μm). Some algae, however, possess huge cells, easily seen by eye. These "giant cells" are especially useful in cell studies because fine needles can be inserted into them for making measurements, and their protoplasm can be removed virtually intact.

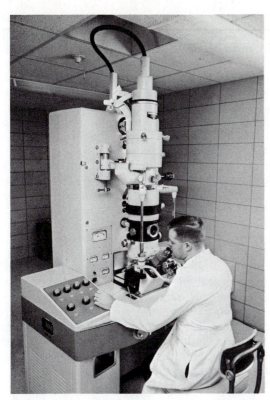

Figure 4-6. An electron microscope. Electrons are emitted from a heated wire filament under vacuum at the top of the column and are accelerated downward as a fine beam by an applied electrical potential. The electron beam is focused by magnetic "lenses" in the walls of the column. The specimen is placed approximately half way down the column. The condenser lens, between the filament and the specimen, focuses electrons on the specimen. Electrons that pass through it are focused by the objective lens located below the specimen holder to give a magnified image. This image is magnified further by a third magnetic lens, which projects the final image onto a screen at the bottom of the column. The screen is coated with a material that fluoresces (gives off light) when bombarded with electrons, enabling the operator to view the much-magnified image. Beneath the screen is film that records a picture of this image when the screen is tipped out of the way. This is how the transmission electron micrographs shown in this book were obtained.

Figure 4-7. Illustration of the size scale of various kinds of cells. Between each successive picture and the next a portion of the previous picture is magnified 10 times. Starting (A) with cells visible to the naked eye, the "giant cells" of the alga *Nitella*, in (B) some mature plant cells and cotton fiber are barely visible, in (C) they can be seen clearly at 100 × magnification, while in (D) (1000 ×) embryonic plant cells can be studied and in (E) (10,000 ×) the structure of bacterial cells can be seen. (D), and especially (E), fall in the range of the electron microscope only.

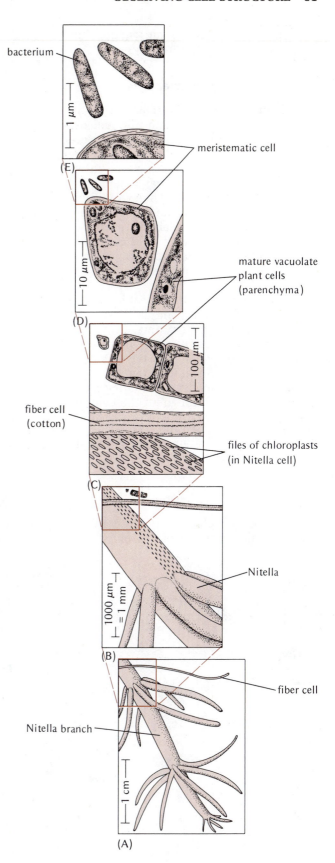

THE FUNCTIONAL ORGANIZATION OF PLANT CELLS

The Cytoplasm

Cytoplasm generally refers to all the protoplasm of the cell apart from the nucleus. Today, however, the term cytoplasm is often used more specifically for the fluid or gel-like, seemingly structureless protoplasmic material in which the cell's organelles and internal membranes are suspended (see Figure 4-A). The cytoplasmic matrix is not just a suspension medium, but the location of many of the cell's metabolic reactions.

Cytoplasm consists of 15–25% protein, about 70–80% water, and small but vital amounts of salts and many other substances involved in the metabolism of the cell.

The cytoplasm is usually at least partly fluid, as demonstrated, for example, by the phenomenon of **cytoplasmic streaming.** Under the microscope the cytoplasm of many plant cells can be seen to flow in regular streams at rates of 2–20 μm/sec or more. Streaming requires energy coming from cell metabolism. Streaming serves to distribute efficiently throughout the cytoplasm substances absorbed from outside the cell and substances released by the nucleus and other organelles. In slime molds, streaming makes locomotion possible.

Ribosomes

As seen by electron microscopy, the cytoplasm contains numerous densely staining regular particles 17–20 nm in diameter (Figure 4-10). These particles are composed of about 50% protein and 50% RNA (ribonucleic acid is related to the DNA of the nucleus); hence the name **ribosomes.** No membrane surrounds individual ribosomes, which thus may be called nonmembranous cell organelles. Ribosomes are vitally important to all cells, both prokaryotic and eukaryotic, because they synthesize proteins and hence produce the protoplasm itself. The evidence for this, as for the functions of other organelles mentioned in the next few pages, comes from experiments in which the organelles have been isolated from cells by the techniques of cell fractionation. Ribosomes so isolated will synthesize proteins when provided with the requisite materials and energy supply.

As illustrated in Figure 4-A, many of the cell's ribosomes (called free ribosomes) are simply suspended in the cytoplasm, while others (membrane-bound ribosomes) are attached to internal membranes. Free ribosomes probably produce proteins needed in the cytoplasm, whereas membrane-bound ribosomes probably synthesize proteins for use in membranes and organelles and for export to the outside of the cell.

The ribosomes of prokaryotic cells are consistently somewhat smaller than the ribosomes of eukaryotic cells; they also differ in functional features, as reflected by the fact that certain antibiotics inhibit the action of prokaryotic but not eukaryotic ribosomes, and vice versa. Prokaryotic ribosomes are referred to as 70S ribosomes and eukaryotic ribosomes as 80S ribosomes, with reference to their different centrifugal sedimentation behavior (see Cell Fractionation) due to their difference in size.

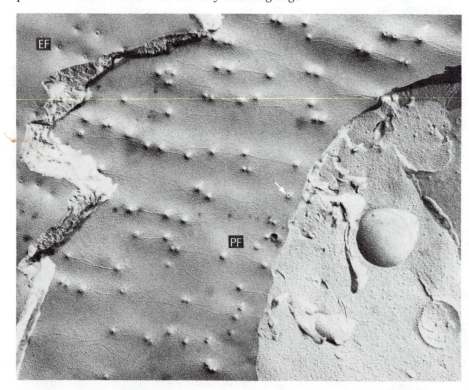

Figure 4-8. Freeze-fracture view of a plant cell plasma membrane. The fracture plane has split open the membrane and exposed the interior surface of each half of the membrane sandwich, the extracellular (or outside) half (EF) and the protoplasmic (or cytoplasmic) half (PF). The cell's cytoplasmic contents show at the right of the electron micrograph. The large globular structures (arrow) on the PF fracture face are plasmodesmata that have been interrupted by the freeze-fracture process. The numerous smaller bumps on both faces are membrane proteins.

Cellular Membranes

The external cell membrane is called the **plasma membrane** or sometimes the **plasma-lemma.** The cell also contains internal membranes, including the nuclear envelope and the membranes delimiting various cytoplasmic organelles. Observed in cross section by electron microscopy, all these membranes look very similar: two dark (electron-dense) layers are separated by a light (electron-transparent) layer. This "sandwich," sometimes called a unit membrane, is about 7–8 nm thick (Figure 4-11).

Membranes are composed primarily of fat-like substances, or **lipids,** which prevent most water-soluble substances from passing through easily. Very small molecules such as water, oxygen, and carbon dioxide, however, can readily cross cellular membranes. For the uptake of most nutrients required from outside the cell, such as sugars and salts, the plasma membrane contains specialized proteins called **carriers,** which transport the needed molecules into the cell. The carrier-mediated uptake of many substances through the plasma membrane is actually an energy-driven pumping process called active transport (Chapter 6).

Proteins embedded within membranes can be seen using a technique called **freeze-fracture,** by which a membrane's surface and interior can be examined in face view, without fixation. Cells are abruptly frozen at the temperature of liquid nitrogen (−196°C), the frozen block is split (fractured) by a blow from a knife, and the fracture surfaces are made visible by coating with a thin layer of metal, which is subsequently examined under the electron microscope. Membranes normally fracture down the middle, that is, along the center line of the unit membrane sandwich, so the interior components of the membrane become visible. Arrays of protein particles occur within most kinds of membranes (Figure 4-8), their size and arrangement often differing among the different membranes of a cell.

Plasmodesmata (singular, plasmodesma) are membrane connections (Figure 4-A) extending between adjacent cells of a plant tissue, through the cell walls. A plasmodesma (Figure 4-9) consists of a tubular, sometimes branched, channel whose bounding membrane is continuous with the plasma membranes of the adjacent cells. Plasmodesmata probably serve to transport materials from the cytoplasm of one cell to that of another without the substances having to cross intervening plasma membranes.

THE INTERNAL MEMBRANE SYSTEM

The **endoplasmic reticulum** (ER) is a system of paired membrane sheets ramifying through the cytoplasm (Figure 4-A), often bearing numerous ribosomes attached to their outer surfaces (Figure 4-10). The two membranes of each pair join to-

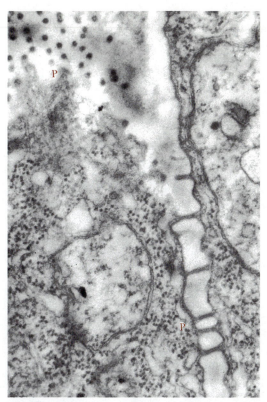

Figure 4-9. Plasmodesmata (P) connecting adjoining plant cells (48,000 ×). Lower right: longitudinal section showing continuity of plasma membranes through the pores that penetrate the cell wall. Upper left: surface view of the plasmodesmata in a fold of the wall.

gether at their edges, forming a flattened sac-like structure called a **cisterna.** The principal known functions of the ER are (1) synthesis of membrane lipids and (2) synthesis, by the ER-bound ribosomes, of membrane proteins and proteins for export from the cell (for example, to become part of the cell wall). Upon synthesis, these proteins are released into the interior cisternal space, which separates them from the cytoplasm, permitting their selective secretion.

Under the electron microscope, ribosome-bearing ER membranes have a rough appearance (Figures 4-A, 4-10) because of the attached particles and, thus, are called rough ER. Some cells contain ER membranes lacking ribosomes, called smooth ER. These membranes can serve for lipid synthesis but not for protein synthesis, which requires ribosomes.

Dictyosomes or **Golgi bodies** are stacks of smooth (ribosome-free) cisternae, usually about 0.25–0.5 μm wide. In cross section, these stacks look like sheaves of wheat (Figure 4-11). The Golgi system functions to secrete protein and carbohydrate products from the cell to its exterior. At their edges the dictyosomal cisternae swell and break up into vesicles that detach into the cytoplasm, and subsequently fuse with the plasma membrane, discharging their contents to the cell exterior, a process called **exocytosis** (Figure 4-12).

Among the products secreted by the Golgi system are cell wall components and the nectar of flowers (which bees collect to make honey). Some kind of connection exists between the Golgi dictyosomes and the ER, because proteins produced by the ER can be secreted via the Golgi system (see Figure 4-12).

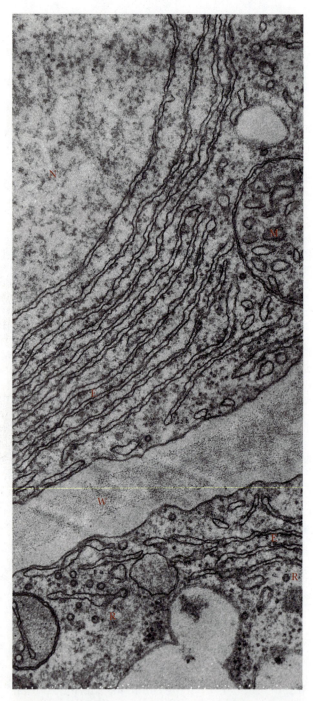

Figure 4-10. Electron micrograph of parts of two plant cells, showing (E) endoplasmic reticulum with attached ribosomes (dark granules), and (M) a mitochondrion. N: nucleus. W: cell wall. (44,000 ×.)

Energy Organelles

MITOCHONDRIA

Mitochondria (singular, mitochondrion) are about 0.2–0.5 μm in length, barely visible under the light microscope. Electron microscopy reveals a very distinctive structure (Figures 4-10, 4-13) consisting of a double membrane (two unit membranes, separated by a space), of which the inner membrane is folded into projections called **cristae** extending into the organelle's interior. Mitochondria are responsible for cellular aerobic respiration, the oxidation of foodstuffs for the production of biologically useful energy for cellular activities such as cytoplasmic streaming, active transport, and protein synthesis (Chapter 6). Thus the mitochondrion has been called the "powerhouse of the cell."

PLASTIDS

Usually the largest cytoplasmic organelles, well-developed **plastids** can reach 3–6 μm in diameter. They are conspicuous under the light microscope not only because of their size but because they often contain pigments (Figures 4-14 to 4-16). The electron microscope shows that plastids, like mitochondria, are enclosed by a double membrane envelope (Figure 4-17).

Chloroplasts (Figure 4-14) are plastids that contain the green pigment chlorophyll and carry on photosynthesis. The structure and function of chloroplasts are discussed in Chapter 7.

Chromoplasts (Figure 4-15) contain yellow or orange pigments called carotenoids, which are related to, and indeed the source of, vitamin A. Chromoplasts develop from chloroplasts in some yellow flowers and in yellow or orange fruits, such as tomatoes and squash, by a modification process involving the loss of chlorophyll and extensive structural changes.

Amyloplasts (Greek *amylon*, starch), also sometimes called **leucoplasts** (Greek *leukos*, white), found in nongreen cells, are colorless plastids that convert sugar into starch granules (starch grains), which are stored as an energy reserve for future energy needs of the plant (Figure 4-17). As obtained from cereal grains and from potatoes and other starchy root vegetables, these starch grains comprise humans' major energy food. Milled cereal seeds release their starch grains plus stored protein to yield flour. The form of the starch grains synthesized by amyloplasts differs noticeably among different food plants (Figure 4-18), permitting the positive identification of components of food products, useful in testing for adulteration and other abuses.

Chloroplasts can grow and divide along with the cells that contain them. However, dividing higher plant

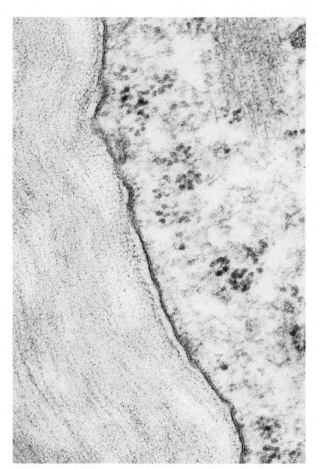

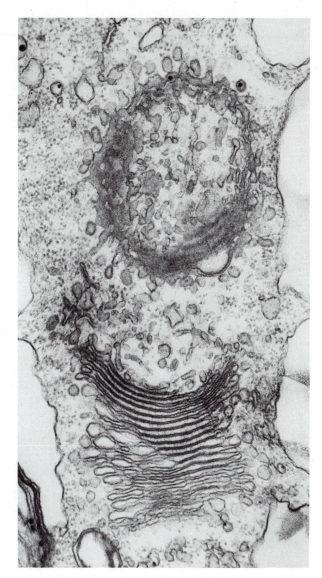

Figure 4-11. Electron micrographs of plant cell membranous structures. (Left) Boundary between cytoplasm (right) and cell wall (left), showing plasma membrane structure (82,000 ×). (Right) A section through two unusually elaborate dictyosomes from the unicellular alga, *Euglena gracilis*. The lower dictyosome is seen in side view, the upper one viewed toward the plane of an individual membrane sac (cisterna). (50,000 ×.)

cells usually do not contain developed plastids, but smaller precursor organelles called **proplastids** (Figure 4-A) of about the same size and form as mitochondria. Protoplastids can grow and develop into plastids such as chloroplasts or amyloplasts when the cell matures.

Partial Autonomy of Organelles

Chloroplasts and mitochondria possess DNA, although very little compared to the amount in the nucleus. Chloroplasts and mitochondria also contain ribosomes. These features indicate that chloroplasts and mitochondria contain genetic instructions and synthetic machinery for producing at least some of their component proteins. Such independence is called genetic **autonomy.** Chloroplasts and mitochondria are only partially autonomous, however, as nuclear DNA actually specifies, and cytoplasmic ribosomes synthesize, many of their components, and the organelles cannot grow or reproduce outside the cytoplasm.

Other Energy Organelles

Microbodies are organelles similar in size to mitochondria, but bounded by only a single membrane. Microbodies contain enzymes for certain special kinds of energy metabolism that can advantageously be kept separate from the cytoplasm, for example, reactions in which hydrogen peroxide, a rather toxic substance, is produced or consumed. Microbodies (see Figures 6-7, 7-15) convert fats into sugar during the germination of fat-storing seeds, and they participate in an important aspect of photosynthetic metabolism called photorespiration.

Other somewhat similar organelles that play a role in energy economy include **liposomes,** which synthesize and store fats, and **aleurone granules,** which accumulate proteins for storage (for example, in seeds). All these organelles probably develop from the internal membrane system of the cell, using lipids and proteins synthesized by the ER, and are not partially autonomous as plastids and mitochondria are.

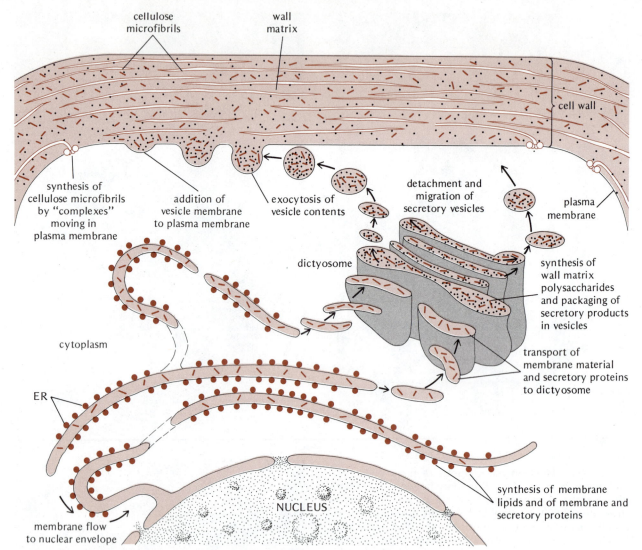

Figure 4-12. Diagram of Golgi secretion and cell wall synthesis in a plant cell. Note that the Golgi system produces and secretes the noncellulosic matrix material of the cell wall, whereas the microfibrils of cellulose are produced at the plasma membrane by their own synthetic system (consisting, according to current research, of a "complex" of membrane-embedded particles, probably proteins which move in the plane of the membrane, spinning out a microfibril as they go). The diagram illustrates the "endomembrane flow" concept, whereby membrane material produced in the ER is delivered, via the Golgi secretion mechanism, to the plasma membrane for its enlargement during cell growth.

Motility Structures

Microtubules (Figure 4-19) are hollow filaments about 25 nm in diameter, built up from protein units called **tubulin**. Microtubules are dynamic, labile structures that can form from their tubulin subunits and then disappear again according to the needs of the cell. Microtubules make up the spindle structure that appears and functions in the division of the chromosomes, a mo-

tility process (see Mitosis). Microtubules are also found as permanent components of the motile appendages or **flagella** that many lower plant cells possess.

Microfilaments are much finer filamentous structures, about 5 nm in diameter and not hollow, made of a protein similar to one of the principal proteins of muscle, actin, which enables muscles to contract and do work. Actin filaments also appear to be responsible for such activities as cytoplasmic streaming.

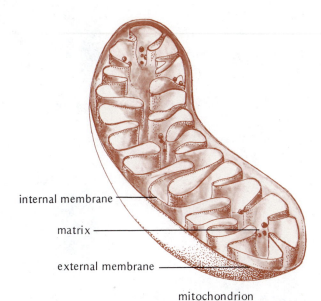

Figure 4-13. Three-dimensional model drawing of a sectioned mitochondrion.

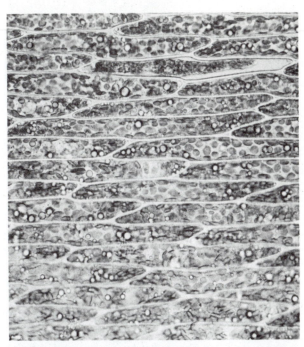

Figure 4-14. Cells containing brilliant green chloroplasts (darker bodies in photograph; the transparent, spherical bodies are oil droplets).

Vacuoles

Unlike animal cells, most plant cells are not largely filled with cytoplasm and organelles, but include more or less extensive water-filled spaces called **vacuoles,** separated from the cytoplasm by a membrane boundary (the vacuolar membrane or tonoplast). In young, dividing plant cells vacuoles may be small and scattered in the cytoplasm (Figure 4-A), but in mature cells a single large vacuole typically occupies most of the cell's volume, surrounded by only a thin peripheral layer of cytoplasm next to the plasma membrane (Figure 4-20) and sometimes traversed by a few **cytoplasmic strands** (Figure 4-3). The contents of vacuoles are often called **cell sap** or vacuolar sap. Unlike cytoplasm, vacuolar sap is not rich in protein and does not seem to participate actively in cell metabolism, although it can contain special-purpose proteins.

Vacuoles function (1) as a means for cells to increase in size without synthesizing an equivalent volume of protein-rich cytoplasm; (2) as storage compartments for reserves of metabolically useful compounds such as sugars; (3) as sites for the breakdown of cell constituents and organelles when this is necessary; and (4) as depots for metabolic end products called **secondary products,** many of which would interfere with cell functions if free in the cytoplasm, but which may play some special role in plant activities, for example, to repel predators or disease organisms as noted in Chapter 1. Illustrative of secondary products are anthocyanins, water-soluble blue or red pigments that occur in vacuoles (Figure 4-16) and are responsible for the colors of many flowers, fruits, and autumn leaves. Crystals, usually of calcium

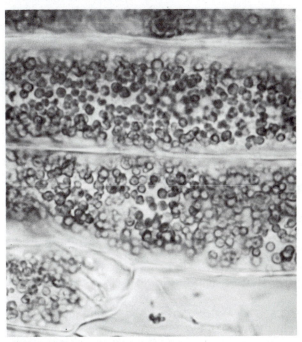

Figure 4-15. Cells from the petals of a yellow flower, containing chromoplasts (orange-yellow).

oxalate, are widespread vacuolar components (Figure 4-21). Oxalate is often regarded as a useless metabolic by-product but it may actually be formed to regulate the acid–base balance of plant cells.

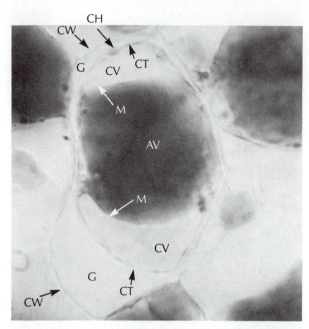

Figure 4-16. Plant cell showing anthocyanin-containing vacuole (AV). This cell, from anther epidermis of staminate catkins of poplar, is unusual in containing several large vacuoles. Only one (AV) of the three vacuoles in the cell pictured contains anthocyanin (bright red in actuality), the other two (CV) being colorless as is the thin peripheral layer of cytoplasm (CT), showing that anthocyanin is restricted to a vacuole and is not found throughout the cell. So the cytoplasm (containing a few chloroplasts, CH) would show clearly, the protoplast was contracted osmotically within its cell wall chamber (plasmolyzed; Chapter 9) by placing it in a strong sugar solution before it was photographed, creating a gap (G) between the plasma membrane and the cell wall (CW).

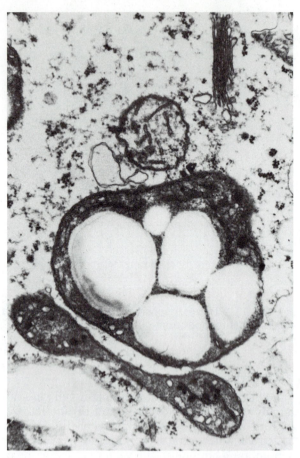

Figure 4-17. An amyloplast, containing starch grains. Note double-membrane envelope, typical of plastids.

The Nucleus

In young dividing cells the nucleus is located centrally (Figure 4-A). In mature cells, in which the vacuole occupies most of the cell's volume, the nucleus is often to one side, surrounded by a thin layer of cytoplasm (Figure 4-20). In some vacuolate cells the nucleus is suspended centrally within the vacuole by cytoplasmic strands connecting it with the peripheral cytoplasm (Figure 4-3). These strands usually exhibit active cytoplasmic streaming for exchange of materials between the nucleus and the peripheral cytoplasm.

In some cases the nucleus is not required for the cell's immediate life and survival. For example, with the nucleus removed, a cell of the alga *Acetabularia* (Figure 14-18) lives for weeks or even months. Such a cell, however, cannot reproduce and eventually dies, because long-term maintenance as well as growth and developmental changes require the nucleus.

Genetic evidence shows that the DNA of the nucleus contains information for the synthesis of most of the proteins and kinds of RNA that the species is capable of making. The nucleus controls cell maintenance and development by instructing the cell's ribosomes to synthesize particular proteins (Chapter 5). The RNA involved in cell function, for example, the ribosomal RNA contributing to the structure of the ribosomes, is produced in the nucleus. However, the nucleus lacks functional ribosomes and therefore cannot synthesize proteins; it imports its proteins from the cytoplasm.

SUBSTRUCTURE OF THE NUCLEUS

The nucleus has a double bounding membrane or **nuclear envelope** (Figure 4-A). Unlike the double envelopes of chloroplasts and mitochondria, however, the nuclear envelope is perforated by numerous openings called nuclear pores, allowing the exchange of materials between nucleus and cytoplasm. The nuclear envelope con-

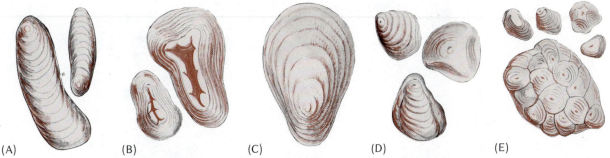

(A) (B) (C) (D) (E)

Figure 4-18. Starch grains. (A) Banana; (B) garden bean; (C) potato; (D) sweet potato; (E) rice (several adhering grains called a compound grain, formed within one amyloplast). The layering within starch grains is caused by the alternate deposition of different molecular forms of starch (see Figure 5-2) having different densities.

nects with cisternae of the ER, perhaps for transfer to the nucleus of proteins synthesized on ribosomes attached to the ER.

Within the nucleus are one or more smaller spherical bodies (Figure 4-A) called **nucleoli** (singular, nucleolus), often large and dense enough to be seen under the light microscope (Figure 4-3). Nucleoli, consisting largely of RNA destined to become cytoplasmic ribosomes, are located at nuclear sites of ribosomal RNA synthesis.

DNA is usually dispersed within the nucleus in its cytoplasm-like fluid called nuclear sap. The DNA plus associated proteins is called **chromatin** (from Greek *chroma*, color, because DNA stains intensely with certain dyes). Chromatin actually consists of a limited number of discrete threadlike structures called **chromosomes** (*chroma*, color, and *soma*, body) which are not recognizable in the nondividing nucleus because they are ex-

tremely extended and intertwined with one another. During cell division the chromosomes contract and become microscopically visible (see Mitosis, below). The chromosomes carry the hereditary units of DNA, or **genes,** which specify the particular proteins of the cell.

DETERMINATION OF ORGANELLE FUNCTIONS

Cell Fractionation

Much information on organelle and membrane functions is gained by isolating different kinds of organelles and membranes to identify their individual components and metabolic capabili-

Figure 4-19. Microtubules in guard cells of the yellow nut sedge, *Cyperus esculentus* L. (A) Transverse section. The microtubules appear as small circles just inside the plasma membranes of two adjacent cells. (B) Longitudinal section, taken in a plane perpendicular to that of (A). The plane of section passes through the microtubules and also through large portions of the cell wall (gray shadow at bottom and right). The microtubules run parallel to adjacent cell wall microfibrils as seen at lower left. [Magnification (A) 55,000 ×; (B) 59,000 ×.]

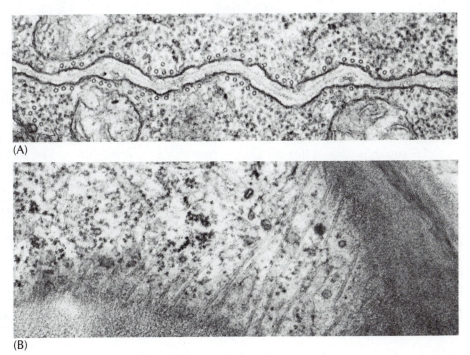

(A)

(B)

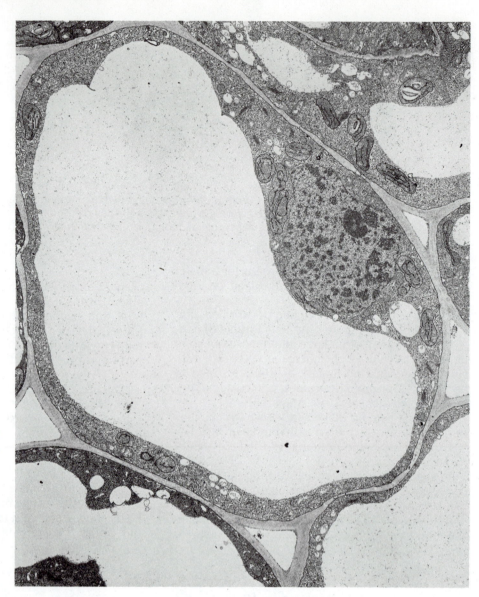

Figure 4-20. Vacuolate cell from a growing pea stem (5,400 ×). CW, cell wall; CY, cytoplasm; D, dictyosome; ER, endoplasmic reticulum; I, intercellular space; MI, mitochondrion; ML, middle lamella; N, nucleus; NL, nucleolus; P, plastid; PM, plasma membrane; V, vacuole; VM, vacuolar membrane (see diagram at right).

ties. This isolation process is done with a **centrifuge** (Figure 4-22), which produces centrifugal force to move subcellular particles through the medium in which they are suspended.

To separate organelles and membranes by centrifugation, the cells must first be broken or **homogenized** so as to release their delicate organelles into the extraction medium with as little disruption as possible. Although homogenization is sometimes done with a household blender, the preferred method for most organelles is to grind the tissue by hand in a mortar and pestle (Figure 4-23) containing an extraction medium whose composition is chosen to resemble that of the cytoplasm itself in concentration, acidity, and other important features. The resulting homogenate is strained through cloth to remove coarse debris such

as large pieces of cell walls, then centrifuged at successively greater speeds (greater centrifugal forces) to sediment or "pellet" successively smaller classes of organelles and membranes fragments, a procedure called **differential centrifugation** (Figure 4-23).

During cell homogenization the plasma membrane, vacuolar membrane, dictyosomes, and ER usually break up into minute, spherical membrane enclosures or vesicles unless special methods are used to keep them intact. These small membranous particles, pelleted only by very great centrifugal forces, are collectively called **microsomes** (meaning "small bodies"). Because the bulk of the microsomal fraction (Figure 4-23) is usually derived from the ER, the term "microsomal" is often used loosely to denote activities, such as lipid and protein synthesis, associated with the ER.

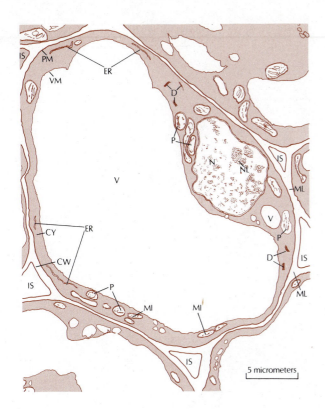

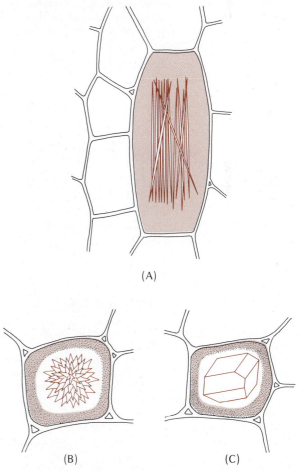

(A)

(B)   (C)

Figure 4-21. Calcium oxalate crystals found in the vacuoles of living plant cells. (A) *raphides*, or needle-like crystals; (B) a *druse*, or star-shaped cluster of crystals; (C) a simple prismatic crystal. (ca. 2,000 ×.)

appear over cell nuclei (Figure 4-25), where most of the cell's DNA is located. By autoradiography it has been shown, for example, that cell wall components are synthesized in the dictyosomes or Golgi bodies and then transported to the cell wall as illustrated in Figure 4-12.

THE CELL WALL

The cell wall, composed mainly of carbohydrate, is a mechanical structure—tough but pliable in the case of thin cell walls, stiff and virtually rigid in the case of thick walls of specialized supporting cells. As noted in Chapter 2, the possession of a cell wall outside the plasma membrane is one of the principal features distinguishing plant cells from animal cells, which are bounded simply by a plasma membrane. The cell walls of plant

However, to distinguish whether an activity is carried out by the ER or by other membranes, it is necessary to separate the different membrane components of the microsomal fraction. One way of doing this is by **density gradient centrifugation** (Figure 4-24).

Autoradiography

Autoradiography is a technique that allows the detection of localized synthetic and transport processes within intact cells. When cells that are synthesizing a particular cell constituent such as DNA or protein are supplied with a radioactive form of one of the building blocks for that constituent, they take up and incorporate the radioactive material. For example, to detect DNA synthesis, one can allow the cells to incorporate radioactive thymidine, a building block of DNA. Then the cells or tissues are fixed, and sectioned if necessary, and in the dark a thin photographic film is placed over them. Atomic radiation coming from the radioactivity in the cells exposes the film, which is then developed like a photographic negative but left in contact with the cells. Wherever radioactive material occurs in the cells, grains of silver appear in the film, easily seen under the microscope. This preparation is called an autoradiograph, meaning a photographic image made by radiation coming from the object itself. In the above-mentioned instance of radioactive thymidine incorporation, almost all the silver grains will

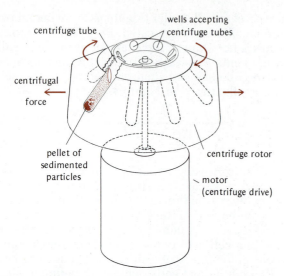

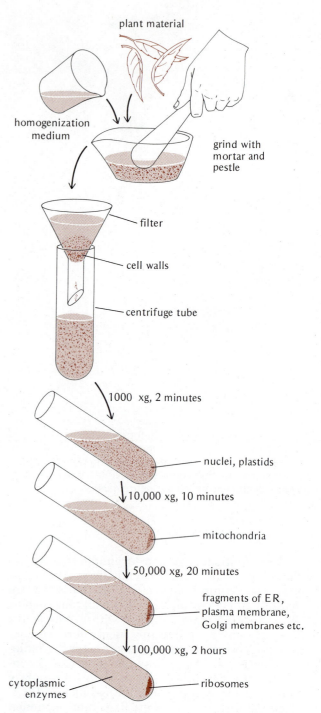

Figure 4-22. Centrifuge, simplified schematic diagram. In most centrifuges for biological work the rotor is mounted inside a refrigerated chamber to keep the samples cold during centrifugation. The drive (i.e., electric motor) is located beneath this chamber and the unit also includes speed controls, a brake for stopping quickly, and often a vacuum system to reduce friction between rotor and air, which would tend to heat up the rotor and samples at high speed (as much as 100,000 rpm in some centrifuges).

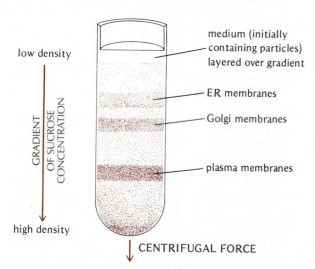

Figure 4-24. Separation of "microsomal" membranes by density gradient centrifugation. Centrifugal force drives each type of membrane down into the gradient to the point at which the medium has the same density as the membrane. At that point, the membrane particles stop moving because they "float." Membranes and organelles of different density (due to different biochemical composition) can thereby be separated.

Figure 4-23. Cell fractionation procedure (differential centrifugation). Each of the pellets is resuspended by stirring carefully with a suitable resuspension medium in order to obtain an organelle or membrane preparation for use in enzyme assays or other biochemical tests. (1000 × g is 1000 times gravity.)

tissues serve the supporting role played by skeletal structures in animals.

The principal carbohydrate of higher plant cell walls is **cellulose,** a chain-like molecule built of sugar units. Cellulose is present in the form of thin filamentous structures called **microfibrils** (Figure 4-26), usually about 4–20 nm in diameter, running within the wall parallel to the cell surface. Cellulose microfibrils are about as strong as a steel thread of equivalent size. Normally they

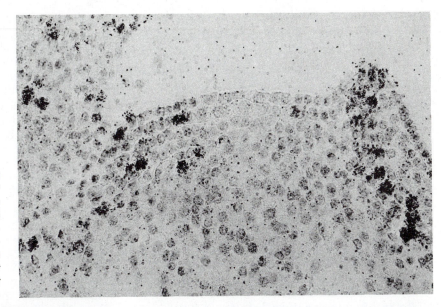

Figure 4-25. Longitudinal section of the shoot apex of sunflower (*Helianthus annuus*), after exposure to ³H-thymidine for 24 hours to obtain incorporation of thymidine into DNA. Subsequent autoradiography shows that only some of the nuclei (where black silver grains occur) were replicating their DNA.

Figure 4-26. Electron micrographs of cell walls, showing cellulose microfibrils. (A) A primary wall (12,000 ×); (B) a secondary wall (18,000 ×). (These pictures show cell walls of the alga *Valonia*, but higher plant primary and secondary walls have somewhat similar structure.)

are embedded within a nonfibrillar material called the wall **matrix,** composed of noncellulosic materials. This composite structure has been compared with reinforced concrete, in which the reinforcing rods are comparable to the cellulose microfibrils.

The **middle lamella,** also called the intercellular cement, is the zone of contact between the cell walls of adjacent cells (Figures 4-A, 4-20). It does not contain cellulose microfibrils but consists primarily of **pectins,** carbohydrates familiar from their use in jellies and jams.

Primary Walls

The thin cell wall that envelops young growing cells (Figure 4-A) is called a **primary wall.** In older cells (Figure 4-20) the primary wall is still present but may have become covered by a specialized secondary wall (see below). The primary wall expands or extends as a cell grows, and is therefore not rigid. It can be defined as a cell's original wall, at least potentially capable of further growth in area. Most primary walls are less than about 0.1 μm thick, but certain specialized growing cells (epidermis, collenchyma) develop primary walls 1 μm or more thick. The primary wall is normally traversed by numerous plasmodesmata, often concentrated in local areas called pit fields.

Electron microscopic observation shows that the structure of the primary wall is relatively loose (Figure 4-26A). Its microfibrils run in varying directions and are separated by a comparatively large amount of matrix material. Cellulose usually comprises less than one-third of the solid matter of these walls; the rest is mainly other carbohydrates, especially pectins, plus a mixed assortment called **hemicelluloses,** somewhat intermediate in chemical nature between pectin and cellulose. Some protein also occurs in primary walls and probably plays a structural role.

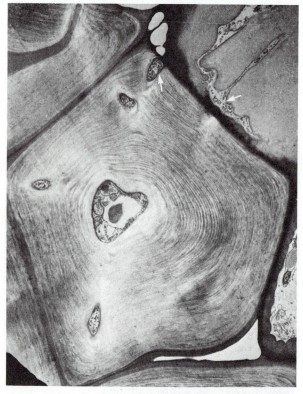

Figure 4-27. Thick, layered secondary wall in a stone cell (cross section) from the fleshy part of a pear fruit. The wall layers are zones of different density due to daily variations in the manner of wall deposition during development of the cell. Secondary wall is not deposited in regions of plasmodesmata (arrow) resulting in the formation of pits, or channels, through the wall connecting the living cytoplasm of adjacent stone cells.

Secondary Walls

Many cells, after completing their growth, deposit to the inside of their primary wall a distinct layer of wall material (Figure 4-27) called a **secondary wall.** When present, it is usually much thicker than the primary wall and is easily seen under the light microscope. During secondary wall formation cells often do not deposit wall material in pit field areas, where plasmodesmata occur. Thus after wall deposition a depression, called a **pit,** occurs in the secondary wall at each of these points (Figure 4-27). Pits permit the continued uptake of materials, via plasmodesmata, by living cells with secondary walls.

The secondary wall usually has a much higher cellulose content than the primary wall (often well above 50%). Its cellulose microfibrils are much more compactly arranged and are aligned almost perfectly parallel in any given layer (Figure 4-26B). Often several layers of secondary wall are deposited, in which the cellulose microfibrils run in different directions, like the layers in a multiple-ply tire. Unlike the primary wall, the secondary wall contains little or no pectin. Its matrix consists of hemicelluloses plus, in most cases, a noncarbohydrate stiffening constituent called **lignin.** Cell walls containing lignin are said to be lignified. Their thickness and other mentioned structural features make secondary walls much stronger and stiffer than primary walls. Because of this, secondary walls cannot extend or grow in surface area, this often being considered a defining characteristic of a secondary wall.

Plants employ secondary walls in mechanical tissues that must carry heavy loads, for example, the wood of trees, and the supporting cells called **fibers** found in tree bark and in the stems of most other plants. Humans exploit tissues with strong secondary walls for structural materials and for rope and textile fibers. Cotton fibers, which are hairs that grow from the surface of cotton seeds, develop unusual secondary walls containing no lignin and more than 90% cellulose. As a result they are strong yet pliable, explaining cotton's usefulness as a textile fiber.

Pulp and Paper

Another important use of secondary wall cellulose is in producing paper. The best-quality paper (rag paper) is made from cotton, traditionally from discarded cotton rags, and thus contains no lignin. Most paper, however, is made from low-grade timber ("pulpwood") by the process of pulping, which consists of chemical treatments to break down the middle lamellae so the fibrous cells of the wood come apart, yielding a mushy material called wood pulp. This pulp can be spread as a thin layer and dried to yield paper and cardboard. Wood pulp is also used extensively in the production of chemically modified cellulose plastics such as rayon, cellophane, and celluloid. The magnitude of the pulping industry is astonishing; the United States alone annually consumes more than 100 million tons of pulpwood. To pulp the wood, lignin must be extracted, because during wall lignification this material is deposited not only in the secondary wall but also in the primary wall and especially the middle lamella, rigidifying it. To remove lignin and other noncellulose components, large volumes of dilute alkalies and extraction liquors containing sulfur (sulfite or sulfide) are used, leading to very unpleasant-smelling, ecologically disastrous pollution of streams. For cheaper grades of paper such as newsprint, the lignin is only partially extracted; as a result, the paper tends to turn gradually yellow or brown from the oxidation of lignin residues. Some of the lignin extracted by the pulping industry is now utilized in manufacturing a variety of synthetic products and as a tanning agent in the production of leather from hides.

Formation of the Cell Wall

The matrix carbohydrates of the cell wall, both pectin and hemicelluloses, are made in the dictyosomes of the Golgi system and secreted into the cell wall by Golgi-derived vesicles, as noted earlier. The dictyosomes or other internal parts of higher plant cells contain no cellulose; during wall synthesis new cellulose appears only in the cell wall, at its surface, next to the plasma membrane. Cellulose, therefore, is probably synthesized in or on the plasma membrane.

In most growing cell walls the cellulose microfibrils are deposited in an oriented fashion— in a highly parallel alignment in secondary walls, as already noted, and with a preferential but not perfect orientation in most primary walls. Microtubules almost always occur in the adjacent cytoplasm, next to the plasma membrane, running in the same direction as that in which cellulose microfibrils are being deposited (Figure 4-19). This and other evidence suggests that microtubules in some manner control the directionality of cellulose deposition, perhaps by directing the movement of the cellulose-making machinery in the plasma membrane.

The formation of lignin in the secondary wall is a process completely different from the deposition of its carbohydrate structure. Lignin formation begins after cellulose deposition is completed and takes place throughout the thickness not only of the secondary wall but also of the primary wall on which the secondary layers have been deposited. Lignin is not manufactured within the cell and then secreted, but appears to arise by polymerization (Chapter 5), directly in the cell wall, of precursor material released by the developing cell.

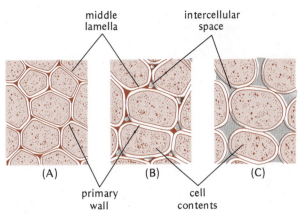

Figure 4-28. Formation of intercelluar gas spaces in a plant tissue by breakdown of the middle lamella. (A) Young tissue without gas spaces. (B) Gas spaces appearing in corners between cells. In many compact parenchyma tissues the process stops at about this point. (C) Expansion of gas spaces by extensive separation of middle lamellas and rounding up of cells. This condition develops in the internal tissues of leaves, and much larger gas channels develop in some tissues specialized for aeration. Very large gas channels can also be formed by breakdown of entire cells or groups of cells.

Intercellular Spaces

At a certain point in the growth and development of many tissues, the middle lamellae weaken or break down, especially at corner points where three or four cells come together. As a result, the walls of adjacent cells separate in these places, creating **intercellular spaces** (Figure 4-28). These merge to create a system of continuous air-filled channels, called the intercellular gas system, that runs around the cells and may open to the outside air. In some tissues entire cells or groups of cells break down, creating very large air channels. The intercellular gas system serves the important function of delivering oxygen for respiration, or carbon dioxide for photosynthesis, from the outside air to the cells deep within a plant organ.

CELL SHAPES AND TYPES

The cell wall gives a plant cell its shape. Cells that remain alive at maturity and have a relatively cuboid or compact shape (Figure 4-29A) are called **parenchyma.** The bulk of most young, soft plant organs is composed of parenchyma cells, which usually have only a thin primary wall. However,

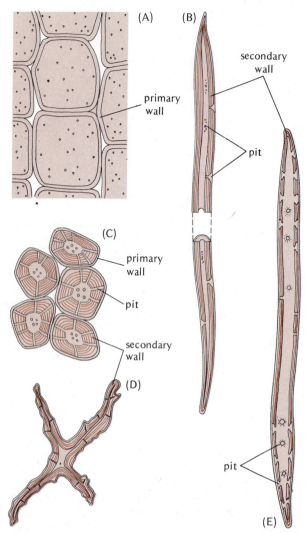

Figure 4-29. Some differently shaped and thickened plant cell types. (A) Parenchyma (cell contents not shown); (B) fiber cell; (C) compact sclereid ("stone cell"); (D) branched sclereid; (E) tracheid.

Some important specialized cell types are distinguished by the production of unique wall components not found generally in plant cells. Cells of the **epidermis,** or surface cell layer of young plant tissues, secrete a waxy layer called the cuticle onto the outer part of their exterior wall, facing the external environment (Figure 9-7). Similarly, **cork cells** comprising the outer layers of the bark of woody plants (Chapter 13) impregnate their walls with layers of another somewhat wax-like material called suberin (Latin *suber,* cork). These specialized cell walls protect the aerial organs of plants against drying out and are important to the ability of plants to grow and survive on land.

CELL REPRODUCTION

A basic corollary of the cell theory holds that every living cell arises from a preexisting cell by cell division. The division of plant cells consists of two main processes: nuclear division (**mitosis**) and cytoplasmic division (**cytokinesis**).

Mitosis

In dividing, the nucleus of a eukaryotic cell undergoes an elaborate process called mitosis, which provides each daughter nucleus with a complete set of all the genes that the parent nucleus possessed. During mitosis the previously elongated chromosomes, all of which have been duplicated, condense into recognizable units. The duplicate chromosome sets are then separated from one another and two new nuclei, the daughter nu-

wood contains compact living cells that are called wood parenchyma but develop a secondary wall.

The different types of specialized cells found in plant tissues (Chapters 8–13) differ mainly in possessing different shapes and kinds of cell walls. Figure 4-29B–E illustrates several of the named cell types. **Fiber cells** (Figure 4-29B) are elongated cells with a very thick secondary wall, and **sclereids** (Figure 4-29C, D) are compact, irregularly shaped, or branched cells with a thick secondary wall. Thick-walled mechanical tissue, including fibers and/or sclereids, is called sclerenchyma. In wood, elongated cells with relatively thin secondary walls are **tracheids** (Figure 4-29E). Most specialized cells that develop secondary walls, in contrast to parenchyma, die after completing wall deposition.

Figure 4-30. Mitosis in a single living endosperm cell of African blood lily (*Haemanthus katherinae*), as seen with the phase-contrast microscope. These cells, which do not possess a cell wall, have been flattened in order to show the mitotic process more clearly. However, the mitotic spindle does not show up well in phase-contrast microscopy and appears here as a zone from which most organelles are excluded (cf. Figure 4-31). The numbers in parentheses represent the time, in minutes, elapsed since the first photograph. (A) Late prophase; (B) early prometaphase (nuclear membrane has just disappeared and chromosome ends are protruding into the cytoplasm); (C) late prometaphase; (D) metaphase; (E) early anaphase; (F) midanaphase; (G) late anaphase; (H) early telophase (cell plate beginning to form); (I) telophase (cell plate nearly complete).

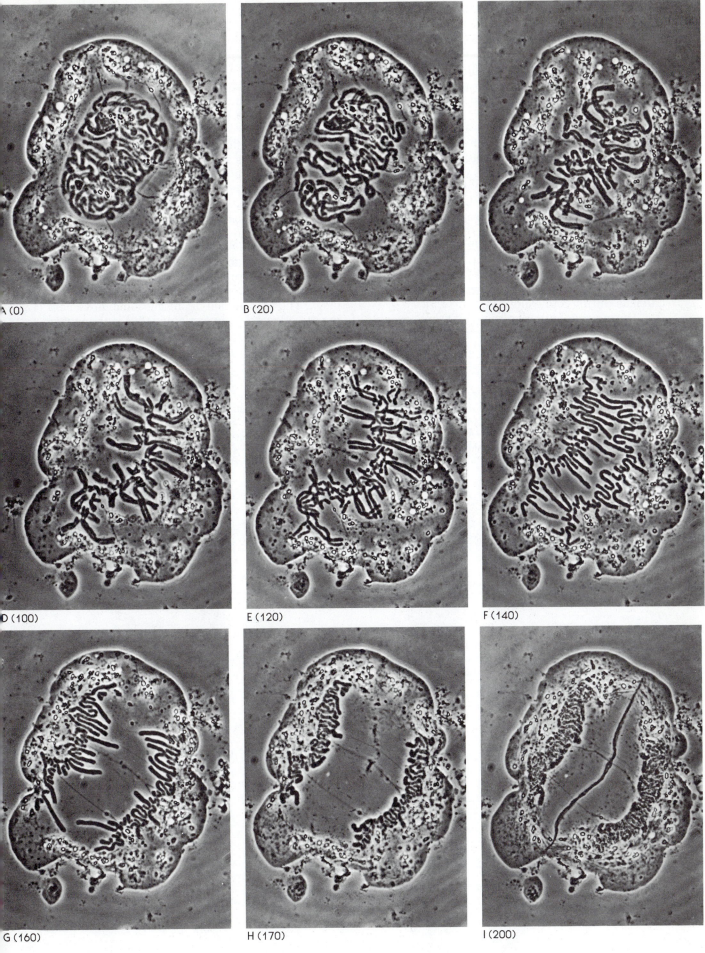

A (0) B (20) C (60)

D (100) E (120) F (140)

G (160) H (170) I (200)

( ) = time (min) elapsed since first shot

clei, become organized around each set (Figure 4-30). Mitosis is arbitrarily divided into four stages called **prophase, metaphase, anaphase,** and **telophase** (Figures 4-30, 4-31). The progression from each of these stages to the next is a continuous process.

During **interphase** (Figure 4-31A), the non-dividing state of the nucleus between one mitosis and the next, the chromosomal DNA becomes copied or duplicated, a process called **DNA replication.** By replication, the DNA of every chromosome becomes represented by two identical copies. Replication cannot be seen by eye but can be detected by (1) staining to determine quantitatively the amount of DNA in the nucleus, which doubles as a result of replication, or (2) supplying radioactively labeled thymidine (a building block of DNA), which becomes incorporated into the nucleus during replication (Figure 4-25).

During prophase (Figure 4-31A) the chromo-somes contract and become microscopically visible. As a result of the preceding replication each chromosome is longitudinally double, comprised of two identical half-chromosomes, or **chromatids** (Figure 4-30B), held together at one point called the **kinetochore** (or **centromere**). The nuclear envelope and nucleoli disappear (Figure 4-30C), and the fibrillar **spindle,** made up of microtubules, forms around the set of chromosomes, which move toward the midline ("equator") of the spindle, their kinetochores lining up on the equator at metaphase (Figure 4-30D).

Anaphase begins when the chromatid pairs separate (Figure 4-30E) and each is pulled toward an opposite pole of the spindle (Figure 4-30F) by microtubules attached to their respective kinetochores. This action ensures that one copy of every chromosome is received by each daughter nucleus. The mechanism of chromosome movement is not yet completely explained, but is thought to

Figure 4-31. Diagrams of the stages of mitosis. Except for the interphase diagram at upper left, these drawings correspond with the photomicrographs in Figure 4-30 bearing the same letters, and give an interpretation of those micrographs. The two copies (chromatids) of one particular chromosome are shown in heavy black and color at prophase, and subsequent diagrams show what happens to these chromatids.

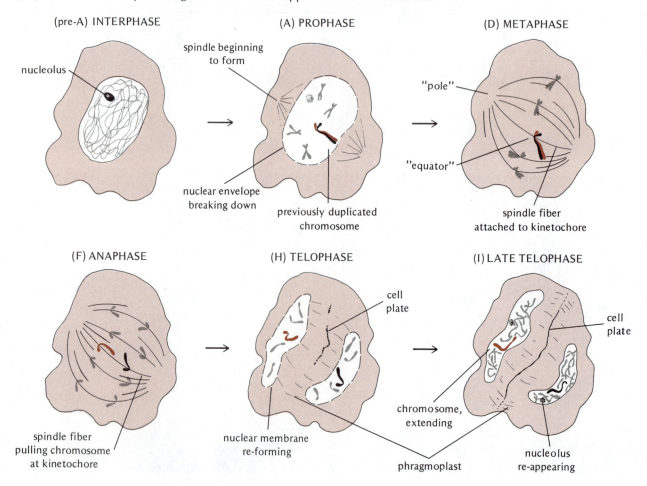

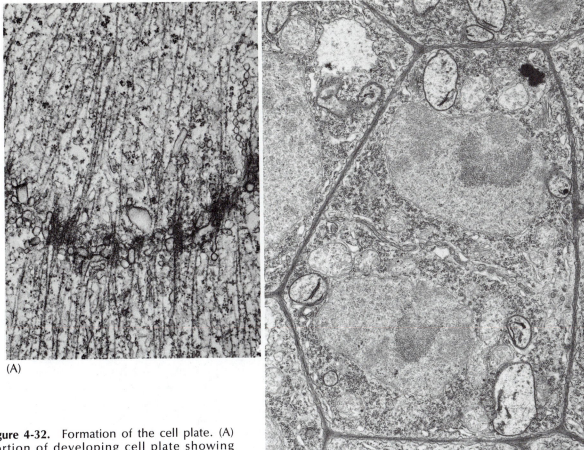

(A)

(B)

Figure 4-32. Formation of the cell plate. (A) Portion of developing cell plate showing phragmoplast microtubules and vesicles that will fuse to form the cell plate (75,000 ×). (B) Cell plate formed from fused vesicles with openings that will become plasmodesmata (20,700 ×).

involve the sliding of the microtubules attached to kinetochores, past other microtubules in the spindle.

When the chromosomes reach the poles of the spindle and cease moving (Figure 4-30G), telophase begins. The chromosomes gradually elongate (Figure 4-30H, I), nucleoli appear, and a nuclear membrane forms around each group of chromosomes (Figure 4-31H). The stages of mitosis can be completed in 1–3 hr, with anaphase normally the most rapid (note time data in Figure 4-30).

Cytokinesis

Division of a plant cell into two daughter cells normally begins in telophase by the formation of a disk-like structure called the **cell plate** in the center of the spindle region, about where the chromosomes were located at metaphase (Figures 4-30H, 4-31H). The cell plate, consisting mainly of pectin, grows outward at its edges until it meets

the parent cell's wall (Figures 4-30I, 4-31I), dividing the cell in two. Each daughter cell deposits a primary wall on its side of the cell plate, which becomes the middle lamella between the adjacent cells.

The cell plate is formed from the contents of many small membrane vesicles (Figure 4-32A) probably originating from the dictyosomes. These vesicles are moved to the midline of the spindle region, perhaps by the microtubules abundant in this area, which run in the same direction as the former spindle microtubules and form a structure called the **phragmoplast** (Figure 4-31H). These vesicles fuse to form the cell plate, bounded on either side by the membrane that previously bounded the individual vesicles (Figure 4-32B). At the outer edge of the spindle region additional microtubules appear, extending the phragmoplast and, in turn, the cell plate toward the wall of the parent cell. The membranes enclosing the cell plate become part of the plasma membranes of the two daughter cells. During the fusion of vesicles to form the cell plate, places where ER membranes pass through the vesicle zone remain open and become plasmodesmata between the daughter cells.

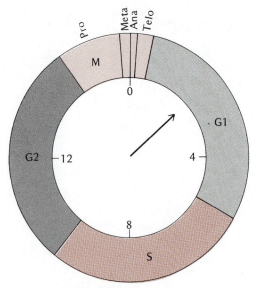

Figure 4-33. The cell cycle, represented as a clock face, measured in hours after the beginning of anaphase (ANA). G1 is the interphase period after completion of mitosis but before DNA replication; S is the period of DNA replication; G2 is the postreplication gap before the beginning of mitosis, M. Duration of the cycle (here 16 hr) and relative lengths of its phases vary with temperature and between different species and cell types.

The Cell Cycle

Mitosis, cytokinesis, and the biosynthetic activities of interphase that double the amounts of all cell constituents and organelles comprise the **cell cycle** by which dividing cells reproduce. The concept of the cell cycle as a recurring sequence of specific events, each of definite duration, is represented in a form analogous to a clock face in Figure 4-33. The duration of the entire cycle is called the cell **generation time.** Because the functions of the cycle depend on metabolism, the generation time varies with the temperature and the nutrient supply; it can be as short as 10 hr in actively growing plant tissues under favorable conditions. Mitosis itself occupies only a small fraction of the generation time (Figure 4-33).

Although some functions, such as respiration and protein synthesis, occur continuously during cell growth, their rates may vary rather dramatically over the course of the cell cycle; for example, protein synthesis and respiration typically fall to a minimum during mitosis. RNA synthesis ceases completely during mitosis. Certain other functions occur mainly or only at a particular point in the cell cycle. The most important example is nuclear DNA replication, which normally does not continue throughout interphase but occurs during a relatively limited time period called the S phase (synthesis period), more or less in the middle of interphase (Figure 4-33). Cells that are in S phase can be identified by

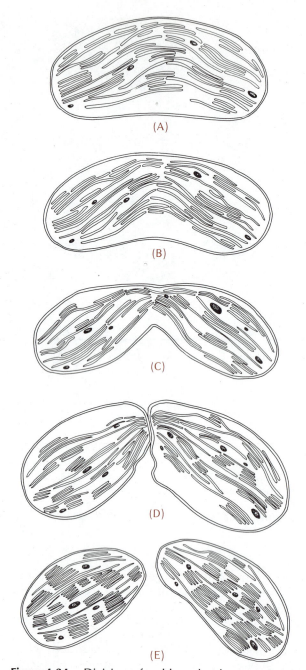

Figure 4-34. Division of a chloroplast by constriction.

their ability to incorporate radioactive thymidine into nuclear DNA (Figure 4-25).

Other organelles also must be duplicated in the course of the cell cycle. Mitochondria and chloroplasts or proplastids grow, replicate their DNA, and divide merely by constriction (Figure 4-34), and dictyosomes sometimes divide; in most cases these events are not synchronized with one another or with the division of the cell itself. In various algal cells with only one or two large chloroplasts or dictyosomes per cell, however, these organelles divide at a particular point in the cell cycle, usually late in interphase, or during mitosis or cytokinesis.

SUMMARY

(1) Protoplasm, or the living substance of cells, is a protein-rich fluid or gel in which metabolism takes place, containing organelles responsible for certain aspects of metabolism.

(2) Eukaryotic cells possess a membrane-bounded nucleus and membranous cytoplasmic organelles. Prokaryotic cells (bacteria and blue-green algae) lack these features. All plants other than the groups mentioned are eukaryotic.

(3) Typical plant cells are 10–50 μm (0.0004–0.002 in) in diameter and can be seen only with the microscope; some unusually large plant cells are visible to the unaided eye. The structure of cell organelles can be discerned clearly only with the electron microscope.

(4) Cellular membranes regulating the composition of cells and their parts include the external or plasma membrane, the membranes of organelles, and the internal membrane system or endoplasmic reticulum (ER). The ER is a site of synthesis of protein and lipid (fat-like) components for cellular membranes. Golgi bodies or dictyosomes are specialized parts of the internal membrane system responsible for the synthesis of cell wall carbohydrates and the secretion of these and of proteins and other materials to the outside of the cell. Plasmodesmata are membrane channels for the transfer of materials between adjacent plant cells.

(5) Ribosomes are particles that synthesize proteins. Free cytoplasmic ribosomes make proteins needed in the cytoplasm; ribosomes attached to the ER make membrane proteins and proteins for export from the cell.

(6) Energy-transforming organelles include mitochondria, which are the sites of cellular respiration, and plastids, which include green chloroplasts responsible for photosynthesis and colorless amyloplasts that synthesize and store starch. Yellow chromoplasts are responsible for the yellow colors of many flowers and fruits. All these organelles have a double membrane envelope. Other, single membrane-bounded organelles are responsible for other aspects of energy metabolism such as the accumulation and breakdown of fats stored as energy reserves.

(7) Fibrillar organelles function in motility processes such as cell division (microtubules) and cytoplasmic streaming (microfilaments).

(8) The nucleus, bounded by a double membrane nuclear envelope, contains the cell's hereditary material (DNA) combined with proteins in the chromosomes, plus one or more nucleoli containing RNA for cytoplasmic ribosomes.

(9) Vacuoles, bounded by a vacuolar membrane, contain nonprotoplasmic contents. The vacuole comprises most of a typical mature plant cell's volume. Vacuoles serve for the storage of soluble reserves such as sugars and for the accumulation of metabolic products such as red and blue anthocyanin pigments and oxalate crystals.

(10) Isolating individual cell organelles and membranes by centrifugation permits their functions to be investigated and determined. Autoradiography allows the occurrence and location of biosynthetic processes to be determined in intact cells by providing them with suitable radioactively labeled compounds.

(11) The usually thin primary cell wall, capable of growth in area, contains a relatively loose array of cellulose microfibrils embedded in matrix material. Between the primary walls of adjacent cells is the middle lamella, also composed of matrix material. As cells grow and mature the middle lamella breaks down in places, yielding intercellular gas spaces for the uptake of oxygen and carbon dioxide from the environment.

(12) A secondary wall, rigid and incapable of growth in area, is deposited upon the primary wall by many cells to provide mechanical support for the cell or the plant. The secondary wall consists predominantly of highly ordered cellulose microfibrils, in successive layers, usually encrusted with lignin.

(13) Plant cell shapes and specializations are due mainly to variations in their cell walls. Parenchyma cells are thin-walled living cells, compact in shape. Sclerenchyma cells have a thick secondary wall and are often elongated in shape. Epidermal and cork cells produce walls containing waxy materials that prevent water loss.

(14) In nuclear division (mitosis) the previously replicated chromosomes shorten (prophase), move to the equator of the newly formed spindle (metaphase), separate into two identical halves that move in opposite directions (anaphase), and become organized into two daughter nuclei (telophase).

(15) Cytoplasmic division (cytokinesis) occurs during telophase by the formation of a cell plate in the midregion of the spindle (phragmoplast).

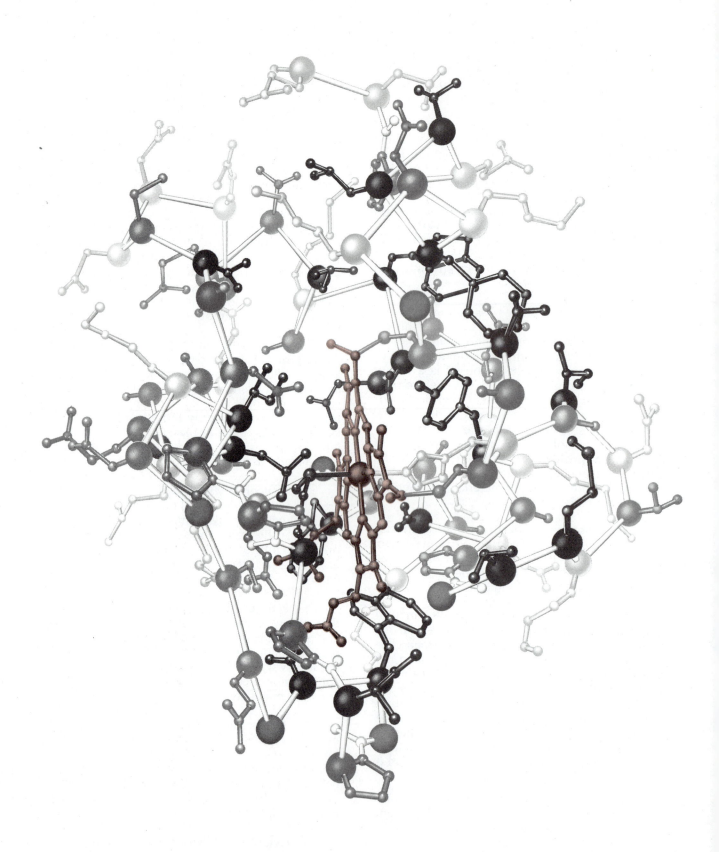

Chapter 5

CELL MATTER AND INFORMATION

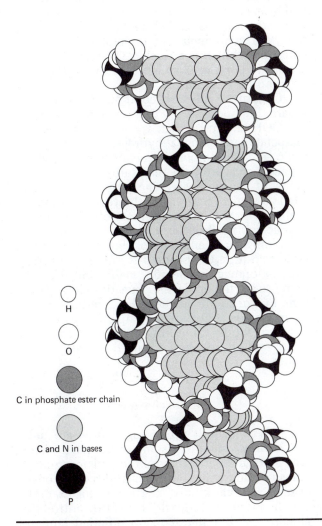

H

O

C in phosphate ester chain

C and N in bases

P

Cell functions, and hence the functions of whole plants, depend on the substances that make up the cell substructures discussed in Chapter 4. To synthesize these substances, detailed information is needed. Plant cells, like all known cells, carry this information in their DNA. This chapter introduces the important plant cell constituents and the pathway of information transfer that makes their production possible. To understand these matters an acquaintance with some of the elementary principles of chemistry is needed, as presented in Appendix I.

Quantitatively, **water** is the most important cell constituent, comprising about 80–90% of most plant cells. It is an essential solvent for many cell substances, allowing them to participate in cellular chemistry or metabolism, and it engages directly in many of the cell's metabolic reactions, including those of photosynthesis. Cells, of course, get water from their environment. For land plants, obtaining water is not just a cellular problem but a function of the entire organism, covered in Chapters 9–11. Inorganic salts from the environment are also important in cell function; their uptake is covered in Chapters 6 and 11. Here we consider the nature, functions, and synthesis of the organic or carbon constituents of cells.

Functionally we can distinguish several kinds of organic components of plants: (1) **Structural molecules** are comparable to the components of buildings in the human economy. (2) **Catalytic molecules,** which facilitate the occurrence of

Figures 5-A and B. Structural models of informational macromolecules. (A, left) A bacterial form of the protein, cytochrome *c*, important in cell metabolism. Its polypeptide chain consists of 82 amino acids. Each amino acid is shown as a sphere (representing its peptide bond-forming parts) plus an outline of its distinguishing "R group" (Figure 5-6), which projects outward from the polypeptide backbone, the chain of larger spheres connected by heavier "struts." The unique amino acid sequence of the protein expresses information that is encoded in the cell's DNA. The roughly circular configuration at left center is a part of the polypeptide chain that is coiled into a corkscrew-like arrangement (α-helix), here seen end on. Similar helical regions can be seen obliquely elsewhere in the molecule. The cytochrome *c* molecule also contains, at its center, an organic ring system (*heme,* shown in color) containing an atom of iron which participates in this protein's electron-carrying function in cell respiration. (B, above) The double helix of DNA (deoxyribonucleic acid), the cell's carrier of genetic information. It consists of two separate polynucleotide chains twisted around one another, visible on the outside of the molecule. Each chain is a backbone of alternating sugar and phosphate units, bearing a specific sequence of nitrogen bases which associate closely in the center of the structure, holding the two chains together and coding the information that the molecule carries.

processes in the cell, are analogous to machinery in our economy. Many of the structural and catalytic molecules of cells are **polymers,** that is, very large molecules called macromolecules (Greek *macro,* large) built up by joining many smaller subunit molecules together more or less repetitively. (3) **Informational macromolecules** are comparable to the designs and blueprints for making machinery and buildings. (4) **Metabolites** are substances that are being changed from one chemical form into another within the cell to supply useful energy or to yield some end product such as a structural or catalytic polymer. Metabolites can be compared with goods in production and energy in generation and transmission. (5) **Storage reserves** are photosynthetic products (or materials derived from them) that can be moved about in the plant or stored in cells, for future use in metabolism. Storage reserves, which can be compared with raw materials and fuels being transported and stockpiled, are sometimes called plant foods because plant cells use them for their energy production and biosynthetic activities much as animals and humans use the food they eat. (6) **Secondary products** are metabolic end products not having an immediate functional or structural use in the cell. While traditionally often regarded as "waste products" comparable to human bodily or industrial wastes, many secondary products in fact have a significant ecological value to plants, as noted in Chapter 1. Secondary products, which result from specialized kinds of metabolism, are discussed in Chapter 6; the present chapter covers only constituents that are common to cells generally.

Basic cell constituents fall into four main chemical groups, mostly familiar from an everyday knowledge of nutrition: (1) **carbohydrates;** (2) **lipids** or fat-like substances including fats themselves; (3) **proteins** and their building blocks; and (4) **nucleic acids** and their building blocks, called nucleotides. Each of these chemical groups includes molecules from several of the functional categories previously mentioned.

The nature and function of important cell constituents can be introduced without going into most of the details of their chemical structure, which become complicated. For reference, however, some of these details are given as structural formulas in the diagrams, which can be regarded as optional material, important to the subject but not essential at the introductory level.

CARBOHYDRATES

Carbohydrates are the sugars and related compounds. The name carbohydrate comes from the fact that common sugars such as **glucose** or dextrose ($C_6H_{12}O_6$) contain carbon, plus hydrogen and oxygen atoms in the same proportion as in water (H_2O; Greek *hydor,* water). Glucose (Figure 5-1A) is perhaps the most basic of all foodstuffs or metabolites, being the starting material in most cells for both energy production and biosynthesis. Glucose is stored both free and as a constituent of various reserve products; for example, cane sugar or **sucrose,** which is a compound sugar or disaccharide (Figure 5-1B), is made by joining one molecule of glucose to one of the slightly different sugar, **fructose.** Sucrose is accumulated in large amounts as a storage reserve by sugarcane and sugar beet, from which it is obtained as commercial table sugar. Both glucose and fructose are six-carbon or **hexose** sugars ($C_6H_{12}O_6$; Greek *hex,* six, and *-ose,* ending that denotes sugars). Sugars with five carbon atoms, called **pentoses** (Greek *pente,* five), are components of important macromolecules including nucleic acids. Other kinds of sugars are important as metabolites in both respiration and photosynthesis.

Organic acids are compounds related to sugars but possessing one or more carboxyl (—COOH) groups, which are acidic because they dissociate, releasing a hydrogen ion (H^+). Organic acids are important metabolites into which sugar is converted in respiration and in some kinds of biosynthesis. They are also accumulated as storage reserves in the vacuoles of many plant cells; the names of most of the common organic acids reflect the scientific names of plants containing large amounts of them, for example, citric acid ($C_6H_8O_6$; Figure 5-1C) from *Citrus* fruits such as orange and lemon, malic acid ($C_4H_6O_5$) from *Malus,* the apple, and fumaric acid ($C_4H_4O_4$) from *Fumaria,* the flower called fumitory. Stored organic acids contribute an important part of the flavor of foods containing them, and as lemon juice, and so on, they are used to flavor many dishes.

Polysaccharides

Polymers built up of sugar units are **polysaccharides,** including some of the most important storage reserves and structural macromolecules of plant cells. Polysaccharides may consist of only a single kind of sugar unit, for example, starch (Figure 5-1D), which is built entirely of glucose; or they may contain different kinds of sugar units hooked together as in pectins and hemicelluloses, structural macromolecules composing the middle lamella and the matrix (nonfibrillar) material of cell walls. Polysaccharides may consist of a straight chain of sugar units (Figure 5-2A, B), or the chain may branch more or less extensively (Figure 5-2C, D).

Starch is a glucose polymer, accumulated in starch granules in amyloplasts and chloroplasts, and consisting of both unbranched and branched-chain components (Figure 5-2B, C). Starch can be readily broken down by the cell into its glucose

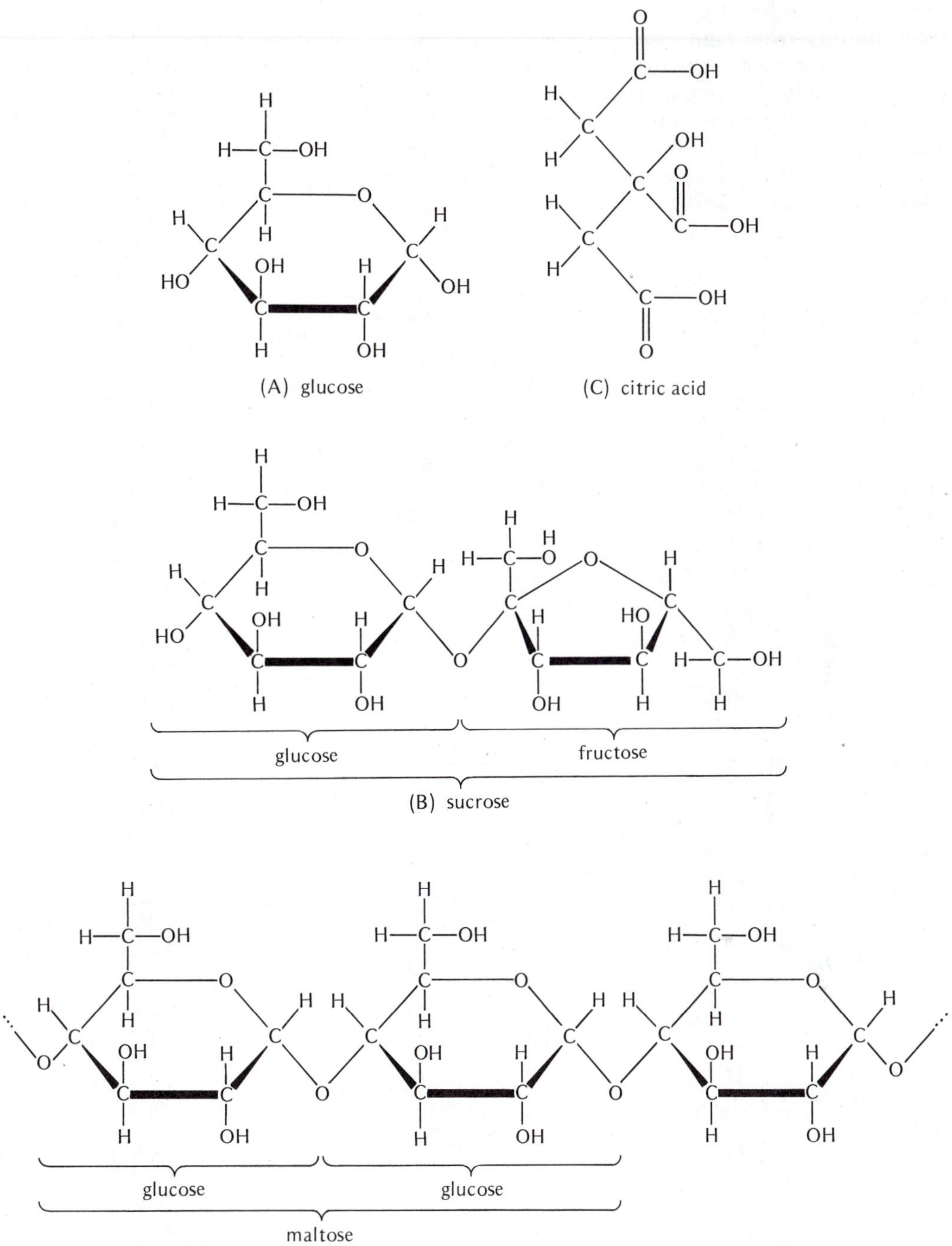

(D) part of a polysaccharide (starch) chain

Figure 5-1. Examples of carbohydrates (A, B, D) and related organic acids (C). (A) A simple sugar; (B) a disaccharide; (D) part of the (polysaccharide) chain of starch, the repeating unit of which is the disaccharide maltose.

subunits for use in metabolism, so it is a compact storage reserve. Starch stored by cereal grain seeds, and by storage roots and tubers such as the potato, comprises the staple energy food of mankind.

Cellulose is a simple chain of glucose units, so linked that the chain tends to lie straight (Figure 5-2A) rather than curling as the chains of starch do. The extended chains of cellulose readily and indeed avidly associate to form **microfibrils** (Figure 4-26), strong cable-like structures composed of some 30 to as many as several hundred cellulose chains side by side. These microfibrils provide the basic skeletal material of most plant cell walls. Since woody plant tissues, consisting predominantly of cellulose, are the most copious biological product on earth, cellulose has been called "nature's most abundant macromolecule."

Although humans use cellulose as a fuel both

directly and indirectly, as coal (fossil carbon derived mainly from cellulose), cellulose does not serve as a storage reserve for the plants that make it, because they cannot usually break it down, like starch, into its glucose subunits. Humans cannot digest cellulose, either, so the cellulose content of foods comprises much of what is called the "crude fiber," of no caloric value.

Chemical modification of cellulose to yield nitrocellulose, cellulose acetate, methyl cellulose, and so on, has made possible the production of rayon fibers, celluloid films and plastics, acetate wrapping materials, pyroxylin lacquers, cellulose gum food stabilizers, thickening and emulsifying agents, and explosives (gun cotton). The success of the industries utilizing these products ushered in the era of plastics and synthetic fibers, which have revolutionized con-

Figure 5-2. Examples of plant polysaccharides. (A–C) Simple polysaccharides, composed only of glucose units (shown as hexagonal rings). (A) Cellulose. (B) The amylose (straight-chain) component of starch: its glucose units are linked such that the chain coils into a helical (corkscrew) configuration. (C) Branched-chain or amylopectin component of starch. Its chains all tend to coil as in (B), but for clarity this feature is omitted. (D) Examples of hemicelluloses and pectins, complex polysaccharides composed of different sugar units. Only representative parts of these macromolecules are shown, consisting of the hexose units G, glucose; GAL, galactose; F, fucose; and R, rhamnose; the pentose units A, arabinose, and X, xylose; and the sugar acid, galacturonic acid (U). N indicates straight runs of up to about seven galacturonic acid units; S is an arabinose-galactose side chain similar to that shown just above.

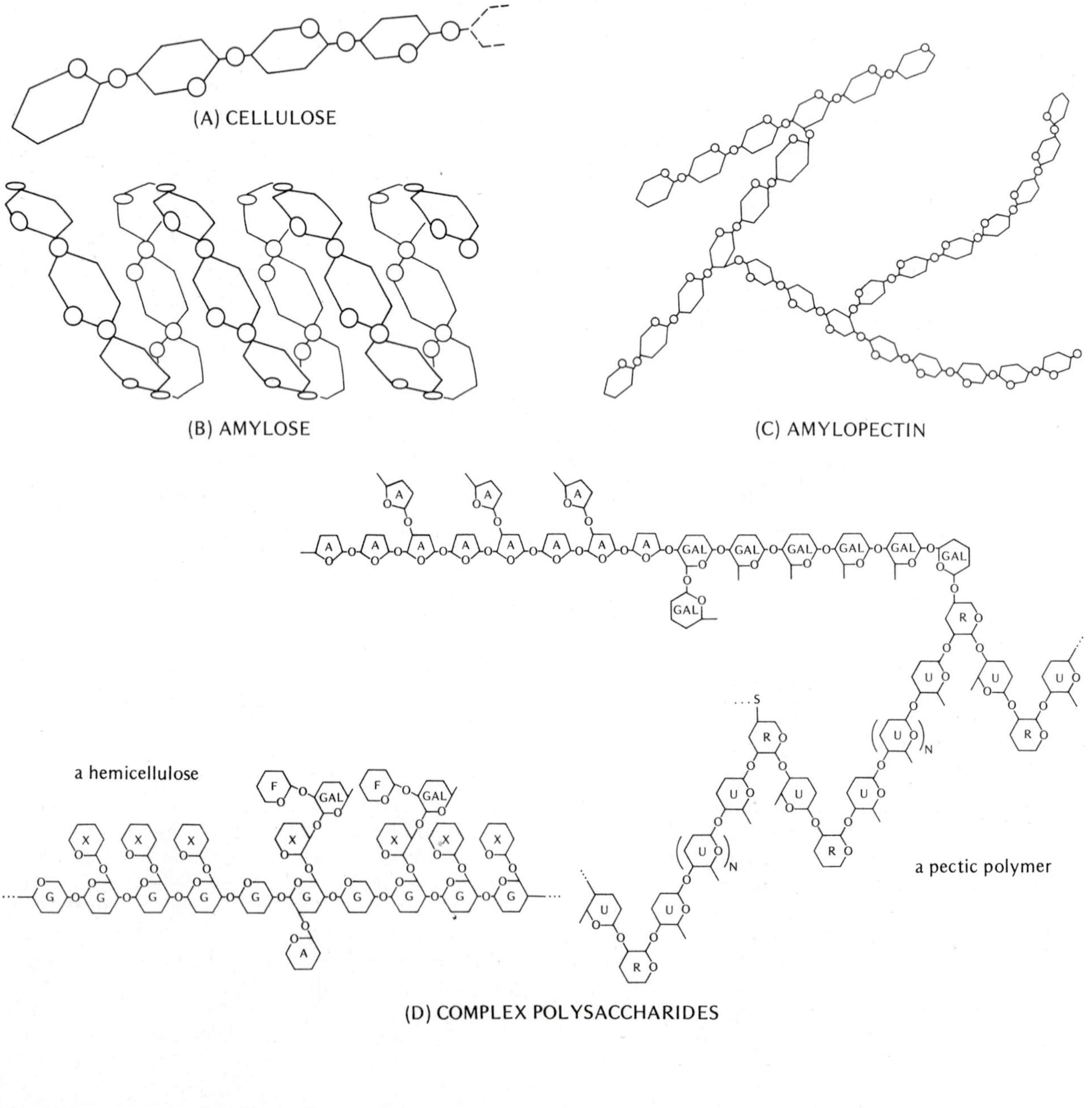

(A) CELLULOSE

(B) AMYLOSE

(C) AMYLOPECTIN

a hemicellulose

a pectic polymer

(D) COMPLEX POLYSACCHARIDES

sumer goods and marketing in the last several decades and are now based primarily on petrochemicals (compounds obtained from petroleum). With the impending exhaustion of the petroleum supply, a resurgence of interest in cellulose-based synthetic fibers and plastics may well occur.

Pectins and hemicelluloses (Figure 5-2D) are the "glue" that binds cellulose microfibrils together in cell walls. Hemicelluloses include a variety of usually branched polysaccharide structures, most of which contain pentose sugars instead of or in addition to hexoses. Pectins are composed primarily of a "sugar acid," galacturonic acid, similar to a hexose (galactose) but possessing an acid (—COOH) group. Hexose and pentose sugars also occur in pectins (Figure 5-2D).

LIPIDS

Lipids are fat-like cell substances that tend to associate with one another and to separate from water as a distinct oily phase, familiar, for example, in the separation of olive oil (lipid) from vinegar (water solution) that occurs in Italian salad dressing. Lipids are hydrophobic ("water fearing"), in contrast to carbohydrates, nucleic acids, and most proteins, which attract water molecules or are hydrophilic ("water loving") and tend to dissolve in water.

Fats

Many cells store fats as energy reserves. Unlike solid animal fats, most plant fats are liquids, called vegetable oils. The commercially important vegetable oils used in cooking, salad dressings, margarine, paint, and so on, are storage reserves obtained from various seeds and one fruit (the olive). The oily or greasy properties of fats and most other lipids are due to the fatty acids that they contain (Figure 5-3A), which are very hydrophobic because their long carbon and hydrogen chains lack oxygen atoms which would attract water. In a fat, fatty acids are attached to a three-carbon carbohydrate, glycerol. Because of their relative lack of oxygen, fats have a considerably higher energy content per unit of weight than carbohydrates do, so they make a very efficient energy reserve.

Membrane Lipids and Membrane Structure

The principal components of cellular membranes are phospholipids. These contain phosphate and some other hydrophilic group (Figure 5-3B), besides glycerol and fatty acids as in a fat. While the hydrophobic fatty acid "tails" tend to associate into a distinct lipid phase, the hydrophilic "head group" (including the phosphate) attracts and associates with water. This favors the formation of a lipid layer two molecules thick, with the fatty acid tails in the interior and the head groups exposed to water at the surfaces (Figure 5-4). This phospholipid bilayer appears to be the basis of cellular membrane structure, apparently corresponding to the three-layered "unit membrane" seen by electron microscopy (Chapter 4), as explained in Figure 5-4. This structure provides, at cell and organelle surfaces, a thin lipid boundary that hydrophilic substances such as sugars and salts cannot readily penetrate and, therefore, cannot easily cross. Thus phospholipids have both a structural and a functional role in membranes.

Cellular membranes also contain special proteins called membrane proteins which penetrate into, or through, the lipid bilayer; the actual structure of a cellular membrane is apparently a "mosaic" or composite of lipid bilayer areas and intercalated membrane proteins or protein complexes (Figure 5-4). Some of these proteins function as trans-membrane carriers for the transport of hydrophilic substances such as sugars and particular inorganic ions needed by the cell or its organelles (Chapter 6).

At normal cell temperatures the lipids of cellular membranes are relatively loosely associated, the membrane being a fluid rather than a solid structure. This fluidity permits membrane lipids and proteins some movement within the membrane, affording a certain degree of permeability to other molecules and enabling transport carriers to deliver needed nutrients to the other side.

Waterproofing Lipids

Cutin and suberin are specialized lipids that, respectively, impregnate the walls of epidermal and cork cells at the aerial surfaces of plants. These lipids contain fatty acids that bear —OH groups permitting the fatty acids to be linked together and polymerized to form a polyester (Figure 5-3C) analogous to the polyester materials now used for synthetic fibers. These lipids do not contain glycerol, but suberin contains other components in a more complicated structure than that illustrated for cutin in Figure 5-3C. In addition to these polymers deposited in the cell walls, plant surfaces are also covered by waxes, which are a mixture of long-chain fatty acids and derivatives thereof, including long-chain alcohols. Evidence indicates that these surface waxes are especially important in protecting against desiccation.

PROTEINS

By far the most versatile of biological substances, proteins are the "action molecules" of cells. Proteins called enzymes bring about the chemical reactions of metabolism, providing the cell with useful energy and conducting the synthesis of cell components. Other proteins bring about motility, including cytoplasmic streaming and chromo-

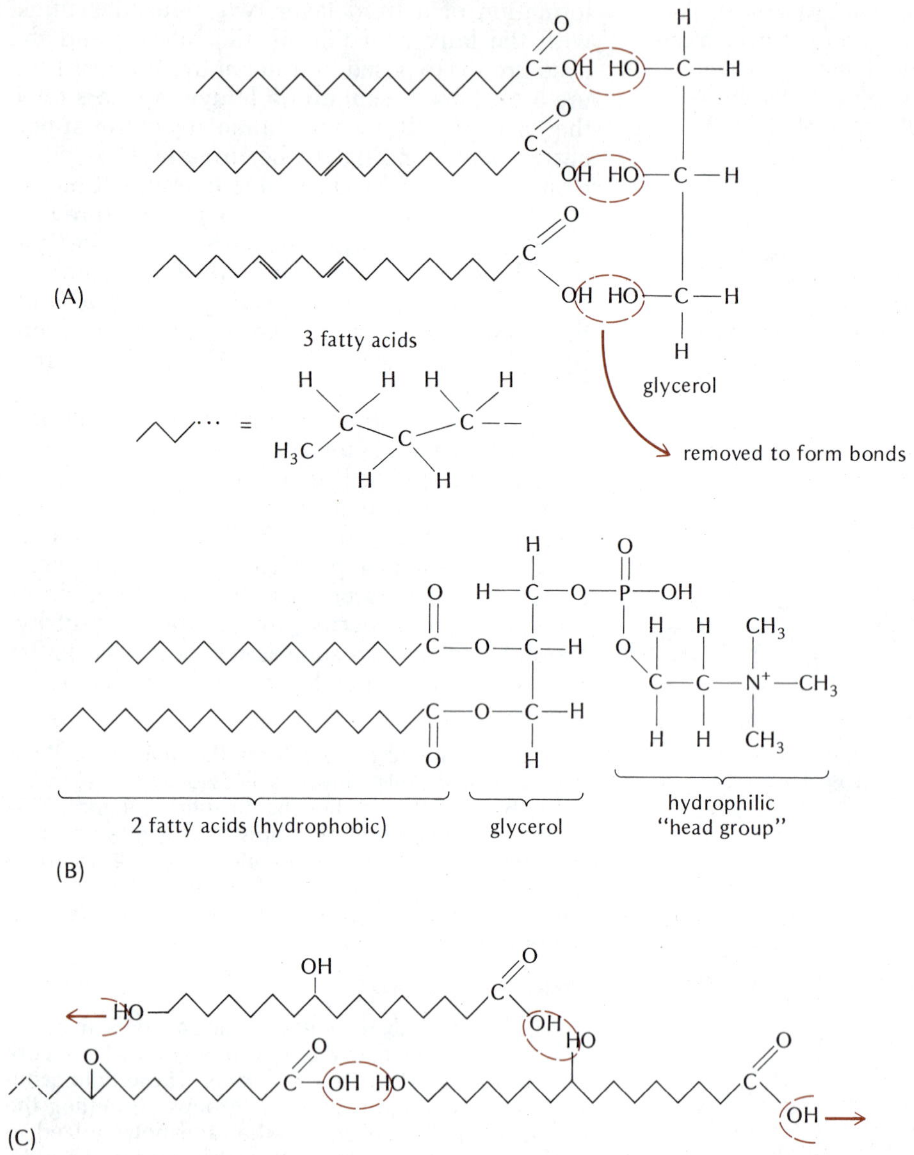

Figure 5-3. Some examples of lipids. (A) How fatty acids and glycerol can be joined, by removal of H₂O, into a triglyceride (fat or vegetable oil). Fatty acid chains (composed of carbon and hydrogen only) are shown conventionally by a zigzag symbol as explained below diagram (A). Double lines represent double bonds in the chain. (B) A phospholipid (lecithin). (C) Hydroxy fatty acids from cutin, showing how they can be joined together. Only part of the third fatty acid chain (lower left), similar to the others, is shown. Arrows indicate how repetition of the linkage between fatty acids can build up the polymer (a polyester).

some movement in mitosis as well as various kinds of locomotion. Still others have a structural role, for example, as components of membranes and of primary cell walls. Special proteins accumulated by seeds serve as storage reserves, used by the embryo plant for its growth during seed germination.

Proteins are polymers built of units called **amino acids,** which contain nitrogen in the form of amino (—NH₂) groups. Proteins are therefore rich in nitrogen, and this is why plants need relatively large amounts of nitrogen as a mineral nutrient. The polymer chains of proteins vary from

about 100 to over 1000 amino acids in length and are called **polypeptide** chains because the linkage between each amino acid unit and its neighbor is known as a peptide bond (Figure 5-5). About twenty different amino acids (Figure 5-6) are linked together in various combinations to form polypeptide chains. Two of these amino acids contain sulfur, so most proteins contain a small amount of sulfur, which is essential to their structure and function and is one of the reasons plants require sulfur as a mineral nutrient.

Every cell contains many different proteins, each kind of protein being a specific, recognizable

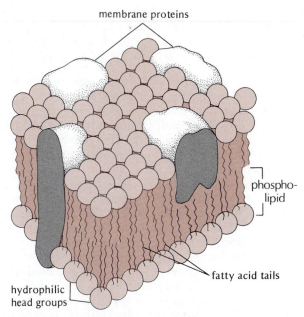

membrane proteins

phospho-lipid

fatty acid tails

hydrophilic head groups

Figure 5-4. Phospholipid bilayer structure of cellular membranes. The two layers of hydrophilic head groups apparently correspond to the electron-dense (dark) outer layers of the "unit membrane" three-layer structure seen by electron microscopy (Figure 4-11A). Membrane proteins are inserted into the lipid bilayer occasionally, as illustrated.

molecule having a particular size and polypeptide chain length and a specific and unique sequence of amino acids running from one end of its polypeptide chain to the other. Most protein molecules are relatively compact and globular, each kind having a unique three-dimensional shape essential to the protein's specific function. This shape arises from the coiling, bending, and folding of the polypeptide chain around and back upon itself in a highly specific way (Figure 5-A). Disruption of the native shape of a protein, called **denaturation**—for example, by heating it to a temperature near boiling—destroys its biological activity. This is why cooking kills cells and organisms.

Many proteins, including numerous enzymes, consist of more than one polypeptide chain or **subunit**, associated together. Proteins of this type may consist of several identical subunits or of two or more different subunits differing in amino acid sequence and shape. The biological function of such a protein may depend completely upon the mutual combination of its subunits or may simply be modified or improved by the association. This feature of protein structure adds another dimension of possible variations in shape and function beyond that due to variations in the amino acid sequences and resultant three-dimensional shapes of single polypeptide chains. Some of the most important proteins in plant function, including the enzyme that captures CO_2 in photosynthesis, have this multi-subunit structure.

Enzymes

Enzymes are proteins that catalyze, that is, speed up, chemical reactions without being permanently altered themselves in the process. Under the temperatures and other conditions prevailing in cells, most of the chemical reactions comprising metabolism would occur at uselessly slow rates if it were not for the enzymes that catalyze them. Enzyme catalysis is highly specific regarding both the kind of reaction catalyzed and the particular metabolite(s) that engages in it, which is called the **substrate** of the enzymatic reaction. Because of this specificity, virtually every different reaction in cell metabolism requires a distinct enzyme; because metabolism involves many different reactions, cells must produce many different enzymes in order to function.

Enzymes are named, conventionally, by adding the suffix -ase to the name of one of the substrates of the catalyzed reaction (or to a term for the type of reaction, e.g., citrate synthase, an enzyme that synthesizes citrate or citric acid). Catalysis of the reaction is indicated by writing the name of the responsible enzyme above the arrow that represents the occurrence of a reaction; for example, one of the reactions in respiration is

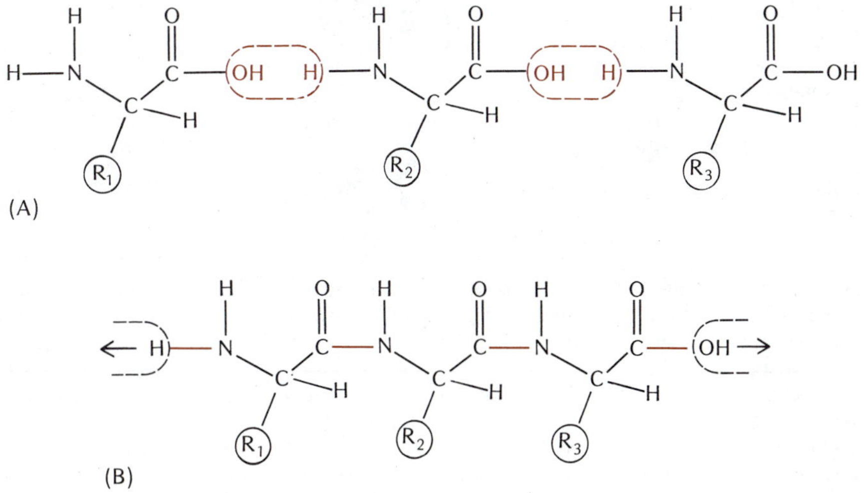

Figure 5-5. Amino acids (A), showing how peptide bonds (color lines in B) can be formed by removing the elements of water from amino ($-NH_2$) and carboxyl ($-COOH$) groups. Arrows in B show how the process can continue indefinitely to build up a polypeptide. R_1, R_2, and R_3 stand for the side chains that distinguish different amino acids from one another (Figure 5-6).

(A)

(B)

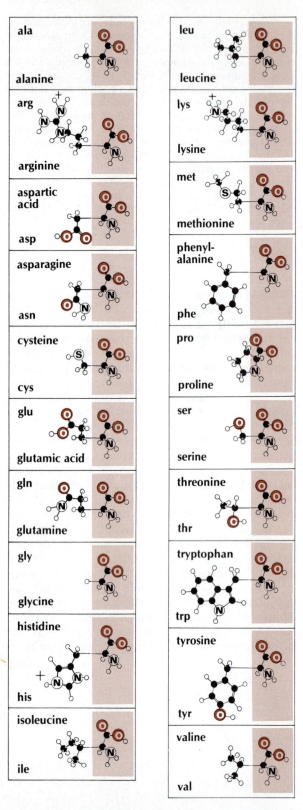

$$C_4H_4O_4 \quad + \quad H_2O \quad \xrightarrow{\text{fumarase}} \quad C_4H_6O_5 \quad (5\text{-}1)$$

fumaric acid   water   malic acid
(substrates)   (product)

Enzymes catalyze reactions essentially by allowing the reacting molecules to interact with one another in a specific way more easily than they would when colliding randomly as in an uncatalyzed reaction. An enzyme combines transiently with one or more of its substrates; for example, fumarase associates with a fumaric acid molecule to form an enzyme–substrate complex, which then undergoes reaction to yield the product, malic acid [Eq. (5-1)]. The formation of the enzyme–substrate complex is made possible by the specific three-dimensional shape of the protein molecule constituting the enzyme: its surface bears a groove, cleft, or pocket called the **active site** (Figure 5-7) into which the substrate fits rather like a hand in a glove. When inserted into the active site, the substrate molecule is slightly bent or distorted in such a way that it undergoes reaction much more readily than it otherwise would. The reaction product, such as malic acid in Eq. (5-1), dissociates from the active site, allowing the enzyme molecule to accept substrate molecules again and again, catalyzing the reaction repeatedly.

COFACTORS

To catalyze a reaction, many enzymes require the help of a specific inorganic ion, often called a **cofactor.** For example, many enzymes of metabolism require magnesium ions (Mg^{2+}); this is one of the reasons magnesium is needed as a mineral nutrient. Many enzymes require some small organic molecule as a cofactor. Other important enzymes contain a metal atom or non-amino-acid organic group as an essential component of the enzyme molecule. Some of the vital enzymes of respiration, for example, contain an atom of iron that participates in the reaction catalyzed by the enzyme.

TEMPERATURE EFFECTS

Like chemical reactions in general, the rate of most enzymatic reactions varies greatly with temperature, rising two- to threefold for every increase of 10°C in temperature. Consequently, cell functions depend greatly on temperature. The me-

Figure 5-6. Names and structures of the 20 amino acids that make up most proteins, along with the three-letter abbreviations commonly used to designate amino acids in representing protein structure. The shaded area covers the parts of each amino acid that form peptide bonds (Figure 5-5). Different amino acids differ from one another in the remainder of their structure, shown outside the shaded area (R groups in Figure 5-5). These differences confer different properties on the respective parts of a protein's polypeptide chain, helping it to fold up into a specific three-dimensional shape (Figure 5-7) and making for local differences in the protein molecule's surface that are important in its biological (e.g., enzymatic) activity. N, nitrogen; O, oxygen; S, sulfur; black circles, carbon; white circles, hydrogen.

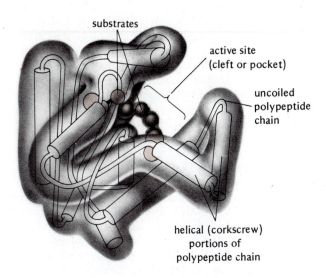

substrates

active site
(cleft or pocket)

uncoiled
polypeptide
chain

helical (corkscrew)
portions of
polypeptide chain

Figure 5-7. Approximate three-dimensional shape of an enzyme, adenylate kinase, showing the active site or cleft into which fit its two substrates [called respectively AMP (left) and ATP (right)]. The path known to be taken by the polypeptide chain of 194 amino acids, in creating the enzyme's specific shape, is shown with a dark line and cylinders within the external outline, the cylinders conventionally representing regions where the polypeptide chain takes a helical (corkscrew) course. Because of projecting amino acid R groups (Figure 5-6) the enzyme's external surface must actually be more irregular than the approximate outline drawn. (Adapted from R. J. P. Williams, Biological Reviews, Vol. 54, pp. 389–437, 1979, Fig. 7.)

tabolism of most cells virtually ceases if they are cooled to 0°C, the temperature of ice water. However, unlike an uncatalyzed chemical reaction such as combustion, high temperatures also stop enzymatic reactions by denaturing the responsible enzyme, thus destroying its specific shape and catalytic activity.

EFFECTS OF ACIDITY OR pH

Acid (H^+) or alkali (OH^-) severely inhibits most enzymes, which show catalytic activity only at H^+ concentrations relatively near neutrality. A logarithmic scale called **pH** (Figure 5-8) is used to express in convenient numbers the low H^+ concentrations that must exist inside cells for them to function properly. This scale spreads out the low H^+ concentration range and displays the fact that any given enzyme typically exhibits a definite **pH optimum** for its activity, usually in the vicin-

ity of neutrality (pH 7). Cells must regulate their internal pH, keeping it relatively near the pH optima of the enzymes that are important in metabolism. Plant cells seem to keep their cytoplasm at about pH 7.1–7.3.

NUCLEIC ACIDS

The nucleic acids, DNA and RNA, are the informational macromolecules that enable cells to construct their specific proteins and thus to carry out the diverse enzymatic reactions of metabolism. Nucleic acids are built of subunits termed **nucleotides.** As discussed in Chapter 6, several nucleotides function also, or instead, as important metabolites or enzyme cofactors, crucially involved in energy metabolism and biosynthesis. A nucleotide (Figure 5-9A) consists of a phosphate group attached to a pentose sugar, in turn attached to a nitrogen-containing ring structure called a **nitrogen base.**

Structure of DNA and RNA

Nucleotides are coupled into nucleic acid polymers by joining the phosphate group of each

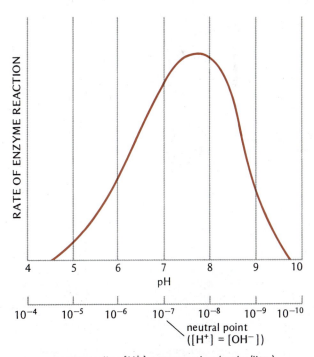

corresponding [H⁺] concentration (moles/liter)

Figure 5-8. Typical dependence of enzyme activity on acidity, or pH. Most cytoplasmic enzymes show maximum activity at pH 7–8 (their "pH optimum"). The numbers below the pH scale show the corresponding H^+ concentrations, in moles/liter, expressed in powers of 10, e.g., 10^{-4} = .0001, 10^{-5} = .00001, etc.

nucleotide to the pentose of the next (Figure 5-9B). This yields a "polynucleotide" backbone of alternating phosphate and pentose groups with nitrogen bases projecting out from the pentoses as side groups. RNA (ribonucleic acid) is built of nucleotides containing the pentose **ribose,** whereas DNA (deoxyribonucleic acid) contains instead **deoxyribose,** with one less oxygen atom than ribose. Four different bases called adenine (A), guanine (G), cytosine (C), and uracil (U) occur in RNA, whereas DNA contains A, G, C, and thymine (T).

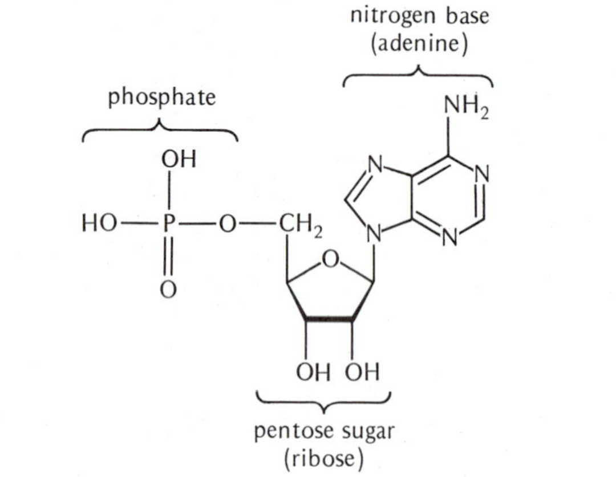

(A)

(B)

Figure 5-9. Nucleotides and nucleic acid structure. (A) A nucleotide, AMP (adenosine monophosphate). (B) Shows how nucleotides are joined together to form a polynucleotide (nucleic acid) chain. In this case AMP is joined (at the dashed line) to a nucleotide containing the nitrogen base *uracil;* arrows show how the process can be repeated to extend the chain indefinitely. Notice that in this and subsequent examples of conventional shorthand depiction of complex organic structures, a carbon atom is understood at each angle of a ring unless the symbol for some other element (N, O, P, etc.) appears there, and H atoms attached to ring carbons are usually omitted.

James D. Watson and Francis Crick discovered that these bases tend to attract or pair with one another in a specific relationship called **complementary pairing:** G pairs with C, and A pairs with T or U (Figure 5-10). This property, due to weak forces called **hydrogen bonds** that occur between the respective bases, is the basis of the hereditary and informational function of nucleic acids. The nature of these functions first became apparent from the structure of DNA deduced by Watson and Crick (Figure 5-B), called a **double helix** because it contains two polynucleotide chains twisted in a helical (corkscrew) fashion around one another. The two chains are not interconnected by chemical bonds, but adhere because of the hydrogen bonds involved in complementary pairing between their bases. At each point where one chain bears A, the other bears T, and vice versa; where one chain bears G, the other bears C, and vice versa (Figure 5-10). Therefore every base has a complementary base on the other chain with which to pair, holding the two chains together.

DNA Replication

During DNA replication, prior to mitosis, the two chains of the DNA double helix separate, and against each a new polynucleotide chain is built (Figure 5-11). The enzyme **DNA polymerase,** which catalyzes this process, will insert into the new chain at any point only a nucleotide whose base pairs with the base already present on the intact strand at that point. The preexisting strands thus act as **templates,** directing the formation of new strands perfectly complementary to each of them; the end result is two DNA double helices that are identical copies of the original one, conserving in every detail the sequence of bases from one end of the structure to the other. This capability for self-directed exact replication allows DNA to serve as a hereditary material.

DNA in eukaryotic chromosomes is associated with specific proteins called **histones** into a complex but precise nucleic acid–protein aggregate, or nucleoprotein. The DNA chains appear to be wound around successive globular aggregates of histone proteins, creating a beads-on-a-string formation, the units (beads) of which are called **nucleosomes** (Figure 5-12). This nucleoprotein is somehow aggregated into an even higher order of structure in forming the microscopically visible chromosomes.

The Genetic Code

Watson and Crick suggested, and it was later proved, that DNA encodes genetic information in the sequence of bases along its chains. A primary function of genes is to specify the proteins that a cell can make. The identity of a protein is due, as noted above, to the amino acid sequence of its

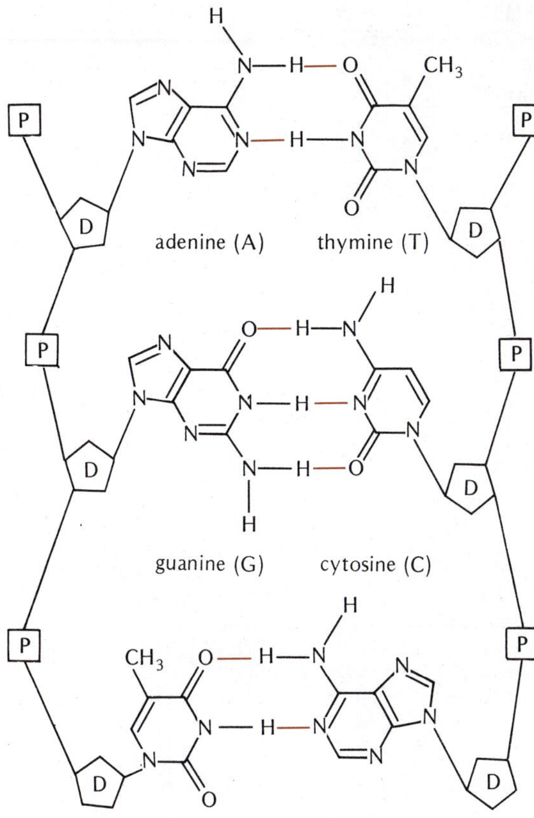

Figure 5-10. Complementary pairing between bases in DNA. The bases of each type of pair specifically interact with one another by weak forces called hydrogen bonds (shown in color) between the indicated chemical groups. Bases are depicted using the conventions mentioned in the legend to Figure 5-9. D indicates the sugar (deoxyribose) and P the phosphate units of the polynucleotide chains. The actual positioning of paired bases in a DNA double helix is depicted in Figure 5-B.

polypeptide chain(s). The base sequence in the DNA of a given gene determines the amino acid sequence of the corresponding polypeptide by means of the **genetic code.** Successive amino acids are specified by successive sets of three bases along the gene's DNA strands. These sets, analogous to three-letter words, are called **codons** (Figure 5-13). Since any one of the four bases can occur at any point in the nucleic acid chain, there are 4 × 4 × 4 or 64 possible three-letter codons that can occur in DNA, but only 20 different amino acids that need to be specified. Each of the 20 amino acids except 2 can be specified by codons ("synonyms" in the genetic "language"), so in fact almost all of the possible codons are utilized (Table 5-1). The genetic code appears to be almost universal among organisms on earth, from the simplest viruses and prokaryotes to the most advanced plants and animals.

RNA Synthesis: Transcription from DNA

The cell's DNA includes genes that code for the specific base sequences of various kinds of RNA, such as the ribosomal RNAs needed for ribosome formation. From the DNA sequences coding for them, these RNA sequences are produced

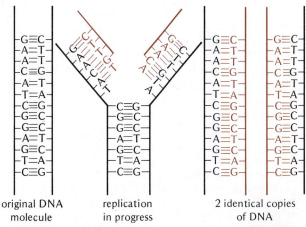

Figure 5-11. Replication of DNA. The complementary strands, here shown as if in one plane (ignoring their helical intertwining), are pulled apart and new strands (shown in color) are assembled by inserting nucleotides according to the rules of complementary pairing. The result is two identical copies. (The actual process is somewhat more complicated because the second new strand is not assembled in exactly the same way as the first.)

TABLE 5-1   The Genetic Code[a]

First position	Second position				Third position
	U	C	A	G	
U	UUU UUC Phe UUA UUG Leu	UCU UCC UCA UCG Ser	UAU UAC Tyr UAA UAG Term[b]	UGU UGC Cys UGA Term[b] UGG Trp	U C A G
C	CUU CUC CUA CUG Leu	CCU CCC CCA CCG Pro	CAU CAC His CAA CAG Gln	CGU CGC CGA CGG Arg	U C A G
A	AUU AUC Ile AUA AUG Met	ACU ACC ACA ACG Thr	AAU AAC Asn AAA AAG Lys	AGU AGC Ser AGA AGG Arg	U C A G
G	GUU GUC GUA GUG Val	GCU GCC GCA GCG Ala	GAU GAC Asp GAA GAG Glu	GGU GGC GGA GGG Gly	U C A G

[a]The codons in the transcribed DNA strand would be complementary to the RNA codons listed in this table. The amino acid abbreviations used here are defined in Figure 5-6.
[b]Codon specifying the end of a polypeptide chain (chain termination).

in the nucleus by the action of **RNA polymerase.** This enzyme forms RNA by copying DNA sequences in the same complementary way as DNA polymerase replicates DNA, except that RNA polymerase uses ribose-containing nucleotides, thus yielding an RNA chain. RNA polymerase copies only one of the two DNA strands (Figure 5-14), producing an RNA base sequence complementary to the copied DNA strand. This is called the **transcription** of DNA information into RNA information. RNA polymerase also transcribes the genes that specify proteins, producing **messenger RNA** (mRNA) molecules in which the DNA base sequence of each gene is again represented by a complementary RNA sequence. After their formation all these kinds of RNA are released from the nucleus into the cytoplasm, where the mRNA can be translated by ribosomes to yield proteins (see below).

Before release into the cytoplasm, newly transcribed RNA undergoes **posttranscriptional modifications** or "**processing,**" including the splitting and removal of certain sections to create the final form of ribosomal and messenger RNAs and the addition of a "poly(A) tail" (an RNA chain containing only adenine as base) to many of the mRNAs. The significance of these modifications is not completely clear. Poly(A) addition, however, allows scientists to isolate mRNA from cells because the poly(A) tails will stick to a matrix bearing poly(dT) chains (artificial DNA chains containing only thymine as base).

Multiple-Copy DNA

The majority of the DNA of plant cells consists of certain base sequences repeated many times, called

multiple-copy DNA. The genes specifying ribosomal and transfer RNAs, and histone proteins, occur as multiple copies and are transcribed in order to form these products. Most of the genes for cell proteins occur as single copies and most of the multiple-copy DNA does not seem to be transcribed; its significance is still a matter of speculation. Some investigators have suggested that it might serve to regulate the function of other genes, but this is still not certain. Others regard it as merely "junk DNA," a kind of genetic excess baggage due to historical accidents or mutations that multiplied some of the cell's DNA for no apparent reason.

Noncoding Sequences in DNA

The DNA base sequence of a simple gene (such as the genes of bacteria or viruses) codes uninterruptedly, from one end of the gene to the other, for its corresponding polypeptide or protein. Various genes of eukaryotic cells, however, have been found to contain occasional intervening stretches of DNA that, although they are transcribed into RNA along with the rest of the gene, do not code for any part of the protein and are called **introns.** For the RNA from the transcription of such a gene to be used as mRNA, the introns must first be clipped out and the remaining (information-carrying) stretches of RNA joined back together; this process, called **splicing,** is another example of a posttranscriptional modification of RNA.

PROTEIN SYNTHESIS: TRANSLATION OF mRNA

Ribosomes combine with mRNAs that appear in the cytoplasm, attaching at a point correspond-

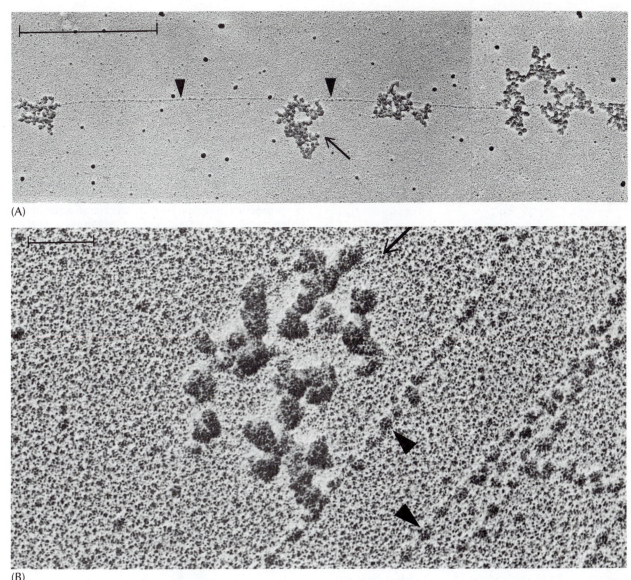

(A)

(B)

Figure 5-12. High-power electron micrographs showing chromatin (DNA-protein) strands, with nucleosome subunits (arrowheads), from maize root-tip cells. Full arrows point to RNA-protein complexes forming as a result of transcription of genes along the chromatin strand (the knob-like proteins in this complex apparently are already present within the nucleus and attach to the mRNA as fast as it is produced by transcription). Bar in (A) equals 1 μm, and in (B), a higher-magnification micrograph, 0.1 μm. The globular particle at the base of the RNA-protein complex in (B), adhering to the chromatin strand, is apparently an RNA polymerase molecule in the act of transcribing one of the DNA strands.

ing to the start of the base sequence coding for a polypeptide and then progressing along the mRNA strand, "reading" the code words sequentially and inserting, into the polypeptide chain being synthesized, the corresponding amino acids in the specified sequence (Figure 5-15). This process is called **translation.**

The key to accurate translation is the class of RNAs called **transfer** (or adaptor) **RNAs** (tRNAs). Each transfer RNA is designed to translate mRNA codons for a particular amino acid, which is carried at one end of the tRNA molecule. In order to translate a codon on the mRNA, a ribosome must combine with a tRNA, take the amino acid from it, and add that amino acid to the polypeptide chain the ribosome is producing. The only tRNA that can combine with the ribosome at any given time is the one designed to translate the mRNA codon then positioned at the appropriate site on the ribosome (Figure 5-16). Thus, the sequence of codons in the mRNA is translated by the ribosome into the sequence of amino acids specified by that mRNA, and hence by the gene from which that mRNA was transcribed.

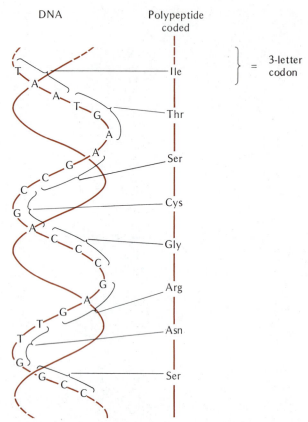

DNA   Polypeptide coded

T — Ile } = 3-letter codon
A
A
T — Thr
G
A
A — Ser
C
G
C
G — Cys
A
C
C — Gly
C
G
G — Arg
A
G
T — Asn
T
G
G — Ser
G
C
C

Figure 5-13. The sequence of bases in DNA specifies the amino acid sequence in a polypeptide, by means of sets of three bases called codons. The abbreviations used here to denote individual amino acids are given in Figure 5-6. Base sequence is shown only for the one DNA strand that is used as information for specifying a polypeptide. Its complementary strand, although essential in DNA replication (Figure 5-11), does not code for a polypeptide. As explained in subsequent figures, the information-carrying strand actually directs polypeptide synthesis via formation of a complementary messenger RNA. The codons in DNA shown here are therefore complementary to the RNA codons listed in Table 5-1.

The translation process and its products can be readily studied in the test tube using an extract of wheat germ (embryos from wheat grains) containing ribosomes, tRNAs, and associated enzymes, fortified with an energy source and a mixture of all the amino acids including one radioactively labeled amino acid. If a sample of poly(A) mRNA isolated from some plant tissue is added to this mixture, the ribosomes translate the added mRNA molecules, producing radioactively labeled polypeptides which can be separated by physical methods and detected by autoradiography.

Organelle DNA Function

The processes just described occur separately, to a more limited extent, in chloroplasts and mitochondria. The organelles' DNA codes for certain protein components of the organelle and also specifies a DNA polymerase and an RNA polymerase, ribosomal RNAs, and transfer RNAs to work with the organelle's ribo-

somes. However, many of the protein components of organelles are produced in the cytoplasm from nuclear information and are imported by the organelles. Also, the organelle DNA is not organized like the nuclear chromosomes but consists, like the so-called chromosome of bacterial cells, simply of a large molecule of DNA with many different genes (specific base sequences) along its length (Figure 5-17). Chloroplast DNA occurs in multiple copies, as many as 50 copies per chloroplast in the alga *Chlamydomonas*.

Posttranslational Modification

After the formation of polypeptide chains by ribosomes, certain polypeptides are modified before becoming active proteins. Some of the amino acids of the polypeptide may be modified; for example, in certain proteins a hydroxyl group is added to proline (Table 5-1), converting it to hydroxyproline. Hydroxyproline-rich protein is a distinctive structural component of primary

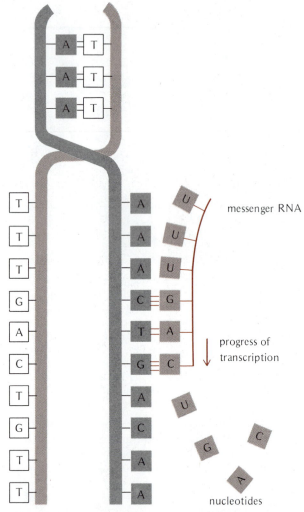

messenger RNA

progress of transcription

nucleotides

Figure 5-14. Transcription. The two DNA strands are locally pulled apart, and one of them is copied as an RNA strand following the rules of complementary pairing, with U substituting for T. After transcription, the DNA strands reassociate into their stable double-helical configuration.

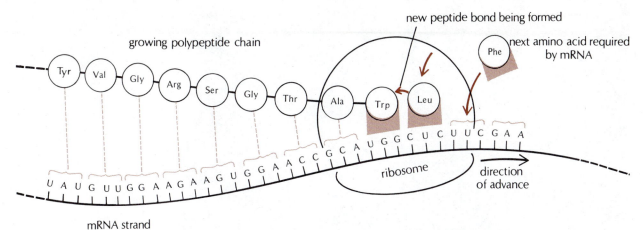

Figure 5-15. Translation. A ribosome progresses along an mRNA strand, adding amino acids to a growing polypeptide chain as called for by the sequence of codons in the mRNA. The correspondence between codons and amino acids is shown by brackets and dotted lines. Figure 5-16 explains the mechanism that specifies which amino acid will be inserted for each codon along the mRNA.

Figure 5-16. "Adapter" or decoding role of transfer RNAs (tRNA) in translating the genetic code during polypeptide synthesis. The tRNAs for each amino acid have an "anticodon" loop with three bases complementary to code words for that amino acid (bases shown by letters at the bottom of the tRNA's helical configuration). The tRNA, carrying its amino acid (i.e., "charged"), pairs with the appropriate codon via the anticodon loop when the ribosome is attached to that codon. The ribosome adds the amino acid to the growing polypeptide chain, releasing unloaded ("uncharged") tRNA, then advances to the next codon and repeats the process. For reuse, each tRNA is "recharged" by a specific enzyme that loads it with the correct amino acid. Only the bases of the anticodon portion of each tRNA are indicated in this diagram; the actual configuration of tRNA molecules is somewhat more complicated than shown here.

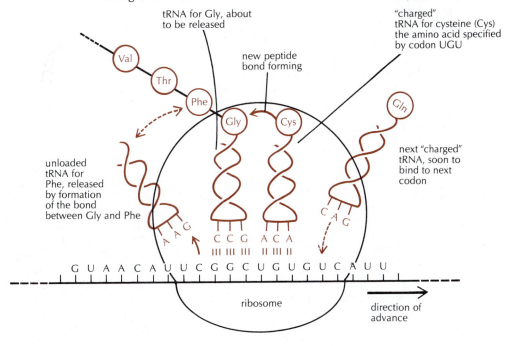

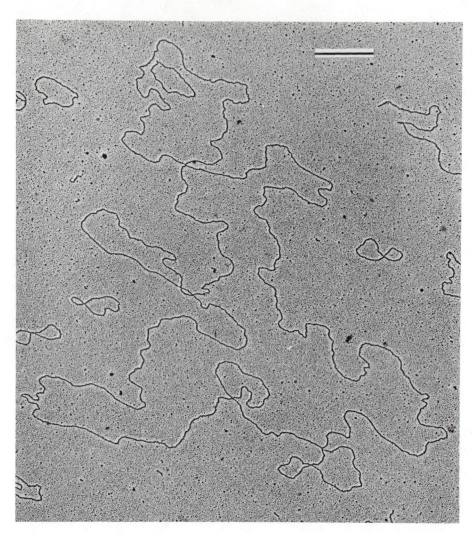

Figure 5-17. Electron micrograph of a circular DNA molecule or "chromosome" from a chloroplast of corn, displayed by spreading the molecules out as a thin film on a liquid surface. The small circular molecules accompanying the large convoluted circle of chloroplast DNA are virus particles of known size, for comparison with the chloroplast DNA. Bar = 1 µm. Although not visible by electron microscopy, the circle is known to consist of a double-stranded DNA helix.

cell walls. Sugar units may be added to certain amino acids of the polypeptide chain, yielding a **glycoprotein.** These seem to be important as membrane proteins, especially at cell surfaces, and the hydroxyproline-rich protein just mentioned is also an example, because sugars are added to many of its hydroxyproline units before this protein is secreted into the cell wall. This addition occurs in the Golgi dictyosomes of the cell.

As formed on the ribosomes by translation of the appropriate mRNAs, the polypeptide chains for various proteins are longer than the completed and active proteins. Conversion of the primary translation product into the final polypeptide can occur by splitting off a specific part of the polypeptide chain from one end. In various cases this splitting accompanies transfer of the polypeptide across a membrane; the peptide portion clipped off in the process is regarded as a "signal sequence" specifying transport of that particular polypeptide through the membrane and is not required for the function of the active protein. Examples are various chloroplast proteins, including one of the subunits of

the photosynthetic CO_2-capturing enzyme, made on cytoplasmic ribosomes and then transferred into the chloroplast through the chloroplast envelope.

Specification of Nonprotein Cell Components

As far as we know, DNA does not contain direct information for the structure of cell components other than RNA and protein, for example, for cellulose, cutin, or membrane phospholipids. The cell's genes specify these materials instead by coding for specific biosynthetic enzymes, for example, an enzyme (cellulose synthetase) that can couple glucose units together in the specific way they are joined in a cellulose chain. While most plants have a gene for this kind of enzyme activity, and thus can form cellulose as a structural material, animals generally do not.

CONTROL OF GENE ACTIVITY

Although by the process of mitosis each cell inherits a complete copy of the hereditary information, or genes, possessed by that individual organism, normally only some of these genes express themselves in any one cell. "Housekeeping" genes, such as those for enzymes of respiration and for RNA, protein, and phospholipid synthesis, must be functioning in all living cells, but genes for specialized proteins (such as enzymes for photosynthesis or for forming specialized nonprotein products such as lignin or cutin) are expressed only in cells specialized for a particular function. Gene activity is, therefore, subject to control by the cell. Such control not only enables each cell to maintain a balanced composition but also enables complex organisms such as higher plants to **differentiate,** that is, develop functionally specialized tissues and organs such as leaves, stems, and roots. There are several levels of possible control over the expression of a gene's effect.

(1) **Transcriptional control,** or control over the production of RNA, especially mRNA, is the most basic kind of regulation and is well demonstrated in bacterial cells. Transcription of mRNAs for certain enzymes is induced by foreign molecules called inducers, and transcription of many other mRNAs is inhibited or repressed by specific cell metabolites. The presence of these small molecules thereby controls how much of a given enzyme the bacterial cell makes.

These control mechanisms operate by means of specific proteins called **repressor proteins,** one form of which can specifically bind to the DNA at the site where RNA polymerase begins to transcribe a given gene, thus blocking its transcription. As explained in Figure 5-18, metabolites that repress act by occupying a specific site on the specific repressor protein; occupancy gives the repressor protein the right shape for binding to the DNA. A specific gene is needed to code for each specific repressor protein, which in turn binds to a specific DNA base sequence near the site for commencement of transcription of the regulated gene. Several genetic elements are involved, therefore, in the control of each regulated gene or series of genes. **Structural genes** are the genes that code for RNA or protein molecules needed for general cell functions, and are subject to regulation. **Regulator genes** are genes that specify control elements like those just mentioned.

(2) **Posttranscriptional and translational controls** involve interference either with the export of transcribed mRNAs from the nucleus to the cytoplasm or with the initiation or completion of the translation process by the ribosomes. Such controls are thought to be important in eukaryotic cells although the nature of these controls is less well understood than transcriptional control.

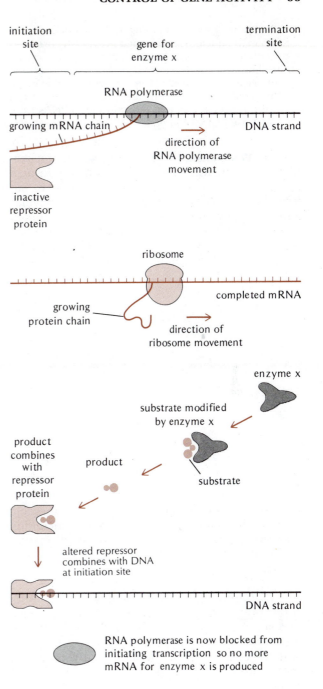

Figure 5-18. Control of gene expression by repression. The product formed by action of a particular enzyme can combine specifically with, and alter the shape of, a repressor protein. When so altered, this protein attaches to the initiation site on the DNA at which transcription of the gene for this particular enzyme commences, thus blocking its transcription. The repressor protein is specified by a separate regulator gene. Note that DNA and RNA messages for actual enzymes have a much longer base sequence than shown in this simplified diagram.

Posttranscriptional modification or processing of mRNAs in the nucleus, mentioned previously, provides a potential means of regulating gene expression by controlling which of the transcribed messages become exportable and translatable. There is evidence that much of the transcribed RNA is broken down before it can leave the nucleus, possibly representing mRNAs from genes whose expression is thereby being suppressed.

(3) **Posttranslational controls** are mechanisms for regulating the activity of already-formed enzymes. One potentially general means, in eukaryotic cells, is by compartmentation of enzymes and substrates into different membrane-bounded compartments or organelles within the cell. The enzyme can then act only as fast as its substrate is made available to it, by transport through some membrane. Another means of control is **effector action,** the inhibition or stimulation of enzyme activity by specific nonsubstrate molecules, called effectors, as well as by nonspecific enzyme cofactors such as Mg^{2+}. This kind of action is important in regulating cell metabolism, as discussed in Chapter 6.

Posttranslational modification of polypeptides, noted above, provides other potential means of regulation. Some enzymes become activated by clipping off a peptide fragment from the end of the original translation product. For others, which function in association with a specific membrane or organelle, posttranslational modifications that aid in placing the enzyme in its proper location within the cell are important determinants of the organelle's or membrane's activity.

SUMMARY

(1) Cells are composed of water, inorganic salts, and organic carbon compounds including structural, catalytic, and informational macromolecules, metabolites, storage reserves, and secondary products.

(2) Carbohydrates include sugars and structural and reserve polysaccharides (e.g., cellulose and starch, respectively) that are built from sugars. Sugars (and some polysaccharides) are named with the ending -ose. Organic acids are important metabolites and storage products, derived from sugars.

(3) Lipids include storage fats, or vegetable oils, and membrane lipids (phospholipids) which form a lipid bilayer structure that blocks the passage of water-soluble substances such as salts and sugars through the membrane.

(4) Proteins are macromolecules (polypeptides) built from amino acids. They include structural molecules, storage reserves, and, most important, enzymes. The biological function of a protein is attributable to its three-dimensional shape, which depends upon the specific sequence of amino acids in its polypeptide chain(s) and how the chain(s) is coiled and folded.

(5) Enzymes catalyze biochemical reactions, making cellular metabolism possible. Each enzyme is specific for one kind of reaction and for one or a few particular substrates. Enzymes are named with the ending -ase following the name of the enzyme's principal substrate and/or kind of reaction.

(6) An enzyme works by combining transiently with its substrate(s) by means of an active site into which the substrate fits. The action of many enzymes requires an inorganic ion and/or a small organic molecule as a cofactor or as an integral part of the enzyme molecule. Enzyme action is sensitive both to cool or cold temperatures and to heat (denaturation) and is strongly affected by acidity or pH.

(7) Nucleic acids (DNA and RNA) are macromolecules built from nucleotides, each of which consists of a nitrogen base, a pentose sugar, and a phosphate group. DNA contains four nitrogen bases (A, G, C, and T) in two polynucleotide chains, which are paired together by hydrogen bonds between complementary bases at opposing points on the two strands (A opposite T, G opposite C). This arrangement arises by complementary synthesis during DNA replication, ensuring exact

copying of the base sequence and enabling DNA to carry hereditary information.

(8) RNA contains the same bases as DNA, although U is present instead of T. RNA strands are made by transcription, or copying DNA strands complementarily (with U substituting for T). Messenger RNA (mRNA) transcribes the genetic information of DNA for amino acid sequences of proteins, encoded by means of three-letter words or codons; each codon, or sequence of three bases, specifies a particular amino acid.

(9) Genetic information in mRNA is translated by ribosomes with the aid of transfer RNAs (tRNAs), which act as adaptors to bring the correct amino acid into position for each codon in order along the mRNA. The ribosome travels along the mRNA chain, incorporating amino acids one by one into a polypeptide in the sequence called for by the sequence of codons along the mRNA.

(10) Gene activity may be controlled at either the transcriptional level (repression of mRNA formation), the posttranscriptional or translational level (mRNA processing or export; initiation or completion of translation), or the posttranslational level (control of enzyme activity).

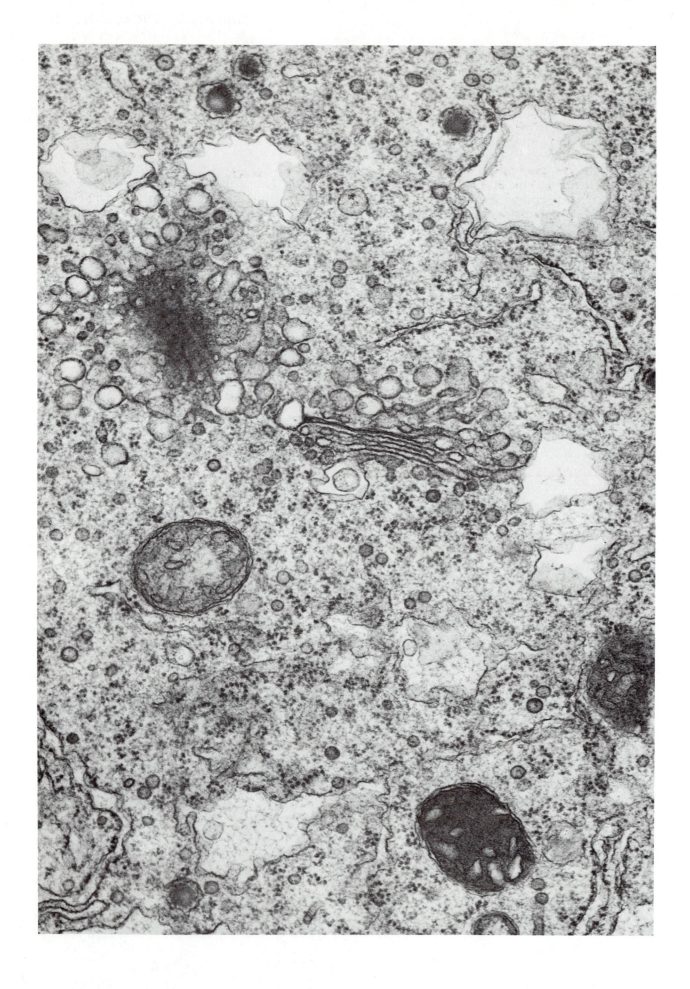

Chapter 6

CELL METABOLISM, ENERGY, AND TRANSPORT

When any complex biochemical molecule such as a protein, nucleic acid, polysaccharide, or lipid breaks down into its component parts, energy is released. It therefore *takes* energy to assemble such molecules from their component building blocks such as amino acids or nucleotides. It also takes energy to produce the required building blocks from simple food materials that are available from photosynthesis or in a heterotrophic cell's food supply. Cytoplasmic streaming, mitosis, and other kinds of motility also require energy. So do the processes called active transport, by which cells take up through their membranes and concentrate inside themselves the nutrients they need.

Although light is a photosynthetic organism's ultimate energy source, light energy is not directly usable, even by a green cell, for most of the general cell functions just mentioned. Photosynthetic and heterotrophic cells alike obtain all or most of the useful energy needed for general cell functions by breaking down "fuel" substances, usually carbohydrate or fat, in the complex energy-yielding metabolic processes of respiration or fermentation. Biochemical studies have revealed that these processes comprise the central metabolic activity of the cell. They are tied to other metabolic functions such as amino acid and lipid biosynthesis, which employ metabolites formed in the course of respiration or fermentation. This chapter provides an introductory look at these processes and how they make possible energy-requiring activities such as biosynthesis and transport.

Figure 6-A. A section of a metabolically active plant cell, showing mitochondria, dictyosomes, rough endoplasmic reticulum, and ribosomes (43,000 ×). Note the outer and inner mitochondrial membranes, with the inner thrown into many folds, or *cristae*. Cristae in section often appear as vesicles within the mitochondrion. The enzymes of the citric acid cycle and respiratory chain are embedded in the mitochondrial inner membrane.

THE CELL'S ENERGY CURRENCY: ATP

Respiration is chemically equivalent to combustion, the oxidation of fuel substances by oxygen from the air, which of course releases heat energy. Respiration is useful to the cell because the energy it releases appears not entirely as heat, but partly in a chemical form, **adenosine triphosphate** or **ATP,** which can be used for energy-requiring cell functions. ATP is a nucleotide, derived from the simple nucleotide adenosine monophosphate (AMP; Figure 5-9A) by adding two more phosphate groups (Figure 6-1). An unusually large amount of chemical energy is associated with a bond between two phosphate groups, which are therefore called **high-energy phosphate groups.**

Energy for energy-requiring functions, such as joining sugar units together into a polysaccharide or coupling amino acids into a polypeptide (Figure 5-5), is supplied by removing one or two of the phosphate groups from an ATP molecule in an enzymatic reaction that proceeds in conjunction with the energy-demanding process and forms either ADP or AMP (Figure 6-1). Similarly, work processes such as active transport and locomotion can be energized by splitting high-energy phosphate groups from ATP. The cell's energy metabolism, that is, respiration or fermentation, restores high-energy phosphate groups to ADP and AMP, regenerating ATP to be used over again as an energy source.

The use of ATP for an energy-requiring biosynthesis is illustrated by the formation of peptide bonds in protein synthesis (Figure 6-2). The reaction of an amino acid with ATP, catalyzed by an "amino acid-activating enzyme," produces a high-energy form of the amino acid (amino acyl-AMP) which can act as an amino acid donor. It transfers the activated amino acid to a tRNA, a coupling that also retains a substantial amount of energy, permitting the amino acid to be joined to another by peptide linkage on the ribosome, ex-

103

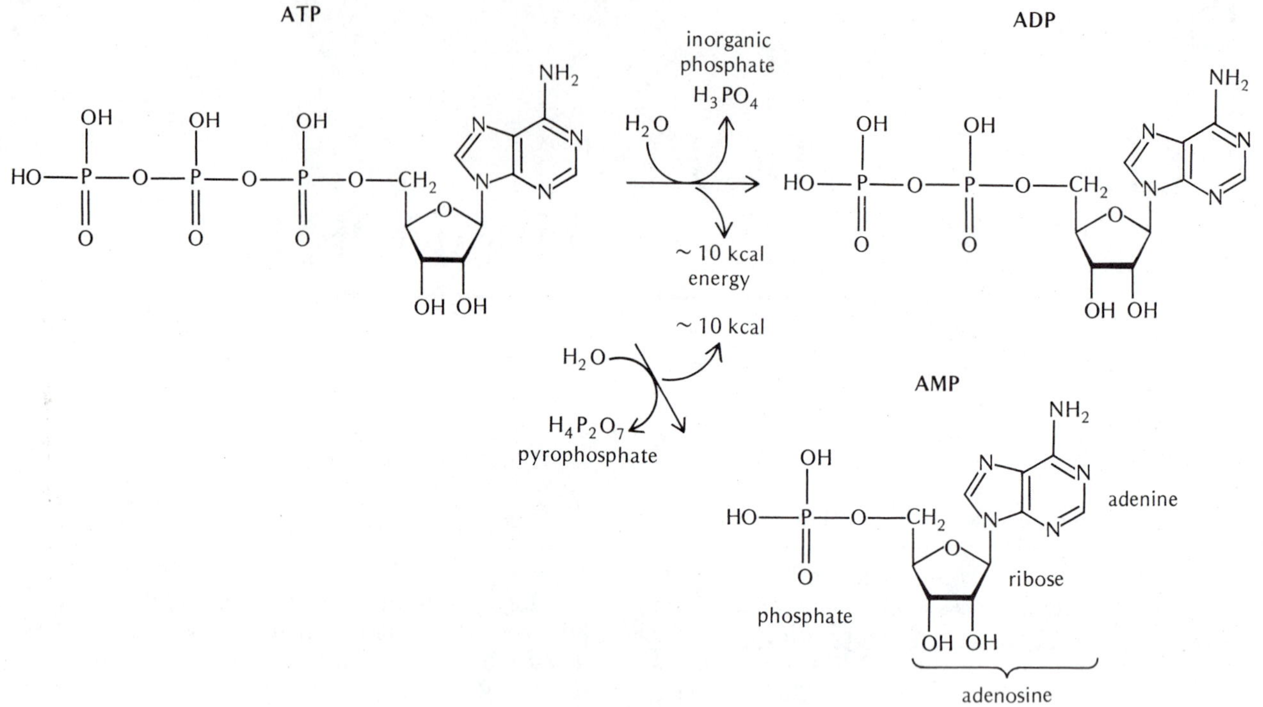

Figure 6-1. ATP (adenosine triphosphate) showing its high energy phosphate groups (color) and how they can be split (hydrolyzed) to release energy and either ADP (adenosine diphosphate) or AMP (adenosine monophosphate, the nucleotide depicted in Figure 5-9). A similar energy-releasing conversion of ADP to AMP is also possible in principle, but does not seem normally to be used as an energy source by cells.

Figure 6-2. How ATP energy (color) is used to "activate" an amino acid and "charge" a tRNA with it, leading to formation of a peptide bond in a growing polypeptide. Ad, Rib, and $\textcircled{P}$ are the adenine, ribose, and phosphate components of ATP or AMP (Figure 6-1).

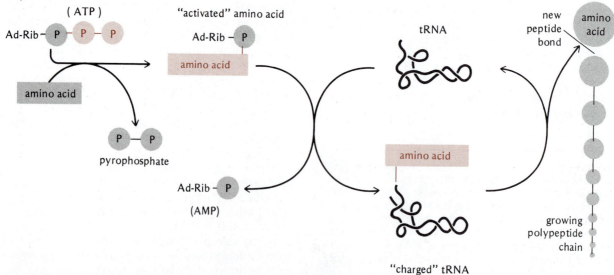

tending the growing polypeptide chain. The tRNA step is actually superfluous from the energetic point of view but is needed for the information processing (translation) aspect of polypeptide synthesis.

Other nucleoside triphosphates besides ATP energize certain biosynthetic processes. UTP (uridine triphosphate) supplies energy for the synthesis of most cell wall polysaccharides, GTP (guanosine triphosphate) is used in addition to ATP in protein synthesis, and CTP (cytidine triphosphate) is used for phospholipid synthesis. These nucleotides are also substrates for RNA synthesis by RNA polymerase (Chapter 5), providing both the building blocks and the energy for this process. The analogous deoxy/nucleoside triphosphates of A, C, G, and T are used by DNA polymerase. Triphosphates other than ATP derive their extra phosphate groups indirectly by transfer of high-energy phosphates from ATP, the high-energy compound produced directly by cellular energy metabolism.

ENERGY METABOLISM

Glucose, free or derived from a storage reserve such as starch or sucrose, is the most universal substrate of energy metabolism. In **respiration** glucose is broken down to CO_2 and water using oxygen:

$$C_6H_{12}O_6 + 6 O_2 \longrightarrow 6 CO_2 + 6 H_2O + 686 \text{ kcal energy.*} \quad (6\text{-}1)$$
glucose

Since this is equivalent to the combustion of glucose (e.g., from the cellulose of wood), it yields the same amount of energy as combustion. Biochemical knowledge about respiration outlined later shows that for every glucose molecule respired, 38 ATP molecules can be formed from ADP. Since about 10 kcal of energy is associated with each high-energy phosphate group, some 380 of the total yield of 686 kcal of energy is conserved as ATP, an energy conversion efficiency greater than 50%, considerably better than that of most man-made devices. The rest of the energy yield is released as heat, as can be detected by placing rapidly respiring plants such as germinating seeds into a thermos bottle along with a thermometer: the temperature inside the bottle will climb as much as 10–20°C (18–36°F) above that of the room.

In the absence of oxygen (anaerobic conditions) cells can conduct only a partial breakdown or **fermentation** of their energy substrate, yielding much less energy than respiration because most of the chemical energy of glucose is still tied up in the fermentation products. Most plant cells un-

der anaerobic conditions ferment glucose to alcohol and CO_2:

$$C_6H_{12}O_6 \longrightarrow 2 C_2H_5OH + 2 CO_2 + 56 \text{ kcal}$$
glucose   ethyl   (6-2)
            alcohol

Corresponding with the much lower energy yield, only two ATP molecules can be made per molecule of glucose fermented, so this kind of energy metabolism is much less effective in meeting cellular energy needs.

Fermentation yields enough energy for plant cells to maintain themselves for limited periods (hours) under anaerobic conditions, but most plants cannot grow without oxygen. The roots of many crop plants and trees are injured or killed if flooded and thereby deprived of sufficient oxygen for more than a day or so. Many trees are affected even by extra soil piled up over their roots, the reason why grading operations during home and building construction frequently kill desirable shade trees. However, yeast cells, which have an extraordinary capacity for fermentation, can grow and multiply anaerobically using fermentation as their energy source, as they do in beer and wine making (Figure 6-3). Many bacteria can also grow anaerobically by means of some type of fermentation (e.g., lactic bacteria which sour milk).

Although fermentation involves no overall oxidation of glucose, oxidation reactions that take place as part of the process are actually the basis for extracting ATP energy in both respiration and fermentation. To comprehend metabolism, it is therefore essential to understand something about biochemical oxidation processes.

Biological Oxidation and Reduction

Oxidation is the removal of electrons from a substance, by oxygen or some other electron-accepting compound, called an oxidant. The addition of electrons to a substance is **reduction**, so an oxidant is reduced when it oxidizes something. In metabolism, the oxidation of organic compounds usually occurs by the removal of hydrogen atoms, two at a time, from the compound. A hydrogen atom (H) is a proton plus an electron, so removing a hydrogen atom means removing an electron, that is, oxidation.* Hydrogen atoms removed from a substrate during metabolism are usually transferred to an organic oxidant or hydrogen acceptor, most often a specialized nucleotide called NAD, containing the B vitamin niacin

---

*A calorie is the energy needed (at 15°C) to heat 1 g of water by 1°C. A kilocalorie (kcal; same as the food Calorie used in nutrition) equals 1000 cal, or the energy needed to heat about 1 liter of water by 1°C.

*In any biochemical writing, one must be careful always to distinguish between a hydrogen *atom* (H), which carries an electron and is involved in oxidation/reduction, and a hydrogen *ion* (H$^+$), which carries no electron. Loss or gain of a hydrogen ion, as when an acid dissociates or when a base reacts with a hydrogen ion to neutralize it, is not oxidation or reduction, but an acid/base reaction.

Figure 6-3. Beer in fermentation. The thick foam produced by escaping carbon dioxide excludes air from the solution being fermented by yeast beneath.

or nicotinamide. The hydrogen transfer reaction, catalyzed by enzymes called **dehydrogenases,** can be represented as follows, where BH_2 represents the compound being oxidized:

$$BH_2 + NAD \xrightarrow{\text{dehydrogenase}}$$

$$\underset{\substack{\text{substrate}}}{BH_2} + \underset{\substack{\text{H acceptor}\\ \text{(oxidant)}}}{NAD}$$

$$\underset{\substack{\text{oxidized}\\ \text{product}}}{B} + \underset{\substack{\text{reduced}\\ \text{acceptor}}}{NADH_2} \quad (6\text{-}3)$$

Some dehydrogenase reactions release a fair amount of energy, enough to permit the formation of ATP. This is the basis of ATP production in fermentation. In respiration, hydrogen atoms from the reduced hydrogen acceptor, $NADH_2$, are oxidized ultimately by oxygen, releasing much additional energy and yielding much more ATP.

Glycolysis

The biochemical study of energy metabolism goes back to 1897 when the brothers Eduard and Hans Buchner, in Germany, serendipitously discovered how to prepare a cell-free extract of yeast in which fermentation, previously viewed as a vital process occurring only inside living cells, took place. Investigation of what happens in yeast juice eventually showed that an orderly sequence of 12 biochemical reactions, each catalyzed by its own enzyme, converts glucose step by step into the final product, alcohol. Such a reaction sequence is called a **metabolic pathway.** Most of this pathway, called **glycolysis,** turned out to be common to fermentation and respiration. Subsequently many other metabolic pathways have been discovered in cells, leading to the conclusion that orderly, stepwise pathways are the basic pattern of metabolism. These pathways, rather like assembly lines with the enzymes as the individual

workers, enable cells to produce efficiently the great variety of highly specific biochemical products that are needed for growth, as well as to extract useful energy in breaking down glucose and other energy sources.

For our purposes it is unnecessary to go into the complicated details of glycolysis or other metabolic pathways (they may be found in any textbook of biochemistry or physiology). The net result of glycolysis is shown in the upper left part of Figure 6-4. Hexose (6-carbon) sugar becomes split into two 3-carbon halves which are oxidized by an energy-releasing dehydrogenase reaction similar to that of Eq. (6-3). This dehydrogenase works in such a way that the energy yield of the reaction is largely conserved by the formation of ATP; a subsequent step allows the formation of a second molecule of ATP. The final products of glycolysis are two molecules of a three-carbon organic acid called pyruvic acid ($C_3H_4O_3$) and the two molecules of $NADH_2$ that were formed from NAD in the dehydrogenase step.

The term glycolysis (glyco-, sugar; lysis, splitting) comes from the fact that sugar molecules are split in half midway through this process. All the enzymes of glycolysis are found in solution in the cytoplasm, that is, not associated with membranes or organelles. Thus, this phase of metabolism always occurs in the cytoplasm.

Fermentation

Under anaerobic conditions, there is no oxygen to oxidize $NADH_2$ back to NAD, so glycolysis would soon stop for lack of the NAD needed in its dehydrogenase step if the cell could not find a way to dispose of the hydrogen atoms of $NADH_2$. Plant cells can use hydrogen atoms from $NADH_2$ to form ethyl alcohol from acetaldehyde, a two-carbon compound obtainable from pyruvic acid

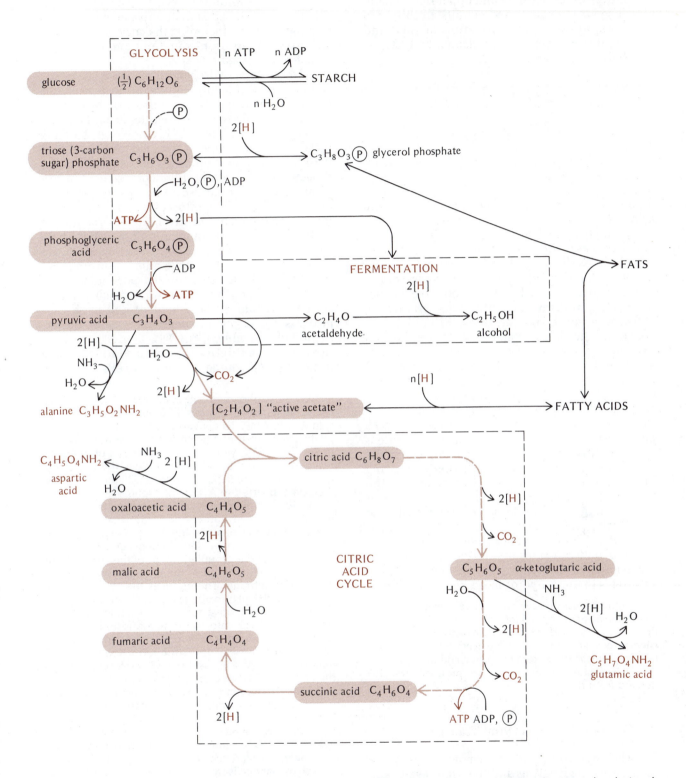

Figure 6-4. Simplified metabolic map of energy metabolism, showing glycolysis, alcoholic fermentation, and the citric acid cycle, with many of the intermediate compounds omitted. Since one of the early reactions of glycolysis splits a hexose (6-carbon sugar) molecule into two trioses (3-carbon sugars), each of which is further metabolized as indicated, for simplicity the diagram is drawn for the metabolism of just one triose (= ½ glucose). The diagram also shows how biosynthesis and breakdown of starch, fats, and certain amino acids (shown in color) ties into the pathways of basic energy metabolism.

by splitting off CO_2 (center and upper right, Figure 6-4). The NAD so generated permits glycolysis to continue, allowing the formation of ATP and causing the accumulation of alcohol and CO_2 as metabolic end products.

RESPIRATION

The Citric Acid Cycle

In cellular respiration, pyruvic acid from glycolysis is further metabolized in the mitochondria. In the 1930s the biochemist Hans Krebs found that citric acid and certain other organic acids are involved in the breakdown of pyruvic acid, which he realized must take place as a cyclical process, the **citric acid cycle,** now often called the **Krebs cycle.** Its net effect is indicated in the lower part of Figure 6-4. Pyruvic acid is oxidized (removing 2H), one carbon atom is split off as CO_2, and the remaining two carbons are added to a four-carbon organic acid to yield the six-carbon acid, citric acid. Citric acid is then oxidized by a series of dehydrogenase reactions and removals of CO_2 until it has been degraded back to the original four-carbon acid, completing both the cycle and (after two pyruvic acid units are gone) the conversion to CO_2 of all six carbon atoms of a glucose molecule. These reactions take place inside the mitochondria, catalyzed by enzymes tightly bound to the inner mitochondrial membrane (Figure 6-A). The reaction converting fumaric to malic acid [Eq. (5-1)] is one of the steps of this cycle.

The citric acid cycle has one step that, like glycolysis, yields ATP. However, the principal yield of ATP in respiration comes from subsequent oxidation, via respiratory electron transport (see below), of the hydrogen atoms removed from substrates in the dehydrogenase steps of the cycle.

The two-carbon fragment "active acetate" derived from pyruvic acid and fed into the citric acid cycle (Figure 6-4) is acetic acid combined with a nucleotide called **coenzyme A** (acetyl-CoA). Active acetate is the key metabolite: fatty acids are formed from it and broken down into it (see later); many secondary products are also derived from it.

Respiratory Electron Transport

The hydrogen atoms removed by dehydrogenases in glycolysis and the Krebs cycle, and transferred to hydrogen acceptors such as NAD, are oxidized by oxygen (yielding H_2O) via a chain of hydrogen or electron carriers called the electron transport chain or **respiratory chain.** The enzymes of the respiratory chain, like those of the citric acid cycle, are embedded in the mitochondrial inner membrane; therefore mitochondria are the sites of respiratory oxygen consumption inside the cell. The most distinctive enzymes of the respiratory chain, including the enzyme that reacts with oxygen, are iron-containing proteins called **cytochromes** (Figure 6-5).

Respiratory hydrogen atoms have an energy content approximately equivalent to that of free hydrogen gas; when they are oxidized by oxygen, much more energy is released than could be conserved in a molecule of ATP. However, handing these hydrogen atoms or the electrons associated with them step by step down the sequence of hydrogen and electron carriers comprising the respiratory chain (Figure 6-5) breaks up the large overall energy yield into several smaller increments, each capable of driving the formation of ATP from ADP and inorganic phosphate. Three electron transfer steps, as shown in Figure 6-5, are apparently coupled to ATP production, so three ATP molecules can be formed per molecule of $NADH_2$ oxidized by the respiratory chain. This conserves about 30 kcal or over half of the approximately 52 kcal of energy released by the oxidation of one mole of $NADH_2$. This ATP production, called **oxidative phosphorylation,** is the principal ATP source of respiring cells.

Whereas phosphorylation converts ADP to ATP, as noted earlier some of the cell's energy-using processes split ATP to AMP. To convert this back to ATP, a phosphate group must first be added to AMP from another molecule of ATP, yielding two molecules of ADP (the reaction catalyzed by adenylate kinase, the enzyme illustrated in Figure 5-7). Both these ADP molecules can then be converted to ATP by phosphorylation.

Oxidative phosphorylation is now generally believed to occur in mitochondria by what is called a **chemiosmotic** mechanism, involving membrane transport. The transfer of an electron between electron carriers at each of the "coupled" steps of the chain pumps a hydrogen ion (H^+) out of the mitochondrion, across its inner membrane. This reduces the H^+ concentration (raises the pH) inside the mitochondrion, generating a "proton motive force" or a tendency for H^+ ions to flow back into the mitochondrion across its membrane. The influx of two H^+ ions, down this energy hill, in some way drives the formation of one ATP molecule from ADP. Thus three ATP molecules are produced per pair of electrons coming down the respiratory chain to oxygen from $NADH_2$. If an acid substance is added that can readily cross the mitochondrial membrane, the H^+ ions that it carries into the mitochondrion destroy the electron transport-generated proton motive force, cancelling out oxidative phosphorylation. Such agents, called **uncouplers,** block biosynthesis and growth in plant cells by depriving them of most of their ATP energy supply, even though respiration continues and indeed is usually stimulated by uncoupling.

The respiratory chain of plant mitochondria differs somewhat from that typical of animal cells or bacteria. Many plant cells possess an alternate pathway, not involving cytochromes, for electron flow from the middle part of the respiratory chain to oxygen. Whereas the cytochrome path is blocked by cyanide (the reason cy-

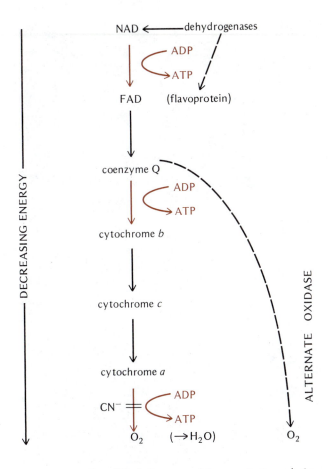

DECREASING ENERGY

NAD ← dehydrogenases

ADP
ATP

FAD   (flavoprotein)

coenzyme Q

ADP
ATP

cytochrome *b*

cytochrome *c*

cytochrome *a*

ALTERNATE OXIDASE

CN^-
ADP
ATP

O_2 ($\rightarrow H_2O$) O_2

Figure 6-5. Simplified diagram of the respiratory chain in plant mitochondria. The arrows indicate transfer of H atoms or electrons between successive electron carriers, with steps coupled to phosphorylation indicated in color. FAD is a nucleotide (flavin-adenine dinucleotide) similar to NAD, but containing the B vitamin, riboflavin. In the respiratory chain, FAD is associated with a specific protein (forming a flavoprotein or flavin enzyme). The cytochromes are iron-containing enzymes, somewhat similar to the oxygen-carrying protein, hemoglobin, of vertebrate blood. Coenzyme Q is the only nonprotein compound illustrated, being a small molecule of the type called a quinone. The iron atoms in cytochromes, and the riboflavin and niacin components of FAD and NAD respectively, as well as the quinone molecule, are all able to accept and then pass on electrons or H atoms, enabling the respiratory chain to hand electrons from respiratory substrates to oxygen. As indicated, cyanide inhibits this last reaction, thus blocking respiration. (Dashed line shows alternate cyanide-insensitive route to oxygen possessed by many plant mitochondria, as noted in text. The complete respiratory chain contains several additional components not listed here.)

anide poisons cells), the alternate path is insensitive to cyanide, so tissues possessing this path continue to respire (consume oxygen) in the presence of cyanide, a phenomenon called cyanide-resistant respiration. The significance of the alternate path for cell function is still not clear.

BREAKDOWN OF ENERGY RESERVES: DIGESTION

Complex storage reserves such as sucrose, starch, and fat cannot be respired directly for energy but must be broken down into units that can be fed into the pathways of glycolysis or the citric acid cycle. This breakdown is sometimes called **digestion** because it is somewhat analogous to the digestion or breakdown of complex foods that animals and heterotrophic plants carry out to provide their cells with absorbable food material. Breakdown of reserves is also called **mobilization,** because it converts insoluble products such as starch and fat, which cannot be transported through the plant, into soluble products that can be.

The first step in mobilization is to split a complex or polymeric storage product into its component molecules by reacting it with water, a type of reaction called **hydrolysis,** catalyzed by enzymes collectively called hydrolases. For example, starch is hydrolyzed to glucose by enzymes called **amylases** (Greek *amylon,* starch). In seeds that accumulate starch, amylases appear at the beginning of germination, increase rapidly in activity, and digest the stored starch for the use of the growing embryo.

Hydrolysis of starch is the objective of **malting,** in which barley grain is prepared for making beer. Yeast cells do not produce amylases, so they cannot ferment ungerminated barley (containing starch but not sugar). Barley seeds are therefore soaked in water and spread out en masse on trays in the malting shed (Figure 6-6). After 2 or 3 days, germination has begun and the seeds have produced copious amounts of amylase which has just started to digest their starch. At this point the sprouting grain is gently dried at a temperature lower than would destroy amylase activity. Lack of water stops hydrolytic digestion of the starch. The dried grain, called malt, can be kept for a considerable time; when soaked in warm water (brewed), the active amylases in the malt quickly digest its starch into sugar, which comes out into the brewing water and is subsequently fermented by yeast to yield beer.

Storage fats can be hydrolyzed to fatty acids and glycerol by enzymes called **lipases.** Glycerol is closely related to one of the intermediate metabolites of glycolysis and goes directly into that pathway. Fatty acids are used by breaking them down, oxidatively, into two-carbon "active acetate" fragments that can be fed into the citric acid cycle just like the two-carbon units derived from the pyruvic acid from glycolysis (see Figure 6-4).

For use by a germinating seed, fatty acids must first be converted into a soluble food for transport from the storage tissue (endosperm or cotyledon) to the growing points of the embryo. To accomplish this, fat-storing seeds develop, during germination, a special metabolic cycle called the **glyoxylate cycle** that allows fatty

Figure 6-6. A malting shed in operation.

acids to be converted into sugar, which readily moves to where energy is needed. The glyoxylate cycle, like the citric acid cycle, involves organic acids as intermediates; it takes place in special organelles of the microbody type (Chapter 4) called **glyoxysomes** (Figure 6-7) which appear in fat-storing tissues during fat mobilization and contain the unique enzymes of this cycle.

BIOSYNTHETIC PATHWAYS

While the sugar building blocks of cellulose and other cell wall polysaccharides are available from general carbohydrates directly or with only minor modification, the amino acid, nucleotide, and fatty acid building blocks of the other major cell constituents must be specially produced. Plant cells possess a metabolic pathway for forming each of the needed constituents; for example, they have metabolic pathways, involving unique sets of enzymes, leading to each of the 20 amino acids required for protein synthesis. This is not true of animals and humans, which lack the enzymes to produce certain of the necessary amino acids and therefore must obtain these (called the "essential amino acids") in their diet. Similarly plants can produce several vitamins required as components

of certain important nucleotides functioning in metabolism, such as NAD, whereas humans cannot.

This contrast in biosynthetic capabilities between plants and animals leads to some important problems in human nutrition. For example, vitamin deficiencies arise when people eat foods such as polished rice that consist only of storage tissue lacking the vitamins that would have to be present if the tissue were metabolically active. Whereas storage proteins in cereal grains serve humans as an important amino acid source, they serve the seeds themselves simply as a nitrogen reserve, which can be used to manufacture any of the 20 needed amino acids. Thus the amino acid composition of its storage protein is not particularly important to the embryo plant: the major storage proteins of wheat, rice, and corn grains actually have an amino acid composition considerably different from that of plant or human cell protein, relatively deficient in several of the dietary essential amino acids (especially lysine and tryptophan; Table 32-2). As a result, cereal grain protein has a nutritional value for humans only about half that of milk or cell proteins such as meat, eggs, and vegetables. This nutritional inadequacy contributes to the serious protein deficiency widespread in countries where people depend mainly upon cereal grains for their protein.

Several of the simpler amino acids needed for making proteins can be produced merely by ad-

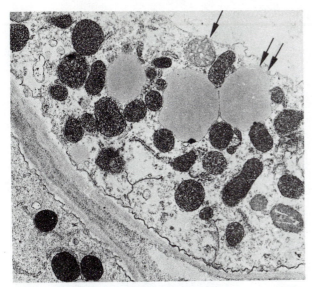

Figure 6-7. Glyoxysomes (dark-staining "microbodies") in the endosperm tissue of castor bean. For comparison, note the mitochondrion (single arrow) and spherosomes (oil droplets, double arrow) (8400 ×).

ding nitrogen, in the form of ammonia (NH_3 or $NH_4{}^+$), to the carbon skeleton of one or another of the metabolites of glycolysis and the citric acid cycle. The amino acid called alanine, for example, comes from pyruvic acid, and aspartic acid comes from the oxaloacetic acid of the citric acid cycle (see Figure 6-4). These simple amino acids can be made by animal as well as plant cells and are non-essential in the animal diet. More complex amino acids, including those "essential" to animals, are formed in plant cells by more lengthy pathways leading from the simple amino acids and/or other

metabolites. Most of these pathways require the use of ATP. The nitrogen bases of nucleotides are formed by similar special pathways.

Regulation of Biosynthetic Pathways

If cells are to grow without accumulating useless, if not actually injurious, surplus metabolites, there must be mechanisms to balance different biosynthetic pathways, ensuring that the amounts of different amino acids and nucleotides that the cell produces are just the amounts required to make the nucleic acids and proteins that gene activity is directing. In bacteria, by the mechanism called repression (Chapter 5) a surplus metabolite turns off the transcription of genes for enzymes of the pathway producing it, gradually shutting down production. It is not clear that repression is important in plant cells, but **end product inhibition** has been demonstrated, in which a given amino acid inhibits the action of an enzyme in its biosynthetic pathway, preventing overproduction (Fig. 6-8).

End product inhibition is an example of an important class of influences on enzyme activity only mentioned in Chapter 5. The activity of many enzymes is affected by one or more specific metabolites, called **effectors,** that are not substrates of the enzyme (do not undergo reaction) but interact with the enzyme at a special regulatory site distinct from the enzyme's catalytic or active site. Positive effectors increase activity; negative effectors (for example, a compound exerting end product inhibition) depress activity. They do this by altering slightly the shape of the enzyme and thus affecting how it interacts with its substrates. This is called an **allosteric effect.** These

Figure 6-8. Concept of feedback regulation of a biosynthetic pathway by negative effector action of its end product, acting on the first enzyme of the pathway. The mechanism prevents excess amounts of the specific end product from being produced by the cell.

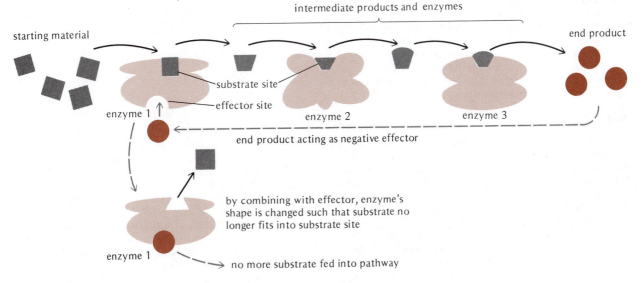

effects regulate metabolism automatically, for example, by preventing the overproduction of a certain metabolite as just mentioned.

Secondary Products

As noted in Chapter 1, many plants produce specialized chemical products not necessary for general cell functions but possibly important ecologically to the plant or important to people who grow and use the plant—spices, stimulants and drugs, rubber, dyes, and so on. The formation of these secondary products requires special, often elaborate, biosynthetic pathways, each involving a unique set of enzymes and usually requiring at least some ATP. Many secondary products (lignin is an important exception) are formed only by specific plants or the members of particular plant groups; indeed evolutionary relationships can be studied in terms of secondary products, the field of "chemotaxonomy." Moreover, secondary products are usually not formed throughout the plant, but by cells or tissues metabolically specialized for the purpose.

There are several important classes of secondary products, each embracing many different compounds having some chemical and biosynthetic similarity.

Phenolics

These are compounds containing six-membered (benzene) rings bearing hydroxyl (—OH) groups, as appear in Figure 6-9. All vascular plants make lignin, the complex phenolic polymer (Figure 6-9) deposited in secondary cell walls for stiffening (Chapter 4), but a huge variety of other phenolics, mostly small molecules rather than polymers, are made by different plants. These include tannins, purple or red (anthocyanin) and yellow (flavonoid) pigments, and many other compounds.

The biosynthesis of many phenolics begins with the amino acid phenylalanine, which contains a benzene ring (Figure 5-6). Removing the nitrogen from it yields a benzene ring with a three-carbon side chain, called a C_6–C_3 unit, a structure found in many phenolic products including lignin (Figure 6-9), which is a polymer of such subunits. Some phenolic rings are derived from carbohydrates via the early part of the pathway to phenylalanine, and others are derived from active acetate (acetyl-CoA; Figure 6-4).

Isoprenoids

These are compounds made up of branched five-carbon units called isoprene (Figure 6-10A). Isoprenoids include the yellow pigments, carotenoids, that are important to all photosynthetic cells (Chapter 7). A huge variety of terpenoids (named after turpentine), volatile aromatic compounds composed of isoprene units (Figure 6-10B), are formed by different plants, yielding many of the flavors and aromas of herbs, spices, resins, and so on. Natural rubber is an isoprene polymer (Figure 6-10C). The isoprene units of all these products are

formed from the active acetate (acetyl-CoA) of respiration (Figure 6-4).

Alkaloids

These are ring compounds containing nitrogen, making them somewhat basic or alkaline (hence their name). An enormous diversity of alkaloids are formed by different plants, for example, the opium poppy alkaloids such as morphine, and the peyote alkaloids such as mescaline. Since many of these have powerful effects on animal nervous systems, their formation by plants may reflect "chemical warfare" against plant-eating animals as noted in Chapter 1. Some of the important plant alkaloids are considered in Chapter 32.

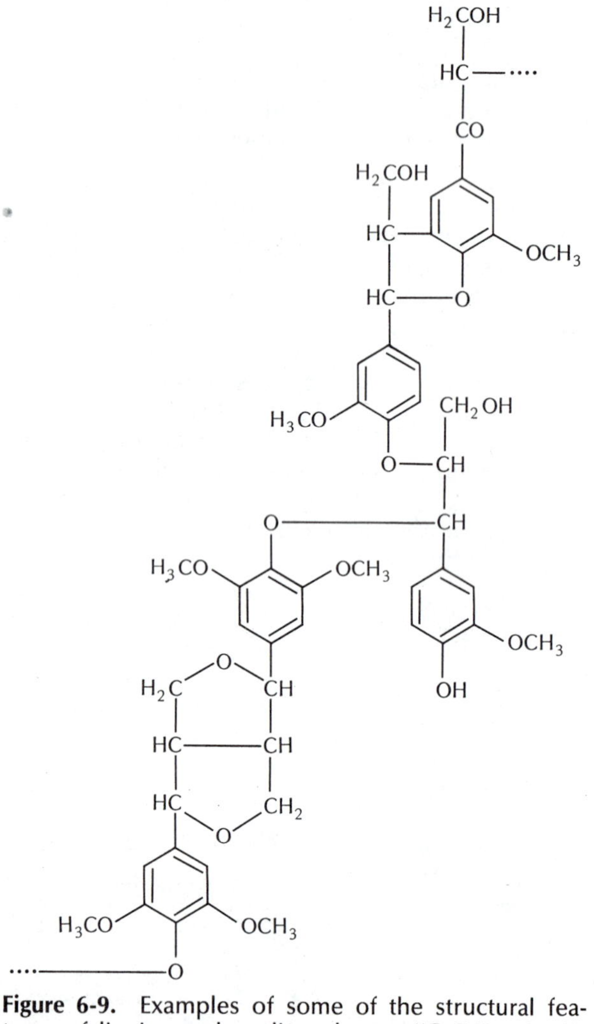

Figure 6-9. Examples of some of the structural features of lignin, a phenolic polymer. "C_6C_3" units (6-carbon benzene ring + 3-carbon side chain) are linked together in a variety of ways, only a few of which are illustrated here. The complete structure of this very complex polymer is not yet known; it is probably a random combination of units with different linkages rather than a unique, precise structure such as individual proteins have. The C_6C_3 unit is also used to build many other phenolic secondary products.

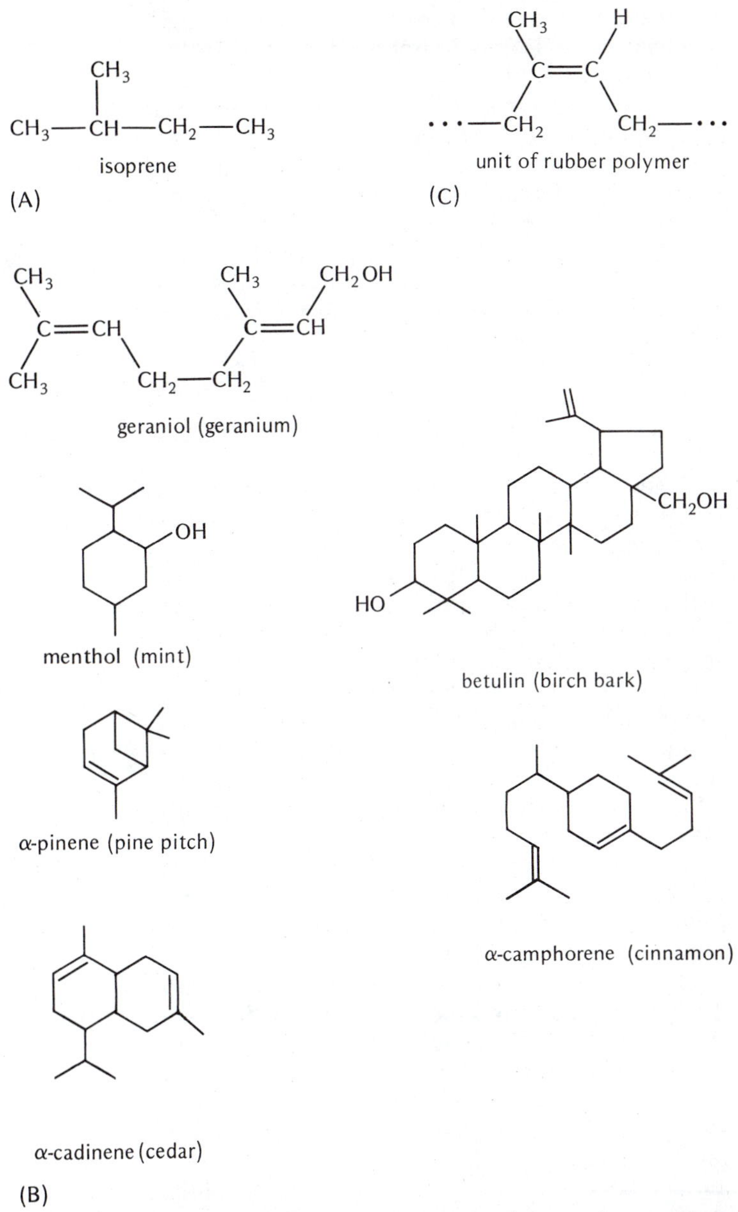

Figure 6-10. (A) Isoprene; (B) examples of isoprenoid (terpenoid) secondary products, indicating one plant source of each. Ring structures are drawn using the conventions given in Figure 5-9; all these structures are built up from isoprene units. (C) Isoprene subunit of rubber, a polymer of some 3000–6000 such units.

TRANSPORT PROCESSES AND ENERGY

Substances that cells need can be taken up from their surroundings either by passive uptake (not requiring work) or by active transport, requiring energy. Passive uptake occurs by **diffusion** resulting from the random movement of molecules (due to their kinetic energy, a reflection of temperature). Passive uptake across membranes is governed by the **permeability** of the membrane and by the difference in concentration of the diffusing substance on opposite sides of the membrane.

Respiratory Oxygen Uptake

Cellular membranes are rather permeable to small molecules such as O_2 and CO_2, so a cell can obtain passively through its membranes adequate amounts of oxygen for its respiration and can easily get rid of the resulting CO_2. However, over longer distances such as from the outside of a stem or root to its interior, diffusion tends to be much slower and would be inadequate to supply the O_2 the interior cells need if it were not for the intercellular spaces or gas channels of a plant tissue (Chapter 4). O_2 diffuses much more rapidly in air than in water or cell sap, so plants are able to meet their respiratory oxygen needs by diffusion through their gas channels and do not need an O_2-carrying circulatory system like that of higher animals. Diffusion suffices thanks to the fact that most of a mature plant tissue's volume is nonrespiring vacuoles rather than cytoplasm. Thus plant tissues have a much lower respiratory oxygen demand than typical animal tissues have.

Another important kind of passive transport is **osmosis** (often regarded as a special case of diffusion), by which cells obtain water. Osmosis is discussed in Chapter 9.

Active Transport

The extracellular concentration of nutrients such as sugars and inorganic salts is often rather low. Furthermore, cellular membranes are not very permeable to these nutrients, which are hydrophilic, water-soluble, and hence do not readily dissolve in membrane lipids. Passive uptake, therefore, tends to be inadequate for a cell's nutrient needs. Cells improve their nutrient uptake through active transport "pumps" which, by using energy, bring nutrients in rapidly even from a dilute source and can accumulate them inside the cell to concentrations much higher than that outside.

Active transport is specific for individual kinds of molecules or ions. It appears that for each kind of substance actively transported, the membrane contains **carrier** proteins having carrier sites, analogous to the active site of an enzyme, into which the transported molecule or ion fits. The carrier is generally thought to change shape within the membrane, moving the bound molecule or ion to the other side and releasing it (Figure 6-11). Input of energy giving the transport directionality may occur at any of the steps mentioned if that step needs energy and can be driven by the splitting of ATP. As a result, the carrier system acts as an **ATPase,** that is, an enzyme that hydrolyzes ATP to ADP (or AMP) as it pumps the transported substance through the membrane.

ATPase activity that depends on the presence of a particular nutrient molecule or ion offers a possible way of recognizing an active transport carrier in isolated organelle or membrane preparations. A potassium- (or sodium-) dependent ATPase widely detectable in membrane preparations from plant tissues is believed to represent a K^+ pump in the plasma membrane.

In the chemiosmotic mechanism of oxidative phosphorylation mentioned earlier, a pH plus electrical potential difference (proton motive force) across the mitochondrial inner membrane constitutes a form of energy that can do work, that is, make ATP. Proton motive force actually offers a form of energy alternative to ATP that can be used for certain other cell processes, particularly some kinds of active transport. For example, active uptake of sugars by plant cells seems to be driven

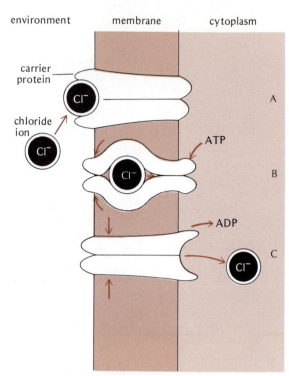

Figure 6-11. Example of how a carrier protein or ion channel located in a lipid membrane may actively transport an ion through the membrane. (A) In its initial or "relaxed" state, the carrier specifically binds a particular kind of ion (Cl^-, in this case) at the outside surface of the membrane. (B) Interaction of the carrier with ATP causes the carrier to change shape, forcing the ion through the channel. (C) The ion is released into the cytoplasm, and ADP is released as a result of the splitting of ATP that caused the carrier to change shape. The carrier then spontaneously reverts to the original conformation (A) by a process that requires no energy.

directly not by ATP but by proton motive force generated across the plasma membrane by actively pumping H^+ ions out of the cell (probably by an ATP-driven H^+ pump).

Another type of energy-dependent transport occurs via membrane vesicles produced by the ER or Golgi cisternae. This method is used especially for the secretion of proteins and polysaccharides to the outside of the cell, by the process of vesicle discharge or exocytosis (Figure 4-12). Since vesicles carry not only the secretory product but also the enclosing membrane material, exocytosis can supply the plasma membrane with phospholipids and membrane proteins from their sites of synthesis inside the cell (the ER; see Chapter 4).

METABOLIC COSTS OF GROWTH AND MAINTENANCE

In order to grow, a cell must respire a certain proportion of the food material available to it (from either its own photosynthesis or an external source)·

to provide the ATP needed to **assimilate** food material, that is, turn it into cell substances (Figure 6-12). This respiration covers the previously explained energy costs of both biosynthesis and active uptake of inorganic and (in a heterotrophic cell) organic nutrients used in growth. From the chemical composition of cells or tissues, plus a biochemical knowledge of the ATP yield of respiration and the ATP costs of biosynthetic pathways and polymer biosynthesis, one can predict that of every gram of photosynthetic products used for growth, between 0.15 and 0.25 g must be respired in order to assimilate the rest. The theoretically maximum possible efficiency of a green plant's growth is therefore 75–85% of its photosynthetic production. The actual efficiency of plant growth estimated for crops and natural vegetation varies from 60–80% for rapidly growing annual crop plants to as low as about 20% for tropical forest trees.

Turnover and the Cost of Maintenance

The growth yields just noted are less than the theoretical value primarily because cells need respiration for maintenance, even if they are not growing. Maintenance respiration provides energy for active transport needed to maintain the cell's ionic composition despite leakage of ions from the cell, and for resynthesizing cell components that are continuously being broken down, or **turned over.**

The occurrence of turnover is shown by radioisotope labeling experiments. If a tissue is supplied with some radioactively labeled amino acids, they soon become incorporated, through biosynthesis, into protein; but over a period of days this radioactivity decreases and finally disappears, due to breakdown and replacement of the protein containing the labeled amino acids. RNA and membrane phospholipids are also subject to turnover. The value of turnover is apparently to maintain the cell's enzymes, membranes, and so on, in a fully active and functional state.

Maintenance respiration is estimated to consume food equal to between about 1 and 5% of a living plant tissue's biomass (weight of organic material) per day, depending primarily upon the temperature, because the higher the temperature, the faster the turnover as well as the necessary maintenance respiration. This mounts up, over an entire growing season, to a substantial consumption of photosynthetic products, greatest under warm tropical conditions (Figure 6-13).

Figure 6-12.   Major aspects of metabolism, showing where energy (color arrows) from respiration goes during plant growth.

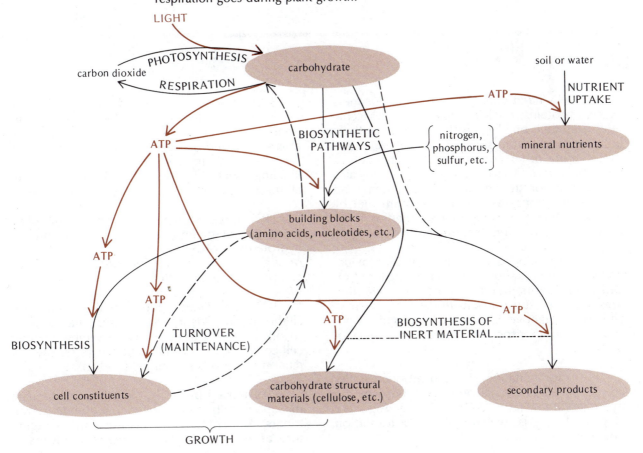

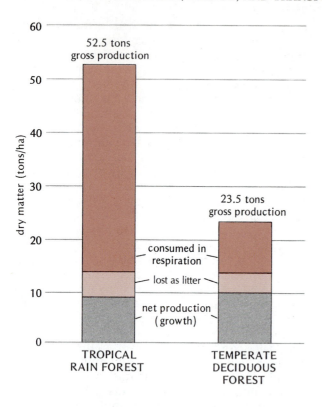

Respiration of plants is minimal under the low temperatures of winter, when growth is not occurring and maintenance requirements are at a minimum, enabling vegetation to pass even the protracted winter seasons of arctic and alpine habitats with relatively little respiratory cost. This is an advantage plants have over warm-blooded animals, which have to pay the full cost of maintaining themselves throughout this difficult period unless they are able to hibernate, reducing their body temperature and respiration to low values more like those of winter vegetation. Other kinds of dormancy similarly involve a much reduced respiration. Dry seeds respire almost undetectably and can survive for long periods with negligible depletion of their stored food supply.

Figure 6-13. Annual photosynthesis ("gross production"), respiratory costs (dark brown), loss of leaves and branches ("litter," light brown), and net incremental growth (grey) of trees in a tropical rain forest (Ivory Coast, Africa) and a temperate deciduous forest (beech woods, Denmark).

SUMMARY

(1) Cells need energy to assemble polymers such as proteins, nucleic acids, and polysaccharides from their building blocks; to produce these building blocks from simple food materials; and to assemble other complex molecules such as phospholipids and secondary products. They also need energy for the active uptake of nutrients and for motility processes. These needs are met by ATP, a high-energy compound formed as a by-product of respiration or fermentation.

(2) Respiration is the energy-yielding oxidation of cell energy substrates such as sugar to CO_2 and water. Fermentation is the energy-yielding breakdown of substrates in the absence of oxygen. The metabolic pathway called glycolysis, involved in both processes, occurs in the cytoplasm.

(3) In fermentation, the end product of glycolysis is reduced to alcohol to dispose of the hydrogen atoms removed by the dehydrogenase step of glycolysis, which makes ATP formation possible.

(4) In respiration, the end product of glycolysis is oxidized to CO_2 by dehydrogenase reactions of the citric acid cycle, and the hydrogen atoms removed are oxidized by oxygen via the respiratory chain, releasing much energy and yielding much ATP (oxidative phosphorylation); all these processes occur in the mitochondria.

(5) Complex storage reserves must be broken down (digested) into their component units by hydrolases in order to be utilized for respiration or as carbon sources for the biosynthesis of cell components (assimilation).

(6) Plant cells have biosynthetic pathways for all the amino acids, nitrogen bases, vitamins, and so on, that they need for the assembly of their cell constituents, as well as pathways for the formation of various secondary products.

(7) Passive uptake, by diffusion, is adequate to supply cells with oxygen but not most nutrients. Active transport, involving the expenditure of energy, absorbs nutrients by means of

specific carriers located in the membrane, which combine with the molecule or ion and move it through the membrane by an ATP-dependent process.

(8) The assimilation of storage reserves or new photosynthetic products into cell components involves a respiratory cost, to provide energy for biosynthesis and for the absorption and processing of nutrients. Living cells also have a respiratory cost for the maintenance of their components (turnover). Because of these costs, plant growth is less than the actual photosynthesis by the plant's photosynthetic cells.

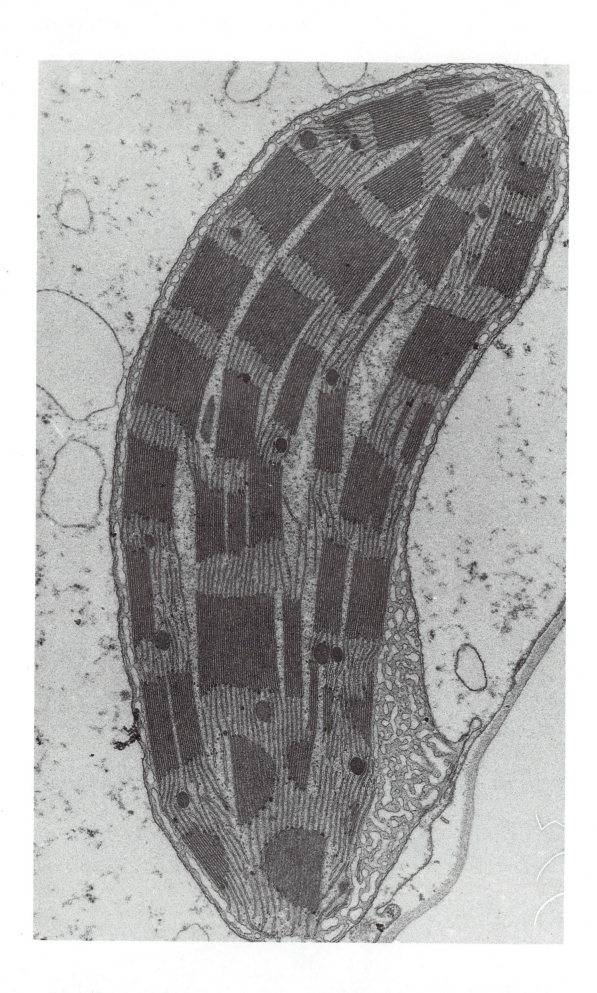

Chapter 7

PHOTOSYNTHESIS: THE CHLOROPLAST

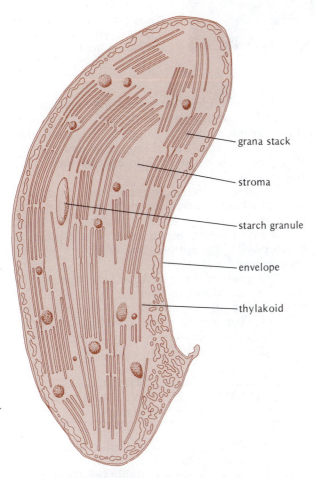

grana stack

stroma

starch granule

envelope

thylakoid

In photosynthesis, plants convert sunlight energy and carbon dioxide into the chemical energy of carbohydrates, releasing oxygen. On sunny days aquatic plants often release visible bubbles of oxygen into the water around them, showing the occurrence of photosynthesis (Figure 7-1). Photosynthesis by a leaf in the air is not directly

Figure 7-A. Chloroplast structure. An electron micrograph and (B, above) accompanying line drawing of a section through a maize leaf cell chloroplast.

visible, but can be detected either by measuring CO_2 uptake or O_2 release with relatively elaborate equipment or simply by testing for the accumulation of starch, a carbohydrate product of photosynthesis (Figure 7-2). The fact that starch accumulates only where light directly strikes a leaf, enabling one to make a starch print like that in Figure 7-2, reflects the fact that photosynthesis is conducted by individual cells rather than being a function of the leaf as a whole. Photosynthesis indeed is a subcellular process, localized to a green cell's chloroplasts. This chapter considers the subcellular process of photosynthesis, which is similar in the green cells of plants as different as unicellular algae and giant redwood trees, while the next chapter takes up the special physiology and attendant structure of photosynthetic leaves.

CHLOROPLASTS

Chloroplasts are conspicuous under the light microscope (Figure 4-14) because of their relatively large size and bright green color. They are the sites of photosynthesis within all green plant cells except blue-green algae, which lack discrete chloroplasts. The photosynthetic function of chloroplasts is indicated both by the fact that chloroplasts contain chlorophyll, which is essential to photosynthesis, and because within the chloroplasts of sunlit leaves, microscopically visible starch granules accumulate as photosynthetic products (Figure 7-A).

From certain leaves such as spinach one can isolate, by centrifugation, chloroplasts that can perform complete photosynthesis, that is, produce carbohydrate from carbon dioxide in the light, proving that they contain all the equipment necessary for photosynthesis. However, isolated chloroplasts are very fragile, ceasing photosynthesis within a few minutes due to loss of some of the enzymes needed for the process.

Figure 7-1. An aquatic plant (*Elodea*) emitting oxygen bubbles in the light.

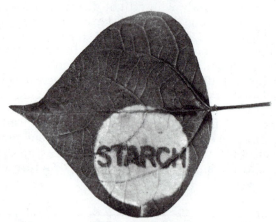

Figure 7-2. A "starch print" produced by photosynthesis in a garden bean leaf. A stencil of the word "starch" was placed over the leaf of a plant that had been kept overnight in the dark to deplete its leaves of starch. The plant was then exposed to light for several hours, after which the stencil was removed and the chlorophyll was extracted from the leaf with alcohol. The leaf was then stained with iodine solution which forms a blue-black complex with starch. Starch was produced where the stencil allowed light to reach the leaf, resulting in the illustrated print.

Chloroplast Structure

The chloroplast (Figure 7-A) is bounded by a double outer membrane or envelope (two membranes separated by a narrow gap) like that of the mitochondrion. The interior of the chloroplast contains multiple layers of membranes. Chlorophyll is located in these membranes. Adjacent membranes curve around and join at their edges, forming a flattened sac called a **thylakoid.** The thylakoids of higher plant chloroplasts usually aggregate into groups resembling stacks of pancakes, called **grana stacks** (see Figure 7-A). Some thylakoid membranes, called intergrana lamellae, extend from one grana stack to another within the chloroplast.

Freeze-fracture electron microscopy (Chapter 4) of thylakoids reveals conspicuous arrays of particles within the membranes (Figure 7-3). These are believed to represent proteins and protein–chlorophyll aggregates involved in energy capture and conversion.

The region of the chloroplast between the grana stacks or outside the thylakoids is called the **stroma.** The stroma appears to contain the enzymes for conversion of carbon dioxide to carbohydrate; these enzymes readily leak out of the chloroplast if its envelope is damaged.

CHLOROPLAST PIGMENTS AND LIGHT ABSORPTION

For light energy to be utilized in any process, such as photosynthesis, it must first be absorbed. This requires a suitable light-absorbing compound, or **pigment.** A pigment appears colored because it absorbs certain colors or wavelengths of light more strongly than others. Chlorophyll absorbs red and blue light more strongly than green, so light passed through chlorophyll looks green.

Evidence for what kind(s) of pigment absorbs light for a light-dependent process may be obtained by determining the effectiveness of different wavelengths of light, called the **action spectrum** of the process, and comparing this with the

Figure 7-3. Particulate structure of thylakoid membranes as revealed by the "freeze-fracture" method (70,000 ×). A thylakoid suspension was quick-frozen, the ice block was split, and a replica of the fractured surface was prepared and examined by electron microscopy. The illustration shows the large particles typically associated with the inner (nearer the thylakoid space) portion of the thylakoid membrane and the smaller particles associated with the outer (nearer the stroma) portion of the thylakoid membrane. The larger particles have been associated with photosystem II activity and the smaller ones with photosystem I activity (see p. 127).

light absorption characteristics of pigments. To determine an action spectrum, light must first be separated into its component colors, for example, by passing it through a glass or quartz prism. This displays the sequence of light wavelengths and corresponding colors called the spectrum, familiar as the colors of the rainbow. In 1882 the German botanist T. W. Engelmann, by projecting a tiny spectrum onto a chain of green algal cells in water containing oxygen-sensitive motile bacteria (see Figure 7-4), found that oxygen was produced most vigorously in red and blue light. Since red and blue are the wavelengths that chlorophyll absorbs most strongly, it appears that light absorbed by chlorophyll supplies the energy for photosynthesis.

Action and Absorption Spectra

A quantitative action spectrum for photosynthesis can be obtained by measuring the rate of photosynthesis under light of different wavelengths but of equal intensities, as in Figure 7-5A. The quantitative light absorption characteristics of a pigment such as chlorophyll are expressed by its **absorption spectrum,** its relative absorption of different wavelengths (Figure 7-5C). The close similarity between the action spectrum for photosynthesis and absorption spectra of chlorophylls found in the leaf (Figure 7-5A, B) shows convincingly that chlorophylls are the principal light absorbers for photosynthesis.

Chloroplast Pigments

Photosynthetic cells contain other pigments besides chlorophyll. Furthermore, higher plants and many others contain more than one kind of chlorophyll. Table 7-1 summarizes various kinds of photosynthetic pigments (cf. Figure 7-6). All green plant cells contain chlorophyll *a*, which appears indispensable to their photosynthesis be-

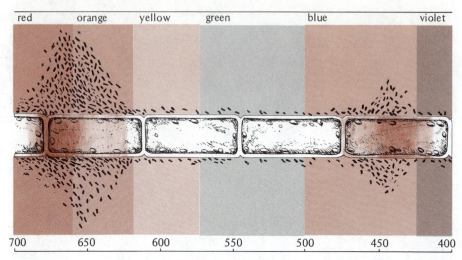

| red | orange | yellow | green | blue | violet |

700   650   600   550   500   450   400

Wavelength, nm

Figure 7-4.   Engelmann's 1882 experiment to determine which wavelengths of light are most effective for photosynthesis. He projected a minute spectrum, as illustrated, onto a filament of the alga *Cladophora* while watching it, under the microscope, in water containing motile bacteria that swim toward a source of oxygen. The bacteria congregated along the filament in the blue and red regions of the spectrum, which the alga's chlorophyll absorbs strongly, as indicated by the dark appearance of the filament in these parts of the spectrum. This result demonstrated that oxygen production is due to light that chlorophyll absorbs.

cause mutants lacking it are invariably nonphotosynthetic. Chloroplast pigments other than chlorophyll *a* are called **accessory pigments.** Action spectrum data indicate that accessory pigments contribute energy, to varying degrees, to photosynthesis. In higher plants, for example, chlorophyll *b* is as effective in contributing energy as chlorophyll *a*, whereas energy absorbed by carotenoids is used only about half as efficiently. In many algae, however, energy absorbed by nonchlorophyll accessory pigments is used with high efficiency.

Although in most photosynthetic cells the energy contribution from carotenoids is unimportant, carotenoids have another function that makes them indispensible. They protect chlorophylls and other components of the photosynthetic machinery against photodestruction. In sunlight the leaves or photosynthetic cells of carotenoid-less mutant plants rapidly bleach white due to loss of chlorophyll and, conse-

quently, lose their photosynthetic activity completely. To prevent this, carotenoids are formed by all normal photosynthetic cells.

PHOTOSYNTHESIS AS AN OXIDATION–REDUCTION PROCESS

Taken in its entirety, photosynthesis is the reverse of respiration; thus the overall equation for the photosynthetic formation of glucose is

$$6 \ CO_2 \ + \ 6 \ H_2O \ \xrightarrow{\text{light}} \ C_6H_{12}O_6 \ + \ 6O_2. \quad (7\text{-}1)$$

As respiration releases energy, so photosynthesis must consume it, and this of course is why light energy is required in the process. It is easier to comprehend the elements of this process if we consider the conversion of one CO_2 molecule to one carbon atom of glucose or other carbohydrate,

TABLE 7-1   Photosynthetic Pigments[a]

Pigment Class	Pigment	Color	Occurrence
Chlorophylls	Chlorophyll *a*	Grass green	All green plants
	Chlorophyll *b*	Olive green	Higher plants, green algae
	Chlorophyll *c*	Yellow-green	Brown algae, etc.
	Bacteriochlorophyll	Blue	Photosynthetic bacteria
Carotenoids	Carotenes[b]	Yellow to red	All photosynthetic cells
	Xanthophylls[b,c]	Yellow to red	Most photosynthetic cells
Biliproteins	Phycocyanin	Blue	Blue-green and red algae
	Phycoerythrin	Red	Red and blue-green algae

[a]The chemical structures of several of these pigments are shown in Figure 7-6.
[b]Many different specific pigments are included in this subclass.
[c]Xanthophylls differ from carotenes in containing oxygen whereas carotenes contain only carbon and hydrogen (see Figure 7-6).

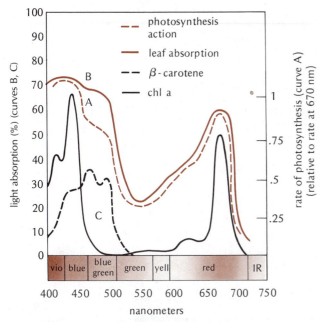

Figure 7-5. Absorption spectra of chloroplast pigments compared with an action spectrum for photosynthesis. The lower curves (C) are absorption spectra of chlorophyll a and of β-carotene, a typical carotenoid absorption spectrum. Above (B) is shown the absorption spectrum of a leaf of the water plant *Anacharis* (*Elodea*), and an action spectrum (A) for photosynthesis by this leaf. The action spectrum clearly implicates chlorophylls as the principal light absorbers for photosynthesis.

represented (CH_2O), dividing the above equation by 6:

$$CO_2 + H_2O \xrightarrow{\text{light}} (CH_2O) + O_2 \qquad (7\text{-}2)$$

As the reverse of respiration, which is the oxidation of carbohydrate to CO_2, in photosynthesis CO_2 must be reduced; that is, hydrogen atoms or electrons must be added to it. Four H atoms are required to reduce one molecule of CO_2 to (CH_2O) (see Figure 7-7, process labeled C).

The Production of Reducing Power: The Hill Reaction

The hydrogen atoms needed for the reduction of CO_2 in chloroplasts must come from water. As indicated in process A in Figure 7-7, two water molecules are needed to supply the necessary four hydrogen atoms; their removal leaves a molecule of oxygen as a by-product. this representation indicates that all of the oxygen produced in photosynthesis should come from water, not CO_2. This fact can be demonstrated by allowing algal cells to photosynthesize in water labeled with the heavy oxygen isotope, ^{18}O: the oxygen produced has the same ^{18}O content as the water supplied.

The H atoms used for carbon reduction are carried by the hydrogen-accepting nucleotide called NADP, which is closely related to (but has one more phosphate than) the H carrier NAD of respiration (Chapter 6). The formation of $NADPH_2$ and oxygen (from NADP and water) requires much energy, because the H atoms of $NADPH_2$ are about equal in energy content to hydrogen gas, an excellent fuel. Light, via chlorophyll, provides the energy for $NADPH_2$ production, which occurs on the thylakoid membranes. The English biochemist Robert Hill discovered, in the 1930s, that chloroplasts isolated from ground-up leaves by centrifugation can produce oxygen and reduce NADP or artificially added hydrogen or electron acceptors in the light, even though they usually lose the ability to reduce carbon dioxide (because enzymes of the carbon reducing system leak from the stroma through damaged chloroplast envelopes). Light-driven production of oxygen and hydrogen atoms by chloroplasts (Figure 7-7A) is called the **Hill reaction,** after its discoverer. The Hill reaction carried out with isolated chloroplasts enables the process by which chlorophyll converts light into chemical energy to be studied in the test tube, contributing enormously to our understanding of how this conversion occurs.

In certain circumstances algal cells produce an enzyme that will release as hydrogen gas (H_2) the H atoms generated by the Hill reaction. This should make possible, in principle, the biological production of hydrogen fuel from water, using sunlight energy. Much interest currently centers on trying to make biological production of hydrogen a practical fuel source.

Photophosphorylation

Associated with the Hill reaction in thylakoid membranes, ATP is formed from ADP, a process called **photophosphorylation** (Figure 7-7B), which conserves some of the energy of the absorbed light as high-energy phosphate, in addition to the large amount of energy in the hydrogen output of the Hill reaction. ATP produced by photophosphorylation is needed as an additional energy source for driving carbon dioxide reduction (Figure 7-7C) and can also be used for certain other energy-requiring processes in green cells.

The Hill reaction and photophosphorylation are often called the **light reactions** of photosynthesis since they are directly driven by the input of light energy from chlorophyll. The reactions of carbon dioxide reduction are called **dark reactions** because they do not directly involve light, but are driven indirectly by the products of the light reactions (Figure 7-7).

Bacterial Photosynthesis

A few groups of bacteria are photosynthetic; unlike green plants, they do not produce oxygen. Purple

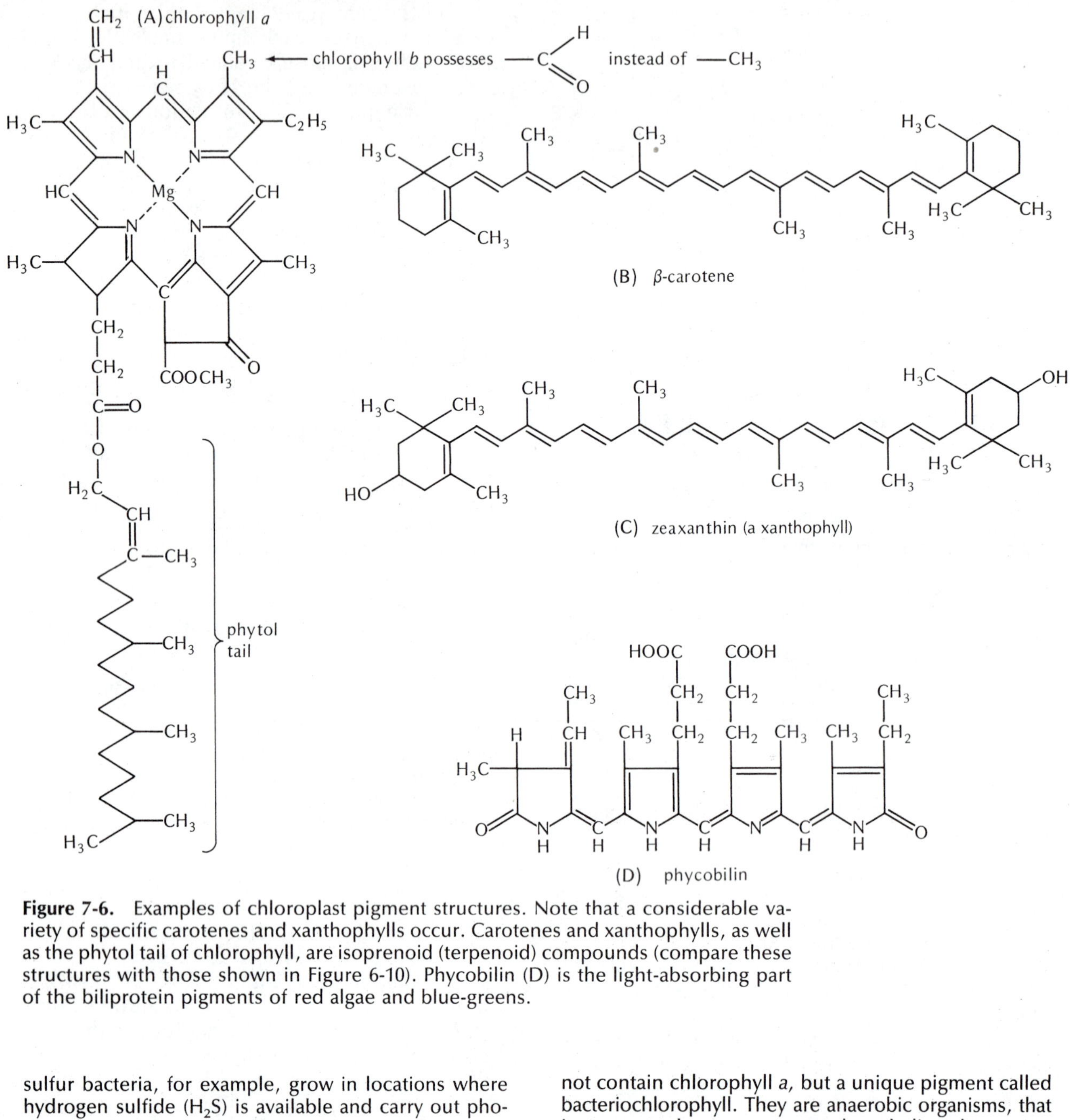

Figure 7-6. Examples of chloroplast pigment structures. Note that a considerable variety of specific carotenes and xanthophylls occur. Carotenes and xanthophylls, as well as the phytol tail of chlorophyll, are isoprenoid (terpenoid) compounds (compare these structures with those shown in Figure 6-10). Phycobilin (D) is the light-absorbing part of the biliprotein pigments of red algae and blue-greens.

sulfur bacteria, for example, grow in locations where hydrogen sulfide (H_2S) is available and carry out photosynthesis according to the equation

$$CO_2 + 2 H_2S \xrightarrow{\text{light}} (CH_2O) + H_2O + 2 S \quad (7\text{-}3)$$

Here obviously 2 H_2S provides the four H atoms needed for CO_2 reduction, and by providing them it becomes oxidized to sulfur (2 S). The striking similarity between Eqs. (7-3) and (7-2) originally suggested that the oxygen produced in green plant photosynthesis must come from water.

Compared with the green plant's production of H atoms from water, rather little energy is needed to obtain H atoms useful for reducing CO_2 from a chemical reducing agent such as H_2S. It appears that the most important role of light in bacterial photosynthesis is to provide, by photophosphorylation, ATP needed in the process of CO_2 reduction. Photosynthetic bacteria do not contain chlorophyll a, but a unique pigment called bacteriochlorophyll. They are anaerobic organisms, that is, cannot tolerate oxygen, and are believed to exemplify the primitive photosynthesis that first evolved on earth, before green plants. Today, photosynthetic bacteria are confined to anaerobic habitats such as the mud on the bottom of ponds and shallow lakes.

LIGHT REACTIONS OF PHOTOSYNTHESIS

Energy Absorption and Transfer

Light energy is absorbed as a discrete unit, or **quantum,** by a pigment molecule, raising the molecule (actually, one of its electrons) to an energized or **excited** state, conventionally indicated by an asterisk (*) after the pigment's name or an ab-

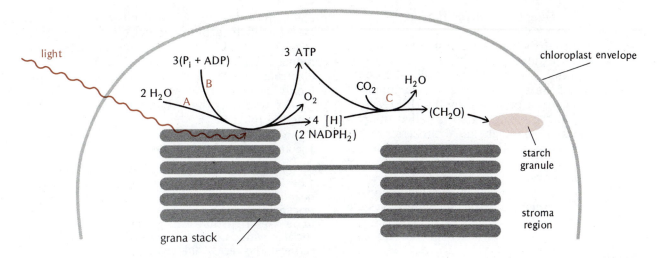

Figure 7-7. Component processes of photosynthesis: (A) Hill reaction. (B) Photophosphorylation. (C) Carbon dioxide reduction to carbohydrate (CH_2O).

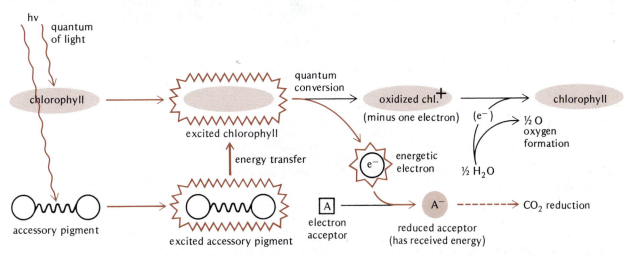

Figure 7-8. Capture and transformation of energy (colored arrows) in photosynthesis. A quantum can be captured either by a chlorophyll molecule directly (upper left), or by an accessory pigment (lower left) followed by energy transfer to chlorophyll. Loss of an energy-laden electron from an excited chlorophyll molecule to an electron acceptor, the process of quantum conversion, provides the reducing power for CO_2 reduction (lower right). The oxidized (electron-depleted) chlorophyll left by quantum conversion becomes restored to its normal state by an electron coming from water, resulting in production of oxygen (upper right; formation of one O_2 molecule results from four such electron transfers).

breviation thereof, for example, chl for chlorophyll (see Figure 7-8). Among the chloroplast pigments, it appears that only chlorophyll a can convert excitation energy into chemical energy to drive the process of photosynthesis. This is why chlorophyll a is essential to photosynthesis. Energy absorbed by accessory pigments can be used in photosynthesis because it can migrate *from* the accessory pigment *to* chlorophyll a (Figure 7-8).

Excitation energy migration between pigments is a very rapid process and depends upon their being packed closely together, as they are in the thylakoid membranes. Excitation energy migrates as a unit (or quantum) from one pigment molecule to another, rather than spreading out diffusely among many pigment molecules. As a result, energy received by chlorophyll a from other pigments is equivalent in photosynthetic value to light quanta absorbed directly by chlorophyll a.

Quantum Conversion

To be used for photosynthesis, the physical energy in an excited pigment molecule must be converted into chemical energy, specifically into the energy of the electrons or hydrogen atoms of the reducing agent produced by the Hill reaction (Figure 7-7A). This conversion is called the **primary process** or photochemical process of photosynthesis, also called **quantum conversion** because it converts a quantum, or unit amount of energy, from physical to chemical form. It occurs by the transfer of an electron from an energized chlorophyll *a* molecule to the biochemical system of the chloroplast:

$$chl^* \longrightarrow chl^+ + e^- \qquad (7\text{-}4)$$

chlorophyll	chlorophyll	electron,
energized by	minus 1	associated
absorption of	electron, i.e.,	with electron
light	oxidized	acceptor

This is somewhat like the operation of a photoelectric cell, as in a photographer's light meter, the light-sensitive unit of which releases electrons generating an electric current. In the chloroplast, electrons released by quantum conversion are transferred to a biochemical electron acceptor (Figure 7-8) which, by accepting them, becomes reduced. From the primary electron acceptor these electrons pass to other electron carriers, emerging (combined with H^+ ions from the simple ionization of water) as the hydrogen output of the Hill reaction ($NADPH_2$).

The Reaction Center and Photosynthetic Unit

Several lines of evidence indicate that only about 1 of every 250–300 chlorophyll molecules can convert quanta. This special form of chlorophyll is called the **reaction-center chlorophyll.** A reaction center appears to be one or possibly two chlorophyll *a* molecules associated with an electron acceptor and other electron carriers. The rest of the chlorophyll *a* plus accessory pigments absorb light energy for delivery, by migration between pigments, to the reaction center chlorophyll. These bulk pigments are therefore called antenna pigments by analogy with a radio or TV antenna, which collects and feeds energy to the instrument. The aggregate of several hundred antenna pigment molecules plus a reaction center is called a **photosynthetic unit** (Figure 7-9). The large number of antenna pigment mol-

Figure 7-9. Concept of a photosynthetic unit. Absorption of a quantum of light by an antenna pigment; transfer of the excitation energy from pigment to pigment until it reaches the reaction center; conversion of excitation energy at the reaction center by release of an electron to an electron acceptor, reducing it and leaving the reaction center chlorophyll positively charged (lacking one electron). Color shows a path of energy migration, ultimately to the reduced electron acceptor. (Diagram shows a photosynthetic unit of photosystem I, the reaction centers of which are a special form of chlorophyll *a* called chlorophyll a_{700} or P700).

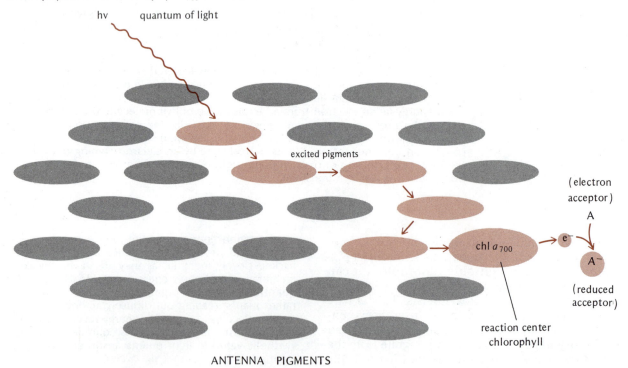

ecules in each photosynthetic unit enables its reaction center to be supplied constantly with quanta of energy, even though at ordinary light intensities any given pigment molecule captures a quantum of light only infrequently.

Chlorophyll occurs in the photosynthetic unit in at least two types of association with proteins. The "light-harvesting chlorophyll protein" carries much of the antenna chlorophylls a and b, while another protein complex called "chlorophyll protein I" bears reaction-center chlorophyll a plus a part of the antenna chlorophyll a.

The Two Photosystems

From detailed action spectrum data for photosynthesis, two kinds of quantum-converting pigment systems, called photosystems I and II, can be distinguished. Photosystem I primarily utilizes energy absorbed by chlorophyll a, whereas system II preferentially receives energy from accessory pigments. The two photosystems appear to act one behind the other in series, like the two batteries of an ordinary flashlight, to boost electrons to a higher energy level (give them more reducing power) than one system could alone (Figure 7-10). Photosystem II takes electrons from water, oxidizing it to oxygen, and delivers these electrons to photosystem I, which raises their energy level further, allowing them to reduce NADP to NADPH₂.

Photosynthetic Electron Transport

As indicated in Figure 7-10, cytochromes and other electron carriers found in thylakoid membranes function in electron transport chains that allow electrons to move (a) from photosystem II to photosystem I; (b) from photosystem I to the ultimate hydrogen acceptor of the Hill reaction, NADP; and (c) from water to photosystem II.

Photosynthetic electron transport, like electron transport in mitochondria, leads to ATP production (photophosphorylation). Electron flow from photosystem II to photosystem I releases energy, as represented by the downhill direction of the arrows in this part of Figure 7-10. Photophosphorylation conserves some of this energy as ATP.

A second type of photophosphorylation called **cyclic photophosphorylation** can occur, represented by the broken arrows in Figure 7-10. Electrons from photosystem I can return to its reaction center via the electron transport chain between photosystems II and I, allowing ATP production independent of the reduction of NADP. Electrons flow in a cycle out of, and back to, photosystem I. Cyclic photophosphorylation supplements the photophosphorylation coupled to the Hill

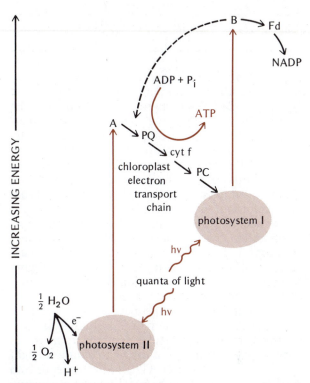

Figure 7-10. Action of the two chloroplast photosystems in series raises electrons to a high enough energy level to reduce NADP. Black arrows show route of electron flow; zigzag and straight colored arrows show, respectively, quantum absorption and conversion by the pigment systems. Names of the electron carriers whose abbreviations appear in the diagram: (PQ) plastoquinone; (PC) plastocyanin (a copper-containing protein); (cyt f) cytochrome f; (Fd) ferredoxin (an iron-containing protein). A and B designate the (not yet conclusively identified) primary electron acceptors of photosystems II and I, respectively (it is a historical accident that the first photosystem in the series is termed system II while the second is termed system I). Broken arrow indicates cyclic route of electron flow from photosystem I, causing cyclic photophosphorylation.

reaction (reduction of NADP) to provide sufficient ATP for the carbon reduction cycle (see below).

Photophosphorylation involves a "chemiosmotic" mechanism, similar to that for mitochondrial phosphorylation. Light-driven flow of electrons between electron carriers, located in the thylakoid membranes, causes H⁺ ions to be transported across the thylakoid membrane to the interior space of the thylakoid. This raises the H⁺ concentration inside, building up a pH difference across the thylakoid membrane, tending to drive diffusion of H⁺ out of the thylakoid. Outward diffusion of H⁺ from thylakoids drives the formation of ATP from ADP (Figure 7-11). The need to produce and maintain a transmembrane pH difference to drive photophosphorylation appears to be the reason thylakoid membranes are arranged as closed sacs.

CARBON DIOXIDE CAPTURE AND REDUCTION

The problem of how carbon dioxide is converted to carbohydrate in chloroplasts was solved by Melvin Calvin and a group of associates at the University of California at Berkeley in the 1950s using the radioactive isotope of carbon, ^{14}C, to detect and identify the compounds in which carbon from CO_2 first appears during photosynthesis: these compounds rapidly become radioactive when $^{14}CO_2$ is fed to green plant cells in the light. From this information they deduced that photosynthesis occurs by a complex cycle of reactions, now often called the **Calvin cycle,** which is sketched in simplified form in Figure 7-12. The key reactions of this cycle are (a) CO_2 fixation, bringing CO_2 into organic combination by joining it to a CO_2 "acceptor" molecule, producing a three-carbon organic acid as the primary fixation product; and (b) reduction of the fixation product to yield a three-carbon sugar, which depends upon and is driven by the products of the photosynthetic light reactions, $NADPH_2$ and ATP, and introduces into the final product the chemical energy associated with carbohydrates. The rest of the reactions of the cycle serve (c) to form six-carbon sugar (e.g., glucose) from the three-carbon sugar (triose) produced in step b, and (d) to generate the CO_2 acceptor for step a from the three- and six-carbon sugars formed in steps b and c. As the cycle turns, driven by $NADPH_2$ and ATP coming from the light reactions, surplus six-carbon sugars accumulate and can be stored by conversion to reserves such as starch and sucrose.

The Calvin cycle was worked out originally by experiments on unicellular algae, but was later found in all groups of photosynthetic plants. Because its initial CO_2 fixation product is a three-carbon acid, this generally occurring type of photosynthetic metabolism has been called **C3 photosynthesis** and plants possessing it, C3 plants, to distinguish them from C4 plants with a specialized kind of photosynthetic metabolism involving four-carbon acids (Chapter 8).

The CO_2-Fixing Enzyme

In chloroplasts the enzyme **ribulose bisphosphate carboxylase** (RuBP carboxylase) adds CO_2 to a five-carbon or pentose sugar (ribulose bisphosphate; RuBP), the addition product splitting to yield two molecules of phosphoglyceric acid (PGA), one of the two bearing a carbon atom from CO_2 (Figure 7-13). RuBP carboxylase is produced by green cells in large quantities, often comprising as much as half the total protein of a leaf, this large amount apparently being needed to obtain an adequate capacity for CO_2 fixation.

The RuBP carboxylase of green plants is a very large enzyme (molecular weight of about 540,000) consisting of two kinds of protein subunits (molecular weights of 13,000 and 54,000, respectively); eight of each kind make up the enzyme molecule. Interestingly, the large subunit is coded for by chloroplast DNA and synthesized by chloroplast ribosomes (Chapters 4 and 5), whereas the small subunit is specified by nuclear DNA, made by cytoplasmic ribosomes, and imported into the chloroplast, where it combines with the large subunit to form the functional enzyme.

Formation and Interconversion of Sugars

Figure 7-13 shows some of the details of how ATP and $NADPH_2$ from the light reactions cause the reduction of PGA (from RuBP carboxylase) to three-carbon sugar (triose) phosphate. Two of these readily combine to form six-carbon sugar (hexose) phosphate, and by complicated interconversions using triose and hexose

Figure 7-11. Chemiosmotic mechanism of photophosphorylation. Photosystems I and II are oriented within the thylakoid membranes in such a way that electron flow from water to NADP picks up H$^+$ ions from the outside and releases them in the interior of the thylakoid, thus building up a higher H$^+$ concentration (lower pH) inside the thylakoid than outside it. Outward diffusion of H$^+$ ions, driven by this pH difference, drives ATP formation by an ATPase located on the thylakoid membrane. Note that only one of the many photosynthetic units and ATPase complexes in the thylakoid membrane are shown. (Inward pumping of H$^+$ ions by the photosystems is somewhat more complicated than represented here, involving the electron transport chain between systems II and I as well as the oxygen-producing and NADP-reducing steps shown in the diagram.)

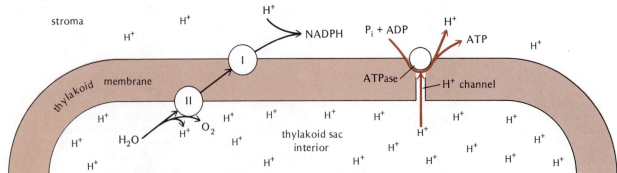

phosphates, pentose phosphate (ribulose-*P*) can be formed. An extra ATP molecule must be used to convert this to RuBP, replacing the RuBP molecule that was previously used by RuBP carboxylase to capture one CO_2 molecule. This explains why three ATP molecules are required per CO_2 molecule captured and reduced in photosynthesis (Figure 7-7C). After six CO_2 molecules have been captured and reduced, and six RuBP molecules have been regenerated, two triose or one hexose phosphate molecule remains in surplus as a net product of photosynthesis.

Photorespiration

The 21% oxygen in normal air depresses the photosynthesis of most plants to about two-thirds the rate observable in air containing only 1 or 2% oxygen. This oxygen inhibition can be seen not only in leaf photosynthesis measurements but also in the total growth and primary production of plants, which increases as much as 30% in a low-oxygen atmosphere (Figure 7-14). High oxygen inhibits net photosynthesis at least in part by inducing, in green cells in the light, a substantial supplemental respiration called **photorespiration,** which "recycles" back to CO_2 part of the CO_2 being captured in CO_2 fixation. Researchers are very interested in the nature and mechanism of photorespiration because, if it could be reduced or prevented in crop plants, their photosynthetic production would increase significantly.

Photorespiration results at least in part from an ability of oxygen to interact with RuBP carboxylase as an alternative substrate, that is, alternative to CO_2. The combination of O_2 with the enzyme's active site not only blocks the useful reaction with CO_2 but splits RuBP oxidatively into PGA and a two-carbon acid, glycolic acid ($HOCH_2COOH$), draining carbon from the Calvin cycle. To salvage and return this carbon to the cycle, an oxidative pathway, the glycolate pathway, operates in organelles of the microbody type (Chapter 4) called **peroxisomes,** which cluster around the chloroplasts of green cells (Figure 7-15) and contain the principal enzymes of this pathway. It converts two molecules of glycolic acid to one molecule of PGA (returning this carbon to the Calvin cycle) and one molecule of CO_2, the photorespiratory CO_2 production.

Certain flowering plants have evolved a specialization called C4 photosynthesis which effectively suppresses photorespiration, giving these plants a significantly higher photosynthetic capacity than C3 plants have. This adaptation, which involves anatomical as well as biochemical specialization within the leaf, is considered in Chapter 8.

PHOTOSYNTHETIC ENERGY EFFICIENCY

The information that photosynthesis research has uncovered about the mechanism of this process enables us to predict the maximum efficiency with which chloroplasts might be able to convert sunlight energy into that of carbohydrates, perhaps the most basic of all limitations to the proliferation of life on earth.

Figure 7-12. Simplified diagram of the carbon dioxide fixation and reduction cycle of photosynthesis. (A) CO_2 fixation, yielding two molecules of PGA (phosphoglyceric acid). (B) Reduction of PGA to triose (3-carbon sugar) phosphate. Both $NADPH_2$ and ATP are required to drive the reduction. (C) Formation of hexose (6-carbon) sugar by coupling two triose phosphate molecules together. (D) Formation of the CO_2 acceptor, RuBP (ribulose bisphosphate), from triose and hexose phosphates by a complex set of reactions that interconvert various sugar phosphates. Details of the most important compounds and reactions of the cycle are given in Figure 7-13.

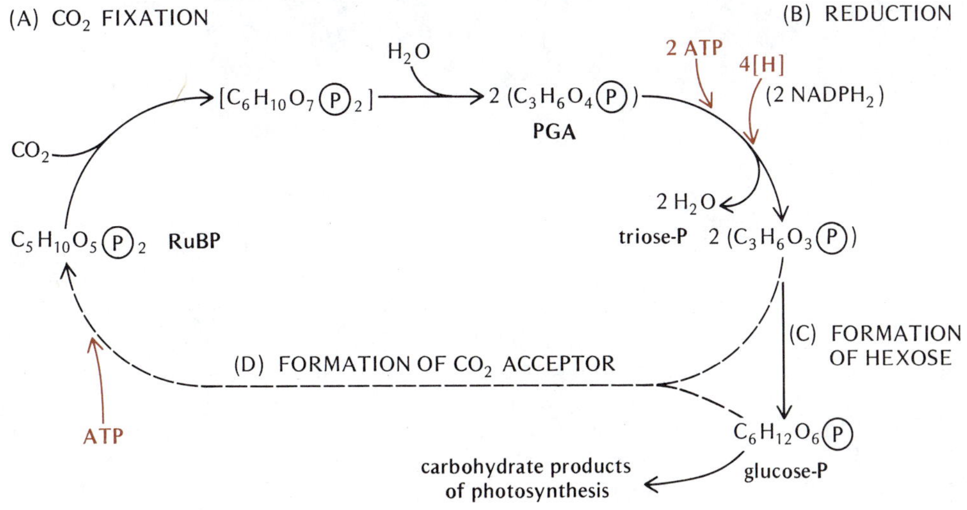

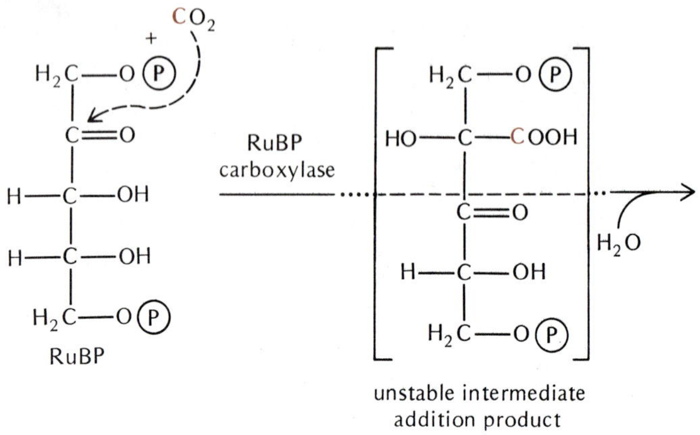

Figure 7-13. Details of chloroplast reactions by which CO_2 is fixed and the fixation product (PGA) is reduced, with energy input from ATP as well as $NADPH_2$. Both PGA molecules yielded by the RuBP carboxylase reaction are reduced in the same way, thus using 2 $NADPH_2$ and 2 ATP per CO_2 molecule fixed. This set of reactions yields six carbohydrate carbon atoms (two triose phosphates) in place of the original five (of RuBP). Hexose phosphate is formed by coupling two triose phosphates together.

Figure 7-14. Inhibitory effect of atmospheric oxygen on plant growth. Bean seedlings with their roots in a nutrient solution were grown, beginning at the stage shown in the upper two photos, in air (left, 21% O_2) or in nitrogen containing 2.5% O_2 and the same CO_2 concentration as normal air (right). The same plants after six days further growth are shown in the lower photos; note the much greater growth not only of the top of the plant in low O_2 but also of its root system, due to more available photosynthetic products. Average growth (increase in dry weight) per plant in low O_2 over the six days was 700 mg, compared with 330 mg for the air controls (average initial dry weight was 210 mg at start of experiment).

The Quantum Requirement

For each H atom delivered by the light reactions (Figure 7-7A) for CO_2 reduction, two quanta of light are needed, one for each of the two photosystems. Since four H atoms are needed to reduce one CO_2 molecule, eight quanta should be required to generate the needed reductant. Of the three ATP molecules needed for each CO_2 molecule converted into carbohydrate in the Calvin cycle (Figure 7-7C), at least two are produced by the photophosphorylation that accompanies production of four H atoms of reductant by the Hill reaction. So it appears there should be at most only a small shortfall of ATP, needed for the Calvin cycle, to be met by additional photophosphorylation using additional quanta.

The highest photosynthetic efficiencies experimentally measured, using algal cells under conditions of maximally efficient light utilization

(chemical reaction scheme)

2 molecules of PGA → (phosphoglycerate kinase, ATP → ADP) → diphosphoglyceric acid → (triose phosphate dehydrogenase, NADPH₂ → NADP, H₂O) → triose phosphate

REDUCTION

(low light intensity, high CO_2 concentration), are equivalent to one CO_2 molecule fixed for every 10–12 quanta of light taken up. With higher plant leaves, photosynthetic yields as high as one CO_2 molecule captured per 12–15 quanta absorbed have occasionally been measured.

The difference between these values and the 8 quanta needed in principle for reductant formation presumably is due (a) to the need for additional ATP beyond that formed in conjunction with the Hill reaction, (b) to inefficiencies in the light-harvesting and -converting pigment system, resulting in the energy of some quanta being lost as heat, and (c) to the occurrence of photorespiration, although photorespiration should be minimal under the high CO_2 concentration used to measure maximum photosynthetic efficiency (see Chapter 8).

Energy Conversion

We can calculate the maximum possible efficiency of photosynthetic energy conversion by taking as the energy input the energy content of quanta of red light; quanta of shorter wavelengths contain more energy but at least 10 of them are still required, so more of their energy is wasted. Red light contains about 40 kcal/mole of quanta, or 400 kcal/10 moles of quanta. Since 1 mole of glucose ($C_6H_{12}O_6$) contains, according to Eq. (6-1), 686 kcal of stored energy, the $\frac{1}{6}$ mole of carbohydrate that could be formed photosynthetically from 1 mole of CO_2 (using 10 moles of quanta) will contain 114 kcal. The maximum efficiency of photosynthetic energy conversion is therefore 114/400 = 28.5%.

It must be emphasized that this is an absolutely maximum value, good only for red light and for completely optimal conditions, including the virtual absence of photorespiration. Conditions in the field almost never correspond to those for maximum photosynthetic yield. Photorespiration substantially reduces photosynthesis by most plants under field conditions, and the costs of growth and maintenance respiration explained in Chapter 6 always reduce net primary production considerably below the plant's actual photosyn-

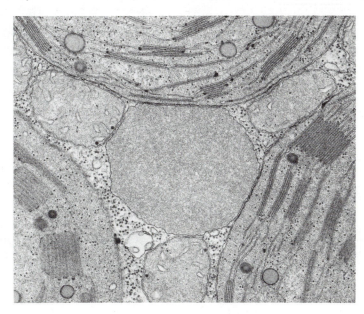

Figure 7-15. A microbody of the peroxisome type (P) in cell of a tobacco leaf (29,000 ×). The peroxisome is surrounded by three mitochondria (M) and chloroplasts (C); note how the outer membrane of these organelles is double while the membrane of the peroxisome is single.

thesis. Agriculture and forestry give energy conversion efficiencies mostly below about 1% in practice.

The Effect of Light Intensity

One of the most important factors that makes practical photosynthesis differ from the laboratory measurement of maximum energy efficiency is that, in the field, photosynthesis normally goes on in full sunlight. The rate of photosynthesis typically depends on the light intensity in the manner shown in Figure 7-16. At low light intensities photosynthesis is starved for energy, or **light-limited;** the system uses most of the quanta the pigments capture and is therefore maximally efficient, but because it gets few quanta, its rate is low and may only slightly exceed the rate of respiration, so net photosynthetic production by the cells is actually very poor. The rate of light-limited photosynthesis of course increases steeply with the energy input, as seen at the lower left in Figure 7-16. As the light intensity is raised to a level that supports substantial net photosynthesis (i.e., well above the rate of respiration), the increase in rate begins to flag (middle part of Figure 7-16), and at higher intensities it reaches a maximum value or plateau, usually at about one-fifth to one-third the intensity of full sunlight. This is called **light saturation** (right-hand part of Figure 7-16).

Light saturation does not result from any limitation in the capacity of chlorophyll to absorb light. It represents, instead, the maximum rate at which the dark reactions of photosynthesis can use energy from chlorophyll. A further increase

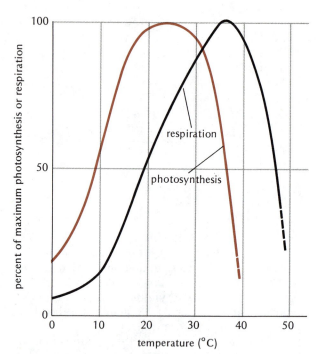

Figure 7-17. Typical effect of temperature on light-saturated photosynthesis, compared with temperature dependence of respiration, as seen in pea seedlings. (The two curves should not be compared vertically; maximum photosynthetic rate is several times greater than the maximum respiration.)

in the energy supply is then superfluous, and the excess energy that chlorophyll absorbs simply becomes converted into heat and wasted. Obviously the overall energy efficiency of photosynthesis falls lower and lower the farther the light intensity is raised above the light saturation point.

The Effect of Temperature

Figure 7-17 shows that the light-saturating rate of photosynthesis depends significantly upon temperature, as expected because the dark reactions of the system, limiting its rate, are temperature-dependent enzymatic processes. Under low light intensity, on the other hand, photosynthesis depends little on temperature because quantum capture and conversion are temperature-independent physical or photochemical processes, and the temperature-dependent activity of the dark reaction enzymes is in excess and does not limit the overall process.

Although light-saturated photosynthesis depends on temperature, photosynthesis is peculiar, compared with most metabolic processes such as respiration, in having a rather low **temperature optimum** (Figure 7-17). Photosynthesis by most plants increases only up to about 20–25°C (68–77°F) or less, leveling out and then actually declining markedly as the temperature approaches

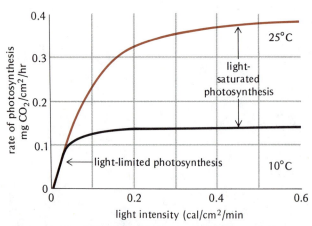

Figure 7-16. Effect of light intensity on rate of photosynthesis by typical higher plant or algal cells at optimum temperature (colored line) and at a lower temperature (10°C, black line). Light intensity (of photosynthetically active radiation, 400–700 nm) at far right corresponds approximately to full maximum summer sunlight.

or exceeds the human body temperature (37°C or 98.6°F), which we of course regard as the physiologically optimum temperature. This characteristic significantly reduces photosynthetic productivity during hot weather and is another important factor reducing practical energy conversion compared with the laboratory value.

The low temperature optimum of photosynthesis seems to be due at least partly to photorespiration. In Chapter 8 we shall pursue this aspect, as well as the dependence of photosynthesis on CO_2 concentration, in comparing C3 and C4 plants, which differ in their responses to these factors as well as to light intensity.

SUMMARY

(1) Chloroplasts, the organelles in which photosynthesis takes place, are bounded by a double membrane envelope and contain a system of internal membranes or thylakoid sacs, often grouped into grana, embedded in a matrix phase or stroma.

(2) Chlorophyll, associated with the thylakoid membranes, absorbs light for photosynthesis, as shown by the action spectrum of photosynthesis. Chlorophyll *a* is the essential photosynthetic pigment of all green plants. Other kinds of chlorophyll (chlorophylls *b* and *c*) and nonchlorophyll pigments (carotenoids, biliproteins), called accessory pigments, also absorb light energy and transfer it to chlorophyll *a* for use in photosynthesis.

(3) Hydrogen atoms (or electrons) needed to reduce CO_2 to carbohydrate are obtained from water using energy from chlorophyll and leaving oxygen as a by-product. This process, the Hill reaction, occurs in the thylakoids.

(4) The Hill reaction is driven by quantum conversion, wherein an excited chlorophyll molecule releases an electron yielding a chemical reducing agent, leaving an oxidized chlorophyll which constitutes a chemical oxidant. By two such photoreactions in series (two photosystems), thylakoids take electrons from water and convert them into the electrons or H atoms of $NADPH_2$, a reductant strong enough to reduce CO_2. This electron flow involves an electron transport chain in the thylakoid membranes, producing ATP (photophosphorylation).

(5) Carbon dioxide is captured in the stroma by RuBP carboxylase; the three-carbon primary fixation product is reduced to a three-carbon sugar by $NADPH_2$ from the Hill reaction, with the help of ATP from photophosphorylation. By means of a cycle of reactions (Calvin cycle), the CO_2 acceptor (RuBP) is produced. Excess three-carbon sugars from the operation of the cycle are converted to six-carbon sugar, from which carbohydrate end products (sucrose, starch) can be made. An oxidative pathway, photorespiration, occurs as a side reaction of the Calvin cycle and reconverts some of the fixed carbon back to CO_2, wasting energy.

(6) From the known features of the photosynthetic mechanism, a minimum of eight quanta of light should be required to convert one CO_2 molecule to one (CH_2O) molecule. Experiments give a yield of one CO_2 molecule fixed per 10–12 quanta absorbed by algal cells under special conditions giving maximum efficiency. This corresponds to about a 28% conversion of red light energy into energy stored in carbohydrate. Energy conversion in nature or agriculture is much less efficient. One reason is that photosynthesis is light-saturated (limited by dark reactions) under full sunlight; another is the low temperature optimum (20–25°C) of normal photosynthesis (due to photorespiration), which depresses photosynthesis in warm weather.

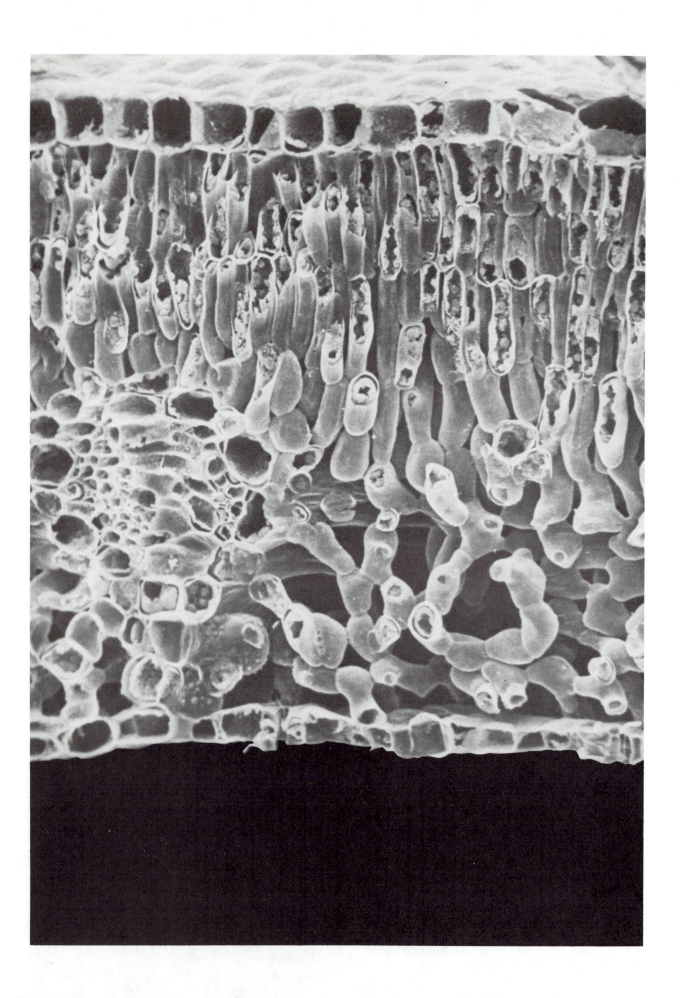

Chapter 8

THE LEAF: STRUCTURE AND PHOTOSYNTHETIC ACTIVITY

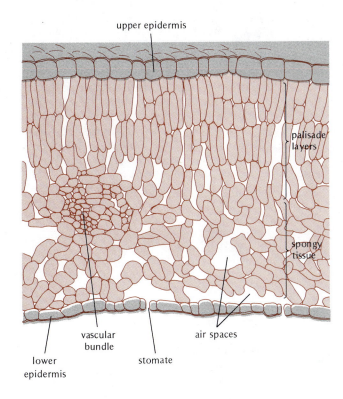

upper epidermis

palisade layers

spongy tissue

vascular bundle

air spaces

lower epidermis

stomate

Figure 8-A. An apple leaf in transverse section, scanning electron micrograph (500 ×) and (B, above) accompanying line drawing. Notice the individual chloroplasts visible in palisade cells that have been torn open by sectioning. Palisade plus spongy layers comprise the leaf's mesophyll tissue.

Leaves are the higher plant's solar energy collectors. Sunlight being a diffuse energy source, leaves expand and proliferate to present a maximum surface area for energy capture, like that of a man-made solar collector. But unlike a man-made solar collector, leaves also have to collect matter, specifically carbon dioxide, from their environment, which is perhaps an even more diffuse source of CO_2 (only 0.03% by volume) than of energy. Leaves have evolved a distinctive structure that enables them to perform both these functions effectively and simultaneously.

THE STRUCTURE OF LEAVES

The principal tissues of a leaf (Figure 8-A) are the (1) epidermis, (2) mesophyll, and (3) vascular bundles.

The Epidermis

The **epidermis** is the external layer of cells covering the leaf surface and is usually one cell thick. The cells usually lack chloroplasts and develop an extra-thick cell wall on their outer surface, onto which a water-repellent layer of cutin and waxes (Chapter 5) is secreted, restricting evaporation of water. For the photosynthetic uptake of CO_2 from the air into the leaf, the epidermis possesses openings called **stomates*** (Figure 8-1). The stomate is a slit-like intercellular space between two specialized epidermal cells called **guard cells** which, with bending movements, can open and close the pore (Chapter 9). During periods of active photosynthesis the stomates are open, allowing CO_2 to enter the leaf.

---

*Also called stomata; singular, stoma (Greek, mouth).

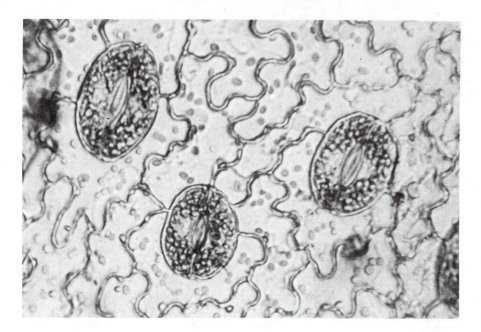

Figure 8-1. Stomates on the lower epidermis of Christmas fern (150 ×). The cells shaped like jigsaw-puzzle pieces are ordinary epidermal cells; a pair of guard cells, each containing many chloroplasts, flanks each stomatal opening. There are approximately 57,000 stomates/in² (9200 stomates/cm²) in the lower epidermis of leaves of this species. (The ordinary epidermal cells of this plant are unusual in having some chloroplasts, visible in the photograph.)

Stomates may occur in both upper and lower epidermis. With the exception of grass leaves and vertically held leaves, however, stomates are usually much more abundant in the lower epidermis, where hundreds of stomates commonly occur per square millimeter of leaf surface (Figure 8-1). Many leaves, including those of most woody plants and arid-climate plants, have stomates only in the lower epidermis.

The epidermis of many leaves bears **hairs,** which arise as outgrowths from epidermal cells, often become multicellular, and occur in diverse forms (Figure 8-2). Hairs help reflect excess light or retard water loss from leaves. "Stinging" hairs repel potential predators, an unpleasantly familiar example being the hairs on the leaves of nettle plants (Figure 8-2).

The Mesophyll

The **mesophyll** comprises all the internal tissue except the vascular bundles (Greek *mesos*, middle, and *phyllon*, leaf). In most leaves, the mesophyll is differentiated (Figure 8-A) into an upper layer of one or more tiers of elongated **palisade parenchyma cells** and a lower layer of more irregularly shaped **spongy parenchyma cells.** Chloroplasts are usually larger and more numerous in the palisade cells. In a sunflower leaf, for example, the palisade layer averages 77 chloroplasts per cell, while the spongy tissue averages only 27 chloroplasts per cell. Thus the palisade layer is the principal photosynthetic tissue of the leaf. The elongated shape of palisade cells probably serves to transmit light down from the upper epidermis efficiently, these cells acting like microscopic "light pipes."

An elaborate system of air-filled intercellular spaces runs between the cells of both the palisade and the spongy layers but is especially obvious in the spongy tissue (Figure 8-A). This system permits rapid diffusion of CO_2 within the leaf from the stomates to the photosynthesizing mesophyll cells, as considered later in this chapter.

The mesophyll of some leaves, such as grasses (Figure 8-5) and conifers (e.g., pine needles), is not differentiated into distinct palisade and spongy layers. On the other hand, vertically held leaves often have a palisade layer on both sides, for example the vertically hanging leaves of *Eucalyptus* trees.

Vascular Bundles

Strands of vascular or conducting tissue run through the mesophyll, transporting material to and from the mesophyll cells and, to some extent, helping to support the leaf. The larger **vascular bundles** appear as the macroscopically visible veins of the leaf (Figures 3-11, 3-12) which connect via the leaf stalk or petiole to the vascular tissue of the stem. The veins of a dicot leaf branch and rebranch into numerous smaller and smaller microscopically visible bundles, forming a very extensive net-like system (Figure 8-3). The vascular bundles of an elm leaf, for example, would stretch more than 200 m (600 ft) if laid out end to end. No mesophyll cell is more than a few cells away from one of the leaf's vascular bundles (Figure 8-4). In monocots most of the vascular bundles run parallel through the leaf, interconnected only by minor laterals.

Vascular bundles transport water into the leaf and photosynthetic products out of the leaf, a

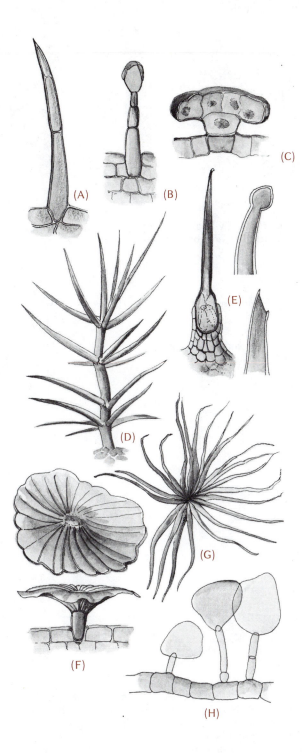

process called **translocation** of photosynthate (Chapter 10). During the day, translocation normally does not keep pace with photosynthesis, so carbohydrate (including starch) accumulates in the leaf (Figure 7-2). Photosynthetic products disappear during the night because of continued translocation.

A tight sheath of parenchyma cells with no intercellular spaces between them, called the **bundle sheath,** completely surrounds the smaller vascular bundles (Figure 8-5). Since everything transferred between mesophyll and vascular tissues must pass through these cells, they can control the exchange between the leaf and the rest of the plant.

ENERGY CAPTURE BY LEAVES

Land plant leaves are much more effective energy collectors than one might guess from the absorption spectrum of chlorophyll (Figure 7-5C), which suggests that leaves should hardly capture green light at all. In fact they absorb it almost as completely as red or blue light (Figure 8-6A) and

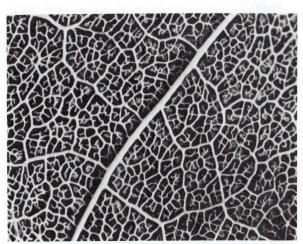

Figure 8-3.   Portion of a dicot leaf with mesophyll mostly removed, showing the larger veins and their branches.

Figure 8-2.   Examples of plant hairs. (A) Simple hair, and (B) glandular hair of geranium leaf (*Pelargonium*). (C) Shield-shaped glandular (secretory) hair of mint leaf (*Mentha piperita*), which produces the aromatic mint oils. Similar glandular hairs, producing a resin containing the drug tetrahydrocannabinol, occur on floral bracts of marijuana plants (*Cannibis sativa*). (D) Branching hair of mullein (*Verbascum thapsus*), giving the leaf a felt-like covering. (E) Stinging hair of nettle leaf (*Urtica dioica*); details show how tip breaks off on contact yielding a hypodermic needle-like end from which toxic contents are ejected by compression of the bulb-like base of the hair. (F) Shield-shaped scale hair of olive leaf (*Olea europaea*), top and side view. (G) Star-like hair of *Eleagnus angustifolia*, top view. (H) Bladder-like hairs of *Atriplex portulacoides*. Hairs of types D and F–H reflect some of the light falling on the leaf, helping to keep it from overheating in bright sunlight.

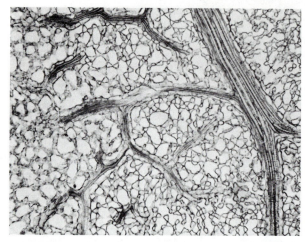

Figure 8-4. Vein branching in a privet leaf (40 ×). Section cut parallel to the leaf surface and through the spongy mesophyll. Branches of the smaller veins end in the spongy tissue.

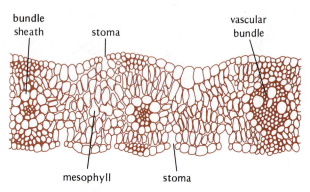

Figure 8-5. A wheat leaf in cross section, showing vascular bundles and bundle sheath cells.

use it quite well for photosynthesis (Figure 8-6B). This is due first of all to the large amount of chlorophyll in a leaf's several cell layers, which compensates to a considerable extent for the weak ability of individual chlorophyll molecules to absorb green light. Utilization of green light is important because the sun's energy is strongest in the green wavelengths; moreover, green light is practically the only useful energy available to a leaf that is shaded by others, since leaves absorb practically all the red and blue light falling on them (Figure 8-6A).

When light falling on a leaf passes through the palisade layers and enters the spongy tissue, the irregular air spaces of this tissue scatter and reflect it back and forth through the tissue several times, on the average, before it escapes from the leaf. This gives pigments in the leaf not just one but several chances to absorb incident light, helping them capture weakly absorbed wavelengths such as green. This light-scavenging function may explain why spongy tissue occurs so generally in leaves.

The Importance of Canopy Structure

As a plant grows it increases its total leaf area available for energy capture and, hence, its total photosynthesis, as long as all its leaves are fully exposed to the sun. But as the leaves grow in size and number, they start to shade one another, reducing the energy they receive. Many plants counter this reduction with an ability to position their leaves for minimal mutual shading, a pattern called a **leaf mosaic** (Figure 8-7). However, once the total leaf area exceeds the area of the land on which the plants are growing, mutual shading is usually unavoidable.

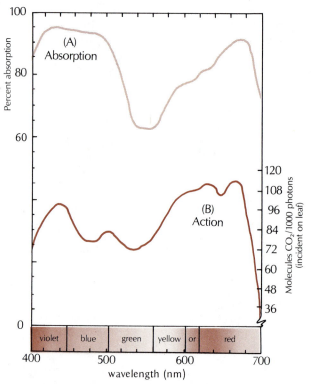

Figure 8-6. Comparison of a leaf absorption spectrum (A) with the action spectrum for photosynthesis (B) of a bean leaf. Note the much greater relative absorption and action in the green part of the spectrum (500–575 nm) by comparison with the very thin leaf of *Anacharis* (Figure 7-5), which lacks internal air spaces to reflect and scatter the incident light within the leaf, making the light pass repeatedly through many of the chloroplasts. (Data of S. E. Balegh and O. Biddulph, *Plant Physiology*, Vol. 46, pp. 1–5, 1970.)

Since, as indicated in Figure 8-6A, leaves usually transmit only about 5–15% of the useful light falling on them, shaded leaves find themselves far down on the "light saturation" curve relating photosynthetic rate to incident light in-

Figure 8-7. A leaf mosaic pattern on a maple branch. The leaf petioles respond to the direction and intensity of light by elongating more rapidly when shaded and bending toward the light, causing the leaf blades to move into less shaded regions.

tensity (Figure 8-12), so they do the plant little good. Photosynthesis by a leaf more than two or three leaves down in the canopy may not even equal its respiration; it is then a liability rather than an asset to the plant. Older, lower leaves therefore tend to die and be shed by the plant (Figure 8-8).

Because the photosynthesis of leaves exposed to full sun is usually light-saturated (Figure 8-12 and Chapter 7), these leaves waste a good part of the energy they absorb, energy that light-starved leaves lower in the canopy could put to good use if they could only obtain it. Consequently total photosynthesis can be increased if a plant tips its upper leaves at an angle to the sun, cutting down the amount of light falling on them (only slightly reducing their light-saturated photosynthesis) and allowing more light to fall on the lower, light-starved leaves, which greatly increases the photosynthesis of these leaves. Various trees such as firs use this principle (Figure 8-9). Plant breeders are using the same principle in breeding high-yield crop strains, finding it advantageous to breed for a tendency to hold leaves more erect so the lower leaves get more light (Figure 8-10).

ADAPTATIONS OF LEAF STRUCTURE AND PHOTOSYNTHESIS

Many departures from the basic type of leaf anatomy described above have evolved as plants have adapted themselves to different conditions of water supply. Some of these variations are given as examples of ecological adaptation in Chapter 19. Here we shall look at a couple of anatomical and physiological adaptations that affect significantly the leaf's actual photosynthetic performance.

Shade Leaves

Ordinary plants, adapted for living in full sunlight, are called **sun plants. Shade plants** are species specially adapted to low-light habitats. Certain plants can adapt physiologically to either high or low light intensities and thus can live in either shady or sunny habitats, forming sun-adapted leaves in high light and shade-adapted leaves or shade leaves under low light. Many trees form sun leaves at the top where they are exposed to full sun and shade leaves on their lower branches.

Shade leaves are usually larger and much thinner than sun leaves, often with fewer palisade cell layers (Figure 8-11), and are built with much less carbohydrate and protein per square centimeter of leaf area. Since under low light the plant has only limited photosynthetic production, it is

Figure 8-8. Leaf "self-pruning" in a dense stand of young slash pine.

Figure 8-9.   Lower (left) and upper (right) branchlets of fir, showing horizontal and vertical leaf positions, respectively.

Figure 8-10.   A corn plant bred for stiffer, more upright leaves (right) compared with a normal plant (left). The upright-leaved plants allow sunlight to penetrate to the lower leaves and so increase photosynthesis. The drooping leaves (left) shade out the lower leaves of the plant.

advantageous to make the products go as far as possible for new leaf construction.

The light saturation curves of sun and shade leaves differ (Figure 8-12). Shade leaves' photosynthesis saturates at a much lower light intensity and at a much lower light-saturated rate than that

of a typical sun leaf, because shade leaves contain fewer photosynthetic enzymes, due to the mentioned economies in their construction. This lower photosynthetic capacity is of no disadvantage to a shade plant because it does not normally receive the higher light intensities that a sun plant could use profitably.

Because shade leaves have less cell material to maintain and contain fewer mitochondria per unit area, they respire substantially more slowly than sun leaves, and it takes less photosynthesis to equal the leaf's respiration rate. Consequently the shade leaf possesses a lower **light compensation point,** or light intensity at which photosynthesis just equals respiration (Figure 8-12). This balance point represents the absolute lower limit of light intensity under which any green plant can indefinitely survive. By their adaptation, shade plants are able to live in places so shady that sun plants simply die, even though the photosynthesis of shade leaves is inherently no more efficient than that of sun leaves.

Shade plants from forest floor habitats (especially tropical rain forests) are commonly used as houseplants, growing successfully in the low light intensities often prevailing indoors which many other ornamental plants cannot tolerate. A number of these tropical forest floor shade plants such as *Peperomia, Monstera,* and *Dieffenbachia* (Figure 8-13) have relatively thick, leathery leaves, in contrast to the shade leaves of temperate-zone plants mentioned above. But compared with the leaves of temperate-zone plants, these tropical leaves tend to live a long time (important for houseplants from the decorative point of view) and thus eventually repay through their photosynthesis the greater investment of carbohydrate in them. They often contain unusually dark chlorophyll pigmentation capable of absorbing virtually all light that falls on them, and their photosynthetic and respira-

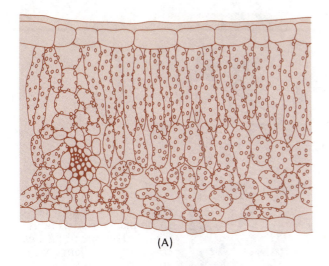

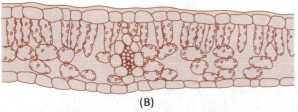

Figure 8-11. Cross sections of sun and shade leaves of sugar maple (*Acer saccharum*). (A) Sun leaf (from the south side of an isolated tree). Note the multiple palisade layers and the thick cuticle over the upper epidermis. (B) Shade leaf (from the center of the crown of an isolated tree).

tory specializations are, if anything, even more extreme than described above.

Many shade plants, called "tender" in horticultural parlance, are actually injured by exposure to direct sun and must be protected if grown in the open. If unprotected, the leaves bleach as a result of photodestruction of chlorophyll and their photosynthetic activity is lost, such shade leaves evidently having an inadequate system for protecting chloroplast components against photodestruction. Leaves developed in the shade by plant species capable of tolerating either shade or sun are also often sun-intolerant and will become "burned" if the plant is placed in direct sun for any period. Such plants recover photosynthetic capacity by forming new, sun-tolerant leaves.

C4 Plants

Although at low light intensities the photosynthesis of corn and sugarcane is no better than that of other plants, it does not fully saturate at higher light intensities, so in full sunlight they achieve a photosynthetic rate as much as 50% higher than that of ordinary plants (Figure 8-12, curve for C4 plant). They might be called supersun plants, with an adaptation opposite to that of shade plants. This adaptation makes corn and sugarcane among the most productive of crop plants, attracting much attention because of the

urgent need to increase crop productivity so as to improve world food supplies.

When $^{14}CO_2$ is fed to photosynthesizing corn or sugarcane leaves, radioactivity appears within seconds in four-carbon organic acids, and only later (after a minute or two) in compounds of the Calvin cycle. Therefore, these leaves first capture carbon in C4 acids, then transfer it to the Calvin cycle. This leads to the term **C4 photosynthesis,** contrasting with ordinary or C3 photosynthesis, where carbon is first captured in a three-carbon acid in the Calvin cycle (Chapter 7). C4 photosynthesis has also been found in various other grasses, mostly of tropical origin, and in a variety of dicot groups, mostly unrelated to one another. It has evidently evolved independently in a number of different flowering plant groups.

The leaves of virtually all C4 plants have a peculiar anatomical structure called **Kranz anatomy** (Figures 8-14, 8-15). They have large bundle sheath cells possessing conspicuous chloroplasts. The bundle sheath is surrounded by a sheath of chloroplast-containing mesophyll cells, and no normal palisade or spongy mesophyll occurs. This arrangement gives the appearance, in cross section, of a green wreath around each vascular bundle and has become known as Kranz anatomy from the German word for wreath. The bundle sheath

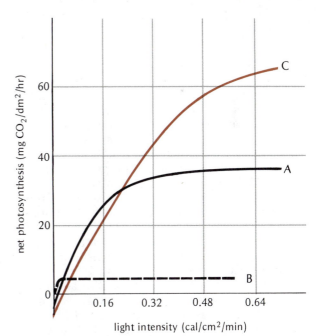

Figure 8-12. Photosynthetic light saturation curves of (A) C3 sun leaf, (B) C3 shade leaf, and (C) C4 leaf. Full sunlight intensity is approximately 0.6 cal/cm²/min. Dark respiration rates are represented by the points at which the three curves intersect the vertical axis *below* 0 (values for net CO_2 *production* in the dark). Light compensation points are the points at which the curves intersect the horizontal 0 line of zero net photosynthesis (actual photosynthesis just equal to respiration). Note that the compensation point of curve (B) is considerably lower than that for (A) or (C).

Figure 8-13. Tropical forest shade plants used as houseplants. (left) *Monstera deliciosa;* (right) *Dieffenbachia amoena.*

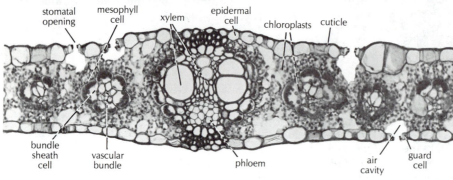

Figure 8-14. Cross section of a C4 leaf, showing Kranz anatomy (corn, *Zea mays*). Although this leaf looks superficially like the wheat leaf in cross section (see Figure 8-5), wheat is a C3 monocot.

cells fit tightly together, whereas intercellular spaces run between the mesophyll cells, which seem to be responsible for CO_2 uptake.

In C4 leaves starch usually appears only in the bundle sheath chloroplasts, not in the mesophyll. This reflects a remarkable specialization in which CO_2 is captured from the air and combined into C4 acids by the mesophyll cells, this carbon then being reduced to carbohydrate by the bundle sheath cells via the Calvin cycle (see Figure 8-16).

C4 plants lack the photorespiration that so substantially reduces the photosynthesis of C3 plants (see Chapter 7); this is presumably why C4 plants have a greater photosynthetic rate under high light. C4 plants also typically show a substantially higher temperature optimum for photosynthesis than C3 plants (Figure 8-17) and can therefore be regarded as adapted to both high light and high temperature, fitting them for tropical as well as warm desert conditions. Their high temperature optimum seems to be due again to the lack of photorespiration, which as might be expected is quite temperature-dependent and cuts into C3 photosynthesis more and more at higher temperatures.

The lack of photorespiration in C4 plants can be explained by the above-mentioned separation of CO_2-capturing and Calvin cycle functions between mesophyll and bundle sheath cells of a Kranz leaf. The vigorous production of C4 acids in the mesophyll plus their import into, and splitting to release CO_2 in, the bundle sheath cells (Figure 8-16) act as a CO_2 pump, building up a much higher CO_2 concentration in the bundle sheath than could be obtained by simple diffusion from the low CO_2 concentration present in air, as in C3 leaves (see below). This high CO_2 competes effectively against O_2 for RuBP carboxylase, suppressing its oxygenase reaction leading to glycolic acid and thus preventing photorespiration (see section explaining photorespiration in Chapter 7).

CO_2 Relations of C3 and C4 Leaves

Because of their photorespiratory CO_2 production, photosynthesis in C3 leaves can deplete the CO_2 concentration within the leaf only down to about 0.005%, called the **CO_2 compensation point.** The CO_2 compensation point of C4 leaves, in contrast, is virtually nil (less than 0.001%); that is, they can remove practically all the CO_2 from the intercellular spaces around their mesophyll cells. Although this has no practical value for photosynthesis in normal air, it enables C4 plants to conserve water better than C3 plants can (see Chapter 9).

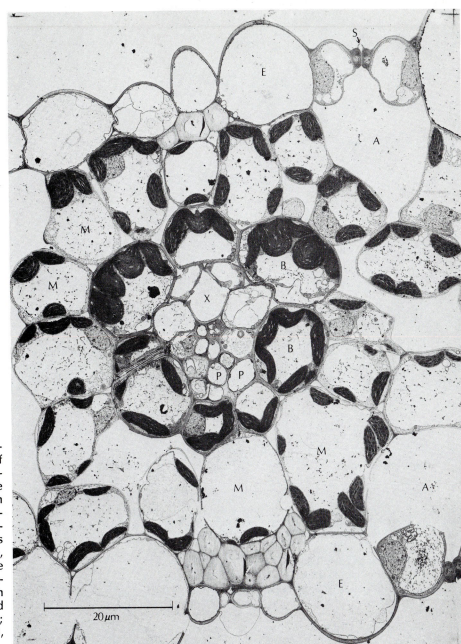

Figure 8-15. Transverse section electron micrograph of C4 leaf of the grass *Echinochloa colonum*, showing the concentric bundle sheath (inner circle, B) and mesophyll (outer circle, M) chloroplast-containing cell layers around a vascular bundle, typical of C4 plants. Note the intercellular gas spaces between mesophyll cells, which are the site of CO_2 uptake and initial fixation. A, air space; E, epidermis; P, phloem; S, stomate; X, xylem.

CO₂ Fertilization

Under high light, neither C3 nor C4 photosynthesis is saturated with CO_2 at the normal concentration of 0.03% CO_2 in the atmosphere, meaning that photosynthesis and, consequently, plant growth can be substantially increased by artificially adding CO_2 to the air. The idea readily arises of increasing agricultural productivity through CO_2 fertilization. This has proved economically profitable in the greenhouse production of vegetables and cut flowers, CO_2 being released into the greenhouse from tanks of liquid CO_2. Although some scientists advocate CO_2 fertilization as a way to solve the world food crisis, the problems in conducting it on an agricultural scale are formidable, especially the difficulties and expense of trying to enclose air over large acreages and keep the enclosures from overheating under summer daytime conditions without losing the added CO_2.

CARBON DIOXIDE UPTAKE FROM THE AIR

Both C3 and C4 leaves obtain CO_2 from the surrounding air by diffusion, which takes place from regions of higher to regions of lower concentration of a given kind of molecule. By consuming CO_2 during photosynthesis, photosynthetic cells lower the CO_2 concentration within the leaf below that in its environment. This concentration difference drives the diffusion of CO_2 into the leaf from the air.

Because of the very low CO_2 concentration (0.03%) in normal air, at most only a small concentration difference is available for the diffusional uptake of CO_2 by leaves—very small, for example, compared with the O_2 concentration dif-

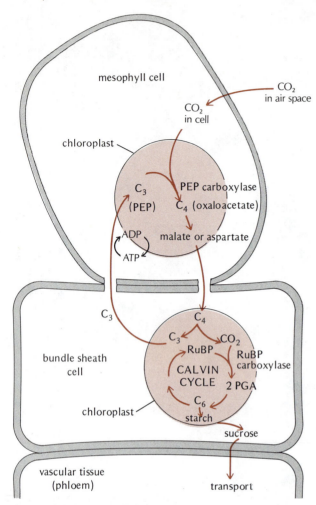

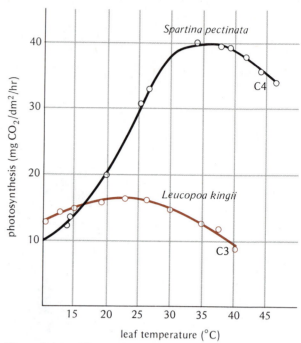

Figure 8-17. The temperature dependence of photosynthesis for a C3 and a C4 plant. Photosynthetic rates were measured under full sunlight intensity.

Figure 8-16. A simplified diagram of the C4 photosynthetic cycle. CO_2 is fixed into four-carbon organic acids, such as malate or aspartate in the mesophyll cells by an enzyme especially abundant in those cells, called phosphoenolpyruvate (PEP) carboxylase. The C4 acids then move into the bundle sheath cell chloroplasts where they are decarboxylated. The released CO_2 is there fixed by RuBP carboxylase into PGA and reduced to carbohydrate by the normal Calvin cycle sequence. The three-carbon acid resulting from decarboxylation is returned to the mesophyll cells where it is converted back to PEP (using ATP from photophosphorylation), which can again be carboxylated to keep the cycle going. Sugars produced by the Calvin cycle can be stored as starch or moved directly into a neighboring phloem cell for transport to the rest of the plant.

ference driving respiratory oxygen uptake in our lungs from the air we breathe, which contains 21% oxygen. Therefore, the rate at which CO_2 can diffuse into leaves tends to limit the rate of photosynthesis by land plants.

The structure of leaves has evolved to minimize this limitation as much as possible. First, leaves are usually rather thin (typically about 0.1 mm or 100 μm thick), which maximizes the rate of diffusion by minimizing the distance, or **path length,** over which diffusion of CO_2 must occur

(the shorter the path length, the faster the diffusion, for a given concentration difference). Second, the extensive system of intercellular air spaces or gas channels (Figure 8-A) greatly improves CO_2 uptake, because CO_2 diffuses about 10,000 times faster in air than through a liquid medium such as cell cytoplasm. Third, the nearly complete separation of a leaf's mesophyll cells from one another by intercellular spaces provides a very large area of mesophyll cell surface (called the **internal surface** of the leaf) for absorption of CO_2 from the gas space system. The internal surface of a lilac leaf, for example, is about 13 times its external surface area.

The impediment that a leaf or other tissue offers to diffusion is termed the **diffusion resistance** of the tissue. The various features of typical leaves just mentioned serve to minimize the diffusion resistance of the leaf for CO_2 uptake.

Stomatal Diffusion

Stomatal openings, the route for entry of CO_2 into the leaf, usually comprise 1% or less of the epidermal surface (Figure 8-1), so the epidermis might be expected to impose a large diffusion resistance upon the leaf. Surprisingly it does not, as long as the stomates are open. This is essentially because the stomatal channels are so short (typically less than 10 μm), compared with the overall path length for CO_2 uptake into the leaf, that their resistance comprises only a minor fraction of the total diffusion resistance of a typical leaf.

The need for very efficient diffusional uptake of CO_2 to support a reasonable photosynthetic rate by leaves creates their most serious problem, the evaporative loss of water, or transpiration. In the next chapter we consider this problem and the plant's solution to it in terms of leaf structure, stomatal behavior, and water absorption activity.

Some investigators of leaf gas exchange think that when it is windy, a significant fraction of the CO_2 absorbed by leaves may enter by bulk flow of air through the stomates, rather than by diffusion. Bulk flow of air into and out of the leaf's gas channels, which would be distantly analogous to human breathing, might be caused by the flapping of a leaf in the wind or by wind-generated pressure differences between opposite surfaces of the leaf. The exact mechanism of CO_2 influx actually makes only a minor difference in considerations of the water cost of photosynthesis. Since diffusion is inevitably responsible for at least part of the CO_2 uptake and water loss (for all of it, in still air), for simplicity in Chapter 9 we consider these exchanges in terms of diffusion only.

SUMMARY

(1) Typical leaves consist of (a) an external epidermis containing stomates for CO_2 intake; (b) internal mesophyll tissue comprising the palisade layer, which is the principal photosynthetic tissue, and the spongy layer with large intercellular spaces for inward diffusion of CO_2; and (c) vascular bundles (veins) for transport of water into and photosynthetic carbohydrate out of the leaf.

(2) Despite chlorophyll's weak absorption in the green part of the spectrum, leaves capture, and can use for photosynthesis, almost all visible light wavelengths, including green. This is due to the large amount of chlorophyll in the leaf (and to light scattering within the leaf, enhancing absorption). Mutual shading by leaves reduces their photosynthetic performance and is significant in productivity considerations.

(3) Photosynthesis by shade leaves saturates at a lower light intensity and lower rate than that by sun leaves and has a lower light compensation point because the shade leaf has a lower respiration, per unit of leaf area. Shade leaves are usually markedly thinner, with fewer cell layers and less biomass per square centimeter than sun leaves. These features enable shade plants to grow at lower light intensities than sun plants can.

(4) In C4 photosynthesis, a specialization found in some important warm-climate plants such as corn (maize) and sugarcane whose leaves have Kranz anatomy, four-carbon acids are formed from CO_2 in the mesophyll cells, while the captured carbon is reduced to carbohydrate via the Calvin cycle in bundle sheath cells. The C4 cycle suppresses photorespiration and thus increases the photosynthetic capacity of C4 plants at high light intensities and high temperatures. C4 plants use CO_2 more efficiently than C3 plants, but photosynthesis of both types can be increased by raising the CO_2 concentration above that in normal air (CO_2 fertilization).

(5) Leaves take up CO_2 from the outside air for photosynthesis by diffusion through the stomates and intercellular gas channels, followed by absorption into the photosynthetic cells through the large internal surface of the leaf. Inward diffusion of CO_2 results from a CO_2 concentration difference between the leaf interior and the outside air due to photosynthetic CO_2 consumption inside the leaf. Diffusion through the stomates occurs much faster than one might expect from the 1% or less of the leaf's surface they make up.

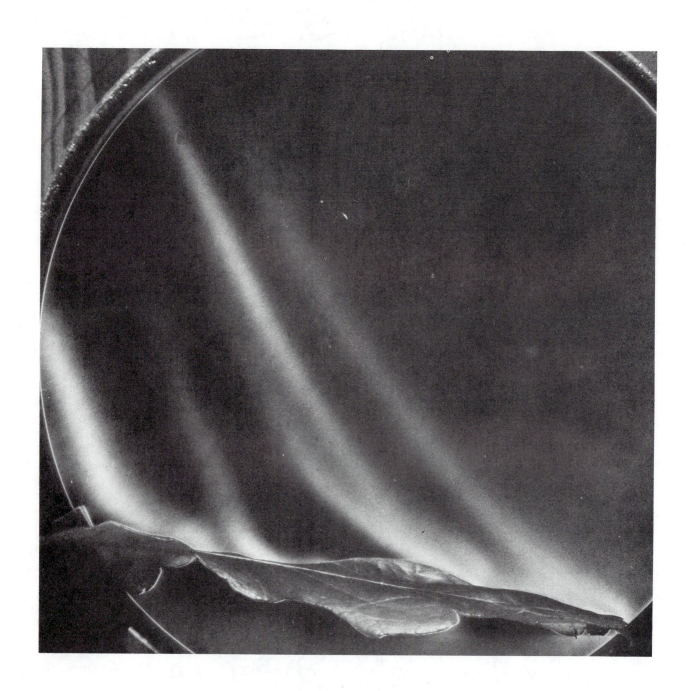

Chapter 9

LEAF TRANSPIRATION AND WATER BALANCE

To carry on photosynthesis, as we learned in Chapter 8, the leaves of land plants obtain carbon dioxide from the air around them by diffusion. At the same time, water evaporates from the leaf into the air, the process known as **transpiration.** This water loss creates a major demand for water to keep the leaves from drying up, a requirement which limits the vigorous growth of land plants to locations where adequate soil moisture is available. It necessitates the irrigation, at vast expense, of large areas of the earth's surface that receive inadequate precipitation, if these areas are to be used productively for agriculture. The purposes of this chapter are to understand the transpirational water cost of photosynthesis, the means by which transpiration can be controlled, and the osmotic forces by which cells can absorb water to maintain water balance despite the occurrence of transpiration.

TRANSPIRATION: THE WATER COST OF CARBON DIOXIDE UPTAKE

The leaf's diffusion pathway allows water vapor to diffuse from the internal air space of the leaf to the air outside just as efficiently as CO_2 can diffuse in. As water vapor diffuses out, water evaporates into the gas channels from the moist internal surface of the leaf, so the leaf loses water (transpiration).

The water transpired by a leaf or branch can be determined by measuring the water consumed by it with an apparatus called a potometer (Figure 9-1) or, more simply, by measuring the decrease in weight of a potted plant whose pot and soil surfaces are sealed against evaporation, for example with a plastic wrapping film. Results given in Table 1-1 (p. 8), obtained by an even simpler method, illustrate that plants normally transpire much more water than they retain in their growth. Data for field crops given in Table 9-1 show that typical land plants transpire several hundred grams of water for every gram of biomass that they accumulate by their photosynthesis. This is called the **transpiration:photosynthesis ratio;** it constitutes the water cost of doing photosynthesis on land. Because of the large quantity of water required, its absorption and transport by a land plant to meet the demands of transpiration are actually the plant's predominant activities, dwarfing the material changes involved in photosynthesis and growth (Figure 9-2).

The reason for the large water cost of photosynthesis on land is easily found from the principles of diffusional gas exchange explained in Chapter 8. Evaporation of water from the moist internal surface of a leaf saturates its gas channels with water vapor. At temperatures favorable for photosynthesis, that is, 25–35°C, air holds 2–4% by volume of water vapor at saturation, the precise amount depending upon the temperature. On a sunny day the open air usually contains not more than about 50% of its saturating water vapor content (i.e., a relative humidity of 50%), or about 1–2% water vapor. Therefore, the concentration difference driving the diffusion of water vapor out of the leaf commonly lies between about $(2-1) = 1\%$ and $(4-2) = 2\%$. The CO_2 concentration difference driving photosynthetic CO_2 uptake, on the other hand, is always less than 0.03%, because that is the CO_2 concentration in the air outside, and may often be about 0.02%, since the gas channels of most leaves seem to contain at least 0.01% CO_2 during even the most active photosynthesis. Since the concentration difference for water vapor diffusion is therefore at least 50–100 times greater than that for CO_2, and the diffusion resistance for water vapor is comparable to that for CO_2, the rate of transpiration must be at least 50–100 times the rate of photosynthetic CO_2 uptake.

Figure 9-A. Photograph of a sunlit oak leaf taken with Schlieren optics to reveal the convective air stream flowing across and away from the leaf. Because of the sunlight it is absorbing, the leaf is warmer than the surrounding air and warms the air, which therefore becomes less dense and appears light in the photograph. The less dense air rises from the leaf surface in streams, carrying away heat and also water vapor released by transpiration.

147

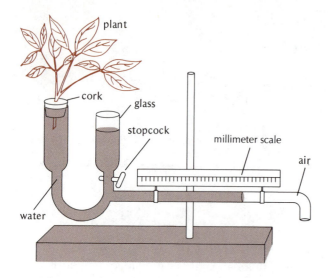

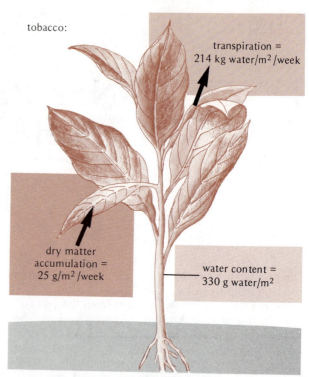

Figure 9-1. A potometer, used to measure water uptake by a transpiring branch. When the leaves transpire water vapor they take up an equal amount of water from the shoot's xylem, in turn drawing water from the potometer through the cut-end of the stem. This water uptake causes the water/air boundary (meniscus) in the tube under the scale to move to the left, allowing the volume of water transpired to be calculated. The water in the reservoir controlled by the stopcock can be used to reset the position of the meniscus on the scale before another measurement.

Figure 9-2. Dry matter accumulation, water content, and transpired water for tobacco plants in the field in one week. Data are given on the basis of one square meter of land planted at a typical density with tobacco. Note that transpiration rate is given in *kilograms* of water, the other values in *grams*.

Experimentally measured transpiration:photosynthesis ratios (Table 9-1) are generally much greater than 100. The reasons for this include the following: (a) the resistance for CO_2 uptake is actually greater than for water vapor; (b) the concentration difference driving CO_2 uptake may often be less than estimated above; and (c) as noted in Chapter 6, part of the photosynthetic output of the leaves is respired during growth and does not add to the dry matter of the plant, on which Table 9-1 is based.

The Water Cost of C4 Photosynthesis

C4 plants such as corn (Table 9-1) transpire appreciably less water, per gram of photosynthetic growth, than typical C3 plants. This is due to the fact that C4 photosynthesis can reduce the CO_2 concentration within the leaf to a lower level than can C3 photosynthesis (Chapter 8). Thus the concentration difference driving CO_2 uptake is greater in C4 than C3 plants and, for a given amount of water transpired, the C4 plant gains more carbon. By conserving scarce water, C4 photosynthesis should be beneficial in arid climates.

The Magnitude of Transpiration

Sunlit leaves commonly transpire 100–200 g of water (100–200 ml or about half a cup to a cup)

per hour, per square meter of leaf surface. A square meter of typical leaves weighs about 100 g; thus a leaf can transpire as much water as its entire weight in 1 hr! Obviously if transpiration were not compensated by transport of water to the leaves, they would very quickly dry up. One square meter is approximately the leaf area of a large herbaceous plant such as a sunflower. Such a plant therefore needs about 1.2–2.4 liters (1.3–2.6 qt) of water over a 12-hr day. A beech tree 40 cm (16 in.) in diameter has about 300 m² of leaf area and thus may transpire about 360–720 liters (100–200 gal) in a 12-hr day, or some 15,000–30,000 gal in the course of a 5-month growing season. An acre of corn may transpire 400,000 gal during its growing season, equivalent to 16 in. of rainfall. Irrigated agriculture is thus a large consumer of water supplies, and forests greatly reduce the amount of water that can be collected from precipitation for human use.

Environmental Influences on Transpiration

Transpiration is increased by environmental conditions that promote the evaporation of water,

TABLE 9-1. Water Use by Field Crops. Data obtained by Shantz and Piemeisel[a] for water consumption relative to dry matter production over the entire growing season at Akron, Colorado.

Crop	Type of photosynthesis	grams water/ gram biomass[b] 1915	1916
Alfalfa	C3	695	1047
Oats	C3	445	809
Barley	C3	404	664
Wheat	C3	405	639
Corn (maize)	C4	253	495
Millet	C4	202	367
Sorghum	C4	303	296

[a]Journal of Agricultural Research 34:1093–1190 (1927).
[b]Data for 1915 and 1916 illustrate approximately the extremes in water use observed over the period 1911–1917, showing that water use is not a constant but varies considerably with environmental conditions: 1915 was a considerably cooler, more moist year than 1916, the water consumption by most of the crops reflecting this. These measurements were made long before it was discovered that corn, millet, and sorghum have a distinct type of photosynthesis.

notably wind, bright sun, high temperature, and low atmospheric humidity (creating a larger leaf-to-air water vapor concentration difference driving diffusion). The water cost of photosynthesis tends therefore to be substantially less for plants in sheltered, shady, and/or moist habitats than in sunny, dry, and/or windy ones, and in any habitat the water cost varies considerably with the weather. Ironically heat, wind, bright sun, and low humidity, which promote transpiration, tend to predominate in habitats and regions where water is least available, such as deserts.

Sunlight has both physical and biological (via the stomates; see below) effects on transpiration. Bright sunlight directly warms the leaf, increasing its transpiration because of the effect of temperature. The sun also heats up the air, which further warms the leaf and increases transpiration. The combination of these effects leads to a marked variation in transpiration rate over the course of a typical day, usually reaching a maximum near midday correlated strongly with the sunlight intensity and air temperature (Figure 9-3A).

Transpiration in Leaf Energy Balance and Temperature

Evaporation consumes heat, as we are aware from our own bodies' cooling through perspiration. Transpiration therefore cools the leaf. In the shade, transpiration can reduce a leaf's temperature appreciably below that of the surrounding air, noticeable when one feels a leaf to distinguish a real plant from a decorator's plastic imitation. Sunlight, on the other hand, warms the leaf, because most of the radiant energy that the leaf absorbs is not utilized in photosynthesis but becomes heat. The temperature that a leaf attains is determined by the balance between cooling processes and the input of solar energy tending to heat up the leaf (its **heat load**). Under mild summer day conditions as much as half of a typical leaf's heat load may be carried by transpirational cooling, so the leaf will not become much warmer than the air around it. During hot, dry weather, which promotes transpiration, the entire heat load may be absorbed by transpiration and the leaf can even be cooled several degrees below the air temperature despite being in bright sun. This is advantageous when the air temperature is above the optimum for photosynthesis, since cooling then increases the rate of photosynthesis and protects the leaf against heat injury that could occur if its temperature rose too high. Transpiration in these circumstances is not just an expense of doing photosynthesis, but is beneficial to the process.

However, transpirational cooling is very expensive in terms of water. As water runs short and stomates close to protect leaves against desiccation, transpirational cooling is eliminated. The leaf temperature then must rise above that of the air until simple cooling by the air (Figure 9-A) plus radiation of infrared heat from the leaf to its environment (Figure 9-4) dissipate the leaf's heat load. If the air is already hot, the leaf may get so hot that it suffers high temperature injury. This is a problem for crop and garden plants during summer "heat wave" conditions.

Enclosed Agriculture

The water cost of crops in arid regions can be considerably reduced by growing them in enclosed greenhouses, in which the plants maintain by their transpiration a high humidity, restricting further transpiration. However, such installations have to be artificially cooled against heating by the sun, to substitute for the plants' natural outdoor cooling processes just discussed. Despite this expense, enclosed agricultural systems are attractive in areas such as parts of the Middle East where outdoor agriculture is impossible for lack of water. It is possible to cool greenhouses evaporatively with salt water and, by condensing the resulting water vapor, recover enough pure water to supply most of the water needed by the crop (Figure 9-5). This technology is surprisingly simple and may see wide future application in coastal desert areas.

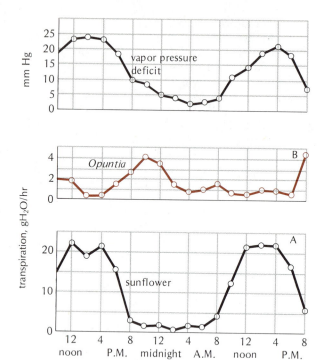

Figure 9-3. Daily variation in transpiration of sunflower and cactus (*Opuntia*) plants in well-watered soil in the summer. (A) Sunflower, a typical C3 plant, shows a maximum transpiration rate near midday and very low transpiration at night, when its stomates are closed. (B) *Opuntia*, a crassulacean acid metabolism (or "CAM") plant, shows a maximum transpiration rate at night. The vapor pressure deficit, plotted at top, shows in pressure units (mm Hg) the extent to which the air was less than saturated with water vapor at different times during the experiment, a measure of the tendency toward evaporation. Because the cactus opens its stomates widely only when the vapor pressure deficit is relatively low (night), it transpires much less than the sunflower does when its stomates are open (day). Data of P. J. Kramer, *American Journal of Botany*, Vol. 24, pp. 10–15, 1937.

STOMATAL CONTROL OF TRANSPIRATION

Stomates (Figure 9-6) are the leaf's principal means of regulating transpiration. Stomates of most plants close in the dark, restricting transpiration at night to values very much lower than in the daytime (Figure 9-3A). Stomatal closure restricts transpiration because evaporation directly from the outer surfaces of the leaf is largely prevented by the cuticle and surface waxes of the epidermis (Figure 9-7).

Evaporation from the epidermal surfaces of the leaf, called **cuticular transpiration,** is for most leaves less than 5% of the transpiration rate when the stomates are open. As noted in Chapter 5, the present evidence is that the surface ("epicuticu-

lar") waxes (Figure 9-7B) secreted by the epidermis are actually more important than the cuticle itself in reducing cuticular transpiration.

Response to Water Stress

Stomates close when a leaf runs short of available water, called a state of **water stress.** By restricting further transpiration, the stomates and cuticle protect the cells of the leaf from desiccation injury or death during the periods of water stress that almost all land plants at least occasionally experience. The epidermal cuticle and stomatal mechanism, found universally among vascular land plants, is evidently one of the most important adaptations permitting them to successfully inhabit the land, with its perpetual threat of water shortage. However, stomatal closure, by blocking diffusional uptake of CO_2, virtually stops photosynthesis. For this reason water stress, such as during a drought or in unirrigated arid areas, severely limits plant growth compared with more moist conditions.

Stomates of some desert plants exhibit a closing response to low atmospheric humidity, even

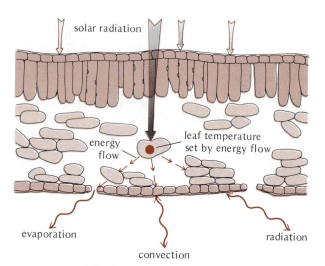

Figure 9-4. Leaf energy exchange mechanisms. Only a small fraction of the leaf's solar radiation input is transformed by photosynthesis into chemical potential energy. The remainder accumulates as heat, tending to warm the leaf. Three principal mechanisms dissipate heat from the leaf: (1) emission of infrared *radiation* to the environment, (2) simple cooling or *convection* (flow of heat to the surrounding air), and (3) *evaporative cooling* (transpiration). A leaf's temperature is determined by the balance between its heat load and the three cooling processes. The double-headed arrows for convection and radiation signify that the energy flux may go in either direction via these mechanisms depending upon whether the leaf is warmer or cooler than its surroundings.

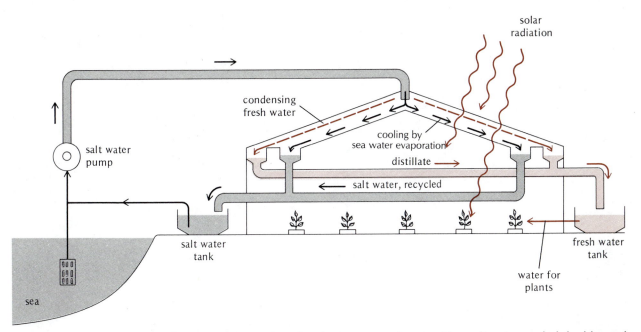

Figure 9-5. An enclosed agriculture greenhouse with a salt water-cooled double roof for condensational recovery of fresh water for irrigating the plants. Sea water flows down the inner roof, evaporating to take up heat from the greenhouse (coming from incident sunlight). Water vapor evaporated from the salt water condenses on the underside of the outer roof, runs down, and is carried away by an eaves trough to a holding tank. Because of the high humidity prevailing inside the greenhouse, the fresh water recovered from salt water cooling in practice more than meets the water needs of the crop.

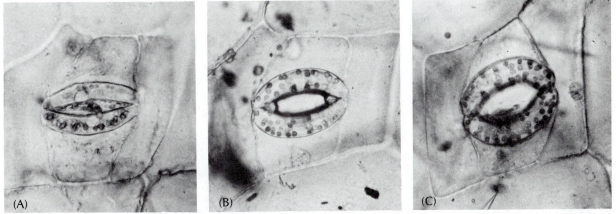

Figure 9-6. Stomatal opening in *Zebrina*. (A) Closed, (B) partially open, (C) fully open. Note that chloroplasts are present only in the guard cells of the stomatal complex, not in ordinary epidermal cells.

when the leaf is not under water stress. This presumably helps these plants conserve water by increasing their gas exchange resistance when conditions are most favorable for transpiration and the water cost of photosynthesis would be highest.

Antitranspirants

The closing response of stomates to water stress is at least in part hormonally mediated by the plant growth regulator, abscisic acid. This compound (Chapter 12) is produced by leaf tissue under water stress and induces stomatal closure. Abscisic acid and

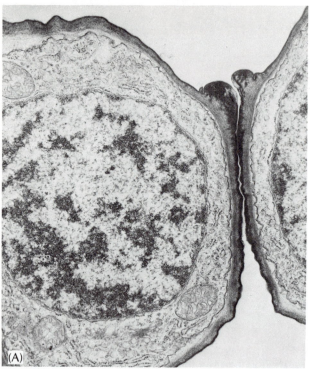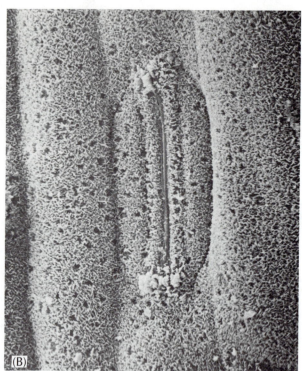

Figure 9-7. The cuticular surface of leaves. (A) Electron micrograph of cross section through guard cells and stomatal pore (closed) of an oat leaf (27,500 ×), showing cuticle on outside surface of their thick outer wall and surfaces of the stomatal opening. Ordinary epidermal cells have a similar cuticle. (B) A scanning electron micrograph of the epidermal surface of a wheat leaf, showing a stomate (closed) and a heavy deposit of wax platelets over the entire surface (2300 ×). The wax platelets are secreted through, and lie on top of, the cuticle.

other compounds that similarly induce stomatal closure are sometimes called **antitranspirants.** When applied to plants, antitranspirants will reduce transpiration even when the plants are not under water stress. There is hope of controlling agricultural water consumption in areas of water scarcity by the judicious application of antitranspirants to cause partial stomatal closure. By restricting CO_2 uptake this will inevitably restrict crop photosynthesis to some extent, so the conservation of water will take place at the expense of maximum productivity.

Feedback Control: Light and CO_2 Responses

Light rather quickly induces stomatal opening (Figure 9-8). This response increases the leaf's capacity for CO_2 uptake when photosynthetic demand for CO_2 is increasing. The stomatal response is partly a positive response of guard cells to light and partly a negative response to CO_2 concentration. Treatment of a leaf with an above-normal CO_2 concentration causes stomatal closure (Figure 9-8), while a low or zero CO_2 concentration stimulates stomatal opening. Thus photosynthetic CO_2 consumption, by depleting the leaf's internal CO_2 concentration, creates a stimulus for stomatal opening. This "feedback" response (Figure 9-9) constantly adjusts the leaf's diffusion resistance

to suit the CO_2 demand. It restricts transpiration at times when the leaf's internal CO_2 supply exceeds its photosynthetic demand, holding down the total water cost of photosynthesis compared with what it would be if unrestricted evaporation from the leaf occurred all the time.

The mechanism by which guard cells open and close stomatal pores and respond to light and

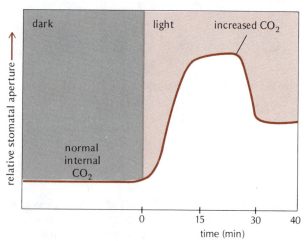

Figure 9-8. Time course of changes in stomatal aperture of a maize leaf in response to light and to high CO_2 concentration (0.04%).

to water stress will be considered after we examine the osmotic behavior of plant cells, upon which stomatal action depends.

CAM Plants: Nocturnal CO_2 Uptake

Cacti and other stem and leaf succulents (Chapter 3) adapted to dry habitats have a mechanism, called **CAM**, for taking up CO_2 at night rather than during the day. CAM stands for "crassulacean acid metabolism" (Crassulaceae are an important family of leaf succulents). The stomates of CAM plants are closed during the day; they open in the evening and stay open most of the night. During the night the plant takes up CO_2 and traps it as carboxyl groups of organic acids. During the day, with stomates closed, the organic acids are broken down again to CO_2 within the plant, and this internal CO_2 is used for photosynthesis. CAM metabolism differs from C4 photosynthesis (Chapter 8) by involving temporal rather than spatial separation of CO_2 capture from carbon reduction.

Because of nocturnal stomatal opening, CAM plants show a daily transpiration rhythm inverted from that of C3 or C4 plants (Figure 9-3B). Since all the environmental conditions that promote transpiration (see above) tend to be at a minimum at night, CAM plants lose much less water, in obtaining CO_2 for photosynthesis,

than do plants that take up CO_2 during the day while conducting photosynthesis. Transpiration:photosynthesis ratios as low as 50 have been measured for cacti and as low as 18 for agave, another CAM plant. CAM enables these plants to conserve their stored water and continue a slow rate of photosynthesis over long periods of drought.

THE WATER BALANCE OF PLANT CELLS

Cells of a leaf or other transpiring organ compensate for evaporative water loss by absorbing water by a passive physical process, **osmosis.** Osmotic water uptake is further important to leaves in stomatal opening and in their mechanical support or rigidity; osmotic principles explained here will also be utilized in subsequent chapters to understand other aspects of plant function.

Osmosis

Osmosis is the movement of water across membranes in response to physical driving forces. The first such force is due to dissolved substances

Figure 9-9.   The feedback response of stomates to light and CO_2. When light-induced photosynthesis depletes the leaf's internal CO_2, the stomates open wider and more CO_2 diffuses in, allowing photosynthesis to proceed. When internal CO_2 increases because photosynthesis is not keeping pace with CO_2 uptake, stomates close, restricting water loss at a time when additional CO_2 is not needed for photosynthesis.

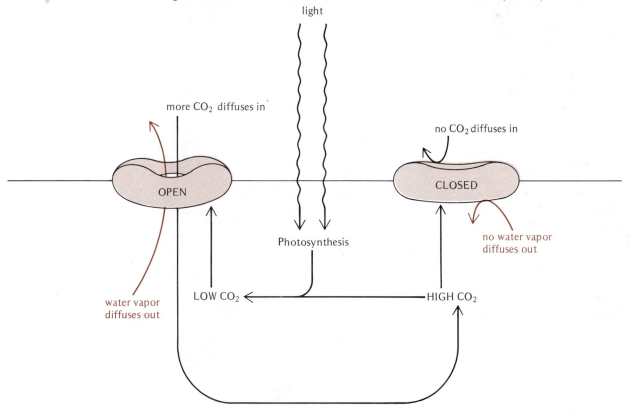

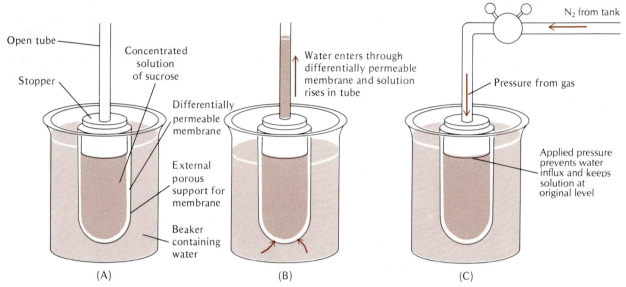

Figure 9-10.   Osmotic water movement and the effect of back pressure in an experimental osmometer. The osmotic movement of water from the beaker into the tube shown in (A) and (B) can be prevented (C) by applying the proper amount of pressure (here, using compressed gas) to the solution inside the tube. The amount of artificially applied pressure that just prevents osmotic influx of water (i.e., that induces osmotic equilibrium) is equivalent to the turgor pressure that would build up inside a plant cell with a total solute concentration equal to the concentration of sucrose shown here, if that cell were placed in, and allowed to come into osmotic equilibrium with, pure water.

(solutes), which in effect reduce the water concentration, resulting in a tendency of water to diffuse toward areas where the solute concentration is higher, that is, where the water concentration is lower. If the solute concentration differs on the two sides of a membrane that is permeable to water, water will tend to move across the membrane toward the side with the higher solute concentration (Figure 9-10A and B).

Since plant cells normally have a higher solute concentration than their environment, they tend to absorb water osmotically and swell up. A plant cell does not swell extensively, however, because of its strong cell wall. As water enters a cell, its expansion (Figure 9-11, B→A) builds up a pressure, called **turgor pressure,** inside the cell, much as pumping air into a tire builds up pressure within the tire. A cell or tissue having substantial turgor pressure is termed **turgid.**

Pressure is the second important force that influences osmosis. As illustrated in Figure 9-10C, the imposition of sufficient pressure will stop osmosis into a solution, counteracting the osmotic effect of the solutes. Water can continue to move osmotically into a plant cell only until turgor pressure becomes great enough to offset the osmotic action of the cell's solutes. This turgor pressure halts further osmosis and imposes a state of osmotic equilibrium or water balance between the cell and its environment.

In isolated cells turgor pressure buildup and osmotic equilibration take place within a few sec-

onds to a few minutes, depending on the water permeability of the cell's membranes. The same phenomenon occurs when one crisps limp celery or carrot sticks by placing them in water, but more slowly because water must pass osmotically through many cells to reach the interior of these relatively massive tissues.

The crisping of limp celery or carrot sticks by soaking in water illustrates how the stretching of plant cell walls by osmotically generated turgor

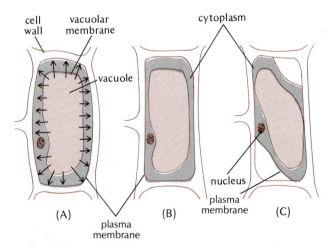

Figure 9-11.   Osmotic states of a plant cell: (A) turgid, (B) flaccid (wilted, no turgor), (C) plasmolyzed. Arrows in (A) denote internal pressure or turgor.

pressure causes firmness or rigidity in soft tissues. This is the principal basis of mechanical support for soft plant parts such as leaves. When the cells of such organs lose turgor, the tissue becomes limp and droops, or wilts.

The turgor pressure that will be reached at osmotic equilibrium between a solution and pure water, called the osmotic pressure or **osmotic potential** of the solution, is approximately 24 kg/cm² times the total molar solute concentration or **osmolarity** of the solution [a 1.0 osm (osmolar) solution contains a total of 1 mole of solute particles per liter]. The solute concentrations of most plant cells range between 0.2 and 0.6 osm, hence turgor pressures of 5–15 kg/cm² [70–200 psi (pounds per square inch)] can develop in them.

Water Potential

The thermodynamic or energy status of water in cells and tissues, which dictates passive water transport processes such as osmosis, is called the **water potential.** Water potential is usually expressed in units of pressure (kilograms per square centimeter, or bars). Passive water transport takes place from higher to lower potential, for example, from pure water (water potential = 0) to a cell with a water potential of -5 bars. For an ordinary vacuolate plant cell the water potential is essentially the sum of its turgor pressure (a positive quantity, tending to raise the potential) plus the osmotic potential due to its osmolarity (a negative quantity, tending to lower the water potential). For example, a plant cell whose solute concentration is 0.4 osm has an osmotic potential of about -9.6 bars ($= -24 \times 0.4$); if its turgor pressure is 5.0 bars, it has a water potential of $5.0 - 9.6 = -4.6$ bars. By osmosis this cell would take water from a cell whose water potential is -3 bars but would give up water osmotically to a cell with -6 bars of water potential.

Water potential indicates numerically the water status or degree of water stress in a plant cell or tissue. A water potential of zero represents complete saturation with water (no water stress), while a water potential of -15 bars or lower would indicate a state of severe water stress.

Plasmolysis

If a plant cell is exposed to a medium of higher solute concentration than its own (called a hypertonic medium), the cell must lose water. It will shrink until it loses any turgor pressure that it had initially and, even then, will continue to lose water. Further contraction of the cell volume pulls the plasma membrane away from the cell wall, leaving a fluid-filled space between the protoplast and the cell wall chamber, a condition called **plasmolysis** (Figure 9-11C). Plasmolysis is usually not lethal; the cell will normally recover if returned to a dilute medium in which it can take up water and deplasmolyze. However, plasmolysis eliminates the cell's ability to participate in water transport functions.

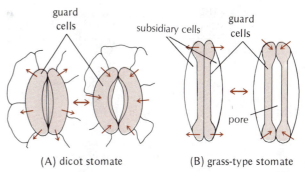

(A) dicot stomate   (B) grass-type stomate

Figure 9-12. Stomatal opening in dicot (A) and grass-type (B) stomates. Arrows indicate direction of water movement. Guard cells of dicots are commonly kidney-shaped, with thicker, less extensible cell walls on the side of the cell lining the stomatal pore. When water moves into the guard cells, the resulting increase in turgor pressure distends the thinner outer walls more readily, causing the cells, tightly attached to each other at both ends, to bow away from each other. This opens the stomate. Water loss from the guard cells decreases their turgor pressure and allows the elastic inner walls to return to their position adjacent to one another, closing the stomate. Stomates of the type developed by grass leaves open by inflation, under turgor, of the thinner-walled bulbous ends of the elongated guard cells. Expansion of these bulbs spreads the thick-walled central part of the guard cells apart from one another, creating a slit-like opening.

Thus excess application of chemical fertilizers, which dissolve in soil water to create a high solute concentration, plasmolyzes the water-absorbing cells of the roots and causes plants to wilt and develop "fertilizer burn."

STOMATAL ACTION

Opening of the stomatal pore is due to osmotic water uptake by the guard cells and their resultant expansion and development of turgor pressure. If their turgor pressure is eliminated by exposing guard cells to a hypertonic medium, the stomate closes. Guard cells possess uniquely thickened cell walls that cause these cells to change shape as turgor pressure builds up in them, separating the jaws of the stomatal pore from one another as illustrated for the two principal kinds of stomates in Figure 9-12. Thus stomates function as pressure-actuated valves to control the gas exchange of the leaf.

When guard cells are exposed to light and/or a low CO_2 concentration, they rapidly develop a very high solute concentration, causing them to take up water. This concentration increase is due primarily to the energy-dependent uptake and accumulation of a high concentration of potassium ions (K^+), apparently obtained from other epidermal cells. In the dark or in high CO_2 concentrations the guard cells lose much of their K^+, their solute concentration declines, they therefore lose water, and the stomate closes.

Because stomatal opening depends upon turgor pressure and turgor pressure declines or disappears under water stress, stomates will inevitably close when a leaf runs short of water. This passive closure constitutes "fail-safe" protection against the water stress hazard. However, as stated previously, there is evidence for an active closing response to water stress, involving the production of abscisic acid which induces a loss of K^+ from the guard cells.

K^+ uptake by guard cells is accompanied by the uptake of chloride (Cl^-) from other epidermal cells and/or by the production of organic acids from starch, which decreases, in the guard cells, in the light. How light or CO_2 actually controls these transport processes and metabolic conversions is not yet known with certainty. The model in Figure 9-13 suggests a possible mechanism for the complementary responses of K^+ transport to light and CO_2. It is based on the fact that, unlike ordinary epidermal cells, guard cells generally contain chloroplasts (Figures 9-6, 9-7A), and action spectra show

that chlorophyll is at least partly responsible for stomatal opening. However, part of the action of light on guard cells appears to involve nonphotosynthetic pigments, and other objections can be raised indicating that the model in Figure 9-13 is at best an oversimplified one. How the guard cells, whose action is so generally important to life on land, actually control stomatal opening is a subject of active current research.

WATER UPTAKE IN RESPONSE TO TRANSPIRATION

When a cell loses water by evaporation it loses some of its turgor pressure. Since the remaining turgor pressure is no longer sufficient to balance the osmotic effect of the cell's solutes, the cell will tend to absorb available water by osmosis. Primitive nonvascular land plants such as lichens, bryophytes, and terrestrial algae maintain water balance essentially in this way, absorbing water when available from the substrate beneath them.

When cells of vascular plant leaves transpire water and decrease in turgor pressure, they can

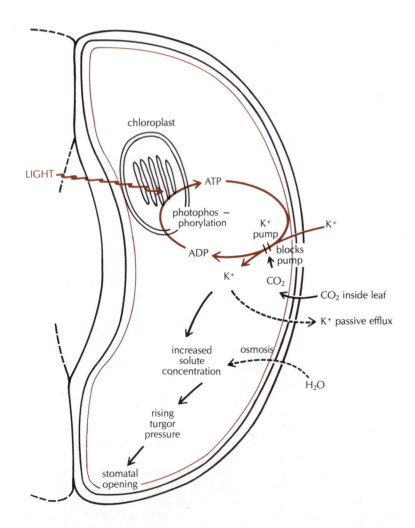

Figure 9-13. Simplified model of guard cell action. Light absorbed by guard cell chloroplasts causes photophosphorylation, producing ATP which drives a K^+ uptake pump, thus increasing the guard cell's solute concentration. This causes osmotic water uptake, leading to high turgor pressure and resultant stomatal opening. CO_2 reverses this by blocking the K^+ uptake pump, allowing the solute concentration to fall by passive leakage of K^+ from the cell.

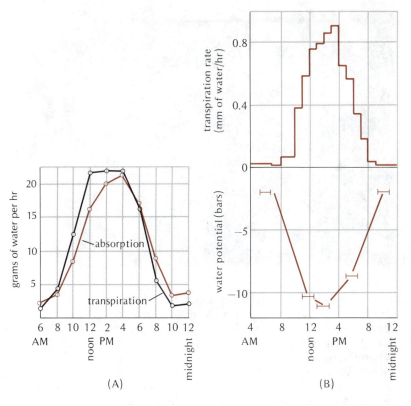

Figure 9-14. (A) Transpiration and water absorption by a sunflower plant out of doors during a summer day and the following evening. (B) Diurnal changes in transpiration rate and water potential of tops of well-watered, field-grown corn plants.

absorb water from the xylem tissue of the leaf's veins. This supplies the transpiring cells with water and sets in motion the process of long-distance water transport through the xylem, eventually causing absorption of water by the roots from the soil. Since rapid absorption cannot occur until the leaves have lost enough water to create a substantial osmotic imbalance and hence a large enough driving force for water transport, there is a definite lag between daily changes in water absorption and in transpiration (Figure 9-14). Because the driving force for water transport and absorption results

from transpiration, it is sometimes called **transpiration pull.** How it causes water transport in the xylem and absorption by the roots is considered in Chapters 10 and 11.

In terms of the water potential concept, transpiration causes the water potential of leaves to fall (Figure 9-14) below the water potential of the soil. The resultant soil-to-leaf water potential difference is the driving force for water uptake by roots and transport to the leaves.

Under midday summer conditions highly conducive to transpiration, leaves may transpire

Figure 9-15. Temporary wilting in squash (*Cucurbita maxima*) leaves. (A) In the late afternoon of a hot day. (B) Early the next morning. The plants have absorbed water and recovered during the night.

(A)

(B)

Figure 9-16.   Guttation from tomato leaves.

fast enough to lose turgor pressure and wilt (Figure 9-15A), dramatically indicating the development of water stress. This brings on "midday stomatal closure," depressing photosynthesis just when the energy supply is greatest. If the soil contains sufficient moisture, water uptake and transport to the leaves continue due to the osmotic driving force in the wilted leaf cells, restoring turgor, stomatal opening, and photosynthetic activity in the afternoon or evening when conditions become less favorable for transpiration (Figure 9-15B). This **temporary wilting** is most often seen when the soil is somewhat depleted of water but not severely dry.

When soil is seriously depleted of moisture, leaves become unable to recover from wilting even at night, a situation termed **permanent wilting.** Their stomates remain closed all the time, and photosynthesis remains shut down. Permanent wilting, which can be relieved only by providing water, reflects soil water-holding properties considered in Chapter 11.

Guttation

At night, when transpiration is not occurring, the leaves of many low-growing plants (but not trees) exude drops of water (Figure 9-16). This phenomenon, called **guttation** (Latin *gutta,* a drop), indicates that the roots are absorbing more water than the leaf cells need at that time. Guttation water emerges from vein endings called **hydathodes,** specialized for the easy outflow of liquid from the vascular system (Figure 9-17). Guttation is another osmotic process, due to the mineral nutrient-absorbing activity of the roots as considered in Chapter 11.

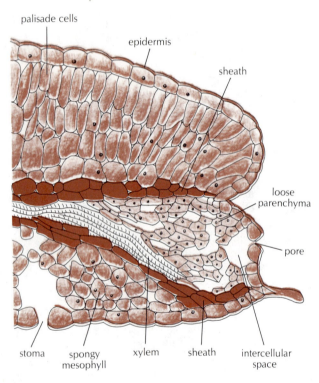

Figure 9-17.   A hydathode at the edge of a leaf of *Saxifraga lingulata.* Xylem sap, when under positive pressure, moves through intercellular spaces in the adjacent loose parenchyma tissue and is forced out through the hydathode's pore, a modified stomate. Water is prevented from moving into other leaf tissues by a compact cellular sheath lacking intercellular spaces.

SUMMARY

(1) The structural arrangements for efficient uptake of CO_2 permit water vapor to diffuse rapidly from the water-saturated intercellular gas spaces of the leaf to the air outside, resulting in the rapid evaporation of water from cells within the leaf, that is, transpiration. The transpiration:photosynthesis ratio, or water cost of photosynthesis, is usually greater than 200:1. (This is because the water vapor concentration difference driving transpiration is usually much greater than the CO_2 concentration difference that drives photosynthetic CO_2 uptake, which is limited by the low CO_2 content of air (0.03%). Certain strategies, namely, C4 photosynthesis and CAM, can moderate this water cost.)

(2) Wind, low humidity, high temperature, and sunlight promote transpiration for physical reasons and, thereby, increase the water cost of photosynthesis. Transpiration cools the leaf, disposing of some of the surplus sunlight energy tending to heat up the leaf. This is beneficial during hot weather when the air temperature is above the optimum for photosynthesis.

(3) Stomates open in response to light and low CO_2 concentrations, and close in darkness or under high CO_2. These responses adjust the diffusion resistance of the leaf relative to its photosynthetic demand for CO_2, holding down the water cost of photosynthesis by comparison with unregulated evaporation.

(4) Stomates close in response to water stress; because of the epidermal cuticle and waxes, closure protects the leaf against desiccation injury. However, stomatal closure inhibits photosynthetic CO_2 uptake.

(5) Due to its internal solute concentration, a plant cell absorbs water by osmosis and builds up against its cell wall a turgor pressure, which stops osmotic uptake. This allows water balance between the cell and a dilute environment and is important for the mechanical support of soft plant tissues, such as most leaves.

(6) Light or a low CO_2 concentration stimulates guard cells to develop a very high solute concentration [primarily by uptake of potassium ions (K^+) and accompanying uptake of chloride (Cl^-) or production of organic acids], causing osmotic water uptake, turgor pressure increase, and swelling of the guard cells, which opens the stomatal pore. The reverse occurs in darkness, under high CO_2, or under water stress, but water stress will cause stomatal closure in any case when the leaf wilts and the guard cells lose turgor.

(7) Evaporation of water from a leaf cell reduces its turgor pressure, bringing on osmotic uptake of water from the vascular system to offset the transpiration loss and acting as a driving force for water transport through the xylem and absorption from the soil.

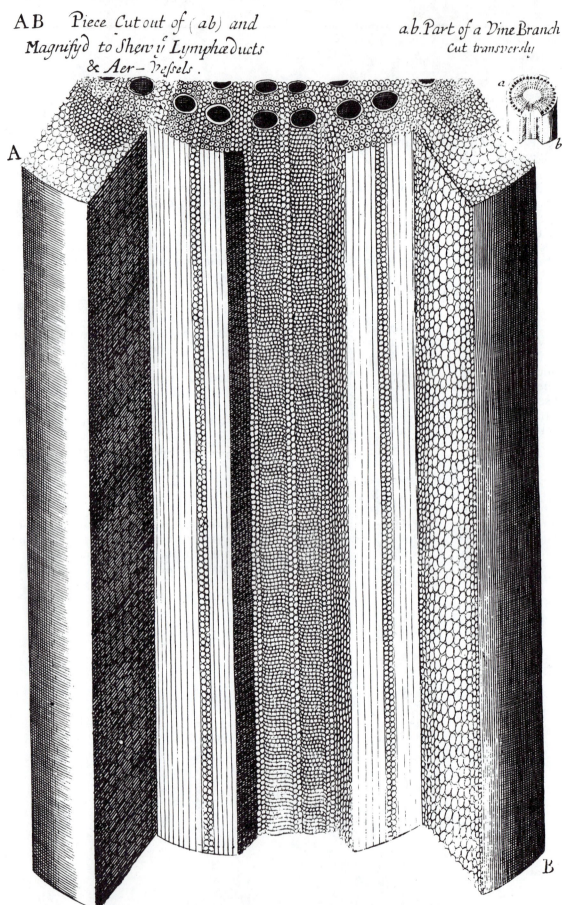

AB Piece Cut out of (ab) and
  Magnify'd to Shew ÿ Lymphæducts
    & Aer-Vessels.

a.b.Part of a Vine Branch
    Cut transversly

A

B

Chapter 10

THE STEM AND VASCULAR SYSTEM

The stem of a plant transports water to the leaves to replace the large amounts they lose through transpiration. It also transports products of photosynthesis (photosynthate) from the leaves to feed the nonphotosynthetic parts of the plant, such as roots, flowers, and fruits, and many non-photosynthetic cells in the stem itself. These transport functions are performed by the plant's vascular system. The stem also supports the leaves, flowers, fruits, and so on, in the air; cells that aid in mechanical support occur both within the vascular system and outside it. The performance of these and other diverse functions requires a relatively complicated structure. In this chapter we look at the general features of organization of stems and at the structure and function of the vascular tissues. More detailed aspects of stem structure are covered in Chapters 12 and 13, where these features are considered in terms of their development.

The plant vascular system consists of two structurally and functionally quite different tissues, the **xylem** (pronounced "zye'-lem") for water transport and the **phloem** (pronounced "flow'-em") for photosynthate transport. These two tissues are always separate but usually run parallel, side by side. A strand of vascular tissue or a vascular bundle, illustrated previously by the veins of a leaf (Figure 8-4), consists of parallel strands of xylem and phloem. The xylem can be recognized by the fact that most or all of its cells possess distinctive lignified secondary walls, described below, and (in flowering plants) by the occurrence of tubular water-conducting structures called vessels (Figure 10-A). Cells of the phloem are usually much smaller, and many or most of them lack secondary walls.

GENERAL FEATURES OF STEM STRUCTURE

Herbaceous Stems

The relatively soft stems of herbaceous (non-woody) plants consist of (1) vascular bundles that run lengthwise through the stem, embedded in (2) soft tissue called **ground tissue** composed mainly of parenchyma and comprising the bulk of the stem's volume, covered on the outside surface by (3) epidermis. The epidermis of stems is basically similar to that of leaves, being a single cell layer that secretes a waxy cuticle onto its outer wall and is perforated by stomates. These are less abundant than on leaf surfaces, in keeping with the more modest gas exchange needs of the stem. Although the ground tissue often contains photosynthetic cells, these are generally much less active than those of leaves.

MONOCOT STEMS

Stems of corn or wheat (Figure 10-1) illustrate the basic anatomical plan found in monocots: numerous vascular bundles scattered in an extensive

Figure 10–A. Internal structure of a grapevine stem as figured by Nehemiah Grew, one of the two "fathers" of plant anatomy, in his book *The Anatomy of Plants* (1682). Grew shows a microscopic view of a small portion removed from the outer part of the stem as in the small diagram at upper right; from the center part of this portion the rind or bark has been stripped off, exposing the xylem (wood). Large tubes cut transversely on the top surface are vessels in the xylem (Grew called them air vessels, erroneously believing that they conduct air within the plant). Semi-circular cross sections of bundles of small cells in the rind at top left and right are phloem (Grew called them lymph ducts, correctly believing that they conduct some kind of sap). Other features of stem structure such as rays, tracheids or fibers, cells of the cortex, and the pattern of pits in the side walls of two exposed vessels can be made out in the drawing, but are represented imperfectly because of the limitations of Grew's microscope and the lack of any previous understanding of plant anatomy to help him interpret what he was observing.

(A)

(B)

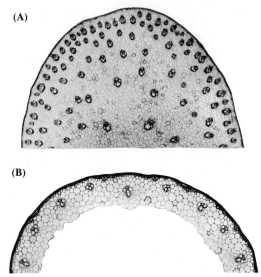

Figure 10–1.   Cross sections of corn (A) and wheat (B) stems, showing the scattered distribution of vascular bundles. The hollow center of the wheat stem arises by collapse of its central parenchymatous tissue during enlargement of the stem in primary growth.

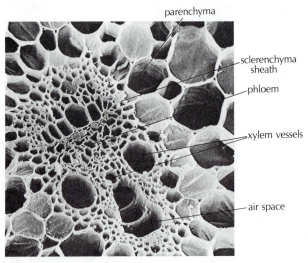

Figure 10–2.   Vascular bundle of corn stem, scanning electron micrograph. The center of the stem is to the lower right.

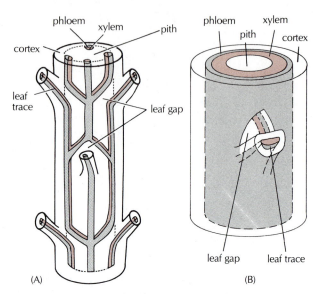

Figure 10–3.   Diagrams showing the nature of the vascular system, leaf gaps, and leaf traces in (A) an herbaceous dicot stem and (B) a young woody stem.

to stiffen the stem both against bending and against injury from outside. This tougher outer part of the stem, including the more numerous, smaller vascular bundles near the epidermis, is sometimes called the "rind."

DICOT STEMS

In a dicot stem the vascular bundles are usually arranged in the form of a cylinder, around a central core of parenchymatous ground tissue called the **pith** (Figure 10-3A). The ground tissue outside the vascular bundles is called the **cortex.** It often includes, just beneath the epidermis, bundles of a thick-walled strengthening tissue called collenchyma. This tissue is important principally in the mechanical support of young, still-elongating parts of the stem and is considered in connection with growth in Chapter 12.

The vascular bundles, like those of monocot stems, always contain phloem toward the outside of the stem and xylem inward from this phloem (Figure 10-4) (in a few families of dicots some phloem also occurs to the inside of the xylem). Fibers with heavy secondary walls commonly develop in the outer part of the phloem of each vascular bundle, forming a prominent "fiber cap" (Figure 10-4) which adds further strength to the stem. The fibers of the flax plant, from which linen cloth is made, are of this type.

Vascular tissue is continuous throughout the root, stem, and leaves of the plant body. At each node in the stem, strands called **leaf traces** depart from the stem's cylindrical array of vascular bundles and extend out into the leaf petiole, becoming the veins of the leaf. Each leaf receives commonly one, three, or five leaf traces, depending upon the kind of plant (most monocot leaves have

ground tissue. The xylem of each vascular bundle (Figure 10-2) is located in the half of the bundle toward the center of the stem, while phloem occurs in the half toward the outside surface of the stem. The bundle is surrounded by a sheath of elongated, heavy-walled sclerenchyma fiber cells, which provide mechanical support for the stem but actually belong to the ground tissue rather than the vascular system. Similar vascular bundles from certain monocot leaves constitute important commercial fibers such as sisal and Manila hemp. In monocot stems heavy-walled sclerenchyma cells also commonly occur at the outer edge of the ground tissue, just beneath the epidermis, serving

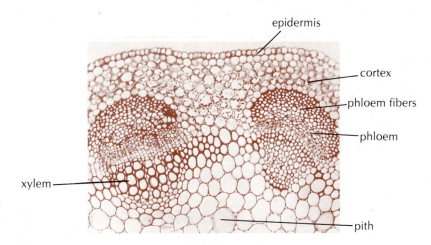

Figure 10–4. Cross section, portion of an herbaceous dicot stem (clover, *Trifolium hybridum*).

many traces). The leaf traces originate by branching of the stem's vascular bundles, usually one or more nodes *below* the node where the trace departs into a leaf (Figure 10-3A). Where leaf traces depart from the stem's vascular cylinder, they leave a parenchyma-filled gap, called a **leaf gap,** in the cylindrical array of vascular strands (Figure 10-3A). Above the node adjacent bundles fill the leaf gap by altering their course and/or by branching. In the nodal region some of the vascular bundles also may interconnect laterally. By such features the vascular system usually acquires a complex pattern (Figure 10-3A) which tends to be distinctive for any given plant species.

Woody Stems

In a young woody stem a continuous vascular cylinder develops, interrupted by leaf gaps where leaf traces depart (Figure 10-3B). These gaps are soon closed by secondary growth, resulting in a

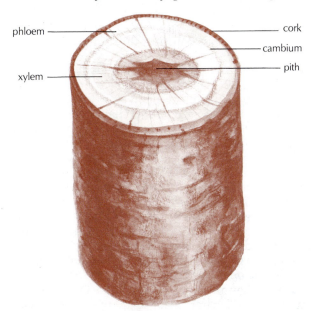

Figure 10–5. Surface view of a cross-sectioned young woody stem.

solid cylinder of xylem and, to the outside of it, phloem (Figure 10-5). Between the xylem and the phloem lies the vascular cambium, a thin sheet of dividing cells, invisible to the naked eye, which causes the growth in diameter of the stem. The xylem portion of the stem, as it becomes more massive by further secondary growth (Figure 10-6), becomes what we recognize as the **wood** of a tree (hence the term xylem, from Greek *xylon*, wood). The tissues outside the wood and the cambial layer are called the **bark** (Figure 10-6). Its inner part comprises the tree's phloem (from Greek *phloios*, bark), while the outer bark consists of multiple layers of a tough protective tissue called **cork** or phelloderm. The development and structure of these tissue layers are covered in Chapter 13.

THE PROBLEM OF LONG-DISTANCE WATER TRANSPORT

In Chapter 9 we learned how transpiring cells can maintain their water content by osmotic absorption if water is available to them. The resistance to osmotic water flow across a single plasma membrane is not large, so a single cell (such as an algal cell growing on the soil surface) can meet its water needs satisfactorily by osmosis, if it has direct access to water. However, for the osmotic movement of water over macroscopic distances, the resistance is much larger because the resistances of the many cellular membranes that must be crossed en route add up cumulatively. Due to this large resistance, osmosis through tissues over macroscopic distances occurs much too slowly to supply the water needs of transpiration. Nonvascular land plants such as mosses and liverworts are consequently restricted to a small size and usually to relatively moist and shady habitats where transpiration is modest. Successful large-scale colonization of the land by living organisms depended upon the evolution of vascular plants with their low-resistance pathway for water transport, the xylem.

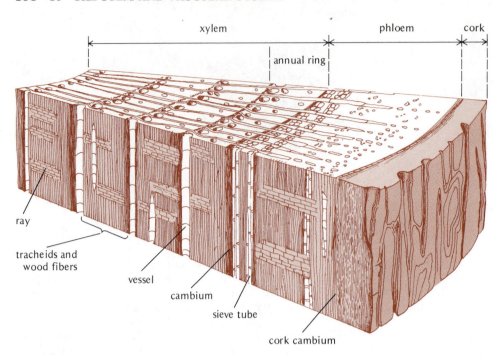

Figure 10–6. Wedge sector of an older woody stem. Only the outermost four of this tree's annual xylem growth rings are shown.

The water-conducting function of the xylem can be demonstrated by placing the base of a cut, transpiring branch in a water solution of a dye or by injecting a dye into a tree trunk. The dye travels rapidly upward in the xylem; microscopic observations show that it moves within cellular structures called tracheids and (if present) vessels (Figure 10-7A, B). The rate of xylem water flow can be measured in an intact plant or tree by certain ingenious methods. For example, if one supplies heat from a small electric heater to a local area of stem or tree trunk, one can measure how long it takes for an increase in temperature, due to flow of warmed water in the xylem, to occur at a known distance above the point of heating. During rapid transpiration typical water flow rates in the xylem are between 10 and 100 cm/min, whereas when atmospheric conditions or stomatal closure (e.g., at night) shut down transpiration, almost no flow occurs.

STRUCTURE OF THE XYLEM

Befitting their function, the water-conducting structures of the xylem (Figure 10-7A, B) are elongated, pipelike cells or cell assemblages. Their most basic specialization is that at functional maturity they die and their cytoplasm, organelles, and cellular membranes break down, eliminating any osmotic resistance to water flow through them. Loss of membranes and cell contents means loss of turgor pressure; to prevent collapse of the conducting cells, their walls must be stiffened and rigidified, which is accomplished during cell development by laying down a prominent lignified secondary cell wall before the mature conducting cell dies.

Tracheids (Figure 10-7A) are elongated single cells whose ends usually overlap with those of other tracheids and whose secondary walls often possess conspicuous pits. These pits, invariably paired up between the overlapping tracheids, provide a route for relatively free water flow from one tracheid to another, through the thin primary walls separating the tracheids to either side of a pit pair (Figure 10-8).

A **vessel** (Figure 10-7B) is a multicellular tube created from the fusion of several cells by breakdown, in one or more places, of the walls between them (Figure 10-9). These openings in the cell walls are called **perforations**; the individual cells fused to form a vessel are called **vessel members** or vessel elements. The lateral secondary walls of vessels often possess pits through which water can pass to or from adjacent vessels, tracheids, or other cells. Within a vessel, the perforations between vessel members eliminate resistance to water flow from cell to cell. Thus the vessel is a more efficient water conduit than a series of tracheids, which impose some resistance to water flow through the primary walls that must be crossed when water flows from one tracheid to another through pits. Vessels are the most advanced water-transporting structures that have evolved in plants.

Vessels are found in most flowering plants (angiosperms) but are lacking in gymnosperms, ferns, and other lower vascular plants (with rare exceptions), xylem transport in these latter plants occurring entirely through tracheids. Angiosperm xylem (Figures 10-6, 13-15) contains both vessels and tracheids, but because the conducting efficiency of vessels is greater, they normally comprise the most important pathway of water transport.

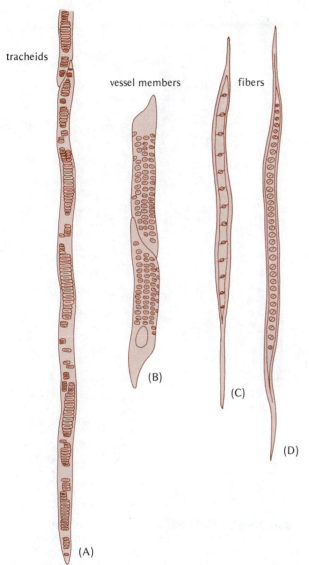

tracheids

vessel members   fibers

(B)

(C)

(D)

(A)

Figure 10–7. Conducting and supporting cells of the xylem.

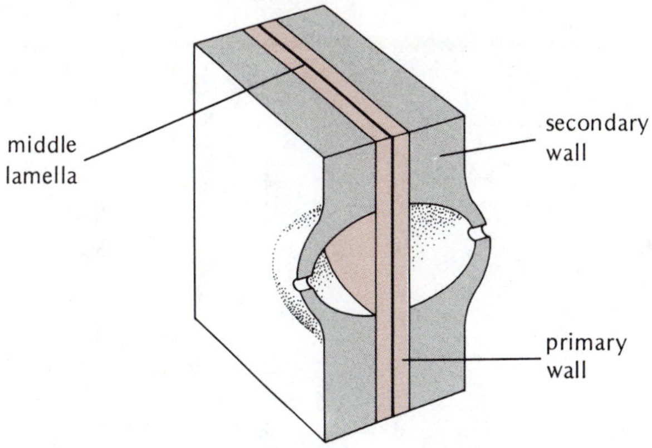

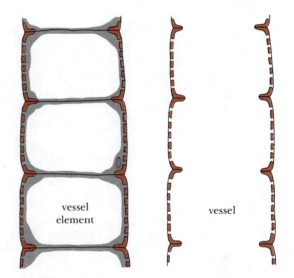

middle lamella

secondary wall

primary wall

Figure 10–8. A bordered pit pair, longitudinal section.

vessel element

vessel

Figure 10–9. Formation of a vessel. (Left) Differentiating vessel elements, arranged in a longitudinal series. The horizontal walls are primary only. (Right) The primary walls have disintegrated.

Variations in Form of Conducting Cells

Long, narrow vessel members with tapering ends and multiple perforations like those of *Magnolia* or of the tulip tree (*Liriodendron*) (Figure 10-10A) are considered least advanced, being rather like tracheids with enlarged pits near the cell tips. The primary walls and middle lamella in these pitlike areas break down at maturity. More advanced vessel members have relatively transverse end walls and simple, single perforations, with a tendency toward a shorter and wider cell shape (Figure 10-10B). This reaches an extreme in the very wide vessels of oak and similar woods (Figure 10-10C). Such vessels are often wide enough to be visible with the unaided eye. Enlargement of the vessel diameter greatly increases transport efficiency, because the resistance to flow through

a tube decreases with the fourth power of its diameter.

The form of secondary wall thickening of tracheids and vessel members varies in different parts of the plant. Xylem cells developed in non-elongating regions such as a tree trunk possess a continuous secondary wall decorated with pits as already described (Figures 10-7, 10-11A). Xylem cells developed in elongating regions of stems and roots and in expanding leaves, flowers, and so on, possess a discontinuous secondary wall laid down either in the form of a helix (Figure 10-11B) or as a series of separate rings (Figure 10-11C), called an annular configuration. Because the primary wall between the secondary thickenings is extensible, helical and annular vessels and tracheids can elongate as the tissue containing them grows, in contrast to pitted vessels or tracheids whose con-

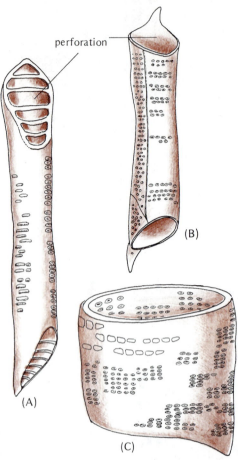

perforation

(B)

(A)

(C)

Figure 10–10. Forms of vessel members: (A) with multiple scalariform (from Latin *scalaris,* "of a ladder") perforations, from wood of tulip tree; (B) with simple perforations, from wood of cottonwood. (C) Barrel-shaped, from wood of oak.

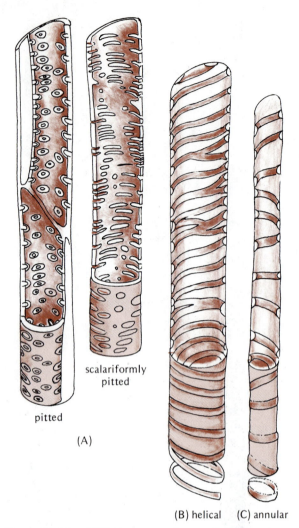

pitted

scalariformly
pitted

(A)

(B) helical   (C) annular

Figure 10–11. Different secondary wall patterns in xylem vessels, from the stem of Dutchman's Pipe (*Aristolochia*). Thickenings are (A) pitted, (B) helical, (C) annular. Vessel types (B) and (C) are formed during early growth, when the stem is still elongating. Type (A) vessels are formed later and are characteristic of secondary xylem, but are also found in the primary xylem formed after organ elongation ceases.

tinuous secondary wall is a rigid enclosure precluding elongation.

Other Types of Xylem Cells

In herbaceous stems, cells specialized to aid in mechanical support occur mainly in the phloem and/or the ground tissues, as noted above, but for a tree its xylem, that is, wood, becomes its principal means of mechanical support. The tracheids of conifer wood (Figure 13-13) serve both for mechanical support of the tree and for water transport. The vessels of angiosperm wood, however, are mechanically weak. For strength, angiosperm wood (Figures 10-6, 13-15) usually contains **wood fiber cells,** similar in shape to tracheids but with heavier walls bearing fewer and smaller pits (Figure 10-7C, D).

The xylem also contains compactly shaped living cells, called parenchyma cells. **Vascular rays** (Figure 10-6) are made up of parenchyma cells in radial rows, that is, running in the direction from the center to the outside of the stem, not only through the xylem but extending into the phloem. Rays apparently serve for transport in the

radial direction, for example, transport inward of photosynthate from the phloem to feed living cells in the xylem and pith. **Wood parenchyma cells** occur in groups running in the vertical direction. Wood and ray parenchyma can also act as storage cells accumulating starch reserves for use, for example, during the spring flush of shoot growth. The various cell types and their pattern of organization in wood are discussed further in Chapter 13.

THE MECHANISM OF XYLEM TRANSPORT

Because tracheids and vessels are dead cells lacking solute-impermeable membranes, they cannot take up or move water by osmosis, as living cells do. Transport of water in the xylem occurs simply by bulk flow, like the flow of water in pipes. The bulk flow of water in the xylem, as in a system of pipes, requires a gradient of pressure to drive it. Therefore, during transpiration a pressure gradient must arise in the xylem, causing the rapid flow rate that can be measured as noted earlier. The living cells in the xylem are not responsible for this, because they can all be killed by steam or poisons without stopping the flow of water in the stem. From the fact that a detached leafy branch lacking roots transports water through its xylem as rapidly as an intact plant does, we can see that the forces needed for water transport must arise principally in the transpiring leaves. Since these forces evidently draw water through the xylem toward the transpiring leaves, the motive force for xylem transport is sometimes called transpiration pull.

Water Transport and Water Potential

Water transport in the xylem, since it is a passive physical process, can most simply be regarded as a flow of water down a gradient of water potential. As explained in Chapter 9, transpiration from the leaves lowers their water potential; water consequently flows toward the leaf cells and evaporating surfaces from the xylem of the leaf veins, lowering the water potential in the xylem. This sets up a water potential gradient within the low-resistance pathway comprising the xylem, down which water flows toward the leaves from the roots, eventually lowering the water potential in the roots and causing water to flow in from the soil. This generalized explanation of water transport, while compact and quantitatively useful, sidesteps the physical nature of the water potential gradient required for water flow in the xylem, which is explained by the cohesion–tension theory of water transport.

The Cohesion–Tension Theory

The removal of water from the xylem by leaf cells and evaporating surfaces may be expected to create a kind of suction in the xylem cells, constituting a pull for moving water through the xylem, toward the leaf (Figure 10-12). The existence of a suction or reduced pressure in transpiring plants was indeed shown by Stephen Hales (1727) in the earliest plant physiology experiments (Figure 1-9). However, the force involved is not a simple suction like that produced by a suction pump, because suction can raise water by at most 10 m or 32 ft (the distance atmospheric pressure can push water), whereas many trees grow taller than 10 m (Figure 10-13) and some, notably the California redwoods (*Sequoia sempervirens*) and certain Australian *Eucalyptus* species, approach and sometimes exceed 100 m in height. A force equivalent to 10 times atmospheric pressure (i.e., 10 bars or 10 kg/cm²) is required to lift water to the top of a tree 100 m tall. The **cohesion–tension theory** of water transport explains this in terms of water's capacity to sustain a negative pressure, or tension (see Figure 10-14).

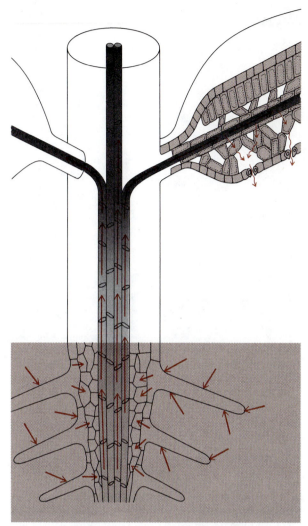

Figure 10–12. Diagram illustrating transpiration pull. Evaporation of water from leaf cell surfaces (wavy arrows) draws water into those cells from the leaf xylem. This sets up a gradient of reduced pressure in the vessel system with the pressure most reduced at the top of the system (increasing color tone). Water flows up the stem (arrows) in the direction of reduced pressure. Because of this pull, water from the soil enters the plant through the root hairs and other surfaces of the young root, then moves through the root cortex into the root xylem.

Because the conducting vessels and/or tracheids of the leaf xylem have rigid walls and are entirely filled with water, cohesion should cause tension to develop in the water within them when water is drawn from them during transpiration. The possibility of developing a tension in the xylem extends the range of pressure differences potentially available to drive water transport, beyond the difference of only 1 bar that could be generated by suction (Figure 10-13). Tension makes it possible, in principle, for transpiration pull to raise water to the tops of tall trees, provided the "tensile strength" or amount of tension that water will in practice withstand within the xylem is greater than the pressure difference physically re-

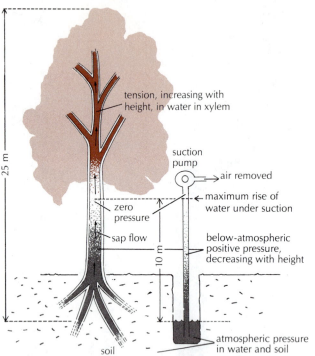

Figure 10–13. Pressure distribution in the xylem of a tall tree, compared to that in a pipe to which suction is applied. Intensity of gray tone indicates amount of positive pressure; intensity of color tone indicates amount of tension (negative pressure); no shading indicates zero pressure. No matter how hard the suction pump is driven, water will rise no higher in the pipe because the suction pump can at most reduce the pressure at its end of the pipe to zero by pulling out all the air in it. (During a high rate of transpiration the tension in the top of the tree would become greater and tension would extend lower in the trunk, because of the increased force needed to drive rapid sap flow through the wood.)

quired to raise water. Calculations from the known strength of interaction (hydrogen bonding) between water molecules indicate that water should, in theory, support tensions of at least 300 bars, much more than is needed to raise water to the tops of the tallest known trees. Experimental measurements of water's tensile strength give much lower values, in the range of 20–30 bars, which is nevertheless still more than adequate.

Although it has thus far proved impossible to measure tensions within the xylem directly, indirect evidence such as detectable shrinkage of tree trunks during the day as compared with the night does suggest that tension builds up in the xylem, and indirect methods of measurement indicate the presence of tensions of the theoretically correct magnitude.

Hazards of the Water Transport Mechanism

Tension is an inherently unstable state. Any disturbance breaking the continuity of water in a

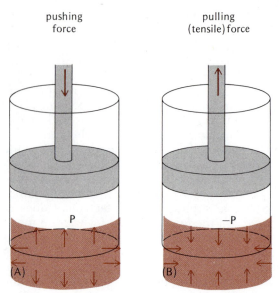

Figure 10–14. Mechanical piston and cylinder model illustrating concepts of pressure, suction, and tension. (A) If a pushing force is applied to the piston it compresses the liquid, generating a pushing force or positive pressure (P) between liquid and cylinder walls, and pushing back on the piston itself. (B) If a pull, or tensile force, is applied to the piston it creates in the liquid a pulling force or tension (negative pressure, −P) against the cylinder walls and the face of the piston. In order for this to be sustained it is essential that the hypothetical piston allow no air to leak between it and the cylinder wall, and that the molecules of the liquid attract one another strongly (**cohesion**) and also interact positively with the walls of the container.

tracheid or vessel, or introducing a gas bubble, will destroy the state of tension and create a **gas embolism,** blocking water transport through that conduit. Fortunately such gas bubbles cannot readily pass through the pits which allow water flow between tracheids or vessels. The pit "membranes" (primary walls; Figure 10-8) act as gas-excluding valves, restricting a gas embolism from spreading throughout the xylem and completely disrupting water transport. An enormous number of conducting cells side by side—redundancy of functional components—is thus the first line of defense against gas embolisms, ensuring that when some conduits become blocked by gas embolisms, there are plenty of others to carry on transport. The second defense is the continual addition of new, water-filled functional tracheids and vessels by secondary growth, replacing conducting elements that may have become nonfunctional.

PHOTOSYNTHATE TRANSLOCATION IN THE PHLOEM

The second part of the vascular system of plants, the phloem, enables the export of photosynthetic products, or photosynthate, from the leaves to the rest of the plant. The necessity of the inner bark for transport of food materials to the roots of trees has long been known from the results of girdling, or removing a ring of bark all the way around a tree. This causes the tissues below the gap to cease growing for lack of food (Figure 10-15), while fruit produced above the girdled point grows larger and sweeter than normal due to the surplus of photosynthetically produced sugar. Eventually the roots die of starvation and, with them, the entire tree. Early colonists in this country used girdling to kill trees to eliminate their shade so they could plant crops on forested land without having to cut down the trees. Foresters sometimes girdle unwanted trees to eliminate them from woods without damaging desirable trees by felling the unwanted ones. Girdling of individual branches, which does not kill the root system, is used in specialized horticultural endeavors to produce extra-large, premium-quality fruits.

Conducting Structures of the Phloem

The unique cells in the phloem have conspicuous sievelike areas on their walls (Figure 10-16), interconnecting these cells with one another (Figure 10-17), suggesting a transport role. In angiosperm phloem these cells occur in longitudinal rows, called **sieve tubes** (Figure 10-16B), consisting of individual cells called sieve tube members

or sieve tube elements, an arrangement analogous to the makeup of vessels in the xylem. The conspicuous sievelike wall between adjacent sieve tube members is called a **sieve plate** (Figure 10-16C). Experiments in which radioactive tracers are fed to leaves and the phloem of the petiole or adjacent stem is examined by autoradiography confirm that sieve tubes are the route by which photosynthate is transported from the leaves.

Sieve tubes differ structurally and functionally from the vessels in the xylem. Sieve tube members have a somewhat thickened cell wall

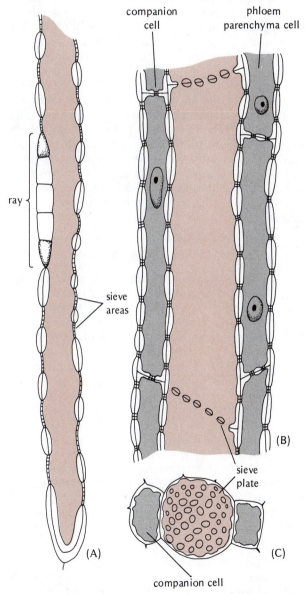

Figure 10–16. Sieve elements and associated cells. (A) Sieve cell of a conifer, *Pinus pinea,* with an associated ray. The diagram shows only about one third of the sieve cell's total length. (B) Sieve tube member of squash, with adjacent cells (diagrammatic). (C) Simple sieve plate and adjacent cells as seen in cross section.

Figure 10–15. Black cherry (*Prunus serotina*), end of second season of growth after girdling (stripping off the phloem). Cambial growth was stopped below the ring but continued above it.

(Figure 10-18), but it is not a lignified secondary wall. Sieve tube members are living cells with an osmotically effective plasma membrane. In contrast to transport in the xylem, killing the cells of the phloem stops the transport of photosynthate. Sieve tube members are peculiar among living plant cells in that they lack a nucleus; the nucleus usually breaks down and disappears during development of the specialized cell. Electron microscopic observations show that sieve tube members also lack internal membranes, ribosomes, and most other organelles (Figures 10-17, 10-18). Usually they contain only a few mitochondria and plastids, often with a degenerate appearance, attached to the plasma membrane. Sieve tubes contain no proper cytoplasm, since the vacuolar membrane disappears, eliminating the normal distinction between cytoplasm and vacuole. The contents of a sieve tube are usually called simply sieve tube sap.

Invariably accompanying and developmentally related to the sieve tube members are nucleated, cytoplasmically dense **companion cells** (Figure 10-16B). These cells possess abundant organelles (Figure 10-18) and communicate with the adjacent sieve tube member via numerous plasmodesmata. Companion cells probably perform the metabolic functions needed for maintenance and repair of the sieve tube member with which they are associated.

The Process of Translocation

Sieve tube sap contains principally sugars. This can be determined by cutting into the conducting phloem and collecting the sap that exudes, primarily from severed sieve tubes. A more ingenious method is to sample the contents of a single sieve tube using the insects called aphids. These troublesome garden pests feed on plants by inserting their tubular feeding apparatus (stylets) into a sieve tube (Figure 10-19). If a large feeding aphid is anesthetized and immobilized with CO_2, the stylets can be severed and the body of the aphid removed, leaving the stylets in place. Sap then flows out of the cut stylets from the sieve tube, often for many hours, and can be collected. Both sampling methods show that simple sugars such as glucose and fructose are not present in sieve tube sap, but compound sugars, especially sucrose, occur at high concentrations (5–20%).

In the majority of plants sucrose is the principal component of sieve tube sap. The sieve tubes of many plants, however, contain the **raffinose** family of compound sugars, which consist of sucrose with one or more galactose units attached. Radioisotope evidence shows that the principal sugar constituents, whether sucrose or the raffinose series, are the main materials translocated. Small amounts of amino acids also occur in sieve tube sap and presumably are translocated.

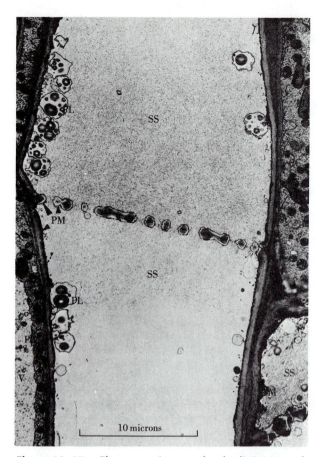

Figure 10–17.   Electron micrograph of adjoining ends of two sieve-tube members of the tobacco plant, showing the large pores that connect the adjacent cells (3300 ×). CC, companion cell; PC, phloem parenchyma cell; M, mitochondrion; PM, plasma membrane; PL, plastid; SS, sieve-tube sap, containing slime protein which shows as a fibrous material; V, vacuole. Note absence of separate cytoplasm and vacuole in the sieve-tube members.

Phloem translocation lacks a fixed directionality, moving sugar either up or down in the plant, depending upon where sugar is needed and where it is being produced. This is called "**source-to-sink**" directionality. Because leaves produce sugar by photosynthesis, while the greatest bulk of nonphotosynthetic tissue, or largest "sink," is the root system, translocation occurs downward toward the roots through most of the shoot system. However, if one feeds radioactive CO_2 to a leaf near the tip of a growing stem, or near a flower or growing fruit, the radioactive products are translocated upward from that leaf toward the nearby sink, as illustrated in Figure 10-20B. If one experimentally removes the upper leaves and feeds radioactive CO_2 to a lower leaf, its products are translocated upward, because the source–sink relationships within the plant have been changed (Figure 10-20C).

The speed with which carbohydrates move in the phloem can be determined in various ways,

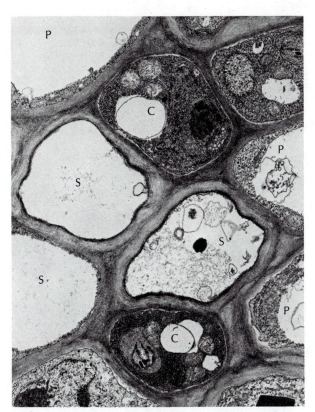

Figure 10–18. Electron micrograph of cross section of phloem, showing sieve tube members (S), companion cells (C), and phloem parenchyma cells (P). Note the dense cytoplasm, abundant organelles, small vacuoles, and presence of a nucleus in the companion cells.

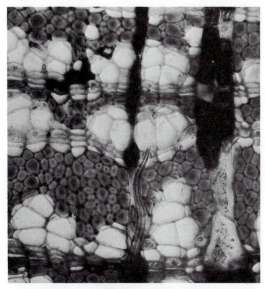

Figure 10–19. (Below) Mature aphid (*Longistigma caryae*) feeding on lower side of a branch of basswood (*Tilia americana*). The aphid's feeding apparatus (stylets) is inserted in the bark and only the sheath can be seen. The aphid is about 6 mm long. (Above) Stylet tips in an individual sieve element. The positive pressure in the punctured phloem cells drives the phloem solution into the stylet and through the aphid's digestive system. Drops of only slightly altered phloem sap (called "honeydew") are exuded from the aphid's anus, as in the lower photograph.

for example, by following the movement of radio-active products through the stem after feeding a leaf radioactive CO_2 (Figure 10-20). Such measurements indicate transport rates of 50–200 cm/hr. This is enormously faster than solutes such as sugars can move by diffusion, so translocation clearly does not occur simply by diffusion of sugar through the sieve tubes.

The Mechanism of Translocation

The most widely believed theory of phloem translocation, the **pressure-flow hypothesis,** holds that the sugar-containing sieve tube sap is driven through a sieve tube toward a sink en masse, through the sieve plates from one sieve tube member to the next by an osmotically generated pressure gradient. As indicated in Figure 10-21, because of loading and unloading of sugar in source and sink areas, respectively, a solute concentration difference tends to arise between the two ends of the sieve tube system. According to the osmotic principles considered in Chapter 9, this should lead to a difference in turgor pressure between the source and sink ends of the system, which should drive sap flow through the sieve tubes toward the sink.

An attractive feature of the pressure-flow theory is that it explains the source-to-sink directionality of translocation without the need for instructions from the plant. It also explains why phloem translocation is nonspecific: any substances that enter the sieve tubes become translocated, including mineral nutrients or dyes applied to the leaves and virus particles introduced by aphids when they puncture a sieve tube; this would be expected if a mass flow of sieve tube sap were oc-

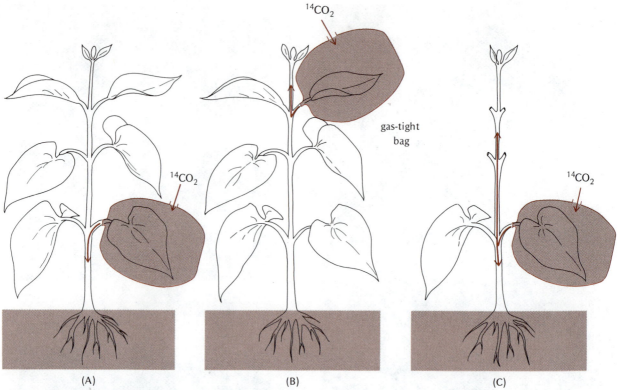

Figure 10–20. Source-to-sink translocation of radioactively labeled sugars in the phloem. When a leaf is fed $^{14}CO_2$, radioactive carbon is fixed into the sugar produced by photosynthesis. The movement of the sugar can be followed using autoradiographic techniques. (A) Lower leaf given $^{14}CO_2$, translocation downward. (B) Upper leaf given $^{14}CO_2$, translocation upward. (C) Upward leaves removed and $^{14}CO_2$ fed to a lower leaf, some translocation in both directions.

curring. According to the pressure-flow theory, during sieve tube sap flow water is taken up osmotically at the source end of the system and given up at the sink end. The amount of water involved would be very much less than is normally transported in the opposite direction in the xylem, but a minor recirculation of upwardly transported water back to the roots via the phloem would be expected.

Although mass flow of sap within a sieve tube certainly occurs when sap is exuding from a severed aphid stylet as described above, and this exudation proves the existence of substantial turgor pressure within the sieve tube, it would not be easy to demonstrate directly a mass flow of sap in an intact sieve tube. There is some experimental evidence for it, and some that has been interpreted as contradicting it. Sieve tubes are very difficult to experiment upon because they cannot be studied in isolation from the other cells and tissues in which they are embedded. Several mechanisms alternative to the pressure-flow model have been proposed for translocation, but no really convincing evidence for any of them has been forthcoming. The pressure-flow model still appears to be the most plausible, with the possible complication that transfer of sugar from one sieve tube to another along the translocation route might be an active process, with passive pressure flow itself as described above operating within each individual sieve tube.

Loading of sugar into the sieve tubes in source areas, which is crucial to the function of the system, appears to occur by active uptake (Chapter 6) because the sugar concentration is often much higher in the sieve tubes than in the photosynthetic cells of the leaf. There is evidence that ATP energy is used in sieve tube loading and that sucrose moves into sieve tubes across the plasma membrane rather than via plasmodesmata. In the small veins of many leaves the cells adjacent to sieve tubes have elaborately infolded plasma membranes (Figure 10-22), suggesting a proliferation of surface area to improve membrane transport into the sieve tubes, because of which these cells have been named **transfer cells.**

Sieve Tube Plugging

Sieve tubes possess mechanisms to block the overflow of sap from them when they are injured, preventing the loss of valuable photosynthate by the plant, much as the clotting mechanism blocks the loss of blood in animals. Plugging mechanisms contribute to the above-mentioned difficulties in experimenting upon functioning sieve tubes. Sieve tube sap contains a fibrillar protein called **P protein** (Figure 10-17) which plugs up the sieve plates during the surge of sap flow that occurs when a sieve tube is cut.

Sieve tubes also have the capacity to form a distinctive glucose polysaccharide called **callose.** Callose

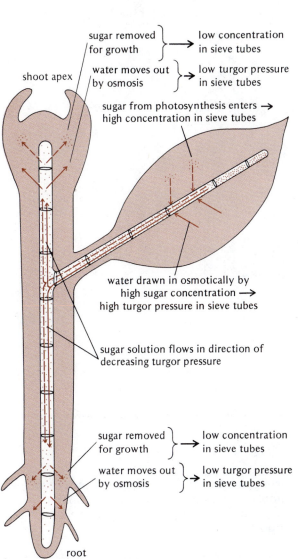

sugar removed | for growth } → low concentration in sieve tubes

water moves out | by osmosis } → low turgor pressure in sieve tubes

sugar from photosynthesis enters → high concentration in sieve tubes

shoot apex

water drawn in osmotically by high sugar concentration → high turgor pressure in sieve tubes

sugar solution flows in direction of decreasing turgor pressure

sugar removed | for growth } → low concentration in sieve tubes

water moves out | by osmosis } → low turgor pressure in sieve tubes

root

Figure 10–21. Model diagram illustrating the pressure-flow hypothesis of photosynthate translocation in sieve tubes.

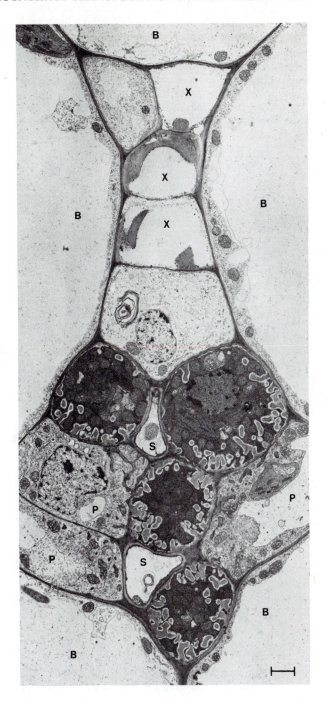

Figure 10–22. Electron micrograph of cross section of a small veinlet in leaf of groundsel (*Senecio vulgaris*), showing transfer cells with elaborate finger-like cell wall ingrowths and plasma membrane infoldings, surrounding two sieve tube members (S). These darkly staining transfer cells are actually companion cells, connected to sieve tube members by plasmodesmata through which the sugar that these transfer cells presumably absorb through their plasma membranes can pass into the adjacent sieve tube. The lighter-staining phloem parenchyma transfer cells (P) have wall and plasma membrane ingrowths facing the companion cells, possibly promoting release of sugar (from photosynthesis) into the intervening cell wall space, for uptake by the companion cells and delivery into the sieve tubes. Above are three xylem elements (X; tracheids or vessels) with annular or helical secondary wall thickenings running obliquely through the section. The large, highly vacuolate parenchyma cells (B) surrounding the vein are the bundle sheath cells. Bar = 1 μm.

becomes deposited on sieve plates, especially of injured or old sieve tubes, and can accumulate to form large callose masses (Figure 10-23). Such callose-plugged sieve tubes are no longer functional but may often be microscopically the most conspicuous structures in the phloem. Because callose is stained specifically by the dye aniline blue, staining of tissue with this dye provides a cytochemical test for the presence of sieve tubes.

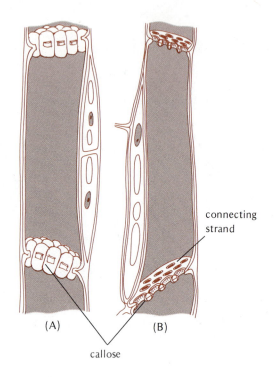

Figure 10–23. Sieve tube members with different degrees of callose deposition on sieve plates. The sieve plates of the older sieve tube (A) have a much larger amount of callose, essentially blocking the connecting strands, as compared with the sieve plate of the younger sieve tube (B). Similar callose deposition can occur rapidly upon injury, helping to block loss of sieve tube sap.

SUMMARY

(1) Herbaceous stems contain, embedded in parenchymatous ground tissue, vascular bundles of xylem plus phloem, in a cylindrical array in dicot stems and scattered more irregularly in monocots. Leaf traces diverge into leaves from this array and become the vein system. Woody stems acquire a solid cylinder of xylem surrounded by a cylinder of phloem and an external protective layer of cork.

(2) Xylem conduits (tracheids and vessels) afford a low-resistance pathway for water transport because at maturity they are dead and lack cellular membranes and, in the case of vessels, also lack cell walls separating the conducting units (vessel members). Most angiosperms possess vessels, whereas most gymnosperms and lower vascular plants have only tracheids for water transport. More advanced or specialized vessels are much larger in diameter, giving them still lower resistance to water flow. The secondary wall of tracheids and vessels is deposited either as a pitted, continuous rigid wall or as a helical band or a series of ringlike bands which allow the con-

ducting cell to stretch during growth. Other cells found in xylem are ray cells, wood parenchyma cells, and (in angiosperms) wood fibers aiding in support.

(3) Water is moved through the xylem by a pressure gradient caused primarily by a reduction of pressure at the leaf end due to the removal of water from xylem cells by transpiration (transpiration pull). Because the xylem cells are water filled, the removal of water can develop a negative pressure, or tension, in them, enabling trees taller than 10 m to raise water to their tops, which is not possible by suction. Since the existence of tension in the water column depends upon cohesive forces between water molecules and between water and the conduit walls, this explanation is called the cohesion–tension theory of water transport.

(4) Photosynthate is translocated in the phloem via living sieve cells or (in angiosperms) sieve tubes, which interconnect through sieve plates in their end walls. Each sieve tube member

lacks a nucleus and most organelles but is accompanied by one or more closely associated, metabolically active companion cells possessing these features.

(5) The principal material translocated in sieve tubes is sugar, commonly sucrose or cane sugar. Sugar moves much faster in the phloem than it could move by diffusion. The pressure-flow theory explains translocation as due to a buildup of sugar concentration and (from osmosis) of turgor pressure in sieve tubes at a "source" (site of sugar production), coupled with a depletion of sugar and resultant loss of turgor pressure in "sink" areas (sites of sugar consumption): the resulting pressure difference between source and sink should drive a mass flow of sieve tube sap toward the sink.

Chapter 11

THE ROOT, THE SOIL, AND PLANT NUTRIENTS

The ability of land plants to grow and produce depends inescapably on the soil, from which their roots absorb the water needed for cell functions including photosynthesis, as well as the mineral elements required for growth and function. Roots also anchor the plant in the soil so it can hold itself upright, which is necessary for most plants to compete successfully for the light essential to photosynthesis. Thus, like the stem, the root performs both mechanical and transport functions and possesses structural features serving each.

THE ROOT SYSTEM

In Chapter 3 we distinguished two common types of root systems, the tap root system, in which a single main root bearing laterals penetrates more or less straight down into the soil, and the fibrous type of root system such as that of grasses, which has no main axis and tends to develop as a dense tangled mass relatively near the soil surface.

The root systems of woody plants, such as trees, usually spread out laterally rather than penetrating deeply into the soil (Figure 11-A). The greatest number of roots is usually found in the uppermost meter (3 ft) of soil. Among different plants the depth of penetration usually varies between about 0.5 and 10 m (1.5–30 ft), although greater extremes occur. In areas where only a thin soil layer overlies hard rock or (as in the arctic) ice, root systems can be restricted to an even shallower depth. The lateral extent of the root system often exceeds a plant's height and the spread of its branches. An aspen tree 8.5 m (25 ft) high, for example, had a main lateral root 16 m (47 ft) long. An oak tree 12 m (36 ft) high had a tap root 4.7

m (14 ft) long, but lateral roots extended out 20 m (60 ft) from the base of the tree.

By repeatedly branching and rebranching, roots proliferate amazingly. To fully appreciate this, one must carefully wash the root system free of soil rather than merely pull up the plant, which breaks off and leaves behind the finer roots. Table 11-1 shows the extent of the root system of a single rye (*Secale cereale*) plant as determined by H. J. Dittmer using this approach. This plant had more than 386 *miles* of root length, with a surface area of about 230 m^2 (2500 ft^2, equal to the floor area of a good-sized house), some 50 times the total area of the shoot system with its leaves. Despite this enormous ramification, the actual dry weight or biomass of the root system is often less than that of the shoot, indeed only about half in the case of most crop plants.

The Root Tip

The root system extends itself through the soil by growth at the tips of its individual branches. The tip of a root (Figure 11-1) is covered by a protective cellular structure called the **root cap** which is completely distinctive for roots, with no counterpart in shoot organs. It serves as a mechanical shield for the root's delicate growing point (apical meristem) as the root pushes between soil particles. Cells toward the periphery of the root cap show a very active Golgi system (Figure 11-2) that secretes a mucilaginous polysaccharide, called "slime" or mucigel, to help the cap and the growing region of the root behind it slide easily through the soil as the root elongates. Cells at the outside of the root cap become rubbed off as the root pushes the cap through the soil, but the addition of new cells to the cap by cell division in the apical meristematic region just behind the cap compensates for this.

As explained in Chapter 12, where the cellular aspects of meristems and growth are considered, elongation occurs only in the first few millimeters of the root tip. Behind this elongation

---

Figure 11-A. The upturned root system of a Giant Sequoia tree.

TABLE 11-1   Dimensions of Shoot and Root Systems of a Single Winter Rye Plant (*Secale cereale*). The plant had grown 4 months in a 30 × 30 × 56 cm container of dark loam soil, from which its roots were carefully removed for measurement by Howard J. Dittmer (American Journal of Botany 24:417–420, 1937).

Organ System/Organs	Number	Average Length (*cm*)	Total Length[a] (*m*)	Surface Area (*m²*)
Shoots				
stems (+ leaf sheaths)	80	50	40	0.13
leaf blades	480	38	182	4.64[b]
total for shoots				4.77
Roots			(*km*)	
main roots[c]	143	45.6	0.065	0.14
secondaries[d]	35,600	14.5	5.18	4.27
tertiaries[d]	2,296,000	7.6	174.9	70.47
quaternaries[d]	11,483,000	3.8	441.9	162.5
total roots[e]	13,815,000		622.0	237.3
root hairs	14.33 × 10⁹	0.074	10,628	401.4
total, roots + root hairs[e]				638.7

[a]Note that total lengths for roots are in kilometers rather than meters.
[b]Total surface area of both sides of leaves.
[c]Fibrous root system with 143 main axes.
[d]Secondaries, tertiaries, etc. are successive branchings from main root axes.
[e]Combined length of all roots equals 386 miles, with a total surface area, including root hairs, of 6,875 square feet.

zone, root hairs (Figure 11-3) arise as tubular outgrowths of many of the root's epidermal cells (Figure 11-4) and can reach lengths of as much as a centimeter. Root hairs further increase the root's surface area (Table 11-1). They also increase by as much as 20-fold the amount of soil the root contacts. These features increase the root's capacity for absorption of water and nutrients and indeed seem to be essential for adequate water uptake from any but the wettest soil. This is shown by the water stress and wilting that plants often experience after being transplanted because, even with the greatest care, it is almost impossible to dig up a plant without breaking off or injuring many of its delicate root hairs.

Root hairs are comparatively short-lived. As the root progresses through the soil, new root hairs form while the older hairs, farther back on the root, collapse and slough away. The area of the root bearing root hairs, therefore, does not change greatly in size or position relative to the root tip as growth proceeds, but moves progressively through the soil. This is important because, except when a soil is saturated, water does not flow freely through soil toward an absorbing root. The roots must continually grow into and tap different areas within the soil to supply the plant adequately with water.

Behind the root hair zone is the region where new lateral (branch) roots are initiated within the tissues of the parent root, and grow out (Figure 11-5).

Techniques of Transplanting

Transplanting injury can be minimized by moving a block of soil with the roots, but usually many root hairs still get broken. Partially shading the transplanted plant, to reduce its transpiration rate, is an important protective measure. If the plant can be protected from desiccation injury for a few days, during which the further growth of surviving root tips generates new root hair-bearing surface and the initiation of laterals from roots with broken-off tips regenerates sites of root hair formation, the transplantation will often succeed.

Some kinds of nursery-grown deciduous woody plants are transplanted **"bare root"** in winter while dormant. The roots are removed from the soil and wrapped in damp peat moss or other material to keep them from drying out. This uprooting breaks off all smaller roots and root hairs, but because there is no transpiration when the plant is leafless, it may be kept for an extended period in this condition. After planting, the ini-

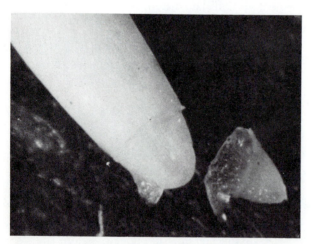

Figure 11-1.   A young root tip with its cone-shaped root cap removed, exposing the apical meristem to view.

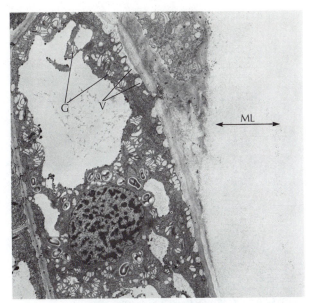

Figure 11-2. Root cap cells from the root tip of a wheat seedling. The enlarged Golgi bodies (G) are giving rise to vesicles (V) which fuse with the plasma membrane adjacent to the outer cell wall, secreting the mucilaginous layer (ML, visible as a finely flocculent material outside the cells) characteristic of this tissue.

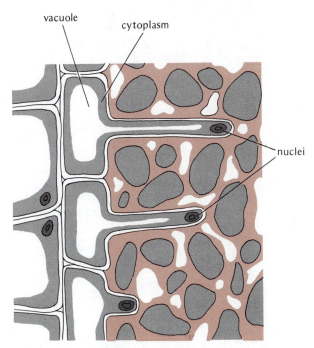

Figure 11-4. Root epidermal cells, showing stages in root hair development and interpenetration of soil particles.

Figure 11-3. Root hairs of young root of radish (*Raphanus sativus*).

tiation of adventitious root tips from older roots provides the absorbing surface needed at the onset of spring growth. Bare root propagation is widely used for roses and deciduous fruit trees.

Most evergreen trees and shrubs such as conifers, rhododendrons, and azaleas do not tolerate bare root transplanting; this is true also of various deciduous trees such as beech, birch, dogwood, and some magnolias, which do not regenerate growing root tips from old roots vigorously enough. Such trees are transplanted by digging up a ball of earth around the roots. The ball is usually wrapped firmly with an open material such as burlap, which will hold the soil in place and can be planted directly in the ground, allowing root tips to grow through it into the surrounding soil. This procedure still involves considerable injury and risk because, except for very small trees, it is never possible to remove a ball of earth nearly as large as the entire root system.

STRUCTURE OF THE ROOT

Within the mature root are three principal anatomical regions: the epidermis, the cortex, and a central vascular cylinder or **stele** (Figure 11-6). The stele is composed of conducting tissues (xylem and phloem) with associated parenchyma cells.

The most prominent cells of the vascular cylinder are the heavy-walled vessels and tracheids

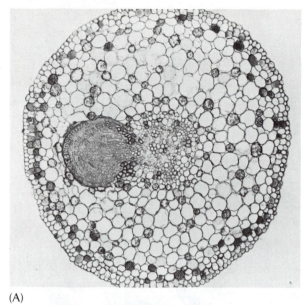

(A)

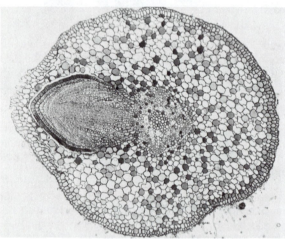

(B)

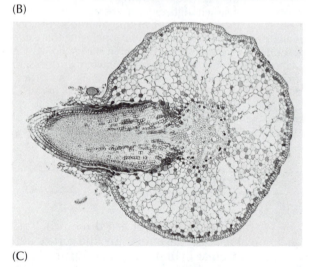

(C)

Figure 11-5. The emergence of a branch root. In this sequence of cross sections of a parent root, the branch root is initiated by cell divisions in the pericycle tissue, becomes organized into a conical root apex, and grows outward, pushing aside or crushing cells of the parent cortex and epidermal tissues as it emerges.

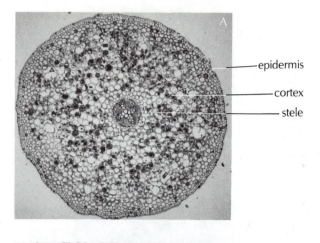

epidermis

cortex

stele

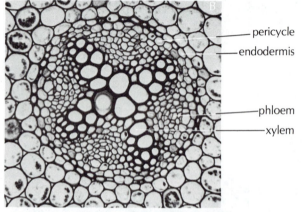

pericycle

endodermis

phloem

xylem

Figure 11-6. Cross-sections of a dicotyledonous root (*Ranunculus*). (A) Entire root. (B) An enlarged view of the stele.

of the xylem. In cross section, the xylem of a dicot root appears as a pattern of arms radiating from the center. The phloem cells occur in groups between the xylem arms. This contrasts with the arrangement of vascular bundles of xylem plus phloem in the young stem (Chapter 10).

The number of xylem arms is fairly constant for a species, with typically 2–4 in dicot roots (Figure 11-6) and as many as 20 in monocot roots (Figure 11-7). In conifers and most dicots the cells in the very center of the root mature into xylem, and there is no pith. In many monocots and some herbaceous dicots, on the other hand, xylem fails to differentiate in the center of the root, where instead a parenchymatous pith occurs (Figure 11-7).

Surrounding the conducting tissues and comprising the outermost tissue of the stele is a narrow zone of parenchyma cells, the **pericycle.** This is a developmentally important tissue, the location at which branch roots may arise.

A distinctive single cell layer, the **endodermis,** separates the stele and the cortex. The endodermis has a specialized waxy thickening on its radial and transverse cell walls, called the **Cas-**

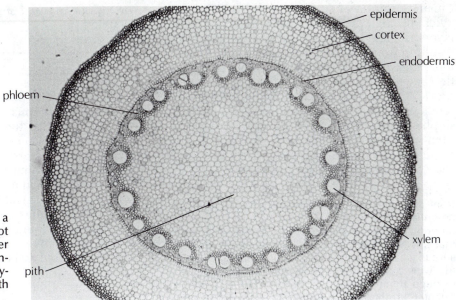

Figure 11-7. Cross-section of a young monocotyledonous root (*Zea mays*). The stele is larger than in most dicot roots, contains many more "arms" of xylem, and surrounds a central pith of parenchyma cells.

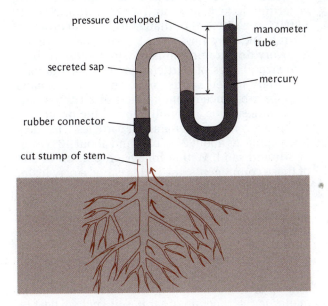

Figure 11-8. Apparatus showing the development of root pressure in the xylem of the root system of a detopped plant, to which a manometer (pressure-measuring device) has been attached. Root pressure, analogous to turgor pressure in a single plant cell, arises by osmotic uptake of water across the living cell layers of the root into its xylem, driven by solutes accumulated in the xylem by active transport of mineral nutrients by the root from the moist soil into the xylem.

parian thickening or Casparian strip, giving these walls a beaded appearance in cross section (Figure 11-16). The endodermis is an important physiological boundary between the stele and the outer tissues of the root. In older monocot roots most of the endodermal cells lay down an extremely thick secondary wall virtually sealing off the stele from the tissues outside it, except for the presence of a few unthickened endodermal "passage cells" opposite the xylem arms.

The root **cortex** consists of relatively undifferentiated parenchyma, usually many cells wide. Much of the bulk of young roots is cortex. These cells expose a large surface of plasma membrane across which mineral nutrients that have entered the cell walls of the cortical region from the soil can be absorbed into the living cells.

The **epidermis,** the outermost cell layer of the root, differs from the epidermis of the shoot in lacking stomates and not producing a cuticle, which would interfere with the absorption of water and mineral nutrients from the soil. The absence of a cuticle, however, makes young roots very sensitive to desiccation if brought out of the soil into the air, in contrast to stems and leaves. This is another frequent source of transplanting injury. The root epidermis is active in absorption, especially through its formation of root hairs as discussed previously.

Roots of dicots and gymnosperms which increase in diameter by secondary growth alter, in this process, their internal structure until it becomes similar to that of a woody stem (Figure 10-5). Secondary growth is considered in Chapter 13.

ABSORPTION OF WATER BY ROOTS

Water absorption by roots is basically an osmotic process. Uptake of water when transpiration is not taking place, leading to guttation (Figure 9-16) or to the development of **root pressure** in a detopped plant (Figure 11-8), occurs because roots transport mineral nutrients into their xylem. This raises the solute concentration in the xylem, causing osmosis of water from the soil into the xylem through the membranes of intervening cell layers (Figure 11-9A). In the case of absorption resulting from daytime transpiration, the driving force for osmotic uptake is not solute concentration but a reduced pressure in the root xylem (Figure 11-9B), due to the removal of water from the root xylem by transpiration pull. This osmotic absorption mechanism enables plants to take up water from soil that appears macroscopically to contain no liquid water and cannot be used as a water source by animals such as humans.

In terms of water potential, the removal of water from the root by transpiration pull reduces the water potential in the root xylem, creating a water potential difference between the soil water and the root xylem that drives inward water flow. Some investigators believe that a significant fraction of the water flows around the living cells of the root via their cell walls rather than through them by osmosis. For reasons explained in Chapter 10, the resistance to water movement across the living cell layers of the root is relatively high compared to that to flow in the xylem. Consequently, as noted above, the root system must have a rather large surface area to compensate for the slowness of uptake

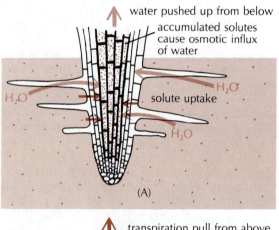

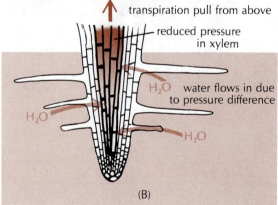

Figure 11-9. Mechanisms of water uptake in roots (schematic longitudinal sections). (A) Water absorption due to accumulation of solutes in the xylem by active solute uptake. (B) Passive water absorption due to transpiration pull. Color shading in the xylem indicates a gradient of pull, or tension, increasing upwards, caused by transpiration in the leaves.

by any one root and provide an adequate total uptake of water.

THE SOIL

To complete the picture of how roots obtain water we must consider the structure and resultant properties of the soil. Soil is the combined product of physicochemical weathering processes by which rocks are degraded and of biological processes by which relatively persistent organic material is formed at the earth's surface. Soils in general contain (1) mineral particles; (2) inorganic ions, including the nutrient ions required by plants; (3) living cells and tissues, notably plant roots, microorganisms including algae, protozoa, and especially fungi and bacteria, and various animals from earthworms to microscopic worms called nematodes; (4) dead organic matter; (5) water; and (6) air.

The mineral particles of soils are classified, according to size, as sand [particles from 2 down to 0.05 mm (50 µm) in diameter], silt (50 down to 2 µm), and clay (less than 2 µm). Coarse-, fine-, and very fine-grained soils in which these respective size classes of particles predominate are accordingly called **sands, silts,** and **clays. Loams** contain a balanced combination of all three particle classes, thus a combination of the properties of the coarser- and finer-grained particles. The ideal loam contains about 20% clay and about 40% each of silt and sand. With a higher proportion of one of these components, the soil is called a clay loam, silt loam, or sandy loam, respectively.

Decay of plant and animal remains by soil bacteria and fungi leaves behind an ill-defined brown or black polymeric organic material called **humus.** Most soils of humid, temperate climates contain only a minor proportion (1–10%) of humus and are called **mineral soils.** However, under swamp or bog conditions, where decay of organic matter is retarded, an **organic soil** such as peat or muck can develop, consisting primarily of organic matter, with a spongy texture very retentive of water.

The humus in a mineral soil encourages its clay and silt particles to aggregate into macroscopically visible clumps called **crumbs** (Figure 11-10) with relatively large spaces called macropores between them, comparable in size to the interstices between grains in a sandy soil. Macropores improve the oxygen supply (aeration) of roots and the penetration of rainfall into the soil, and by lightening the soil, crumb structure facilitates root growth. Thus large benefits for plants derive from the relatively small humus content of most mineral soils. Similar benefits are believed to result from the activity of earthworms burrowing through soil, creating new macropores which improve soil structure.

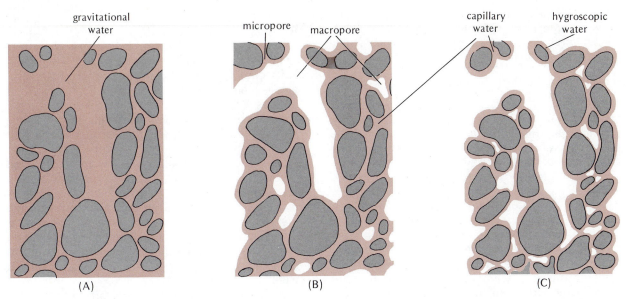

Figure 11-10. A "structured" soil (containing crumbs) under three conditions of hydration. (A) Water-saturated, all pores filled with water. (B) Field capacity, macropores air-filled and micropores water-filled. (C) Permanent wilting point, pores air-filled, hygroscopic water films around soil particles, small amounts of capillary water at points of particle contact.

Soil Water

The pores, or interstices between soil particles, permit soil to absorb and retain water. As illustrated in Figure 11-10, the size of the soil particles, plus the existence of crumb structure, determines the number and range in size of a soil's pores. Because of forces called capillary attraction, the smaller the size of a pore, the more strongly it holds water.

Water in the largest pores (the macropores, those more than about 50 μm wide) drains freely under the force of gravity, and is called **gravitational water.** Except in soils with impeded drainage or where the water table reaches the surface as in marshy land, gravitational water is present only just after rains or irrigation, soon draining into the subsoil and leaving the macropores filled with air for good root aeration. When present, gravitational water is freely available to plant roots, but it blocks the air space of the soil and interferes with the aeration and healthy root growth of most crop plants. This is why good drainage is so important in agricultural land.

Capillary water is found in the smaller soil pores (micropores, less than about 50 μm wide; Figure 11-10B). Micropores retain their water against gravity, yet hold most of it weakly enough that roots can absorb it. In a well-drained soil just after rainfall or irrigation, when gravitational water has drained away, capillary water remains stored in the soil for future plant use. The finer textured the soil, the greater the amount of micropore space and, hence, the more useful water can be stored (see Table 11-2). When the micropore space is fully

TABLE 11-2   Typical Soil Moisture Percentages

Soil Type	Field Percentage	Permanent Wilting Percentage	Storage of Available Capillary Water
Sand	6	3	3
Sandy loam	10	5	5
Silt loam	20	10	10
Clay	40	20	20

charged with water, as after a rain, the soil is said to be at its **field capacity** for water.

Hygroscopic water is water that humus and the surfaces of the soil's mineral particles bind by forces such as hydrogen bonding. Hygroscopic water and the free water in the smallest micropores are held too strongly for roots to absorb and thus are unavailable to plants. Fine-textured soils with a high clay content contain a very large particle surface area and thus retain more unavailable, hygroscopic water than do coarser soils (Table 11-2).

When evaporation and removal by roots deplete a soil's capillary water, leaving only water that roots cannot take up, the soil is at its **permanent wilting point** (Figure 11-10C). Plants must wilt and stay wilted, even if they are not transpiring, until more water is added to the soil, because the osmotic forces in the plant are not great enough to remove hygroscopic water to a significant extent. The permanent wilting point represents the lower limit of soil moisture that will support plant activity.

In terms of water potential (Chapters 9 and 10), the permanent wilting point is the water content at which a soil's water potential becomes as negative as the water potential of leaf cells when they have no turgor pressure, normally about −10 to −15 bars, depending upon the leaf cells' osmolarity. Wilted leaf cells cannot bring about water uptake from the soil when its water potential is as low as theirs, so they stay wilted. The water potential of most soils decreases so steeply with water content at this point that permanent wilting is reached at almost the same soil water content, regardless of how high the osmolarity of the leaf cells may be. The permanent wilting point scarcely depends, therefore, on the kind of plant involved.

Soil Conductivity

The small size of the micropores of fine-textured soils, that is, silts and especially clays, offers considerable resistance to water flow. Therefore, infiltration of water into the soil during a rain tends to be slow and much of the precipitation can be lost as runoff, whereas most of the rain falling on a sandy soil usually soaks into it be-cause its macropores conduct water freely. Except for very heavy rain, this is usually also true of loams having good crumb structure, despite a loam's high content of fine particles. Thus loam soils combine the advantages of high water storage and easy infiltration.

As water is removed from a soil the larger pores, which hold water least strongly, become emptied first. Since the smaller the pore, the greater its resistance to water flow, a soil's conductivity for water drops markedly as it becomes depleted below its field capacity, and water in it can no longer flow significantly toward an absorbing surface from elsewhere in the soil. Continual growth of roots through the soil to tap new water supplies thus becomes important, as noted previously, and root hairs, by contacting many more micropores than the unaided root would, aid enormously in water uptake.

Heavy, unstructured soils significantly impede the growth of roots, making it more difficult for plants to avoid water stress. Furthermore many

TABLE 11-3   Mineral Elements Required by Green Plants

Element	Approximate Level Needed[a]	Role in Plant Function
Macronutrients		
Nitrogen (N)	15	In proteins, coenzymes, nucleic acids, etc.
Potassium (K)	5	Activates certain enzymes in glycolysis; important in cell membrane potentials
Calcium (Ca)	3	Structure and permeability properties of membranes; structure of middle lamella
Phosphorus (P)	2	In nucleic acids, coenzymes, ATP, metabolic substrates
Sulfur (S)	1	In proteins; in coenzymes for carbohydrate and lipid metabolism
Magnesium (Mg)	1	In chlorophyll; Mg^{2+} a required cofactor for many enzymes
Micronutrients		
Iron (Fe)	0.1	In enzymes of electron transfer chain (cytochromes, ferredoxin), nitrogenase, etc.; essential for chlorophyll synthesis
Boron (B)	0.05	Unknown (possibly in cell wall formation in meristems and/or in sugar translocation)
Manganese (Mn)	0.01	Formation of O_2 in photosynthesis; cofactor for various enzymes
Zinc (Zn)	0.001	In several dehydrogenases of respiration and nitrogen metabolism
Copper (Cu)	0.0003	In respiration (cytochrome oxidase) and photosynthesis (plastocyanin; ribulose bisphosphate carboxylase)
Molybdenum (Mo)	0.0001	In nitrate reductase; in nitrogenase (N_2 fixation)
Cobalt[b] (Co)	<0.00001	For symbiotic N_2 fixation; for blue-green algae and (as part of vitamin B_{12}) for various other algae
Chlorine (Cl)	0.05	Cl^- activates O_2-producing system of photosynthesis
Sodium[b] (Na)	0.05	Required by *Atriplex* (saltbush) and probably by other saline-habitat plants, sugar beet, and blue-green algae; role unknown
Silicon[b] (Si)	0.1	Essential for *Equisetum* (horsetail) and probably for rice and some other grasses and sedges; required by diatoms for cell wall formation
Iodine[b] (I)	0.001	Required by certain brown and red algae; role unknown

[a]Millimoles per liter of nutrient solution. These figures are extremely approximate and vary considerably with the species.
[b]These elements are not as yet known to be required *generally* by plants, as are the other elements listed in the table.

soils, especially clays, harden as they dry out, further impeding root growth just when it is most needed. For these various reasons plant growth is generally at an optimum when the soil is at field capacity, providing both water and oxygen freely to roots. Below field capacity the risk of water stress steadily increases, whereas above field capacity, for example, when the soil is waterlogged, root growth can be reduced and plants can be injured by inadequate aeration of the roots.

MINERAL NUTRIENTS

A soil's ability to supply mineral nutrients, that is, required chemical elements other than carbon, hydrogen, and oxygen, is the principal factor in its fertility. To determine the elements specifically required from the soil, plants can be grown with their roots suspended in aerated water containing known chemical salts, a technique called **hydroponic culture.** If the omission of a particular chemical element results in poor growth and in disease, which can be cured only by supplying the omitted element, that element can be considered an **essential element** for plant growth. This technique enabled physiologists to establish the list of essential elements in Table 11-3. An important feature of these findings is that most of the known essential elements—with a few exceptions noted in the table—are required quite generally by higher plants and, in most cases, also by lower plants such as algae and fungi. Another

Figure 11-11. Tobacco plants showing mineral deficiency symptoms. The plant marked "Ck" received all essential elements; the others were supplied with all essential elements except the one indicated on the label. All plants are the same age. (Minus N) Plants generally stunted, older leaves pale and yellowed ("chlorotic"), younger leaves turn pale later since N in older leaves is mobilized and transported to the younger growing parts of the plant as needed. (Minus K) Characteristic mottled chlorosis of older leaves, spreading later to the younger ones (because K, like N, is highly mobile), tips and margins of leaves often curl under, more bushy growth habit due to reduction of stem growth. (Minus Ca) Deficiency most striking in young tissues since most of the calcium in the plant is immobile once absorbed; in meristematic areas, incomplete cell division or mitosis without the formation of new cell walls completely stunts growth, and the tips of leaves that do manage to grow at all are hooked or withered. (Minus Mg) Deficiency symptoms usually appear in older leaves first since magnesium is mobile; chlorosis characteristically develops *between* the veins of older leaves. (Minus S) General chlorosis and yellowing of leaves, beginning with the younger ones, unlike N deficiency. (Minus P) General stunting of growth, older leaves become chlorotic first but more slowly than in N deficiency.

striking feature is that the symptoms of mineral element deficiency, illustrated by the plants in Figure 11-11, tend to differ in deficiencies of different elements and to be similar in deficiencies of the same element in different plants. A knowledge of typical deficiency symptoms often enables one to infer, by inspection, what nutritional problem a sick houseplant or garden plant may have.

Classes of Essential Elements

Macronutrients are elements required in relatively large amounts, typically 50–200 kg/ha (hectare). **Micronutrients,** or minor or trace elements, are needed only in minute amounts, ranging from a few parts per million to as little as one part per billion of a culture solution, or a few grams per hectare. Despite the small amounts needed, the requirement for most micronutrients is just as absolute as for macronutrients, and severe disorders result from a deficiency (Figure 11-12).

Roles of Essential Elements

Elements such as nitrogen, phosphorus, sulfur, iron, and magnesium have many functions in metabolism, too numerous to list in Table 11-3. Certain functions are specific to plants, such as the roles of magnesium in chlorophyll, manganese in photosynthetic oxygen formation, and molybdenum in nitrogen utilization. However, many functions of essential elements are common to animals and plants. Most of the trace metals, for example, are needed as components of enzymes that function in respiration or other metabolic processes.

The symptoms of particular deficiencies reflect some of the functional roles of essential elements. Because chlorophyll contains nitrogen and magnesium, a deficiency of either of these causes **chlorosis,** the failure to produce normal amounts of chlorophyll in the leaves, which consequently acquire a yellow color (due to their carotenoids, which contain neither nitrogen nor magnesium and can still be formed). However, iron deficiency also causes severe chlorosis, even though chlorophyll does not contain iron. Iron is evidently required for the formation of chlorophyll or, perhaps, of the chloroplast structure that develops concomitant with chlorophyll biosynthesis.

Internal Recycling of Nutrients

Several of the deficiency symptoms appear at a characteristic location within the plant (Figure 11-11). Nitrogen, potassium, or magnesium defi-

Figure 11-12.   Zinc deficiency in potatoes (*Solanum tuberosum*) in Washington State. The plants in the background received zinc and phosphorus fertilizers; the dying plants (arrow) in the foreground received phosphorus but no zinc.

ciency symptoms appear first and most markedly in the older leaves. Iron or calcium deficiency symptoms appear instead at the shoot tip and in the youngest leaves. The old leaves formed before the beginning of an iron deficiency remain bright green, while the new leaves are chlorotic. These features reflect the plant's ability or inability to mobilize nutrient elements internally. Nitrogen, potassium, magnesium, and phosphorus can be withdrawn from existing leaves and transported to the growing points for use in new growth during a deficiency. On the other hand, iron and calcium that have been moved into a leaf cannot later be withdrawn from it to meet the needs of new growth (see Figure 11-13).

Due to the frequently marginal supply of, and competition for, nutrient elements in nature, it is important for plants to mobilize and reuse essential nutrients internally. When a leaf grows old it turns yellow, reflecting the breakdown of chlorophyll as well as most of the leaf protein, the nitrogen of which is mostly recovered by the plant before the leaf is shed. The export of nutrients from leaves, like the export of photosynthate, occurs in the phloem. A similar phenomenon occurs prior to autumn shedding of the leaves of deciduous trees, during the period of autumnal coloration. In this case, the recovered nutrients are stored in the stem or roots for use in next year's spring flush of growth.

Utilization of Newly Absorbed Nutrients

From the fact that the nutrient needs of plants can be met by supplying inorganic salts, it is clear that plants absorb their nutrients as electrically charged ions (into which salts dissociate, in water). Most of the essential metals are absorbed as their simple ions (K^+, Mg^{2+}, Fe^{3+}, etc.) and utilized either as such or by incorporation of the ion into an organic compound, for example, Mg^{2+} in chlorophyll and iron (Fe^{3+}) in the cytochrome enzymes of respiration. Phosphorus, sulfur, and nitrogen are available in the soil largely or entirely in the form of oxidized ions (phosphate PO_4^{3-}, sulfate SO_4^{2-}, nitrate NO_3^-). Phosphate is utilized as such (mainly via the formation and use of ATP), but sulfur and nitrogen must be in a reduced form, combined with hydrogen, to be utilized in most of the organic compounds (proteins, etc.) for which they are needed. Plants reduce nitrate and sulfate by means of a complex of several enzymes called a **reductase system,** one reductase system being specific for nitrate and another system for sulfate. The nitrate reductase system (Figure 11-14) reduces nitrate to ammonium (NH_4^+), the form of nitrogen that can be used directly to form amino acids and other organic nitrogen compounds. If NH_4^+ is available in the soil or in a hydroponic culture solution, plants can use it as a nitrogen source and, of course, then do not have to reduce it.

Plants grown with only ammonium as a nitrogen source lack nitrate reductase activity. The reductase system is inducible; that is, supplying the substrate (nitrate) induces the plant to make the enzymes of the nitrate reductase system. Inducible control of enzyme activity is common in bacteria, whereas the striking inducibility of nitrate reductase is one of the few examples of this phenomenon known in plants.

The nitrate reductase system occurs in roots. In many plants, if nitrate uptake is not too rapid, incoming NO_3^- becomes completely reduced in the roots to NH_4^+, which is then utilized to form amino acids or related organic nitrogen compounds. This organic nitrogen is delivered into the xylem and moved in the transpiration stream to the top of the plant for use in its growth. Some plants, however, transport nitrate in the xylem even at modest rates of fertilization (e.g., sunflowers). Leaves can form a very active nitrate reductase system and thus make available for protein synthesis in the top of the plant the nitrate that reaches them. Very heavy nitrogen fertilization, however, can overload the nitrate reductase system even of the leaves, resulting in nitrate accumulation in the top of the plant, occasionally to an extent that has made produce toxic to human infants.

Nitrogen Fixation

The principal form of nitrogen on earth, the nitrogen gas (N_2) that makes up almost 80% of the atmosphere, cannot be used directly by higher plants. It is, however, the ultimate source of the nitrogen in the soil and the biological world, by means of processes that bring nitrogen into combination with other elements. In nature the most important such process is biological **nitrogen fixation,** carried out by certain bacteria and other prokaryotic organisms that occur in water and in soil, and also in association with the roots of certain higher plants, especially the legumes. N_2-fixing organisms have enzymes that can reduce N_2 to NH_4^+. Nitrogen fixation is discussed in Chapters 22 and 25.

MINERAL NUTRIENT UPTAKE

Plant cells can accumulate many of their nutrient ions to concentrations much greater than those in the plant's source of nutrients (Figure 11-15). To achieve ion accumulation, a cell must move ions from lower to higher concentration, which means moving them energetically uphill by an active transport process, requiring an energy input. Passive transport processes such as diffusion and bulk flow are also involved, however, in the absorption of nutrient ions by the root and their movement to the shoot.

Ion Uptake by Roots

By exposing roots to a solution containing radioactively labeled ions, such as phosphate labeled with radioactive phosphorus (^{32}P), and measuring the uptake of radioactivity by the roots, one finds that ions diffuse rapidly into roots via the hydrated cell walls of the cortex, since the wall system is not separated from the external medium by any impermeable barrier such as a cuticle (Figure 11-16). This portion of the tissue, to which solutes from the environment have free access, is called its **free space.** Because of this access, all cells of the cortex can participate in the uptake of ions from the root environment as noted earlier.

Radiolabeled ion transport experiments indicate that root cells take up different nutrient ions such as K^+ and Mg^{2+} by separate transport mechanisms, specific for each of the important ions. The transport of a given ion behaves as if the plasma membrane possesses a limited number of specific sites, called **carriers,** capable of transporting that ion. Carriers are believed to be specific proteins embedded in the membrane (see Figures 5-5, 6-11), although no ion carriers from higher plant cells have yet been isolated and conclusively identified. Carriers for nutritionally important ions have a high affinity for the specific ion; that is, they combine with and transport it even at very low concentrations. This enables a root to take up nutrient ions effectively even when their concentration outside the root is quite low.

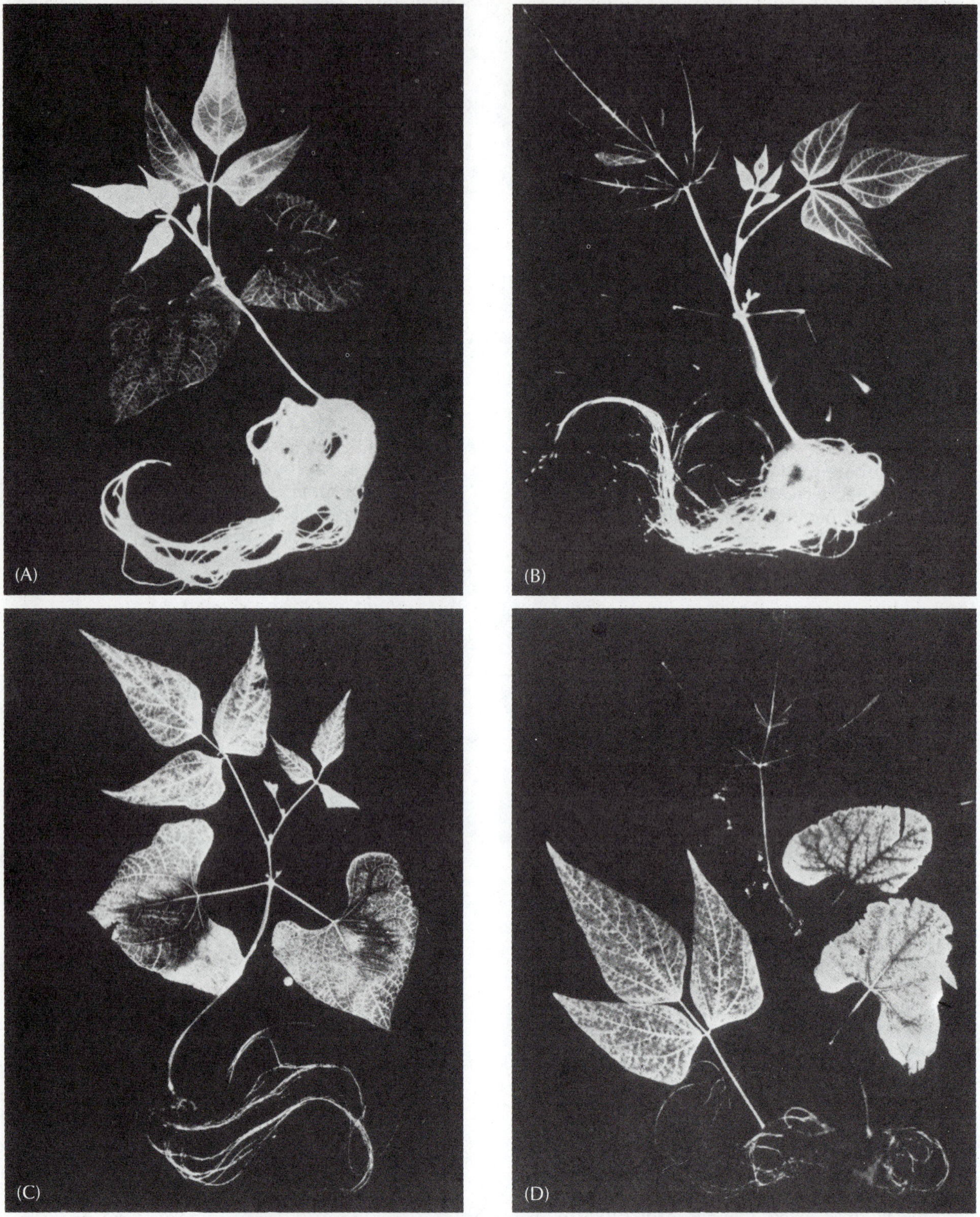

Figure 11-13. Autoradiographs showing internal redistribution of ^{32}P (A, B), and not of ^{45}Ca (C, D) in bean plants. (A, B) The plant in (A) was grown for a short time in a solution containing radioactive ^{32}P; the ^{32}P is distributed throughout the plant. The plant in (B) was grown in the same solution, then removed and grown in nonradioactive solution; the ^{32}P has been concentrated in younger leaves. (C, D) The plant in (C) was grown for a few hours in a solution containing radioactive ^{45}Ca. The plant in (D) was grown in the same solution, then in nonradioactive solution; ^{45}Ca has *not* been redistributed from the older leaves to the younger. (The technique of autoradiography is discussed in Chapter 4.)

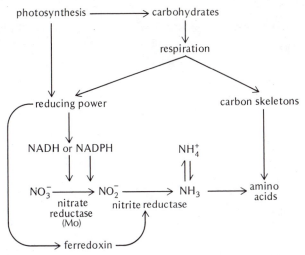

Figure 11-14. Utilization of nitrate by plants. The reduction of nitrate to ammonium requires two enzymes: *nitrate reductase,* to reduce nitrate to nitrite, and *nitrite reductase,* to reduce nitrite to ammonium. Nitrate reductase contains molybdenum as a component essential to its catalytic activity, and this appears to be the reason Mo is required by plants. In Mo-deficient plants nitrate cannot be reduced and a nitrogen shortage results.

Poisons of respiration, such as cyanide, and chemicals called uncouplers, which disrupt the ATP formation normally coupled to mitochondrial respiration, block the uptake of most nutrient ions by root cells. This indicates that ion carriers transport ions through the plasma membrane by an ATP-dependent process.

Ion uptake is complicated considerably by the fact that most plant cells have distinct cytoplasmic and vacuolar compartments separated by an internal membrane, the vacuolar membrane or tonoplast, across which ion transport also occurs. Across either the tonoplast or the plasma membrane, or both, an electrical potential difference like the membrane potential of animal cells may exist, affecting the flow of ions because of their electrical charge. Like animal cells, the interior of most plant cells is electrically negative by 50–150 mV (millivolts) compared with the external environment. Although the complications created by these facts must be understood in order to deal properly with ion transport, they lie beyond the scope of this discussion and are usually covered in courses and books on physiology.

Transport into the Xylem and Shoot

By following the movement of radiolabeled ions fed to the roots of an intact plant, it has been found that nutrient ions travel upward to the shoot in the xylem, being carried along passively by the transpiration stream (Chapter 10). Supplying the mineral nutrient needs of the shoot thus becomes a question of how roots deliver into their xylem some of the ions they absorb.

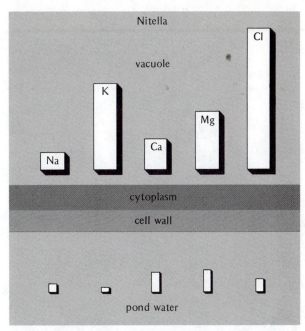

Figure 11-15. Diagram showing relative concentrations of several ions in the vacuolar sap of *Nitella* cells and in the pond water in which they are cultured. All ions shown reach a much higher concentration in the vacuoles of *Nitella* than in the external solution.

Since the tracheids and vessels of the xylem lack impermeable cellular membranes, ions transported into the xylem would tend to be lost by diffusion back into the soil via the free space if not for the endodermis that surrounds the stele of a root (Figure 11-6B). Its Casparian thickenings block the free space pathway between the stele and the cortex (Figure 11-16).

The prevailing view of nutrient transport into the stele is that ions taken up by cells of the epidermis and cortex diffuse from cell to cell via the plasmodesmata interconnecting these cells and, by the same means, pass through the endodermal cells into living parenchyma cells (pericycle, etc.) within the stele. From these cells, they either leak out into the free space of the stele or are pumped into it, hence reaching by diffusion the vessels and tracheids of the xylem (Figure 11-17). The system of cell cytoplasms interconnected by plasmodesmata that this theory visualizes as the pathway of ion movement to the stele is called the **symplast.** There is still no rigorous proof of the symplast theory of ion uptake by roots.

MYCORRHIZAS

The roots of many plants contain fungi that grow on the surface or penetrate into the interior without killing the roots. This association, called a **mycorrhiza** (*myco,* fungus; *rhiza,* root), improves the plant's ability to take up nutrient ions, as explained in Chapter 25.

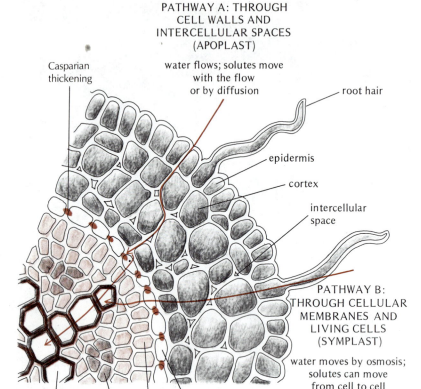

PATHWAY A: THROUGH
CELL WALLS AND
INTERCELLULAR SPACES
(APOPLAST)

Casparian
thickening

water flows; solutes move
with the flow
or by diffusion

root hair

epidermis

cortex

intercellular
space

PATHWAY B:
THROUGH CELLULAR
MEMBRANES AND
LIVING CELLS
(SYMPLAST)

water moves by osmosis;
solutes can move
from cell to cell
via plasmodesmata

xylem   phloem   endodermis

pericycle

Figure 11-16. Diagrammatic cross section of a root, showing alternative pathways for uptake of water and nutrients. For simplicity the cortex is shown as being only a few cells thick, although in typical roots it is much thicker. The thickness of cell walls of living cells is exaggerated in the diagram. Note how the Casparian thickenings in cell walls of the endodermis block off pathway A all the way around the vascular core of the root.

MINERAL NUTRIENTS IN THE SOIL

The mineral nutrients directly available to plant roots are the ions dissolved in the soil's water, or **soil solution.** Most soils contain additional nutrient ions, called **exchangeable nutrients,** bound to soil particles, principally clay and humus. These particles bear negative charges (Figure 11-18) which attract and hold positively charged ions (cations) such as K^+ and Mg^{2+}; this binding is termed **cation exchange.** Because cation exchange is reversible, exchangeably bound nutrients become released into the soil solution as roots absorb and deplete nutrients from the soil solution. Cation exchange thus provides a substantial reserve of nutrients for plant use over and above the nutrient ion content of the soil solution. Cation binding furthermore retards the leaching or washing away of mineral nutrients from the soil during rainfall.

Because a soil's cation-exchange sites are due mainly to its clay and humus, the cation-exchange capacity of soils varies widely with their composition, in much the same way as the field capacity does (Table 11-2). Loams, clays, and mucks, with a substantial to high clay and/or humus content, have a high cation-exchange capacity and tend to be fertile. Sands, silts, and sandy loams, espe-

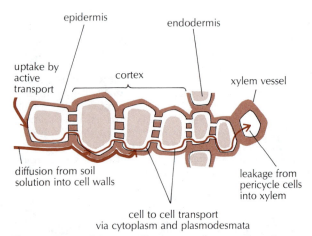

epidermis

endodermis

uptake by
active
transport

cortex

xylem vessel

diffusion from soil
solution into cell walls

leakage from
pericycle cells
into xylem

cell to cell transport
via cytoplasm and plasmodesmata

Figure 11-17. Diagram illustrating the symplast theory of ion uptake by roots. Ions diffusing into the cell walls of epidermal or cortical cells can be actively taken up into the cytoplasm through plasma membranes. These ions can then pass from the cytoplasm of each cell to that of the next, through plasmodesmata, without crossing any membranes until the ions are released into the xylem. Plasmodesmata are represented much larger, relative to cell size, and fewer in number, than in actuality.

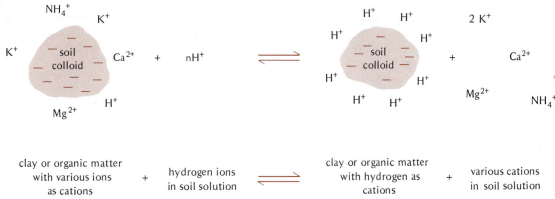

Figure 11-18. Cation exchange with soil particles. Arrows show that cation binding is reversible.

cially if low in organic matter, have a low cation-exchange capacity and therefore contain little nutrient reserve, are very subject to nutrient loss by leaching, and tend to be rather sterile. They are agriculturally useful only if regularly fertilized.

Soil Nitrogen

Because soil particles do not bear positive charges, they do not bind negative ions, such as nitrate, to a significant extent. Therefore nitrate, the principal available form of nitrogen in most soils, is leachable and tends to be depleted by heavy rains. This not only is costly to farmers, but is an important cause of nutrient pollution from agricultural fertilization, making for excessive growth of algae in streams and lakes (eutrophication).

By the microbial decay of organic residues and by ammonia fertilization (see below), nitrogen enters the soil as ammonium ions (NH_4^+). Not only is NH_4^+ directly useful to plants as a nitrogen source as noted above but, being positively charged, NH_4^+ binds quite well to the soil's cation-exchange sites and hence is resistant to leaching. This ordinarily has little practical benefit because of soil bacteria that rapidly oxidize NH_4^+ to nitrate, a process called **nitrification,** which is why nitrate is the principal form of available nitrogen in soils. This and other microbial transformations of soil nitrogen important to soil fertility are considered in Chapter 22.

Besides available nitrogen (NO_3^- plus NH_4^+), a good soil contains an insoluble reserve of nitrogen, sometimes as high as 0.5% of the soil dry weight, in its organic matter or humus. Soil microorganisms slowly break down humus and release NH_4^+ from it, gradually making the soil's insoluble nitrogen available to plant roots. This is called **mineralization** of nitrogen. Because of the importance of humus in both the nitrogen economy and the cation-binding capacity of soil, hu-

mus-rich soils such as black loams are among the most fertile natural soils.

FERTILIZERS AND FERTILIZATION

Crop removal, leaching by rainfall or irrigation, and soil erosion deplete the soil of its supply of essential nutrients and, if not compensated for, lead to infertility. Experience has proved that some kind of fertilization, the intentional addition of sources of mineral nutrients, is needed for sustained agricultural production on most soils. In one experiment in which corn was grown continuously without fertilization for 70 years in an Illinois plot, the yield dropped gradually from about 100 bushels an acre to only 23. One application of fertilizer increased the yield nearly sixfold, to better than the original yield. An increased supply of fertilizers is the single most important requirement for increasing food production in many underdeveloped areas of the world to meet the needs of a growing population.

There are two general kinds of fertilizer materials, the inorganic or "chemical," and the organic. Organic fertilizers are biological residues such as manure, fish scraps, cottonseed meal, and blood meal. The value of manuring was known and exploited in ancient times, and the placing of fish in hills of corn by American Indians is proverbial. As noted in Chapter 1, the ancients mistakenly thought of manure as food for plants in the same sense as our food, that is, as providing body substance, which perhaps accounts for the popular term "plant food" for fertilizers, still used today.

The teachings of Justus Liebig, a German scientist internationally prominent in the mid-1800s and the "father of agricultural chemistry," and the work of J. B. Lawes in England, who in 1843 established the Rothamsted Experiment Station to study the use of fertilizers, brought about general recognition that the value of organic fertilizers lies in the mineral elements they contain. These elements become released (mineralized) by microbial decay of the fertilizer material after it has entered the soil and, thereby, become available for absorption by plant roots. These findings led to the development and introduction of inorganic chemical fertilizers, which were less expensive to obtain and distribute and, pound for pound, much more potent in their beneficial effects on crop growth than the traditional organic fertilizer materials.

Most chemical fertilizers are soluble salts prepared from natural inorganic materials, for example, potassium chloride and sodium nitrate from brines and salt deposits or superphosphate from phosphate rock. An agriculturally very important nitrogen fertilizer, liquid ammonia (NH_3), is produced by reacting nitrogen gas from the air with hydrogen at a high temperature and pressure in the **Haber process.** When absorbed by soil (Figure 11-19), ammonia becomes ammonium ion (NH_4^+) as noted above.

Chemical fertilizers directly supply absorbable nutrient ions and hence are faster acting than organic fertilizers, which soil microorganisms must break down before plants can utilize them. Chemical fertilizers have various other advantages, including the ease with which balanced combinations or **mixed fertilizers** containing all the important nutrients can be prepared, in contrast with organic fertilizers, which are often rich in one nutrient element and deficient in another.

In terms of quantities needed (Table 11-3) the most important nutrients, and the ones that most often become deficient in the soil, are nitrogen (N), phosphorus (P), and potassium (K). Therefore, the potency of fertilizer materials is conventionally rated numerically as an N:P:K ratio, the numbers standing for the percentage of the material's weight that is N, P, and K, respectively. This NPK value is called the fertilizer "grade." (Traditionally, the P and K values have actually represented the percentages by weight of phosphorus oxide and potassium oxide that would correspond to the fertilizer's phosphorus and potassium contents, but this complication in the meaning of fertilizer grade is being eliminated.)

Organic Gardening

Followers of the philosophy of organic gardening advocate the exclusive use of organic fertilizers, believing that chemical fertilizers exhaust the soil and yield produce that is nutritionally inferior or even harmful to health. However, some of the fields at the Rothamsted Experiment Station in England have now received only chemical fertilizers for more than a century, yet are still fertile and produce high-quality crops. Since a plant takes up essential elements in the same form, inorganic ions, whether they are supplied as such or as part of biodegradable organic compounds, there is no basis for believing that the food value of plant produce should depend upon the form in which the essential elements are supplied. Fear of harmful effects of chemical fertilizers on health

Figure 11-19.   Applying anhydrous ammonia (NH_3) for corn in Illinois. The gas is forced into the soil under high pressure at a depth of 30–45 cm (12–18 in).

tends to confuse mineral nutrients, which even our own bodies need, with chemical pesticides commonly used in agriculture, about which concern regarding human health is certainly warranted.

The real advantage of organic fertilizers over chemical ones is that they help maintain the humus content of the soil, which is beneficial to its ability to hold water and nutrient ions, helps control erosion, and provides a reserve of nitrogen that slowly becomes available to plants by mineralization. Since cultivation encourages the breakdown of soil organic matter, agriculture carried on exclusively with chemical fertilizers gradually depletes the humus content and, with it, the quality of the soil (see Chapter 33). Besides the use of organic fertilizers, other practices that help maintain the humus content are now being increasingly recommended, including turning under the stubble and other nonharvestable remains of crops and growing and turning under a seasonal "cover crop" such as winter rye or barley, called **green manuring.** On the scale of the home garden, a properly tended compost pile produces a valuable soil conditioner for maintaining or improving the humus content even if the compost does not necessarily have the high fertilizer value that enthusiasts often ascribe to it.

Liming

A time-honored soil-conditioning practice is the addition of limestone (calcium carbonate; $CaCO_3$) or lime (calcium oxide; CaO). Although calcium is an essential element, most soils contain an abundance of this element; the benefit of liming is ordinarily due not to calcium itself but to neutralization of soil acidity by lime or limestone.

Under cultivation soils tend, especially in moist climates, to become acid, that is, to accumulate H^+ ions, which have several unfavorable effects. Hydrogen ions combine with the cation-exchange sites, allowing nutrient cations to be readily leached away. Acid pH also makes excessively available a soil's manganese and aluminum, which can become toxic to plants. Regular neutralization of soil acidity with lime prevents these changes and thus helps significantly to maintain soil fertility. Liming was known and practiced even in ancient Greece and Rome, although the reasons for its benefits were not then understood.

Intensive Fertilization

In recent decades industrialized countries such as Japan, the countries of western Europe, and the United States have planted crops more and more densely and fertilized them more and more heavily, especially with nitrogen, thus obtaining per-acre yields far above the traditional ones (Figure

11-20). Although this accomplishment is spectacular, it has negative consequences. Once the rate of fertilization has reached what may be called the luxury level, much additional fertilizer is required for each further gain in yield (Figure 11-21). The data in Figure 11-20, for example, show that the per-acre yield of corn in Illinois doubled over a period in which the rate of nitrogen fertilization was increased 20-fold. This means that only a minor part of the nitrogen now being applied to the soil is going into crops. The rest is disappearing primarily through (1) leaching, causing an accumulation of nitrate in underground waters, making them hazardous for drinking; and (2) rainfall runoff, leading to the eutrophication of streams and lakes, which spoils them for recreation and as domestic water sources.

Intensive fertilization is also entangled with the energy problem. The current high rates of nitrogen fertilization are economically feasible only because of the availability of liquid ammonia. But the Haber process of ammonia production is extremely energy-expensive. Estimates show that fertilizer production creates a very substantial share of the energy consumption associated with agriculture, which as practiced today in developed countries does not repay, in the energy value of the foods that reach our tables, the energy cost of producing, distributing, and cooking them.

Alternative Methods of Fertilization

Essential elements are sometimes supplied directly to the leaves, usually as a foliage spray. That this can be effective seems surprising in view of the water-impermeable nature of the cuticle and surface waxes of leaves (Chapter 9), but the cuticle evidently contains channels that allow nutrients to be absorbed. In certain pineapple-growing regions of Hawaii, where iron is unavailable due to soil conditions, a weak solution of iron

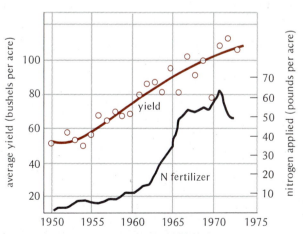

Figure 11-20. Average corn yield and nitrogen fertilizer use over 23 years in Illinois. The increase in yield during this period can be attributed largely to increased fertilization and the denser planting thereby made possible, because high-yielding hybrid corn was already being planted on virtually all Illinois farmland by 1950.

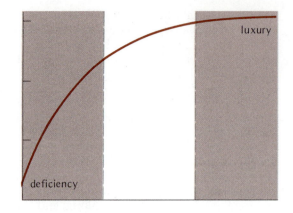

Figure 11-21.   A schematic fertilization/yield curve, indicating decreasing effects upon yield of continuing increases in nitrogen fertilization.

Figure 11-22.   Growth of tomato (*Lycopersicum esculentum*) in fertile soil (A), in nutrient solution (B), and in pure sand irrigated with a nutrient solution (C). All plants have shown excellent growth and set large numbers of fruits.

sulfate is sprayed at frequent intervals directly upon the pineapple leaves. The small amount absorbed is sufficient to maintain normal growth. Iron is often similarly supplied to azaleas and pin oaks, and zinc (as zinc sulfate) to citrus, tung, and pecan trees. Orange trees are sprayed with copper and manganese salts, and zinc–copper sprays are used on avocados.

Although foliar sprays are mostly used for special cases of micronutrient deficiency, nitrogen in the form of urea (H_2N–CO–NH_2; the principal nitrogen excretion product of mammals) is readily absorbed by the leaves of various plants. Since urea can be broken down by plants to NH_4^+, which serves as an excellent nitrogen source, this method of nitrogen fertilization is now being practiced occasionally, most notably in apple orchards. However, the application of nutrients to the soil will undoubtedly remain the principal method of fertilization.

Growth of plants with their roots in a mineral nutrient solution (hydroponic culture), or in sand or gravel watered with such a solution, has been developed for commercial use, particularly in greenhouses. Under favorable conditions high yields of premium-quality produce can be grown by this method (Figure 11-22). But because it is expensive and energy-intensive compared with conventional agriculture, hydroponic culture is unlikely to provide more than a minor supplement to conventional agriculture, except in special situations where conventional agriculture is impossible. As noted in Chapter 9, in certain desert areas, notably in the Middle East, intensive greenhouse-enclosed hydroponic culture systems have been introduced in which vegetable plants are grown with their roots directly in desert sand, watered with a balanced nutrient solution.

SUMMARY

(1) Because of frequent branching and development of numerous root hairs, root systems are usually much more extensive than shoots in both linear extent and surface area. The penetration of roots through soil is aided by the root cap. Disturbance of the root system, as in transplanting, breaks root hairs and smaller roots, seriously impairing the roots' absorptive ability until these are regenerated.

(2) A young root possesses an epidermis, an extensive cortex, and a central vascular cylinder or stele, separated from the cortex by the endodermis, which forms a boundary between vascu-

lar and nonvascular regions. The stele contains strands of xylem and phloem (sometimes with a central pith) surrounded by a cell layer called the pericycle, where branch roots originate.

(3) Roots absorb water by osmosis when transpiration pull reduces the pressures in the root xylem. Roots can take up a limited amount of water in the absence of transpiration, because of solute accumulation in the xylem due to mineral nutrient uptake. This causes root pressure and guttation.

(4) Soil consists of mineral particles of different sizes (sand, silt, and clay) and organic matter or humus. The finer textured the soil, the more capillary water it can hold against gravity, in micropores, available for plant use. Fine-textured soils also retain more hygroscopic water bound to particle surfaces, unavailable to plants. Fine-textured soils tend to have a much lower conductivity for infiltration of water during rain or irrigation, because of their relative lack of large pores (macropores). Loam soils, combining a wide size range of particles that, aided by humus, often aggregate into a crumb structure, contain both macropores and micropores in relative abundance, so infiltrate readily yet store relatively much water for plant use. Sandy soils, having little micropore space, store little water and readily become depleted.

(5) About 13 mineral elements, normally obtained from the soil, are generally required by plants. Many of these have one or more essential functions in metabolism. Deficiency symptoms reflect not only functions of essential elements but also whether the plant is able to redistribute the element from older tissues to growing points.

(6) Mineral nutrients are absorbed as ions and used either as such or, in the case of nitrogen and sulfur, after reduction by a specific reductase system (e.g., the nitrate reductase system for nitrate) in order to incorporate them into organic compounds.

(7) Root cells absorb ions by respiration-dependent active transport, using carriers specific for each of the important nutrient ions.

(8) The root delivers nutrients to the shoot by transporting them into the xylem, in which they are carried upward by the transpiration stream. The endodermis of roots prevents loss of nutrients from the xylem by diffusion through the free space back into the soil. Transport of nutrients from absorbing cells in the outer part of the root, to the xylem, is believed to occur by inward diffusion from cell to cell via plasmodesmata (symplast theory) followed by release into the xylem from living cells within the stele.

(9) A reserve of positively charged nutrient ions, available to roots, is held in the soil by cation exchange. Negative ions, such as nitrate (NO_3^-), are not held by exchange sites and are therefore leachable. Soil humus comprises a reserve of insoluble nitrogen that mineralization slowly makes available to plants.

(10) Chemical fertilizers, which include chemical salts and liquid ammonia, are more efficient sources of mineral nutrients than organic fertilizers, which must decay to be utilized by plants. Organic fertilizers, however, help maintain the humus content of soil. Liming counteracts the development of acid pH in soil. Fertilization is both a major socioeconomic necessity and a major source of ecological and energy problems in today's world.

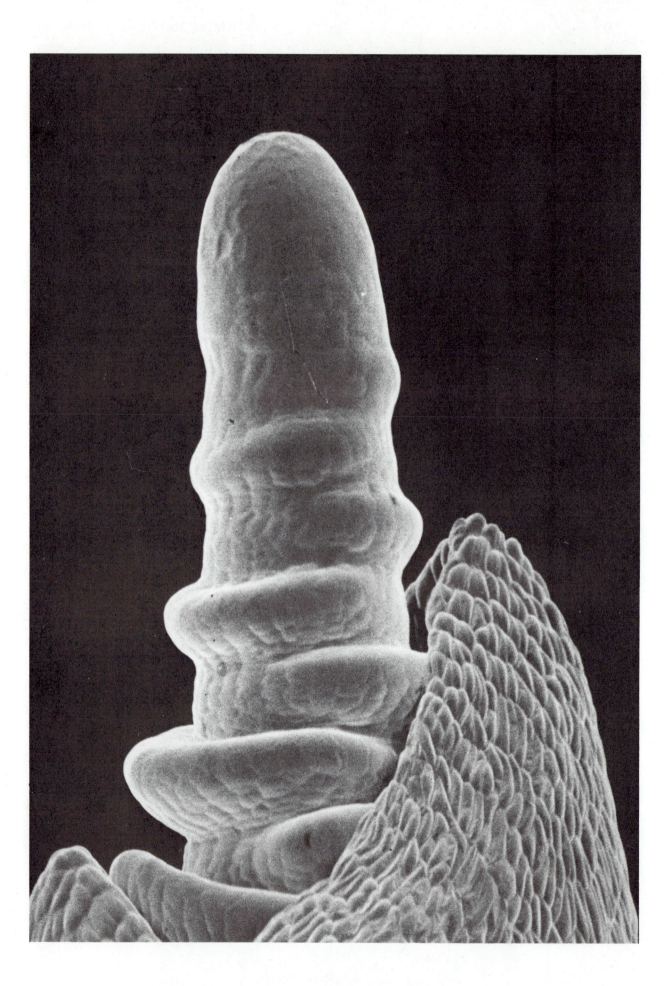

Chapter 12

PRIMARY GROWTH AND ITS HORMONAL CONTROL

Once past the stage of seed germination the body of a seed plant grows only in highly localized growth zones, in contrast to the growth of animals, which typically occurs throughout the body. Accompanying growth, or permanent increase in size, in these zones are **morphogenesis,** or development of the specific shape and form of cells and organs, and **differentiation,** or development of structural and functional specializations in cells and tissues. As noted in Chapter 3, we distinguish two types of growth in plants: **primary growth** or an increase in length, which usually takes place at the tips of roots and shoots; and **secondary growth** or growth in diameter or thickness without elongation, which occurs in the older parts of stems and roots. This chapter considers the nature of primary growth and the hormones by which plants are able to regulate it.

The activity of a growth zone depends upon cell multiplication in a usually even more restricted area of cell division, the meristem. Meristematic cells are usually not specialized for different functions as many cells of mature tissues are. Most meristematic cells are small, with thin walls, a prominent nucleus, and only small vacuoles (Figure 12-1A). As they grow in size they divide, keeping the average cell size more or less constant in the most active part of a meristem, despite the fact that individually all the cells in it are constantly enlarging by their growth.

On the periphery of a meristem, cell division slows down and finally stops. However, after ceasing to divide, the cells normally continue to grow, often more rapidly than within the meristem itself. As a result the cells in this zone become larger and larger, typically growing to 10–20 times the size of meristematic cells (Figure 12-1B). Thus the zone in which cells grow without dividing is often called the **cell enlargement zone** or, in elongating organs, the zone of cell elongation, somewhat misleading terms because the process of cell enlargement also occurs, as just noted, in meristems. But whereas in a meristem, enlargement is attended by the production of all cell components and maintenance of the meristematic type of cell structure, in a cell enlargement zone, increase in cell volume is due mainly to uptake of water into vacuoles, which enlarge and eventually fuse, giving rise to the large central vacuole typical of mature plant cells (Figure 12-1B). Since the final size of plant organs is due to growth in cell enlargement zones, plants achieve their size with much less biosynthesis of cytoplasmic components and attendant consumption of mineral nutrients such as nitrogen and phosphorus than would be needed if mature plant cells were filled with cytoplasm, as are animal cells. However, cytoplasmic changes do occur during the cell enlargement phase, for example, the development of proplastids into chloroplasts in cells of young leaves and the outer part of stems.

PRIMARY GROWTH OF THE SHOOT

The Apical Meristem

The shoot **apical meristem,** located at the very tip of the stem (Figures 12-A, 12-2, 12-3), makes possible the continued and potentially unlimited growth of the shoot. The apical meristem not only produces new cells that are added to the stem beneath it, but also produces new leaves (called **leaf primordia;** singular, primordium) and frequently new lateral buds. As new leaf primordia appear (Figure 12-4), new stem internodes are delimited.

Cell division is by no means confined to the apical meristem proper, which refers to the dome or cone of dividing cells at the very apex, on the

Figure 12-A. Scanning electron micrograph of the shoot apical meristem of Arawa wheat (200 ×). The tip of the apex consists of apical initial cells which divide, adding to the shoot; leaves are initiated on the side of the apex, as ridges. Cell division and expansion increases the size of the ridge, leading to the shape of a leaf.

197

(A)   (B)

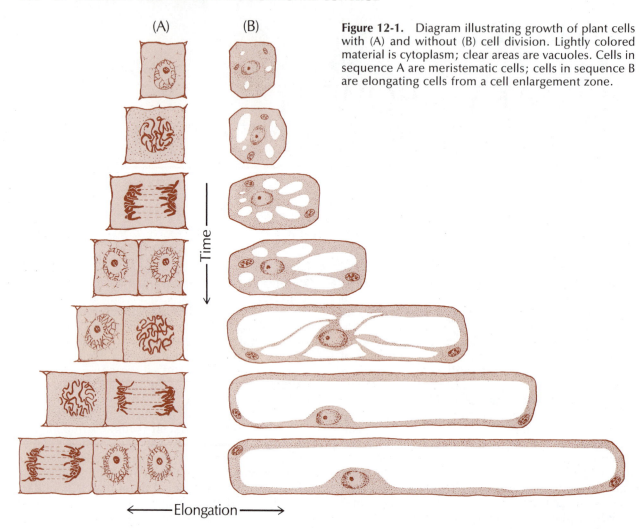

Time →

← Elongation →

Figure 12-1. Diagram illustrating growth of plant cells with (A) and without (B) cell division. Lightly colored material is cytoplasm; clear areas are vacuoles. Cells in sequence A are meristematic cells; cells in sequence B are elongating cells from a cell enlargement zone.

flanks of which leaf primordia arise. Cell division continues in each newly formed internode and leaf primordium. The apical meristem is primarily a region of initiation of new meristematic organs and is not where most of the new cells added to the shoot in primary growth arise, as sometimes thought. Most of the new cells actually come from cell division in the young meristematic leaf primordia and internodes.

In cellular structure the apical meristem shows some degree of organization (Figure 12-5). One or more regular cell layers (called the **tunica**) cover its surface (in flowering plants) because cells in these layers divide in the plane of the surface. Divisions in the interior of the meristem (called the **corpus**) occur in various directions. Cell division activity is not uniformly high throughout the apical meristem. Its central region, sometimes called the **quiescent center,** is much less active than the periphery where new leaves are initiated and the zone just below the apex where young stem tissue is developing.

The shoot apex of ferns and many other lower vascular plants is organized around a single much-enlarged **apical cell** at the top center of the apical meristematic region. The apical cell divides regularly in different di-

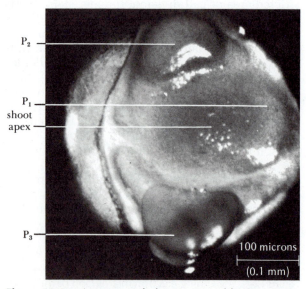

P_2

P_1
shoot
apex

P_3

100 microns

(0.1 mm)

Figure 12-2. An exposed shoot apex of lupine (*Lupinus albus*) viewed from above (180 ×). The apex itself is a dome-like mound in the center and is surrounded by the three most recent leaf primordia it has produced: P_1 (youngest), P_2, and P_3.

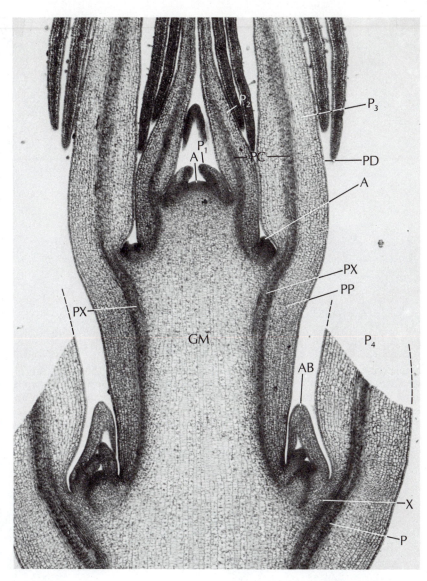

Figure 12-3. Longisection of lilac (*Syringa*) vegetative shoot tip. A, apical meristem; AB, young axillary bud; P$_1$, P$_2$, etc., successively older leaf primordia; PC, procambium (provascular tissue); PD, protoderm; PX, protoxylem at point where its differentiation is beginning; PP, protophloem; GM, ground meristem; P, primary phloem; and X, primary xylem, after considerable vascular differentiation has occurred. Dark structures at top are sections of leaf primordia that are folded in the bud such that the plane of section cuts across them.

rections, cutting off smaller daughter cells, which then multiply, giving rise to leaf primordia and new internodes in much the same way as do cells in the apical meristem of seed plants.

Growth of Stem Internodes

Figure 12-6 diagrams the changes occurring in an internode after its initiation from cells of the apical meristem (Figure 12-6A). In young internodes close to the apex (Figure 12-6B) active cell division accompanies elongation, building up longitudinal files or ribs of cells, because of which this subapical meristematic region is sometimes called a rib meristem. In contrast to the apical meristem, the subapical meristematic region is a strictly temporary meristem since the internode will eventually become part of the mature non-growing tissue of the plant.

Rather close to the apex, differences between cells start to develop in the relation between cell growth and cell division, such that strands of relatively elongated cells called **procambium** appear (Figure 12-6A, B), destined to differentiate later into vascular bundles. In lower (therefore older) regions cell division in all tissues slows behind the rate of growth, so the average cell size starts to increase (Figure 12-6B). Enlargement has already become conspicuous by the time cell division ceases completely and the so-called cell enlargement phase of growth begins (Figure 12-6C). Here differentiation of xylem and phloem cells is beginning in the procambium. At the end of the cell enlargement phase (Figure 12-6D) all the cells have become much longer. Because extensive tissue differentiation occurs in this region, it is called the **zone of maturation.**

In typical shoots the terminal region of primary growth represented in Figure 12-6A–D involves several internodes and extends as much as 10 or even 15 cm (6 in) behind the shoot apex. An exception to the pattern of strictly terminal primary growth is found in grasses

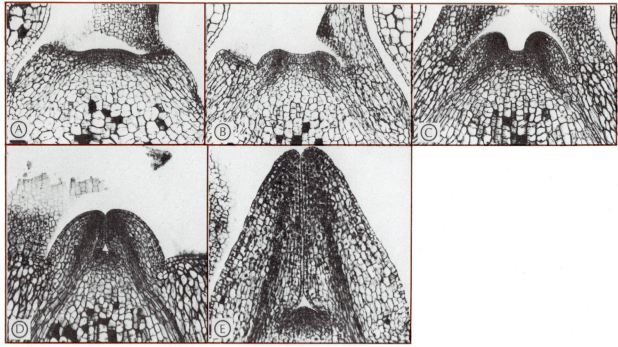

Figure 12-4.   The formation of a pair of leaf primordia by the shoot apical meristem of *Kalanchoë* (135 ×). By stage E, in which the leaf primordia are 0.4 mm long, the next pair of primordia are just beginning to differentiate from the shoot apex at right angles to the pair shown. The new primordia cannot be seen in E, however, because they are out of the plane of section for that photomicrograph. The darker colored, more elongated cells in the primordia at stages C, D, and E are provascular tissue.

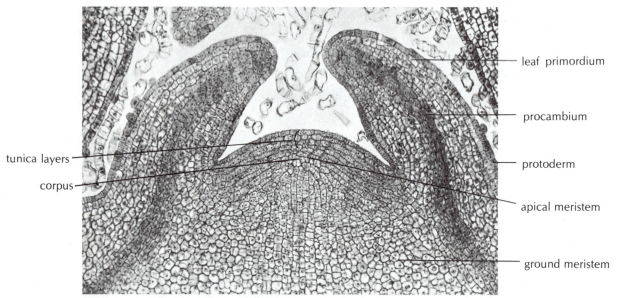

Figure 12-5.   Median longitudinal section of shoot apex of white ash (*Fraxinus americana*).

and certain other plants, whose internodes retain for some time a capacity to elongate in a basal zone of **intercalary growth** involving an intercalary meristem (Figure 12-7) derived from the subapical meristem by persistence of cell division activity in the intercalary growth zone.

Origin and Growth of the Leaf

The leaf originates with an accelerated division of cells in a localized area along the flank of the apical meristem. The first extra divisions usu-

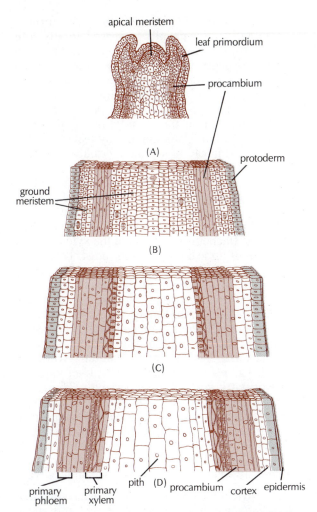

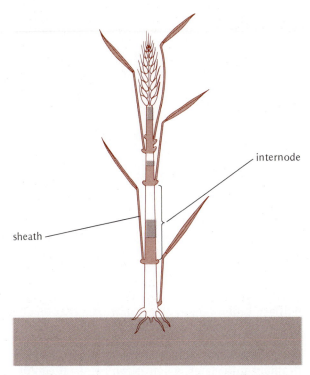

Figure 12-6. Developmental cellular changes in internodes from the shoot apex downward. Diagrammatic longitudinal sections. (A) Shoot apex, showing apical meristem, youngest leaf primordia and youngest internode. (B) Subapical meristematic zone (rib meristem region). (C) Cell enlargement zone, showing beginning of xylem and phloem differentiation at inner and outer edges of the procambium, respectively. Epidermal cells are synthesizing a heavy outer cell wall. (D) Maturation zone, showing further differentiation of xylem and phloem tissues, and the results of cell enlargement.

Figure 12-7. Diagram of intercalary growth zones in a rye plant. Intercalary meristems are shown in color, and zones of cell enlargement in grey. Ordinarily, in a grass such as rye, the base of each leaf wraps completely around the stem, "sheathing" it for a distance of at least one internode. Leaf sheaths are only partially represented here as extending upward from each node and terminating where leaf blades diverge from them.

ally take place beneath the outermost regular or tunica cell layer, but soon the cells of the surface layer also begin to divide more rapidly than other cells of the apical meristem. The stimulated tissue grows out to form first a bump, then a fingerlike or peglike projection, the **leaf primordium** (Figures 12-2–12-4). The lower part of the peg is the embryonic petiole; the upper part is the embryonic midrib or axis of the leaf. A zone of cells on either side of the midrib portion now begins to divide and grow laterally to form the blade, the expanded portion of the leaf.

Later enlargement involves growth throughout the leaf, with cell division continuing for some time but slowing down and eventually stopping before the leaf has grown to full size so, as in the stem, the cells of the mature leaf become much larger than the dividing cells of the leaf primordium. The last few cell divisions in the epidermis give rise to the guard cells and (if present) subsidiary cells of the stomatal apparatus, these cells differentiating structurally as the leaf is completing its enlargement. During the final stages of leaf enlargement cells of the epidermis grow more than the mesophyll cells within the leaf, allowing the mesophyll cells to separate from one another, forming the extensive intercellular spaces of the mature leaf.

The apical meristem initiates leaf primordia in a specific, regular pattern around its circumference. This pattern determines what the pattern of leaf arrangement (**phyllotaxis**) will be in the developed shoot, and it therefore differs among plants with different leaf arrangements. Phyllotactic patterns are quite precise and have mathematical properties that have interested developmental botanists and mathematicians for over a century. There are two principal classes of patterns: (1)

the opposite and whorled patterns, where two or more leaf primordia are initiated simultaneously on different sides of the apical meristem (Figures 12-3, 12-5); and (2) the spiral or "alternate" patterns, in which new primordia appear one by one in a regular but often complex spiral progression around the apical dome (Figure 12-2). These patterns are evidently related to how the plant controls the initiation of new leaf primordia (see below).

Origin of Axillary Buds

Within the subapical meristematic region, soon after the emergence of a leaf primordium, accelerated growth and division of cells in the angle between the leaf primordium and the adjacent internode begin to form an axillary bud (Figure 12-3). The extent of its further development during primary growth varies widely among plants; some axillary buds remain small and macroscopically invisible, whereas others become large and conspicuous. In either case, however, the axillary bud normally stops growth at some point and becomes dormant until later, when it can be reactivated to form a branch.

Differentiation of Primary Tissues

The tissues that differentiate from cells produced during primary growth are called **primary tissues;** collectively they comprise the primary body of a plant. The phloem is primary phloem; the xylem, primary xylem. The epidermis, the cortex and pith of stems, and the mesophyll of leaves are likewise primary tissues. The forerunners of these tissues can be distinguished just below the apical meristem as the primary meristematic tissues (Figure 12-6B): (1) the **protoderm,** which forms the epidermis; (2) the **procambium,** which gives rise to the primary vascular tissues; and (3) the **ground meristem,** which becomes the ground tissues, that is, the pith and cortex of the stem, and mesophyll of the leaf.

The procambium consists of elongated cells developing in strands, which in dicot and conifer stems are arranged in a cylindrical pattern, appearing as a circular array in cross section (see Figure 12-8A). Some of these strands are the precursors of leaf traces, extending out into developing leaf primordia (Figures 12-3, 12-5). The outer cells of the procambium gradually differentiate into phloem; the inner cells, into xylem. If the procambial strands remain distinct while this is happening, differentiation leads to distinct vascular bundles, separated by interfascicular (between-bundle) parenchyma in stems (Figure 12-8B, C) or by mesophyll cells in leaves.

As development proceeds the early-formed procambial strands usually increase laterally both through cell division within the procambium and by recruiting cells from adjacent ground meristem tissue. In many woody plants this expansion soon leads to a complete ring or cylinder of procambium, except where leaf traces depart (Figure 12-8B'). Differentiation then results in a cylinder of primary xylem and phloem, cut by leaf gaps (Figure 12-8C'), instead of separate vascular bundles.

As shown in Figure 12-8, vascular differentiation is progressive, that is, the cells do not all differentiate simultaneously. In seed plants the earliest sieve tubes (called **protophloem**) differentiate at the outer edge of the procambium, while the earliest tracheids or vessels of the xylem (called **protoxylem**) appear at the inner edge. Later conducting cells differentiate in sequence from these points toward the center of the procambium; the vascular cells formed after elongation ceases are called metaphloem and metaxylem elements.

In xylem differentiation cells deposit a heavy, lignified secondary wall. It is essential not to confuse the terms primary and secondary wall, defined in Chapter 4, with the deceptively similar terms primary and secondary growth. The growing cells in both primary (lengthening) and secondary (laterally enlarging) growth regions possess primary walls capable of enlargement, and in the zones of maturation associated with both primary and secondary growth, xylem cells and others deposit a heavy, inextensible secondary wall.

Protoxylem tracheids and vessels, which differentiate within the primary growing zones of stems, roots, and leaves, deposit ringlike (annular) or corkscrewlike (helical) secondary wall thickenings (Figure 10-11B, C). These patterns allow the xylem cells to stretch as the tissue grows, despite their tough secondary wall thickenings. The earliest-formed xylem elements often eventually collapse or become torn apart by subsequent elongation (Figure 12-6D). After elongation ceases, metaxylem cells differentiate with a continuous secondary wall bearing pits, like the walls of secondary xylem that may later be formed.

Supporting Tissues of Young Shoots

In the cortex of elongating stems and the petiole and midrib of enlarging leaves, there commonly develop strands of **collenchyma** cells with exaggerated primary wall thickenings, usually heaviest at the corners of the cell (Figure 12-9). Collenchyma strands are mechanically strong, as illustrated by the "strings" of celery, which are collenchyma. Their thick primary cell walls, unlike lignified secondary walls, retain a capacity for enlargement. Thus collenchyma cells can elongate with the growing stem or leaf they are helping to support.

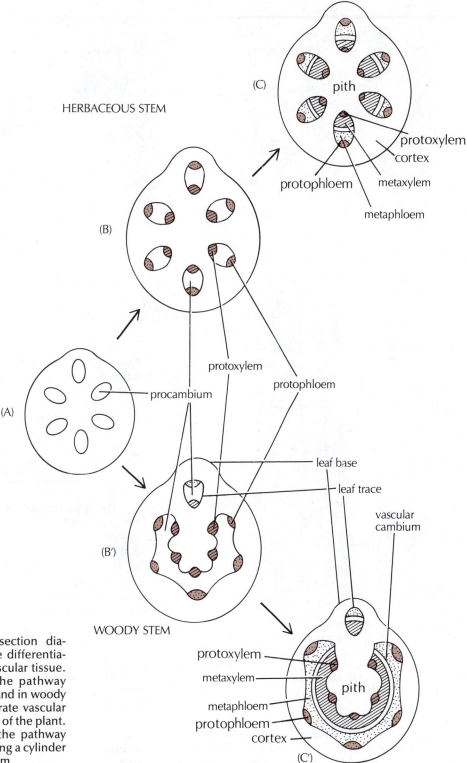

HERBACEOUS STEM

(C) pith
protoxylem
cortex
protophloem
metaxylem
metaphloem

(B)

protoxylem
procambium
protophloem

(A)

leaf base
leaf trace
vascular cambium

(B')

WOODY STEM

protoxylem
metaxylem
pith
metaphloem
protophloem
cortex

(C')

Figure 12-8. Stem cross section diagrams, showing progressive differentiation of the cambium and vascular tissue. The sequence A–B–C is the pathway found in herbaceous stems and in woody stems that begin with separate vascular bundles in the primary body of the plant. The sequence A–B'–C' is the pathway found in woody stems forming a cylinder of primary xylem and phloem.

After stem elongation has ceased, **fiber cells** with thick, lignified secondary walls often differentiate in the outer part of the primary phloem (Figure 10-4), where the first sieve tubes previously differentiated. Fibers may also develop outside the phloem, in the cortex.

Seasonality of Primary Growth

In a herbaceous shoot the developmental progression described above occurs continuously as long as the shoot grows vegetatively. In a woody shoot, as noted in Chapter 3, primary growth usu-

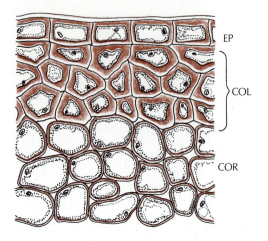

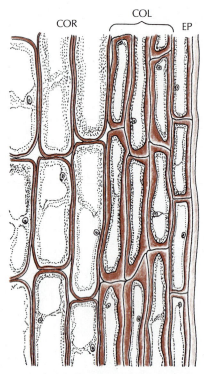

Figure 12-9. Collenchyma cells (COL) as seen in transverse (top) and longitudinal (bottom) sections of a young, elongating stem. Also shown are epidermal cells (EP) with thick outer walls, and thinner-walled cortical parenchyma cells (COR) to the inside of the collenchyma tissue layer.

ally occurs as a brief, rapid "flush" utilizing internode and leaf primordia previously formed in the resting terminal bud, often during the preceding season. In some woody plants all the organs that appear after bud burst are preformed within the bud, whereas in others, additional leaf primordia, stem internodes, and sometimes flower primordia initiated by the apical meristem during the growth flush may enlarge and differentiate at that time, as they do in herbaceous stem growth.

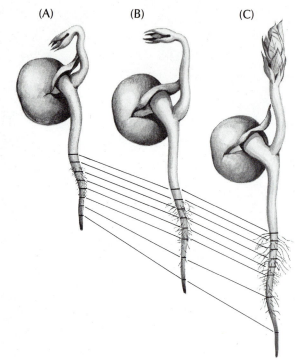

Figure 12-10. Region of elongation in a pea (*Pisum sativum*) root. (A) Ink marks, 1 mm apart, are placed on a growing root. (B, C) The distribution of the marks over the next several hours shows that most of the growth occurs near the root apex.

PRIMARY GROWTH OF THE ROOT

By applying marks such as spots of India ink along a root tip and periodically remeasuring the distance between them (Figure 12-10), one finds that primary growth in roots is considerably more localized than is typical for shoots. The growth zone ranges from 1 mm in timothy grass (*Phleum pratense*) roots to as much as 10 mm (about 0.4 in.) in corn roots.

The Meristematic Zone

The apical quarter or third of the elongating portion of the root tip, behind the root cap, consists of small meristematic cells of relatively uniform size (Figures 12-11, 12-12). The cells elongate and divide in the direction of the root axis, building up obvious cell files. In most root meristems the cells also grow to some extent in the transverse direction, so the diameter of the young root increases with the distance from its tip (Figure 12-12). The meristematic cell files converge toward the tip of the root to a focal point just behind the root cap (Figure 12-12). From this region, the apical meristem, originate the cells that multiply into extensive files farther back in the

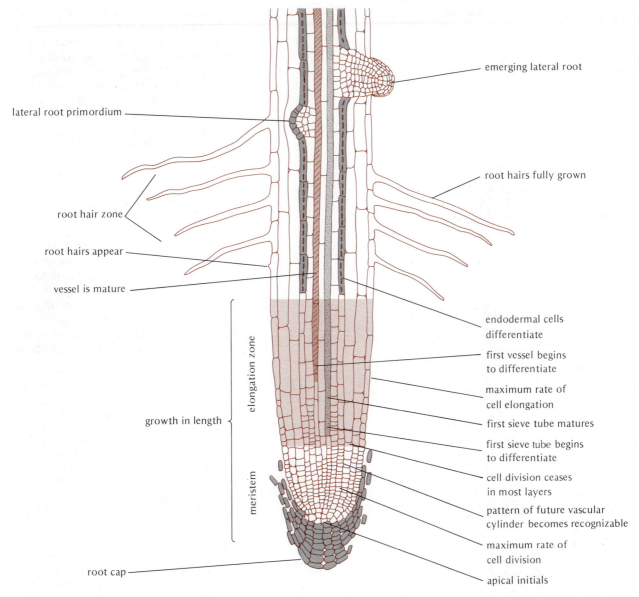

Figure 12-11.   Diagram of primary growth zone of a root as seen in longitudinal section. The number of cells is much greater in typical roots than in this diagram, which has been made as simple as possible. Differentiation of only the first-formed sieve tube and vessel is shown. Branch root initiation is shown at the top of the figure.

subapical meristematic zone. Cells that add to the root cap also originate in the apical meristem or in a narrow zone just ahead of it. Noticeably fewer mitotic divisions occur in the central part of the apical meristem (the quiescent center) than in the subapical meristematic cell files, where most of the new cells added to the root arise.

In the subapical meristematic region primary meristematic tissues (protoderm, ground meristem, and procambium) become distinguishable, which will later give rise, as in the stem, respectively to the epidermis, cortex, and central vascular cylinder.

The Elongation Zone

Farther back in the subapical meristematic region, the rate of cell division begins to decline whereas growth in length accelerates, running ahead of cell division so the cells begin to become larger. Behind this point cell division soon ceases completely, but the rate of growth rises to a maximum. Successive cells down any cell file therefore become progressively and markedly longer (Figure 12-11), reflecting the fact that each successive cell down the file has been located within the elongation zone longer and has therefore

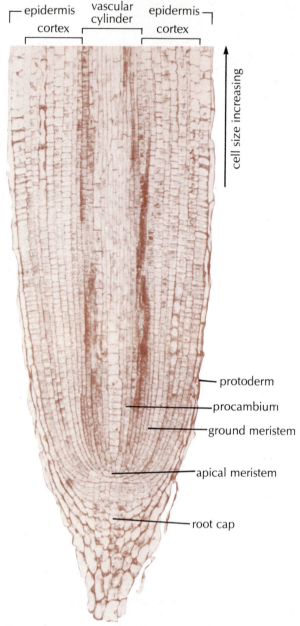

epidermis   vascular   epidermis
cortex   cylinder   cortex

cell size increasing

— protoderm

— procambium

— ground meristem

— apical meristem

— root cap

Figure 12-12. The root tip of a dayflower (*Commelina*) in longitudinal section.

undergone more enlargement. At the basal end of the elongation zone growth in length ceases, so cells beyond that point show no further increase in average length. This is the **root hair zone,** where root hairs arise by outgrowth from epidermal cells (Figure 12-11). This could not occur earlier or the young root hairs would be sheared off by movement of the growing part of the root past soil particles.

Differentiation

The first conducting cells of the xylem and phloem differentiate within the elongation zone.

The earliest sieve tubes may indeed differentiate even within the meristematic region. Differentiation of vascular tissue and other specialized cells continues progressively and reaches completion in the root hair or maturation zone. In contrast to the stem, not only the protophloem sieve tubes but also the protoxylem tracheids or vessels of the root differentiate at the outer edge of their respective regions of procambium, further differentiation taking place progressively inward toward the center of the root (Figure 12-13).

Initiation of Lateral Roots

Behind the root hair zone, small groups of cells within the pericycle become stimulated to renewed growth and cell division, forming new root apical meristems. Their activity leads to the outgrowth of lateral roots through the cortex (Figure 12-11), a mode of branching radically different from that of the shoot from externally located axillary buds formed in the subapical meristematic zone and therefore representing persistent apical meristematic tissue.

GROWTH REGULATORS AND HORMONES

Growth regulators are specific chemical substances that in relatively minute amounts specifically influence certain developmental processes, not merely as nutrients but by a specific regulatory action. A **hormone** is a biochemical regulator produced in one tissue and transported to a different tissue where it exerts a physiological effect, serving as a chemical "messenger" between the two tissues. Most of the known types of plant growth regulators qualify as hormones because, at least in certain plants or circumstances, they seem to be produced in organs different from those they affect.

Although each type of hormone or growth regulator exerts certain distinctive actions that we regard as characteristic, all the important kinds of growth regulators cause a multiplicity of effects on different plant organs and different plants. This multiplicity means that a relatively small number of hormones can regulate a large variety of developmental phenomena. The regulatory possibilities are further increased, as discussed below, by interaction between different hormones having either antagonistic or complementary effects upon a given tissue or process. Hormone effects provide a means of **growth correlation,** enabling different organs or growth centers to regulate one another's growth, ensuring the balanced development of the whole plant. Several examples of growth correlation systems are given below.

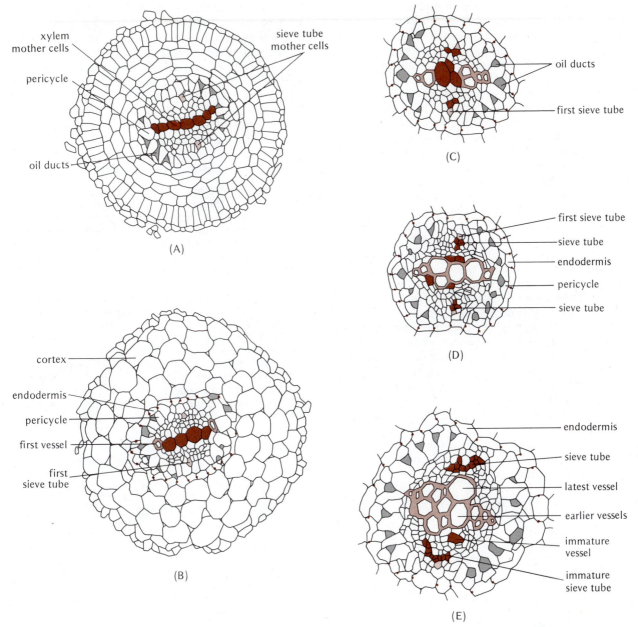

Figure 12-13. Progressive tissue differentiation as seen in transverse sections of the young root of the carrot plant. Sections A–E were cut successively farther back from the tip of the root and hence represent successively later stages of differentiation. The cortex and epidermis, although present, are not shown in diagrams C–E. Mature xylem and phloem cells are indicated with heavy walls. Xylem and sieve tube "mother cells" are undifferentiated procambial cells whose position and size show that they will subsequently differentiate into conducting cells. Oil ducts are specialized structures formed in tissues of carrot and certain other plants. (A) Within meristem, 0.2 mm from tip of root. (B) In elongation zone, about 1.5 mm from tip of root. (C–E) Successively farther back into the root hair zone. In (E) primary differentiation is complete and secondary growth (which gives rise to the "carrot") is about to begin.

Auxin

The type of growth regulator called auxin was discovered through the study of plant growth responses to light and gravity, especially using the shoot sheath or coleoptile (Figure 3-35A) of grass seedlings, most often the oat (*Avena*) seedling. Study of these growth responses suggested that the tip of the coleoptile produces a substance that stimulates elongation in the coleoptile tissue below the tip. In the late 1920s a young Dutch botanist, Frits Went, devised a method for detecting and assaying this substance, later named auxin, by incorporating it into small blocks of agar and applying these to one side of detipped coleoptiles (Figure 12-14). The coleoptiles grow more on the side that receives auxin and hence develop a curvature, proportional to the amount of auxin that the agar block contains. This *Avena* curvature test is an example of a **bioassay,** a procedure by which a biologically active chemical substance can be measured by making use of a biological response to it. The Avena curvature assay enabled biochemists to isolate auxin and identify it as indoleacetic acid (IAA). **Auxin** is now used as a generic term meaning IAA or any related substance with an action similar to that of IAA.

Auxins promote elongation of stems and stemlike organs of higher plants generally (the coleoptile is actually a modified leaf, but behaves like a stem). Demonstrating this usually requires depriving elongating tissue of its own auxin, for example, by removing the coleoptile tip or, in the case of a leafy shoot, its young leaves, which are its principal sites of auxin production. Since auxin is produced at a specific location (coleoptile tip or young leaves) and transported to a different site (elongating part of coleoptile or elongating stem internodes, respectively) where it exerts its physiological action, auxin acts as a hormone. It enables a plant to regulate the growth of young internodes in proper relation to the growth of young leaves at the shoot tip.

Plants translocate auxin from sites of production to sites of regulatory action by means of a special active transport mechanism specific for auxin, called **polar transport** because movement occurs in a polarized fashion, from the distal or apical end of a stem or coleoptile toward the basal end (Figure 12-15). Because of polar transport, a growth curvature develops along the entire length of a coleoptile when auxin is applied to one side of the decapitated tip (Figure 12-14). Polar transport is not restricted to vascular tissue, but also occurs in nonvascular parenchyma.

Indoleacetic acid (see Figure 12-16) is related structurally to the amino acid tryptophan (Figure 5-6), from which IAA is formed. Although only a few naturally occurring compounds other than IAA have auxin activity, the structural requirements for activity are relatively loose, so many synthetic ring compounds bearing a short acidic side chain will exert a biological action similar to that of IAA. A number of synthetic auxins have important applications in agriculture and gardening, including use as herbicides (weed killers). Plants cannot break down the synthetic auxins as easily as they can IAA, so when sprayed with a synthetic auxin such as 2,4-D (see Figure 12-16), they readily develop a hormonal overdose that deranges development and metabolism, leading to death. An especially valuable feature of synthetic auxin herbicides is their selectivity: broadleaf plants tend to be much more sensitive to

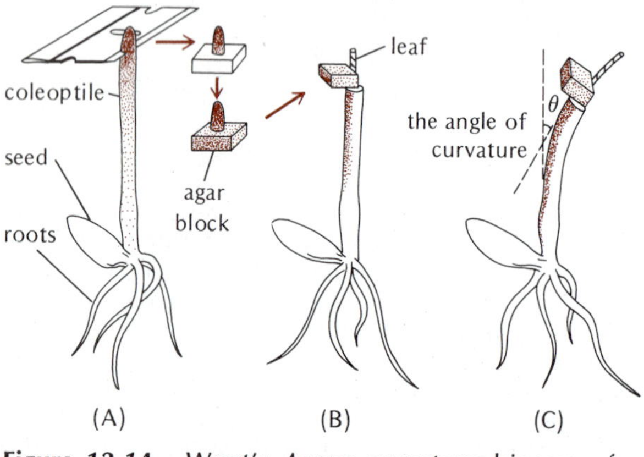

Figure 12-14. Went's *Avena* curvature bioassay for auxin. (A) A 3-day-old oat seedling is decapitated, and the leaf inside is pulled up. (B) An agar block containing auxin (here illustrated as obtained by placing the coleoptile tip in contact with the block for about an hour) is applied to one side of the leaf so that it contacts the cut surface of the coleoptile. (C) Auxin from the block becomes transported down the coleoptile and causes that side to grow faster than the opposite side, resulting in curvature. The angle of the curvature, θ, can be used to determine the amount of auxin in the agar block.

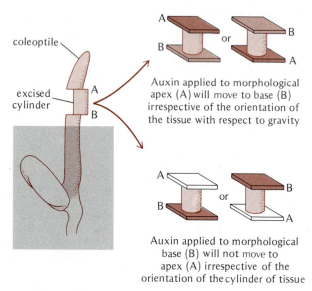

Figure 12-15. The polar transport of auxin as seen in a polar transport test. A known amount of auxin is applied to one end of the excised coleoptile section in an agar block, called the "donor" block (darker color). A second block of plain agar, the "receiver" block (lighter color), is applied to the other end of the section. The amount of auxin moving into the receiver block during the period of the transport test can be determined by, for example, the *Avena* curvature bioassay for auxin (see Figure 12-14).

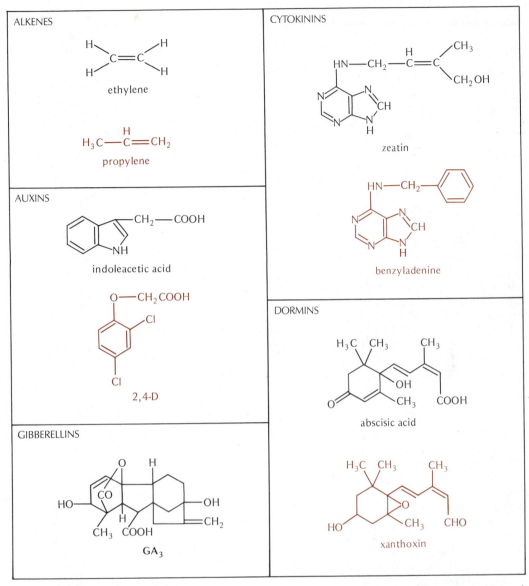

Figure 12-16. Chemical structures of plant growth regulators and hormones. Natural hormones are shown in black; synthetic analogs having similar activity are shown in color (xanthoxin is an artificial breakdown product of carotenoids, rather than a synthetic analog).

them than grasses are, so these herbicides are used extensively to kill broadleaf weeds in grain crops such as wheat, rice, and corn, as well as in lawns.

AUXIN IN CELL DIVISION

The elongation of coleoptile and stem internode tissue that auxin promotes is typically cell enlargement without cell division. Auxins also stimulate growth, however, in a number of situations in which cell division occurs, that is, meristematic growth. One very important such situation in research and in applied botany is **tissue culture.** Cells from many kinds of plant tissue, if placed on a suitable medium in a culture tube or flask, can be induced to grow and multiply into a disorganized mass of dividing cells called a **callus** (Figure 12-17). Besides nutrients and often vitamins, in most cases an auxin must be supplied in order for growth and cell division to take place. This suggests that auxin also probably plays a role in normal meristematic growth at the shoot apex.

INHIBITORY EFFECTS OF AUXIN

In striking contrast to growth promotion in shoots, even small amounts of auxin strongly in-

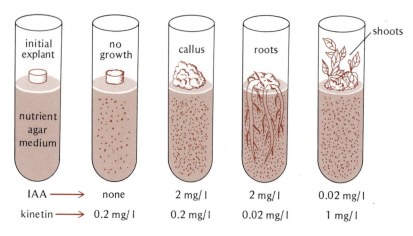

Figure 12-17. Growth responses of a plant tissue culture to auxin and kinetin. The initial explant is a small piece of sterile tissue derived from the pith of tobacco stem and placed on a nutrient agar medium, as shown at left. After several weeks, the kinds of growth illustrated occur on media supplemented with the indicated levels of growth factors.

hibit the primary growth of roots, whose cells are in this sense physiologically the opposite of stem cells. Auxins do not even stimulate all kinds of shoot cells. The growth of young leaves is not much affected by auxins, but this may be merely because leaves produce their own auxin. However, auxins *inhibit* the growth of axillary buds. This may be the means by which dormancy of axillary buds, noted above and in Chapter 3, is controlled, the phenomenon of **apical dominance** (Figure 12-18).

If a shoot's terminal bud is injured or removed, one or more lateral buds promptly grow out, replacing the lost primary growth region. This shows that an active terminal bud inhibits, or dominates, the axillary buds behind it. If a source of auxin (such as lanolin paste containing IAA) is applied to a shoot tip immediately after the terminal bud has been removed, this substitutes for the apex and inhibits the outgrowth of lateral buds below (Figure 12-18). Apparently the terminal bud dominates over the axillaries simply by producing auxin, which moves down to the axillary buds by polar transport. Some investigators think, however, that inhibition of axillary buds is not due directly to auxin, but to an inhibitory growth regulator whose production auxin stimulates in the vicinity of the axillary bud.

The various effects of auxin noted above illustrate the important principle of **target tissue specificity.** Different kinds of affected (target) tissues give different responses to the same hormone, depending upon their respective states of differentiation. Some of the auxin effects mentioned are also good examples of growth correlation mechanisms referred to at the beginning of this section.

Gibberellins

As noted in Chapter 1, **gibberellins** were discovered by Japanese scientists as products of a disease fungus, *Gibberella fujikuroi,* that causes rice seedlings to grow abnormally tall. Later, gib-

berellins were found in higher plants and shown to be involved in normal growth. Indicating this are genetic dwarf strains of various plants, such as corn and peas, which grow in a very stunted manner but, when supplied with gibberellin, will grow just like "normal" plants (Figure 12-19). These dwarfs evidently have a defect in the gibberellin system that controls normal growth. Normal growth cannot be restored by giving them auxin. Clearly, even though both gibberellin and auxin promote stem elongation, gibberellin is a different type of hormone from auxin. Apparently both hormones must be present for rapid shoot elongation to occur. Gibberellins, in contrast to auxins, do not inhibit root growth, again showing that gibberellins are physiologically distinct from auxins.

An important use plants make of gibberellins is to regulate rosette shoot growth. As noted in Chapter 3, many herbaceous plants grow in rosette form, with nonelongated internodes, during part of their development or annual cycle, then "bolt" by elongation of the shoot apex, usually just prior to flower formation. Treatment with gibberellin will cause the rosette to elongate and bolt prematurely (Figure 12-20). In its normal development a plant appears to switch from rosette to elongated growth by turning on gibberellin production.

Gibberellins comprise a family of more than 45 closely related compounds with the kind of ring structure shown in Figure 12-16. The structural requirements for gibberellin activity are considerably more stringent than in the case of auxins; highly active synthetic analogues comparable to the synthetic auxins have not been found.

Cytokinins

Cytokinin was discovered as a factor needed for cell division (cytokinesis) of cells in tobacco pith tissue cultures, which would enlarge but not divide if supplied only with auxin. Cytokinin proved subsequently to be required for cell divi-

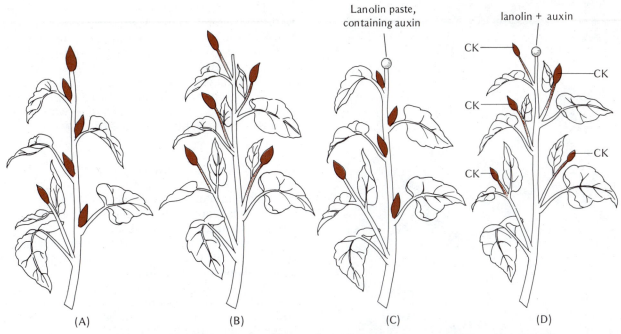

Lanolin paste, containing auxin

lanolin + auxin

CK

CK

CK

CK

CK

(A)   (B)   (C)   (D)

Figure 12-18. Apical dominance and the effect of auxin and cytokinin. In (A) all lateral buds except the lowest are inhibited by the shoot tip, since when the shoot tip is removed (B) the lateral buds begin to grow. The lowest lateral bud in (A) previously escaped from inhibition because it is so far from the tip. (C) If the shoot is removed but immediately replaced by auxin, lateral bud growth is prevented as if the tip were still present, although this auxin does not inhibit the lateral branch that was already growing. (D) If a cytokinin (CK) is added directly to buds inhibited as in (C), the cytokinin can overcome the dominance exerted by auxin, and bud outgrowth occurs.

Figure 12-19. Effect of gibberellin (GA) on normal and dwarf corn plants. From left to right: normal, normal + GA, dwarf-5, dwarf-5 + GA. Plants were treated during growth with gibberellic acid (gibberellin A_3, total treatment 280 μg per plant). Note that the dwarfs responded much more markedly to GA than did the normals, gibberellin causing the dwarfs to grow as rapidly as the normal plants. Dwarf-5 is a single-gene recessive mutant impaired in normal gibberellin metabolism.

Figure 12-20. Gibberellic acid (GA) causes bolting of cabbage. (Left) Untreated. (Right) Sprayed several times with 0.1% GA.

sion in various kinds of tissue cultures and is believed to be important in cell division generally. Cytokinin is often bioassayed by measuring the growth of soybean callus, which depends completely on cytokinin for its growth. In most cases, auxin is needed for cytokinins to exert their growth and cell division effects (Figure 12-17).

Cytokinins are a family of compounds, most of which contain the purine base adenine (familiar in DNA and RNA) bearing one or another substituent group attached to its amino group (see Figure 12-16). Besides occurring free in plants, cytokinins also occur as minor components of transfer RNA molecules (see Chapter 5), but it is not clear that this occurrence has anything to do with the hormonal activity of free cytokinins. Various synthetic compounds exert cytokinin activity, most of them artificially made adenine derivatives (example in Figure 12-16).

In contrast to auxin, cytokinins *promote* the outgrowth of axillary buds (Figure 12-18), in this situation opposing the action of auxin rather than complementing it as in the promotive action of both hormones on cell division. This antagonistic control of axillary bud growth may be significant when a shoot produces laterals while its terminal bud is still active.

The reverse control is suggested for the initiation of lateral roots. Auxin promotes this and cytokinin inhibits it. Cytokinin, coming from the root tip, would tend to prevent lateral root initiation, whereas auxin, coming down from the shoot by polar transport, would tend to cause lateral initiation. The balance between these opposing effects would determine the locations where new laterals appear behind the root tip.

Dormins: Abscisic Acid (ABA)

The name dormin was coined for a chemical factor from woody shoots that suppressed shoot elongation and induced the formation of resting buds resembling dormant winter buds. This factor proved to be identical to a compound that had been isolated with the aid of a leaf abscission (leaf fall) bioassay and was named abscisic acid (ABA), a somewhat unfortunate (but now established) name because ABA seems to be much more significant, generally, in dormancy than in leaf abscission. A few other compounds with similar action and structure are known and it seems reasonable to call the class of growth regulators exerting this type of action **dormins.**

The inhibitory action of ABA on growth is remarkable in being nontoxic, in contrast with most other growth-inhibiting substances, for example, inhibitors of energy metabolism or of RNA or protein synthesis, which act as poisons and eventually kill the affected tissue. The nontoxicity of ABA is an important reason for thinking that its growth-suppressing effect must be physiologically significant.

Dormins such as ABA, and gibberellins, are mutually antagonistic in bud growth, gibberellins tending to reverse the dormancy imposed by dormins, and dormins tending to suppress the stimulation of growth by gibberellins. The formation of resting buds, which are important in the seasonality of primary growth in woody plants, may be controlled by the balance between dormins and gibberellins in the shoot.

Ethylene

Early in this century the development of certain abnormalities in houseplants was traced to the gas **ethylene,** a minor component of the illuminating gas in widespread use at that time. Much later, sensitive methods for detecting ethylene revealed that plant tissues themselves generally produce small amounts of ethylene. Ethylene exerts effects at such low concentrations, as low as a tenth of a part per million (1 ppm = 1 ml of ethylene/1,000,000 ml, or 1000 liters, of air), that the small amounts of ethylene plant tissues produce can be physiologically significant.

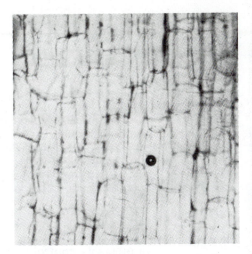 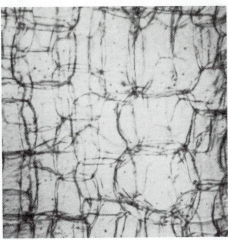

Figure 12-21. Ethylene effect on lateral expansion of cortical parenchyma cells in stems of pea seedlings (longitudinal sections). (Left) Seedlings untreated. (Right) Seedlings treated 24 hours in air containing only 0.5 μl ethylene per liter of air (0.5 parts per million).

Ethylene inhibits elongation in most growing tissues. The affected cells typically continue growth instead by lateral expansion, that is, in the direction transverse to the normal axis of elongation (Figure 12-21). This makes the growing portion of the stem or root become abnormally thick and stocky. In various tissues and organs ethylene promotes **senescence** or aging, and it promotes fruit ripening; these effects are of enormous practical significance in the storage and handling of agricultural and horticultural produce for market (see Chapter 17).

Plant cells produce ethylene, rather surprisingly, from the amino acid methionine [CH_3—S—CH_2—CH_2—$CH(NH_2)COOH$]. The responsible enzyme system first couples methionine to ATP, then converts the methionine skeleton into a ring compound that becomes enzymatically split, releasing as ethylene (CH_2 = CH_2) the —CH_2—CH_2— that had been next to the sulfur atom of methionine.

REGULATION OF COMPONENT PROCESSES IN PRIMARY GROWTH

Cell Enlargement

A plant cell is confined within its cell wall, a structure strong enough to resist the turgor pressures of 3–5 kg/cm^2 (45–75 psi) that normally exist inside a cell. However, for a cell to grow in size, its wall must enlarge. As previously noted, the capability of increasing in surface area is a characteristic distinguishing the primary cell wall of growing cells from the secondary wall that many differentiated cells possess.

How does the growing primary wall enlarge? The classical botanical theory held that the primary wall enlarges by intercalation of new wall material, a process called "intussusception." Although autoradiographic evidence shows that some wall components (pectins, hemicelluloses) do become introduced into the interior of a growing primary wall, the wall clearly does not grow simply by accretion of new material because during cell enlargement it may become substantially thinner than it was initially. Furthermore, plant cell growth requires turgor pressure and stops, for example, if tissue becomes wilted, even if carbohydrate for biosynthesis is available. Turgor force is needed to stretch the cell wall in order for it to enlarge during growth.

Growth in cell size therefore has a mechanical or biophysical aspect seemingly rather different from the simple accumulation of cell biomass often considered as growth. The property of the growing wall enabling it to undergo irreversible or plastic extension under turgor forces is called **plasticity** (Figure 12-22).

Turgor pressure, the driving force for plastic expansion of the cell wall, is due of course to osmosis resulting from the cell's solute concentration or osmolarity. Osmosis has a second role in cell enlargement, namely, to bring water into the cell and increase its volume as its cell wall stretches plastically. In a cell enlargement zone this water goes mainly into vacuoles, which increase in volume correspondingly. Water uptake by a meristematic cell, on the other hand, expands the volume of its cytoplasm, which is maintained in normal physiological condition by the concomitant biosynthesis of cytoplasmic components such as proteins and membranes. In meristems wall expansion and water uptake must be tightly controlled, therefore, relative to protein biosynthesis, whereas growth in cell enlargement zones is not under a comparable restraint. Biosynthesis and deposition of cell wall polymers, however, are important in both types of growth because if the wall becomes, by its expansion, too thin to withstand the cell's turgor pressure, the cell will burst, an accident that indeed sometimes happens to growing plant cells, notably root hairs.

An agent can influence plant cell growth by affecting either wall plasticity or turgor pressure. Mineral nutrients and photosynthesis can influence turgor pressure by providing solutes, raising the cell's osmolarity. Auxin, on the other hand, does not cause an increase in cell osmolarity, but induces wall plasticity. The development of wall plasticity requires metabolic effort on the part of

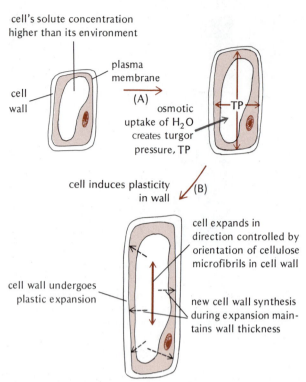

Figure 12-22. Physical principles of plant cell enlargement. During actual growth, processes (A) and (B) occur together, with (A) continuously driving the plastic expansion of the primary wall involved in (B).

the cell, being blocked, for example, by inhibitors of respiration. What exactly causes wall plasticity is still not clear, but it is popular today to think that a subtle breakdown of wall polymers must occur in response to growth hormones. The nature of the auxin effect is considered further in Chapter 14.

Cell Division

Growth in cell enlargement zones shows that cell division does not necessarily accompany cell growth and must be regulated separately. The fact that meristem cells normally divide every time they double in size indicates that meristematic growth involves a mechanism coupling cell division tightly to the occurrence of cell enlargement. Because of their effects on cell division, both auxins and cytokinins offer possible means of regulating cell division in meristems. However, it is not at all clear at present how the complex patterns of growth and cell division in meristematic regions described above are actually controlled.

Morphogenesis

A cell may acquire a specific shape through either division or enlargement in a specific direc-

tion or directions, or by a combination of these processes. Plant cells seem to control these processes using the same organelles that function in the mitotic spindle, the microtubules.

CONTROL OF THE DIRECTION OF DIVISION

Division in a given direction consists of positioning the cell plate, at the end of mitosis, in that direction so that the new cell walls form there. Development of the stomatal complex (guard cells and subsidiary cells) in the epidermis of a growing leaf (Figure 12-23) illustrates how specifically oriented cell divisions can create cells with particular shapes and arrangement. In this and other situations in which the positions of cell division are precisely controlled, prior to prophase of the division in question, a band of microtubules appears in the peripheral cytoplasm next to the plasma membrane (Figure 12-24), just where the future cell plate will attach to the old cell wall.

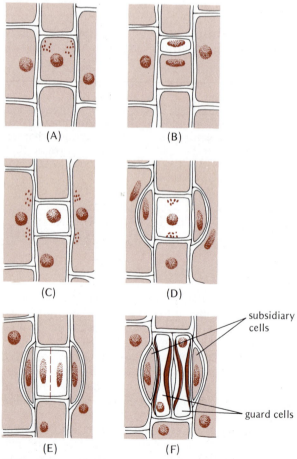

Figure 12-23. Development of the stomatal complex in wheat leaf epidermis, illustrating the role of preprophase bands of microtubules (colored dots) in determining positions of mitosis and cell plate formation that bring about the specific morphogenesis. Note that between each successive cell division all cells lengthen as a result of elongation growth.

This **preprophase band** (Figure 12-23) disappears prior to the onset of mitosis, so it apparently indicates the development of a specialized patch of plasma membrane to which the phragmoplast (in which the cell plate forms; see Chapter 4) can later be pulled into its correct position.

DIRECTIONALITY OF CELL ENLARGEMENT

The shape of growing organs, as well as of most of the cells composing them, comes about by a definite pattern of enlargement in three dimensions. If cells grew equally in all directions, they and the organs they make up would become spheres (Figure 12-25A). An elongated axis such as a stem or root, or an elongated cell such as a root hair, arises by the preferential growth of cells in one direction.

Cell elongation can occur in two quite different ways. One is to restrict cell wall expansion to the tip end (or both ends) of a cell, as illustrated in Figure 12-25C. This **tip growth** is the mode of morphogenesis of tubular cells such as root hairs and is involved in elongation of the tips of some fiber cells and tracheids.

The second type of cell elongation occurs in an elongating stem or root, whose cells grow throughout their length or surface area (**surface growth**). These cells elongate because of the oriented deposition of strong cellulose microfibrils, building a cell wall that can extend predominantly in but one direction (Figure 12-25B).

Close to the plasma membrane, microtubules run in the same direction as the microfibrils being deposited in the adjacent cell wall (see Figure 4-19), apparently controlling the direction of deposition. This is shown, for example, by treating an elongating cell with the drug colchicine, which disrupts microtubules. The treated cell continues to make and deposit cellulose, but in a disoriented manner as in Figure 12-25A. As a result it grows into a sphere. It is not understood how microtubules control the direction of microfibril deposition.

Organ Initiation

Not only in the formation of lateral roots but also in the initiation of adventitious roots and buds that occurs in various developmental circumstances (Chapter 3), new centers of cell division, growth, and morphogenesis arise from cells that had stopped dividing or growing and had differentiated as mature tissue. Shortly after Went discovered auxin, researchers found that auxins induce stem tissues to initiate adventitious roots. This discovery is much used today in the propagation of plants by cuttings (Chapter 15), even though it may seem incongruous that auxins, which inhibit root growth, stimulate root initiation. In callus tissue cultures, also, treatment with high

levels of auxin induces root formation, whereas cytokinins tend to induce bud formation instead (Figure 12-17), these hormones opposing one another's actions here as in axillary buds. These "morphogenetic" effects of hormones enable plants to initiate adventitious organs in response either to accidents or to special functional needs noted in Chapter 3. However, they are of little help in explaining morphogenesis itself since the hormone seems to act merely as a trigger, starting growth activities which then regulate themselves internally to create their respective morphogenetic patterns.

Leaf initiation at a shoot tip is an important example of internal regulation by a growth center. It is often said that a leaf primordium is initiated by accel-

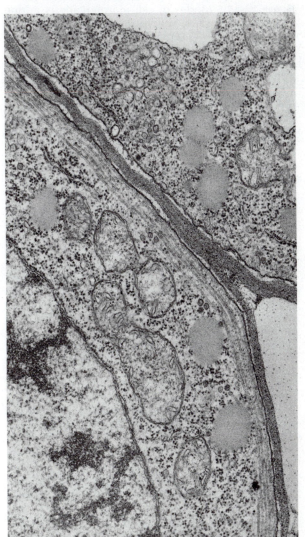

Figure 12-24. Preprophase band of microtubules next to the plasma membrane in a cell of a young wheat leaf. The nucleus, partly visible at lower left, was about to begin mitosis, with its mitotic spindle oriented in the direction perpendicular to the page. After mitosis the cell plate would attach to the old cell wall along the line of the preprophase microtubule band. The preprophase band is represented in cross section in the diagrams of Figure 12-23.

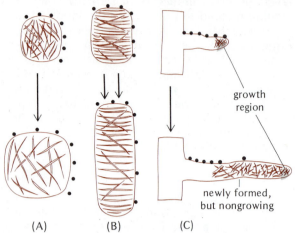

(A)   (B)   (C)

Figure 12-25. Modes of cell morphogenesis during growth. Cellulose microfibrils in the cell wall are represented by lines. Markers placed on the cell surface during growth (dots) indicate regions of cell wall expansion. (A) Isodiametric cell enlargement; randomly oriented microfibrils. (B) Cell elongation by directional surface growth; microfibrils with preferred orientation in one direction strengthen the cell wall preferentially in that direction. The growing wall thus expands primarily at right angles to the principal direction of the microfibrils. (C) Tip growth of a tubular cell (root hair); randomly oriented microfibrils. Elongation is due to wall expansion occurring only at the tip of the cell.

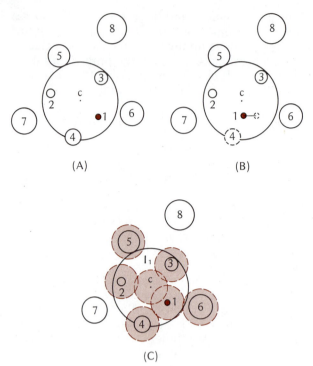

Figure 12-26. Pattern of leaf initiation around the apical meristem. (A) A normal apex with the primordia numbered in succession, beginning with the most recently formed (in color), grouped around the center of the apex c. (B) A comparable apex in which primordium 4 (P_4) was killed shortly after its initiation. P_1 has arisen closer to the position of P_4 than in the normal apex. (C) The interpretation of leaf positioning based on the existence of inhibitory fields (light color) around each primordium and the apical meristem. The next incipient primordium (I_1) will arise where the interacting inhibitory fields are weakest (i.e., where the largest gap in the color is).

erated cell division on the shoulder of the apical meristem, but the actual outgrowth of a primordium requires stimulated enlargement, and it is difficult to prove whether enlargement or division really comes first. As noted previously, leaf primordia arise in a regular pattern around the apical meristem, but this is not a fixed property of the system. If a young primordium is punctured, the next primordium will arise much nearer to the eliminated primordium's position than if this primordium had not been destroyed. This implies that young primordia inhibit growth in the area around them, perhaps by producing and releasing an inhibitory hormone (e.g., ABA) into surrounding tissues. These areas of negative control, called **fields**, around already-initiated primordia probably determine the positions at which new primordia will arise (Figure 12-26).

Differentiation

One might expect that tissue differentiation and formation of the complex yet precise anatomical patterns in shoots and roots would be induced and controlled by internal regulatory agents such as hormones. There is some evidence for hormonal control at least of vascular differentiation. For example, if a vascular bundle of the common houseplant, *Coleus*, is severed by notching one corner of the stem, the plant will differentiate a new vascular bundle from parenchyma cells in the pith, connecting the ends of the severed bun-

dle (Figure 12-27A). Physiological experiments indicate that this differentiation is due mainly to a stimulus coming down the severed bundle from the leaves above the cut. The effect of the leaves may be replaced by applying auxin above the cut (Figure 12-27B), so auxin evidently can induce parenchyma cells to differentiate into xylem and phloem elements. One can visualize that streams of auxin coming from young leaves near the shoot apical meristem could induce the development of procambial strands in the young leaves and the stem below them, leading to the vascular pattern of the stem.

In growth, morphogenesis, and differentiation, we are dealing with the processes by which the hereditary information or genes of a plant express themselves in specifying its form and structure. In Chapter 14 we consider how the genes and the environment interact to influence and determine a plant's growth and development. In that connection we shall also consider the molecular nature of plant hormone action.

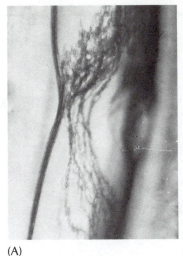
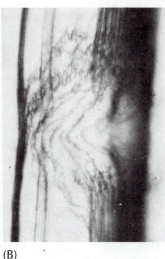

Figure 12-27. Regeneration of xylem elements around a wound in the stem of *Coleus blumei*. A notch-shaped wound was made to sever a major vascular bundle in one corner of the square stem of the plant. After seven days, the internode was stained to make the xylem visible. (A) Normal regeneration around a wound, all leaves left on the plant. Pith parenchyma cells have differentiated into xylem elements in a pattern circumventing the wound, reconnecting the ends of the severed bundle. If all leaves above the wound and the shoot apex of the plant are removed, no xylem regenerates, but (B) if auxin is added, xylem regenerates as in (A).

(A)   (B)

SUMMARY

(1) Growth and division in meristems produce new cells for addition to the plant body. Enlargement of these cells in so-called zones of cell enlargement or elongation transforms the typically small, cytoplasm-rich meristematic type of cell into the relatively large, highly vacuolate mature plant cell.

(2) The shoot apical meristem initiates leaf primordia and new internodes of the stem, which grow with extensive further cell division in the subapical meristematic region where the primary meristematic tissues—protoderm, procambium, and ground meristem—become distinguishable. They complete their growth in a zone of cell enlargement, accompanied by progressive tissue differentiation that becomes completed after growth ceases (zone of maturation). These processes may occur concomitantly in a herbaceous shoot but are often staggered seasonally in woody shoots.

(3) The root's apical meristem, just beneath the root cap, is the source of cells for the root cap and for a very active subapical meristematic zone behind the apical meristem, where most of the cell multiplication that adds new cells to the root occurs. Behind this, division stops and the cells undergo elongation, then stop growing. Differentiation of primary tissues is completed in the root hair zone (or maturation zone), where root hairs arise. Differentiation and lateral branch formation take place differently in roots and in stems.

(4) The hormones recognized as important in controlling primary growth are auxins and gibberellins, which are both necessary for normal elongation; cytokinins, which in addition to auxin are needed for cell division; dormins, including abscisic acid, which reversibly inhibit shoot growth and may be involved in resting bud formation and dormancy; and ethylene, which induces lateral expansion of elongating cells and could be involved in morphogenesis.

(5) Cell enlargement requires the force of turgor pressure, tending to stretch the cell wall, and cell wall plasticity, making the wall yield to this force and extend irreversibly. This allows water to be absorbed by osmosis, increasing the cell's volume. Auxin stimulates growth by causing the cell to increase the plasticity of its cell wall.

(6) Morphogenesis of individual cells can be due either to specifically oriented cell division(s), as in the formation of stomatal complexes, or to oriented and/or localized cell wall expansion, as in tip-growing tubular cells such as root hairs. Organ morphogenesis is due to localized and/or oriented cell enlargement. Directionality is given to cell enlargement by directionally reinforcing the cell wall with cellulose microfibrils, whose direction of deposition is controlled by cytoplasmic microtubules. In an elongating cell, microfibrils are deposited in the transverse direction, reinforcing the wall against lateral expansion.

(7) Balanced and compensatory growth of different organs is achieved by hormonal growth correlation mechanisms. Examples include stimulation of internode growth by auxin from growing leaves, inhibition of axillary bud growth by auxin from the terminal bud (apical dominance), induction of vascular differentiation and of adventitious root formation by auxin, and stimulation of axillary bud growth or adventitious bud formation by cytokinins.

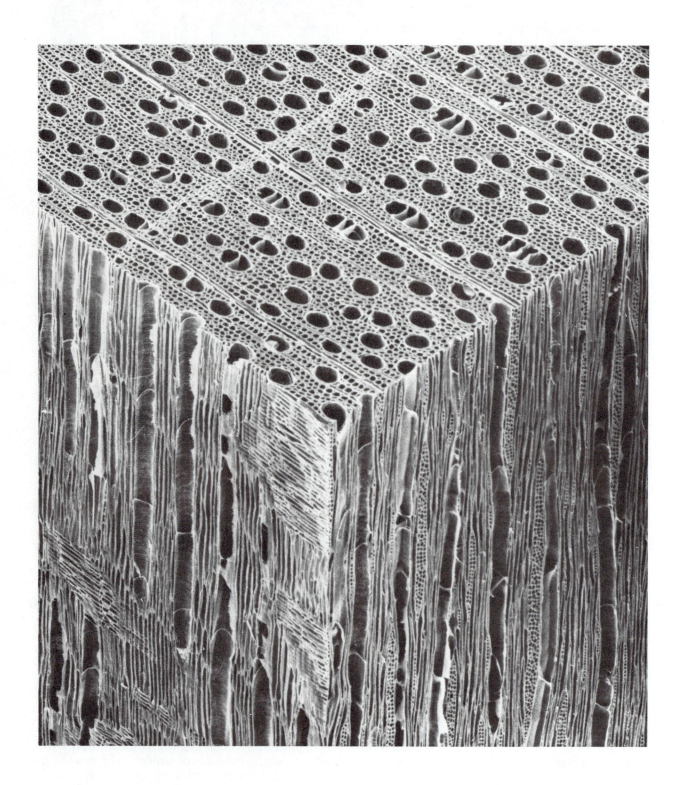

Chapter 13

SECONDARY GROWTH: FORMATION AND STRUCTURE OF WOOD AND BARK

Growth in thickness, or secondary growth, of stems and roots produces the bulk of the plant biomass in forested areas of the earth's surface and is the source of many important useful plant products such as lumber, paper, fuelwood, cork, tannins, pitch or oleoresin, rubber, and root vegetables. Secondary growth is due to two kinds of lateral meristems or cambia, which give rise to **secondary tissues** collectively comprising the secondary body of the plant. The **vascular cambium** produces secondary xylem and secondary phloem. The **cork cambium** produces a protective layer of cork on the external surface of woody stems and roots.

ORIGIN AND ACTIVITY OF THE VASCULAR CAMBIUM IN STEMS

In woody stems the vascular cambium usually becomes active before the primary vascular tissues have fully differentiated, whereas in herbaceous stems showing secondary growth the cambium usually begins its activity more slowly. It originates by the division of procambial cells located between the primary phloem and xylem of the young vascular bundles (Figure 13-1A). This activity soon extends across the parenchyma tissue between vascular bundles (Figures 13-1B, 13-2), creating a complete cylinder (a ring, as seen in cross section) of vascular cambium.

In woody stems that form a complete ring of primary phloem and xylem (Figure 12-8C'), the vascular cambium develops in the form of a continuous ring or cylinder from its inception, except where the vascular cylinder is interrupted by leaf gaps. A cambium arises in each leaf gap and, through its activity, eventually closes this gap in the cylinder.

Division of cambial cells adds secondary xylem to the stem on the inner side of the cambium and secondary phloem on the outer side (Figure 13-1B, C). As a result, secondary xylem is laid down outside the primary xylem, and secondary phloem is added to the inner side of the primary phloem. In a twig, by the end of the growing season the first growth, or annual, ring of xylem has been formed (Figure 13-1C). This consists of a small amount of primary xylem next to the pith, the remainder of the ring being secondary xylem. In woody plants the cambium forms secondary xylem and secondary phloem, year after year, during the life of the plant. The older phloem gradually becomes crushed and/or torn apart by the pressure of new tissues formed to the inside, but the secondary xylem accumulates (Figure 13-1D), eventually becoming the bulk of the plant body, that is, the wood of the trunk and branches of a tree, made up of a succession of growth rings.

In many herbaceous dicots secondary growth occurs only to a limited extent, so primary tissues constitute the bulk of the plant body. In stems of some herbs and vines the cambium between vascular bundles produces only parenchyma, so separate vascular bundles persist after considerable secondary growth. In others, cambial activity is limited to the vascular bundles (e.g., clover, *Trifolium*). In a few dicots (e.g., buttercups, *Ranunculus*) a cambium is absent, as it is in most monocots, whose tissues are therefore entirely primary in origin (the peculiar type of secondary growth that occurs in a few monocots is explained later). Since a herbaceous stem, in temperate climates, lives for only a few months between spring and autumn, its cambium functions for only one growing season. Even so, in some herbs such as sunflower the stem becomes hard and woody. In others the amount of secondary tissue formed is small and the stem remains relatively soft.

Figure 13-A. Scanning electron micrograph of a small block of red maple wood, seen in oblique view (105 ×).

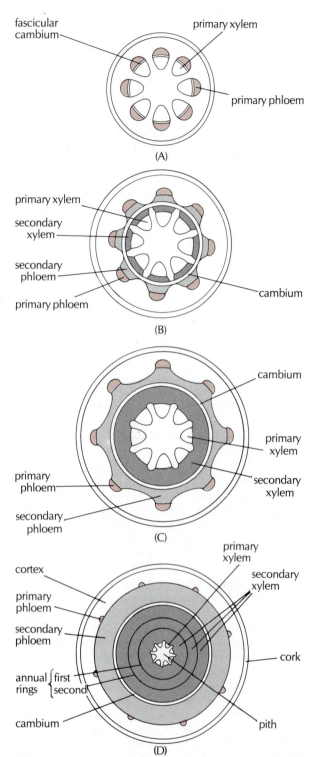

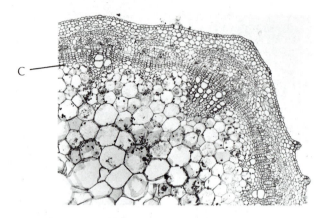

Figure 13-2. Cross section of a portion of soybean (*Glycine max*) stem, showing an early stage in vascular cambium development. The noticeably smaller, thin-walled cells running through the middle of each vascular bundle are cambium tissue. Cambial activity (C) is also evident in between the vascular bundles so that a ring of cambium now circles the stem.

Figure 13-1. Diagrammatic cross sections, showing differentiation of vascular cambium and start of secondary growth in a young woody stem. (A) Primary body differentiated cambium initiated within vascular bundles. (B) Cambial differentiation has extended between vascular bundles, secondary xylem and phloem have been produced within the bundles. (C) End of first year's growth, complete cylinders of secondary xylem and phloem have been produced. (D) Three-year-old stem, shown at reduced magnification, with annual rings of secondary xylem.

Seasonality and Control of Cambial Activity

In temperate-climate trees and shrubs, secondary growth occurs most actively in the spring. However, unlike the spring flush of primary growth discussed in Chapter 12, secondary growth usually continues for most of the "growing season," gradually slowing down but sometimes not stopping until frost. The boundary between one year's increment of secondary xylem and the next, which delimits the growth ring or annual ring, is distinguishable because of structural differences between the cells of the early-season and late-season wood (see below).

In the spring, cell division activity in the cambium begins in a tree's twigs and moves gradually downward as a wave of activity spreading into the larger branches and trunk, suggesting that something stimulatory to cambial activity comes from the buds as they begin to grow, and moves down the stem. In agreement with this, the removal of buds from a twig stops its cambial activity. The application of auxin (for example, in lanolin paste) to the tip of such a branch can restore cell division activity in its cambium. Therefore, it appears that the production of auxin by buds as they commence growth, and the polar transport of this auxin down the branch system, controls cambial activity. This is another example of a valuable growth correlation mechanism, enabling buds to control, relative to their growth, the amount of secondary growth that occurs for both conduction to and mechanical support of new shoots.

Trees of continuously warm climates such as the tropics often form indistinct growth rings which are related to the more or less annual flushes of primary growth that occur either spontaneously or in response to alternating wet and dry seasons. Such trees may produce more than one growth ring in a year, making the rings unreliable for their traditional use in determining the age of the tree. "False" annual rings can be generated in temperate-climate trees by a variety of stresses, such as defoliation by insects or a late spring frost.

The width of individual growth rings can vary greatly from year to year, depending upon the tree's environment, a favorable season yielding a wide growth ring and an unfavorable season a narrow one. In moist climates, temperature is an important determinant, but the amount of secondary growth that can occur is also profoundly affected by the shading of a tree (Figure 13-3) because this reduces its photosynthesis. Summer drought, which depresses photosynthesis by causing stomatal closure, can also reduce the year's increment of secondary growth. In semiarid and arid climates, precipitation tends to be the main determinant of ring width.

Beginning in the 1920s, A. E. Douglass found that the conifer trees over a wide area in Arizona and New Mexico show the same complex sequence of narrower and wider rings, which correlate impressively with annual variations in regional precipitation. Earlier parts of the ring sequence seen in present-day trees could be matched with the outer part of the ring sequence seen in previously cut trees or pieces of lumber, making it possible to determine when these trees were felled (Figure 13-4). By taking wood cores from timbers in older and older Indian buildings and ruins, and finding portions of their ring sequence that matched part of the sequence in younger wood, Douglass was able to build up a "tree ring chronology" extending back some 1900 years before the present. By this means he dated many of the Indian pueblos and cliff dwellings and inferred features of the climate that affected their occupancy. Severe droughts in 840, 1067, 1379, and 1632 were indicated by very narrow growth rings. A calamitous 23-year drought starting in 1276 may have caused the abandonment of most of the ancient villages during that period. The oldest still-inhabited southwest Indian pueblo is the Hopi village of Oraibi in northern Arizona, which has been occupied since at least 1150 A.D. according to the oldest timbers that have been found there.

(A)

(B)

Figure 13-3.   Stem cross sections showing growth rings, heartwood and sapwood. (A) White oak (*Quercus alba*). The dark area in the center is the heartwood; the lighter part, the sapwood. Growth rings were much broader when the tree was young, perhaps before over-topping by adjacent trees. (B) Ponderosa pine, showing the increased width of the growth rings following the removal of adjacent trees. A count of the narrow rings shows that the growth rate was retarded for about 54 years.

SECONDARY GROWTH OF ROOTS

In gymnosperm and dicot roots a vascular cambium arises first in the parenchyma located between the arms of the primary xylem and the inner edge of the primary phloem (Figure 13-5A). This cambium forms secondary xylem to the inside and secondary phloem to the outside, as in the stem. By resumption of cell division in the pericycle the cambium extends around the ends

of the xylem arms and begins to form secondary tissues there also. At this stage the outline of the cambium as seen in cross section is lobed, conforming to the outline of the primary xylem arms (Figure 13-5B). As growth continues the lobed outline disappears, because secondary xylem is produced more rapidly between the xylem arms.

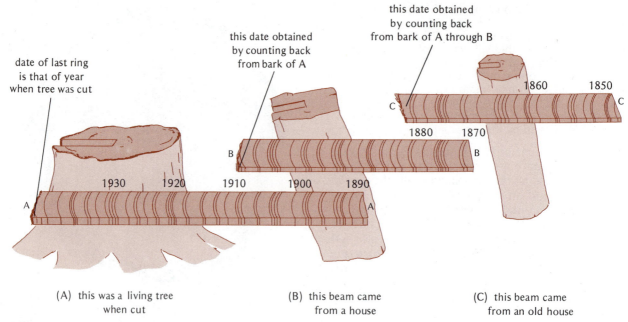

date of last ring
is that of year
when tree was cut

this date obtained
by counting back
from bark of A

this date obtained
by counting back
from bark of A through B

1860   1850

C   C

1880   1870

B   B

1930   1920   1910   1900   1890

A   A

(A) this was a living tree   (B) this beam came   (C) this beam came
when cut   from a house   from an old house

Figure 13-4. Tree-ring dating. An old beam or tree trunk can be dated by matching tree-ring patterns in wood cores from progressively older wood from the same growing area.

The cambium then exhibits a circular outline (Figure 13-5C). Further secondary growth of the root is essentially like that in stems.

In plants with edible storage roots, such as beet, carrot, turnip, parsnip, horseradish, sweet potato, and cassava, the greater part of the storage organ is composed of secondary tissues. However, the cambium does not produce thick-walled xylem tissue like that of the stem, but a few relatively thin-walled vessels, usually in small groups, scattered in a mass of thin-walled xylem parenchyma (Figure 13-6), which makes such roots edible: if thick-walled vessels and fibers were produced, the roots would be woody and not usable as food.

STRUCTURE AND FUNCTION OF THE VASCULAR CAMBIUM

Repeated division of cambial cells in the radial (inward to outward) direction in a stem or root builds up conspicuous radial cell files (Figure 13-7A), giving the secondary xylem and phloem derived from these cells a more or less regular structure of radial cell rows. During periods of little or no cambial activity, for example, autumn and winter, the cambial zone of meristematic cells between the xylem and phloem may be only one to a few cells wide (Figure 13-7A), but during rapid secondary growth it can increase by cell multiplication to as much as 20 cells in width. As these cells grow and divide, those toward either side of the meristematic zone will become added one by

one to the xylem or the phloem. For this reason, eventually there can remain within the cambial zone itself only the progeny of one particular cell layer now located at some point in the interior of the cambial zone. These ultimate progenitor cells, from which, over the long term, the rest of the cambial zone and, from it, the secondary xylem and phloem arise, are called the **cambial initials.** The initials structurally resemble the dividing cells on either side of them, which are called, respectively, **xylem mother cells** and **phloem mother cells** (Figure 13-8) because their progeny eventually all become xylem or phloem cells. The mother cell zones are analogous to the subapical meristem or rib meristem zones in shoot and root tips (Chapter 12), all cells of which eventually become part of the primary body.

In many textbooks the cambium is held to refer strictly to the single layer of true cambial initials, and the complete meristem of initials plus mother cells is termed the cambial zone. Since the initials often are not distinguishable in structure from the mother cells to either side of them, and may not be physiologically different either, some authorities consider that the entire meristem may be called the cambium.

Cambial cells do not conform to the usual concept of meristematic cells as small, nonvacuolate cells. Cambial cells typically have large vacuoles. Although not apparent in cross-sectional view (Figures 13-7A, 13-8), cambial initials are of two kinds (Figures 13-7B, 13-9): (1) greatly elongated cells tapering at both ends, termed **fusiform initials** (from Latin *fusus*, spindle, and form); and (2) much smaller cells, the **vascular ray initials.**

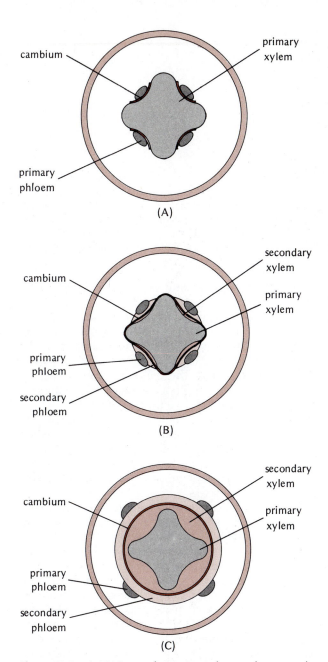

Figure 13-5. Initiation and progress of secondary growth in a dicot root. (A) Primary structure, with vascular cambium initiated only between "arms" of xylem. (B) Early stage in secondary growth: the completed cambium has formed a small amount of secondary phloem and xylem between the xylem arms, and has extended around their ends. (C) Later stage, showing circular cambium and a xylem cylinder.

Parenchyma cells derived from the ray initials become the vascular rays (Figure 13-10). The mother cells derived from fusiform initials give rise to the vertically elongated cells of the xylem and phloem, including the conducting cells (Figure 13-7B).

Cell division in the fusiform cambial cells is remarkable. The phragmoplast that arises around the telophase mitotic spindle grows gradually up and down toward the tips of the cell (Figure 13-7B), eventually splitting the already narrow fusiform cell into two even narrower daughter cells just as long as the parent cell.

Usually, more derivatives are formed on the xylem than on the phloem side of the cambium. In fast-growing conifers (Figure 13-7B), more than 10 xylem cells may be produced for each phloem cell. As a trunk or branch increase in circumference, the number of cambial initials around its circumference must increase. This happens by the occasional division of fusiform initials at right angles to their normal direction of division, that is, in the circumferential direction. New groups of ray initials are formed when one or more fusiform initials divide longitudinally into several relatively block-shaped cells. This initiates a new vascular ray, which will extend outward in the wood only from that point of initiation, in contrast to the oldest rays, which extend out from near the center of the stem.

Cell Enlargement and Differentiation

On either side of the cambial zone is a narrow zone of cell enlargement where cells derived from the cambium enlarge to their mature widths, which, in the case of ray cells and the large vessels in angiosperm wood, can be many times the width of the cambial initials (Figure 13-10). Then comes a narrow zone of differentiation where derivative cells become either phloem or xylem cells, depending upon their position, by undergoing structural changes. On the xylem side not only do the prospective tracheids, vessels, and wood fiber cells (if present) deposit a secondary wall, but also the ray parenchyma and wood parenchyma cells (which, unlike the conducting cells, remain alive after completing differentiation) deposit one, giving the wood a uniformly hard consistency.

Vertical rows of wood parenchyma in dicot wood (Figure 13-15), or of phloem parenchyma (Figure 13-7B), arise when a fusiform xylem or phloem mother cell divides longitudinally into several box-shaped cells prior to differentiation. This resembles the origin of vascular rays by subdivision of fusiform initials, described above.

STRUCTURE OF WOOD

The wood of most tree trunks and larger branches consists of two distinguishable zones, the **sapwood** and the **heartwood** (Figure 13-3). The heartwood is often darker in color. Microscopic examination reveals that in the heartwood the cells that remain alive after the differentiation of xylem (wood ray cells and wood parenchyma, if present) have died; the boundary between sapwood and heartwood is the point at which they are dying. As they die, tannins that they contain often become oxidized to dark-colored products,

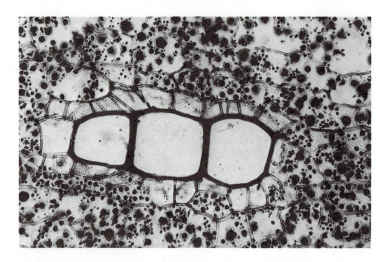

Figure 13-6. Cross section of a portion of fleshy root of cassava (*Manihot esculenta*). Note the group of vessels surrounded by xylem parenchyma cells containing many starch grains, here stained black with an iodine stain for starch.

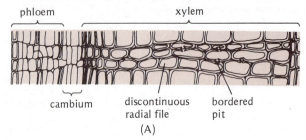

(A)

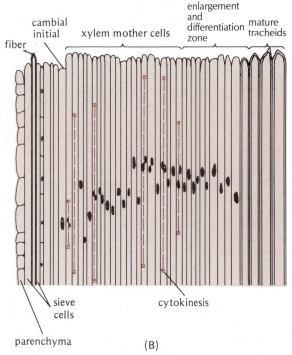

(B)

Figure 13-7. Vascular cambial region of western red cedar (*Thuja occidentalis*). (A) Transverse section in late summer, showing cell files, narrow cambial zone. (B) Longitudinal section in spring, showing wide cambial zone with divisions occurring in xylem mother cells derived from the cambial initial (phragmoplasts participating in cytokinesis are shown in color). Mother cells and derived tracheids, as well as some of the phloem cells, are longer than the cambial initial because they elongate at their tips after being formed by division of the cambial initial.

and other chemical and structural changes may occur.

If a dye is injected into the trunk of a tree to follow the movement of water in it, one finds that water is not transported in the heartwood. The boundary between heartwood and sapwood works its way outward in the trunk or branch year by year, while new sapwood is added to the outside of the xylem by the vascular cambium.

In some woods such as oak the vessels become blocked, during the conversion of sapwood into heartwood, by obstructions called tyloses (Figure 13-11). These form by ingrowth into the vessel cavity of adjacent ray or wood parenchyma cells before they die.

The wood of conifers, called **softwood,** has a much simpler structure than that of dicot trees, called **hardwood.** Dicot wood tends to be harder because it contains more thick-walled cells, plus cell arrangements that make it more difficult to split. Because of its thicker cell walls, hardwood is generally heavier than softwood. Softwood is more easily worked in sawing, planing, drilling, nailing, and so on, than hardwood and is therefore the wood predominantly used in building construction. Hardwood is preferred, on the other hand, for furniture manufacture not only because its hardness gives it a more durable surface, but because its more crooked grain and complex cell structure confer upon finished surfaces a much more handsome appearance than can be obtained from most softwoods.

Some hardwoods, especially oak, have become important in recent years for making pallets or portable platforms for the prestacked or "palletized" handling of goods, the most common way that goods are now moved in industrialized countries. Because of their toughness, oak pallets hold up much better than softwood pallets under heavy use. Not all "hardwoods," however, are so hard. Poplar and basswood are softer than the "softwoods" hemlock and yellow pine. Balsa wood, produced by a tropical broad-leaved tree (*Ochroma*), is familiar as the softest and lightest of available kinds of wood. Although very flimsy by comparison with other woods, balsa is remarkably strong relative

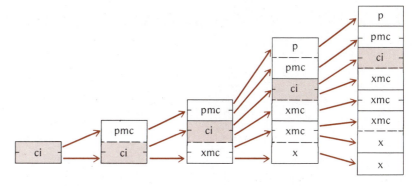

Figure 13-8. Diagram showing division of cambial initials, forming xylem and phloem mother cells. These in turn divide, their derivatives enlarging and differentiating into cells of secondary xylem and phloem. ci, cambial initial; pmc, phloem mother cell; xmc, xylem mother cell; x, xylem; p, phloem.

to its extremely light weight and, for good reason, is preferred for weight-critical construction such as model airplane building.

Because of its radial cell files and rays, and its circumferentially running growth rings, wood looks quite different both macroscopically and microscopically when viewed in the different planes of section shown in Figure 13-12A. In transverse section (plane 1, Figure 13-12A) its conducting elements are visible in cross section. In the two longitudinal planes of section they can be seen from the side, but in radial section (plane 2) one observes the rays from the side, running horizontally, whereas a tangential section (plane 3) cuts across the rays and allows their structure to be seen end-on.

Softwoods

Conifer wood is composed mainly of tracheids in regular radial rows (Figure 13-13), each row arising from one cambial initial. A typical softwood tracheid is about 4 mm long. The tapering ends of each tracheid overlap the tracheids above and below it by about one-quarter of their length. This affords an extensive area of contact across which water can flow from one tracheid to the next through bordered pits (Figure 10-8) in the walls between them. In addition to their water-conducting function, tracheids serve as the supporting or skeletal system of coniferous trees, holding up huge pines, firs, and redwoods and resisting the large stresses that wind and snow add to that of their weight. The rays in conifer wood are relatively simple sheets of radially elongated cells (Figure 13-13).

Growth rings in a softwood are due to variations in tracheid structure. Tracheids formed during the rapid secondary growth in the spring are wider and thinner-walled than those formed in summer as secondary growth is slowing down (Figure 13-7A). The summer- or latewood is thus denser and darker. The boundary between one annual growth increment and the next results from an abrupt change to the formation of springwood

tracheids as growth resumes at the beginning of a new season.

The wood, and also the bark, of many conifers contains **resin canals** or resin ducts, which are vertically or horizontally running channels (Figure 13-14) lined by living parenchyma cells that secrete into the duct cavity a resinous material called pitch or **oleoresin**. If the tree is injured this pitch exudes, presumably helping to discourage invasion of the tree by insects or fungi. Pine pitch or oleoresin yields the turpentine and rosin of commerce.

Hardwoods

In dicot wood (Figure 13-15) different cell types are specialized for water conduction and support. Besides tracheids, most dicot woods contain vessels for water transport. The presence of vessels allows one to distinguish hardwood from soft-

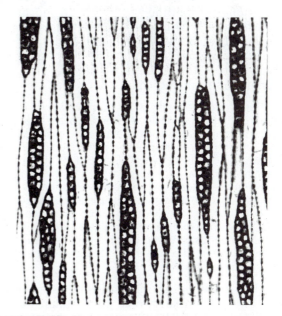

Figure 13-9. Dormant cambium of pear (*Pyrus communis*), tangential view. The elongated cells are the fusiform initials; the rounded cells in clusters are the vascular ray initials.

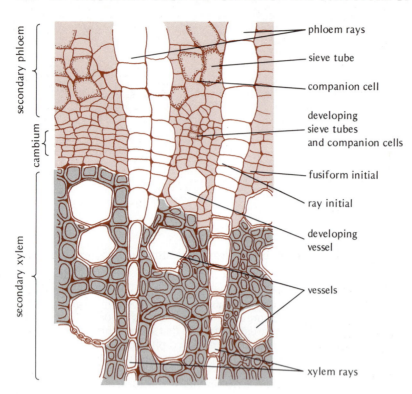

Figure 13-10. Diagram of dicot cambium and adjacent cell enlargement and differentiation zones (transverse section). Notice that large vessels and sieve tubes arise by enlargement of smaller mother cells.

wood visually, since vessels are often wide enough to be seen as pinpricks or channels with the unaided eye or a simple magnifying glass. The vessels of different dicots vary widely in size and in structural features such as the shape and number of perforations between vessel members, as illustrated in Figure 10-10.

For mechanical support, many hardwoods contain **wood fibers,** which may be regarded as evolutionarily modified tracheids. Wood fibers range from cells with somewhat thicker walls and smaller pits than tracheids (fiber tracheids; Figure 10-7C) to extremely heavy-walled cells (libriform fibers; Figure 10-7D) similar to the heavy-walled fibers found in the phloem.

Hardwoods generally differ from softwoods in the nature of their rays. While softwood rays are only one cell wide, many hardwoods contain also, or instead of such simple rays, more massive rays several to many cells wide and many cells high (Figures 13-9, 13-15). The larger rays may, as in oak, be visible to the unaided eye when the wood is cut parallel with the rays. This confers a striped appearance to oak that is valued in furniture making. Hardwoods almost always contain vertical rows of wood parenchyma cells, formed as described previously. Wood parenchyma cells may be clustered around the vessels or separated from them in different species.

The growth rings of hardwoods consist of springwood (or "earlywood") and summerwood (or "latewood") with differences similar to those described above for softwood. Two important types of hardwoods are recognized on the basis of vessel distribution. **Diffuse-porous** hardwoods have vessels of similar width throughout most of each growth ring (Figures 13-16A, 13-A). **Ring-porous** hardwoods contain very wide vessels in the springwood, grading down to much narrower ones in the summerwood (Figure 13-16B). The wide springwood vessels are much longer, apparently extending uninterrupted almost the entire height of the tree, as compared with diffuse-porous woods, whose individual vessels are usually from 1 to about 5 m long. These features of the springwood vessels give ring-porous wood a much greater efficiency for water conduction than most diffuse-porous woods.

Measurements of typical daytime sap flow as described in Chapter 10 show rates of several centimeters per minute in diffuse-porous trees, but as much as half a meter to a meter per minute in ring-porous trees. This mirrors the greater conductivity of ring-porous wood just noted, but is peculiar because ring-porous trees do not transpire appreciably more water, for a given-sized tree, than do diffuse-porous trees. It appears that in ring-porous trees only the outer part of the sapwood, perhaps only the most recently formed growth ring, functions in water transport; in this narrow zone the sap must flow with a very high velocity to supply the water demands of the leaves.

Reaction Wood

Trees produce an anatomically distinct type of wood on the lower or upper side of their larger

Conifers produce reaction wood called **compression wood** on the lower side of inclined branches or trunks. Its tracheids differ in size, shape, and wall structure from ordinary tracheids. An inclined dicot stem produces reaction wood called **tension wood** on its upper side. Tension wood contains smaller and fewer vessels than normal hardwood and has conspicuous thick-walled wood fibers that help support the tension caused by the weight of the branch. Tension and compression wood seem to be formed in response to the lateral movement of auxin induced by gravity (see Chapter 14).

The formation of reaction wood can move and reorient a branch. When the main shoot or "leader" of a conifer is cut off or killed, a lateral branch even several years old will often bend upward toward the vertical at its base, replacing the lost leader. This bending can occur in a matter of weeks and is due to vigorous production of compression wood at the base of the branch.

Knots

A **knot** is the base of a branch. As long as the branch is alive, its cambium and resulting growth rings are continuous with that of the trunk (Figure 13-18A). Once the branch stops growth and dies, the advance of the trunk's cambium by further secondary growth buries the base of the branch, bark and all, in the wood (Figure 13-18B). A board sawed from the part of a tree containing such buried branches will usually cut them transversely (Figure 13-19). If the plank came from wood formed after the branch died (cut b in Figure 13-18B), the knot will be loose (Figure 13-19B) and may fall out. Even if the knot is tight (cut a in Figure 13-18B; Figure 13-19A), the reaction wood in the branch makes the knot harder and more brittle than normal wood and makes it shrink differently when lumber is dried, causing cracking and trouble in sawing or nailing into the wood in or around the knot. The rustic appearance of knots, on the other hand, is valued for decorative "knotty pine" paneling.

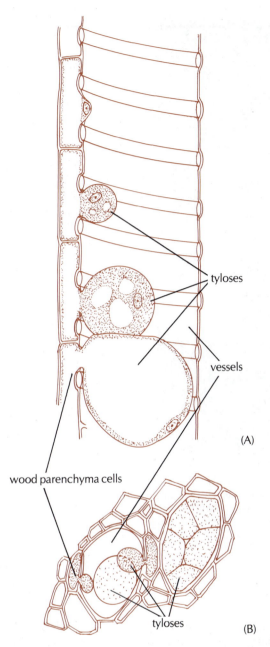

Figure 13-11. Tyloses developing in xylem vessels of black locust (*Robinia pseudoacacia*). (A) Longitudinal section; (B) transverse section. The tyloses develop as outgrowths from wood parenchyma cells into the vessel cavity.

inclined or horizontal branches, or of trunks when they become inclined away from the vertical. This wood is called **reaction wood** because it is formed in reaction to gravity. It serves to stiffen inclined stems against being broken off or bent downward. Reaction wood is produced more vigorously than normal secondary growth, often causing inclined trunks or the bases of large branches to acquire a noticeably elliptical cross section or even a wing-like form (Figure 13-17).

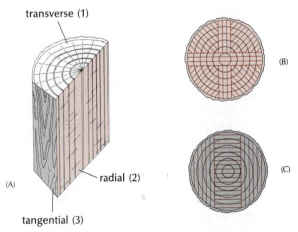

Figure 13-12. Planes of section of a tree trunk segment. (A) Three-dimensional diagram showing transverse, radial, and tangential planes. *Quarter-sawed* lumber is produced by cutting logs mainly in the radial direction (B). *Plain-sawed* lumber is produced by cutting logs mainly in the tangential section (C).

Figure 13-13. Scanning electron micrograph of conifer wood (*Pinus echinata*) (45 ×). Top, lower left, and lower right are respectively transverse, tangential, and radial sections. Wide tracheids comprising the spring (or early-season) wood are visible at lower left and upper right, the narrower summer wood tracheids in between defining a growth ring. The large channels seen in cross section in the upper center are resin canals. A pair of rays are visible in longitudinal section along the radial face of the block at lower right, and on the tangential face many cross-sectioned rays can be seen between the tracheids.

STRUCTURE AND GROWTH OF BARK

Bark, comprising all tissue outside the vascular cambium of a tree, consists of inner bark or phloem, plus outer bark that contains layers of cork. In some cases tissue derived from the cortex of the primary body remains between these two.

Phloem or Inner Bark

The phloem's most characteristic components are the sieve cells (in gymnosperms) or sieve tubes (in angiosperms), specialized for the transport of photosynthates. However, secondary phloem usually also contains bands of elongated sclerenchyma fibers or **phloem fibers** (Figure 13-20), which develop a heavy, lignified secondary wall contrasting sharply with the unlignified primary walls of sieve tubes, companion cells, and unspecialized phloem parenchyma cells that are usually present. Secondary phloem also often contains short heavy-walled sclerenchyma cells called **sclereids,** which help toughen the bark.

In the differentiation of both primary and secondary phloem, a sieve tube member and its associated companion cell(s) always develop from a single progenitor cell or mother cell, which divides transversely into two unequal cells, the larger becoming the sieve tube member and the smaller becoming one or (after further division) several companion cells (Figure 13-21). The mother cells for sieve tube members and companion cells of the secondary phloem are cambial derivatives on the phloem side of the vascular cambium, as previously indicated.

Figure 13-14. Resin canal in transverse section of *Pinus palustris* wood (225 ×). The living parenchyma cells which line the canal are partially collapsed from sectioning and preparation for scanning electron micrography. The photo also shows a close-up view of the narrow, thick-walled summer wood tracheids of a growth ring.

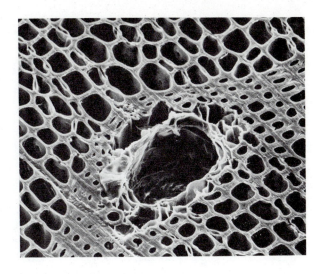

The functional life of sieve tube members is usually short, commonly only a single growing season. The functioning secondary phloem is therefore a narrow zone, usually less than a millimeter wide, immediately outside the vascular cambium. However, in some trees such as linden, sieve tube members can function for 5 to 10 years. Even more striking is the case of long-

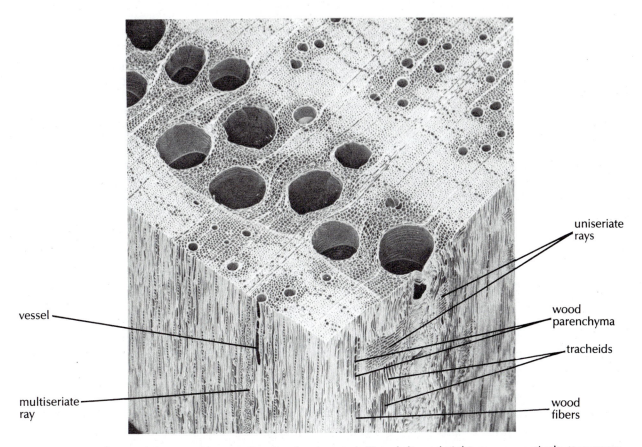

Figure 13-15. Scanning electron micrograph of red oak wood. Top, left, and right are respectively transverse, tangential, and radial planes of section. (100 ×) Huge vessels showing in the spring wood just above the annual ring, and much smaller ones in the summer wood beyond. *Uniseriate* rays are one cell wide; *multiseriate* rays are many cells wide.

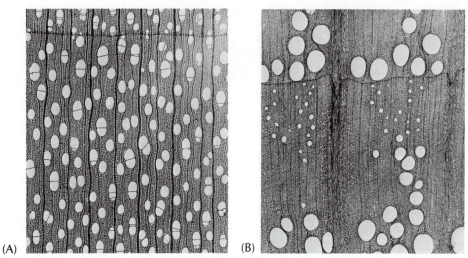

Figure 13-16. Hardwood vessel distribution, transverse sections. (A) Diffuse-porous wood of river birch (*Betula nigra*). (B) Ring-porous wood of red oak (*Quercus rubra*).

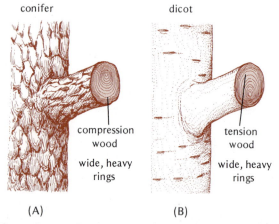

Figure 13-17. Reaction wood in cross sections of horizontal branches. (A) Conifer, showing compression wood below. (B) Dicot, showing tension wood above.

lived monocots, which, since they have no cambium and form no secondary phloem, depend for their whole life upon the original primary sieve tubes in even the oldest portions of the trunk. In one palm it was shown that sieve tube elements at least 50 years old were still functional, and in other species it has been estimated that the functional life span must be more than 100 years.

Vascular rays are prominent in the phloem, derived from the same ray initials that give rise to the wood rays. Phloem rays and wood rays are therefore continuous with one another across the vascular cambium, via the ray initials (Figure 13-20). Phloem rays often bring about circumferential growth of the phloem as the trunk or branch enlarges by secondary growth. Cells in some of the phloem rays begin to elongate and divide in the circumferential direction (arrow in Figure 13-20).

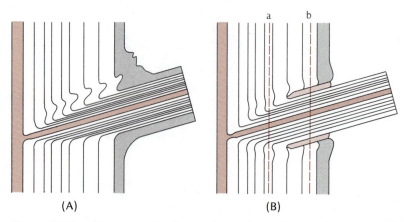

Figure 13-18. Diagrams of radial sections of wood containing knots. (A) Young trunk with living branch, continuous cambium and growth rings. (B) Older trunk after death of branch, showing buried bark of branch and most recent growth rings of trunk discontinuous with branch. Colored lines show cut through tight knot (a); cut through loose knot region (b). Light color shows pith of trunk and branch.

The progress of this circumferential growth with time results in the ray becoming wider and wider with distance away from the vascular cambium, creating what is called a **flaring ray** (Figure 13-20). This growth keeps the inner bark from splitting as it becomes stretched circumferentially by the growth of the cylinder of wood which it encloses.

Latex Ducts

Various plants develop specialized tubelike cells or cell systems called **latex ducts** or **laticifers,** producing internally a usually milky juice called latex which contains secondary products, often with a sticky consistency. Latex ducts may occur in any tissue of the shoot or root, but are commonly found in the phloem (Figure 13-22). Latex exudes from the latex ducts when the organ is cut or injured. Familiar examples are the milky sap seen when one breaks the stem or root of dandelion, milkweed (hence the name), or lettuce (*Lactuca;* from Latin *lactis,* of milk, because of this characteristic). Latex from the rubber tree, *Hevea brasiliensis,* is the source of natural commercial rubber. The exudation of latex, because of its unpleasant consistency and/or taste, and sometimes because of poisonous constituents, may help plants defend themselves against insects and herbivorous animals by discouraging feeding.

There are two principal types of latex ducts. (1) Articulated ("jointed") laticifers are formed by many individual cells fusing together in series at their ends to create a tubular system, analogous to the formation of xylem vessels except that, unlike vessels, laticifers remain alive after fusion and the laticiferous ducts often branch. The laticifers in the bark of the commercial rubber tree (Figure 13-22) are of this type. (2) Nonarticulated laticifers are single cells which, by long-continued tip growth, develop into an extensive, branched tubular system that lacks the "joints" or sites of cell–cell fusion seen along an articulated latex duct. Laticifers of milkweed, spurge (*Euphorbia,* a common weed), and oleander (*Nerium,* an ornamental shrub much used in warm climates) are of this type.

Formation of Cork

Woody plants generally begin to form a layer of cork, their external protective tissue, at or near the surface of the young stem or root during its first season. Some herbaceous plants or plant parts also form cork, an example being the "skin" of a potato. The cork cambium, which functions like the vascular cambium but is much less complex, usually arises by renewed division of cells in the outer part of the cortex, just beneath the epidermis (Figure 13-23). Cork cambial initials are more or

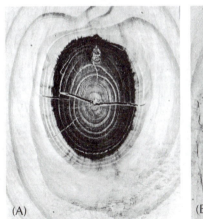

Figure 13-19. Hard pine boards showing knots. (A) Tight knot (cut a in Figure 13-18B). (B) Loose knot (cut b in Figure 13-18B).

(A)

(B)

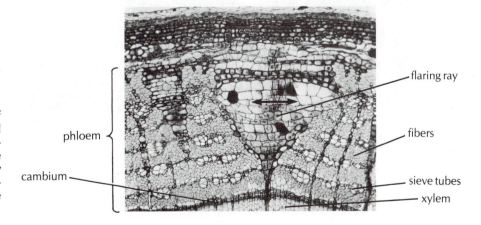

Figure 13-20. Portion of cross section of basswood (*Tilia americana*) stem, showing the flaring rays in the phloem. The colored arrow indicates the direction of circumferential growth in the flaring ray.

phloem

cambium

flaring ray

fibers

sieve tubes

xylem

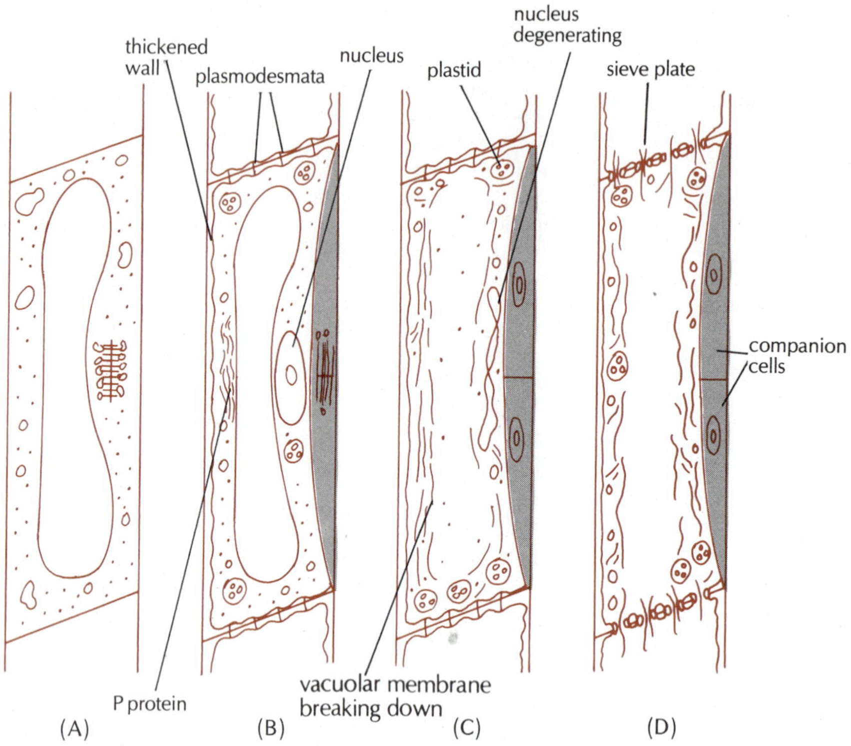

thickened wall

plasmodesmata

nucleus

plastid

nucleus degenerating

sieve plate

P protein

vacuolar membrane breaking down

companion cells

(A) (B) (C) (D)

Figure 13-21. Development of a sieve tube member and companion cells from a sieve tube mother cell. (A) Dividing sieve tube mother cell. (B) After unequal longitudinal division into young sieve tube member and precursor of companion cells, now undergoing mitosis. (C) During degeneration of nucleus, internal membranes, and organelles, and development of sieve plate pores. (D) Fully differentiated sieve tube member and companion cells.

less block shaped, not elongated like the fusiform initials of the vascular cambium. Their divisions build up, outside the cork cambium, radial rows of block- or slab-shaped cells. These derivatives differentiate into cork cells by depositing a suberin-containing wall, and then die, while the cork cambial initials beneath them continue to divide and add more cells to the cork layer. The cork cambium may also add inwardly a thin layer of unspecialized parenchyma cells.

Suberized cork tissue protects the living tissues of the inner bark and vascular cambium against desiccation and injury. Being impervious, cork would cut off oxygen uptake for respiration of living cells inside the stem, if it were not for specialized gas exchange structures called **lenti-**

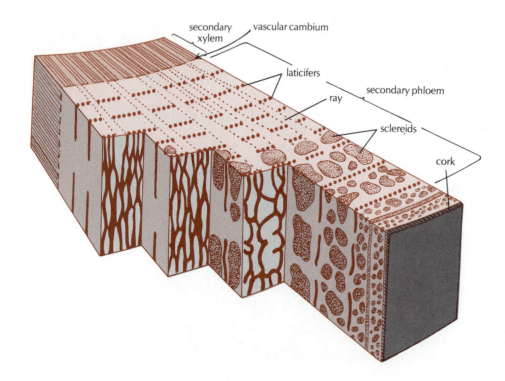

secondary xylem

vascular cambium

laticifers

ray

secondary phloem

sclereids

cork

Figure 13-22. Block diagram of the para rubber tree (*Hevea brasiliensis*), showing zones of laticifers. To obtain rubber a cut is made into the bark, severing many laticifers and allowing the latex they contain to flow out and be collected.

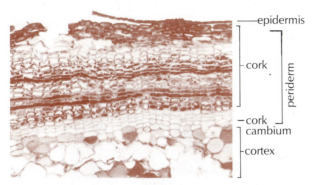

Figure 13-23. Young cork cambium and cork layer in the stem of *Erodium macradenum*.

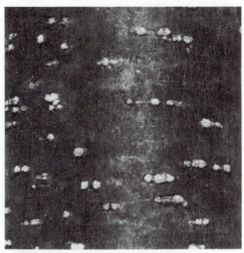

Figure 13-24. Lenticels of alder (*Alnus rugosa*). This stem was 10 cm in diameter.

cels which develop here and there in the cork, often looking like raised patches or horizontal streaks on the surface of the bark (Figure 13-24). Lenticel tissue (Figure 13-25) contains intercellular gas-filled spaces that permit oxygen to diffuse into the trunk through the cork.

In some trees such as beech, birch, and some species of cherry, the original cork cambium remains active, like the vascular cambium, for the life of the stem. In most trees, however, the cork cambium has a limited life span. After it has produced a cork layer of a certain thickness, the cork cambium dies and is replaced by a new cork cambium which arises by renewed division of cells in the secondary phloem located inward from the first cork cambium. This new cork cambium proceeds, like the first one, to build up a layer of cork to its outer side. Thus a series of cork layers is produced in the outer bark, which becomes a composite of cork layers sandwiched between layers of dead secondary phloem (Figure 13-26).

The outer bark becomes gradually but, in time, severely stretched in circumference as the trunk or branch grows. Since cork tissue is dead, it cannot accommodate this stretching by circumferential growth as the secondary phloem can. In some trees such as birch the outer layers of older cork merely stretch elastically until they break and peel away, leaving a smooth layer of newer cork at the surface. More commonly, multiple cork cambia are initiated in the secondary phloem producing scales or plates of cork, which pull away from one another as the stem's circumference increases (Figure 13-26), and are ultimately shed. This produces the scaly or furrowed outer bark of many trees.

Healing of Wounds

Besides causing growth in size and replacement of conducting, mechanical, and protective tissues, secondary growth also serves for wound repair. When primary tissues are wounded, for example, if a potato is cut in half, cell division activity begins in cells just be-

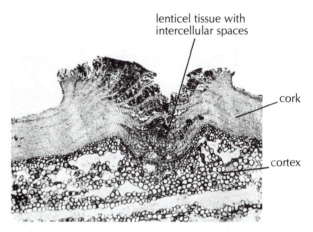

lenticel tissue with intercellular spaces

cork

cortex

Figure 13-25. Lenticel of black cherry (*Prunus serotina*), as seen in cross section of a young stem.

neath the cut within a few days. This activity initiates a cork cambium, producing cork that seals up the wound. When secondary tissues such as a woody branch or root are cut, the healing process is not so simple because most of the cut surface (the wood) contains only scattered living cells (or none at all, in the heartwood), which cannot be recruited for wound repair. Instead, active proliferation of dividing cells begins at the edges of the wound, from the vascular cambium and secondary phloem. This builds up a callus of undifferentiated cells, from which a cork cambium soon develops in the outer layers and produces a covering of cork. A small wound may become covered by callus, which internally differentiates vascular cambium in continuity with that to either side of the wound, allowing normal secondary growth to proceed in the injured area. Larger injuries, however, must be closed gradually by an exceptionally vigorous secondary growth that begins from the marginal callus, in which a vascular cambium differentiates. The production of xylem and phloem forms a roll-shaped extension of bark-covered wood that progressively extends over the

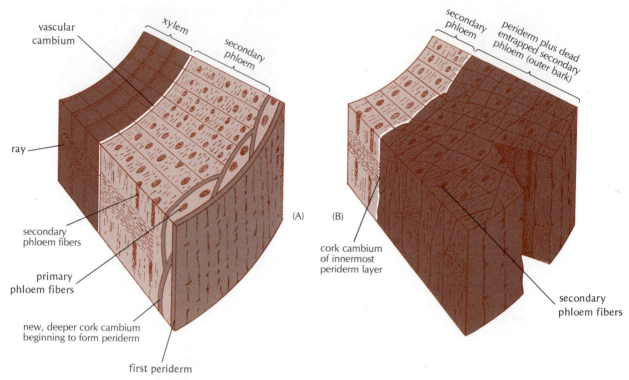

Figure 13-26.   Block diagrams showing the origin and structure of furrowed outer bark. The term *periderm* in these diagrams refers to the layers of cork and parenchyma tissues produced by a cork cambium. (A) An early stage in the development of cork. A second cork cambium has differentiated from parenchyma cells of the secondary phloem to the inside of the first periderm. Between the second and the first cork cambia remain several layers of cortex and primary phloem tissue. (B) Fully differentiated outer bark. Successive cork cambia have differentiated from secondary phloem tissue, each new cork cambium arising to the inside of the previously active one and separated from it by several layers of secondary phloem cells. Each new cork layer cuts off contact between cells of secondary phloem and previous cork cambium to the outside of it, and living tissues of the inner bark. The isolated tissues die, but remain sandwiched between successive cork layers. Only the innermost cork cambium is alive and active.

Figure 13-27.   Healing over in progress, four years after a wound was created by cutting off a large branch from a red oak trunk. Active cambial growth from the right side has nearly closed halfway over the wound, but injury to the bark on the left has retarded healing on that side.

open cut and closes it (Figure 13-27). Objects such as nails or signs attached to the cut surface may be buried by the advancing layers of secondary xylem, phloem, and cork.

Growth in Thickness in Monocots

Certain monocots that have attained tree size carry on secondary growth by a method basically different from that of dicots. Examples are the "dragon tree"

(*Dracena*), a commonly planted ornamental in warm climates, and the picturesque Joshua tree (*Yucca brevifolia*) of southwestern deserts (Figure 13-28), as well as other yuccas. The stems of these plants have a cambium near the outside, which adds new cells inwardly to the stem. These cells differentiate into an array of vascular bundles, each containing both phloem and xylem, with a limited amount of parenchyma between the vascular bundles. Thus the secondary body of these plants resembles the primary body of a monocot stem with its scattered vascular bundles (Figure 10-1) and

differs markedly from the concentric structure of woody dicot or conifer stems.

In contrast, the tallest of the tree monocots, palms, achieve their size without secondary growth. The large diameter of a palm trunk arises by extensive lateral expansion and cell multiplication just behind the shoot apex (Figure 13-29), in a zone sometimes called a **primary thickening meristem.** The thick stems of corn and sugarcane, and of various herbaceous dicots, arise by the action of somewhat similar primary thickening meristems, as does the potato tuber.

Figure 13-28.   Joshua trees (*Yucca brevifolia*), showing (right) development of corky bark and gradual increase in stem diameter by secondary growth of these monocots.

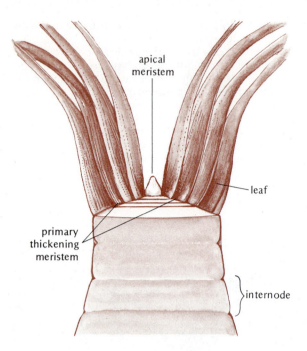

apical
meristem

leaf

primary
thickening
meristem

internode

Figure 13-29. Palm trunk mode of growth, showing the location of the primary thickening meristem (longitudinal section).

SUMMARY

(1) In stems of most plants having secondary growth, a vascular cambium arises by renewed or continued cell division in a thin layer of procambial cells between the primary xylem and the primary phloem, and in the parenchyma between vascular bundles if the primary vascular system is composed of bundles. Cells produced by cell multiplication in the cambium differentiate, on the outer side of the cambium, into secondary phloem (inner bark) and, on the inner side, into secondary xylem which accumulates as a cylinder of wood composed of yearly growth increments or annual rings.

(2) Secondary growth activity is controlled by auxin coming from growing buds and can be severely reduced by low temperature, water stress, and shading, which depress photosynthesis.

(3) Secondary growth in roots begins with the initiation of a vascular cambium at the outer edge of the xylem and becomes similar to secondary growth in stems once the sinuous outer boundary of the root's primary xylem has been converted, by the addition of secondary xylem, into a cylindrical outline.

(4) The cambial zone consists of a single layer of cambial initials plus up to several meristematic cells, called phloem and xylem mother cells, lo-

cated, respectively, just outward and just inward from each initial and ultimately derived from that initial. Division of the mother cells produces most of the new cells added to the phloem and the xylem, as well as creating particular cell shapes and arrangements such as that of phloem and xylem parenchyma and sieve tube members with associated companion cells. Elongated fusiform initials give rise, via similarly shaped mother cells, to the elongated conducting and mechanical cells of the xylem and phloem; compact ray initials give rise to radially elongated vascular rays.

(5) Differentiation of secondary xylem normally involves secondary wall deposition by all the cells derived from the cambium, including wood ray and wood parenchyma cells that remain alive after differentiation (in many storage roots, secondary xylem parenchyma remains thin walled, making the tissue nonwoody and edible). In the phloem, heavy secondary walls are deposited by phloem fibers (and, if present, sclereids).

(6) The more recently formed layers of wood, which contain living ray and (if present) parenchyma cells, are called the sapwood; at a certain distance from the cambium they die, the wood becoming heartwood, usually darker in color than the sapwood. Water transport is restricted to the sapwood, in some cases only to its outermost layers.

(7) Conifer wood, or softwood, contains tracheids, rays a single cell layer thick, and usually resin canals. Growth rings are visible because cells formed later in the season (summerwood or latewood) are narrower and thicker walled than those of the springwood or earlywood.

(8) Dicot wood, or hardwood, contains tracheids, vessels, rays up to several cells in thickness, wood parenchyma, and usually wood fibers, which can be very thick walled. Growth rings are due to differences in cell width and wall thickness, as in softwoods, plus differences in the distribution of large vessels in many cases (ring-porous woods contain large vessels only in the earlywood).

(9) As a stem or root grows in diameter, its bark must grow in circumference. In the secondary phloem, circumferential growth and cell division often occur in the rays, producing flaring rays which, by their growth, keep the inner bark from splitting.

(10) Layers of protective cork are produced outside the secondary phloem by a cork cambium that usually arises first in the cortex of a young stem, usually being replaced later by cork cambia arising successively inward in the secondary phloem.

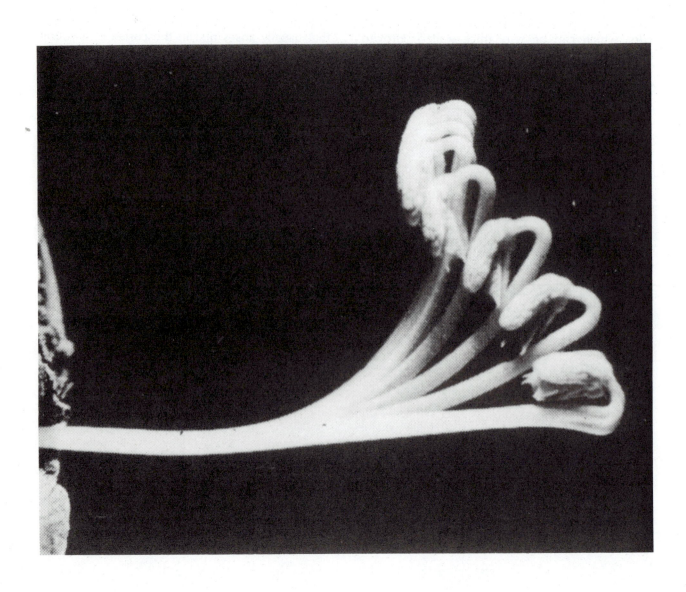

Chapter 14

ENVIRONMENTAL AND GENETIC CONTROL OF DEVELOPMENT

From little acorns grow great oaks, not pine trees or flowers: development is the means by which the unique hereditary complement of any species expresses itself, creating the body form and structure characteristic of that species. Viewed in this light, the hormones and other growth-regulatory substances involved in aspects of plant development are simply a part of the machinery by which genes achieve their expression.

We tend to think, out of relative familiarity with the precision of human development, that heredity specifies body form (and thus development) within narrow limits. The progress and outcome of development in plants, however, is strongly subject to environmental influences. For example, one cannot predict the shape of an oak tree in advance; its form depends to a considerable extent on what happens to it during its development. One reason for this is that the tree is directly exposed to the environment during a much more important part of its development than are most animals. Another reason is that, in contrast to that of animals, plant development serves not only to produce the functional organism but also to enable it to respond and adjust to favorable and unfavorable environmental factors, compensating to some extent for a plant's inability to move about and improve its situation. Environmental factors affect, adaptively, the expression of hereditary information for a plant's development.

The purposes of this chapter are to consider the developmental responses of plants to the environment and to examine the important problem of how intrinsic as well as environmentally mediated aspects of development rest upon genetic information.

Kinds of Developmental Responses

Two principal kinds of developmental responses to environmental stimuli can be distinguished. **Morphogenetic responses** are changes in the rate or the nature of developmental processes (for example, the appearance of flowers on a previously vegetative shoot), not oriented relative to the direction of an inducing stimulus. **Tropisms** are bending or curving responses toward or away from (or sometimes at a specific angle to) a directional stimulus such as light or gravity. Tropic curvatures develop by differential growth, that is, faster elongation on one side of an organ than on the other side, in a primary growth zone.

Many plants display bending responses that result not from growth but from reversible changes in size of certain cells, due to an increase or decrease in turgor pressure. These responses, which include stomatal opening and the daily opening and closing of many flowers, are called **turgor movements.** Although they are not developmental changes, turgor movements share a number of important features with developmental responses so the two are usually considered together.

Because plants need light for photosynthesis, and because light provides useful cues for the occurrence of other changes in a plant's environment, many responses to light (photoresponses) have arisen. Plants can also respond to gravity (Figure 14-A) as already noted, to temperature, and in many cases to touch or contact with foreign objects.

TROPISMS

If placed at an angle to the vertical, the primary growing zones of shoots and roots will curve upward (shoots) or downward (roots) in response to gravity, a response called **geotropism** (also **gravitropism;** Figure 14-A). Shoots will also bend

---

Figure 14-A. Bean seedling responding negatively to gravity (negative geotropism). Composite photo with successive exposures made every 40 minutes.

Figure 14-1. Positive phototropic bending of a radish seedling. Composite photograph with successive exposures made every 40 minutes.

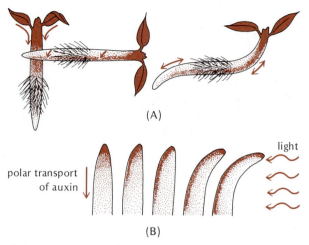

Figure 14-2. Auxin relations during geotropism (A) and phototropism (B) of seedling plants. Auxin-producing regions are shown in color and the distribution of auxin in shoot, coleoptile, and root is shown by color dots. (A) When the seedling is turned horizontally, auxin is transported to the lower side. As shown by arrows, auxin promotes elongation of cells in the shoot so that the lower side grows more rapidly and causes upward curvature; in the root, auxin inhibits cell elongation so that the upper side grows more and causes downward curvature. (B) When light is directed at one side of a grass seedling coleoptile, a difference in auxin concentration on the two sides arises at the tip (the light-sensitive site) and moves down the coleoptile because auxin is transported toward the base. As a result, curvature begins just below the tip and progresses toward the base of the coleoptile.

toward light, a response called **phototropism** (Figure 14-1). These curvatures arise by differential growth of the cells on opposite sides of the responding organ. Study of these responses in the coleoptile or first leaf sheath (Figure 3-35A) of grass seedlings first revealed the existence of the plant growth hormone later named auxin.

In the 1930s, Went and his colleagues used the *Avena* curvature bioassay (Figure 12-14) to measure the amounts of auxin transported down the two halves of a coleoptile or a shoot subjected to lateral gravity or light. They found more auxin in the lower than in the upper half of a horizontal shoot and more auxin on the side of a shoot away from the light than on the side toward the light (Figure 14-2). Since auxin stimulates cell elongation, the difference in auxin levels explains why the growth rate increases on one side, resulting in the observed curvature. Some mechanism causes a lateral transport of auxin in response to lateral gravity or light.

Geotropism

Under gravity auxin molecules do not simply fall to the lower side of the stem. Molecules are much too small to fall any appreciable distance. Instead, auxin seems to be moved laterally in stems by a deflection of the polar transport mechanism that normally transports auxin through a shoot toward its base (Figure 12-15).

Gravitational force is probably perceived by the displacement of heavy starch-containing amyloplasts (often called starch **statoliths**) that almost always occur in geotropically responsive tissues and fall, under gravity, to the lower side of the cells containing them (Figure 14-3).

How the displacement of statoliths causes lateral auxin transport is an unsolved problem, but an asymmetric distribution of cytoplasm and/or organelles in the cell as a result of statolith displacement could conceivably shift toward one side of each cell the normally lengthwise polarity of auxin transport. Some experiments suggest, however, that the pressure of the statoliths against the cell surface is responsible for their effect on the polarity of transport.

Geotropism of Roots. Roots, as noted above, respond to gravity in a manner opposite to that of shoots, bending downward (Figure 14-4) whereas shoots bend upward. The root cap possesses cells containing large amyloplasts that are displaceable within each cell by gravity; if the root cap is removed, the root loses its ability to respond to gravity, even though it continues to elongate. Apparently a hormonal stimulus is transmitted from the gravity-sensitive statolith-containing cells of the root cap to the primary growth zone of the root.

In contrast to its effect on cells of the stem, auxin *inhibits* the growth of root cells. Thus, lateral transport of auxin in the direction of gravity would reduce elongation on the lower side of a horizontally oriented root, tending to make it curve downward. However, recent research indicates that inhibition of root cell elongation by abscisic acid, released and translocated from the root cap, may control geotropism, in addition to or

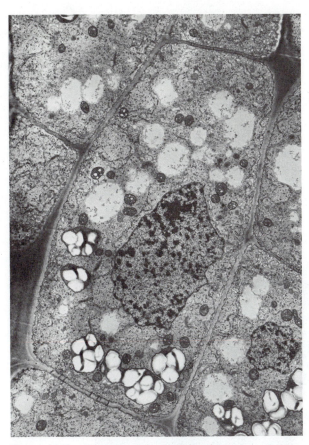

Figure 14-3. Electron micrograph of gravity-sensitive cells from a root cap, showing how amyloplasts containing large starch grains (white globules) become displaced to the lower side of the cell by gravity. If the organ is turned in a different direction, the amyloplasts fall to the new lower side of each cell within a few minutes. Note that neither the nucleus (large body near center of cell) nor mitochondria (smaller, darkly staining organelles with internal membrane sacs) are displaced to the lower side of the cell, because their density is similar to that of the cytoplasm, whereas starch is much denser.

instead of auxin. The actual mechanism of geotropic curvature in roots is still uncertain.

Phototropism

To be perceived by an organism, light must be absorbed by some kind of photoreceptor pigment. The photoreceptor for phototropism clearly is not chlorophyll, because red light (in fact all wavelengths above about 500 nm) is completely inactive in causing phototropism (Figure 14-5), whereas chlorophyll absorbs red light strongly (Figure 7-5). Because blue light most effectively induces phototropism, its receptor is often called the **blue light photoreceptor.** The blue light photoreceptor mediates a number of other light responses in plants besides phototropism. The blue light photoreceptor is probably a flavoprotein (a

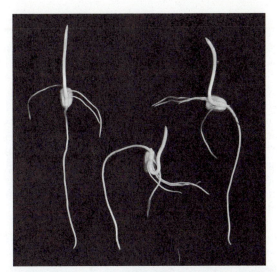

Figure 14-4. Negative and positive geotropism in germinating corn (*Zea mays*). The grain was germinated in either a vertical position (left), a horizontal position (right), or inverted (center). In all, the shoot has grown upward (negative geotropism) while the root has grown downward (positive geotropism).

protein containing the B vitamin, riboflavin). How the blue light photoreceptor causes lateral auxin transport, leading to phototropism, is still unknown.

The blue light photoreceptor evidently absorbs not only blue but also near-ultraviolet radiation (wavelengths of about 360 nm). Plants respond phototropically with a high sensitivity to these ultraviolet wavelengths, which the human eye cannot detect. This may not be important to the plant, since in nature it receives much more blue than ultraviolet radiation, but is a valuable clue to the identity of the blue light photoreceptor, and is one of the principal lines of evidence that the photoreceptor must be a flavoprotein rather than a carotenoid pigment. As yet, however, the blue light photoreceptor has not been isolated or conclusively identified.

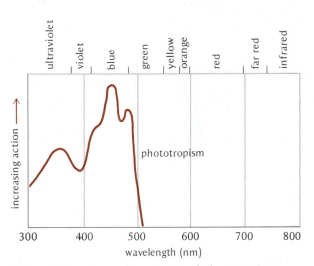

Figure 14-5. Spectral sensitivity of phototropism.

Phototropism of Leaves. In keeping with their function as solar energy collectors, most leaves orient their blades perpendicular to the prevailing direction of principal illumination, so as to receive the maximum light intensity. A familiar example is the way a houseplant kept next to a window positions its leaves to face the window (Figure 14-6). Many plants can adjust the location of their leaf blades by curvature and elongation of their leaf petioles to minimize mutual shading. The leaves of certain plants make daily "solar tracking" movements to follow the sun. These are turgor movements and are considered below.

Thigmotropism

Tendrils, the grasping organs of climbing plants, grow more or less straight until they touch some foreign object. Contact induces the tendril to coil, by differential growth on its two sides, so as to wrap itself around the object and attach the plant to it (Figure 14-7). The response, **thigmotropism** (from Greek *thigma*, touch), can be demonstrated by rubbing an uncoiled young tendril; some tendrils will curve to make a complete circle or more within 5 or 10 min after stimulation. A similar but slower response is shown by the stems

Figure 14-6. Phototropism in *Coleus*. The stem is inclined toward the light, and the leaf blades are approximately at right angles to the direction of the light.

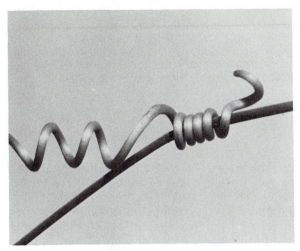

Figure 14-7. A tendril of *Smilax* wrapped around the stem of another plant.

of twining plants such as morning glory and bindweed (*Convolvulus*).

Treatment with auxin induces tendrils to coil in the absence of contact stimulation, but the role auxin plays in the response to contact is not clear. The epidermis of tendrils contains specialized cells that appear to be sensitive to contact, but how they induce differential growth in the tendril is also unclear. Some tendrils curve toward the side that has been stimulated by contact, but others curve in an anatomically predetermined direction regardless of which side receives stimulation. In the latter case the response is not strictly thigmotropism, but **thigmonasty** (nastic movement refers to a bending or curving response that is independent of the direction of the inducing stimulus; see below).

Thigmomorphogenesis. Plants tend to grow shorter and stockier, and to deposit thicker cell walls in supporting cells such as collenchyma and fibers, when frequently stimulated by contact or shaken (as, for example, by the wind) than when kept still. This type of response, which improves the plant's mechanical support when it is subject to external disturbance, is termed thigmomorphogenesis.

TURGOR MOVEMENTS

Sleep Movements of Leaves

Perhaps the most widely occurring turgor movements are nocturnal changes in the orientation of leaves, or "sleep movements." Familiar examples are the evening "closure" of the three leaflets of sorrel (*Oxalis*) and the nocturnal elevation of the paired leaves of the "prayer plant" (*Maranta*) into a configuration resembling a pair of hands in prayer (Figure 14-8), responsible for the common name of this widely grown houseplant. Sleep movement is known technically as **nyctinasty** (from Greek *nyktos*, night, and *nastos*, pressed together), referring to the frequently

Figure 14-8. Sleep movements (nyctinasty) of leaves of the "prayer plant" (*Maranta*), in daylight (above) and at night (below).

prominent juxtaposition of different leaflets or leaf blades at night.

Nyctinasty and other turgor movements are due to the expansion and contraction of specialized, enlarged cells called **motor cells** which gain or lose turgor pressure in response to an external stimulus. Leaf movements result from the bending or straightening of a local joint-like structure called a **pulvinus** (Figure 14-9), containing motor cells. In the light, the motor cells on one side of the pulvinus develop a higher turgor pressure than those on the other side, bending the pulvinus and moving the leaf blade into a horizontal position. In the dark, the reverse happens, allowing the pulvinus to straighten (Figure 14-9) or bend in the opposite direction, until the blade or leaflet assumes a relatively vertical position.

In the examples of nyctinasty that have been investigated, the pulvinus senses light not by chlorophyll or by the blue light photoreceptor active in phototropism, but by means of the photomorphogenic receptor pigment, phytochrome, considered below. The direction of illumination does not affect the position the leaf assumes.

When stimulated to enlarge by phytochrome and light, motor cells increase their internal solute concentration by absorbing K^+ and other ions from adjacent tissues; this increase draws water into the cells by osmosis, increasing their turgor pressure. They contract by releasing their abnormally high K$^+$ content, reversing the osmotic water gain that previously inflated the cells. Thus the mechanism of sleep movement is similar to that of stomatal opening and closure (Chapter 9), except that phytochrome does not serve as a photoreceptor in stomatal guard cells, so far as we know.

Although invariably influenced by light, sleep movement in a number of plants will continue to take place for several to many days under a constant temperature and continuous darkness or low-intensity light, evidently driven by a spontaneous physiological rhythm within the plant, called a **circadian rhythm.** This remarkable phenomenon is considered in Chapter 16 in connection with time measurement by plants.

The value of the widely occurring nyctinastic responses of plants is not entirely clear. One apparent function is to reduce the nighttime cooling of leaf blades by exposing them less to the cold night sky and reducing (via their mutual juxtaposition) their exposed surface area for losing heat. Another function, in the case of leaves that show daytime closure in cloudy weather, may be to cause less water to be trapped on leaf surfaces when it rains, so more rain reaches the soil for later absorption and use by the plant.

The day/night opening and closing of many flowers is a light response similar to nyctinasty, except that the motor cells, located at or near the base of each petal, are usually not organized into an obvious pulvinus like that responsible for leaf movements. Light-induced flower opening allows bees and other daytime pollinators to gain access to the flower in good weather but protects it against nighttime dew and from damage when rain threatens.

Solar Tracking

The leaves of certain plants follow or "track" the sun in the course of each day, their blades facing the sun's rays at all times (Figure 14-10) and thereby maximizing the light intensity received. Common examples of solar-tracking plants are clover, soybeans, sunflower, cotton, and lupine.

Solar tracking is due to the action of a pulvinus in the leaf petiole, similar to that involved in sleep movements. But the solar tracking response is highly oriented relative to the direction of illumination perceived by the leaf blade, which transmits some kind of stimulus to the pulvinus, influencing the behavior of its

Figure 14-9. Action of a pulvinus. Photomicrographs of sections of a leaf axis (below) bearing two pulvini at the bases of individual leaflets of the nyctinastic leaf of the "silk tree" (*Albizzia julibrissin*). The motor cells are the large parenchyma cells in the cortex (outer part) of the swollen basal region to either side of the vascular strand running up into each leaflet. In the daytime position (left) the motor cells on the inner side of each pulvinus are expanded and the motor cells on the outer side are partially collapsed, allowing the two leaflets to spread apart. In the night position (right) the outer motor cells expand and those on the inner side contract, pressing the leaflets together. These changes result from osmotic water uptake or loss, caused respectively by uptake or loss of solutes (mainly K$^+$ and Cl$^-$ ions) by the motor cells.

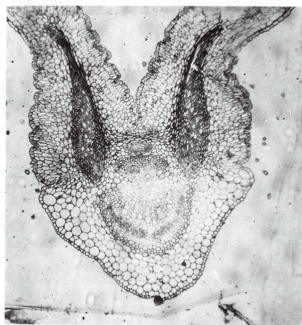
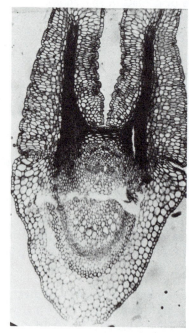

Figure 14-10. Daily cycle of leaf blade movement by a solar tracking plant, western wood sorrel (*Oxalis oregana*). Left, 8 AM; center, noon; right, 4 PM. The potted plant, with sticks supporting some of its leaf petioles, was placed indoors beneath a light bulb mounted on a clock-driven arm, the base of which shows as a bright bar pointing in the direction of the light. The bulb moved slowly above the plant from directly left at 6 AM to directly right at 6 PM, simulating the daily motion of the sun (the narrow parallel lines are markers on the background to record the time of day by the position of the bar). Note how the leaflets incline so as to virtually face the light throughout the day, by means of a pulvinus located at the top of each petiole.

motor cells. Auxin from the leaf blade can affect the turgor pressure of the motor cells, but it is not yet clear whether auxin is the stimulus that controls these cells in responding to the direction of illumination.

Turgor Movements in Response to Touch or Shaking

The leaves of the familiar "sensitive plant" (*Mimosa pudica*; Figure 14-11) possess pulvini that "close" their leaflets together if the plant is shaken or the leaf is stimulated by touching or cutting it. The pulvini work like those involved in sleep movements, but the motor cells on one side of the pulvinus become permeable to K^+ and other ions when mechanically stimulated. Thus they lose ions and, as a result, lose water by osmosis, leading to loss of turgor, bending of the pulvinus, and closure of the leaf. The value of this kind of response, which also occurs in a few other plants, may be to make the plant's foliage become difficult to see when an animal starts to feed on it, diverting the animal's attention and feeding to some other nearby plant.

Leaflet closure also increases the exposure of sharp prickles located along the leaf axis, tending to discourage herbivores or other intruders by giving them a painful experience.

Somewhat similar touch-sensitive turgor movements function in the leaves of some "carnivorous" plants such as the Venus flytrap (Figure 19-23D) and as pollination-controlling devices in various flowers (Chapter 16). The most interesting feature of the sensitive plant and the Venus flytrap is their capacity to transmit a stimulus from touch-sensitive cells to motor cells located elsewhere (Figure 19-23E). In the sensitive plant the closing stimulus can be transmitted from pulvinus to pulvinus within one leaf at a rate of about 0.5 to 1 cm/sec; after a severe stimulation such as local burning, a closing stimulus is transmitted from leaf to leaf along the stem. This superficially resembles nerve transmission in animals, except that it is much slower and does not involve cells specialized for this specific purpose. As in nerves, however, both electrical and chemical stimuli seem to be involved in the transmission.

Figure 14-11.   The leaf closure response of the "sensitive plant" (*Mimosa pudica*). (Left) Open; (center) beginning to close; (right) closed.

MORPHOGENETIC LIGHT RESPONSES: PHYTOCHROME

Plants grown in the dark develop a syndrome of symptoms called **etiolation.** Etiolated plants, for example, the "bean sprouts" used in Chinese cooking, are pale yellow or white due to lack of chlorophyll, they elongate much more than a light-grown plant, and their leaves do not expand (Figure 14-12). This form of growth is extremely valuable to a buried shoot or a seed germinating deep in the soil, because it raises the shoot tip toward the surface as rapidly as possible, giving the plant the greatest chance of reaching photosynthetically useful light before the plant's store of food reserves is exhausted.

An etiolated plant lacks chlorophyll partly because the conversion of a precursor compound, called photochlorophyll, into chlorophyll is a light-requiring reaction. The other symptoms of etiolation, involving an altered pattern of development, are manifestations of a class of morphogenetic responses in which light is detected by a unique photoreceptor pigment called **phytochrome.** Phytochrome absorbs red light (wavelengths of about 680 nm) strongly; thus etiolation can be stopped and the "normal" pattern of growth

Figure 14-12. Etiolated and normal potato (*Solanum tuberosum*) plants. The etiolated plants on the left were grown in darkness; the normal plants on the right, in the light. The plants were the same age.

restored by exposing plants to red light (even though the plants will not bend phototropically to red light). Young leaves expand rapidly and turn green; the stem stops its excessive elongation and begins to thicken. The direction from which light comes does not affect these responses.

Phytochrome occurs in two forms, P_R and P_{FR}, which light can interconvert. P_R is the red-absorbing form, occurring in dark-grown plants. Light absorption by P_R converts it into P_{FR}, the active agent in phytochrome effects such as reversal of etiolation. But P_{FR} can absorb "far-red" wavelengths of about 700–730 nm (barely visible to the human eye), which convert it back to the biologically inactive form, P_R:

$$P_R \underset{\text{far-red light}}{\overset{\text{red light}}{\rightleftarrows}} P_{FR} \longrightarrow \text{response.}$$

Consequently, treating plants with far-red light after exposing them to red light (or white light, which has the same effect as red) characteristically blocks phytochrome-mediated effects. For example, a few minutes of red light exposure will induce substantial subsequent expansion of the leaves of an etiolated plant, but if this plant is treated with far-red light for a few minutes just after the red light, expansion fails to occur (Figure 14-13). Because the interconversion of P_R and P_{FR} is reversible, after repeated alternating doses of red and far-red light, the ensuing response is simply to the last kind of light given: if red, the leaves expand; if far-red, they do not. This property of far-red reversibility constitutes a diagnostic test for the involvement of phytochrome in a light response.

Besides leaf expansion and stem elongation, there are diverse other phytochrome-mediated responses. They range from the control of sleep movements in leaves, leaf fall, and shoot dormancy to the induction of flowering (Chapter 16), seed germination (Chapter 17), secondary product synthesis, and pigment formation in flowers, fruits, and leaves.

Detection and Isolation of Phytochrome

By the sensitive measurement of changes in light absorption of plant tissues after exposure to red and far-red light, the interconversion of P_R and P_{FR} can be observed directly, and extracts of tissues can be assayed for the amount of phytochrome they contain. This assay enabled researchers to purify phytochrome. It proved to be a blue-green pigment closely related to the photosynthetic pigment, phycocyanin, of cyanobacteria and red algae.

Phytochrome absorbs light because the protein bears a pigment ring system (phycobilin) similar to that

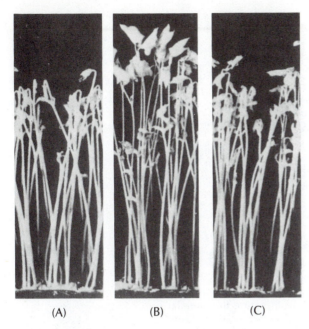

(A)   (B)   (C)

Figure 14-13. Phytochrome-mediated morphogenetic responses (leaf expansion, and plumular hook opening) in dark-grown bean seedlings. (A) Untreated controls (no light). (B) Brief dose of red light (670 nm). (C) Brief dose of red light, followed immediately by short exposure to far-red light (730 nm).

shown in Figure 7-6D. P_R and P_{FR} apparently differ in the configuration around one or more of the double bonds in the phycobilin ring system. Absorption of light by the phycobilin converts or isomerizes it into the alternative configuration, thus changing P_R into P_{FR} and vice versa. The conformation (shape) of the attached protein is also modified, which is believed to be responsible for the resultant biological effects of P_{FR}.

Although the basic morphogenetic photoresponses to white light are due to the conversion of P_R to P_{FR}, certain situations in nature provide a relative abundance of far-red light, causing the reverse photoreaction and, thus, an opposite response. This is ecologically significant in seed germination and is explained in that connection (Chapter 17). Moreover, a number of morphogenetic responses require much higher doses of light than those necessary just to convert P_R to P_{FR} and are also sensitive to blue and ultraviolet light. Either phytochrome or the blue light photoreceptor system may induce these "high-irradiance responses," but how they work is not clear at present.

Chloroplast Development

An important intracellular morphogenetic response partly under phytochrome control is the development of chloroplasts, during cell enlargement and differentiation, from small precursor organelles called proplastids (Chapter 4) which occur and multiply in meristematic cells. When a plant is grown in darkness, chloroplasts develop into only partially differentiated organelles called

etioplasts that contain no chlorophyll and no thylakoid membrane system, possessing instead a lattice-like tubular membrane system called the prolamellar body (Figure 14-14A). Etioplasts lack the enzymes of photosynthetic electron transport and of carbon metabolism, such as RuBP carboxylase (Chapter 7). Exposure to light induces the prolamellar body to extend and transform into thylakoids, which then stack up into grana stacks, yielding the typical structure of mature chloroplasts (Figure 14-14B, C). Accompanying these changes, chlorophyll formation and a dramatic production of photosynthetic enzymes occur.

DORMANCY AND LEAF ABSCISSION

The annual cycle of many plants, especially those of temperate and colder latitudes, involves dormancy, or shutdown of growth, and **abscission** (pronounced ab-sizh'-un), or leaf fall, in preparation for winter. Plants of arid regions may become dormant and shed their leaves in the dry season. Dormancy is not just growth cessation but usually involves the development of specialized dormant terminal and/or lateral buds ("winter buds" of winter-dormant plants), often possessing specialized organs such as bud scales not developed by an ordinary growing bud.

As noted in Chapter 3, winter dormancy and leaf abscission typically set in by late summer or early autumn, well before the onset of the freezing weather that these responses help plants tolerate. Plants typically enter dormancy and undergo leaf abscission not in response to cold temperature itself, but to the shortening of the day length that occurs in late summer and fall, providing a reliable cue that winter is approaching. Such a response to the length of the daily light and dark periods (Figure 14-15) is called **photoperiodism** and is considered further in Chapter 16 in connection with flowering, in which it is very important. The photoperiodic responses of plants involve the detection of light by phytochrome.

The Mechanism of Dormancy

Bud dormancy probably involves the action of growth-inhibitory hormones or dormins, such as abscisic acid (Chapter 12), whose formation may be stimulated by the short-day conditions that induce dormancy. Whether abscisic acid is the active agent, as the earlier work in this field suggested, has been challenged, but the artificial application of abscisic acid to growing woody shoots does inhibit their growth in a manner similar to naturally induced dormancy. Gibberellin can reverse natural or abscisic acid-imposed dormancy, suggesting that the internal balance be-

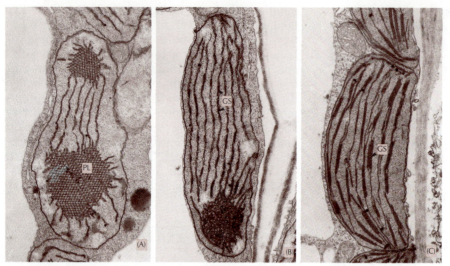

Figure 14-14. Light-induced chloroplast development in leaf of oat seedling. (A) Etio-plast as found in dark-grown leaf; PL, prolamellar body. (B) Leaf after two hours of exposure to light; note disorganization of PL, from which is proliferating the lamellar membrane system with grana stacks (GS) beginning to form. (C) Differentiated plastid after 24 hours of exposure to light (21,000 ×).

Figure 14-15. Photoperiodic control of shoot dor-mancy in Douglas fir (*Pseudotsuga menziesii*). Plants were exposed for one year to daily photoperiods of 12 hours (left) ("short days"), 20 hours (right) ("long days"), and 12 hours plus a one-hour interruption near the middle of the dark period (center). Short-day treat-ments with no interruptions of the night by light in-duced bud dormancy (left) and shoot growth ceased.

tween gibberellin and dormins may actually con-trol dormancy.

Most winter-dormant plants become released from dormancy by a type of temperature response that is curious but widespread in plants inhabit-ing regions with cold winters. A certain amount of exposure to *cold* is required to break bud dor-mancy. Temperatures slightly above freezing, for a period varying from weeks to months depending on the kind of plant, are effective. As might be expected, plants adapted to colder latitudes re-quire a longer period of cold. The cold require-ment ensures that, under favorable temperatures, buds will break dormancy and begin to grow only after experiencing an extended cold period, typi-cal of an entire winter, and not during brief warm spells in the middle of winter. The cold treatment probably influences the hormone balance, for ex-ample, by inhibiting the synthesis of a dormin such as abscisic acid while some destruction of it con-tinues.

If the winter does not satisfy a dormant tree's cold requirement, the tree will leaf out in spring only feebly and erratically, if at all. This is frequently a problem, for example, with deciduous fruit trees such as apples and cherries when planted at relatively warm latitudes. Alternatively, exposure to long days can release from dormancy the buds of some trees such as birch, red oak, and larch. This ensures that the trees will even-tually leaf out in spring even if a particular winter fails to fully satisfy their cold requirement. Some trees re-quire exposure to both cold and long days before dor-mant buds will begin to grow. Because of their demand for sufficiently long days, these trees may remain leaf-less for a considerable time after temperatures favor-able for growth arrive in early spring.

Leaf Abscission and Senescence

Detachment of the leaf, called abscission, occurs by the development of a weakened layer of cells (the **abscission zone**) across the leaf petiole, usually at its base (Figure 14-16). This development is inhibited by auxin from the leaf blade and, on the other hand, can be stimulated dramatically by ethylene. Under long days the leaf evidently produces enough auxin to prevent its abscission. Short days apparently alter the hormone balance such that abscission is no longer inhibited.

In preparation for abscission the leaf undergoes **senescence** or the breakdown of its cellular constituents and membranes, leading ultimately to cell death. This breakdown enables plants to recover nitrogen and other mineral nutrients from the leaves before they are shed. One of the nitrogenous constituents broken down is chlorophyll; this is the first cause of the autumnal change in color of foliage, since the carotenoid accessory pigments of the leaf remain, making it appear yellow.

The second cause of autumnal coloration is the formation of red or purple anthocyanin pigments in the leaves, which in combination with the carotenoids can color the leaf various shades from orange to deep red. The functional value, if any, of anthocyanin formation during senescence is not known, but anthocyanin formation is an example of a high-irradiance phytochrome response and is therefore stimulated by sunny weather; it is also stimulated by cool temperatures. Therefore, the most spectacular autumnal coloration develops in years when cool clear weather prevails in autumn.

A remarkable effect of cytokinins is to prevent or retard leaf senescence. A drop of cytokinin solution applied to a senescing leaf will keep that spot green while the rest of the leaf turns completely yellow (Figure 14-17). Thus a reduced level of cytokinin production by the plant might contribute to autumnal leaf senescence. Ethylene, on the other hand, not only stimulates senescence but is produced in extra large amounts by senescing tissue. Since ethylene stimulates development of the abscission layer, senescence would be expected to promote the ultimate detachment of the leaf.

Because of its name, it might seem that abscisic acid (ABA) should be important in leaf abscission. Although ABA stimulates leaf abscission in cotton, this effect has not been found generally with other plants, so at present a major role for ABA in leaf abscission is not recognized.

GENETIC BASIS OF DEVELOPMENT

The controlling role of nuclear hereditary material in development can be demonstrated with the single-celled but morphologically com-

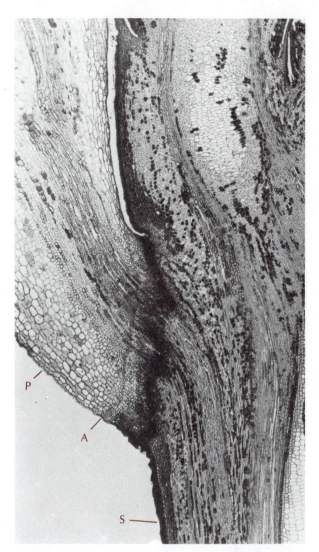

Figure 14-16. Longisection of the base of a leaf petiole (P), showing development of the abscission zone (A) along which the leaf will detach from the stem (S) during autumnal leaf fall.

plex alga *Acetabularia*. As illustrated in Figure 14-18, different species of *Acetabularia* develop a parasol-like "cap" of distinctively different form. The plant normally contains a single nucleus, located in the basal part of the plant body. By experimental manipulations this nucleus can be removed and replaced by a nucleus taken from a different *Acetabularia* species. The cap that develops after such a nuclear transplantation has the morphology characteristic of the species from which the nucleus was obtained. This result shows that the nucleus dictates the morphogenetic processes by which the specific form of the cap is created.

Totipotency

The fertilized egg or other reproductive cell from which a plant or animal develops must contain genetic instructions for developmental processes leading to every kind of cell and organ in the mature plant or animal body. A fertilized egg is developmentally **totipotent;** that is, it has the potentiality for all kinds of development of which the species is capable. Although a possible

Figure 14-17. Prevention of senescence in local spots on a detached leaf by spot applications of a cytokinin. (Left) A cytokinin has been applied to the areas designated by the dark circles. (Right) Two weeks later, while the rest of the leaf has senesced, the two treated spots remain green.

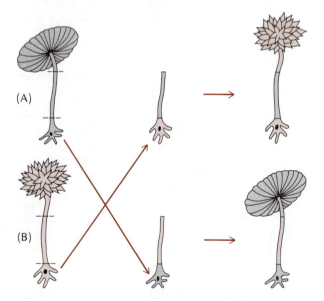

Figure 14-18. Transplantation experiments using two species of *Acetabularia* which differ in cap shape, *A. mediterranea* (A) and *A. crenulata* (B). In the upper sequence, a stalk from *A. mediterranea* is grafted to a base (with nucleus) from *A crenulata*. In the lower sequence, the reciprocal graft is shown. Similar results are obtained by transferring just the nucleus from one species to an enucleated stalk of the other species.

means of channeling the development of the cells derived from a fertilized egg into different directions would be to parcel out different genetic instructions to different cells, this does not seem to be how development is controlled, at least in plants. When grown under suitable conditions in test tube culture, individual cells obtained from specific vegetative tissues such as a carrot root can develop into entire plants (Figure 14-19). Although they were part of the differentiated plant body, these cells are clearly totipotent and must possess, as does a fertilized egg, all the genetic instructions of the species. The formation of adventitious buds on roots, and of adventitious roots on shoots (Chapter 3), shows that totipotent cells exist throughout the plant body. Therefore, specific kinds of development appear to be specified in general not by parceling out, but by controlling the expression of, genetic instructions. Different kinds of cells make use, in their development, of different parts of the genetic information that all the cells possess.

Developmental Programs of Gene Expression

As indicated in the descriptions of shoot and root development in the preceding chapters, development typically follows a precise, relatively invariant, and often complex course in giving rise to any specific kind of tissue or organ. Some kinds of changes occur early, and others much later, in the developmental sequence. We infer the existence of a developmental program, an orderly regulation of gene action over time and as a function of cell position within the organism, as a basis for developmental changes. An excellent example of a temporal program of specific gene expression is found in the development of the slime mold *Dictyostelium*, which reproduces by developing a stalked, spore-containing structure called a sporangium, as illustrated in Figure 14-20. In the course of sporangial development, several new enzymes appear and increase dramatically in activity, in an undeviating order corresponding to a sequential expression of the genes coding for these various enzymes. Each of these enzymes may bring about a particular step in the growth and construction of the sporangium.

The first step in gene expression is transcription to produce a messenger RNA, which is then translated by the ribosomes of the cell to yield the specific protein (enzyme, etc.) for which that gene codes (Chapter 5). A primary means of controlling development is therefore regulation of the processes of transcription and translation. In *Dictyostelium* (Figure 14-20), prior to the appearance of each new enzyme in the developmental sequence, there is a period during which its mRNA is produced; the developmental program involves a sequential turning on of the transcription of different genes.

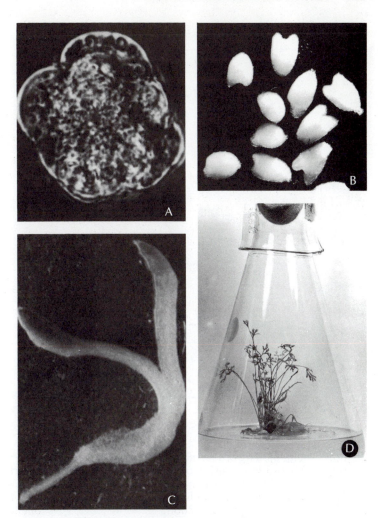

Figure 14-19. The development of a carrot plant from a single cell. (A) Globular embryoid resulting from growth of a single cell of a suspension of embryo origin. (B) Embryoids showing cotyledon primordia from a suspension of leaf petiole origin. (C) Fully developed embryoids. (D) Carrot plant grown from an embryoid in a 1-liter flask.

A delay of some hours occurs, however, between the transcription of each of the controlled genes and the appearance of the corresponding enzyme, indicating that the translation of each mRNA is also being regulated.

The best-understood kind of regulation of gene transcription, illustrated in Figure 5-18, involves the action of a specific regulatory protein (called a repressor) that potentially can bind to the DNA of a gene, blocking its transcription. The repressor can interact with a specific small molecule that, depending on the case, either causes the repressor to bind to the DNA (as in Figure 5-18) and block transcription, or prevents the repressor from binding (Figure 14-21), inducing transcription, in which case the small molecule is called an **inducer.** In these ways, as yet understood in detail only with bacteria, transcription can be either switched off or switched on.

GROWTH REGULATOR ACTION

In plants the principal agents known to influence gene expression developmentally are the hormones and other growth regulators. In princi-ple, growth regulators can affect gene expression and development in several possible ways, including (1) regulating gene expression directly by controlling either the transcription or the translation of mRNA; (2) influencing the activity of gene products (enzymes); and (3) regulating transport through membranes, indirectly influencing either (1) or (2). Although the mode of action of none of the plant hormones and growth regulators is yet known in complete detail, we have apparent examples, at least, of actions falling into categories (1) and (3).

In principle, any hormone or growth regulator that exerts a specific developmental effect must interact, in the affected cells, directly with some specific "target" process or structure, influencing it with consequences resulting eventually in the observable developmental effect. This direct interaction and influence is called the **primary action** of the hormone or growth regulator. It is generally expected that the primary action of a hormone should result from its binding to a hormone-specific **receptor** site, most likely on a specific receptor protein. A receptor site is analogous to an effector site on an effector-regulated enzyme (Figure 6-8) or on a repressor protein (Figures 5-18, 14-21). The binding of the hormone to the receptor site should influence positively or negatively the biological activity of the re-

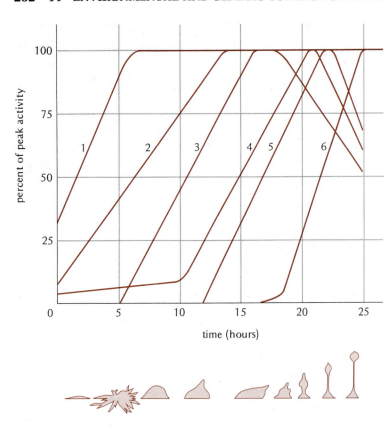

Figure 14-20. Slime mold (*Dictyostelium*) sporangium development. The diagrams along the bottom of the figure indicate the time of appearance of developmental stages through which the slime mold passes as it forms a mature sporangium (far right). Curves 1 through 6 represent the activities of some of the enzymes produced in the slime mold at specific times during the production of a sporangium, the order of enzyme appearance corresponding to a sequential expression of the genes coding for them. (See Chapter 24 for explanation of the slime mold's development and reproduction.)

ceptor protein. Binding of several hormones to binding sites (possibly receptor sites) in plant cells has recently been detected. Current research seeks to discover the direct consequences of this binding, which would constitute the hormone's primary action.

In a number of instances plant growth regulators influence development by affecting the production of, or sensitivity to, another growth regulator or hormone. Auxin, for example, exerts some of its effects by stimulating the formation of ethylene. Phytochrome affects growth in certain cases by controlling ethylene production and in others (including control of stem elongation) apparently by affecting the formation of gibberellin or the responsiveness of cells to this hormone.

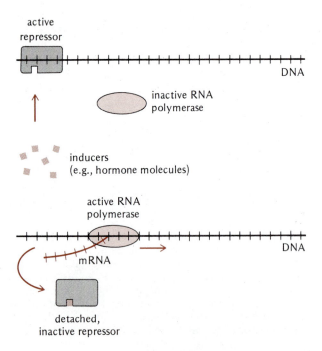

Figure 14-21. Concept of induction of mRNA synthesis (transcription) by a regulatory molecule such as a hormone. For the regulated gene there is a specific repressor protein that tends to bind to the site on the gene (DNA) where transcription must begin, blocking the access of the enzyme responsible for transcription (RNA polymerase) to this initiation site. The repressor has a specific binding site for the regulatory molecule. When this site is occupied it modifies the shape of the repressor, resulting in loss of its tendency to bind to the gene, thus giving RNA polymerase access to the initiation site and allowing transcription to begin. Translation of the resultant mRNA leads to the appearance of the protein for which the gene codes. Although this simple mode of gene activity induction is demonstrated in bacteria, hormonal induction in eukaryotes seems generally to be more complicated. In many cases the hormone apparently acts indirectly on transcription (i.e., as a result of other changes the hormone causes), rather than interacting directly with a repressor as represented here.

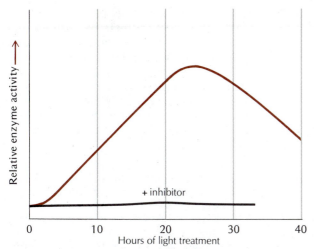

Figure 14-22. Phytochrome-mediated induction of phenylalanine ammonia lyase (PAL) in mustard seedlings exposed to light (colored line). When the seedlings are given light in the presence of an inhibitor of either RNA synthesis or protein synthesis, PAL activity does not increase (black line).

In these cases, to understand the developmental effect fully, we must determine the primary action of the intermediating hormone.

Control of Gene Activity

All the growth-stimulatory plant hormones stimulate RNA and protein synthesis at least over the longer term (hours or days), and since sustained growth and cell division require general protein synthesis, it is perhaps not surprising that hormonal stimulation of development is almost invariably suppressed by agents that block RNA or protein synthesis. It is much more difficult, however, to show that hormones stimulate growth or cell division by acting directly on transcription or translation, for example, by acting as inducers in the sense shown in Figure 14-21.

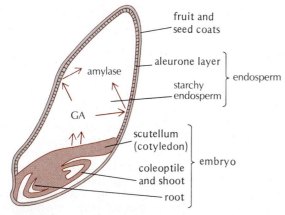

Figure 14-23. Barley seed in longisection, showing the thin aleurone layer (actually 2–3 cells thick) surrounding the extensive starchy endosperm. Gibberellin (GA) released from the embryo at the beginning of germination induces the aleurone cells to produce α-amylase and secrete it into the starchy endosperm.

When converted to the P_{FR} form, phytochrome induces the synthesis of certain enzymes (Figure 14-22) by a process that requires RNA synthesis (presumably the formation of mRNA). These enzyme increases explain the phytochrome stimulation of anthocyanin pigment and phenolic secondary product formation. In stimulating leaf abscission, ethylene induces cells of the abscission zone (Figure 14-16) to form enzymes (pectinase, cellulase) that weaken cell walls by degrading their components, so that the petiole tissue eventually breaks along the affected zone. These are examples of growth regulator control of gene expression, but it is not clear that the primary action is directly on transcription in any of these cases.

GIBBERELLIN INDUCTION OF ENZYMES IN ALEURONE CELLS

The best-understood case of hormonal control of enzyme formation in plants is the induction, by gibberellin, of enzyme synthesis in seeds of grasses such as wheat and barley. During germination the outer cell layer of the storage tissue or endosperm of the seed, called the **aleurone layer** (Figure 14-23) because its cells contain storage protein (aleurone) granules, produces large amounts of amylase. This enzyme breaks down starch stored in the endosperm, mobilizing this energy reserve for uptake and use by the embryo. The aleurone cells fail to produce amylase if the embryo is removed from the seed before the seed is wetted, indicating that some message from the embryo controls the phenomenon, enabling the embryo to induce the mobilization of its nutrients when it needs them. During germination the embryo releases gibberellin, and gibberellin actually induces the aleurone cells to produce amylase (Figure 14-23). Supplying gibberellin to an embryo-less seed or to isolated aleurone tissue will induce the aleurone cells to produce amylase in the absence of the embryo (Figure 14-24). RNA synthesis in the aleurone cells is necessary for amylase production in response to gibberellin.

By testing its ability to stimulate amylase synthesis by isolated ribosomes from wheat germ (Chapter 5), assays of mRNA from aleurone cells indicate that these cells turn out mRNA for amylase during the first few hours after exposure to gibberellin, before enzyme synthesis starts. However, gibberellin evidently also controls the translation of amylase mRNA, because if the tissue is deprived of gibberellin after amylase production is underway, enzyme synthesis stops even though the mRNA for amylase is definitely present. Although important features of this hormone regulation phenomenon are thus recognized, the primary action of gibberellin even in this response still remains to be identified.

Membrane-Level Regulation

Auxin stimulates growth, and phytochrome converted to P_{FR} affects the sleep movements of leaves, within just a few minutes, that is, much

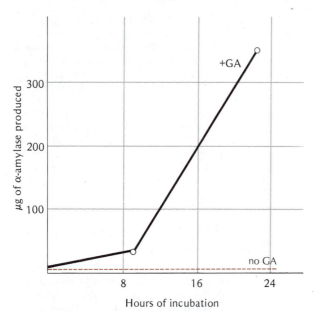

Figure 14-24. Induction, by gibberellin (GA), of α-amylase production by isolated barley aleurone layers floating on a dilute aqueous medium. Most of the produced amylase is extruded by the aleurone cells and appears in the culture medium. If no GA is supplied virtually no amylase is produced (dotted line at bottom of graph), whereas with GA rapid amylase production occurs after a lag of about 8 hours. If an inhibitor of RNA synthesis is added to the medium at the start of GA treatment, no amylase is produced (results are just like "no GA"), whereas if the same inhibitor is added more than 8 hours later it does not prevent amylase formation because by then the aleurone cells have already produced the mRNA for amylase in response to GA.

more rapidly than the recognized inductions of enzyme formation (Figures 14-22, 14-24). It thus appears that auxin and phytochrome may exert these and certain other actions by influencing some preformed biological system in the cells rather than by inducing new production of mRNA and proteins. Because the action of the motor cells in sleep movements is due to ion transport through their

membranes, it seems likely that phytochrome exerts at least some of its effects by influencing membrane transport, although how it does so is not yet clear. In promoting senescence, ethylene also apparently works by influencing membranes, causing them to become more permeable. Vacuolar and cytoplasmic constituents mix and the cell loses solutes, eventually causing it to lose turgor, or wilt.

As explained in Chapter 12, auxin promotes cell enlargement by causing a weakening of the cell wall and an increase in its plasticity. Recent evidence indicates that auxin does this by inducing cells to actively pump H^+ ions out into their cell walls; an acidic pH in the cell wall causes it to become more plastic so it expands, enlarging the cell (Figure 14-25). It remains to be shown how an acidic pH in the cell wall increases its plasticity; this may be because acidic conditions promote the breakdown of certain wall polymers or of bonding forces between them.

DIFFERENTIATION

In the specialization of cells for different functions during development, as noted previously, different parts of an organism's hereditary information become selectively expressed in different cells. Among the most important problems in biology is how this selective gene expression is achieved.

Genes that control the transcription of other genes (for example, by specifying repressor proteins) are called regulator genes (Chapter 5). Because multiple-copy DNA base sequences are characteristic of eukaryotic cells capable of differentiation, it has been suggested that these sequences might comprise regulator gene networks, capable of selectively turning on many different combinations of genes. This remains largely speculation. Although little concrete explanation of gene expression in differentiation can be given as yet, some general features pertinent to plant cell differentiation are noted below.

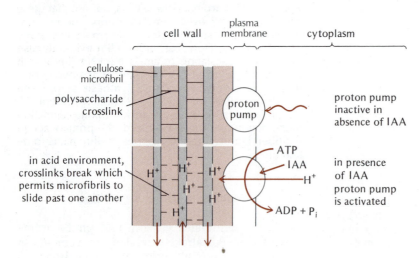

Figure 14-25. Hydrogen ion extrusion theory of auxin action on cell enlargement. The cell wall contains structural polymers, here represented as matrix polysaccharides crosslinking the microfibrils of the wall, preventing wall extension. The wall becomes acidified by extrusion of H^+ ions from the cell by a proton pump activated in the presence of IAA. Acidic conditions in the wall weaken its structural crosslinks, allowing its microfibrils to slide and the wall to extend under the forces exerted by the cell's turgor pressure, resulting in irreversible enlargement (growth) of the cell.

Determination

In the first step of differentiation, **determination,** an embryonic or meristematic cell becomes committed to a particular developmental program that will lead eventually to the second step, overt differentiation, during which the cell acquires its specialized structure and composition. An example of determination is seen in the development of leaves. With certain plants it is possible to remove young leaf primordia at different times after their emergence from the flank of the apical meristem and to grow the isolated primordia on a culture medium. Leaf primordia older than about the last five or six to be initiated will usually grow and differentiate into recognizable leaves (Figure 14-26). These primordia therefore have already been determined to become leaves, even though they do not have the morphology or anatomical structure of a leaf at the time they are excised. With fern leaves, younger leaf primordia behave more erratically, sometimes developing into leaves and sometimes not, whereas the smallest leaf primordia that can be grown successfully in isolation (the second and third youngest) usually develop into stem-like organs rather than leaves and evidently have not yet been determined. With flowering plants that have been investigated, even the smallest primordia that can be grown successfully (the second youngest) develop into leaves and are already determined at that early stage of their development.

The steps of differentiation, at least of xylem and phloem cells, are under hormonal control, as noted in Chapter 12. In experimental test systems the involved cells divide before hormonally induced differentiation occurs. It seems that the act of division may be necessary to set in motion the developmental program for differentiation.

Polarity

When daughter cells from a cell division are going to differentiate for different functions, such as into a sieve tube member and a companion cell (Figure 13-21) or into a stomatal guard cell complex and an ordinary epidermal cell (Figure 12-23), the division that gives rise to the differently determined daughters is generally an unequal one, as illustrated in the figures cited. This reflects the acquisition of some kind of directional organization or **polarity** by the parent cell prior to its division, often indicated by a dif-

Figure 14-26. Leaf primordia of cinnamon fern (*Osmunda cinnamomea*) grown *in vitro.* (A) A frond resulting from the growth of an older (P5) primordium (4 ×). (B) A very small (P2) primordium may grow into a whole plant instead of a leaf; shoot apex (arrow) and first leaf (30 ×).

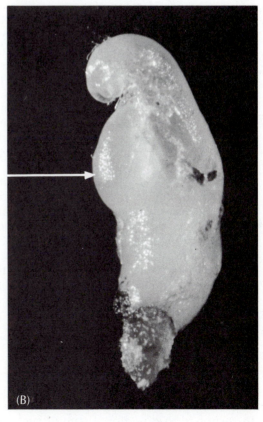

ference in cytoplasmic density and organelle distribution between the two halves of the cell. Therefore, when the cell divides, the daughter nuclei find themselves in different cytoplasmic environments, which may cause them to enter different developmental programs. Thus the process leading to determination may begin even before the cells that are eventually to differentiate arise.

The plant system best suited to studying experimentally the development of cell polarity is the fertilized egg of the seacoast alga, *Fucus* (Figure 23-40). This cell is initially symmetrical, but in response to directional illumination or certain other stimuli, it becomes polarized, leading to outgrowth on one side and an asymmetric division that separates the cells that will give rise to the shoot and the root-like parts of the plant (Figure 14-27). Delicate electrical measurements show that during the acquisition of polarity (prior to polar outgrowth or division), a minute electric current develops in the egg, caused by output of ions (especially Ca^{2+}) from one side of the cell and inflow of ions on the other. This type of current, which has also been found in other polarized plant cells, may help bring about the asymmetric distribution of cytoplasmic components responsible for the subsequent differentiation of the daughter cells. It has been suggested that electrical polarity is a determining principle in plant cell differentiation and morphogenesis, but the generality of this hypothesis remains to be demonstrated.

Overt Differentiation

The specialized functions of many differentiated animal cells are attributable to distinctive proteins, such as the oxygen-carrying protein (hemoglobin) of red blood cells, and the contractile proteins actin and myosin of muscle cells. Thus overt differentiation is often equated with the activation of genes for "cell-specific proteins."

Plant tissues that specialize for particular metabolic functions, such as photosynthesis, deposition of cutin and cuticular waxes, and formation of secondary products, produce the special enzymes required for these biosynthetic pathways. Examination of the protein composition of plants shows differences between different organs such as leaves, stems, roots, and flower petals, and there is some evidence for protein differences between different tissues in the same organ. However, there is almost no evidence on the actual significance of these protein differences in the different development and structure of different cell

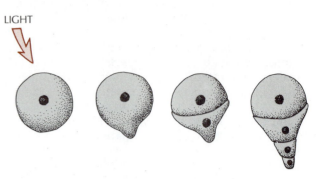

LIGHT

Figure 14-27. Acquisition of polarity and early development of a fertilized *Fucus* egg. At far left, a fertilized egg is subjected to a directional light cue. A protuberance grows out from the side of the cell directly opposite the light. The first (asymmetric) division produces a smaller cell which will give rise, on the darker side, to the root-like portion of the plant, and a larger cell which will produce the shoot on the lighted side.

types, apart from the need for special enzymes for specialized metabolism.

The most common and conspicuous method of cell differentiation in plants, the formation of a secondary wall, produces a variety of cells such as tracheids, vessel elements, fibers, and sclereids that differ importantly in structure and function not because their walls differ in composition, but because their cell shapes differ and because their walls have been deposited to different thicknesses and in different patterns. Differentiation of particular cell types is a matter of specifying how cellular metabolic products will be distributed and oriented, a question of morphogenesis. Morphogenesis, although clearly under genetic control (Figure 14-18), is probably a good deal more subtle than simply turning on genes for the formation of particular products. This aspect of the genetic control of development in plants is not yet understood in concrete terms.

During the maturation or differentiation of many plant tissues, the chromosomes in each cell's nucleus replicate up to several times without the occurrence of mitosis, yielding cells with multiple chromosome sets (**polyploid** cells; Chapter 20). This increase in the number of copies of the cell's genes (gene amplification) may help enhance the formation of messenger RNA for cell-specific proteins when large amounts of these must be synthesized during differentiation. However, gene amplification does not seem to be a generally essential feature of plant cell differentiation, because in various plants normal differentiated cells develop without becoming polyploid.

SUMMARY

(1) Tropisms are oriented growth curvatures in response to directional stimuli such as light (phototropism) and gravity (geotropism). In shoots, these curvatures result from the stimulus-induced

lateral transport of auxin. (In roots, geotropism may be controlled instead by abscisic acid.) Light for phototropism is detected by a pigment called the blue light photoreceptor, probably a flavoprotein. Surface contact induces coiling (thigmotropism) of the tendrils and twining stems of climbing plants.

(2) Turgor movements are reversible, usually bending or curving motions, of plant organs caused by changes in turgor pressure. Sleep movements (nyctinasty) change the position of many leaves and flower petals on a day–night cycle, which is sensitive to light detected by phytochrome. Some leaves show daily solar tracking movements, which keep them facing the sun for maximum solar energy capture at all times of day. [Certain leaves, such as those of the sensitive plant and the Venus flytrap, undergo turgor movements in response to touch (contact) or shaking.] In these responses, an increase in turgor seems to be due primarily to K^+ uptake, increasing the cells' solute concentration and resulting in osmotic water uptake, while a decrease in turgor results from leakage of K^+ and other ions from the cell, causing osmotic water loss.

(3) Morphogenetic responses are developmental changes that are not oriented relative to the direction of a stimulus. Light induces diverse and profound morphogenetic responses (photomorphogenesis), exemplified by the induction of leaf expansion and inhibition of excessive internode elongation when dark-grown (etiolated) plants are exposed to light. In these responses light is detected by a protein pigment, phytochrome, which absorbs red light strongly (wavelengths of about 670 nm), resulting in conversion of the pigment to a biologically active form (P_{FR}) capable of absorbing far-red light (730 nm). Far-red light reconverts P_{FR} to the biologically inactive red-absorbing form (P_R), and will therefore block photomorphogenesis if a dose of far-red light is given immediately after a photomorphogenically active dose of white or red light. The far-red reversibility of a light response provides specific evidence that it is phytochrome mediated. The development of chloroplasts from etioplasts, precursor organelles derived from proplastids in the dark, is an intracellular phytochrome-mediated photomorphogenic response.

(4) The formation of dormant winter buds and abscission (shedding) of leaves in autumn are phytochrome-mediated morphogenetic responses to day length (photoperiodism). Buds normally become released from dormancy in response to extended exposure to cold temperatures, usually experienced in the course of winter (in some cases, exposure to long days breaks dormancy instead). The balance between dormins and gibberellins may be how the occurrence of dormancy is controlled. Leaf abscission may be controlled by the balance between auxin (inhibits) and ethylene (stimulates) [the senescence preceding or accompanying abscission may involve cytokinins (inhibitory) and ethylene (stimulatory)].

(5) Nuclear control of development is demonstrated, for example, by nuclear transplantation in the alga *Acetabularia*. A developmental program of selective gene expression is illustrated by sequential enzyme changes leading to sporangium formation in the slime mold *Dictyostelium*. The genetic complement of nuclei is not altered during developmental selective gene expression, because complete plants can be regenerated from vegetative plant cells either in cell and tissue cultures or via adventitious bud and root initiation from differentiated organs; nuclei of differentiated plant cells are thus demonstrated to be totipotent.

(6) Environmental factors that influence development, and hormones or growth regulators that mediate environmental effects and internal controls of development, affect gene expression either directly or indirectly. In aleurone cells of grass seeds such as barley, gibberellin turns on the synthesis of mRNA for the enzyme amylase, leading to massive production and excretion of amylase from the cells. Various phytochrome, ethylene, and auxin effects also apparently involve the production (or activation) of specific enzymes. Certain hormone or growth regulator actions, such as that of auxin on cell enlargement, occur quickly and are thought to result from an effect on transport through membranes.

(7) Differentiation, a special case of selective gene expression, consists of determination, or commitment to a particular developmental program, plus overt differentiation, or development of distinctive cell structure and function. The steps to differentiation, especially of vascular cells (phloem and xylem), can be influenced or controlled by hormones, particularly auxin. Cell differentiation frequently follows the development of cytological asymmetry or polarity in a plant cell and its unequal division into daughter cells that then follow different developmental programs. (Developmental polarity, in eggs of the alga *Fucus* and certain higher plant cells, is indicated by an electrical polarity, or directed flow of ion current through the cell.) Overt differentiation involves not only the formation of cell-specific products but also their specific organization within or around the cell, that is, morphogenesis, as illustrated by the different types of plant cells that differentiate by forming a secondary wall deposited in different specific patterns.

Chapter 15

VEGETATIVE REPRODUCTION

Higher plants have basically two ways of reproducing their kind: sexual reproduction, resulting from the combination of hereditary material from different cells or individuals and leading, in flowering plants, to seeds; and asexual or vegetative reproduction, which propagates a given individual without a sexual process. Since vegetative reproduction is simpler and, in most cases, is merely an extension or specialization of the ordinary vegetative (nonreproductive) development discussed in the preceding chapters, we consider it first.

Vegetative reproduction is of enormous importance both in the propagation of cultivated plants and in the competition among plants in nature. Its first advantage is that, by producing what are essentially offshoots of itself, a plant can give rise to progeny that are much more robust, vigorous, and likely to succeed on their own than are small, delicate seedlings. The second advantage is that, because vegetatively generated offspring are produced essentially by mitotic cell division, which passes on all the cell's genes to its daughters (Chapter 4), vegetatively formed progeny are genetically identical to their parent, as human identical twins are to one another. Vegetative progeny inherit any and all genetic advantages the parent may possess. This enables a genetically superior individual to spread and multiply its advantageous genes rapidly without mixing them up with others in the course of sexual reproduction. The collection of genetically identical individuals produced by the vegetative reproduction of a single individual is called a **clone.**

Sexual reproduction has its advantages, however, including the ability to create advantageous new combinations of genes from different individuals (Chapter 18) and the possibility of long-distance dispersal by specialized seeds or fruits (Chapter 17). Most plants that reproduce vegetatively, therefore, also engage in sexual reproduction. Few if any annual plants have a means of vegetative reproduction, instead relying entirely on seeds for their propagation. Natural vegetative reproduction is almost universal among herbaceous perennial plants but is less common, though by no means lacking, in woody plants, especially trees. Because of the usefulness of vegetative reproduction in horticulture and agriculture, humans have devised various artificial methods of propagating plants vegetatively, considered later in this chapter.

ORGANS OF VEGETATIVE REPRODUCTION

Chapter 3 discussed and illustrated many of the types of modified stems and roots, and the phenomena of adventitious bud and root formation, that make vegetative reproduction possible. Specialized underground shoots such as rhizomes, tubers, and bulbs are often basically organs for winter or dry-season survival and comprise the permanent axis of many perennial herbaceous plants. Because most such underground shoots can branch, however, they inevitably give rise to new individuals when different branches are separated from one another by senescence and decay of the older parts. Many plants with a localized principal vegetative axis also develop outwardly growing rhizomes or other organs specialized for vegetative propagation of the individual (Figure 15-1).

The extent to which plants can spread by means of **rhizomes** is impressive. Figure 15-2 shows the extensive growth and large number of new, potentially separate plants arising in one growing season from rhizomes produced by a single seedling of cattail (*Typha*), a common marsh plant. A blueberry species native to the southeastern United States spreads about 30 cm/year by sending out rhizomes. One of these plants, with

Figure 15-A. Water hyacinths in Brazil. The extremely rapid rate of vegetative reproduction of water hyacinths has made the plant a major problem in many countries where it covers and clogs inland waterways and lakes.

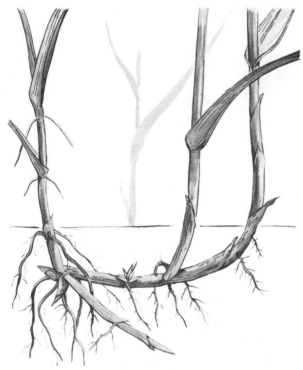

Figure 15-1. Elongated horizontal rhizome growing out from the base of a vegetative plant. New shoots that have developed from buds along the rhizome are shown to the right.

hundreds of aerial shoots all connected by an underground rhizome system radiating from a center, covered an area nearly half a mile in diameter and was estimated to be 1000 years old. Rhizomatous grasses, including many of the prairie grasses, similarly spread extensively. Genetic analysis of plants of the grass *Holcus mollis* in Sweden revealed, for example, that particular individuals had developed, by vegetative reproduction, into clones covering areas as much as half a mile wide.

Rhizomes exhibit **diageotropism** (Greek *dia,* away from), a growth curvature response that orients the rhizome tip approximately perpendicular to the direction of gravity. (Lateral branches in an aerial shoot system also show diageotropic behavior.) Some rhizomes have a light-avoidance response (negative phototropism) which further helps them to keep their position suitably below the surface of the ground. Auxin seems to be involved in these responses, as in ordinary phototropism and geotropism, but the mechanism of diageotropism is relatively little understood.

Plants that overwinter by means of **bulbs** reproduce vegetatively by "division" of the bulb. This results from the development of axillary buds within the parent bulb, which form their own bulb scales and become separate bulbs (called **offsets**) once the parent bulb's outer scales have died and

Figure 15-2. (Above) Cattail (*Typha latifolia*) plant grown from a single seed in 6 months. (Below) The same plant with leaves cut back and roots removed. The star in the center indicates the original shoot.

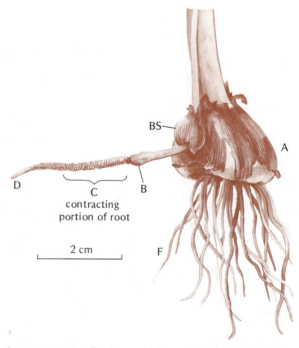

Figure 15-3. Bulb dispersal by contractile root action in the wild onion, *Allium neapolitanum.* A, parent bulb with fibrous roots (F); B, offset (laterally formed bulblet) bearing horizontal root whose growing tip (D) is anchored in the soil, allowing contraction of the root's basal region (C) to pull the offset away from the parental bulb, and out of its outer bulb scales (BS) located where the offset was formed next to the parent bulb.

weathered away. The "cloves" within a garlic bulb illustrate this process partway toward completion. An offset is initially attached to its parent bulb, but the offsets of certain plants begin independent life by sending out a strong horizontal contractile root that pulls the offset some distance away from the parent bulb (Figure 15-3), enabling the offset to colonize new ground. The depth of bulbs and corms below ground is also maintained by the yearly production of contractile roots as noted in Chapter 3.

Some plants produce aerial organs for vegetative reproduction. The **runner** (Figure 15-4), as found in the strawberry, is an elongated, horizontally growing shoot, often with reduced leaves; at its nodes it develops adventitious roots enabling the developing axillary buds to become independent plants.

A **bulbil** is a small bulb-like bud, formed on the shoot, which detaches by means of an abscission zone and, in contact with the ground, develops adventitious roots and grows into a new plant. In different plants bulbils are formed either on leaves (Figure 15-5) or on the aerial stem, often where flowers would otherwise occur (Figure 15-6).

Many plants reproduce vegetatively by forming adventitious buds and shoots ("suckers") on their roots (Figure 3-18). Weeds that have this habit, such as bindweed, wild blackberry, and Canada thistle, are almost impossible to get rid of by cultivation because this merely breaks up the roots, each piece subsequently sprouting a new plant. A number of shrubs and trees also spread vegetatively by producing adventitious buds from their

Figure 15-5. Upper and lower surfaces of leaves of *Kalanchoë diagremontianum*. The margins of the leaves bear young plantlets.

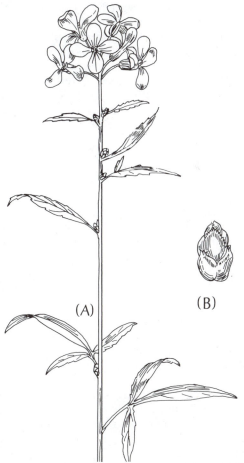

(A)   (B)

Figure 15-6. (A) Shoot of the mustard-like wildflower *Dentaria bulbifera*, bearing bulbils [magnified view, twice natural size, in (B)] in axils of leaves. The bulbils drop off and sprout new plants.

runner

adventitious roots

Figure 15-4. Vegetative reproduction in wild strawberry (*Fragaria virginiana*).

Figure 15-7. Groves of aspen trees in spring, in a mountain habitat. The uniformity with which all the aspens in the middle foreground are leafing out, in contrast with other groves on the hill just beyond where the buds have only begun to break, indicates that the closer grove may be a single clone, all the trees derived from one original individual by suckering from adventitious buds forming on the roots, a mode of multiplication known to lead to aspen colonies.

roots, for example, the common beach plum (*Prunus maritima*) of the northeastern United States coast and other species of wild plum and wild cherry. The aspen tree (*Populus tremuloides*) common in the Rocky Mountains is especially given to this kind of reproduction; whole aspen groves (Figure 15-7), covering more than 200 acres and consisting of as many as 47,000 trees, are derived from a single original individual. A record for the known age of this type of clone is an aspen grove covering several acres in Minnesota, which has probably existed since the end of the last ice age some 8000 years ago.

Another method of vegetative spreading is seen in weak-stemmed, sprawling shrubs such as wild blackberries and raspberries. Their stems root wherever they fall on the ground, and thereby become new individuals extending the area occupied by the clone. This type of reproduction is termed layering (see Figure 15-10).

ARTIFICIAL VEGETATIVE PROPAGATION

Vegetative propagation is widely practiced by gardeners and commercial plant growers to reproduce desirable genetic strains or cultivars and to multiply plants more quickly or conveniently than by germinating seeds. A robust, fully developed or produce-bearing plant can often be obtained much more quickly by vegetative propagation than via seed. Many lilies and orchids, for example, grow very slowly from seed, several years frequently being required before new plants bloom, whereas new plants obtained by vegetative propagation often flower within one or two seasons.

Table 15-1 lists some economically important plants that are usually propagated vegetatively.

Some of these, including the cultivated banana, pineapple, navel orange, Marsh grapefruit, and Thompson seedless grape, entirely lack seed reproduction and can be propagated only vegetatively.

Over time, useful new genetic traits or combinations of traits occasionally appear in individuals of a cultivated plant, either as a result of genetic mutation or when genes from different individuals come together in sexual reproduction or hybridization to yield a new variety with advantageous characteristics. These desirable features can be reproduced unchanged if the individual can be propagated vegetatively. All the trees of any named cultivar of apple, such as McIntosh or Golden Delicious, comprise a vegetatively propagated clone, as do individual cultivars of strawberries, peaches, roses, potatoes, and many other vegetatively propagated plants. Because the

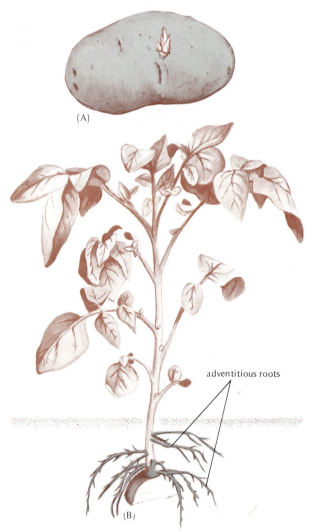

Figure 15-8. Potato propagation from cuttings from tubers. (A) Uncut seed potato, showing eye. (B) Eye sprouted into a new shoot from which have arisen adventitious roots.

TABLE 15-1   Vegetative Propagation of Horticultural Plants
Information from H. T. Hartmann and D. E. Kester, *Plant Propagation, Principles and Practices*, 3rd ed., Prentice-Hall, 1975

Method	Ornamentals	Crop plants	Method	Ornamentals	Crop plants
Rhizomes[a]	orchids	banana		trumpet vine	red raspberry
	iris	asparagus		oriental poppy	blackberry
	day lily	artichoke	Leaf cuttings	begonia	
	calla lily	mint		african violet	
	aster			Peperomia	
	shasta daisy			Sansevieria	
Tubers	Caladium	white potato	Layering	magnolia	black raspberry
Stolons (runners)	red osier	strawberry		Dieffenbachia	gooseberry
	dogwood			holly	currant
	bugle (*Ajuga*)			rhododendron	filbert
Corms	crocus	taro		azalea	litchi
	gladiolus			lilac	apple rootstocks
	Freesia			rubber plant (*Ficus elastica*)	
	cyclamen				
Bulbs	tulip	onion	Stem grafting	camellia	grapevine
	daffodil, etc.			rose	walnut
	lily				apple
Offsets or "stem suckers"[b]	century plant	date palm			olive
		pineapple	Root grafting[d]	rhododendron	pear
Tuberous roots	Dahlia	sweet potato		wisteria	apple
	Ranunculus	manioc (tapioca)	Budding	crabapple	apple, pear
Stem cuttings	azalea	sugar cane		rose	peach, plum, almond
	camellia	grapevine			apricot, cherry
	rhododendron	blueberry			walnut, pecan
	mulberry	currant			grapevine
	quince	fig			citrus fruits
	geranium	pineapple			avocado
	chrysanthemum				para rubber tree
Root cuttings and root	poplar	apple rootstocks	Tissue culture	orchids	asparagus
"suckers"[c]	black locust	horseradish		chrysanthemum	sugar cane

[a]Including dividing the "crown," the complex of rhizomes or basal shoots of a herbaceous perennial at or below ground level.
[b]Lateral shoots arising near the base of the stem.
[c]Shoots arising from adventitious buds on the roots.
[d]Grafting a stem (scion) onto a root (stock).

sexual process involved in seed formation reshuffles or recombines genes, a cloned cultivar derived in the above manner usually does not "breed true" from seed; for example, most of the trees that could be grown from the seeds in a Golden Delicious apple would probably bear apples more like crabapples than like their parent, the Golden Delicious. Therefore, many fruit and "root crop" plants, which have traditionally been propagated vegetatively, must be so propagated in order to retain the desirable produce to which we have become accustomed. This is not true of traditionally seed-propagated domesticated plants such as cereal grains and most garden vegetables, the cultivars of which are not clones.

Artificial vegetative propagation methods exploit, where possible, the modified stems or roots that plants have evolved for natural vegetative reproduction as considered in the previous section. In addition, over the centuries growers have discovered various techniques for improving a modest tendency of some plants toward vegetative reproduction, techniques enabling many other plants that do not naturally reproduce by vegetative means to be so propagated. The more important of these techniques are examined in the following sections.

Cuttings

Vegetative plant parts removed and encouraged to form adventitious roots are **cuttings.** They are made from stems, roots, and even leaves, but the most common are stem cuttings, often called slips, especially if nonwoody. The basal part of a section of stem 7–20 cm or more in length, bearing several nodes and lateral buds, is placed in moist sand or soil, sometimes in a warmed greenhouse bench and sometimes with a misting arrangement to keep it from drying out. If successful, adventitious roots grow out from the lower end and shoots from the upper buds. Prominent among the food plants propagated by cuttings from aerial stems are sugarcane, pineapple, and cassava. Sugarcane is propagated by planting sections of stalks, each bearing one or more nodes and buds. Roots spring

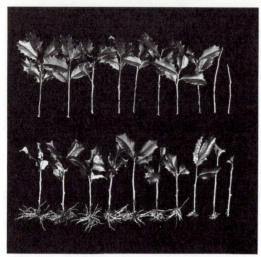

Figure 15-9. Induction of adventitious root formation in cuttings of American holly (*Ilex opaca*) by auxin. (Top row) Without auxin; (bottom row) cuttings treated with auxin.

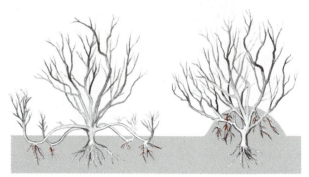

Figure 15-10. Vegetative reproduction by layering. (Left) In "simple layering," branches that are bent down and partially covered with soil form adventitious roots where covered and may then be cut from the parent plant and grown independently. Raspberries and Rhododendrons, among many other horticultural plants, are propagated this way. (Right) In "mound layering," a plant is pruned back forcing outgrowth of new shoots from the base. These form adventitious roots where they are buried by hilling soil around the plant, and may then be separated as independent plants. Currants, gooseberries, and certain apple and quince cultivars are propagated this way.

from the region just above the nodes, and buds grow into shoots.

The white, or Irish, potato is propagated by cuttings of its tubers (Figure 15-8). The tuber is cut up so that every "seed piece" bears at least one node or "eye," from which a shoot develops, producing adventitious roots from its underground portion. Later, rhizomes develop from lateral buds located at underground nodes and expand at the tips into new tubers (Figure 3-28).

AUXIN EFFECT ON ROOTING

Not long after the discovery of the plant growth hormone, auxin, it was found that treatment with auxin stimulates adventitious root initiation by cuttings (Figure 15-9). Since adventitious roots arise by the development of a meristem of dividing cells from mature cells of the stem that would not otherwise have divided again, this is another effect of auxin on cell division besides those mentioned in Chapter 12. This discovery greatly broadened the range of plants that can be propagated by cuttings. The basal end of a cutting is dipped in a dilute solution or a powdered formulation of the hormone, then planted in soil. Cuttings thus treated produce roots more rapidly and in greater numbers than untreated cuttings (Figure 15-9). Such plants as yew, holly, and camellia, difficult to root from cuttings, respond to this treatment. Commercial hormone preparations under a variety of trade names are now widely used by amateur gardeners as well as by professional nursery workers.

Layering

If they contact the soil, the stems of many plants produce adventitious roots. This fact is uti-

lized in the horticultural method called **layering** (Figure 15-10) for propagating raspberry, rhododendron, rose, currant, gooseberry, and various other shrubs and trees. Some difficult-to-root woody plants can be propagated by "air layering," wounding a branch and surrounding it at that point with a ball of damp moss, perhaps dusted with a root-inducing hormone. If this induces adventitious roots to grow out into the moss, the branch can be cut off and planted.

Root Propagation

Many garden ornamentals, such as phlox, delphinium, columbine, oriental poppy, and dahlia, and some root crops such as the sweet potato (Figure 15-11) are propagated either by root cuttings or by division of the roots. In the latter case, the new growth comes from adventitious buds formed either at the base of the old stem or on the roots themselves. Dahlias, widely grown for their attractive autumn flowers, form a clump of fleshy roots at the base of the stem. These produce adventitious buds at the junction of the root and stem. The clumps are stored over the winter, then divided into individual roots for planting in the spring.

Propagation by Leaves

The leaves of a considerable number of plants (24% of 1204 species tested) will produce adventitious roots and shoots under laboratory conditions. In general, however, only fleshy or leathery leaves are used for propagation on a commercial

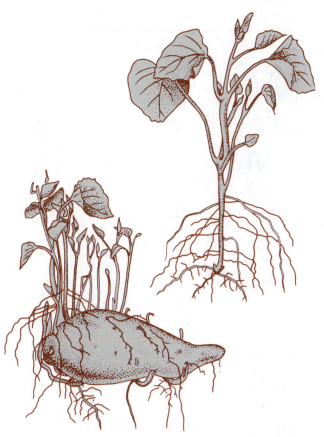

Figure 15-11.   Vegetative reproduction in sweet potato (*Ipomoea batatas*). (Left) Adventitious shoots growing from swollen storage root. (Right) Single shoot, called a "slip" or "draw," removed and ready for planting.

scale, since such leaves possess food and water reserves sufficient to keep them alive until a new plant is established. Among the plants regularly propagated by leaves are the snake plant (*Sansevieria*), African violet (Figure 15-12), gloxinia, and several begonia species. In some cases the entire leaf is planted; in others leaf cuttings are made.

Meristem Culture

Greenhouse orchids may be propagated vegetatively by division of the rhizome, but a more rapid method involves tissue culture. Buds are removed from the bases of the leaves (Figure 15-13) and the apical meristem of each bud, including several leaf primordia, is dissected out under sterile conditions. The excised piece (Figure 15-13) is then planted in sterile nutrient agar, where it grows into a small rounded body, several millimeters in diameter, called a **protocorm.** If a protocorm is cut into several pieces, buds develop on each piece and grow into new protocorms. Each protocorm is capable of forming a new plant. As long as the protocorms are cut into pieces they continue to

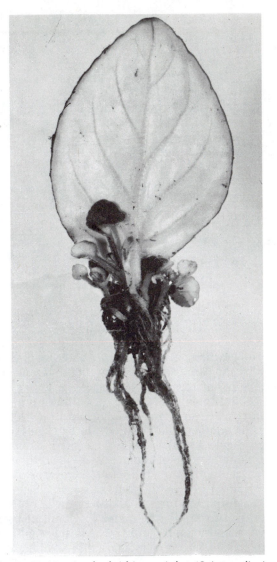

Figure 15-12.   Leaf of African violet (*Saintpaulia ionantha*) bearing adventitious plants at the base. These were produced when the petiole was set in soil.

yield new protocorms, but when cutting is stopped each protocorm produces a new plant. This technique can be regarded as a modification of the method of taking cuttings, but involves only the apical meristem.

Meristem culture is also used to obtain virus-free stock of commercially valuable plant varieties, including the potato and strawberry. In some geographic regions certain viral infections are widespread, and the ordinary means of vegetative propagation, for example, by tuber pieces or stolons, result in infected progeny due to virus carried in the tissues used for propagation. The actively growing meristems of infected plants are the only reliably virus-free parts. Meristems are excised under sterile conditions and grown into virus-free plants by placing the meristematic tissue on a sterile medium containing the appropriate nutrients and keeping it under favorable growth conditions. Such virus-free stock can then be used to multiply the variety by vegetative propagation.

Figure 15-13. Propagation of *Cymbidium* orchids by meristem culture. Apical meristems are removed from within lateral shoots (projection at upper right in A) borne by the plant's basal bulb, and (B) cultured first in tubes (right) and later in bottles (left), proliferating and eventually producing numerous plantlets (C) which can be raised in the greenhouse.

Grafting and Budding

In **grafting,** a horticultural practice discovered by the Chinese as early as 1000 B.C. and described in ancient writings such as Aristotle's and the Bible (Romans 11:17–24), a cutting from one plant (called the **scion**) is attached to a root or stem of another plant, called the **stock** or **rootstock** (Figure 15-14). Tissues of the stock and scion fuse along the line of contact, forming a firm graft union, which permits the scion to grow forth as a shoot, supported by the roots of the stock.

Budding is a form of grafting in which a single bud, surrounded by a small amount of bark, serves as the scion (Figure 15-15). Peaches, plums, pears, citrus, and perhaps half of the commercial stock of apples for planting are produced by budding.

In making a graft it is important that the cambial region of the stock be closely associated with that of the scion. Under the protection of a wrapping or the grafting wax, cells of the cambial zone of both the stock and the scion form a mass of unspecialized wound tissue, or callus. A cambium arises in this callus, uniting the cambium of the stock with that of the scion and producing new layers of wood and bark running continuously between the stock and the scion. These transport water and minerals from the stock to the scion, and photosynthetic products from the scion (after it produces leaves) to the stock, providing for the growth of the whole plant.

Figure 15-14. Whip, or piece-root, grafting of apple (*Pyrus malus*). (A) Scion of the desired variety. (B) Stock; a piece of seedling apple root. (C) Clefts in scion and stock forced together. (D) Graft completed and wrapped. (E) Established graft showing callus.

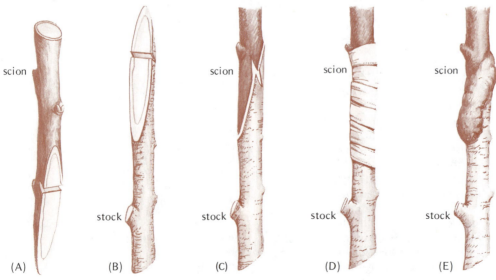

scion   scion   scion   scion

stock   stock   stock   stock

(A)   (B)   (C)   (D)   (E)

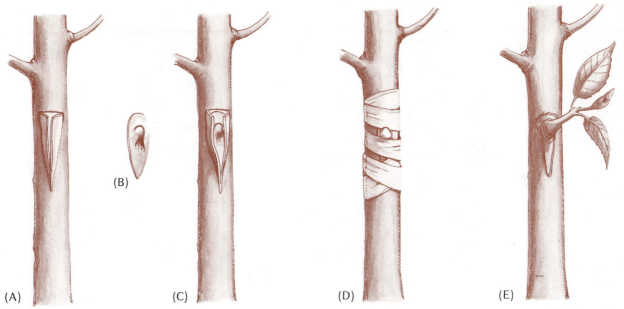

(B)

(A) (C) (D) (E)

Figure 15-15. Budding of a peach (*Prunus persica*). (A) The growing seedling stock is cut to receive a bud. (B) A bud with a shield of bark, cut from the variety to be propagated. (C) Bud inserted in stock. (D) Bud wrapped to protect and hold it in place. (E) Outgrowth of the grafted bud.

Grafting is most effective between related plants, such as different cultivars of the same or closely related species. Grafts between species of different but related genera, however, are not uncommon, as when a pear (*Pyrus communis*) scion is grafted onto a quince (*Cydonia vulgaris*) rootstock. Plants are said to be **compatible** if they can be grafted together. If they are incompatible, a graft union between them will not develop; that is, the graft will not "take." Tomatoes can be grafted on potatoes, olives on lilac, and watercress on cabbage. This is very different from the grafting capabilities of vertebrate animals and humans, in which tissue or organ grafts or "transplants" can usually be made only between very closely related individuals unless rather extreme measures are taken to suppress rejection mechanisms.

Grafting and budding are widely employed in the propagation of woody plants that do not breed true from seeds and are not easily propagated by cuttings. The technique is also used to combine two kinds of plants, as when a weeping mulberry is grafted onto an upright mulberry trunk to form an umbrella-type tree, or when several cultivars of apples are grafted onto the same stock for use in gardens. Another important application is to enable cultivars with pest-, disease-, or cold-susceptible roots (or merely feeble roots) to be grown as vigorous plants under environmental conditions unfavorable to the cultivar on its own roots. Many important cultivars of fruit and nut trees including walnuts, pears, peaches, and apples are widely grown on "wild" rootstocks for this rea-

son, and so are many cultivars of roses. European grapevines (*Vitis vinifera*) had been grown on their own roots, propagated by cuttings, for thousands of years until, in about 1864, a "root louse" or aphid (*Phylloxera*) to which their roots are susceptible was accidentally introduced into France from America, spread rapidly, and destroyed most of the vineyards of France and other European countries within a decade. In the 1870s and 1880s the vineyards were replanted by budding the traditional vine cultivars onto *Phylloxera*-resistant rootstock of wild American grape species such as *Vitis riparia*, an essential method of propagation ever since.

Because the scion develops entirely from cells of the bud or stem that was originally grafted onto the stock, it seems natural that the scion should retain its genetic identity and individual characteristics independently of the stock upon which it is grafted. A Bartlett pear grafted onto a quince rootstock, for example, produces pears untainted by any taste or appearance of quince. This fact, however, enormously intrigued horticulturalists of the Renaissance and Georgian era, when modern grafting techniques were first being developed and enthusiastically applied for setting out new orchards and gardens. Still imbued with the Aristotelian concept that the substance of plants comes from "nutritive juice" brought up by the roots (Chapter 1), people in 1700 found it almost miraculous that the sap from a quince rootstock would, when it passed into a pear scion, support the formation of pears rather than the quinces that it would have given rise to had it remained in a quince branch.

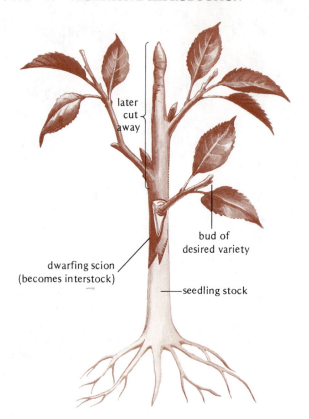

Figure 15-16. Interstock grafting for dwarfing, in apple.

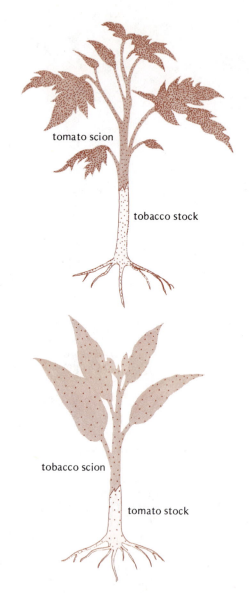

Figure 15-17. Tomato/tobacco intergrafts, showing nicotine accumulation with dots. (Above) Tomato shoot grafted onto tobacco rootstock, with nicotine accumulating in the tomato branches. (Below) Tobacco shoot grafted onto tomato root stock, with little nicotine accumulation in the tobacco leaves.

Although the scion and stock are genetically distinct, the fact that they depend upon one another physiologically leads to significant effects of rootstocks and scions upon one another, some of which are exploited horticulturally. One important example is the use of **dwarfing rootstocks** to produce dwarf fruit trees, desired for home gardens because of their convenience and the relatively small space they occupy. Particular rootstocks have a dwarfing effect upon scions grafted onto them. For example, dwarf pear trees can be produced by grafting certain pear cultivars onto quince rootstocks, and dwarf peach trees by grafting onto apricot.

Dwarf apple trees are produced using **dwarfing interstocks.** A piece of stem of a dwarfing stock is first grafted, as the scion, onto a standard (nondwarf) rootstock. Later, most of this scion's shoot is cut off (Figure 15-16) and the desired cultivar such as Golden or Red Delicious is budded onto the remaining short section of the former (dwarfing) scion, which becomes the interstock and remains between the rootstock and the fruit-bearing top. The short piece of dwarfing wood in the trunk makes the tree semidwarf and early bearing.

Many instances in which different rootstocks affect the size, quality, or yield of fruits, or affect characteristics of the scion's shoots such as winter hardiness,

are known. These and the dwarfing effects noted above presumably result from the transfer (or lack of transfer) of materials such as hormones and nutrients between the stock and the scion. Transfer of a species-specific compound between grafting partners is demonstrated by grafts between tobacco and tomato. Tobacco leaves, of course, contain nicotine, and tomato plants do not, but when tomato shoots are grafted onto tobacco rootstock, considerable nicotine appears in the tomato branch (Figure 15-17). If a tobacco shoot is grafted onto tomato roots, very little nicotine appears in the tobacco leaves. In normal tobacco plants nicotine evidently is produced mainly in the roots, moving from there to the leaves where it accumulates.

SUMMARY

(1) Vegetative reproduction permits a given genetic stock to propagate itself genetically unchanged, yielding a clone; to produce progeny that are initially much sturdier and more vigorous than seedlings are; and to spread over considerable areas from the initial location of one individual.

(2) Natural vegetative reproduction utilizes (a) underground overwintering organs (rhizomes, tubers, bulbs, corms) that branch or "divide," yielding new individuals; (b) under- or aboveground organs specialized for vegetative reproduction (rhizomes, runners, bulbils); or (c) the formation of adventitious buds (suckers) on roots or of adventitious roots on stems (layering).

(3) Artificial vegetative propagation of clonal cultivars is effected by means of the above-mentioned organs and processes as well as by cuttings (using auxin stimulation of adventitious root initiation), leaf propagation, meristem culture, grafting, and budding. The limits of compatible grafting in plants extend to relatively distantly related individuals as compared with the limits of tissue grafting in higher animals.

Chapter 16

SEXUAL REPRODUCTION: THE FLOWER

In contradistinction to vegetative or asexual reproduction, sexual reproduction not only perpetuates and multiplies a species but also enables it to combine, in its offspring, favorable genetic characteristics from different individuals. Flowering plants reproduce sexually in flowers, yielding seeds. Since flowering plants lack locomotion, they depend upon environmental agencies, mainly animals and wind, for the transfer of sexual cells, or pollen, from the flowers of one individual to those of another. Hereditary material can be exchanged between different individuals only if they come into bloom at the same time. Also, if reproduction is to succeed, the formation of flowers must begin early enough in the growing season for seed development to be completed before the onset of harsh environmental conditions such as frost, injurious to the delicate cells involved in sexual reproduction. On the other hand, reproduction is best not begun until a plant becomes large and vigorous enough to produce a strong crop of seeds, since few and weak seeds yield few and weak offspring. These needs dictate the evolution of mechanisms controlling and timing the occurrence of flowering and of flower behavior and structure promoting or compelling cross-transfer of pollen between different individuals. This chapter considers these biological aspects of flower reproduction, as well as the nature of the sexual processes taking place in a flower. Chapter 17 completes the picture by examining the development of a new embryo plant and the seed and fruit that enclose it, the mechanisms for dispersal of seeds, and their germination.

THE FLOWER

Although most people today take the existence of sex in flowers almost for granted, the sexual function of flowers is by no means obvious. It

Figure 16-A. A sunflower being pollinated by a bee.

was missed by the astute ancient observers Aristotle and Theophrastus even though they knew that the Egyptians and Assyrians pollinated date palms artificially to ensure the production of dates; it remained unknown through Medieval and Renaissance times. Nehemiah Grew, the codiscoverer of plant anatomy, seems to have been the first to suggest publicly (in 1676) that flowers are in general sexual and that pollen is their "male" element. Between 1691 and 1694 a German botanist, Rudolf Camerarius (1665–1721), proved by careful experiments on several kinds of flowers that seeds develop only if pollen is placed onto the **pistil** of the flower (Figure 16-1A), which came to be regarded as its "female" part. This discovery led to the making of hybrids between plants and, eventually, to the science of genetics and the modern techniques of plant breeding and crop improvement (Chapter 18).

A typical flower (Figure 16-1A) consists of four types of organs: (1) **sepals,** which enclose the flower when in bud and are mainly protective in function; (2) **petals,** usually brightly colored and showy; (3) **stamens,** the pollen-producing organs; and (4) one or more pistils. Most flowers contain only one pistil, composed of two or more hollow subunit structures called **carpels,** fused together (Figure 16-1). In certain flowers, such as those of blackberry and raspberry, the carpels are not fused together, so the flower contains several pistils.

Many plants develop their flowers in a specialized branch system called an **inflorescence,** familiar examples being the flower clusters of lilac and the flowering stalks of iris. An inflorescence often bears modified leaves called **bracts,** which are usually much smaller and simpler than foliage leaves (Figure 16-2). Flowers occur at the ends of branches of the inflorescence and/or in the axils of bracts. This and other evidence suggests that flowers are modified shoots and their sepals, petals, stamens, and carpels are much-modified leaves.

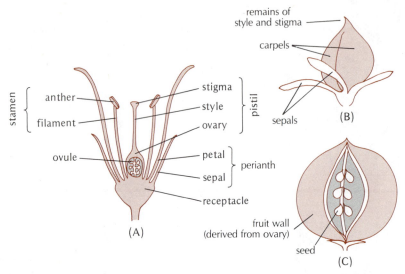

Figure 16-1. The flower and its development into a fruit. (A) Structure of a simple flower, as seen in longisection. (B) Enlargement of ovary after pollination, external view showing the two carpels making up this ovary. The other flower parts except sepals have been shed. (C) Fruit, formed by further growth of ovary, sectional view showing seeds (derived from ovules) inside. Shrivelled sepals persist at the base of this fruit.

Sexual reproduction depends on the fusion of two sex cells or **gametes,** a sperm and an egg, bringing their hereditary material together in a fusion cell, or **zygote.** This act of sexual fertilization occurs within the pistil some time after pollination, or the transfer of pollen grains from an **anther,** the terminal, pollen-producing part of a sta-

men, to the pistil's pollen-receptive surface or **stigma** (Figure 16-3). Fertilization initiates the development of the fertilized egg or zygote into an embryo plant within a seed and, concomitantly, the growth of the pistil (Figure 16-1B) into a fruit (Figure 16-1C) enclosing the seed or seeds.

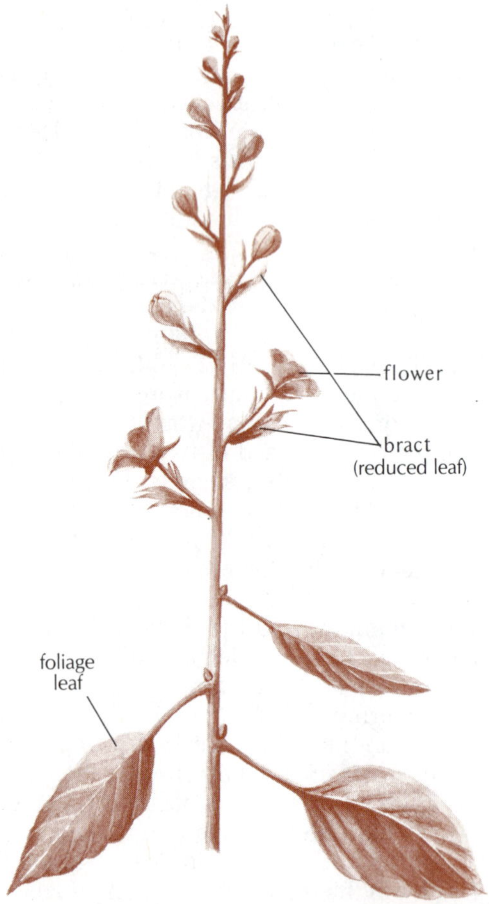

Study of how gametes are produced within flowers, first carried out successfully using the improved light microscopes and fixation and sectioning techniques of the late nineteenth century, revealed that anthers and pistils are not really male and female organs, as popularly stated. They are instead modified spore-producing organs that give rise to, and house and nourish, a microscopic sexual plant generation that actually produces the sperms and eggs for fertilization. To understand the nature of sexual reproduction in a flower, we must first consider the cytological nature of sexual cycles in general and the peculiar type of sexual cycle, called alternation of generations, that plants have evolved.

THE SEXUAL CYCLE

Each of the gametes (egg and sperm) participating in sexual reproduction carries a single set of chromosomes, that is, just one representative of each of the kinds of chromosomes that the species possesses. This is termed the **haploid** condition and is symbolized n, standing for the basic number of chromosome types characteristic of the species—for example, 10 in corn and 23 in humans. When a sperm fuses with an egg in fertilization, the resulting zygote nucleus receives two complete sets of chromosomes; this is called the **diploid** condition, designated 2n. Since the zygote develops into the macroscopic organism in the case of both a human and a flowering plant, the cells of the organism are diploid (20 chromosomes in corn, 46 in humans).

Figure 16-2. An inflorescence, showing flowers, bracts, and foliage leaves.

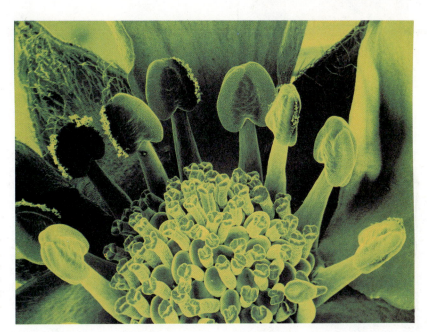

Figure 16-3. Scanning electron micrograph of a strawberry blossom, showing many pistils and stigmas surrounded by nine stamens (40 X). Although only one stigma has a pollen grain on it, many pollen grains are visible on the margins of the anthers.

To carry out sexual reproduction, a diploid organism must produce haploid gametes from its diploid cells. This is accomplished by **meiosis** or reduction division, a sequence of nuclear divisions that reduces the chromosome number from diploid to haploid. The details of meiosis are important in understanding the inheritance of genetic characteristics during sexual reproduction and are considered in Chapter 18. For the present purpose it suffices to note that meiosis always involves two successive divisions (Figure 16-4), normally yielding from the original diploid nucleus a set of four haploid nuclei or cells, called a **tetrad** (from Greek *tetras*, four). The appearance of tetrads in an organ or tissue gives an indication that meiosis has occurred there.

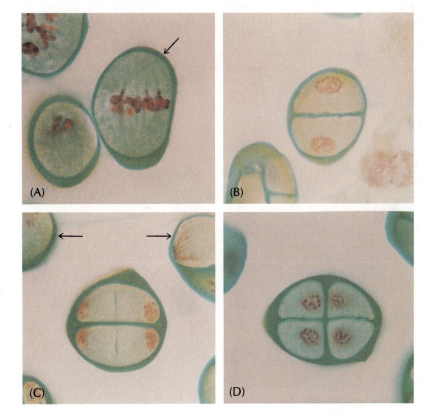

Figure 16-4. Meiosis in the formation of pollen grains from pollen mother cells (microsporocytes) in the anthers of a flower. (A) Metaphase of the first division (arrow). (B) First division completed (interphase). (C) Anaphase (arrows) and telophase (center of picture) of the second division. (D) Tetrad of haploid cells after completion of the second division. Each of these cells will become a pollen grain.

Alternation of Generations

In the sexual cycle of animals, diploid cells of the body give rise by meiosis to haploid ga-

metes, that is, sperms and eggs, which fuse in fertilization to yield a diploid zygote that develops into a new individual (Figure 16-5A). For reasons of evolutionary history most plants have a more complicated sexual cycle, involving distinct diploid and haploid individuals or generations occurring in an obligatory alternating sequence. Haploid gametes are produced not from a diploid plant body, but by ordinary mitosis from the haploid type of plant, called a **gametophyte** (gamete + Greek *phyton*, plant). After gamete fusion in fertilization the resulting diploid zygote develops into a diploid plant called a **sporophyte.** The sporophyte, as its name implies, produces spores; these are formed by meiosis and are therefore haploid. Each spore grows into a new haploid gametophyte, continuing the cycle (Figure 16-5B).

In most lower plants, up to and including ferns, the gametophyte and sporophyte generations are separate, independent, free-living plants (Chapters 23–29). In flowering plants, however, through evolution the gametophyte or haploid generation has become reduced to just a small group of haploid cells, enclosed within the flower and completely supported by the parent diploid sporophyte, which comprises the macroscopic body of the plant. The gametophyte generation consists of separate female or egg-producing gametophytes contained within the pistil and male or sperm-producing gametophytes that comprise the pollen grains (Figure 16-5C). In the following section we trace the origin and activity of these stages.

GAMETE FORMATION AND FERTILIZATION IN THE FLOWER

The actual sexual processes within a flower are microscopic, often very transient, and embedded within many layers of tissue, making them difficult to observe. Only by painstaking fixation and sectioning of floral organs at successive developmental stages, followed by careful microscopic study of the sections, was it possible to determine what takes place. The phenomena thus

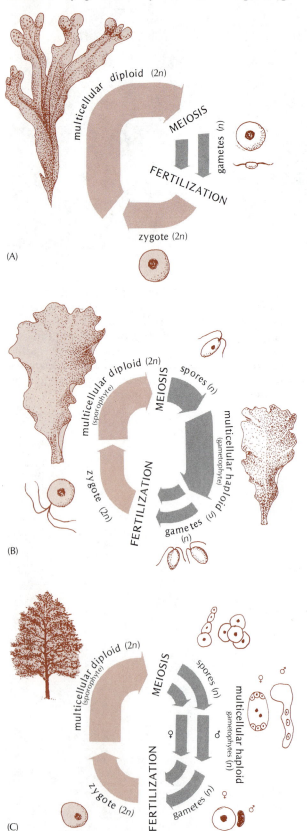

Figure 16-5. Reproductive cycles. (A) Rockweed (*Fucus*), a marine alga, which produces haploid gametes by meiotic division of diploid cells, as animals do. (B) Sea lettuce (*Ulva*), another alga, which reproduces by an alternation of separate haploid and diploid generations. The gamete-producing haploid (gametophyte) generation arises from haploid spores produced by meiosis in the diploid (sporophyte) generation. (C) A flowering plant, which reproduces by a similar cycle, but with the haploid generation reduced to a microscopic, nonphotosynthetic haploid phase developing within (and parasitic on) the floral organs of the diploid (sporophyte) plant body.

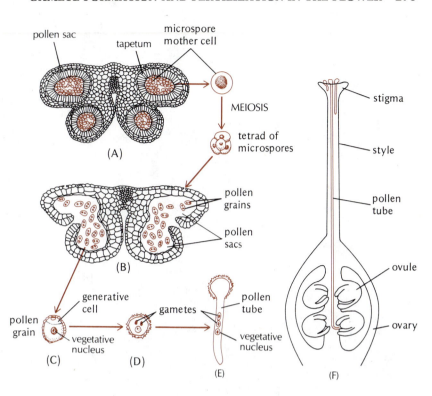

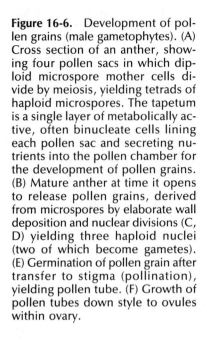

Figure 16-6. Development of pollen grains (male gametophytes). (A) Cross section of an anther, showing four pollen sacs in which diploid microspore mother cells divide by meiosis, yielding tetrads of haploid microspores. The tapetum is a single layer of metabolically active, often binucleate cells lining each pollen sac and secreting nutrients into the pollen chamber for the development of pollen grains. (B) Mature anther at time it opens to release pollen grains, derived from microspores by elaborate wall deposition and nuclear divisions (C, D) yielding three haploid nuclei (two of which become gametes). (E) Germination of pollen grain after transfer to stigma (pollination), yielding pollen tube. (F) Growth of pollen tubes down style to ovules within ovary.

revealed are surprisingly complex and vary significantly among different groups of flowering plants. The following description covers the most typical developmental sequence in general outline, leaving out most of the finer details now known from modern research using electron microscopy. (Some of the most interesting variations from the usual pattern are noted in Chapter 31.)

The floral organs, like the sporophyte plant body as a whole, are diploid. Each pollen-producing organ or anther (Figure 16-3) contains four chambers (pollen sacs) in which many diploid cells divide by meiosis (Figure 16-6), each yielding a tetrad of haploid cells called **microspores** (so called because they are much smaller than the megaspores discussed below). The haploid nucleus of each microspore then divides by mitosis, producing a "vegetative" and a "generative" haploid cell (Figure 16-6); the enclosing microspore wall becomes thickened and often elaborately sculptured (Figure 16-7). This haploid structure, a pollen grain, is the partially developed male gametophyte or microgametophyte. The generative nucleus eventually divides again, yielding two sperms that are merely haploid nuclei with a little associated cytoplasm. At some point before or after this division the anther opens, releasing the pollen grains for transport to a stigma.

The pistil (Figure 16-6F) consists of an expanded structure called the **ovary,** capped by a projection called the **style,** which terminates in a stigma. In one or more chambers within the ovary, microscopic globular structures called **ovules,** the progenitors of seeds, develop. Within each ovule

one diploid cell enlarges and divides by meiosis (Figure 16-8), yielding four large haploid cells called **megaspores** (Greek *megas,* mighty). Only one of these develops further, growing and undergoing three successive mitotic nuclear divisions to yield an even more enlarged sac-like supercell containing eight haploid nuclei (Figure 16-8). This structure, the female gametophyte or megagametophyte, is also called the **embryo sac** because the embryo plant later develops within it. Six of the haploid nuclei within the embryo sac segregate into distinct membrane-enclosed cells (without cell walls); one of these becomes the **egg.**

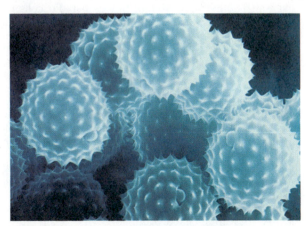

Figure 16-7. Scanning electron micrograph of ragweed pollen grains, showing elaborate ornamentation. The pore out of which the pollen tube grows is visible in some of the grains.

Pollen Tube Growth and Fertilization

A pollen grain reaching the stigma **germinates,** sending forth a tubular extension called the **pollen tube,** which penetrates the stigma and grows down through the style into the ovary, toward an ovule. As it grows, the pollen tube carries along the two sperm nuclei from the pollen grain (Figure 16-6). The germinated grain with its pollen tube is the fully developed microgametophyte. Upon reaching an ovule the pollen tube grows into it, usually via the pore (micropyle) between the outer tissue layers (integuments) of the ovule, and discharges its sperms into the embryo sac. One sperm nucleus fuses with the egg, effecting fertilization. The fertilized egg or zygote subsequently begins dividing and developing into an embryo plant.

Many kinds of pollen can germinate and grow pollen tubes in an artificial liquid medium (Figure 16-9), making it possible to study the nature and requirements of pollen tube growth. The requirements are often quite simple: sugar and small amounts of mineral salts. If sugar is omitted or removed, the pollen tubes soon burst at their tips, probably because they constantly need sugar to build new cell wall polysaccharides to compensate for the very rapid apical extension and thinning of the cell wall occurring in these strictly tip-growing cells (see tip growth, Chapter 12). If an ovule is placed into a suitable pollen tube culture, the tubes curve and grow toward it due to a chemical stimulus given off by the ovule, a type of growth response called chemotropism. This explains how pollen tubes grow toward and enter ovules in the ovary. Experiments indicate that the chemotropic stimulus released by ovules is simply calcium ions (Ca^{2+}).

An interesting detail of the events leading to fertilization within the ovule, as revealed by electron microscopy, is that the pollen tube actually discharges into one of the two haploid cells, called **synergids,** lo-

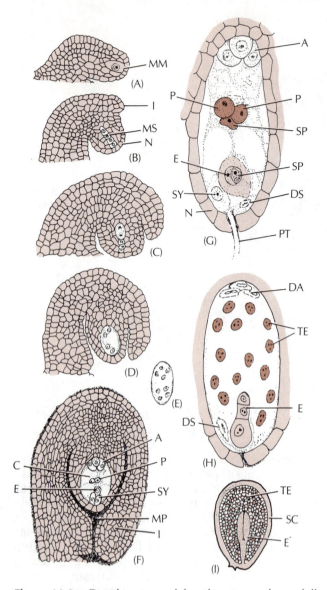

Figure 16-8. Development of female gametophyte, followed by double fertilization and its consequences, as seen in longisections of a developing ovule within the ovary of a flower. The ovule is attached to the ovary wall by the stalk at the lower left in each diagram. Diploid tissues color-tinted, haploid cells white, triploid cells shown with heavily colored nuclei (color dots, in I). (A) Young ovule containing enlarged diploid megaspore mother cell (MM) about to divide by meiosis, yielding (B) Four haploid megaspores (MS); a fold of tissue called an integument (I) is beginning to develop around the megaspore-enclosing nucellus tissue (N). (C) Three megaspores are degenerating while the fourth is enlarging and dividing. Further divisions yield a four-nucleate (D, other megaspores now completely crushed) and eight-nucleate (E) female gametophyte or embryo sac. (F) The eight haploid nuclei segregate into distinct cells and differentiate, yielding three antipodal cells (A), two synergids (SY), one egg (E), and two polar nuclei (P) occupying the large central cell (C) of the gametophyte; integument (I) closes around nucellus leaving only a narrow opening, the micropyle (MP). (G) Enlarged view of female gametophyte at time of double fertilization. After penetrating nucellus (N), pollen tube (PT) has discharged sperm nuclei (SP) into one of the synergids, which is degenerating (DS). This allows sperm to enter and fuse respectively with the egg (E) yielding a zygote, and with the polar nuclei (P) in the central cell yielding a triploid nucleus (primary endosperm nucleus). (H) Zygote has divided into three-celled young embryo (E); triploid nuclei (TE) derived from primary endosperm nucleus are multiplying by mitosis in central cell; remaining haploid synergid (DS) and antipodal (DA) cells are degenerating. (I) Much less magnified schematic diagram of seed, with mature embryo (E) embedded within triploid endosperm tissue (TE), surrounded by seed coats (SC) derived from nucellus and integuments of ovule.

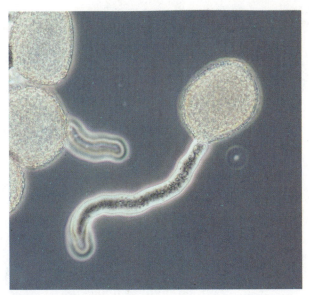

Figure 16-9. Pollen tube growing from an *Impatiens* pollen grain, germinated *in vitro*.

Double Fertilization: The Endosperm

Along with fertilization of the egg, yielding the zygote, a remarkable phenomenon occurs. This event is totally unlike anything in animal reproduction or even in lower plants: the second sperm nucleus from the pollen tube fuses with two other haploid nuclei (**polar nuclei**) of the embryo sac to yield a **triploid** (3n) nucleus. The phenomenon of zygote and triploid nucleus formation by simultaneous nuclear fusions is called double fertilization. The triploid nucleus immediately begins to multiply by mitosis, soon populating the embryo sac with a large number of triploid nuclei (Figure 16-8), while the zygote begins to develop into the embryo. Later, cell walls form between the triploid nuclei, creating a triploid tissue called the **endosperm** in which the diploid embryo is embedded.

The endosperm serves initially to produce nourishment and hormones that stimulate the growth of the embryo. Later in the development of many seeds the endosperm stores up reserve products (starch, storage protein, and/or vegetable oils) for subsequent use in seed germination, as described in Chapter 17. In most cases the remaining haploid nuclei of the original female ga-

cated next to the egg, and this cell invariably degenerates at about the time of discharge (Figure 16-10). The synergid cell is, in effect, expended as a vehicle for allowing entry of the sperm nuclei into the embryo sac.

Figure 16-10. Electron micrographs showing pollen tube discharge into the embryo sac (female gametophyte), and double fertilization, in spinach flowers. (A) Apex of the embryo sac, with part of the egg cell (EC) to the left, and of the one synergid cells in the center, into which a pollen tube (PT) has grown and discharged via the pore that has formed at its tip. Large starch grains (S) found within the synergid but characteristic of the pollen tube show that the pollen tube contents have been injected into the synergid, which is now degenerating (sperm nuclei are not seen in this section). Pollen tube has entered the synergid by growing through a specialized area of wall ingrowths called the filiform apparatus (FA). (TC denotes diploid cells of the ovule surrounding the haploid female gametophyte.) (B) Double fertilization: upper part of egg cell (EC/ZY), above and to left of view in (A), one sperm nucleus (SN 1) fusing with egg nucleus (N; prominent dark bodies are their respective nucleoli), yielding the zygote. Above, in the central cell (CC/EN) of the embryo sac, the two polar nuclei (PN 1 and 2) are fusing with the second sperm nucleus (SN 2) to form the primary endosperm nucleus. The degenerating vegetative nucleus (VN) from the pollen tube is also visible.

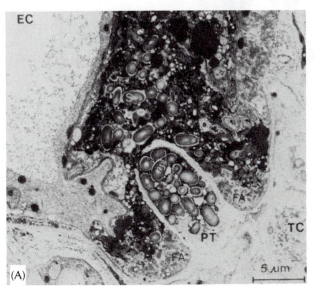

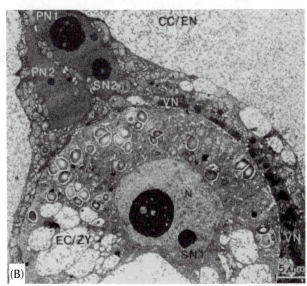

metophyte degenerate and disappear long before the seed has completed development, so hardly a trace of the gametophyte generation remains in the mature seed.

As the endosperm and embryo grow, the ovule that housed the original embryo sac also grows. The outer layers (integuments) of the ovule eventually become tough layers, the **seed coats,** which protect the seed after it is released into the environment. The basic flowering plant seed thus consists of the embryo or new sporophyte generation plant; the seed coats, tissue from the parental sporophyte generation; and the triploid endosperm (belonging to neither sporophyte nor gametophyte generation), storing reserve products for use in germination. This triploid tissue, containing the bulk of the food stores of cereal grains, comprises the principal staple food of mankind.

One or more seeds are formed, in the manner described above, within each fruit, which as previously noted represents the enlarged ovary of the flower's original pistil (Figure 16-1). The way in which the fruit, the embryo, and other parts of the seed develop and the important variations in their

structure are discussed in Chapter 17. Figure 16-11 summarizes the mechanism of sexual reproduction in flowering plants as explained above.

POLLINATION MECHANISMS

Obviously the embryo plant receives hereditary contributions from two parents only if the pollen that reaches the stigma comes from a flower on a different plant. This is called **cross-pollination. Self-pollination** occurs when the pollen comes from the same flower or from another flower on the same plant. Flowers of the majority of plant species have features that favor or ensure cross-pollination.

Pollination by Animals

The shapes and colors of a flower's showy structures (usually petals; see Figure 16-1) attract prospective animal pollinators to the flower. So

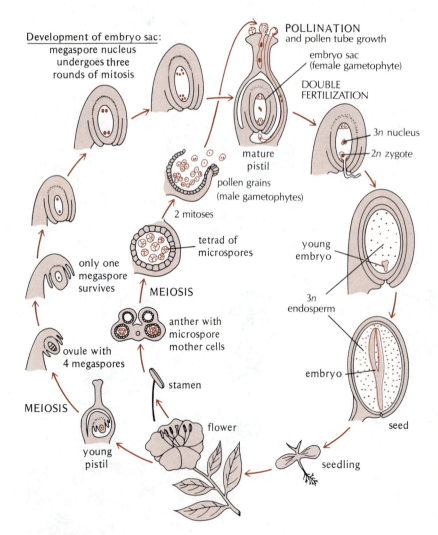

Development of embryo sac: megaspore nucleus undergoes three rounds of mitosis

POLLINATION and pollen tube growth

embryo sac (female gametophyte)

DOUBLE FERTILIZATION

3n nucleus

2n zygote

mature pistil

pollen grains (male gametophytes)

young embryo

2 mitoses

tetrad of microspores

3n endosperm

only one megaspore survives

MEIOSIS

embryo

anther with microspore mother cells

ovule with 4 megaspores

stamen

seed

MEIOSIS

young pistil

flower

seedling

Figure 16-11. Summary of flowering plant reproductive cycle. Diploid tissues are marked with continuous color and triploid tissues with color stippling; haploid cells are white. Ovule is drawn much larger, relative to ovary and pistil, than it is in reality. The fruit surrounding the seed during its maturation is ommitted.

Figure 16-12. Aggregate flowers (inflorescences). (A) Dogwood. (B) Poinsettia. (C) Daisy. (D) *Embothrium grandiflorum,* a member of the family Proteaceae. In this family, the colored stamen filaments are often the principal "showy" organs.

do odor or fragrance and the sugar secretion called **nectar,** which many flowers produce in specialized glands called nectaries. Nectar provides a carbohydrate energy source as a reward for the pollinator, motivating it to visit the flower. Bees, for example, collect nectar for making honey. Pollen, which is rich in protein, is collected by many pollinators as a protein food source, but during this activity (or during the collection of nectar) they often brush pollen against the stigma and pollinate the flower.

Some so-called "flowers" are actually entire inflorescences whose showy organs are either brightly colored, relatively large bracts (Figure 16-12A, B) or specialized individual flowers (Figure 16-12C, D). Despite their fundamentally different morphological nature, these structures are equivalent to ordinary flowers in their ability to attract pollinating animals and thereby produce seed.

Many flowers are specialized in structure, color, and odor for pollination by one or another

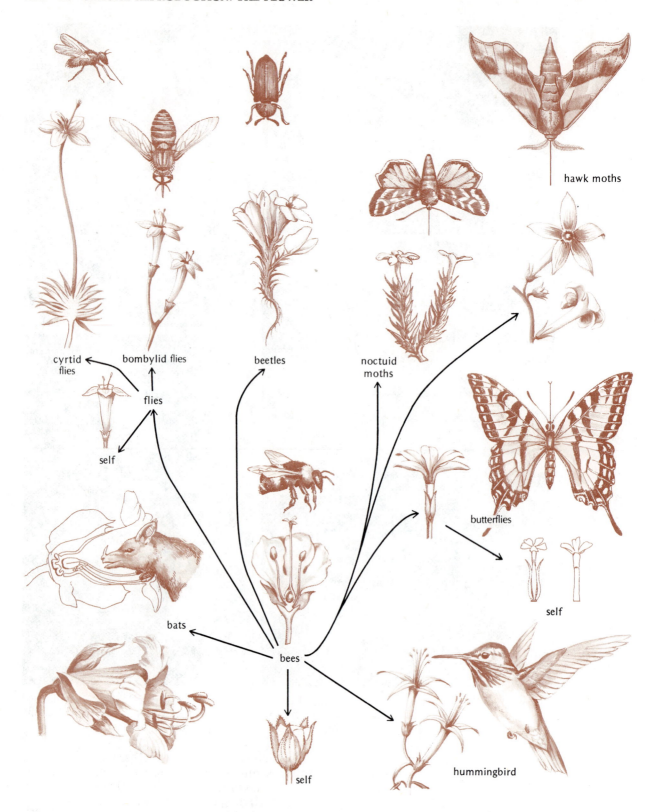

Figure 16-13. Morphological diversification of flowers in different genera and species of the phlox family (Polemoniaceae) for pollination by different pollinators, and for self-pollination which has arisen in several branches of this family. This is a rather extreme example of pollinator diversity among closely related plants. (Note that the fly-pollinated flowers in this group attract special types of flies that somewhat resemble bees in appearance and behavior, rather than the carrion-feeding flies for which commonly recognized fly-pollinated flowers are specialized.)

kind of animal, including a wide variety of insects as well as hummingbirds and even bats (Figure 16-13). The structural and behavioral features of both flowers and their pollinators present strong evidence of mutual evolutionary changes: pollinating animals have specialized to obtain food by this activity, and different flowers have evolved to exploit most efficiently the pollinating capabilities of different animals. This is called **coevolution** (Chapter 19). The structural evolution of flowers and floral organs is considered explicitly in Chapter 31; here we mention only a few of the structural variations important in pollination mechanisms.

Many flowers present their pollinator with a specific task to perform in order to get pollen, nectar, or some other reward. Flowers are often so designed that, in performing its task, the pollinator gets pollen on part of its body, which then comes into contact with the stigma of the next flower the insect enters. Sometimes flower parts are arranged so as to move when the pollinator enters, ensuring the specific deposition of pollen on it or even the obligatory transfer of pollen only to the stigma of a different flower (Figure 16-14). These arrangements promote cross-pollination.

Bees, butterflies, and moths seek nectar and/or pollen, are attracted by bright colors and sweet fragrances, and are the pollinators of most showy flowers. Many kinds of flies, on the other hand, are attracted by decaying material. Fly-pollinated flowers such as skunk cabbage (Figure 16-15A) and jack-in-the-pulpit therefore tend to produce carrion-like odors; some make a display resembling pieces of flesh, sometimes even replete with fur (Figure 16-15B).

Bee Pollination

Bees are the most generally important pollinators. They can see near-ultraviolet light (wavelengths of about 390 nm), invisible to the human eye. Many flowers that appear plain white or yellow to us possess ultraviolet-absorbing pigments in their petals, often in the form of spots or lines that give the flower a distinctive pattern when photographed under ultraviolet light (Figure 16-16). Bees can see these patterns, which probably help guide them toward the nectar source and also help them distinguish between flowers of different plant species, a demonstrated ability of bees that makes them especially effective in cross-pollinating the flowers of any particular species chosen for nectar collection that day.

Bee pollination is of great economic importance; over 50 kinds of crop plants grown in the United States are insect-pollinated, mainly by bees. Important to this type of agriculture is the provision of bees from commercially-maintained hives, which are transported from place to place. Because bees do not forage for food under inclement weather, weather during the flowering season can be a significant agricultural hazard for

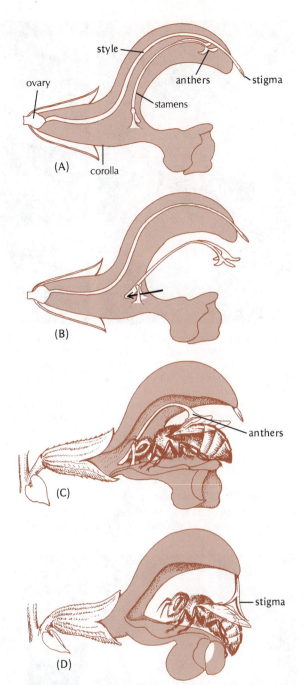

Figure 16-14. Flowers of sage (*Salvia*), showing specific structural arrangements for transferring pollen via its pollinators (bees). (A) Longisection of flower, showing stamens attached to the tubular corolla (fused petals) by a hinge bearing a basal projection which, if pushed (B, arrow), tips the stamens downward. (C) A bee entering the flower to reach the nectar at its base pushes against the projection, causing stamens to swing down and deposit pollen on bee's abdomen. (D) Alighting on a flower whose stigma is exposed and receptive to pollen, the pollen-dusted part of the bee's abdomen contacts stigma and pollinates it.

bee-pollinated crops, most notably fruits trees such as apples, cherries, plums, and almonds. By keeping bees inactive, rain can cause the loss of much or all of a

Figure 16-15. Fly-pollinated flowers. (A) Skunk cabbage (*Symplocarpus foetides*) in which a purplish modified leaf (spathe) encloses a spike-like inflorescence (spadix) bearing many small flowers which emit a putrid, fly-attracting aroma responsible for the plant's name; the tip of the spadix can be seen within the opened spathe to the right. (B) *Stapelia hirsuta,* a cactus-like stem succulent of the milkweed family, whose furry and blotchy red flowers simulate a killed small animal, a common attraction for flies. (C) *Stapelia variegata,* whose very foul-smelling flowers may resemble pieces of decaying meat.

year's crop due to failure of pollination during the rather brief period when the flowers are receptive to pollen.

Hummingbird Pollination

Hummingbirds are specialized for collecting nectar as their principal or exclusive energy source. Quite a few flowers are specialized for pollination by these animals. Hummingbird flowers are usually red (which hummingbirds, but not bees, can see) and tend to have a very elongated, tubular form with nectar glands down in the base of the tube (Figure 16-17). The hummingbird reaches the nectar by hovering in front of the flower

Figure 16-16. Plain yellow flowers of evening primrose (left) and lesser celandine (right) photographed under visible light (above) and under near-ultraviolet light (below). The human eye would see these flowers as they appear above, but the bee eye would see the intricate patterns below, caused by the distribution of UV-absorbing pigments.

Figure 16-17. Hummingbird-pollinated flowers. (A) Large scarlet *Salvia*. (B) *Fuchsia*. (C) *Fuchsia* being visited by a hummingbird.

Figure 16-18. Wind-pollinated flowers, lacking petals. (A) Beech, staminate flowers in catkins, pistillate flowers located singly (spherical green bodies to upper right, with projecting stigma). (B) A grass (*Dactylis glomerata*) with pendulous stamens and minute greenish feathery stigmas visible among the stamens.

and inserting its long beak and tongue into the tube (Figure 16-17C). Hummingbirds, in contrast to bees, have a poor sense of smell and apparently find their nectar sources entirely by sight. Hummingbird flowers, therefore, usually lack fragrance, but they produce unusually copious amounts of nectar in keeping with the need to support a very active animal much larger than a bee. The tubular part of the flower is often so long and thin that insects cannot get to the nectar; only a hummingbird, with its long beak and tongue, can reach it.

Wind Pollination

Wind-pollinated flowers, which include the flowers of grasses and most temperate-zone deciduous trees, do not need (and usually lack) showy organs (Figure 16-18), fragrance, and nectar. Wind pollination is a much less efficient method than pollination by animals, because animals seek out flowers and thus carry pollen directly from one to another, whereas wind disperses pollen everywhere. Therefore, in order to reproduce successfully, wind-pollinated plants must produce a great deal more pollen than animal-pollinated plants. The pollen grains of wind-pollinated plants are relatively small, so in the air they do not fall rapidly. They also have dry surfaces, so they do not stick together, in contrast to the stickiness of pollen from animal-pollinated flowers.

The stamens of wind-pollinated flowers usually project out prominently and often dangle from the flower, so the slightest breeze will sway them and cause pollen release. Many wind-pollinated trees develop flowers in dangling tassel-like clusters called **catkins,** which the wind moves easily (Figure 16-19). The stigmas of wind-pollinated flowers are often highly branched and feather-like, increasing their chances of catching any pollen grains that come their way through the air.

Figure 16-19. Catkins of paper birch (*Betula papyrifera*).

Hay Fever

Hay fever is an allergic reaction to specific kinds of pollen grains occurring in substantial numbers in the air. People differ in the pollens to which they are sensitive and in the strength of their allergic reactions. Insect-pollinated flowers such as goldenrod, roses, and sunflowers are sometimes blamed for hay fever, but actually their pollens do not occur in the air to a significant extent. Although wind-pollinated plants cause nearly all hay fever, not all wind-borne pollens cause hay fever. Cattails (*Typha*), for example, are wind-pollinated and release large amounts of pollen, but are not allergenic; this is also true of most kinds of conifer trees, even though all conifers are wind-pollinated.

In many parts of the United States there are three hay fever "seasons," corresponding to the principal flowering periods of major kinds of wind-pollinated plants. The early spring season (March to May, depending on the latitude) is caused by wind-pollinated deciduous trees such as elms, box elders, poplars, birches, oaks, hickories, ashes, and sycamores. The late spring–early summer season (May–July) is due mainly to the flowering of wild and pasture grasses; most cereal crops, which are also grasses and therefore wind-pollinated, do not contribute significantly, except for rye (*Secale*). The late summer and fall season (August–September) is due primarily to ragweeds (*Ambrosia*; Figure 16-20), wind-pollinated herbaceous dicots whose pollen seems to be the most highly allergenic of any. Where ragweed does not occur abundantly, as in the drier southwest areas of the United States, this season may be unimportant or lacking.

Figure 16-21.   Corn plants in flower. White arrows indicate pistillate flowers (in "ears"), colored arrows indicate staminate flowers (in "tassels").

Separation of Sexes

Many wind-pollinated plants, including most wind-pollinated trees and corn, develop separate pollen-producing (staminate, or "male") and pis-

Figure 16-20.   Inflorescence of great ragweed (*Ambrosia trifida*).

til-producing (pistillate, or "female") flowers (Figure 16-18A), making it impossible for a flower to pollinate itself. Usually both kinds of flowers are borne by each plant, a condition termed **monoecious** (pronounced "mahn-ee'-shus"). The staminate flowers of corn are grouped into the well-known terminal "tassel," while its pistillate flowers are borne laterally, in the "ears" (Figure 16-21).

In some species, notably willow trees, each individual is genetically either "male" or "female," bearing only staminate or pistillate flowers, respectively, a situation analogous to the separate sexes of animals. This condition, termed **dioecious** (pronounced "die-ee'-shus"), completely ensures cross-pollination, but at the cost of allowing only half of all plants to produce seed (since, as in animals, usually about half of all individuals are of each "sex").

The monoecious and dioecious specializations are much more common among wind-pollinated than animal-pollinated flowers, doubtless because pollen shed into the air is much more likely to fall on the stigma of the flower shedding it (if that flower has a stigma) than is the sticky pollen of animal-pollinated flowers. Squash, however, illustrates the unusual situation in which an insect-pollinated species is monoecious (Figure 16-22).

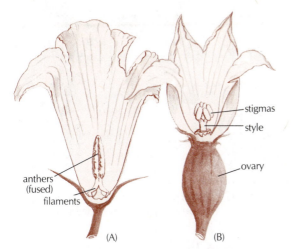

Figure 16-22. Monoecious flowers of squash. (A) Staminate flower. (B) Pistillate flower. The two kinds of flowers are both produced by each plant; a fruit (squash) develops from the ovary of a pistillate flower after pollination by pollen from a staminate flower.

Physiological Cross-Pollination Mechanisms

Many flowers possessing both stamens and pistil develop stigmatic receptivity to pollen either before or else after the given flower sheds its pol-

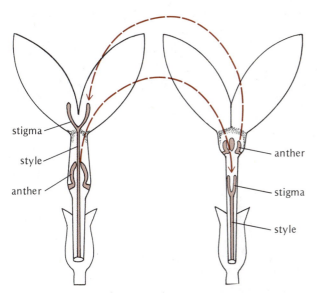

Figure 16-23. Heterostylic flowers of bluet (*Houstonia caerulea*), an attractive eastern spring wildflower. Flowers of the type on the left called "pin," and right, called "thrum" (shown split in half lengthwise), are borne by separate and genetically different plants in each local population. Because of the reciprocal positioning of anthers and stigmas in the two flower types, pollen deposited on a pollinator's proboscis inserted into one of the types tends to be transferred to the stigma of the other type when the pollinator visits it, and vice versa (arrows). In addition, the pollen of each flower type is compatible only with the stigma of the other genetic type.

len, preventing any one flower from pollinating itself. This behavior, like the monoecious flower morphology described above, encourages but does not compel cross-pollination, because flowers at different stages of development on the same plant can pollinate each other.

SELF-INCOMPATIBILITY

Some species are constituted genetically so that pollen from the same flower, or from other flowers on the same plant, cannot cause fertilization. This behavior, termed **self-incompatibility,** ensures that seed production results only from cross-pollination. To yield seed, a pollen grain must be genetically different from the flower it pollinates.

Many wild plant species are self-incompatible; most cultivated plants are not, except for certain fruit and nut trees such as sweet cherries, almonds, and some apple, pear, and plum cultivars. Since these trees are propagated vegetatively, all trees of a given variety are a genetically equivalent clone and cannot successfully pollinate one another. To obtain fruit, one must interplant at least two cultivars.

Self-incompatibility genes prevent fertilization by blocking the growth of the pollen tube in the style. This blockade occurs if the pollen grain carries a self-incompatibility gene that also occurs in the stigma, as is inevitably the case if the pollen comes from the same plant. If the pollen comes from a genetically different individual, it usually carries a self-incompatibility gene different from those found in the first plant, so the growth of this pollen tube will not be blocked.

Heterostyly

Heterostyly (Figure 16-23), which occurs sporadically in different insect-pollinated plant families, is a genetically determined combination of morphological and physiological mechanisms promoting cross-pollination. The two types of flowers have a complementary form: each tends to deposit pollen on the part of the pollinator's body that comes into contact with the stigma of the opposite type of flower. The two genetic forms are cross-compatible, but pollen of each type is usually unable to fertilize the same or other individuals of that type. Populations usually consist of about equal numbers of the two genetic types. The situation is somewhat analogous to separate sexes except that each individual is capable of producing offspring.

Self-Pollination

Most cultivated plants, and many (but probably a minority of) wild ones, commonly or normally reproduce by self-pollination. Such flowers lack self-incompatibility mechanisms and release pollen when their stigmas are receptive. Some plants, including many weeds and crop plants (such as peas, beans, and tomatoes), have become specialized for self-pollination. The tendency of the flower to attract pollinators is reduced by its small size (Figure 16-13), inconspicuously colored (or no) petals, and lack of fragrance or nectar. Anthers and stigma usually remain within the flower, in contact with one another, ensuring self-pollination. In some cases the flower never opens [a condition called **cleistogamy** (from Greek *kleistos*, closed)], eliminating any possibility of cross-pollination.

Self-pollination obviously reduces or prevents genetic interchange between individuals or populations of a species. This has special significance in genetic adaptation and evolution (Chapter 19).

Apomixis

Certain plants such as the common dandelion circumvent the sexual processes normally involved in seed formation, developing seeds without gamete formation or fertilization. This process is called **apomixis** (Greek *apo-*, away from, and *mixis*, mingling, referring to the lack of fertilization). The embryo may develop, for example, from a diploid cell of the ovule rather than from a zygote. From the genetic point of view, apomictically formed seeds are equivalent to vegetatively produced offspring.

CONTROL OF FLOWERING

Most plants will not flower until they have attained a certain minimum size or age, which may be a few weeks for small annuals, a couple of months or more for large annuals such as many vegetable crops, 1 year or more for many perennial plants, and 5 to 20 or 30 years or more for many trees and for monocarpic perennials such as the "century plant" (*Agave*) which flower only once, after many years of vegetative development. "Ripeness to flower" denotes the state of readiness or competence for flower intiation eventually reached as a plant grows and matures. Internal controls governing ripeness to flower ensure that a plant does not produce seeds until it has achieved sufficient vigor and accumulated enough storage reserves for good reproductive performance.

Where flowering requires no specific environmental signal, as in many crop plants grown for their fruits or seeds, the time required to achieve ripeness to flower determines how soon a crop can be produced, often given on seed packets as "days to maturity." Of course this is only an approximate figure, since the rate of growth and progress toward ripeness to flower, as well as of flower and fruit development once flower buds have been initiated, depends on the temperature and other physical and nutritional factors.

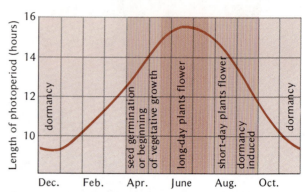

Figure 16-24. Yearly photoperiod cycle, showing flowering periods for long-day and short-day plants.

Control by Day Length: Photoperiodism

Once ripeness to flower has been attained, flowering in many plants depends on the day length, a response called **photoperiodism.** Day length varies in the course of the seasons as shown in Figure 16-24. Besides synchronizing the flowering of different individuals of the same species to take place at the same time of year, photoper-

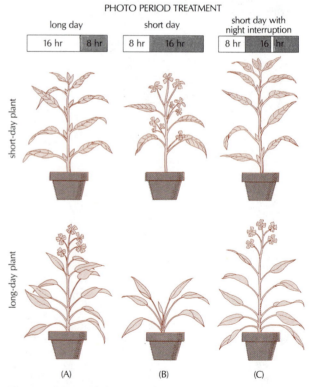

Figure 16-25. Photoperiodic response of a short-day plant (upper row) and a long-day plant (lower row) to (A) long day/short night photoperiodic cycles, (B) short day/long night cycles, and (C) short days with long nights interrupted in the middle with a few minutes of light. Most long-day plants, like that illustrated, grow as a rosette under noninducing (short-day) conditions, and when induced bolt prior to flowering. This is not true of most short-day plants.

iodism also helps ensure that flowers begin to form early enough in the growing season to complete seed development before the onset of unfavorable autumn or winter conditions.

Two basically opposite types of photoperiodic response occur. **Long-day plants** initiate flower production when the day length exceeds a certain minimum value, or critical day length. **Short-day plants** initiate flowers when the day length becomes less than a characteristic critical day length (Figure 16-25). The critical day lengths of most long-day and short-day plants fall in the same range, about 12 to 14 hr. Thus long-day plants bloom in spring and early summer, when the days are relatively long, and short-day plants bloom in late summer and autumn, when the days become shorter (see Figure 16-24). Examples of long-day plants are hollyhock, radish, garden beet, spinach, iris, timothy grass, and red and sweet clover. Short-day plants include soybean, chrysanthemum, ragweed, most asters and goldenrods, and poinsettia. Plants whose flowering is insensitive to day length are termed **day-neutral.** Examples include peas, garden beans, tomatoes, nasturtiums, roses, snapdragons, and many common weeds.

Some species show **obligate** photoperiodism, meaning that the plant *must* be exposed to the photoperiodically effective ("inductive") photoperiod to

Figure 16-26. Photoperiodic response in chrysanthemum (*Chrysanthemum morifolium*), a short-day plant. The flowering plant on the left received 8-hour days and 16-hour nights. The vegetative plant on the right received 16-hour days and 8-hour nights.

initiate flowers; in a noninductive photoperiod the plant will continue vegetative growth indefinitely. Soybeans, for example, will grow vegetatively for an unlimited time under long days and will flower only if exposed to short days. Certain stonecrops (*Sedum*), which are long-day plants, have continued to grow vegetatively for as long as 9 years when given less than 12 hr of light per day; they flowered only when finally placed under 15-hr days. On the other hand, in **facultative** photoperiodism, which is probably more common, an inductive photoperiod stimulates flower initiation, but the plant will eventually form flowers even in a noninductive photoperiod: an inductive photoperiod merely causes it to flower earlier. Still, this can quite effectively synchronize the flowering of different individuals.

Knowledge of photoperiodism is utilized in horticultural practice for "forcing" flowers into bloom at times of year other than their normal flowering season. Fall-flowering short-day plants such as chrysanthemums are brought into bloom in spring or summer by shielding the plants morning and evening with frames covered with black cloth or black paper, to reduce the number of hours of light they receive (Figure 16-26). Long-day plants may be brought into bloom in winter in the greenhouse by extending the short winter day length using supplementary electric light.

Various aspects of vegetative development are controlled photoperiodically, as noted in Chapter 14. Most research on the mechanism of photoperiodic responses of plants, however, has been done with flowering.

THE NATURE OF THE PHOTOPERIODIC RESPONSE

Since the ability to form its specific reproductive structures comprises an important part of the hereditary makeup of any species, inductive photoperiodic stimulation apparently causes genes to become expressed that until then were not being utilized. Photoperiodic induction switches on the previously unexpressed developmental program for flower formation. Apical and/or axillary meristems cease initiating vegetative shoot organs and begin to develop flowers or an inflorescence bearing flowers.

Although the developmental response occurs in buds, experiments in which only part of a plant is exposed to an inductive photoperiod show that photoperiod is actually perceived by the leaves. In some cases only a single leaf needs to be exposed to an inductive photoperiod in order for the plant to flower. This implies that some kind of flower-inducing stimulus passes from the leaves to the meristems. A leaf or branch from a photoinduced plant, if grafted onto an uninduced plant, will induce the latter to flower (Figure 16-27). This suggests that induced leaves produce and release a flowering hormone, given the name **florigen.** In certain cases where a short-day and a

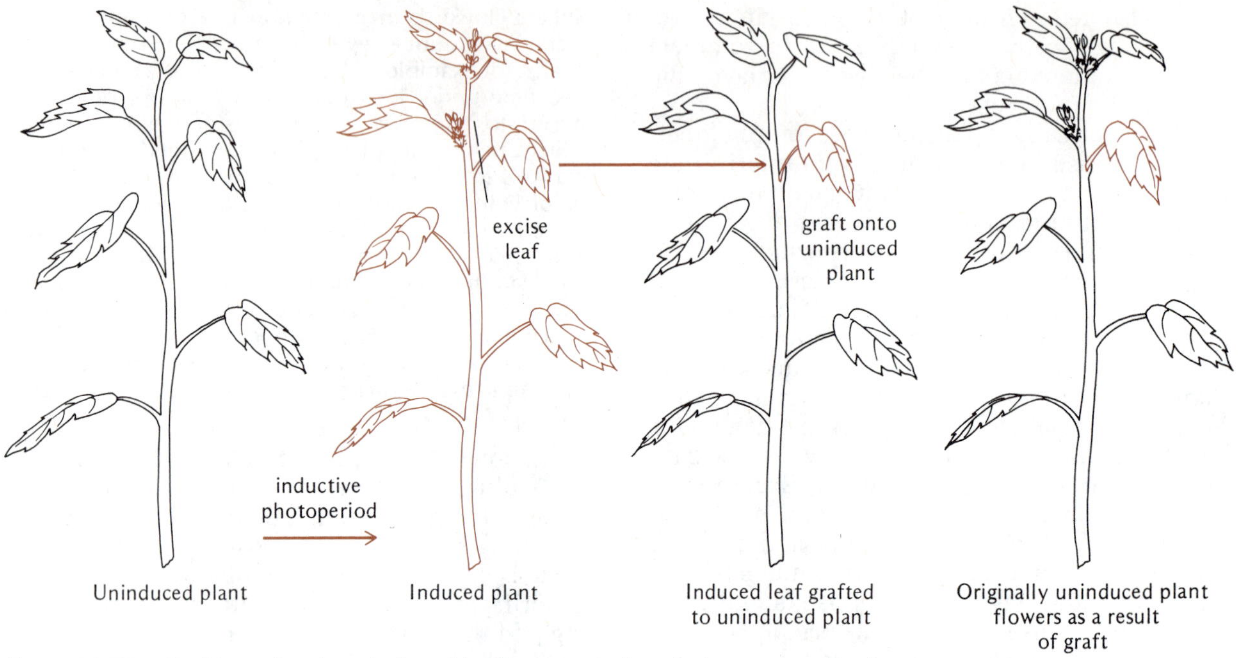

excise
leaf

graft onto
uninduced
plant

inductive
photoperiod

Uninduced plant

Induced plant

Induced leaf grafted
to uninduced plant

Originally uninduced plant
flowers as a result
of graft

Figure 16-27. Grafting of a photoinduced leaf to an uninduced plant causes the latter to flower.

long-day plant can be grafted together, if one is induced, it will induce the other to flower, indicating that the florigen of both long- and short-day plants is the same. However, no chemical substance that will generally substitute for a photoperiodic stimulus and induce flowering has yet been convincingly detected, isolated, or identified, so the reality of florigen remains in doubt.

A number of long-day plants can be induced to flower under short days by treatment with a gibberellin (Figure 12-20). These and many other long-day plants grow vegetatively as a rosette and elongate or bolt prior to flower formation (Figure 3-3). Some long-day rosette plants bolt but do not produce flowers when treated with gibberellin, and short-day plants generally do not flower in response to gibberellin. Thus gibberellin is not florigen, but seems to be involved in the response of long-day rosette plants, being responsible for the bolting that precedes flower formation. In cases where bolting is the limiting step in flower initiation, gibberellin causes flowering.

PERCEPTION OF PHOTOPERIOD

Exposure each night to light for a brief time, such as 5 min, halfway through its normal dark period (a "night interruption") causes a long-day plant on a short-day cycle to flower and *blocks* flower induction in a short-day plant held on a short-day cycle (Figure 16-25). Red light (wavelengths of about 670 nm) is most effective for a night interruption, and its effect can be com-

pletely reversed by exposing the plant for a few minutes to far-red light (730 nm) immediately after it has received a night interruption with red (or white) light (see Figure 16-25). The opposing effects of red and far-red light show that phytochrome is the photoreceptive pigment by which plants detect light in the photoperiodic response.

The results of night interruptions indicate that photoperiodism involves an ability of the plant to measure the length of an uninterrupted dark period. This implies that the plant possesses a time-measuring capability, or **biological clock.** It is still not known exactly what a biological clock is. But as early as the 1930s the German botanist Erwin Bünning inferred that organisms can measure time using their innate rhythms of physiological activity or circadian rhythms, which proceed in the absence of any external stimuli with a period or cycle duration of about 24 hr and are typically almost independent of temperature. A manifestation of circadian rhythms is the daily sleep movement rhythm of many leaves (Figure 16-28), noted in Chapter 14. If a light stimulus occurs during the part of a plant's circadian cycle that normally takes place at night, the plant sees the photoperiod as a long day (short night), whereas if no light stimulus occurs in the nocturnal part of the cycle, the plant interprets it as a short day (long night). We have no idea as yet, however, how this turns on the production of a flowering stimulus causing genetic information for flower formation to become expressed.

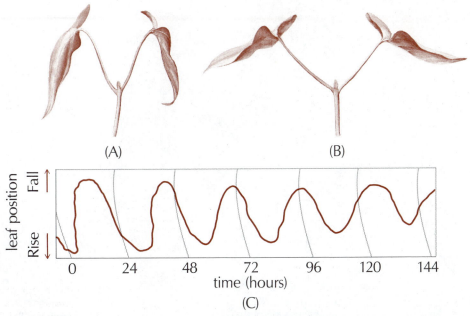

Figure 16-28. Circadian leaf movement rhythm as recorded in bean seedlings. (A) Nocturnal leaf position; (B) diurnal leaf position; (C) record of rise and fall of leaf petiole attached to a recording device, while the plant is maintained under continuous dim light at constant temperature. The leaf continues to move up and down on a cycle with a slightly longer than 24-hour period, an example of an endogenous (internally driven) rhythm.

Temperature Responses: Vernalization

Many plant species of temperate and cold-temperate climates are biennials: they grow vegetatively for the first year, then flower, go to seed, and die during the second year. This pattern ensures that the plant will begin reproduction only after a full season of vegetative growth and accumulation of storage reserves. Flowering in these plants is induced by a certain amount of exposure to cold temperature, normally received during the winter, inducing the plant to flower during the following warm season. Flower induction by exposure to cold is called **vernalization** (from Latin *vernalis*, of spring, since the plant usually flowers in the spring of its second season). Many root crops of temperate climates, such as beets, carrots, and turnips, are vernalizable biennial plants that store photosynthetic products in their edible storage roots for use in reproduction during their second season, following vernalization (Figure 16-29). Such roots are useless to us as food or fodder once mobilization of their storage reserves begins, and must be harvested before vernalization occurs. If continuously protected from cold temperatures, biennials will continue to grow vegetatively year after year.

Winter wheat is a crop that depends on vernalization. Planted in the autumn, the young plants become vernalized during the winter and flower promptly in the spring, producing an early crop. This type of wheat is especially suited to the "wheat belt" plains states, where the hazard of summer drought makes an early spring wheat crop advantageous. Winter wheat cultivars cannot be planted successfully in spring because, without receiving a cold stimulus, they grow vegetatively for much too long and usually fall over before producing a crop. Vernalizability is a genetic characteristic; strains of wheat lacking it are called **spring wheat** because they can be planted in spring to yield a summer wheat crop.

Vernalization has something in common with other inductive effects of cold on plants, such as the induction of growth in dormant winter buds. However, vernalization involves (as does photoperiodism) not just the resumption of growth but also a modified expression of genetic information. As in photoperiodism there is evidence from grafting experiments for a flowering stimulus that is produced upon the exposure of vernalizable plants to cold, but this stimulus has not been identified chemically and how it causes the expression of new genetic information is a mystery that may be solved by future research.

Other plants show a variety of different temperature responses or requirements in their flowering. Flowering of tropical and subtropical plants, especially, may require specific temperatures either above or below certain relatively high values. Various plants express a photoperiodic response only over a certain temperature range. Water supply can also have an important effect on flowering,

(A)   (B)

Figure 16-29. Vernalization response in henbane (*Hyoscyamus*). (A) Plant kept continuously under warm temperatures grows vegetatively, as a rosette. (B) After a period of exposure to low temperature, when returned to warm temperatures plant bolts and flowers.

especially in desert annuals, where water shortage hastens flowering, ensuring some seed reproduction when there is too little water for a normal amount of vegetative growth. We cannot even mention all the kinds of environmental responses known to affect reproduction; those described in the preceding sections, however, are recognized as the most generally important.

SUMMARY

(1) Flower reproduction can combine hereditary material from different parents in the offspring by the occurrence of cross-pollination, effected either by animals or by wind. Animal-pollinated flowers often possess specialized features to attract and reward a particular kind of animal as pollinator and often have a structure specifically conducive to the transport of pollen from one flower to another by the pollinator. Wind-pollinated flowers must produce a relative abundance of pollen and usually possess features promoting the shedding of pollen when the wind is blowing and preventing the accidental pollination of the same flower, such as separate staminate ("male") and pistillate ("female") flowers. Some plants prevent self-pollination by genetic self-incompatibility. Others practice frequent or normal self-pollination, and some are specialized for this.

(2) The sexual cycle of higher plants involves the fusion of haploid gametes to yield a diploid zygote that grows into a diploid plant or sporophyte. The sporophyte produces, by meiosis, haploid spores that grow into a haploid gametophyte, which in turn produces gametes. This cycle is referred to as an alternation of generations. In flowering plants, the gametophyte generation is microscopic, contained inside the flower.

(3) Flowers consist of sepals and petals (protective and showy organs); stamens that bear anthers, the pollen-producing structures; and pistils that contain ovules, the progenitors of seeds.

(4) Pollen grains are two-celled male gametophytes derived from haploid microspores, produced by meiosis in the anthers. Transferred to the stigma of the pistil (pollination), the pollen grain forms a pollen tube that delivers two sperms to the ovule.

(5) In each ovule four haploid megaspores are formed by meiosis, and one of these grows into a female gametophyte or embryo sac containing eight haploid nuclei. The embryo sac receives the two sperms from the pollen tube. Double fertilization then occurs: one sperm fuses with the egg nucleus, yielding the zygote; and the other fuses with two other haploid nuclei of the embryo

sac, yielding a triploid nucleus that multiplies, giving rise to triploid endosperm tissue within the developing seed. The zygote develops into the embryo, and the outer tissue of the ovule develops into the seed coats. The ovary portion of the pistil meanwhile grows into a fruit.

(6) Flowering is controlled and timed by an internal governor called the development of ripeness to flower and by environmental stimuli, especially photoperiod and temperature.

(7) Long-day plants are stimulated to flower when the day length exceeds a certain critical value; short-day plants flower when the day length is less than their critical day length. The light regime is detected by phytochrome in the leaves, resulting in the formation of a presumably hormonal flowering stimulus ("florigen") moving to the buds, a stimulus that can be transferred from one plant or plant part to another plant by grafting them together.

(8) Vernalization, found in biennials (and winter wheat), is the induction of flowering by exposure of the plant to cold temperatures. Other kinds of reproductive temperature responses are found in various plants, especially tropical and subtropical plants.

Chapter 17

EMBRYO, SEED, AND FRUIT

Embryonic development, occurring (in higher plants) within the growing seed, creates a new individual from a fertilized egg. Besides reproducing a species, seeds are also the principal means higher plants have for long-distance dispersal or migration to colonize new territory or to find an environment permitting survival when local conditions change unfavorably. Seed production and dispersal thus play for plants one of the important roles that locomotion plays for animals. Plants have evolved various means of promoting seed dispersal, most often utilizing fruits specialized for this purpose.

Seed germination, like birth or hatching in animals, is a critical step in a plant's life cycle. Because of the limited size of the embryo and the limited food reserves that can be stored in a typical seed for use in its germination, most seedlings are relatively frail. Yet they must be able to withstand environmental hazards such as storms, temperature extremes, and herbivorous animals, and the new seedlings must be able to compete with older plants in order to become established. Plants have therefore evolved environmental responses in seeds helping to assure that they germinate under the most advantageous environmental conditions and at the most favorable times of year, and responses in their seedlings helping them overcome some of the difficulties they often encounter. This chapter considers these phenomena as well as the development of the embryo plant and the typical structure of seeds and fruits, completing the account of sexual reproduction in flowering plants begun in Chapter 16.

EMBRYO DEVELOPMENT

Figure 17-1 shows the stages through which the development of a dicot embryo typically pro-

gresses, from the zygote in the embryo sac to the embryo within the seed. The zygote undergoes a series of mitotic divisions, producing a short chain or filament of cells called the **proembryo** (Figure 17-1A–C). From the beginning the proembryo exhibits a directionality or polarity which becomes amplified in further development. The more basal cells of the proembryo, located toward the tip of the embryo sac, form a separate structure called the **suspensor,** which does not become part of the eventual embryo plant, while the more apical cells multiply and enlarge into a globular cell mass comprising the embryo proper. The suspensor, through enlargement of its cells, pushes the young embryo deeper into the embryo sac. It apparently also absorbs nutrients for the embryo, which is nonphotosynthetic and completely heterotrophic at this stage.

From the globular stage the embryo progresses to a "heart" stage (Figure 17-1G) in which the two cotyledons begin to emerge at the pole away from the suspensor. The shoot apical meristem develops between the cotyledons while the opposite pole, next to the suspensor, becomes the root apical meristem. The embryo elongates into the "torpedo" stage (Figure 17-1H) and differentiates a vascular system. Further development often involves a disproportionate growth of the cotyledons, which may comprise most of the bulk of a full-term embryo.

In its later growth the embryo of many seeds becomes curved or bent double within the confines of the seed coats (Figure 17-1I). We mentioned in Chapter 3 that besides the cotyledon(s), a fully developed embryo consists of organs termed (1) the radicle or primordial root; (2) the hypocotyl or stem-like axis between the radicle and the first node, where the cotyledon or cotyledons attach; and (3) the plumule or primordial shoot apex, including any foliage leaf primordia that have been initiated (Figure 17-1I).

The development of monocot embryos involves proembryo and globular stages comparable to those of dicots, but because only a single cotyledon develops, there is no "heart" stage. The mature embryo, however, possesses organs com-

Figure 17-A. Dandelion "seed" head showing individual dandelion fruits, some still attached to the head and some being blown free to float on the wind with the help of their parachute-like portion, the pappus.

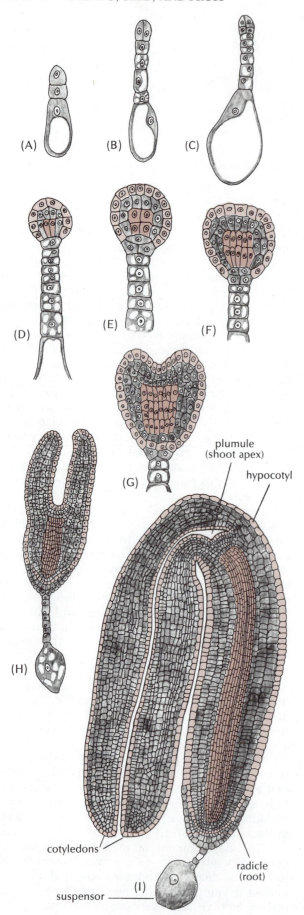

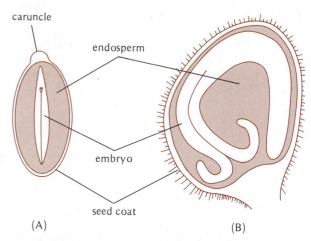

Figure 17-2. Diagrams of endosperm-containing dicot seeds in longisection (note embryo's two cotyledons). (A) Castor bean, twice natural size. The caruncle is a soft external appendage which apparently promotes seed dispersal by providing food for animals which disperse the seed. (B) Tomato, about 15 × natural size.

parable to those of the dicot embryo (Figure 17-3A). Embryos of grasses develop unusual sheathing structures around the shoot and root tips, called the coleoptile and coleorhiza, respectively (Figure 17-3B).

Embryoids from Vegetative Cells

Under tissue culture conditions embryo-like structures, called **embryoids** (meaning "embryo-like"), sometimes develop from cells derived from vegetative tissues (Figure 14-19). When suitably transplanted, embryoids can grow into complete plants. Embryoids develop polarity and pass through morphological stages similar to those just described for zygotic embryos, often even passing through a filamentous proembryo-like stage. This fact indicates that the developmental program for embryogenesis can be called forth in cells that would normally never utilize it, reflecting their genetic and developmental "totipotency" (Chapter 14). The

Figure 17-1. Embryo development during seed formation in shepherd's purse (*Capsella*). (A) Chain of cells (proembryo) formed by two divisions of zygote in embryo sac. (B) At the 7-cell stage the terminal cell has divided transversely; yielding two cells that will develop into the embryo proper; the rest of the proembryo including the enlarged cell at the base is the suspensor. (C–E) Further cell divisions and growth yield a globular embryo (E) atop a further-developed suspensor. (F–G) Formation of "heart" stage (G) by initiation of cotyledons. Beginning of vascular differentiation (procambium) in center of embryo. (H) "Torpedo" stage formed by elongation of cotyledons, hypocotyl, and radicle (root). (I) Fully developed embryo, bent double by elongation to twice the length of the seed coat chamber within which it has developed. Suspensor persists at root tip.

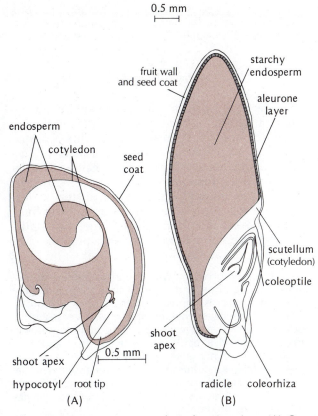

Figure 17-3. Monocot seeds in longisection. (A) Onion; (B) wheat. A "seed" of wheat and other grasses is actually an entire fruit (called a *grain* or *caryopsis*), the wall of which is tightly fused to the coat of the enclosed single seed. The aleurone layer, which is the outermost part of the endosperm, is usually 2–3 cells thick rather than uniseriate as in diagram.

principal conditions found to favor embryoid initiation are (1) a relatively high nitrogen supply and (2) transfer of the cells from an auxin-containing medium to one that lacks or contains only a low level of auxin (a hormonal "step-down").

Even more remarkably, in immature anthers placed in test tube culture the microspores will sometimes divide repeatedly, giving rise to embryoids rather than to pollen grains. Since the microspores are haploid, the resultant embryoids are haploid. This shows that the developmental program for embryo formation does not really depend on the diploid state, even though the program normally begins with the attainment, at fertilization, of diploidy. From haploid embryoids entire haploid vegetative plants can be grown. Embryoid and haploid embryoid induction has important uses in genetic studies and plant breeding (Chapter 18).

Role of the Endosperm

The endosperm of most developing seeds is at first noncellular, or liquid, consisting of the embryo sac cytoplasm populated by rapidly multiplying triploid endosperm nuclei. Possibly famil-

iar examples are the liquid contents of corn kernels in the "milk" stage and the milk of coconuts, which is a portion of the endosperm of the coconut seed that remains liquid after the outer part of the endosperm has become cellular. In the earlier, frequently liquid, stages of its development, the endosperm apparently serves for the biosynthesis of nutrients and hormones for the growing embryo. Liquid endosperm such as coconut milk is very rich in such materials: coconut milk is widely used as a nutritional and hormonal supplement to make possible or improve the growth of plant cells and tissues in test tube culture.

Embryo Culture

The dependence of early embryo growth upon exterior nutrients and hormones is shown by experiments in which embryos are removed from embryo sacs and cultured on agar-solidified media in culture tubes or flasks. Whereas embryos beyond the "torpedo" stage can continue development on very simple media containing only sugar and mineral salts, younger and younger embryos show a progressively greater dependence upon hormones, vitamins, and sometimes amino acids such as are found in liquid endosperm. In one investigation, for example, it was found possible to grow globular embryos of *Capsella bursa-pastoris* (shepherd's purse) into full-sized embryos similar to those in seeds if, in addition to sugar, mineral salts, and vitamins, the medium also contained auxin, a cytokinin, and the purine base, adenine.

Final Development and Structure of the Seed

Later in the development of many seeds, after the endosperm (if at first liquid) has become cellular through the ingrowth of cell walls between its nuclei, the endosperm begins to convert available nutrients into insoluble reserves such as starch, storage proteins, and vegetable oils, storing up these products for the future use of the embryo during seed germination. In such seeds at maturity the seed coats enclose an embryo plus endosperm (Figures 17-2A, 17-3). In the development of certain seeds, however, the later growth of the embryo occurs at the expense of endosperm tissue, which becomes broken down and disappears while the embryo's cotyledons enlarge, synthesize and store up food reserves, become thick and fleshy, and eventually occupy most of the volume of the seed, which in this case contains little or no recognizable endosperm at maturity (Figure 17-4). Examples of this type of seed, which is common in dicots, include beans and peas, peanuts, squash and sunflower seeds, and oak acorns. Although its final structure is simpler than that of endosperm-containing seeds, the developmental route to the endosperm-less condition actually involves an additional step (endosperm digestion and ov-

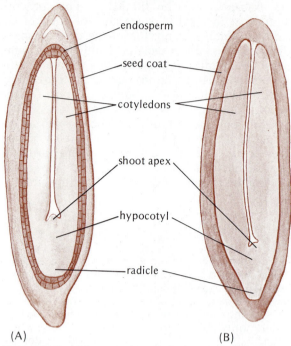

Figure 17-4. Dicot seeds with little or no endosperm. (A) Lettuce (about 20 × natural size), endosperm only two cells thick (B) Squash (about 8 ×), endosperm lacking. The cotyledons are the principal reserve-storing organs in both.

quently have features or associated structures that promote dispersal in the same way that many fruits promote seed dispersal in flowering plants (see below).

Embryo Dormancy

The final important feature of embryo development within the seed is the cessation of growth when the embryo attains full size. This is a temporary, physiologically imposed dormancy, somewhat analogous to winter bud dormancy (Chapter 14). If it fails, the embryo continues to grow and the seed germinates within the fruit (or, for example, corn kernels germinate within the husk while the ear is still attached to the corn stalk), the seed becoming useless either as produce or as a means of reproduction or dispersal.

Young embryos grown in embryo culture do not stop growing or enter dormancy, indicating that in the seed, dormancy is imposed on the embryo from outside it. At the stage when embryo growth ceases in the developing seed, a large increase in the inhibitory growth regulator, abscisic acid, occurs in the fruit and seed (Figure 17-6). Since adding abscisic acid to a culture medium makes a cultured embryo stop growth and become dormant, the production of abscisic acid may be the means by which a maturing fruit imposes dormancy on the seeds it contains.

erdevelopment of cotyledons) compared with the development of seeds containing endosperm.

The final size attained and the amounts of reserve products stored by seeds vary enormously among different plants—from the small seed, 1–2 mm in diameter, produced by many weeds and some crop plants (lettuce, tomato), through relatively large seeds, 0.8–2 cm across, such as peas, beans, corn, and lima beans, to giant seeds including the coconut and the even larger seeds of certain other palms (Figure 17-5). At the other extreme, orchid seeds are almost microscopic, contain little or no food reserves, and require organic nutrients from outside in order to germinate.

The seed's coats, derived from the outer layers or integuments of the ovule, vary from papery to stony among different seeds and sometimes show specializations such as hairs, spines, or mucilage layers that may have a role in dispersal or germination.

Seeds of conifers and other gymnosperms have a fundamentally different structure from those of flowering plants. The nature of gymnospermous seeds is explained in Chapter 30. Although different in nature, the behavior of the gymnosperm seed and its embryo in dispersal and germination is similar to that of flowering plant seeds. The subsequent discussion of these aspects of seed reproduction therefore applies to both types of seeds, except that gymnosperm seeds are not borne in or dispersed by a fruit; however, they fre-

Figure 17-5. Fruit and seed of the Seychelles nut palm or "double coconut," *Lodoicea seychellarum* (native to the Seychelles Islands), the world's largest known seed, taking five years to mature and weighing up to 18 kg (40 lb). Man is holding one of the seeds, after removal of the fruit husk, showing the midline groove leading to the name "double coconut."

Vivipary

A few plants show incomplete embryo dormancy as their normal condition, germination beginning while the seed is still attached to the parent plant. The mangrove tree, *Rhizophora mangle,* which grows in intertidal mud along tropical seacoasts, produces viviparous seeds with a heavy, spear-like root (Figure 17-7) which tends to impale itself in the mud below when the seed finally falls. This "instant planting" improves the seedling's chances of not being washed out to sea with the next tide.

FRUIT DEVELOPMENT

Pollination, and the ensuing fertilization within the ovules, stimulates the ovary enclosing them to begin growth into a fruit. Within a few days after successful pollination the ovary begins to swell perceptibly. This is called **fruit set.** If pollination or fertilization fails, the fruit fails to set, and the ovary or entire flower soon falls away.

Applying the hormone auxin or gibberellin to the ovaries of flowers can in many cases induce a fruit to develop without pollination or fertilization. This is called a **parthenocarpic fruit** and is, of course, seedless (Figure 17-8).

Pollen is one of the richest natural sources of auxin, so it is thought that the initial stimulus to fruit growth comes simply from the auxin to which the pistil is exposed as a result of pollination. In some plants the ovary starts to grow even before the pollen tubes have reached the ovules or fertilization has occurred. Later growth of the fruit also depends upon auxin and gibberellin, this requirement being satisfied by hormone production in the developing seeds, first by the endosperm and later by the embryo.

In a number of horticultural situations the hormone level provided by pollination or subsequent seed development seems to be marginal, so a significant number of flowers fail to set fruit or the partly grown fruits later drop off ("June drop"). In such cases, it is advantageous to spray the plants with auxin, which increases the initial fruit set and decreases the loss by June drop, significantly increasing the yield of fruit. This is common practice today in tomato and soft fruit (plums, peaches, etc.) horticulture.

Some important horticultural fruits such as the banana, pineapple, navel orange, and seedless grape are naturally parthenocarpic, thus seedless. In some cases the pollen is sterile, but still provides enough auxin to induce fruit development. In other cases fertilization occurs but the embryo aborts during development, so the fruit becomes seedless.

Fruit Ripening

When its seeds are ready for dispersal, a fruit "ripens." Some aspects of fleshy fruit ripening are familiar: changing from green to some bright color,

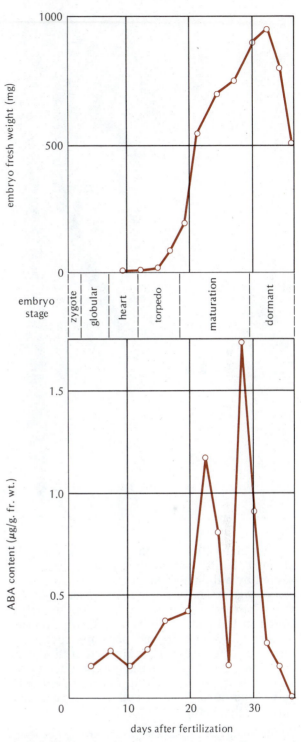

Figure 17-6. Embryo growth in weight (above) and abscisic acid (ABA) content (below) during bean seed development.

sweetening, and softening. A fruit's color changes due to chlorophyll breakdown and to the formation either of yellow, orange, or red carotenoid pigments in chromoplasts (banana, orange, tomato) or of red or purple vacuolar anthocyanin pigments (red apples, raspberries, grapes). Sweetening is due to the accumulation of sugar, which

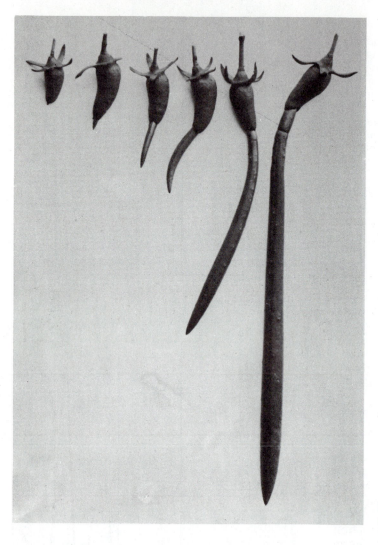

Figure 17-7. Stages in germination of viviparous seeds of mangrove (*Rhizophora mangle*).

reaches high concentrations (over 20% by weight) in some fruits. Sweetening is usually accompanied by a dramatic decrease in the fruit's content of organic acids, which make green fruits taste sour. Conversion of these acids, and of the starch stored in the green fruit, into sugar is often important in sweetening and accounts completely for the sweetening of fruits ripened off the plant, a common practice in marketing produce today. The softening of fruit tissue during ripening involves the breakdown of cell wall components, particularly pectin.

Ripening can be induced prematurely by exposing fruit to air that contains only a few parts per million of the gaseous plant growth regulator, ethylene (Chapter 14). Ripening can be delayed by storing fruits in air kept as free of ethylene as possible and/or supplemented with extra CO_2, which antagonizes the physiological action of ethylene. This technique of **controlled-atmosphere storage,** now extremely important in the produce distribution industry, takes advantage of the fruit's own method of inducing ripening. The onset of ripening is preceded by a sharp increase in the production of ethylene by the fruit itself (Figure 17-9), which apparently induces the observed ripening changes. If the accumulation or action of ethylene is prevented by the techniques of controlled-atmosphere storage, ripening can be delayed and the storage life of the produce greatly extended.

In fruits such as bean pods that become dry at maturity ("dry fruits"), ripening is seemingly different, amounting simply to senescence and drying out of the fruit tissues, the fruit often splitting open to release the seeds. These features are explained further in the next section. However, since the ripening of a fleshy fruit such as an apple or pear involves senescence and (when overripe) death of the fruit cells, the two kinds of ripening have important features in common. Both are stimulated, for example, by ethylene.

Types of Fruit

Development of the ovary into a fruit creates a wide variety of structural specializations yielding extremely different fruit forms and structures, mostly designed to promote seed dispersal in one way or another. In some instances parts of the

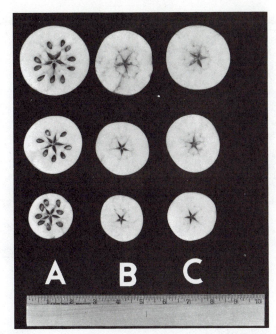

Figure 17-8. Seeded and parthenocarpic apple fruits in section view. (A) Seeded fruits from pollinated flowers; seeds removed and placed on cut surface. (B) Seedless fruits produced by treating unpollinated flowers with lanolin paste containing gibberellic acid. (C) Seedless fruits produced by treating unpollinated flowers with a lanolin paste containing a paste of embryonic apple seeds.

flower other than the ovary, or parts of the stem on which it is borne, become fused to the ovary during its development and form part of the mature fruit.

As indicated above, there are two main categories of fruits. (1) **Fleshy fruits** include the soft, sweet, brightly colored fruits referred to by the everyday use of the word fruit, as well as soft but nonsweet fruits that are colloquially called vege-

tables, such as squashes, bell peppers, eggplants, and tomatoes. (2) **Dry fruits** become more or less hard and dry by the time the seeds are ready to shed. The pods of peas and beans and the seed pods of many familiar garden flowers fall into this category, as do many nuts, consisting of an edible seed enclosed by ovary tissue that has become stony through the deposition of heavy secondary walls by its cells.

Fruits may be either many-seeded, such as peas, beans, squashes, and melons, or single-seeded, such as most edible nuts and many soft fruits including peaches, plums, cherries, and avocados. Fruits become single-seeded either because the ovary contains only one ovule to begin with or because all but one of the several ovules that the ovary contains abort during postpollination development. Rarely, more than one of the ovules grows into a seed, explaining the occasional two-seeded almond or hazel nut. One-seeded dry fruits are usually mistaken by the nonspecialist for simple seeds. Examples are thistle, dandelion, sunflower, and lettuce "seeds." Particularly important examples are the "seeds" of grasses including cereal grains, which actually consist of a seed fused tightly to its enclosing ovary (fruit) wall.

The diversity of fruit forms and structures can be understood more basically in terms of their floral, and sometimes extrafloral, tissue makeup, as explained in Table 17-1 and Figures 17-10 and 17-11 for some of the more important named fruit types.

SEED DISPERSAL

As noted previously, fruits very often serve for seed dispersal. In some cases other structures

Figure 17-9. Respiratory CO_2 production (in ml) and ethylene formation (in μl, multiplied by 50 for convenience in plotting) by banana fruits during their ripening. The respiratory maximum observed in fruits when they ripen is called the respiratory "climacteric." A dramatic burst of ethylene production precedes the climacteric, initiating the ripening process.

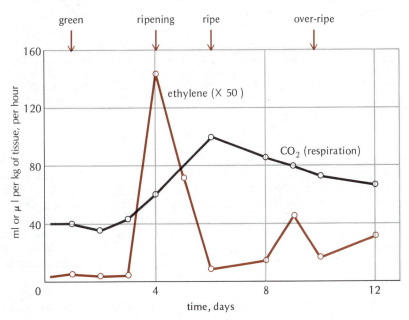

DRY FRUITS

FLESHY FRUITS

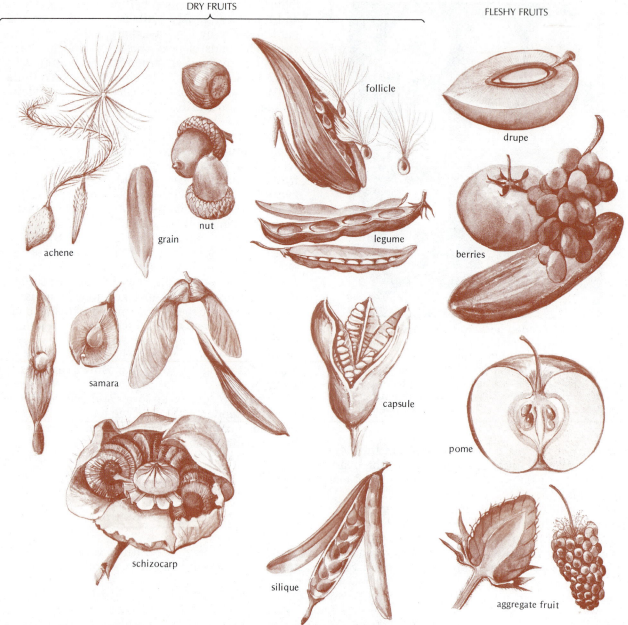

follicle

drupe

nut

grain

achene

legume

berries

samara

capsule

pome

schizocarp

silique

aggregate fruit

Figure 17-10. Principal types of fruits, illustrated by examples named in the accompanying table (page opposite) which lists the defining features of various fruit types. The samaras and the plume-bearing dandelion achenes are examples of wind-dispersed *fruits,* while wind-dispersed *seeds* are illustrated by the plume-bearing seeds escaping from the milkweed follicle.

such as sepals, or modified leaves called bracts that surround the fruit, aid in dispersal. In other cases seeds themselves bear structures that promote dispersal after the seed is released from the fruit. The term **dispersal unit** refers to any detached plant part serving as a vehicle for seed dispersal.

Dispersal by Animals

The bright colors and agreeable flavors of fleshy fruits lure animals and humans to gather and eat them; as a result the seeds are moved from place to place, especially when resistant to the digestive process of an animal's alimentary tract and eliminated after the animal has moved on. Thus plants exploit the food-seeking locomotion of animals for their own migration. Some seeds can remain viable in the intestinal tracts of migratory shore birds long enough to be transported several thousand miles.

Another common means of dispersal by animals is through dispersal units with hooks, bristles, spines, or adhesive secretions that stick in

TABLE 17-1   Fruit Types and Terms[1]

SIMPLE FRUITS: formed from a single pistil (composed of one or more carpels)

Dry fruits: ovary wall (**pericarp**) becomes dry upon ripening
  Indehiscent fruits: remain closed when ripe
    Achene: small, single-seeded dry fruit with relatively thin pericarp, separate from the enclosed seed (left, *Clematis,* and right, dandelion)
    Grain (caryopsis): similar to achene, but pericarp tightly fused with enclosed seed's coats (grass "seeds")[2]
    Nut: large, stony-walled single-seeded fruit (hazel nut, above, and oak acorns, below)
    Samara: winged, usually single-seeded fruit (left to right: tree-of-heaven (*Ailanthus*), elm, maple,[3] ash "seeds")
    Schizocarp: multiple-seeded fruit that splits into several single-seeded portions when ripe (mallow)
  Dehiscent fruits: split open when ripe, releasing seeds
    Follicle: derived from a single carpel that splits down one side when ripe (milkweed)
    Legume: also from a single carpel, but splitting when ripe into two halves along dorsal and ventral lines (locust, above, and pea, below)
    Capsule: derived from pistil composed of two or more carpels, and opening along two or more lines or pores when ripe (iris)
    Silique: particular type of capsule, with outer surfaces of two carpels separating from a central partition when ripe (mustard)

Fleshy fruits: ovary wall and/or associated tissue soft and fleshy when ripe
    Drupe: fruit wall consists of outer fleshy layer (mesocarp) and inner stony layer (endocarp, or "pit") surrounding usually one seed (peach)
    Berry: one to many seeds embedded in entirely fleshy pericarp (tomato, grape, cucumber)[4]
    Pome: like a berry but with outer, main fleshy layer derived not from ovary wall but from surrounding receptacle (stem) tissue (apple)

AGGREGATE FRUITS:[5] a cluster of several separate carpels, from a single flower (strawberry,[6] blackberry)
MULTIPLE FRUITS:[5] derived from several to many flowers, whose parts more or less fuse (pineapple, Figure 17-11)

[1]Examples pictured in Figure 17-10 on opposite page are noted in parentheses.
[2]Structure of grain is shown in Figure 17-3A.
[3]Maple fruits consist of two winged, one-seeded parts that separate and disperse independently upon ripening.
[4]Note the considerable difference between botanical and colloquial meanings of "berry." The botanical term is not restricted to small, sweet fruits as is colloquial usage, and some colloquial "berries" such as the strawberry, raspberry, and blackberry (see aggregate fruits) are not berries in the botanical sense.
[5]These classes include both dry and fleshy types, as with simple fruits, but we include examples only of the more familiar fleshy representatives.
[6]Fleshy tissue is receptacle (stem), bearing on its surface numerous achenes ("pips") derived from the flower's numerous pistils.

animal fur or human clothing (Figure 17-12). Rather than motivating a potential dispersing animal with a food reward, as fleshy fruits do, these types of dispersal units are successful despite creating a nuisance for the nonconsenting animal or human partner in the dispersal process. In trying to get rid of the burrs and stickers picked up while moving through fields or vegetation, the animal or human unwittingly scatters the seeds in places distant from the location of their parent plants.

Dispersal by Wind

Many small seeds (such as milkweed) and dry fruits have feathery plumes or bristles (for exam-

ple, dandelion and thistle achenes) or wings (samaras), enabling transport of the dispersal unit by wind. This is perhaps the most effective means of long distance seed dispersal. Another form of wind dispersal occurs in the "tumbleweeds," common in open, arid terrain. These are annual plants whose tops break off at seed maturity and are blown across the landscape by the wind, scattering seeds as they go (Figure 17-13).

Dispersal by Water

Because the large seed ("nut") of the coconut palm can float and resist salt water, coconut palms have spread over water to many remote oceanic

(A)

(B)

Figure 17-11. The pineapple, an example of a multiple fruit, shown attached to its parent pineapple plant. The fruit is a condensed inflorescence; the flowers which have become fused into a compact mass during development are responsible for the diamond-shaped pattern on the pineapple's surface. Above the fruit the apex of the inflorescence produces a small leafy shoot, which can be rooted to propagate the plant vegetatively. Other less familiar multiple fruits are the fig, mulberry, breadfruit, and osage orange.

islands in tropical latitudes. This type of dispersal is important chiefly to ocean-shore plants.

Forcible Discharge

Some fruits are specialized to propel their seeds away from the parent plant (Figure 17-14). For example, *Impatiens*, or "touch-me-not," is so called because its ripe fruits "explode" on the slightest disturbance. They actually split apart and curl up suddenly due to turgor forces, scattering the seeds. In many legumes and some capsule fruits the drying seed pod tends to twist, generating a stress that suddenly lets go when the pod splits open, hurling the seeds away from the plant. Dwarf mistletoe (*Arceuthobium*) and "squirting cucumber" (*Ecballium*) seeds are blown as much as 10 m (30 ft) by the discharge of turgor pressure from within the ripe fruits. The woody fruit segments of the tropical sandbox tree (*Hura crepitans*) strain like bows as they dry out, finally splitting apart and slinging their heavy seeds reportedly as far as 100 m, in an explosion said to be dangerous to bystanders.

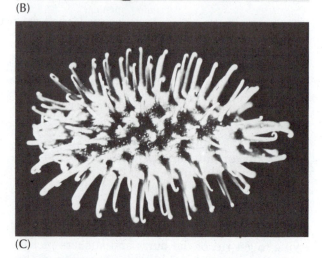

(C)

Figure 17-12. Examples of animal-dispersed fruits or other dispersal units, bearing barbs or hooks that catch in fur, feathers, or clothing. (A) Fruits of Queen Anne's lace, (B) Fruit cluster of "sticktight seed," each fruit detaching separately from the head and bearing two barbed bristles. (C) Fruit head of cocklebur (*Xanthium strumarium*), an entire inflorescence with bracts modified as hooks, enclosing two seeds.

Figure 17-13. Tumbleweed (*Salsola kali*). The plant on the right is living; the one on the left is dead, dried, and detached.

SEED DORMANCY AND CONTROL OF GERMINATION

As it completes its development and becomes ready for dispersal, the seed generally becomes dehydrated, reaching a water content sometimes as low as 5–15% of its weight, and its embryo enters a dormant state. Germination begins with the physical uptake of water by the dry seed, called **imbibition,** followed by resumption of growth by the embryo at temperatures favorable to growth. Some seeds, such as those of willows, poplars, and silver maple, begin to germinate immediately after shedding, as soon as they reach moist, warm soil. Other seeds, exemplified by wild oats (*Avena barbata*), will not germinate when they are shed, but lose their dormancy and become able to germinate after a limited time (usually a few weeks). This is called **afterripening.** Many seeds possess a more persistent dormancy that often can be lifted only by environmental agencies, responses that help germination to occur at favorable times of year and under conditions favorable to the success of the seedling.

The seeds of various wild plant species are dormant merely because they possess tough seed coats impervious to water or oxygen or mechanically preventing growth of the embryo. After dispersal in nature, exposure to the elements gradually breaks down and weakens the seed coats, eventually permitting germination. This behavior tends to space out widely the germination of different seeds from the same year's crop, depending on the accidents of exposure, enabling the species to take advantage of chance opportunities for seedling establishment whenever they occur. When seeds of this type are gathered and stored indoors for horticultural use, they remain dormant but can be induced to germinate by mechanically abrad-

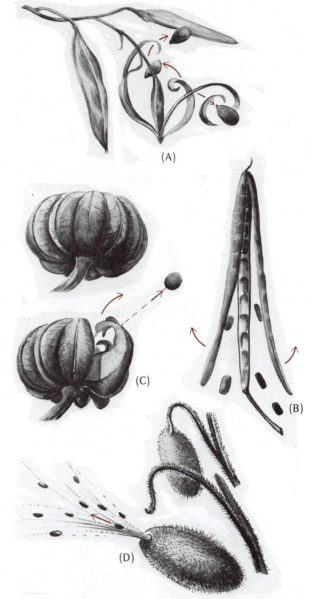

Figure 17-14. Examples of forcibly discharging fruits. (A) Fleshy fruits of touch-me-not (*Impatiens*), the segments of which curl up when they split apart, due to tension between inner and outer fruit wall layers caused by turgor pressure in the cells. (B) Dry fruits (siliques) of bitter cress (*Cardamine*), the halves of which separate and curl up explosively due to a bending stress created by drying. (C) Woody segments of the fruit of the sandbox tree (*Hura crepitans*) discharge their seeds by a sling action when the segments, severely stressed by drying out, split loose. (D) Squirting cucumber (*Ecballium*) fruits eject their seeds by discharge of internal turgor pressure when the fruit separates from its stalk.

ing or nicking the seed coats (**scarification**), or by softening them by soaking the seeds temporarily in a corrosive medium (such as sulfuric acid) before planting. These procedures are widely used by nurseries to obtain seedlings of ornamental plants.

Other seeds show a "constitutive," or self-imposed, dormancy of the embryo itself, which can be terminated by a specific environmental signal such as cold temperature, light, dry heat, or leaching by sufficiently heavy rainfall. Heat-inducible seed germination occurs in plants adapted to colonizing ground opened up by forest or brush fires. Leaching-dependent seeds of desert annuals ensure that germination will occur only after enough rain has fallen to permit seedlings to complete their life cycle even if it does not rain again that season. Cold- and light-inducible germination occurs much more widely.

Cold Requirement

Seeds of continental-climate plants often require an extended (weeks or months) exposure, after imbibition, to temperatures near freezing before they will germinate at temperatures favorable to growth. This requirement, normally satisfied by the temperatures of winter, ensures that the seeds will germinate only in spring, with a whole growing season ahead of them, rather than in late summer or autumn when harsh weather would soon arrive. The cold requirement is basically similar to that for breaking winter bud dormancy (Chapter 14) and, like the latter, probably involves growth inhibitors (dormins) such as abscisic acid.

Seeds of many ornamental plants possess a cold requirement; to be germinated in nurseries, they must be exposed artificially, under moist conditions, to near-freezing temperatures for an extended period before being put into a warm place for germination. This horticultural procedure is termed **stratification** because the cold treatment is often given to the seeds after spreading them, in flats, between layers (strata) of moist sand or some other planting medium. This is necessary because exposure of seeds to cold when they are dry is not effective in breaking dormancy.

Another seed type is induced to germinate not by any specific temperature but by a sufficiently large fluctuation in temperature. A day–night temperature difference of about 15°C (27°F) seems to be most often optimal for this response, found in many weed seeds. There is as yet little information on how a temperature fluctuation might act to induce germination.

Light Requirement

Small seeds, such as lettuce, commonly require exposure to light for germination. This helps

Figure 17-15.   The control of lettuce seed germination by red (R) and far-red (I) light. Seeds are moistened, then exposed to red light (for 1 min each exposure) and far-red light (for 4 min each exposure) in the sequences indicated. If the last exposure is to red, most of the seeds germinate; if to far-red, they remain dormant.

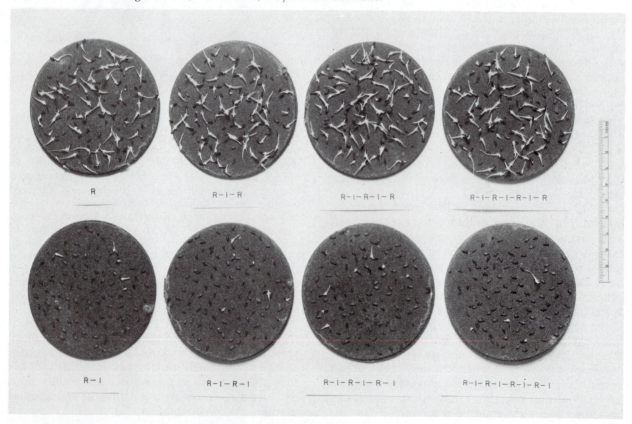

ensure that a seed will not germinate when buried too deeply for the seedling shoot to reach the surface of the ground. The germination of many weed seeds is light sensitive. When land is tilled, seeds buried in the soil are brought to the surface, exposed to light, and thus induced to germinate. This enables weeds to reappear quickly after tillage and persist in agricultural fields despite cultivation.

The light requirement has been studied especially thoroughly with lettuce seeds; indeed such work led to the original discovery of the photomorphogenic pigment, phytochrome. Red light (wavelengths of about 660 nm) most effectively stimulates germination: after the exposure of imbibed seeds to light for a few minutes, they will germinate in complete darkness. However, "far-red" light (730 nm) given immediately after a red or white light exposure cancels the effect of the red or white light and blocks germination (Figure 17-15). This indicates that phytochrome is the light receptor, as explained in Chapter 14.

The opposing responses to red and far-red light enable light-sensitive seeds to respond oppositely, depending on whether they are exposed directly to the sun or are shaded by other plants. Because chlorophyll absorbs red light strongly but does not absorb far-red light, sunlight that has passed through green leaves contains far-red but not red light (Figure 17-16) and therefore *inhibits* the germination of light-sensitive seeds. Consequently, they are inhibited from germinating when heavily shaded by vegetation, a situation that would severely disadvantage the seedling in competing for light for photosynthesis. Seeds are induced to germinate at a time of year when they receive direct sunlight or if the overlying vegetation is removed. This response improves the seedling's chances of becoming established.

Longevity of Seeds

Seeds retain their vitality and ability to germinate, called **seed viability,** for greatly varying lengths of time. The seeds of some plants of the moist tropics are viable for only a few weeks. In the dry tropics and in temperate and colder latitudes, seeds usually must survive either a lengthy dry season or winter and thus must possess greater longevity. The longevity of temperate-zone crop seeds, which lack any dormancy mechanism, is increased by storage under cool-dry conditions. Corn seeds, whose viability normally lasts only about 5 years under farm-storage conditions, have remained viable for more than 20 years when stored at near-freezing temperatures under low moisture. One sample of wheat showed 99% germination after 32 years of cool-dry storage. On the other hand, the seeds of some tropical crops, such as cacao and coffee, rapidly lose viability unless kept moist. Furthermore, the longevity of temperate-zone seeds possessing a dormancy mechanism also

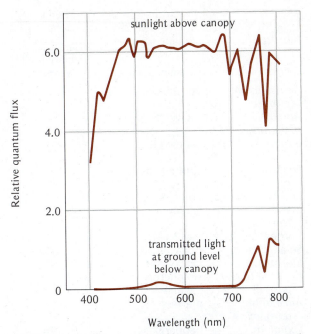

Figure 17-16.   The spectral quality of sunlight passed through green leaves. Notice the preponderance of far-red light (greater than 730 nm) in light transmitted through a forest canopy (lower line) compared to sunlight above the canopy, which contains at least as much red (660 nm) as far-red light.

seems to be considerably greater under cool-moist than under dry conditions.

An experiment begun in 1879 by W. J. Beal of Michigan State University has determined the longevity of weed seeds when buried in the soil. Beal filled 20 bottles with sand holding 50 seeds of each of 20 kinds of herbaceous plants, mostly weeds—a total of 1000 seeds in each bottle—and buried the bottles 45 cm (18 in.) below the soil surface. Bottles have been dug up at intervals, and germination tests made, ever since. The seeds of only three species survived longer than 50 years (Table 17-2). By 1970, after 90 years of burial, only one species, *Verbascum blattaria* (moth mullein), survived, still giving 20% germination.

Germination of some seeds at much greater ages has been recorded. For example, seeds from a herbarium specimen of the "silk tree," *Albizzia julibrissin*, collected in 1795 and stored in the British Museum, germinated in 1942, when 147 years old. Viable weed seeds were excavated from cool, low-oxygen sediments in several Danish archeological sites dated as 600–1200 years old. Unfortunately there is no proof that these seeds were really as old as the sediments in which they were found, and the same must be said for an even more remarkable record of viable lupine seeds recovered from frozen ground in Alaska, 3–7 m below the surface, geologically estimated to be more than 10,000 years old. The record for proved longevity seems to be some viable water lily (*Nelumbo nucifera*) seeds, dated by the radioisotope method (Chapter 21) to be about 1040 years old, which were found in a peat layer (from a former swamp bed) in Manchuria.

TABLE 17-2   Results of Beal's Buried-Seed Experiment
Data from A. Kivilaan and R. S. Bandurski, *American Journal of Botany,* Volume 60, p. 141, 1973.

Species	Common name	Viability limit (years)[a]
Agrostemma githago	corn-cockle	<5
Ambrosia artemisiaefolia	ragweed	<5
Bromus secalinus	brome grass	<5
Euphorbia maculata	spotted spurge	<5
Trifolium repens	creeping clover	5
Malva rotundifolia	mallow	20
Anthemis cotula	mayweed	30
Setaria glauca	foxtail grass	30
Stellaria media	chickweed	30
Capsella bursa-pastoris	shepherd's purse	35
Amaranthus retroflexus	pigweed	40
Lepidium virginicum	peppergrass	40
Portulaca oleracea	purslane	40
Brassica nigra	black mustard	50
Polygonum hydropiper	knotweed	50
Oenothera biennis	evening primrose	80
Rumex crispus	dock	80
Verbascum blattaria	mullein	>90[b]

[a]Last occurrence of seed germination. Tests were made every 5 years until the 40th year, and every 10 years thereafter. <5 indicates no germination even after the first 5 years.
[b]In the 1970 test these seeds still gave 20% germination.

The Seed Pool in the Soil

Because of persistent seed dormancy, many of the seeds shed each year do not germinate in that or the next growing season, and some may not germinate for many years, even though still viable. Thus a pool or "bank" of buried seeds builds up in the soil whose numbers often greatly exceed the number of growing plants in the area. As many as 1000 to 5000 viable seeds/m² have been found under various kinds of vegetation and 20,000 to as many as 80,000 weed seeds/m² of cropland. This seed pool enables the regeneration of plant populations after catastrophic events such as fire, flood, drought, and unseasonable frosts. Similarly, weed seed pools in agricultural soils ensure the continual regrowth of weeds despite efforts to control them.

SEEDLING GROWTH AND RESPONSES

The general phenomena of seed germination and the basic distinction between the hypogeous and epigeous modes of germination were explained in Chapter 3. Here we shall examine features of growth and behavior important in helping germinating seedlings establish themselves in their environment.

After swelling of the seed due to the purely physical imbibition of water, the first visible sign of germination is usually elongation of the radicle or primordial root and its protrusion through the seed coats into the environment. This is a metabolic growth process and requires a favorable temperature. Early growth involves little or no cell division, which begins only gradually in the apical meristems of the root and shoot. Respiration, which is virtually undetectable in a seed, begins and increases rapidly as the embryo begins to grow. Almost from its outset the growth of most seedlings is strictly dependent on oxygen for respiration; the seedlings are injured or killed if flooded and thereby deprived of adequate oxygen, a crop injury frequently observable in low places after flooding occurs due to heavy spring rains. Certain seeds such as rice, however, are adapted to germination under water; their seedlings can make some growth even when completely deprived of oxygen.

Soon after emergence the young root and shoot acquire the capacity for geotropism, or growth curvature response to gravity, ensuring that the root grows downward and the shoot upward regardless of how these organs were pointed when

Figure 17-17.   Orchid seedlings growing on sterile nutrient agar. The seeds were sown on the agar and germinated there. The seedlings may now be transferred to soil.

Figure 17-18. Mobilization of insoluble storage reserves (mainly starch and protein) during germination of a grass seed such as barley. Diagrams A–E depict conditions within the seed at the times that appear directly above on the graph. (A) Seed seen in longisection, showing release of gibberellin (GA) from embryo into endosperm after the seed is wetted. (B) Aleurone cells respond to GA by producing α-amylase, protease, and other hydrolytic enzymes (hydrolases, shown by black dots), excreting these into the starchy endosperm. (C) Hydrolases break down starch, protein, and other molecules in endosperm, releasing soluble nutrients (color dots), which (D) the embryo's cotyledon (scutellum) absorbs and delivers to the shoot and root for growth. As coleoptile appears above ground its tip opens, and the enclosed first foliage leaf emerges (D). (E) By the time storage reserves are depleted, seedling's first foliage leaf has expanded and begun photosynthesis. Although the formation of hydrolases to mobilize insoluble storage reserves occurs in seeds generally, control of hydrolase formation by GA seems to be found largely or exclusively in seeds of grasses.

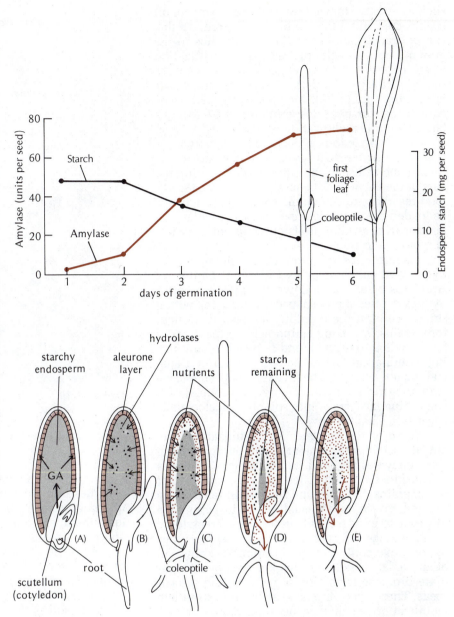

the seed came to rest on or in the ground. The shoot develops phototropic sensitivity, helping it find its way toward the light around obstacles such as stones.

Although light may trigger germination and helps orient seedling growth, photosynthesis is not necessary for early seedling development, which occurs at the expense of the food reserves stored in the seed's endosperm or the embryo's cotyledon(s). The larger the seed, the more reserves it contains and the longer and farther it can grow before using them up and becoming totally dependent on light for photosynthesis. Thus large seeds such as beans may be planted much deeper than small seeds such as carrots and lettuce.

The germination of orchid seeds, which as noted above are extremely small and rudimentary, involves extraordinary nutritional features. These seeds are not capable of germinating on their own in nature, but grow only when they become invaded by a particular kind of fungus, which appears to supply nutrients to the seedling for its early, very slow, nonphotosynthetic growth. This represents symbiosis between a lower and a higher plant, other more generally important examples of which are considered in Chapter 25. Orchid germination may be hastened considerably, and made more reliable, by planting the seeds on a nutrient agar medium containing sugar and vitamins as well as mineral nutrients, similar to the culture media used for many plant tissue cultures. In this case the seedling lacks the usual fungus, at least until it is transplanted out into the soil. This technique of seed culture (Figure 17-17) is now widely used for the horticultural propagation of orchids.

Mobilization of Reserves

Soon after outgrowth of the embryo, breakdown or **mobilization** of insoluble stored reserves such as starch, storage proteins, and vegetable oil begins, converting these into soluble products that can be translocated from the storage sites to the growing points of the root and shoot. Utilization of reserves stored in the endosperm also requires that the embryo *absorb* the soluble products; this is the first function of the cotyledon(s) in seeds containing storage endosperm (Figure 17-18). Uptake of soluble mobilization products such as sugars by the cotyledons appears to be an active, energy-requiring transport process. In endosperm-less seeds such as beans and squash, mobilization occurs in the cotyledons themselves, where the seed's food reserves are stored; the soluble products can be loaded directly into the embryo's vascular system for translocation to its growing points, without an uptake step.

Insoluble reserve products are mobilized by hydrolytic enzymes or hydrolases (Chapter 6). Amylases break starch down to sugars. Proteases hydrolyze reserve protein, usually occurring within storage cells in the form of granules called aleurone grains, releasing amino acids. Lipases split fats or oils into glycerol and fatty acids, and during germination fat-storing seeds develop a special metabolic cycle (the glyoxylate cycle), occurring in unique organelles (glyoxysomes), for converting fatty acids into soluble, translocatable sugar (see Chapter 6).

Some of the enzymes for mobilization may already occur within the dry seed, but more often these enzymes are produced by protein synthesis only as the seed germinates. In the germination of cereal grains, the hormone gibberellin released by the embryo induces specialized endosperm cells to produce amylase and protease for breakdown of the storage reserves (Chapter 14). Other types of hormonal control of mobilization are known in other seeds; for example, cytokinins appear to regulate protease formation in germinating squash seeds. These controls enable embryos to stimulate mobilization when they need nutrients for germination.

Positioning Responses

As long as it remains underground the seedling shoot grows, because of lack of light, in the etiolated form, elongating excessively and without leaf expansion. This behavior maximizes the shoot's chances of reaching photosynthetically useful light before its food reserves are used up. Once the shoot emerges from the soil, light (via phytochrome) induces leaf expansion and inhibits further hyperelongation of the stem.

In epigeous seedlings the cotyledons usually expand and turn green when brought into the light, becoming the seedling's first photosynthetic organs. This developmental switch in function represents a remarkable biochemical transformation

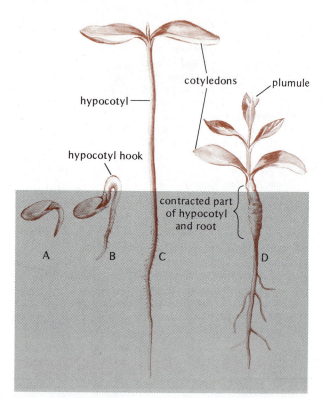

Figure 17-19. Epigeous germination of a dicot seedling, showing formation and later straightening of the hypocotyl hook, and subsequent contraction of primary root and hypocotyl to lower the cotyledons to near ground level.

in the cells of the cotyledons, breaking down the enzymes that were important during the preceding mobilization phase and producing new photosynthetic enzymes and organelles (chloroplasts and peroxisomes).

As noted in Chapter 3, most seedling shoots grow at first in a hook-shaped configuration, which helps them work their way through the soil more effectively. The first part of the shoot to emerge from the soil is the elbow or topmost part of the shoot hook (Figure 17-19). The first response of such shoots when they emerge is straightening of the hook. That hook straightening is a light response becomes apparent when seeds are germinated in a darkroom or dark closet: the etiolated shoots continue growing far into the air with a hook-shaped terminal region. Hook straightening is a typical phytochrome-mediated response, showing red light sensitivity and far-red light reversibility (Chapter 14). Hook straightening and the anti-etiolation responses mediated by phytochrome serve to position the new leafy shoot correctly above the soil surface regardless of the depth at which the seed was planted.

In epigeous seedlings (Figure 17-19), just after hypocotyl hook straightening the seedling can be damaged rather easily because its emergent hypocotyl is relatively long and weak. In many seedlings this instability is soon corrected by a contractile action of the primary root, pulling the hypocotyl back down into the ground and the cotyledons closer or down to the soil surface. Contractile root action is a special kind of growth in which cells of the root enlarge in the transverse direction while contracting in length.

Grass seedlings, which instead of a hook-shaped shoot use their coleoptiles to penetrate the soil (Figure 3-35A), also possess light responses that properly position the young shoot. Its initial growth is due to elongation of the section of stem called the mesocotyl, extending from the base of the coleoptile (inside which the shoot apex is located) to the seed and primary root (Figure 3-35A). The mesocotyl's elongation is very sensitive to inhibition by light. As the shoot tip approaches the soil surface, the small amount of light penetrating the soil stops further mesocotyl elongation, and the coleoptile (whose growth is not very sensitive to inhibition by light) then elongates until it emerges. This behavior positions the shoot apex of the grass just beneath the soil surface and ensures that the seedling's first adventitious roots, which grow from the node at the base of the coleoptile (Figure 3-35A) and eventually become the plant's main root system, begin to develop near to but not above the soil surface.

SUMMARY

(1) In flowering plant seed formation the fertilized egg, within an ovule, develops into an elongated proembryo comprised of basally located suspensor cells plus tip cells that multiply into a globular cell mass which becomes the embryo proper, composed of the primordial root (radicle), the leaf or leaves [cotyledon(s)], the stem axis (hypocotyl) between the cotyledon(s) and the root, and an embryonic shoot (plumule), the development of which, even in the mature seed, often has not progressed beyond the formation of a shoot apical meristem. (Artificially induced embryonic plants or embryoids, developing from vegetative plant cells or from microspores in culture, pass through similar stages.)

(2) Early development of the embryo requires nutrients and hormones from outside, apparently produced by the growing endosperm. In many seeds the endosperm then stores reserve products (starch, vegetable oil, storage proteins) for subsequent use during germination, but in the development of other seeds embryo growth digests the endosperm, and reserve products for use in germination are stored in the cotyledons instead.

(3) Pollination and subsequent fertilization stimulate the ovary to grow into a fruit, dependent on hormones (especially auxin) from the pollen and from the developing seeds. Parthenocarpic, seedless fruits can be induced to develop by hormone treatment, without pollination. (Fruit ripening is controlled through stimulation of fruit tissue senescence by ethylene produced by the fruit.)

(4) Fruits are often specialized to assist in dispersal of the enclosed seed(s). Fleshy fruits gatherable as food, and dry fruits bearing hooks or spines that catch in fur and clothing, promote seed dispersal by animal and human activity. Small, usually one-seeded fruits bearing wings or feathery bristles enable the wind dispersal of many seeds. Seashore plants, such as the coconut palm, have floating, seawater-resistant fruits or seeds adapted for dispersal by water. Some fruits forcibly propel their seeds away from the parent plant. These include various pod- or capsule-like dry fruits that open upon ripening, releasing their seeds. Such seeds often have specialized dispersal adaptations similar to those mentioned for fruits.

(5) As the seed matures prior to shedding, its embryo enters a hormonally imposed dormant state. The embryos of some seeds lose their dormancy and are able to germinate upon or shortly after shedding, but other seeds exhibit a protracted dormancy due either to tough and impervious seed coats or to an internally imposed growth inhibition that can be lifted by some environmental stimulus such as light, cold, or heat. Light-sensitive dormant seeds detect light by means of phytochrome. Whereas the seeds of tropical plants often show only a very limited longevity, those of temperate-crop plants may survive for a few decades if stored under optimum conditions, and the dormant seeds of some wild plants and weeds have longevities of many decades to hundreds of years.

(6) Germination begins with water uptake, or imbibition, followed by temperature- and respiration-dependent growth of the embryo, whose radicle emerges initially from the seed, followed by the hypocotyl and then the plumule. Storage reserves in the cotyledon(s) or endosperm are broken down into soluble products and translocated to the embryo's growing points [after uptake by the cotyledon(s), in the case of endosperm reserves] to supply substrate for growth.

(7) Seedling roots orient their growth through geotropism, and shoots through geotropism and phototropism. Rapid shoot growth toward the soil surface and positioning of the young leafy shoot just above the soil surface are achieved by phytochrome-mediated shoot hook straightening, leaf expansion, and light inhibition of the stem etiolation that occurs in the dark.

Chapter 18

INHERITANCE

The ability of organisms to produce offspring like themselves is fundamental to life. Not only do species reproduce their kind, but distinctive individuals tend to produce offspring with similar distinctive features. The science of genetics seeks to understand the laws and mechanisms governing the transmission of hereditary characteristics from parents to offspring. The application of genetics to the breeding of useful plants and animals has greatly modified agriculture, horticulture, and animal husbandry. This chapter is devoted to the principles of plant genetics and some of its practical applications, while Chapters 19 and 20 cover genetic aspects of ecological adaptation and organic evolution.

Mendel

The modern concepts of inheritance were first introduced in 1866, with the publication of the work of Gregor Mendel (Figure 18-1), an Austrian monk. Prior to Mendel's time much plant and animal breeding had been done, but no one had been able to formulate any clear laws of heredity that could account for or predict the results of breeding experiments. Inheritance was generally thought merely to blend the characteristics of parents in their offspring, like mixing two fluids, as epitomized by the still popular expression "blood" for hereditary material (as when people claim to "have some of George Washington's blood" in their veins).

For 8 years Mendel carried out hybridization or crossing experiments, mostly with garden pea plants differing in a limited number of striking hereditary characteristics. He found that these characteristics did not become blended in the off-

spring but were inherited as units, in a peculiar but regular pattern. Rather remarkably, without any useful prior concepts of the subject to guide him, Mendel devised what is now recognized as a correct and essentially complete hypothesis explaining the genetic makeup of his experimental plants and how their hereditary units (now called genes) were transmitted during sexual reproduction. In retrospect, one reason other breeders had not recognized that hereditary characteristics are transmitted as units was that their crosses involved the simultaneous inheritance of many genetic differences between the parents, rather than the single well-defined genetic characteristics with which Mendel chose to work. Unfortunately,

Figure 18-1. Gregor Mendel pictured in his garden making crosses between pea plants.

Figure 18-A. A tulip flower being pollinated with pollen from a genetically different tulip, to make a cross or hybridization. The experimenter is using a fine brush to transfer pollen from anthers (held in his palm at right) taken from a flower of the intended pollen parent, to the stigma of a flower of the intended pistillate parent, the anthers of which have been removed so the flower cannot self-pollinate.

his contemporaries did not grasp the importance of Mendel's results and of his hypothesis regarding inheritance. His work was ignored and remained forgotten for several decades, to be rediscovered and appreciated only posthumously, at the turn of the twentieth century, when his hypothesis was found to be true for animals as well as plants. Advances in the science of genetics then came rapidly. The phenomena of inheritance of discrete hereditary units, which Mendel discovered, are called Mendelian inheritance in his honor.

LAWS OF MENDELIAN INHERITANCE

Dominant and Recessive Characters

Mendel cross-pollinated pea plants showing a large, well-defined hereditary difference, such as tall (2 m in height) versus dwarf (0.3 m in height). Since the pea is normally self-pollinated, Mendel accomplished cross-pollination by opening a bud-stage (before shedding of pollen) flower borne, for example, by a tall plant, removing its stamens, and pollinating its stigma with pollen from a dwarf plant (Figure 18-A shows this operation in a more easily visualized form, with a tulip flower).

When Mendel planted seeds from this cross, the resulting hybrid plants were not intermediate between the two parents, as might have been expected; instead all grew to about 2 m tall, like the tall parent. Mendel made separate crosses involving six other pairs of characters and found that in every instance the hybrid resembled one of the parents and not the other. He termed **dominant** a character such as tallness, which prevails in the hybrid, and **recessive** the opposite type of character, such as a dwarf habit, which does not express itself in the hybrid. He found round seeds to be dominant over wrinkled seeds, yellow cotyledons to be dominant over green, and flowers distributed along the stem to be dominant over terminal flowers.

Genotype and Phenotype

Mendel permitted the above-mentioned hybrid plants, called the **F₁** or first filial generation (from Latin *filius* or *filia*, son or daughter), to self-pollinate, forming seeds that produced an **F₂** (second filial) generation. Whereas the F₁'s had uniformly resembled their dominant parent, among the F₂ offspring from each cross, some of the plants resembled the dominant parent while others resembled the recessive parent (Figure 18-2). Mendel saw that the F₁'s must have carried, but not expressed, the recessive hereditary factor such as that for dwarfness: although their physical ap-

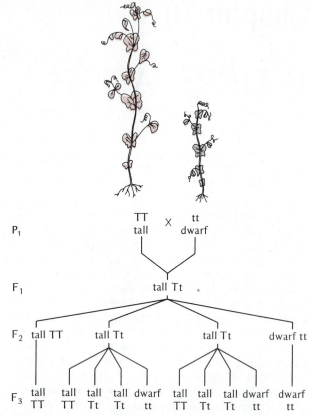

Figure 18-2. Typical results of Mendel's crossing of tall with dwarf peas, listing the phenotypes and inferred genotypes of the parents (P₁), the first generation hybrids (F₁ or first filial generation), and the second and third generations of offspring (F₂, F₃) derived by self-pollinating the F₁ and then the F₂ plants. Where all offspring from a given plant are alike, only one phenotype and genotype is listed, while if the offspring differ in phenotype, the typical proportions of different phenotypes and genotypes are represented by the relative number of individuals listed.

pearance was the same as that of the dominant parent, the genetic constitution of the F₁'s was different, because the parent strain when self-pollinated yielded only plants of the parental type and therefore did not carry the recessive hereditary factor, as did the F₁'s. Thus we must distinguish between the **genotype** or genetic constitution of an individual and its **phenotype** or physical expression of hereditary characteristics.

Whereas the genotype is an inherent feature of the individual from the moment it begins life as a zygote, its phenotype is the result of its development and is the product of both its genotype and its environment. As noted in Chapters 14 and 16, during plant development environmental factors can influence gene expression profoundly. Two apple trees are never identical, even when they have identical genotypes as a result of having been propagated vegetatively from the same parent tree. And a genetically "tall" pea plant may be as stunted as a genetic dwarf if the "tall" plant receives too few nutrients or too little water during its growth. The existence of biological and environmentally caused variation in phenotype for a given genotype must always be kept in mind when evaluating the genetic basis of a characteristic.

Symbols and Terms for the Genotype

Because the F_1's in his crosses must have carried both a dominant and a recessive genetic factor, Mendel concluded that any individual carries two genes of a given kind, such as for tall or dwarf growth, one derived from its female parent and one from its male parent. We now know (although Mendel did not) that this is because each individual is diploid, having received one set of gene-bearing chromosomes from each of its parents. Specific genes are usually designated by letters, a capital letter for a dominant gene and a small letter for the corresponding recessive form of the gene, such as T for tallness and t for dwarf. The genotype of the tall parent, which carried only the gene for tallness, would be TT; that of the F_1 hybrid would be Tt; and that of the dwarf parent would be tt: it is dwarf because it carries no dominant gene to obscure the effect of the recessive t gene on its phenotype.

The different forms of a given gene, such as T and t, governing a particular character, such as height, are termed **alleles** (shortened from the original term allelomorph, from Greek *morphé*, form, and *allelos*, of one another).

Individuals carrying only one allele, exemplified by the genotypes TT and tt, are termed **homozygous** (Greek *homos*, same, plus zygote). Individuals carrying different alleles, exemplified by the genotype Tt, are **heterozygous** (*heteros*, other).

Genetic Segregation

Mendel found that in each of his crosses involving a dominant versus a recessive character, about one-quarter of the F_2 offspring showed the recessive phenotype. As illustrated in Figure 18-2, these individuals would always "breed true," that is, when self-pollinated they would yield only (recessive) offspring like themselves, indicating that they were homozygous for the recessive allele (i.e., were tt in the F_2 of tall crossed with dwarf peas). Among the approximately three-quarters of the F_2 offspring showing the dominant phenotype, Mendel also found individuals that bred true (Figure 18-2) and evidently carried only the dominant allele (i.e., were TT). These again comprised about a quarter of the entire F_2 population. It became apparent to Mendel that during sexual reproduction the genes contributed to an individual (the F_1 hybrid) by its parents become separated from one another in many of its offspring. This is now called genetic segregation or **Mendelian segregation.** The regular phenotypic and genotypic ratios resulting from segregation and seen in the F_2 and subsequent generations after a cross are called Mendelian ratios.

Mendel's observations indicated that the characteristic 3:1 ratio of dominant to recessive phenotypes seen in the F_2 generation occurs because the heterozygous and two parental homozygous genotypes segregate in the ratio $1\ TT : 2\ Tt : 1\ tt$. Mendel was able to explain this by supposing that (1) in the formation of both male and female gametes, the two alleles become separated from one another, yielding equal numbers of gametes carrying one or the other of the respective alleles (equal numbers of T and of t gametes); and (2) in fertilization, gametes combine randomly, without regard for their genotypes. These suppositions quantitatively predict a 1:2:1 segregation of genotypes, as can be seen by considering, with the help of the diagram in Figure 18-3 (called a Punnett square), the different kinds of combinations that can occur during fertilization between the male and the female gametes produced by the F_1 generation. These principles of genetic segregation are called Mendel's first law of inheritance.

Mendel discovered and formulated some other important principles of hereditary transmission, but before introducing them we consider how the chromosome theory of inheritance explains Mendel's observations on genetic segregation.

THE CHROMOSOMAL BASIS OF MENDELIAN INHERITANCE

In Mendel's time the role of the nucleus and chromosomes in heredity was unknown, so Men-

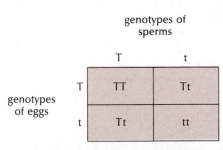

Figure 18-3. Punnett square showing the genotypes that result from mating of sperms and eggs produced by a Tt individual. Because the sperms combine at random with the eggs, and each include equal numbers of T and t cells, the combinations shown in the different boxes should all appear with equal frequency. Since the combination Tt can occur in two different ways, this genotype should appear twice as often as the other genotypes.

del could not point to any physical basis for the observed segregation of genes during reproduction. The consequent abstractness of his hypothesis was probably one of the reasons it was not grasped and appreciated by other scientists of his day. By 1900 the behavior of chromosomes in mitosis and meiosis had been carefully described. It was then realized that Mendel's laws of inheritance would be accounted for if the hereditary units or genes were located on the chromosomes. By now there is compelling evidence of several kinds showing that DNA carried by the chromosomes of a cell's nucleus comprises the cell's principal hereditary material.

The place on a chromosome where a given type of gene, such as the *T* or *t* alleles for tallness, occurs is called a genetic **locus** (plural, loci; from Latin *locus*, place). One can refer to a particular locus with the letter used for the alleles that occur there; for example, in the foregoing we are dealing with the *T* locus. From evidence explained later we know that any given chromosome carries a specific set of loci in a particular linear order, as illustrated in Figure 18-11.

Meiosis

The chromosome behavior accounting for Mendelian inheritance occurs during meiosis or reduction division, which forms haploid from diploid cells and leads to the production of haploid gametes. As noted in Chapter 16, meiosis always involves two successive nuclear divisions, called meiosis I and meiosis II (Figure 16-4).

Because a diploid cell contains two sets of chromosomes, one set from the original egg and one set from the sperm that fertilized this egg, the cell contains two of each type of chromosome that the species possesses; for example, cells of the pea plant contain two chromosomes bearing the *T* locus for tallness. These two chromosomes of a given genetic type are called **homologous chromosomes** or **homologs.**

During or prior to prophase of meiosis I the chromosomes replicate, as they do in mitosis; but in meiotic prophase homologous chromosomes associate or **pair** together, side by side, forming as many chromosome pairs as there are different types of chromosomes in the cell (Figure 18-4). Corre-

Figure 18-4. Diagrams of meiosis showing the behavior of two pairs of homologous chromosomes, a given chromosome and its homolog being shown in different colors. (A) Early prophase, homologs becoming paired (each chromosome has already undergone replication but this is not yet visible). (B) Late prophase, chromosomes have shortened. (C) Metaphase I, homologs tending to separate but not yet detached from one another; each can be seen to consist of two chromatids, resulting from the preceding replication. (D) Late anaphase/early telophase I. (E) Late telophase I or interphase. (F) Metaphase II, chromosomes consisting of two chromatids aligned on the equator of the spindle in each daughter cell. (G) Late anaphase/early telophase II, chromatids separated and distributed to opposite poles. (H) Post-meiotic tetrad of haploid cells (chromosomes would have elongated into an interphase condition rather than remaining short as shown here).

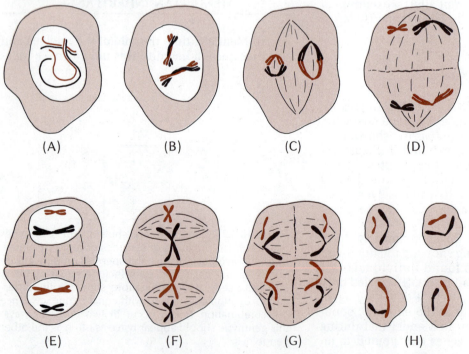

(A)   (B)   (C)   (D)

(E)   (F)   (G)   (H)

sponding loci on the respective homologs come together at each point along the length of the two chromosomes in each pair. At metaphase I these pairs line up on the equator of the spindle. In anaphase I the members of each pair separate from one another, one homolog going to one pole of the spindle and the other to the other pole (Figure 18-4D). This explains why, if the parental plant is heterozygous (e.g., *Tt*), half of the haploid cells resulting from its meiosis carry one of the alleles (e.g., *T*) and half of them carry the other (*t*) allele.

As a result of anaphase I each spindle pole or daughter nucleus receives half the number of chromosomes found in the original diploid nucleus, but because of the preceding chromosome replication each of these chromosomes actually consists of two chromatids, like the chromosomes in prophase of mitosis. Meiosis II resembles mitosis: at metaphase II the chromosomes, each consisting of two chromatids, line up on each spindle. In anaphase II these chromatids split apart and move to opposite poles, so each of the four final daughter nuclei receives a simple haploid set of chromosomes.

The most important difference between meiosis and mitosis is the pairing of homologs during prophase I and their separation during anaphase I (rather than the splitting apart and separation of chromatids as in mitosis): this is what reduces the chromosome number to half and segregates into separate haploid nuclei the different alleles for which the parental nucleus was heterozygous.

Explanation of Mendelian Segregation

Figure 18-5 illustrates how the separation of homologous chromosomes during meiosis segregates alleles into equal numbers of gametes and how the random fusion of gametes carrying the respective chromosomes results in the observed 1:2:1 segregation of genotypes in the F_2 generation. This process provides a complete explanation of typical Mendelian segregation.

In most animals haploid gametes are formed directly from diploid cells, as indicated in Figure 18-5, but in plants the formation of gametes is more complicated. Meiosis actually yields haploid spores, which develop into gametophytes producing sperms or eggs. Since, in flowering plants, each female gametophyte

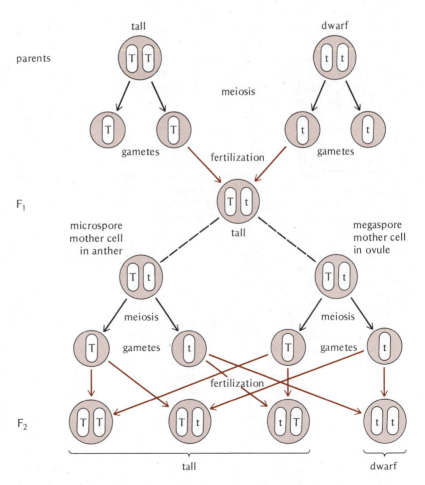

Figure 18-5. Diagram showing the chromosomal explanation of the segregation and recombination of alleles during sexual reproduction represented in Figure 18-3. Only the behavior of the homologous chromosomes bearing the *T* locus is shown. Random mating of sperms and eggs yields the 3:1 phenotype proportion or 1:2:1 genotype proportion in the offspring.

produces one egg and each male gametophyte (pollen grain) produces two sperms (Chapter 16), any alleles for which a plant is heterozygous still appear in equal numbers both in eggs and in sperms. Therefore, the simplified depiction of gametogenesis in Figure 18-5 is serviceable for flowering plant genetics, even though developmentally inaccurate. It is not serviceable for lower vascular plants such as ferns or for most algae and fungi, which have an independent haploid generation, each gametophyte usually producing many sperms and/or eggs (Chapters 23–29). Genetic segregation patterns in these types of organisms differ from those explained in the present chapter.

The Role of Chance in Genetic Segregation

The appearance of 1:2:1 segregation of genotypes in the F_2 generation of a single-character cross depends not only on the formation of equal numbers of gametes carrying each allele, due to meiosis, but also on these gametes' combining at random during fertilization. Randomness means that the element of chance enters into the results of inheritance to an important degree.

The situation is similar to that involved in tossing a coin. Although there is an equal chance of getting either heads or tails, equal numbers of heads and tails usually do not occur if the coin is tossed just a few times, because the outcome of any throw is completely independent of all previous throws. Commonly three or even four heads (or tails) may come up in a row, by chance. Only if the coin is tossed many times do the numbers of heads and tails begin to approach the expected ratio of 1:1. Similarly, the phenotypic segregation ratio for a one-character cross will usually deviate appreciably or substantially from the ideal Mendelian 3:1 value if the number of offspring is not very large. For example, among 1064 F_2 pea plants from the cross tall × dwarf, Mendel obtained 787 tall and 277 dwarf, a ratio of 2.84:1; a 3:1 ratio would have been 798 tall to 266 dwarf. The difference between the observed and the ideal ratios is actually no greater than might be expected on the basis of chance.

By examining the segregation of characters that are expressed in the pea *seed* (determined by the genotype of its embryo), Mendel was able to score a much larger number of F_2 progeny—7324 from round (dominant) × wrinkled (recessive) seeds and 8023 from yellow (dominant) × green (recessive) seeds—obtaining segregation ratios of 2.96:1 and 3.01:1, respectively, which closely approach the ideal ratio. At the other extreme, if we test only four F_2 progeny, it is simply not reasonable to expect that they will necessarily include one *TT*, two *Tt*, and one *tt*, genotypes, as Figures 18-2 and 18-5 might erroneously be thought to indicate. There is actually a greater than 50–50 chance of failing to find one of the homozygous genotypes (*TT* or *tt*) among only four F_2 individuals.

The role of chance and, therefore, of the total number of progeny in a Mendelian segregation ratio can be seen by counting the number of pigmented and

Figure 18-6. Examples of Mendelian segregation and recombination as seen in corn seeds. Each seed in an ear has its own phenotype reflecting its particular genetic constitution. (Left) F_2 generation from the cross smooth blue × wrinkled white gives seeds of four phenotypes, namely, smooth blue, wrinkled blue, smooth white, and wrinkled white, approximately in the ratio 9:3:3:1. White and wrinkled are recessive, independently assorting traits. (Right) Blue × white F_1 backcrossed to white (homozygous recessive) gives seeds of blue and white phenotypes in an approximately 1:1 ratio, showing that the F_1 was heterozygous for the color alleles.

nonpigmented corn grains in each row of the ear shown on the left in Figure 18-6 and comparing the ratios found in different rows to the ratio for the total of grains in all four rows. We suggest you try this to see what actual genetic segregation data look like (ignore the kernels in any part of a row where their phenotypes are not clear in the photograph).

FURTHER MENDELIAN PRINCIPLES

The Backcross and the Test Cross

Mendel discovered another way, besides using the phenotypes of progeny obtained from selfing, to determine the genotype of a hybrid individual having a dominant phenotypic character, namely, to cross it with the homozygous recessive parental type (Figure 18-7). Crossing a hybrid with one of the parental types is called a **backcross**; this particular kind of backcross is called a **test cross** since it reveals very plainly the genotype of

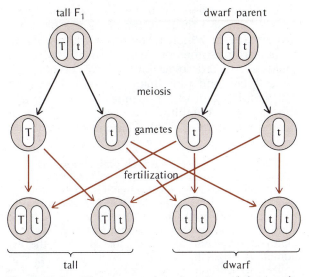

Figure 18-7. Chromosomal explanation of the results of a backcross between an F$_1$ heterozygous at the *T* locus, and its homozygous recessive (*tt*) parent. Random fertilization among the available gametes yields a 1:1 ratio of genotypes and phenotypes in the offspring of the backcross.

the tested individual. If that individual is homozygous dominant, for example, *TT*, then the offspring all show the dominant phenotype (e.g., tall) because all of them are heterozygous (*Tt*) just like an F$_1$ hybrid. If, however, the tested individual is heterozygous, then, as illustrated in Figure 18-7, 50% of the offspring show the dominant phenotype because they are heterozygous (*Tt*), while the other 50% show the recessive phenotype, being homozygous recessive (*tt*). This 1:1 segregation ratio results because the heterozygous individual produces equal numbers of gametes bearing the dominant and the recessive allele, whereas the gametes with which these are mated, coming from a homozygous recessive parent, carry the recessive allele only (Figure 18-7). An example of this kind of segregation is shown in Figure 18-6 (right).

Independent Assortment: Recombination

The previously described crosses involving a single genetic character at a time are called **monohybrid crosses**. Mendel also mated peas differing with respect to two genetic characters, called a **dihybrid cross**. For example, plants with round, yellow seeds were crossed with plants having wrinkled, green seeds. As noted previously, round (R) is dominant over wrinkled (r), and yellow (Y) over green (y), so the F$_1$ seeds were all round/yellow (genotype *RrYy*). In the F$_2$ generation, the wrinkled/green and round/yellow characters from the respective parents did not remain together during segregation, but segregated completely independently of each other, each at a 3:1 ratio. Thus, of all the round F$_2$ seeds, three-fourths

were yellow and one-fourth were green, and likewise for the wrinkled F$_2$ seeds. An analogous case of independent assortment of two phenotypic characters in corn seeds is shown in Figure 18-6 (left).

Independent segregation of different genetic characters, called **independent assortment** or Mendel's second law, occurs when the loci for the two characters are carried on different (nonhomologous) chromosomes. As illustrated in Figure 18-8, in the F$_1$ hybrid of genotype *RrYy*, independent separation of the different chromosome pairs bearing the *R* and the *Y* loci produces four different gamete genotypes, in equal numbers. To predict the outcome when these gametes combine with one another at random during fertilization, one can construct a Punnett square (Figure 18-9), accounting for the various types of fertilizations that should occur, all at an equal frequency. Adding these up, the resulting nine F$_2$ genotypes should fall into four phenotypic classes occurring in a ratio of 9:3:3:1, as Mendel indeed observed. The fewest F$_2$'s are the wrinkled green homozygous double recessives, *rryy*.

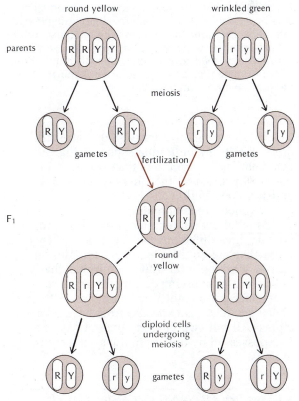

Figure 18-8. Genetic segregation in a dihybrid cross involving two loci (*R* and *Y*) located on different (nonhomologous) chromosomes. Upper part of diagram shows derivation of the F$_1$'s genotype. Lower part shows how the two alternative possibilities for chromosome separation during meiosis in the F$_1$'s cells produces gametes of four genotypes. Because either type of segregation is equally probable, the four gamete genotypes are all produced in approximately equal numbers. The recombination among these four classes of gametes during fertilization is shown in Figure 18-9.

male gametes of F₁ (from anther)

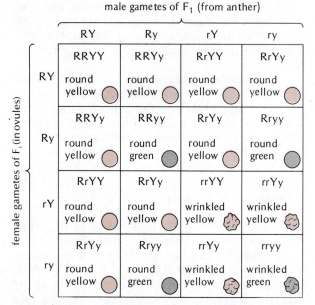

Figure 18-9. Recombination pattern for gametes produced, as shown in Figure 18-8, by the F₁ from a dihybrid cross, indicating the genotypes and seed phenotypes (from the cross round yellow seeds × wrinkled green seeds) that are expected to appear in equal numbers by random mating of gametes all occurring in equal numbers. The dominance/recessivity relations and the formation of some genotypes by two or more gamete combinations lead to a 9:3:3:1 ratio of double dominant : dominant/recessive : recessive/dominant : double recessive phenotypes.

It is evident in Figure 18-9 that a dihybrid cross results in individuals possessing different combinations of characters than occurred in either parent, in this case the phenotypes round green and wrinkled yellow. This illustrates the principle of genetic **recombination.**

Mendel even attempted a trihybrid cross, involving three different genetic characters. This was more than he could adequately handle, but scientists working after 1900 confirmed that a trihybrid cross involving dominant versus recessive characters gives eight F₂ phenotypes in the ratio 27:9:9:9:3:3:3:1, as expected from Mendel's principles. All but two of these phenotypes (i.e., the two possessed by the original parents) are recombinants.

LINKAGE AND CROSSING-OVER

Genes at different loci on the same chromosome are physically connected, or **linked,** and do not show independent assortment. If the two loci are very close together (tightly linked), alleles at the respective loci usually remain together through meiosis, so the gametes and therefore the offspring receive only parental combinations of these alleles, and no new combinations. However, some recombination does occur between alleles at most linked loci, as a result of the phenomenon called **crossing-over,** which takes place when homologous chromosomes are paired in prophase I of meiosis. At a crossover point the two homologs in effect break and interchange strands (Figure 18-10), resulting in chromosomes containing new (nonparental) combinations of alleles, which appear in the gametes as these chromosomes are separated by the meiotic divisions.

During prophase I each chromosome consists of two strands or chromatids (Figure 18-10A). Most crossovers occur between one chromatid of each of the two paired homologs, leaving the other chromatid of each homolog unaffected (Figure 18-10C). As a result, beyond the crossover point, the chromatids of each homolog differ genetically at any loci at which the two homologs originally carried different alleles, as symbolized by different colors in Figure 18-10C, D. This difference is carried along as the chromosomes separate in anaphase I (Figure 18-10E). To segregate the parental genes completely and obtain genetically haploid cells, each carrying just one allele at every locus, it is necessary to separate the chromatids from one another. This is the function of the second division of meiosis (Figure 18-10F–H).

Genetic Maps

Crossing-over generally occurs more or less randomly along the length of each chromosome pair. Consequently, the greater the distance between different linked loci, the greater the frequency of crossing-over between them, and the more often nonparental combinations of their alleles appear in the offspring. By measuring the recombination frequency between alleles at several mutually linked loci, one can arrange the loci into a linear sequence called a **genetic map,** showing the order and relative distance between the loci on a particular chromosome (Figure 18-11). Fairly extensive genetic maps are now available for the chromosomes of corn, peas, and various other organisms.

The Genome and the Karyotype

The total of all genetic loci encompassed within the linkage groups or different chromosome maps of a species comprises its haploid **genome** (strictly speaking, its nuclear genome, because additional genes occur in the cytoplasm as noted below). If the cell is diploid, it possesses two haploid genomes; if it is triploid, three; and so forth. A hybrid individual possesses one genome derived from each parent.

The genetically different chromosomes comprising a species' haploid chromosome set or haploid genome often differ from one another in total length, relative length of the two arms on either side of the centromere, and pattern of constric-

Figure 18-10. Details of crossing-over and subsequent segregation of genes located on one homologous pair of chromosomes during meiosis. (A) Early prophase; both chromosomes have been replicated into two strands (chromatids), and are becoming paired. (B) Still during early prophase, one strand of each of the paired homologs breaks and (C) rejoins to the opposite chromosome strand, creating an interchange of genes between the two. (The chromosomes are actually much more extended than shown here, at this point in the process.) (D) Metaphase I. (E) Anaphase or early telophase I, and (F) interphase, showing how genes on the distal part of each chromosome, beyond the crossover point, have not yet segregated. (G) telophase II; chromatids of each chromosome have separated, segregating genes on the interchanged segments. (H) Chromosome constitution of the four cells formed by meiosis. All four are different.

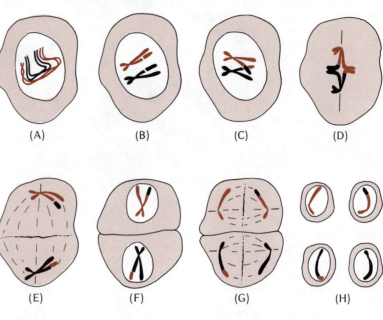

(A)   (B)   (C)   (D)

(E)   (F)   (G)   (H)

tions and swellings (Figure 18-12). These distinctive features of the haploid chromosome set comprise the **karyotype** of a species. The karyotype is the physical form of the species' linkage map or nuclear genome. The karyotype usually differs between species (and sometimes between variants within a species).

ACTION OF MENDELIAN GENES

As explained in Chapter 5, we now know that the basic type of Mendelian gene, called a structural gene, acts by coding for the structure of a specific polypeptide (an enzyme or other specific protein, or a subunit thereof) called its gene product. If the cell's nucleus contains a given structural gene, the nucleus will be able to transcribe this gene's information into messenger RNA molecules, and the cell's ribosomes will translate those mRNAs, yielding molecules of the specified gene product. Thus the cell function that depends on this gene product will take place if one copy of the gene occurs in the cell and will not necessarily benefit further if two copies are present, just as two copies of a given book's information are of no greater use than one copy. Such an allele is dominant, since its phenotypic effect (Figure 18-13) is apparent whether heterozygous (present in one copy) or homozygous (present in two copies).

In contradistinction, a recessive allele in the simplest case is an informationally defective form of the structural gene, failing to specify the gene product on which the phenotype governed by this genetic locus depends (Figure 18-13). For example, white color in flowers is typically a recessive character, caused by failure of the cells to produce an enzyme needed for flower pigment formation.

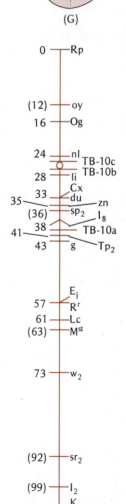

Figure 18-11. Genetic map of chromosome no. 10 of maize. The letter symbols to the right of the cross lines stand for known genetic loci, located at the distances (in "map units") from one end shown by numbers to left (one map unit essentially represents 1% recombination between the respective loci). Circle on chromosome axis shows position of kinetochore.

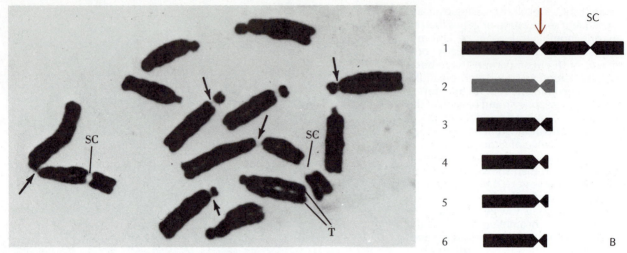

Figure 18-12. (A) Mitotic metaphase spread of the broad bean's diploid set of 12 chromosomes, each consisting of two chromatids (T) resulting from preceding replication. Arrows mark kinetochores (centromeres) of several chromosomes. Chromosome 1 has another ("secondary") constriction (SC), the place at which the nucleolus forms during telophase. (B) Conventional haploid karyotype diagram for this species (*Vicia faba*) showing relative size and morphology of the six different chromosome types (two homologs of each type in the diploid set), numbered in order of size and lined up by their kinetochores (arrow).

The converse character, colored flowers, is dominant, because if the cells contain an allele coding for the pigment-synthesizing enzyme, they produce the enzyme and therefore make pigment, regardless of whether the inactive or "silent" (recessive) allele is present. Only if the silent, recessive allele is the only one present—that is, is homozygous—does its functional inactivity control the phenotype. For many genes, such as the gene controlling flower distribution along the pea shoot, we do not necessarily know for what polypeptide the dominant allele codes, but we can presume that there generally is one.

Many genetic loci possess different alleles neither of which is dominant over the other: each allele exerts an effect on the phenotype, even when heterozygous. Such **codominant** alleles both presumably specify functional, but mutually different, gene products.

In some cases the effects of dominant and recessive alleles differ quantitatively rather than qualitatively. This may be because the recessive allele specifies a gene product with a weaker biological activity than the dominant allele's product, rather than a totally inactive (or no) gene product. For example, in the case of recessive dwarfism in peas, the *T* allele's gene product evidently functions in the gibberellin hormone system of the plant, because as noted in Chapter 14, *tt* dwarfs can be made to grow as tall as the tall genotype (*Tt* or *TT*) merely by supplying them with gibberellin

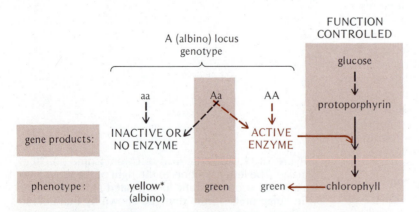

*Due to carotenoids, present also in green phenotype.

Figure 18-13. Chlorophyll formation, an example of the influence of dominant and recessive alleles on the phenotype. Seedlings with the phenotype "albino," which are blocked for formation of chlorophyll, are nonphotosynthetic and survive only as long as their seed storage reserves last (an example of a "lethal" phenotype).

(cf. Figure 1-2). It was originally thought that *tt* plants must be blocked in gibberellin biosynthesis, but this proved not to be the case. It seems that *tt* plants are relatively insensitive to their own gibberellins and may possess an altered gibberellin receptor protein.

The gene product of a given locus may affect the phenotype directly, as in the examples given above, or indirectly, by controlling the activity (gene product formation) of other loci. The latter type of gene is called a **regulatory gene** (see Chapter 5). A locus sometimes affects two or more seemingly quite different characteristics, such as leaf shape and time of flower formation. This may be because its gene product is required for the development of both of the respective features or because it is a regulatory locus influencing the expression of more than one structural gene.

Multiple Alleles

Certain genes are represented by several to many alleles, each with a different, distinguishable phenotypic effect. One example in plants is the gene system controlling self-incompatibility in flowers (Chapter 16). In clover, as many as 100 different *s* alleles (self-incompatibility alleles) can be distinguished. Each allele probably specifies a slightly different protein.

In principle, every gene is probably represented by multiple forms, brought about by mutation (see below), but unless the proteins they specify differ enough to affect the phenotype, these forms are not externally recognizable as different alleles. These cryptic (hidden) alleles may be detected, however, by examining the organism's proteins by high-resolution methods such as electrophoresis (Chapter 20).

POLYGENIC OR QUANTITATIVE INHERITANCE

Mendel himself encountered an important kind of exception to simple Mendelian inheritance when he crossed white-flowered with purple-flowered beans. Instead of a regular segregation of parental phenotypes in the F_2 and later generations, he obtained progeny with a graded series of flower colors between purple and white. Mendel surmised that flower color in this case might be the cumulative effect of genes located at a number of separate loci. We now know that many characteristics of plant and animal size, shape, form, and color are controlled in this manner rather than by single loci with large effects like those Mendel originally analyzed. In this kind of inheritance the genes individually have relatively small effects, but these effects are additive, so large differences can result from several to many such genes acting together, and almost any degree of

quantitative intermediacy between the extremes can be specified genetically. This type of inheritance is therefore called **quantitative inheritance** and is also termed **polygenic inheritance** because of the participation of many different genetic loci. This phenomenon explains the popular view of heredity as a blending of parental characteristics.

In quantitative inheritance the alleles at each of the several to many loci influencing the given character segregate individually in simple Mendelian fashion, but their multiplicity and small individual effects lead to a continuous, "bell-shaped" segregation pattern in the F_2 generation of a cross (Figure 18-14), rather than a Mendelian ratio of phenotypes. Because of quantitative inheritance, genetic variation among individuals in a population commonly takes the form of a seemingly continuous range of forms, rather than the limited number of discrete phenotypes that would be specified by simple Mendelian genes.

Modifiers

How great an effect a given Mendelian gene has on the phenotype can be conditioned by "modifier" genes at other loci. Sometimes there are many modifier loci with individually small effects, quantitatively altering the expression of a particular allele of the basic Mendelian locus (Figure 18-15). In view of the genetic similarity between modifiers and polygenic inheritance described above, a polygenic system may simply be a set of modifiers for some structural gene whose gene product is required for the character in question.

Modifiers and polygenes may, therefore, be sets of regulatory genes. For example, modifiers affecting the extent of flower pigmentation (Figure 18-15) might control which cells of a petal turn on the transcription of a structural gene for a pigment-forming enzyme and/or how much of its mRNA transcript (and therefore of enzyme) these cells make. Extensive regulation of gene action seems to be needed to bring about development and differentiation in higher organisms. Regulatory genes in bacteria act by coding for a gene product that controls the transcription of one or more structural genes (Chapter 5), and we know of cases in which transcription is regulated in higher organisms (Chapter 14). Alternatively, modifiers may influence structural gene expression by acting at any of the following levels: (a) primary RNA transcript processing into functional mRNA, (b) mRNA translation, or (c) posttranslational processing or intracellular transport of the gene product; or they may influence intracellular conditions or membrane properties affecting the activity of the gene product. Probably most or all of these levels of regulation are employed in genetic control of development, but as noted in Chapter 5 we have little information at present about the actual mechanisms of regulation at post-transcriptional levels.

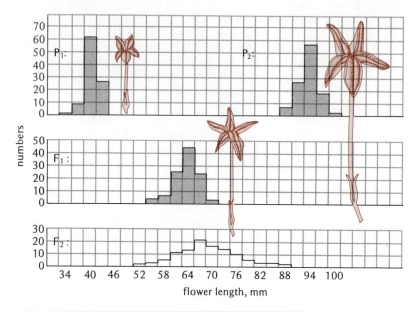

Figure 18-14. Quantitative inheritance of flower length in tobacco (*Nicotiana*). Parental strains P₁ and P₂, with flower lengths and forms indicated, were crossed to give an F₁ (average length and form also shown). The distribution of flower lengths among 441 F₂ offspring obtained by inbreeding the F₁ generation is shown in the lowest graph. Note that no F₂ individuals had flowers as long or as short as the respective parents, as would have been the case if only 1–3 genetic loci were involved in the difference. (Data from E. M. East, *Genetics*, Vol. 1, p. 164, 1916.)

CYTOPLASMIC INHERITANCE

Because genes contributed by the egg and the sperm are equally effective in determining the phenotype of an offspring, in a cross it ordinarily does not matter which of the partners is the pollen parent and which is the female or seed parent. Certain characters, however, are transmitted preferentially or exclusively through the female parent, a situation called **maternal inheritance.** This is due to certain genetic factors being carried in the cytoplasm: the pollen of the male parent usually contributes only the sperm nucleus to the offspring; the cytoplasm of the zygote is the cytoplasm of the egg, derived from the female parent. Thus, maternal inheritance is also called **cytoplasmic inheritance.**

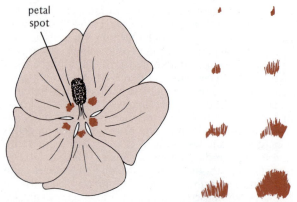

Figure 18-15. Example of modifier genes in cotton (*Gossypium*). Presence of anthocyanin pigment spot on petal (left) is determined by a single dominant Mendelian gene. In the presence of this gene, however, the extent of development of the spot varies widely depending upon the modifier genes present, the patterns at right illustrating the range of spot size (at least 22 degrees of spot development, corresponding to at least that many modifier genotypes, can be distinguished).

Many heritable variations in the form and the extent of chloroplast development are cytoplasmically inherited (Figure 18-16). These characters are due to chloroplast DNA, which comprises a minor but essential fraction of a plant's DNA. Cytoplasmically inherited chloroplast variants include a number of the variegated or mottled-leaved horticultural plant forms considered attractive as ornamentals and houseplants (Figure 18-17). Chloroplast genes code for the chloroplast ribosomal RNAs and transfer RNAs and for at least several chloroplast proteins (enzymes, ribosomal proteins, thylakoid membrane proteins). Cytoplasmically inherited chloroplast defects are due to mutation of some of these chloroplast genes.

The egg normally does not contain differentiated chloroplasts, but its cytoplasm includes proplastids (Chapter 4) carrying chloroplast DNA. By multiplication and differentiation, the chloroplasts of the next generation arise from these proplastids.

Many chloroplast characters, such as the formation of chlorophyll itself, are inherited in a strictly Mendelian fashion, being controlled by nuclear genes. Chloroplasts therefore develop by the cooperation of nuclear and chloroplast genes. An especially remarkable cooperation is involved in the formation of the key chloroplast enzyme RuBP-carboxylase, the CO_2-fixing enzyme of photosynthesis (Chapter 7), consisting of two kinds of subunit polypeptides, one of which is coded by a nuclear gene and the other by a chloroplast gene.

Other cases of cytoplasmic inheritance involve mitochondria, which, like chloroplasts, contain DNA coding for certain of their components [including cytochromes of the respiratory chain (Chapter 6) and mitochondrial transfer and ribosomal RNAs]. Mitochondrial genes are responsible for **cytoplasmic male sterility,** encountered in a number of crop plants. The pollen of an individual carrying the male-sterile genetic factor aborts or cannot carry out fertilization. The sterility seems to be due to defective function of mitochondria in the pollen. Certain nuclear genes, called fertility restorers, can compensate for the mitochon-

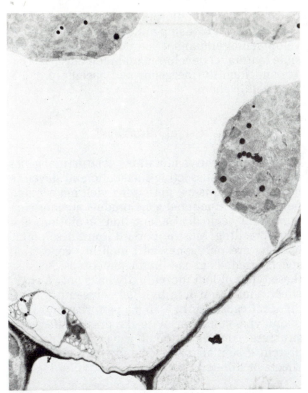

Figure 18-17. Variegated pattern of leaf color in a plant of *Hosta* containing mutant and normal chloroplasts. The lighter patches each contain many cells that developed from a cell that happened to receive only mutant (nongreen) plastids in the cell division that formed it.

Figure 18-16. Electron micrograph of a cell of *Hosta undulata* (a lily family plant) containing both normal chloroplasts (top and right) and a mutant chloroplast with drastically different structure, lacking grana stacks. That this type of chloroplast occurs in the same cell with the normal ones shows that its altered structure is not due to a nuclear gene or environmental influence.

drial defect, making the pollen fertile despite its carrying the cytoplasmic male sterility factor.

Chloroplast and mitochondrial DNA is not organized into microscopically visible chromosomes like those of the nucleus, but occurs as closed circular double-stranded DNA molecules. These resemble the bacterial cell's DNA, called the bacterial chromosome (Chapter 22). Although chloroplasts normally contain only one type of "chromosome," multiple copies of it occur (as many as 50 copies per chloroplast).

A third kind of cytoplasmic inheritance is due to self-replicating pieces of DNA, called **plasmids**, which can occur in the cytoplasm. Plasmids are found extensively in bacterial cells and in some lower eukaryotes, such as yeast (see Chapter 22).

MUTATION

A **mutant** is an individual possessing an altered form of a particular gene. Mutants arise spontaneously as novel heritable variants ("sports") in a population. Various agents accelerate mutation, including (a) high-energy radiation such as ultraviolet light, X rays, and atomic radiation; and (b) certain chemicals, called mutagens, which react with DNA or affect its replication. Spontaneous mutation may be due either to natural high-energy radiation (ultraviolet and cosmic radiation, and

natural radioactivity), to artificial or naturally occurring mutagens in the environment, or to mistakes in copying DNA during normal replication. Mutation is essentially a random alteration in the base sequence of the DNA comprising a gene. Mutations of nuclear, chloroplast, and mitochondrial genes can and do occur. The normally occurring form of the gene, prior to mutation, is called the **wild type.**

A mutant gene may specify a gene product with a different amino acid sequence and, consequently, with properties and functional activity quantitatively different from those of the wild type. In other cases the mutant gene may be "silent," specifying a functionally inactive gene product or no gene product at all. The molecular differences between these kinds of mutations are considered in Chapter 20. The mutation of a structural gene to give a functionally inactive allele, which is normally recessive for reasons given previously, is the type of mutation observed and studied most frequently.

Since the loss of a function is usually detrimental, when homozygous many mutations are detrimental, and some are lethal. Deleterious and lethal recessive mutations can persist and be propagated, however, because they do not express themselves when heterozygous with the wild-type allele.

Mutants as Analytical Tools

Because a structural gene mutation yields a highly specific alteration of a cell or organism's functional equipment, mutants are of enormous value in efforts to analyze and understand life processes. Any function that depends on the action of one or more enzymes or other proteins specific to that function can be specifically blocked by mutation of any of the genes specifying those proteins. For example, by obtaining mutants blocked in the biosynthesis of a certain molecule such as a particular amino acid, one can work out its pathway of biosynthesis and obtain information about the enzymes involved, because (a) each step in the pathway requires a different specific enzyme; (b) each of these enzymes is specified by a different genetic locus; and (c) the wild-type allele at each locus can be mutated into a defective allele, which when homozygous will specifically block the biosynthetic pathway at the step catalyzed by that particular enzyme (Figure 18-18).

George Beadle and Edward Tatum, who analyzed mutants of the red bread mold, *Neurospora*, at Stanford University in the 1940s, were among the first scientists to see clearly that different gene loci in general code for different specific enzymes. Many plant functions, from the biosynthesis of chlorophyll and flower pigments to the electron

transport chain of chloroplasts have subsequently been dissected by using mutants. By obtaining developmental mutants it should eventually be possible to unravel developmental programs (Chapter 14) and identify the genes and specific proteins involved in them.

Useful Mutations

While defective mutations of structural genes are useful for dissecting metabolic and developmental mechanisms, mutations yielding a modified but still functional gene product are more important in ecological adaptation, evolution, and plant breeding. Most important from these points of view are polygenes and modifier genes that specify graded changes in the phenotype. Because these types of loci individually have only a small effect, a mutation at individual polygenic or modifier loci causes only a slight perturbation of the phenotype, compared with the drastic effect of a structural gene mutation altering or eliminating a protein essential to the phenotype. Because the effects of different polygenic or modifier loci are

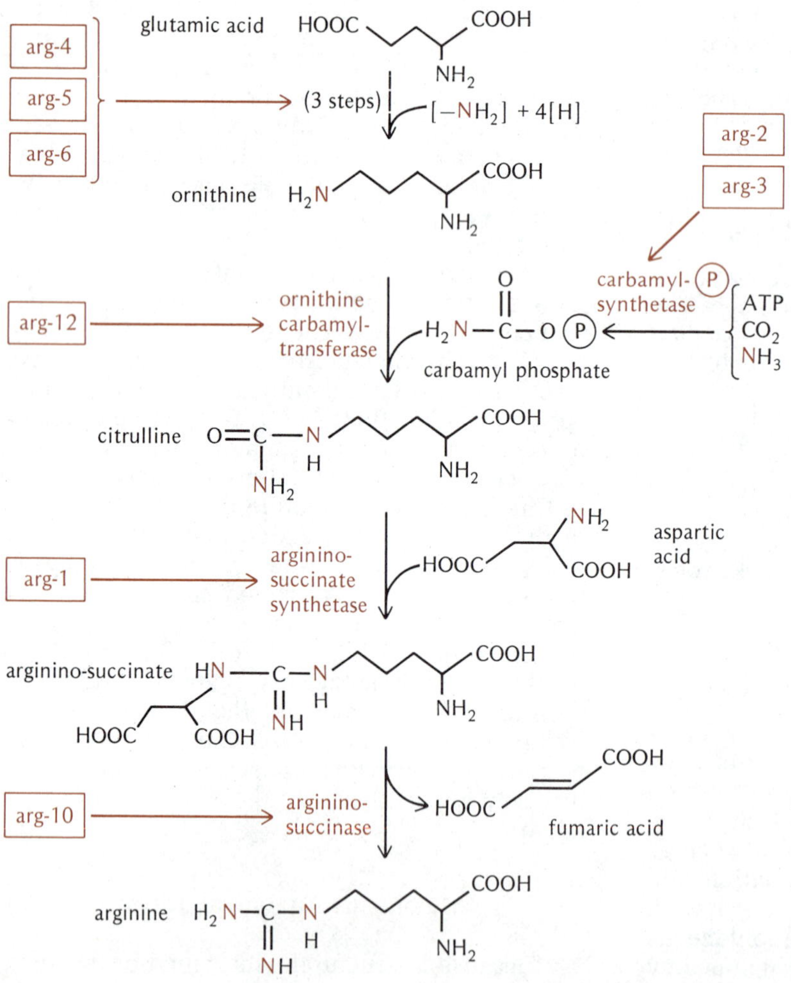

Figure 18-18. Metabolic pathway for synthesis, in the fungus *Neurospora*, of the complex amino acid arginine from the two simpler amino acids glutamic and aspartic acid. Names of metabolites are in black, and of enzymes in color. The zigzag line indicates the carbon backbone of glutamic acid, which is converted intact into the arginine backbone, with the addition of three nitrogen atoms (color) and one carbon. In boxes are the names of the known genes specifying enzymes of this pathway. Mutations in any of these genes cause the fungus to require arginine in its medium. For simplicity the first four steps of the pathway are not given in detail. (Note that arg-2 and arg-3 specify the two polypeptide subunits of an enzyme needed to produce carbamyl phosphate, a substrate essential for the carbamyl transferase step of the pathway.)

additive, more and more mutations exerting effects in the right direction can be found and combined, ultimately improving the phenotype significantly. This is the basis of most kinds of genetic improvement of plants by breeding.

PLANT IMPROVEMENT

Plant breeding is one of the most important applications of the genetic principles discussed in the preceding sections. A discussion of some of the techniques and accomplishments of modern plant breeding follows.

Selection

The original, at least partly unconscious method of plant improvement by humans was simply selecting desirable variants in a crop, year by year, for use in planting the next season's crop. Because of the variety of genes occurring in a wild species, it is possible by selection to derive strains differing considerably from the wild type. The cole crops (Figure 20-11), all of which belong to one species, *Brassica oleracea*, are perhaps the best example of morphologically diverse cultivars derived from a wild plant species by human selection.

Selection reaches a limit when any given genetic strain is selected so long and intensively that it becomes virtually homozygous at all genetic loci. Further improvement, including most of the modern improvement techniques, depends on hybridization and/or mutation. Certain of the cole crop vegetable cultivars are known to have appeared suddenly during historical time, probably as a result of mutation.

Hybridization

Hybridization can bring together desirable genes from two or more individuals, cultivars, or even species. Crosses are made between selected genetically different parents, and the F_1 hybrids are grown and then propagated by self-pollination (or by cross-pollination among them) for several generations, an operation called **inbreeding.** In the F_2 and subsequent generations, new gene combinations appear by the processes of recombination considered above. In each generation, forms with the desired phenotype are selected as the source of seed for the next generation. Because of genetic segregation, some of the selected individuals are homozygous for the traits of interest. Because genetic segregation continues to occur in each successive generation, the proportion of the progeny homozygous for the traits being selected continues to rise, generation by generation. This can be appreciated by considering what the genetic makeup of successive generations will be in Figure 18-2 if, in each generation, we select for propagation only the plants with a tall phenotype. Eventually an inbred or "purebred" line or genetic strain, homozygous for the traits being selected for, is obtained from the original cross (Figure 18-19A). This requires as many as 10 or more generations, depending on how many different loci are involved in the traits we wish to combine. If combinations of traits from more than two individuals or cultivars are wanted, more complex programs of successive hybridizations and selections can be used.

Hybridization and selection enable us (a) to obtain the best combinations of the genes already present in the parent stocks and (b) to incorporate and make use of beneficial mutant genes when they appear. To improve our wheat and other crops, botanists of the United States Department of Agriculture (USDA) have searched the world for genetic forms that could be adapted to the United States or crossed with United States crops to improve them in specific ways. For example, in 1900 Mark Carleton introduced several valuable cultivars and strains of hard red winter wheat from Russia to the United States. Many of our present-day cultivars of this type of wheat were developed from these introductions.

Frequently, incorporating one or a few new characters from a "foreign" genetic strain into an already highly improved cultivar is desirable. This introduction of new genes is accomplished by repeated backcrossing of hybrids with plants of the parental cultivar whose improvement is desired. From the first and successive backcross progeny generations, individuals carrying the desired traits from the foreign strain are selected for use in the next backcross. After several generations this yields a cultivar genetically and phenotypically comparable to the original one, except for the selected traits derived from the foreign parent (Figure 18-19B). Backcrossing is especially useful in developing disease resistance in crops, an important aspect of current plant breeding. Crossing a disease-resistant plant of otherwise poor qualities with a disease-susceptible but otherwise valuable plant, followed by repeated backcrossing to the susceptible parent type, accompanied in each generation by selection for disease resistance, yields a new disease-resistant strain of the previously susceptible cultivar.

With vegetatively propagated plants including fruit trees, potatoes and other root crops, and many ornamental flowers such as the traditional spring bulb flowers and orchids, new genotypes resulting from hybridization or mutation can be propagated as clones (Chapter 15). Therefore, inbreeding and selection are not needed for producing new cultivars of this type (Figure 18-19C). However, since these cloned hybrids and mutants are highly heterozygous, they do not breed true and must be propagated vegetatively rather than by seeds to obtain offspring true to type.

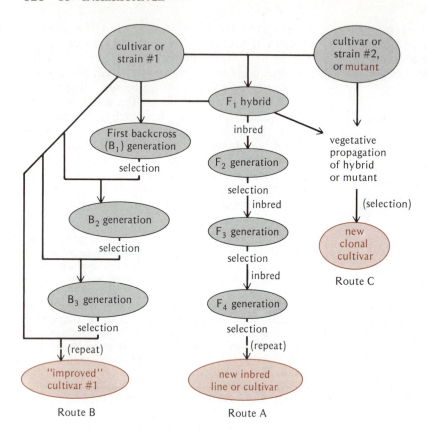

Figure 18-19. Major routes for plant improvement by plant breeding. Horizontal line indicates hybridization; vertical line descending from this indicates offspring of the cross. In each "selection" step, progeny possessing desired traits are picked out for use in further propagation.

Hybrid Seed

An important improvement in the productivity of corn (maize) and certain other crops has resulted from the introduction of hybrid seed. Hybrid corn seed is produced by controlled hybridization between selected inbred lines. Inbred strains are derived, as noted above, by self-pollination plus selection over five or more generations. Since corn is monoecious, it is self-pollinated by dusting pollen from the staminate inflorescence or "tassel" of the plant onto the silks (stigmas) of the same plant (Figure 16-21). To prevent cross-pol-

lination the silks are covered before selfing and again immediately after the pollination.

As a result of genetic segregation during inbreeding, deleterious recessive alleles occurring at many genetic loci in a traditional cross-pollinated maize cultivar become homozygous and therefore express themselves. Thus the size and vigor of the strain decline in the course of inbreeding, as illustrated in Figure 18-20. The weaker offspring from the inbreeding programs are discarded, selecting against some of the undesirable recessive genes all cultivars carry.

Figure 18-20. From left to right, corn plants representative of successive generations of self-pollination (inbreeding). The reduction in vigor generation by generation is evident, but seems to be leveling out after generations of inbreeding, making the plants homozygous at most genetic loci.

Figure 18-21. Hybrid vigor (heterosis). Hybrid corn is shown at center; inbred parent strains at left and right.

To produce seed for crop production, inbred lines are crossed by interplanting two selected lines in the field and detasseling the plants of the intended seed parent, so pollen for fertilization of its pistillate flowers must come from the tassels of the other genetic strain (Figure 18-23). Plants grown from such hybrid seed are much taller and sturdier, and bear much larger ears with more grain, than their inbred parents (Figure 18-21). They can yield as much as 35% more produce than the original cross-pollinated cultivars because of the previous selection against unfavorable genes during the inbreeding program and because of a phenomenon called hybrid vigor or **heterosis.**

Heterosis is associated with the high degree of heterozygosity occurring in the progeny of a cross between strains that differ from one another at many loci. Heterotic vigor is due in part to the fact that genes for vigor are often dominant. If the crossed inbred lines differ genetically at many loci, chances are that each line will carry dominant genes for vigor at different loci than the other line does. When combined in the hybrid, these dominant genes will complement one another, yielding exceptionally vigorous offspring. Besides this, however, it seems that at some loci heterozygosity in and of itself is conducive to greater vigor. Despite the importance of heterosis in crop plant yield improvement programs, a full explanation of heterosis is not yet available.

Commercial use of hybrid corn seed in the United

States corn belt began in the early 1930s and became adopted by almost all Iowa farmers by 1940 (and by all United States farmers within another 10 years). Over this period the average yield of corn per acre rose by about 60%, with no increase in the rate of fertilizer application (Table 18-1). Besides affording larger plant and ear size, the hybrids can be planted three or four times more densely than the traditional open-pollinated varieties, if enough fertilizer is supplied. Instead of the traditional 7000–8000 plants per acre, growers now plant as many as 20,000–30,000, with very heavy fertilization. Development of this practice has contributed importantly to the further increase in yield of corn per acre since 1940. A large hybrid corn seed-producing industry has developed to provide seed for the roughly 30 million hectares (ca. 70 million acres) in the United States, and growing acreages in other countries, currently planted with hybrid corn each year.

For a decade or so prior to 1970 the detasseling labor needed to make crosses for hybrid corn seed production was eliminated by using as the female parent a line carrying cytoplasmic male sterility, and as the male parent a line carrying a (nuclear) fertility restorer gene enabling the resulting hybrid crop to produce pollen for its own fertilization. This practice was dropped when a new form of the southern corn blight disease appeared, to which corn plants carrying the type of cytoplasmic male sterility factor then in use are susceptible. In future hybrid seed production it may be possible to eliminate detasseling by employing other known male sterility factors, but because of the serious loss of United States corn production in the blight epidemic of 1970, breeders are reluctant to allow seed production to again become generally dependent upon any one gene. Male sterility is used, however, for hybrid seed production with certain other crops such as onions (Figure 18-22).

High-Yielding Wheat and Rice: The Green Revolution

In the 1950s a group of plant geneticists and plant pathologists led by Norman Borlaug began a program to improve the wheat cultivars grown in Mexico. Mexican wheat was chronically afflicted with wheat rust disease (Chapter 26), substantially reducing its yield. Furthermore, the traditional cultivars, traditionally grown without fertilization on relatively nutrient-poor soils (which were another factor in their poor yield), responded to fertilization by growing too tall and often falling over (lodging) under the extra weight of the increased amount of grain, which would consequently be lost. Borlaug's group therefore aimed to incorporate genes for resistance to rust and for improved response to fertilization into wheat cultivars adapted to tropical latitudes. Useful genes were sought from numerous other strains of wheat and incorporated by hybridization, followed by backcrossing accompanied by selection for the desired characteristics.

A favorable response to fertilization was achieved primarily by introducing genes for a semidwarf growth habit (Figure 1-13), derived

Figure 18-22.   Commercial hybrid onion seed field in California. The light rows are the female parents, which are male-sterile. The dark rows are inbred onions producing normal pollen. This is carried by insects to the male-sterile plants, which produce the hybrid seed. Hybrid onions have advantages similar to those of hybrid corn.

principally from the Norin 10 wheat cultivar from Japan. The dwarfing genes confer a short, stiff growth habit, enabling the plant to support much heavier grain loads under fertilization. By tolerating fertilization this genotype permits substantially more plants to be grown per acre than is practicable with traditional varieties. By holding down vegetative growth, the dwarfing genes also cause a larger proportion of the plant's photosynthetic carbohydrate to become stored in the grain. These improvements, along with a resistance to the prevalent kinds of wheat rust, enabled Mexican farmers to produce substantially more grain on a fixed amount of land. Within 7 years after the first of the semidwarf varieties was introduced commercially in Mexico in 1962, the average national wheat yield per hectare had doubled. This transformed Mexico from a chronically grain-short, wheat-importing country to an important exporter of wheat on the world market.

The usefulness of the new semidwarf wheat varieties became apparent in other countries as far away as India and Pakistan, where within a few years their use began to increase explosively, resulting in major changes in long-stagnant agricultural economies. The advances with wheat inspired similar breeding programs with rice, carried out especially at the International Rice Research Institute in the Philippines. This effort also led to semidwarf varieties with greatly improved yields, the use of which has spread through the rice-growing lands of Southeast Asia, bringing similar improvements in the public food supply and in the economic position of farmers. The complex of improved food supplies and accelerated economies that has resulted from adoption of the new "miracle" wheat and rice strains, along with the technologies of fertilization and irrigation necessary to take proper advantage of them, became known as the Green Revolution.

TABLE 18-1   Average yields of maize in Iowa for sample years between 1930 and 1965
Data of G. F. Sprague, in *Change in Agriculture*, ed. A. H. Bunting, Praeger, 1970. Data years chosen are those with generally higher yields, representing more favorable weather conditions.

Year	Hybrid seed used (%)	Fertilizer nitrogen (kg/Ha)	Grain yield (ton/Ha)[a]
1930	0	0.09	2.3
1935	6	0.01	2.5
1938	52	0.03	3.1
1940	90	0.07	3.5
1946	100	1.4	3.8
1952	100	9.3	4.2
1958	100	20.8	4.4
1962	100	52.6	5.2
1965	100	91.9	5.5

[a]In metric tons per hectare; 1 T/Ha equals 890 lb./acre or 14.8 bushels/acre.

Genetic Hazards in Plant Breeding

The explosion in worldwide use of high-yielding grain cultivars and hybrids developed by plant breeding methods has brought some important hidden dangers to prominence. Highly selected, genetically uniform crops are being planted over vast areas where formerly many traditional cultivars, carrying much genetic diversity, were grown. We call such genetically uniform populations **monocultures.** "Never before," as USDA scientist Louis Reitz put it, "have so few genes been responsible for filling so many mouths with food." The situation is like the proverbial having all of one's eggs in one basket. What will happen if an environmental change occurs that these particular genotypes cannot tolerate?

The most immediate hazard is from plant diseases. Genes of a disease organism can mutate, yielding a new strain no longer controlled by the resistance genes carried by a particular crop cultivar. If this cultivar is grown extensively, the new disease strain can multiply and spread widely within a short time, creating a disease epidemic that would severely threaten the food supply. A serious example close to home was the Southern corn blight epidemic in the United States corn belt in 1970, mentioned above and explained more fully in Chapter 26.

Another danger in widespread planting of the new high-yielding cultivars is that they displace, and are leading to the disappearance of, many traditional crop cultivars comprising the pool of crop genetic diversity from which breeders can get new genes for disease resistance or other needed characteristics. Thus the success of the highly bred high-yielding cultivars is compromising the genetic potential of crop species by threatening to reduce it to the very limited base of genes contained in these varieties alone. Recognizing the seriousness and irreversibility of this loss of genes, various organizations such as the USDA are trying to build up and maintain collections of traditional crop cultivars, to preserve a diversity of genetic stocks for future plant breeding needs. This is expensive, however, and has so far not been adequately supported, so the genetic impoverishment of our important crop species is at present only being slowed down.

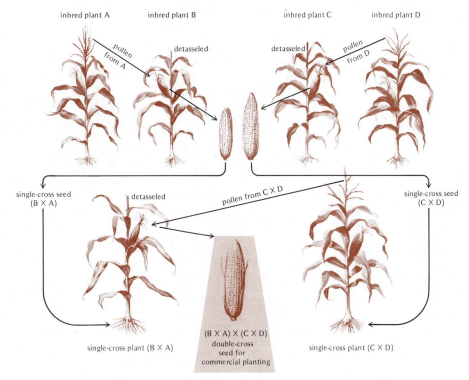

Figure 18-23. "Double cross" method for producing hybrid corn seed, combining genes from four inbred lines. The F₁'s from two separate crosses between different inbred strains are crossed the next season, to yield seed supplied commercially to farmers. The second crossing step multiplies greatly the amount of hybrid seed, and yields seed of especially high quality because the F₁'s that serve as pistillate parent in the second cross are very vigorous due to heterosis. Sometimes the pollen parent used in the second cross is another inbred line rather than a hybrid, in which case the commercial seed contains genes from three inbred lines rather than four.

inbred plant A   inbred plant B   inbred plant C   inbred plant D

pollen from A   detasseled   detasseled   pollen from D

single-cross seed (B × A)   detasseled   pollen from C × D   single-cross seed (C × D)

single-cross plant (B × A)   (B × A) × (C × D) double-cross seed for commercial planting   single-cross plant (C × D)

Seed Protein Improvement

Cereal grains, which on a worldwide basis provide the bulk of human food, have too little protein relative to carbohydrate to constitute a properly balanced diet. Moreover, the protein they do contain is deficient in certain dietarily essential amino acids (Chapter 6) to such an extent that their nutritional value is less than half the actual protein content. This leads to chronic dietary protein deficiency in many areas where people depend mainly on grain for their protein, and reduces significantly the value of feed grains for raising animals. Therefore, besides increasing the yield, another important need in cereal breeding is to increase the protein content and nutritional quality of cereal seeds.

Considerable attention has been given to mutations that increase the content of lysine, the amino acid most deficient in cereals. Two genes in corn, opaque-2 and floury-2, increase the lysine content of corn protein substantially by depressing the seed's production of a major class of storage proteins particularly deficient in lysine. Because these genes reduce the yield of the plants, give the grain an abnormal appearance, and unfavorably affect the cooking characteristics of flour made from it, high-lysine corn has thus far seen little use in areas such as Latin America where corn is important in human diets, and chronic protein deficiency is widespread. It may be possible to do better in the future by introducing mutations substituting lysine for other amino acids at certain points in the polypeptide chains of major storage proteins, without altering the content or properties of these proteins significantly.

Genetic Engineering Techniques in Plant Breeding

Radical new methods for gene transfer between plants, called "genetic engineering," are based on the use of plant tissue and cell cultures, as well as on recombinant DNA technology (recombination between or within DNA molecules in the test tube, plus introduction and integration of such DNA into cells utilizing techniques of molecular genetics).

Haploid Plants As noted in Chapter 17, by culturing isolated anthers of certain crop plants in the test tube, one can obtain haploid embryoids which can be grown into haploid plants. Such plants should make searching for useful mutations much easier and quicker, since recessive mutations, which are not expressed when heterozygous in a diploid, must express themselves in a haploid. Selection for desired mutations in haploid cell cultures followed by regeneration of plants from the selected cells seems to be a potentially powerful technique.

Another important potential use of haploids is to bypass the five or more generations of inbreeding or backcrossing and selection required, in conventional plant breeding, to obtain an inbred line or a new true-breeding cultivar following a cross. Since a haploid possesses only one allele at every locus, if its chromosome number is doubled (which treatment with certain drugs can accomplish) the resulting diploid will be homozygous at all loci. Therefore, by obtaining haploid plants from anthers of a hybrid and doubling their chromosome number, it should be possible to obtain new true-breeding recombinant lines in one generation.

Somatic Cell Genetics The available pool of genes that can be used to improve a given crop is at present limited to the species with which it can be hybridized sexually. Many valuable genes for resistance to diseases, better performance under environmental stresses, nutritionally superior seed proteins, and so on, might be exploited if chromosomes from more distantly related or even quite unrelated plants could be introduced into the genomes of crop plants by transfer between vegetative ("somatic") cells. This may be achievable using protoplasts, naked cells stripped of their cell walls by treatment with cellulose-degrading enzymes. As noted in Chapter 1, protoplasts from quite unrelated species (even carrot protoplasts and human cells) can be induced to fuse (Figure 1-18). In certain cases hybrid protoplasts can divide (Figure 18-24), and more rarely they can grow into cultures from which embryoids and eventually whole plants can be obtained (Figure 1-19). The hope is to transfer particular genes or chromosomes from the somatic hybrid to the crop whose improvement is desired, through further sexual or somatic backcrossing steps.

Further possibilities for genetically engineering higher plants emerge from aspects of prokaryotic cell genetics and are discussed in Chapter 22.

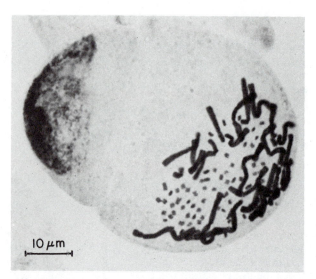

Figure 18-24. Cell undergoing mitosis, derived by somatic hybridization (protoplast fusion) between soybean (*Glycine max*) and vetch (*Vicia hajastana*). The large chromosomes are derived from vetch and the small ones from soybean, proving that the cell is a hybrid.

SUMMARY

(1) Diploid organisms such as higher plants carry two homologous chromosomes (homologs) of each genetic type characteristic of the species. Each homolog bears certain specific genetic loci, each of which in turn affects a particular characteristic. One of the two homologs of each type is derived from each of the individual's parents.

(2) At each locus different forms, or alleles, of the gene located there may occur. One allele is often dominant over another, which is termed recessive. A dominant allele expresses its effect on the phenotype, or physical manifestation of hereditary makeup or genotype, whether or not a recessive allele is present; that is, whether the organism is heterozygous (dominant allele on one homolog, recessive allele on the other) or homozygous (dominant allele on both homologs) for that trait. A recessive allele affects the phenotype only when homozygous. The F_1 or first-generation hybrids in a cross between individuals homozygous for a dominant gene and its corresponding recessive allele, respectively, are heterozygous, so they have the dominant phenotype.

(3) During sexual reproduction alleles at a given locus segregate into different gametes by the process of reduction division or meiosis, in the first division of which homologous chromosomes initially pair together and then separate from one another to opposite poles of the spindle. Gametes fuse at random during sexual fertilization. This results in a 3:1 segregation of dominant and recessive phenotypes (1:2:1 segregation of homozygous dominant, heterozygous, and homozygous recessive genotypes) in the F_2 generation derived by self-fertilization or intercrossing of F_1 hybrids. This is the simplest example of Mendelian segregation.

(4) Because different chromosome pairs behave independently in meiosis, genes located on different chromosomes segregate independently of one another during reproduction (independent assortment), resulting in recombination or the appearance of new combinations of parental characteristics, at predictable ratios, in the F_2 generation of a cross.

(5) Genes located on a given chromosome are genetically linked and do not assort independently. Recombination between linked loci occurs by crossing-over between them, due to the random interchange of strands between paired homologs during prophase of the first meiotic division. The crossover frequency between loci is used to construct genetic maps showing the relative location of the different genes carried on a chromosome.

(6) Many characteristics of size, form, and color are controlled by genes with small additive effects, located at several to many independently assorting loci (polygenic or quantitative inheritance). In this case phenotypic segregation in F_2 and subsequent generations of a cross does not yield Mendelian ratios, but rather a range of phenotypes usually intermediate between those of the parents.

(7) Certain characters are maternally or cytoplasmically inherited because they are due to chloroplast or mitochondrial genes contained in the DNA of these organelles.

(8) A typical Mendelian gene acts by coding for a specific polypeptide called its gene product, often an enzyme (or a subunit thereof). Different alleles specify slightly different gene products or, in the case of "silent" recessive alleles, a defective (functionally inactive, or missing) gene product.

(9) Gene mutation, due to alteration of the DNA comprising a gene, converts the gene into a different allele, often one yielding a defective gene product. Defective mutants are valuable for analyzing biochemical, physiological, and developmental mechanisms because such mutants usually represent highly specific blocks in function due to the absence of a single polypeptide essential to the function in question.

(10) Crop improvement by plant breeding is achieved by (a) selection for desirable traits in a genetically mixed crop plant population or cultivar; (b) hybridization between genetically different individuals or cultivars (sometimes, species) followed by inbreeding or backcrossing accompanied by selection for desired traits; and (c) crossing of inbred lines to produce hybrid seed, which yields exceptionally vigorous and productive plants due to heterosis. [Genetic engineering methods involving somatic (vegetative) crossing and the use of cell and tissue cultures for mutation, gene transfer, and selection are being introduced into plant breeding and are expected to yield important advances in the future.]

Chapter 19

ECOLOGICAL ADAPTATION

Between extremes of arctic cold, humid tropical warmth, dry desert heat, and waterlogged swamp the earth presents a dramatic diversity of environments. Yet apart from polar ice, there are almost no places on the earth's surface where some kinds of green plants do not manage to grow and support a community of animals and microbes. Ecology is the science of the interactions of organisms with their environment, how they adapt to its disadvantages and exploit its advantages. The particular environment in which an organism lives is its **habitat**, and the set of organisms occupying a given habitat is a **community.** An **ecosystem** is the composite of a community and its physical environment.

The members of a community interact with and influence one another. They may also affect their physical environment to some degree—for example, trees cool the ground by their shade and moderate the wind by their bulk, factors that are exploited in planting shade trees and windbreaks. This chapter examines the relations of plants, as individuals, with their environment. These relationships, which involve both physiological and genetic characteristics, are the unit processes of ecosystems. The makeup of plant communities and the functioning of ecosystems as a whole are considered in Chapters 33–34.

The principal kinds of ecological variables to which a plant must adapt are (1) **climatic,** pertaining to the aerial environment (including sunlight) and its effects on conditions at the earth's surface; (2) **edaphic** (from Greek *edaphos*, ground), pertaining to the plant's substratum (water or soil); and (3) **biotic,** pertaining to the effects of other organisms of all kinds. A summary of important specific ecological factors is given in Table 19-1, which indicates the chapter(s) in which the effects of these factors are discussed. In an ecosystem the effects of different factors are interconnected to a considerable degree. For example, climate influences the nature of the soil that develops in a given site, while the nature of the soil (water-holding capacity, humus content, cation exchange capacity, and so forth) in turn influences the mineral nutrient supply and the effects of rainfall on plants.

STRATEGIES OF ADAPTATION

The different complexes of behavior, physiology, and body structure by which organisms cope with the physical and biotic factors of their habitats are often called **adaptive strategies.** Although the word strategy implies conscious choice or intent, it must be kept in mind that except in humans and perhaps a few other animals, adaptive strategies represent genetically determined physiological and behavioral features developed as a result of evolution rather than matters of conscious choice.

For a species to succeed and persist in any habitat, it must (1) survive the most extreme or severe conditions occurring in that habitat, and (2) grow and reproduce as well as or better than other species under the more favorable conditions occurring in the habitat. Because of the seasonal nature of climatic extremes in most habitats, these **performance** and **stress-resistance** aspects of adaptive strategies are usually temporally and physiologically separate, at least to some degree. Deciduous trees, for example, carry on photosynthesis and growth in summer when conditions are mild, but in winter they shed their leaves and pas-

Figure 19-A. The bristlecone pine, *Pinus longaeva,* a plant adapted to a harsh habitat, tolerating chronic drought, extreme cold, and severe winds: the subalpine timberline zone of high mountain peaks in the Great Basin. The very slow growth of this tree is not disadvantageous because (as can be seen in this picture) it has no significant competitors in this habitat. Wind injury eventually kills off the exposed side of most of these trees but they continue to grow on the leeward side, yielding very asymmetrically shaped specimens which can survive as long as 4900 years, the longevity record for individual organisms on earth.

TABLE 19-1 Ecological Factors Important for Plants

Numbers in parentheses indicate other chapters in which the effects of the factor and methods by which plants adjust to or deal with it are discussed. Aspects of some of the most important factors are considered in the present chapter.

Climatic Factors	Edaphic Factors	Biotic Factors
Solar radiation (light, heat) (8, 9, 14)	Mineral nutrients (11)	Parasites (26)
Temperature (5, 8, 9, 14, 34)	pH (11)	Predators (herbivores) (1, 20)
Precipitation, water supply (11, 34)	Salinity (33)	Pollinators (16)
Humidity (9)	Soil water retention (11)	Beneficial symbionts (25)
Wind (9)	Soil hardness or porosity (11)	Competitors
Fire* (34)	Aeration of roots (11)	Organisms modifying the
Air pollutants* (33)	Toxic materials in soil* (11, 20)	physical environment (33)
Carbon dioxide† (8, 33)		Land use and abuse by
		humans (33)
		Allelopathy‡ (34)

*Due in part to human activity.
†Potentially important in aquatic habitats, in enclosed agriculture (Chapter 8), and outdoors because of human modification of the atmosphere (Chapter 33).
‡Negative chemical effect of a species on its competitors.

sively tolerate cold without functioning photosynthetically or developmentally. During the summer the tree must be able to carry on photosynthesis and grow faster than its competitors if it is not to become shaded out, whereas during the winter what counts is ability to resist the most severe frosts that occur in the location.

The distinction between performance and stress-resistance adaptations is often blurred. Obviously, to perform under certain conditions an organism must first be able to tolerate them, but the converse is often not true, as in the example just given. Some specializations contain elements of both stress-resistance and performance adaptation, as in many of the adaptations to different kinds of water supply, which will be considered later. Nevertheless, the conceptual distinction between performance and stress-resistance adaptations helps to clarify some important contrasting aspects of adaptive strategies.

STRESS RESISTANCE

Resistance adaptations may be regarded as adaptations that help an organism survive environmental stresses but usually do not contribute directly to its photosynthetic performance or growth. From the biological or evolutionary point of view, stress resistance of course has value ultimately because it permits or improves subsequent growth performance, especially eventual reproductive performance that perpetuates the species. Obviously, an individual that succumbs to an environmental hazard does not reproduce further. However, a stress such as frost, drought, disease, or attack by herbivorous animals often injures rather than kills the plant, reducing its fitness for performance because tissues are lost or functionally impaired. Photosynthate that could

have been used for new growth or reproduction has to be expended instead to replace or repair organs. Resistance adaptations that reduce these costs and consequences thus contribute indirectly but importantly to total performance.

Resistance adaptations are of two general kinds—**tolerance,** or ability to withstand severe conditions (Figure 19-A), and **avoidance,** or behavior that evades the problem. Herbaceous perennials, for example, avoid extremes of winter cold by dying down to underground organs which, because of the insulating effect of soil and snow, are not exposed to the minimum temperatures that occur in the air above (Figure 19-1). Leaf fall in deciduous trees is a kind of avoidance, but their stems and buds must be able to tolerate the full winter cold. In evergreens of cold-winter climates the leaves must also be frost-tolerant. Ability to tolerate subfreezing temperatures, called **frost hardiness,** varies from tropical plants with no frost tolerance at all, through subtropical species tolerant of mild frosts, to plants of cold-temperate and high latitudes able to withstand winter temperatures of −50°C or below (Figure 19-2), at which temperature most of their tissue water freezes solid.

Many tropical plants and warm-summer annuals, called **chilling-sensitive,** cannot even tolerate cool *above*-freezing temperatures below about 10°C (50°F; see Figure 19-2). Chilling injury, such as the blackening of bananas placed in the refrigerator, seems to be due to metabolic imbalance or changes in membrane properties at cool temperatures—changes for which the chilling-tolerant plants of cooler climates are able to compensate.

Tolerance Range

An individual or genotype can survive variations in environmental conditions only within

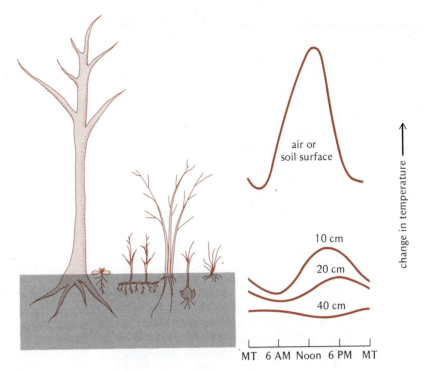

Figure 19-1. Relative diurnal temperature changes at increasing soil depths, compared with the soil surface or the air above. Underground plant organs experience much milder temperature extremes than aerial organs are exposed to. Low-growing herbs and small shrubs close to the soil surface are also protected considerably from temperature extremes by snow cover, often present during the coldest weather, acting as an insulating blanket. Snow consequently holds underground temperature fluctuations to even smaller values, compared with surface or air temperature changes, than indicated in these graphs which are for bare soil.

specific limits, called its **tolerance range.** This capacity limits the geographical range over which the species or genotype can occur. As with cold tolerance, resistance to high temperatures again illustrates how the modest tolerance range of plants not adapted to a particular kind of stress can be increased by adaptations of the vegetative body to that stress and can be extended much further by specialized structures, dormancy states, or

avoidance strategies that allow the plant to survive conditions its functioning cells could not tolerate.

One of the most heat-sensitive plant functions is photosynthesis, which in many plants is irreversibly damaged when leaf temperature rises only slightly above 40°C (104°F). Strong transpirational cooling during hot dry weather (Chapter 9) allows the leaf to avoid injury by keeping its

Figure 19-2. Killing temperatures for various plants during their growing season, and when winter-acclimated. Note that apple and red osier, although very winter-hardy, are sensitive to relatively modest frosts during the growing season, whereas stone pine (*Pinus cembra*) an alpine species, can withstand a fairly severe frost even in midsummer.

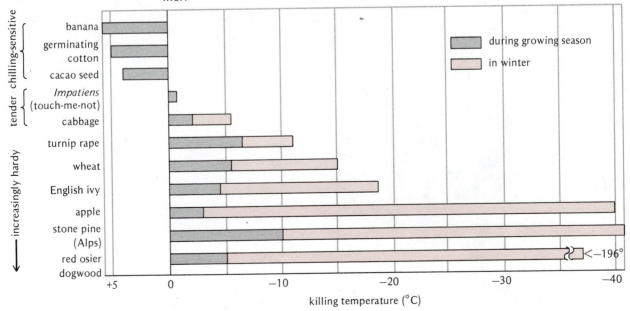

temperature below the critical level. However, this avoidance option is excluded when water is scarce, as in deserts. In plants that tolerate high leaf temperatures the critical temperature for photosynthetic breakdown is usually high, apparently because their chloroplast thylakoid membranes are more temperature-stable owing to a modified lipid composition.

The most heat-tolerant photosynthetic systems are found in certain **thermophilic** ("heat-loving") blue-green algae that grow in hot springs and geyser basins at temperatures up to almost 80°C. Thermophilic organisms possess enzymes that are extraordinarily stable to high temperatures compared with normal enzymes, and are obligately restricted to hot spring environments because they cannot function effectively at lower temperatures (Figure 19-7).

The trunks of many trees withstand ground fires with temperatures above the kindling point. This is actually a type of heat avoidance; the heat-sensitive inner bark and cambial tissues are insulated from external heat by the outer bark's dead cork layers. The thicker the outer bark, the longer or hotter a fire the tree can survive. The western redwoods (Figure 3-10) produce extraordinarily thick, fire-retarding, asbestos-like outer bark, giving them unusual fire resistance. Perennial grasses and many shrubs adapted to fire-prone habitats lose their above-ground organs in a fire but regenerate from surviving underground stems or roots not exposed to the fire's full heat (Figure 19-3), another avoidance strategy. However, most plants and trees whose tops are destroyed by fire survive only as seeds, especially those sheltered from heat in the soil. The seeds of some fire-adapted plants show considerable direct tolerance to dry heat, in some cases surviving temperatures as high as 120°C.

Figure 19-3. "Crown sprouting" of new shoots from the roots of a fire-adapted shrub after a brush fire has killed its top. Chaparral vegetation in the mountains of southern California.

Limiting Tolerances

By the weakest-link principle, it might appear that the distribution of a species should be limited by the tolerance range of the most stress-sensitive stage of its developmental and reproductive cycle. The most sensitive stage in a seed plant's life cycle is usually the seedling, which because of its limited root system and delicate shoot is ill-equipped to survive hazards such as water stress, mechanical damage, or fire. However, the geographical range of perennial plants is not strictly limited by the seedling's tolerance range because perennials can occupy and persist in an area even if their seedlings succeed only in occasional more favorable years. Seedlings of the bristlecone pine (Figure 19-A) in the White Mountains of California, for example, apparently succeed only in about one out of every 50–100 years, but because the tree lives for thousands of years once it is established, this is not really a handicap. Even annuals, by producing seeds that remain dormant for several years, can persist in habitats where their seedlings fail completely in some years. However, a species obviously cannot occupy a habitat in which its seedlings never succeed, unless the plant reproduces and invades vegetatively.

Various annuals and perennials mitigate the problem of seedling sensitivity to environmental hazards by producing two morphologically and physiologically different types of seeds ("seed polymorphism"), which differ in their germination delay or requirements. By causing seeds to germinate at different times, this adaptation improves the chances that some of the seedlings will find a stress-free "window" in which to succeed. Seeds of Douglas fir trees are reported to germinate faster when lying on one side than on the other, spreading out temporally the germination of seeds from a given crop simply because they come to rest on the soil in different positions.

Many plant species seem to be limited, at least on some sides of their range, not by tolerance of physical extremes but by biotic factors—for example, competition with other plants, attack by herbivorous animals, or plant diseases. In milder and more humid climates, disease organisms are often much more successful in infecting potential hosts, so resistance against diseases is an important aspect of adaptation. In environments that greatly restrict photosynthesis, such as arid climates, consumption by herbivores is especially costly to a plant because it has so little photosynthate available for use in replacing lost tissues. Perhaps for this reason, spines, thorns, and obnoxious secondary products tend to be especially frequent as defenses against herbivorous vertebrates and insects in desert and semidesert plants.

COMPETITIVE PERFORMANCE

Specializations enabling a plant to achieve better growth and reproductive performance un-

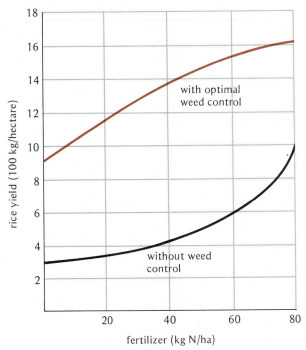

Figure 19-4. Effect of weed infestation on yield of rice in India at different levels of nitrogen fertilization. The weed competed so effectively for nitrogen that only a very heavy fertilization was able to give a rice yield in weedy plots equal to the unfertilized yield in weeded plots.

der conditions prevailing in its habitat help it compete with its competitors. Competition between photosynthetic plants amounts to competition for light, water, mineral nutrients, and the living space supplying these amenities. [Carbon dioxide, usually a marginal resource (Chapter 8), is normally not significant in plant competition on land because of nearly continuous renewal of the carbon dioxide supply by air movement, but competition for CO_2 can limit growth of aquatic plants.] Competitive performance is also influenced by how effectively or efficiently a plant makes use of whatever resources it captures.

Competition is illustrated by the negative effect of weed infestation on the growth and yield of crop plants (Figure 19-4). The example shown here illustrates competition for mineral nutrients, but competition between weeds and crop for light and water is often significant too. Previous chapters have discussed various kinds of performance adaptation likely to be important in competition, such as adaptations of photosynthesis to light intensity and temperature (Chapter 8), specializations for efficient absorption and transport of water (Chapter 10) and mineral nutrients (Chapter 11), and adaptations that reduce the water cost of photosynthesis (Chapter 9).

Developmental responses, especially to light (Chapter 14), are probably important in competition

also. For example, far-red light selectively transmitted through green plant tissues, as explained previously in relation to seed germination (Figure 17-18), stimulates stem elongation by reversing the phytochrome inhibition of elongation. When a plant is surrounded or shaded by other vegetation, the preponderance of far-red in the filtered light stimulates its elongation, helping it to outgrow its competitors and reach direct sunlight. This response represents "shade perception."

Environmental Optima

The optimum temperature, light intensity, and other conditions for a plant's growth can be determined by growing it in controlled environment chambers (Figure 19-5) under different climatic regimes; its photosynthetic optimum temperature can be determined using leaf chambers (Figure 19-6). Plants inhabiting cool-summer climates such as arctic, seacoast, or mountain areas have a markedly lower photosynthetic optimum temperature for growth (Figure 19-7) or photosynthesis (Figure 19-17) than plants of warm-summer climates. The roots of arctic plants, which often occur in soils kept cold throughout the summer by underlying permanent ice (permafrost), have temperature optima for growth as low as 5°C, a temperature at which ordinary roots are virtually inactive. Such adaptations presumably require enzymes and membranes that operate more efficiently at low temperatures than do normal ones. This may be achieved at the cost of metabolic maladjustments or outright instability at higher temperatures, because cool-climate plants usually cannot compete in climates with a warmer growing season.

Figure 19-5. Interior of a large controlled-environment plant growth chamber, in which environmental effects on growth and flowering of chrysanthemums are being investigated. Banks of incandescent and specially shaped high-intensity fluorescent lamps line the ceiling; sensors for detecting and regulating the temperature and humidity of the air with which the chamber is continually ventilated are located on wall behind man at left [the late Harry Borthwick, one of the discoverers of phytochrome (Chapter 14)].

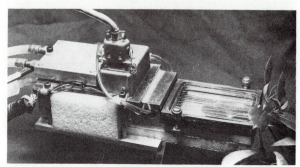

Figure 19-6. Leaf chamber for measuring photosynthesis of a leaf attached to an intact plant. The leaf, from plant at right, is being held between fine wires inside the glass-covered chamber, with a soft seal around the leaf petiole where it passes out of the chamber. Light from above falls on the leaf, and air from one of the hoses at left passes over it, returning via the other hose to an instrument that measures CO_2 concentration, determining how much CO_2 is being removed from the air by the leaf. Temperature and humidity of the air passing by the leaf are controlled by equipment in the left-hand part of the chamber.

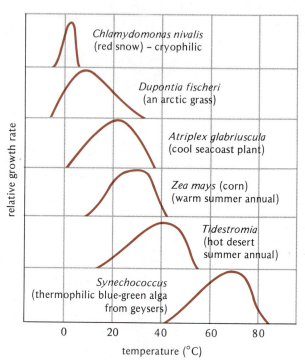

Figure 19-7. Examples of temperature optima and limits for plants adapted to various climates and habitats.

Because of photorespiration, as noted in Chapter 8, plants with ordinary or C3 photosynthesis apparently cannot evolve a temperature optimum much above about 30°C, which seems to place C3 plants at a disadvantage in areas of prevailingly hot (above 35°C) summer days. C4 photosynthesis has apparently evolved in part as an adaptation to such conditions, since it confers a considerably higher temperature optimum (Figure 8-17). Certain desert C4 plants have a photosynthetic temperature optimum of close to 50°C (122°F). Although many plants of hot-summer climates are C4 plants, most plants in such areas are in fact C3 plants that are environmentally adapted by means of high-temperature tolerance, but are photosynthetically inactive in the hottest weather. Most of their photosynthesis apparently takes place at cooler times of the year and/or during the cooler morning and evening hours in summer.

EDAPHIC OPTIMA

Plants can evolve differing quantitative requirements for mineral nutrients in adapting to different soils (Figure 19-8). One broad class of performance optima concerns soil pH (Figure 19-9). Many plants grow best in near-neutral or slightly acid soils, at pH values of about 5.0–6.5, but some, called **basophilic** ("base-loving"), are specifically adapted to the alkaline (pH above 7) and high-calcium conditions of soils in limestone (calcium carbonate) areas. In alkaline soils the essential elements iron and manganese tend to be scarce or unavailable (because their hydroxides and carbonates are very insoluble), so basophilic plants must have a more effective uptake mechanism for these nutrients or must utilize small amounts more efficiently. **Acidophilic** ("acid-loving") plants are adapted instead to acidic soil conditions (pH 3–4) that occur in many areas or habitats. As noted

in Chapter 11, an acid pH makes iron, manganese, and aluminum abnormally available, so ordinary plants often experience aluminum and manganese toxicity in acid soils. Acidophiles such as rhododendrons and azaleas not only tolerate these conditions but require them for optimal growth; they often suffer an iron deficiency if grown in ordinary neutral soils unless extra iron or an acidic fertilizer is supplied.

Ecological Optima and Fitness Range

The environment in which a species achieves its greatest success in nature and apparently prefers to live, based on field observations, is not necessarily its climatic and edaphic optimum but a compromise between its performance optima and tolerance range on the one hand and the negative influences of competitors, predators, and parasites on the other. The ecological fitness range of a species is essentially the range of environments over which it can survive and perform better than its potential competitors. Tolerance to conditions that weaken or exclude competitors, and at least limited performance adaptation to these conditions, are thus key features in the adaptive strategy of many plants. For example, acidophilic and basophilic plants are often restricted much more narrowly to acidic or alkaline soils, respectively, than their tolerance range requires (Figure 19-9). This restriction probably results because they tolerate the more extreme pH conditions better and perform better in such soils than do ordinary plants,

Figure 19-8. Example of a striking edaphic effect on vegetation, a boundary between serpentine and nonserpentine soils. The oak woodland at left occurs on an ordinary soil with a normal calcium content, while the wildflower-carpeted meadow to the right consists of species specially adapted to the calcium-poor soil derived from serpentine rock, a soil on which the oaks cannot grow. A few species are able to live on both serpentine and nonserpentine soils, by having evolved genetically different ecotypes adapted to the respective calcium conditions. Pacific Coast Range near Palo Alto, California.

whereas the latter compete better in normal neutral soils and thus usually exclude acidophiles and basophiles from neutral-soil habitats.

OPPORTUNISTS

A strategy alternative to competitive adaptation is specialization for temporary colonization of unoccupied habitats that have been made available by any kind of disturbance or catastrophe (such as floods, landslides, wind blow-downs, fires, volcanic eruptions, disease epidemics, or plowing of fields). In such openings there is a complete or relative absence of competition for space and resources. The opportunist strategy can thus be regarded as one of competition avoidance.

An opportunist or colonizer (Figure 19-10) must have (1) an efficient method of dispersal from one disturbed site to another and (2) prolific seed reproduction, which improves its chances of reaching new openings and enables its population to expand rapidly and occupy such openings. Opportunists must, of course, have performance adaptations sufficient for adequate growth under the physical conditions of the habitat, and they must be able to survive its extremes. Many do so simply by avoidance, in the form of seeds, because they are annuals. The premium on vigorous and rapid seed reproduction and lack of emphasis

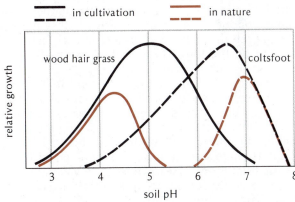

Figure 19-9. Influence of soil pH on wood hair grass (*Deschampsia flexuosa*), an acidophilic species, and coltsfoot (*Tussilago farfara*), a basophilic species, showing their intrinsic optima and tolerance ranges as compared to their range of occurrence in nature. They are both excluded from the middle range of soil pH by competition from neutrophilic plants.

Figure 19-10. Herbaceous opportunist species growing and blooming prolifically on the floor of a forest after a forest fire has killed the trees. The colonizers will eventually be displaced by a new growth of trees (succession, Chapter 33).

on longevity favors the annual habit for colonizing species.

ALLOCATION OF RESOURCES

The development of stress-resistance arrangements such as dormancy, overwintering organs, leaf fall and replacement, or thorns and insect-repelling secondary products requires resources and time that would otherwise be available for growth to keep ahead of competitors. A plant's optimal balance between these alternative activities and investments clearly depends on the relative importance of competition compared with environmental stresses for survival and success in a given habitat. It would be very different, for example, in a frost- and drought-free, humid tropical climate compared with a harsh arctic or desert environment. The relative commitment, or **allocation,** of its resources that a plant makes to different vegetative and reproductive tissues also affects, in principle, its potential competitive performance, because total plant growth is tied to photosynthesis, and photosynthesis in turn is correlated with leaf area. The more of its photosynthate and mineral nutrients a plant can invest in additional leaves, the greater its photosynthetic capacity and consequently its ultimate growth or reproductive output.

Actually, in most habitats plants of greatly different growth form such as trees, vines, perennial herbs, and annuals grow side by side, using widely different investment strategies. Trees grow upward as rapidly and as far as possible to preempt a large share of the habitat's sunlight. In order to succeed at this they must invest at least half of their photosynthetic production in nonphotosynthetic structural tissues of the stem and roots rather than in new leaf area. Moreover, the respiratory cost of maintaining extensive mechanical and transport tissues makes trees inherently less efficient in growth than many herbaceous species. Understory plants in a forest, which are at a disadvantage in obtaining light energy, compete by various special strategies in addition to the opportunist strategy described previously, in which plants take advantage of accidents to avoid competition for light.

Climbing vines, utilizing the supporting structures provided by trees, can allocate more of their limited photosynthate toward stem elongation to raise their leaf surfaces up toward the light. Herbaceous plants also allocate comparatively less carbon to mechanical tissues and more to new leaf area. **Shade plants** (Figure 19-11) utilize the weak or spotty light missed by the tree canopy, allocating their photosynthate to new leaf more efficiently than trees do. Because they have a lower respiratory cost for leaf maintenance, they can grow

Figure 19-11. Summer condition of a deciduous broadleafed forest. The herbaceous plants on the forest floor are shade plants that allocate a larger proportion of their resources to leaf production and produce leaves with considerably less investment of carbon per unit of leaf area than do the sun plants making up the overlying tree canopy.

under light intensities that would not support sun plants. **Ephemerals** such as early-spring annuals and bulb plants in deciduous woods, on the other hand, conduct most of their yearly photosynthesis during the limited spring season of high light intensity on the forest floor, before the trees have come fully into leaf (Figure 19-12).

Perennial herbaceous plants tend to allocate a relatively high proportion of their photosynthetic carbon to underground organs of survival and vegetative reproduction. Annuals, on the other hand, allocate relatively little to roots and a disproportionate amount to seeds compared with perennial herbs and trees, whose survival and competitive ability do not require yearly success in seed reproduction.

Timing Strategies

To achieve the greatest total growth and reproductive output a plant should initially invest

Figure 19-12. Spring condition of a deciduous broad-leafed forest. Many of the plants that are conspicuous on the forest floor at this time are ephemerals that conduct most of their yearly photosynthesis before the trees' leaves expand and cut down the light intensity reaching the ground. As this happens many ephemerals die down to dormant underground survival organs, or (if annual) die and persist as seeds through the rest of the normal growing season, thereby avoiding the costs of maintaining leaves and other physiologically active organs when the plant is not getting enough light to permit rapid photosynthesis.

For weeds, and for desert annuals that receive unreliable rainfall, the advantages of delaying investment in reproduction are usually outweighed by the risk of premature death—by the plow or the hoe or by running out of water, respectively. Many of these plants have a growth habit that involves continuous flower initiation accompanying vegetative growth, so that even if short-lived they leave some offspring, and the longer they live the more seeds they produce.

Although trees and other perennials can delay reproduction, they cannot afford to delay the timely development of resistance adaptations that anticipate yearly environmental extremes. The timing of emergence from dormancy and initiation of growth is also very important. These timing decisions involve a critical tradeoff between a plant's total period for photosynthetic production and growth, and the risks of injury by unseasonable extremes such as late spring or early autumn frosts. Thus important photoperiodic and temperature-sensing mechanisms have evolved that control leaf fall and the induction and breaking of dormancy (Chapter 14).

Most trees concentrate their shoot growth in a brief spring (and sometimes also midsummer) growth **flush** that utilizes previously stored photosynthate (Chapter 3) rather than continuously increasing their leaf area by investment of new photosynthate. Deciduous trees can thereby generate a complete leaf canopy quickly in spring, after which a further increase in leaf area seems superfluous because practically all the useful light is already being intercepted. The growth-flush pattern of allocation to shoot development also helps trees control insect pests. Many plant-eating insects feed on soft growing leaves and cannot effectively attack mature leaves toughened by continued cell-wall deposition. Available insects usually cannot consume more than a small fraction of a tree's young leaves during the brief period of a growth flush, and because the insects are deprived of this resource the rest of the time, they cannot multiply into a large population. This "herbivore avoidance" value may explain why tropical trees commonly grow by flushes even though there seems to be no climatic need for this.

GENETIC AND DEVELOPMENTAL BASIS OF ADAPTATION

Although physiological, anatomical, and behavioral adaptations obviously all have a genetic basis, they also depend to some degree upon the developmental responses of the genotype to its environment. The tolerance range and competitive adaptability of a species, which determine its geographical range, comprise two aspects, the tolerance range and adaptability of any given individual, and the range of genotypes with different tolerance, avoidance, and performance characteristics found in different populations of the species as a whole.

its photosynthate only in vegetative organs that are directly or indirectly essential to photosynthesis, expanding its photosynthetic capacity as rapidly as possible and only later allocating photosynthate to reproduction. Thus many plants conduct strictly vegetative growth for an entire growing season (biennials) or part of a season (annuals), and then abruptly begin to allocate all their photosynthate and storage reserves to reproductive effort. Ripeness-to-flower, photoperiodic, and vernalization controls (Chapter 16) help these plants time this change to occur after substantial vegetative growth yet early enough to complete reproduction before environmental conditions deteriorate. Trees usually delay reproduction for 10–20 years or more by passing through a lengthy juvenile phase.

Plasticity

The breadth of a given genotype's ability to tolerate and adjust successfully to different kinds of habitats is termed its **phenotypic plasticity** (Figure 19-13). Plasticity may be narrow or wide depending on the species. Since the climatic, edaphic, and biotic factors in a given habitat are usually relatively similar from year to year, one might expect that evolution should favor not plasticity, but optimum adaptation to the conditions of a particular kind of habitat. Opportunists or colonizer species are exceptions, because for them there is a premium on the success of their seeds in any newly opened habitat regardless of its microenvironment (local environmental conditions). Therefore, colonizer species have evolved a relatively broad plasticity. Once humans started to disturb the environment and alter it from year to year, colonizer species with broad plasticity such as the dandelion were greatly favored in human-occupied areas. They are the common weeds.

Phenotypic plasticity includes relatively passive adjustments as well as active, adaptive responses to environmental variables. An example of a passive adjustment is the growth forms of plants that tolerate chronically windy habitats such as seacoasts or mountaintops (Figure 19-A). By means of "wind pruning" the plant's top develops a streamlined or "windswept" form that extends in the direction of the prevailing winds and results in smoother air flow that does less (or no) damage to the shoot system than would be done to a plant or tree of more normal shape.

ACCLIMATION

Many performance and stress-resistance adaptations develop in response to environmental cues or conditioning. For example, most "hardy" plants actually cannot tolerate frost during the summer months while they are growing (Figure 19-2) but gradually acquire frost tolerance during autumn and early winter (Figure 19-14), first by a photoperiodic response to the short days of autumn and then in response to progressively lower prevailing temperatures, a response called "hardening." The ability to be hardened and the degree of cold that can be tolerated after hardening (Figure 19-2) are genetically determined, but the hard-

Figure 19-13.   Example of phenotypic plasticity in a clone of cinquefoil (*Potentilla glandulosa*) native to an elevation of 1400 m (4500 ft) in the Sierra Nevada mountains (California). Individuals derived by vegetative propagation of this clone were grown outdoors (A) at near sea level (30 m elevation) in coastal California, (B) at 1400 m, and (C) at 3050 m (10,000 ft) in the High Sierra, near timberline (see Figure 19-16). In the latter location the plant developed a diminutive form similar to that of species native to high altitudes. At sea level the plant grew less vigorously than at its native elevation, but survived. Clones of this species native to low altitudes, in contrast, usually failed to survive at timberline, and clones native to high altitudes often failed to survive at sea level, the climatic differences between these elevations being greater than the tolerance range of these respective genotypes.

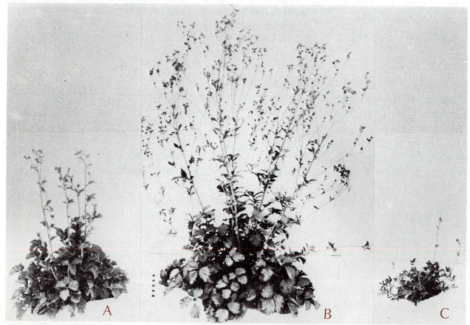

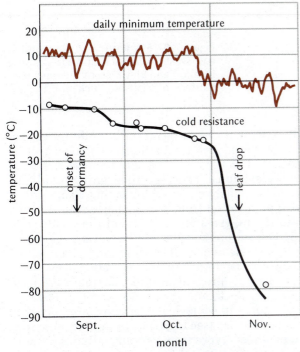

Figure 19-14. Progress of cold acclimation (development of frost tolerance) of red osier dogwood (*Cornus stolonifera*) during autumn in Minnesota. "Cold resistance" shows the lowest temperature survived by branches of the shrub removed and tested on the indicated dates. The early part of the acclimation (down to about −20°C cold resistance) takes place before occurrence of frost and is induced by short days; the subsequent deep acclimation (down to below −196°C cold resistance from December onward) is induced by exposure to freezing temperatures. (Data of van Huystee, Weiser, and Li, *Botanical Gazette*, Vol. 128, p. 200, 1967.)

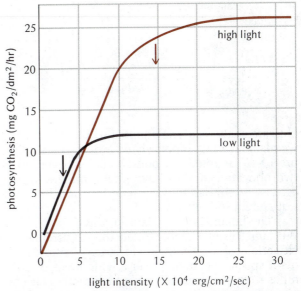

Figure 19-15. Dependence of photosynthetic rate on light intensity for leaves of a clone of goldenrod (*Solidago virgaurea*) grown under either low or high light. Arrows on the respective curves indicate intensity under which the plants had been grown. This clone was from an ecotype capable of acclimating to either low- or high-light habitats. Obligate shade and sun ecotypes of the same species have photosynthetic characteristics similar to the respective curves shown but do not change substantially when grown under the opposite light regime.

ening process itself is a physiological response to the environment. Plants whose photosynthetic mechanism can adapt seasonally to high temperatures respond to a warmer environment by an upward shift in the temperature that is critical for photosynthetic breakdown, and plants that inhabit climates with cool and warm phases in the growing season may show seasonal changes in their photosynthetic temperature optimum similar to the differences shown in Figure 19-17. Such adaptive adjustments of tolerance or performance in response to environmental variations are called **acclimation.**

The physiological or biochemical changes responsible for acclimation are little understood, although in the case of shifting temperature optima it seems probable that quantitative or qualitative changes in enzyme complements are necessary. The hardening process is generally accompanied by accumulation of sugars in the cells and progressive decrease in their water content, but it is not clear how these or other changes allow cells to tolerate the desiccation that occurs when cell water is pulled out by freezing into ice outside the cells.

The changes in many acclimation responses resemble genetically determined differences between species or genotypes that have a narrower degree of phenotypic plasticity—that is, a greater degree of specialization for particular environmental conditions. For example, some plants are *facultative* shade plants: under low light intensity they develop shade-type leaves, and under high light they produce sun-type leaves that have a much higher light-saturated photosynthetic rate (Figure 19-15; cf. Figure 8-11). In some cases the shade leaves, if later exposed to full sun, can acclimate themselves by producing more photosynthetic enzymes, thus transforming themselves into sun leaves. The ability to acclimate to different light intensities is a special genetic characteristic not found in genetically obligate sun or shade plants.

Genetic Adaptation: Ecotypes

Wild plant species occurring over any considerable climatic or edaphic range have usually adapted to different conditions by evolving genetic strains with different tolerance ranges and performance adaptations. Such strains are called **ecotypes.** For example, the goldenrod, *Solidago virgaurea*, in Sweden consists of three ecotypes, an obligate sun ecotype adapted to open fields, an

Figure 19-16. Vertically exaggerated east-west profile of California showing location and elevation of the transplant gardens of Clausen, Keck, and Hiesey.

obligate shade ecotype adapted to the interior of woods, and a facultative shade ecotype (as in Figure 19-15) adapted to woodland/field borders (the two obligate ecotypes have photosynthetic characteristics similar to the shade leaf and the sun leaf shown in Figure 19-15).

The botanists Jens Clausen, David Keck, and William Hiesey demonstrated the occurrence of climatic ecotypes in a classic series of studies on plant species occurring over a wide range of altitudes from the seacoast to the high Sierra Nevada mountains of California. They set up "transplant gardens" at warm coastal, cool middle-elevation, and cold high-elevation sites (Figure 19-16), where they planted clones (genetically uniform, vegetatively reproduced populations) of a given species collected at the same elevation and at the other two elevations. They recorded growth and survival year after year in each planting. Plants grown at the elevation from which they came did best; after a few years few or none of the plants from high elevations survived in the warm coastal habitat, and vice versa. By hybridizing plants of the same species from different elevations and growing their F_2 progeny in the transplant gardens, Clausen, Keck, and Hiesey showed that ecotypic differences in ecological adaptation are polygenic characteristics (see Chapter 18). Among other differences, ecotypes from varying altitudes differ in their photosynthetic temperature optima (Figure 19-17).

Ecotypes have also been demonstrated in species occurring over a wide range of latitudes. One important example is photoperiodic adaptation. At higher latitudes, autumn frosts come earlier, but daylength is longer during summer and early fall (Figure 19-18). Therefore, if short-day responses such as flowering of short-day plants or bud dormancy and leaf fall in deciduous trees are to be completed before frost, the plant's critical daylength must be considerably longer at high latitudes. Thus ecotypes with appropriately different critical daylengths have evolved at different latitudes (Figure 19-19). Transplanted to the wrong latitude, they are badly adapted: when grown north of their native latitude they do not respond photoperiodically soon enough for that latitude and are damaged or killed by early frosts, while if grown south of their own latitude they cease growth too early and lose out in competition with other plants. Crossing ecotypes of different latitudes and determining the distribution of critical daylengths in the resulting F_2 generation has shown that critical daylength, like other ecotypic differences, is polygenically determined.

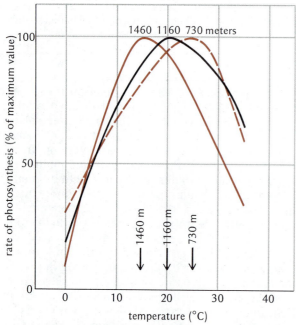

Figure 19-17. Temperature dependence of light-saturated photosynthesis by seedlings of balsam fir (*Abies balsamea*) from elevations of 730 m (2400 ft), 1160 m (3800 ft), and 1460 m (4800 ft) in the White Mountains of New Hampshire. Arrows on abscissa show respective temperature optima. All the seedlings had been raised under identical conditions, so the differences in their photosynthetic characteristics are evidently genetic. The decrease of about 2.7°C in optimum temperature per 305 m (1000 ft) of increase in elevation corresponds rather closely to the typical environmental temperature decline of about 3°C per 305 m of elevation.

Edaphic ecotypes also occur, even within a local area. Certain grass species include acidophilic and basophilic ecotypes occurring on acidic and alkaline soils, respectively. Certain plants normally found on ordinary soils have evolved ecotypes adapted to serpentine soils (Figure 19-8) or to heavy-metal-polluted soils near mines and smelters (Chapter 20). Other ecotypes have also evolved in response to human activities. Certain common urban weeds, such as annual bluegrass (*Poa annua*) and plantain (*Plantago major*), consist of distinct ecotypes with either a prostrate growth form that tolerates mowing and occurs in

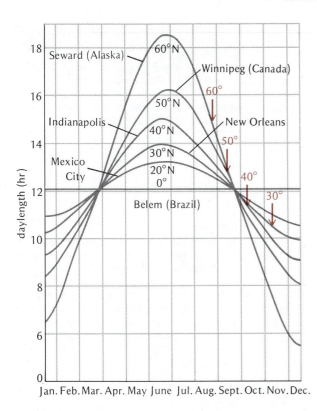

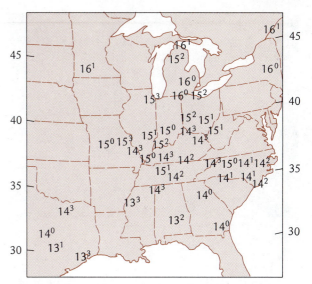

Figure 19-19. Differences in photoperiodic critical day length between populations of cocklebur (*Xanthium strumarium*, a short-day plant) in the eastern and central United States and adjacent Canada. Figures show critical daylengths in hours and quarters of an hour (superscripts), i.e., 14^3 means 14¾ hours.

Figure 19-18. Annual cycle of changes in photoperiod at different latitudes. Color arrows show approximate date and photoperiod of first killing frost at the locations cited (Mexico City and Belem experience no frost).

lawns, or an upright growth habit adapted to competing with rank grass and other weeds in untended vacant lots and roadsides. Ecotypic differentiation in plant species thus seems to be rather general.

ADAPTATIONS TO ENVIRONMENTAL WATER STATUS

Plants have evolved diverse behavioral and morphological adaptations to the wide variations in water supply and accompanying variations in temperature that occur in different areas. According to their adaptations to water supply plants are grouped as (1) **mesophytes,** adapted to moist but not wet habitats (*mesic* habitats, from Greek *mesos,* middle or intermediate); (2) **hydrophytes,** adapted to wet habitats, in which plants grow partly or completely submerged in water; (3) **xerophytes,** adapted to habitats with seasonal or persistent drought (*xeric* habitats, from Greek *xeros,* dry); and (4) **halophytes** (from Greek *halos,* salt), adapted to salty (*saline*) habitats, where water is osmotically unavailable because of high salt concentration. Because of intermediate degrees of adaptation

to flooding, drought, and salinity, these named categories are not completely distinct, but their extremes differ so greatly in form and are so commonly encountered that the terms are widely used.

Mesophytes

The more familiar vascular plants are mesophytes; they usually have relatively large, thin leaves and lack the specializations described below. Mesophytes range from species that require a consistently moist environment and/or tolerate some flooding, to flooding-intolerant species that are capable of surviving occasional droughts and grade ultimately into xerophytes. These gradations, combined with the previously mentioned variations in growth habit that are adaptive toward different temperature regimes, are the principal components of the major variations in land vegetation considered in Chapter 34.

Phreatophytes

Plants that obtain their water from the **water table** of permanently water-saturated soil or rock, usually located some distance below ground, are called **phreatophytes** (pronounced "free-at'-o-fite," from Greek *phreatos,* well, + *phyton,* plant) because their method of water procurement is analogous to drawing water from a well. Phreatophytes such as willows, alders, sycamores, and cottonwoods comprise the distinctive vegetation of stream banks (Figure 19-20) and lake shores, where the water table is close to the surface.

Figure 19-20. Cottonwoods (*Populus fremontii*), an example of a phreatophyte, growing in a desert creek bed or "wash." By means of deep roots these trees obtain water from the underground water table when the surface ground is dry. Thus they are able to achieve a lush growth contrasting with the low, sparse scrub vegetation below the cliffs in the distance, which is all the area's direct precipitation will support.

(text continued from page 345)
Their roots, unlike those of most land plants, can tolerate the low-oxygen conditions of submergence. Thus phreatophytic trees commonly occupy bottom land areas subject to seasonal flooding. Phreatophytes usually cannot compete with ordinary mesophytic vegetation in upland areas where the soil often becomes somewhat dry, and where the water table is too far beneath the ground surface for their roots to reach it.

Figure 19-21. Some submerged or partially emergent hydrophytes. (A) Marsh crowfoot (*Ranunculus sceleratus*) has morphologically different submerged and (where the shoot grows up out of the water) aerial leaves. (B) Tape grass (*Vallisneria americana*) has thin, ribbon-like leaves. (C) Water milfoil (*Myriophyllum humile*) has finely pinnately divided leaves. (D) A water lily (*Brasenia schreberi*) has surface-floating leaves with stomates on their upper surface in contact with the air. (E) Bladderwort (*Utricularia intermedia*) has finely divided submerged photosynthetic leaves, plus bladder-like leaves (lower branch) which capture small aquatic animals as a mineral nutrient source for the plant (as do the leaves of "insectivorous" plants; Figure 19-23).

A   B   C   D

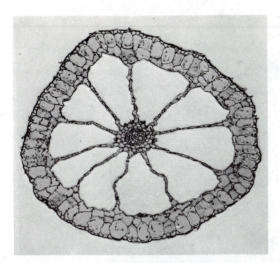

Figure 19-22. Cross section of stem of a hydrophyte (*Ceratophyllum demersum*) showing air passages and reduced vascular system.

Hydrophytes

Hydrophytes are vascular plants adapted to living in swamps, bogs, lakes, ponds, rivers, and shallow seacoast waters. Some hydrophytes are free-floating plants; others grow submerged (Figure 19-21). Still others (**emergent** hydrophytes) grow up into the air from roots or rhizomes anchored in mud or sand beneath the water.

Submerged hydrophytes usually have minimal roots or none. Their leaves commonly lack a cuticle, stomates, and internal gas channels and are often either thin and ribbon-like or dissected (Figure 19-21). This structure improves diffusion of carbon dioxide from the water into the leaf for photosynthesis in the absence of intercellular gas channels (cf. Chapter 9). Mechanical tissues and xylem are often reduced in amount or almost absent. The shoots of some hydrophytes can grow either submerged or, as the water level falls or the shoot gets longer, in an emergent manner. Typically, they then develop leaves with a mesophytic morphology quite different from that of their submerged leaves (Figure 19-21A). Other submerged hydrophytes, such as water lilies, have leaves that float (Figure 19-21D). These contain gas channels for carbon dioxide diffusion and possess stomates only on their upper surface, through which they obtain carbon dioxide from the air above the pond.

The bottom mud in which the roots and rhizomes of emergent hydrophytes such as bulrushes and cattails (Figure 15-2) are located is usually anaerobic. Emergent hydrophytes possess a well developed aeration system consisting of greatly enlarged intercellular air spaces (Figure 19-22) extending throughout the plant body. These supply oxygen to the underwater nongreen organs by diffusion from the air above, the plant's emergent stems or leaves acting as "snorkels." Certain emergent hydrophytes develop specialized snorkel roots (Chapter 3). The root "knees" of swamp cypresses in the southern states are usually considered to be aerating organs, although this has been questioned.

Swamps and bogs commonly have an acid pH and severe mineral nutrient depletion due to leaching by the flowing or percolating water. To solve this problem certain bog plants have evolved the ability to capture and digest insects (Figure 19-23). Although popularly called "carnivorous plants," insect-trapping plants are fully photosynthetic; their growth does not benefit from insect capture except under soil nutrient deficiency. Insect capture seems to be simply a means of obtaining nitrogen, phosphorus, potassium, and other mineral nutrients from a source other than the soil.

Xerophytes

Xerophytes are characteristic of desert and semidesert vegetation, which grows in areas receiving either little precipitation or more substantial precipitation restricted to part of the year, so that the vegetation must regularly endure a seasonal drought. Xerophytes also occur in moister regions in places where the available water is frequently deficient, as in sandy areas or on cliffs. At least four growth-form strategies are successful in xeric environments: (1) **ephemerals,** short-lived annuals that die as or before the dry season begins; (2) **drought-deciduous** perennials and woody plants; (3) **succulents;** and (4) **xeromorphs** (Greek *xeros,* dry, + *morphe,* form), woody plants with tough, drought-resistant leaves or other morphological adaptations for endurance of drought. Many xerophytes produce thorns or spines and/or obnoxious secondary products for reasons noted earlier.

EPHEMERALS

Desert annuals are often not regarded as xerophytes because they survive drought by avoidance, as seeds, and usually do not have morphological specializations against water stress. However, to be successful, desert annuals need physiological adaptations to ensure that their seeds will germinate only after sufficient rainfall and that a seedling will produce at least some seeds even when the water supply runs out unusually early. In such years desert annuals can be very minute ("belly plants") compared with their appearance in years of better precipitation. In years when precipitation fails to satisfy the seeds' germination requirements (e.g., at least 15 mm of precipitation in the Mojave Desert of California/Nevada), few or no ephemerals are seen, whereas better years bring forth an abundance of annuals yielding the spectacular spring flower displays for which deserts are famous (Figure 19-24).

Figure 19-23.   Some "insectivorous" plants of bogs and marshes. (A) Pitcher plant (*Sarracenia purpurea*), growing in a peat bog (*Sphagnum* moss in lower foreground), has pitcher-shaped leaves with downward-pointing bristles inside, which force entering insects down into digestive juice in base of pitcher. (B) Sundew (*Drosera intermedia*) has leaves (C) with tentacle-like projections each tipped with a drop of sticky secretion which catches an insect like fly paper; adjacent tentacles then quickly bend to engulf it in their drops of digestive secretion. (D) Venus fly-trap (*Dionaea muscipula*) has leaves consisting of a photosynthetic blade tipped by a folded, open-book-like trap which rapidly closes and traps any insect that contacts the touch-sensitive hairs on its upper surface (E, F); the sides of the trap later press together (G), compressing the insect and secreting digestive fluid onto it; the products then released are absorbed by the leaf. Trap closure is due to a sudden change in size of specialized cells located along the midrib of this segment of the leaf.

Figure 19-24. Desert ephemerals (annuals) in bloom after a winter with enough rainfall to stimulate substantial seed germination, leading to a spring flower display. The permanent vegetation of this area consists of scattered xeromorphic shrubs; the base of one is visible in the right foreground.

DROUGHT-DECIDUOUS PLANTS

As illustrated by the spectacular ocotillo (Figure 19-25) of the Sonoran Desert, drought-deciduous plants survive drought by avoidance: they shed their leaves as soon as their water supply runs short, avoiding further substantial water loss and entering a state of drought dormancy that can last for many months, until sufficient rain falls again. They then rapidly become active and unfold a new crop of leaves. Dryland bunch grasses behave somewhat similarly: their leaves and aboveground stems die when drought begins, while the plants survive by means of underground rhizomes.

SUCCULENTS

Succulents tolerate drought by storing and retaining water in fleshy, parenchymatous, water-storage tissue located in either their stems (e.g., cacti, Figure 3-24) or leaves (Figure 19-26, 3-19).

Stem succulents usually have extremely reduced leaves or none. The ribbed or pleated surface structure of succulent stems provides for a bellows-like action of the tough external tissue layers, allowing the water-storage tissue within to greatly expand and rapidly store large amounts of water when rain occurs (Figure 19-27). By means of a very thick epidermal cuticle and by conducting carbon dioxide uptake at night (as described in Chapter 9), succulents conserve their stored water over long periods of drought, continuing photosynthesis and avoiding dormancy. In certain succulents, photosynthesis has been measured after more than a year without water. However, the succulent strategy does not permit very rapid photosynthesis, so succulents usually cannot compete in habitats where there is adequate moisture for mesophytes.

Because they keep their stomates closed during the day and hence lack transpirational cooling, succulents tend to get very hot in the full desert sun and must be capable of tolerating high temperatures. The ribs of large cacti like the saguaro (Figure 3-25) increase these plants' surface area for cooling by the air around them, somewhat reducing their overheating problem. "Window-leaf plants" (Figure 19-28) and "living stones" (Figure 19-26D) are succulents specialized to develop largely underground. The leaves of window-leaf plants possess a somewhat translucent or transparent patch ("window") at their tops, where they emerge from the soil, which admits light for photosynthetic cells in the subterranean, probably cooler, part of the leaf.

Many succulents have only shallow, relatively weak root systems (Figure 19-29A). These rapidly gather water for storage in the water-storage organs when surface soil layers are wetted by rain but are almost inactive the rest of the time.

Because of their exotic appearance and geometrically pleasing shapes and surface patterns, as well as for their handsome flowers, many succulents are prized as house plants. The fact that they tolerate infrequent watering is an advantage for many people on busy schedules. One of the principal weaknesses of cacti as indoor plants is indeed their sensitivity to overwatering, which allows the stem to become diseased and rot, usually below the ground line. Increasing use is being made

Figure 19-25. Ocotillo (*Fouquieria splendens*), an example of a drought-deciduous xerophytic shrub occurring in the deserts of the American Southwest. (A) Habit of shrub. (B) In flower, at end of wet season, with leaves beginning to fall. (C) Completely leafless dormant thorny branch during dry season.

Figure 19-26. Some leaf-succulent xerophytes. (A) *Crassula* (jade plant). (B) *Aloe*, a succulent monocot. (C) *Mesembryanthemum* (ice plant). (D) *Lithops* ("living stone").

Figure 19-27. Young saguaro cacti (*Carnegia gigantea*) illustrating how the bellows-like expansion of their ribbed surface permits the stem to expand greatly and store much water. (A) After a long dry season, stem is contracted in girth, with its ribs folded closely together. (B) After rain, stems have swollen, spreading ribs apart (this young cactus has already developed a two-branched stem).

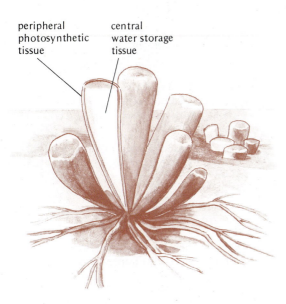

peripheral photosynthetic tissue

central water storage tissue

Figure 19-28. *Fenestraria,* a South African "window leaf" succulent, with one leaf cut in half to show structure. The translucent leaf tip, protruding through the soil, and transparent central water-storing tissue of the leaf, provide light for photosynthetic cells located in the outer part of the underground portion of the leaf.

of succulents for outdoor landscaping in the "sun belt" states where water for conventional landscaping is becoming increasingly short. Large old specimens now fetch high prices for landscaping purposes, and some species of cacti are becoming endangered by rampant collecting and removal from their native habitats.

XEROMORPHS

Xeromorphs are mostly shrubs or subshrubs that possess one or more leaf specializations for

Figure 19-29. Root systems of different kinds of xerophytes. (A) Cactus, with shallow superficial roots utilizing water from even light rains. (B) Deep-rooted xeromorphic shrub, tapping water falling on a much wider land surface area than that covered by the plant's foliage. (C) Extremely deep-rooted phreatophyte (mesquite), obtaining water from a water table far below the surface.

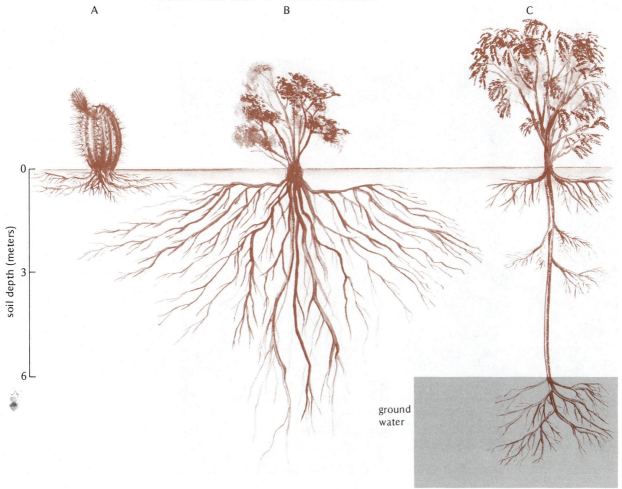

A

B

C

soil depth (meters)

0

3

6

ground water

TABLE 19-2   Xeromorphic Specializations of Leaves

A xeromorphic leaf often possesses several but not necessarily all of the specializations listed.

Feature	Specialization	Probable Effects
Size	Small (or lacking)	Improve cooling of leaf by air; reduce transpirational surface (if lacking, stems are photosynthetic)
Shape	Narrow, needle-like, or rolled up	
Stomates	On one side only; often sunken below surface in pits or larger crypts	Increase gas exchange resistance, reducing stomatal transpiration, especially in wind
Cuticle	Very thick and waxy	Prevent transpiration when stomates are closed
Epidermis	Sometimes more than one cell thick	
	Bears reflective hairs or scales	Reduce solar heating (hairs, scales may also reduce stomatal transpiration)
Leaf position	Often oblique or parallel to sun's rays	
Mesophyll cells	Several compact palisade layers, often on both sides of leaf	Reduce internal gas exchange.
	Little or no spongy tissue	
Internal gas channels	Small and few	
Leaf texture	Tough (leathery or hard), often with spiny margins	Prevent wilting and leaf damage under water stress; discourage consumption by herbivores and insects
Mechanical tissue	Abundant sclerenchyma, as bundle extensions or beneath epidermis	

Figure 19-30. Shoots of xerophytes, showing xeromorphic leaf features. (A) Thick, leathery, spiny leaves (toyon, *Heteromeles arbutifolia*). (B) Narrow, rolled, gummy-surfaced leaves (bush monkey flower, *Diplacus aurantiacus*). (C) Needle-like leaves (chamise, *Adenostoma fasciculatum*). (D) Tough, minute leaves (microphylls) (buckthorn, *Rhamnus crocea*). (E) Thick, small leaves with hairy stomatal crypts (white pattern) on lower surface (buck brush, *Ceanothus cuneatus*). (F) Dissected, narrow, hairy leaves (California sage, *Artemisia californica*).

Figure 19-31. Manganita (*Arctostaphylos*), a xeromorphic shrub whose vertically held leaves reduce the incident light intensity and resultant solar heating.

tolerating severe drought stress, as listed in Table 19-2. The frequently small leaf size and narrow outline (Figure 19-30) helps air to cool the leaf when it is heated by intense sunshine and transpirational cooling is lacking because of stomatal closure due to water stress. Xeromorphs with flat leaves also often avoid overheating by turning the leaf vertically (Figure 19-31), reducing its exposed surface and thus its midday heat load.

The surface structure of xeromorphic leaves (Figure 19-32) seems designed to impede gas exchange and slow down transpiration even when the stomates are open, especially in the windy conditions often prevalent in deserts. Nevertheless, xeromorphs transpire and photosynthesize fairly vigorously when their water supply is adequate and their stomates are open; their powerful ability to restrict transpiration comes fully into action only when water stress develops in the leaves and their stomates close. Rolling of the leaf

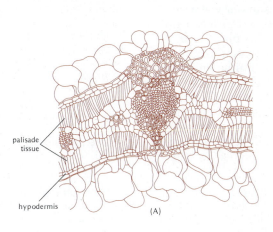

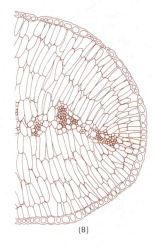

palisade
tissue

hypodermis   (A)

(B)

Figure 19-32. Examples of xeromorphic leaf structure. (A) Flat leaf blade of saltbush (*Atriplex canescens*), with swollen light-reflective hairs and palisade layer on both sides. (B) Cross section of compact leaf of *Greggia camporum*, having low surface-to-volume ratio and internal tissue in all palisade cells. (C, D) Cross sections of needle-like leaves, (C) of Russian thistle (*Salsola kali*), with a single palisade layer outside its internal water storage tissue, and (D) *Hakea platysperma* (an Australian xeromorphic shrub) similarly with external palisade layer, extremely thick cuticle, and sunken stomates.

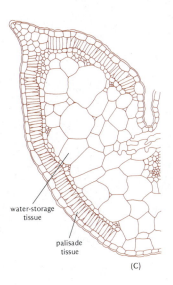

water-storage
tissue

palisade
tissue

(C)

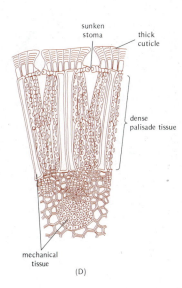

sunken
stoma

thick
cuticle

dense
palisade tissue

mechanical
tissue

(D)

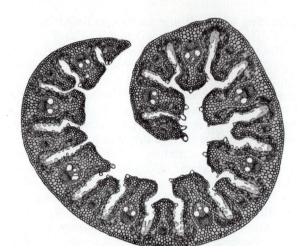

Figure 19-33. Cross section of a xerophytic grass leaf that rolls up under water stress. Stomata are absent from the lower surface, but are numerous along the sides of the furrows on the upper surface which becomes enclosed when the leaf rolls up.

to enclose the stomatal surface, as occurs, for example, in xeromorphic grasses under water stress (Figure 19-33), further increases the leaf's transpirational resistance. When the plant is under only moderate water stress, a slight, transient stomatal opening, which allows a limited amount of photosynthesis, may occur in the early morning and sometimes in the late afternoon (Figure 19-34B), the daylight hours least favoring transpirational water loss. In the hottest, dryest parts of the year xeromorphs must keep their stomates closed

almost all the time, and they become virtually inactive metabolically. To survive they must tolerate considerable desiccation, a capability termed **drought hardiness.** For example, leaves of the creosote bush (*Larrea tridentata*), one of the southwest desert's most abundant and widespread xeromorphs, lose three fourths of their total water during the summer drought. Over half of the shrub's leaves actually die and are shed during the dry season, making it intermediate between a xeromorph and a drought-deciduous xerophyte.

The tough texture and mechanical tissue of many xeromorphic leaves (Table 19-2, Figure 19-32) prevent the leaf from wilting or collapsing under water stress, which under windy conditions would cause damage to the leaf. The tough, sclerenchyma-rich type of xeromorphic leaf is called a **sclerophyll** (pronounced "sklair'-uh-fill"). The heavy wall thickenings of sclerophylls may permit tension to develop in leaf cells under water stress, like the tension normally present in the xylem of mesophytes (Chapter 10). This tension greatly increases the force (water potential) that the plant can exert for removing water from the soil compared with mesophytes. Xeromorphic leaves under water stress may even take up significant amounts of water from the air, either from fog or dew that condenses on them, or from water vapor at night when the air is relatively humid. Certain sclerophyllous shrubs of "fog desert" vegetation, for example, in coastal Chile, actually transport water from their leaves into the dry soil around their roots when the leaves are wetted by fog. The shrub can later absorb this soil water to cover its transpiration costs during dry weather, enabling these shrubs to exist in areas where rain almost never falls.

Figure 19-34. Examples of daily course of photosynthesis and transpiration for a xeromorphic shrub (left) under well-watered conditions, and (right) under water stress, its stomates opening only transiently in the early morning and late afternoon to allow a limited amount of photosynthesis when the temperature, light, and humidity conditions are least conducive to transpiration.

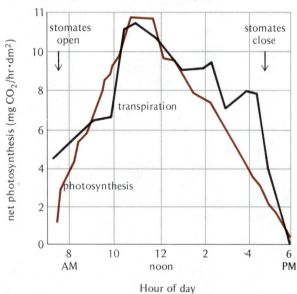

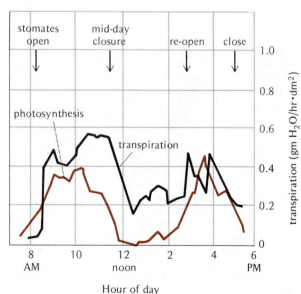

Figure 19-35. Xeromorphs with reduced or ephemeral leaves and a green, photosynthetic stem. (A) A branch of *Ephedra* ("mormon tea"), a shrub bearing nonphotosynthetic scale leaves (globular structures in picture are the cone-like, ovule-bearing female reproductive structures of this gymnosperm; see Chapter 30). (B) Palo verde (*Cercidium floridum*), after shedding of its short-lived foliage leaves. The green stems support the tree photosynthetically for most of the year.

The leaves of some xeromorphs have been reduced to virtually nonfunctional scales; the photosynthetic tissue is located instead in the stems, which are green (Figure 19-35A). Other xeromorphs, such as the well-known palo verde ("green stick"), a tree of the arid southwest, have green photosynthetic stems bearing small short-lived photosynthetic leaves that drop off in the dry season (Figure 19-35B).

In contrast to succulents, xeromorphs are usually very deep-rooted and tap as much soil volume as possible for their water supply. Roots of the widely spaced desert bushes spread out and collect precipitation falling on a much larger area than that covered by the plant's top (Figure 19-29B); the plant thus invests a considerably larger fraction of its photosynthate in its root system than do most mesophytes.

Some desert and semidesert xeromorphs, including the widespread mesquite, are phreatophytes, which reportedly send down roots as deep as 54 meters (175 feet) to reach a deeply located water table (Figure 19-29C).

Evergreen Xeromorphs of Cold-Winter Climates. Conifers, with their needle-like or scale-like leaves (Figures 30-19, 30-21), are xeromorphs. Although conifers such as the pinyon pine and the junipers of the Great

Figure 19-36. Examples of halophytes. (A) Pickleweed (*Salicornia bigelovii*), which has a succulent photosynthetic stem, with leaves reduced to small scales, occurs widely in salt marshes. (B) Saltbush (*Atriplex canescens*), with narrow xeromorphic leaves covered with a light-reflecting layer of scales, is common on saline soils in the Great Basin (winged structures at top of shoot are the shrub's fruits).

Basin or the pines that inhabit sandy "pine barrens" in the eastern states are xerophytes—that is, they occur in xeric habitats—many conifers are confined to mesic regions such as mountain ranges and higher latitudes. The ability to tolerate water stress is actually important for evergreens of cold-winter climates because their leaves continue to transpire to some extent even when the soil and often the xylem are frozen, thus preventing water uptake and transport, a situation called "winter drought." Doubtless for the same reason, the leaves of evergreen broad-leaved plants of cold-winter climates, such as rhododendrons and mountain laurel (*Kalmia*), have xeromorphic features. Excessive winter transpiration caused by warm windy weather while the ground is frozen can, however, injure evergreens, causing "winter burn."

Halophytes

Salt-adapted plants occur in seacoast salt marshes and around salt lakes, "dry lakes," and salt flats such as those in the Great Basin. Ordinary mesophytic and xerophytic plants cannot obtain water from these sources because these habitats contain salt at a higher concentration than the osmotic concentration of normal roots. Halophytes can take up salt to very high concentrations and can tolerate these salt concentrations in their cells. Thus they can obtain water from these salty habitats by normal osmotic means. Most halophytes have either a succulent or a xeromorphic form (Figure 19-36), reflecting a need for water conservation. Some of them possess **salt glands** that excrete excess salt from their shoots.

Although ordinary plants (mesophytes) apparently do not require sodium as a mineral nutrient (Chapter 11), sodium benefits the growth of halophytes even under nonsaline conditions. Certain halophytes, notably the Australian shrub *Atriplex vesicaria,* absolutely require Na^+. The amount of sodium needed as a nutrient, however, is much less than the amounts accumulated by these plants when growing in saline habitats. A high salt concentration is probably as injurious to the cytoplasm of halophyte cells as it is to ordinary cells; cells of halophytes apparently accumulate sodium chloride mainly in their vacuoles and produce special organic solutes at an equally high concentration in their cytoplasm to prevent its being dehydrated by the osmotic action of vacuolar sodium chloride.

SUMMARY

(1) Ecological factors to which plants adapt and which determine the characteristics of different habitats include climatic factors (light and aspects of the aerial environment), edaphic factors (aspects of the soil or aquatic medium), and biotic factors (effects of other organisms, both beneficial, competitive, and harmful). The effects of individual factors frequently interact.

(2) Adaptations for stress resistance, which allow a plant to endure environmental hazards and stresses, include avoidance, or evasion of environmental stress, and tolerance, or ability to withstand the stress. Persistence through unfavorable climatic conditions as seeds (annual plants) or underground organs (herbaceous perennials) are examples of avoidance. Direct endurance of high temperatures, frost (frost-hardiness), or drought (drought-hardiness) are examples of tolerance. Resistance to diseases and defenses against herbivores are other important stress-resistance adaptations. Resistance adaptations contribute indirectly to photosynthetic performance and growth by reducing or preventing costly losses of fitness caused by deleterious environmental factors.

(3) Adaptations that improve photosynthesis, growth, or reproductive performance under particular environmental conditions allow a plant to compete better with others, either by capturing more resources (light, water, mineral nutrients) or by using a limited resource more efficiently. Examples of optimized performance relative to environmental variables are specialized temperature optima for photosynthesis and growth in cool and warm climates and special pH optima for root function in acidic and alkaline soils (acidophilic and basophilic plants). Shade adaptation illustrates a more efficient use of a resource, the limited amount of photosynthate available in dim light.

(4) Opportunist strategy, or specialization for colonizing temporarily unoccupied habitats, may be regarded as competition avoidance. Because it places a premium on high reproductive output plus efficient dispersal and not on individual longevity, many colonizers are annuals.

(5) Different patterns of resource allocation both in space and in time are the basis of different

competitive and survival strategies. Annuals emphasize investment in reproduction (seeds), herbaceous perennials allocate resources preferentially to vegetative survival organs, and trees invest disproportionately in mechanical tissues in order to compete. The timing and extent of commitment to survival structures and behavior and/or to reproductive effort, as against investment in additional photosynthetic capacity, affect a plant's total potential growth and represent a tradeoff between its competitive performance and its preparedness for environmental hazards.

(6) Phenotypic plasticity denotes the ability of a genotype to specify varying body form, behavior, and physiological characteristics as adaptive responses to different environments. Acclimation is adaptive adjustment of performance optima or tolerance limits as a response to exposure to different conditions—for example, the optimum temperature for photosynthesis is lowered when the plant is kept at a cool temperature. Phenotypic plasticity tends to be greatest in opportunists such as weeds.

(7) Ecotypes are genetic strains of a species adapted to different local habitats or to the different environments of various geographical areas. By evolving ecotypes a species can adapt to and occupy a wider range of habitats than its phenotypic plasticity would allow. Altitudinal and latitudinal differences in temperature response or tolerance, and genetic adaptation to sun and shade habitats or to mowed and unmowed areas, are examples of ecotypic adaptation.

(8) Major differences in growth form and anatomical structure enable higher plants to adapt to habitats of different water status with their associated temperature, mineral nutrient, and aeration regimes. Hydrophytes are adapted to aquatic, often low-oxygen habitats, submerged hydrophytes differing greatly from bottom-rooted, emergent hydrophytes in their specializations for gas exchange and mechanical support. Mesophytes are plants of ordinary moist (mesic) habitats and are often intolerant of flooding. Halophytes have adapted to salty (saline) habitats by accumulating and tolerating large concentrations of salt in their cells and by possessing xeromorphic features.

(9) Xerophytes, which are adapted to seasonally or habitually dry (xeric) habitats and regions, embrace several different strategies. Ephemerals (annuals) and drought-deciduous perennials survive drought by avoidance. Succulents retain water in fleshy, water-storage tissue and conserve this water for long periods by specialized photosynthetic metabolism that involves stomatal opening and carbon dioxide uptake at night instead of during the day. Xeromorphs tolerate drought by possessing leaves specialized in various ways to minimize transpiration under water-stress conditions, to improve cooling by the air when transpirational cooling is prevented by stomatal closure, and to tolerate protracted periods of wilting without leaf damage. Some xeromorphs are leafless and conduct photosynthesis in their stems.

Chapter 20

EVOLUTION

The concept of organic evolution has probably altered our view of nature and of ourselves more profoundly than any other scientific idea. The theory of evolution holds that despite the faithfulness with which individual species reproduce their kind, organisms in the course of the earth's history acquire new and different forms and functions, and give rise to new and different species that did not previously exist. These changed features often enable the new species to perform better in its environment or to adapt to and exploit a different environment. Step by step, simple organisms evolve into more complex ones with improved or entirely new capabilities.

Although the idea of organic evolution can be traced back to ancient times, it first became influential in the nineteenth century through the writings of Lamarck and especially of Darwin. In works that appeared between 1801 and 1835 the French zoologist Jean Baptiste de Lamarck argued that animal and plant species should be modified in the course of many generations because each individual transmits to its offspring characteristics and capabilities that it acquires in adapting to its environment. Lamarck's concept of evolution was faulty because, as was demonstrated later, acquired characteristics generally are not inherited.

Darwin

Not until 1859, when the English naturalist Charles Darwin (1809–1882) (Figure 20-1) published his book, *The Origin of Species*, was a compelling case made for the hypothesis that not only has evolution occurred, but natural forces ensure that it *must* occur. Darwin's book became one of the most influential works of all time, not only in science but also in philosophy and in popular thinking, since acceptance of its reasoning obliged people to revise their traditional religious belief

Figure 20-1. Portrait of Charles Darwin as a young man. This likeness was made not long after he returned from round-the-world biological explorations as naturalist aboard the H.M.S. Beagle, making many observations important in leading him to his concepts of organic evolution.

Figure 20-A. Natural selection in action. This Eucalyptus forest in Australia was recently burned by a forest fire. The trunks and larger branches of some of the trees have survived the fire and are producing a new crown of leaves by outgrowth of adventitious buds that have formed in the bark ("epicormic regeneration"). Other trees have been killed and will not reproduce further. Many seedling eucalypts, having germinated after the fire, are sprouting up and will reproduce some of the killed trees. Fire stress, which occurs frequently in these woods, selects in favor of characteristics for withstanding, or for vigorous seed germination after, a fire.

in the doctrine of *special creation*—the idea that they and the earth's plants and animals were created separately and independently in their present forms.

Both Darwin and another English naturalist, Alfred Russell Wallace, realized at about the same time the profound significance of the fact that only some of the offspring produced by a species normally survive life's competitive forces and reproduce. The genes of successfully reproducing individuals are inherited by the following generation, whereas the genes of unsuccessful individuals are not. Through competition, genes that help individuals to survive and reproduce in their environment are automatically selected for transmission from one generation to the next. This is the principle of **natural selection,** often paraphrased as the "survival of the fittest" (Figure 20-2). According to the principle of natural selection, the genetic makeup of a species should change with every generation in a direction that enables its individuals to compete better, on the average, in their environment. Over many generations, Darwin and Wallace reasoned, the inexorable repetition of this process would genetically change a given species so substantially as to create other, fundamentally different species.

Evidence of Evolution

Darwin and subsequent biologists have detected many indications of evolution by comparing the form and structure of different organisms and by examining their geographical occurrence. In plants relationships have been noted among species with a basically similar pattern of growth and form of reproduction. These similarities, together with the fact that groups of species sharing such similarities often occur only in a particular part of the world, can be understood by assuming that groups of species have descended by evolution from common ancestors. In subsequent chapters there are numerous illustrations of the principle of morphological similarity between evolutionarily related plant groups.

The study of **fossils** provides a second important line of evidence indicating the past occurrence of evolution. Fossils are the remains or traces of plants and animals in rocks representing ancient sediments deposited during successive periods of the earth's history. The sequence of these fossil-bearing sedimentary beds, worked out by geological surveying techniques, shows unmistakably that many different animal and plant forms have marched across the stage in the course of the earth's history.

A third line of evidence for evolution is found in the structure of specific RNAs and proteins, the genetic products directly specified by genes. This chapter examines the genetic and molecular as-

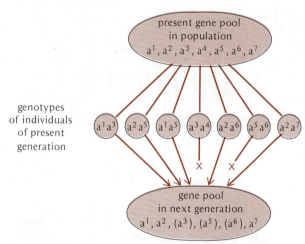

Figure 20-2.   Natural selection of genes in a population, represented in very simplified form. Suppose that in a small population, presently containing seven alleles at locus a (top of diagram), the particular genotypes shown in the middle of the diagram happen to occur in the present generation, and that an individual must carry either the a^1 or a^2 allele in order to reach maturity and reproduce. The genes transmitted to the next generation will then be as indicated; ( ) means occurring in the next generation at a reduced frequency. Note that allele a^4 disappears entirely from the gene pool in the example given. All other alleles except a^1 and a^2 will continue to decline in frequency, or disappear, in subsequent generations. (More realistically, the *number* of successful offspring each individual produces, rather than its mere survival and reproduction, is what must be considered, as shown later in the chapter.)

pects of evolution, while subsequent chapters consider the extensive geological and morphological evidence of the evolutionary diversification of plants.

Component Processes in Evolution

As Darwin and Wallace recognized, evolutionary genetic change in a population or a species is inevitable if genetic variation occurs among individuals and if some genotypes are more capable than others of dealing with their environment and producing successful offspring. The sources of genetic variation within a species are (1) **mutation,** which alters the genetic material, and (2) **recombination,** which creates new and different gene combinations or genotypes from among the genes occurring in a population. **Natural selection,** the third major component process in evolution, acts upon the resulting genetic variation.

MUTATION

Mutation, or heritable alteration of an organism's genetic elements, provides the raw material for evolution. As noted in Chapter 18, mutations can be caused either by mistakes during DNA replication or by environmental factors such as ultraviolet light, atomic radiation, or cosmic radiation, which create essentially random changes in the structure of a cell's DNA or in the arrangement of its genome. Two general kinds of mutations are (1) **gene mutations,** which are alterations of or within individual genes, and (2) **chromosomal mutations,** which are alterations in the location and arrangement of genes on chromosomes.

Gene Mutations

Under natural conditions most genes mutate very infrequently, on the order of only one to ten mutations per million new genes. Because each individual possesses 10,000 or more genes, however, the chances that it carries a new mutation at some genetic locus or other are considerable (on the order of 1–10%). Among the millions to hundreds of millions of individuals that may comprise the total population of a species, mutation may be expected to take place in almost every gene, in one individual or another in the population, in every generation. Continuing generation after generation, mutation in the long run has an enormous potential for creating genetic variation in a species.

The simplest type of gene mutation is a **base substitution,** in which the base that normally occurs at one point in the gene's DNA base sequence is replaced with a different base (Figure 20-3A). This alters one code word (codon) in the mRNA transcribed from the gene, which can result in one wrong amino acid being substituted in the amino acid sequence of the gene's gene product. (Some base substitutions in the third position of a codon still specify the same amino acid because of the redundancy of genetic code words. See Table 5-1.) If the mutant gene product containing one wrong amino acid functions just as well as the normal protein, the mutation will not be detectable in the phenotype. However, many amino acid substitutions alter the gene product's functional (e.g., enzymatic) activity rather drastically, sometimes inactivating it completely. Certain amino acid substitutions may, on the other hand, actually improve the performance of the gene product relative to particular needs or environmental conditions. For example, enzymes of blue-green algae (cyanobacteria) that are specialized for living at temperatures near boiling, such as in geyser and hot spring waters in Yellowstone and Mt. Lassen National Parks, show unusual stability against denaturation at high temperatures compared with the same enzymes of thermally normal species.

The second type of gene mutation is **deletion** or **insertion** of bases. Deletions or insertions of three bases or small multiples thereof, resulting in deletion or insertion respectively of one to a few amino acids in the gene product's amino acid sequence (Figure 20-3B), have the same range of possible effects as simple amino acid substitutions, depending upon how important the affected part of the polypeptide's amino acid sequence is to its biological activity.

A small deletion or insertion of *other than* three bases or a multiple thereof, on the other hand, can completely destroy the functional activity of the gene product. This occurs because during mRNA translation such an insertion or deletion shifts the ribosome's recognition and reading of *all* codons beyond the point of mutation (Figure 20-3C), so the *entire* amino acid sequence beyond that point is altered. This is called a **frameshift mutation,** altering the "reading frame" for the codons beyond the point of insertion or deletion.

Large deletions, which eliminate a substantial part of the gene's base sequence and therefore of the gene product's amino acid sequence, are almost always functionally defective. So are base substitutions or deletions located at or just before the start of the gene, either at the site where its transcription into mRNA is initiated or at the site where translation of its mRNA begins. Such mutations result, respectively, in no transcription of mRNA or in no translation of the transcribed mRNA.

Since mutations yielding a defective gene product will be lethal, when homozygous, if the gene product is essential to metabolism or development, such mutations in general are likely to be much less important in evolution than mutations specifying an altered but still functional gene product.

The existence of gene mutations within populations of a species can be detected by examining a specific protein (such as a particular kind of enzyme) by the method of **electrophoresis** (Figure 20-4). This method shows the presence, in different individuals, of different alleles specifying slightly different proteins of the same type (these proteins usually differ from one another by a single amino acid substitution).

Molecular Evolution

When the amino acid sequence of the same protein is compared in different species, it is generally found that the more distant the taxonomic relationship between the species, the greater the number of amino acid substitutions, deletions, and insertions distinguishing the respective proteins (Figure 20-5). By comparing amino acid sequences of a given protein in diverse species, an "evolutionary tree" can be constructed showing the kinds of mutations that have accumulated in the gene specifying this protein during the evo-

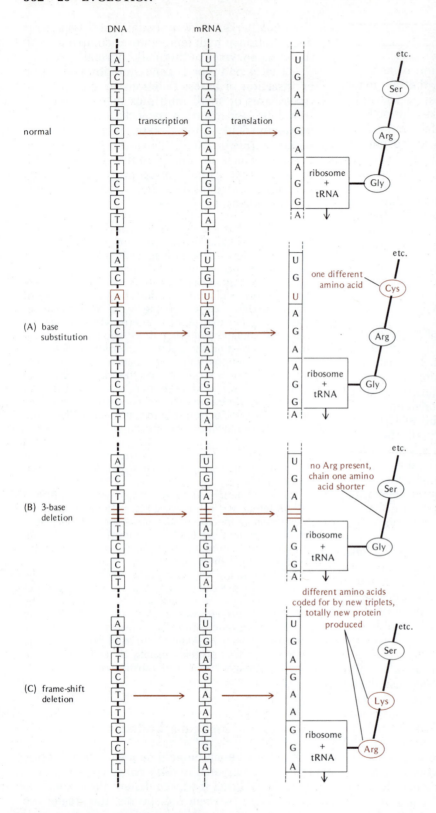

Figure 20-3. Types of gene mutations. Top diagram shows transcription of part of a gene's base sequence and its translation as a sequence of three-base code words (codons) specifying particular amino acids in a polypeptide ("gene product"). (A) A *base substitution* mutation, in which one base in the DNA has become replaced by a different one, changes one codon, resulting in substitution of a different amino acid at one point in the gene product. (B) A *deletion* of three bases deletes one codon and thus one amino acid from the gene product but does not change its amino acid sequence as a whole. (Similarly an *insertion* of three bases inserts one extra amino acid into the polypeptide chain without changing the entire amino acid sequence). (C) A deletion or insertion of one or two bases, or any number other than a multiple of three, is a *frame-shift* mutation because it alters the "reading frame" for the entire base sequence beyond the point of the mutation. The identity of all amino acids beyond the mutation is altered, usually resulting in complete loss of the gene product's biological activity. Examples of base substitutions, insertions, and deletions can be seen in comparing the ribosomal RNA sequences shown in Figure 20-7 (in this case the gene product is simply the RNA, rather than a polypeptide).

(*Text continued from page 361*)
lutionary divergence of different groups of organisms (Figure 20-6).

The DNA sequences that code for the functional RNAs (ribosomal and transfer RNAs) of the cell's protein synthesis machinery (see Chapter 5) are similarly subject to mutational changes during evolution. However, mutations that alter the structure of ribosomal or transfer RNAs endanger the translation of all the genetic messages of the organism, which would be lethal. Ribosomal and transfer RNA gene sequences have therefore evolved only very slowly. Thus pea and wheat,

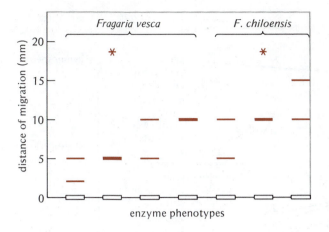

Figure 20-4. Genetic variation in complement of an enzyme (peroxidase) in different individuals from two species of wild strawberry (*Fragaria*) in California, as shown by gel electrophoresis. Tissue extracts from different individuals are placed in slots cut into a slab of starch gel, along a line corresponding to the bottom of the diagram. A high voltage is applied across the gel for a period of time, the (+) side being the top of the diagram and the (−) side being below the 0 line. Peroxidase molecules, being negatively charged, are pulled toward the (+) side. Enzyme molecules with different electric charge (due to one or more different amino acids in their polypeptide chains) migrate at different rates, separating into distinct bands (heavy horizontal lines) as detected by applying a substrate that peroxidase converts into a colored product. As seen, four different forms (allozymes) of peroxidase occur among the individuals studied. Individuals with 2 peroxidase bands are heterozygous (possess alleles for 2 allozymes). An (*) indicates the phenotype most frequently encountered in each species. Data from J. F. Hancock, Jr., and R. S. Bringhurst, *American Journal of Botany,* Vol. 65, p. 795 (1978), and Vol. 66, p. 367, 1979.

(Text continued from page 360)

which are almost as distantly related as any two flowering plants can be (the one a dicot, the other a highly advanced monocot), have identical phenylalanine transfer RNA sequences.

The 5S ribosomal RNA sequences of various flowering plants differ only slightly (Figure 20-7) compared

Figure 20-5. Portions of the amino acid sequences of cytochrome *c* from various higher plants, showing the results of molecular evolution. The one-letter symbols used for amino acids are defined below the diagram using the standard three-letter abbreviations of amino acid names given in Figure 5-6. The large omitted portions of the approximately 111 amino acid sequences are largely identical among all the plant cytochromes *c*. Sequences for representative monocots and dicots are placed in order of similarity respectively above and below the sequence for *Ginkgo*, a representative gymnosperm (primitive seed plant). The sequences do not fall, however, into a single progressive series, but represent the end points of a branching phylogenetic tree as shown in Figure 20-6. [Amino acid symbols (D or N) and (E or Q) are given in parentheses when the methods used did not distinguish between the members of these respective pairs, the most probable identity being given. Data mostly of D. Boulter et al., from M. O. Dayhoff, *Atlas of Protein Sequence and Structure,* 1972, 1973, and 1976. Amino acids not listed among the symbols happen not to occur in these sections of the cytochrome *c* molecule.]

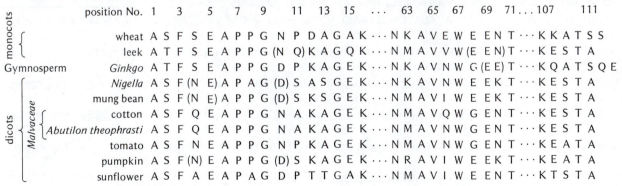

symbols   A ala   D asp   E glu   F phe   G gly   I ile   K lys   M met   N asn   P pro   Q gln   R arg   S ser   T thr   V val   W try

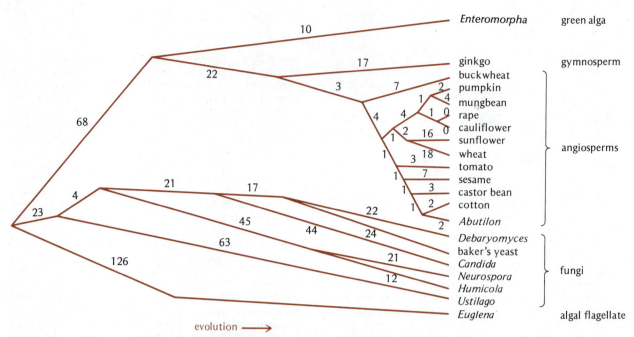

Figure 20-6. Apparent phylogenetic tree for evolutionary divergence of cytochrome c sequences in plants and fungi. Numbers between forks in diagram show calculated probable number of base substitutions in the cytochrome c gene accounting for divergence of the sequences between different groups and species. The deduced pattern conforms with general taxonomic relationships, such as placing seed plants, fungi, and green algae in separate Subkingdoms. Note that fungi, which have existed on earth much longer than seed plants (Chapter 21), are much more diverse in their cytochrome c sequences than are seed plants. From G. W. Moore et al., *Journal of Molecular Biology,* Vol. 105, p. 15, 1976, and data of J. A. M. Ramshaw, et al., *New Phytologist,* Vol. 71, p. 773, 1972.

Figure 20-7. Base sequence of 5S ribosomal RNA of the green alga, *Chlorella,* and of six angiosperms, showing the extensive homology between them, and the base substitutions, deletions, and additions involved in their modification through evolution. In listing the sequences, those containing deletions relative to longer sequences have been spread apart at deletion points so the matching portions of all sequences appear above one another. Base substitutions and additions, relative to the sequence next above, are shown in color. (The sequences did not necessarily evolve in the order given: "deletions" in some cases may instead represent insertions into the shorter type of sequence to yield the longer one. Total length of these rRNAs is about 120 nucleotides. Data from V. Erdmann, *Nucleic Acids Research,* Vol 6(1), r29, 1979.

Chlorella	AUGCUACGAUCAUACACCACGAAAGCACCCGAUCCCAUCAGAACUCGGAAGUUAAACGUGGUUGGGCUC
wheat	GGAUGCGAUACCAUCAGCACUAAAGCACCGGAUCCCAUCAGAACUCCGAAGUUAAGCGUGCUUGGGCGA
rye	GGAUGCGAUACCAUCAGCACUAAAGCACCGGAUCC–AUCAGAACUCCGAAGUUAAGCGUGCUUGGGCGA
tomato	GGAUGCGAUACCAUCAGCACUAACGCACCGGAUCC–AUCAGAACUCCGAAGUUAAGCGUGCUUGGGCGA
sunflower	GG–UGCGAUACCAUCAGCACUAAUGCACCGGAUCC–AUCAGAACUCCGCAGUUAAGCGUGCUUGGGCGA
common bean (*Phaseolus*)	GG–UGCGAUACCAUCAGCACUAAUGCACCGGAUCC–AUCAGAACUCCGCAGUUAAGCGUGCUUGGGCGA
broad bean (*Vicia faba*)	AGG–UGCGAUACCAUCAGCACUAAUGCACCGGAUCC–AUCAGAACUCCGCAGUUAAGCGUGCUUGGGCGA

with differences in the gene for cytochrome c among the same plants (Figure 20-6). In Chapter 22 the more substantial differences in ribosomal RNA sequences in the most basic, ancient evolutionary groups of prokaryotes and eukaryotes are described.

Chromosomal Mutations

High-energy radiation passing through cells can break chromosomes. Cells possess mechanisms to repair this damage by linking broken chromosome segments together. In this repair process chromosome segments are sometimes linked to one another in the wrong relationship, yielding novel gene arrangements or chromosomal mutations called **duplications, deletions, inversions,** and **translocations** (Figure 20-8). Duplications and deletions can also result from mistakes in crossing-over during meiosis (Figure 18-10), and entire chromosomes can be deleted or duplicated by accidents in meiosis or mitosis.

Deletion of any chromosome or a substantial part thereof usually eliminates many functions and is lethal. Duplications, inversions, and translocations, by changing the position of structural genes relative to regulator genes, can affect the degree to which particular genes are expressed in the phenotype. By affecting the meiotic pairing relations between chromosomes, chromosomal mutations disturb meiosis and can drastically affect gene recombination during reproduction; chromosomal mutations are among the most important factors that permit several distinct species to arise from one common ancestor species, as we shall see below.

A local repeat, or **gene duplication** (repeating the complete base sequence of one or more genes), creates an important potential for protein evolution. When the haploid genome contains two copies of a gene, a mutation of one of these copies yielding a functionally defective gene product will result in no ill effects on the organism. Over a period of time, further mutations

can occur and accumulate in this gene until the gene product it specifies happens by chance to result in a new and beneficial function, such as to catalyze a new type of enzymatic reaction. In this way organisms can evolve entirely new functional capabilities.

As noted in Chapter 5, an intriguing peculiarity of eukaryotic nuclear genomes, especially those of plants, is that a large proportion of the genome consists of multiple-copy DNA, or particular base sequences several hundreds of bases long repeated hundreds to many thousands of times in the genome. Most of the multiple-copy DNA is not genetically active (exceptions are rRNA, tRNA, and histone genes, which occur in multiple copies). Most structural genes have unique base sequences (single-copy DNA, i.e., represented only once in the haploid genome). Multiple-copy sequences apparently arise by repeated duplication of certain base sequences and then gradually diverge from one another by base substitutions and other mutations. It is not yet known whether multiple-copy sequences have a regulatory role controlling structural gene activity, as has been proposed, or are genomic "baggage" possibly utilized occasionally in evolution when a mutation in a sequence just happens to code for a potentially useful gene product.

RECOMBINATION

The Role of Sex in Evolution

Sexual reproduction, by recombining genes from different individuals, increases the amount of genetic variation exposed to agencies of natural selection. Consider, for example, a population in which, over the course of time, a mutation occurs in each of 100 different genes. If the population reproduces only asexually there will be 101 different genotypes, namely, the original type plus the 100 different mutants. In a sexually reproducing diploid population, on the other hand, each mutant gene g and its wild-type allele G can be combined in the three (heterozygous and homozygous) combinations gg, Gg, and GG. Each of these

---

GACUAGUACUGGGUUGGAGGAUUACCUGAGUGGGAACCCCGACGUAGUGU

GAGUAGUACUAGGAUGG–G––UGACCUCC–UGGGAA–GUCCUCGU–GUUGCAUCCUC

GAGUAGUACUAGGAUGG–G––UGACCUCC–UGGGAA–GUCCUCGU–GUUGCAUCCU

GAGUAGUACUAGGAUGG–G––UGACCCCC–UGGGAA–GUCCUCGU–GUUGCAUCCU

GAGUAGUACUAGGAUGG–G––UGACCCCC–UGGGAA–GUCCUCGU–GUUGCA–CCU

GAGUAGUACUAGGAUGG–G––UGACCUCC–UGGGAA–GUCCUCGU–GUUGCA–CCUUU

GAGUAGUACUAGGAUGG–G––UGACCUCC–UGGGAA–GUCCUUGU–GUUGCA–CCUU

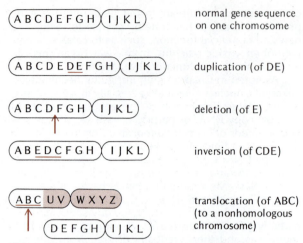

Figure 20-8. Types of chromosomal mutations. The letters designate successive genetic loci along the chromosome, and the constriction represents its centromere. Note that a translocation can move genes to another location on the same chromosome, as well as to a different chromosome as shown here.

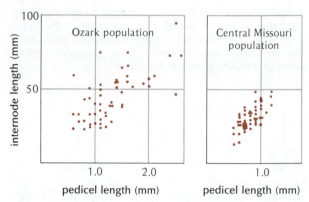

Figure 20-9. Example of variation pattern in plant populations. The mean length of the lowest eight internodes of the inflorescence of different individuals of two populations of *Camassia scilloides,* a spring wildflower in Missouri, is plotted against the length of the individual stalk (pedicel) of the lowest flower in the inflorescence of the same individuals (these measurements reflect the degree of elongation and of lateral expansion, respectively, of the inflorescence). Note that these characters vary continuously in each population although the range and average values differ between the populations. After R. O. Erickson, *Annals of the Missouri Botanical Garden,* Vol. 28, 293–298, 1941.

combinations can in turn be combined with any one of the similar three combinations that will occur at all of the other 99 loci carrying a mutation, so the total number of potential genotypes that could be formed is 3^{100}. This is an astronomically large number, amounting to the figure 51,537,754 followed by 40 zeros. Obviously, in any real population only a small fraction of these genotypes would actually occur. But since estimates indicate that the total number of genetic loci in a higher organism is in the tens of thousands, if different alleles occur at even a small fraction of a species' genetic loci, probably every individual in the population (except for identical twins and asexually produced offspring) will have a unique genotype and will offer a genetically unique opportunity for natural selection.

Because most characteristics of structure and function in higher organisms are polygenic (quantitatively inherited) traits (see Chapter 18), genetic recombination in a sexually reproducing population tends to create a virtually continuous phenotypic variation among its member individuals rather than an assortment of markedly different phenotypes as in the classic Mendelian type of cross. This kind of variation, for example, variation in leaf size and shape, height of plant, and so on (Figure 20-9), is what agencies of natural selection normally work on. Exceptionally, discrete phenotypes such as different flower forms (Figure 16-23) or colors, determined by simple allelic differences among individuals in a population, are encountered. Variation of this type is called **polymorphism.**

The Breeding System

The way in which different alleles become distributed through a population, and the chances

of obtaining any particular gene combination, depend on the breeding system, that is, the pattern of genetic interchange between individuals during reproduction. In an **inbreeding** population, individuals reproduce mainly by self-fertilization. An **outbreeding** or **outcrossing** population reproduces mainly by cross-fertilization (cross-pollination, in seed plants). Genes are exchanged extensively among members of an outcrossing population, resulting in the greatest chance of obtaining new and unusual combinations of rare genes that appear by mutation or by introduction from another population.

Although recombination in an outcrossing species can create special gene combinations of adaptive value, it also tends to break up these combinations during sexual reproduction of individuals possessing them. Many plant species possess a means of asexual or vegetative reproduction (Chapter 15), which enables an unusually well-adapted genotype to multiply and spread, greatly increasing the abundance of its genes in the population.

In annual plants, which do not reproduce vegetatively and have but one season in which to leave successful offspring, there is a premium on transmitting advantageous gene combinations with high frequency through the individual's seeds. One solution to this problem, found in some outcrossing annual species, is a greatly decreased chromosome number, which maximizes the genetic linkage tending to hold particular gene combinations together through sexual reproduction (some of these plants have as few as two chromosomes in the haploid set). Much more commonly, annual species tend to specialize for self-pollination (inbreeding). Compared with outcrossing, self-pollination greatly increases the chances that a parental gene combination will reappear in its progeny. As noted in Chapter 18, inbreeding during a number of generations renders most loci homozygous, ensuring the transfer of a parent's gene combinations to most of its progeny. Occasional accidental crossings between individuals of different inbred lines can yield bursts of segregation and recombination, providing a variety of new genotypes that are subject to natural selection. Thus, genetic variation and evolution are not necessarily static in a normally inbred species.

NATURAL SELECTION

Genes that increase the chances that individuals bearing them will reach reproductive age or that increase the average number of successful offspring produced by an individual will be transmitted to succeeding generations in greater numbers than other genes and thus must increase in relative abundance, or **gene frequency,** in the population in successive generations. This is the meaning of natural selection, a more subtle principle than just "survival of the fittest." What is called the selective **fitness** of an individual is a measure of its reproductive performance relative to that of others, including both survival and a surviving individual's **fecundity** or quantitative production of offspring. An individual's fitness, in the evolutionary sense, increases in proportion to that individual's contribution to the gene pool of the next generation, as simplistically represented in Figure 20-10.

Darwin was much impressed by how selection by humans had created extremely different

varieties or breeds of domesticated plants and animals out of ancestral wild species (Figure 20-11). This evidence convinced him that natural selection of genes for adaptive characteristics should lead to new and different morphological and physiological types.

The contemporary occurrence of natural selection is evident, for example, in plant species that have colonized land adjacent to mines and metal refineries where, due to pollution, the soil has acquired a toxic load of heavy metals such as lead or copper. Tests show that over a relatively short period of time the plant populations growing in these situations have been selected for an ability to tolerate levels of heavy metals that are poisonous to plants of ordinary populations of the same species (Figure 20-12).

The nature and strength of environmental factors tending to select for or against various traits is called the **selection pressure** of the particular environment. Different physical and biotic environments exert selection pressures that differ both in intensity and direction, as evident in metal-polluted and adjacent normal soil. The fitness of any genotype is meaningful only in relation to the selection pressure of its particular environment. A genotype of superior fitness in one kind of habitat may be of substandard fitness in another; in other words, in another environment it will be selected *against.* Heavy metal tolerance, for example, though obviously advantageous in metal-polluted habitats, evidently has a lower fitness than the normal (intolerant) phenotype on ordinary soils; populations of the same plant species as that in Figure 20-12, growing on normal soils adjacent to metal-polluted sites, remain metal-intolerant despite the nearby presence of metal-tolerant geno-

Figure 20-10. Illustrations of quantitative selective fitness. In (A), all offspring have an equal chance (about 30%) of succeeding, but one genotype (color) produces more offspring than do others (black). In (B), all genotypes produce the same number of offspring but a greater proportion of those of one genotype (color) than of others (black) become successfully established in the next generation. In either case the relative abundance of the "color" gene must increase in the population in successive generations, that is, it confers greater fitness than the alternative genes. Actual cases are usually a combination of both effects, since more vigorous offspring are more likely to be successful, to grow larger and in turn produce more offspring. (Cases A and B correspond roughly to the extremes of r-selection and K-selection, respectively. For simplicity, the distinction between heterozygous and homozygous genotypes in the population is omitted in the diagrams.)

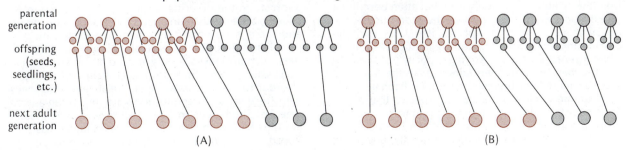

parental generation

offspring (seeds, seedlings, etc.)

next adult generation

(A)   (B)

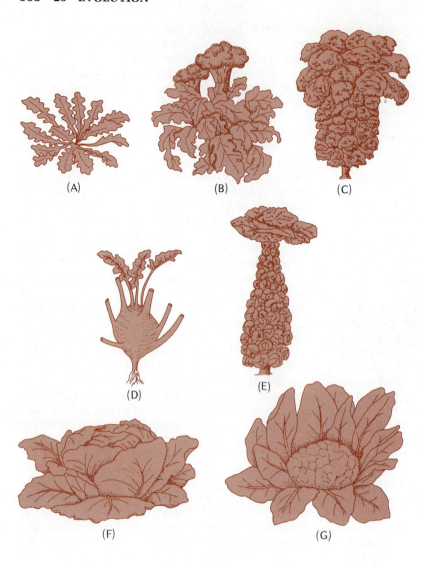

Figure 20-11. The subspecies or varieties of *Brassica oleracea*, the cabbage group of vegetables, illustrating the degree of morphological change that selection can bring about. (A) Wild cabbage, (B) broccoli, (C) curly kale, (D) kohlrabi, (E) Brussells sprouts, (F) heading cabbage, (G) cauliflower. Most of these varieties evolved from wild cabbage during historic time as a result of selection by humans.

(A)   (B)   (C)   (D)   (E)   (F)   (G)

types spreading their pollen and their seed into the adjacent unpolluted pasture land.

r- Versus K-Selection

Depending upon the kind of habitat and the competitive strategy that a species adopts, the survival and fecundity aspects of evolutionary fitness can have quite different relative importance. Selection for maximum reproductive rate is called **r-selection,** while that for maximum survival ability is called **K-selection.** Derived from the symbols in a widely used equation describing population growth, r represents the reproductive potential or fecundity of the species, its relative multiplication rate when unrestrained by limited resources, while K represents the "carrying capacity" of the habitat, the maximum number of individuals that can exist there as limited by the resources of the habitat.

Extreme r-selection is seen in species that act as opportunists, colonizing temporarily unoccupied territory (Chapter 19). An opportunist can be successful only if it can expand its population rapidly, beginning with one or a few immigrants and taking over the available terrain and resources before others do. Opportunists therefore are under selection pressure for a maximum rate of reproduction, producing uniformly well-adapted, uniformly successful offspring. This type of selection pressure favors an annual habit, with production of a maximum number of small seeds, and this in turn favors self-pollination and inbreeding for reasons given in the previous section. These adaptations appear in many weeds, which are typical opportunists or r-selected species.

In a fully developed canopy of vegetation such as a forest, on the other hand, all available space and resources are already being used by existing individuals, and there is no opportunity for rapid expansion of a population. In order to leave any progeny at all under these circumstances a new individual must compete successfully with others for scarce light and nutrients in order to become established; relatively large seeds with substantial storage reserves are thus often advantageous. The longer an individual survives, the greater are the chances that the seeds or seedlings it has produced will be on hand when the rare opening permit-

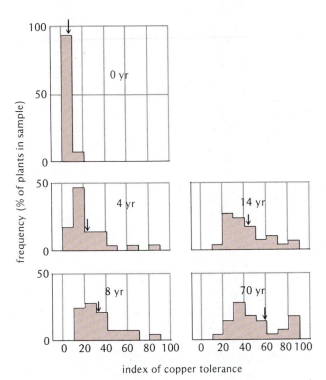

Figure 20-12. Evidence for progressive natural selection of genes for copper tolerance in the populations of the grass *Agrostis stolonifera* occurring in a copper-smelting area in Lancashire, England, the soils of which have been contaminated with 10–20 parts per million of available copper for known lengths of time as indicated (0 years = populations from nearby uncontaminated soils). "Index of copper tolerance" is relative mean growth of roots of the given grass clone when grown in a nutrient solution containing 0.5 parts per million of copper, as a percentage of root growth in copper-free solution (100 = no copper inhibition; 0 = complete inhibition); 30 clones from each population were tested. Arrows show the mean copper tolerance index for each population. Roots of plants from uncontaminated soils hardly grew at all in the copper tolerance test, while roots from copper-stressed populations of increasing ages grew progressively better. The differences have been shown to be polygenic. From L. Wu and A. D. Bradshaw, *Nature*, Vol. 238, 167–169, 1972.

ting progeny to develop to maturity appears. K-selected species therefore tend to be long-lived. Similarly, although there may be much unoccupied ground in a harsh environment such as a desert, alpine mountain-top, or arctic tundra, rapid expansion of a population is hardly possible. Fitness here is mainly a matter of being able to survive the severe conditions. These examples illustrate extremes of K-selection. There are many species between these extremes of selective influences, but the extremes show that the behavior of the organism itself affects the nature of the selective pressures that determine its fitness.

SPECIES FORMATION

In a species inhabiting a wide geographical range, different selection pressures in different areas can result in physiologically and morphologically distinguishable populations called **ecotypes** (Chapter 19), **races,** or **subspecies** (Figure 20-13). Individuals from different races or subspecies are usually capable of interbreeding to produce healthy offspring, so the races will tend to exchange genes and intergrade where they overlap geographically. However, a gap in distribution may eventually separate the races, so that they do not exchange genes at all. This is called **spatial isolation** (or geographical isolation). When spatially isolated, genetic differences may evolve that reduce or even completely prevent the possibility of successful mating between individuals of the different races. The different populations have then become **reproductively isolated,** and can move into and inhabit common territory without exchanging genes with one another or losing their respective recognizable characteristics. The different populations from this point onward comprise separate genetic entities that thereafter normally evolve independently of one another and can be identified as distinct **species.** The **biological definition of a species,** therefore, is "a group of actually or potentially interbreeding populations that are reproductively isolated from other such groups."

The gradual evolution of a new species by the process explained above has never been witnessed directly and evidently usually occurs over a time scale that is long compared with the few thousand years of recorded human history but very short compared with the geological time scale of millions of years.

The "biological" definition of a species as a reproductively isolated system of populations is not rigidly applicable to every situation. In various plant groups strikingly and consistently different plant forms, which are traditionally recognized as different species, can be crossed to give fertile hybrids—for example, the sycamore trees of the eastern United States (*Platanus occidentalis*) and those of the eastern Mediterranean (*Platanus orientalis*) (Figure 20-14). The continued existence of separate but interfertile species may be due in different instances to geographical or ecological isolation (different habitats), or to reproductive behavior such as flowering at different times of year, using different pollinators, or adopting strict self-pollination.

Mechanisms of Reproductive Isolation

Various kinds of mechanisms have arisen to reproductively isolate different species. Some genetic incompatibility mechanisms operating between species are analogous to self-incompatibility mechanisms (Chapter 16): the pollen does not

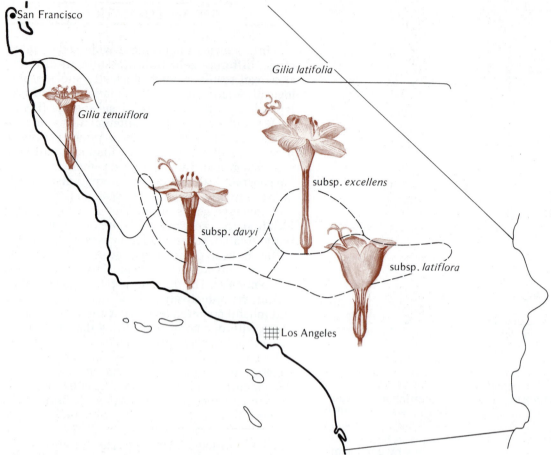

Figure 20-13. Ranges and distinguishing flower features of geographical races (subspecies) of a California wildflower, *Gilia latifolia,* and of a closely related species, *G. tenuiflora,* which probably was formerly a subspecies of *G. latifolia* but has become a distinct species as shown by its overlap with *G. latifolia* in one area without intergradation. From A. and V. Grant, *El Aliso,* Vol. 3, 203, 1956. For simplicity, three other subspecies of *G. latifolia* with narrower distributions have been omitted.

germinate or the pollen tube does not grow in the style of the foreign species, so fertilization never occurs. Or fertilization may occur, but the zygote fails to complete embryonic development owing to either specific genetic incompatibility factors or a general imbalance between the genomes of the two parents.

In many cases the genomes of related species are compatible and yield a viable or vigorous hybrid that, however, is sterile. Again, this may be due to specific genetic factors but is often caused by chromosomal differences. If the chromosomes of the two parental species differ by inversions or translocations or are simply different enough not to pair together during prophase I of meiosis, an unequal or haphazard separation of chromosomes takes place during meiosis. As a result, few if any gametes receive a complete haploid set of chromosomes, and most of the gametes either die or cannot yield viable offspring.

Although hybrid sterility blocks gene exchange between the respective species, it does not always entirely block evolutionarily significant recombination between their genes. In fact, it can create a curious route for the sudden appearance of a new species via a chromosomal mutation that results in polyploidy, as described below.

New species can arise in other ways than through the gradual divergence of spatially isolated races by natural selection. For example, if a local population happens to become very small during a particular generation, chromosomal mutations can become established in the genome of that population purely by chance (as explained in Chapter 18). The population thereby becomes reproductively isolated from others of the same species, and if it proves to be well enough adapted, it can subsequently expand and overlap these other populations without interbreeding with them and losing its own identity; it has become a new species. Any genetic characteristics distinguishing members of the small original population from the principal populations of the parental species will be retained as the new species

native to
eastern
United States

native to
eastern
Mediterranean

×

*Platanus
occidentalis*

*Platanus
orientalis*

× *Platanus
acerifolia*
(hybrid)

Figure 20-14. An example of interspecific hybridization. Two species of sycamore (*Platanus*) which occur naturally on different continents can be crossed to give a vigorous hybrid (called × *Platanus acerifolia* to designate it as a hybrid), commonly planted as a street tree under the common name "London plane tree."

multiplies, enabling it to be distinguished morphologically from the parental species. The chances for such a sudden or catastrophic type of species origination are probably increased if the parental species is under severe stress that tends to reduce the size of its populations. The chance combination of a physiologically superior genotype with a chromosomal mutation restricting interbreeding between it and other members of the parental species would then have greatly increased selective fitness.

CHROMOSOMAL EVOLUTION

Since chromosomal mutations are often involved in reproductively isolating different species from one another, a progressive change in karyotype often accompanies the evolution of a diversity of species within a genus (Figure 20-15).

Polyploidy

If a hybrid between two species is sterile it obviously cannot be evolutionarily selected, even if it is especially well adapted to a particular environment. However, through the mutation called chromosome doubling, which multiplies by two the number of chromosomes per cell, a sterile hybrid may be rendered fertile. Possession of multiple (four or more) chromosome sets is called polyploidy (Greek *poly*, many).

Chromosome doubling can take place when chromosome replication preparatory to mitosis or meiosis is followed by failure of division to continue to completion, resulting in the daughter chromosomes all remaining within one nucleus. If the original cell was diploid (2n), the doubled cell possesses 4n chromosomes and is called tetraploid ("four-ploid"). Further doublings can produce cells or individuals that are octoploid (8n), 16-ploid, and so on. Chromosome doubling is a rather common mutation. When polyploidy occurs in a nonhybrid individual, it usually leads to partial or complete sterility, but in an already sterile hybrid between two species the results can be dramatically opposite.

As noted earlier, if the two parental haploid chromosome sets in an interspecific hybrid are functionally too different for normal meiotic chromosome pairing, the chromosomes will separate erratically at meiosis (Figure 20-16), resulting in sterile gametes. If chromosome doubling occurs in this hybrid (producing a tetraploid if both parents were diploids), each chromosome will be represented twice and will therefore have an identical mate to pair with during meiosis (Figure 20-16). The tetraploid will therefore have perfectly regu-

Figure 20-15. Diagrams of haploid chromosome sets (karyotypes) of different species of the herbaceous dandelion-like plant, *Crepis* (sunflower family). Chromosomal evolution is thought to have proceeded from left to right, that is, with reduction of chromosome number and conversion of relatively symmetrical chromosomes (with arms of nearly equal length) into asymmetric ones with a subterminal or terminal centromere. These changes are presumably the result of translocations.

C. kashmirica   *C. sibirica*   *C. conyzaefolia*   *C. capillaris*

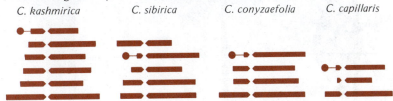

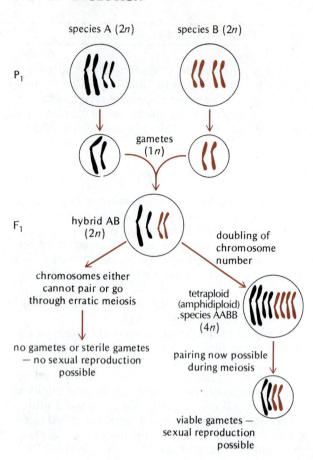

species A (2n)   species B (2n)

P₁

gametes
(1n)

hybrid AB
(2n)

doubling of
chromosome
number

chromosomes either
cannot pair or go
through erratic meiosis

tetraploid
(amphidiploid)
species AABB
(4n)

F₁

no gametes or sterile gametes
— no sexual reproduction
possible

pairing now possible
during meiosis

viable gametes —
sexual reproduction
possible

Figure 20-16. Formation of a fertile allopolyploid (amphidiploid) new species by chromosome doubling of cells of a sterile F₁ hybrid between two species.

lar meiosis, and its gametes will be fertile. It will be unable to exchange genes readily with either of the parent species because such hybrids would have unpaired or mispaired chromosomes during meiosis and thus a low fertility. The tetraploid hybrid and its offspring thus constitute a reproductively isolated population, a distinct new species. Such a hybrid polyploid species is called an **allopolyploid,** or amphidiploid.

New allopolyploid species have appeared by chromosome doubling following artificial hybridization between different species (Figure 20-17). Many plant genera contain species with multiple chromosome numbers (Figure 20-18), indicating that polyploidy has developed many times in the course of plant evolution. From data collected on chromosome numbers it is estimated that between a quarter and a third of all higher plant species are allopolyploids. This has been proved in the case of certain wild species such as the common weed "hemp-nettle" (*Galeopsis tetrahit*), which has 32 chromosomes, because it was possible to create an essentially identical plant by crossing

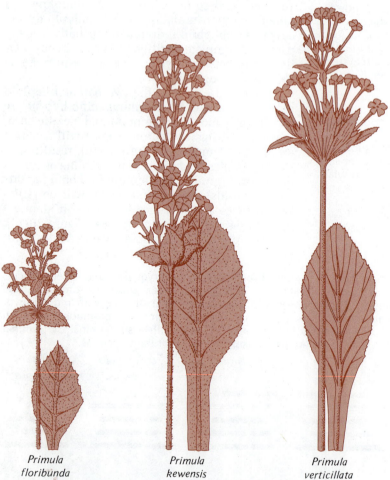

*Primula
floribunda*

*Primula
kewensis*

*Primula
verticillata*

Figure 20-17. An allopolyploid species, *Primula kewensis,* that arose in historic time from a cross between *P. floribunda* and *P. verticillata.*

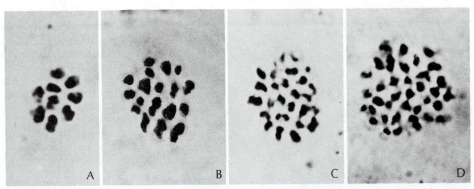

Figure 20-18. Chromosomes of a polyploid evolutionary complex, illustrated by Australian species of saltbush (genus *Atriplex*). Pictures show photomicrographs of metaphase I of meiosis in pollen mother cells (microsporocytes), the usual means for determining a plant's chromosome number: each darkly stained body is a chromosome *pair* (see Chapter 18); the total number of pairs equals the plant's basic (*n*) chromosome number. (A) *Atriplex angulata*, a diploid species, *n* = 9; (B) *A. vesicaria*, a tetraploid species, *n* = 18; (C) *A. cinerea*, a hexaploid, *n* = 27; (D) *A. nummularia*, octoploid, *n* = 36. A hexaploid can arise by chromosome doubling of a hybrid between a diploid and a tetraploid species; an octoploid can arise similarly from a cross between two tetraploids or between a diploid and a hexaploid.

two wild diploid species of *Galeopsis* (with 16 chromosomes each) and picking out a hybrid whose chromosomes had spontaneously doubled.

Contemporary records show that some wild allopolyploid plant species have arisen and spread within historic times. Cord or marsh grass (*Spartina townsendii*), which is now abundant along the coast of the English Channel, is one example. This species was first noticed in southern England about 1870. It apparently arose as a result of hybridization between *Spartina maritima* (2*n* = 60), a native European species of cord grass, and *Spartina alternifolia* (2*n* = 62), which had been accidentally introduced from North America. These two species are capable of crossing to give a sterile F₁ hybrid. The allopolyploid (2*n* = 122) proved to be very vigorous, and it spread rapidly, covering thousands of acres along the south coast of England by 1906. It also crossed the English Channel and now grows extensively along the coast of France.

Allopolyploidy has also played an important role in the evolution of cultivated crop plants, for example, wheat (Chapter 32). Among horticultural plant groups there are also important examples of **autopolyploids,** formed by chromosome doubling in a nonhybrid individual, and **triploids** (three chromosome sets), formed by a combination of an ordinary haploid gamete with a diploid gamete resulting from failure of the first division of meiosis. These aberrations may be unusually vigorous vegetatively and horticulturally useful but they are usually sterile; in such a case they can be propagated only vegetatively.

COEVOLUTION

Interacting organisms tend to evolve behavioral and structural features that help each to exploit the other's capabilities; in the course of evolution they often become progressively more dependent on one another. Coevolution is defined as evolutionary change in two or more species resulting from their action as selective forces on one another. Symbiotic associations, in which a particular plant lives in close nutritional association with a different kind of plant or with a microorganism or animal, are consequences of coevolution and are considered in Chapter 25. Coevolution also occurs between independent, free-living organisms, as seen in the striking structural and behavioral relationships between flowers and their pollinators (Chapter 16) and in some of the mechanisms plants have evolved to employ animals for seed dispersal (Chapter 17).

Coevolution can involve a highly species-specific relationship, for example, the pollination of the fig by a particular kind of wasp that inhabits the fig flower, or of yucca flowers by the Pronuba moth (Figure 20-19). In these cases both the plant and the animal partner have become strictly dependent on each other for their reproduction. More often, a coevolutionary relationship develops between groups or kinds of plants and animals, for example, between the hummingbird family and the numerous species or genera of plants that have specialized for pollination by these birds (Chapter 16). An example of a very broad coevolutionary relationship is that between grazing mammals and range grasses. Under the selection pressure created by the hoofed herbivorous mammals that first appeared in the post-dinosaur period of the earth's history (Chapter 21), grasses evolved as a plant form that tolerates grazing better than other types of vegetation. Due to this tolerance, grasslands increased under the activities of grazing animals, and this spread not only encouraged the multiplication and diversification of grazing mammals but also selected for adaptations to grass by animals, such as a special tooth struc-

Figure 20-19. Flowers of "Spanish bayonet" (*Yucca*), opened to expose within them the Pronuba moths specialized for pollinating these flowers. A female moth collects pollen from the anthers (A) of one flower, flies to another flower, and forces this pollen into the stigma (S), thereby pollinating the flower. Then it lays its eggs in the ovary (O) of that flower. The moth's larvae feed on some of the developing seeds, but allow the rest to mature. Both moth and plant depend completely on one another for their reproduction.

ture and growth pattern that compensates for the abrasive action on the grazers' teeth of the silica granules deposited by grasses in their leaf cells.

As the foregoing examples illustrate, positive coevolutionary relationships between plants and animals usually involve the plant's providing food for the animal or its offspring, while the animal furnishes some kind of help in the plant's reproduction or dispersal. Other interesting cases involve defense against predators. Some plants develop sugar-secreting glands called **extrafloral nectaries** on their stems or leaves. These attract ants and wasps, which collect the sugar as food, often take up residence on or near the plant, and aggressively attack and repel plant-eating insects such as beetles or cutworms that trespass on the defended plant. Certain trees and shrubs called "ant plants" develop special stem (Figure 20-20) or leaf (Figure 3-23B) struc-

tures inhabited by ant colonies that defend the host plant. Although as regards animal behavior these insects are defending a territory and a food source, from the point of view of the plant the insects are in effect being recruited and paid (with food and shelter) to protect the plant from predators. Plants of this type have evidently become dependent on their defenders because in their absence the plants are decimated by predators.

Plant diseases and the interaction between plants and herbivorous insects are examples of negative coevolutionary relationships. As noted in previous chapters, many plants have evolved chemical defenses against insect predation by producing special chemical compounds called secondary products that deter or repel insects. These defenses have forced herbivorous insects in turn to evolve highly specific tolerances to, or even preferences for, certain secondary products, and some insects are restricted to food plants containing particular products. For example, the cabbage group of vegetables produces mustard oils that deter most insect pests, but the cabbage butterfly has evolved an attraction for these products, and its larvae feed exclusively on vegetables of the cabbage group.

The evolution of chemical defenses by plants against insect pests and the tolerance or preference of insects for them has been called an "evolutionary arms race." Certain insects have turned the race to their advantage by using disagreeable-tasting secondary plant products to make themselves unpalatable to other predators, almost reversing the predator protection relationship described previously. The Monarch butterfly, for example, lays its eggs on milkweed (*Asclepias*) plants, which produce latex containing distasteful or poisonous cardiac glycosides. The larvae feed on the plant and accumulate the secondary product, making the adult butterflies very distasteful to, and consequently avoided by, birds that would otherwise eat them.

Coevolution Under Domestication

When prehistoric humans undertook agriculture and plant domestication they started coevolutionary relationships that led, even by ancient times, to the almost total dependence of civilized humanity on domesticated crop plants for its food and well-being. Crop plants, as noted previously, have undergone substantial evolution under human selection pressure since prehistoric times. New species have arisen or been created; most kinds of mutations and natural selection phenomena that are recognized in the evolution of wild plants have contributed to the diversification and improvement of cultivated species, sometimes to an exaggerated degree. During some 5000 years of evolution under domestication in North and South America, for example, corn has become so different from related wild grasses that experts cannot agree on any existing wild species as its progenitor.

Human activity in disturbing, clearing, and cultivating land has also helped various "opportunist" species succeed by becoming weeds. Weed specialization took place during prehistoric times,

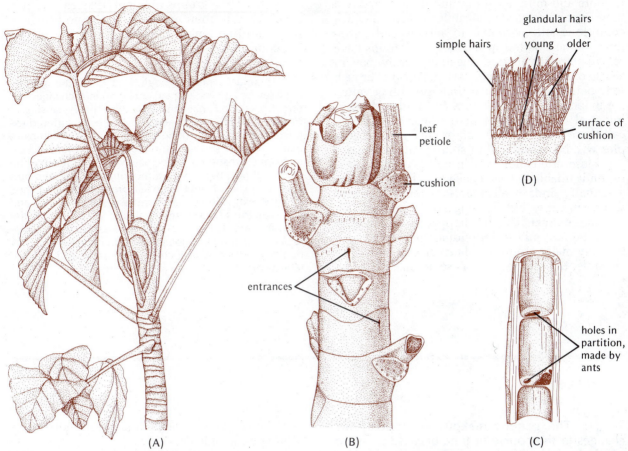

Figure 20-20. *Cecropia,* a Central and South American tropical "ant tree." (A) shows one of the coarse, jointed shoots of the tree. (B) shows a larger view of the subapical part of the stem with entrances for ants into its hollow internodes (section in C), and the specialized cushions, below each leaf base, bearing glands called "Müllerian bodies" (D) which provide food for the ants inhabiting the internodes.

as shown by the widespread occurrence of common weed seeds in archeological deposits.

Weeds and, perhaps even more dramatically, cultivated plant species have come to depend largely or entirely on humans for their survival and growth. Most cultivated plants cannot compete with wild vegetation without human help in clearing the land, planting and cultivating the crop, protecting it from predators and diseases, and harvesting and replanting it. Crops growing in many semi-arid areas could not possibly exist there without irrigation. Certain cultivated plants, such as corn and some seedless fruits (bananas, seedless grapes), cannot propagate themselves, even on land free of competition, without human help.

Although from the point of view of natural selection, specialized domesticated plants might be regarded as genotypes of poor fitness, disadvantaged in competitive and reproductive abilities, these genotypes obviously have enjoyed spectacular success in the modern world, multiplying enormously and taking over a substantial part of the earth's entire land area. The actual high degree of fitness of these genotypes is

due to their ability to produce crop products wanted by humans. From the coevolutionary standpoint, desirable crop plant genotypes have in effect recruited humans to work for their growth, protection, and multiplication at the expense of other genotypes, both domesticated and wild. Such survival and multiplication are the essentials of evolutionary fitness, as emphasized previously. Furthermore, by allowing the human species to multiply to the point where it depends totally on them, these domesticated plant species have ensured their survival and success for as long as humans remain important on earth and continue to depend on them.

EXTINCTION

Under the changing selection pressures of a changing physical or biological environment, mutation and recombination either are or are not able to provide new genotypes of sufficient fitness to permit a species to perpetuate itself. If a species fails to perpetuate itself successfully, it declines in more and more localities and is restricted to a narrower and narrower geographical range in which conditions still permit it to survive and reproduce. Some existing species or plant groups have remained relatively unchanged in morphology, in a state of apparent evolutionary stagnation, through long periods of geological time, providing valuable evidence of the structure and probable physiological capabilities of plants of past ages. We call these **relict** species and groups; many examples will be considered in subsequent chapters.

More often, however, a species whose evolution is unable to keep pace with changing times eventually finds itself unable to reproduce anywhere and dies out, leaving no descendants—it becomes **extinct.** This has happened over geological time not only to individual species but also to many major groups of plants and animals—for example, the dinosaurs. It is sometimes said that such species or groups died out because they had specialized in their evolution to the point where they could no longer adjust adequately to changing conditions.

Many species of both plants and animals are unable to adjust rapidly enough, if at all, to the increasingly severe and widespread environmental changes and direct predation that the human species is causing in nature. A number of species have already suffered extinction at the hands of humans, and, as is well known, many more are seriously threatened. Environmental catastrophes resulting from major disturbances of the earth's atmosphere (e.g., a monumental volcanic outburst or a gigantic meteorite fall) are sometimes blamed for mass extinctions such as those of the dinosaurs and many of the plants comprising the vegetation of the dinosaur era (Chapter 21). In the perspective of geological time the enormous effects on nature exerted by humans today may well also appear as an environmental catastrophe.

SUMMARY

(1) The genetic makeup of populations changes in the course of time because different genotypes have different degrees of reproductive success in their environment. Those individuals that reproduce more extensively inevitably transmit their genes to the next generation in greater numbers than do individuals that reproduce less or not at all. This constitutes natural selection, first visualized clearly by Charles Darwin and Alfred Russell Wallace.

(2) Gene mutation occurs spontaneously at a low rate but over a period of time creates new alleles at virtually all genetic loci. Mutant alleles yield either an inactive or a missing gene product (defective mutants), or a modified but still functional gene product. A mutant protein often differs from the normal one by amino acid substitution at one point in its polypeptide chain, due to a base substitution in the DNA, which alters one codon. Deletions or insertions of bases in the DNA sequence usually alter the gene product more severely. In time, these kinds of mutations accumulate in genes for specific gene products, so that the same enzymes or RNAs in different species often differ in structure (molecular evolution).

(3) Chromosomal mutations, due to breakage of chromosomes and rejoining of the fragments in different configurations, alter the genetic map and disturb meiosis in ways that tend to block gene recombination between mutant and wild-type individuals.

(4) Recombination, during sexual reproduction, brings together mutations or alleles that occur in different individuals, greatly increasing the diversity of genotypes in a population. The breeding system (the amount of outcrossing or inbreeding) determines the extent and pattern of gene recombination in a population.

(5) Selective or evolutionary fitness denotes an individual's capacity to produce successful progeny, thus transmitting its genes to the next generation in preference to genes of other individuals. Fitness includes both the ability to survive to and beyond reproductive age, and the quantitative productivity of offspring (fecundity). The nature and intensity of natural selection for or against different kinds of genotypes or phenotypes is called the selection pressure of the given environment.

(6) New species can be formed by the selection of genetically different races in response to different selection pressures in different parts of a species' range. Spatial isolation of the races may follow, and chromosomal or gene mutations may appear in the separate races, resulting in reproductive isolation (inability to exchange genes) of the races, which thereby become separate species.

(7) New species can also be formed by crossing two species to produce a sterile hybrid that undergoes chromosome doubling, yielding a fertile tetraploid (amphidiploid) that is distinct from either parental species. (The sudden appearance of a distinct new species from a previously existing species by other means is also possible.)

(8) Coevolution is evolutionary change in two or more species that results from their action as selective forces on one another, and is exemplified by the relationship between particular flowering plants and particular animals that serve as pollinators. Domestication of plants by humans has involved the coevolution of domesticated plants and at least the cultural coevolution of the human species to the point where humans and domesticated plants are mutually dependent on one another.

(9) If the selective fitness of an entire species decreases sufficiently, due to the environment changing more rapidly than the species can adapt by evolution and/or migration, it will leave fewer and fewer offspring each generation and eventually become extinct. Extinction can occur in a short time due to an environmental catastrophe, including perturbation by humans.

Chapter 21

PLANTS THROUGH TIME: ORIGIN, EVOLUTIONARY HISTORY, AND CLASSIFICATION

Well before the time of Darwin and Wallace in the mid-nineteenth century, sedimentary rocks were known to contain fossil remains of animals and plants totally unlike any known on earth. This knowledge led to a suspicion that life on earth had not always existed in its present form, as the then-prevailing doctrine of special creation maintained. Darwin's theory of evolution made these fossil remains understandable. Ever since Darwin's theory appeared, morphological research into both fossil and modern plants and animals has flourished, with the goal of working out the evolutionary history, or **phylogeny,** of living organisms, and classifying them accordingly. In recent decades both fossil and chemical evidence has been found that extends our knowledge about primitive living organisms even further back in time, and has led to concepts of the probable origin and early evolution of living cells. This chapter considers the origin of life and the early evolution of photosynthetic cells, provides a brief outline of the subsequent evolutionary history of multicellular plants, and describes the evolutionary or phylogenetic classification of plants as background for the following chapters on major plant groups.

Figure 21-A. Fossil leaves, stems, and shoots from the Carboniferous period or coal ages, the remains of fern-like and horsetail-like plants that do not exist on earth today. Study of fossil remains provides one of the important lines of evidence for progressive evolution of plant and animal life on earth.

GEOLOGICAL EVIDENCE

How Fossils Are Formed

Plant and animal remains on the earth's surface are subject almost universally to microbial decay and to destruction by weathering, eventually leaving no trace of the former living organism. Only under special conditions existing in localities that are relatively rare on the modern earth do organic remains last long enough to leave their mark in the geological record. These localities are usually places in which sediment is being deposited and where lack of oxygen retards microbial decay. Lakes and river deltas, where silt and sand transported by rivers are dropped, often under anaerobic conditions, and swamps, where organic remains pile up faster than they are destroyed by decay, are important potential sites of fossil formation.

Plant remains deposited in sediments can give rise either to surface impressions that display only superficial features of the organism (Figure 21-1A) or to fossils with more thickness and substance that sometimes preserve some anatomical structure (Figure 21-1B). In the type of fossil called a **petrifaction,** such as petrified wood (Figure 21-1C), mineral matter has infiltrated the plant specimen cell by cell, often preserving exquisite anatomical details that may be detected by sectioning and other techniques used in microscopic examination of the rock (Figure 21-1D).

379

Figure 21-1.   Different types of fossils. (A) Impression (part of a fossil seed fern leaf), showing the matching halves of a rock split along the plane of the impression. (B) Cast (a fossil seed), exposed by chipping away the enclosing rock matrix. (C) Petrified trunk of the extinct conifer *Araucarioxylon*, of Mesozoic age, exposed by weathering of the soft sediments in which it was embedded ("petrified forest" near Santa Cruz, Patagonia). (D) Section of a petrifaction (fossil wood), photomicrograph showing cellular details preserved in the fossil.

The Geological Time Scale

The sedimentary rock record indicates that the earth's history has consisted of lengthy intervals of relative stability, during which mountains eroded extensively and the eroded material was deposited as sediment; these stable periods were interrupted by periods of vigorous mountain building called geological revolutions. In reconstructing the earth's geological history geologists have named the major intervals of sediment deposition **periods** (see Table 21-1). Each period differed in important ways from those that preceded and followed it in the kinds of animal and plant life that were prevalent. The periods are grouped into four large **eras,** which are distinguished by especially major differences in the fossil animals and plants. The periods of the most recent era, the Cenozoic, are subdivided into named **epochs** (Table 21-1) that reflect the relatively detailed evidence available from more recently deposited sed-

iments. The older geological periods are now also often subdivided into named episodes (not listed in Table 21-1).

The historical time assigned to the eras and periods (see Table 21-1) is based on estimates of the age of sedimentary rocks that are arrived at by several methods. The most definitive method, radioisotope dating, makes use of the known rate of radioactive "decay" or disappearance of radioactive atoms deposited in the sediments when they were laid down. Measuring the relative amounts of different kinds of atoms occurring in a rock, usually in trace amounts, can determine the extent to which certain radioactive atoms have disappeared, and thus the age of the sediments composing the rock.

The fossil life found in sedimentary rocks begins with very simple forms in the pre-Cambrian era and develops into more and more complex animals and plants from the Cambrian period onward.

TABLE 21-1. Geological Time Scale

Geological sequences are conventionally shown with the youngest or most recent at the top, because this is how sedimentary rock strata are deposited.

ERA	Period	Epoch	Duration in millions of years (approx.)	Millions of years ago (approx.)	Major biological milestones
CENOZOIC	Quaternary	Recent	Approx. last 5,000 years		Civilization
		Pleistocene	2.5	2.5	Glaciation, rise of humans
	Tertiary	Pliocene	4.5	7	Diversification of mammals and angiosperms
		Miocene	19	26	
		Oligocene	12	38	
		Eocene	16	54	
		Paleocene	11	65	
MESOZOIC	Cretaceous		71	136	Rise of angiosperms Eventual extinction of dinosaurs
	Jurassic		54	190	World-wide gymnosperm forests; heyday of dinosaurs; first birds
	Triassic		35	225	Spread of gymnosperm forests and dinosaurs; first mammals
PALEOZOIC	Permian		55	280	Rise of conifers, reptiles Extinction of swamp forest trees
	Carboniferous		65	345	"Coal age" swamp forests; first reptiles, insects; early seed plants
	Devonian		50	395	Primitive land plants & amphibians
	Silurian		35	430	Earliest land plants
	Ordovician		70	500	Early fishes Abundant, diverse invertebrates
	Cambrian		70	570	Macroscopic algae (seaweeds) Hard-shelled invertebrates
PRE-CAMBRIAN	Ediacarian		~130	~700	First macroscopic soft-bodied invertebrate animals
	(Proterozoic)			≧3400	Unicellular prokaryotes, plants, animals
	(Archean)			≧4600	First unicellular life Pre-biotic "primordial soup" develops Origin of earth

THE BEGINNINGS OF LIFE

The age of the earth is estimated to be about 4.5 billion years. The oldest known sediments on earth, rocks in Greenland dated as about 3.8 billion years old, contain indications that life was already beginning to develop, if indeed it did not already exist, at that time. These indications include traces of organic compounds, which are typical products of living cells but apparently were first produced by nonbiological processes.

Origin of Life

The atmosphere of the primordial earth, in marked contrast to that of the modern earth, appears to have contained no oxygen. Its major constituents were probably nitrogen, carbon dioxide, carbon monoxide, and water vapor, and it may also have contained at least small amounts of hydrogen, methane (CH_4), and/or ammonia (NH_3). (A widely believed but disputed theory holds that the latter constituents predominated.) Experiments in which mixtures of these components are

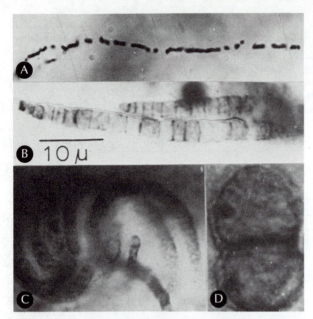

Figure 21-2. Microfossils of primitive stages in the evolution of life, seen in thin sections of pre-Cambrian rocks. (A) Chain of bacterium-like cells from 3.5 billion year-old sediments in northwestern Australia that contain some of the oldest microfossils currently known. (B, C) Filaments of cyanobacteria (blue-greens) from the Bitter Springs formation, central Australia, about 0.9 billion years old. Similar but less well preserved fossils have been found in sediments as old as 2.2 billion years. (B) resembles modern *Oscillatoria*; (C) is similar to modern *Anabaena*, the larger cell in center of picture possibly being a heterocyst (cell specialized for nitrogen fixation, see Chapter 22). (D) Probable eukaryotic cells from the Bitter Springs formation. These two cells seem to be daughters from a recent cell division; each contains a protoplasmic remnant resembling a nucleus.

subjected to forms of energy such as electrical discharges or ultraviolet radiation, which must have been abundantly available on the primordial earth, show that organic compounds such as amino acids and nitrogen bases can be formed from these inorganic components and can then condense into polymers. Such "pre-biotic" synthesis of organic compounds apparently occurred on the primitive earth, giving rise to a random mixture of organic carbon compounds dissolved in surface waters, called the **primordial soup.** In shallow ponds this mixture may have attained a concentration of organic solids as high as 1%, comparable with a weak bouillon. Eventually, living cells evidently arose from the contents of the primordial soup.

Artificially produced random polypeptides ("proteinoids") show in feeble form a number of the catalytic activities of modern enzymes; such products may thus have acted as primitive catalysts for the beginning of biochemical activity. The surfaces of clay particles may also have acted as catalysts or templates for polymer formation in the primordial soup. Polypeptides and certain other polymers can condense, forming micro-

scopic vesicular (bladder-like) structures, and any lipids that appeared would have tended to form membrane-like films on the surfaces of such structures. This suggests how cellular organization could have begun in the primordial soup, yielding organic aggregates called **protocells.**

Any molecules of the nucleic acid type that were accidentally formed in the primordial soup would have tended to multiply because of the inherent tendency of a nucleic acid base sequence to dictate its own duplication by complementary pairing. Some of these primitive self-replicating molecules would have been included accidentally in protocellular structures. The essential step toward the evolution of an autonomous self-multiplying cellular system would have been the appearance, presumably by chance, of a catalyst that tended to polymerize amino acids in some definite relationship to the base sequences of the nucleic acids that were present. This relationship was no doubt rather loose at first, but any improvements that arose in it would have improved the tendency of such a protocell, called a **progenote,** to perpetuate itself and multiply. Presumably this early development led in time to the elaborate, precise system of genetic coding and its translation possessed by modern organisms. The fact that all modern organisms share one genetic code language suggests that they all descended from a single progenote stock.

Before early cells could develop very far there had to be a mechanism for obtaining energy from the components of the primordial soup for use in synthetic activities. Fermentation was probably the original kind of energy metabolism, since no oxygen was available for aerobic respiration. The organisms existing at this stage would have been prokaryotes, and possibly were comparable with some present-day anaerobic bacteria.

Earliest Fossils

Recognizable fossil remains of bacteria-like cells and chains of cells have been found in a northwest Australian geological formation about 3.5 billion years old and in South African rocks that are nearly as old. These fossils (Figure 21-2A) indicate that a considerable diversity of unicellular life already existed on earth at that time, suggesting that the first cells must have arisen well before the earth was a quarter of its present age.

EVOLUTION OF UNICELLULAR LIFE

Origin of Photosynthesis

The growth of primitive organisms eventually depleted the organic compounds in the primordial soup, giving an advantage to any organisms that could produce organic compounds from inorganic materials such as carbon dioxide. The carbon isotopes found in the Australian and South African sediments mentioned above indicate that photosynthesis was already occurring as early as

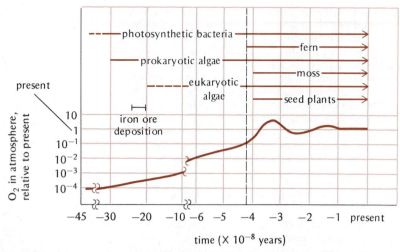

Figure 21-3. Possible history of earth's atmospheric oxygen content, along with biological changes responsible for, or permitted by, atmospheric evolution. Note that O_2 concentration is plotted on a log scale, as a fraction of present ($=1.0$ or 21% by volume O_2) oxygen content. The vertical dashed line indicates the time (early Silurian period) when, according to this interpretation, the air's oxygen concentration reached 10% of the present concentration (2% by volume), permitting an ozone layer to form and shield the land surface from lethal ultraviolet radiation. Note that development of land vegetation apparently led to a rapid increase in atmospheric O_2 content, reflecting the much greater total photosynthetic capacity of land vegetation compared with the oceans.

3.5 billion years ago. Since geological evidence indicates that the atmosphere at that time still contained no oyxgen, this photosynthesis apparently was the kind found in modern photosynthetic bacteria, which do not produce oxygen (Chapter 7).

Oxygen-producing photosynthesis probably arose with the evolution of blue-green algae or cyanobacteria (Figure 21-2B, C), which, like other bacteria, are prokaryotes. Fossils that are unmistakably identifiable as blue-green algae occur in the Gunflint Formation north of Lake Superior, which is 1.9 billion years old, and in South African sediments 2.2 billion years old.

As explained in Chapter 7, bacterial photosynthesis utilizes light energy absorbed by a pigment system (bacteriochlorophyll) to raise electrons to a higher energy level; the electrons are obtained from an externally available souce of reducing power such as hydrogen sulfide (H_2S). This photoprocess gives the electrons enough reducing power to reduce carbon dioxide efficiently, and enables the organism to produce ATP (needed for carbon dioxide reduction) at the expense of light energy (photophosphorylation). Oxygen-producing photosynthesis resulted from the development, by cyanobacteria, of an additional photosystem (photosystem II) which is able to utilize electrons from water (releasing oxygen) and deliver them to the kind of photosystem possessed by photosynthetic bacteria (photosystem I). Since water is ubiquitous, this development freed cyanobacteria from dependence on habitats supplying H_2S or some other usable natural reductant and enabled them to proliferate worldwide, doubtless greatly increasing the earth's photosynthetic productivity compared with that in the preceding era of bacterial photosynthesis.

When cyanobacterial photosynthesis arose, the presence on the earth's surface of materials such as ferrous iron (Fe^{2+}) that are capable of reacting with free oxygen at first prevented the oxygen produced by cyanobacteria from accumulating in the air. Large iron ore deposits, such as that in northern Minnesota, were laid down all over the earth during the period between 2.2 and 1.9 billion years ago and represent "fossil oxygen" from the oxidation of Fe^{2+} to insoluble ferric oxides and hydroxides by the oxygen then being released by photosynthesis. Once most of the available iron was oxidized, considerable amounts of oxygen could accumulate in the air (Figure 21-3). By about 1.8–1.9 billion years ago its concentration apparently reached 1% of the present value, a concentration high enough to support respiration in single-celled organisms.

For the photosynthetic and other anaerobic bacteria present at the time, the appearance of oxygen in the air was a catastrophe, because anaerobic organisms do not have biochemical equipment for coping with free oxygen and are poisoned by it. A desperate process of evolution for tolerance to oxygen must have then occurred, and organisms that were unsuccessful in adapting to oxygen were forced to retreat to the more and more restricted anaerobic habitats that persisted, where many still flourish today. Organisms that evolved the process of respiration, on the other hand, began to enjoy a much superior means of producing useful energy (Chapter 6); this new breed of organisms included aerobic bacteria and eventually also eukaryotic cells.

Another important nutritional advance that probably occurred in the pre-Cambrian era was the development of nitrogen fixation, which allowed cells to use atmospheric N_2 as a nitrogen source, freeing them from dependence on combined forms of nitrogen. Nitrogen fixation was probably occurring at least as early as 2.2 billion years ago because filaments of fossil blue-green algae apparently containing heterocysts [cells specialized for nitrogen fixation (see Figure 21-2C and Chapter 22)] occur in rocks that are that old. Since certain modern anaerobic bacteria and photosynthetic bacteria are capable of nitrogen fixation, it seems likely that this process actually arose prior to oxygen-producing photosynthesis, but exactly when is not known.

Origin of Eukaryotic Cells

Relatively large fossil cells apparently representing single-celled eukaryotic algae (green algae) occur in Australian sediments 0.9 billion years old (Figure 21-2D). Large, probably eukaryotic cells have been found in rocks as old as 1.4 billion years. It is difficult to be sure when eukaryotic cells originated, but scientists now generally assume that it was between 1.5 and 2 billion years ago. In view of the basic differences in ribosome structure between all modern eukaryotes on the one hand, and modern prokaryotes on the other (Chapter 5), eukaryotic cells seem to represent a line of descent that diverged from that of bacteria and cyanobacteria at a rather early stage of evolution.

Many unicellular eukaryotes, or Protista (Chapter 2), are **flagellates**—that is, cells that are motile by means of thin whip-like projections termed flagella (Figure 21-4); this was probably an early feature of the Protista. Many Protista differ from prokaryotes in their ability to engulf foreign particles as food (phagotrophic nutrition). This too was probably a primitive characteristic of Protista, for reasons discussed below. However, probably rather early, some Protista began to specialize for saprotrophic nutrition, eventually leading to the evolution of fungi. Others became more and more completely phagotrophic, leading to the unicellular animals (Protozoa) and eventually to higher animals. Still others specialized for autotrophic nutrition by acquiring chloroplasts, leading to the evolution of plants (see Figure 2-13).

One of the most important innovations of the Protista for subsequent evolution was sexual reproduction—meiosis and sexual cell fusion. Fossil cell tetrads possibly indicating the occurrence of meiosis have been found in sediments as much as 1 billion years old.

Endosymbiotic Origin of Organelles

As noted in previous chapters, the mitochondria and chloroplasts of eukaryotic cells contain their own DNA and their own ribosomes, the RNA and some proteins of which are coded by the DNA of the organelle rather than by nuclear DNA. In size, biochemical properties, and RNA and protein components, organelle ribosomes resemble

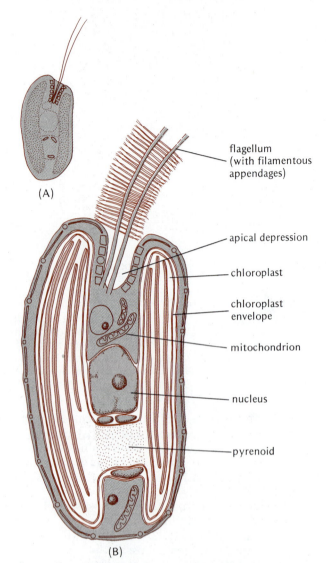

Figure 21-4. Cell of a photosynthetic flagellate, *Cryptomonas*, (A) as it appears with the light microscope, and (B) ultrastructure as seen by electron microscopy. The cell has one large, lobed choroplast, and two flagella decorated with filamentous appendages, characteristic of certain algal groups (Chapter 23). (The cell contains various specialized cytoplasmic structures not labelled in this diagram; for explanation of pyrenoid, see Chapter 23.)

ribosomes of bacteria and blue-green algae rather than the cytoplasmic ribosomes of eukaryotic cells. Organelle DNA also resembles bacterial DNA rather than the chromosomes of the eukaryotic nucleus. This and other evidence suggests that mitochondria and chloroplasts arose through the engulfment of prokaryotic cells by primitive phagotrophic eukaryotic cells. The prokaryotic cells then became **endosymbionts,** or foreign cells residing symbiotically inside a host cell. A small group of present-day flagellates called Glaucophyta contain cyanobacterial endosymbionts called cyanelles (Figure 21-5), which probably illustrate this stage of eukaryotic cell evolution. The endosym-

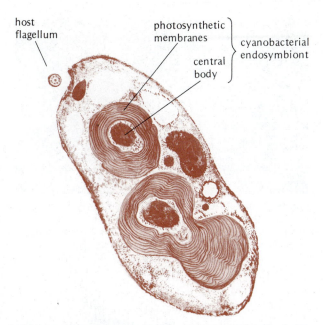

Figure 21-5. Electron microscope view of cell of the flagellate *Cyanophora paradoxa*, containing photosynthetic endosymbiotic cyanobacteria (blue-greens) called cyanelles. One of the endosymbionts is just completing cell division (fission).

biotic prokaryotes apparently became integrated eventually into the host cell as a permanent and essential part of its organization. The double membrane envelope characteristic of mitochondria and chloroplasts probably resulted from the envelopment of the plasma membrane of the prokaryotic cell by a host cell membrane in the process of engulfment (see Figure 21-6).

Mitochondria presumably arose from prokaryotes that were similar to aerobic bacteria (bacteria capable of respiration), which became the respiratory organelles of the eukaryotic cell. The fact that all eukaryotes possess similar mitochondria suggests that this endosymbiosis occurred very

early in eukaryotic evolution and that primitive eukaryotes were phagotrophic.

Compared with prokaryotes and the nuclear genome of eukaryotes, mitochondria apparently use a somewhat simplifed version of the genetic code, translating both UGG and UGA as tryptophan and both AUG and AUA as methionine (compare with Table 5-1). This suggests that mitochondria may have arisen from prokaryotes that were different from any existing bacteria. Alternatively, the genetic code of a bacterial endosymbiont might have become simplified after the integration of the symbiont into the eukaryotic host cell.

MULTIPLE ORIGIN OF CHLOROPLASTS

It seems likely that photoautotrophic flagellates arose more than once during early eukaryotic evolution by the acquisition of different prokaryotic endosymbionts, leading to several independent lines of photosynthetic plant evolution. This theory is suggested to account for differences in chloroplast components between different plant groups. The chloroplasts of the subgroup of algae called red algae, containing biliprotein accessory pigments (phycocyanin and phycoerythrin) like those of cyanobacteria, probably originated as cyanobacterial endosymbionts, similar to the endosymbionts in Glaucophyta (Figure 21-5). The green algal and higher plant type of chloroplast, possessing chlorophylls *a* and *b* and lacking biliproteins, probably originated from a different prokaryotic group represented by a recently discovered organism called *Prochloron* (Figure 21-7), which is now known only as a symbiont inside the bodies of marine animals called tunicates. The third principal type of chloroplast, occurring in the golden- and brown-colored algae, which are sometimes called chromophytic algae (Greek *chromos*, color; *phyton*, plant) and include the brown algae or kelps, contains chlorophyll *c* and distinctive carotenoid accessory pigments. This type may have arisen from a now unknown type of prokaryote.

Figure 21-6. Concept of chloroplast origin by endosymbiosis. (A) Phagotrophic engulfment of photosynthetic prokaryote by eukaryotic flagellate; (B) retention of the autotroph as a walled or (C) wall-less endosymbiont; (D) integration of endosymbiont into cytoplasm, the former endosymbiont's plasma membrane plus surrounding host cell vacuolar membrane becoming the double membrane envelope of the plastid.

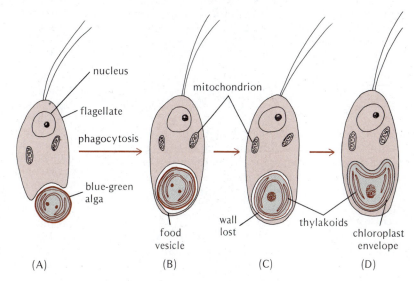

(A)   (B)   (C)   (D)

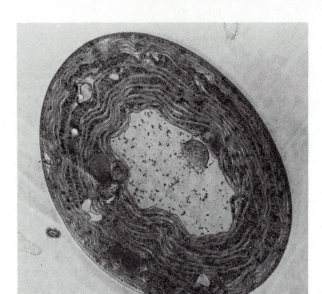

Figure 21-7. Ultrastructure of *Prochloron,* a prokaryotic unicell possessing the green algal photosynthetic pigment system.

Figure 21-8. Reconstruction of an early Paleozoic seascape, showing abundant invertebrate animals, primitive fish, and several kinds of macroscopic algae known from fossils, including a massive colonial form, a coral-like (calcium carbonate-encrusted) form, and kelp-like forms of increasing size.

The chromophytic chloroplast might have been derived instead by loss of biliprotein pigments from the type of chloroplast found in cryptomonad flagellates (*Cryptophyta,* Figure 21-4). Their chloroplasts contain both chlorophyll *c* and biliproteins and are therefore intermediate between red algal and chromophytic chloroplasts.

Most chromophytic chloroplasts and certain others, notably those of the green flagellate *Euglena,* are surrounded by multiple membranes (Figure 21-4) rather than the double-membrane envelope typical of red algal and green plant chloroplasts (Chapter 7). Multiple-envelope chloroplasts may be the result of secondary endosymbiosis, in which a colorless flagellate acquired as an endosymbiont a photosynthetic, chloroplast-containing eukaryote, the product of an earlier primary endosymbiosis of a prokaryote with a flagellate. During subsequent integration, the endosymbiotic eukaryote would lose its cytoplasm and nucleus; its chloroplast and sometimes its plasma membrane, plus the host membrane involved in its engulfment (as in Figure 21-6), would remain as the photosynthetic system of the eventually evolved cell.

RISE OF MACROSCOPIC LIFE

Throughout most of the vast stretch of pre-Cambrian time, spanning more than 80% of the earth's known geological history, no multicellular organisms more complex than the chains of cells comprising filamentous fungi and simple algae appeared, so far as revealed by the geological record.

The first known fossil macroscopic organisms, representing soft-bodied invertebrate animals, appeared about 700 million years ago near the close of the pre-Cambrian era. By the beginning of the Cambrian period, about 600 million years ago, macroscopic life had greatly diversified, and the seas teemed with invertebrate animals. Macroscopic green seaweeds (specialized green algae) of complex structure evolved and became the dominant plant forms in the Cambrian oceans (Figure 21-8). There is some evidence supporting the occurrence of macroscopic brown algae (kelps) and especially of red seaweeds in the Cambrian, and definitely in the ensuing Ordovician and Silurian periods. Thus, the major groups of modern seaweeds (though not the modern species thereof) evolved during the early Paleozoic.

Similarly, most of the major modern groups of invertebrate marine animals had appeared by the end of the Cambrian and evolved further during the Ordovician and subsequent periods. Vertebrates, primitive jawless fish somewhat analogous to modern lamprey eels, apparently existed during the Cambrian and Ordovician, and became diversified and prominent in the Silurian. Bony fish (the most important modern group of fish) appeared and became dominant in the seas during the Devonian, the time when plants were colonizing the land.

There has been much conjecture, still largely inconclusive, attempting to explain the apparently rather sudden appearance of macroscopic life in the late pre-Cambrian and its rapid diversification during the Cambrian. Among the suggested reasons are (1) a rise in atmospheric oxygen concentration, due to photosynthesis, to a level that shielded the earth's surface from injurious ultraviolet solar radiation (via formation of the stratospheric ozone layer) and allowed an organism of macroscopic size to obtain enough oxygen for its respiration; (2) the evolution of meiosis and sexual reproduction, which permitted much more rapid and extensive gene recombination than was possible in asexually reproducing organisms; (3) the emergence of continental land masses from the ocean, which improved the mineral nutrient supply in the adjacent seas and

Figure 21-9.   Reconstruction of a mid-Devonian landscape. Primitive leafless land plants of the types that occurred in the early Devonian, such as the low-growing stems shown at the near shore in the center of the picture and the taller, branching stems in the center right foreground, were now accompanied by more complex plants such as the horsetails left and right of center and some tree-like forms such as the tree ferns shown at left and right.

thus the algal growth and food supply for animals there; and (4) the action of newly evolved predatory animals as agents of natural selection for the evolutionary adaptation and specialization of organisms preyed upon (both animals and plants). Some of these reasons can be disputed, but it is certain that once macroscopic creatures appeared on earth, evolution proceeded much more rapidly than it ever had in the earlier pre-Cambrian.

EVOLUTION OF LAND PLANTS

Although on land cyanobacteria and simple green algae may have grown on rocks and at the soil surface in the Paleozoic as they do today, up to the Silurian period all macroscopic life was apparently aquatic, primarily marine. Silurian sediments yield enigmatic plant fossils consisting of tubular cells (including some thick-walled tracheid-like tubes) in a flat sheet covered by a waxy cuticle, suggesting that these plants were terrestrial and possibly grew flat on the ground or on rocks. By the late Silurian, 405 million years ago, primitive spore-bearing plants had appeared on land. These plants had an erect branching stem covered by a cuticularized, stomate-bearing epidermis, and contained a strand of xylem and phloem. Their successors in the Devonian period developed roots. These features, as explained in Chapters 9–11, are vital to the success of land plants in an aerial environment. The early vascular plants, which were probably confined at first to wet, low-lying habitats (Figure 21-9), became more complex and greatly diversified during the Devonian period. By providing a terrestrial food supply they allowed primitive land animals (amphibians) to evolve from fishes by the end of the Devonian.

Modern vascular plant groups seem to be descended from the primitive Devonian vascular plants. Because all modern vascular plants have photosynthetic pigments that are basically similar to those of green algae, it appears that land plants evolved from some group related to green algae (see Chapter 23).

Bryophytes (mosses and liverworts), the most primitive modern land plants, largely lack the land-adaptive features of the Devonian vascular plants and therefore seem physiologically even more primitive than these early land plants. However, although fossils probably representing bryophytes occur in Devonian and possibly Silurian rocks, the fossil record reveals no clear stage of land colonization by bryophyte-like plants prior to early vascular land plants. If bryophytes preceded the late Silurian and early Devonian vascular land plants, either they failed to be fossilized or their fossils have eluded discovery. The nature of known Silurian and Devonian early land plants and of the subsequent coal-age plants mentioned below is considered in some detail in Chapter 28.

The Coal Ages

By the late Devonian, land plants evolved secondary growth based on a vascular cambium, which made possible the development of tree-like but still rather soft-bodied plant forms and the appearance of the first forests. By the advent of the Carboniferous period, or "coal ages," some 350 million years ago, low-lying land areas were covered by a luxuriant vegetation of giant tree-like plants up to 30 m (100 ft) tall, bearing small scale-like or wedge-shaped leaves (Figure 21-10) and reproducing mainly by spores. Ferns were also abundant. This growth led to the deposition, in the extensive swampy lowlands that were present at that time, of enormous quantities of dead plant material that became fossilized as coal and pro-

Figure 21-10. Reconstruction of Carboniferous swamp forest, showing giant tree-like spore-bearing vascular plants, ferns, horsetails (low plants in foreground and small tree to right), and early gymnosperms (Cordaites, the massive, tapered trunks in background).

vides the principal modern source of this fuel (some coal is of more recent origin).

Rise of Seed Plants

Undoubtedly the most important evolutionary advance in the success of land plants during the less humid geological periods following the Carboniferous was the seed. The seed enables a plant to provide its offspring with food materials to help it become established, assistance that is not possible with a microscopic spore. The kind of sexual reproduction leading to seed formation (Chapter 16) also frees the plant from the dependence on direct contact with water for sperms to swim to eggs for fertilization. In the development of plant reproduction, the evolution of seeds is somewhat comparable to the acquisition by higher animals of the capacity to bear live young (vivipary).

The fossil record shows that evolutionary changes enabling the development of the seed took place during the latter part of the Devonian period; at least one seed-like structure has been found in rocks of the late Devonian. Several groups of seed-bearing plants appeared and diversified during the Carboniferous, including fern-like plants called **seed ferns,** which bore seeds on their leaves, as well as other members of the important modern seed plant group of gymnosperms or naked-seeded plants, which include the conifers.

Age of Gymnosperms

During the succeeding Permian period most of the dominant Carboniferous plants died out, while seed plants steadily rose in importance. Some of the Carboniferous seed plant groups such as the seed ferns also largely died out in the Permian and ultimately became completely extinct, whereas the conifers diversified and, along with several new groups, including palm-like gymnosperms called **cycads,** became the dominant forest vegetation of the Mesozoic era (Figure 21-11), which began about 225 million years ago. This was the "age of dinosaurs," but from the point of view of vegetation it was the age of gymnosperms.

The gymnosperm forests of the Mesozoic dominated the earth for about 100 million years, into the Cretaceous period. They then declined and by the mid-Cretaceous had been replaced over much of the earth by flowering plants. Although like the dinosaurs most of the Mesozoic gymnosperms died out by the end of the Cretaceous, the conifers evolved further and persisted as modern evergreen forests, primarily of colder latitudes and higher altitudes.

Advent of Flowering Plants

Fossils show that flowering plants (angiosperms) arose during, or at least by, the early Cretaceous period, which began 136 million years ago

Figure 21-11. Reconstruction of a mid-Mesozoic landscape with fern understory beneath gymnosperm forest of conifers, cycads (palm-like trees), and ginkgos (foliage at upper left).

and lasted 70 million years. By the mid-Cretaceous, angiosperms had greatly diversified, and by the end of this period they had become the dominant vegetation over much of the earth. During the early epochs of the Cenozoic era, or age of mammals, which began about 65 million years ago, most of the modern botanical families of angiosperms appeared. By the Miocene epoch, some 25 million years ago, essentially the modern range of diversity in angiosperms had evolved, although many present-day species and even a few now prominent families were yet to arise.

The principal changes in vegetation revealed by the fossil record during the Tertiary period are major migrations of tropical and temperate forest communities toward lower latitudes, indicating a general cooling of the earth's climate. It was so warm in the late Cretaceous that many forest trees such as oaks, sycamores, and magnolias grew in Greenland, which is now barren arctic terrain. The climate was still warm enough in the Eocene that walnuts grew in Alaska, while subtropical moist forests including palms occurred as far north as central Oregon. In the time span of 50–60 million years from the Eocene to the Pliocene (see Table 21-1) the belts of tropical, subtropical, temperate hardwood, and subarctic coniferous forests moved southward in North America more than 2000 miles (Figure 21-12). This migration took place over many plant generations, presumably by seed dispersal and the gradual replacement, in any one locality, of the formerly dominant plants by immigrant species from the north that proved better adapted in the face of a cooling climate.

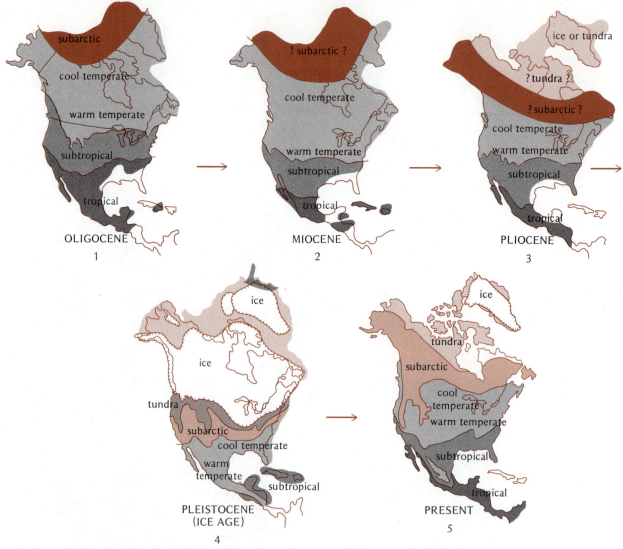

Figure 21-12. Progressive migration of climatic zones in North America during the Cenozoic Era, as estimated from plant fossils and geological evidence of glaciation. In the Paleocene and Eocene epochs warm climates extended even farther north, as noted in text. "Subarctic" denotes boreal conifer forest; "tundra" refers to treeless arctic vegetation.

Pleistocene Glaciation

The Tertiary cooling trend culminated in extensive and repeated spread of glacial ice from the polar regions toward the lower latitudes during the Pleistocene epoch, or "ice ages," which began about 2.5 million years ago (Figure 21-12). Glaciations had taken place several times previously in the earth's history, as far back even as the pre-Cambrian, but this latest glaciation occurred so recently that its effects upon the earth's landscape are still prominent. Some northern plants survived the advance of ice by persisting on occasional peaks and mountain tops projecting above the ice sheets, as in New England. During retreats of the ice (interglacial intervals), vegetation re-colonized areas previously laid bare, such as the north central and northeastern United States, only to be destroyed by a subsequent advance of ice, the last of which ended only about 10,000 years ago. Since some of the interglacial intervals of the Pleistocene were longer than 10,000 years, the present era may actually be only another interglacial period, to be followed by a renewed advance of ice.

PLANT CLASSIFICATION

Plants were being classified long before the concept of evolution motivated scientific classification. The human mind needs some form of classification merely to deal with and make sense of the otherwise bewildering number and diversity of plants. Virtually all cultures, primitive or industrialized, implicitly or explicitly classify plants according to their human uses (see Chapter 32). Such utilitarian groupings are examples of **artificial classifications,** groupings made without regard for evolutionary relationships, in contrast to **natural** or **phylogenetic classifications,** which seek to group organisms according to their apparent genetic or evolutionary affinities. The concept of natural classification had actually developed prior to Darwin's theory of evolution, which, when it appeared in 1859, strikingly accounted for natural "relationships" that had previously been perceived but not understood.

Pre-Darwinian Classification

The story of plant classification prior to the theory of evolution helps to explain the way we view plants and why the theory of evolution found such ready acceptance by scientists when it appeared in 1859.

Theophrastus, the "father of botany" (ca. 300 B.C.), classified plants as trees, shrubs, subshrubs, and herbs. Although ecologically and practically important and of course much-used even today, these categories are artificial because there are many examples of similar and obviously rather closely related herbs, shrubs, and trees. Moreover, there is no sharp dividing line between these plant forms.

Through the Middle Ages botanical knowledge was nurtured mainly through interest in the medicinal uses of plants, set down in books called **herbals,** which named and illustrated various plants and listed for each its known or supposed therapeutic value. The later herbalists, especially in the sixteenth century, acquired a detailed understanding of plant forms and depicted them very artistically (Figure 21-13), but they dealt at most with only a few hundred plants and felt no need to classify them in more detail than had Theophrastus. As the herbal era drew to a close around 1600, however, botanists acquired a growing enthusiasm for finding, describing, and naming new kinds of plants, cultivating them in botanical gardens, and building collections (called herbaria, singular herbarium) of pressed, dried plant specimens for scientific study. By 1623 the Swiss botanist Caspar Bauhin (Chapter 2), one of the last herbalists, was able to list some 6000 known plant species, more than ten times the number known to herbalists of a century earlier. Thereafter the number grew rapidly (and now approaches 400,000); without a classification system this diversity would soon have become incomprehensible.

Bauhin introduced the first element of modern classification, the distinction between genus and species. Subsequent botanists created higher categories of classification, at first mainly expanding on Theophrastus' classification by gross body form but gradually departing more and more from it as they realized that reproductive structures, especially flowers, reveal broader and apparently more fundamental similarities between plants (Figure 21-14).

According to the concept of natural classification that botanists since the time of Bauhin had been de-

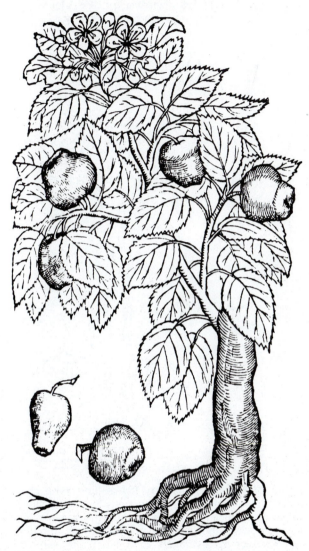

Figure 21-13. Woodcut illustrating the apple tree, from one of the famous herbals, Rembert Dodoens' *Herbarius oft Cruydt-bock* ("herbal or plant-book"), published in Leyden (Netherlands) in 1608. The drawing is at least partly from the artist's imagination since it pictures the tree as bearing both flowers and fruits, which do not occur on an apple tree at the same time of year.

veloping and that Linnaeus (1707–1778) considerably elaborated, the sum total of characteristics shared between species indicates an "affinity" or basis for natural groupings, which were felt to be fundamental even though their significance could not be scientifically explained at that time. Later botanists, especially the Frenchman A. L. de Jussieu (1748–1836) and the Genevan Swiss A. P. de Candolle (1778–1841), advanced the concept of natural classification, improving and adding to the natural "orders" proposed by Linnaeus, which eventually became many of the important botanical Families in modern usage. They arranged these into higher ranks, culminating in the basic subdivisions

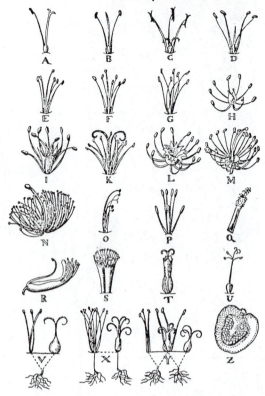

Clariſ: LINNÆI.M.D.
METHODUS plantarum SEXUALIS
in SISTEMATE NATURÆ
deſcripta

Lugd. bat: 1736

G.D. EHRET. Palat-heidelb
fecit & edidit

Figure 21-14. Contemporary figure illustrating characteristics of the classes in Linnaeus' "sexual system" of plant classification. Diagrams show the stamens, and in some cases the pistil, of flowers, mostly omitting other flower parts. Classes V, X, and Y represent, respectively, monoecious, dioecious (see Chapter 16), and "polygamo-dioecious" plants whose individuals bear perfect (bisexual) flowers plus *either* staminate *or* pistillate flowers. Class Z represents plants such as the fig, in which many small flowers are fused into a compound structure (yielding a multiple fruit). After a plate drawn by the famous botanical artist G. D. Ehret in 1736.

their sexual reproduction was obscure compared with that of seed plants. As optical microscopes and preparative techniques such as sectioning and staining improved during the nineteenth century, it became possible to investigate critically the reproduction of both lower and higher plants. These investigations revealed that most cryptogams actually have very definite sexual reproduction and that their morphology, development, and reproduction are much more diverse than those of seed plants, implying several distinct lines of evolutionary descent. When studied microscopically, the reproduction of seed plants was discovered to be a modification of that of spore-bearing lower vascular plants such as ferns, which had previously been assigned to the cryptogams. The sexual stages of flowering plants proved to be much smaller and more obscure than those of ferns and other lower vascular plants.

These comparative morphological results revolutionized concepts of plant relationships and classification. The spore-bearing and seed-bearing vascular plants were seen to comprise a single natural evolutionary unit, whereas the lower (nonvascular) plants comprise an extensive series of distinct major groups, not just one or two.

Although botanists in many countries took part from about 1850 onward in the crucial comparative morphological studies that led to modern concepts of ma-

of monocots and dicots. These classifications began to reveal an overall rationality in the diversity of flowering plants; they prepared the way for, and were impressively explained by, Darwin's theory of evolution. Since then, natural relationship has meant phylogenetic relationship, and this concept has dominated efforts to classify plants as well as other organisms.

Development of General Plant Classification

Until about the middle of the nineteenth century lower plants were hardly understood at all, and were thrown into a few catch-all groups collectively called "cryptogams" (Greek *kryptos*, hidden; *gamos*, union) in the mistaken belief that

Figure 21-15. Wilhelm Hofmeister (1824–1877), the German botanist who discovered the nature of the reproductive cycles of vascular plants.

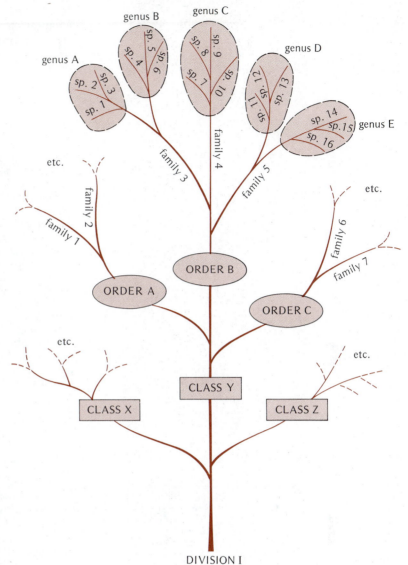

Figure 21-16. Concept that the successively higher or larger categories of a phylogenetic classification reflect successively older and more distantly related branches of the tree of evolutionary descent of living organisms. Species, the most closely related and recently separated lines of descent, are represented only for one of ten orders, and orders and families are shown for only one of the classes within this division. (Most classes contain many more orders, families, and species than shown here). Actual classifications undoubtedly are only imperfect approximations of the concept represented by this diagram.

jor plant groups, the most important pioneering efforts were made by a series of admirably thorough German researchers, including Wilhelm Hofmeister (1824–1877), who studied vascular plants; Anton de Bary (1831–1888), who studied primarily fungi; and N. Pringsheim (1823–1894), who investigated both algae and fungi. Hofmeister (Figure 21-15) discovered the remarkable alternation of diploid and haploid generations in the sexual cycle of ferns and other lower vascular plants and clearly showed that the reproduction of gymnosperms and flowering plants is a modification of this type of sexual cycle. This was one of the most significant concepts clarifying our understanding of plants, as will become evident in the discussion of vascular plant groups (Chapters 28–31).

CONCEPTS OF EVOLUTIONARY CHANGE USED IN CLASSIFICATION

In the hierarchy of formal taxonomic categories used today (Chapter 2), the largest categories or divisions correspond to the lowest (i.e., the oldest) branches of the "tree" of evolutionary descent. Successively smaller subdivisions such as classes, orders, families, and genera should correspond to more and more recently established lines of descent from more and more recent common ancestors, that is, to a closer and closer genetic relationship between species (Figure 21-16).

From the kinds of comparative morphological studies mentioned in the previous section, several principles of evolutionary change have emerged that play an important role in phylogenetic interpretation and in efforts to devise phylogenetic classifications.

Organ Modification: Homology

Evolutionary advance often occurs by modification of the form and structure of organs to serve modified or different functions. Examples are the

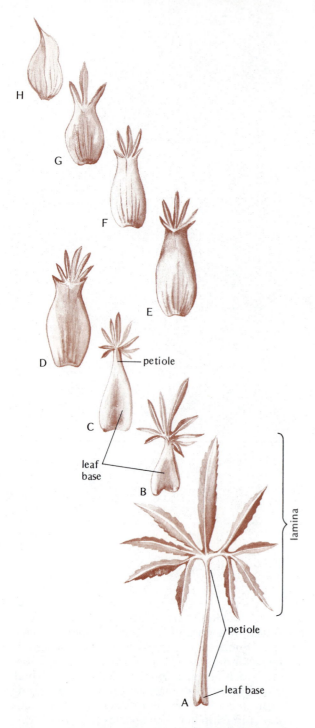

Figure 21-17. Successive foliar (leafy) organs up the stem of the "Christmas rose" (*Helleborus*) showing a gradual transition from photosynthetic foliage leaves (A) to scale-like floral bracts (H) by expansion of the leaf petiole and reduction of the palmately lobed leaf blade (lamina). These organs are obviously homologous, representing different degrees of modification of the developmental program for leaf growth. A developmental program might similarly become gradually modified genetically during evolution to transform one organ into a homologous organ of different form and function, such as a foliage leaf into a bud scale, a spine, or a flower part.

leaf, stem, and root modifications discussed in Chapter 3. Different organs that appear to represent modifications of the same original type of organ are said to be **homologous** (Figure 21-17).

The concept of homology, although used at least implicitly in almost all evolutionary discussions, and applied not only to organs but also to intracellular structures such as chloroplasts and mitochondria, has some important limitations. Although we commonly refer to structures evolving and becoming modified, it is actually not the structure but the genetic program for its development (Chapter 14) that evolves and becomes modified. Agencies of natural selection act, of course, at the phenotypic level—that is, at the level of the already developed structure, the performance of which influences the fitness of the individual. Presumably, altered genetic programs specifying useful modifications of development and function are selected evolutionarily in this way. It is an interesting, but still moot, question how much genetic information is actually shared between the developmental programs for homologous structures in different plants, especially plants that are only distantly related.

Genetic homology can be seen clearly at the molecular level. The sequence of amino acids in proteins and of bases in ribosomal and transfer RNAs shows conclusive evidence of the progressive modification, by mutation, of the genes specifying these gene products in the course of evolution of different species.

Reduction

During evolution, organs or tissues may become reduced in size or even lost. The scale leaves of some flowering plants (Figure 21-17) are examples of reduction. A more profound series of reductions has occurred during the evolution of the sexual reproductive cycle in land plants (Chapters 27–31).

Evolutionary reduction can occur for different reasons. One is specialization of function. For example, reduction of leaf size in many xerophytes (Chapter 19) serves to reduce the transpirational surface area and also the tendency of leaves to become overheated by the sun when not transpiring because of stomatal closure. Another reason for reduction is lack of need for the organ or structure. If the organ's function has become superfluous, mutations that reduce the extent of its development can be tolerated and may in fact be selected for because they spare needless investment of photosynthate. This may account, for example, for the loss of petals by wind-pollinated flowers (Chapter 16).

Evolutionary Conservatism

Vegetative structures such as leaves and stems are apparently more readily modified in the course of evolution than are reproductive structures. Thus,

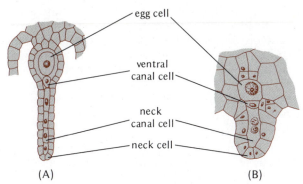

egg cell

ventral
canal cell

neck
canal cell

neck cell

(A)   (B)

Figure 21-18. Example of an evolutionarily conservative feature, the egg apparatus or *archegonium* of land plants (Embryonta). (A) Archegonium of a liverwort, which belongs to the Bryophyta, the most evolutionarily primitive division of the Embryonta. (B) Archegonium of a fern, representing the most evolutionarily advanced class of the Embryonta other than seed plants. The function of the archegonium is explained in Chapters 27–29. A much-reduced archegonium can still be recognized even in the gymnosperms or more primitive seed plants (Chapter 30).

larger phylogenetic groups are recognized mainly through basic similarities in their reproductive mechanisms, as noted previously. For example, the flower is fundamentally similar in all angiosperms, despite enormous variety in growth, form, leaf morphology, and so forth. A character that evolves relatively slowly and thus provides a superior indication of relationship is called conservative (Figure 21-18).

One possible reason for evolutionary conservatism is that the more complex the developmental program for a given organ (the greater the number of genes involved), the more difficult it is to modify that organ evolutionarily, and/or the greater the morphological or anatomical traces of homology that remain after the modification. Another possible reason is that the more

vital a given function is to the organism's evolutionary fitness, the fewer the mutations or genetic variations that may produce a useful modification and be tolerated. This may explain why chloroplast pigmentation and structural characteristics have persisted through a great deal of vegetative and reproductive evolution of major plant groups.

Evolutionary conservatism can also be recognized at the molecular level, and for a similar reason. The very conservative evolution of ribosomal and transfer RNA base sequences (Chapter 20) and ribosomal proteins is one example. Another is the histone proteins associated with eukaryotic chromosomes (Chapter 5), which are very similar in organisms as far apart evolutionarily as seed plants and mammals.

Convergent Evolution

In adapting to a given kind of environment, different lines of evolutionary descent may evolve similar forms and structures. The close resemblance in form between the desert-dwelling cacti of the New World and the unrelated cactus-like Euphorbia-family plants of the South African deserts (Figure 21-19) is one spectacular example, as is the very similar Kranz leaf anatomy of many different flowering plant groups that have independently evolved C4 photosynthesis (Chapter 8).

Different kinds of convergent evolution may be distinguished: (a) the acquisition, by different (nonhomologous) organs, of similar form and function—for example, leaf succulents compared with stem succulents (Chapter 3), and (b) the evolutionarily independent acquisition, in different plant groups, of similar specialized form and function in homologous organs, as in the examples of stem and leaf structure cited at the end of the previous paragraph. Convergent evolution of the first type is often relatively easy to detect because differences in morphology or anatomy usually persist between nonhomologous but functionally analogous organs—for example, between the different types

Figure 21-19. Example of convergent evolution of Xerophytic shoot morphology. These plants belong to different, unrelated botanical families and have evolved their similarities independently. (A) *Stapelia*, of the milkweed family (Asclepiadaceae). (B) *Euphorbia brightoni*, of the spurge family (Euphorbiaceae). (C) *Myrtillocactus* of the cactus family (Cactaceae).

of thorns or spines described in Chapter 3. Convergent evolution of the second type may be considerably more difficult to establish, and is often a subject for debate in evolutionary discussions.

Reticulate Evolution

If a group derives parts of its genome from two or more species, its evolution is reticulate ("net-like"), in contrast to the branching pattern of evolutionary divergence from a single common ancestor (Figure 21-16). One commonly occurring kind of reticulate evolution is allopolyploid species formation from an interspecific hybrid (Chapter 20). This is possible only in relatively closely related species, which can cross to give a viable hybrid.

More radical gene transfer between unrelated species, for example, by means of viruses, may be significant in the evolution of bacteria (Chapter 22) as well as possibly making a rare contribution to that of higher organisms, although this remains to be proved. The endosymbiotic origin of organelles, which is now widely accepted, constitutes reticulate evolution, since the genes for organelle (former endosymbiont) components, whether remaining in the organelle or integrated into the nucleus, become part of the eukaryote's genome.

CONTEMPORARY CLASSIFICATIONS

Although numerous examples of the kinds of evolutionary change listed in the previous section can be demonstrated by comparative morphology, and although many evolutionary relationships are clear beyond any reasonable doubt, decisions about evolutionary relationships are nevertheless always interpretations of evidence that in many cases leaves room for differences of opinion. For example, a particular similarity between groups may be due to evolutionary conservatism or to convergent evolution. In theory, the fossil record could settle many of these issues, but it is actually very incomplete, a minuscule sampling of past life biased in favor of organisms growing in habitats most conducive to fossilization of their remains, and of organisms and organs most amenable to preservation as fossils. As a result, the fossil record of all plant groups contains huge gaps unlikely to be adequately filled by future fossil discoveries, and many questions about phylogenetic relationships remain unsolved by geological evidence.

These problems and limitations have the strongest effect on our interpretation of the relationships between the most distantly related groups, the highest categories in a phylogenetic classification. Because of these uncertainties, there is little point in insisting dogmatically on any given classification scheme. For the purposes of this book, however, we need a working classification that displays the major features of plant diversity in at least a roughly or possibly phylogenetic pattern and helps the student to understand and remember the diversity of plants and the most salient evolutionary trends.

Current plant classifications, as elaborated in many recent botanical textbooks, recognize an ever increasing number of principal divisions or major groups (often 20 or more), creating a bewildering impression compared with the few and simple late nineteenth century groupings listed in Chapter 2, many of which are now regarded as artificial categories that are usable only as common names. Proliferation of major divisions is due primarily to the interpretation of more and more already known subgroups as substantially independent lines of evolution. At the other extreme is the now popular five-kingdom classification (Figure 2-13), which lumps all eukaryotic algae into one kingdom, Protctista.* This classification does not display graphically the diversity among algae, which comprise at least several distinct lines of evolution probably as ancient as the lines leading to the animal, fungal, and higher plant kingdoms or subkingdoms. The working classification given below seeks a compromise between the oversimplification in the five-kingdom classification and the multiplicity of many other current classifications.

Working Classification Used in This Book

Figure 21-20 shows a classification modified from the earlier five-kingdom classification shown in Figure 2-13. It separates the eukaryotic algae into three named subkingdoms, corresponding to three basic lines of photosynthetic eukaryote evolution as reflected in the three types of chloroplasts discussed earlier (Multiple Origin of Chloroplasts): (1) the **Bilonta** or biliprotein-containing eukaryotes; (2) the **Chromonta** or golden- and brown-colored algae (chromophytes); and (3) the **Chloronta** or algae with green chloroplasts that are relatively similar to those of land plants (**Embryonta**), which presumably descended from some member(s) of the Chloronta. The members of each of these groups share features of chloroplast structure and function that often make it useful to refer to them as a group. The slime molds, which have apparently evolved as a special line distinct from

---

*The original five-kingdom classification recognized the kingdom **Protista** as comprising unicellular eukaryotes other than unicellular fungi, and placed multicellular algae (reds, browns, and multicellular greens) with higher plants in kingdom **Plantae,** an artificial group that plainly included multiple lines of descent (cf. Figure 21-20). A later revision remedied this by combining Protista with multicellular algae in a new kingdom named **Protoctista,** restricting kingdom Plantae to land plants (subkingdom Embryonta in our classification).

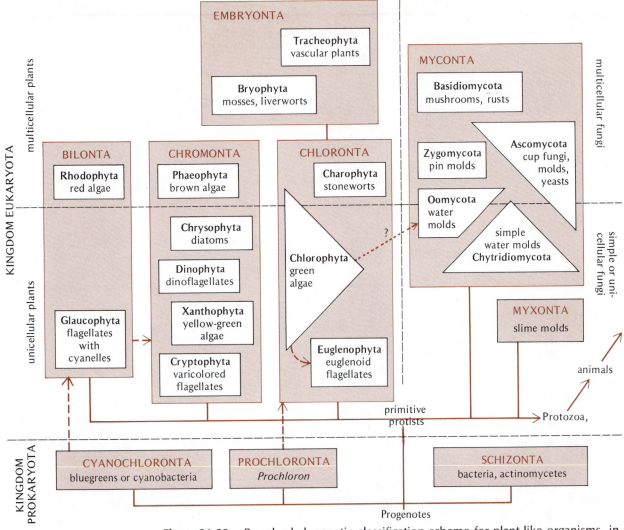

Figure 21-20. Rough phylogenetic classification scheme for plant-like organisms, involving division of the eukaryotes other than animals into six subkingdoms (color rectangles) mostly containing several divisions (black rectangles and triangles), and division of the prokaryotes into three subkingdoms. Organisms to left of vertical dashed line are mostly photoautotrophs, and those to its right, heterotrophs. Generally, the higher the position of a group on the diagram, the greater the size and complexity of its members. Triangles are used for certain divisions containing both unicellular and muticellular organisms; — — —> indicates possible or probable origin of chloroplasts by endosymbiosis; · · · ·> indicates possible origin of one division of fungi from algae, independent of the evolution of other fungi.

either the fungal (**Myconta**) or animal lines, are also given their own subkingdom (**Myxonta**).

The major groups (divisions) belonging to each subkingdom are indicated in Figure 21-20. These divisions will be considered in subsequent chapters.

The ending -*onta* used for subkingdom names in Figure 21-20 comes from Greek meaning "things that exist." The names are patterned after **Cyanochloronta,** "blue green ones," used by Bold and Wynne (1978) for the cyanobacteria. Thus, Chloronta means roughly "green ones," Chromonta "colored ones," Myxonta "slimy ones," Myconta "fungal ones," Bilonta "ones (that contain) biliproteins,'' and Embryonta ''ones (that produce) embryos'' (which is equivalent to the Embryophyta of some classifications). Our Chromonta are equivalent to the group Chromophyta used by certain authors. **Schizonta** (''fission ones'') for bacteria derives from Schizophyta, the term formerly used for this group. [The ending -*phyta* (''plants'') is now by convention reserved for photosynthetic plant divisions.]

We group the subkingdoms simply into two kingdoms, Prokaryota and Eukaryota, representing a very ancient evolutionary divergence, as noted previously. Unicellular and multicellular animals may be classified as additional subkingdoms of the kingdom Eukaryota. It should be emphasized that the classification represented by Figure 21-20 is intended as a working compromise for the display of major features of plant evolutionary diversification rather than as a strictly phylogenetic diagram, for which it would undoubtedly be an oversimplification.

SUMMARY

(1)   Fossilized cells in pre-Cambrian rocks show that prokaryotic life, probably including photosynthetic bacteria, existed at least 3.4 billion years ago, having presumably arisen from a ''primordial soup'' of organic compounds at the earth's surface. This mixture of organic compounds was produced by chemical reactions between components of the earth's primitive oxygen-less atmosphere. Oxygen-producing photosynthetic cyanobacteria arose at least 2.2 billion years ago, leading to oxygen accumulation in the air and thereby to respiration in living organisms.

(2)   Eukaryotic cells probably appeared 1.5–2 billion years ago, evolved meiosis and sexual reproduction, and apparently acquired mitochondria by engulfment, endosymbiosis, and subsequent integration of aerobic bacterial cells; chloroplasts were acquired by endosymbiosis of prokaryotic photoautotrophs such as cyanobacteria. Different types of chloroplasts found in the major lines of photosynthetic plant evolution may have arisen by endosymbiotic uptake of different types of prokaryotic (or photosynthetic eukaryotic) cells into the cells of primitive eukaryotic flagellates.

(3)   According to the fossil record, macroscopic organisms first appeared in the late pre-Cambrian, about 700 million years ago. Macroscopic algae (seaweeds) developed in the Cambrian period. In the late Silurian, about 405 million years ago, primitive vascular plants began to colonize the land. They diversified during the ensuing Devonian period, and by the Carboniferous, 350 million years ago, dense forests of spore-bearing and early seed plants covered extensive low-lying, swampy terrain. This type of vegetation declined and largely disappeared before the end of the Paleozoic era, 225 million years ago.

(4)   Gymnosperms (conifers and more or less related forms) were dominant through most of the subsequent Mesozoic or dinosaur era, up to about 100 million years ago (middle Cretaceous period), when angiosperms (flowering plants) diversified and soon became predominant over most of the earth. Through the Cenozoic era (65 million years ago to the present) the earth's climate steadily cooled, driving tropical and subtropical vegetation toward the equator from high latitudes where it formerly existed. This trend culminated in the Pleistocene glaciations, which destroyed much vegetation at higher latitudes and altitudes.

(5)   Phylogenetic classification of plants, which seeks to represent evolutionary relationships among plant groups, is based primarily on evidence from comparative morphology, ultrastructure, and fossils. Recognized trends useful in judging relationships are (a) evolutionary modification of organs or structures into functionally different but homologous structures; (b) evolutionary reduction; (c) convergent evolution; and (d) evolutionary conservatism of particular features, such as reproductive structures and chloroplasts and their pigment systems. Reticulate evolution is the contribution of genes from different species to an evolutionary line, as in endosymbiotic acquisition of organelles.

(6) Major ancient botanical evolutionary complexes, classified here as subkingdoms, are (a) slime molds (Myxonta), (b) fungi (Myconta), (c) biliprotein-possessing algae (Bilonta), (d) golden- and brown-pigmented algae (Chromonta), (e) green-pigmented algae (Chloronta), and (f) land plants (Embryonta), which probably descended from certain of the Chloronta.

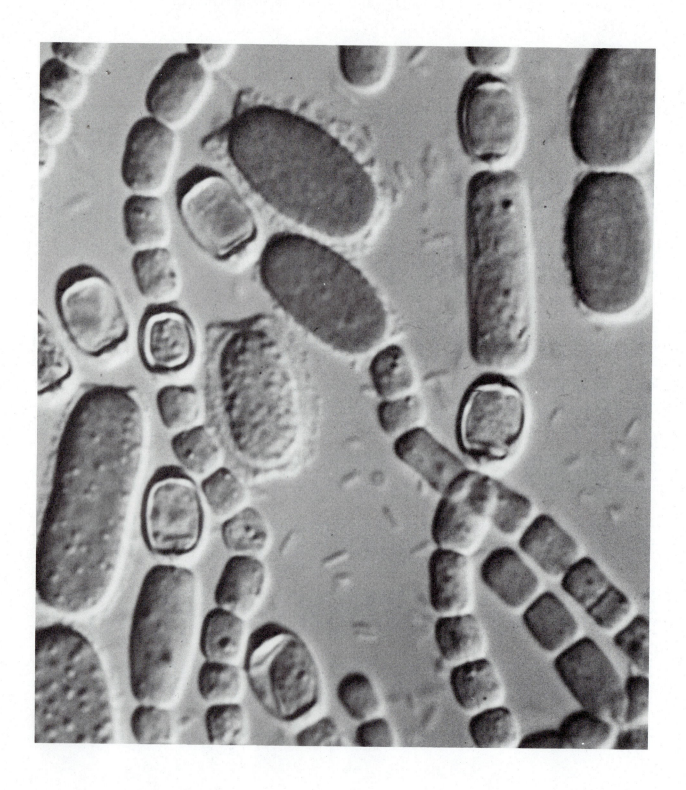

Chapter 22

PROKARYOTES

Prokaryotic organisms represent the oldest type of living cell, and the simplest in terms of cell structure. Despite their relative simplicity, prokaryotes exhibit an astounding diversity of means for adapting to different environments, and at least some kinds of prokaryotes occur almost everywhere on earth. Some species inhabit environments as extreme as hot springs with temperatures above 90°C, acid mine wastes containing essentially dilute sulfuric acid, salt brines nearly saturated with sodium chloride, or habitats devoid of oxygen such as waste-polluted waters, swamp mud, sea water trapped in submarine basins such as deep fjords, and the depths of the Baltic and Black Seas, which are completely anaerobic.

Although invisible to our unaided eyes, bacteria surround us in enormous numbers, not only on the surface of our bodies but in the air we breathe, in the water we drink, and on almost everything we normally touch. A gram of average soil contains over 10^9 prokaryotic cells. As causes of infectious diseases, bacteria have killed more humans than all wars combined. Yet the number of pathogenic (disease-producing) bacteria is greatly exceeded by the number of nonpathogenic, beneficial forms that are useful or essential to humans and other animals. For example, if soil bacteria and other microorganisms did not decompose dead plant and animal bodies, essential mineral elements would soon become completely tied up in organic compounds, leaving none available for plant growth and food production. Different bacteria possess biochemical pathways for degrading almost any organic compound. Bacteria are the ecosystem's principal defense against waste accumulation and pollution.

The subject of microbiology, which deals mainly with prokaryotes, is a vast one, requiring entire textbooks just to give a general survey and others to cover its medical, agricultural, and industrial aspects. This chapter, necessarily restricted in scope, presents some basic general characteristics of prokaryotes, considers the more plant-like prokaryotes (blue-greens or cyanobacteria), and discusses other bacteria whose activities are particularly important to green plants. We include an elementary discussion of gene transfer and recombination in bacteria, since this concept is basic to current and prospective work in plant genetics and genetic engineering, which are repeatedly mentioned elsewhere in this book.

Most prokaryotes fall into three main groups: (1) **Cyanobacteria** (blue-greens) exhibit an oxygen-producing, green plant type of photosynthesis. (2) **Actinomycetes** are heterotrophic prokaryotes that grow in a filamentous, fungus-like manner. (3) **Eubacteria,** or "true bacteria," among which are counted the large majority of prokaryotes, are mostly unicellular, nonphotosynthetic organisms; those few that are photosynthetic do not produce oxygen, as noted in Chapter 7. Several other smaller groups of prokaryotes are also recognized, including the Archaebacteria mentioned later and the Mollicutes or mycoplasmas, which are considered along with other plant parasites in Chapter 26.

GENERAL FEATURES OF PROKARYOTES

Cell Size and Shape

Prokaryotic cells are among the smallest living things (Figure 22–1). A single drop of liquid can contain as many as 50 million bacteria. Cells of the smallest bacteria (about 0.2 μm in diameter) can just be detected under the light microscope, while those of the larger prokaryotes (in the range of 5 μm in diameter) overlap the sizes of the smallest eukaryotic unicellular organisms. Typical mitochondria of eukaryotic cells are about the same size as small bacterial cells, which is not just a coincidence in view of the theory that mitochondria originated from endosymbiotic bacteria (Chapter 21).

Figure 22-A. *Cylindrospermum,* a filamentous blue-green alga (cyanobacterium) containing heterocysts, extremely thick-walled cells which lack a nucleus and in which nitrogen fixation occurs. The larger, more cylindrical cells with what appear to be lumpy contents are resting spores (akinetes). Some of the filaments have fragmented, releasing individual heterocysts and resting spores.

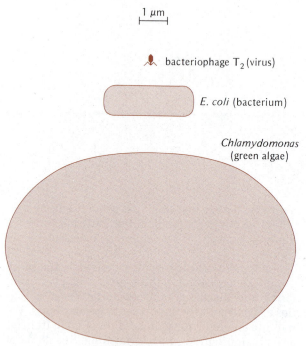

| ⊢ 1 µm ⊣

🜋 bacteriophage T₂ (virus)

E. coli (bacterium)

Chlamydomonas
(green algae)

Figure 22-1. Comparative sizes of a prokaryotic cell (*E. coli*), a eukaryotic unicellular microorganism (*Chlamydomonas*), and a bacterial virus.

Figure 22–2 shows the three most common types of cell form among eubacteria. These are referred to respectively as (1) **cocci** (singular, coccus), (2) **rods** (formerly called bacilli), and (3) **spirals** (formerly called spirilla).

Growth

Generally, the smaller organisms are, the more intense their metabolism. Thus prokaryotes usually have more rapid growth rates and shorter cell cycles (generation times), given an optimum environment, than larger organisms with which they may be competing for nutrients—an obvious ad-

vantage, at least for the short term. *Escherichia coli*, a bacterium found universally in the human gut and much used in molecular biological and genetic studies, has a doubling or generation time of about 20 min under optimal growth conditions. Bacteria with a generation time of several hours are regarded as slow growing, whereas, as noted in Chapter 4, meristematic higher plant cells with a cell cycle of about a dozen hours are considered fast growing.

Reproduction

Prokaryotes usually multiply by **fission,** cell division by simple septation due to ingrowth of the peripheral cell wall in the midregion of the enlarged cell (Figure 22–3). A few species multiply by **budding,** or formation of a new cell as a local outgrowth from the parent cell, a type of reproduction that also occurs in some eukaryotes (e.g., yeasts, Figure 24–25).

Prokaryotes lack the eukaryotic type of sexual reproduction involving fusion of haploid gametes and meiosis. However, bacteria do have several mechanisms for transferring genetic information between cells, discussed in a later section.

SPORE FORMATION

One group of eubacteria forms dormant survival spores called **endospores.** During this process, a part of the cell contents including the DNA becomes surrounded by a very thick spore coat (Figure 22–4). The spore, released on disintegration of the original cell wall, has a low water content and very low metabolic activity and is quite resistant to heat, disinfectants, drying, and other environmental stresses. Subsequently, it can germinate into a new vegetative cell after tiding the organism over a period of unfavorable conditions.

Because bacterial endospores can survive boiling, in the canning process food is heated under pressure

Figure 22-2. Three common bacterial shapes. (A) Approximately spherical cells, termed cocci; (B) cylindrical or rod-shaped cells; and (C) spiral (actually helical) forms. Bars in (A–C) are 5, 5, and 10 µm, respectively.

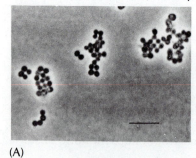

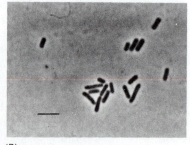

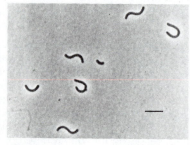

(A)   (B)   (C)

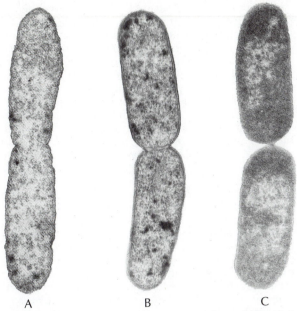

A   B   C

Figure 22-3. Cell division in a rod-shaped bacterium (*Escherichia coli*). Septum formation occurs by ingrowth of the peripheral wall near the midpoint (A, B) of the cell. Daughter cells of many species then separate when septum formation is complete (C). In some species, daughter cells remain together to form a chain or filament (see Figure 22-7).

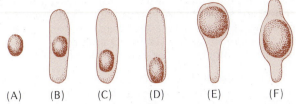

(A)   (B)   (C)   (D)   (E)   (F)

Figure 22-4. Endospores (A) of various bacilli are formed centrally (B), subterminally (C), or terminally (D), in positions characteristic of the species. In some species, spore diameters are larger than those of the vegetative cells (E, F), resulting in a swollen appearance.

to obtain temperatures above boiling. If food is contaminated by spores of the botulism bacillus, *Clostridium botulinum,* that are not killed by sufficiently high temperatures, the vegetative cells that can grow from them within the can or jar will produce the deadly botulism toxin. This bacterium also requires for growth a pH above 4.5 and the anaerobic conditions that occur in cans or tightly sealed containers. Only acidic foods, such as tomatoes, apples, and rhubarb (pH less than 4.5), can be safely canned by simple boiling. Commercially canned, less acid foods are processed at temperatures of 110°–120°C (230°–248°F) for periods of 12 min to 1½ hr, depending on the volume of the cans, the pH, and the nature of the foods.

Since one bacterial cell differentiates into only one endospore, endospores are not a means of reproduction, in contrast to the reproductive spores of actinomycetes and of fungi, algae, and other spore-bearing plants (Chapters 23–29), in which one cell usually produces several to many spores. Endospores are instead a modification of the vegetative cell for survival during stress.

PROKARYOTE NUTRITION AND METABOLISM

Every organism requires a carbon source that provides carbon for synthesis of its organic molecules; an energy source to provide energy needed for biosynthesis, membrane transport, and other cell functions; and sources of elements other than carbon, hydrogen, and oxygen that are needed to form vital organic molecules, such as nitrogen and sulfur for proteins, nitrogen and phosphorus for nucleic acids, and so on (see Chapters 5 and 6). According to the carbon and energy sources they use, organisms can be grouped into four nutritional categories based on the distinction between autotrophs, which use inorganic carbon (CO_2), and heterotrophs, which must have an organic carbon source (Chapter 2). Organisms that use light energy are designated by the prefix *photo-*, while *chemo-* indicates organisms that gain their energy from chemical sources (Table 22-1). Thus, photoautotrophs, such as green plants and the photosynthetic bacteria, use light energy and fix CO_2. Certain bacteria are photoheterotrophs, obtaining energy from light but requiring an organic carbon source. Chemoautotrophic bacteria, as noted in Chapter 2, fix CO_2 using energy obtained by oxidizing inorganic chemicals obtained from their environment. Chemoheterotrophs, including most bacteria and all animals, usually obtain both carbon and energy from organic carbon compounds.

A unique kind of photoheterotrophic metabolism occurs in the halophilic ("salt-loving") bacterium, *Halobacterium halobium.* It grows in waters containing very high salt concentrations, such as evaporating seawater ponds, and possesses a carotenoid-containing purple protein pigment called bacteriorhodopsin, which is remarkably similar to the human eye pigment rhodopsin, responsible for vision. Bacteriorhodopsin occurs in *H. halobium's* cell membrane and absorbs light energy, resulting in protons being pumped out of the cells. The light-induced proton gradient between the

TABLE 22-1   Types of Prokaryotic Energy and Carbon Nutrition

Energy Source	Carbon Source	
	CO_2	*Organic Compounds*
Light	Photoautotrophs	Photoheterotrophs
Chemical compounds	Chemoautotrophs	Chemoheterotrophs

inside and the outside of the *H. halobium* cell drives conversion of ADP to ATP, much as in mitochondrial oxidative phosphorylation (see Chapter 6). This enables *H. halobium* to obtain ATP for its energy needs by a mechanism other than respiration or fermentation.

Heterotrophic bacteria are extremely diverse with respect to the organic substrates they can utilize. As noted above, almost every organic compound, natural or synthetic, can be attacked and utilized as a carbon or energy source by at least a few kinds of bacteria. Organisms called **prototrophs** need only a carbon and energy source; from it plus inorganic forms of such mineral nutrients as nitrogen, phosphorus, or sulfur they can make all their cell constituents. Wild-type strains of *Escherichia coli*, for example, are prototrophs. Many bacteria, however, require organic sources of nitrogen or sulfur. Some, called **auxotrophs,** need specific dietary requirements such as particular amino acids or vitamins, like most animals.

Aerobes and Anaerobes

Aerobes are organisms that require oxygen and depend on respiration for their energy production; **anaerobes** can live and grow in oxygen-free habitats (Figure 22-5), usually obtaining their energy from fermentation or from a process called anaerobic respiration, which will be considered later under the heading Denitrification. Some are **facultative** anaerobes—able to grow anaerobically but growing also (usually better) under aerobic conditions. Many bacteria, however, are **obligate** anaerobes that are unable to use oxygen and are actually poisoned by it. Obligate anaerobes, presumably resembling primitive prokaryotes that existed on earth before its oxygen-containing atmosphere evolved (Chapter 21), exploit permanently anaerobic habitats but are also abundant in aerated soil for reasons explained later.

Heterotrophic anaerobic bacteria obtain energy by a wide variety of means of fermentation. Some ferment carbohydrate to ethyl alcohol, like yeasts and higher plants (Chapter 6), but many more produce lactic acid ("lactic bacteria," including those responsible for souring milk and producing fermented milk products such as yogurt, buttermilk, and cheese), or a mixture of lactic acid and alcohol. Various other fermentation products, often with distinctive or foul odors, are produced by other bacteria; identification of their specific fermentation products is important in recognizing many bacterial groups or species. Bacterial fermentation often involves metabolic pathways more complicated than the classic glycolysis pathway explained in Chapter 6 as the basis of alcoholic fermentation and aerobic respiration.

Obligate anaerobes mostly lack the cytochrome proteins of the aerobic electron transport chain and therefore cannot conduct respiration. Oxygen is toxic to obligate anaerobes mainly because oxygen initiates oxidation reactions that form small amounts of the very strong oxidants hydrogen peroxide (H_2O_2) and superoxide (O_2^-), which attack and damage proteins and membrane lipids. Peroxide and superoxide are formed from oxygen in cells of aerobes as well as of anaerobes, but aerobes are able to handle the problem because they possess the enzymes **catalase** for decomposing peroxide ($2\ H_2O_2 \rightarrow 2\ H_2O + O_2$) and **superoxide dismutase** for decomposing superoxide ($2\ O_2^- + 2\ H^+ \rightarrow H_2O_2 + O_2$), thus preventing these oxidants from building up inside the cell.

Autotrophic Bacteria

Most chemoautotrophic bacteria are aerobic, requiring oxygen for the oxidation reaction by which they obtain energy from an inorganic chemical source. Examples are iron bacteria (Figure 2–1) and sulfur bacteria, which oxidize sulfide or hydrogen sulfide (H_2S) occurring in natural waters or produced by microbial decomposition of proteins in the soil. This action is beneficial to plants because H_2S in the soil would be toxic to roots. "Hydrogen bacteria" oxidize hydrogen gas (H_2) derived from certain bacterial fermentations or produced as a by-product of the nitrogenase system of nitrogen-fixing soil organisms. The nitrifying bacteria, another important group of aerobic chemoautotrophs, will be considered later. However, certain chemoautotrophs (the methane producers) are obligate anaerobes, deriving their energy from an oxidation-reduction process not involving oxygen.

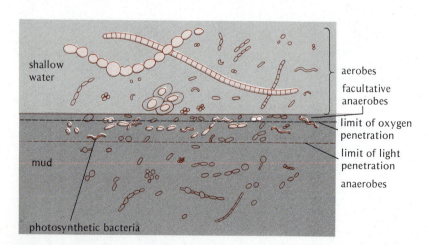

shallow water

mud

photosynthetic bacteria

aerobes

facultative anaerobes

limit of oxygen penetration

limit of light penetration

anaerobes

Figure 22-5. Zonation of aerobic, facultatively anaerobic, and obligate anaerobic microorganisms in and above the mud at the bottom of a pond, puddle, or ditch. Photosynthetic organisms are shown in color. Oxygen production by photosynthetic cyanobacteria and eukaryotes in the water keeps it aerobic. Photosynthetic true bacteria, which are anaerobes, occur in the upper, illuminated layers of mud below the limit of oxygen penetration.

TABLE 22-2.   Comparison of Types of Photoautotrophic Bacteria and of Two Eukaryotic Groups.

	Purple Bacteria	Green Bacteria	Cyano-bacteria	Eukaryotic Green Algae and Plants
Primary carbon source	CO_2 or organic compounds	CO_2 or organic compounds	CO_2	CO_2
Reductants for CO_2 assimilation	H_2, H_2S, or organic compounds	H_2, H_2S	H_2O	H_2O
Principal photoactive pigments	bacterio-chloro-phyll a or b	bacterio-chloro-phyll c, d, or e	chlorophyll a, phycobili-proteins	chlorophylls a and b
Photosystem				
I (produces no oxygen)	+	+	+	+
II (oxygen-producing)	−	−	+	+

Photoautotrophic Bacteria

The cyanobacteria, or blue-greens, are of course aerobic, since they produce oxygen and must be able to tolerate its presence. They are described in the final section of this chapter. In contrast, the few kinds of photoautotrophic eubacteria are mostly obligate anaerobes, growing in oxygen-free habitats such as lake- and pond-bottom mud. They use an externally obtained chemical reductant such as H_2S or H_2 for photosynthetic reduction of CO_2 (Chapter 7), oxidizing this reductant in the process and producing no oxygen, which would be toxic to them. The two groups of photoautotrophic eubacteria, the **green bacteria** and the **purple bacteria,** differ, as their names imply, in the types of chlorophyll and photosynthetic accessory pigments they possess (Table 22–2) and also in certain features of intracellular structure.

The chlorophylls of photosynthetic eubacteria are a family of pigments called **bacteriochlorophylls,** which differ structurally from the cyanobacterial and plant chlorophylls (Chapter 7) and differ from one another in smaller ways analogous to the differences between plant chlorophylls a and b (Figure 7–6). Besides differences in their bacteriochlorophylls (Table 22–2), green bacteria also differ from purple bacteria in their carotenoids, which are responsible for the often red or purple color of cells of purple bacteria.

Culturing Bacteria

The recognition in the 1800s that bacteria cause diseases created a major impetus for isolating, identifying, and studying bacterial species. To distinguish different kinds of bacteria that often look similar microscopically it was found necessary to isolate each in **pure culture** and to use physiological criteria for identification in addition to cell morphology. A pure or **axenic culture** (Greek *xenos*, foreign, plus *a-*, not) consists of a single kind of organism growing on a nutrient me-

dium (Figure 22-6A) under "sterile" conditions (i.e., without contaminating foreign organisms). The nutrient medium or environment must include an energy source plus sources of carbon, nitrogen, and other essential elements appropriate for the organism being cultured, plus any special auxotrophic requirements such as vitamins. Isolation of pure cultures is accomplished by spreading onto a sterile, solid nutrient medium a suspension of microbial cells, much diluted to assure separation of individual cells from one another on the surface of the medium (Figure 22–6C). These cells grow and multiply until the accumulated progeny of each original cell form a mass of cells known as a **colony** that is large enough to be visible with the naked eye (Figure 22–6B). A colony is equivalent to a clone of eukaryotic cells or organisms; all cells of the colony have arisen by asexual reproduction of one cell and are genetically identical except for mutation.

Colonies of different types of bacteria often have different characteristics, such as color, texture, size, and margin appearance (Figure 22–6B), which aid in distinguishing kinds of bacteria. With a sterile needle, cells from one isolated colony can be transferred to a fresh nutrient medium for further growth (the process of **subculturing**). By transferring cells periodically to fresh sterile nutrient media, a pure culture of one kind of bacterium can be maintained indefinitely if the appropriate growth conditions are provided. The solidifying agent normally used in laboratory media is **agar,** a polysaccharide (obtained from certain red algal seaweeds) that very few microorganisms can break down.

Enrichment Cultures

In many cases, the key to isolating a desired kind of prokaryote is to use a **selective medium** and selective growth conditions, taking advantage of particular nutrients needed (or not needed) and specific, unusual conditions needed or tolerated by the desired bacte-

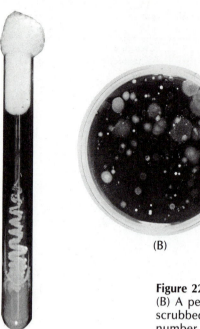

(A)

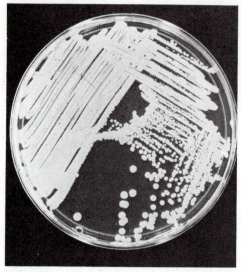

(B)

(C)

Figure 22-6. (A) A pure culture of a bacterium growing on a nutrient agar medium. (B) A petri dish filled completely with solid medium was inverted against a recently scrubbed floor, then removed. After several days of incubation, adhering cells of a number of species have multiplied, resulting in visible colonies. (C) Colonies originating from single cells can frequently be obtained by streaking dilute suspensions of bacteria on a solid medium. This is one procedure used in obtaining pure cultures.

rium to the exclusion of most others. This is called an **enrichment culture.** For example, organisms that fix atmospheric nitrogen can be isolated on culture media lacking combined nitrogen because other organisms cannot grow without an added nitrogen source. Similarly, to isolate soil organisms capable of breaking down organophosphate pesticides, small amounts of the pesticide are incorporated into a culture medium lacking any other phosphorus source, so that only organisms able to degrade the pesticide and obtain phosphorus from it can multiply.

CELL STRUCTURE

Cell Walls and Surface Structures

Bacteria were early classified into two major groups based on a staining procedure called the Gram stain. **Gram-positive** bacteria retain the purple color of a dye used in the procedure, while **Gram-negative** cells do not and can be distinguished by subsequently staining with a second dye of a different color. Gram-positive bacteria have cell walls of relatively simple structure, composed largely of a polymer called **peptidoglycan,** which contains both carbohydrate and amino acid subunits. Gram-negative bacteria have a more complex, multilayered wall containing a number of components besides peptidoglycan, including lipopolysaccharides, which are complex macromolecules composed of sugar units attached to lipid molecules. Many bacteria also secrete mucilaginous or slime-like polymers outside their cell walls, forming an external sheath often called a capsule (Figure 22-7).

Peptidoglycan is an interesting polymer composed of polysaccharide chains cross-linked into a molecular network by short polypeptide chains (Figure 22–8). This provides the semirigid framework of most prokaryotic cell walls, just as cellulose does in most plant cell walls. Since cross-linked peptidoglycan chains apparently extend all the way around a bacterial cell, the bacterial wall has been called a "bag-shaped macromolecule." In order to grow, the bacterial cell must break its enclosing peptidoglycan chains at certain points, introducing new chains into the structure.

By marking the bacterial cell wall with radioactive or fluorescent tracers it can be shown that the cells of cocci grow in a local band around their equator (Figure 22–9B). Some rods seem to elongate by growth in local patches or bands (Figure 22–9A), unlike most elongating higher plant cells, which grow throughout their length (Chapter 12).

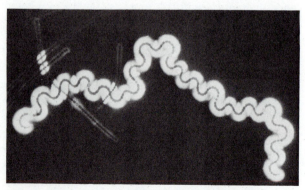

Figure 22-7. Thick mucilage surrounds this chain of cells of a cyanobacterium (*Anabaena spiroides*). Suspending the cells in a dilute solution of ink allows the normally invisible sheath to be seen with the light microscope. (80 ×)

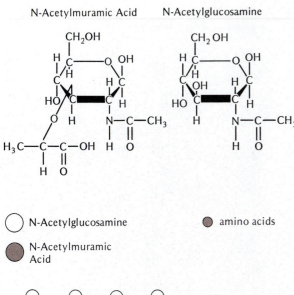

N-Acetylmuramic Acid   N-Acetylglucosamine

○ N-Acetylglucosamine   ● amino acids

● N-Acetylmuramic Acid

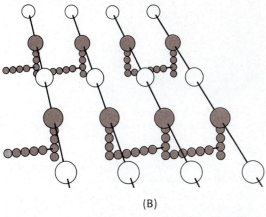

(B)

Figure 22-8. The structure of peptidoglycan, present in all walled prokaryotes except the Archaebacteria, is diagrammed in (B). The two amino sugars shown in (A) alternate to form the backbone chains of the molecule in (B). Each N–acetylmuramic acid unit usually has attached a chain of four amino acids (tetrapeptides) shown as small vertical circles in (B). Other amino acids can form cross-bridges between some of these tetrapeptides, forming a very large molecule that can extend in a layer around the cell. *E. coli* has one such layer while most Gram-positive bacteria have many layers of peptidoglycan molecules in their walls.

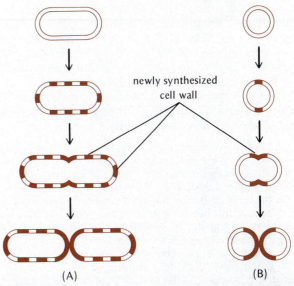

newly synthesized cell wall

(A)   (B)

Figure 22-9. Modes of cell growth demonstrated by different kinds of bacteria: (A) by insertion of new cell wall components in patches or bands about the cell, or (B) by insertion in a midzone region of the cell, which will become the poles of the daughter cells following septum formation. (After E. Nester et al., *Microbiology* 2nd ed., 1978.)

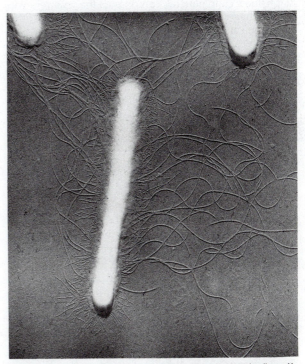

Figure 22-10. A rod bacterium having both flagella (long, wavy, hair-like structures) and numerous short pili surrounding the cells. Cell movement caused by the flagella is an energy-requiring process. (20,000 ×)

FLAGELLA AND OTHER MOTILITY STRUCTURES

Most motile bacteria have whip-like structures called **flagella** (singular, flagellum) projecting from the cell surface (Figure 22–10), anchored in the cell wall and to the plasma membrane by a specialized hook-like region. Flagellated bacteria have one to many flagella arranged at specific locations, such as at the ends of rods or spirals or distributed around the cell surface. By a rotary motion, driven by ATP from within the cell, flagella propel the cell. A few eubacteria and all cyanobacteria that move do so by **gliding** rather than by flagellar activity. Cells of gliding species do not have any apparent cell appendages associated with the property of motility, although most secrete mucilage and leave slime trails.

Pili (pronounced "pilly"), hair-like structures protruding from the cell surface, are even smaller than flagella and can be seen only with the electron microscope (Figure 22–10). They occur mainly on Gram-negative bacteria, including the enteric bacteria that inhabit the animal and human gut, such as *E. coli*. Special pili that aid the transfer of genetic material occur in conjugating bacterial species (see below).

Flagella and pili consist of protein subunits arranged in a helical manner about a hollow center. The proteins, however, differ in flagella of different species and in different kinds of pili. Prokaryotic flagella are simpler and fundamentally different from those of eukaryotic cells.

Cell wall and surface components are very important in human defense mechanisms against pathogenic bacteria. Many, if not most, of the antibodies the human body forms after contact with bacteria are directed against capsular, flagellar, or cell-wall molecules. Many of the antibiotics used against pathogenic bacteria act by interfering with the synthesis or condensation of the bacterial cell wall, thereby stopping growth or causing cell lysis (disintegration). For example, penicillin blocks peptidoglycan synthesis and is particularly effective in stopping growth of Gram-positive species such as staphylococci. In general, compounds that interfere with the synthesis of unique cell-wall polymers or with other biochemical pathways found only in prokaryotes are the preferred types of therapeutic agents used in medicine to treat bacterial infections and diseases because such compounds do not interfere with the metabolism of the host's cells. That is, such antibiotic compounds are selective or specific in their action against the prokaryotic cell.

Internal Structure

The DNA of prokaryotic cells is typically located in a central region of the cell, sometimes called a **nucleoid,** that appears less dense in the electron microscope and can be seen as fibrils if special preparative techniques are used (Figure 22–23B). As emphasized in Chapter 4, the nucleoid is not bounded by a membrane. Prokaryotic cells also have no endoplasmic reticulum; all ribosomes are free in the cytoplasm surrounding the nucleoid region. Prokaryotic ribosomes differ from eukaryotic ones in being smaller (about 15 nm in diameter compared with about 20 nm in eukaryotic cytoplasm) and in having protein and RNA components that are structurally different from their eukaryotic counterparts. Further differences occur in their mechanisms of protein synthesis.

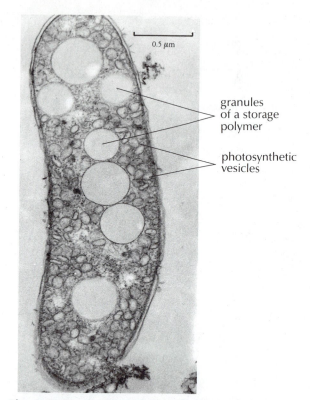

granules of a storage polymer

photosynthetic vesicles

Figure 22-11. A thin section of a photosynthetic bacterium showing the numerous intracellular membranes that contain light-absorbing pigment molecules. These invaginations from the cell membrane appear as rounded or elongated vesicles when sectioned for examination in the transmission electron microscope.

MEMBRANES

Prokaryotic cells lack mitochondria; the respiratory electron transfer chain of aerobic bacteria is located in the cell's plasma membrane. From this membrane, according to the theory of endosymbiotic origin of mitochondria in eukaryotes (Chapter 21), arose the mitochondrial inner membrane bearing the mitochondrial electron transfer chain components.

The cytoplasm of prokaryotic cells sometimes contains membranous structures. Certain species with high respiration rates, such as nitrifying bacteria (Figure 22–23), have plasma membrane infoldings carrying additional respiratory enzymes. Cyanobacteria and other photosynthetic bacteria (Figure 22–11) have internal membranes, thus increasing their photosynthetic pigment-carrying surface. These infoldings and intrusions of the plasma membrane into the cytoplasm may look like vesicles or cisternae, depending on the species, but in most species they are actually connected with the plasma membrane at some point.

Storage Reserves

Like eukaryotic plant cells, prokaryotic cells can accumulate in their cytoplasm granules of storage reserves such as the polysaccharide glycogen ("animal starch," a glucose polymer similar to plant starch). However, these reserve granules are not formed within membranous organelles as are the storage reserves of plant cells.

Many bacteria store globules of a lipid polymer, poly-β-hydroxybutyrate, a polyester chemically analogous to the extracellular cutin polyester of the higher plants' epidermis (Chapter 5). Some prokaryotes store granules of polyphosphate (Figure 22–12), a polymer of inorganic phosphate units. This apparently serves not only as a phosphate reserve but also as an energy store, because the bonds between the phosphate groups of polyphosphate are high-energy phosphate bonds, and phosphate groups can be transferred from the polymer to ADP, forming ATP. Polyphosphate is considered a primitive high-energy compound that may have been used in energy metabolism by early forms of life before the evolution and perfection of ATP as an energy currency.

PROKARYOTE CLASSIFICATION AND EVOLUTION

Because bacteria exhibit so many different combinations and permutations of cell and biochemical characteristics, it has proved difficult for microbiologists to perceive or agree on well-defined natural groups within the eubacteria. The traditional microbiological approach to classification is, therefore, largely pragmatic and, compared with botanical classification (Chapter 21), not greatly concerned with attempts to express phylogenetic relationships. Eubacteria are classified into formally named Families analogous to botanical Families, but above the Family level they are conventionally lumped into a series of about 16 informal, probably mostly artificial, numbered but not formally named groups. These are defined by combinations of physiological and morphol-

ogical characteristics such as cell shape, Gram-staining properties, motility, endospore formation, aerobic or anaerobic habit, and type of nutrition (photoautotrophic, chemoautotrophic, etc.)

Recently, progress in clarifying evolutionary relationships among bacteria has been made by comparing ribosomal RNA and protein sequences, which are coded by evolutionarily conservative genes. The considerable divergence in these sequences found among bacteria, as represented in simplified form in Figure 22–13, suggests that many of the bacterial genera or even species separated evolutionarily from one another very long ago. The separation between Gram-positive and Gram-negative eubacteria, and between these and cyanobacteria, appears very ancient (Figure 22–13). The ribosomal RNA sequence results also agree with the theory that plant chloroplasts were derived from cyanobacteria or related primitive photoautotrophic prokaryotes (Chapter 21).

Archaebacteria

By their ribosomal RNA sequences a special subgroup of bacteria, named Archaebacteria, has been recognized that differ as greatly from the remaining bacteria (called Eubacteria proper) as the eubacteria differ from eukaryotes (Figure 22–13). The Archaebacteria also differ fundamentally from other prokaryotes and eukaryotes in certain other characteristics, including their RNA polymerases and membrane lipids, and they differ from other prokaryotes in lacking a normal peptidoglycan in their walls (some have an analogous but chemically different polymer). Therefore, as their name implies, Archaebacteria are thought to be an extremely ancient group that separated evolutionarily from other prokaryotes even earlier than eukaryotes did—according to some investigators perhaps as long as 4 billion years ago. The several chemoautotrophic members of the Archaebacteria differ from chemoautotrophic and photoautotrophic eubacteria and cyanobacteria in fixing CO_2 not by means of the Calvin cycle (Chapter 7) but by a cycle involving organic acid intermediates, suggesting that autotrophy originated independently in the Archaebacteria and in the other prokaryotes. It has been proposed that Archaebacteria should be recognized as a separate kingdom, distinct from both the Eukaryota and the rest of the Prokaryota.

Archaebacteria grow in restricted, specialized habitats, some of which may resemble conditions that existed on earth more than 3 billion years ago, such as acid hot springs and extremely saline environments (e.g., *Halobacterium*, mentioned previously). Many of the archaebacteria are anaerobes. The best known are the chemoautotrophic methane producers, which derive their energy anaerobically by fermenting methanol or oxidizing hydrogen gas, released by other bacteria, with CO_2 to produce methane (CH_4):

$$CO_2 + 4\,H_2 \rightarrow CH_4 + 2\,H_2O + \quad 32\ \text{kcal}$$
$$\text{(energy yield)}$$

This is the basis for production of methane fuel by microbial decomposition of garbage and other wastes, a process of considerable current interest as a possible energy source.

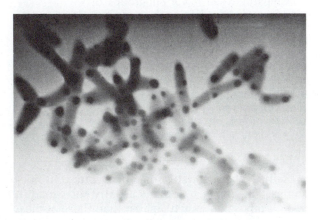

Figure 22-12. Polyphosphate granules are accumulated by a number of bacteria (*Corynebacterium diphtheriae* shown), and are visible under a light microscope when stained with basic dyes or as electron-dense regions in cells viewed in an electron microscope.

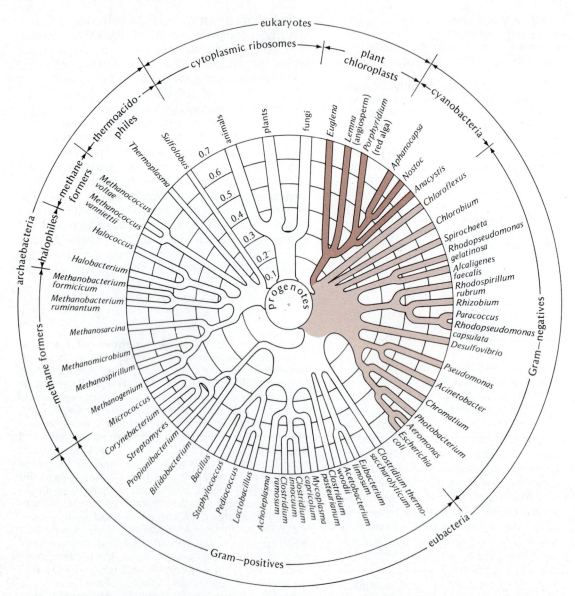

Figure 22-13. Diagram showing possible evolutionary relationships between major subgroups of prokaryotes (and between them, plant chloroplasts and eukaryotic cells) based on analysis of ribosomal RNAs. The sequence similarities of the ribonucleic acid bases were compared among the small ribosomal subunits ("16S" ribosomal RNA of prokaryotes and plastids and "18S" RNA of eukaryotic cell ribosomes). The circular scale from 0.1 to 0.7 is an index of sequence similarity (1.0 = identical sequences). It has been estimated that sequences of similarity 0.6 probably diverged in the early Paleozoic (500–600 million years ago), whereas those of similarity less than 0.3 probably diverged more than 3.4 billion years ago, that is, prior to the earliest known fossil prokaryotic cells. These latter divergences probably go back to the time of "progenotes" or primordial organisms (Chapter 20). Lighter color indicates lines of nonoxygenic (bacterial) photosynthesis; heavier color indicates the lines possessing oxygen-producing photosynthesis. (After O. Kandler, *Naturwissenschaften* **68**:185, 1981)

PROKARYOTE GENETIC MATERIAL AND GENETICS

Prokaryotic genetic material consists of circular double-stranded DNA molecules, termed chromosomes even though they lack the complex structural organization, with associated histone proteins, characteristic of eukaryotic chromosomes. The entire genome is usually contained in one chromosome, much smaller than those of most eukaryotic organisms because the smaller cell size and generally lower degree of complexity of prokaryotes require less genetic information. The *E. coli* chromosome contains 4.5×10^6 nucleotide pairs, compared with 1.5×10^{10} pairs in the genome of corn (*Zea mays*).

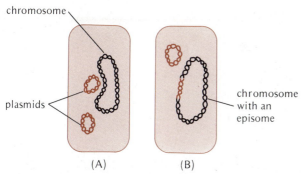

Figure 22-14. (A) Many bacterial strains contain plasmids, extrachromosomal DNA that is self-replicating and contains additional genetic information. One or a number of "copies" may be present. (B) Integration of plasmid DNA into the chromosome allows replication with the chromosome, and the incorporated DNA is termed an episome.

Because prokaryotic cells normally contain just one type of chromosome, they are genetically haploid. Other genetic elements in addition to the chromosome can occur in prokaryotes. These include nonchromosomal replicating DNA fragments known as **plasmids,** which are circular double-stranded DNA molecules usually much smaller than the bacterial chromosome (Figure 22–14A). Some plasmids have short DNA base sequences (insertion elements) complementary to sequences in the bacterial chromosome, whereby they can be incorporated (integrated) into the chromosome. They are then replicated along with the chromosomal DNA. Such plasmids are called **episomes** (Figure 22–14B). Other plasmids remain independent of the chromosome in the bacterial cell. Much current interest centers on plasmids because they can be used to transfer genes from one species to another for genetic engineering.

The plasmids found in prokaryotes in nature usually confer some selective advantage on the bacterium containing them. Some plasmids or episomes contain genes bestowing drug resistance on the host bacterium. For example, the pathogenic bacterium *Shigella dysenteriae,* which is responsible for bacillary dysentery, can acquire an episome rendering it insensitive to several antibiotics, including streptomycin and tetracycline.

Genetic Recombination in Prokaryotes

Certain bacteria have ways of obtaining one of the advantages of sexual reproduction, genetic recombination. These mechanisms probably operate infrequently in nature, since they depend mostly on close contact of large populations of compatible strains of bacteria. Perhaps specialized habitats such as the human gut, with its enormous numbers of bacterial cells, meet the conditions for gene transfer between bacteria. We shall consider these mechanisms because they form the basis of various proposals for improving the productivity of higher plants by genetic engineering.

There are three ways of transferring genetic material between bacterial strains: (1) **transformation,** in which cells take up "foreign" DNA from their external environment; (2) **transduction,** in which a bacterial virus (bacteriophage) transfers a fragment of a bacterial chromosome to another bacterium; and (3) **conjugation,** in which DNA passes directly from one bacterial cell to another. In all these mechanisms genetic information is transferred in one direction, from a donor genome to a recipient cell.

TRANSFORMATION

In transformation, free DNA released or extracted from one organism is taken up by cells of another organism and incorporated into its genome; a strand of the absorbed foreign DNA replaces a homologous segment (a segment containing similar genetic loci) of the recipient chromosome (Figure 22–15). For this replacement to occur the base sequence of the introduced DNA fragment must be similar to that of a segment of the recipient's chromosome. Thus transformation is most easily accomplished between different strains of a single species having many common DNA sequences, but successful transformations between different bacterial genera have been made.

Expression of the introduced genes may result in a phenotypic change in the recipient. In the first transformation experiments, the genes substituted into the recipient bacterial strain changed the colony appearance and the virulence (disease-causing ability) of that strain.

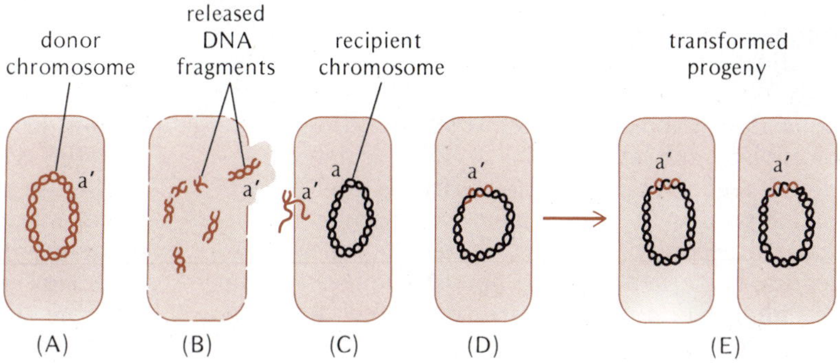

Figure 22-15. Transformation. DNA from a donor strain (A, B) is taken up by a recipient cell (C). Substitution of a fragment of donor DNA in the chromosome of the recipient (D) changes the genetic constitution (gene *a* is replaced by gene *a'*). The "new" genes appear in progeny cells (E).

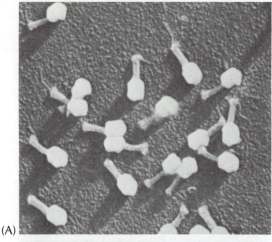

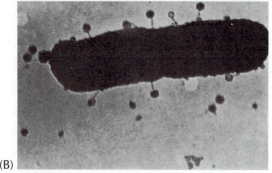

Figure 22-16. (A) Bacteriophage T₂ of *Escherichia coli*. Several of the well-studied bacteriophages have a characteristic tadpole-like shape with a polyhedral protein-coated head containing DNA, and a cylindrical protein tail ending in fibers which attach it to the host bacterial cell. (B) Bacteriophages (T₅ of *E. coli*) infecting a host cell. The viral DNA is transmitted through the protein sheath of the phage tail into the bacterium.

Transformation of higher plant cells with DNA from other plants has been attempted repeatedly, but a convincing demonstration of the integration and action of foreign genes has been achieved in few, if any, instances.

BACTERIAL VIRUSES (PHAGES)

To understand the second method of bacterial genetic recombination, transduction, we must first consider the nature of a bacterial virus. Viruses are self-replicating entities that use living cells to supply the energy and building blocks for their replication. They are generally smaller than the smallest bacteria (Figure 22–1) and can be seen only with the electron microscope. Most viruses consist of a sheath of protein surrounding a core of nucleic acid, either DNA or RNA, which carries the viral genetic information. Viruses infect all the major groups of organisms; plant-pathogenic viruses are considered in Chapter 26. Viruses infecting bacteria are called **bacteriophages** or simply **phages** (Figure 22–16A) (from Greek *phagein*, to eat, a misnomer).

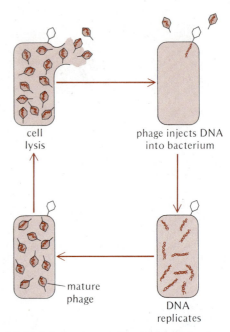

Figure 22-17. Infection cycle of a bacteriophage. Viral genes injected into the bacterium (upper right) direct the synthesis of coat proteins and their own replication. Following a period of synthesis, many new viruses are assembled and the host cell lyses, releasing new infective bacteriophages.

After attachment of the protein coat of a phage to the bacterial host cell (Figure 22–16B), the phage's nucleic acid is transmitted into the bacterium. Once inside the cell, viral genes direct the host's biochemical machinery to synthesize many copies of the viral DNA and the viral coat proteins. From these copies new phage particles assemble, and the host cell bursts (lyses, Figure 22–17), enabling the released phage particles to infect other bacterial cells.

Some phages do not immediately replicate and lyse their host cell following infection. The DNA of latent or **temperate** phages is replicated along with the bacterial chromosome, like plasmid DNA described above, and is carried along in the progeny of the bacterium until the phage is **induced** (either spontaneously or by some agent such as certain drugs or ultraviolet light) to multiply and lyse the host cell.

TRANSDUCTION

Certain temperate phages mediate transfer of genes from their host bacterium to a recipient bacterium, a process called transduction. This gene transfer is made possible by the integration of temperate phage DNA into the bacterial chromosome, acting essentially like an episome. When it subsequently becomes de-integrated, the phage DNA can carry with it genes from the host chromosome, transferring these to a cell infected by the released phage particle (Figure 22–18). The

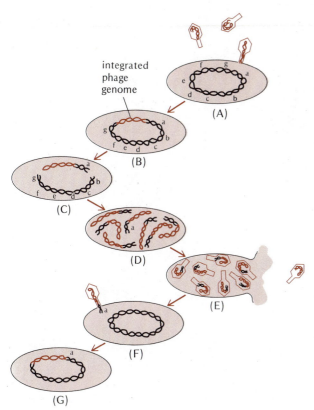

integrated
phage
genome

(A)

(B)

(C)

(D)

(E)

(F)

(G)

Figure 22-18. Transduction by temperate bacteriophages. (A) Bacteriophage infects bacterium and becomes latent (B). Viral genome is carried along and replicated with host chromosomal genes for varying time periods, through a number of cell generations. Induction causes deintegration of phage DNA, which may carry along a portion of the bacterial chromosome (C), and replication of the viral genome and attached host DNA fragment (D). Incorporation of the DNA into coat proteins and lysis of donor cell (E) releases infective phages. Infection of a recipient strain (F) introduces donor genes (*a*), resulting in recombination (G).

transferred genes may become integrated permanently into the new host's chromosome.

De-integration of phage DNA from the host chromosome occurs by excision (splitting out) of the phage base sequence by cuts made at its ends, where it is joined to the chromosomal DNA strand. Occasionally, the phage DNA is not excised from the host chromosome exactly, some adjacent genes of the host chromosome being included in the excised DNA segment (Figure 22-18). Upon induction of phage multiplication as described above, these genes are replicated and packaged into phage heads along with the phage DNA to which they are attached. Following lysis and infection of new host cells, these genes may replace homologous genes in the recipient bacterial cells.

CONJUGATION

Certain plasmids, termed conjugative plasmids, can be transferred from one bacterial cell to another by conjugation, or temporary cell fusion. Conjugative plasmids contain genes that cause conjugation between their host cells and other cells not carrying the plasmid. Either the plasmid alone is transferred from host to recipient cell (Figure 22-19A) or, if the plasmid has been integrated into the host chromosome, part or all of the chromosome may be transferred (Figure 22-19B), making possible extensive recombination between genes of donor and recipient strains.

A conjugative plasmid that has appeared on the medical scene with serious consequences is responsible for the occurrence of penicillin resistance in *Neisseria gonorrhoeae,* the Gram-negative coccus that causes gonorrhea. The widespread application of penicillin therapy in the late 1940s brought this venereal disease under control temporarily, but the rapid spread of resistant strains of *N. gonorrhoeae* since 1976 has again made gonorrhea a major public health problem. Genes providing resistance to many antibiotics, by coding for enzymes that act upon the drugs, are frequently carried by plasmids, as noted above. Resistance plasmids (also called R factors) have been found in a number of different pathogenic bacterial genera, and many carry resistance genes to several antibiotics at once, making the bacterium particularly difficult to combat.

Conjugative plasmids contain genes coding for production of "sex pili" in a host strain carrying the plasmid. When a "donor" bacterial strain carrying a conjugative plasmid is mixed with a non-plasmid-containing strain, pili on donor cells form bridges to cells of the other (recipient) strain. The cells then move together, probably by retraction of the pili, and plasmid DNA is transferred into recipient cells. Replication of the plasmid DNA in the donor cell accompanies transfer, so after conjugation both cells contain the plasmid (Figure 22-19A).

Conjugative plasmids are integrated into the host chromosome infrequently (in about 1 cell in 10,000). In episome form the conjugative plasmid can still induce its host cell to form sex pili and to conjugate. When mixed with recipient cells, the chromosome of the donor cell breaks upon conjugation and the plasmid genes carry along the chromosome (Figure 22-19B). Transfer of the entire chromosome into the recipient cell usually takes more than an hour. If the conjugating cells are agitated during the transfer period, they break apart after only a portion of the donor chromosome has been transferred. After conjugation the recipient cell will have received all or, more often, only a part of the donor chromosome. A portion of the introduced chromosome may infrequently integrate into and replace genes in the recipient cell's chromosome, accomplishing recombination between genes of the donor and recipient strains.

Gene Transfer by a Plasmid into a Plant Cell

A case in which a bacterial gene carried by a plasmid can be transferred to, and acts in, a higher plant cell is the disease **crown gall,** a tumor-like growth of plant tissue induced by infection with the bacterium *Agrobacterium tumifaciens* (Figure 22-20). Apparently the bacteria cause tumorous

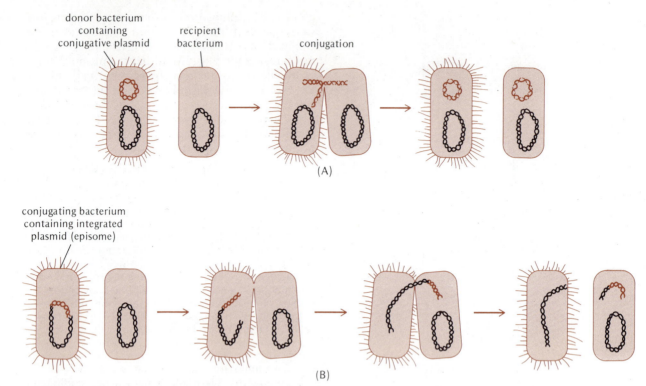

Figure 22-19. (A) Conjugation with transfer of the conjugative plasmid. One DNA strand is transferred and generally is replicated at the same time (dotted line). Cells with conjugative plasmids first contact recipient cells via pili on the cell surface. (B) Transfer of some or all of the donor cell's chromosome can occur during conjugation of cells containing integrated conjugative plasmids.

(callus-like) growth by inducing the plant cells to produce growth-stimulating quantities of auxin. The pathogenicity of *A. tumifaciens* strains is correlated with the presence of a large plasmid called Ti (tumor-inducing plasmid) in the bacterial cells. Tumor tissue can be grown in laboratory culture, and the causal bacteria can be eliminated by heat or antibiotic treatments without killing the plant cells, which continue to grow in tumor form and to contain the Ti plasmid.

The Ti plasmid can carry bacterial genes that cause the tumor tissue to synthesize certain unusual nitrogen-containing compounds (opines) not normally found in plant tissues. Thus, in crown gall a bacterial plasmid transfers genes to higher plant cells, and that genetic information becomes expressed in the host cells. Genetic engineering researchers hope to use *Agrobacte-*

Figure 22-20. Tumorous growth (a gall) of plant tissue caused by infection with a plasmid-carrying common soil bacterium, *Agrobacterium tumefaciens*. These galls occur on the aerial or crown portions of many plant species, frequently near the juncture of stem and root.

rium strains carrying Ti or similar plasmids to transfer desired genes, such as those for nitrogen-fixing enzymes, into plant cells.

NITROGEN-FIXING BACTERIA

Among the most important ways in which prokaryotes affect the nutrition and activities of plants, thereby indirectly affecting the entire ecosystem, is by interconverting different chemical forms of nitrogen occurring in the environment. Nitrogen fixation is second only to photosynthesis in importance for maintaining life on earth, because nitrogen is essential to all life and is usually the limiting nutrient for terrestrial and marine productivity. Although nitrogen gas (N_2) makes up 78% of the atmosphere, most organisms cannot use it as a nitrogen source. Only certain bacteria and blue-greens can convert atmospheric nitrogen to combined nitrogen that cells can use in forming nitrogen-containing organic molecules.

Nitrogen-Fixing Organisms

Nitrogen-fixing prokaryotes include free-living nitrogen fixers and those that fix nitrogen in a symbiotic association—for example, symbiosis between legumes such as clover and alfalfa and their root nodule bacteria, members of the genus *Rhizobium*. Symbiotic nitrogen fixers (Chapter 25) are the most important agriculturally, but in certain kinds of vegetation and in the open ocean, free-living nitrogen fixers are often ecologically significant (Chapter 33). The process of nitrogen fixation is similar in both; it will be discussed here for free-living eubacteria and (later in the chapter) for cyanobacteria.

Free-living, nitrogen-fixing prokaryotes enrich the soil or water with nitrogen when they die, and their accumulated organic nitrogen is released by decomposition. Two common genera of free-living, heterotrophic soil bacteria that include nitrogen-fixing species are *Azotobacter* and *Clostridium*. *Azotobacter* is aerobic, growing in well-aerated, neutral or slightly alkaline soils. *Clostridium* species, on the other hand, are obligate anaerobes that are responsible for nitrogen fixation in waterlogged soils and acidic bogs and swamps, where anaerobic conditions prevail. Clostridia, however, are commonly present even in well-aerated, cultivated soils because the interior of soil particle aggregates (granules or crumbs, see Chapter 11) can be anaerobic due to depletion of oxygen by the respiration of microorganisms decomposing organic matter within the granules. Several of the anaerobic photosynthetic bacteria also fix nitrogen for their own growth, as do certain chemoautotrophic bacteria such as the ferrous iron oxidizer, *Thiobacillus ferrooxidans*.

The Nitrogen-Fixing Process

Nitrogen fixation occurs by conversion of N_2 to ammonia (NH_3, or ammonium ion, NH_4^+), the form of combined nitrogen that cells can use to make amino acids and other necessary organic nitrogen compounds. Conversion of N_2 to NH_3 involves reduction, that is, addition of hydrogen atoms. Nitrogen-fixing prokaryotes use a complex enzyme called **nitrogenase** to reduce nitrogen to ammonia. The nitrogenase enzyme molecule appears to be similar in all nitrogen-fixing organisms. It is composed of two different protein subunits, one containing molybdenum and both containing iron atoms, which are essential for the enzyme's function. Nitrogenase action requires ATP energy; at least 12 ATP molecules are used for each N_2 molecule reduced to NH_3 (Figure 22–21), making nitrogen fixation a very energy-expensive process.

To conserve a nitrogen-fixing organism's energy, nitrogen fixation must be closely regulated. Nitrogenase activity is inhibited by accumulation of ammonia or amino acids derived from nitrogen fixation, shutting down nitrogen fixation when the cell already contains

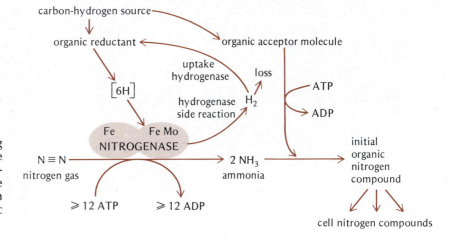

Figure 22-21. The nitrogen-fixing reaction catalyzed by the enzyme nitrogenase. The initial organic nitrogen compound formed is the amino acid glutamine, formed from the acceptor molecule, glutamic acid.

sufficient combined nitrogen. Agricultural-fertilization therefore depresses fixation of nitrogen by soil micro-organisms.

Oxygen inactivates nitrogenase and also re-presses its synthesis. Therefore, all aerobic nitro-gen-fixing organisms have an oxygen problem in fixing nitrogen. In *Azotobacter*, a very high res-piration rate, combined with the unusually large cell size of this prokaryote, keeps the intracellular oxygen concentration low enough to allow nitro-gen fixation.

Another problem with nitrogen fixation is a defect in the nitrogenase system that allows reducing power and energy to be lost through production and release of hydrogen gas (H_2). This is a side reaction, catalyzed by nitrogenase, that essentially represents failure of some of the reducing power input (electrons) to be added to nitrogen molecules (Figure 22–21). It is somewhat anal-ogous to the oxygenase side reaction of RuBP carbox-ylase in photosynthesis that leads to photorespiration and involves nonproductive wastage of energy and re-ducing power (Chapter 7). More efficient strains of ni-trogen-fixing prokaryotes possess an "uptake hydro-genase" that enables them to recapture, with energy expenditure, some of the reducing power lost owing to the hydrogenase side reaction of nitrogenase (Figure 22–21).

Increasing Biological Nitrogen Fixation

Because little or no symbiotic nitrogen fixation ac-companies growth of major cereal grain crops (Chapter 25), modern agriculture uses large amounts of fertilizer nitrogen to increase crop yields. It is estimated that biological nitrogen fixation currently furnishes only about one-half to two-thirds of the nitrogen utilized

worldwide by agricultural crops, most of the remainder being supplied by industrial nitrogen fixation, a petro-leum-dependent process. Because industrial nitrogen fixation is becoming increasingly expensive, there is great interest in increasing the amount and efficiency of biological fixation.

Many of the efforts to improve nitrogen fixation are directed toward symbiotic associations (Chapter 25). Reducing the hydrogenase side reaction and increasing the uptake hydrogenase's effectiveness are among the goals. A number of investigators, however, are hoping to modify higher plants for nitrogen-fixing capability directly by introducing genes for nitrogenase (called *nif* genes) from bacteria. Some of the requirements visu-alized for development of such a genetically engi-neered plant include (1) a portion of a prokaryotic chromosome carrying *nif* genes must be transferred into plant cells; (2) the genes must be stable in the foreign cell environment; (3) the genes must be ex-pressed, that is, transcribed and translated by the host plant cells to yield active enzyme molecules; and (4) the plant cell or tissue must provide the required en-vironment for productive nitrogenase enzyme activ-ity—for example, a low enough oxygen concentration, large amounts of ATP, and organic acceptor molecules for the fixed nitrogen.

BACTERIA TRANSFORMING SOIL NITROGEN

Ammonifiers and Nitrifiers

As noted previously, many bacteria and fungi act as decomposers, breaking down the remains of dead plants, animals, and other microbes. De-

Figure 22-22. Biological nitrogen transformations. Prokaryotes play prominent roles in transformations of fixed nitrogen in soil. Encircled reactions indicate the organisms involved.

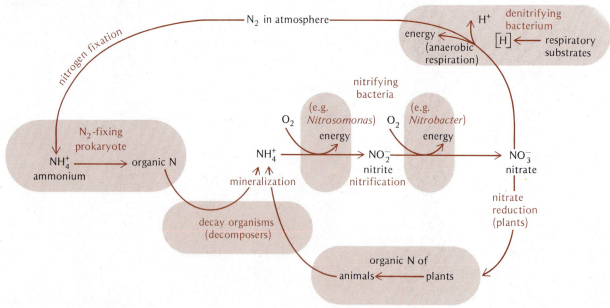

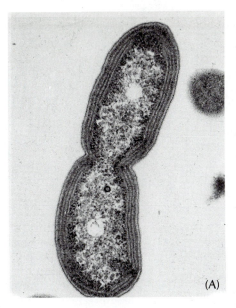

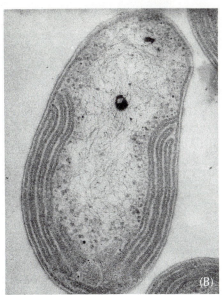

Figure 22-23. Electron micrographs of sections of (A) *Nitrosomonas,* an ammonia oxidizer, and (B) *Nitrobacter,* a nitrite oxidizer. Note the characteristic peripheral cell membranes that contain enzymes catalyzing these energy-yielding reactions. In (B) the prokaryotic nucleoid region with DNA fibrils is prominent.

composers release organically bound nitrogen (mostly from protein) as NH_4^+, a process known as **ammonification** (also called **mineralization** of nitrogen) (Figure 22–22). Ammonium, however, normally does not accumulate in the soil because other soil bacteria called **nitrifiers** rapidly oxidize it to nitrite, then nitrate, a process known as **nitrification** (Figure 22–22). *Nitrosomonas* and *Nitrobacter* (Figure 22–23) are common nitrifying bacteria in soils, but a number of other bacteria also carry out nitrification. *Nitrosomonas* and *Nitrobacter* are chemoautotrophic. They do not need an organic carbon source; from oxidation of ammonia or nitrite they obtain energy and reducing power for fixing and converting CO_2 to organic compounds. Nitrification requires oxygen and therefore occurs only in aerated soils; in flooded or poorly drained soils, nitrification is inhibited and much or all of the soil's inorganic combined nitrogen occurs as NH_4^+. This is also true in acid soils because nitrifying bacteria are not acid-tolerant.

Nitrate, although readily used by many higher plants as a nitrogen source, has the disadvantage of being readily lost from the soil. This occurs first by leaching and second by a microbial process called denitrification. Because nitrates are very water-soluble, they are rapidly leached from the soil by water flow during rain or flooding, entering ground water or streams fed by runoff. Ammonium, by comparison, is relatively immobile because it binds to the cation exchange sites (negative charges) on soil particles (Chapter 11).

Denitrification

Under anaerobic conditions, certain bacteria called **denitrifiers** can shift from utilizing oxygen as an electron acceptor for respiration to utilizing nitrate or nitrite as a substitute for oxygen. In this process, **denitrification,** nitrate and nitrite are reduced by addition of electrons (Figure 22–22), the reverse of the oxidative nitrification process discussed above. Denitrification, however, converts nitrate or nitrite to gaseous forms of nitrogen (N_2, NO, or N_2O), which rapidly escape from the soil, depleting its nitrogen content. Denitrifying ability is possessed by various facultative anaerobic bacteria and is of major importance in the nitrogen balance of the environment. The amount of combined nitrogen lost from soils by denitrification is estimated to range from 20% of applied fertilizer nitrogen in agricultural soils to as much as 100% of biologically fixed nitrogen in some nonagricultural soils. Denitrification can occur in well-aerated agricultural soils for the same reason that nitrogen fixation by anaerobes is possible in these soils—namely, because the interior of soil granules is often depleted of oxygen.

Denitrification is an example of **anaerobic respiration,** a type of energy metabolism that is an alternative to ordinary oxygen-consuming respiration. In anaerobic respiration a chemical oxidant (such as nitrate or sulfate) substitutes for O_2 as an ultimate acceptor for hydrogen atoms or electrons removed from a cell's energy substrate during energy-yielding enzymatic oxidation (Figure 22–22). Electron transfer from substrates to the electron acceptor (NO_3^-, SO_4^{2-}, etc.) occurs by means of a shorter electron transfer chain than that leading to O_2 in respiration and produces less ATP than does aerobic electron transfer to O_2 but much more than can be formed during fermentation (Chapter 6). Therefore, for organisms like denitrifiers that can conduct it, anaerobic respiration is a kind of anaerobic energy metabolism superior to fermentation because it permits faster growth under anaerobic conditions than fermentation.

Because it is cheaper to produce and apply on a large scale, ammonia is now the most widely used agricultural nitrogen fertilizer. Because nitrogen losses

TABLE 22-3. Increase in Crop Yields and Protein by Inhibiting Nitrification of Applied Nitrogen (Ammonium-Nitrogen) Fertilizer.[1]

Crop	Percent Increase in Yield	Percent Increase in Protein
Field corn	22	17 (in grain)
Cotton	10	N.D.[2]
Sugar beets	9	12 (in leaves)
Potatoes	6	50 (in tubers)
Rice	10	8 (in grain)
Wheat	18	N.D.

[1]Data averaged from figures cited by D. M. Huber *et al.*, BioScience 27:523. 1977.
[2]N.D. = not determined

from soil by leaching and denitrification are due to conversion of ammonia to nitrate, these losses could be greatly reduced if nitrification could be prevented in agricultural soils. Several chemicals that inhibit microbial nitrification are now being added to commercial fertilizers. One of these, nitrapyrin, is specifically toxic to *Nitrosomonas* and has other desirable features such as a short half-life in soil and a relatively modest cost. Table 22–3 gives data illustrating the effects of a nitrification inhibitor on fertilizer nitrogen utilization. Application of the inhibitors at the time of spring planting can delay nitrification of the applied ammonium long enough for the crop plants to absorb most or all of the supplied nitrogen before it is converted to nitrate and subjected to leaching and denitrification losses. Reducing soil nitrogen losses by this means can eliminate the need for multiple applications of nitrogen fertilizer during the growing season, saving time and fuel energy used for production and application of fertilizer. Another benefit is to decrease the fertilizer-derived nitrogen pollution of streams and lakes from runoff water from agricultural land, which results in excess algal growth (eutrophication, Chapter 33).

ACTINOMYCETES

As the suffix -*mycetes* (from Greek, *mykes*, fungus) implies, these "filamentous bacteria" were once thought to be related to fungi because they grow in a branching filamentous pattern somewhat like the filaments of fungi (Figure 22-24; see also Figure 24-A, 24-2), and, like fungi, many actinomycetes produce asexual spores for dispersal in chains at the tips of aerial branches (Figure 22-24C). Electron microscopic observations, however, show prokaryotic cell structure in actinomycete filaments, in contrast to fungi, which are

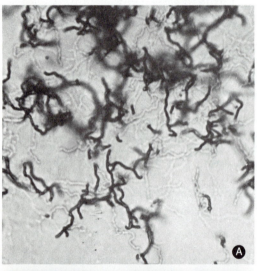
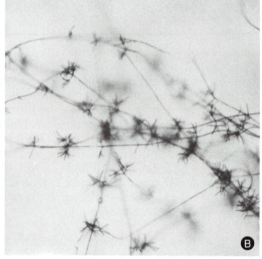
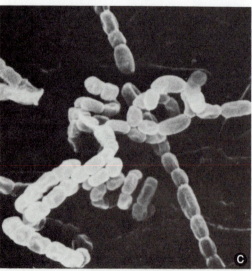

Figure 22-24. Light micrographs of two genera of actinomycetes (A, B) show the filamentous form of growth of cells. (C) is a scanning electron micrograph of spores formed by segmentation of some aerial branches of *Streptomyces*, a genus cultured for the production of a number of important antibiotics.

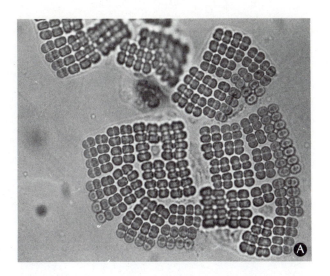

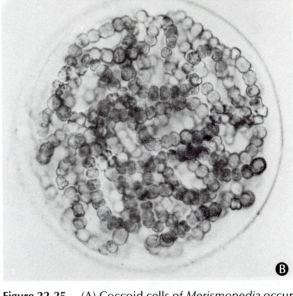

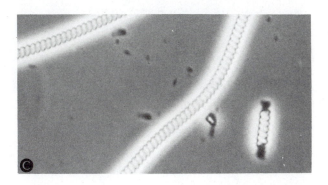

Figure 22-25. (A) Coccoid cells of *Merismopedia* occur in plate-like colonies. (B) Unbranched filaments of the bead-like cells of *Nostoc commune* occur in gelatinous aggregates that often reach macroscopic size. (Filaments of cylindrical cells are shown in Figure 22-26). Certain *Nostoc* species occur in some lichen and higher plant symbioses (Chapter 26) and contribute fixed nitrogen to their partners. (C) The helical cells of *Spirulina* occur in both marine and fresh waters and have been harvested from several lakes in Africa and Mexico for use as a human dietary protein supplement.

eukaryotes. The small cell diameter of actinomycetes (0.5–1.2 μm) is similar to that of most other prokaryotic cells and unlike that of fungi. The cell walls of actinomycetes also contain the peptidoglycans unique to bacteria.

Actinomycetes are largely nonmotile and restricted to soil, where they are among the most abundant organisms: in numbers of cells per gram of soil, actinomycetes almost equal all other kinds of bacteria combined. Certain species are responsible for the earthy smell characteristic of soil. Actinomycetes contribute importantly to decomposition and recycling processes in the environment. Various species can attack a wide variety of plant residues, as well as some rather exotic substrates such as rubber, paraffin, and cresol. Actinomycetes are important decomposers in straw and compost piles, which may become quite hot during rapid microbial degradation, sometimes even catching fire. *Thermoactinomyces* is a common thermophilic actinomycete growing in such compost heaps.

Many actinomycetes produce antibiotics that lyse and kill bacteria, a form of "chemical warfare." The majority of well-known antibiotics used medicinally against infectious bacteria are obtained from actinomycetes. Members of the genus *Streptomyces* synthesize streptomycins, actinomycins, and tetracyclines, among others. Antibiotics are produced commercially

from mass cultures of these organisms in a manner similar to the production of penicillin and other fungal products (to be described in Chapter 24).

A few actinomycetes cause diseases of humans and other animals; several *Streptomyces* species cause plant diseases such as potato and sugar beet scab. Actinomycetes of the genus *Frankia* form nitrogen-fixing nodules with certain nonleguminous higher plants such as alder (Chapter 25).

CYANOBACTERIA (BLUE-GREENS)

As noted previously, the photosynthetic prokaryotes known as blue-greens differ from other photosynthetic bacteria in producing oxygen during photosynthesis by a process similar to that used in chloroplasts of eukaryotic algae and higher plant cells (Chapter 7). Because of their plant-like mode of nutrition and a degree of resemblance to some of the simpler green algae, blue-greens have traditionally been called **blue-green algae.** However, an increasing recognition that in their prokaryotic cellular and biochemical features they are much closer to eubacteria than to any algal group has led to a growing and now predominant use of the name cyanobacteria.

The same types of cell shapes seen in eubacteria occur among cyanobacteria (Figure 22–25),

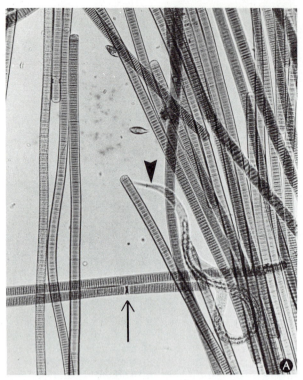
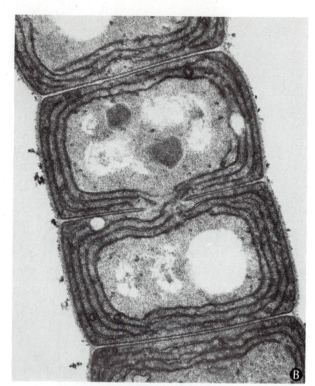

Figure 22-26.   Trichomes of *Oscillatoria* species are comprised of cells that are usually wider than long, in contrast to the elongated cylindrical cells of rod-shaped eubacteria. (A) Arrow indicates a potential fragmentation point in the trichome where the segment to the right, after separation, could glide away to populate another area. The arrowhead points to a roundworm or nematode, common soil animals comprising thousands of species, some parasitic on plants and on animals. (B) An electron micrograph of a longitudinal section through a trichome shows one cell in the process of division by fission. Thylakoid membranes are located about the cell peripheries; storage granules and the nucleoid region are also prominent.

but several of the most common cyanobacteria form long chains or filaments of cells. The type of filament developed by *Oscillatoria* (Figure 22–26A), consisting of squat cylindrical cells adhering tightly together by their end walls (Figure 22–26B), is called a **trichome** ("hair").

Trichomes of some species can move by gliding. No cyanobacteria have flagella; their gliding motion works only when they are in contact with a solid surface. Many blue-green species produce mucilaginous or slime sheaths. This gummy material sometimes holds dividing unicellular species in aggregates (Figure 22–25B) or forms thick, rigid tubes about filamentous species. A gliding filament leaves its sheath behind as a microscopic slime trail.

Like other bacteria, most cyanobacteria reproduce asexually by fission. Trichomes can break into multicellular segments called **hormogonia**, usually where certain cells have died (Figure 22–26A); the hormogonia can then glide off to a new location. In some filamentous species, thick-walled cells (**akinetes**) filled with food reserves serve as resting spores.

Photosynthetic Apparatus

The light-harvesting chlorophyll and carotenoid pigments of cyanobacteria, as in all photosynthetic organisms, occur in membranes. In blue-greens, these membranes consist of flattened sacs called **thylakoids,** which are like those in eukaryotic chloroplasts (Chapter 7). However, blue-green thylakoids do not occur in stacks (grana) and are not themselves enclosed by any envelope except the cell's plasma membrane. They occur commonly around the periphery of the cell in whorl-like or wavy conformations (Figure 22–26B) attached to the plasma membrane at certain points.

Among the various chemical forms of chlorophyll (Chapter 7), cyanobacteria contain only chlorophyll *a*, lacking both bacteriochlorophyll and the chlorophyll *b* found in green algae and higher plants. Like all photosynthetic organisms, they contain carotenoid accessory pigments. Their distinctive class of light-harvesting pigments is the **phycobiliproteins,** especially a group of blue pigments, **phycocyanins,** which are responsible for the usually bluish-green color of cyanobacteria.

Phycobiliproteins occur in small granules (**phycobilisomes**) attached to the outer surface of the thylakoid membranes of cyanobacteria (Figure 22-27). Phycobiliproteins have a light-absorbing group similar to that shown in Figure 7-6D. Three types of phycobiliproteins, called phycocyanins, allophycocyanins, and phycoerythrins, occur in cyanobacteria. Phycocyanins and allophycocyanins, occurring in all cyanobacteria, are both blue pigments but differ somewhat in their absorption spectra. If large amounts of phycoerythrins, which are orange-red, are present, they can make the cells appear red, purple, or brown instead. These accessory pigments extend the range of visible light wavelengths that are strongly absorbed by the cell for photosynthesis (Figure 22–28), compared with the wavelengths absorbed strongly by chlorophyll *a*.

Figure 22-29. Gas vesicles in a cell of *Nostoc* prepared for the electron microscope by freeze-etching.

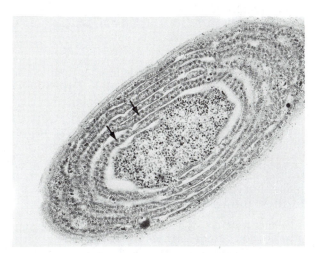

Figure 22-27. Phycobilisomes appear as granules (arrow) on the surface of concentric thylakoid membranes in this slightly oblique section of a cyanobacterial cell viewed in the electron microscope. This blue-green (*Synechococcus lividus*) grows at temperatures over 60°C in hot springs in Yellowstone National Park.

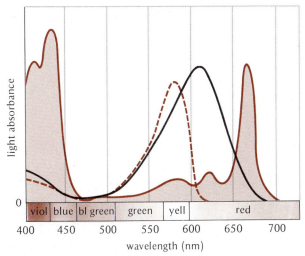

Figure 22-28. Light absorbance by photosynthetic pigments of cyanobacteria. The main phycobiliprotein pigment is phycocyanin (solid line), while the red phycoerythrin pigment (dotted line) absorbs slightly shorter wavelengths. All blue-greens also contain, besides chlorophyll *a* (shaded), β-carotene, which absorbs strongly between 450 and 500 nm where these other pigments absorb little radiation.

Some cyanobacteria differ photosynthetically from higher plants in being able to adapt their photosynthesis to anaerobic conditions, when encountered, by shifting to a non-oxygen–producing form of photosynthesis using H_2S as a hydrogen source for CO_2 reduction, like that of photoautotrophic green and purple bacteria (Table 22–1). This is a further indication of a relationship between cyanobacteria and eubacteria. The ability to adapt photosynthesis to anaerobic conditions apparently helps such blue-greens continue to grow when trapped in anaerobic mud at the bottom of a pond, as happens seasonally in certain ponds.

Other Intracellular Structures

Storage granules that can develop in vegetative cells of blue-greens include glycogen, from surplus photosynthate, and polyphosphate, which also occurs in eubacteria, as previously mentioned. In nitrogen-rich environments blue-greens can form granules containing a unique nitrogen storage polypeptide, **cyanophycin,** composed of two amino acids, arginine and aspartic acid, in equal amounts. Cyanophycin differs from ordinary polypeptides not only in its simple amino acid composition but also in the fact that ribosomes do not synthesize it.

Gas vesicles are hollow, air-filled cylinders with conical ends (Figure 22–29) found in vegetative cells of certain aquatic blue-greens and a few other prokaryotes. The surface "membrane" of these structures consists of protein subunits and contains no lipid, in notable contrast to typical biological membranes. Gas vesicles simply contain in gaseous form whatever gases are dissolved in the cell's cytoplasm rather than any specially secreted gas. One function of gas vesicles is flotation, bringing cells close to the water surface where there is maximum light intensity for photosynthesis. Flotation sometimes causes large numbers of cyanobacterial cells to accumulate at the surface of ponds and lakes, forming a "bloom," an often disagreeable scum.

Nitrogen Fixation by Blue-Greens

Nitrogen-fixing blue-greens occur ubiquitously in aquatic habitats, and on land they frequently form films on damp soil surfaces. Many investigators believe that they are responsible for most nonsymbiotic nitrogen fixation in temperate regions. They play a major role in some tropical areas. In Asia, growth of blue-greens in flooded rice paddies has enabled rice cultivation to be continued in the same fields with little apparent change in soil fertility for hundreds of years without application of nitrogen fertilizers.

HETEROCYSTS

In nitrogen-fixing filamentous blue-greens, enlarged, heavy-walled cells called heterocysts occur at intervals along the cell chains, an example of cell differentiation in prokaryotic organisms (Figure 22–30). It has recently been shown that heterocysts are specialized for nitrogen fixation; apparently most free-living, ecologically important nitrogen-fixing blue-greens are heterocyst-forming species.

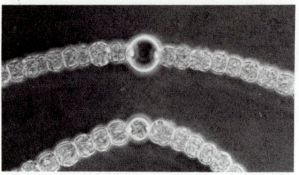

Figure 22-30. The larger heterocyst cell in a chain of *Anabaena* cells is differentiated for the function of fixing atmospheric nitrogen.

Nitrogenase, as noted previously, is sensitive to oxygen, a property incompatible with photosynthetic oxygen production by vegetative cyanobacterial cells. When isolated, heterocysts are found to contain nitrogenase and to lack the oxygen-producing and CO_2-fixing parts of the photosynthetic apparatus. Their thick wall apparently restricts oxygen uptake, enabling respiration to deplete their internal oxygen concentration to a low enough level to allow nitrogenase to function.

Since heterocysts do not fix CO_2, their carbon supply must come from neighboring vegetative cells. Thus there is a division of labor among the cells of a filament, vegetative cells supplying photosynthate to heterocysts and heterocysts supplying fixed nitrogen to vegetative cells. Heterocysts use imported photosynthate as a source of reducing power (hydrogen atoms) for reduc-

Figure 22-31. Exchanges of carbon and nitrogen compounds between a heterocyst (center cell in diagram) and the flanking vegetative cells of a cyanobacterium. Energy for nitrogen fixation by the heterocyst comes from cyclic photophosphorylation (photosystem I): no CO_2 is fixed nor O_2 produced. Both photosystems function in the adjacent vegetative cells which can supply photosynthate (sugar) to the heterocyst; fixed nitrogen can flow in both directions from the heterocyst, although arrows are not shown for all possible exchanges between the two cell types.

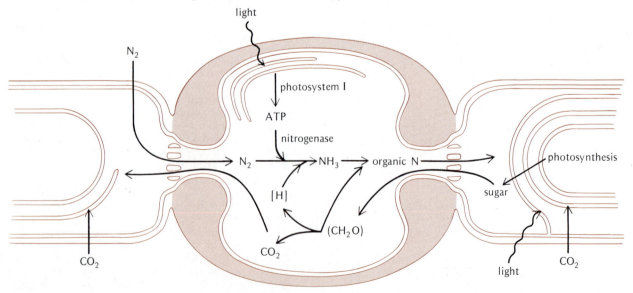

tion of N_2 by nitrogenase and for respiration to provide some of the ATP needed in this process. However, heterocysts apparently get most of the energy needed for N_2 fixation from light, because although they lack photosystem II, the oxygen-producing photosystem of photosynthesis, they still contain photosystem I and can use it to carry on cyclic photophosphorylation without producing oxygen (see Chapter 7). Figure 22–31 shows the reactions supporting nitrogen fixation that are believed to occur in heterocysts and the interchanges between them and vegetative cyanobacterial cells.

SUMMARY

(1) Important groups of prokaryotes are the eubacteria, the blue-greens (cyanobacteria), and the actinomycetes. Eubacteria and actinomycetes show great versatility in utilizing organic and inorganic substrates as carbon and energy sources and are found almost everywhere on earth. Practically all organic molecules can be degraded by at least one kind of bacterium. Some eubacteria are phototrophic (i.e., obtain energy from light), including both photoautotrophic and photoheterotrophic species, none of which produce oxygen. Other eubacteria are chemoautotrophic, but most are chemoheterotrophic. Some can tolerate environmental extremes (temperature, pH, lack of oxygen) better than most eukaryotes; spore-formers are extremely temperature-resistant.

(2) Prokaryotic cells have no membrane-bounded organelles and are typically enclosed in a peptidoglycan-containing cell wall. They are generally smaller than eukaryotic cells and have a smaller genome. The prokaryotic genome consists of a single chromosome, a closed double-stranded DNA molecule. Many prokaryotes contain other replicative DNAs, such as plasmids and viruses (phages), that may become integrated into the chromosome for varying time intervals. When free in the cell (not integrated) a phage multiplies rapidly, then causes the host cell to lyse, releasing phage particles capable of infecting other bacterial cells. Plasmids can be transferred to other cells by temporary cell contact (conjugation).

(3) Genetic recombination can occur in prokaryotes by transformation, conjugation, or transduction. Genetic engineering utilizes these processes to transfer genes between organisms.

(4) Nitrogen-fixing bacteria include both free-living and symbiotic genera. All nitrogen-fixing prokaryotes use a similar nitrogenase system to reduce N_2 to NH_3. They have varying mechanisms for protecting the enzyme from inactivation by oxygen. Large amounts of cell energy (ATP) are required for N_2 fixation. The importance of nitrogen in food production has generated interest in transferring genes for nitrogen fixation to agricultural plant cells.

(5) Ammonifying bacteria release NH_3 or NH_4^+ from organic nitrogen compounds during decay. Chemoautotrophic nitrifying bacteria in soil oxidize NH_4^+ to NO_3^- as an energy source, making nitrate the principal form of inorganic nitrogen available in aerated neutral soils. Nitrate tends to be lost from such soils by leaching and by denitrification (a type of bacterial anaerobic respiration in which nitrate substitutes for oxygen as an oxidant for respiratory substrates).

(6) Actinomycetes, or "filamentous bacteria," are heterotrophic prokaryotes that frequently grow in branching cell chains; some produce aerial spores. Many well-known antibiotics are produced by certain of these ubiquitous soil-dwellers, which are important decomposers of animal and plant remains.

(7) Cyanobacteria (blue-greens) produce oxygen in photosynthesis in the same way as eukaryotic plants. This and other similarities in the photosynthetic process link them evolutionarily to green plants. Heterocysts, specialized cells capable of fixing nitrogen, occur in some filamentous species, fixing ecologically and agriculturally significant amounts of nitrogen.

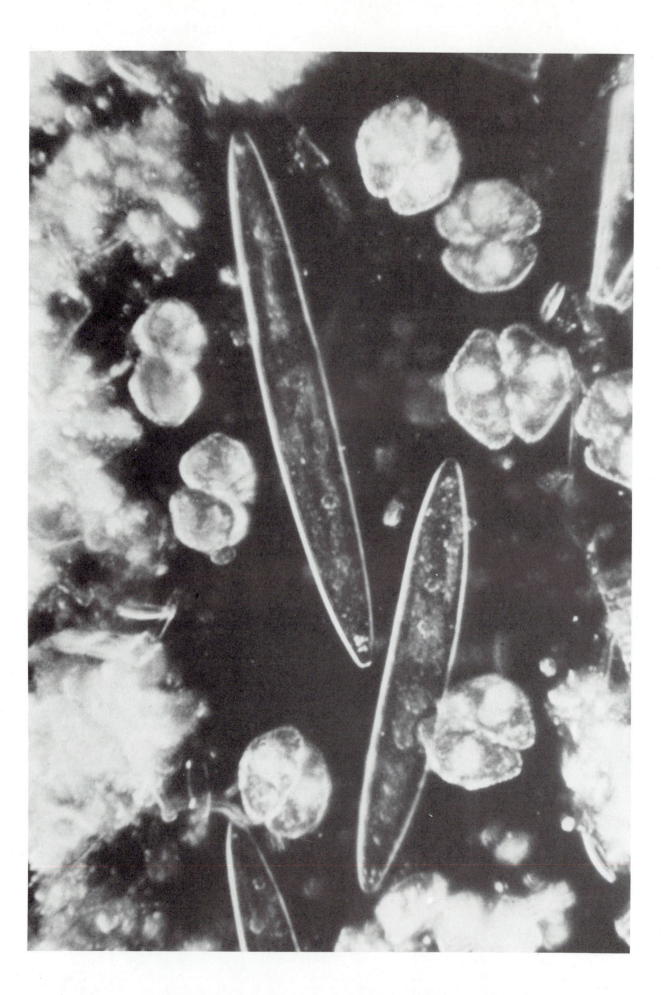

Chapter 23

THE ALGAE

The algae were the first true plants. The oldest pre-Cambrian fossils that are thought to be eukaryotic cells appear to have been unicellular algae. They were preceded, as noted in Chapter 21, by oxygen-producing photosynthetic prokaryotes, the blue-greens.* Geochemical evidence indicates that the earth's oxygen concentration was still low, however, at the time of these microfossils. Early algal photosynthesis, therefore, probably contributed importantly to building up the earth's oxygen supply, eventually making possible the evolution of higher forms of life. Readily recognizable multicellular marine algal fossils date back to the Cambrian period, still several million years before the appearance of the vascular plants.

Algae are nonvascular plants and consist of a thallus, a plant body with little or no tissue differentiation. They range from unicellular, through multicellular but still microscopic cell filaments, to large and massive macroscopic plants (seaweeds) composed of millions of cells (Figure 23-1). Although seaweeds are obviously plant-like in form and habit, growing attached to rock surfaces or other substrates, at the other extreme many unicellular algae are motile, free-swimming forms, moving about by means of flagella, such forms being called **flagellates**. Although algae are usually regarded as simple organisms, it is perhaps not surprising, considering their long history, that they have evolved many complex and often beautiful variations in structure at both the cellular and the multicellular levels, and differ from one another in processes as basic as photosynthesis, mitosis, and cell division (cytokinesis).

---

*The prokaryotic blue-greens (cyanobacteria), formerly called blue-green algae, are now classified with the bacteria and are described in Chapter 22.

Figure 23-A. These freshwater unicellular green algae with rigid cellulosic walls are called desmids. The cells of many species are composed of two mirror-image halves, or semi-cells.

THE GREAT GROUPS OF ALGAE

Phycologists, or specialists on algae (Greek *phykos*, seaweed), recognize at least eight major groups or Divisions of algae. These differ in form, in certain of their important photosynthetic pigments, and, as a result, in their usual coloration. The differences in prevailing coloration between most of the Divisions have led to common names for them that are listed, along with the formal Division names, in Table 23-1. For simplicity we can group the Divisions into three broad Subkingdoms possessing the three major types of photosynthetic pigment systems and associated chloroplast structural features.

All photosynthetic algae possess chlorophyll *a*, the basic energy-converting plant pigment (Chapter 7); the differences lie in the additional or **accessory** pigments present. The Chloronta, or green-colored algae, have the same kind of photosynthetic pigment system as higher plants, namely chlorophyll *b* in addition to chlorophyll *a*, plus a relatively modest content of yellow carotenoids. The Chromonta, or yellowish to brown algae, possess larger amounts of carotenoid pigments, often of distinctive structure; they lack chlorophyll *b* but usually have a different form of chlorophyll called chlorophyll *c*.* The Divisions within the Chromonta differ in more specific features of photosynthetic pigmentation and resultant coloration, for example, the brown algae (brown seaweeds, kelps) and diatoms contain a distinctive carotenoid called fucoxanthin, which is not found in most other algae. The Bilonta, or biliprotein-pigmented algae, consisting essentially of seaweeds known as red algae, possess the biliprotein pigments phycoerythrin (red) and phycocyanin (blue), but neither chlorophyll *b* nor chlorophyll *c*. The light-absorbing characteristics of accessory pigments, to be explained later in connection with the relations of algae to their light environment, lead to a definite ecological role for these pigments.

---

*Actually two closely related forms, chlorophylls c_1 and c_2.

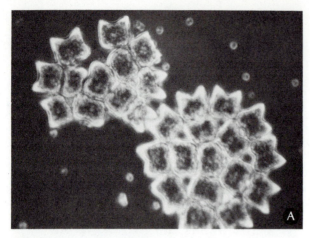

Figure 23-1. Several thallus types seen in algae. (A) A flat colonial form (*Pediastrum*) with a fixed number of cells per colony. (Several cells have been lost from one of the colonies.) (B) Filament of cells (*Oedogonium*). (C) A large brown seaweed (*Macrocystis*). (A) and (B) are microscopic. A small gas-filled bladder at the base of each "leaf" in (C) keeps the meters-long fronds floating on the surface for maximum light.

The differences between the algal Subkingdoms and Divisions will be explained more fully in the second half of this chapter after we have considered some important general features of algal life, form, and structure.

ALGAL HABITATS AND ADAPTATIONS

Algae occur almost everywhere in the environment. On land, small or unicellular algae commonly grow on moist soil, rocks, and wood surfaces, and on the bark of trees. Algae are much more important, however, in the water. They are the principal primary producers in the oceans, lakes, and most other aquatic habitats. The smaller, usually unicellular aquatic algae, suspended in marine and fresh waters (Figure 23-A), are known as **phytoplankton** (Greek *phyton*, plant; *planktos*, wandering). Phytoplankton serve as food for microscopic and nearly microscopic floating or swimming aquatic animals (zooplankton), often

called phytoplankton "grazers," for example, copepods (Figure 23-2), and for larger animals such as clams and certain fish that strain or filter plankton from the water (filter-feeders). Many phytoplankton are flagellates; although motile, they are so small that they are carried about by water currents along with nonmotile phytoplankton species.

Algae growing attached to rocks or other substrates, such as corals, at the bottom of bodies of water are termed **benthic.** These algae are usually filamentous or macroscopic. The most conspicuous benthic algae are the seaweeds, which grow attached to rocks along ocean shores in the intertidal zone and below low tide (Figure 23-3). Freshwater benthic algae, mainly green algae, are mostly much smaller and less complex.

The Algal Light Environment

Since light usually penetrates natural waters to a maximum depth of only about 200 m (650 ft),

TABLE 23-1.   Major Groups of Algae

Subkingdom	Division	Common Names
Chloronta (green-pigmented)	Chlorophyta Euglenophyta Charophyta	green algae euglenoids stoneworts
Chromonta (yellow-green, yellow-brown, & brown pigmented)	Xanthophyta Chrysophyta Dinophyta Phaeophyta	yellow-green algae diatoms, etc. dinoflagellates brown algae
Bilonta (red pigmented)	Rhodophyta	red algae

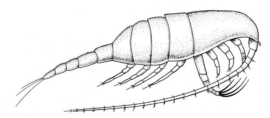

Figure 23-2. A copepod, an important component of the zooplankton.

Figure 23-3. Vertical zonation of benthic algae on New England coastal rocks in early spring. From top to bottom (left) are blue-greens and a green comprising the crust in the spray zone and upper intertidal, with small reds below. The "leafy" seaweeds are brown algae (fucoids) and the lowermost zone of red algae appears almost black.

both planktonic and benthic algae can grow only in this surface region, called the **euphotic zone** of seas and lakes. Benthic algae are obviously restricted to waters shallow enough that the bottom falls within the euphotic zone. Elsewhere, in deeper lakes and seas, the bottom is somewhat like a desert, without plants but inhabited by specialized animals subsisting on the "rain" of phytoplankton and other debris falling from the overlying surface waters (Figure 23-8).

ROLE OF ACCESSORY PIGMENTS

Most of the photosynthetic accessory pigments strongly absorb light of wavelengths different from those absorbed strongly by chlorophyll *a* (Figure 23-4), transferring this energy to chlorophyll *a* for use in photosynthesis. Efficient absorption of the widest range of wavelengths is especially important for deep-water algae, which grow where the light intensity is low. Furthermore, because water selectively absorbs the longer wavelengths (red and yellow), the light that penetrates the greatest depth consists mainly of blue and green wavelengths (Figure 23-5). Accessory pigments that strongly absorb the blue-green to green range of wavelengths, as do phycoerythrin and some carotenoids such as fucoxanthin (Figure 23-4), help algae that possess them to grow at greater depths than can the green-colored algae, which depend largely on absorption by chlorophylls. Therefore, brown seaweeds (which possess fucoxanthin) extend to greater depths than the green algae, and red algae (with phycoerythrin) grow at the greatest depths of any algae.

Light Responses

Dictated by their photosynthetic need for light, algae have orientation and movement responses to light that are analogous to the phototropic and developmental responses of higher plants (Chapter 14). Algal flagellates generally show positive **phototaxis,** or ability

Figure 23-4. Light absorption by the major pigments occurring in green algae (A), and brown algae and diatoms (B). Major accessory pigments in red algae are phycobiliproteins, particularly phycoerythrin, whose absorption spectra are shown in Figure 22-31.

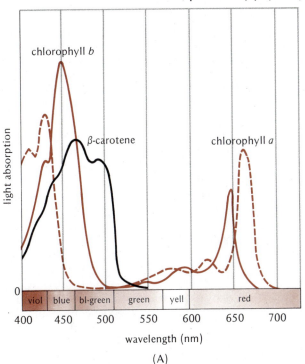

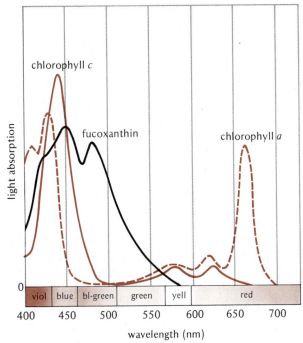

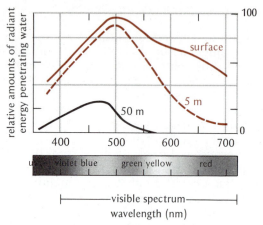

Figure 23-5. Light composition at different depths of clear oceanic water.

to swim toward a light source. This helps keep flagellated phytoplankton from sinking below the euphotic zone. The motile reproductive cells of many nonmotile algae are also phototactic, helping them settle in a well-illuminated spot. Nonmotile algal cells may show chloroplast orientation responses. For example, if a spot of light is shone on a part of the tubular thallus of *Vaucheria* (Figure 23-33), chloroplasts from elsewhere in the cell congregate in the illuminated area within a few minutes, improving their photosynthesis. Each cell of the green alga *Mougeotia* (Figure 23-9D) has a single, flat, strap-shaped chloroplast that under dim light conditions rotates until it faces the light, giving it the maximum possible light intensity. Under very bright light, on the other hand, the chloroplast rotates edge-on to the light, protecting itself from the photodestructive effects of too much light. Chloroplast photo-orientation in *Mougeotia* is governed by phytochrome, whereas the response in *Vaucheria* is controlled by a blue light photoreceptor similar to that involved in higher plant phototropism (Chapter 14).

Water Balance and Salinity

Permanently submerged algae obviously have no evaporative water loss problem like that of higher plants. Intertidal algae, when the tide is out, and terrestrial algae do have a desiccation problem since they lack a waxy cuticle restricting evaporation. Certain features that help algae tolerate periods of emergence will be mentioned later. Terrestrial algae either must be able to tolerate and recover from severe dehydration of their cells or are restricted to almost constantly moist habitats.

Most nonmotile algal cells maintain water balance with their surroundings essentially as higher plant cells do, namely by developing an internal osmotic concentration somewhat higher than that of the medium around them, causing water to enter the cell by osmosis and building up a turgor pressure against the cell wall, which stops further water uptake (Chapter 9). However, many

algal flagellates, especially those lacking a cell wall, possess an alternative, active water balance mechanism involving a **contractile vacuole** (Figure 23-23), such as that occurring in unicellular animals (protozoa). The contractile vacuole takes up water that has entered the cell and expels it through a pore in the cell surface periodically.

Freshwater algae generally cannot tolerate the osmotic water loss (and other effects of high salt concentration) caused by salt water. Conversely, marine algae, adapted to a high-salinity habitat, generally cannot survive in fresh water. Thus, freshwater phytoplankton and benthic algae consist of entirely different species from those occurring in the ocean. Indeed, relatively few algal families contain both freshwater and marine species, and some even larger groups, including certain of the Divisions (Table 23-1), are restricted or virtually restricted to either salt or fresh water. Both habitats, however, have phytoplankton with a similar range of morphology (in different species or genera).

A number of unicellular algae can survive and grow in brackish waters, where the salinity varies. Such waters occur in estuaries where fresh water flows in from rivers or streams during outgoing tides but salt water enters and mixes to some extent when the tide comes in, and in ponds and lagoons as they dry up during drought. The ability to tolerate changing osmotic conditions depends on a response called osmoregulation.

Algae that osmoregulate adjust their internal osmotic concentration to match that of the external solution, so that water is not extensively lost from nor gained by the cell as the salinity and therefore the osmotic strength of its environment changes. For example, the golden-colored flagellate *Ochromonas* osmoregulates when subjected to high salinity by rapidly breaking down a high molecular weight storage polysaccharide into a low molecular-weight compound called isofloridoside. This greatly increases the cell's solute (osmotic) concentration, preventing it from losing water (Figure 23-6). When shifted to a low-salinity medium the cell quickly reverses the process, converting isofloridoside back into the storage polysaccharide, thus reducing its own osmotic concentration so that it does not gain water and swell pathologically under these conditions.

Seasonality and Temperature

Aquatic habitats in general have much smaller global and seasonal temperature extremes than occur on land, the lowest temperature normally experienced in bodies of fresh water being 0°C and in the ocean −1.8°C, the freezing point of sea water; the warmest temperature is probably usually under 40°C (except in the local thermal environment of hot springs and geyser basins). Algae are affected by the equator-to-pole temperature gradient in the oceans, the species composition of benthic and planktonic marine algal flora of trop-

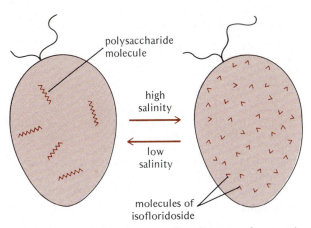

Figure 23-6. *Ochromonas* cells adjust to changes in external osmotic concentrations by the interconversion of isofloridoside and its polysaccharide polymer.

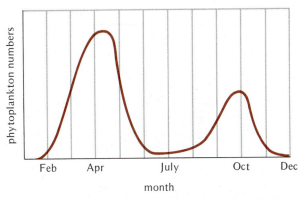

Figure 23-7. In temperate regions total phytoplankton numbers typically show a maximum in spring and a smaller peak in the fall. Each peak is the sum of smaller peaks for many different species. Combinations of environmental and biological factors such as temperature, light, nutrients, and predation control the population of each species.

ical waters being different from that of temperate and subpolar oceans. Algal growth is of course slowed by low temperatures, but certain algae (diatoms) are able to photosynthesize and grow even on the underside of the Arctic Ocean ice pack, at −1.8°C.

The very low light intensities that penetrate the water in winter when the sun is low in the sky seem to be as important as cold temperatures in causing a die-off or disappearance of most of the phytoplankton in lakes and oceans during the winter season in temperate and colder latitudes. In spring the rapidly increasing intensity and duration of sunlight, plus the gradually rising water temperature, greatly accelerate the growth of phytoplankton. Phytoplankton populations typically go through a series of spring and autumn **blooms** or proliferations of different species or groups of species (Figure 23-7). The early blooms tend to consist of relatively few, fast-growing species. Later in the season slower growth but a greater species diversity is usually found.

Most benthic algae of temperate and cold-temperate lakes and streams, and filamentous marine benthic algae of colder latitudes, are annuals. They usually overwinter by means of heavy-walled, cold-resistant dormant spores or "resting spores," commonly produced by sexual reproduction. Both annual and perennial seaweed species occur. In some perennials a small part of the thallus persists through the nongrowing season and initiates new growth when favorable environmental conditions return, whereas the entire thallus of some of the larger seaweeds survives and continues to grow for a number of years. These differences are analogous to the different higher plant growth forms considered in Chapter 3. In warmer, subtropical or tropical waters where temperature is never limiting and there is sufficient sunlight year round, algal growth can occur more or less continuously, and some annual algae can produce several generations per year.

In seasonal climates the reproduction of annual and also perennial algae must occur at an environmentally favorable time of year, prior to harsh conditions

that prevent growth or kill vegetative plants. Cool temperatures are probably a reproductive signal for many algae, but in at least some cases autumn reproduction is controlled photoperiodically as a short-day response, somewhat similar to photoperiodism in higher plants (Chapter 16).

Nutrition

Algae require much the same mineral nutrients as higher plants (Chapter 11). The primary productivity of fresh and sea waters, therefore, depends strongly on their mineral nutrient content. As phytoplankton in these waters die and sink below the euphotic zone, the nutrients they contain are taken out of circulation for photosynthetic organisms because the temperature gradient that exists in most waters "stratifies" the water or prevents any mixing of deeper layers with those near the surface. In lakes of seasonal climates, thermal stratification breaks down once a year during autumn cooling and the water "turns over," bringing nutrients to the surface that permit algal blooms to develop the next spring. In most parts of the open ocean turnover never occurs, the surface waters consequently become and stay depleted of nutrients, especially nitrogen and phosphorus, and phytoplankton populations are held to a low level. The productivity of tropical and temperate open oceans is therefore very low, comparable to that of deserts on land (Table 34-2).

In subpolar regions, where mingling of cold and warmer water masses causes some mixing of surface and subsurface water layers, the nutrient supply and productivity are much better. Along seashores, where nutrients in rivers and streams and from shoreline erosion are entering the ocean, algal populations and productivity are relatively

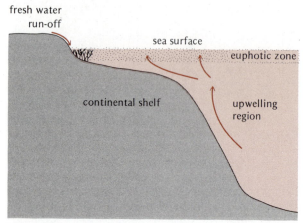

Figure 23-8. The concentrations of marine algae are highest in coastal regions where nutrients are supplied by run-off from the land and by seasonal upwellings in temperate and polar latitudes. The occurrence of phytoplankton (dotted areas) and benthic algae is also restricted by such factors as light and temperature.

high. Along certain coasts underwater ocean currents come to the surface, a process called upwelling, bringing abundant nutrients up from the depths (Figure 23-8). These waters permit extensive growth of phytoplankton populations and have among the highest primary productivities on earth. This production supports large fish populations and is the basis of some of the most important world fisheries, such as the anchovy catch off the Peruvian coast.

Certain algae require special mineral nutrients not known to be generally needed by plants. Some brown and red algae require iodine; kelps are notable for accumulating iodine to some 10,000 times its concentration in sea water. Before a cheaper source of this element was found, kelps were harvested commercially to obtain iodine to add to salt for preventing iodine deficiency (goiter) in humans. Diatoms, one of the principal groups of phytoplankton, require silicon in the form of silicate ($Si_2O_3^{2-}$) because their cell walls are made of silica (SiO_2). The silicate concentration of natural waters is never very high, and its depletion by diatoms, limiting their growth but not affecting most other kinds of phytoplankton which do not require silicon, probably contributes to the seasonal succession of phytoplankton species in lake and sea waters.

Heterotrophy and Facultative Heterotrophy

Most algal Divisions, but especially the various groups of flagellates, include colorless, heterotrophic species resembling the photoautotrophic members of the same group except for lack of functional chloroplasts. Certain heterotrophic forms have colorless plastids (leucoplasts) or give other evidence of having arisen from photosynthetic forms by mutation. Some of the colorless flagellates that resemble photosynthetic ones may, on the other hand, be forms whose ancestors never acquired plastids by endosymbiosis (Chapter 21). Many of the related photosynthetic flag-

ellates can grow heterotrophically in the dark, either by absorbing soluble carbon sources such as glucose, acetic acid, or ethanol (saprotrophy), or by ingesting particulate matter (phagotrophy). Some do both. For example, the photosynthetic flagellate *Ochromonas* can ingest particulate food and can also be grown in laboratory culture in the dark by feeding it soluble sugars. Certain facultatively heterotrophic algae are important inhabitants of sewage-treatment ponds. In very murky sludge-filled liquids, where poor light penetration limits photosynthesis, they can grow by consuming organic compounds released by bacterial fermentations, yet when light reaches them they oxygenate the suspension, speeding its aerobic decomposition.

Auxotrophy

Many photosynthetic algae, including at least some species or strains from every algal Division, are auxotrophic for certain vitamins, that is, they cannot make them and require an external supply. Vitamin B_{12}, thiamine (B_1), and biotin are the vitamins most commonly required. In nature the needed vitamins come from other algae or bacteria that synthesize and release them into the water. Vitamin production and dependency may be another significant factor in sequential algal species blooms.

ALGAL CELL STRUCTURE

Chloroplasts

Depending on the group, algal cells may contain just one or two large chloroplasts each, or smaller, more numerous chloroplasts. Chloroplasts occur in various forms—ribbon-like, cup-shaped, lobed, disk-shaped, spiral, or reticulate (net-like), a much greater diversity than is encountered in higher plants (Figure 23-9).

In internal structure algal chloroplasts are also much more diverse. Local sets of thylakoids stacked into grana, like those of higher plant chloroplasts (Chapter 7), occur only in a few members of the Chloronta. In most algae the thylakoids traverse the entire length or breadth of the chloroplast (Figure 23-10) and occur either singly (red algae) or in groups of two, three (in most of the *Chromonta*; Figure 23-10), or occasionally more (*Chloronta*).

One or more dense, nonpigmented bodies called **pyrenoids** (Figures 23-9, 23-41) occur within or projecting from the chloroplasts of many algae. There is evidence that a high concentration of RuBP carboxylase, the key enzyme for photosynthetic carbon dioxide fixation, occurs in pyrenoids. A carbohydrate end product of photosynthesis such as starch frequently accumulates near the pyrenoid, either inside (green algae, Figure 23-23) or outside the chloroplast. These observations suggest that pyrenoids are structures specialized for performing the Calvin cycle of photosynthesis (Chapter 7).

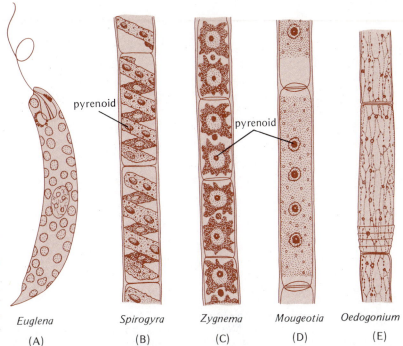

Figure 23-9. Chloroplast shapes. Illustrated are (A) disc, (B) spiral, (C) stellate, (D) ribbon, and (E) net forms. Other cell organelles are not shown and the organisms are not drawn to scale. All are green-pigmented algae (Chloronta), but similar shapes occur in other algal groups. A cup-shaped chloroplast is illustrated in *Chlamydomonas* (Figure 23-23).

Euglena (A) *Spirogyra* (B) *Zygnema* (C) *Mougeotia* (D) *Oedogonium* (E)

Many flagellates and motile algal reproductive cells have one or more small orange-pigmented spots, known as an "eyespot" or stigma, which help the cell detect the direction from which light is coming and respond phototactically. The stigma is usually located within a chloroplast and is a local area occupied by lipid droplets containing orange carotenoid pigments.

Flagella

All major groups except the red algae include species that either are flagellated or produce flagellated cells at some time during their life cycles. Algal flagella, like those of all eukaryotes, are structurally more complex than the bacterial flagella described in Chapter 22. Figure 23-11A shows the arrangement of microtubules in an algal flagellum as seen in cross section. A circle of nine pairs of microtubules surrounds two central ones, all enclosed in a membrane that is actually continuous with the cell's plasmalemma. This arrangement, frequently called a "9 + 2" structure.

Many variations in external features of flagella occur in algae. Different species may have spines, scales, or fine hairs in various arrangements covering their flagella (Figure 23-11B, C). These ornamentations, and the number and place of insertion of flagella in the cell,

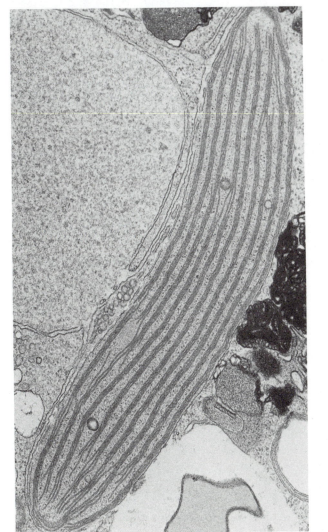

Figure 23-10. Chloroplasts of algal groups have distinctive ultrastructural features. Brown algal chloroplasts have thylakoids in groups of three (note at left end of chloroplast), and have a total of four membranes surrounding the plastid, the outermost one being continuous with the outer nuclear membrane.

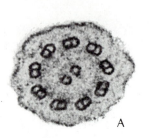

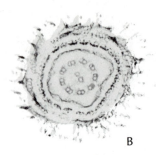

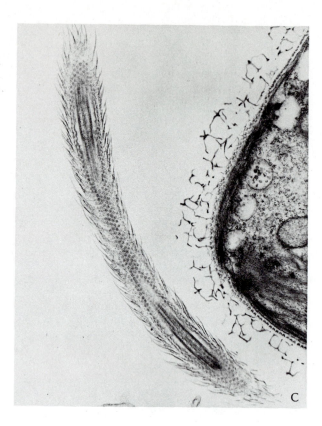

Figure 23-11. (A) Cross section of a typical eukaryotic flagellum, present in many algae, with the "9 + 2" arrangement of microtubules inside a membrane. A number of algal species have various ornamentations (hair, spines, or scales) on the outer surface of the flagellum; that of a green unicell (*Pyramimonas*) is shown in cross section (B) and longitudinally (C), with grazing cuts revealing the interior microtubules in two places. (A segment of the sectioned cell, also covered with scales, shows at the right.)

TABLE 23-2. Cellular Characteristics of Major Algal Divisions

| Division | Chloroplast Features | | | Storage Products |
	Distinctive Accessory Pigments	Number of Envelope Membranes	Thylakoids in Groups of	
Chlorophyta	chl *b* lutein[1]	2	2–many	starch
Euglenophyta	chl *b*	3	usually 3	paramylon, lipids
Charophyta	chl *b*	2	many	starch
Xanthophyta	chl *c*	4	3	chrysolaminarin (?), lipids
Chrysophyta	chl *c* fucoxanthin[1]	4	3	chrysolaminarin, lipids
Dinophyta	chl *c* peridinin[1]	3	3	starch lipids
Phaeophyta	chl *c* fucoxanthin[1]	4	3	laminarin, mannitol, lipids
Rhodophyta	chl *d*	2	1	glycogen-like starch

[1]Distinctive carotenoid pigments (all algae contain at least some types of carotenoid accessory pigments).

are conservative features that aid in classification of algae (Table 23-2).

Cell Walls and External Coverings

All the multicellular algae have cell walls, mostly containing cellulose and other polysaccharides, sometimes in combination with proteins (glycoproteins). Although some flagellates as well as nonmotile unicellular algae have cell walls, their composition varies more widely. Many photosynthetic flagellates entirely lack a cell wall and are simply naked cells. Many flagellates and other unicells are covered by scales, plates, or other structures of highly specific form (Figure 23-12), remarkable considering their minute size. The nature and composition of surface structures, either cell walls, scales, or plates, and so on, tends to be characteristic of or even diagnostic for particular algal groups.

Electron microscopy shows that the surface scales of many flagellates are synthesized inside the cell, within its Golgi apparatus, and then are secreted onto the outside surface by fusion of Golgi-derived vesicles or cisternae with the plasma membrane (exocytosis). Thus the formation of these structures resembles that of genuine cell walls, in which the Golgi apparatus participates importantly (see Chapter 4).

Many algae secrete mucilages (i.e., gelatinous or slimy polysaccharides) outside their cells. In scale-covered flagellates such secretions stick the plates to the cell surface. More generally, mucilage might be regarded as a protective coating, somewhat like the "capsule" of many bacteria (Chapter 22). Since mucilages hold many times their own weight of water, they help retard desiccation of terrestrial algae and of intertidal marine algae during exposure to the air at low tide. The dense polysaccharide surface coating found on many algal cells or thalli is sometimes called a cuticle, but it apparently is never waxy and therefore is not comparable to the cuticles of land plants. A multilayered "cuticle" on the external surface of certain red and brown seaweeds gives their thalli an impressive iridescent appearance (Figure 23-42; the red seaweed *Iridaea* is named for this feature) by physical interference between light waves reflected from the successive layers.

Components of Cell Covering		Flagella		Division
Microfibrillar	*Non-microfibrillar[2]*	*Number*	*Position on Cell*	**Division**
cellulose, or mannan, or none	protein or $CaCO_3$ in some	1–many	apical	Chlorophyta
none	protein ("pellicle")	1–2	apical	Euglenophyta
cellulose	$CaCO_3$ in some	2	subapical	Charophyta
cellulose or none		1–many	apical	Xanthophyta
cellulose or none	SiO_2 in diatoms, $CaCO_3$ in some	1–3 or absent	apical	Chrysophyta
cellulose plates or none		2	lateral	Dinophyta
cellulose	alginates	2	lateral	Phaeophyta
cellulose or xylan	galactan-SO_4 (agar, carrageenan); $CaCO_3$ in some	absent	—	Rhodophyta

[2]Cell walls of algae typically contain amorphous polysaccharides in which the microfibrillar component is embedded. Only additional unique or characteristic components are listed.

Figure 23-12. Many unicellular algae have unique cell coverings. Diatoms (A) have overlapping valves largely composed of silica. Coccolithophorids are covered by calcium carbonate scales (coccoliths) (B). Many dinoflagellates (C) have cellulosic plates beneath the plasma membrane. Desmids (D) often have cellulosic walls in elaborate shapes.

REPRODUCTION

Vegetative Versus Asexual Reproduction

In lower plants, **vegetative reproduction** refers to multiplication of unicellular organisms by cell division, and of multicellular ones by fragmentation of the thallus, each fragment giving rise to a separately growing plant. The latter is comparable with some types of vegetative reproduction conducted by higher plants (Chapter 15). **Asexual reproduction** refers to the formation, by mitotic divisions, of **spores,** single cells that can germinate to produce new individuals. Frequently, algal spores are flagellated, free-swimming cells without a cell wall called **zoospores** (Greek *zoion,* animal, referring to their animal-like motility). A vegetative cell may form one or, by division, a number of spores that are released by rupture of the parent cell wall. Spores are to be distinguished from **gametes,** which are cells that must fuse with another gamete before a new individual can develop.

Sexual Reproduction

In the simplest type of sexual reproduction in algae, two gametes of similar size and form, either flagellated or nonflagellated, fuse together into a zygote (Figure 23-13A). This is called **isogamous** reproduction (Greek *isos,* equal). In other species, the gametes that fuse differ noticeably in size but still have a similar form, for example, both of them may be flagellated (Figure 23-13B). This is called **anisogamous** reproduction (Greek *an-,* not; plus *isos*). The larger of the two gametes is commonly called the "female" gamete, and the smaller the "male." In **oogamous** sexual reproduction (Greek *oion,* egg) the larger gamete is nonmotile and is called the egg, while the much smaller gamete is termed the male gamete or sperm (Figure 23-13C), and is usually motile, but may be nonmotile, for example, in red algae.

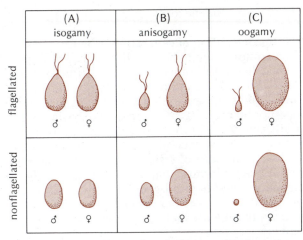

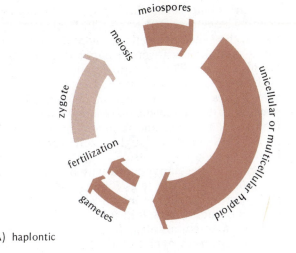

(A) haplontic

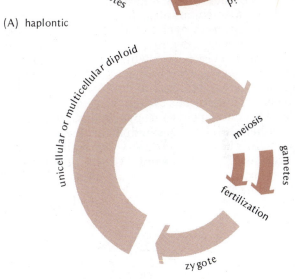

(B) diplontic

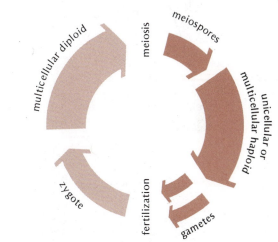

(C) alternation of generations

Figure 23-13. Types of sexual reproduction. Some nonflagellated gametes migrate by ameboid movement to fuse with gametes of the opposite mating types. Anisogamous reproduction involving nonflagellated gametes (B column, lower half) occurs rarely if ever in algae, but is a type seen in other organisms such as fungi. In oogamy, the gametes differ not only in size but in form. In oogamy, the gametes differ not only in size but in form. In red algae, where oogamy with nonflagellated male gametes occurs, the female cell typically differentiates a receptive hair-like structure to which the male gametes attach.

In most algal species, to be compatible and capable of fusion, the two gametes (whether isogamous, anisogamous, or oogamous) must come from different plants, that is, from genetically distinct individuals. Such a species is termed **heterothallic** (Greek *heteros*, other). If gametes produced by the same thallus can fuse successfully, the species is termed **homothallic** (Greek *homos*, same). This is similar to self-fertility or self-compatibility in some higher plants.

A heterothallic species possesses a self-incompatibility or mating-type genetic locus, usually with two alleles designated + and −. Gametes can fuse only if they are of opposite mating type. As discussed below, in many algae the vegetative plants that produce the gametes are haploid, and thus are either of genotype + or genotype − and can produce only gametes of that one genotype. This arrangement ensures that all zygotes are derived from cross-fertilization between different individuals.

It was noted above that temperature and light conditions are important in controlling reproduction. Another important control is nutrition, especially in phytoplankton and in filamentous algae that compete as colonizers or "opportunists." A shift to nutritionally unfavorable conditions, such as lack of a nitrogen or phosphorus source, induces sexual reproduction. This can help the organism escape locally unfavorable conditions by dormancy or dispersal.

Figure 23-14. Basic life cycle types.

Algal Life Cycles

Like all other basic features of algae, their reproductive cycles are fundamentally much more diverse than those of higher plants, which reproduce by a stereotyped alternation of diploid and haploid generations (Chapters 16 and 27-31). The sexual reproduction of algae generally conforms with one of the three following types of cycles, diagrammed in Figure 23-14:

1. **Haplontic cycle,** or exclusively haploid vegetative stage (Figure 23-14A). In haplontic organisms, the vegetative cells are haploid and reproduce sexually by forming haploid gametes by simple mitosis. Their fusion product, the zygote, is the only diploid cell in the life cycle. The zygote divides by meiosis, yielding haploid cells that may be called **meiospores** (spores produced by meiosis) if they become independent of one another. These haploid daughter cells become new haploid vegetative plants.

Haploid vegetative thalli may be unicellular, as in many flagellates (e.g., Figure 23-15), or each meiospore may divide mitotically a number of times, producing a haploid multicellular plant as in many filamentous algae (e.g., Figure 23-16).

2. **Diplontic cycle,** or exclusively diploid vegetative stage (Figure 23-14B). In this type of life cycle the vegetative thallus is diploid and produces haploid gametes by meiosis. Since these fuse immediately, yielding a diploid zygote, they are the only haploid cells of the life cycle. The zygote develops into a new vegetative thallus, which may be either unicellular (as in diatoms, Figure 23-17) or multicellular (as in the widespread seashore "rockweed" *Fucus,* Figure 23-38C). This type of reproductive cycle is universal among animals, including humans, but in plants is encountered only rarely, indeed only in a small minority of algae.

3. **Alternation of generations** (Figure 23-14C). This complex life cycle, similar to that of higher plants, involves both haploid and diploid vegetative thalli. The haploid generation is called the **gametophyte** because, like the haploid thallus in a haplontic cycle, it reproduces by releasing (haploid) gametes. The resulting diploid zygotes divide by mitosis, becoming diploid vegetative individuals or **sporophytes,** so-called because they reproduce by forming haploid meiospores. These grow into new haploid gametophyte thalli, completing the cycle. Either or both generations may also reproduce asexually, by zoospores, or vegetatively.

In contrast to the flowering plants, in which the gametophyte generation is much reduced and dependent on the sporophyte (Chapter 16), in practically all algae both generations are independent and free-living. In some algal species the two generations are vegetatively virtually identical in size and form, for example, the common seashore green alga *Ulva,* called "sea lettuce" (Figure 23-18). This is termed an **isomorphic** alternation of generations (Greek *isos,* equal, *morphē,* form). In others, such as kelps, the haploid and diploid generations differ in size and form

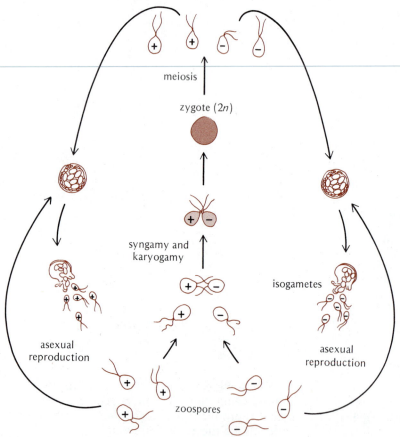

meiosis

zygote (2n)

syngamy and karyogamy

isogametes

asexual reproduction

asexual reproduction

zoospores

Figure 23-15. *Chlamydomonas* life cycle, a haplontic type.

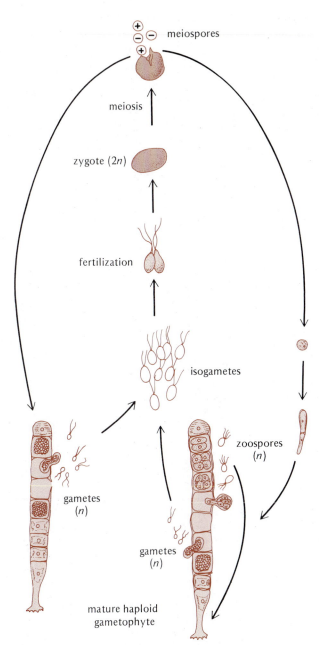

Figure 23-16. *Ulothrix* life cycle, a haplontic type.

mechanisms, for example, the production and precisely positioned excretion of surface scales or plates (Figure 23-12). **Colonial** forms, aggregates of independent or virtually independent cells, are of two types. One constitutes cells, indefinite in number, that multiply by mitosis, while the colony multiplies by fragmentation. This type has representatives in most algal groups. The other type of colony, a **coenobium** (Greek *koinos*, in common; *bios*, life), is restricted to green algae, where many species have colonies with a fixed number of cells established when the colony is formed (Figures 23-1A, 23-24). This precise number of cells and their species-specific arrangements demonstrate the existence of complex morphogenetic controls over cell division.

Filamentous Forms

Perhaps the most elementary type of morphogenesis is that found in the very common, unbranched filamentous algae that develop, in the simplest cases, merely by the continued elongation and division of cells in one direction (Figure 23-20A). In some filamentous algae, such as *Oedogonium* (Figure 23-26), cell division is more or less restricted to particular cells within the filament, a beginning stage in the progression toward local zones of growth and division, or meristems, such as those typical of higher plants.

The next step in morphogenetic complexity is development of a **branching** filament system by controlled outgrowth and cell division at an angle to the filament axis at certain points (Figure 23-20B). This allows an individual to develop a much denser cell mass with a definite two- or three-dimensional form. Since the interdigitating or interlocking branches in such a plant body would probably often be bent or broken if the filament axes were to elongate behind the points where branches have arisen, in branching filament systems cell growth and division usually become restricted to the apical part of each filament, sometimes to just the tip of its terminal cell.

Macroscopic Thallus Forms

A simple two-dimensional morphogenetic pattern is formation of a sheet of cells by repeated division in all directions within one plane (Figure 23-20C). This creates a leaf-like organ that is rather effective for energy capture; it occurs for example in *Ulva* (Figure 23-18), in which two cell layers engage in cell division, generating an extensive cell sheet two cells thick. More complex three-dimensional body forms such as those of many brown and red algae (Figure 23-20D) develop by precise combinations of directionally oriented cell divisions.

(Figure 23-19), a **heteromorphic** alternation of generations (Greek *heteros*, other). In certain genera of golden-brown algae (Chrysophyta), a unicellular diploid stage alternates with a filamentous haploid one.

BODY FORMS AND MORPHOGENESIS

Unicellular algae might be thought to be the simplest morphologically, but the internal structure and external form of many unicellular species is elaborate and complex, indicating some highly evolved intracellular morphogenetic

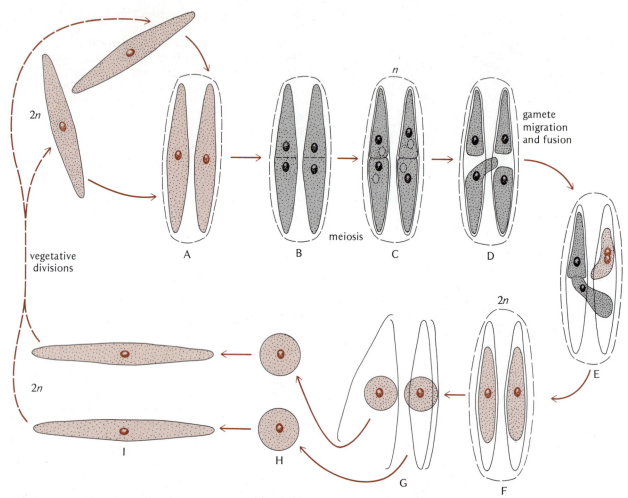

Figure 23-17.   Sexual reproduction in a diatom. The sexually induced diatoms surround themselves in a jelly-like mucilage (A) and each undergoes meiosis (B, C). After the first nuclear division of meiosis (B), each cell divides into two; after the second meiotic division (C) one of the resulting nuclei degenerates (dotted circles), leaving each shell with two haploid gametes. One gamete from each diatom then migrates through an opening into the other cell and fuses with the stationary gamete (D, E). The zygotes (F) resulting from nuclear fusion shed the old cell coverings (G), increase in size (H), and eventually form new shells (I).

Coenocytes

A basically different method of thallus development is found in various, mostly green, algae. The initiating cell (spore or zygote) grows and its nucleus divides mitotically, but cytokinesis does not occur, so the cell becomes multinucleate (Figure 23-20E). By continued growth it can even become macroscopic (Figure 23-21). Such a multinucleate one-celled organism is called **coenocytic** (pronounced "seen-o-sit'-ic", from Greek *kytos*, chamber; *koinos*, in common). Chloroplasts multiply, along with the nuclei, into astronomical numbers within the coenocyte. Its basic body form is usually tubular, created by tip growth similar to that in root hairs or pollen tubes. However, complex two- and three-dimensional, fern-like or tree-like forms (Figure 23-21A, B) can be developed by precisely controlling the branching and the direction and extent of branch, compared with main axis, growth. By developing a dense felt of interlocking coenocytic filaments, a massive, macroscopic plant body of species-specific form can also be created (Figure 23-29).

Although the vegetative thallus of coenocytes may have no cell divisions, coenocytes are not incapable of cytokinesis; cell divisions occur during formation of reproductive organs and cells.

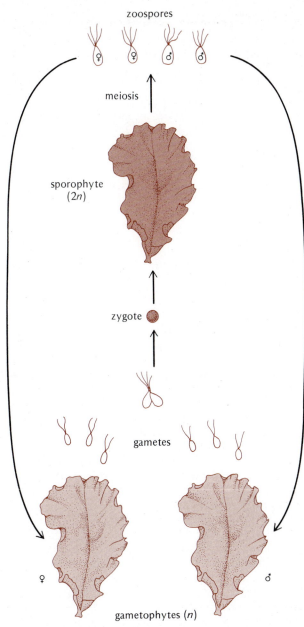

Figure 23-18. The *Ulva* life cycle illustrates an isomorphic alternation of generations.

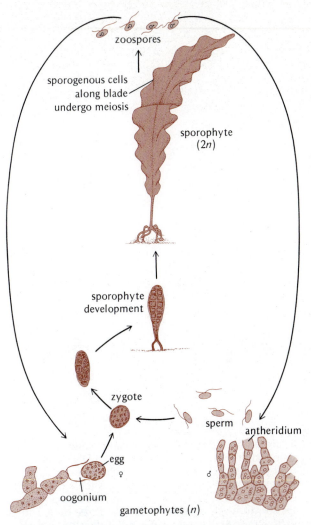

Figure 23-19. *Laminaria* life cycle, in which the large sporophyte generation alternates with a microscopic gametophyte generation.

The following sections present additional information on the morphology and biology of specific algal groups. This will represent only a very limited selection from the great diversity of existing algae.

SUBKINGDOM CHLORONTA, THE GREEN-COLORED ALGAE

This Subkingdom includes, as noted already, those algae having photosynthetic pigments similar to those of higher plants: chlorophylls *a* and *b*, plus typical "higher plant" carotenoids such as

β-carotene. Because of these and certain other similarities with higher plants, such as formation of starch as the storage reserve product and the presence of granal structures (thylakoids stacked into local aggregates, Chapter 7) in the chloroplasts of some of the Chloronta, higher plants (Subkingdom Embryonta) are thought to have descended from the Chloronta.

Among the approximately 500 genera of Chloronta can be found most of the kinds of body form discussed above, ranging from flagellated and nonflagellated unicells and simple colonial forms probably similar to pre-Cambrian algal life, up to multicellular macroscopic plant bodies of highly specific form and morphogenesis. Curiously, out of this wide range of forms there are none resembling at all closely the plant bodies of Bryophytes or of primitive vascular plants. Therefore, the origin of the Embryonta cannot be identified easily with any particular subgroup of Chloronta.

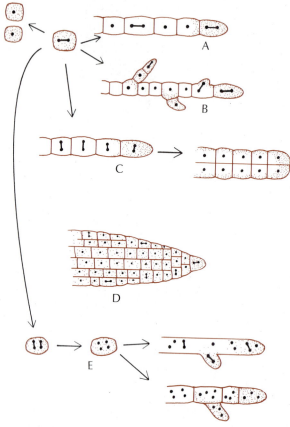

higher plants (Chapter 4): in the central part of the telophase mitotic spindle, membranous vesicles fuse into a cell plate, formed at right angles to the direction of the spindle microtubules. This structure (i.e., the residual spindle microtubules plus the cell plate cutting across them) is called a **phragmoplast** (Figure 23-22A). In many green algae, however, the spindle microtubules disappear at telophase and a new set of microtubules forms between the daughter nuclei at right angles to the former spindle microtubule direction (Figure 23-22B, C). Within this new microtubule array the cell plate, running parallel to the microtubules, forms by vesicle fusions or by infurrowing of the plasma membrane from the cell surface. Green algae dividing by this mechanism, called a **phycoplast,** are thought to comprise an evolutionary line separate from that which led to higher plants, only those green algae with phragmoplast division being possibly related to the higher plant evolutionary line. These include the stoneworts (Charophyta) and a few members of the Chlorophyta (see below).

Although some of the Chloronta are marine, as a whole the green algae are much more prominent and diverse in fresh waters. It seems likely, however, that the green algae originated in the pre-Cambrian oceans and only later diversified in fresh waters, a habitat that was not invaded significantly by some of the other major algal groups such as the brown and the red algae. Among the reasons for the presumption of a marine origin is the discovery of *Prochloron*, a marine prokaryote with a photosynthetic pigment system of the green algal type (Figure 21–7). *Prochloron* is thought to be a surviving member of the prokaryotic group from which the green algal type of chloroplast evolved by endosymbiosis.

In numbers and diversity the **Division Chlorophyta,** to which the common name "green algae" is usually restricted, is the principal group of Chloronta. The preceding comments thus pertain largely to the Chlorophyta. Following are a few representative examples selected from the wide range of body forms and modes of reproduction found among the Chlorophyta. After discussing these we shall compare the two much smaller and more specialized groups included in the Chlo-

Figure 23-20. Directed growth and planes of cell division determine thallus forms. (A) Divisions of attached cells in only one plane result in a uniseriate row of cells. Branching filaments (B) or sheets of cells (C) arise when additional planes of division occur. (D) A section through a growing tip of a three-dimensional thallus reveals complex patterns of division and elongation of cells. (E) Coenocytic cells or thalli arise when cell divisions are not coupled to nuclear divisions.

The range of Chloronta that may be related to the higher plants' progenitors is narrowed considerably by the recent discovery, using electron microscopy, of unusual ultrastructural features of cytokinesis (cell division) in these organisms. After mitosis, some of the Chloronta undergo cytokinesis in a manner similar to

Figure 23-21. Some coenocytic thallus forms illustrated by marine green algae: (A) a tubular feather-like form (*Caulerpa sertularoides*), (B) the parasol form of *Acetabularia,* and (C) a sac-like thallus (*Valonia*). These three are tropical species and fall within a 2–10 cm height range.

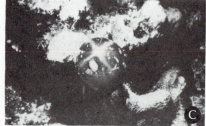

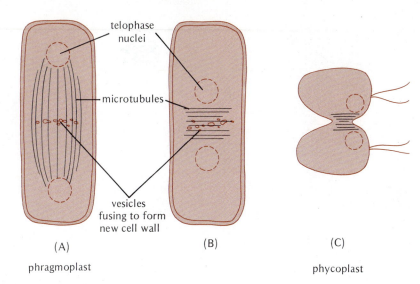

Figure 23-22. Types of cell division found in green algae. The phragmoplast type (A) is also seen in cells of land plants. The phycoplast type, seen in many green algae, is thought to characterize a separate evolutionary line. Green algae without cell walls (C) show an infurrowing parallel to the microtubules of the phycoplast.

ronta (the euglenoids, and the stoneworts or charophytes).

Unicellular and Colonial Green Algae

CHLAMYDOMONAS

This alga is a unicellular flagellate (Figure 23-23). It occurs chiefly in stagnant pools, but some species are specialized for growing in the meltwater around ice crystals in mountain snow banks that persist through the summer, causing the phenomenon of "red snow." Large amounts of carotenoid pigments, responsible for this unusual coloration in green algae, probably help protect the cells' chlorophylls from photodestruction under the intense sunlight to which they are exposed at the surface of snowbanks.

Chlamydomonas, as previously noted, has the simple haplontic type of sexual reproductive cycle (Figure 23-15). It is excellently suited to genetic studies. Mutations are easily detected because of the haploid condition of the vegetative cells. The characteristics and hereditary transmission of mutations can be readily studied using *Chlamydomonas* cell populations grown in laboratory culture.

Chlamydomonas cells (Figure 23-23) lack cellulose but produce a glycoprotein wall. Two flagella arise from the anterior end of the cell; a single cup-shaped chloroplast with a large central pyrenoid occupies much of the cell's internal volume. This chloroplast divides just after the nucleus, immediately preceding cell division. An eyespot occurs near the flagellated end of the cell. The cell contains two contractile vacuoles located near the bases of the flagella.

Asexual reproduction in *Chlamydomonas*, which is more specialized than in most flagellates, involves division of a vegetative cell into as many as eight daughter cells (Figure 23-15). After forming cell walls and new flagella these progeny are liberated by breakdown of the parent cell's wall.

A structurally similar but nonmotile green alga, *Chlorella*, is also easily grown in laboratory culture and is used extensively in studies on photosynthesis; the

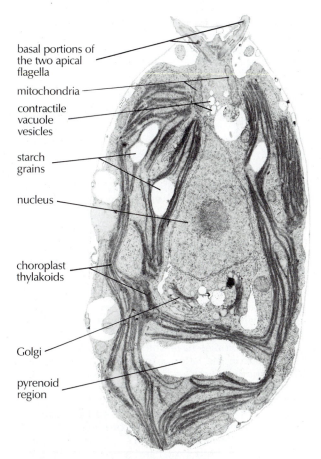

Figure 23-23. An electron micrograph of a section of the flagellate, *Chlamydomonas*. A peripheral cup-shaped chloroplast surrounds the eukaryotic cell organelles.

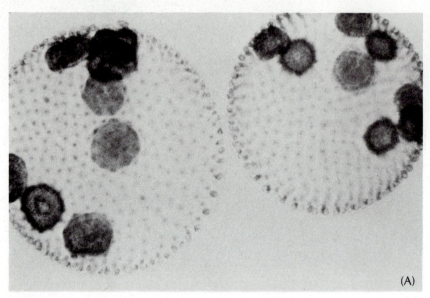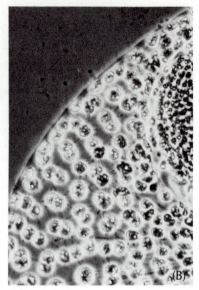

Figure 23-24. *Volvox* colonies seen under the light microscope. A number of darker daughter colonies are evident inside two mature colonies (A). Dark lines that are the cytoplasmic connections between individual cells comprising the sphere are visible in (B).

Calvin cycle (Chapter 7) was discovered mainly using this organism. Because *Chlorella* is an efficient protein producer, mass culture of it and similar unicellular algae as a human food source or protein supplement is a technology with possibly important future applications, including cultivation in space stations or space vehicles.

VOLVOX

This curious organism, found in ponds and lakes, illustrates a motile colonial stage of plant body evolution. The spherical colonies (Figure 23-24A), large enough to be visible to the naked eye, contain 500–20,000 flagellated cells that form a single-layered hollow sphere. A gelatinous sheath (perhaps a type of cell wall) separates each cell from its neighbors, which are sometimes interconnected by cytoplasmic strands (Figure 23-24B). By the combined action of the outwardly directed flagella of all its cells, the colony rolls slowly through the water.

Volvox colonies are coenobia, the colonies of each species having a characteristic, unvarying number of cells. Individual *Volvox* cells are haploid and much like those of *Chlamydomonas.*

Volvox colonies have both sexual and asexual reproduction. In the latter, a new colony arises from a precise number of nuclear and cell divisions in selected vegetative cells that enlarge and push into the central cavity of the parent sphere (Figure 23-24A). New vegetatively produced colonies are released by disintegration of the parent colony. Sexual reproduction involves a haplontic cycle similar to that of *Chlamydomonas,* but it is oogamous. Both sexual and asexual reproduction in *Volvox* show rather dramatic specialization or differentiation in what is commonly regarded as a simple colonial organism.

DESMIDS

The cells of many of these aquatic, mostly unicellular green algae are nearly divided into symmetrical halves (Figure 23-25A). They can develop elaborate and beautiful shapes that show the occurrence of complex species-specific morphogenetic controls involving precisely regulated local growth of their cellulosic cell walls (Figure 23-25B).

Desmids lack flagella, but some of them move by extruding a mucilage through small pores in the cell walls, a mechanism analogous to gliding motility in bacteria and blue-greens (Chapter 22). There are numerous species of desmids, and a wide variety of specific cell shapes. They are particularly common phytoplankton in freshwater ponds and lakes, and sometimes are the principal primary producers in these habitats. The reproduction and life cycle of desmids are somewhat similar to that of *Spirogyra,* considered below.

Multicellular Green Algae

ULOTHRIX

This common filamentous green alga (Figure 23-16) is a simple haplontic organism with asexual and sexual reproduction mechanisms similar to those of *Chlamydomonas. Ulothrix* filaments are differentiated only to the extent that their lowermost cell, called a **holdfast** cell, is specialized for attachment to a substratum and does not have the capacity for reproduction, as do all other cells of the filament.

In asexual reproduction the protoplast of an individual cell in a filament converts itself into one or two

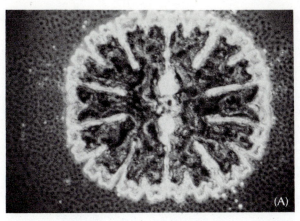
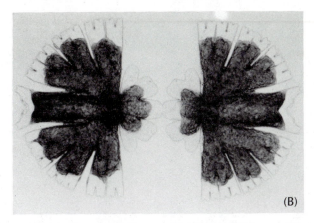

Figure 23-25.   A desmid (*Microsterias*) with one of the most intricate cell forms (A). Each semi-cell of the parental cell forms a new half during cell division (B).

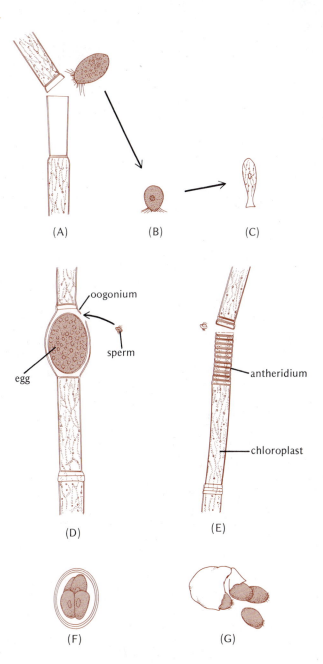

(A)   (B)   (C)

oogonium

egg   sperm

antheridium

chloroplast

(D)   (E)

(F)   (G)

motile zoospores. These resemble free-living flagellates like *Chlamydomonas* and are released and swim away from the parent filament to settle in another location and grow into a new filament. In *Ulothrix's* isogamous sexual reproduction, cells of a filament divide into several small, flagellated gametes, which, after their release, fuse in pairs to yield zygotes. The zygote germinates by meiosis, as in *Chlamydomonas,* to release four motile haploid zoospores, each of which can grow into a filament (Figure 23-16).

OEDOGONIUM

The filaments of this common alga (Figure 23-1B; pronounced "ee-doe-go'-nee-um") form a green fuzz attached to stones or aquatic higher plants in most freshwater ponds. Although haploid like *Ulothrix,* and similarly reproducing asexually by flagellated zoospores (Figure 23-26A–C), *Oedogonium* shows a more highly evolved, oogamous type of sexual reproduction (Figure 23-26D–F). It produces eggs individually in enlarged cells of the filament. Specialized small cells divide to form sperms. Upon release, the motile sperms swim to an egg, attracted by a chemical attractant it releases. The resulting zygote becomes a dormant resting spore through formation of a heavy cell wall; it germinates by meiosis, usually after overwintering, to release haploid zoospores that can grow into new filaments (Figure 23-26F–G).

Although some species of *Oedogonium* are bisexual, each individual producing both eggs and sperms, other species show additional sexual differentiation into

Figure 23-26.   *Oedogonium* asexual reproduction: (A) Escape from cell of filament of a zoospore with characteristic crown of flagella. (B) Settling, and (C) germination of zoospore. Sexual reproduction: (D) Filament showing oogonium at time of fertilization by a sperm from (E), a filament containing antheridia. (F) A zygote after meiosis. (G) Haploid zoospores released at germination of the zygote.

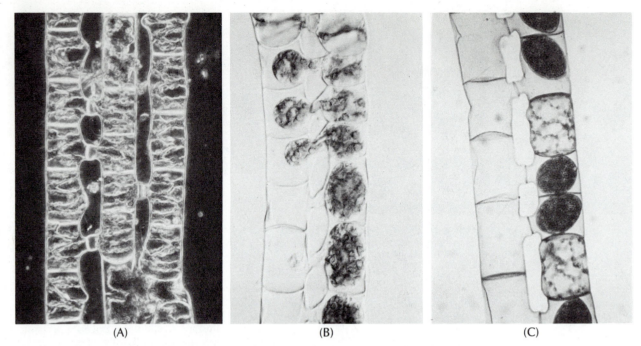

(A)   (B)   (C)

Figure 23-27. Sexual reproduction in *Spirogyra* occurs by fusion of nonflagellated gametes. Conjugation tubes form between cells of adjacent filaments of different mating types (A). Gametes of the donor filament migrate through the conjugation tubes (B) and fuse with other gametes. The resulting zygotes form thick cell walls and become dormant (C).

separate male and female individuals (dioecious condition). In some species the male individuals are much reduced dwarfs that grow attached by their holdfasts to female individuals, presumably improving the male's chances of successfully fertilizing a female. Thus, among these haploid organisms we encounter behavioral, developmental, and genetic sexual specializations analogous to some that have evolved in higher forms of life.

SPIROGYRA

The stout filaments of *Spirogyra* commonly form green masses ("pond scum") on the surface of quiet freshwater ponds and streams. The filaments feel slimy owing to excreted mucilage. Each cell possesses one or more remarkable, helically shaped ("spiral"), bright green chloroplasts (Figure 23-9), which are responsible for the genus' name; these are located in a peripheral layer of cytoplasm surrounding a large central vacuole, as in many higher plant cells.

Spirogyra spreads vegetatively by fragmentation but has no specialized method of asexual reproduction. It carries out sexual reproduction by **conjugation,** the fusion of two similar nonflagellated cells (Figure 23-27). Since the participating gametes (protoplasts of cells from separate filaments) are not morphologically different, the process is isogamous, but the two gametes participating in each conjugation do differ physiologically since one moves (in an ameba-like manner) through the conjugation tube between the two cells to join the other. The zygote develops into a heavy-walled, overwinter-

ing spore, as in *Oedogonium,* and it similarly germinates by meiosis to yield a new haploid filament, a haplontic life cycle.

STIGEOCLONIUM AND CLADOPHORA

Stigeoclonium, a common freshwater green alga (Figure 23-28A), illustrates the development of a branching filamentous body form that allows the individual to build a dense mass of filaments and capture much more light in a local area than a single filament can. *Cladophora* (Figure 23-28B), an unusual genus because it occurs both in fresh waters and in the ocean, is similarly densely branched but illustrates a condition intermediate between cellular and coenocytic body construction. Its large cells are multinucleate (cf. Figure 23-20E).

Although the reproductive cycle of *Stigeoclonium* is not understood completely, the organism is probably haploid, with a mode of reproduction somewhat like that of *Ulothrix. Cladophora,* on the other hand, reproduces by an isomorphic alternation of generations.

ULVA

The macroscopic thallus of "sea lettuce" (Figure 23-18) is common in intertidal and shallow subtidal zones of seacoasts worldwide. Seldom more than about 30 cm long, it grows attached to rocks and other objects by means of a holdfast at

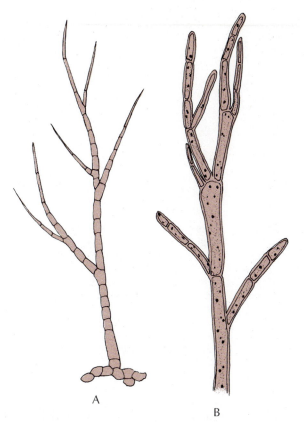

Figure 23-28. Two common branched filamentous green algae: (A) *Stigeoclonium* and (B) *Cladophora*.

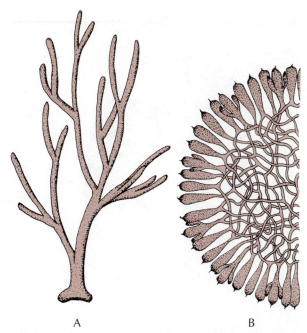

Figure 23-29. The green seaweed *Codium fragile* has a rope-like thallus (A) composed of numerous interwoven coenocytic filaments (B). The enlarged exposed tips contain the numerous disc-shaped chloroplasts for photosynthesis. (After G. Smith, Cryptogamic Botany I, 2nd ed., McGraw-Hill, 1955.)

the basal end. *Ulva* shows, as noted previously, an alternation of isomorphic (vegetatively identical) gametophyte and sporophyte generations.

In *Ulva* evidently the same developmental program of morphogenesis expresses itself in both the haploid and the diploid states. The sexual process is isogamous, as in *Ulothrix*. Ulva gametophytes and resulting gametes are of + and − mating types, which must come together to achieve fertilization.

Coenocytic Green Algae

Although the coenocytic body plan occurs in other algal groups, it has been elaborated most extensively in the Chlorophyta. The seemingly simplest form is found in sac-like coenocytes like the tropical marine green alga *Valonia* (Figure 23-21C). This simple appearance is deceptive, however, because *Valonia* achieves and sustains its form by means of an elaborate cell wall of multiple cellulose microfibril layers that is somewhat comparable to a multiple-ply tire. The formation of this wall is evidence of highly precise morphogenetic controls.

Other, tubular coenocytic algae develop stem- and leaf-like forms (Figure 23-21A). One of the most remarkable algae of this group, the tropical marine *Ace-*

tabularia, which produces a parasol-like photosynthetic "cap" (Figure 23-21B), delays any nuclear divisions until after the cap is formed and the plant is ready to undergo reproduction. Prior to this the entire macroscopic body has only one nucleus, a fact that has allowed biologists to conduct rewarding nuclear transplantation experiments to study the role of the nucleus in development (Chapter 14). The marine coenocytic *Codium fragile* (Figure 23-29) attains a massive macroscopic body form through coordinated growth of many tubular coenocytic filaments. *Codium* and other green coenocytes have a diploid thallus, reproducing by the diplontic type of sexual cycle (Figure 23-14B).

The Euglenoids (Division Euglenophyta)

These organisms are motile flagellates (Figure 23-30), exemplified by the large green flagellate *Euglena*, which is studied in many biology laboratories and has species widespread in both fresh and salt waters. In acid or nitrogen-rich waters it often grows to excess, forming objectionable scums or blooms.

Euglenoids differ from the flagellates included within the green algae (Chlorophyta) by producing a glucose storage polysaccharide different from starch, called paramylon, and in not possessing a cell wall. Instead the cells have a flexible complex surface layer called a **pellicle** that permits the cells to change their shape in a somewhat ameba-like fashion. This, plus the existence

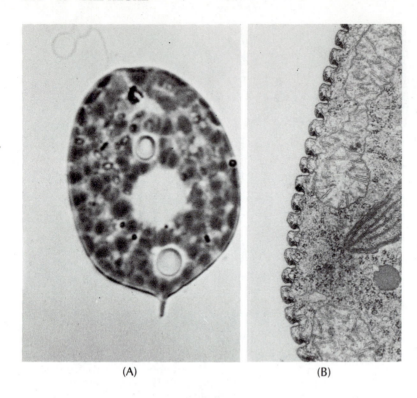

(A)   (B)

Figure 23-30. (A) Light micrograph of a euglenoid (*Phacus* sp.) of ovoid shape with a terminal spine. Note the anterior flagellum and the numerous dark discoid chloroplasts throughout the cell. (B) Electron micrograph of a section of a *Euglena* cell, showing the ridged surface of the pellicle. The ridges run in a helical pattern around the cells. Portions of mitochondria and a chloroplast can also be seen.

of colorless, saprotrophic or phagotrophic euglenoids, causes some biologists to regard euglenoids as protozoa.

Because euglenoid chloroplasts are surrounded by multiple membrane envelopes, it has been suggested (see Chapter 21) that they arose by secondary endosymbiosis, that is, integration of a chloroplast-containing endosymbiotic eukaryote, in this case presumably a unicellular green alga, into the cells of a colorless flagellate. In that event the basic organism might well be a protozoan, its evolutionary connection with the rest of the Chloronta being only in the origin of the chloroplasts of the green euglenoids.

The Stoneworts (Division Charophyta)

The relatively few but widespread species of the Charophyta comprise one of the most interesting and distinct of all plant groups, and some specialists doubt whether they should even be regarded as algae. These plants, which fall mainly into two genera, *Nitella* and *Chara*, grow on the bottom of clear freshwater ponds and streams. They are called stoneworts because when growing in calcium-rich (hard) waters many of them deposit calcium carbonate (limestone) outside their cells, forming a rough stony coating. Their body form (Figure 23-31A) superficially resembles a higher plant, consisting of a "stem" with "internodes" and "nodes" bearing whorls of "leaves," all arising from an apical growth zone analogous to a shoot apex. However, the body consists of gigantic

coenocytic cells that are structurally similar to the coenocytic Chlorophyta. Each *Nitella* "internode" is a single cell measuring up to about 1 mm in diameter and as much as 40–50 mm long, containing row upon row of chloroplasts and innumerable nuclei. The "leaves," similarly, are single coenocytic cells, or, in most *Chara* species, both "leaves" and "internodes" consist of a single large axial cell covered by a single layer of smaller tubular coenocytic cells (Figure 23-31B). The "shoot apex" is a single apical cell that by division cuts off cells that grow into "nodes" or, by enlarging prodigiously, "internodes."

The Charophyta have a highly developed oogamous sexual reproduction in which, uniquely among algae but similar to bryophytes and vascular plants, the sperms and eggs are produced within multicellular structures, that is, surrounded by a jacket of "sterile" or nonreproductive cells (Figure 23-31B). The chloroplasts of Charophyta resemble those of higher plants rather closely in granal structure, unlike the chloroplasts of most green algae. These features, plus the fact that the Charophyta have the phragmoplast type of cell division, have encouraged the idea that they are related to the ancient progenitors of land plants. However, the Charophyta apparently have a haplontic type of life cycle (Figure 23-14A), unlike the obligatory alternation of diploid and haploid generations found in all land plants, and their coenocytic body construction is totally unlike that of any known land plants. One proposal is that the Charophyta are a much modified and reduced, secondarily aquatic, evolutionary offshoot of an early terrestrial plant group that evolved into vascular plants in the Paleozoic, leaving no primitive remnants at the algal level other than these specialized relatives.

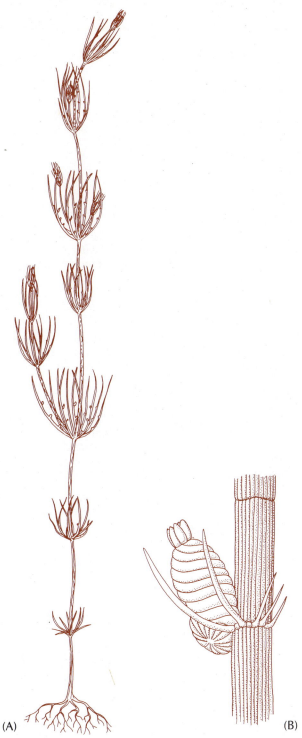

(A)   (B)

Figure 23-31. Stoneworts or charophytes are composed of long internodal cells alternating with short nodal cells where whorls of branches occur (A). The spherical, sperm-producing antheridia and the elliptical oogonia (B), formed by nodal cells, are both covered by a jacket of sterile cells. The calcified oogonia, preserved in many fossil deposits dating back as far as the Upper Silurian, are readily recognized by the spiral indentations on their surfaces. ((A) After R.D. Wood, Charophytes of North America; (B) after R.F. Scagel et al, An Evolutionary Survey of the Plant Kingdom, Wadsworth.)

The "giant cells" of *Nitella* and *Chara* are advantageous material for experimentally investigating mechanisms of cell function by manipulating individual cells or parts thereof—for example, the mechanism of cell enlargment (involving experimental manipulation of the cell wall), membrane transport (inserting electrodes into the cell to measure its electrical potential, and manipulating ion and ATP concentrations within the cell), and cytoplasmic streaming (involving isolation of actin microfilament bundles, cf. Chapter 4). These giant cells are prominent in current research in these and other areas of cell physiology. *Valonia,* among the coenocytic Chlorophyta (Figure 23-21C), is another "giant" algal cell similarly used advantageously in physiological experiments.

SUBKINGDOM CHROMONTA, THE YELLOWISH TO BROWN ALGAE

Except for the brown algae (Phaeophyta), the algal Divisions grouped in the Chromonta are predominantly or entirely unicellular or colonial, many of them being among the most important oceanic and freshwater phytoplankton. They lack chlorophyll *b* and possess (in most cases) chlorophyll *c* and abundant, often group-specific, carotenoid pigments such as fucoxanthin that confer a yellowish, golden, or brown cast to their cells. These algal Divisions also share certain basic cell structural features: (1) their chloroplast thylakoids generally occur in groups of three (Figure 23-10); (2) their photosynthetic storage products form outside the chloroplast rather than within it as in the Chlorophyta, Charophyta, and higher plants; and (3) their chloroplasts are enclosed by a multiple (three or four) membrane envelope, the outer membrane(s) of which are sometimes called "chloroplast endoplasmic reticulum" (chloroplast ER). However, the Divisions within the Chromonta differ from one another in other fundamental aspects of cell and body structure and modes of reproduction; they clearly represent very ancient separate evolutionary lines. Grouping them together as the Chromonta (formerly called Chromophyta) simplifies considerations of algal diversity and is mainly justified by the probably similar evolutionary origin of their chloroplasts.

As noted already for euglenoids, extra chloroplast envelopes suggest that the chloroplasts of the Chromonta were derived by secondary endosymbiosis. Because of the diversity in cell and chloroplast structural features encountered among the Chromonta, it appears that different but related golden- or brown-pigmented flagellates may have been acquired as endosymbionts independently by rather unrelated host cells ancestral to different Divisions of the Chromonta. As noted in Chapter 21, a small group of photosynthetic flagellates, the Cryptophyta, has chloroplasts containing biliproteins plus the pigments characteristic of the Chromonta. This suggests that the Chromonta's chloroplasts may have originated from primitive flagellates related or belonging to the biliprotein-containing algal Subkingdom (Bilonta, see below). If so, the biliprotein pigments were lost during further evolution of the Chromonta.

Among the wide variety of photosynthetic flagellates falling within the Chromonta many phycologists now recognize several Divisions besides those listed in Table 23-1 and considered below. In the interests of simplicity these are lumped with the more traditional Divisions (mainly with the Xanthophyta and Chrysophyta) in this discussion.

The Divisions of the Chromonta differ biochemically from one another and from other algae mainly (1) in their storage carbohydrates (Table 23-2), which are mostly glucose polysaccharides of the **laminarin** family (named after the kelp *Laminaria*) that are structurally different from starch, and (2) in the structure of their major xanthophylls, details of which are too complex for inclusion in Table 23-2.

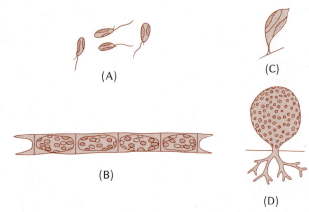

Figure 23-32. Some representative forms of Xanthophyta. (A) A flagellate, *Heterochloris;* (B) filamentous *Tribonema;* (C) a stalked unicell, *Characiopsis;* (D) a sac-like coenocyte, *Botrydium,* that grows on damp soil, attached by rhizoids.

The Yellow-Green Algae (Division Xanthophyta)

The Xanthophyta have plastids of a yellowish-green color due to large amounts of yellow carotenoids. The group includes unicellular flagellates and colonial, filamentous, and tubular coenocytic members (Figures 23-32, 23-33), and is found primarily in fresh waters.

The common and conspicuous coenocyte, *Vaucheria* (Figure 23-33), although a rather atypical member of this group, is an interesting organism frequently studied in botany laboratories, so we shall note some of its features. Some *Vaucheria* species are aquatic, while others are terrestrial, growing in dense felt-like masses

on the banks of ponds and streams or as thin films on soil and greenhouse flowerpots. Their tubular, occasionally branching thalli possess a dense layer of chloroplasts (plus nuclei) next to the cell wall. Terrestrial thalli sometimes form small colorless branches called **rhizoids** that penetrate the soil and attach the thallus to it. *Vaucheria* reproduces asexually by large, coenocytic, multiflagellated zoospores (Figure 23-33A), each produced singly in a terminal cell or sporangium. (These remarkable zoospores may be equivalent to a mass of many individual biflagellate zoospores that fail to separate by cytokinesis.) *Vaucheria* is haplontic and has a conspicuous oogamous sexual reproduction (Figure 23-33C) that leads to new thalli in much the same way as in *Oedogonium.*

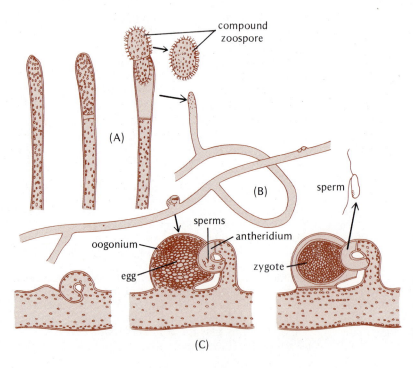

Figure 23-33. *Vaucheria* asexual reproduction: terminal sporangia are delimited by cross walls and each produces a single multiflagellate zoospore (A). Sexual reproduction: antheridia produce numerous sperms capable of fertilizing the egg in the adjacent oogonium (C). A thick-walled dormant zygote results from fertilization.

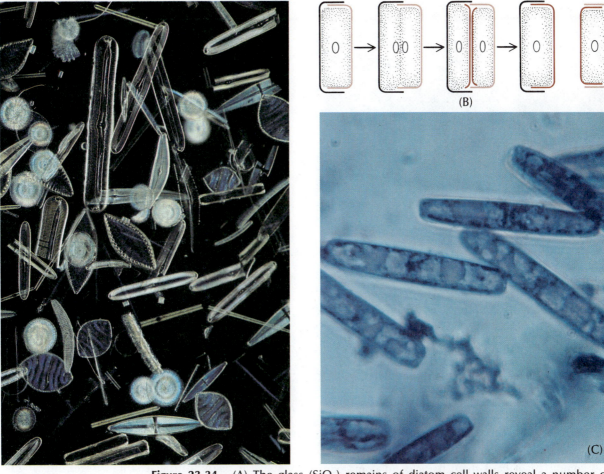

Figure 23-34.   (A) The glass (SiO$_2$) remains of diatom cell walls reveal a number of shapes seen in various species. **Centric** types have circular valves, while elongated valves occur in **pennate** species. (B) Vegetative cell division results in two cells, each having one parental valve and one newly synthesized, resulting in one progeny cell being smaller than the other. (The valves diagrammed are from a side view, whereas (A) and (C) show top or bottom views of valves.) (C) A colorless variant of a pennate diatom. Absence of plastids allows the large oil droplets to be seen in these stained cells.

The Golden-Brown Algae (Division Chrysophyta)

These are mostly motile unicellular or colonial forms, including the diatoms, one of the most important groups of marine and freshwater phytoplankton. Abundant xanthophylls and β-carotene mask the chlorophylls in these algae.

DIATOMS

As noted earlier, diatoms (Figure 23-34A) produce a unique type of cell wall composed of silica (SiO$_2$, a component of glass). Most diatom species are free-floating (plankton), but some live attached to larger plants and other objects. There are both unicellular and colonial forms. Diatoms are extremely abundant in both fresh and marine waters in the cooler latitudes, usually comprising the majority of the phytoplankton. Off the American east coast, for example, there may be up to half a million diatoms per liter of sea water.

The diatom cell wall is composed of two halves or "valves," one overlapping the other like the top and bottom of a petri dish. When the cell divides, the valves separate, and each daughter cell produces a new half complementing the one it inherited (Figure 23-34B). The valves are often marked by elaborate ridges and grooves and pierced by minute pores that allow contact between the protoplast and its environment. When resting on the bottom some of the elongated (pennate) diatoms can move in a gliding manner by extruding mucilage through the slits and pores in the lower valve. This enables them to move back into the light if they are shaded.

Figure 23-35. Thick deposits of white diatomaceous earth near Lompoc, California, are mined for industrial usage.

Figure 23-36. A coccolithophorid, a common component of open ocean phytoplankton populations.

A diatom cell contains one to several golden-brown plastids. Reserve food is stored as a liquid form of laminarin and as an oil, microscopically visible in the cells as large round droplets. As diatoms fall to the ocean bottom and are trapped in sediments, petroleum may be formed from this oil. Their indestructible cell walls accumulate in marine sediments without limit, leading to the gigantic deposits of diatomaceous earth that occur in many parts of the world; these represent former marine sediments that were later raised above sea level. The largest beds in the United States, some 450 m (1400 ft) thick, are located near the California coast (Figure 23-35). Chemically inert, indestructible, light, porous, yet composed of hard sharp particles, diatomaceous earth has many uses in construction and industry and is mined extensively for these purposes.

Diatoms reproduce mainly by vegetative means, namely by simple cell division (Figure 23-34B). Since their silica wall cannot enlarge, sexual reproduction or a special type of division, not described, is needed to restore to full size the daughter cells that have received the smaller half of the parental wall during normal divisions and have consequently become reduced in size. Centric diatoms have oogamous sexual reproduction involving motile sperms; others show conjugation of nonmotile isogametes (Figure 23-17). In either case, the resulting naked zygotes form new valves. The vegetative diatom cell is diploid, and its sexual cycle is diplontic.

Other Important Chrysophyta

The **silicoflagellates** have an intracellular silica skeleton in contrast to the extracellular silica covering of diatoms. Other Chrysophyta are decorated with surface scales or plates of silica or calcium carbonate. The **coccolithophorids,** bearing calcified plates called "coccoliths" (Figure 23-36), are abundant in warm waters and are often the predominant phytoplankton in tropical oceans.

The Dinoflagellates (Division Dinophyta)

This Division (also called Pyrrhophyta) includes a great variety of unicellular, motile flag-
ellates, whose chief common characteristic is the presence of two grooves or furrows in the cell surface, each containing a single flagellum (Figure 23-37A). One furrow runs transversely and completely encircles the cell; the other is longitudinal and extends only along one side of the cell. The combined action of these flagella makes the cell rotate as it swims forward, hence the name Dinoflagellate (Greek *dinos*, to rotate). Armor-like cellulosic plates, closely joined together, occur under the cell's surface membrane in many species. In some the plates have striking horns, spines, or other elaborate ornamentations (Figures 23-12, 23-37). Many dinoflagellates are photosynthetic, with the usual Chromonta type of pigment system including an excess amount of carotenoids that give the cells a reddish, brownish, or yellowish color. Many others are colorless, nonphotosynthetic flagellates that have been classified as protozoa.

Many dinoflagellates are bioluminescent, that is, they produce light by an oxidative enzymatic reaction. This reaction is frequently responsible for sea water "phosphorescence" that is sometimes visible when waves break or when boats, ships, or animals disturb the water.

Although dinoflagellates are usually secondary to diatoms in importance as phytoplankton, they sometimes occur in such great numbers that they color the ocean yellowish- to reddish-brown over great areas, a phenomenon called the "red tide." In one affected area the sea was estimated to contain more than 150 million dinoflagellate cells per liter. Several of the species re-

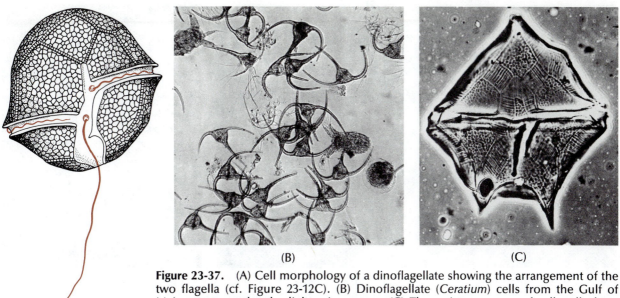

Figure 23-37. (A) Cell morphology of a dinoflagellate showing the arrangement of the two flagella (cf. Figure 23-12C). (B) Dinoflagellate (*Ceratium*) cells from the Gulf of Maine, seen under the light microscope. (C) The resistant nature of cell wall plates allowed this dinoflagellate (magnified about 1350 ×) to be preserved in the fossil record in late Cretaceous rock sediments (Red Bank formation in New Jersey). Many of the algal species now fossilized existed for relatively short geological time periods, and once the ages of the microfossils are established, recognition of their distinctive features, such as the number, shape, and ornamentation of plates or scales, aids geologists in dating sedimentary rock layers containing them. Individual plates and the bands with parallel markings that join them contain sporopollenin, a carotenoid derivative also found in the resistant walls of higher plant pollen grains.

sponsible for red tides produce toxins that become concentrated by shellfish feeding on the plankton. These toxins can paralyze other animals who eat the shellfish, including humans. For this reason, shellfish consumption by humans is not allowed during the summer months when dinoflagellates are abundant in the ocean.

Vegetative reproduction in dinoflagellates occurs by a longitudinal splitting of the cell, following a very unusual type of mitosis, which is accomplished without breakdown of the nuclear envelope or nucleolus or formation of a normal mitotic spindle. Each daughter cell inherits one half of the parent cell wall and builds the other half anew, as do diatoms and desmids. The nucleus of a dinoflagellate cell during interphase has a unique, so-called mesokaryotic structure in which the chromosomes remain permanently condensed, much as they are only during mitosis in most eukaryotes. Certain dinoflagellate cells possess two nuclei, a typical mesokaryotic one plus a more ordinary eukaryotic nucleus. The latter seems to be part of a relatively recently acquired photosynthetic endosymbiont whose chloroplast has become the chloroplast of the dinoflagellate, illustrating an early stage in secondary endosymbiosis. All in all, the dinoflagellates are a most bizarre group, probably becoming a separate evolutionary line early in pre-Cambrian eukaryote evolution and acquiring chloroplasts independently by endosymbiosis several times. Certain modern species are themselves endosymbionts (Chapter 25).

Isogamous sexual reproduction occurs in at least some dinoflagellates, both haplontic and diplontic life cycles being reported in different species.

The Brown Algae (Division Phaeophyta)

The most familiar seashore seaweeds are the larger brown algae, or kelps (Figures 23-1C, 23-38A, B) and "rockweeds" (Figure 23-38C). The brown algae are almost exclusively marine and seem to be basically adapted to cold waters, being prolific along temperate and colder coastlines but relatively unimportant in the tropics. The browns include the largest and most massive of all algae, some becoming as much as 100 m long. However, they range from these largest forms down to microscopic filamentous forms, although no unicellular brown algae are known. Their characteristic brown or olive color is due to large amounts of the distinctive xanthophyll, fucoxanthin, which also occurs in the Chrysophyta.

Attached to rocks, the common rockweed *Fucus* (Figure 23-38C) and related genera grow profusely along seashores in the intertidal zone, along with smaller brown algae. Most of the browns, however, grow from about the average low tide line to about 10 m of depth, often forming extensive jungle-like kelp beds or "kelp forests" in shallow offshore waters (Figure 23-38A). This dense growth captures most of the light falling on it and is highly productive, supporting a rich community of marine animals.

Competition for light is keen in a kelp bed. Most kelps develop expanded, leaf-like organs ("blades") for light capture, often along with gas-filled floats to buoy

Figure 23-38. Some common brown algae. (A) Kelp beds off the central California coast. (B) The attachment of kelps by their holdfasts to rock substrata is so tight that storms frequently wash up these algae with pieces of rock still attached to their bases. (C) Fucoids in and around a tidepool.

them toward the surface (Figure 23-1C). These floats contain some carbon monoxide (CO), a gas unusual in organisms. The floats and blades are borne on a stem-like axis or **stipe,** which extends down and attaches to the bottom by a massive, disc-shaped branch system, the **holdfast** (Figure 23-38B). Kelps thus exhibit very def-

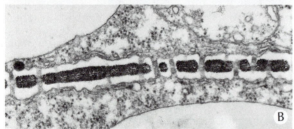

Figure 23-39. Light micrograph (A) of trumpet filaments (named for the cell shape) comprising the central part of a kelp blade. An electron micrograph (B) shows the sieve-plate–like structure of the cross walls between cells.

inite organ differentiation, belieing the standard definition of thallophytes as lacking differentiation. However, the kelp plant body is nevertheless by convention still called a thallus.

Because of the strong light absorption by a kelp bed canopy, the stipes and holdfasts may not get enough light to support themselves photosynthetically and can be fed by the blades. The large kelps develop within their stipes chains of cells called sieve tubes or trumpet filaments (Figure 23-39), which have expanded, sieve-like end walls resembling the sieve plates of vascular plants' sieve members (Chapter 10). Radioactive tracer experiments indicate that the kelp translocates carbohydrates (especially mannitol) toward its holdfast through these specialized filaments, which comprise a primitive vascular tissue evolved independently of that in higher plants.

Because many kelps and rockweeds are subjected to relentless, often severe, wave action in their habitat, they must be tough yet flexible and resilient. Fragmentation of the thallus by breakage does not usually lead to vegetative reproduction, as in so many other algae, because the surf quickly casts detached plants or branches up on the shore, where they soon die and decompose. An exception is the Sargasso Sea, an area in the Atlantic where large detached, floating masses of brown *Sargassum*, related to *Fucus*, grow and accumulate.

Besides cellulose, kelp cells produce and secrete distinctive gelatinous wall polysaccharides called fucoidin and alginic acid, whose mechanical properties are conducive to flexibility of the thalli. Alginic acid, and its salts called alginates, are extracted from commercially harvested kelps for extensive use as thickeners and stabilizers in the cosmetic and food industries. Their ability to hold particles in suspension and increase product viscosity is advantageous in the preparation of many products, including chocolate milk and ice cream. This is the principal economic importance of brown algae, although some are used as human food,

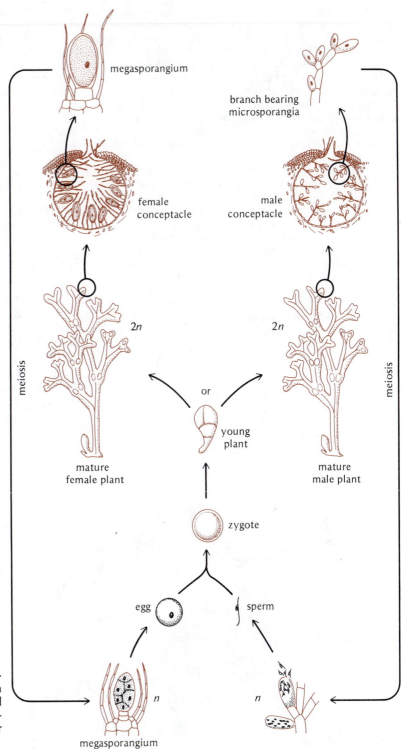

Figure 23-40. *Fucus* life cycle. The diplontic life cycle, relatively uncommon in algae, is characteristic of all fucoid brown algae (*Fucus, Sargassum, Ascophyllum,* and related genera in the order Fucales).

for example, in Japan and China, and kelps are sometimes employed as cattle feed in some northern coastal countries.

REPRODUCTION AND DEVELOPMENT

Sexual reproduction in brown algae varies from essentially isogamous in simple filamentous forms to oogamous in *Fucus* and kelps. The reproductive cycle of most browns is an alternation of generations, isomorphic (like *Ulva*) in simple browns but highly heteromorphic, as described earlier (Figure 23-19), in kelps such as *Laminaria*. The macroscopic kelp plant is the diploid (sporophyte) generation, which reproduces by releasing motile meiospores in enormous numbers, these cells being an important food source for filter-feeding animals and zooplankton.

The rockweed *Fucus* and several related genera are unusual among browns in reproducing by a simple diplontic life cycle (Figure 23-40). The diploid vegetative plant produces haploid eggs and sperms by meiosis within reproductive chambers called conceptacles located near the thallus tips, shedding the gametes into the water where fertilization occurs.

The fertilized egg of *Fucus* or of a kelp first attaches itself to the bottom by producing during germination a system of single-celled, root-like outgrowths or rhizoids, as illustrated previously for *Fucus* (Figure 14-28). Rhizoid outgrowth involves a delicate control by light of the polarity of development, as discussed in Chapter 14. From the rhizoid portion of the germling, the massive multicellular holdfast eventually develops. As the shoot portion of the germling enlarges, it eventually differentiates one or more local meristematic growing zones, somewhat analogous to the meristems of higher plants. The growth zone is apical in some species such as *Fucus*, which has a single apical cell engaged in cell division at the tip of each branch of its thallus. Other species have a subapical growth zone or, in some cases, an intercalary growth zone located far back from the tips of the blades or branches (e.g., at base of blade in Figure 23-38B). This intercalary meristem generates new photosynthetic tissue ahead of it, replacing older, distally located organs being eroded away by wave action and herbivorous animals.

The sporophytes of *Laminaria* and many other kelps are perennial. *L. hyperborea* of arctic seacoasts, for example, can live as long as 18 years. In the arctic, kelp sporophytes begin to grow during late winter when light for photosynthesis is minimal. At this time the plants grow largely at the expense of stored reserves (laminarin), subsequently replenishing these under summer conditions favorable for photosynthesis. An advantage of this curious cold-season growth cycle may be the opportunity to capture and use the greater amounts of mineral nutrients available in sea water when growth of other algae is not competing for the nutrient supply.

SUBKINGDOM BILONTA, THE BILIPROTEIN-PIGMENTED ALGAE

The only major eukaryotic group possessing the blue and red biliprotein pigments, phycocyanin and phycoerythrin, is the red algae (Division Rhodophyta). Red algal chloroplasts consistently differ from those of both the Chloronta and the Chromonta because their thylakoids occur singly and bear distinctive biliprotein-containing particles called **phycobilisomes** (Figure 23-41). These features resemble the internal structure of cyanobacterial cells, from which the red algal chloroplast probably originated by endosymbiosis (Chapter 21). Complete lack of any motile, flagellated stages also sets the red algae apart from the other algal subkingdoms.

Besides the red algae, one small flagellate group, the Cryptophyta, possesses phycocyanin and phyco-

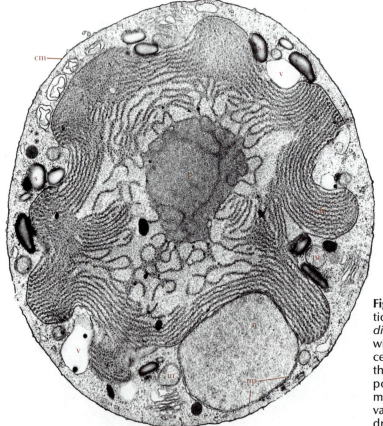

Figure 23-41. Electron micrograph of a section through a one-celled red alga (*Porphyridium cruentum*). A single stellate chloroplast with a central pyrenoid occupies much of the cell volume. Phycobilisomes can be seen on the single thylakoids. n, nucleus; np, nuclear pore; er, endoplasmic reticulum; cm, cell membrane; ch, chloroplast; p, pyrenoid; v, vacuole; s, starch-like bodies; m, mitochondrion (14,850 ×).

Figure 23-42. Two types of small red seaweeds: thalli of a coralline species with calcified segments are seen at the corners of the photo around a foliose irridescent species (*Fauchea*). Thalli of both are several centimeters in length.

Figure 23-43. *Porphyra* ("nori") beds ready to harvest at Sendai, Japan.

erythrin pigments, and therefore might also be included in the Subkingdom Bilonta. But in addition to having flagella and lacking phycobilisomes the Cryptophyta possess the pigments and chloroplast structure characteristic of the Chromonta, and therefore are more reasonably placed in that Subkingdom.

The Red Algae (Division Rhodophyta)

Like the browns, the red algae are largely confined to the oceans, and their more than 4000 species exceeds the number of known species of brown and marine green algae combined. Their geographical prominence contrasts with that of the browns and marine greens, the former being most important in colder waters while the reds are most prominent and diverse in tropical oceans. Like the browns, they are mostly benthic multicellular organisms, although a few unicellular red algae are known. Because of the capacity of their pigment system to absorb strongly all wavelengths of visible light, reds are the algae best adapted, as noted earlier, for inhabiting deeper waters; they sometimes extend as deep as 200 m, considerably deeper than any other algae. However, reds are not confined to deep waters, and many species occur up to or even above the low-tide line.

Although some red seaweeds develop a relatively massive, blade-like thallus that superficially resembles some of the smaller kelp, the largest reds rarely exceed 1 m in length, and most of them are much smaller, more delicate organisms with a wide variety of often beautifully intricate body forms (Figure 23-42).

Many small red algae grow as **epiphytes,** that is, attached to other plants (usually brown or other red algae) or even to animals. Some are **obligate** epiphytes, restricted to growth on specific "host" seaweeds, and a number have evolved into parasites of other red algae.

Red algae produce and secrete around their cells cellulose and massive amounts of distinctive noncellulosic polysaccharides. One of the commercially important red-algal polysaccharides is **agar,** obtainable from the genus *Gelidium* and various others. Its property of forming solid gels even at low concentrations (1–2%) and of being resistant to degradation by most microorganisms makes it the substrate of choice for microbiological culture media (Chapter 22), and this use accounts for a considerable share of the current demand. Agar is also extensively used as a thickener in the food preparation and pharmaceutical industries. A similar but somewhat weaker gelling red-algal polysaccharide called **carrageenan,** usually obtained from Irish moss (a mixture of the red algae *Chondrus crispus* and *Gigartina stellata*), is extensively employed for thickening dairy products, and is also used in puddings, salad dressings, and other food and cosmetic products.

Certain red algae called **coralline** reds deposit a crust of calcium carbonate (lime) about themselves, giving them a coral-like appearance (Figure 23-42). Although true coral is formed from the activities of minute sedentary animals, lime-secreting red algae are sometimes more important in the formation of coral reefs than the coral animals themselves.

Some red algae are important food plants of Asiatic countries, where they not only are gathered from bays and lagoons but have been cultivated for hundreds of years (Figure 23-43). One of the most widely used, *Porphyra*, contains protein and carbohydrates that are digestible by humans, in contrast to the Irish moss polysaccharides, which pass unaltered through the digestive tract.

REPRODUCTION

Red algae reproduce by an alternation of isomorphic to extremely heteromorphic generations, depending on the species. The known reproductive cycles include considerable complications and variations that are beyond the scope of this introduction. We note below a few of the unusual and at least relatively general features of red algal reproduction.

Red algal gametes are nonmotile. Gametophytes release small "male" gametes called **spermatia,** which are carried about passively by water currents. Game-

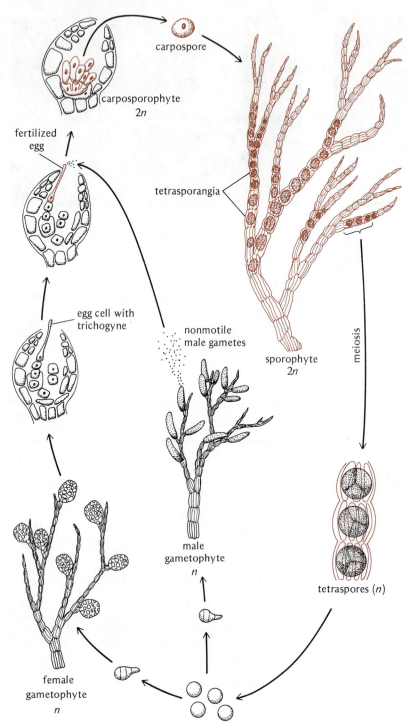

carpospore

carposporophyte
2n

fertilized
egg

tetrasporangia

egg cell with
trichogyne

nonmotile
male gametes

sporophyte
2n

meiosis

male
gametophyte
n

tetraspores (n)

female
gametophyte
n

n

Figure 23-44. *Polysiphonia* life cycle. Many red algae have two diploid stages alternating with the haploid generation.

tophytes also produce egg cells bearing an elongated receptive projection called a **trichogyne** (Figure 23-44), to which spermatia of the same species will stick upon contact, apparently by a species-specific cell surface recognition mechanism. The adhering spermatium fuses with the trichogyne, and its nucleus travels down and fertilizes that of the egg. This leads to a unique feature of red algal reproduction: the diploid zygote usually multiplies in place, still attached to the parent gametophyte, to form a transient diploid generation called the **carposporophyte,** which by mitosis produces and releases a number of diploid spores (carpospores), each

of which potentially can grow into an independent diploid plant. The carposporophyte is essentially a way of multiplying the results of each sexual fertilization, giving it a number of chances of yielding a successful independent offspring. The independent diploid plants derived from the carpospores are comparable with ordinary sporophytes. They eventually reproduce by forming haploid meiospores, which, since they are formed in tetrads (groups of four) has led to the name **tetrasporophyte** for this generation. The meiospores grow into gametophytes, completing the cycle (Figure 23-44).

SUMMARY

(1) Algae are primarily aquatic thallophytes. Although they include the oldest, and usually considered the simplest, eukaryotic green plants, they are very diverse in form, and many are quite complex. They range from unicellular forms, often with elaborate internal structure or external form, through colonial and filamentous types with some degree of cell differentiation and definite morphogenetic controls, to leaf-like and branch-like macroscopic bodies with organ differentiation and modes of development somewhat like those of higher plants. Most of these larger algae are benthic, living attached to a substratum by a holdfast, while many of the unicellular and colonial forms are motile by means of flagella. Phytoplankton are motile and nonmotile algae living suspended in marine and fresh waters.

(2) Algae can be grouped into three broad assemblages or Subkingdoms according to their photosynthetic accessory pigment systems, associated chloroplast structural features, and types and locations of photosynthetic storage products. These Subkingdoms are (a) the Chloronta, usually green in color, with pigments similar to those of higher plants, including chlorophyll *b;* (b) the Chromonta, usually yellow-green to brown in color, with abundant and often structurally distinctive carotenoid pigments, and with chlorophyll *c* but not chlorophyll *b;* and (c) the Bilonta, usually red in color, with biliprotein pigments (phycoerythrin and phycocyanin) but neither chlorophyll *b* nor chlorophyll *c.* Each Subkingdom contains several Divisions differing in cell structure and in details of pigments and storage reserves (except the Bilonta, which contains only one Division, the red algae). These Divisions seem to comprise relatively clear, separate evolutionary lines within the algae, while the members of each broad Subkingdom as a whole may have little in common evolutionarily except chloroplasts of similar endosymbiotic origin.

(3) Because of their need for light, algae are limited to the euphotic zone of lakes and oceans. In marine ecosystems the benthic algae may be found in zones corresponding to some extent with depth in the order greens, browns, and reds, according to the effectiveness of their accessory pigments in supplementing the light absorption by chlorophylls. Fresh- and salt-water algae are almost always of different species or major groups. Brown and red algae are significant only in the sea, while greens are much more important in fresh waters. Phytoplankton of fresh and salt waters belong mostly to the green algae and to several Divisions of the Chromonta. Only a few algae (mostly flagellates) can adapt to both fresh and salt waters.

(4) The makeup of both benthic and planktonic marine algal floras varies with water temperature: red algae and dinoflagellates (plus coccolithophorids, etc.) predominate in the tropics, while brown algae and diatoms predominate in temperate and subpolar waters. In these colder waters, algal growth is highly seasonal. Spring and autumn blooms of temperate freshwater phytoplankton are apparently controlled by yearly variation in the light and mineral nutrient supply, as well as by temperature. Phytoplankton growth in the open ocean is very limited, owing to poor mineral nutrient supply caused by lack of mixing between surface and deep waters; the ocean's productivity is therefore very low except along the coasts or where upwelling brings nutrients to the surface from below.

(5) Most algae reproduce both vegetatively or asexually, and sexually. Sexual reproduction is isogamous, anisogamous, or oogamous in different algae. Most algae possess genetic mating-type systems that ensure cross-fertilization.

(6) Among different algae are found all the biologically known types of sexual reproductive cycles, including (a) the haplontic type, in which the vegetative body is haploid and produces gametes yielding a zygote that germinates by meiosis, restoring the haploid condition, (b) the diplontic type, which involves a diploid vegetative body producing haploid gametes by meiosis, and (c) alteration of haploid (gametophyte or gamete-producing) and diploid (sporophyte or meiospore-producing) generations, each generation reproducing by giving rise to the other. Many of the simpler algae show an isomorphic alteration of generations (of identical vegetative form), whereas many of the more complex brown and red algae have a more highly evolved, heteromorphic alternation of generations, with the diploid or sporophyte generation usually being larger, different in form, and more complex than the gametophyte, as in higher plants.

(7) An evolutionary progression in vegetative body and reproductive features can be seen within many of the algal Divisions, examples being given for the green algae (Chlorophyta). Some other important algae are the diatoms (in the Chrysophyta or golden-brown algae) and the dinoflagellates (Dinophyta), which are major phytoplankton groups belonging to the Chromonta. The kelps or larger brown algae (Phaeophyta) and some of the red algae (Rhodophyta) are commercially important as sources of algal polysaccharides, notably alginic acid and agar, used in the food and microbiological industries. The stoneworts or Charophyta are of interest as the members of the Chloronta that are perhaps most closely related to the progenitors of land plants.

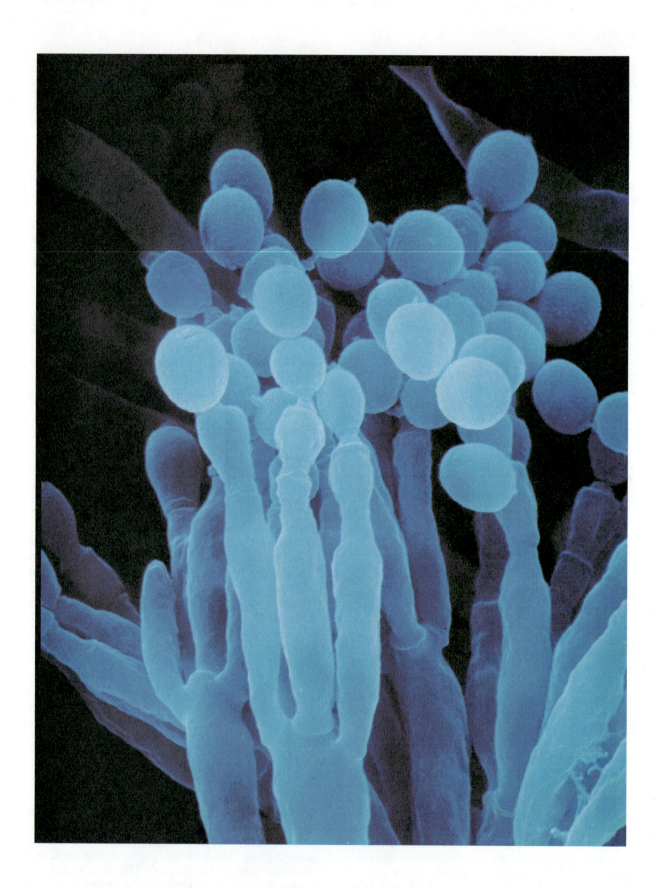

Chapter 24

FUNGI

Fungi are heterotrophic eukaryotes traditionally included in the plant kingdom primarily because they do not move about. Although nutritionally comparable to animals, fungi differ from animals not only in usually lacking locomotion but also in (1) possessing cell walls, (2) reproducing by spores, and (3) obtaining food by absorbing soluble food materials from outside rather than engulfing or ingesting food as animals do. Because most fungi have cell walls that are different in composition from those of green plants and are thought to comprise an ancient evolutionary complex separate from photosynthetic eukaryotes, fungi are treated as a separate Kingdom in the five-kingdom classification (Figure 2-13). In our classification (Chapter 21), grouping all eukaryotes together as one Kingdom, fungi are classed as a Subkingdom (Myconta).

Whatever their evolutionary relationship to plants, fungi interact with green plants in several ways. (1) Fungi and bacteria are nature's primary recycling agents; by releasing mineral nutrients as they decompose plant and animal remains, fungi play an important role in green plant nutrition. (2) Some fungi contribute directly to plant nutrition through mutualistic association with the roots of many higher plant species (mycorrhizal associations). (3) Parasitic fungi are the most important causal agents of plant diseases.

Fungi also interact with humans in economically and socially important ways. Scientists in many agricultural experiment stations and university and industrial laboratories work to solve problems caused by fungi, discovering methods for protecting wood, fabrics, and other materials against fungal deterioration, developing horticultural and agricultural practices to combat fungal diseases, perfecting methods for protecting produce against fungal spoilage, and finding methods for prevention of and for inspection and quarantine against the spread of detrimental fungi. A few fungi cause diseases of humans and other animals, control of which demands research effort. On the other hand, fungi also are exploited by humans for use in making bread, some other foods, and alcoholic beverages; mushrooms, of course, are used directly as a food. Fungi are now utilized for production of many valuable special products such as antibiotics, organic acids, and other compounds used in medicine and industry.

This chapter considers the characteristics, diversity, and economic significance of fungi except for plant diseases and their causal fungi, which are covered mainly in Chapter 26. The other important ecological roles of fungi mentioned above are considered in Chapters 25 and 33. The slime molds are briefly covered at the end of this chapter; they are allied by name with true molds but differ so fundamentally that we classify them in a separate Subkingdom (Myxonta).

THE NATURE OF FUNGI

Vegetative Growth

Some fungi are unicellular, but most are composed of branching filamentous (thread-like) structures termed **hyphae** (singular, hypha). These usually develop into an interlaced network collectively known as a **mycelium** (Figure 24-1). Hyphae elongate by tip growth (Figure 12-25B). **Septate** hyphae are divided by cross walls into cellular units, whereas **nonseptate** hyphae have few or no cross walls (Figure 24-2), except where reproductive structures are formed.

In most fungi the chief microfibrillar component of the cell wall is **chitin,** the same polysaccharide that makes up the exoskeletons of insects. Cellulose occurs in the cell walls of some of the lower fungi but is absent in most others. Like the walls of algae and higher plants, fungal walls include a matrix phase composed of polysaccharides that are different from the microfibrillar polymers.

Figure 24-A. *Penicillium chrysogenum,* a mold that can cause food spoilage but also produces the antibiotic penicillin. This scanning electron micrograph shows the brush-like chains of spherical spores that characterize the genus and form the blue-green macroscopic "mold" frequently seen on various food products.

Figure 24-1. A cottony mass (mycelium) of fungal hyphae growing on and causing decay of a squash.

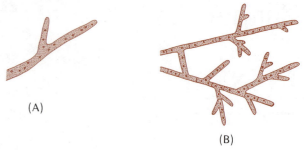

Figure 24-2. Mycelial hyphae representative of "lower" (A) and "higher" (B) fungi.

Fungi are either saprotrophic or parasitic. Some species, including most rusts and powdery mildews, are **obligate parasites,** which are restricted to this mode of nutrition. Others, including many of the large fleshy mushrooms, are **obligate saprotrophs.** Certain fungi are **facultative parasites**—that is, they can grow either saprotrophically or parasitically, and some go through separate parasitic and saprotrophic phases in their life cycle. Saprotrophic fungi and bacteria occur almost everywhere in the environment as agents of de-

cay. They thrive in the soil, in dead remains of plants and animals, and on other organic materials such as food and leather. In the wet tropics, mold can cover shoes and clothing almost overnight. Molds obtain soluble food from these sources by degrading them with enzymes such as proteases, cellulase, amylase, and pectinase, which are produced and released by the hyphae into their surroundings.

The mycelial mode of fungal growth gives fungi an advantage over bacteria in saprotrophically exploiting solid substrates such as wood. The tip growth of fungal hyphae, which is accompanied by secretion of degradative enzymes, allows hyphae to penetrate rapidly and

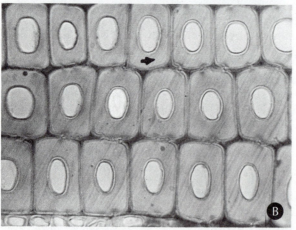

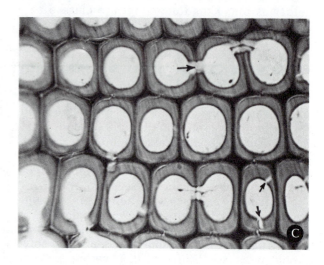

Figure 24-3. (A) Wood-rotting fungi invading early heartwood cells of a conifer (microscopic view of a longitudinal section through the wood). (B and C) Cross-sections through tracheids of uninvaded (B) and partially decayed (C) pine wood. The dry weight of (C) has decreased by two-thirds; arrows indicate where hyphae have grown through cell walls digesting channels, in addition to decreasing overall wall thickness.

spread through solid substrates (Figure 24-3), gaining possession of many of the nutrients that are releasable by degradative enzymes before less invasive saprotrophs such as bacteria can reach these nutrients. Many bacteria, however, can grow without oxygen and thus can proliferate in anaerobic habitats that exclude most fungi.

Although most fungi are aerobic (require oxygen), a few, including a number of yeasts, are facultative anaerobes, obtaining sufficient energy for growth from fermentation. Fungi nearly always require free water, or high humidity, for growth. In general, they tolerate a more acid pH than bacteria, many growing best at or below pH 6. This fact is frequently used in the selective isolation of fungi from natural sources. Like bacteria, most fungi (except obligate parasites) can be grown in pure culture on nutrient media. Some are prototrophs and grow on simple laboratory media containing only a carbon source, such as the sugars glucose or sucrose, and mineral salts, including a nitrogen source. From these they synthesize their own proteins and other growth compounds, including vitamins. Other fungi such as the pink mold *Neurospora* (an important organism in genetic investigations; Chapter 18) are auxotrophic, requiring biotin, thiamine, and/or other vitamins.

Reproduction

The universal occurrence of fungi and their success in colonizing virtually all available habitats are due to their efficient mechanisms of dispersal by air- or water-borne spores. Almost all fungi produce copious numbers of spores—an average mushroom releases an estimated 1 billion spores when mature. Molds, like those that grow on fruits or on bread, likewise produce thousands of spores on each square centimeter of moldy surface. These spores (Figures 24-A and 24-10) are carried about by air currents and are ready to germinate into a new thallus whenever they land on a suitable substratum with the right temperature and moisture conditions. Some of these air-borne spores cause about 15% of all human respiratory allergies. Some aquatic and soil fungi release **motile** spores (**zoospores,** Figure 24-4) into water or the surface films of water in wet soils. These zoospores can swim to fresh nutrient sources, where they germinate. Dispersive spores usually are formed asexually (nuclei arise by mitotic division) and are relatively short-lived. Filamentous

Figure 24-4. The dispersive spores of water molds are flagellated. Members of the Chytridiomycota have a single posterior flagellum that propels the zoospore forward. In contrast to the vegetative stages, these motile cells have no rigid cell wall and the cell organelles (spherical mitochondria clumped about the nucleus) can be seen in this spore, which was dried on a film and shadowed with a thin film of metal for viewing in the electron microscope.

TABLE 24-1 Major Groups of Fungi

Characteristics Distinguishing Lower from Higher Fungi		Divisions	Characteristics of Divisions
Lower Fungi	Multinucleate cell, or mycelium with few septa Asexual spores formed within sporangium Aquatic or terrestrial	1. **Chytridiomycota** (simple water molds) 2. **Oomycota** (water molds, downy mildews) 3. **Zygomycota** (pin molds)	Mostly aquatic or in wet habitats[a] Form motile cells[a] Mostly terrestrial; no motile cells
Higher Fungi	Septate mycelium (rarely unicellular) Asexual spores formed externally Mostly terrestrial	4. **Ascomycota** (molds, powdery mildews, cup fungi, yeasts) 5. **Basidiomycota** (mushroms, puff balls, rusts) 6. **Deuteromycetes** ("imperfect" molds)	Meiospores formed within sac-like cell (ascus) Meiospores formed externally on club-like cell (basidium) No meiospores formed (no sexual reproduction)

[a]These characteristics apply to both Divisions 1 and 2 (water molds and related forms). Distinctions between Divisions 1 and 2 are explained in the text.

fungi can also reproduce vegetatively by fragmentation, and the unicellular yeasts reproduce by budding (see later discussion in this chapter.)

Certain kinds of fungal spores serve as **survival units.** These usually have a thick, resistant, dark cell wall and an obligatory dormancy period, similar to algal zygospores such as those of *Spirogyra* (Chapter 23). Survival spores, often called "resting spores," are frequently derived either from sexual cell fusion or by meiotic division subsequent to sexual fertilization (**meiospores**). Although the asexual cycle, which is important for colonization and extension of the species, is often repeated several or many times each season by means of asexual dispersive spores, the production of meiospores following sexual fusion occurs only once a year in many fungi.

The Kinds of Fungi

A brief overview of the major fungal groups is given in Table 24-1. The three Divisions of Myconta listed in the upper part of Table 24-1 are considered "lower fungi" (formerly called Phycomycetes) because they have characteristics generally considered to be more primitive or evolutionarily older than those seen in the "higher fungi." The characteristics of lower fungi include usually nonseptate hyphae and production of a number of asexual spores within a specialized cell or **sporangium** (external production of spores is discussed later in this chapter). The other three Divisions, called higher fungi (Table 24-1), usually have septate hyphae and a more complex structure, morphology, and life history. Many higher fungi produce their meiospores within complex macroscopic, spore-producing bodies (**sporocarps,** colloquially called "fruiting bodies," e.g., mushrooms), which are composed of large aggregates of hyphae. Higher fungi are grouped into two main Divisions, Ascomycota and Basidiomycota, according to the type of cell in which meiosis occurs, yielding meiospores (discussed below). Certain higher fungi apparently have evolutionarily lost their meiospore-producing or "perfect" stage and engage only in asexual reproduction. These fungi consequently cannot be classified in either of the natural Divisions; they are lumped together in an artificial category, Deuteromycetes, also called Fungi Imperfecti.

Since the number of fungal species probably exceeds 250,000 (roughly equivalent to the number of higher plant species), the variety of morphology, details of life histories, and other aspects of fungi are staggering. Generalizations in the following discussion of major groups often refer to characteristics of common species or of particular species studied in laboratory investigations. They may apply to many but usually not to all members of the group.

LOWER FUNGI

Because many members of the first two groups of lower fungi listed in Table 24-1 are aquatic, these fungi are often called water molds. They are the only fungi producing flagellated motile cells. The motile cells of the two groups differ consistently in number, kind, and position of flagella. Because of these and other important differences, the two groups of water molds seem unrelated.

Chytridiomycota

This group takes its name from that of the **chytrids** (pronounced "kit'-rids"), which are simple unicellular aquatic fungi (Figure 24-5A). The Chytridiomycota or "chytridiomycetes" are the

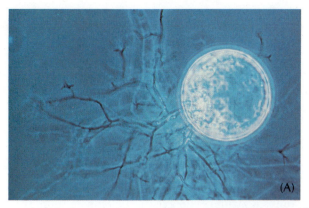 

Figure 24-5.   (A) A chytrid (*Rhizophydium*) grown in laboratory culture, showing the spherical vegetative cell with branching rhizoids which normally grow into a substratum. (B) Mycelium of *Allomyces*, an advanced member of the Chytridiomycota.

most primitive fungi, a group that is basically at the protistan level (Chapters 2, 21) and is probably ancestral to all other fungal groups except the Oomycota. In contrast to the Oomycota (see below), all these fungal groups have fundamentally a haploid vegetative body and chitin-containing cell walls.

The motile cells (zoospores and gametes) of all chytridiomycetes have a single, posteriorly attached flagellum (Figure 24-4). As a group the chytridiomycetes range from the simple, sac-like, unicellular but multinucleate chytrids (Figure 24-5A) to fungi that develop nonseptate, coenocytic hyphae forming a mycelium (Figure 24-5B). Both the sac-like cell and the mycelium

usually arise from a cluster of thin, branching hyphae or **rhizoids** penetrating the organism's substrate (food source) for nutrient absorption. Many chytrids are parasites that attack either algae, aquatic higher plants, or other aquatic fungi. A few parasitize the roots of land plants growing in moist soils. Many other chytrids and most of the mycelial chytridiomycetes are aquatic saprotrophs.

For asexual reproduction the sac-like chytrid cell becomes a sporangium, its multinucleate cytoplasm dividing into uninucleate zoospores that are released into a moist environment, swim to a new substrate, and germinate there, dispersing and propagating the species. Mycelial chytridiomycetes form sporangia at their hyphal tips (cf. Figure 24-6), producing similar zoospores.

Figure 24-6.   (A) Mycelium of an oomycete bearing many terminal sporangia (white tips). (B) A *Saprolegnia* sporangium at higher magnification; when the zoospores formed inside are ready for release, the knob at the tip will dissolve, leaving an exit pore.

 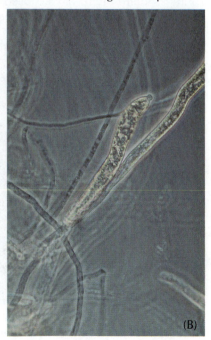

Sexual reproduction of chytridiomycetes usually occurs by fusion of two singly-flagellated motile gametes. Typically, the zygote formed by gametic fusion becomes a thick-walled survival spore. This is called a **meiosporangium** because meiosis occurs in it after a dormancy period, resulting in formation of motile haploid meiospores that upon release swim to a suitable substratum and germinate to produce a new haploid vegetative thallus.

Oomycota

As illustrated by the common water mold *Saprolegnia* (Figure 24-6), most Oomycota or "oomycetes," in contrast to the unicellular chytrids, develop an extensive nonseptate mycelium with few or no cross walls except where sporangia or sexual structures (gametangia) are cut off. Their cell walls contain cellulose, like most green plants, and usually not chitin, as found in other fungi. Within their sporangia oomycetes produce zoospores bearing two laterally attached flagella, in contrast to the singly flagellate motile cells of chytridiomycetes.

Saprolegnia and related saprotrophic oomycetes are easily obtained in water culture by submerging a dead fly or other insect in a bowl containing a bit of soil from a stream or other wet area and some fresh water from a stream or pond (not chlorinated tap water). Within a few days a halo of coarse mycelium surrounds the insect (Figure 24-7).

Some *Saprolegnia* species are important parasites of fish in streams, aquaria, and hatcheries. Other oomycetes, especially the downy mildews, are among the most destructive of all plant pathogens; certain of the resultant plant diseases, such as potato bight, have caused enormous economic losses, human misery, and social upheaval (Chapter 26). Other parasitic oomycetes are among the "damping-off" fungi in the soil, frequently attacking and killing newly germinated plant seedlings—the bane of nurserymen and home gardeners everywhere.

SEXUAL REPRODUCTION

Oomycetes reproduce sexually by nonmotile gametes. Usually several large eggs are produced inside a spherical **oogonium** (responsible for the name oomycetes, Figure 24-8). Separate hyphal branches function as male gametangia or **antheridia,** growing toward and into the oogonium and fusing with each egg, which thus becomes a zygote. The zygote develops a thick wall and becomes a survival spore, the **oospore.** This spore subsequently germinates into a new mycelium.

Recent evidence indicates that the vegetative mycelium contains diploid nuclei; these undergo meiosis within oogonia and antheridia, yielding haploid gamete nuclei prior to sexual fusion. This contrasts with other fungal groups, in which typically only haploid nuclei occur in the vegetative thallus.

The coenocytic hyphae of oomycetes, although not green, resemble some of the coenocytic algae. Oomycetes also resemble algae in possessing cellulose in their cell walls and in certain other biochemical features. These similarities, a diplontic life cycle, and basic differences in flagellation between oomycete and chytridiomycete motile cells, suggest that oomycetes comprise an evolutionary line separate from other fungal groups and possibly derived from certain algae by loss of photosynthetic capability.

Zygomycota

The Zygomycota or "zygomycetes" are basically terrestrial fungi with a nonseptate mycelium; they reproduce by nonmotile spores. Their sporangia (Figure 24-9A) are borne on aerial hyphae; the sporangial wall ruptures or disintegrates at maturity, releasing the spores for dispersal by air currents to fresh substrata.

Some zygomycetes are parasitic on insects, and several have been implicated as causal agents

Figure 24-7. Dead insects, such as this beetle, and larvae frequently serve as food for water molds.

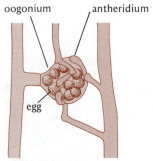

Figure 24-8. Sexual reproduction in *Saprolegnia* and related oomycetes involves the fusion of nuclei from antheridia with egg nuclei in an oogonium. Each resulting zygote becomes a heavy-walled survival spore.

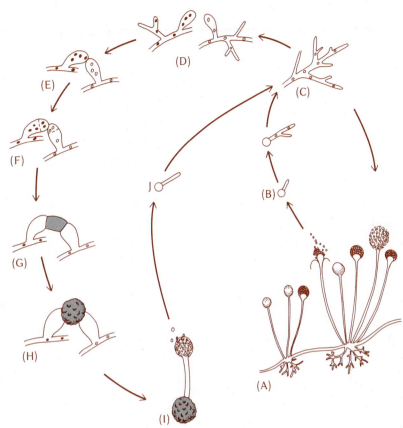

Figure 24-9. (A–C) Asexual cycle of the black bread mold *Rhizopus* with dispersal of sporangiospores by air currents; (C) vegetative nonseptate mycelium; (D) aerial branches of (+) and (−) mating types grow toward each other; (E–G) progametangia contact and fuse. Cross walls form (F), delimiting developing zygosporangium; (H–I) thick-walled zygosporangium which undergoes dormancy period; (J) following meiosis, germinating zygosporangium forms a germ sporangium that releases haploid spores.

of uncommon human respiratory diseases. Most, however, are saprotrophic, for example, the common black bread mold, *Rhizopus stolonifera* (Figure 24-9). One important group forms mycorrhizal associations with higher plant roots (Chapter 25).

Saprotrophic zygomycete molds occur in the succession of different fungi growing on decaying plant and animal matter in the environment. Fungal ecologists call these species sugar fungi because they rapidly utilize as carbon sources the free sugars and simpler carbohydrates in their substratum. Among the fungi appearing in succession on dung, for example, coarse cottony zygomycete mycelia with black sporangia typically appear early, utilizing readily available carbon sources. More resistant sources, such as cellulose, lignin, and other plant cell-wall components making up the bulk of plant residues in the dung are broken down more slowly by enzymes that come mainly from ascomycetes and basidiomycetes, the sporulating structures of which appear much later in the decay succession.

SEXUAL REPRODUCTION

In *Rhizopus* and similar saprotrophic zygomycetes, sexual reproduction resembles in several respects that of the green alga *Spirogyra* (Chapter 23). Two gametangia, formed on separate

mycelia at hyphal branch tips by ingrowth of cross walls (Figure 24-9D–F), fuse together (conjugate). The resulting fusion cell (technically not yet a zygote because its nuclei have not yet fused) differentiates into a thick-walled survival structure or **zygosporangium** (thus the name zygomycetes). After a protracted period of dormancy and eventual fusion of haploid nuclei from the two parents followed by meiosis, the zygosporangium germinates, producing a germ sporangium resembling the mold's asexual sporangia. The released haploid spores germinate into new mold mycelia (Figure 24-9I–J).

In many zygomycetes, sexual self-fertilization is genetically prevented by **mating-type** or self-incompatibility alleles, designated **plus** and **minus**. Any haploid vegetative mycelium has either the *plus* or the *minus* allele. Conjugation can occur only between *plus* and *minus* mycelia. This genetic arrangement is termed **heterothallic,** contrasting with **homothallic** (self-compatible). In the zone of contact between *plus* and *minus* mycelia, diffusible sex "hormones" produced by opposite mating types induce the formation of gametangia. The behavior of the nuclei in the zygosporangium before and after meiosis varies with the genus. In the simplest type only one haploid meiotic nucleus survives, then divides mitotically a number of times so that all spores from a germ sporangium are of the same genotype, either + or −.

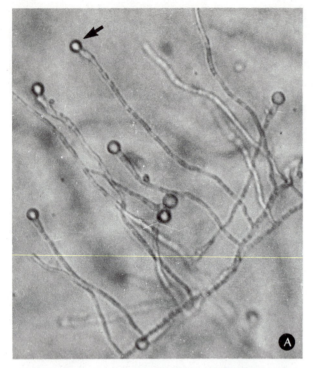

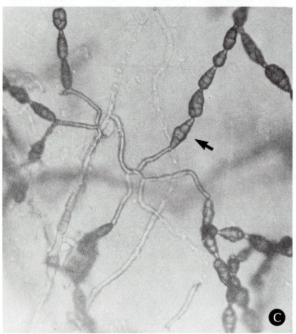

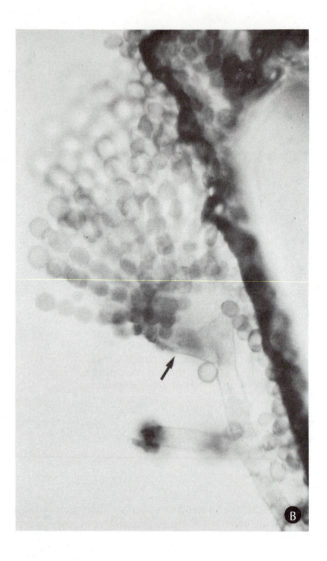

Figure 24-10. Conidia may arise in various ways from the spore-producing branches. Three types are illustrated here. (A) The initial conidium (arrow) of a chain has been produced; (B) many chains of conidia produced on an enlarged end of a conidiophore (arrow) are typical of the common genus *Aspergillus;* and (C) chains of dark-pigmented multicellular conidia are formed by the genus *Alternaria,* one of the mold genera responsible for fungal spore allergies.

ASCOMYCOTA

The Ascomycota ("ascomycetes," or sac fungi) are a very large, mainly terrestrial group that includes most of the molds common on spoiled fruits and vegetables. Some delicious edible fungi, the morels and truffles, are ascomycetes, as are many yeasts. Among the parasitic members of the group are the causal agents of Dutch elm disease, chestnut blight, peach leaf curl, apple scab, and corn blight. Most of the fungi associated with algae in lichens (Chapter 25) are ascomycetes.

As noted previously, the hyphae of ascomycetes are septate, in contrast to the mycelia of lower fungi. However, each "cell" usually contains several haploid nuclei. Electron microscopy shows that the cross walls separating the cells usually contain a small pore (or pores), making the cytoplasm continuous between cells; organelles can pass through the pores in some species.

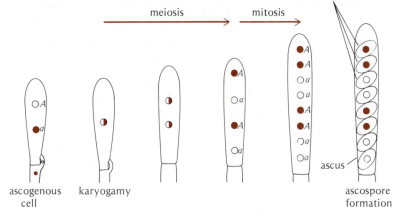

Figure 24-11. Sexual reproduction in ascomycetes results in the formation of ascospores. Karyogamy and meiosis occur in an ascogenous cell, formed following fusion of the two mating types. A mitotic division of the four resulting nuclei frequently occurs, resulting in an ascus containing eight haploid ascospores.

Ascomycete hyphae are generally narrower and more uniform in diameter than those of lower fungi (Figure 24-2). Also, fusion between adjacent vegetative hyphal branches occurs frequently in ascomycetes (and also in basidiomycetes) but rarely or never in lower fungi. These connections appear to facilitate movement of reserve materials or organelles through the mycelium.

Many ascomycetes reproduce asexually by a type of nonmotile spore called a **conidium** (plural, conidia), formed externally on a specialized aerial hypha (**conidiophore**) singly or in chains or clumps (Figures 24-A and 24-10). Most conidia are light and easily air-dispersed. They germinate by outgrowth of a hypha, which grows into a new mycelium (Figure 24-13A).

In all ascomycetes sexual reproduction leads to the occurrence of meiosis in a sac-like cell called an **ascus** (plural, asci; from Greek *askos*, bladder), from which the group takes its name. Within the ascus the resultant meiospores or **ascospores** differentiate (Figures 24-11, 24-13B). Except in yeasts, asci develop on or within a multicellular, often macroscopic sporocarp, of which there are several types (Figure 24-12) characteristic of different subgroups of ascomycetes. Features of the sporo-

carp, the asci, and ascospores are used in classifying and naming ascomycetous fungi.

The Ascomycete Life Cycle

Ascomycetes develop vegetatively as haploid organisms, with only a very transitory diploid phase during sexual reproduction (except in yeasts, which are considered later). The haploid mycelium, which is derived from germination of an ascospore, can grow and, if it produces conidia, reproduce itself asexually indefinitely as long as nutritional and environmental conditions remain favorable (Figure 24-13A). Sexual reproduction often occurs only when conditions start to deteriorate, for example, near the end of a host plant's growing season in the case of many plant pathogens such as powdery mildew. Fusion between sexual cells, the nature of which varies widely among different ascomycete groups, leads to development of a sporocarp containing asci; meiosis takes place in each ascogenous cell, yielding haploid ascospores that begin the cycle over again (Figure 24-13B).

Figure 24-12. Vertical sections of sporocarps, containing asci, in ascomycetes. (A) Apothecium, showing fertile layer composed of asci and sterile hyphae (B) in a palisade-like arrangement. (C) Perithecium of apple scab fungus (*Venturia inequalis*) buried in dead tissues of leaf. (D) Cleistothecium of a powdery mildew (*Microsphaera*).

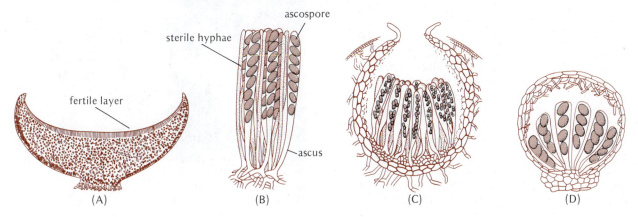

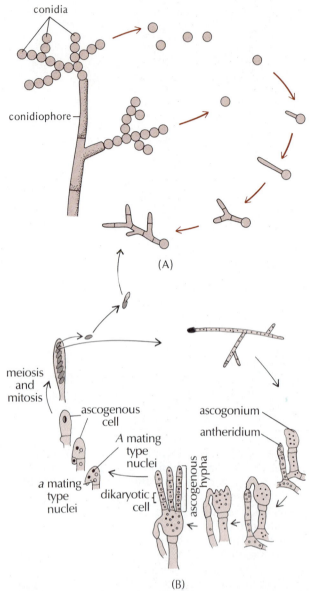

Figure 24-13. (A) Asexual reproduction in an ascomycete. Haploid mycelium (mating type *A* or *a*) forms conidia which disperse the fungus, producing new haploid colonies of identical genotype. (B) Sexual reproduction in an ascomycete.

Unlike most kinds of sexual reproduction, haploid nuclei of the fused ascomycete sexual cells normally do not unite immediately but remain distinct and multiply separately side by side during a limited amount of hyphal growth that leads to formation of asci within the developing sporocarp. Thus each ascus initially contains two haploid nuclei, one derived from each parent of the cross. During ascus development these two nuclei fuse, yielding a diploid nucleus (the only diploid nucleus in the entire life cycle), which immediately divides by meiosis into four haploid

nuclei (Figure 24-11). These usually divide once by mitosis, yielding a total of eight haploid nuclei within the ascus that become packaged into eight ascospores.

Many ascomycetes are hermaphroditic, each mycelium being able to produce both "male" and "female" cells. The word "female" refers here to a cell that *receives* one or more haploid nuclei after sexual fusion from another ("male") cell. To prevent self-fertilization many ascomycetes are heterothallic, possessing two mating-type alleles usually called *A* and *a;* sexual fusion is impossible between male and female cells of the same mating type (as would be produced by any given haploid mycelium). Some ascomycetes, however, are homothallic and can self-fertilize.

In certain ascomycetes sexual reproduction is initiated simply by fusion of vegetative hyphae of opposite mating types, with passage of nuclei from one into the other. In other ascomycetes each mycelium develops specialized male and female hyphae (gametangia), those of opposite mating type being able to fuse (Figure 24-13B). In a third type of sexual fusion, exemplified by *Neurospora*, conidia act as "male" cells, fusing upon contact with a "female" gametangium of opposite mating type (spermatization). During the development and fertilization of a female cell in any of these ways and its post-fertilization outgrowth to yield asci, adjacent haploid hyphae from the parental mycelium grow up and around the female structure, giving rise to the sporocarp housing the asci that result from this sexual union (see Figure 24-12).

CUP FUNGI AND MORELS

Among the ascomycetes cup fungi and morels produce the largest and most conspicuous sporocarps, called **apothecia** (singular, apothecium; Figure 24-12), some rivaling mushrooms in size and complexity; indeed, they are often regarded as mushrooms by nonbotanists. Cup fungi, as their name implies, typically produce cup-shaped apothecia (Figure 24-14). These are quite diverse in different species, ranging in size from a few

Figure 24-14. A cup fungus (*Aleuria*) which has the fertile asci-bearing layer on the exposed upper surface of the saucer-like apothecium.

(A)

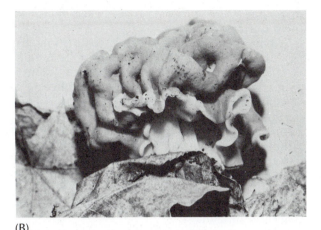

(B)

Figure 24-15. (A) A true morel (*Morchella esculenta*) has a pitted fertile layer bearing the asci. The stalked apothecia of toxic false morels, such as *Gyromitra* (B), have a wrinkled or convoluted fertile layer.

millimeters (less than 0.5 in) to over 12 cm (5 in) in diameter, in form from disc-shaped to vase-shaped, and in color from black or dull brown to bright yellow, red, orange, or purple. When mature, their asci develop a high turgor pressure, enabling them to discharge their ascospores forcibly. Any disturbance, such as a puff of wind, causes the tips of many asci to rupture simultaneously in miniature explosions that throw a visible cloud of ascospores into the air above the apothecium.

Most of the apothecial fungi are saprotrophs, growing on soil or dead wood. Unlike most other ascomycetes, they generally lack asexual reproduction (conidia), reproducing and dispersing themselves solely by means of their ascospores. Their sporocarps tend to appear abundantly in early spring, when conditions are most favorable for ascospore germination.

Morels and their relatives have stalked apothecia (Figure 24-15). "True morels," which are edible and highly regarded for their flavor, are among the most easily identified "mushrooms" because of the pitted or sponge-like structure of the spore-bearing surfaces (Figure 24-15A) that is different from that of all real mushrooms (basidiomycetes). Some related "false morels" are poisonous but can be easily distinguished because they have at most a folded spore-bearing surface (Figure 24-15B).

The similarly much-esteemed **truffles,** renowned in French cuisine, are related to the morels. These notoriously but understandably expensive, tuber-like sporocarps grow entirely underground from mycelia associated mycorrhizally with certain trees, particularly oaks. Rodents and insects burrowing underground eat truffles and disperse their spores, which pass through the digestive tract and are left behind in the animal's dung. In France truffles are found and harvested commercially with the aid of dogs or pigs specially trained to locate them by smell but not to eat them. The growth of truffles is encouraged in southern France by making plantations of the appropriate oak trees, inoculated as seedlings with strains of truffle-forming ascomycete mycelium. Thus far it has proved impossible to obtain truffles from truffle fungi in culture.

Perithecial and Cleistothecial Ascomycetes

These fungi, which include many of the common saprotrophic molds as well as serious parasites such as powdery mildews and the fungus causing Dutch elm disease (Chapter 26), produce their asci in minute globular sporocarps (perithecia and cleistothecia, Figure 24-12C, D) usually less than 1 mm in diameter. Most of these fungi vigorously reproduce asexually by conidia; their sexual, ascospore-producing stage is infrequently encountered.

Neurospora, or pink bread mold, forms **perithecia,** flask-shaped enclosures so constructed that when the asci inside discharge their ascospores, they are shot out of the perithecium into the environment through the neck-like opening at the top (Figure 24-12C). **Cleistothecia** (Figure 24-12D) are similar small sporocarps but have no external opening; they liberate their ascospores by breaking down the sporocarp wall. For example, cleistothecia of many plant parasites, formed on their host plant at the end of its growing season, break down during the winter.

Some species of the extremely common molds *Penicillium* and *Aspergillus* (Figures 24-A and 24-10B) are the conidial or "imperfect" stages (formerly classified as Deuteromycetes) of certain cleistothecial ascomycetes whose perfect stages are rarely seen. They produce conidia in prodigious numbers, giving characteristic colors to colonies of their different species. Because these spores are very readily distributed through the air, all foods are exposed to them, leading to everyday spoilage of bread, fruits and vegetables, and other foods. Certain *Aspergillus* species cause diseases in animals and humans.

Although mold infection is usually regarded as spoilage, certain *Penicillium* species produce much-desired flavors in cheeses such as blue cheese (blue because of *Penicillium* conidia), Roquefort, Camembert, Gorgonzola, and Stilton.

Because *Penicillium* and *Aspergillus* conidia are constantly being dispersed through the air, these molds frequently contaminate laboratory cultures of other microorganisms. In 1929, Alexander Fleming observed inhibition of bacterial growth near a colony of *Penicillium notatum* that was accidentally contaminating a culture of bacteria; this led him to discover penicillin. The usefulness of this first microbially produced antibiotic led to a search for others in the late 1940s, a search that blossomed as more and more medically useful microbial products were discovered. Production of antibiotics is not necessary for growth of the fungus, so these metabolites can be regarded as secondary products (cf. Chapter 5). However, production of biologically active secondary metabolites such as antibiotics may afford a competitive advantage to the producer fungus in the natural environment, an example of "chemical warfare" between organisms.

Certain *Aspergillus* species, utilizing sugar as a carbon source, form citric acid, which is used in soft drinks, and other organic acids used commercially. Industrial processes have been developed for producing microbial secondary products such as organic acids and antibiotics by large-scale cultivation of molds and other microorganisms in huge tanks of liquid media. Secondary products accumulate mainly after active growth of the fungus has slowed or ceased (Figure 24-16), sometimes in very large amounts secreted into the surrounding medium. Eventually production of the secondary metabolite slows and stops. To maximize the amounts obtained, commercial producers strive to delay this decline as long as possible, so much current research is directed toward mechanisms by which the fungal cells control the biosynthesis of secondary products.

Compounds called **mycotoxins** produced by some fungi are poisonous to humans or other animals, for example, the mycotoxins of poisonous mushrooms (see later discussion). **Aflatoxin,** produced by certain *Aspergillus* species, is one of the most potent poisons and

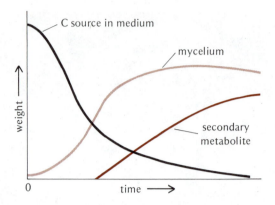

Figure 24-16. Production of secondary metabolites by fungi in culture. Time course comparison of growth (increase in weight) of fungus and the accumulation of products such as antibiotics in the culture medium or mycelium. The carbon source declines as it is used for growth, maintenance, and energy by the fungus.

carcinogens (cancer-causing compounds) known. When grain with a high moisture content is stored, aflatoxin contamination may occur due to growth of *Aspergillus flavus*. Such an incident in 1960 wiped out some 100,000 turkeys and other domestic fowl in England, due to feed containing contaminated peanut meal. Because of its extreme toxicity and carcinogenicity, no detectable amount of aflatoxin is allowed in food sold for human consumption in the United States. The aflatoxin hazard is increased by the ability of certain *Aspergillus* species to infect grain and produce toxin in the field, particularly after insect damage.

YEASTS

Fungi that grow as single rounded cells (Figure 24-17) rather than as filamentous mycelia are called yeasts.* When growing on solid substrates, yeast cells form compact colonies like those of bacteria, in contrast to the fluffy or cottony, rapidly spreading mycelial colonies formed by molds. Certain yeasts multiply like bacteria by fission (dividing in two), but most yeasts form new cells by a process of **budding**—following cell growth and mitosis, a bud forms as a local protrusion of the cell wall (Figure 24-17). As the bud increases in size, one of the daughter nuclei from mitosis migrates into the cytoplasm of the bud. When the bud has grown to about the size of the parent cell, it can detach and proliferate independently.

Although all three major groups of higher fungi contain species that grow as yeast forms, most yeasts, like common baker's and brewer's yeast,

---

*Single-celled chytrids are not called yeasts because their cells are usually anchored by rhizoids, and proliferate by producing and releasing zoospores rather than by budding or fission.

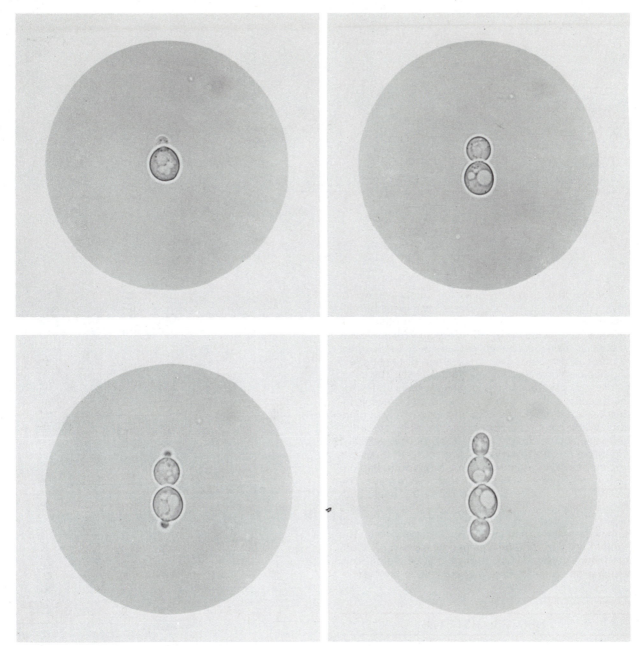

Figure 24-17. Yeasts are unicellular forms of fungi. Most reproduce asexually by budding, as shown here.

Saccharomyces cerevisiae, are ascomycetes. Some filamentous fungi can be induced to grow in a yeast-like form under certain conditions, such as decreased oxygen concentration. Several human pathogenic fungi grow in yeast form in the body at 37°C but change to a filamentous mode of growth when cultured on laboratory media at lower temperatures.

Because of their high vitamin B and protein content, certain yeasts such as "torula" (*Candida utilis*) have been used as food supplements for many years. Yeast species that can utilize certain industrial wastes as nutrients (e.g., citrus peels or the sulfite liquors from paper manufacturing) are being studied as potential low-cost protein sources for livestock feed or human food.

In contrast to most other fungi, many yeasts can ferment sugars to alcohol and carbon dioxide (see Chapter 6), yielding energy for growth. This property is exploited in brewing beer and making wine. In bread making, the carbon dioxide formed by yeast fermentation of sugar in the dough causes it to rise, giving the porous texture of the baked loaf.

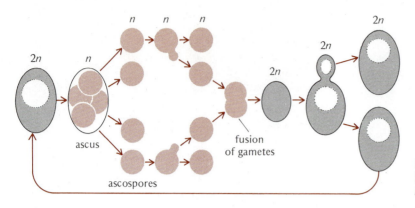

Figure 24-18. Sexual reproduction in baker's and brewer's yeast, *Saccharomyces cerevisiae.*

Sexual reproduction in ascomycetous yeasts invovles the formation of ascospores in an ascus but differs in several ways from reproduction in other ascomycetes. Several life-cycle patterns occur in these yeasts. In baker's yeast (*Saccharomyces cerevisiae*), in contrast to other ascomycetes described earlier, distinct diploid and haploid generations occur (Figure 24-18). The large, ellipsoidal, vegetative cells ordinarily seen under the microscope and used in the baking and brewing industries are diploid. They multiply extensively by budding. When conditions become unfavorable for continued vegetative activity, the yeast reproduces sexually. Any diploid cell can function directly as an ascus: its nucleus undergoes meiosis and forms four haploid ascospores (Figure 24-18), two of each mating type. Following release from the ascus these spores begin budding and reproduce vegetatively for a time, yielding a haploid generation of cells, which are spherical and somewhat smaller than the ellipsoidal diploid cells. The haploid cells are potential gametes; any two of opposite mating type can fuse to form a large diploid cell, which proliferates by budding, continuing the cycle.

DEUTEROMYCETES

Deuteromycetes, also known as Fungi Imperfecti, comprise over 10,000 species for which a "perfect" or sexual stage is not known, as noted previously. Many deuteromycetes cause plant diseases, and many others are saprotrophic. Most of the skin- or hair-inhabiting human "dermatophytes" (Figure 24-19), such as those causing ringworm and athlete's foot, are deuteromycetes.

Genera of deuteromycetes are based largely on the morphology (form) of their conidiophores and conidia, that is, the asexual or **anamorph** form. *Penicillium, Aspergillus,* and *Alternaria* are common anamorph genera. When a sexual stage, the **teleomorph,** is discovered, the former deuteromycete species is renamed and transferred to the fungal Division (Ascomycota or Basidiomycota) to which the perfect state of the fungus belongs.

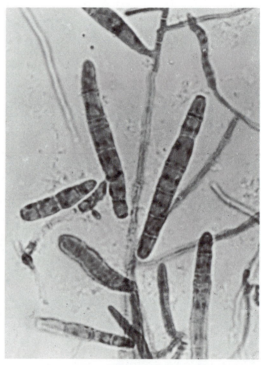

Figure 24-19. Multicellular conidia of the dermatophyte *Trichophyton mentagrophytes,* a common cause of athlete's foot.

BASIDIOMYCOTA

The Basidiomycota, or "basidiomycetes," include the largest and most complex fungi, among them the familiar field and woodland mushrooms. These structures are sporocarps that develop from a vegetative mycelium composed of septate hyphae growing in the ground or other substratum. Basidiomycetes are so named because they produce meiospores of a special type called **basi-**

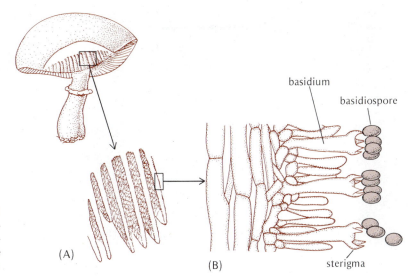

Figure 24-20. The basidiospores in gilled mushrooms are formed by terminal hyphal cells (basidia) that occur on the surfaces of the gills (A). Mature spores are forcibly ejected from the sterigmata (B).

diospores, which are borne externally on a club-shaped hyphal tip or **basidium** (Figure 24-20) and are usually developed (in large numbers) in or on a sporocarp. Among the basidiomycetes are rusts and smuts, which are responsible for economically important plant diseases. These fungi do not produce conspicuous sporocarps like the mushrooms but nevertheless form basidiospores. Some of the rusts have the most complex life cycles of any fungi (see Chapter 26).

This chapter focuses only on basidiomycetes that have macroscopic sporocarps. Unlike most other fungi, these basidiomycetes usually do not reproduce by asexual spores but multiply and disperse themselves by means of their meiotically produced basidiospores, which are formed in enormous numbers by each sporocarp.

The Gilled Fungi

The gilled fungi, which include the majority of mushrooms, are mostly saprotrophic, but many species associate with the roots of angiosperms or conifers as mycorrhizae (Chapter 25). The commercial mushroom sold in food markets is the sporocarp of *Agaricus bisporus*, a saprotrophic species that is commercially cultivated (Figure 24-21) in specially designed mushroom houses. Features of the sporocarp of the closely related meadow and field mushroom, *A. campestris*, are shown in Figure 24-22: a central stalk (**stipe**) bears an umbrella-like **cap.** On the underside of this cap, sheet-like **gills** radiate from the stipe and produce the basidiospores (Figure 24-20).

The *Agaricus* sporocarp begins underground as a local proliferation of hyphae somewhere on the fungus' vegetative mycelium. The swelling in-

creases in size and breaks through the soil surface as a small ball, the **button,** consisting of the unexpanded stipe, cap, and gills. The stipe elongates and the cap expands, breaking the thin **veil** that enclosed and protected the delicate gills in the preceding button stage (Figure 24-22). Remnants of the veil usually persist on the upper part of the stipe as a ring, the **annulus** (Figure 24-22). Some species do not have a veil, so the presence or absence of an annulus on the stipe of the expanded mushroom is one of the important features used to identify different species. Another identifying

Figure 24-21. A commercial crop of *Agaricus* sporocarps growing on bins of compost in darkness and under controlled temperature, humidity, and aeration.

Figure 24-22.   A young (left) and mature (right) sporocarp of the field mushroom (*Agaricus campestris*).

feature is the color of the basidiospores (Figure 24-23).

The Mushroom Life Cycle

Each basidium normally produces four haploid basidiospores (Figure 24-24) that result from meiosis in the basidium. (In *A. bisporus* each basidium forms only two basidiospores, hence its specific name.) Upon germination each basidio-

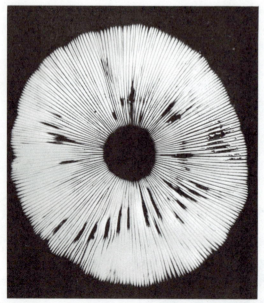

Figure 24-23.   Masses of microscopic basidiospores discharged by a mature mushroom produce a spore print visible to the unaided eye (above, a white spore print of *Amanita verna*). Determination of spore color, usually required for identification of unknown mushroom species, can be made by removing the stipe and placing the cap, gills down, on paper for several hours.

spore grows into a mycelium containing haploid nuclei. Most mushrooms possess self-incompatibility or mating-type genes that act to ensure cross-fertilization in sexual reproduction. For sporocarps to be formed, two haploid mycelia carrying different mating-type alleles (i.e., derived from different and compatible basidiospores) must meet and fuse.

As in ascomycete sexual reproduction, upon contact and fusion of compatible mushroom hyphae their respective haploid nuclei do not fuse but multiply side by side during further hyphal growth, yielding **dikaryotic** (two-nucleate) cells containing one haploid nucleus derived from each parent. The dikaryotic mushroom hyphae grow vegetatively into a new mycelium called a **dikaryon,** which initiates sporocarps as described above. Thus the mushroom sporocarp consists of dikaryotic tissue (Figure 24-24), and each basidium initially contains two haploid nuclei, one derived from each of the haploid mycelia that originally fused to initiate the dikaryon. These two nuclei fuse within the basidium, and the resulting diploid nucleus immediately divides by meiosis, yielding four haploid nuclei that become the nuclei of the four basidiospores. Germination of these yields haploid mycelia, beginning the cycle anew (Figure 24-24).

Establishment of a mushroom's dikaryotic phase does not immediately lead to nuclear fusion and meiosis as it does in ascomycetes. Instead, the dikaryon can continue to grow and spread through the substratum, becoming the principal vegetative mycelium and producing numerous sporocarps during favorable conditions. In soil and leaf litter, these frequently appear in the form of a rough ring representing the current zone of outwardly advancing mycelium (Figure 24-25). This contrasts with the reproductive cycle of ascomycetes, in

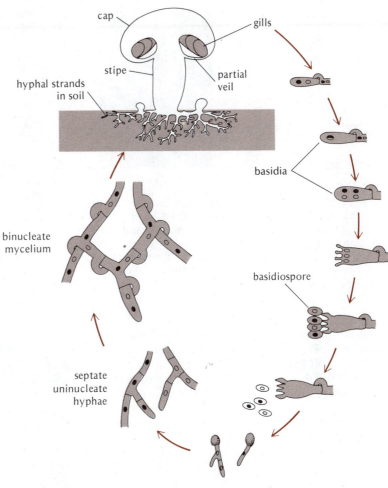

Figure 24-24. Sexual reproduction in a gilled mushroom.

Figure 24-25. A portion of a large meadow fairy ring. All of the sporocarps are formed by a mycelium that began growth at the approximate center of the ring and grew radially outward.

which each sporocarp arises from a separate sexual event.

Although the two haploid nuclei in a dikaryotic cell remain separate except in the basidium, genes in each of these nuclei can express themselves in the cell, so

a dikaryon is genetically equivalent to a diploid. Mushrooms have therefore become functionally diploid organisms, even though they lack diploid nuclei except just before meiosis. With continued growth over many years the dikaryon can become quite extensive, tapping a large area of nutrients. For example, an *Agaricus* ring 60 m (180 ft) in diameter and approximately 250 years old was reported in a Colorado grassland. Estimating conservatively that this dikaryon may have produced over its lifetime an average of at least 90 sporocarps per year, each of which releases some 10^9 spores, this one individual would have given rise to some 2×10^{13} potential offspring by the time its measurements were recorded.

At the growing hyphal tips of a dikaryon the two complementary haploid nuclei divide synchronously, furnishing each new cell with a nucleus of each genetic type. In many dikaryotic basidiomycete hyphae, mitosis and distribution of daughter nuclei into daughter cells involves formation of a lateral hyphal loop termed a **clamp connection** (Figure 24-26).

Edible vs. Poisonous Mushrooms

Many people in many countries collect wild mushrooms to eat. Edible and poisonous species can be very similar and frequently occur in the

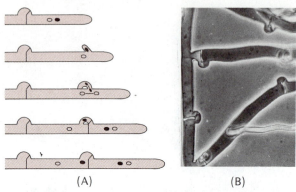

(A)   (B)

Figure 24-26. Maintenance of the dikaryotic condition in many basidiomycetes is accomplished by the formation of clamp connections during mitosis and partitioning of a new cell at a hyphal tip. One daughter nucleus migrates through the loop as shown (A). Occurrence of clamp connections identifies a mycelium as belonging to a basidiomycete fungus (B).

same genus. Because of distinctive characteristics, a few edible species can be identified by casual inspection, but this is not true for most mushrooms. Correct identification usually requires careful study of the mushroom with attention to detail. Unfortunately, contrary to some popular myths, *there are no simple rules of thumb for distinguishing harmless from poisonous species. If the mushroom cannot be positively identified, it should not be eaten.* Several excellent guidebooks on mushrooms are available to help in identifying wild mushrooms, indicating which species are edible and which are known to be poisonous.

More than 70 species of gilled fungi cause poisoning in humans. Some of the most poisonous are members of the genus *Amanita*—several relatively common amanitas are fatally poisonous. Amanita sporocarps (Figure 24-27A) have at the base of the stipe a cup of tissue (**volva**) that remains from a "universal" veil completely enclosing the developing sporocarp in the "egg" stage (Figure 24-27B). This volva is a key feature in identifying members of the genus. Amanitas contain toxic peptides called amatoxins and phallotoxins. One of the amatoxins, α-amanitin, powerfully inhibits messenger RNA synthesis in animal cells, a likely component of the mushroom's extreme toxicity.

Probably the best-known amanita is the worldwide *A. muscaria,* or fly amanita (Figure 24-27C), traditional in fairy tale illustrations with its colorful red cap covered by prominent white scales, remnants of its universal veil. Although poisonous, this species is not as extremely toxic as others in the genus and was reportedly used in some primitive cultures as a hallucinogen.

Pore Fungi

Other common basidiomycetes with the same type of life cycle described for gill fungi include those producing sporocarps with tubes or pores rather than gills as their basidia-bearing surfaces

Figure 24-27. Various growth stages of *Amanita* (A), showing the emergence from the enclosing universal veil of the "egg" stage (B), which remains as the volva at the base of the stipe. Remnants of this veil leave characteristic white patches on the cap (C) of some species such as *A. muscaria.* A partial veil (cf. Figure 24-22) is also present and leaves an annulus around the stipe.

(Figure 24-28). Two groups of basidiomycetes, not closely related, have sporocarps with this structure—the **boletes,** which are fleshy like the gilled fungi, and the woody "bracket fungi."

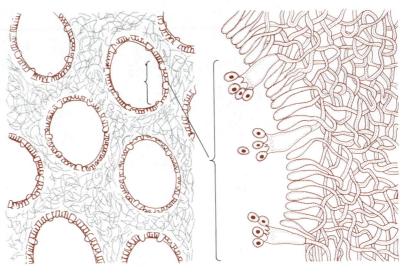

Figure 24-28. Reproduction in a pore fungus. (Left) Cross section, small portion of sporocarp; the openings are cavities of the tubes. (Right) Portion of one tube, greatly enlarged, showing the basidia.

Boletes look superficially like gilled mushrooms, but the undersurface of the cap has round or irregularly shaped pores rather than gills (Figure 24-29). Most boletes associate with forest trees as mycorrhizae. Many, such as *Boletus edulis*, are sought-after edible species.

The bracket fungi, or "polypores," are economically important because many are serious pathogens that attack and kill forest trees (Chapter 26), and others cause dry rot and decay of wood in homes, boats, and other structures, resulting in serious economic losses. Bracket or shelf fungi are so-called because the sporocarps of many species grow in the form of a shelf or bracket (or "horse's hoof") that emerges laterally from the attacked tree, log, or piece of timber (Figure 24-30).

Wood-rotters attack and penetrate wood by secreting cellulase, which breaks down cellulose, yielding sugar for the nutrition of the fungus and allowing progressive invasion of the entire timber or log by the advancing hyphae (Figure 24-3). Some wood-rotters also break down and utilize the lignin of wood they invade. Degradation of cellulose and lignin weakens the wood and will eventually cause a wooden structure to disintegrate or collapse.

Other Fleshy Fungi

Among a number of other basidiomycetes with different types of sporocarps the following relatively often

Figure 24-29. Boletes have pores rather than gills on the underside of the caps.

Figure 24-30. A bracket fungus, a saprotrophic wood rotter.

Figure 24-31. (A) Puffballs; (B) earthstars; (C) a coral fungus.

encountered groups will serve as examples. Puffballs and earthstars (Figure 24-31A, B) have a sheath of one or several mycelial layers enclosing the spore-bearing tissue. As the basidiospores mature, much of the inner tissue breaks down, leaving a mass of powdery yellow or brown spores intermingled with branched hyphal threads. In some species a pore forms in the sheath for spore escape, while in others the sheath breaks away, allowing the spores to escape. Coral fungi or clavarias (Figure 24-31C) somewhat resemble branching marine coral colonies. They usually grow on decaying stumps and logs or in rich soil in woods, varying in color from white, yellow, or pink to all shades of brown and gray. The basidia are borne in a continuous layer over the surfaces of the erect branches.

SLIME MOLDS (MYXONTA)

The slime molds, as noted in Chapter 2, are organisms variously regarded as either simple animals or of distant fungal affinity. In their vegetative or food-gathering stage they lack cell walls, are motile, and feed phagotrophically upon bacteria or solid organic matter, as protozoa do. However, in reproduction they resemble fungi, forming walled spores in sporangia that are often stalked (Figure 24-32) like many fungal sporocarps. Slime mold sporangia are often brightly colored and, though small, can be readily found in moist habitats.

Slime molds, although among the simplest of eukaryotic organisms, undergo elementary differentiation during formation of their sporangia. Certain slime molds can be grown and manipulated easily in laboratory cultures and are widely used as simple "model" systems for experimental study of aspects of cell and developmental biology (Chapter 14). There are two major types of slime molds, the *Myxomycetes* (from Greek *myxos*, slime, and *mykos*, fungus), or plasmodial slime molds, and the *Acrasiales*, or cellular slime molds.

Myxomycetes

These slime molds in their vegetative stage consist of a multinucleate protoplasmic mass or

plasmodium (Figure 24-33), which moves slowly through soil, litter, or over wood in damp environments, engulfing food particles such as bacterial cells. The plasmodium usually has a core of branching or net-like strands of protoplasm, streaming at rates as high as 1 mm/sec and fanning out to form the broad moving front of the plasmodium where food is being picked up. The flow can be seen with the microscope to stop and reverse direction about every minute, bringing ingested food particles back to the central protoplasmic strands and resulting in a net forward

(A)

(B)

Figure 24-32. Stalked slime mold sporangia: *Stemonitis* (A) is about 15 mm in height; (B) *Diachea* is only a few mm.

Figure 24-33. Several yellow plasmodia of *Physarum* after extended feeding on bacteria present in leaf litter and soil.

locomotion that is much slower than 1 mm/sec, more on the order of millimeters per hour.

Plasmodial streaming appears to be due to a filamentous contractile protein, actin, which is similar to the contractile protein of animal muscles (see Chapter 4). Unfavorable environmental conditions cause locomotion to cease, terminating the plasmodial stage. The plasmodium may convert itself into a hard-surfaced resting structure or **sclerotium,** composed of separate compartments formed by cleavage of the formerly continuous cytoplasm about its many nuclei and deposition of walls around these "cells." The dormant sclerotium can survive adverse conditions for a protracted period and will regenerate a vegetative plasmodium when favorable weather returns.

Starvation may induce a plasmodium to develop into sporangia, within which numerous spores are formed from the nuclei and portions of the plasmodial cytoplasm. Meiosis apparently occurs within the developing spores, three of the resulting four haploid nuclei disintegrating, yielding uninucleate haploid spores (Figure 24-34). The spore germinates into a haploid "ameba" (a single cell moving by means of protoplasmic streaming), which may in some cases become flagellated. The amebae in many species occur as different mating types. Compatible amebae can fuse to yield a diploid zygote. Its nucleus divides by mitosis without cytokinesis, giving rise to the diploid plasmodium, completing the life cycle.

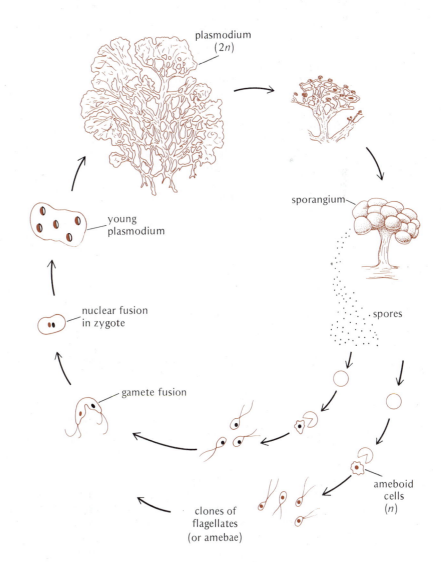

Figure 24-34. *Physarum* life cycle.

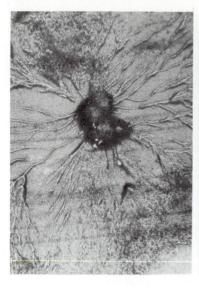 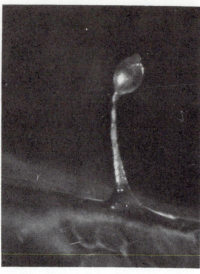

Figure 24-35. Stages in *Dictyostelium* asexual spore formation. (A) Amebae stream into a central mound, which will form a sorocarp (B). The aggregation stage (A) can be seen with a dissecting microscope, whereas the sorocarp is just visible with the unaided eye.

Acrasiales

The cellular slime molds are soil organisms that are not normally seen in the field because of their nearly microscopic size, even when aggregated into a sporangium composed of many thousands of cells. They occur widely, however, and can be isolated relatively easily on laboratory culture media containing either bacteria or appropriate organic nutrients as food. These organisms multiply vegetatively as populations of mitotically dividing, individual ameboid cells that feed phagotrophically. For reproduction these cells aggregate (Figure 24-35A) to form a slowly motile mass of cells resembling a minute slug, which develops into a sporangium or **sorocarp,** as illustrated in Figure 24-35B. This process, in the cellular slime mold *Dictyostelium discoideum*, has been analyzed by extensive biochemical and molecular biological studies, as explained briefly in Chapter 14.

The ameboid vegetative cells of *Dictyostelium* grow and multiply as long as they have sufficient food. Starvation or removal from their food supply induces the precise sequence of developmental changes leading to

a sorocarp (Figure 14-20). Cell aggregation into groups of about 10,000 cells is governed by a hormonal signal (cell attractant), cyclic AMP, which is produced from ATP by a local group of cells that excrete it into the surrounding medium, creating a chemical focal point toward which amebae migrate from the surrounding area (Figure 24-35A). (Certain other species of cellular slime mold utilize a different attractant.) None of the cells fuse during this aggregation process.

Aggregation leads to an elongated slug-like mass of individual cells, or **pseudoplasmodium** (Figure 14-20), which migrates toward light. When it reaches a location providing overhead light, it stops migrating and forms a sorocarp (Figure 24-35B), a process called culmination. Grafting experiments reveal that cells in the front third of the pseudoplasmodium form the stalk of the sorocarp while the remaining cells move up and differentiate into the spores. During culmination the cells, unlike most fungal cells, synthesize cellulose-containing cell walls, both around the spores and for structural support of the stalk. The cells remain haploid throughout the entire process.

Recently, the mycologist Kenneth B. Raper (who originally isolated *D. discoideum*) and his associates have obtained evidence that these organisms go through a sexual cycle that is entirely different from their well-known asexual reproduction by sorocarp formation.

SUMMARY

(1) Fungi are heterotrophic organisms, usually with a filamentous body (mycelium) having chitinous cell walls. Some fungi, especially the yeasts, are unicellular. Fungi reproduce and disperse by either asexual spores or spores produced as part of a sexual cycle or both. Saprotrophic fungi are decomposers, recycling important nutrients. Parasitic fungi, some of which are obligate parasites, cause the majority of economically important plant diseases. Many fungi feed by symbiotic association with a green plant (my-

corrhizae, lichens). Characteristics of major fungal groups are summarized in Table 24-2.

(2) Lower fungi include diverse groups that produce asexual spores within sporangia and have unicellular thalli or nonseptate, coenocytic mycelia. Water molds, which are aquatic and soil-inhabiting organisms, include (a) chytridiomycetes, which usually have a simple unicellular thallus and produce zoospores and gametes with a single posterior flagellum, and (b) oomycetes,

TABLE 24–2 Divisions of the Myconta

Division	Common Types	Thallus Structure	Predominant Life Cycle Phase	Asexual Spore Type	Sexual Reproduction by	Location of Meiosis
Chytridiomycota	Water molds	Many uni-cellular, some mycelial	Haploid	Uniflagellate zoospores, in sporangium	Fusion of motile gametes in most	Zygote (meiospo-rangium)
Oomycota	Water molds, downy mildews	Mainly mycelial	Diploid	Biflagellate zoospores in sporangium	Fusion of gametangia	Gametangia
Zygomycota	Pin molds	Mycelial	Haploid	Nonmotile, in sporangium	Fusion of gametangia	Zygospore
Ascomycota	Yeast, molds, powdery mildews, cup fungi	Mycelial, or yeasts	Haploid	Conidia	Fusion of gametangia, hyphae, or hypha and conidium	Ascus
Basidiomycota	Rusts, smuts, mushrooms, puffballs	Mainly mycelial	Dikaryotic	Uncommon in mushrooms	Hyphal fusion	Basidium
Deuteromycetes[a]	Molds	Mycelial or yeasts	Haploid	Conidia	Absent	Absent

[a]Artificial category (includes "imperfect" species derived from either ascomycetes or basidiomycetes).

which are usually filamentous, produce biflagellate zoospores, and have cellulosic cell walls (in contrast to the chitin-containing walls of other fungi). Certain oomycetes cause serious higher plant diseases (e.g., potato blight); many chytridiomycetes parasitize algae and other aquatic plants.

(3) Zygomycetes such as black bread mold (*Rhizopus*) are terrestrial, usually filamentous lower fungi producing nonmotile, air-dispersed asexual spores. Many are fast-growing, invasive saprophytes that rapidly utilize the simple carbohydrates in plant and animal remains. Their sexual reproduction occurs by fusion of gametangia formed on different mycelia, resulting in a thick-walled survival structure, the zygosporangium.

(4) Ascomycetes, the largest group of fungi, are mainly filamentous with a septate haploid mycelium; they reproduce (a) asexually by conidia and (b) sexually by fusion between cells of different haploid mycelia, usually leading to a sporocarp containing asci in which meiosis occurs, yielding haploid ascospores. Different ascomycete groups develop different types of sporocarps, namely apothecia (cup fungi, morels), perithecia (e.g., *Neurospora* or red bread mold), or cleistothecia (common molds including *Penicillium* and *Aspergillus*, which are spoilage organisms and also industrial sources of valuable products such as penicillin). Many perithecial and cleistothecial ascomycetes are plant pathogens. Yeasts are a small but economically important unicellular group, commonly multiplying asexually by budding.

(5) Deuteromycetes have no known sexual stages. Most reproduce asexually by conidia. They include frequently encountered molds, serious plant pathogens, and causal organisms of a few human and animal diseases.

(6) Basidiomycetes include many saprophytic fungi that produce large fleshy sporocarps (mushrooms, puffballs, coral fungi, and so on) as well as some important plant pathogens (rusts, smuts, some bracket fungi). Basidiomycetes reproduce by haploid basidiospores formed externally on basidia, which are modified hyphal tips within which meiosis occurs; they are located in sporocarps of fleshy fungi. Sporocarps develop from a dikaryotic septate mycelium that is developed following sexual fusion between different haploid mycelia derived from basidiospores.

(7) There are no reliable simple rules for distinguishing edible from poisonous mushrooms. Identification of the particular species with the aid of a competently prepared mushroom handbook containing reliable information on edible and poisonous species is the only way to ascertain independently whether a given wild mushroom can be eaten without danger.

(8) Slime molds, which represent a Subkingdom separate from fungi, possess a motile, ameboid vegetative stage lacking cell walls and developing, in the case of plasmodial slime molds, into an extensive, streaming coenocytic mass of protoplasm (plasmodium). In cellular slime molds the ameboid cells remain distinct even after aggregating into a motile "slug" (pseudoplasmodium) prior to reproduction, which in both types of slime molds occurs by walled, nonmotile spores formed in a sporangium.

Chapter 25

SYMBIOTIC ASSOCIATIONS

In most ecosystems organisms interact with one another in a variety of ways, ranging from merely modifying one another's environment (Chapter 19) to much more specific interdependencies reflecting coevolution (Chapters 16, 17, 20). The ultimate in coevolutionary specialization is the condition in which two or more organisms develop physically interlocked in an intimate nutritional relationship, a state of **symbiosis** (from Greek *syn-*, together; *bios*, life). In the case of lichens, for which the term symbiosis was first proposed, coevolution has produced what are essentially compound organisms (Figure 25-A) consisting of an alga and a particular fungus, for which this is a functionally obligatory combination. The fungi and in some cases the algae are never encountered independently of the other partner. Other specific and obligatory symbioses are now known, many of which involve either an alga or a fungus, as well as less obligatory but nonetheless striking and important symbioses such as the nitrogen-fixing association between plants and certain soil bacteria (Chapter 22), organisms that can and commonly do grow independently.

Symbiotic relationships range from **parasitism,** in which one partner benefits nutritionally from the other, to **mutualism,** in which each partner benefits from the other in some way, with all possible gradations in between. In everyday usage, symbiosis often means mutualistic symbiosis. This chapter deals with important examples of symbiosis involving microorganisms (algae, fungi, and bacteria), both with one another and with plants or animals. Chapter 26 will consider parasitism of plants by their major disease-causing organisms, a similarly important, if more sinister, kind of symbiosis.

Figure 25-A. Certain fungi and algae associate and grow together to form a morphologically distinct lichen.

NATURE OF SYMBIOTIC INTERACTION

In symbiosis at least one of the partners typically gains some form of nourishment from the other. In associations involving a green plant as one of the partners, the plant almost invariably supplies photosynthesized organic carbon to the nongreen, heterotrophic partner. This can be shown by experiments in which radioactive $^{14}CO_2$ is supplied and becomes fixed photosynthetically by the green cells: radioactive carbon subsequently appears in organic compounds of the heterotrophic partner (Figure 25-1). Thus nutritionally, the heterotroph is parasitizing the autotroph. Similarly, in symbioses between different heterotrophs, one partner normally provides the other, the fed partner, with its basic energy and carbon source. We regard the relationship as mutualistic if the part-

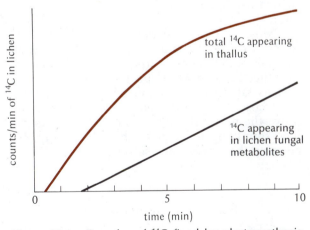

Figure 25-1. Transfer of ^{14}C fixed by photosynthesis (upper curve) by the alga from $^{14}CO_2$ supplied to the lichen can be followed by the appearance of radioactivity in products formed only by the fungal symbiont, such as mannitol or arabitol (lower curve). (After D. C. Smith, *The Lichen Symbiosis*, Oxford Biology Reader #42.)

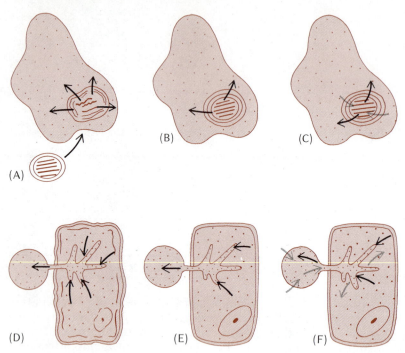

Figure 25-2. Possible stages in evolution of mutualistic symbiosis between (A–C) a predator and its prey, and (D–F) a parasite and its host. (A) Predator eats (engulfs) prey and kills it, taking up its food content. (B) Predator delays or does not kill engulfed prey; takes up food materials that former prey (now symbiont) produces and releases. (C) Former predator (now a mutualistic symbiont) furnishes nutrients that benefit growth and/or productivity of former prey; receives food materials from this symbiont in exchange. (D) Parasite invades host, absorbs nutrients vigorously, kills host, then must move on to new host. (E) Parasite reduces its uptake of nutrients from host, allowing parasitized host to survive and support parasite indefinitely. (F) Parasite absorbs and supplies nutrients from the environment to the host, improving the latter's growth and productivity, thus better supporting the former parasite (now a mutualistic symbiont).

ner supplying food receives some kind of nutritional benefit of its own from the fed partner (such as a vitamin or mineral nutrient supply), or some other advantage. For example, the fed partner may simply protect the food-supplying partner, helping it exist in habitats it could not occupy on its own. The distinction between food-supplying and fed partner becomes convoluted in some cases, as in the well-known symbiosis between ruminant herbivorous animals (e.g., cows) and gastrointestinal bacteria, in which the animal, by eating pasturage or hay, supplies the bacteria with cellulose; the bacteria digest and ferment this raw material, releasing products useful as food to the animal (which cannot by itself digest and utilize cellulose).

Most mutualistic symbioses probably evolved from an initially predatory or parasitic relationship between the prospective partners. Mutations in either the exploited species or the consumer species favoring survival of the prey (or parasitized host), despite its being exploited, may improve the multiplication of both the fed and the food-supplying member of the relationship, increasing their respective selective fitness (Figure 25-2).

Specificity and Recognition

Specificity of interaction between symbiotic partners indicates the existence of cell **recognition** mechanisms by which different kinds of cells can identify one another. Recognition is usually ascribed to interaction between species-specific molecules located on the respective cell surfaces,

either the cell wall or the plasma membrane. Interactions useful for recognition should have an appropriate degree of molecular specificity, exemplified by the combination of an antibody (from the immune system of higher animals) with a specific protein or carbohydrate (antigen) on the surface of a foreign cell. Recognition in plant symbiosis may involve the interaction of specific complementary surface molecules somewhat analogous to the antigen-antibody interaction (see below.

Recognition leads to responses by the prospective symbiotic partners. In a parasitic symbiosis there may be defensive responses by the host and responses by the parasite that help it overcome host defenses and invade host tissues. In a mutualistic symbiosis, recognition elicits expression by both partners of genetic programs for development of the special structural and physiological features of the symbiosis. Symbiosis thus can involve some degree of genetic or at least developmental interaction between partners, sometimes leading to phenotypic expression of genetic information by one or both partners that is not expressed in the absence of the other partner, as examples given below will illustrate.

LICHENS

Lichens are associations of algae and fungi, usually a single species of each, forming thalli that occur worldwide on soil, rock surfaces, tree bark and leaves, dead wood, and manmade structures such as stone buildings, walls, fences, and wood

Figure 25-3. Three common forms of lichens. The entire thallus of the crustose type (A) adheres tightly to the substrate surface, whereas in foliose lichens (B) the leaf-like tips of the thalli are usually free. The stalked cup structures (apothecia) of the fruticose type shown in (C) bear the fungal ascospores. A pendant fruticose lichen and another foliose type are shown in (D).

or slate roofs. Among lichens there are three major growth forms: (1) crust-like (crustose), (2) leaf-like (foliose), and (3) shrub-like (fruticose) (Figure 25-3). The crustose growth form is the most common, but others are very abundant in certain areas. For example, the fruticose lichen *Cladonia rangiferina* ("reindeer moss"), which superficially resembles a gray moss, covers vast areas in the arctic. In arctic and subarctic regions lichens are important pasturage for wildlife such as musk oxen and caribou and for the semidomesticated reindeer.

Fungal hyphae compose most of the lichen thallus, with the algae usually forming a thin layer just beneath the surface (Figure 25-4) and comprising 10% or less of the lichen's dry weight. Over 18,000 lichen species have been described; almost every lichen species contains its own unique species of fungus. The fungal associate (**mycobiont**) frequently forms sporocarps on the lichen surface; lichen classification, like that of free-living fungi (Chapter 24), is based on these structures. Most lichen fungi are ascomycetes, either apothecial or

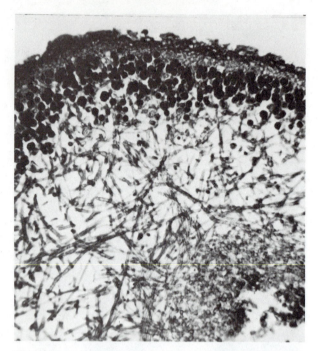

Figure 25-4. A section of a lichen thallus viewed with the light microscope shows algal cells (dark spheres) beneath a surface layer of tightly compacted fungal hyphae. Much looser hyphae occur between the algal region and the hyphae forming the lower surface (lower right).

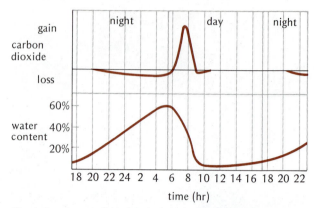

Figure 25-5. A desert-inhabiting lichen studied in Israel absorbs dew during nighttime hours (lower half of graph) and respiration becomes detectable (CO_2 loss, upper graph). With the onset of daylight, photosynthesis begins and CO_2 is fixed for a short time (CO_2 gain) before the lichen dries out and metabolism halts until hydration occurs during the following night. (After D. C. Smith, *The Lichen Symbiosis*, Oxford Biology Reader #42.)

perithecial, but a few are deuteromycetes or basidiomycetes. Thus it appears that the lichen type of symbiosis has evolved independently more than once (convergent evolution).

In contrast to the thousands of mycobiont species, only about 30 algal species (**phycobionts**) occur in lichens. Three genera probably comprise about 90% of all phycobionts. The most common is *Trebouxia,* a unicellular green alga similar to *Chlorella* (Chapter 23). *Trebouxia* is rarely seen in a free-living state, but the other two common genera of lichen algae, the green filamentous *Trentepohlia* and the blue-green (cyanobacterial) *Nostoc,* have free-living members. Both a blue-green and a green alga are present in a few lichen species. The phycobionts in three-membered associations are physically separate: the hypha-entwined cyanobacteria occur in a distinct layer in the thallus or in bumps on the upper thallus surface.

Activities of Lichens

Lichens are notable for their ability to grow in inhospitable habitats and survive adverse conditions for long periods, as the prominence of lichens in arctic and alpine areas illustrates. In forested regions the noticeable growth of lichens on trunks and branches of trees reflects their ability to tolerate severe desiccation. They carry on photosynthesis whenever hydrated by rainfall or dew; during dry spells, or even daily in some climates, they dry out and become metabolically dormant (Figure 25-5). Lichens have survived in the laboratory for more than a year without any water. Tolerance of both water stress and high temperature is important in the ability of lichens to colonize and inhabit bare rock surfaces, one of their most important ecological functions (Chapter 33). Certain species grow on rocks whose temperatures in the sun reach 50°C (122°F) or more.

Rock surfaces and tree branches are mineral-poor habitats. Most rocks lack available (soluble) nutrients and provide usable nutrients only by slow weathering of their insoluble mineral constituents. Lichens apparently obtain some nutrients from rocks by promoting rock weathering, but most of their nutrients seem to be obtained from dust and rainfall. To glean nutrients from such lean sources lichens have very strong nutrient-absorptive capabilities. This leads to the principal handicap for lichens in the modern world, their extreme sensitivity to air pollution. Lichen thalli accumulate sulfur dioxide to thousands of times its concentration in their environment, leading to toxicity from air pollution levels that most other plants can tolerate. As a result, a virtual "lichen desert" has developed in the center of most industrialized areas and extends a considerable distance in the direction of prevailing winds, carrying pollutants outward from their sources (Figure 25-6). Indeed, lichen growth (or disappearance) can be used as an indicator of average air quality in different areas.

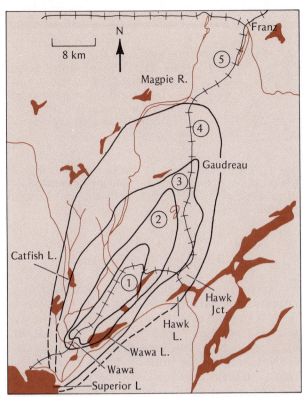

Figure 25-6. Zones of damaged vegetation resulting from air pollutants (mostly sulfur dioxide) from an iron-sintering plant at Wawa in north-central Ontario, Canada. Pollutants carried by prevailing southwesterly winds have resulted in the death of all lichens and bryophytes in zone 1, with gradually increasing numbers surviving in zones 2–4 at greater distances from the source. In zone 5 where "normal" flora occurred, more than 30 lichen species were observed on and about the trunks of each tree. Similar depauperate zones occur in and about urban areas where automobile exhaust and industrial air pollutants affect lichen growth. (After D. N. Rao and F. LeBlanc, *The Bryologist 70*:141, 1967.)

Another unfortunate consequence of the ion-accumulating capabilities of lichens is the buildup of radioactive strontium and cesium, isotopes hazardous to animal and human health, from nuclear weapons testing fallout. In the 1960s worrisome amounts of radioactive Sr and Cs were accumulated by arctic lichens from fallout and passed on to the reindeer grazing on them, and then to the Eskimos and Laplanders who ate the animals or drank their milk.

In three-membered lichens consisting of a fungus, a green alga, and a cyanobacterium, the cyanobacterium is usually a nitrogen-fixing species (see Chapter 22), making such lichens independent of a nitrogen source. Although advantageous, the advantage is not enormous because such lichens are just as dependent as others on the scanty supply of other important mineral nutrients, such as phosphorus and sulfur, in their environment.

TABLE 25-1. Growth Rates of Representative Lichens

Species	Average Increase in Radius (mm/year)
Foliose species	
Parmelia caperata	4.0
P. olivacea	0.7
P. saxatilis	0.5–4.0
Umbilicaria cylindrica	0.01–0.04
Fruticose species	
Cladonia alpestris	3.4
C. rangiferina	2.0–5.0, 4.1
Evernia prunastri	2.0
Ramalina reticulata	11.0–90.0 (7 months)
Crustose species	
Lecanora alphoplaca	0.67–1.40
Rhizocarpon geographicum	1.0
Rinodina oreina	0.57

Modified from M. Hale, *The Biology of Lichens.*

Growth and Metabolism of Lichens

The lichen mode of life, in which survival adaptations predominate, is suited to cells with a very low rate of metabolism that can sink to negligible values for long periods without harming the cells. The growth of lichen thalli is therefore typically very slow (Table 25-1). Slow growth is also enforced by the lichen's usually limited mineral nutrient supply. Lichen metabolism has, however, become evolutionarily adapted to this situation, and lichens will not grow rapidly even when artificially supplied with an abundance of mineral nutrients.

Radioactive tracers show that glucose formed during photosynthesis by *Nostoc* phycobionts is released and utilized by the fungal associate. In lichens containing green algae, sugar alcohols such as sorbitol or ribitol, not glucose, seem to be the forms of carbon transported (Figure 25-1). Most lichen fungi require some vitamins for growth, and the algae presumably also provide these. Some 40–60% of the carbon photosynthetically fixed by the alga seems to be transferred to the mycobiont. When removed from the association in the laboratory, the phycobiont stops releasing photosynthetic products from its cells, suggesting that in the lichen the mycobiont induces release of algal products.

Benefits for the alga from the mycobiont are more difficult to demonstrate conclusively. When isolated and cultured apart from the fungus, the phycobionts usually grow slowly but frequently faster than they do in the lichen association, as might be expected from the facts about photosynthate transfer just mentioned. The fungus, because it is in direct contact with the environment, probably absorbs and supplies the mineral nutrients required by the alga. The fungal mantle probably helps protect the phycobiont against excessive light intensities that frequently occur in lichen habitats and are injurious to the photosynthetic machinery. The fungal layer may also reduce desiccation of phycobiont cells during dry periods. These are examples of bene-

Figure 25-7. Two unique lichen products found in a number of species: usnic acid (A), a yellow pigment, and lecanoric acid (B), a purple aromatic ester used in preparing litmus dye.

usnic acid

lecanoric acid

fits that allow the alga to exist in habitats that it otherwise could not occupy, for example, exposed rock surfaces.

oped, however, because of the slow growth rates of the producer lichens and the difficulty or impossibility of growing many of the lichen species in laboratory culture.

Lichen Products

Mycobionts in the lichen association synthesize a number of unique secondary products, some of which are useful for identifying different species. The most common lichen compounds, which often comprise 5–10% of a lichen's dry weight, are weak phenolic acids, called "lichen acids" (Figure 25-7). A role in the weathering of rocks to help lichens gain mineral nutrients has been postulated for these acids, which can etch the surface of limestone and other alkaline rocks, including gravestones and statuary (Figure 25-8).

Many of these lichen products are unique pigments, responsible for vivid red, orange, yellow, and purple colors of various lichens in nature. **Litmus,** formerly much used as an acid-base indicator in chemical laboratories, was prepared from some of these pigments. Purple, red, and brown lichen pigments were formerly used by the Scots to dye the yarns for clan tartans.

A number of lichen compounds have antibiotic properties, perhaps explaining why bacteria and fungi rarely parasitize lichens. Some of the antibiotic compounds have been isolated and identified and have been shown to be effective against certain human bacterial pathogens. Little commercial application has devel-

Reproduction of Lichens

Lichen reproduction is still somewhat of an enigma. Vegetative reproduction is possible by several means: dried thalli are quite brittle, and fragments carried about by wind, water, or animals may grow into new thalli. Many species, especially foliose or fruticose types, form specialized structures for dispersal. These include **soredia,** which are small, powdery, readily dispersed fragments consisting of fungal hyphae surrounding a few algal cells, extruded through cracks or pores in the thallus surface (Figure 25-9B).

Spores, commonly ascospores, are produced by many lichen fungi. (Apothecia can be seen in Figures 25-A and 25-3C.) Some lichens probably arise in nature from contact between a new mycelium from a germinated spore and some suitable algal cell(s), but this process is not readily observable because the characteristic lichen morphology arises only after formation of the association.

Figure 25-8. The growth rates of some lichen species have been estimated by comparing the diameters of individual colonies growing on man-made stone structures such as old tombstones (A). The age of the colony with the largest diameter for a given species can be no older than the date on the tombstone (1788 for the one in the photo). The colony radius divided by the tombstone age gives a maximum value for the annual growth rate. Erosion rates of statuary caused by lichens (B) are now far exceeded by acid rain.

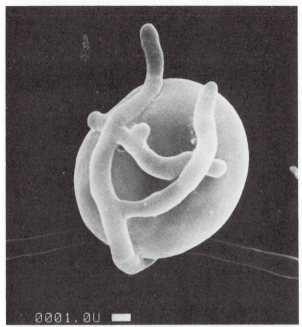

(A)

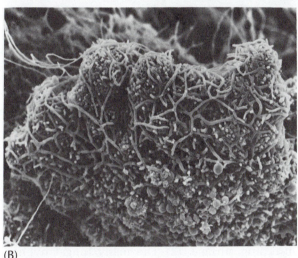

(B)

Figure 25-9. The algal-fungal association in lichen symbiosis: a resynthesis using the natural mycobiont and phycobiont cultured from *Cladonia cristatella* ("British soldiers"). Initial contact (A) shows fungal hyphae encircling a single green algal cell; (B) shows a later stage in thallus development where numerous encircled algal cells are being bound together by a network of hyphae. Soredia consist of such intermingled symbionts in bits small enough to be dispersed by wind, splashing raindrops, etc. Scale in (A) is 1 μm; (B) is approximately 1/10 that magnification.

Separation and Reassociation of Symbionts

Lichens have been resynthesized in a few cases from pure cultures of the fungus and the alga, isolated from the association (Figure 25-9). Both the alga and the fungus are changed morphologically in the lichen compared with their form when growing apart. Algal cells usually have a much thinner cell wall and mucilage

sheath in the lichen association than when growing apart, and cells of some species change in size or shape. Isolated lichen fungi in culture produce only slow-growing mycelial mats without the characteristic form of the lichen thallus; they normally form no reproductive structures. The difficult resynthesis of a lichen requires combined culture of the fungus and the alga under low nutrient levels and is promoted by alternate wetting and drying. Sometimes natural substrates, for example, sterile soil or wood, promote reassociation.

FUNGUS–HIGHER PLANT SYMBIOSIS: MYCORRHIZAE

The majority of higher plants have fungal hyphae closely associated with their growing roots (Figure 25-10), an association called a mycorrhiza (from Greek *mykos*, fungus; *rhiza*, root). Although this kind of association was discovered almost 100 years ago, its mutualistic nature has become generally accepted only relatively recently. Various trees, for example, certain pine species, are difficult or impossible to establish in non-native areas unless soil fungi associated with that tree species in its native habitat are also introduced.

Figure 25-10. Ectomycorrhizae of loblolly pine (*Pinus taeda*) formed by three different fungal associates (two 5-cm lateral root segments are shown for each). All three fungi are basidiomycetes.

Figure 25-11. Cultivated citrus species show a marked dependency on an endomycorrhizal association for adequate growth. Shown here are rough lemon seedlings after 6 months, with and without mycorrhizae, and with weekly applications of 0, ½, or full-strength nutrient solution without phosphate.

Function of Mycorrhizae

In mycorrhizae, photosynthetic products are transferred from the higher plant to the fungus partner (mycobiont). The mycobiont can benefit the plant by improving its mineral nutrient supply. In general, mycorrhizal associations are most frequent in nutrient-poor soils. A number of comparative studies show that under nutrient stress, mycorrhizal plants frequently grow faster and are more vigorous than nonmycorrhizal ones (Figure 25-11), whereas the same plant species growing in rich soil shows little or no difference between fungus-infected and nonmycorrhizal individuals. Analyses of mycorrhizal plants growing in nutrient-deficient environments show greater accumulation of phosphorus and in some cases of nitrogen compared with nonmycorrhizal control plants.

The mycobiont in a mycorrhiza comprises much more mycelium than just that growing around or in the root. It is the mycelium extending into the soil from the infected root that enables the fungus to assist the plant's mineral nutrition. Experiments with tracers such as radioactive phosphate show that free hyphae of mycorrhizal fungi absorb mineral nutrients very efficiently and transfer these nutrients to the host root (Figure 25-12). Two features contribute to this efficiency: (1) The individual hyphae are very small compared with even a small root; the surface area of the much-branched mycelium extending outward from a mycorrhiza is many times the root's own absorbing surface. (2) The many fungal hyphae extending out from the root can explore and exploit for mineral nutrients a much greater soil volume than can the root system itself.

There are two major types of mycorrhizal association, **ectomycorrhizae,** or sheathing mycorrhizae (from Greek *ektos,* outside); and **endomycorrhizae,** or internal mycorrhizae (from Greek *endon,* inside). These differ in the kind of fungi involved and the way they invade the root (Figures 25-13, 25-14) and affect its morphology, but the nutritional relationship between host plant and fungus is apparently similar in both types. Many of the endomycorrhizal fungi are apparently obligate mycorrhizal symbionts because it is difficult or impossible to grow them saprotrophically in culture. Many ectomycorrhizal fungi have some saprotrophic capability, but in nature they seem to be restricted to existence as mycorrhizal symbionts, probably by inferior saprotrophic competitive ability.

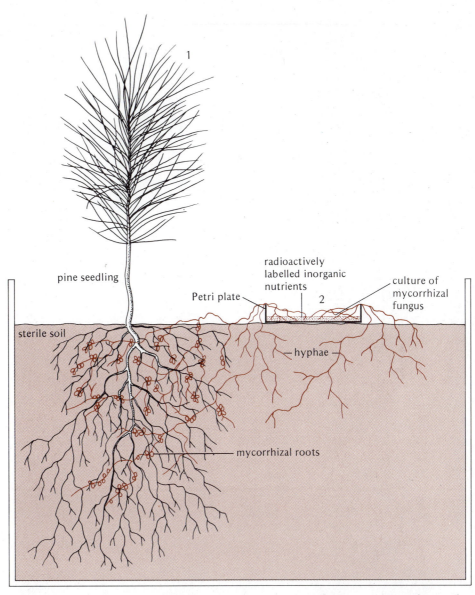

Figure 25-12. Transfer of nutrients can be shown by allowing a pure culture of a mycorrhizal fungus (in Petri plate at right) to grow out of its container into the sterile soil containing an aseptically grown seedling and to form mycorrhizae. $^{32}PO_4$ added at (2) can be traced into plant tissues via hyphae and mycorrhizae, and $^{14}CO_2$ applied to photosynthesizing leaves at (1) eventually results in ^{14}C-labeled cell constituents in the fungus. (After J. L. Harley, *Mycorrhiza*, Oxford Biology Reader #12.)

Ectomycorrhizae

In this type of symbiosis the hyphae form a compact mantle (Figure 25-13) about the young, actively growing lateral roots of the host plant in the upper soil layers. The hyphae grow between cells of the epidermis and of the root cortex but do not grow into the cells and do not extend beyond the endodermis into the stele (vascular core of the root). This hyphal ingrowth is called a **Hartig net** (Figure 25-13C). Infected rootlets are stubby, often branched, and do not develop root hairs (Figure 25-10). The fungal mantle makes up about 40% of the dry weight of beech tree mycorrhizae. Whereas nonmycorrhizal branch rootlets function as absorptive or-

gans for only one season or less, mycorrhizae normally remain functional for more than one growing season.

Most plants forming ectomycorrhizae are woody (often trees), including most conifers, oaks, willows, and some *Eucalyptus* species, among many others. The fungi are usually gilled or pore basidiomycetes and include many common woodland mushroom species such as boletes, *Amanita*, and some puffballs (Chapter 24). A tree may have several different fungal species as mycobionts. Some fungi are nonspecific with respect to the association; for example, *Amanita muscaria* forms associations with conifers, aspens, and hardwoods. Other fungal species are very specific: *Lactarius torminosus* grows only with birch, and the pore fungus *Suillus grevillei* only with larch.

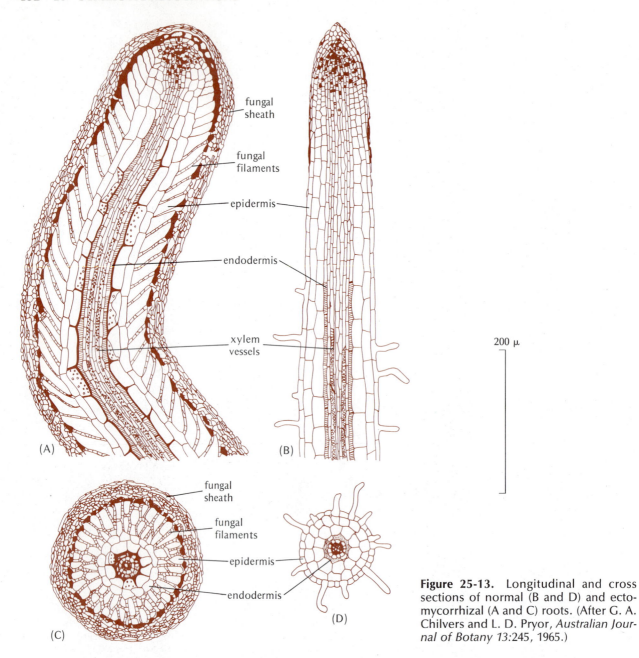

fungal
sheath

fungal
filaments

epidermis

endodermis

xylem
vessels

(A)

(B)

200 μ

fungal
sheath

fungal
filaments

epidermis

endodermis

(C)

(D)

Figure 25-13. Longitudinal and cross sections of normal (B and D) and ecto-mycorrhizal (A and C) roots. (After G. A. Chilvers and L. D. Pryor, *Australian Journal of Botany 13*:245, 1965.)

Many ectomycorrhizal fungi can be grown in pure culture, but they usually grow slowly and do not yield sporocarps (mushrooms), making it difficult to identify the isolated fungus. Many ectomycorrhizal fungi have been presumptively identified by observing that their sporocarps arise only in proximity to a certain forest tree species. To confirm this identification, seedlings of the presumed host species are grown aseptically in sterilized soil and a pure culture grown from spores of the suspected mycobiont (obtained from its sporocarps in nature) is added to see whether mycorrhizae then form.

Figure 25-14. Endomycorrhizae in root cells. Arrow points to cell containing arbuscules. (Dark-staining hyphae are on root surface.)

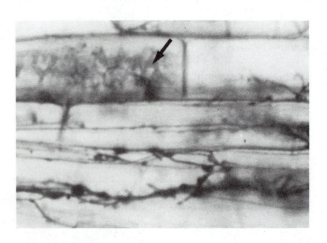

Endomycorrhizae

In these, the most common root-fungal associations, hyphae penetrate inside the root cells, in contrast to the purely extracellular hyphal location in ectomycorrhizae. Hyphae invade mainly epidermal cells but may also grow into cells of the root's cortex. These associations are also called **vesicular-arbuscular** or **V-A mycorrhizae** because the hyphae commonly form vesicles or arbuscules (tree-like shapes) within the penetrated host cells (Figure 25-14). These structures are somewhat comparable with the cell penetrations called haustoria by which many pathogenic fungi parasitize host cells (Chapter 26). The entering hypha does not penetrate the host cell's plasma membrane but pushes it inward, invaginating and convoluting it. Unlike ectomycorrhizae, the V-A mycorrhizal association does not result in obvious morphological changes in the invaded rootlets; instead of a hyphal mantle covering the root tip, only microscopic "threads" of hyphae extend outward from the root surface.

Endomycorrhizal fungi are zygomycetes belonging to the family Endogonaceae. More than 30 species of mycobionts have been described. In general, these mycobionts are not host-specific, even though many seem to be obligate mycorrhizal symbionts, as noted above.

Most herbaceous plant species and many tropical and a few temperate woody species, such as the familiar maples and the tulip tree (*Liriodendron*), form endomycorrhizae. Indeed, it appears that the majority of angiosperm species develop V-A associations. This type of mycorrhiza is also found in lower vascular plants and is essential for development of the gametophytes of some species (e.g., club mosses—see Chapter 28).

Mycorrhizal "Saprotrophs" and "Epiparasites"

A number of flowering plant genera in the orchid family and the heath family (Ericaceae) are nongreen, chlorophyll-less plants, for example, the white "Indian pipe" of eastern forests (Figure 25-15) and the brilliant red "snow plant" of western mountain conifer forests. Although commonly called "saprotrophs," these plants actually are totally dependent (parasitic) on mycorrhizal fungi for their nutrition. In certain cases the nonphotosynthetic angiosperm may be an indirect saprotroph, if its fungus partner is a litter- or wood-decomposing saprotroph. However, the fungi associated with some achlorophyllous angiosperms, such as the Indian pipe, are also associated mycorrhizally with trees or other photosynthetic plants, from which the Indian pipe apparently obtains organic nutrients. If radioactive tracers such as ^{14}C-glucose are injected into the green plant, radioactivity passes through the mycorrhizal fungi to the Indian pipe, and if the fungal connection to the green plant is cut, the Indian pipe fails to grow. Therefore, such achlorophyllous mycorrhizal angiosperms are apparently indirect parasites ("epiparasites") of photosynthetic plants.

Practical Value of Mycorrhizae

Certain studies indicate that native mycorrhizal fungus species may not always give the best

Figure 25-15. White flowering stalks of "Indian pipe" (*Monotropa uniflora*) are about 17 cm high.

possible plant growth. Therefore, an important practical goal is to discover or to create, by genetics or genetic engineering, mycobionts that would enhance plant growth in varying soil types. Particularly, mycobionts are desired that would improve the generally low efficiency with which most crop plants utilize applied fertilizer nutrients, reducing the amounts of fertilizer that need to be applied agriculturally and cutting down the amounts of unused fertilizer nutrients that cause nutrient pollution of surface and ground waters.

NITROGEN-FIXING PROKARYOTE–GREEN PLANT SYMBIOSIS

For agricultural and probably for many natural ecosystems the most important kind of symbiosis is that between specific bacteria and higher plant roots leading to nitrogen-fixing capability. The responsible bacteria occur free in the soil but are able to fix significant amounts of nitrogen only after they infect a host root, inducing it to develop a swelling or specialized structure called a **root nodule** (Figure 25-16), within the cells of which the bacteria occur as endosymbionts (intracellular symbionts) producing the nitrogen-fixing enzyme **nitrogenase** (Chapter 22).

Using isotopic nitrogen ($^{15}N_2$) it can be shown that bacterial root nodules take up N_2 from the air and convert it to combined forms [first NH_4^+ (ammonium), formed from N_2 by nitrogenase, and then amino acids, as explained in Chapter 22]. This mechanism provides the nitrogen needs not

Figure 25-16.   Nitrogen-fixing nodules on the root of a legume.

Figure 25-17.   Nodulated (right) and control soybean plants. The plant on the right was grown in sterilized soil from seed inoculated with specific nodule-forming bacteria; that on the left was grown from uninoculated seed.

which do not involve root nodulation, seem to fix at best only limited amounts of nitrogen compared with specialized nitrogen-fixing root nodules, associative symbioses have attracted attention recently because of a desire to improve nitrogen fixation in forest and range land and to extend nitrogen-fixing capability to major cereal grain crops that are not nodulated.

only of the bacteria but also of the host plant, because nodulated plants are able to grow in the total absence of combined nitrogen (NH_4^+ or NO_3^-), whereas uninfected, nodule-less plants of the same species fix no N_2 and can grow only with an external supply of combined nitrogen (Figure 25-17). Therefore, nitrogen fixed in root nodules is transported to the rest of the host plant.

There are three main types of nitrogen-fixing symbiosis: (1) nodules on roots of plants from the legume family (Leguminosae, such as beans, peas, clover, and alfalfa), containing bacteria of the genus *Rhizobium*, (2) nodules on roots of a variety of nonleguminous flowering plants (i.e., belonging to several botanical families other than the Leguminosae), containing filamentous bacteria (actinomycetes), and (3) specialized chambers in roots or shoots of a few land plants, housing nitrogen-fixing cyanobacteria (blue-greens). Besides these independently evolved kinds of definitive symbiosis, there are loose associations between certain nitrogen-fixing soil bacteria (e.g., *Azotobacter, Azospirillum*) and certain plant roots, including grasses, as well as significant nitrogen-fixing activity in bacteria associated with leaf surfaces. Although these ''associative symbioses,''

Rhizobium-Legume Root Nodules

Nitrogen-fixing nodules of most legumes are small, pink, spherical or elongated lateral structures scattered along the plant's younger roots (Figure 25-16). The nodule (Figure 25-20), apparently a much-modified lateral root, consists of an outer cortex of uninfected root cells surrounding a central core of root cells containing numerous modified *Rhizobium* bacteria called **bacteroids,** which differ in form and size from free-living *Rhizobium* cells. Bacteroids occur within host cells in membrane-enclosed groups (Figure 25-18), the enclosing membrane being derived from the host's plasma membrane. Both the enclosing membrane and the bacteroid plasma membrane and cell wall seem to be modified for increased permeability compared with normal membranes and rhizobial walls, promoting exchange of materials between the host cell and the symbionts. Between the bacteroid-containing core and the cortex of the nodule are strands of vascular tissue connected with the vascular system of the parent root; they supply photosynthate from plant to nodule and promote removal of nitrogen fixation products from nodule to host plant.

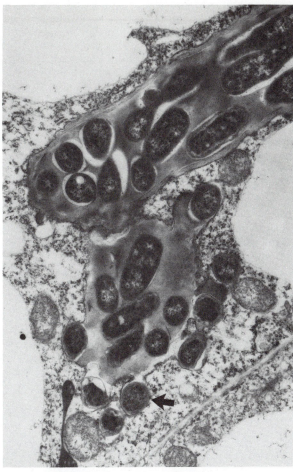

Figure 25-18. An electron micrograph showing sections of a number of bacteroids (dark structures) inside infection threads that have invaded a legume root nodule cell. The arrow indicates one bacteroid, surrounded by plant plasma membrane, being released into the host cell cytoplasm.

Leghemoglobin formation illustrates genetic interaction between symbiotic partners: the protein is coded by the legume plant's DNA but is not produced by plant cells except within the nodule. The bacteroids contribute by producing the iron-containing heme group needed to complete the leghemoglobin molecule.

SIGNIFICANCE OF LEGUME NITROGEN FIXATION

Symbiotic nitrogen fixation by legumes is so much more effective than that of free-living soil bacteria that in agricultural land where legume crops are planted they are responsible for over 90% of the biological nitrogen fixation, adding to the soil as much as 200 kg of nitrogen per year per hectare (about 200 lb/acre). Globally, agricultural legume nitrogen fixation appears to contribute as much as 25% of all the nitrogen fixed biologically on land, the total contribution of legumes being even greater because wild legumes are important nitrogen fixers in many kinds of forest and shrub vegetation. Legumes used in agriculture include those grown for seeds (e.g., peanuts, soybeans, peas, and beans) and those grown for animal forage (clover, alfalfa). Some of these legume crops, as well as others such as vetch, are frequently planted as cover crops to improve soil nitrogen content, and are not harvested but plowed under prior to planting a subsequent nonleguminous crop. Sometimes legumes are mixed with grasses in pastures to improve grass growth for cattle.

Artificial fertilization of soybeans, the most important legume crop, is necessary under agricultural conditions because symbiotic fixation usually does not furnish enough nitrogen for an economically profitable yield. One reason for this is the high energy or ATP cost of the nitrogenase reaction, noted above, which is met by bacteroid respiration of photosynthate from the host plant. Nitrogen fixation in nodules thus competes for photosynthate with growth of the plant (Figure 25-19). Vegetative growth and seed development, by creating strong sinks for photosynthate, tend to deprive nodule bacteria of adequate energy for nitrogen fixation. One of the goals of current plant breeding and genetic engineering is to develop soybean and *Rhizobium* strains with an improved balance between nitrogen fixation and other uses of photosynthate so that more of the crop's nitrogen needs can be met by symbiotic fixation, thus requiring less fertilizer nitrogen.

Symbiotic nitrogen fixation needs a large amount of phosphorus, which is used in the respiratory metabolites and ATP that provide energy for nitrogen fixation. Nodule development and activity frequently appear to be limited by a shortage of phosphorus, and in phosphorus-poor soils they can improved by mycorrhizal fungi that, when associated with roots, aid in phosphate uptake as described above and improve bacterial nodulation. This has potential value both in agriculture and in wasteland reclamation. Mycorrhizal, nodulated plants of both legume and nonlegume groups, seem-

The bacteroid-containing core of the nodule is the site of N_2 fixation, because it is the bacteroids whose DNA codes for and that produce nitrogenase. The bacteroids have very active respiration. Because nodule tissue is very compact and completely lacks the intercellular air spaces for efficient gas exchange possessed by normal roots, bacteroid respiration depletes the O_2 concentration within the nodule core to a low level. This is vital for effective N_2 fixation, because nitrogenase is inhibited by oxygen (Chapter 22). However, the bacteroids require O_2 for the respiration needed to generate the large amounts of ATP necessary for nitrogenase action (see Chapter 22). This need is met by the host cells, which produce a red, oxygen-binding protein called leghemoglobin (because it occurs in legumes) that is similar to the hemoglobin carrying oxygen in vertebrate blood. Leghemoglobin gives legume root nodules their pink coloration. It also catalyzes diffusion of O_2 from the nodule cortex into the bacteroid-containing core, maintaining throughout the core a low O_2 concentration sufficient for the bacteroids' rapid respiration but not high enough to inhibit nitrogenase activity.

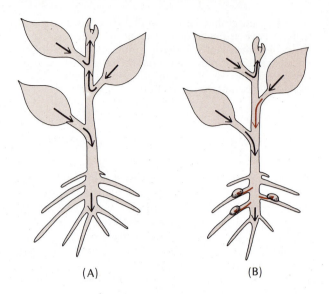

Figure 25-19. Nitrogen-fixing symbionts compete for the photosynthetic carbon and energy resources of the host plant. (A) Non-nodulated plant can allocate about two-thirds of its photosynthate to shoot growth, the shoot comprising about two-thirds of the plant's biomass. (B) In nodulated plant, use of carbohydrate for respiratory energy production for N_2 fixation in nodules creates an additional major "sink" in the root system, diverting photosynthate that would otherwise be available for growth.

(*Text continued from page 495*)
ingly the most nutritionally independent of higher plants, are especially adapted to act as colonizers of impoverished soils such as coal spoils and strip-mined areas (discussed further under nonlegume root nodule symbiosis below). Selection and improvement of these doubly symbiotic plants for use in re-vegetation is an important practical goal.

Figure 25-20. Nodule development in a legume such as clover.

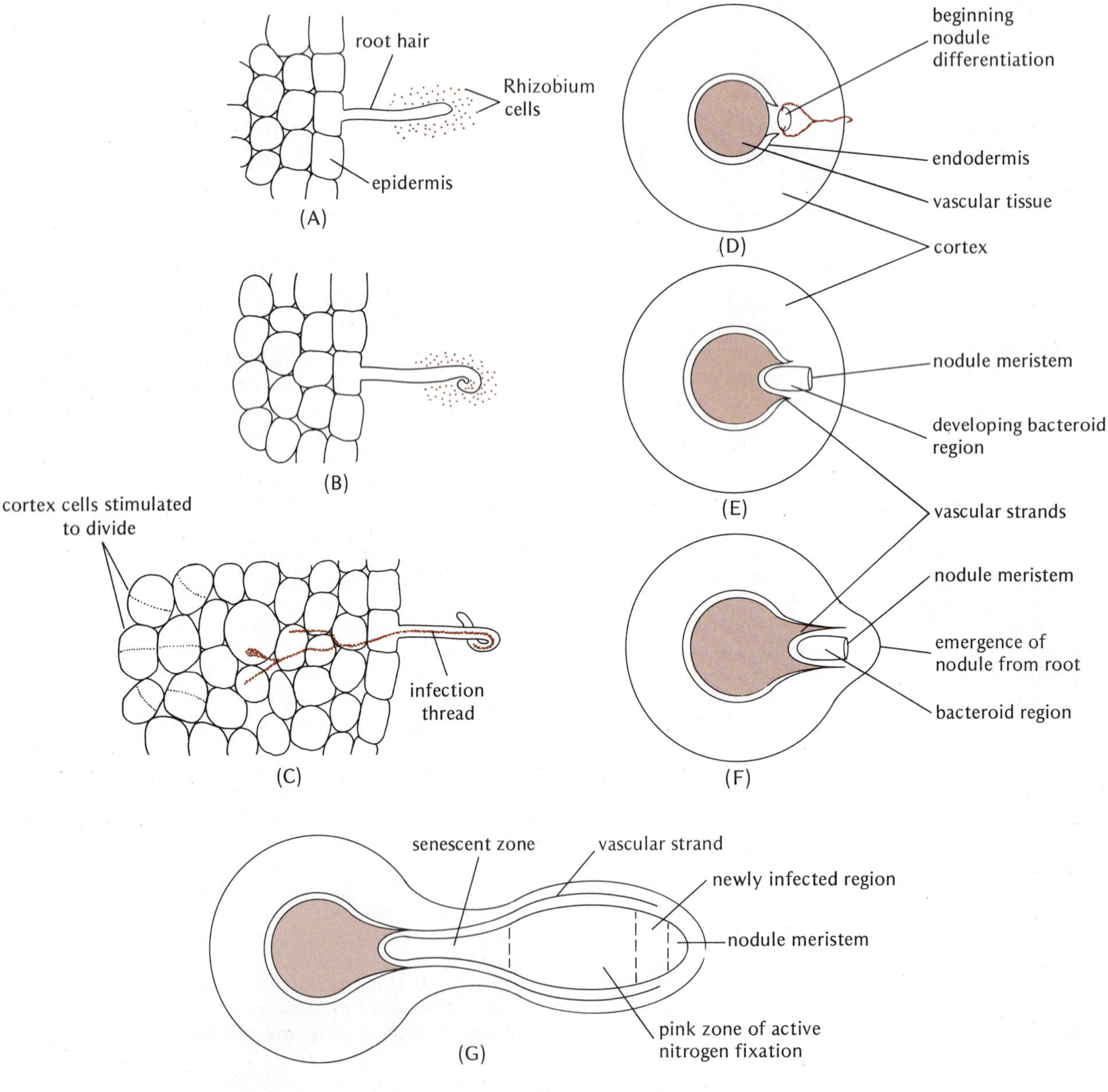

root hair

Rhizobium cells

epidermis

(A)

(B)

cortex cells stimulated to divide

infection thread

(C)

beginning nodule differentiation

endodermis

vascular tissue

cortex

(D)

nodule meristem

developing bacteroid region

(E)

vascular strands

nodule meristem

emergence of nodule from root

bacteroid region

(F)

senescent zone   vascular strand

newly infected region

nodule meristem

pink zone of active nitrogen fixation

(G)

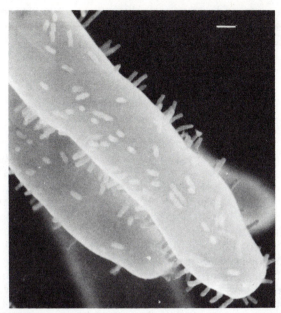

Figure 25-21. Rod-shaped *Rhizobium japonicum* cells attach specifically, by cell ends, to soybean root hairs. (Scale = 2μ)

NODULE DEVELOPMENT

A nodule arises when a root hair becomes infected by *Rhizobium* cells present in the soil around the root (Figure 25-20A–C). As the bacteria multiply, the "infection thread" containing bacteria enclosed by host-cell plasma membrane grows inward and branches (Figure 25-20C), stimulating uninfected cortical cells to begin dividing (probably by releasing auxin). This results in a nodule meristem from whose activity a nodule arises. Cells derived from this meristem receive groups of bacterial cells, enveloped by host-cell plasma membrane, from branches of the infection thread (Figures 25-18, 25-20D); these bacteria differentiate into bacteroids. If the nodule meristem ceases activity relatively early, a spherical nodule develops, whereas if the meristem remains active for a longer period, the nodule becomes more elongated (Figure 25-20F, G).

When nodule cells eventually senesce, their bacteroids apparently usually die, implying that association of *Rhizobium* cells with the legume eliminates rather than improves the fitness of the bacterial cells. This symbiosis thus may have evolved past the point of mutualism to a state in which the host plant effectively enslaves the bacterium. The roots of legumes somehow specifically stimulate the growth and proliferation in nearby soil of *free* rhizobial cells of the appropriate host-specificity group (see below), perhaps explaining how capacity for symbiosis confers an overall evolutionary fitness to *Rhizobium*.

SPECIFICITY

The specificity of the *Rhizobium*-legume interaction is of practical importance for agricul-

tural nitrogen fixation and is also interesting as an indication of the genetic basis of symbiotic interaction. According to the range of legume species that different strains of *Rhizobium* can infect and nodulate, these strains can be grouped into at least eight major host-specificity classes, designated by different specific names such as *Rhizobium leguminosarum* (nodulates peas, sweet peas, vetch, lentils, broad beans) or *R. japonicum* (nodulates only soybeans). Most legumes can be nodulated by *Rhizobium* strains from only one host-specificity group.

Ability to infect a host legume depends first on the ability of *Rhizobium* cells to bind to the legume's root hairs (Figure 25-21). This binding shows a pattern of host and *Rhizobium* species specificity that is comparable to that for nodulation, suggesting that binding is due to genotype-specific surface molecules that can interact, binding the cells to one another.

The *Rhizobium* side of this recognition reaction seems to involve certain polysaccharide molecules present externally on the *Rhizobium* cell wall and localized at polar sites, causing an "end-on" binding of the cells (Figure 25-21). *Rhizobium* mutants lacking these carbohydrate polymers cannot bind to or infect a root hair. Different *Rhizobium* species apparently differ in certain features of polymer composition. Root hair surfaces of a legume seem to bear molecules that can recognize and combine only with unique surface polymers of a particular *Rhizobium* strain, conferring specificity to the interaction. It is now thought that these recognition molecules of the plant are proteins called **lectins,** which have an ability to bind particular carbohydrate groups with high specificity and affinity in a binding reaction that is somewhat analogous to that between vertebrate antibodies and antigens.

Different genetic strains within a *Rhizobium* species can differ greatly in the nitrogen-fixing effectiveness of the nodules they form on different host species or cultivars within the host range of that *Rhizobium* group. Ineffective nodules can be deficient, for example, in leghemoglobin. These variations indicate that a delicate genetic interplay between host and bacterium is needed for full development of the nitrogen-fixing symbiosis. This information is important commercially in choosing *Rhizobium* strains for mixing with seeds of particular crops to ensure good nitrogen fixation in the field.

Nonlegume Bacterial Root Nodules

Nitrogen-fixing nodules on nonleguminous plants, as noted above, contain an endosymbiotic filamentous bacterium (actinomycete) named *Frankia*. This nodulation, although it begins by an infection similar to that by *Rhizobium* cells in legumes, develops into a quite different form, with proliferation of lateral roots from the infected region producing either a ball of stubby nodule roots (Figure 25-22) that sometimes reach baseball size or a cluster of upward-growing infected roots. The

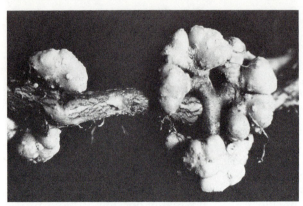

Figure 25-22. A developing ball-like cluster of alder root nodules.

branched filaments resembling fungal hyphae, the ends of which terminate in enlarged vesicles near the periphery of the root cells (Figure 25-23). Indirect evidence implicates these vesicles as the regions of active nitrogenase activity.

SIGNIFICANCE OF NONLEGUME SYMBIOTIC FIXATION

None of the nonlegume nodulated plants are used agriculturally, but alder trees (*Alnus*) are sometimes harvested for hardwood. In the Pacific Northwest, red alder (*Alnus rubra*) is a good "nurse species" for young conifers such as Douglas fir growing in close proximity, presumably because nitrogen fixation by the alder improves their growth. Other nodulated nonlegumes also seem to play a significant role in the nitrogen balance of nonagricultural lands. Several are significant colonizers of lands devastated naturally or by humans, such as eroded slopes and strip-mined or deglaciated areas. These pioneer plants provide increased soil nitrogen and organic matter for subsequent colonization by nutritionally less independent plants.

known *Frankia* hosts are mainly woody (mostly shrubby) species, from several rather unrelated botanical families (Table 25-2).

Within cells of *Frankia*-induced root nodules, actinomycete cells elongate into profusely

TABLE 25-2.   Nodulated, Symbiotically Nitrogen-Fixing Non-Legume Genera

Family[a]	Genus	Common Name(s) (Occurrence)[b]	Growth Form	Known Nodulated Species
Myricaceae (wax-myrtle family)	*Myrica*	sweet gale, wax myrtle	shrubs, trees	26
	Comptonia	"sweet fern"	low shrub	1
Betulaceae	*Alnus*	alder	trees	32
Rosaceae (rose family)	*Dryas*	mountain avens	matted shrubs	3
	Cercocarpus	mountain mahogany	large shrubs	4
	Purshia	antelope brush	shrubs	2
Rhamnaceae (buckthorn family)	*Ceanothus*	buckthorn, wild lilac	shrubs, many thorny	31
	Discaria	(S. America, Australia)	thorny shrubs	2
	Colletia	(S. America)	thorny shrub	1
	Trevoa	(S. America)	shrub	1
Eleagnaceae (oleaster family)	*Eleagnus*	oleaster, Russian olive	shrubs	19
	Hippophaë	sea buckthorn (Europe)	thorny shrub	1
	Shepherdia	soapberry, buffaloberry	shrubs	2
Casuarinaceae	*Casuarina*	"Australian pine"	trees	24
Coriariaceae	*Coriaria*	tanner's bush	shrubs	13
Ulmaceae (elm family)	*Parasponia*	(Java)	shrub	1[c]

[a]The Ericaceae (heath family), usually included in such lists because of one reportedly nodulated species, is omitted because nodulation has not been found widely nor confirmed to fix nitrogen, as with species of all genera listed here.
[b]If no location is given, the genus occurs in North America, and in most cases elsewhere also. (*Casuarina* is native to Australia but has been widely introduced in the warmer parts of North America.)
[c]The only non-legume currently known to be nodulated by *Rhizobium*. All others listed here are nodulated by *Frankia*.

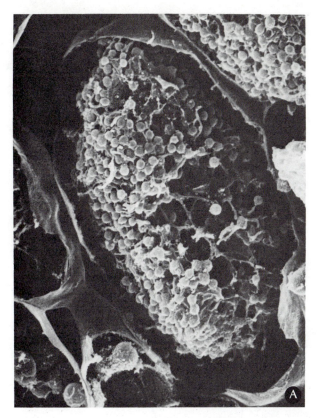
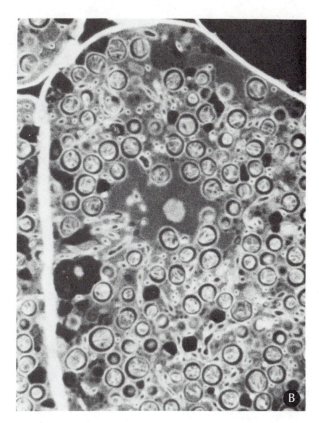

Figure 25-23. A scanning electron micrograph (A) of a host *Eleagnus* root cell filled with *Frankia* filaments. The terminal spherical vesicles, seen in section (B), contain many septa visible by special staining.

The fossil record shows that one nodulated shrubby species, the European sea buckthorn (*Hippophaë rhamnoides*), which can grow in raw sand, probably colonized large areas of Scandinavia after the last glaciation. Creeping mats of the attractive rose family plant, *Dryas octopetala* (Figure 25-24), colonize glacially denuded areas in Canada and Alaska. Nearly impenetrable thickets of pioneer alder trees often precede development of conifer forest in the far north (Chapter 33). In the western United States two nodulated shrubs, *Purshia* ("antelope brush") and *Ceanothus* ("buckthorn" or "deer brush"), seem to be important to the nitrogen supply of semiarid range and forest lands, where few or no legumes occur, and are also able to sustain deer and other wildlife for long periods as their only food source.

Methods for successful isolation and culture of *Frankia* from certain nonlegume nodules and for infecting seedling plants, using these cultures, to yield normal nitrogen-fixing root nodules, have only recently been discovered. These findings will undoubtedly enable investigators to explore and develop the biological principles and applications of the *Frankia* nitrogen-fixing symbiosis as they have those of *Rhizobium* symbiosis.

Figure 25-24. Mats of this tiny angiosperm (*Dryas octopetala*) fix nitrogen in arctic regions.

Figure 25-25.   The small floating water fern, *Azolla* [the specimen in (A) is about 1 cm long], contains nitrogen-fixing cyanobacteria in leaf cavities (B).

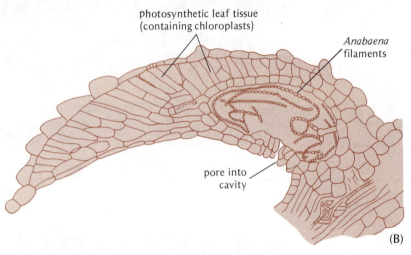

photosynthetic leaf tissue
(containing chloroplasts)

Anabaena
filaments

pore into
cavity

(B)

Symbiosis with Blue-Greens

The most widespread and familiar instance of nitrogen-fixing symbiosis between cyanobacteria and land plants involves the miniature aquatic fern *Azolla* (Figure 25-25A), which grows floating on the surface of ponds and other quiet fresh waters. The upper lobe of its bilobed leaves contains a cavity holding filaments of the nitrogen-fixing blue-green *Anabaena azollae* (Figure 25-25B). Because of the blue-green, the fern requires no combined nitrogen and can grow readily in many waters, sometimes becoming a problem weed.

In Asia, nitrogen fixation by *Azolla* has been utilized for centuries as a nitrogen source for rice cultivation. In Thailand and Vietnam, commercially available *Azolla pinnata* plants are scattered in the field when rice is transplanted into flooded paddies; the fern grows alongside the rice for several months. Increasing temperatures in the spring kill off the *Azolla*, which sinks and decays, releasing its nitrogen to the rice. Recent field trials utilizing a native *Azolla* (*A. filiculoides*) in the

Sacramento Valley of California show that the fern can provide about three-fourths of the nitrogen required by a rice crop. Its use may, therefore, become economically important in the United States.

A few genera of liverworts, primitive nonvascular land plants (Chapter 27), contain nitrogen-fixing cyanobacteria of the genus *Nostoc* in cavities within their thalli (Figure 25-26A), similar to the *Anabaena* chambers of *Azolla* leaves. *Nostoc* and *Anabaena* also inhabit intercellular compartments in specialized coral-like surface roots of cycads (Figure 25-26B), primitive gymnosperms that were major components of land vegetation in the Mesozoic era before flowering plants evolved (Chapter 21). These symbioses were probably important to the nitrogen balance of land ecosystems prior to the advent of rhizobial and *Frankia* root nodules (all of which involve flowering plants); cycad nitrogen fixation still seems to be significant in certain forests in Australia and South Africa where these gymnosperms are an important part of the vegetation.

Gunnera, a genus of herbaceous (sometimes giant) plants of the southern hemisphere and Polynesia (some species are used as ornamentals in the United States),

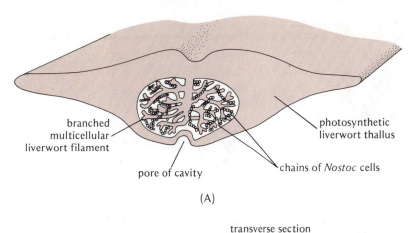

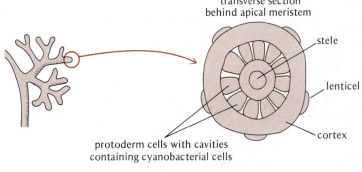

Figure 25-26. Two symbioses between plants and cyanobacteria. (A) Liverwort (*Blasia*) thallus; (B) cycad coralloid root. [(A) After G. A. Rodgers and W. D. P. Stewart, *New Phytologist* 78:441, 1977. (B) After J. I. Sprent, *The Biology of Nitrogen-fixing Organism*, McGraw-Hill, 1979.]

is the only group of flowering plants known to contain symbiotic blue-greens. In *Gunnera,* unlike the other blue-green associations mentioned, *Nostoc* filaments occur as endosymbionts within specialized cells of the plant's stem, analogous to the condition in bacterially nodulated roots. The blue-green symbiont can furnish its host with its entire nitrogen requirement.

LOWER PLANT–ANIMAL SYMBIOSES

Widely different animals have evolved symbioses with algae or with saprotrophic heterotrophs (bacteria or fungi), aiding either in the animal's general food supply or in more specific aspects of its nutrition.

Algae Symbiotic with Invertebrates

Various invertebrates, from some unicellular protozoa to reef-building corals, several sea anemones, and a giant clam, have symbiotic algal associates (phycobionts). Most of these symbioses occur in marine animals, but there are several freshwater associations: the green ciliate *Paramecium bursaria* (Figure 25-27A) and the green hydra, *Hydra viridis* (Figure 25-27B), both containing *Chlorella* cells, are well known to biology students. Among other invertebrate phycobionts are diatoms and other Chrysophyta, blue-greens (cyanobacteria), and the unique *Prochloron*, the

prokaryotic "green alga" (Chapter 21) found in certain sea squirts (ascidians).

In most algal-invertebrate associations algal cells occur as endosymbionts, inside vacuoles within cells of the animal (Figure 25-28), but in some the phycobiont is extracellular. In large invertebrates such as green sea anemones (Figure 25-29A), algae occur only in subepidermal tissue, where they receive the light necessary for photosynthesis. Small animals like *Hydra* (Figure 25-27B), on the other hand, contain algae in cells lining the digestive tract, where because of the animal's transparency there is plenty of light. In the giant clam (*Tridacna*) cells containing dinoflagellate phycobionts occur in the animal's outer fold of tissue or mantle, which is exposed to light when the mostly buried mollusk opens its shell (Figure 25-29B).

FUNCTION OF THE SYMBIOSIS

Feeding experiments with radioactive CO_2 show transfer of photosynthesized products, usually just one or two compounds such as simple sugars or amino acids, from phycobionts to the host animal. The nutritional relationship thus resembles that in lichens, except that some kinds of host animals normally retain their feeding mechanism and do not depend wholly on the phycobionts for carbon, as does a lichen fungus. In many cases the phycobiont apparently merely supple-

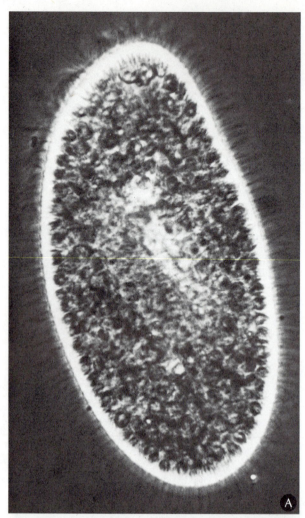

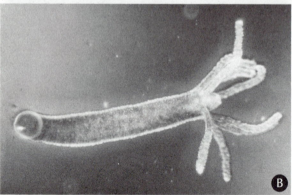

Figure 25-27. A protozoan, *Paramecium bursaria* (A), contains numerous spherical green algal cells. Similar intracellular algal cells make the freshwater hydra, *Hydra viridis*, a bright green, as its species name implies.

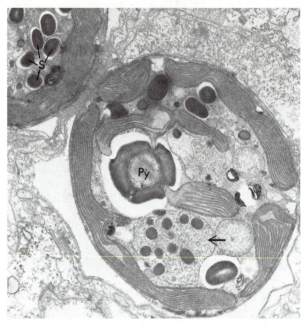

Figure 25-28. Reef-building corals invariably contain endosymbiotic dinoflagellates. The chloroplast structure and pyrenoid (Py) and the nucleus with condensed chromosomes (arrow) are identifying ultrastructural features of these algal symbionts. Starch grains (S) are present in the cytoplasm.

ments the animal's food supply and may be important mainly in tiding the animal over periods when it cannot catch prey. Green *Hydra* and sea anemones, for example, survive starvation better than nongreen (phycobiont-free) individuals of the

same species, provided they are exposed to light, but they cannot grow under these conditions and may eventually die of starvation despite their phycobiont's photosynthesis. At the other extreme is the small marine flatworm, *Convoluta roscoffensis*, which turns the sandy beaches of Brittany (northwestern France) green at low tide. This worm cannot develop to sexual maturity except in symbiosis with a green flagellate. Mature worms, containing intercellular algae, do not feed at all and require only light and the inorganic nutrients in sea water; by virtue of their phycobionts they have become autotrophic animals!

After separation from their host animal, cultured invertebrate phycobionts, like lichen algae, rapidly lose the property of releasing photosynthesized compounds to their surroundings. As in the lichen association, this indicates that the fed partner stimulates specific transport out of phycobiont cells.

The ability of algae to assimilate mineral nutrients may aid the host animal in acquiring or recycling some of its noncarbon nutrient requirements, especially nitrogen, a major limiting nutrient in tropical waters, where algal-invertebrate symbioses are particularly prevalent. Aquatic invertebrates excrete nitrogen largely as ammonia, which plant cells can use very effectively as a nitrogen source (Chapter 11). By consuming ammonia that is metabolically produced by the host animal, phycobionts can scavenge and recycle waste nitrogen that the animal otherwise has to excrete and lose. Experi-

Figure 25-29. (A) One of several species of "green" sea anemones that contain endosymbiotic algae, usually a dinoflagellate species. (B) The giant clam, *Tridacna*, which contains algae in the exposed mantle tissue.

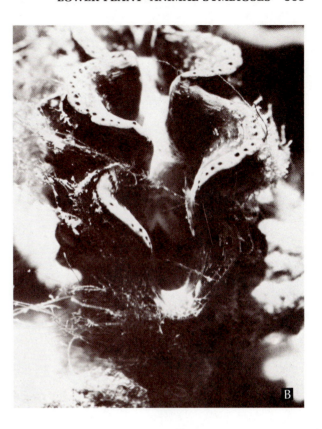

ments with the worm *Convoluta* and with reef corals, which always contain dinoflagellate phycobionts (Figure 25-28), indicate such a recycling of nitrogen. Furthermore, there is a correlation between the rate of phycobiont photosynthesis and the rate of coral calcification (deposition of reef structural material, Figure 25-30), suggesting a contribution by the algae to the calcium metabolism of these animals.

TRANSMISSION OR FORMATION OF THE ASSOCIATION

Hydra viridis and many corals incorporate phycobiont cells into their fertilized eggs before these are shed. In species with no mechanism for transmitting phycobionts to the next generation, such as the flatworm *Convoluta*, or in animals artificially freed of their natural phycobionts, renewed symbiosis usually arises by engulfment of algae by animal cells during phagotrophic feeding. These associations indicate further that symbiosis can arise by selection for restraint in digesting ingested prey. A potential host animal digests other ingested algae but not its phycobiont species, indicating recognition between host and phycobiont cells, the mechanism of which is unknown. Furthermore, the number of endosymbiotic algal cells per host animal cell is kept within a set range in any given symbiosis, indicating some means by which the host can control the phycobiont's multiplication.

A given phycobiont species can usually inhabit a range of host animal species, although different hosts may prefer different strains of the phycobiont. Many corals apparently contain the same species of dinoflagellate (often called *Gymnodinium microadriaticum*).

When free of algae the worm *Convoluta* will accept algal species other than its usual phycobiont, *Platymonas convolutae*, but if this flagellate is subsequently introduced it will displace other phycobionts from the worm.

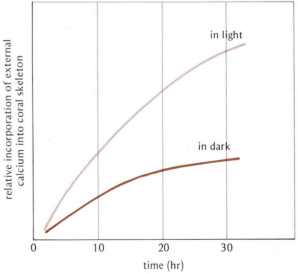

Figure 25-30. Deposition of skeletal material (calcium carbonate) which accompanies growth in reef-building corals is enhanced by photosynthesis of the endosymbiont algae, as shown by measuring radioactive ^{45}Ca incorporation in the light and in the dark. Specific chemical inhibitors of photosynthesis eliminate the enhancement of calcification shown in the light. (After T. F. Goreau, *Biological Bulletin* 116:59, 1959.)

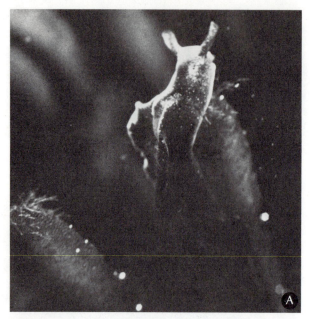

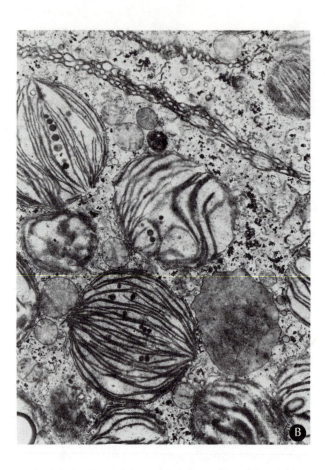

Figure 25-31. (A) A sea slug (*Elysia viridis*) on its food seaweed *Codium*. (B) *Codium* chloroplasts inside *Elysia* cells, showing varying degrees of digestion by the animal cells.

A curious variant of endosymbiosis with algae occurs in some marine slugs (sacoglossans), which feed by sucking out cell contents from siphonous green algae such as *Codium* and *Caulerpa*. Algal chloroplasts engulfed by the sea slug's digestive cells remain intact, at first enclosed in a vacuole, but eventually they become free in the cytoplasm of the animal cells (Figure 25-31), turning the relatively transparent animal green. Experiments with $^{14}CO_2$ show that these chloroplasts can remain capable of photosynthesis for over two months, the carbon they fix appearing in animal cell products. This is not symbiosis because the chloroplasts do not multiply or survive indefinitely, but it clearly represents utilization of plant biochemical machinery by animal cells and might be a step toward evolution of an autotrophic animal.

Fungus-Animal Associations

Among insects, the largest and most diverse group of animals, there are some in which fungi have come to serve a nutritional role analogous to that of the cellulose-decomposing bacteria in the digestive tract of ruminant mammals. Fungus-insect associations are actually much more diverse and important, at least to insects themselves, than most people realize, spanning virtually the entire range of nutritional and physical interactions discussed under the nature of symbiotic interactions

at the beginning of this chapter. Various fungi have specialized as parasites of insects and are now receiving attention as possible specific biological control agents for insect pests such as mosquitoes. Various insects, on the other hand, preferentially use fungi for food or depend on fungus-modified substrates such as rotten wood for their habitat and food source—for example, fungus beetles [belonging to insect families with names like Mycetidae ("fungus ones"), Mycetophagidae ("fungus eaters"), and Mycetobiidae ("fungus living")], and fungus gnats [Mycetophilidae ("fungus lovers")]. Certain insects have developed much closer and more specific associations with particular fungi, including several kinds of specialized symbiosis.

Fungus-Gardening Ants

The best known fungus-insect association is the cultivation of fungi by tropical leaf-cutter or attine ants. These animals live by planting and raising fungi in subterranean burrows or chambers (Figure 25-32), on leaf fragments that the worker ants cut and carry to the fungus garden from nearby vegetation. The ants feed exclusively on the resulting fungal growth, the fungi in effect acting as a digestive agent to convert cellulosic leaf material into food usable by the ants. To promote fungus growth the ants tend the fungal culture with complicated husbandry procedures, including fertilizing the culture with their feces, recycling waste nitrogen within the ant colony. Each new queen ant carries a fungal pellet from the garden with her to start a new garden when she leaves the nest. The fungi involved, mainly basidiomycetes, have also become obligatorily dependent on the association, not being found free-living outside the fungus gardens. Ant fungi are highly species-specific, a distinct fungus species for each species of ant. This association is often called a symbiosis but is really more a coevolutionary **domestication** of the fungus than a symbiosis because the partners, although interdependent, do not develop in intimate physical contact. However, the association includes many of the features found in symbiotic interactions considered previously.

Symbiotic Fungi

Various wood-boring insects have evolved symbioses with deuteromycete fungi that serve somewhat the same digestive role that ant fungi serve for attines. These insects, which include both beetles and wasps, possess body chambers called **mycetangia** ("fungus enclosure," Figure 26-3) for housing and feeding the fungi

with a glandular secretion so they are always available for inoculation of the primary food substrate (wood). When the insect drills into wood, its mycetangia are stimulated to exude secretion, smearing fungal cells or conidia onto the walls of the excavated passage. The fungus invades and breaks down the wood, and the insect feeds on the fungus. The association is obligate for the fungus because it has no means of gaining access to wood as a food source other than by the host insect's boring capability. Again, a high degree of fungus species specificity is involved, although a limited substitution of fungus partners is possible in artificial rearing of some of these insects.

In these associations the insect appears to depend on the fungus not only for its carbon and energy source but also for production of specific auxotrophic dietary requirements, because in the absence of the fungus the insect can be reared only using a complex, highly enriched food mixture and even then may not be able to complete its reproduction. Many other insect species, including various aphids and beetles, carry yeasts or yeast-like fungi within their bodies in special locations or organs, often endosymbiotically within specialized cells called **mycetocytes** ("fungus cells," Figure

Figure 25-32. Fungus-growing ants from Panama on their "garden" of basidiomycete mycelium. These gardens occur in soil "nests" under logs or rocks.

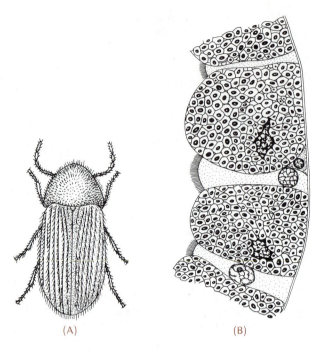

(A) (B)

Figure 25-33. (A) The "drugstore beetle" (*Stegobium panicea*), a pest that infests dried plant materials such as those formerly kept in drugstores for use in remedies (scale bar = 1 mm). Its midgut bears special outgrowths whose lining (B; gut cavity to left) includes "mycetocytes," enlarged cells containing numerous endosymbiotic yeast cells located within vacuoles (larger cells in B). The yeasts produce B vitamins and steroids essential to the beetle's nutrition.

25-33). These fungi are the fed partners in the association, depending completely upon the insect as a food source. However, the fungus seems to be beneficial to the insect; special anatomical arrangements often exist to ensure transmission of fungus to the next generation in the insect's reproduction. The contribution of the fungus seems to be as a prototroph, producing and supplying the host insect with vitamins and other dietary requirements to supplement the nutritionally low-grade food source, such as detritus or phloem sap, on which the insect feeds. Recycling of host nitrogen is also a probable contribution.

A final kind of fungal symbiosis, in which the fungus is the fed partner but apparently does not contribute nutritionally to the insect at all, but instead serves for its protection, is encountered among scale insects. These are sedentary animals that position themselves on a plant stem or leaf and feed like aphids by obtaining sap from the phloem. Certain scale insects are penetrated and parasitized but not killed by a specialized fungus that grows over an entire colony of the insects, enclosing them in a mat-like, mycelial sheath. This covering apparently prevents birds from preying on the scale insects, thus improving their chances of survival.

SUMMARY

(1) Symbiosis includes parasitism, a one-way nutritional relationship that is disadvantageous to the parasitized partner, and mutualism, in which the fed symbiont reciprocates with some nutritional or ecological benefit for the food-providing symbiont. Mutualistic symbiosis, with which this chapter deals, probably arises from a parasitic or predator/prey relationship by evolution of restraints, beneficial to both partners, on the parasite's or predator's exploitation of its food source. It involves recognition between cells of different species and genetic or developmental interactions between them.

(2) Lichens consist of fungi obligately associated with particular algae as their carbon source, usually a distinct fungus species in each species of lichen. The algae involved are either greens or blue-greens, or sometimes both, the blue-greens often being nitrogen fixers. Lichens are specialized by survival adaptations to inhabit and tolerate nutrient-poor, physically hostile environments.

(3) Mycorrhizae are vascular plant roots associated symbiotically with fungi, which are fed by the host plant. Mycorrhizal fungi improve mineral nutrient uptake by the host plant root systems, helping plants to colonize nutrient-poor soils and to compete with soil microorganisms for available nutrients in richer soils. Many common forest and woodland mushrooms are mycorrhizally associated with particular forest trees.

(4) Legume root nodules containing endosymbiotic *Rhizobium* bacteria that actively fix nitrogen can supply the host plant with its entire nitrogen needs, and are responsible for a large part of the available nitrogen supply in nature and agriculture. The association is specific between particular species or genera of legumes and particular genetic strains (named species) of *Rhizobium*. *Rhizobium* bacteria exist free in the soil and cause root nodules to arise by infecting legume roots. Significant nitrogen fixation by *Rhizobium* occurs only in the symbiotic state and results from genetic and physiological interplay

between the partners that satisfies the special conditions (high energy supply, low oxygen concentration, removal of combined nitrogen) needed for vigorous formation and activity of nitrogenase.

(5) Certain nonleguminous woody flowering plants, including alder trees and several shrubs or subshrubs of the rose and buckthorn families, develop nitrogen-fixing root nodules containing an endosymbiotic actinomycete (filamentous bacterium), *Frankia*. These plants are significant colonizers or reclaimers of nutrient-poor soils, including waste areas such as strip-mined and eroded land.

(6) A few lower land plants have evolved symbiotic associations with nitrogen-fixing blue-greens (cyanobacteria), which may have been important to terrestrial nitrogen balance in former geological eras before the evolution of flowering plants and their root nodule symbioses.

(7) Many tropical marine invertebrates and some freshwater ones possess symbiotic algae that supplement their food supply and probably also recycle some of the animals' waste nitrogen.

(8) Insects form diverse symbiotic associations with fungi, some using fungi in effect as digestive agents or as sources of supplemental dietary nutrients and others depending obligately on fungi as their sole food.

Chapter 26

PATHOGENS AND PLANT DISEASES

Arable land must be planted with high-yielding food crops to support the world's increasing human population. Of the thousands of higher plant species, only several dozen provide most of the food consumed by humans and domesticated animals (Chapter 32). It is little wonder then that a single pathogenic fungus, which may be capable of attacking only a specific host or even a specific variety, can cause millions of dollars worth of damage and potential starvation for thousands of people. At least 10% of all agricultural commodities produced worldwide are estimated to be lost through plant diseases. The losses in the United States for major crops such as corn, soybeans, and wheat are in the range of 14% and reach 20% for some crops, despite our advanced technology. A constant battle continues in an effort to prevent and control these losses.

NATURE OF DISEASE AND DISEASE ORGANISMS

Disease refers to a disruption of normal plant functions or metabolism, resulting in decreased growth and reproduction. A variety of infectious microorganisms, parasitic higher plants, parasitic animals (such as wire worms or nematodes), and deficiencies or imbalances of environmental factors (such as lack of required mineral nutrients or the presence of air pollutants) can cause plant diseases. This chapter covers diseases caused by most infectious microorganisms, namely, fungi, viruses, bacteria, and mycoplasmas.* The term

---

*Diseases caused by nematodes (round worms; cf. Figure 22-26) are not covered, but are discussed in zoology books.

Figure 26-A. Corn smut is characterized by large galls, developed when kernels are infected by a basidiomycetous fungus (*Ustilago maydis*). The galls ultimately become filled with masses of powdery, dark brown fungal spores.

pathogen applies to these disease-producing agents. Insects, mites, slugs and snails, birds, and other animals through their grazing activities can also impair plant growth or reproduction but are considered plant pests rather than plant disease agents.

Some animals are important as **vectors** of infectious diseases, that is, animals are the means by which the infective agents are transmitted between plants. Viruses, for example, can be transmitted to healthy plants by insects such as leaf hoppers that have fed on diseased plants. Humans pruning or cultivating healthy plants with tools that were previously in contact with diseased plant parts may transfer infective microorganisms or their propagative units.

Most diseases caused by pathogens are named or classified by the symptoms produced, for example, wilts, rots, blights, smuts, cankers, spot diseases, and so on (Figures 26-A, 26-1).

Nutrition of Pathogens

At first consideration the terms parasite and pathogen might seem equivalent because a parasite obtains at least some of its nutrition from another living organism, thereby robbing the host of resources that could be utilized for its growth. However, although mycorrhizal fungi parasitize various plant species (Chapter 25), they are not normally considered pathogens because the infected host plants frequently grow better, at least under adverse soil conditions, than do uninfected plants of the same species. In other words, a parasite does not necessarily cause harm or damage to the host organism, whereas a pathogen does. All pathogens are parasites, but not all parasites are pathogens.

Growth of some pathogens in a plant rapidly kills the invaded cells as well as surrounding tissues. These **necrotrophic** pathogens (Greek *nekros*, dead body, and *trophē*, nourishment) absorb

Figure 26-1. (A) Maple leaf spot, caused by the deuteromycete, *Phyllosticta minima*. (B) Heart rot of birch caused by a bracket fungus (*Fomes fomintarius*). (C) Rye grains, infected with the ascomycete *Claviceps purpurea*, are known as ergots. Flour, containing ergots and made into bread, caused serious outbreaks of ergotism or "St. Anthony's fire" in Europe during the Middle Ages. Lysergic acid and related alkaloids were responsible for the gangrene of the limbs and other symptoms. (D) Downy mildew (*Plasmopara viticola*) on fruits of grape.

their nutrients from dead tissue. Examples include the fungi causing brown rot of peaches and apricots, and the bacteria causing soft rot of cabbage, cucumber, and lettuce. Necrotrophic path-

ogens are usually **facultative** pathogens, that is, most are capable of growing saprotrophically as well. The "honey mushroom" fungus (*Armillariella mellea*, Figure 26-2), for example, can grow saprotrophically but frequently invades living roots of many woody species. The infection may result in death of the host, particularly if the tree is subjected to additional stresses such as crowding or too much water. The pathogenic capabilities of *A.*

Figure 26-2. Sporocarps of *Armillariella mellea* are attached by mycelial strands to a buried root of a dead tree which perhaps was killed by this fungus.

mellea are responsible for another of its common names, "the oak root fungus."

Necrotrophic infectious organisms are regarded as less specialized than **biotrophic** pathogens, which grow on nutrients from the living tissues they invade. These parasites are usually **obligate** parasites and can grow only on living hosts. All viruses, as well as many fungi, belong in this category.

Obligate parasites cause problems for plant pathologists trying to establish a causal relationship between the organism and a particular set of disease symptoms. Many obligate parasites cannot be cultured on sterile media, a requirement for proving they are the agents of disease by the prescribed steps of infecting healthy plants from pure cultures of the isolated presumed pathogen, observing the development of symptoms of the disease, and reisolating the same kind of infectious organism from the diseased plants.*

Some pathogens produce disease symptoms not primarily through obtaining nutrients at the plant's expense but by producing toxins or causing the plant to produce toxic compounds. Toxins are frequently carried by the vascular system of the plant and can cause symptoms far from the site of infection and location of the pathogen. Some wilt diseases are examples; another is Victoria blight of oats caused by *Helminthosporium victoriae*. Strains of this fungus that do not produce the toxin victorin are not pathogenic.

Infection and Disease Development

PROPAGULES AND ENTRY

Infection of the plant with parasitic microorganisms begins with entry into the plant by a propagule. Sources of infection by fungi are usu-

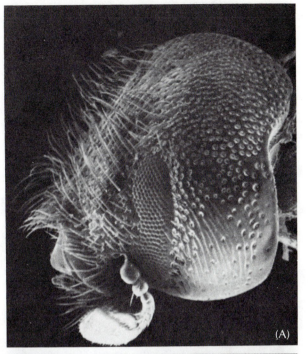

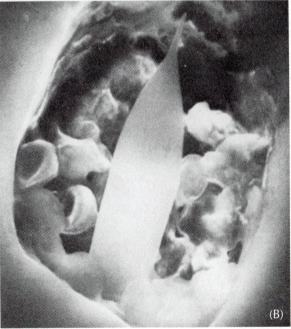

Figure 26-3. The fir-engraver beetle transports pathogenic fungal spores in pits (mycetangia) covering the head of adults, as seen in this scanning electron micrograph (A). An enlargement of one of the pits (B) shows the oily secretion and spherical fungal spores (collapsed by the treatment for the electron microscope) within; the pointed blade-like seta is a part of the pit structure.

ally spores, overwintering structures such as certain thick-walled zygotes or groups of vegetative cells, or mycelia in dormant buds, plant tubers, or plant litter. Wind or insects (Figure 26-3) can disperse these infective units. The spores of many pathogenic water molds are motile (Figure 26-4) and swim through soil water to the roots of the

---

*These formalized steps required to prove the causal relationship between an infectious agent and a disease were first set down by the German bacteriologist Robert Koch, and are known as Koch's postulates.

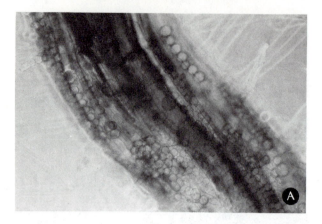

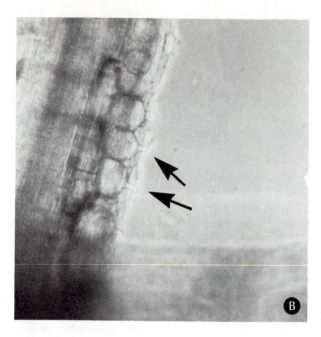

Figure 26-4. Spherical thalli of *Olpidium brassicae* (A) can frequently be seen infecting roots of plants of the cabbage family when they are examined microscopically. Zoospores are released from mature sporangia via the discharge tubes visible in (B).

potential hosts, or, as in *Phytophthora infestans*, the causative agent of potato blight, spores can be dispersed by splashing rain drops. Some facultative pathogens can invade host plants from their nonliving substrates. For example, *A. mellea* can invade living tree roots by outgrowth from its location in fallen timber. Root grafting or root contact can permit transfer of obligate pathogens from one host to another. Streets lined with closely spaced elm trees, for example, have allowed transfer of Dutch elm disease from tree to tree via root contacts.

Fungal and bacterial cells or spores may gain entry through wounds or natural openings such as stomates (Figure 26-5) or lenticels; in some cases fungi grow directly into the epidermal cells (Figure 26-6). Without aid viruses and mycoplasmas cannot invade healthy plant surfaces with intact cell walls, and must be transmitted by animal vectors, commonly insects or nematodes that feed on plant tissue, or by humans during cultural practices such as pruning.

INFECTION AND DEVELOPMENT OF SYMPTOMS

Plant diseases vary in severity depending on three factors: (1) the resistance or susceptibility of the host plant, (2) the virulence of the pathogen, and (3) the environment (chiefly the amount and frequency of rains or heavy dews, relative humidity, and temperature). These factors interact in determining the development of symptoms by the plant in any given year in any locality. The following sections will consider examples of plant diseases and pathogens and possible controls of plant diseases based on the above factors.

DISEASES CAUSED BY FUNGI

Fungal–plant interactions are stressed in this chapter because the vast majority of plant diseases are caused by fungi, followed by viruses and bacteria. All major divisions of fungi have representatives that cause plant diseases. Several important diseases will provide examples of modes of transmission, development, and spread of fungal pathogens.

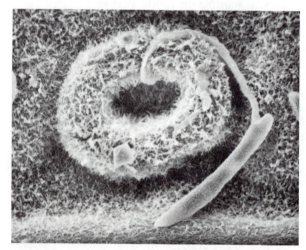

Figure 26-5. A hypha from a germinating fungal spore (lower right) can be seen in this SEM photo to be growing through a stoma of a Scotch pine needle. This ascomycetous fungus (*Scirrhia acicola*) causes brown spot needle blight in a number of pine species in the eastern United States.

Figure 26-6. Germinating powdery mildew (*Erysiphe graminis*) spore on wheat leaf. Arrows indicate where hyphae have penetrated epidermal cells.

Among the many diseases caused by ascomycetes is **Dutch elm disease** (Figure 26-7). The causative agent (*Ceratocystis ulmi*) is one of many pathogens that depend on insects for transmission. The fungus or its sticky spores are carried from tree to tree by elm bark beetles. Female beetles lay their eggs under the bark of dead or diseased trees where mycelia and conidia are also present. When the eggs hatch, the larvae tunnel beneath the bark and ultimately emerge as young adult beetles carrying spores in or on their bodies in a manner similar to the fir-engraver beetles (Figure 26-3). The young beetles seek out living elm trees. Where the beetles feed on and damage the bark, they deposit spores that germinate; the

Figure 26-7. Diagram of Dutch elm disease cycle. (After G. Agrios, *Plant Pathology*, 2/e. Academic Press, 1978.)

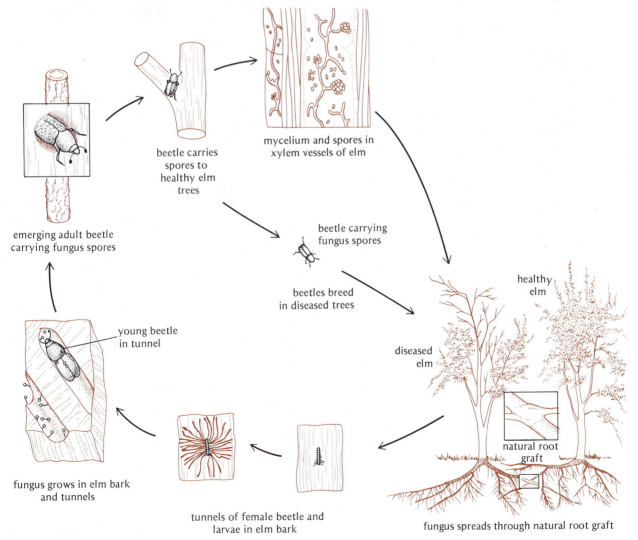

emerging adult beetle carrying fungus spores

beetle carries spores to healthy elm trees

mycelium and spores in xylem vessels of elm

beetle carrying fungus spores

beetles breed in diseased trees

healthy elm

diseased elm

natural root graft

young beetle in tunnel

fungus grows in elm bark and tunnels

tunnels of female beetle and larvae in elm bark

fungus spreads through natural root graft

resulting mycelium grows into the injured region and spreads through the bark and the wood.

When the mycelium reaches xylem vessels, a second type of conidia is produced, and these conidia are carried upward and distribute the fungus through the sapwood. Terminal twigs and branches wilt and die as their water-conducting tissue becomes blocked with fungal mycelia and with tyloses and metabolic products deposited in the xylem cells by the plant in response to the infection. These phenolic compounds are responsible for an outer brown ring of xylem that is seen when infected stems are sectioned. The dead branches or trees provide the sites in which future generations of female bark beetles lay their eggs.

The symbiotic association between the fungus and the bark beetle is a means of providing suitable nutrients and habitat sites for both associates, to the detriment of the host. The fungus is transported to fresh nutrient sources by the beetle, and subsequent growth may result in the death of the tree providing dead wood shelter for beetle eggs and for larval development. Eggs laid in the fall overwinter under the elm bark.

The Powdery Mildews

These ascomycetes are obligate pathogens that take their name from the white powdery or mealy appearance of the fungus on the leaves, petioles, young stems, buds, and even flowers of infected plants. A large number of angiosperm species are susceptible, including cereals, forage and lawn grasses, and numerous garden, orchard, and ornamental plants. In the midwestern and eastern regions of the United States the characteristic white mycelial blotches can usually be seen in summer on the leaves of any lilac bush, and on the West Coast the *Calendulas* or pot marigolds common in spring flowerbeds are particularly susceptible as the host plants age.

Some powdery mildew fungi are capable of parasitizing only one species of host plant, whereas others can infect over 350 kinds of plants in widely scattered plant families. Some of these fungal species are subdivided into a number of **physiological races,** each capable of attacking a limited number of host plants.

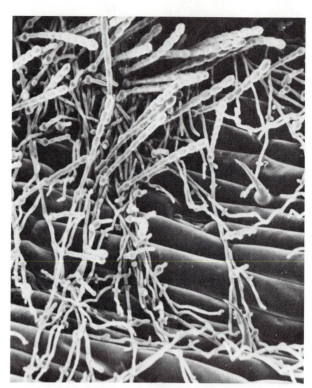

Figure 26-9. A scanning electron micrograph of the conidiating mycelium of *Erysiphe graminis* on wheat.

The fungus is superficial in its growth; most of the hyphae remain on the leaf or stem surface. These masses of tangled hyphae obtain their food from short branches (**haustoria**) that grow into the living epidermal cells. The haustoria may be simple knobs projecting into the cells or they may be elaborately branched (Figure 26-8). The surface mycelia produce enormous numbers of conidia (Figure 26-9) that are carried by wind to other parts of the same plant or to neighboring plants, where they germinate.

Cleistothecia (Figure 26-10), the sexual fruiting structures, are formed toward the end of the growing season and are barely visible to the naked eye as black dots in the mildew patches on the leaf surface. They bear appendages that are commonly curled or branched at the tip in a manner characteristic of each genus. The cleistothecia, containing one to several asci, survive the winter, and in the spring both cleistothecia and asci absorb water and swell. The cleistothecium breaks open and the asci burst, discharging the ascospores. In warm climates many species never form the sexual stage, reproducing solely by conidia.

Phytophthora Root Rots

Fungi of the water mold type (Figure 26-11), that is, fungi with motile dispersive spores, are responsible for a number of diseases that have caused enormous economic losses in the past and others of great significance today. The potato blight

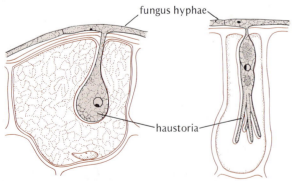

Figure 26-8. Haustoria of powdery mildews in leaf epidermal cells. These structures are commonly produced by biotrophic fungal pathogens.

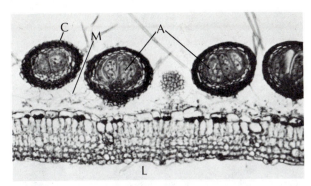

Figure 26-10. Cleistothecia (C) containing asci with ascospores (A) of powdery mildew of willow. The fungal mycelium (M) is on the surface and in epidermal cells of the leaf blade (L).

in Ireland in the 1840s is estimated to have caused the death of one and a quarter million people from starvation or disease associated with malnutrition.

The plant diseases caused by these fungi are most prevalent where the climate is warm and humid, or when the growing season is particularly wet (as was the case in the potato blight epidemic) or the soil in which host plants are growing is wet or waterlogged. Root rots are caused by some species. One of these, the oomycete *Phytophthora cinnamomi*, is a serious pathogen of commercial avocado stands in the United States and infects roots of many other species as well. In Australia it has slowly been spreading through *Eucalyptus* forests for some fifty years and is destroying whole forest communities, including an estimated 5% of Australia's commercial timber resources.

Such wholesale destruction of native plants indicates that the pathogen was recently introduced into Australia from outside, but this is not known with certainty. Because *P. cinnamomi* is a facultative pathogen it presents formidable problems in controlling its spread; the fungus can grow saprotrophically in soil, and its survival spores can remain viable for years, ready to produce zoospores during wet periods. Infection of a range of host plants is possible, and such infections can be transmitted to adjacent healthy plants by root contact. Drainage and irrigation ditches can help the spread of the pathogen.

Rusts

Rusts are among the most widespread and common plant diseases. The name "rust" comes from the reddish, yellow, orange, chocolate-brown, or black powdery appearance of the spores produced in pustules on the leaves and stems of the host plant (Figure 26-12). More than 2000 rust species are known, all parasitic on seed plants or ferns. Rusts parasitize many of the most important crop, forest, and ornamental plants, and attempts to control these diseases have been made for centuries.

The Romans, who called wheat rust (stem rust of wheat) rubigo because of the red lesions, had an annual festival, the Rubigalia, on April 25th at about the time the dog star Sirius was thought to be influencing the first appearance of the rust. During the festival a red dog was sacrificed and red wine was offered to the wheat god Robigus in an attempt to persuade him to control the dog star and diminish the destruction of the wheat crop. Nineteen hundred years later, wheat rust epidemics still occurred; in the United States during 1906 and 1908, rust decreased the annual wheat crop by 2 million bushels. Plant breeding for rust resistance has since then become the modern way of attempting to combat this fungal disease, but such methods are not wholly successful.

Rusts are basidiomycetes, and many have complex life cycles that may involve two host plants and as many as five spore forms. Certain

Figure 26-11. The potato blight fungus (*Phytophthora infestans*) produces on the underside of potato leaves dispersive sporangia (A) that are carried by wind to other plants. In damp, cool weather sporangia produce zoospores (B) that swim in rain water or dew before settling down and forming hyphae that penetrate leaf tissue through stomata, or potato tubers if the zoospores are washed into the soil. At warmer temperatures (above 15°C) a sporangium may germinate directly (C), producing an invading hypha.

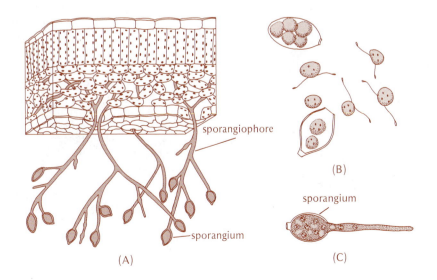

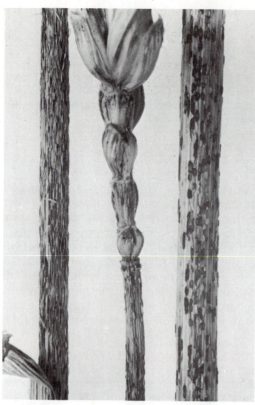

Figure 26-12. Lesions of wheat rust (*Puccinia graminis*) are reddish-brown or rust colored during most of the host growing season. Pustules appearing late in the season are similar in appearance but black, because a different spore type is produced.

Figure 26-13. One of the dispersive spore stages (aecial stage) of the white pine blister rust on white pine (*Pinus strobus*).

spores are produced on one host and others appear on a second or **alternate host.** For example, crown rust (*Puccinia coronata*) of oats and other grasses requires buckthorn (*Rhamnus*) species for the completion of the life cycle, and the white pine blister rust (*Cronartium ribicola*) (Figure 26-13) needs currant or gooseberry bushes as an alternate host. Rusts with only a single host include those parasitizing bean, hollyhock, snapdragon, rose, carnation, chrysanthemum, blackberry, and flax.

Figure 26-14 illustrates the stages of the life cycle of wheat stem rust caused by *Puccinia graminis*. This rust has five spore types and alternates between wheat and barberry to complete its life cycle. The aeciospores and uredospores (Figure 26-14G and H) infect one host (wheat), but only the basidiospores can infect the alternate host, barberry (Figure 26-14C). The teliospores (Figure 26-14A) are nonparasitic resting spores that produce the basidia. All of these spores are dispersed by wind, whereas the spermatia (Figure 26-14D) function only as gametes and are dispersed by insects that are attracted to a sweet fluid produced and released along with the spermatia.

The brick-red rust lesions (**uredia**) occurring on wheat leaves and stems during the late spring and summer produce thousands of asexual uredospores in each pustule (Figure 26-14H). These one-celled dikaryotic spores, which have a spiny, rust-colored surface, are carried by air currents to fresh host plants, where they can germinate, penetrate through a stomate, and produce a new infection. The mycelium at the new infection site takes 1–3 weeks to grow and produce a uredium capable of releasing a new generation of uredospores. Since each pustule produces 50 thousand to several hundred thousand spores, appropriate warm and moist weather can result in four or five asexual generations during the summer months and the spread of the disease over thousands of miles.

The reduced yield of rusted plants is due in large part to the shriveled and shrunken grains formed. The hyphae that grow out from the uredospores (Figure 26-14H) destroy photosynthetic tissue and utilize large quantities of food that should be available for storage in the seed. In addition, the destruction of the epidermis at the site of pustule formation permits increased transpiration. Rapid use of food by the fungus weakens the stem and roots of the plant as well as reducing the filling of the grain. Rusted plants lodge (fall down) readily.

There are several subspecies of *P. graminis* that are very similar morphologically but differ in their ability to infect different plants, that is, they are physiological races. The subspecies designation is based on the host species; thus *P. graminis tritici* subspecies infects wheat (*Triticum* species), *P. graminis secalis* subspecies attacks rye (*Secale cereale*), and so on.

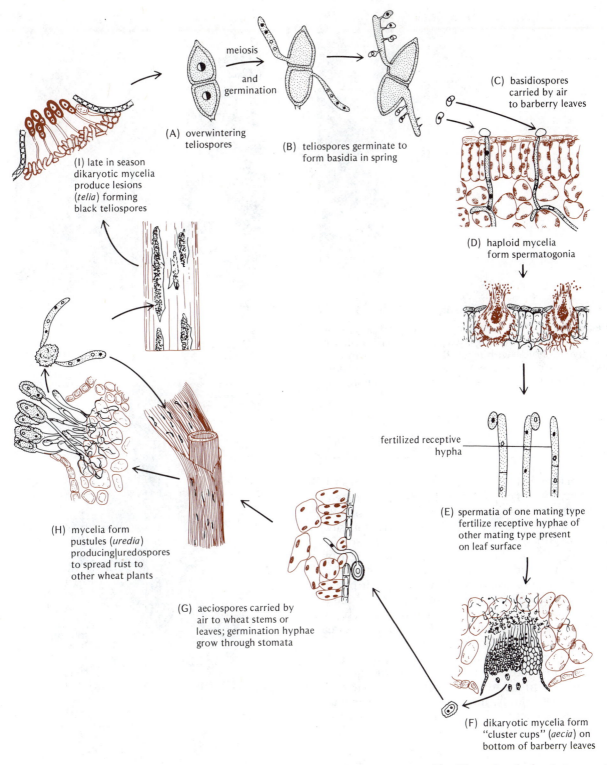

(A) overwintering teliospores

(B) teliospores germinate to form basidia in spring

meiosis and germination

(C) basidiospores carried by air to barberry leaves

(D) haploid mycelia form spermatogonia

fertilized receptive hypha

(E) spermatia of one mating type fertilize receptive hyphae of other mating type present on leaf surface

(F) dikaryotic mycelia form "cluster cups" (aecia) on bottom of barberry leaves

(G) aeciospores carried by air to wheat stems or leaves; germination hyphae grow through stomata

(H) mycelia form pustules (uredia) producing|uredospores to spread rust to other wheat plants

(I) late in season dikaryotic mycelia produce lesions (telia) forming black teliospores

Figure 26-14. The life cycle of wheat stem rust.

VIRUSES

Viruses, specifically bacteriophages, were briefly described in Chapter 22. A number of viruses multiply in plant cells; the first virus to be described and later crystallized was the tobacco mosaic virus. Like many other viruses, it has more than one host species; it causes large economic losses of tobacco and tomato but also infects many other crop plants and weeds.

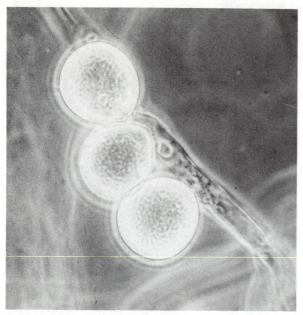

Figure 26-15. Three sporangia of a chytrid (*Rhizophydium graminis*) infecting a wheat root hair (note absorptive rhizoids within hair). Zoospores formed in sporangia on virus-infected plants can transfer viral particles to healthy roots during fungal infection.

Figure 26-16. Rembrandt tulip, showing variegated color, the result of virus infection.

VIRUS DISEASES

Vectors that transmit plant viruses include almost any type of organism that parasitizes or feeds on living plants such as insects, nematodes, parasitic plants such as dodder (Figure 2-A), and several zoospore-forming lower fungi (Figure 26-15). The majority of plant viruses are transmitted by insects that feed on plant juices such as aphids and leafhoppers. Many of the viruses multiply in the insect as well as in the plant host cells. Smokers can transmit tobacco mosaic virus simply by handling susceptible plants after using cigarettes or other forms of infected tobacco.

Mechanical transmission of viruses can result when animal movements or wind causes parts of an infected plant to rub against other parts or against other plants. Infected plants of some species pass viruses to their offspring via seeds, and pollen may carry viruses to healthy plants during pollination.

Viral Infection and Symptoms

Some viruses cause serious plant disease while others cause little apparent damage, and several even result in desirable horticultural traits such as "tulip break," which produces variegated tulip flowers (Figure 26-16). Necrotic ring spot diseases and mosaic diseases are common plant diseases

Figure 26-17. Tobacco mosaic produces a yellow and green mottling, especially of young leaves. Plants are usually not killed, but the quality and yield of the crop are reduced.

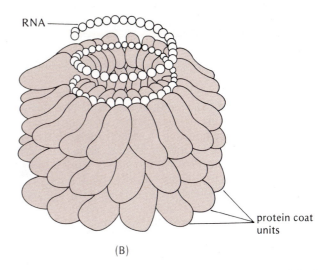

(B)

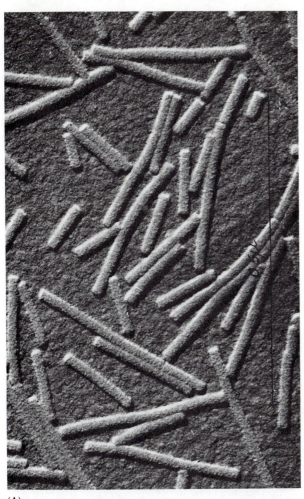

(A)

Figure 26-18. Tobacco mosaic virus (TMV), the first to be purified and crystallized, is a rod-shaped particle as seen in the electron microscope (A). The rods are composed of a helix of RNA surrounded by a protein coat (B).

caused by viruses. Necrosis refers to development of local patches of dead tissue. Mosaic refers to foliage mottling produced by absence of or abnormal pigment production, frequently combined with wrinkling or distortions of leaf surfaces (Figure 26-17); it includes diseases such as cucumber mosaic, alfalfa mosaic, and turnip yellow mosaic, all caused by different viruses.

Persistent infection, in which the viral nucleic acid is replicated in controlled fashion in the host cells with transmission to daughter cells and without releasing free viral particles, occurs frequently in plant viral infections as well as in bacterial and animal hosts. These **latent viruses** are a particular problem in plant species in which vegetative means of propagation are commonly used, such as potatoes and strawberries.

Meristem culture (Chapter 15) and nonlethal heating of infected plant parts will sometimes yield uninfected material for further propagation. Until recently most potato cultivars in the United States contained

one or more latent or mild viruses. Many cultivars had no obvious disease symptoms but gradually over a period of years became less productive and vigorous. Meristem culture of some of these cultivars has provided virus-free stocks and has increased crop yields by 5–60%, depending on the cultivar and virus(es). Among fruit trees, latent virus-infected scions have at times caused great economic losses in the grafted stocks.

Nature of Plant Viruses

Most plant viruses are RNA viruses, but some have DNA instead. Many of the RNA viruses are rod-like, and, as in all viruses, the nucleic acid is at the core of the structure, surrounded by protein that probably serves to protect the viral genetic information while the particles are outside host cells (Figure 26-18). The nucleic acid, either RNA or DNA, contains the genetic information necessary for its own replication as well as for coat proteins and other molecules such as enzymes that are necessary to alter or regulate host cell metabolism. Sizes of viral genomes are usually 1/100–1/1000 those of bacteria, that is, molecular weight = $1.0 \times 10^6 - 1.0 \times 10^7$.

The genome of certain RNA viruses consists of two or more discrete parts carried in separate viral particles of different sizes. Infection by each of the particle types may be necessary for virus replication and appearance of disease symptoms. For example, tobacco rattle virus, whose genome is in two parts, resembles tobacco mosaic virus, but the virus rods occur in two lengths.

In recent years small molecules of RNA have been found to be associated with and causative of several plant diseases such as potato spindle tuber disease, chrysanthemum stunt disease, and cucumber pale fruit disease. The replicating RNA molecules found in some of these diseases are only one-tenth or less the size of viral nucleic acids and have no protein coats. These smallest known pathogenic entities are called **viroids.** The most common method of spread of viroids is mechanical transmission by humans using contaminated tools during cultural procedures.

MYCOPLASMAS

Mycoplasmas are prokaryotic organisms without cell walls. The first mycoplasma described, around 1900, was responsible for pleuropneumonia in cattle, and since that time a number of mycoplasmas causing other animal diseases have been identified. Only recently a number of diseases of plants, insects, and fungi, some of which were attributed to viruses, have been shown to be caused by mycoplasmas. Similar wall-less prokaryotes have been recovered from soil, sewage, and compost, but it is not certain whether these forms are truly saprotrophic or only persist from animal or plant remains.

Mycoplasmas are composed of cytoplasm surrounded by a membrane and containing prokaryotic ribosomes and a DNA chromosome. Genome sizes range from the smallest known size for walled bacteria to about half that value (molecular weight = $0.5 \times 10^9 - 1.0 \times 10^9$). Like viruses, most mycoplasmas can pass through filters that retain walled bacteria; unlike viruses, mycoplasmas are susceptible to treatment with certain antibiotics such as tetracyclines and chloramphenicol (but are resistant to penicillin). Lack of a rigid wall makes possible highly variable cell forms ranging from coccoid shapes to more or less branching filamentous networks.

Helical, motile mycoplasmas comprise the genus *Spiroplasma*. Several species have been cultured and shown to cause plant diseases. The first *Spiroplasma* to be described was *S. citri* (Figure 26-19), a mycoplasma causing citrus stubborn disease in sweet orange trees in the southwestern

Figure 26-20. Excessive proliferation of shoots, called witches' broom, is seen in this plant with peach yellows disease.

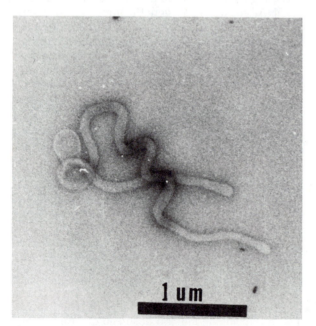

Figure 26-19. *Spiroplasma citri* as seen in the transmission electron microscope.

United States and the Mediterranean regions and infecting various other plants. This is a yellows-type disease, characterized by stunting of the plants and gradual yellowing or reddening of the leaves. Another yellows disease (lethal yellowing disease of palms) apparently caused by a mycoplasma-like organism has killed thousands of coconut palms in southern Florida and the Caribbean during the last decade.

Mycoplasma-like organisms, commonly called MLOs and usually seen in phloem cells using an electron microscope, have been described in a number of diseased plants, especially those with yellows diseases and witches' brooms (Figure 26-20). Most of these suspected plant pathogens appear to be obligate parasites because it has been impossible to culture them despite numerous attempts.

Plant-feeding insects, particularly leafhoppers, are the natural vectors of spread of mycoplasmas and MLOs between plants. Transmission can also occur by grafting and by transfer via parasitic plants. Mechanical transmission by inoculation of infected plant sap or cultured organisms onto abraded plant surfaces has not been successful in investigative studies, probably because the pathogen must reach phloem cells to cause infection.

Figure 26-21. Lima beans (left) resistant to downy mildew (*Phytophthora phaseoli*). The cultivar at right was killed by the fungus.

PLANT DISEASE CONTROL

Plant Resistance

Wild plants are largely resistant to most pathogens, because otherwise the species would have been eliminated during the evolutionary past; only disease-resistant members of a species would be expected to survive repeated epidemics or exposure to surrounding pathogens. However, because wild plants are not usually inbred and are genetically heterogeneous, members of any given population will vary greatly, and there will be a complete gradation between the extremes of complete resistance and susceptibility to any given pathogen. (Also, the production and potential crop yield will, of course, vary.) Thus, in any given generation a relatively high proportion of a wild plant population may show some susceptibility to a given disease, but the genetic variability makes it probable that some individuals will survive and reproduce, no matter what the environmental challenges. Such resistance is sometimes termed polygenic or **horizontal resistance.**

This natural resistance, resulting from centuries of evolution with pathogens common to the same geographic locality, has been overcome in a number of relatively recent notable epidemics when human activities introduced foreign pathogens into a native population. Prominent examples are the chestnut blight in the 1930s that essentially eliminated the American chestnut tree, once a dominant species in eastern hardwood forests, and the Dutch elm disease that is still progressing today.

This complete or nearly complete susceptibility occurs in many crop plant epidemics because plant breeding for genetic uniformity to assure high yields and the monoculture of such varieties result in the possibility of susceptibility of large populations of a single crop to newly virulent races of pathogens. Cultivated plants are bred to include resistance to the current dominant or prevalent pathogen(s) (Figure 26-21). However, genetic variability is a characteristic of plant pathogen populations as well as of wild plant populations, and sooner or later a combination of genes determining virulence in a potential pathogen usually arises. Under favorable environmental conditions the new virulent race multiplies in a crop population that is uniformly susceptible.

One of the more recent examples occurred with southern corn blight. In 1969 a virulent mutant strain of the corn blight fungus *Helminthosporium maydis*, a pathogen that had previously been a relatively minor problem, appeared. This mutant was pathogenic specifically on corn plants carrying "Texas cytoplasm" for male sterility, which was at that time the principal kind of male sterility used in hybrid corn seed production in the United States (in the manner explained in Chapter 18). Most of the United States corn crop therefore was susceptible to the new blight strain, which spread widely through the corn belt in the summer of 1970, causing severe disease in many areas and reducing the national corn harvest by some 17%. This already serious dislocation of the corn supply threatened to become extreme in the following year because no practicable control measures were available. The prospect sent waves of alarm through the farming and scientific communities, the public, and the government. Crash programs were mounted to develop and multiply new hybrids based on cytoplasmic male-sterile genes other than Texas cytoplasm, but because this takes at least two generations there was no possibility of supplying new hybrid seed in significant quantity for the 1971 crop. The feared disaster did not strike in 1971 or 1972, thanks mainly to weather conditions unfavorable for development of the disease in those summers, but the experience sensitized additional people to the real hazards of widespread use of a single plant genotype.

In the past 50 years plant breeders in the agriculturally developed countries have concentrated on producing new crop varieties resistant to each new emergent race of pathogen. Such plant resistance to a particular pathogen race or species is termed **vertical resistance** and is oligogenic (controlled by one or only a few genes). Figure 26-22 illustrates this succession of new resistant plant cultivars and new virulent pathogen races in oats. When many different varieties of a given crop are being grown, replacement of a cultivar that becomes susceptible to a new disease strain is a relatively manageable problem of finding and breeding in genes for resistance to the new strain of disease organism. It is important to discover each new disease strain early enough that the process of searching for and breeding in genes resistant to it can be well under way before the disease has become a severe problem. This is one of the reasons why sustained effort by plant pathologists and plant breeding stations throughout the world

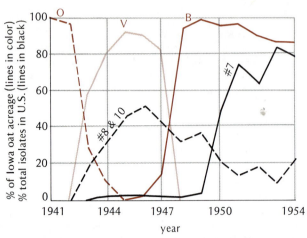

Figure 26-22.   Succession of stem rust race explosions, following the introduction of new oat cultivars specifically (vertically) resistant to particular rust races. In the early 1940s, older, rust-susceptible oat varieties (curve O) were replaced by the rust-resistant variety Victoria (V) to combat rust losses. The dashed black line shows the concomitant rise in stem rust races #8 and #10, pathogenic for Victoria. By 1948 Victoria had to be replaced by the Bond cultivar (B) resistant to rust races #8 and #10, which therefore declined; but after 1949, rust race #7 (solid black line), to which Bond is susceptible, increased to become the predominant race. Similar replacements have occurred after each subsequent introduction of new cultivars with additional vertical resistance genes. (After A. R. Browning and K. J. Frey, *Annual Review of Phytopathology* 7:355, 1969.)

Figure 26-23.   Peach leaf curl can be controlled by interrupting the life cycle of the causative fungus at the overwintering stage.

is now so important for the continued prosperity of agriculture.

GENERAL MECHANISMS OF DISEASE RESISTANCE

Plants have a battery of defenses against both diseases and pests. First, there are **structural features** for avoiding infection. These include surface impregnations of the epidermis with suberin and waxes and the deposition of layers of dead tissue (outer bark) over the surfaces of perennial structures. The presence of lignin in plant cell walls makes them impervious to degradation by most fungi. Abscission layers (Chapter 14) and corky wound tissue also are formed to act as barriers in response to invasion by pathogens.

Pre-formed inhibitors or toxic compounds that are active against microorganisms are present in plant cells. These substances include widely distributed phenolic compounds that prevent germination or growth of many pathogens. One of the best known examples is the occurrence of the phenolic compounds catechol and protocatechuic acid in the outer dead scales of red onion bulbs. These substances inhibit the germination of spores of the onion smudge fungus (*Colletotrichum circinans*), whereas other onion varieties lack these phenols

and are susceptible. Some plants store toxic compounds in inactive forms that then are converted to toxic forms through enzyme action following infection. Pear trees contain arbutin, a complex ring-structured molecule that can be split to release a phenolic portion that is toxic against the fire blight bacterium. Other such compounds release cyanide. The characteristic odor of cabbage family members is due to a volatile oil that is toxic to downy mildew.

Many plants synthesize toxic or inhibitory compounds in response to cell damage by infection or by mechanical or chemical injury. These plant stress metabolites are called **phytoalexins.** A wide variety have been described and more are being discovered all the time. Table 26-1 lists some representative phytoalexins.

It may be possible to apply purified phytoalexins to plants for protection against diseases. Preliminary studies show that spraying capsidol (Table 26-1), which is obtainable in appreciable quantities from green pepper fruits, onto tomato plants protects them from late blight [caused by infection with zoospores of *Phytophthora infestans*, the same fungus that causes late blight in the potato (see p. 515)].

Much recent work centers on the molecules in or produced by fungi that cause the accumulation of phytoalexins. These inducers or elicitors of phytoalexin accumulation have been isolated from several fungi and at least partially characterized. Such compounds, usually cell wall components, may be useful in studying the mechanisms that trigger the induction of plant response to infection. In at least some cases avirulent as

TABLE 26–1   Examples of Phytoalexins

Plants belonging to the same family generally synthesize the same structural type of phytoalexin molecule, as indicated below.

A. Isoflavonoid compounds from Leguminosae

Pisatin, from pea (*Pisum sativum*)

Phaseolin, from common garden bean (*Phaseolus vulgaris*)

B. Terpenoid derivatives from Solanaceae

Rishitin, from potato (*Solanum tuberosum*)

Capsidiol, from sweet pepper (*Capsicum frutescens*)

C. Fatty acid derivatives (polyacetylenes) from Compositae

$$HOCH_2CHOHCH=CH(C=C)_3CH=CHCH_3$$

Safynol, from safflower (*Carthamus tinctoria*)

well as virulent races of a pathogen produce the elicitor molecules that cause phytoalexin synthesis. In comparing the plant–pathogen interactions between such races, the relative speed of colonization by the fungus and the speed of the host response may determine whether a disease becomes established. In infections with virulent races, the pathogen may grow through host tissue before the phytoalexin levels become inhibitory, or the virulent races may degrade the phytoalexin. With avirulent races, growth of the fungus is slowed and stopped owing to rapid plant accumulation of phytoalexin (perhaps coupled with a greater sensitivity to the substance on the part of the avirulent fungus).

Controlling Initial Infection

Measures aimed at controlling the initiation of infection are largely preventative in nature, and many of them are also useful in controlling the spread of infectious organisms. The primary goal, however, is to prevent the pathogen propagules from infecting plants. In temperate regions such control measures are frequently directed at interrupting the pathogen life cycle at the survival spore or overwintering stage.

CHEMICAL MEANS

For diseases such as peach leaf curl (Figure 26-23), in which the spores germinate in the spring and infect young leaves, so-called dormant sprays are used. Oily suspensions of fungicides are applied to twigs and buds just before they open, to kill the germinating spores.

Soil sterilization is sometimes used to eliminate pathogens surviving there. Steam sterilization, used in greenhouses, is impractical for large volumes of soil, but outdoor soils can be fumigated (treated with volatile chemicals such as methyl bromide) at relatively great expense. Fungicides are sometimes applied to soils before sowing seeds to discourage seedling blights and damping-off.* Seed treatment is another measure designed for the same purpose. The chemicals used for dusting or soaking the seeds diffuse into the soil region around them after they are planted, protecting the seeds from attacks by soil-dwelling

*In damping-off, seedlings are invaded at or near the soil surface by any of several soil-borne fungi. Infection causes disruption of the vascular connections between the root and shoot, and the young shoot falls over and dies.

Figure 26-24. Seed treatment of sorghum (*Sorghum vulgare*). (Left) Seed treated with a fungicide (captan); (right) untreated seeds are covered with fungal mycelium.

pathogens during germination (Figure 26-24). Bulbs, corms, tubers, and roots can be similarly treated with these compounds.

SANITATION

For fungi known to overwinter in fallen leaves or fruit (for example, the causative agents of apple scab and brown rot of peaches and apricots), removing the potential spring inoculum by clearing away the orchard debris is a preventive measure. Removing dead or infected elm wood to prevent overwintering of the Dutch elm disease fungus and its insect larval carrier is another sanitary measure based on the pathogen's life cycle.

CROP ROTATION

Some soil pathogens that are restricted to a particular host species and do not have long-lived spores can be eliminated by crop rotation. Potato scab, caused by an actinomycete, can be controlled by alternating potatoes with another crop such as soybeans, and the fungal agent causing take-all disease on cereal grains usually disappears if another type of crop is planted for only one year.

ALTERNATE HOST ERADICATION

An effective control measure for rusts that require two hosts is elimination of the alternate host. For example, in stem rust of wheat discussed earlier, the basidiospores formed from the overwintering teliospores cannot germinate on the economically important host, wheat, but can infect only barberry. Eradication of barberry was used as a preventative measure even before the causative agent of wheat rust was identified in 1865. Farmers had long observed that there was more wheat rust in the vicinity of barberry bushes than else-

where. In 1755 the colony of Massachusetts enacted a law requiring property owners to destroy all barberry bushes growing on their lands; this occurred about 100 years after a similar French law for barberry eradication was instituted. Alternate host eradication effectively eliminates early season infections in much of the United States; however, the disease has not been eradicated by this control attempt. The uredospores that cannot overwinter in northerly regions survive and grow on winter wheat in southern Texas and Mexico. This provides a source of asexual spores that are blown northward in late spring and re-establish the disease each year. Weather conditions determine the length of the growth cycle between successive uredospore waves and the distance travelled by the northward migration of wind-dispersed spores. Figure 26-25 shows a time sequence required for the northward spread.

Plant breeding for early maturation has resulted in the greatest gain in wheat yield in recent years. Varieties that develop ripe grain 10 days earlier than formerly planted varieties allow harvest before the disease becomes epidemic.

INSPECTION AND QUARANTINE

Inspection and quarantine procedures are aimed at avoiding the introduction of a disease from one geographic region into another, usually virgin territory. The devastating results in the United States following the introduction of the pathogens causing Dutch elm disease and chestnut blight were cited earlier. Every part of the world has its own examples. Dutch elm disease was widespread in England in the 1930s and then gradually declined, leaving some apparently resistant or recovered trees. Another outbreak occurred in the 1960s, caused by a more virulent fungal strain introduced in elm timber imported from Canada for boat building.

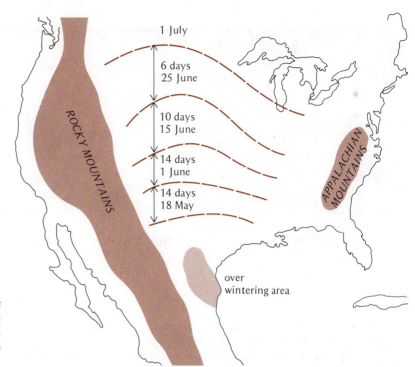

Figure 26-25. The increasing rate of spread of wheat stem rust northward in the 1923 epidemic can be noted from the dates of first appearance up the Great Plains from Texas and Mexico.

The downy mildew fungus, an oomycete that infects and kills grape leaf and fruit tissue (Figure 26-1), is native to North America but was introduced into France, probably in the rootstock of American grapevines used for grafting European wine grapes. This grafting practice, which is still used in the vineyards of Europe and California, had become a necessity because the roots of the European species (*Vitis vinifera*) are highly susceptible to another unwelcome American export, a root aphid. A native American grapevine (*V. riparia*) is relatively resistant both to the aphid and to the downy mildew fungus (a good example of natural selection through evolutionary time), but does not produce grapes suitable for making wine. *V. vinifera* was extremely susceptible to the introduced fungus, which quickly caused great devastation in the French vineyards.

Most countries and some states have laws that attempt to regulate the entry of pathogens and pests on imported products or merchandise. Currently over 500 plant diseases not found in the United States are considered significant and a potential threat. With increased international travel and the movement of millions of carriers between countries and continents, the probability of permanent isolation of potential host plants from foreign pathogens is very low, even with continued vigilance on the part of the United States Animal and Plant Inspection Service. This makes the development of other effective control measures even more imperative.

Controlling Pathogen Spread

USE OF CHEMICALS

Application of chemicals lethal or inhibitory to pathogens is a commonly used control practice

(Figure 26-26). Since most of these chemicals are not absorbed and moved through the plant and since most plant pathogens grow internally, the effectiveness of applied toxic compounds lies largely in preventing the spread of fungal or bacterial propagules to new individuals. These fungicides and bacteriocides are usually applied to aerial plant parts and must cover the plant surface and remain (or be reapplied) through rain, wind, and other external forces, to be effective.

It was in connection with downy mildew of grapes that the Bordeaux mixture, one of the first fungicides used in the control of plant disease, was formulated in France in 1882. Professor Millardet, a mycologist at the University of Bordeaux, noted in a local vineyard that the grapevines close to the road, which were covered with a blue substance, were much healthier than those further from the road, which were not so covered. Investigation revealed that the farmer had placed a combination of copper sulfate (which gave the blue coloration) and lime on the boundary vines to give the appearance of a poison that would discourage poachers from picking his grapes. This discovery led to the formulation and widespread usage of that mixture of compounds, which became known as the Bordeaux mixture. Today other fungicides that are more effective and less toxic to plants have largely replaced this spray or powder mixture.

The term **systemic fungicide** applies to several of the newer fungicides such as benomyl that are absorbed and translocated in the plant. These hold promise in treating established plant infections by means analogous to curing animal diseases with medicines or antibiotics. An additional advantage is that these compounds persist in the host for relatively long periods. Unfortunately, some pathogens have become resistant to the ac-

Figure 26-26. Chemical spraying on a commercial scale to prevent plant disease in a citrus orchard.

tion of these chemicals, in contrast to the older protectant type compounds, such as captan and the Bordeaux mixture.

VECTOR CONTROL

For pathogens transmitted by plant-feeding organisms such as insects or nematodes, control can be directed at the vector. This is particularly applicable to many virus- and mycoplasma-caused diseases. Chemical insecticides, while widely used, have resulted in the emergence of insects resistant to some of these. Biological control of vectors is generally more desirable, but relatively few pathogens of plant disease vectors have been investigated for large-scale production and usage. Commercial dusting preparations containing bacterial (*Bacillus thuringrensis*) spores are active against some plant pests, for example, caterpillar larvae of "cabbage moths." For most of these potential biocontrol organisms, however, there is a lack of needed information about interactions between them and competing organisms in the natural arena or the effects of external physical factors such as dry spells, high rainfall, frost, and so on, on their effectiveness as biocontrol agents.

Recent attempts to control Dutch elm disease have focused on trapping and elimination of male bark beetles using **sexual pheromones,** chemical attractants secreted by females to attract mates. A number of these compounds, which are usually specific for a given species, have been identified chemically. The volatile insect pheromones are the best-studied kinds, but other organisms also produce sexual pheromones, including some algae and water-molds that synthesize water-soluble types. Synthetically produced insect pheromones hold promise as a potential control means for various pests, if the chemicals can be synthesized at low cost. Generally, they are nontoxic, specific, and effective at very low concentrations.

INDUCED RESISTANCE

Evidence in a few plant genera indicates that contact with certain fungi, bacteria, and particularly viruses induces a resistance response. That is, treatment of the plants with heat-killed or avirulent strains of a pathogen results in some resistance of these plants against virulent strains when compared with untreated control plants. For example, when tobacco plants are treated with avirulent strains or heat-killed cells of *Pseudomonas tobacci*, the plants are protected from subsequent exposure to virulent strains of this bacterium that ordinarily would cause the disease known as wildfire. (The symptoms include lesions on leaves and stems, with the leaves becoming distorted and ragged.) Another example occurs when seedlings of watermelon or cantaloupe plants are inoculated with avirulent strains of the fungus (*Colletotrichum lagenarium*) causing anthracnose. The plants later show fewer and smaller lesions when infected with virulent strains than do control plants without exposure to the avirulent strains.

The "immunizing" of plants by application of nonvirulent strains or, in the case of viruses, by inoculation with related, less pathogenic forms is known as **cross-protection.** Plants do not produce circulating antibodies in response to infectious diseases as do animals. The induced resistance response described has been compared with the production of interferon stimulated by viruses in animals. Whatever the mechanism may be, such induced resistance may have at least limited potential application in agriculture. Currently, tomato plants are protected against one of their viral pathogens by spraying with an inactive form of the virus.

INTERFERENCE

Many examples of biological antagonism among microorganisms are known and provide the basis of antibiotic selection, isolation, and usage. Recently, interest has centered on some saprotrophic fungi that protect plants by inhibiting growth of a pathogen (interference). Practical application of this antagonism has been made in pine forests by spraying or painting newly cut stumps with spore suspensions of *Phlebia gigantea*. This saprotrophic basidiomycete grows into the wood and prevents invasion by the bracket fungus *Heterobasidium annosum*, which could cause heart rot of adjacent living trees by growing into them via root grafts.

Some other fungi inhibitory to soil-borne pathogens are being investigated for possible large-scale utilization in preventing root infections of vegetable crops.

CULTURAL PRACTICES

In commercial and agricultural operations sanitation aids in eliminating the spread of pathogens. For example, failure to sterilize pruning implements can result in the spread of fire blight from infected to healthy trees in apple, pear, or peach orchards. Soil pathogens can be spread by other implements or by re-use of unsterilized pots in greenhouses. (Transfer of infected soil on tires and other parts of vehicles used in road construction has been blamed for the spread of *Phytophthora cinnamomi* to new regions in Australia.)

Although it is not usually possible to exert much control over the environment except in greenhouses, one can influence to some extent the spread of disease by cultural practices that alter the environment of individual plants. Proper density of plants may allow air movement and help avoid high localized humidity conditions that encourage many pathogens. Likewise, good drainage reduces infection by soil pathogens such as *Phytophthora* and *Phythium*.

PRESCRIPTION OF CONTROL MEASURES

We have discussed means of plant disease control on the basis of the infectious process, but these disease control measures can equally well be grouped by the nature of the treatments or agents, and they are summarized in this manner in Table 26-2.

A holistic approach to disease control is currently advocated by many plant pathologists. Disease forecasting is used to aid the decision-making process concerning which treatments and how

TABLE 26-2   Methods for Controlling Infectious Plant Diseases

A. *Biological Level*
   Plant breeding for resistance
   Immunization and cross-protection
   Interference
   Parasitism of pathogen[a]
   Heat treatment of infected plants[a]
   Meristem culture[a]
   Quarantine
B. *Cultural Level*
   Alternate host eradication
   Crop rotation
   Sanitation
C. *Chemical Level*
   Seed treatment
   Soil sterilization or treatment[a]
   Dormant sprays[a]
   Foliar/systemic sprays or powders[a]

[a]These items are, or can be, therapeutic measures for eliminating the presence of the pathogen, whereas the other items listed are primarily preventative.

much of each to use for some diseases. Accumulation of data related to disease epidemics over many years and detailed knowledge of pathogen life cycles have allowed computer analysis of various influencing factors to predict the rate of disease spread. The plant pathologist can use this information to determine effective concentrations of fungicides and when to apply them, for example, thus avoiding the expense and other disadvantages of possible overapplication. The further development of this type of forecasting, and continued research on biocontrol measures, plant resistance mechanisms and their genetic control, new treatments, and the improvement of more traditional methods of control will all be required to decrease the inevitable losses of plants by infectious diseases.

SUMMARY

(1) Plant pathogens include fungi, bacteria, viruses, mycoplasmas, and nematodes. Dispersal usually occurs by wind, splashing rain or soil water, animals (mainly insects), and humans. Pathogens may penetrate the plant directly (some fungi and bacteria), or they may enter through natural openings such as stomata, or through wounds.

(2) Diseases range in severity from those greatly disrupting growth and reproduction to latent infections, in which plants show few symptoms but production or crop yields may be diminished. Major factors influencing plant disease are resistance or susceptibility of the plant, virulence of the pathogen, and the environment.

(3) Natural resistance to diseases in a plant population may be only partial and highly variable among individuals and is termed horizontal. Many genes generally determine horizontal resistance. Plant breeding of cultivated species has aimed at increasing resistance to particular strains of pathogens, usually by altering one or a few host genes, yielding vertical resistance.

(4) Structural features, preformed inhibitors, phytoalexins, and immunity factors are among the mechanisms plants utilize to resist infection.

(5) Knowledge of a pathogen's life history may be essential for disease control. Various biological, chemical, and cultural means (Table 26-2) can be used to protect plants from infection or to control spread of the pathogen.

Chapter 27

BRYOPHYTES: NONVASCULAR LAND PLANTS

There is little doubt that the land plants descended from algae-like plants once flourishing in ancient seas, probably green algae (Subkingdom Chloronta, Chapter 23). The decendants of plants that migrated from the water to the land apparently evolved along two different lines. One group developed an efficient system of conducting and supporting cells, a vascular system composed of xylem and phloem. These became the vascular plants, or Tracheophyta, which will be considered in the following chapters of this book. In the other evolutionary line, no such specialized system developed. These plants, lacking an effective mechanism for translocating water, essential mineral elements, and food over any distance and without any specialized supporting elements, remained small and generally inconspicuous.

The plants that come from this second line are the liverworts, hornworts, and mosses, known collectively as the **bryophytes.** Although it is likely that these plants have undergone considerable change since plants invaded the barren land, they are probably not the ancestors of complex modern plants. They therefore constitute an independent line in evolution.

Although individual bryophytes are usually inconspicuous, they are noticeable and even attractive when growing in masses. In general, they are poorly adapted to conditions of terrestrial life and, for the most part, are plants of moist, shady environments. Certain bryophytes, however, grow in places subject to severe drought.

Bryophyte Reproduction

The life cycle of the bryophytes exhibits a pattern of alternation of generations that is found in many algae and is also characteristic of the vascular plants, including the flowering plants. A haploid gametophyte generation alternates with a diploid sporophyte generation. But unlike the condition in the vascular plants, the most prominent feature of the life cycle is the gametophyte. The sporophyte generation is less conspicuous, is always attached to the gametophyte, and is never truly independent, although it may carry on some photosynthesis.

Besides certain vegetative features showing greater specialization of internal structure and body form, the bryophytes differ from their algal progenitors and resemble the vascular plants in two important ways. (1) The sex organs, **archegonia** and **antheridia,** which contain the eggs and sperms, respectively, are multicellular and are covered by an outer layer of protective cells. These features, rare among the algae, also occur in the lower vascular plants but have been partially or completely eliminated in the seed plants. (2) After fertilization the developing embryo remains within the archegonium, and the tissues of the parent gametophyte nourish the young sporophyte. In bryophytes this dependence lasts for the life of the sporophyte, but in vascular plants the sporophyte becomes an independent or free-living plant.

Many bryophytes bear both kinds of sex organs on the same gametophyte. Others have the two kinds of sex organs (male and female) on different gametophytes. This difference is genetically determined. In some species a pair of sex chromosomes has been identified.

Because of similarities in life cycle and in general level of structural organization, most botanists consider the bryophytes a natural group, the Division Bryophyta. The three major groups (classes) within the Division, however, represent separate evolutionary lines, which have been distinct for many millions of years. The evolutionary relationships among the three groups are far from clear. These major groups, usually treated as classes, are the Hepaticae or liverworts, the Anthocerotae or hornworts, and the Musci or mosses.

Figure 27-A. Capsules of moss sporophytes. Characteristics of the capsule are important in the identification of different moss species.

Figure 27-1. *Marchantia,* a thallose liverwort.

THE LIVERWORTS (HEPATICAE)

Some 8000 liverwort species have been described. Most are plants of moist environments. They grow prostrate, or nearly so, on the ground, on tree bark, or on rotted wood and often occur on moist rocks and soil along woodland streams. The plant body of a liverwort exhibits several adaptations to a land habitat, such as a cuticle covering an epidermis and thick-walled spores adapted to dissemination by air.

The liverworts are commonly classified in two groups: the **thallose** liverworts (Figure 27-1) and the **leafy** liverworts (Figure 27-12). In both groups the plant body is **dorsiventral** in form; that is, it is differentiated into an upper, or dorsal, surface and a lower, or ventral, surface. The body, growing close to the soil, is attached to the soil by numerous thread-like rhizoids, comparable in function to root hairs.

Thallose Liverworts

The thallose liverworts are conspicuously lobed. Each time the thallus divides, it forms two more or less equal branches. Growth occurs through the activities of one or more apical cells located in the notch at the tip of each branch of the thallus. The fancied resemblance of the lobed thallus to the human liver is responsible for the names "Hepaticae" and "liverworts" applied by early botanists.

Some of the best known and most attractive of the thallose liverworts are members of the widely distributed genus *Marchantia* (Figure 27-2). *Mar-*

chantia polymorpha is a plant of damp ravines and other moist, shady situations. It invades areas in which other vegetation has recently been destroyed by fire, provided sufficient moisture is present. Under such conditions the plants may

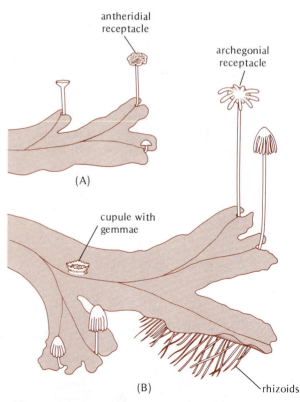

Figure 27-2. *Marchantia polymorpha.* (A) Male plant; (B) female plant.

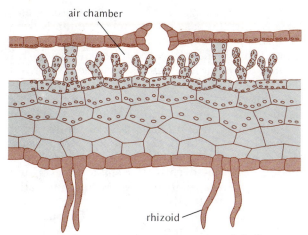

Figure 27-3. *Marchantia*, cross section of thallus.

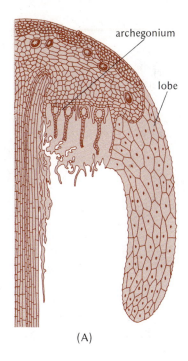

(A)

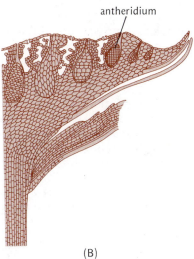

(B)

Figure 27-4. *Marchantia.* (A) Archegonial receptacle cut longitudinally to show position of archegonia. (B) Similar section of an antheridial receptacle.

flourish as a dense matted growth for several years, gradually being replaced by mosses, grasses, and the seedlings of woody plants. The thallus of *Marchantia* and certain closely related liverworts forms a broad, branching ribbon that spreads over the ground, anchored by numerous rhizoids (Figure 27-2B).

The surface of the thallus is marked by diamond-shaped plates, which indicate the position of internal air chambers. A section through the thallus shows these air chambers occupying the upper portion of the thallus and covered by an epidermis (Figure 27-3). Each chamber opens to the outer air by a chimney-like pore, analogous to a stoma except that the opening possesses no guard cells. From the bottoms of the air chambers arise chains of cells, each cell containing numerous chloroplasts. The basal portion of the thallus is composed of colorless, closely packed cells that frequently contain starch grains.

REPRODUCTION

In *Marchantia* sexual reproduction involves separate male plants that bear antheridia and female plants that bear archegonia. Both types of sex organs are located on stalked **receptacles** that are raised above the thallus (Figure 27-2). The stalks bearing the receptacles are vertical branches of the thallus. The archegonial receptacle is expanded like an umbrella, with finger-like lobes around the margin, usually nine in number (Figure 27-4A). The egg-containing **archegonia** are borne in rows between these lobes. The antheridial receptacle is disk-like with scalloped edges, and the sperm-producing **antheridia,** sunken in chambers in the upper or dorsal surface, are connected by narrow canals to the surface (Figure 27-4B). The sex organs basically resemble in structure and function those of most lower vascular plants.

The flask-shaped archegonium contains the egg in its enlarged base (Figure 27-5A). Just above the egg, like a plug holding it in place, is the **ventral canal cell.** The neck of the archegonium is filled with a linear series of **neck canal cells.** The antheridium is an oval structure with a wall that is one cell layer in thickness (Figure 27-6A). This wall encloses a mass of minute cells that develop into sperms or **antherozoids.**

Fertilization takes place prior to the elongation of the stalks of the archegonial receptacles. The mature antheridia burst at the apex (Figure 27-6B) and the sperms escape, splashed by raindrops to the vicinity of archegonia on neighboring plants. They are also motile through the action of their two flagella. Simultaneously, the neck canal

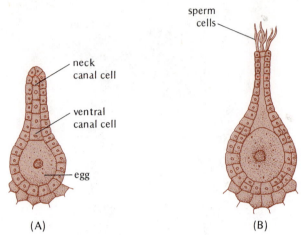

Figure 27-5. *Marchantia.* (A) Young archegonium; (B) archegonium at the time of fertilization.

cells and the ventral canal cell in the archegonium degenerate, forming a protoplasmic mass that extrudes from the archegonial tip (Figure 27-5B). The sperms become entangled in the mucilaginous extrusion and make their way down the neck. One sperm finally penetrates the egg.

SPOROPHYTE GENERATION

After fertilization, the stalks of the archegonial receptacles lengthen, and extensive growth takes place in the dorsal region, which inverts the portion that bears the archegonia. Thus, although originally dorsal and upright, the archegonia and resultant sporophytes eventually become suspended, pointing downward. The zygote formed by fertilization of the egg inside the archegonium

divides repeatedly, forming a multicellular embryo within the base of the archegonium. During the development of the embryo, a cylindrical sheath grows from the base of each archegonium and surrounds the embryo. A sheet of tissue also grows downward on either side of a row of archegonia.

The embryo is at first spherical, but soon a basal portion, the **foot,** grows into the tissues of the receptacle and functions as an absorbing organ (Figure 27-7). The bulk of the embryo forms a **capsule,** which is separated from the foot by a zone of undifferentiated cells, the **stalk.** The capsule contents differentiate into **spore mother cells,** grouped into vertically arranged rows or clusters, and into **elaters,** slender elongated threads with spirally thickened inner walls (Figure 27-8). Each spore mother cell then undergoes meiosis, forming four spores, the stalk elongates, the enlarged archegonium is ruptured, and the capsule is pushed downward (Figure 27-9). The capsule then dries and opens, setting free a loose, cottony mass of spores, which are disseminated by wind. The escape of the spores is aided by the elaters, which are hygroscopic. They coil and twist as they dry, undergoing jerking movements that loosen the spore mass. Spore germination completes the life cycle (Figure 27-10).

In its earlier stages of development, the sporophyte generation of *Marchantia* is entirely dependent nutritionally upon the tissues of the gametophyte. Later, however, the stalk, capsule wall, elaters, and even the foot of the sporophyte become green. The cells of these tissues are abundantly supplied with chloroplasts and possess a considerable capacity for photosynthesis. Food manufactured by the sporophyte thus supplements the supply received from the gametophyte, and so the sporophyte is not entirely dependent upon the gametophyte for food as it is for water and minerals.

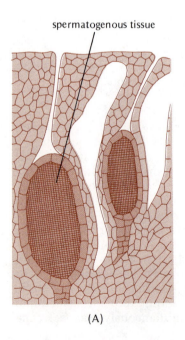

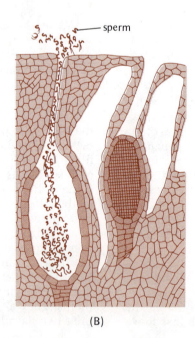

Figure 27-6. *Marchantia.* (A) Antheridium showing the cells which will develop into sperms (spermatogenous tissue) surrounded by a wall one cell layer in thickness. (B) Antheridium discharging sperms.

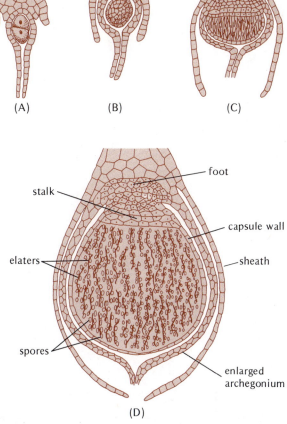

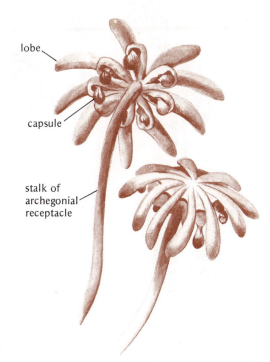

Figure 27-9. Archegonial receptacles with mature, protruding capsules.

Figure 27-7. Stages in the development of *Marchantia* sporophyte. (A) First division of zygote. (B) Many-celled spherical embryo. (C) Older embryo; the foot has begun to penetrate the tissues of the gametophyte, and stalk cells are differentiating. (D) Maturing sporophyte with spores and elaters.

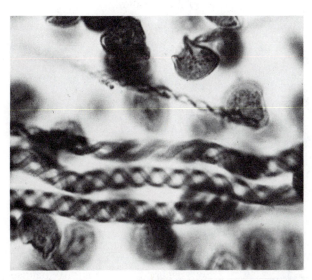

Figure 27-8. Elaters (with spirally thickened cell walls) and spores inside the capsule of a *Marchantia* sporophyte.

Other Thallose Liverworts

Many variations occur in the structural details of both gametophyte and sporophyte in thallose liverworts. The thallus is often much less complex in structure than that of *Marchantia*. Often the sex organs are borne directly on the upper surface of the thallus without any special receptacles. In some thallose liverworts such as *Pellia epiphylla* (Figure 27-11A), a species widely distributed on stream banks and in shady woods in North America, the thallus has a simple internal organization and lacks air chambers and pores. In a few species the thallus consists of a central midrib with lobed, wing-like margins, suggesting a possible relationship to the leafy liverworts. The sporophyte, however, does not necessarily correspond in simplicity, and that of *Pellia* may develop a stalk more than 10 cm long. On the other hand, in such genera as *Riccia* and *Ricciocarpus,* the sporophyte consists of no more than a capsule, and neither foot nor stalk is present (Figure 27-11B). As the spores ripen, the capsule breaks down, leaving them free in the cavity of the enlarged base of the archegonium. They are finally released by the decay of the thallus, and they germinate the following season.

Leafy Liverworts

The leafy liverworts are the largest group of the liverworts. In moist climates, especially the humid tropics, they grow profusely as mats or carpets on rotten logs, on damp soil, and as epiphytes on the trunks and branches of trees. The plant body is typically dorsiventral and is composed of an axis bearing leaf-like expansions (Figure 27-

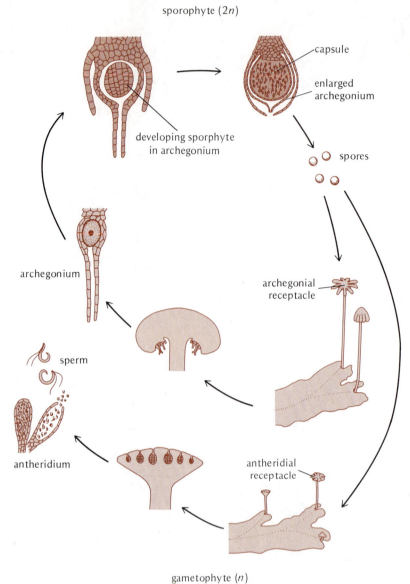

sporophyte (2n)

capsule

enlarged archegonium

developing sporphyte in archegonium

spores

archegonium

archegonial receptacle

sperm

antheridium

antheridial receptacle

gametophyte (n)

Figure 27-10. Life cycle of *Marchantia*. Spores begin the haploid gametophyte generation, which includes all stages up to the fusion of sperm and egg in the archegonium. The diploid sporophyte generation begins with the zygote and ends with meiosis in the spore mother cells in the capsule.

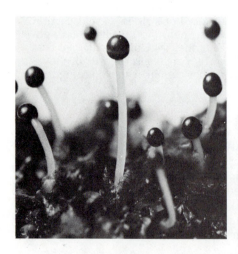

Figure 27-11. Various thallose liverworts. (A) *Pellia,* thallus with sporophytes; (B) *Ricciocarpus,* thallus.

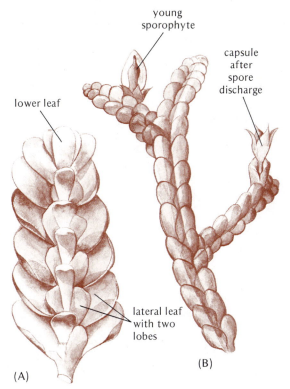

Figure 27-12. A leafy liverwort (*Porella*). (A) Lower surface of axis bearing leaves. (B) Upper surface of plant with attached sporophytes.

12A). There is little or no internal differentiation of tissues, and stomata are absent. The "leaves" are borne in three ranks. Two of these are lateral, extending outward on either side of the axis. The lateral leaves are sometimes divided into two parts. The third rank of leaves, absent in some genera, arises from the lower surface. The leafy liverworts are sometimes confused with the true mosses, but they can be distinguished by anyone who observes their vegetative structures carefully. The mosses are **radially symmetrical;** that is, the leaves are attached all around the stem, in contrast to the liverwort pattern just described. In addition, the moss leaf (with some exceptions, such as the peat mosses and certain others) possesses a midrib, which is absent in the leafy lobes of the liverworts.

The sex organs of the leafy liverworts are borne upon the leafy gametophyte plant. The antheridia are borne in the axils of the leaves, whereas the archegonia are terminal in position, at the apex of the main shoot or of its branches. The sporophyte possesses a foot, a stalk, and a capsule, which opens by four valves (Figure 27-12B).

The young sporophyte is usually enclosed within a protective sheath, from which it emerges as it nears maturity. Investigations have shown that the immature sporophytes of certain leafy liverworts are green and

contain starch grains. Although they are not nutritionally independent of the gametophyte, they probably do produce a considerable portion of their own food supply.

Vegetative Reproduction

In addition to sexual reproduction, many species of liverworts reproduce vegetatively. In *Marchantia* and other thallose liverworts, the older parts gradually die away as the plant grows over the soil. After the death of the thallus at the base of a fork, the two branches become separate plants. Several liverworts, including *Marchantia,* also produce specialized structures, known as **gemmae,** which bring about vegetative reproduction. The gemmae are borne in cup-like structures (**cupules**) on the upper surface of the thallus (Figure 27-2). In *Marchantia* the cupules are bowl-shaped, and the gemmae are minute, lens-shaped bodies attached by a short stalk to the bottom of the cupule. Gemmae are washed free of the thallus by rain water and may be carried some distance from the parent plant. When the gemmae lie flat on the soil, rhizoids develop from the lower surface, and a new thallus is produced. Some leafy liverworts also produce gemmae or deciduous propagative leaves that detach and give rise to new plants.

HORNWORTS (ANTHOCEROTAE)

The hornworts are a small group related to other bryophytes but sufficiently different to justify placing them in a separate class. *Anthoceros* is the best-known genus, and its species are relatively common on the edges of streams or lakes or frequently along damp ditches, paths, and roadsides. The dark green flat gametophyte is small, lobed, and approximately circular in outline, with little internal differentiation of tissues. The cells of the gametophyte usually contain only a single large chloroplast, which includes a pyrenoid, similar in structure and probably in function to the pyrenoids of many algae (Chapter 23). In most species archegonia and antheridia are borne on the same thallus, deeply embedded in its tissues.

The sporophyte that develops from the fertilized egg is a cylindrical, spike-like capsule, slightly tapering toward the apex (Figure 27-13A). It is usually about a centimeter in length, but in one California species it attains a length of 5–6 cm (over 2 inches). The base of the capsule is surrounded by a sheath of gametophyte tissue. The capsule base extends downward as a foot, an organ of attachment and absorption, embedded deeply in the tissue of the gametophyte (Figure 27-13B).

Cells of the sporophyte contain chlorophyll, making it largely independent of the gametophyte for food although still dependent on it for water and minerals. Stomata similar to those in vascular plants occur in the cutinized epidermis. Intriguingly, the gametophytes of several species of *Anthoceros* and other hornworts have

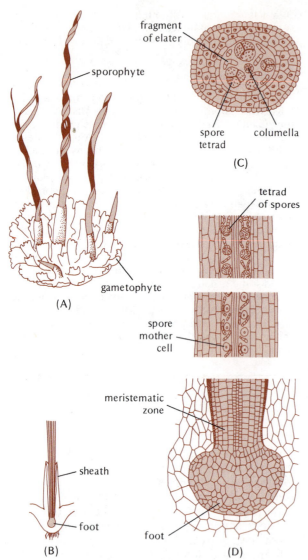

Figure 27-13. *Anthoceros,* a hornwort. (A) Sporophytes and gametophytes. As the capsule ripens from the apex down, it splits into two halves which often twist around each other. (B) Longitudinal section through the base of a sporophyte showing the foot embedded in gametophyte tissue. (C) Cross section of mature region of a capsule. (D) Longitudinal section of a sporophyte at three levels along its length: (top) mature with spores ready to be shed; (center) immature with spore mother cells; (bottom) meristematic zone just above foot.

structures resembling the stomata of the sporophyte. These have been interpreted as nonfunctional stomata homologous with those of the sporophyte. Nowhere else among the land plants do stomata, even nonfunctional stomata, occur in the gametophyte generation.

The structure of the capsule of *Anthoceros* resembles, in certain aspects, the capsule of moss plants, a condition that may be viewed as an example of convergent evolution. A cross section of a mature capsule (Figure 27-13C) reveals a small group of sterile cells, the **columella,** in the center. Surrounding the columella is a hollow cylinder containing elaters and tetrads of spores. The columella and cylinder of spore-pro-

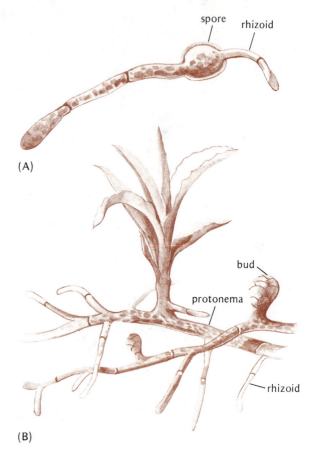

Figure 27-14. Moss gametophyte. (A) Spore germinating and forming a protonema. (B) Protonema bearing rhizoids, buds, and a young, erect leafy stem.

ducing tissue extend vertically through the length of the capsule. Lying outside these is a zone of sterile cells, and this zone in turn is covered by a well-developed and cutinized epidermis, interrupted by stomata. Each of the cells of the sterile zone contains one or several chloroplasts, the number varying with the species. At maturity, the capsule wall splits and the spores are set free.

After an early period of growth the capsule elongates by the activity of a meristematic zone at its base—a developmental feature unique in the bryophytes. This zone gives rise to the cells of all kinds found in the mature capsule—sterile as well as spore-producing tissue (Figure 27-13D). Thus, while spores ripen and are shed from the upper part of the capsule, new spores are continually produced below. In some species the capsule continues to grow and produce new spores as long as the gametophyte lives.

MOSSES (MUSCI)

The mosses are noticed more readily than the liverworts because they grow in more accessible places and usually are more conspicuous. As noted previously, the mosses are radially symmetrical, that is, their leaves arise from all sides of a central axis. In many species the stems are erect. Others have prostrate stems that give rise to erect branches

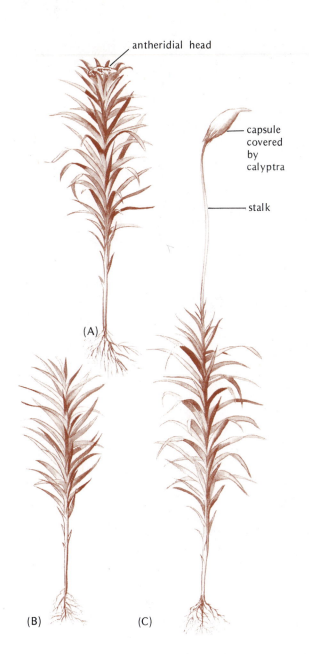

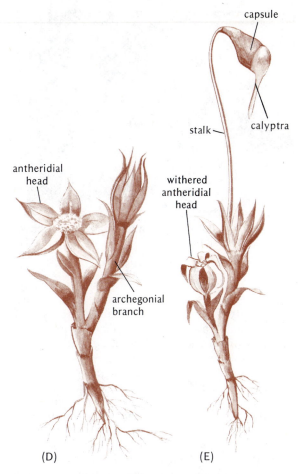

Figure 27-15. (A–C) Male and female plants of pigeon-wheat moss (*Polytrichum*). (A) Male plant with cluster of modified leaves around tip where antheridia are borne. (B) Female plant. (C) Female plant with mature sporophyte after fertilization. (D–E) The cord moss, *Funaria hygrometrica*. (D) Antheridia and archegonia are produced on different branches of the same plant. (E) The mature sporophyte develops on an archegonial branch after fertilization.

and in other species the plant is entirely prostrate and creeping.

Although the lateral appendages of the moss gametophyte are called leaves, they are not homologous with the leaves of the sporophytes of vascular plants. They are more leaf-like than the lobed thallus of the leafy liverworts, and in the true mosses they usually possess a definite midrib.

The mosses include some 14,000 species, most of which belong to one order, *Bryales,* the true mosses. These are the most common mosses. The sphagnum, bog, or peat mosses, *Sphagnales,* form a second order, and the rock mosses, *Andreaeales,* a third. The rock mosses are arctic and alpine plants, dark in color and small in size. The true mosses and the bog mosses will be described in more detail.

Many small plants growing on soil, rocks, or trees or even in the water are popularly known as "mosses."

Such plants, however, technically are often lichens, algae, or even flowering plants. Even plants bearing the common name "moss" may not be mosses. "Sea moss" belongs to the red algae. "Iceland moss" and "reindeer moss" are both lichens, while "Spanish moss" is an angiosperm. The "club mosses" are neither mosses nor related to mosses.

The True Mosses

A moss spore germinates to form a **protonema,** a much-branched, thread-like structure divided into cells by cross walls (Figure 27-14). Branches of the protonema spread over the soil surface. The cells of these branches contain chloroplasts. Non-green rhizoids growing from the protonema penetrate the soil. The form of the protonema resembles that of certain green algae, and may be a holdover from an ancestral algal type.

Figure 27-16. *Polytrichum* moss, male plant, showing heads with antheridia.

The leafy moss plants originate on the protonema as pear-shaped buds. These buds, through the activity of a single apical cell, give rise to the leafy stems. The erect stems develop their own rhizoids in turn and become independent of the protonema, which soon disappears. (In a few mosses, such as *Buxbaumia*, the protonemal stage is persistent and long-lived; the leafy stems are greatly reduced in size.) The leafy moss plants are usually only a few centimeters high, but they vary in size from minute forms that are just visible to the naked eye to aquatic and tropical species 30 cm (1 ft) or more in height. The leaves of the mosses are green and sometimes possess specialized photosynthetic tissues. But in most species they are only one cell in thickness, except along the midrib.

REPRODUCTION

Sex organs are produced at the tips of the leafy stems. In some mosses both archegonia and antheridia are borne on the same plant and sometimes on the same apex; in others, such as *Polytrichum* (Figure 27-15), the haircap or pigeon wheat mosses, they are borne on different plants.

The leaves surrounding the sex organs are frequently modified. This is particularly true of the terminal leaves on stems or branches bearing antheridia. Such leaves are commonly short and may be pale pink or rose. They form a cluster or rosette, superficially resembling a miniature flower (Figures 27-15, 27-16). The antheridia, usually stalked and somewhat club-shaped, consist chiefly of a jacket of cells with one or several cap cells at the apex surrounding a mass of sperm-producing tissue (Figure 27-17). The archegonia

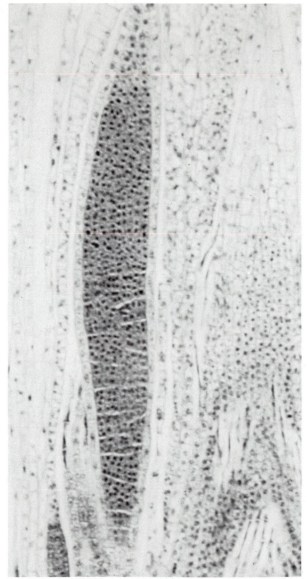

(A)

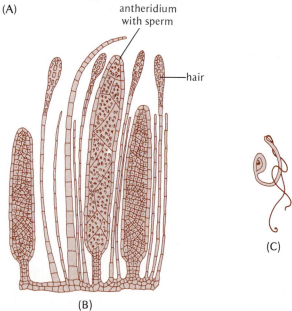

antheridium with sperm

hair

(B)

(C)

Figure 27-17. (A) Longitudinal section of moss male shoot tip, showing antheridium. Shoot apex is at lower right. (B) Longitudinal section of an antheridial head from *Polytrichum*, showing several antheridia with hairs scattered among them. (C) A single sperm.

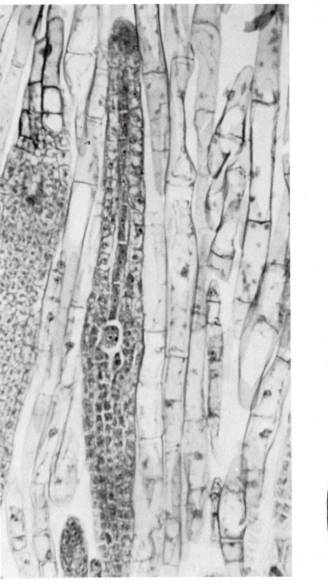

Figure 27-18. (A) Longitudinal section of moss female shoot tip, showing archegonium. (B) Archegonium of a moss (*Polytrichum*). The neck canal cells have broken down in preparation for fertilization.

(A)   (B)

are also stalked and greatly elongated (Figure 27-18). The wall at the base is several cells thick, and the neck contains a considerable number of neck canal cells, which disintegrate as the archegonium matures. Scattered among the sex organs are modified hair-like structures, which contain chloroplasts. These hairs may aid in the retention of water around the archegonia and antheridia.

Preparatory to fertilization, the antheridium absorbs water and swells. The cap cells are forced off, and the sperms emerge (Figure 27-17C). Water is essential for fertilization. In the larger mosses, the sperms are splashed by raindrops from antheridial to archegonial heads. In smaller species, the sperms may spread from the antheridia to the archegonia through a film of water that often cov-

ers the tops of the plants. The biflagellate sperms swim to the apex of the archegonium and down the narrow passage that leads to the egg. One sperm fuses with the egg cell, fertilizing it. Fertilization may occur in more than one archegonium, but usually only one sporophyte develops.

SPOROPHYTE GENERATION

The fertilized egg develops rapidly into an elongated, spindle-shaped embryo (Figure 27-19A). From the embryo grows an elongated stalk, or **seta,** with a spore-producing capsule at the upper end. At the lower end of the stalk is the foot, buried in the tissues of the leafy gametophyte plant (Figure 29-19B, C).

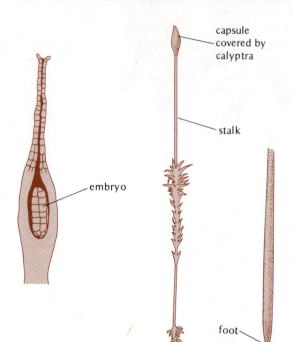

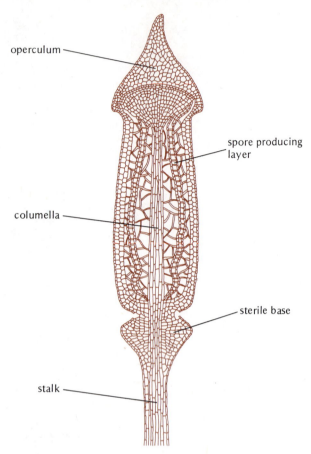

Figure 27-19. Development of the moss sporophyte (*Polytrichum*). (A) A young embryo inside an archegonium. (B) Sporophyte with stalk and capsule attached to gametophyte. (C) Lower part of stalk enlarged, showing foot.

Figure 27-20. Longitudinal section of an immature moss capsule (*Polytrichum*).

The spores are produced within the capsule in a cylindrical zone that surrounds a column of sterile tissue, the columella (Figure 27-20). In some species, the spore-producing cylinder extends the length of the capsule. In others, spore production is limited to the upper portion of the capsule. The cells of the spore-bearing area develop into spore mother cells, which, after meiosis, give rise to spores.

The foot, stalk, and capsule are produced by two apical cells, one located at each end of the spindle-shaped embryo. The capsule, when mature, may be erect, pendulous, or bent at right angles to the stalk. Characteristics of the capsule are important in the identification of the various species of mosses. As the embryo increases in size, the lower part of the archegonium enlarges and elongates and is converted into a **calyptra,** or cap. As the sporophyte lengthens, the calyptra is ruptured at the base and is borne upward, remaining for a period of time (which varies with the species) as a thin brownish or greenish cap or hood covering the capsule (Figure 27-15).

The embryo develops chloroplasts while very young and still enclosed within the archegonium. In many genera photosynthesis is facilitated by the presence of stomata in the epidermal tissues of the capsule. When the sporophyte is fully ripe, the chlorophyll fades and the capsule turns a pale or rich brown. The photosynthetic capacity of the sporophyte probably permits the stalk and capsule to be largely independent of the gametophyte generation for food materials, but the spo-

rophyte is still dependent upon the leafy plant for some food and for water and essential mineral elements.

As the capsule matures (Figure 27-21) and the spores within it ripen, a lid, or **operculum,** be-

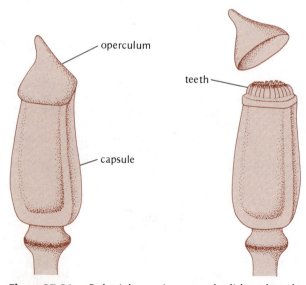

Figure 27-21. *Polytrichum,* ripe capsule, lid, and teeth.

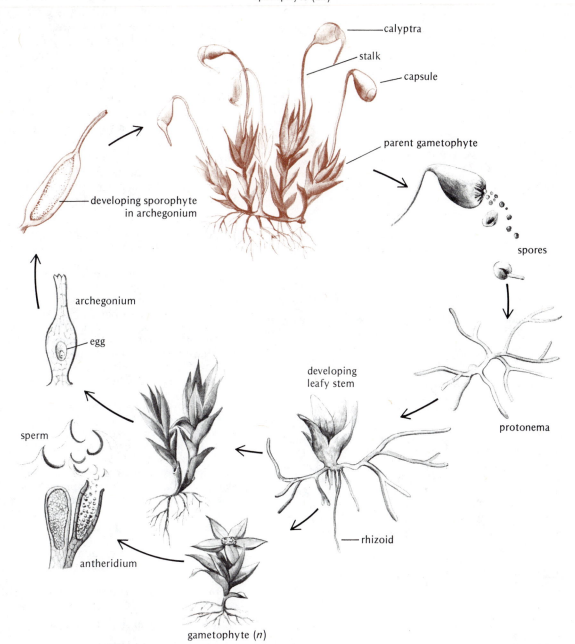

Figure 27-22. Life cycle of a moss. The gametophyte generation begins with the formation of spores and ends with the fusion of egg and sperm in the archegonium. The zygote initiates the sporophyte generation, which ends when meiosis occurs in the spore mother cells.

comes evident at its apex. Just within the mouth of the capsule and under the operculum, one or several rows of teeth also develop. After the capsule dries, the operculum falls away, and the spores are released and are carried away by currents of air. The release of the spores is facilitated by the circles of teeth around the mouth of the capsule.

These teeth are hygroscopic and control the release of the spores. They bend outward when the air is dry, and the spores are released by wind shaking the capsule. When the air is humid, the teeth return to their former position, preventing the escape of the spores. The complete life cycle of a moss is shown in Figure 27–22.

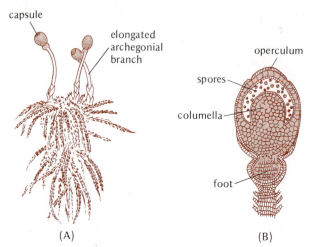

Figure 27-23. *Sphagnum.* (A) Leafy plant bearing capsules on elongated archegonial branches. (B) Longitudinal section of mature capsule.

Conducting Tissues in the Mosses. The presence of vascular tissue, xylem and phloem, is one of the important features distinguishing the tracheophytes from the bryophytes. Nevertheless, some mosses have a conducting system in both the leafy gametophyte and the seta of the sporophyte. Although this system is not composed of xylem and phloem, its cells do show some similarities to those of true vascular tissue. It is best developed in some of the larger mosses such as *Polytrichum*, the hairy cap moss, or the tropical *Dawsonia*, which may be more than 30 cm high. Often in the gametophyte axis there is a central strand of elongated, dead, empty cells known as **hydroids,** which resemble tracheids but lack a secondary lignified wall. Around this there may be another zone of living elongated cells known as **leptoids,** which appear to have an altered cytoplasm but contain a nucleus. This zone of leptoids, however, is often lacking even when hydroids are present in the center. A similar organization may also occur in the seta of the sporophyte, and conducting tissues in some cases are present in the sporophyte even though absent from the gametophyte. Mosses vary widely with regard to the presence of these tissues.

Their shape and cytological characteristics, as well as direct experimental evidence, indicate that the hydroids and leptoids are conducting cells. The hydroid column functions as a preferred pathway of water movement among the tissues of the gametophyte and sporophyte. Similarly, labeled organic compounds move preferentially in the leptoids. However, any similarity to true vascular tissue is usually regarded as an example of convergent evolution not indicating a close relationship to the tracheophytes. The moss conducting system is as characteristic of the gametophyte as of the sporophyte; in contrast, in the tracheophytes vascular tissue occurs only in the sporophyte.

The Peat, or Bog, Mosses

These mosses are represented by a single genus, *Sphagnum*, which contains a large number of species (about 350). Peat mosses are characteristically plants of wet, boggy areas with an acid pH. In such places the plants grow in thick, soft masses, sometimes covering great areas. The stems branch extensively as they grow (Figure 27-23A). The older parts die and partly decay, and the terminal parts grow on from year to year. *Sphagnum* is worldwide in distribution, but it is most abundant in the colder temperate zones where wet soils occur together with high relative humidity and comparatively low temperatures. *Sphagnum* is usually an important plant of moor, bog, and tundra. Either alone or combined with the partly decayed remains of small vascular plants, *Sphagnum* forms the organic material known as peat, which may accumulate in beds of considerable thickness.

Sphagnum growth creates extremely acidic conditions and also releases a phenolic compound, sphagnol, which has antiseptic properties. These conditions favor the preservation of other plant materials in bogs, especially pollen grains; this property has been most valuable in reconstructing vegetational changes in many parts of the northern hemisphere since the last glacial retreat. Remarkably, several hundred well-preserved bodies of Iron Age people, some more than 2000 years old, have been recovered from peat bogs in northern Europe. These bodies are so well preserved that some were originally thought to be recent victims of accident or foul play.

Sphagnum or peat moss has a sponge-like capacity to hold water, up to 20 times its dry weight, which is exploited in a variety of ways. When dried and sterilized, the moss has been used as a substitute for gauze in absorbent surgical dressings, in which its antiseptic qualities were also of value. It is added to heavy or sandy soils as a mulch to improve their texture. It is often used as a medium for starting seeds and rooting cuttings. Great quantities of peat moss are used by nurserymen as packing, in transplanting, and in other operations with plants. In many parts of Europe peat is a valuable fuel both for domestic heating and in industry, and modern mechanical methods of extracting it from bogs are replacing the laborious handcutting techniques used for centuries. Its importance will probably increase as world energy supplies dwindle.

Sphagnum Morphology and Reproduction

The leaves of *Sphagnum* (Figure 27-24) are minute and crowded. They are one cell thick and lack a midrib. The cells of the leaf are of two types, large and small, which alternate with each other, forming a net-like pattern. The smaller cells contain chloroplasts, as do the larger cells when young. But as the leaf matures, the chlorophyll and protoplasm disappear from the larger cells. The cell walls develop spiral or ring-like thickenings, and circular openings appear on either the upper or lower surface. Because of their structure, these large empty cells are able to absorb large quantities of water by capillarity.

The life cycle of *Sphagnum* is like that of the true mosses, with some variations in detail. The spore, upon germination, produces a protonema that is broad and flat rather than thread-like. The leafy plants arise from this protonema and are without rhizoids. The sex organs are borne on side branches just below the apex of the main axis. From the fertilized egg a globular capsule (Figure 27-23) develops, attached by a foot to the tissues of the gametophyte. When young, the capsule contains chloroplasts and is only partly dependent upon the gametophyte for food. As maturity approaches, the chlorophyll disappears, and the capsule turns from green to brown or black. A stalk is present but does not elongate. When the capsule is mature, the upper portion of the archegonial branch elongates and carries the capsule upward. Internally, the capsule consists of a dome-shaped mass of spores that overlies a massive columella. The capsule, unlike that of the true mosses, has no teeth to aid in spore dissemination, but the operculum is discharged explosively, and spores are ejected for distances up to 10 cm.

Vegetative Reproduction

Many mosses, perhaps all, can spread by vegetative reproduction as well as by spores. By this method individual plants can form dense masses extending over considerable areas.

Mosses have three general methods of vegetative reproduction:

1. Branching of the leafy stems. These may branch by the formation and development of buds at the base of the stem. Ultimately, these branches may be detached by the decay of the basal parts, becoming independent plants. The branches may also take the form of stolons, leafless or with only small leaves, which creep on or beneath the soil and grow upward as leafy stems.

2. Formation of detachable propagules. These may be deciduous branches, more or less modified as bulbils, or they may be gemmae. Gemmae, however, are less common than they are among the liverworts.

3. Protonemata. The original protonema may continue to grow for some time, producing several to many leafy stems. More important is the production of secondary protonemata, which, like the original, give rise to leafy plants. The secondary protonemata are produced most abundantly as branches of the rhizoids at the base of the leafy stems, but they may also grow directly from the stems and even from leaves.

EVOLUTION AND RELATIONSHIPS OF THE BRYOPHYTES

The fossil record of the bryophytes reveals little about their evolutionary history except that they are an ancient group. Their oldest definitely identifiable fossils are from the Upper Devonian. These appear to be thallose liverworts that are much like present day forms. The earliest undoubted moss fossils are of Permian age, but there are probable specimens from the late Carboniferous Period.

The bryophytes, like the vascular plants, are believed to have arisen from green algal ancestors, which probably had a life cycle with two independent or free-living generations that were similar in form, that is, an isomorphic alternation of generations. During the transition to terrestrial life, both generations may have undergone some elaboration in size and complexity along with modifications of the reproductive system,

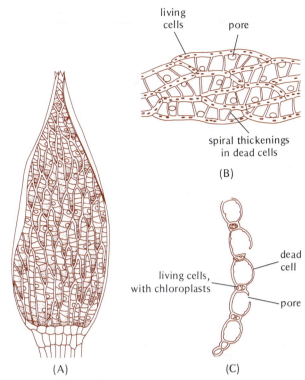

Figure 27-24. The leaf of *Sphagnum*. (A) Entire leaf, surface view. (B) Portion of (A) much enlarged. (C) Cross section of portion of leaf.

including the development of antheridia and archegonia, the retention and nurture of the embryo after fertilization, and the formation of cutinized spores. But very early in the emerging tracheophytes the emphasis was on the sporophyte generation, which became the typical vascular plant with a conducting system of xylem and phloem. In the developing bryophytes, on the other hand, the gametophyte became the conspicuous phase of the life cycle, and the sporophyte was reduced to dependence on it.

The photosynthetic capacity of the sporophyte generation of mosses, liverworts, and hornworts and the presence of stomata in the capsule of mosses and *Anthoceros* are thus interpreted as ancestral features, retentions from the time when the sporophyte was an independent plant. The sporophyte of *Anthoceros*, because of its complex structure and its relatively long life, is regarded as the least reduced of any bryophyte. Stages in the reduction of the sporophyte among the liverworts range from the condition in the leafy forms, where a foot, stalk, and capsule are present, to that in *Ricciocarpus*, in which only a spherical capsule is found.

The gametophyte of the bryophytes has generally remained small and has not developed the kind of terrestrial adaptations characteristic of the conspicuous sporophytic generation of the tracheophytes. The reasons for this particular evolutionary development of the ancestral bryophytes are unclear. It can only be deduced that they became adapted to particular ecological niches with adequate moisture in which this strategem was successful while the vascular plants exploited more truly terrestrial habitats. Bryophytes represent the culmination of a line that became stranded

Figure 27-25. Gametophyte of *Takakia*, a leafy liverwort with radial symmetry which may indicate a relationship to the mosses.

Figure 27-26. Moss on a dead pine tree, Olympic National Park, Washington.

in the backwaters of the main evolutionary stream and gave rise to no higher forms of land plants.

Regarding the relationship among the bryophyte groups there do appear to be intermediate forms between the typical liverworts and the mosses. *Takakia* (Figure 27-25) is a leafy liverwort whose gametophyte possesses slender, erect branching axes with radial symmetry, bearing small leaf-like organs in groups of two or three. This might resemble an ancestral form from which both the leafy liverworts and the mosses could have evolved by different modifications. The existence of some thallose liverworts with lobed leaf-like wings suggests that these forms may have been derived by reduction from ancestral leafy liverworts.

EXPERIMENTAL STUDIES ON BRYOPHYTES

Bryophytes are useful for cytological, genetic, and physiological investigations because they are easily grown on artificial media, and their relatively uncomplicated structure offers advantages in the study of development and its control.

Certain bryophytes exhibit photoperiodism—in some the sexual reproductive structures are initiated under long days and in others under short days. A number of thallose liverworts, such as *Marchantia polymorpha* and *Conocephalum conicum*, and certain leafy liverworts, fall in the long-day category. Thus, it is possible to obtain reproductive material of *Marchantia polymorpha*, widely used for instructional purposes, during the winter months by adding artificial light to in-

crease the daylength. Species of *Riccia*, *Anthoceros*, and *Sphagnum* have been found to be short-day plants.

Light also influences development in a different way in several moss species. The protonemal stage of the gametophyte can grow successfully in darkness if adequate nutrition is provided. However, in the absence of light no upright, leafy shoots are formed, nor do they develop in blue light. Red light most effectively promotes the initiation of shoot buds. Hormones of the cytokinin type increase the formation of leafy shoot buds in the light and moreover, cause their initiation even in the dark or in blue light. Auxins, on the other hand, inhibit bud formation, and cytokinin counteracts auxin inhibition. At least one moss species has a hormone-mediated apical dominance in the leafy shoots of the gametophyte. Removing the apex promotes branching of the axis and formation of secondary protonemata, and the application of auxin to the decapitated shoot prevents these developments, that is, substitutes for the excised apex. Thus the relatively simple moss gametophyte apparently develops under hormonal controls resembling those in the sporophyte generation of vascular plants.

ECOLOGY OF THE BRYOPHYTES

As we have seen, the liverworts and mosses grow on soil, on damp sand, on rocks, and on the trunks and limbs of standing and prostrate trees (Figure 27-26). They also grow in water and among other plants of fields and meadows, bogs, and marshes. Some species flourish in a variety of habitats; others are restricted to specific, limited environments. Certain mosses, for example, have specific requirements with respect to soil pH and are confined either to acid or to alkaline soil. Ecologists use such species as indicators of particular environmental conditions.

In general, mosses are more tolerant of shade than are higher forms of plant life, and this in part accounts for their ability to invade lawns and replace grass in shady spots. The mosses include many xerophytic forms as well as mesophytes and hydrophytes. Drought-resistant bryophytes differ from most xerophytic vascular plants in tolerating extreme drying out rather than resisting water loss. They become inactive when dry, but quickly revive after a rain shower and use available moisture. Bryophytes are likewise adapted to great extremes of temperature, for they range from the arctic zone to the tropics and grow in the vicinity of hot springs. They reach their greatest development in cool, moist forests, such as those of the Pacific northwest and the mountains of the tropics.

Some moss and liverwort species apparently are extremely sensitive to atmospheric pollution by such toxic substances as sulfur dioxide. This sensitivity was first recognized because of their disappearance from expanding urban areas and from the vicinity of certain industrial operations and has been verified by direct experiment. In general, the most sensitive bryophytes are epiphytes. Along with lichens, these sensitive bryophytes are used as indicators in attempts to monitor atmospheric pollution.

The bryophytes play a significant role in the economy of nature. This is partly the result of prolific vegetative propagation, which forms great masses or carpets covering the soil. *Sphagnum* or peat mosses form the basis for whole bog plant communities. Mosses are also significant as soil formers following lichens or other lower forms of plant life on bare rock surfaces (Chapter 33). Retention of water by masses of leafy liverworts and mosses growing on fallen trees and other organic material hastens the processes of decomposition and hence the organic enrichment of the soil. Absorbing but little water from the substratum, they do not dry out the soil but protect it from desiccation. As a result of their ability to retain water, natural beds of mosses undoubtedly act as seed beds for herbaceous and woody flowering plants and for conifers.

Bryophytes also help retard erosion. Carpets or felt-like masses of moss plants slow down rapid runoff of rain water and melted snow. In addition, dense stands of moss hold particles of soil. Insignificant as the individual plants of this group may appear, they play a part, together with other and more advanced forms of plant life, in making and changing the environment.

SUMMARY

(1) The bryophytes include the liverworts, hornworts, and mosses. They are an ancient group of land plants that arose from green algal ancestors and evolved independently of the vascular plants.

(2) All bryophytes have a clear alternation of generations in which the haploid gametophyte is more prominent than the diploid sporophyte.

(3) The gametophytes of liverworts and hornworts are dorsiventral and may consist of a branched thallus or of a branching "leafy" axis. The gametophytes of mosses are radially symmetrical and consist of a leafy axis that arises from an initial, usually filamentous, protonema.

(4) The bryophytes are advanced in their methods of reproduction compared with the algae, but their motile sperms still require water in order to effect fertilization. Multicellular sex organs—archegonia and antheridia—are produced. After fertilization the developing embryo sporophyte within the archegonium is nourished by the gametophyte. In most bryophytes the sporophyte is photosynthetic but does not become totally independent. The sporophyte may be simple, consisting only of a capsule, or relatively large and complex with a foot, stalk, and capsule.

(5) In addition to sexual reproduction, the bryophytes reproduce vegetatively by several methods. Vegetative reproduction accounts in large part for the occurrence of liverworts and mosses in large masses.

(6) Bryophytes are useful experimental organisms because of their simple structure, ease of cultivation, and responsiveness to physical and chemical treatments.

(7) Bryophytes are important ecologically because they retain water, hasten decay processes, and retard erosion. Many bryophytes are sensitive to environmental conditions and are valuable as indicators of atmospheric pollution.

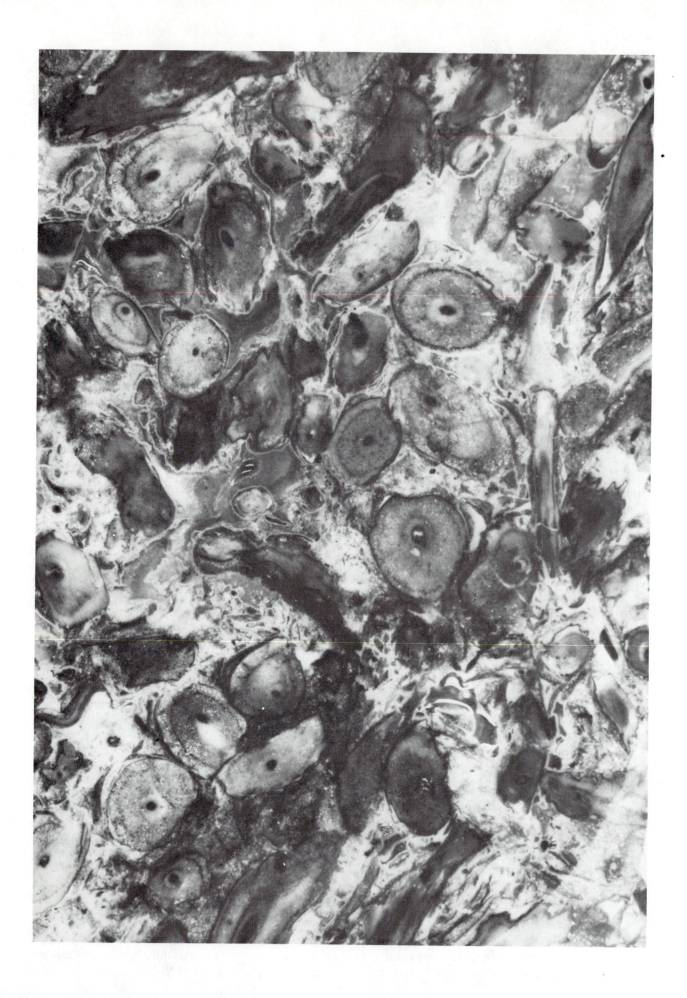

Chapter 28

EARLY VASCULAR PLANTS: EVOLUTION AND MODERN SURVIVORS

The migration of plants from the seas to the dry land surfaces of the earth more than 400 million years ago ranks among the most important events in the history of the biological world, because until plants were present to convert the sun's energy through photosynthesis, no organisms could permanently colonize the land. Since its beginnings more than 3 billion years ago, life had been confined to the watery environment in which it arose, and the fossil record reveals that by the Cambrian and Ordovician periods the oceans teemed with an ever-growing diversity of living forms. As indicated by a scanty fossil record, land plants first appeared in the latter part of the Silurian period and, by early Devonian time, were well established in many parts of the world. The fossil record of this remote age is particularly fascinating, because it contains all that we know, or can hope to learn, of the stages by which the once-barren land was first conquered, opening the way for the evolution of the earth's diverse terrestrial flora.

The transition from an aquatic to a terrestrial mode of life was undoubtedly gradual, and unfortunately the transitional forms between ancient algal ancestors and fully established land plants have not been preserved, or at least are not recognized, in the fossil record. Many changes in body form were necessary to provide protection against water loss, absorption of water and nutrients from the soil, transport of materials within the body, and mechanical support, before these new land plants could prosper. Reproduction also had to be adapted to limited water conditions. Even a cursory examination of almost any land surface demonstrates that the vascular plants, the Division Tracheophyta, adapted most successfully, becoming the overwhelmingly dominant land plants. The story of the land's conquest is largely theirs. The bryophytes, as the previous chapter showed, are also true land plants, but they have had only limited success in comparison with the vascular plants.

This chapter traces the origin and early diversification of vascular plants in the Silurian and Devonian periods as revealed by the fossil record. It also follows some of these types as they became dominant groups in the subsequent Carboniferous period or coal age and examines the few remaining modern descendants of these groups. The living survivors give valuable clues to features of these formerly dominant plant groups that cannot be ascertained from fossils.

THE FIRST VASCULAR PLANTS

The rocks of the Upper Silurian, and especially the early Devonian, yield an assemblage of simply constructed yet surprisingly diverse primitive vascular plants. Knowledge of these ancient plants has grown rapidly during the last two decades; many new forms have been described and other, previously known types have been reinvestigated and often reinterpreted. Clearly, this flora was widespread, and essentially similar forms have been found in many parts of the world. Because of their simple construction, these plants were long regarded as a closely related group, the first truly vascular plants from which all later forms evolved. Now, however, these early land plants are generally considered to fall naturally into several distinct groups of great significance in the interpretation of subsequent vascular plant evolution. A few examples illustrate their natures.

Rhynia and *Cooksonia*

A Lower Devonian fossil deposit near the village of Rhynie in northern Scotland contains some of the best-preserved fossil plants ever found. The excellently fixed cellular detail of the petrified

---

Figure 28-A. A polished piece of Rhynie Chert showing the stems of fossil plants in cross section.

547

sporangium

Figure 28-1. *Rhynia major* is the larger of the two species of *Rhynia* found in a Lower Devonian deposit at Rhynie. Note the absence of leaves and roots and the terminal sporangia.

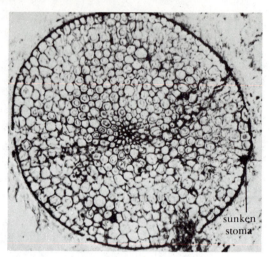

sunken stoma

Figure 28-2. Cross section, stem of *Rhynia*. A solid rod of xylem and phloem appears in the center.

elements of a vascular land plant were present, but in an extremely simple and primitive form.

An even smaller and simpler similar plant type is *Cooksonia*, found in both Europe and North America in beds of uppermost Silurian age (Figure 28-3). This plant was much more slender than *Rhynia*, and many of its dichotomously branching upright axes terminated in small, globose sporangia. Whether these axes arose from rhizomes is

material suggests that the growing plants were submerged by mineral-containing water (Figure 28-A). Two similar *Rhynia* species are among the common plants in this deposit, one about 50 cm (20 in) tall and the other perhaps 22 cm (9 in) (Figure 28-1). Both plant species consisted of an underground rhizome from which erect, leafless stems arose and had no roots, absorption apparently occurring through hair-like rhizoids borne directly on the rhizome. The upright axes branched **dichotomously,** that is, divided into two approximately equal parts at each branching. Some aerial branches were sterile or vegetative, while others terminated in elongated sporangia containing numerous spores with cutinized walls. The spores usually occurred in tetrads and, when separated, showed lines or ridges marking the planes of contact with other spores of the tetrad. The presence of tetrads demonstrates that meiosis occurred in the sporangia of *Rhynia* and that the plants belonged to the sporophyte generation. Stem cross sections show a distinct epidermis with cuticle and stomata and, inside this, a broad cortex of parenchyma, which was probably the plants' photosynthetic tissue (Figure 28-2). The vascular system was a slender solid rod of xylem tracheids surrounded by a cylinder of phloem. All essential

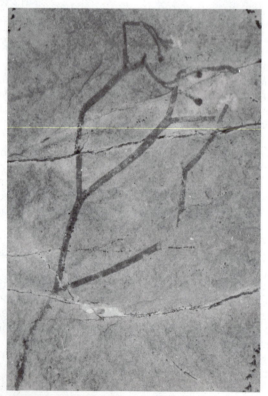

Figure 28-3. *Cooksonia hemispherica,* from the Upper Silurian of central Bohemia. The natural size of this fragment was about $2\frac{1}{4}$ inches.

sporangia

Figure 28-4. *Zosterophyllum* bore sporangia in a lateral position and collected in a terminal spike.

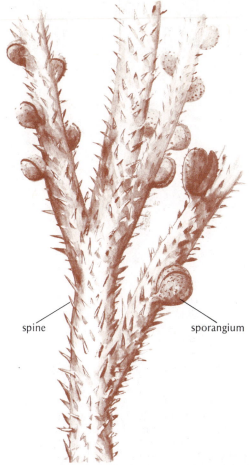

spine

sporangium

Figure 28-5. *Sawdonia* was another Lower Devonian plant which bore sporangia in a lateral position. It was also covered with spine-like emergences.

uncertain. At least two *Cooksonia* species have been recognized, and this plant also extended into the Devonian period. Present in very late Silurian rocks, *Cooksonia* is the oldest known vascular plant. A variety of cutinized spores found in early Silurian deposits, however, extend hope for finding even older vascular plants.

Zosterophyllum and Sawdonia

The Early and Middle Devonian fossil beds have revealed another similarly leafless and rootless, yet consistently different plant type represented by *Zosterophyllum*, a widespread plant with several described species. *Zosterophyllum* was a small, dichotomously branched plant with both horizontal and upright axes and no leaves or true roots (Figure 28-4). It bore sporangia on certain shoots, but, unlike those of the plants previously

described, its sporangia were located in a lateral position, that is, on the side of the axis rather than at the tip. In *Zosterophyllum* the sporangia were collected in a terminal spike, but in other probably related forms they were scattered along the stem. *Sawdonia*, a Lower Devonian fossil from the Gaspé region of Quebec, bore stalked lateral sporangia along the stem and was more or less covered with stout spines or emergences (Figure 28-5). These had no vascular supply and may be compared to rose prickles.

The Silurian and early Devonian land plants such as *Cooksonia* and *Rhynia*, which bore sporangia in a terminal position, that is, as the modified tips of ultimate branches, represent the Class **Rhyniopsida**. Those such as *Zosterophyllum* and *Sawdonia*, which produced sporangia located laterally on the aerial axes, represent the Class **Zosterophyllopsida**. These two groups are significant because they are believed to point the way to the major lines of later vascular plant evolution.

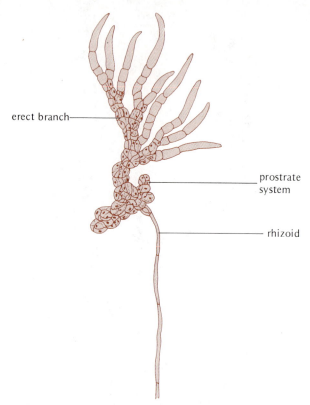

erect branch

prostrate system

rhizoid

Figure 28-6. *Fritschiella* is a green alga found in moist terrestrial environments. Erect branches arise from a prostrate system which also bears rhizoids. Cell divisions in several planes build a three-dimensional body suggestive of the land plants. Although probably not closely related to the land plant ancestors, algae such as this provide useful clues to the nature of the ancestral body form.

MIGRATION TO THE LAND

The earliest known vascular plant fossils reveal organisms already well adapted to terrestrial life. Nevertheless, information from studies of living algae, extinct early vascular plants, and persisting primitive vascular plants permits the development of a hypothesis for the evolutionary events during the land's invasion and the nature of the land plants' aquatic ancestors.

The Aquatic Ancestors

Green algae were probably the immediate progenitors of land plants, both bryophytes and tracheophytes, for reasons which have been discussed in Chapter 23. The living green algae, more than 400 million years after the migration to the

land, do not reveal the exact nature of the actual ancestors of terrestrial plants. Rather it is necessary to identify important features in several different green algal groups and postulate their occurrence together in a hypothetical ancestral form.

From the point of view of body form the most promising characteristics are found among the filamentous green algae. Some of these have a prostrate filament system from which erect, often branched filaments arise. A few have even more elaborate bodies, such as the flattened thallus of the sea lettuce *Ulva* and the limited three-dimensional development of *Fritschiella* (Figure 28-6). Since these green algae are not among the relatively few which share a common pattern of cytokinesis with the higher plants (Chapter 23), they may not be closely related to the ancestors, but they do suggest the nature of the ancestral body form. These tendencies were presumably emphasized in the land-plant ancestors, probably before they left the water, so that the erect filaments developed into three-dimensional, branched shoots, while the lower, prostrate filaments developed as structures to anchor the plant. Extensive three-dimensional growth allowed internal tissue specialization, particularly the formation of elongated cells in a rudimentary conducting system. The concentration of growth in a system of apical cells, as in the Charophyta, probably initiated a scheme of unlimited growth by means of meristems as in the higher plants.

The hypothetical algal ancestor presumably existed in two forms, occurring in a pattern of alternating haploid and diploid generations. Although many existing green algae undergo reduction division in the zygote without forming a distinct sporophyte body, others, like *Ulva*, do show such an alternation of generations, and the two generations are often morphologically identical or isomorphic (Chapter 23). Thus postulating such a condition in the hypothetical land-plant ancestors is not unreasonable (Figure 28-7).

In contrast to this view of possible algal ancestors, it has recently been suggested that the land plants may have evolved from unicellular green algae such as exist today in great numbers and diversity in moist soil. This interesting hypothesis postulates that the transition to a terrestrial existence was achieved at a simple level of organization and that the structural complexity of the higher plants evolved entirely on land.

The Transition to Land

During the migration to land, plants may have lived in intertidal zones or in brackish pools formed as previously submerged land was exposed in periods of receding oceans. They would then have been gradually or periodically exposed to the drying effects of sun and air. As this continued,

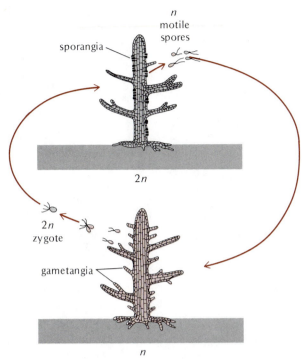

Figure 28-7. Diagram showing the hypothetical nature of the algal ancestors of vascular land plants with isomorphic alternation of generations.

the substrate and became a rhizome system, probably with hair-like outgrowths or rhizoids functioning in absorption. The spores became nonmotile and enclosed in a cutinized wall, a form suited for aerial dispersal. Some of the earliest vascular plant fossils reveal the presence of endophytic fungi in what is believed to be a symbiotic association such as occurs in the majority of land plants today. Thus it appears that from very early in their evolution land plants have been assisted by fungi in the process of absorption.

At this stage an evolutionary process of great importance in the vascular plants probably began. Spores had become effective dispersal agents on land, a great advantage to the colonizing plants. On the other hand, because of the necessity to fuse in pairs, gametes could not be similarly enclosed and dispersed. Sexual reproduction could occur in the moist habitats to which the plants were restricted, but it could not contribute to the plants' spread to new habitats. Increased sporophyte size and productivity would therefore confer advantage on a plant, whereas the gametophyte generation would tend to remain small and ultimately to undergo reduction, lying close to the ground where moisture for fertilization would be more readily available. Thus began a process of gametophyte reduction that can be followed throughout vascular plant evolution (Figure 28-8), and contrasts with the course of evolution in the bryophytes.

At some stage the female gamete became a nonmotile egg retained within the gametangium, where it was fertilized. This structure then evolved into the protective archegonium, which also provided nurture for the developing embryo after fertilization. The male gametangium similarly developed into the antheridium, which, however, continued to produce and release numerous motile male gametes or sperms.

perhaps for several million years, modification of body structure and of reproduction must have enabled survival in this environment and favored further encroachment of dry land by the most successful variants. A cutinized epidermis with stomata developed on the exposed surfaces, and elongated central cells became more efficient in conduction, ultimately giving rise to xylem and phloem. The basal portions of the plants invaded

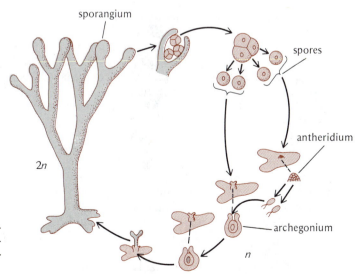

Figure 28-8. Hypothetical early land plant showing reduction of the haploid gametophyte generation. The gametangia have become archegonia and antheridia.

How do the Silurian and early Devonian land plants fit into this scheme of evolutionary origins? Plants such as *Cooksonia* and *Rhynia*, though far removed from their algal ancestors, show a simple plant body without specialized organs and with a central vascular column of small size and uncomplicated structure that could have arisen by the steps just outlined. One troubling question is the nature of the gametophyte generation of these plants. If the gametophytes were small and delicate, perhaps they were not preserved as fossils, but the quality of preservation in the deposits from which *Rhynia* was unearthed is so good that it has been difficult to accept this interpretation. Moreover, at this early stage of their evolution, the gametophytes might be rather similar to the sporophytes. One suggestion is that perhaps some of the plants described are actually gametophytes. In particular, it has been argued that the smaller of the two species of *Rhynia* may be the gametophyte stage, because no sporangia have ever been found actually attached to it and what may be archegonia have been observed in some specimens.

EVOLUTION IN THE DEVONIAN

Following the appearance of the first vascular plants a rapid expansion of plant types occurred in the Devonian period, so that the contrast between the land vegetation near the close of the period and that characteristic of its beginning is dramatic. Plants appeared that seem to be the fore-runners of most of the major groups of the later Paleozoic: the ferns, the gymnosperms, the lyco-pods or club mosses and scale trees, and the sphenopsids or horsetails. These groups became very important in the ensuing Carboniferous period, as will be discussed later in this chapter and in subsequent chapters.

The Trimerophytes

Psilophyton, a plant of Lower and Middle Devonian age, was the first early land plant to be scientifically investigated, and was described in 1859 by the Canadian geologist Sir William Dawson from the Gaspé region of Quebec. *Psilophyton* differs from *Rhynia* in its larger size, up to 90 cm (3 ft) in height, but primarily in its more complex branching pattern (Figure 28-9). Instead of equal branching or dichotomy, there was a main axis that branched unequally with the larger branch overtopping or dominating the smaller one. The larger branch thus contributed to the continuation of the main axis, and the smaller one came to oc-cupy a lateral position. The laterals tended to re-peat the same kind of branching pattern, which, however, became more equal near the tips. The sporangia were terminal in position as in *Rhynia*; but repeated branching of the lateral shoots re-sulted in rather large sporangial clusters. This type of plant body organization probably arose from

Figure 28-9. *Psilophyton* was a more robust plant than *Rhynia* in which unequal branching led to the forma-tion of a main axis with smaller side branches. The sporangia were terminal.

the plan represented by *Rhynia*. *Psilophyton* and several other similar plants are placed in Class **Trimerophytopsida** in recognition of their evolutionary advancement. The name is derived from *Trimerophyton*, a plant that closely resembles *Psilophyton* except that its lateral shoots appeared to branch in threes.

The evolutionary significance of these plants lies in the intermediate position they occupy between plants of the *Rhynia* type and the more advanced ferns and seed plants. In both ferns and seed plants the evolution of leaves as flattened and expanded portions of branched axes is particularly important (Figure 28-10). The first step in this evolutionary process is believed to have been the overtopping described in *Psilophyton* that led to the appearance of subordinate branch systems on what then became a main axis. At first the subordinate branch systems were three-dimensional, but they became progressively flattened into two-dimensional systems. Finally, by the process called **webbing**, the leaf blade developed with the formation of continuous tissue between the branches. Leaves that have developed

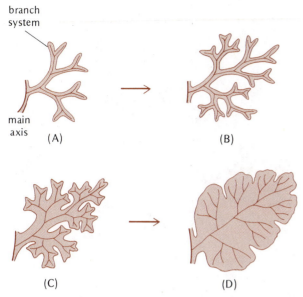

Figure 28-10. Stages in the probable evolutionary origin of a megaphyll from a subordinate branch system.

evolutionarily in this way are called **megaphylls** to distinguish them from microphylls, which had a very different origin (see below). The fossil record supports the view that the leaves of ferns and seed plants arose as megaphylls, but this does not mean that leaf origin has been identical throughout these diverse groups. In the process just described, it was possible for different portions of the subordinate branch systems to become leaves. Thus, in some cases whole lateral branch systems may have evolved into fern-like leaves, while in others only the ultimate divisions became the true foliar organs.

Ferns

Plants resembling ferns were apparently well established before the end of the Devonian period, but they cannot be placed in any of the modern fern groups. These ancient ferns had large frond-like leaves that are often difficult to distinguish from the stems or axes. In some cases the fronds were three-dimensional in organization and were not always webbed or laminated. Their reproductive structures were sporangia borne either terminally on the ultimate leaf segments or marginally or superficially on the leaves. The marginal and superficial positions were probably derived from the terminal as the leaves became progressively more webbed. Internally, as the plants became larger, they developed complex primary vascular systems in both fronds and stem, but with few exceptions, they did not add secondary tissues with the development of a cambium. These

early ferns could have been derived from the *Psilophyton* type of plant body.

The Progymnosperms

One of the most exciting paleobotanical discoveries of recent years has been the recognition that many fossils of the Middle and especially the Upper Devonian, which had long been thought to be ancient ferns, were in fact the forerunners of the gymnosperms (the conifers, etc.). These progenitors are now designated as the Class **Progymnospermopsida**. Like the fern ancestors, the progymnosperms had frond-like branching systems and were also probably derived from the *Psilophyton* type of plant body, but unlike the early ferns, they developed cambial activity and produced extensive secondary tissues. One of the best known of the progymnosperms, and the first to be recognized as such, is *Archaeopteris*. This was a tree of perhaps 15 m (50 ft) in height with a trunk as wide as 1.5 m (5 ft) in diameter near the base (Figure 28-11B). The plant probably had the general aspect of a modern conifer tree, and the secondary xylem was of the conifer type, but the branches were large, frond-like structures in which it is difficult to ascertain precisely what is leaf and what is stem (Figure 28-11A). Although they are believed to be ancestral to the diverse groups of gymnosperms, the progymnosperms were not seed plants. Some are known to have released spores from sporangia borne on their leaves. However, the discovery of fossilized seeds of late Devonian age shows that some plants had attained the seed habit before the end of this period.

Forerunners of the Lycopods

The same fossil beds in Scotland that contained *Rhynia* also included another plant of great interest because it represents an early stage in the evolution of the Class **Lycopsida**, the group that includes the modern club mosses. *Asteroxylon mackeii* was probably less than 60 cm (2 ft) high. Its erect, branching stems were covered with small, closely appressed leafy outgrowths about 4 mm long (Figure 28-12A). These aerial stems arose from leafless rhizomes, small branches of which penetrated the substrate and probably served as anchoring and absorbing organs. These downward penetrating branches are believed to illustrate the way in which root systems evolved in primitive plants by a modification of the basal portions of the originally unspecialized plant body. The sporangia of *Asteroxylon* were large, somewhat kidney-shaped bodies attached laterally by short stalks to the axis and interspersed among the leaves (Figure 28-12B). Other plants of the same age show more fully developed lycopsid characteristics. In

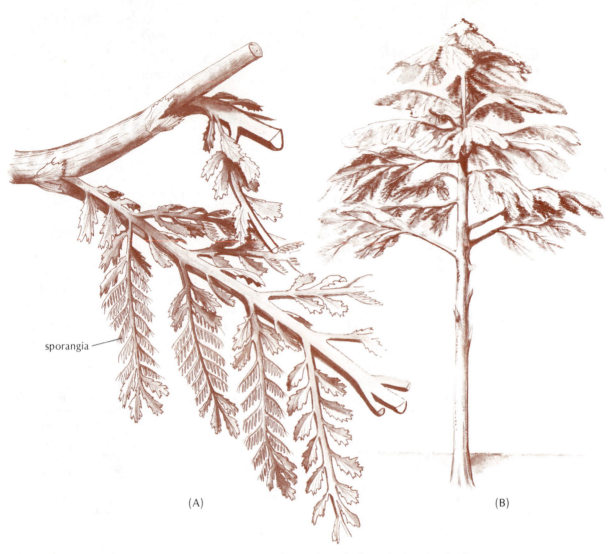

sporangia

(A)   (B)

Figure 28-11. *Archaeopteris*, a progymnosperm. (A) Portion of a branch showing both vegetative and spore-bearing appendages. (B) Reconstruction showing the tree habit.

Baragwanathia (Figure 28-13) the plant body is more robust, the leaves are larger and sporangia are located in the axils of some of the leaves.

These plants reveal the evolutionary origin of the Lycopsida, beginning with the Zosterophyllopsida. Some of the latter were spiny, but none could be said to have true leaves. The more leaf-like appendages of *Asteroxylon* resemble small outgrowths from the stem, or **enations,** and it is believed that they have arisen in this way. They are feebly supplied with vascular tissue by a leaf trace that extends from the central vascular column only to the base of the leaf but not into it. In *Baragwanathia* the trace extends the full length of the leaf as in the later lycopods. Such leaves have had a very different origin from that of the megaphylls of the ferns and seed plants. The term

microphyll is applied to leaves that originated as enations rather than as portions of the branching system, and they are one of the distinguishing features of the Class Lycopsida (Figure 28-14).

THE PSILOPSIDA

Among the living vascular plants, there is a small group that, because of its apparent primitiveness in both structure and reproduction, has long invited comparison with the earliest land plants known only as fossils. Two genera, *Psilotum* with two or three species and *Tmesipteris* with ten, constitute the Family Psilotaceae and the Class **Psilopsida.** *Psilotum* has a wide tropical

sporangium

Figure 28-12. *Asteroxylon mackeii,* an early lycopod. (A) Habit showing aerial stems covered with leafy outgrowths arising from leafless rhizomes; small, root-like branches penetrate the substrate. (B) Portion of an aerial axis with laterally placed sporangia.

(A)

(B)

(C)

Figure 28-13. *Baragwanathia.* (A) Compression of portion of leafy shoot showing the long, simple leaves. (B) Terminal portion of shoot, showing a fertile zone with large kidney-shaped sporangia. (C) Portion of sporangium-bearing region of (B), enlarged.

as the "whiskfern." It is easily cultivated in greenhouses. The aerial stem is frequently and dichotomously branched (Figure 28-15A). Photosynthesis is carried on by the green branches, which have many stomata. The stem contains a slender strand of primary vascular tissue; in the rhizome this is devoid of a pith, but in the aerial stem the center of the strand is composed of sclerenchyma tissue.

The aerial stems bear two kinds of appendages: (1) sterile, small and scale-like and containing no vascular tissue (Figure 28-15B, 28-16A); and (2) fertile, larger and more complicated struc-

and subtropical distribution; *Tmesipteris* is confined to the South Pacific region. These plants, although anomalous in several ways, resemble the Rhyniopsida in some ways. They are slender, rootless, terrestrial or epiphytic plants with erect or pendulous stems. *Tmesipteris* is about 20 cm (8 in) long; *Psilotum* varies from 15 cm to as much as 45 cm in height. The plant body (sporophyte) consists of a rhizome bearing aerial branches and covered with rhizoids. *Psilotum* bears small, scale-like appendages, which some consider rudimentary leaves, on its aerial axes; *Tmesipteris* bears larger leaves, each with a vascular trace.

A consideration of one species, *Psilotum nudum,* is sufficient to indicate many of the important features of the group. This species occurs in the southeastern United States, where it is known

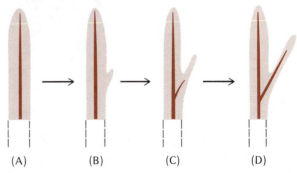

(A) (B) (C) (D)

Figure 28-14. Evolutionary origin of microphylls. It is believed that originally leafless stems (A) gave rise to small outgrowths or enations (B) which subsequently were supplied with vascular tissue in the form of a leaf trace (C and D).

Figure 28-15. *Psilotum.* (A) Dichotomously branched stem. (B) Detail of aerial branches showing sporangia and smaller, scale-like appendages (lower portion of photograph).

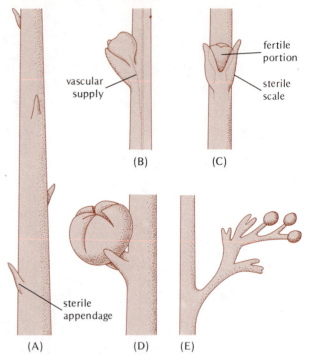

Figure 28-16. *Psilotum.* (A) Stem with sterile appendages. (B) Lateral view of young fertile appendage showing vascular supply to fertile portion. (C) Face view of young fertile appendage showing the forked sterile scale. (D) Mature fertile appendage. (E) Hypothetical ancestral sporangial branch from which the modern *Psilotum* fertile appendage may have been derived by condensation.

tures (Figure 28-15B, 28-16B, C, D). Each fertile appendage is composed of two parts: (1) a fertile portion, consisting of a very short stalk, or branch, vascularized by a branch of the vascular system of the stem and bearing a single three-lobed and three-chambered sporangium (Figure 28-16B), and (2) a sterile outgrowth of the fertile branch, ap-

pearing as a forked, scale-like structure just beneath the three-lobed sporangium. It contains no vascular tissue.

The sporangium dehisces through three clefts, one in each of the sporangial chambers, and the numerous spores are disseminated by wind. They fall upon the soil and give rise to gametophytes. The mature gametophyte (Figure 28-17) is nongreen and subterranean, growing 1 cm or more beneath the surface. It is brownish, elongated, and cylindrical; may branch irregularly or dichotomously; and develops rhizoids that penetrate the surrounding soil. The gametophytes are small, varying in length from 1 to 18 mm and in diameter from $\frac{1}{2}$ to 2 mm. Numerous archegonia and antheridia (containing motile sperms) are scattered over the surface. A feature of interest is the occasional presence, in some of the larger gametophytes, of a slender thread of vascular tissue extending back several millimeters from the apical region.

It is evident that the gametophytes of this seemingly primitive plant show a certain resemblance to the sporophytes in structure. When they occur in the soil it is often difficult to distinguish them from small pieces of sporophyte rhizome or from the rhizomes of young sporophytes. It is necessary to observe the sex organs to be sure. This similarity in a primitive plant supports the theory that originally in the vascular plants the two

Figure 28-18. A stand of *Lycopodium* on Isle Royale, Michigan.

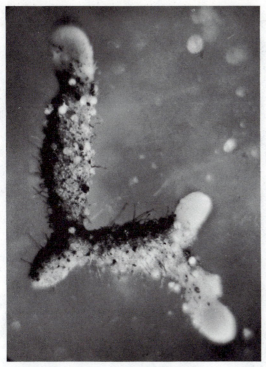

Figure 28-17. *Psilotum nudum,* mature gametophyte bearing sex organs.

generations were similar or identical (isomorphic) in structure and that the once more elaborate gametophytes have undergone an evolutionary reduction.

Shortly after germination the young gametophyte is invaded by hyphae of soil fungi. The gametophyte undoubtedly obtains food from the soil through these associated hyphae, and the relation between the gametophyte and the fungi may be mutualistic. After fertilization, an embryo develops. This is composed only of an absorbing foot and an embryonic stem; neither an embryonic leaf nor a root is present.

Because of the apparently undifferentiated nature of the plant body, the Psilotaceae are thought by many students of phylogeny to be related to the first vascular plants. They may perhaps constitute a branch of the Rhyniopsida or of the Trimerophytopsida that has persisted from the Devonian to the present.

Whether the Psilotaceae are really primitive, however, is disputed. The simplicity of these plants may be due to reduction. For example, the sporangium of *Psilotum* may be the result of the fusion of three sporangia. Further, the fertile appendage probably represents a condensed branch system, reduced from a larger and more complex system of fertile and sterile appendages (Figure 28-16E). Even the sterile appendages of *Psilotum* may be reduced rather than rudimentary structures. Detailed studies of certain primitive living ferns and comparison of various aspects of their structure and life cycles with those of the Psilotaceae have led to a proposal that the Psilotaceae should be placed in the Filicopsida (ferns). Pending general acceptance of this suggestion, however, it seems wiser to retain

them in a separate class, Psilopsida, at least for the time being.

THE LYCOPSIDA

The club mosses are plants with a long history, extending about 400 million years from the Devonian to the present. Although they do not constitute an important component of the earth's vegetation today, in the past many now extinct species were among the dominant plants of the great swamp forests that produced much of the world's supply of coal.

The living Lycopsida include five genera and approximately 1000 species, most of which are grouped in the two genera *Lycopodium* and *Selaginella*. The 200 or so species of *Lycopodium* are known as the club mosses, and the approximately 700 species of *Selaginella* are often referred to as either the small club mosses or the spike mosses. Again, common names are misleading, because these plants are not mosses or even related to the mosses. Two major structural features distinguish the Lycopsida from other groups of vascular plants. (1) The leaves are microphylls that originated evolutionarily as outgrowths or enations from the stem, as described above. The term microphyll is perhaps unfortunate because these structures are not necessarily small. In fact, in some extinct members of the Lycopsida they were as much as 45 cm long. (2) Sporangia are regularly produced either on the adaxial or upper surfaces of certain leaves, known as **sporophylls,** or in their axils. The appearance of this distinctive characteristic in some of the earliest representatives of the Lycopsida has already been mentioned.

Lycopodium

The club mosses (Figure 28-18) are found in much the same habitats as ferns. Most species are

Figure 28-19.   Species of *Lycopodium*, showing a variety of forms. (A) *L. alopecuroides,* with trailing stems bearing erect branches, and small awl-like leaves. (B) *L. lucidulum,* with erect stems and narrow but flattened leaves in the axils of which sporangia are borne. (C) *L. complanatum,* with a habit similar to (A), but its erect branches form cedar-like sprays with scale leaves and bear clusters of spore-producing strobili (cones). Approximately $\frac{2}{3}$ natural size.

The club mosses native to temperate North America are terrestrial plants of varied habit and size (Figure 28-19). The aerial stems may be clustered and rooted at the base, or trailing and creeping, with erect branches arising from horizontal stems that grow over or under the surface of the soil. The erect branches vary from 5 or 7.5 cm to 15 or 20 cm in height. Some species, in which the erect parts are bushy, are known as "ground pine" or "ground cedar." The horizontal stems are elongated by terminal growth, often forming long runners. As the older parts die, the younger branches continue growth, a single plant thus giving rise to a number of separate plants. This vegetative method of reproduction frequently produces large colonies in woodlands and meadows.

Most temperate-zone species of *Lycopodium* are evergreen perennials and are sometimes—unfortunately from the point of view of conservation—sold as Christmas decorations. The leaves of the club mosses are numerous, small, and in some species so scale-like and inconspicuous that the branches may be mistaken for leaves. As is typical of microphylls, they are supplied by a single leaf trace that extends the length of the leaf. The branching is dichotomous, the paired branches

tropical or subtropical, but some flourish in temperate climates, especially in the cool moist forests of the northern hemisphere. Some grow rooted in the soil, but many of the tropical species are epiphytes, with their roots in the crevices of the bark of the trunks or branches of trees.

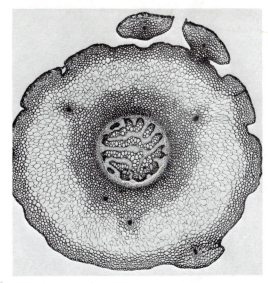

Figure 28-20. *Lycopodium*, cross section of stem.

in some species being equal and in others unequal so that an apparent main axis with subordinate side branches results.

In adult club mosses all roots are adventitious, arising from the base of the stem or the underside of the prostrate parts of the shoot. A cross section of a club moss stem (Figure 28-20) reveals an epidermis, a wide cortex, and a pithless vascular cylinder composed of xylem and phloem. The xylem is deeply and irregularly lobed, with the phloem lying between the lobes of the xylem. The leaf traces are small in diameter and attach to the tips of the lobes with little or no disruption. Unlike most other vascular plants, the structure of the roots is almost the same as that of the stems except for the absence of leaf traces.

THE LIFE CYCLE

The club moss plant is the diploid sporophyte in a life cycle involving alternation of generations (Figure 28-21). The sporophyte produces

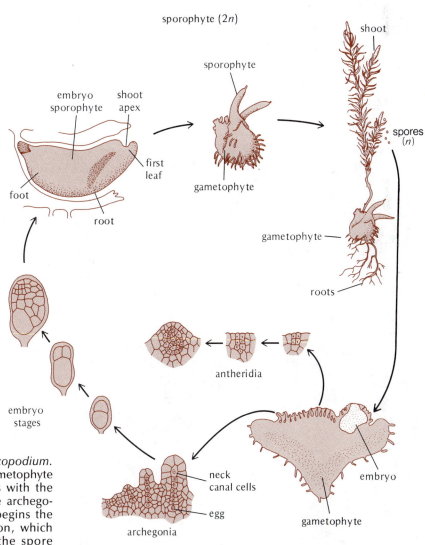

Figure 28-21. Life cycle of *Lycopodium*. Haploid spores initiate the gametophyte generation which culminates with the fertilization of the egg in the archegonium. The resulting zygote begins the diploid sporophyte generation, which continues until meiosis in the spore mother cells leads to the formation of haploid spores.

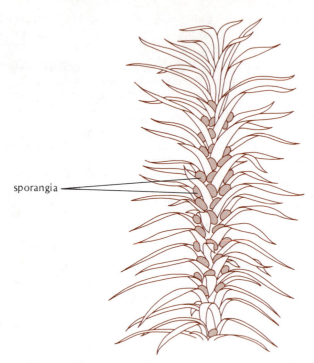

sporangia

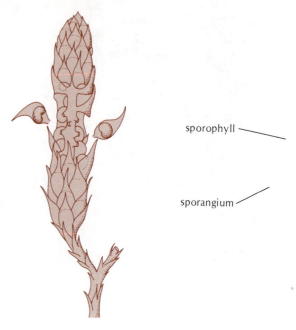

sporophyll

sporangium

Figure 28-22. *Lycopodium lucidulum,* a primitive species with sporangia borne in the axils of sporophylls which are similar to vegetative leaves. These occur in fertile zones along the axis.

Figure 28-23. Strobilus of *Lycopodium* with several sporophylls removed. Each sporangium is borne on the upper (adaxial) surface of a sporophyll.

spores in kidney-shaped sporangia borne singly in the axils of the sporophylls or on their upper (adaxial) surfaces. In one primitive group of species, the sporophylls are similar to the vegetative leaves in size, shape, position, and capacity for

photosynthesis (Figures 28-19B, 28-22). The sporangia are scattered along the leafy shoot, although they tend to form fertile zones that alternate with vegetative zones.

In a second, more advanced group of club mosses, the sporophylls are grouped into cone-like structures called **strobili** (Figure 28-19C). These are found at the ends of leafy branches or raised

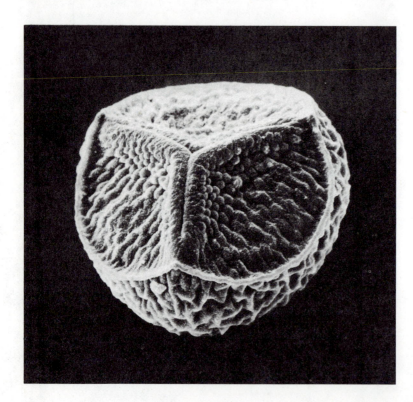

Figure 28-24. Scanning electron micrograph of *Lycopodium inundatum* spore, originally part of a tetrad. Note the contact faces with the other three spores.

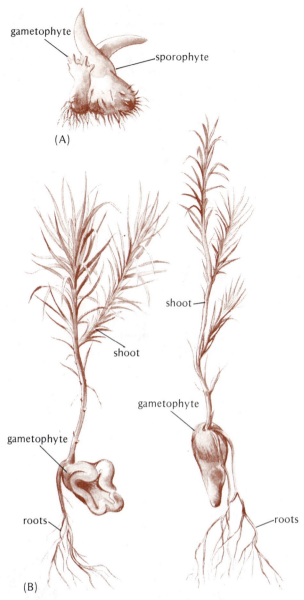

gametophyte

sporophyte

(A)

shoot

shoot

gametophyte

gametophyte

roots

roots

(B)

Figure 28-25.   Gametophytes of *Lycopodium*. (A) Green surface-living gametophyte with a young sporophyte attached. (B) Two forms of subterranean gametophytes with older sporophytes still attached.

on long stalks. The sporophylls are greatly reduced and are scale-like (Figure 28-23) and yellowish. The compact appearance of the strobili, together with the small leaves and the general habit of the plants, suggests the common name of club moss.

The spores are formed from spore mother cells after meiosis. They are tetrahedral and usually have radiating lines on one side (Figure 28-24). The walls have thickened ridges laid down in various patterns. The mature spores escape through a slit in the wall of the sporangium and are disseminated by air.

When a patch of club mosses in which the spores are ripe is disturbed, as by someone walking through it, a yellowish cloud composed of millions of spores arises. The spores contain a high percentage of oils and are thus little affected by water. They have been used in the past as a dusting for pills to prevent sticking together even in damp weather. The spores are also highly inflammable and were formerly used in fireworks and in creating stage effects in theatrical productions.

The Gametophyte.   The spores of the club mosses germinate on or beneath the soil surface and eventually give rise to mature gametophytes bearing the sex organs, archegonia and antheridia (Figure 28-21). The gametophytes vary considerably in size and form in different species, but all are relatively small and tuber-like. The sex organs are borne on the upper surface, sunken in the tissues. The biflagellate sperms reach the archegonia by swimming through a film of free water on the surface of the gametophyte. After fertilization the 2n zygote produces an embryo that grows into a leafy plant. More than one egg may be fertilized, and several sporophytes may arise from the same gametophyte.

Lycopodium gametophytes fall into various categories. In one type, found in many tropical and a few temperate zone species, the gametophyte is a surface-living, erect, ovoid or cylindrical structure with green, photosynthetic aerial branches arising from a yellowish-white basal region (Figure 28-25A). The basal region, which is buried in the soil, is early invaded by intracellular fungus hyphae that appear to be necessary to the continued development of the gametophyte. The gametophytes are small (2–3 mm in diameter) and ordinarily persist only for a few months.

Another type of gametophyte is that found in most species of club mosses of the temperate zone. This type has no chlorophyll and develops entirely underground, buried at depths of 2.5–10 cm (Figure 28-25B). As with the aerial type, fungus hyphae are associated with these underground structures, occurring chiefly in a zone near the periphery of the gametophyte. These hyphae make food available to the gametophyte from the surrounding organic matter of the soil. The relationship between fungus and gametophyte may be mutualistic, but if so, the advantage to the fungus in the association is not known. The subterranean gametophytes vary considerably in size and form but are generally much larger (from 1–2 cm in length or width) than the green type. Depending on the species, they are cylindrical, carrot-shaped, disk-like, or lobed and convoluted.

The green, surface-living gametophytes are relatively easy to find in nature, but the subterranean types of most temperate-zone species are seldom discovered and are considered rare. They are usually located by finding young sporophytes and tracing these downward to the attached gametophytes. They often occur in large numbers at a particular site. Whether they are truly rare or are merely difficult to find does not appear to have been established. It is generally believed that vigorous vegetative reproduction of the sporophyte of

Figure 28-26. Two forms of *Selaginella*. (A) *Selaginella apoda*, "meadow spikemoss," a creeping form. (B) An erect tropical species.

club mosses is important for the survival and spread of these species.

Selaginella

The small club mosses are highly diverse in form (Figure 28-26). Some are creeping and flat; others form tufts or mats; still others grow erect. The genus is much more conspicuous in the moist tropics than in the temperate zone. A number of tropical species branch freely, and the erect forms may attain a height of a foot or more. The delicacy and beauty of the foliage of some tropical species has resulted in their extensive cultivation in conservatories and by florists. The temperate zone species are mostly small and moss-like. They occur over a wide range of habitats, from moist shady banks to dry, rocky slopes. A few species thrive in dry prairie or semidesert habitats where prolonged drought may be the rule. One such species, *Selaginella densa* of the Great Plains, is reported to have revived and begun new growth after being stored for 2 years and 9 months totally dry on a laboratory shelf (Figure 28-27). The "resurrection plant" or "bird's-nest moss" of Texas and Mexico (*Selaginella lepidophylla*) forms a dense green rosette during the growing season, but in dry periods the stems curl inward until the plant assumes the form of a compact ball. With the return of moisture, the stems unroll and growth resumes.

The leaves, or microphylls, are small and overlapping and are usually arranged in four vertical rows along the slender stems (Figure 28-28). The leaves of the two rows on one side, the upper side in creeping species, are distinctly smaller than those of the other two rows. In about 50 species, however, the leaves are all the same size and are spirally arranged.

Each leaf characteristically bears a small and relatively inconspicuous, tongue-like appendage in a pit on its upper surface near the leaf base. This is the **ligule**, a structure found only in certain members of the Lycopsida. Each leaf has a single, unbranched trace extending from the vascular column of the stem through the length of the leaf. The vascular tissues of the stem show diverse arrangements ranging from a single column that may or may not have a pith and is often ribbon-shaped, to a more complex form in which two or more such columns may be present. The xylem of some species contains true vessels. The roots arise in an unusual position in that they originate at branchings of the stem. In upright species they may extend downward for some distance before entering the soil.

Figure 28-27. The drought-resistance of *Selaginella densa*. (A) Deprived of water for nearly three years. (B) Revived after return to moisture.

Figure 28-28. Portion of a shoot of *Selaginella*. Two rows of leaves on the upper side of this creeping species are distinctly smaller than the two rows on the lower side.

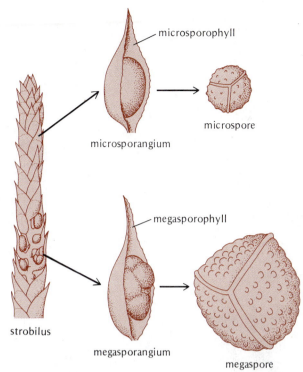

Figure 28-29. *Selaginella* strobilus, sporophylls, sporangia, and spores.

THE LIFE CYCLE

As in *Lycopodium*, the sporangia are borne on the upper surfaces of sporophylls near their axils. The sporophylls are always clustered in a terminal strobilus that is usually four-sided and may range in length from 0.5 to 6 cm depending on the species. *Selaginella*, however, is very different from *Lycopodium* in one important respect. It forms two kinds of spores—small and numerous (up to 600 in a sporangium) **microspores,** borne within a **microsporangium,** and much larger thick-walled **megaspores,** commonly four in number, borne within a **megasporangium** (Figure 28-29). The megasporangium is usually much larger than the microsporangium and is locally distended or lobed by the growth of the spores within. The four large megaspores are formed from a single spore mother cell. A number of spore mother cells are originally present, but all, with the exception of one, degenerate before meiosis, and the degenerating cells contribute to the nutrition of the developing megaspores. This condition, in which spores of two distinct kinds are produced, is called **heterospory,** and *Selaginella* is said to be **heterosporous** in contrast to *Lycopodium* and *Psilotum,* which are **homosporous** (produce only one kind of spore). The obvious difference in size between microspores and megaspores is an important aspect of heterospory, but other differences appear when the spores germinate.

The sporophylls are termed **microsporophylls** or **megasporophylls** according to the kind of sporangium that each bears in its axil. Otherwise they are alike, green, and somewhat smaller than the vegetative leaves. Both kinds of sporophylls and their sporangia are usually borne on the same strobilus. The megasporangia may occur at the base and the microsporangia in the upper portion, or the two kinds may be intermingled.

Male and Female Gametophytes. Germination of the spores, which takes place before the spores are released by the opening of the sporangium, reveals a major feature of heterospory. The two kinds of spores give rise to two kinds of gametophytes that differ both in size and in the kind of gametes produced (Figure 28-30). The microspores produce male gametophytes, whereas the megaspores give rise to female gametophytes. Furthermore, the gametophytes of both types develop mostly or entirely within the spore wall. The microspore first divides to form two cells, one of which is small and lies close to the spore wall (Figure 28-30A). This is called the prothallial cell, and it is thought to be a vestigial structure. The other, larger cell develops into an antheridium consisting of a sterile jacket of cells surrounding many biflagellate sperms. At some stage in this process, varying with the species, the microspores are shed, and development is continued on the soil. Ultimately, the sterile cells degenerate, the

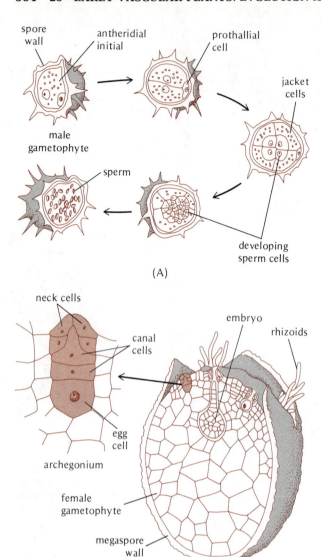

Figure 28-30. *Selaginella*, development of gameto-phytes. (A) Male gametophyte within the microspore wall. (B) Female gametophyte inside the megaspore wall.

spore wall cracks open, and the sperms are released.

The female gametophyte is considerably larger and is not entirely contained within the megaspore wall (Figure 28-30B). First the megaspore nucleus divides repeatedly without wall formation, and the multinucleated protoplast is later partitioned by the formation of walls. In some species this process is never completed, and only the uppermost portion of the gametophyte (i.e., the region in which the spore will open) is truly cellular. At some time during this development, the megaspores are shed and development continues on the soil. Then the spore wall cracks open and the gametophyte protrudes. Archegonia and rhizoids are formed on the exposed portion, but the

gametophyte apparently never becomes green, even when exposed to light. Its development, and the early stages of sporophyte development, must be supported entirely by food reserves stored in the spore. If microspores and megaspores fall in close proximity, the sperms swim to the egg in a film of water, fertilization occurs, and an embryo sporophyte develops. The young plant remains attached for some time to the female gametophyte, still held within the spore wall (Figure 28-31).

Significance of Heterospory

The phenomenon of heterospory is considered to represent an evolutionary advance over homospory, but it is reasonable to ask what advantage heterospory as it occurs in *Selaginella* confers upon the plant. It can be argued that the occurrence of separate male and female gametophytes is significant because it guarantees cross fertilization. But this does not require the spores to be of different sizes. It is perhaps more significant that heterospory is associated with the reduction of the gametophyte and its almost complete retention within the spore wall. Under these conditions, the nutrients required for the development of the gametophytes are stored in the spore, and the gametophytes do not develop vegetative bodies of any consequence. This procedure apparently minimizes the risks to which self-supporting gametophytes are exposed and favors the rapid production of gametes, a factor that could be important when the availability of surface moisture is limited. Since the demands upon the female gametophyte, which must support the early development of the new sporophyte generation after fertilization, are much greater than those upon the male, the difference in size of the spores that produce the two types of gametophytes can be looked upon as an evolutionary process resulting in increased efficiency.

The fossil record shows that the original vascular plants were homosporous, but it also reveals that heterospory made an early appearance in certain groups before the middle of the Devonian period. Most of the living lower vascular plants have remained homosporous, including the Psilotaceae, the club mosses, the horsetails, and most of the ferns. The kind of heterospory found in *Selaginella*, with large, free megaspores and small microspores, occurs in relatively few other groups of living vascular plants, chiefly the water ferns and the quillworts (*Isoetes*). It was, however, present in a number of groups now extinct. On the other hand, the seed plants are heterosporous, but, as will be seen in the following chapter, their reproductive system is considerably modified from that found in *Selaginella*.

Evolution of the Lycopsida

Although they constitute only a minor part of the worldwide flora today, for a period of nearly 70 million years the Lycopsida were a major, if not the dominant, component of the earth's vegetation. The earliest lycopods, of which a number from the Devonian period have been found and described, were much like members of the modern genus *Lycopodium* that do not have their spo-

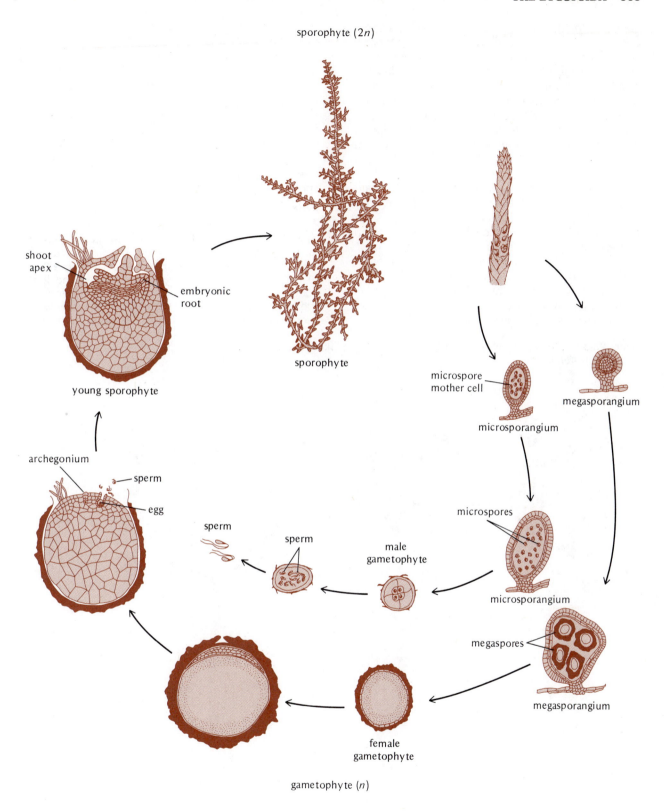

sporophyte (2n)

sporophyte

shoot apex

embryonic root

young sporophyte

microspore mother cell

microsporangium

megasporangium

archegonium

sperm

egg

sperm

sperm

male gametophyte

microspores

microsporangium

megaspores

megasporangium

female gametophyte

gametophyte (n)

Figure 28-31. Life cycle of *Selaginella*. In this life cycle there are separate male and female haploid gametophytes which arise from small microspores and large megaspores, a phenomenon known as heterospory. Both kinds of gametophytes are almost completely retained within the spore walls.

rophylls localized in cones; as far as is known, they were homosporous. *Lycopodium* probably represents a persistence of this ancient lycopod stock, and fossils very much like the modern genus (often assigned to the genus *Lycopodites*) are found in the rocks of the Carboniferous. Carboniferous fossils of *Selaginella* are also known, indicating that a line of heterosporous and herbaceous lycopods had developed, presumably from the same basic stock. This line, of course, has also persisted.

Perhaps the most intriguing of all the lycopods are the giant tree club mosses *Lepidodendron* and *Sigillaria* (Figure 28-32). Arborescent forms of this group had begun to appear before the end of the Devonian, but these tree forms reached their full development, both in diversity and in numbers, in the Carboniferous. They were among the most abundant and conspicuous plants of the vast swamps that became the source of much of the world's coal reserves. Although they were heterosporous like *Selaginella*, these trees are believed to represent a separate line of evolution, another divergence from the basic lycopod stock. Most of them became extinct at the end of the Carboniferous or in the succeeding Permian period, and only a few much reduced forms persisted into later times. Presumably the herbaceous forms (*Lycopodium* and *Selaginella*) that survived, and may be as abundant today as they have ever been, were better able to adapt to changing environmental conditions than were the highly specialized tree forms.

Carboniferous Lycopods

The lepidodendrons were large trees, frequently 30 m or more high and 90 cm or more in diameter at the base (Figure 28-32A). The trunk branched dichotomously, and large cones, in some species more than 30 cm in length, terminated the branches. The lepidodendrons were heterosporous, with both kinds of spores produced in the same strobilus. The leaves were deciduous and narrow, and varied in length from a few

Figure 28-32. Reconstruction of the Carboniferous tree-like lycopods (A) *Lepidodendron* and (B) *Sigillaria*.

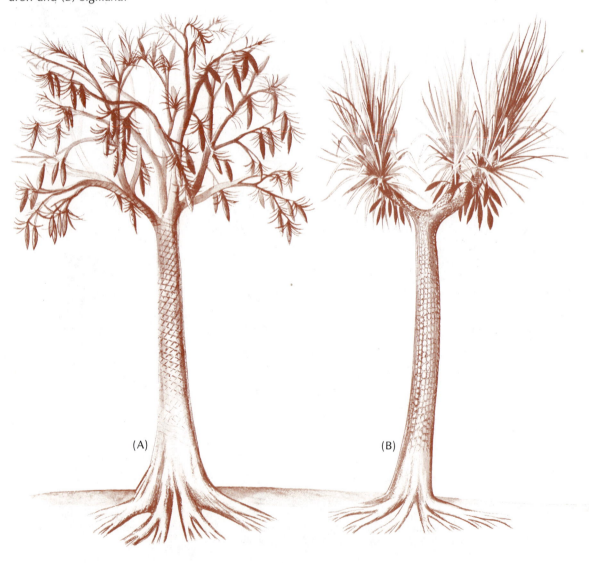

(A)   (B)

centimeters to 60 cm or more. The trunk contained some secondary wood, but the greater part was composed of a thick layer of periderm, produced by a secondary meristem. This periderm, unlike the cork of modern trees, consisted of living tissue. The main trunk and branches were covered with spirally arranged **leaf cushions,** diamond-shaped in outline. The leaf cushion was composed of the enlarged basal part of the leaf, which underwent abscission slightly above the base, leaving a leaf scar in the center or upper part of the leaf cushion (Figure 28-33A). As the tree aged, the increase in thickness of the periderm layer led to the sloughing off of the leaf cushions at the base of the trunk, which became rough and ridged. Impressions of the trunk showing leaf cushions are among the most common and characteristic fossils in Carboniferous strata. At its base, the trunk divided into four large root-like structures that in turn branched repeatedly. These prop-like structures bore many lateral rootlets in a regular arrangement and must have been of great value to tall trees growing in the unstable swamp soil. Botanists are still puzzled about whether these organs were truly roots or modified stems of some sort.

The sigillarias were also large trees, but they differed from the lepidodendrons in several ways. The trunk was sparsely branched, and the top of each branch bore a large cluster of elongated narrow leaves that were sometimes 90 cm (3 ft) long. The trunk of *Sigillaria* also bore leaf cushions, but less prominently raised, and most of the surface of the cushion was covered by a leaf scar (Figure 28-33B). The leaf cushions of *Sigillaria* were spirally arranged, as in *Lepidodendron,* but because of the presence of longitudinal ribs or ridges they appear to be aligned in vertical rows, a feature that distinguishes impressions of the bark from those of *Lepidodendron. Sigillaria* was also heterosporous, but the strobili were usually borne in whorls on the main axis, just below the apex of the trunk.

In spite of their size and arborescent form, these giant lycopods were very unlike modern trees. The small amount of secondary vascular tissue they produced contrasts sharply with the massive woody structure of a pine or an oak. The decline of the tree lycopods may have been caused by the inability of this limited conducting system to meet the demands imposed by the drier climate that succeeded that of the Carboniferous. Most students of the giant lycopods now agree that they were really overgrown herbs. They may have developed very rapidly, produced a massive crop of cones at one time, and then died.

Although the tree lycopods declined and largely disappeared at the end of the Carboniferous period, they did not become entirely extinct, because scattered remains of what were almost certainly reduced and modified forms of these plants are found in several places in the later fossil record. Many botanists believe that a small group of modern lycopods, the quillworts (*Isoetes*), may represent a still surviving remnant of the giant forms.

Isoetes

The much compressed axis of the quillwort body (Figure 28-34) contains a cambium and produces secondary tissues; this is the only instance of secondary growth in a living lycopod. The central axis bears leaves on its upper portion that are produced by an apical

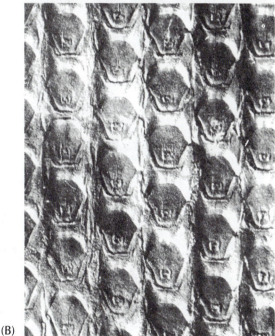

Figure 28-33. (A) Surface of bark of *Lepidodendron,* showing leaf scars and diamond-shaped leaf cushions. (B) Leaf scars of *Sigillaria.*

meristem, and roots borne in a regular arrangement on its lower portion. Moreover, the internal structure of the roots is remarkably like that of the rootlets that are borne on the prop-like basal structures of the lepidodendrons. The axis may represent the very much compressed and reduced plant body of a once arborescent plant, the ultimate in evolutionary reduction. If true, this is a remarkable example of survival through reduction.

Figure 28-34. *Isoetes montana,* in shallow water.

THE SPHENOPSIDA

The living **Sphenopsida** include some 20 to 25 species, all grouped in the single genus *Equisetum,* commonly known as horsetails. Like the lycopods, this limited assemblage gives little suggestion of a group that was a major component of

Figure 28-35. *Equisetum* species with different growth habits. (A) *E. arvense,* the common horsetail, with un-branched, nongreen, nonphotosynthetic "fertile" shoots bearing strobili and separate highly branched green photosynthetic "sterile" shoots, both arising from an underground rhizome. (B) *E. telemateia,* a species with green, unbranched photosynthetic stems bearing stro-bili at their tips.

the earth's vegetation in past geological ages. Members of the genus *Equisetum* are widely distributed, but they are more characteristic of temperate regions than tropical, and are especially abundant in the northern hemisphere. Most horsetails grow in wet or swampy environments such as shallow ponds, marshes, or stream banks, but some thrive in the comparatively dry soil of pastures, roadsides, and even on barren railroad embankments. Occasionally they may become troublesome weeds, and some species contain toxic substances that are harmful to livestock. In general, horsetails vary in height from a few centimeters to about a meter, but several species are taller. One tropical species attains a height of more than 10 m, but its slender stem is supported by surrounding vegetation.

The *Equisetum* Plant Body

The plant is composed of perennial underground rhizomes from which aerial branches arise. In some species the aerial stems are evergreen and perennial; in others they are annual, appearing early in the growing season. The stems (Figure 28-35) are cylindrical, conspicuously jointed, and marked with longitudinal ridges. They may be branched or unbranched. The scale-like leaves, fused by their margins into a cylindrical sheath, are borne in whorls at the nodes. The branches, if present, also arise in whorls at the nodes. Their bushy symmetry suggests the common name of these plants. Photosynthesis is carried on chiefly by the stems and branches, for the leaves are minute and in many species devoid of chlorophyll. Spore-producing strobili are borne terminally on the main stem or the branches. In some species, all the stems

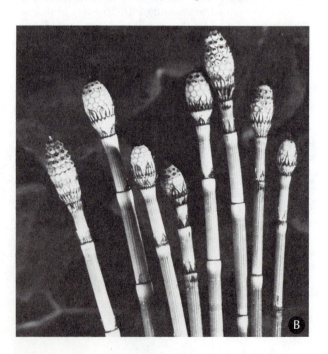

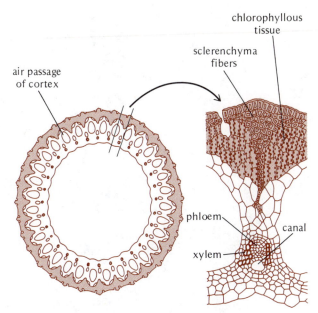

Figure 28-36. Cross section of a stem of *Equisetum* (left) and a small portion more highly magnified (right).

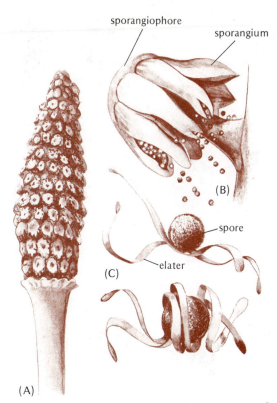

Figure 28-37. *Equisetum.* (A) Strobilus. (B) Sporangiophore with sporangia shedding spores. (C) Spores with elaters.

produce strobili (Figure 28-35B); other species, such as the widely distributed *Equisetum arvense*, are dimorphic, with two kinds of aerial stems, sterile and fertile (Figure 28-35A). In this species the fertile shoot lacks chlorophyll and has but a brief existence, dying to the ground after the spores are shed. The sterile shoots continue to grow vigorously, becoming bushy plants that persist until autumn.

Internally the horsetail stem is hollow except at the nodes, where plates of pith tissue are present. Around the hollow center is the vascular system, a ring of small bundles (Figure 28-36). Each bundle consists of a group of phloem cells with a small mass of xylem on either side and a characteristic canal or cavity formed by the tearing out of xylem cells that matured while the stem was still elongating. True vessels have been observed in the xylem of the rhizomes of five species of *Equisetum*. The wide cortex contains photosynthetic tissue, air passages or canals that extend from one node to the next, and sclerenchymatous fibers that are chiefly responsible for mechanical support of the stem. Both these fibers and the epidermal cells are strongly impregnated with silica, which is so abundant in one species that it was formerly used for scouring pots and pans and received the common name "scouring rush."

The Sphenopsid Life Cycle

In the *Equisetum* life cycle the sporophyte is homosporous, and the gametophytes are free-living, supported by their own photosynthesis. The cone or strobilus of the sporophyte consists of a short, thick axis bearing a succession of whorls of **sporangiophores** that in turn bear the sporangia

(Figure 28-37). The sporangia are elongated, saclike structures attached to the inner side of the stalked, umbrella-shaped sporangiophores. Each sporangium contains a large number of spores formed following meiosis in the spore mother cells. When the spores are ripe, the sporangium opens by a longitudinal slit on the inner side. The axis of the strobilus then elongates, separating the sporangiophores and leaving intervening spaces allowing the spores to be carried away by air currents.

The spores of the horsetails contain numerous chloroplasts, a rather unusual feature among the lower vascular plants. The outermost wall layer is deposited in the form of four spirally arranged strips, which later split away as four elongated appendages that have flattened tips and a common point of attachment. These spore appendages, or **elaters** (Figure 28-37C), are hygroscopic, coiling and uncoiling rapidly with changes in the humidity of the air. The resulting movements within the spore mass apparently contribute to the expulsion of the spores from the sporangium. In addition, the elaters probably play a role in dispersal. Since they are expanded when dry, they may serve as "wings" that aid in spore dispersal by light winds.

On moist soil the spores germinate almost immediately (Figure 28-38). Indeed, they must germinate quickly or die because they lose their vi-

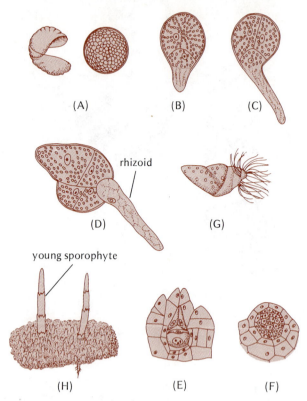

Figure 28-39. *Calamites,* pith cast. The pith disappeared and was replaced by sedimentary material that hardened into rock. The markings are the impressions of the inner surface of the wood.

Figure 28-38. *Equisetum.* (A–D) Stages in spore germination. (E) Archegonium. (F) Antheridium. (G) Motile sperm. (H) Gametophyte with young sporophytes developing after fertilization.

ability rapidly, ordinarily in less than two weeks. Spore germination results in the development of gametophytes, each consisting of a basal cushion, which is attached to the soil by rhizoids and from which arise erect green lobes. The total effect is that of a small pincushion, ranging from 1 mm to 1 cm or more in diameter. Both antheridia and archegonia are borne on the upper surface of the basal cushion, and antheridia may also be produced on the upright lobes (Figure 28-38). The sperms released from the antheridia are motile (Figure 28-38) and swim to the archegonium if water is available. Several sporophytes may arise from the same gametophyte if more than one egg is fertilized.

Equisetum gametophytes are apparently not rare in nature, but they are easily overlooked and for this reason are not often found. At least in some species, gametophytes are initially either male or female. The initially male gametophytes only rarely give rise to archegonia as they age, but the initially female gametophytes regularly become bisexual sooner or later.

Evolution of the Sphenopsida

Fossil plants of Middle Devonian age suggest that the Sphenopsida, like the ferns and seed plants

but independently of them, have arisen from *Psilophyton*-like ancestors among the Trimerophytopsida. It is believed that both the leaves and the unique sporangiophores of the Sphenopsida have evolved through the reduction and modification of small, subordinate branches into sterile and fertile appendages.

The subsequent history of the Sphenopsida is very similar to that of the Lycopsida. In terms of diversity, size, and abundance, the Sphenopsida reached their zenith in the Carboniferous period and constituted an important part of the flora of the great coal swamps. The calamites (*Calamites*) (Figure 21-10) closely resembled the modern horsetails except that they were tree-like in size, sometimes attaining a height of 15 m or more. Unlike *Equisetum,* they had well-developed secondary vascular tissue, and their leaves, usually several centimeters long, are believed to have been photosynthetic. By contrast, the sphenophylls (*Sphenophyllum*) (Figure 21-10) were slender, vine-like plants that may have been climbers. They too had secondary tissues and photosynthetic leaves. These two ancient groups of sphenopsids both became extinct, the calamites barely surviving the end of the Carboniferous and the sphenophylls persisting into the early part of the Mesozoic era. *Equisetum* is also very ancient, and probable fossils (*Equisetites*) have been described from the early Mesozoic. They may even have lived contemporaneously with *Calamites* in the Carboniferous.

attained a diameter of more than 30 cm. The primary vascular system was remarkably like that of *Equisetum,* but unlike the modern herb, this system of primary bundles was embedded in secondary vascular tissues produced by a cambium. The leaves (Figure 28-40) were narrow and sometimes needle-like, longer than those of *Equisetum,* and apparently were the chief organs of photosynthesis. They were borne in whorls, the number in a whorl varying from 4 to 40, depending on the species. The leaves were not fused, or only slightly so, contrary to the condition in the modern horsetails. In some, many of the nodes of the main stem bore large whorls of branches. The sporangia were borne on sporangiophores that were grouped in cones, but unlike the cones of *Equisetum,* these contained whorls of sterile bracts alternating with the fertile whorls. Some became rather complex in structure. Most species were homosporous, but a few were heterosporous. The spores of many species bore elaters.

The sphenophylls (*Sphenophyllum*) clearly represent an evolutionary line distinct from that which gave rise to the calamites and the modern horsetails. The whorled leaves borne at the nodes of the slender stems are wedge-shaped, that is, narrower at the base than at the tip—hence the name *Sphenophyllum.* This group gave its name to the Class Sphenopsida because it was formerly believed to represent the primitive basic stock of the assemblage. This belief was based in part on the ancient origin of the sphenophylls, which extend back into the Devonian period, and in part upon anatomical studies. The primary vascular system of the stem contains a simple, solid core of xylem that is more primitive in form than the system of bundles found in *Calamites* and *Equisetum.* It is now thought that the sphenophylls evolved on a line parallel with that of the extinct and living horsetails and were themselves rather highly specialized plants. They produced enormously complex cones, one of which is said to be the most complicated reproductive structure ever produced by a vascular plant.

FOSSIL FUEL—COAL

Coal is one of the most important of all the materials obtained from the earth's crust. This abundant "mineral" fuel made possible the industrialization of the Western world and is of fundamental importance to our technology and industrial productivity. By its aid we make iron and steel, tools and machines. Converted into electricity, it lights our cities. From coal are obtained innumerable products used in the chemical industries—dyes, medicines, resins, plastics, synthetic rubber, solvents, and perfumes. Products obtained from coal are used to surface our roads and preserve our wood products from decay. This significant modern resource resulted because green leaves trapped the energy of sunlight, hundreds of millions of years ago. The rapid depletion of oil and gas reserves, which is now causing worldwide concern and many economic and political problems, makes it certain that coal will continue to have great importance.

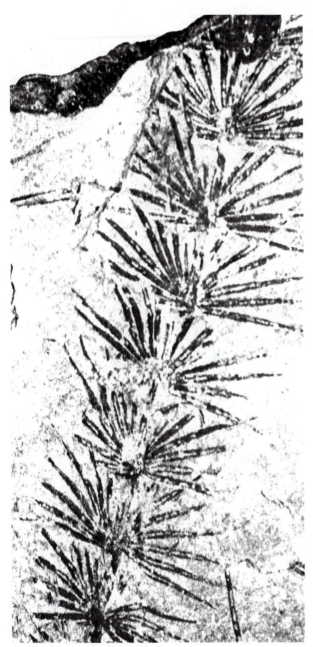

Figure 28-40.   Compression of whorls of leaves of *Calamites.*

Carboniferous Sphenopsids

Fossils of the giant horsetails, *Calamites,* are among the commonest finds in deposits of Carboniferous age and show that many species were present. Petrifactions, casts, and compressions of these plants indicate that they attained their maximum development in the Upper Carboniferous. In spite of their tree-like dimensions, these plants are considered to have been overgrown herbs. Stem fossils of *Calamites* showing nodes and longitudinal ridges strikingly similar to those of *Equisetum* are very common. Such fossils do not, however, represent the surface of the trunk, which was relatively smooth, but are casts of the pith cavity (Figure 28-39). The grooves represent the vascular bundles that projected into the pith. The pith cavity in some species

Coal has been found in all geological strata from the Precambrian to the Pleistocene. There are extensive beds in the Permian, Triassic, Jurassic, and Upper Cretaceous. For the earth as a whole, however, the most extensive and most valuable deposits date from the Carboniferous age (sometimes called the "Age of Coal") and are more than 300 million years old.

Coal is derived from partly decomposed and consolidated plant materials. The Carboniferous plants concerned in coal formation were numerous in kind, but the most important groups were (1) the lepidodendrons, sigillarias, and their relatives; (2) the ferns and seed ferns; and (3) the primitive, conifer-like Cordaitales (Chapter 30). These plants, growing in low swampy forests, died and partly decayed where they lived (Figure 21-10). Their roots, stems, leaves, and spores formed beds of peat, similar to those being formed today in many parts of the world. Decay was only partial, for the stagnant waters of the swamps in which the plants grew excluded oxygen. Wood-destroying fungi did not live under such conditions, and anaerobic bacteria disintegrated wood slowly. And so, year by year, the peat accumulated; this process could proceed year round, for the climate of the Carboniferous, in central Eurasia, western Europe, and North America, where the coal beds were laid down, was generally uniformly warm.

The vast swampy areas of peat and swamp forests—far more extensive than any in the world today—gradually subsided and were covered by the sea. The peat deposits were buried by clays; sand and silt washed in from the surrounding lands; and these sediments eventually became hard rock. Pressure and heat transformed the trapped peat into **lignite,** an immature, or low-grade, coal. The organic material was greatly compressed and underwent many physical and chemical changes during which much of its hydrogen and oxygen were driven off. This eventually changed the lignite into **bituminous,** or "soft," coal, having an average carbon content of about 66%.

Intermittent fluctuations in sea level resulted in successive layers of coal seams. As the seas receded and left vast inland swamps, swamp forests again became established and peat formed, to be covered in turn by sediments as the sea level rose. As many as 100 seams of coal have been found, one above the other, separated by layers of sedimentary rock.

The final stage in coal formation—**anthracite,** or "hard," coal—was produced by even greater pressures and temperatures applied to the coal layers. These pressures resulted from the upward and lateral thrusts of rock strata in periods of mountain building. Where the folding was greatest, the coal beds were altered from bituminous to anthracite coal. Volatile matter was driven out, and the carbon content increased to more than 80%. Most of the anthracite mined in the United States is found in eastern Pennsylvania. It was formed when the folding of upthrusting rocks resulted in the building up of the Appalachian Mountains. Since anthracite is found only in folded strata, it is much less abundant and more difficult to mine than bituminous coal.

SUMMARY

(1) The first vascular plants appear in the fossil record near the end of the Silurian period and especially in the Devonian. The Rhyniopsida (*Rhynia* and *Cooksonia*) were simple, dichotomously branched, rootless plants with terminal sporangia. The equally simple Zosterophyllopsida (*Zosterophyllum* and *Sawdonia*) bore sporangia in a lateral position.

(2) The first land plants probably arose from green algae which had an alternation of generations. During the migration and establishment of vascular plants on land, the gametophyte gradually became much reduced while the sporophyte became more elaborate and eventually constituted the dominant plant body.

(3) Plants such as *Psilophyton* (Trimerophytopsida) are believed to have evolved from *Rhynia*-like ancestors by the development of unequal branching and overtopping, resulting in a main axis with lateral branches. The megaphylls of ferns and seed plants are thought to represent flattened and expanded portions of such branched axes. The progymnosperms appear to bridge the gap between the trimerophytes and the diverse groups of gymnosperms (see figure). The Sphenopsida, or horsetails, may also have been derived from trimerophytes.

(4) Plants like *Asteroxylon* appear to represent the beginnings of the Lycopsida and suggest derivation from the Zosterophyllopsida (see fig-

Relationships among early land plants

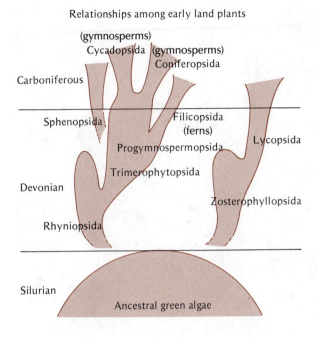

(8) *Selaginella* is heterosporous. Microsporangia and megasporangia are produced on sporophylls, which are always grouped in cones. The microspores and megaspores produce male and female gametophytes, respectively, largely within the spore wall. After the microspores are shed, sperms are produced. The megaspore produces a multicellular female gametophyte with accumulated food. After it falls to the ground, the spore wall splits, the gametophyte protrudes, archegonia are formed, and motile sperms bring about fertilization. The young sporophyte remains attached to the female gametophyte and derives nutrients from it.

(9) The Lycopsida were far more important elements of the earth's vegetation in the past than at present. The giant club mosses, *Lepidodendron* and *Sigillaria*, were large trees, conspicuous in Carboniferous forests. The modern club mosses (with the possible exception of the quillworts) apparently did not evolve from the giant club mosses and are not in the same evolutionary line.

(10) The horsetails (*Equisetum*) are the only living representatives of the Sphenopsida. The stems are hollow and jointed, and the reduced and fused leaves are borne in whorls at the nodes. The branches, when present, are also whorled.

(11) *Equisetum* is homosporous. The sporangia are borne on shield-shaped sporangiophores that are aggregated into terminal cones. The spores give rise to green gametophytes with erect lobes. Both kinds of sex organs are produced on the same gametophyte. Motile sperms bring about fertilization.

(12) The calamites, which were related to the modern horsetails, were the most conspicuous representatives of the Sphenopsida in the Carboniferous.

(13) Coal is composed of buried plant remains that have been subjected to heat and pressure. The major coal deposits of the world are of Carboniferous age and were formed in vast, low-lying swamps.

ure). Lycopod leaves are microphylls and have evolved as outgrowths or enations of the stem rather than as branches.

(5) There are two living genera of the Psilopsida, *Psilotum* and *Tmesipteris*. Although they have many apparently primitive features, their evolutionary origin is unknown.

(6) The chief living representatives of the Lycopsida are the species of *Lycopodium*, the club mosses, and *Selaginella*, the small club mosses.

(7) *Lycopodium* is homosporous. The sporangia are borne in the axils or on the surface of sporophylls, which may be similar to vegetative leaves or modified and grouped in cones. Both male and female sex organs are borne on the same gametophytes, which are small tuberous bodies. The motile sperm swims to the archegonium and effects fertilization, thereby initiating the sporophyte generation.

Chapter 29

THE FERNS

Ferns do not constitute a major part of the earth's vegetation, either in abundance or in number of species, and probably have never done so in spite of the widespread belief that the Carboniferous period was an "Age of Ferns." Yet since the latter part of the Devonian, about 370 million years ago, there have been ferns on this planet; their 10,000 or more species widely distributed today in many habitats show that the ferns are far from being a remnant nearing extinction. There are more than ten times as many species of living ferns as there are of living gymnosperms, but the ferns never form vast stands comparable to the coniferous forests. Because of their great antiquity the ferns are often considered a primitive group. Indeed, in certain aspects of body form and in their reproductive system they have preserved primitive features that are of great interest to students of plant evolution. On the other hand, modern ferns are in many respects highly specialized, and there is no reason to suppose that their evolutionary development has ceased.

Ferns are widely distributed from the moist tropics to beyond the Arctic Circle. The greatest number of species is found in the moister parts of the tropics, especially at higher elevations. But ferns also flourish in temperate regions, where they are familiar in woodlands and meadows and along roadsides and stream banks. They range in size from tiny, floating water ferns (Figure 29-27) to the tree ferns of the tropics, which attain heights of 6 to more than 18 m and bear crowns of huge leaves often 2 to 4 m long (Figure 29-25). They are generally found in moist and shaded habitats,

but some are able to grow in strong sunlight and even in markedly xeric environments. Adaptations to dry environments may be evident in stiff leathery leaves, sometimes with a coating of wax or overlapping scales. Most ferns of the temperate zone are terrestrial, growing on soil or rocks. Some ferns are climbers, and many—largely confined to the tropics—are epiphytes that live at various levels from just above the ground to the canopy of the forest. A remarkable adaptation to this mode of life is found in the bracket epiphytes (*Drynaria* and *Platycerium*) in the tropic and subtropic regions of the world. These are low-level epiphytes that grow on tree trunks. In addition to the large green leaves, specialized short sessile bracket leaves develop on the surface of the trunk. These grow over the rhizome, collect humus, and protect the roots from desiccation (Figure 29-1). The bracket leaves are persistent, and the strong veins that remain, even after the mesophyll decays, hold the humus as in a woven basket.

One of the most successful of the ferns is the bracken (*Pteridium aquilinum*) (Figure 29-2), which occurs in many parts of the world in several varieties. This plant successfully invades pastures and abandoned agricultural land in massive stands that are difficult and expensive to eradicate or even to control. As a result, bracken is classified among the world's important weeds, an unusual distinction for a fern.

Apart from the weedy bracken, which of course must be counted on the negative side, ferns cannot be said to have a great deal of economic significance. They are widely cultivated as ornamentals, both indoors and out, and have an enthusiastic following of amateurs in many parts of the world. The young coiled leaves, or fiddleheads (Figure 29-3), of several ferns are collected in early spring and cooked like asparagus. The fiddleheads of at least one species, the ostrich fern (*Pteretis pensylvanicum*), are harvested commercially in eastern Canada to be canned or frozen. The rhizomes and leaf stalks of two species of *Dryopteris* yield a medicinal product that has long been used in expelling tapeworms, but its use has declined in competition with synthetic compounds. Finally, the matted fibrous roots of several species of *Osmunda* (royal fern, cinnamon fern, interrupted fern)

Figure 29-A. Under side of the pinnately compound leaf of the fern *Pellaea viridis* (green cliff-brake). Many ferns have leaves with intriguing geometry and texture making them attractive as houseplants or outdoor ornamentals. This view reveals the regularly forked (*dichotomous*) venation characteristic of most fern leaves, in contrast with leaves of flowering plants which usually have parallel (monocots) or net (dicots) venation. Along their margins these leaflets bear numerous sporangia, appearing only as small dots at this magnification, which release the microscopic spores that reproduce the species.

Figure 29-1. Staghorn fern, a bracket epiphyte (*Platycerium* sp.).

are used as a growing medium for orchids and other epiphytic ornamentals.

Figure 29-2. Bracken fern (*Pteridium aquilinum*).

from year to year, and most ferns are therefore perennial. In temperate climates, the leaves usually die in the autumn, although certain species are evergreen. Although the stems of some ferns are upright, even becoming tree-like in form, in

EXTERNAL FORM

Like most other vascular plants, the ferns have roots, stems, and leaves. The stems ordinarily live

Figure 29-3. (A) "Fairy ring" of cinnamon fern fiddleheads. (B) Frozen fiddlehead greens.

Figure 29-5. Cinnamon fern *Osmunda cinnamomea*, with short stem and clustered leaves.

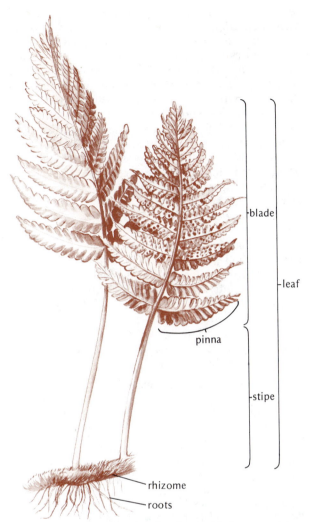

Figure 29-4. Plant body of a fern with leaves arising from a slender rhizome. Upper surface shown in frond on left, lower surface in frond on right.

most they are horizontal, sometimes creeping or climbing over the surface of rocks or tree trunks but usually buried in the soil. If underground, they are correctly referred to as rhizomes. The horizontal stems are slender and elongated with scattered leaves (Figure 29-4), or short and compact with clustered leaves (Figure 29-5). The parenchyma cells of the stems are usually filled with starch, thus providing efficient storage. The root system consists of numerous wiry or hair-like adventitious roots arising from the stem. The trunks of tree ferns are often clothed with a thick mantle of such roots, which contribute considerable support to the slender stem.

Many ferns reproduce by vegetative means, and for those growing outside the moist tropics this is often the major method of propagation. The creeping rhizomes branch as they grow through the soil or leaf mold. If the rhizomes die at the base, the branches persist as independent plants. This is usually the way in which colonies com-

posed of numerous individuals are formed. The manner of growth of several species of ferns, in which the rhizome is short and robust and the leaves form compact clusters, has attracted considerable interest. The branching rhizomes gradually radiate outward from a common center, the site of the original plant, and as older, central parts die and decay, a ring-like group of plants is formed (Figure 29-3A). Similar rings formed by gilled fungi are termed "fairy rings," and the same term may be applied to the ferns. One of these rings, formed by a cinnamon fern, had a diameter of more than 4 m and an estimated age of more than 300 years.

Fern leaves, commonly called **fronds,** consist of two parts, the stalk, or **stipe,** and the **blade** (Figure 29-4). The blade may be simple (undivided) but more commonly is pinnately compound. Each of the larger divisions of the blade of such a compound leaf is termed a **pinna,** and the pinnae in turn may be divided or dissected in various ways. The fern leaf grows in a striking manner. In the leaf of a flowering plant there is a brief initial phase during which growth is concentrated at the apex of the primordium, but this is quickly replaced by more generalized growth throughout the leaf. In the fern leaf growth at the apex continues for an extended period, and the tip of the growing frond becomes tightly coiled like a watch spring. When the frond expands it uncurls because there is more elongation on the inner side of the leaf than on the outer (Figure 29-3A). Uncurling fronds, the fiddleheads referred to previously, are among the attractive spring sights in woodlands and swamps. This pattern of leaf development is one of the distinctive features of ferns and is rarely found in other groups of vascular plants.

INTERNAL STRUCTURE

Internally the ferns are constructed mainly of the same types of cells and tissues that are found in seed plants. With only one exception [*Bot-*

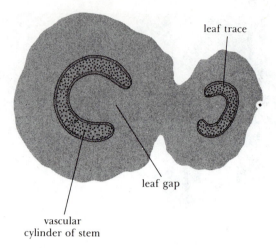

Figure 29-6. Leaf trace and leaf gap in cross section of rhizome of maidenhair fern (*Adiantum pedatum*).

rychium, the grape fern [Figure 29-22)] the living ferns, even the large tree species, are without a cambium or any development of secondary tissues. This is one of the main characteristics that from very early times has distinguished the fern line of evolution from the parallel seed plant line. Instead, the ferns have developed a diversity of patterns in the primary vascular system. Fortunately for students of evolution, the living ferns have preserved a wide range of types from the simplest and most primitive to the most complex and specialized. In some ferns the vascular tissues of the stem are arranged with a solid rod of xylem in the center surrounded by phloem. In others the xylem forms a cylinder surrounding a central pith with phloem lining the outside of the cylinder and often the inside as well (Figure 29-6). A conspicuous feature of such a vascular system is the presence of large leaf gaps associated with the departure of each leaf trace. A simple form of leaf trace, such as that found in the maidenhair fern (Figure 29-6), has the form in cross section of a horseshoe-shaped band of conducting tissue.

In other species two or even several traces connect the vascular tissues of the stem with each leaf, but all traces from one leaf are associated with a single gap. In many ferns the leaf gaps are so extensive, extending below the leaf trace attachment as well as above, that they overlap, and the cylinder is broken into a ring of strands or bundles if observed in cross section (Figure 29-7A), or a network if viewed in three dimensions (Figure 29-8). Elaborations beyond this point include, in different species, the development of bundles of xylem and phloem extending through the pith (Figure 29-9), or even of additional concentric cylinders inside the basic one.

The structure of the blade of the fern leaf, as seen in cross section, is very similar to that of the leaf of a flowering plant. But because most ferns grow in the shade, the leaf is usually thin and contains large intercellular spaces. The distinction between palisade and spongy tissue, so typical of many flowering plants, is not always evident. The roots of most ferns are thin and hairlike or wiry, and lack the complexity of the stem even in species in which the stem shows elaborate vascular patterns. The root cortex often contains sclerenchyma tissue and the vascular system is usually a solid core.

Regardless of the complexity of the vascular system of the adult plant, the juvenile or sporeling stage of most ferns begins with a simple core of xylem surrounded by phloem. As the plant enlarges, with increased diameter of successively formed regions of the stem and formation of larger leaves, this simple vascular system gives way to more complex patterns. This fact suggests that the vascular system of ferns may be rather flexible and subject to modifications in its development, and this has attracted the interest of experimental botanists. For example, is the development of the large leaf gaps characteristic of ferns caused by the leaves that are associated with them? In several species of ferns the apical meristem of the shoot was exposed and a sequence of leaf primordia was destroyed or suppressed as it appeared so that it did not develop. The segment of stem formed during the period of leaf suppression almost always lacked leaf gaps (Figure 29-7B). As soon as leaves were allowed to develop again, the newly formed stem had leaf gaps in its vascular cylinder. Thus the leaf does appear to bring about the formation of the gap associated with it.

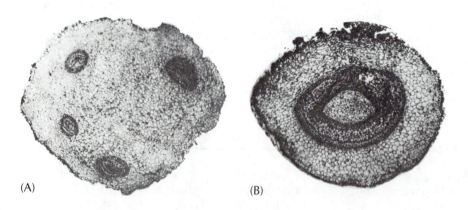

(A)　　　　　(B)

Figure 29-7. (A) Cross section of rhizome of *Onoclea*, showing ring of vascular bundles separated by leaf gaps. (B) Similar *Onoclea* rhizome, after suppression of leaf primordia. No leaf gaps are visible.

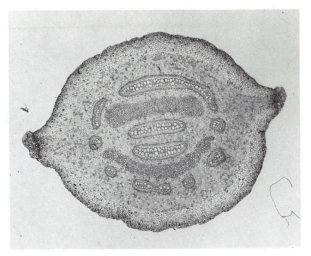

Figure 29-9. Cross section of bracken fern (*Pteridium*) stem with additional vascular bundles in pith.

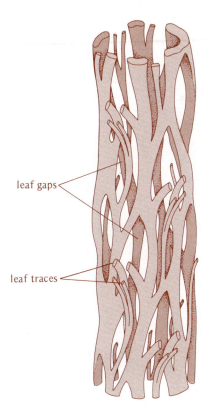

leaf gaps

leaf traces

Figure 29-8. Three-dimensional view of a fern stem vascular system. Large leaf gaps result in a network which appears as a ring of bundles in cross section.

The outer tissues of the stem, the cortex, and the pith often contain an abundance of thick-walled sclerenchyma cells in addition to parenchyma. These cells contribute to the rigidity of the stem and may compensate in part for the lack of secondary tissues. The sclerenchyma tissues often form complex patterns in association with the vascular system. Sections of the trunks of some tree ferns, suitably polished and laquered, make interesting and attractive ornaments, or they may be carved into vessels of various sorts. Strands or bands of sclerenchyma also may accompany the leaf trace or traces in their course through the stipe of the leaf.

THE LIFE CYCLE

Ferns, like other lower vascular plants considered in Chapter 28, reproduce sexually in a cycle involving independent diploid (sporophyte) and haploid (gametophyte) generations (Figure 29-10). The conspicuous fern plant we have been discussing above is the sporophyte, producing haploid spores by meiosis. They develop into small green gametophytes which lack leaves, roots, or vascular tissue and contrast strikingly with the sporophyte. These gametophytes produce sperms and eggs which, after fertilization, develop into new sporophytes. These processes are described in the following paragraphs.

Sori and Indusia

The sporophyte produces haploid spores in **sporangia,** and it is during the development of the sporangia that meiosis occurs. With a few exceptions that will be discussed later, the ferns are homosporous—that is, the spores and the sporangia are all of one type. The sporangia commonly develop in clusters, each of which is termed a **sorus** (Figure 29-11A), or, more popularly, a "fruit dot." Usually these are borne on the lower surface of a leaf and can be seen with the naked eye or a hand lens as small brown or black clusters. The sori occur in various arrangements—near the leaf margin, scattered over the surface, or in two rows, one on either side of the midrib of the pinna; if the pinna is subdivided, they may be found on either side of each subdivision. In some ferns, instead of individual sori, the sporangia occur in large masses covering considerable areas of the leaf surface or in linear bands near the leaf margin. Conversely, some ferns bear their sporangia individually.

Leaves that bear sporangia are known as **sporophylls** or fertile leaves in contrast to the sterile leaves that produce no sporangia and are entirely vegetative in function. This condition, in which two kinds of leaves, fertile and sterile, are borne on the same plant, is termed **dimorphism.** Often there is little or no difference in form between the two types of fronds. In other species, the photosynthetic tissue of the fertile leaf is reduced and may be lost completely in extreme cases. The fertile leaf may then bear little resemblance to a frond. There are also species, such as the interrupted fern, in which only certain pinnae are modified as spore-producing structures.

Associated with the sori of many ferns are thin, membranous structures known as **indusia**

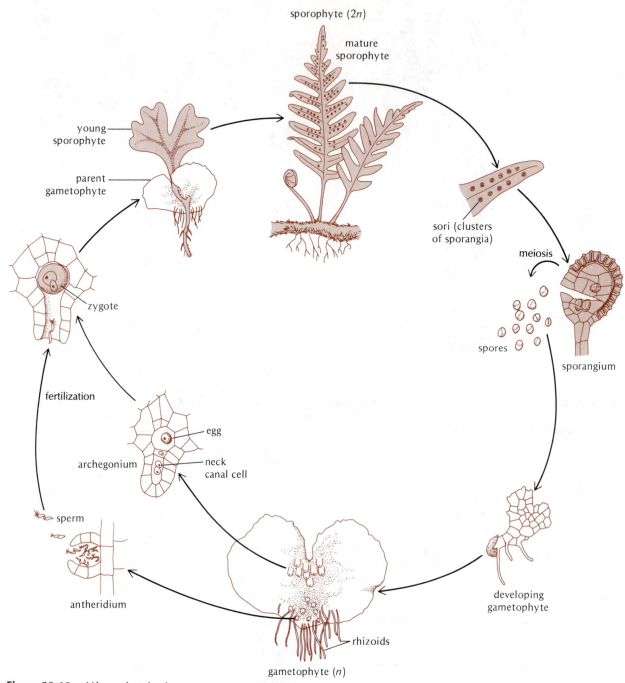

Figure 29-10. Life cycle of a fern.

(Figure 29-11B), which are produced as out-growths of the leaf surface. The form of the in-dusium varies in different species and genera and is a feature of great value in identification and classification of ferns (Figure 29-12). It may be cir-cular or kidney-shaped and attached to the leaf by a stalk. It may be elongated and attached along one side to form a sort of hood, or it may form a cup around the sorus. The indusium commonly covers the entire sorus, at least when young, and it is assumed to be protective in function. In some species a structure called a false indusium is formed by the inrolling of the leaf margin.

The Sporangium

Sporangia are so minute that they must be studied under the microscope to see the details of

Figure 29-11. Fern sori. (A) Without indusium, polypody fern (*Polypodium vulgare*). (B) With indusium, marginal shield fern (*Dryopteris marginalis*). The indusium, or shield, is the light membranous structure extending over the dark sori.

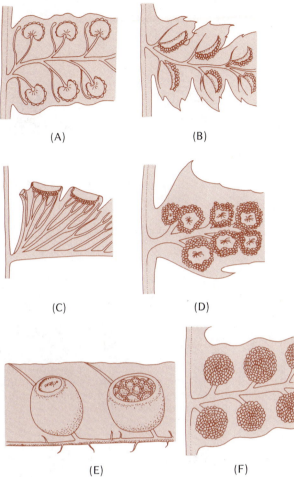

(A) (B)

(C) (D)

(E) (F)

Figure 29-12. Sori and indusia of ferns. (A) Marginal shield fern (*Dryopteris marginalis*). (B) Lady fern (*Athyrium filix-femina*). (C) Maidenhair fern (*Adiantum pedatum*); false indusium. (D) Christmas fern (*Polystichum acrostichoides*). (E) Tree fern (*Cyathea*). (F) Common polypody (*Polypodium virginianum*); indusium absent.

their structure. The sporangium (Figure 29-13A) consists of a hollow lens-shaped case on a short stalk. Flattened cells with thin walls form the sides and enclose a number of spores, commonly 64. Connecting the sides and arching over about two-thirds of the sporangium of many advanced fern species is a row of cells with peculiarly thickened walls. This is the **annulus,** which functions to open the sporangium when the spores are ripe and to discharge them. On one side the annulus is replaced by thinner-walled cells extending to the stalk. Two of these cells are called lip cells.

The Mechanism of Spore Discharge

The mechanism involved in the discharge of fern spores is remarkable. The sporangium breaks open, and the spores are violently discharged, falling near the parent plant or being carried away by air currents. Two stages are involved. In the first stage (Figure 29-13B),

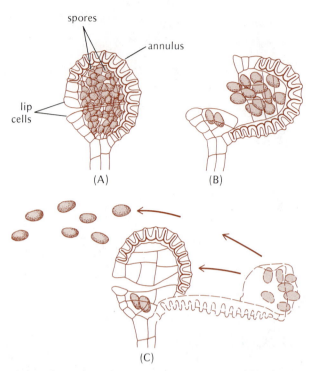

Figure 29-13. The fern sporangium and spore discharge. The top of the sporangium is bent backward by the annulus (A and B), carrying most of the spores with it. Then the annulus suddenly snaps forward (C), discharging the spores.

the sporangium opens slowly, carrying the majority of spores with it. In the second stage, the annulus, acting like a spring, suddenly snaps forward again, and the spores are thrown into the air (Figure 29-13C).

The opening of the sporangium is brought about by the annulus. The outer and side walls of the cells of the annulus are thin and flexible, but the inner tangential and radial walls are greatly thickened. When the sporangium is mature, each cell of the annulus is full of water. If the air is dry, this water tends to evaporate through the thin outer and side walls. This evaporation loss cannot be replaced, and the volume of water in the cells of the annulus decreases. The remaining water goes into a state of tension (negative pressure) comparable to that found in the xylem. The thinner walls of the annulus are drawn in, thus pulling the thick radial walls backward. As a result, the sporangium splits open between the lip cells. The lateral walls of the sporangium are torn, and the annulus slowly straightens out and bends backward. As it does so, it carries the bulk of the spores with it in a cup formed by the side walls of the sporangium.

Finally, as water continues to evaporate from the cells of the annulus, a point is reached when the water within the cells can no longer withstand the increasing tension. The water breaks in one cell, and the resulting shock causes a bubble of water vapor to form suddenly in each cell of the annulus. This releases the tension on the elastic radial walls, which then spring back into their first position. As a result, the top of the sporangium recoils with a speed that the eye cannot follow,

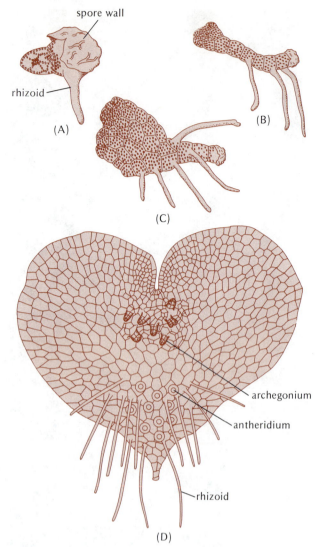

Figure 29-14. Development of the fern gametophyte. (A) Spore germination. (B and C) Stages in development showing the transition from a filament to a plate-like body. (D) Fully developed prothallus, lower surface.

and the spores are catapulted an inch or two into the air. The fern sporangium was used in some early attempts to measure the tensile strength of water in relation to the cohesion-tension theory of water movement in plants (Chapter 10).

The Gametophyte

Under suitable conditions the spore absorbs water, its wall breaks, and the first cell of the gametophyte appears. Cell division soon results in a green filament that becomes attached to the soil by the formation of several colorless rhizoids, the first of which emerges from the spore simultaneously with the filament (Figure 29-14A). After a usually brief period of filamentous growth, a change

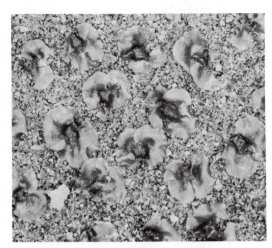

Figure 29-15.   Fern gametophytes growing on soil.

in the pattern of cell division at the apex results in the development of a flat, green, commonly heart-shaped body, some 6–12 mm in diameter, often called the **prothallus** (Figures 29-14B, C, D; 29-15). Most of the gametophyte is only one cell thick, but the central part just below the apical sinus or indentation forms a cushion several cells thick. Growing by means of a meristem in the sinus, the gametophyte bears the sex organs and rhizoids on its lower surface adjacent to the soil and is attached to the soil by the rhizoids (Figure 29-14D). Gametophytes are usually found in damp, shaded spots because, lacking roots or xylem for water transport or an effective epidermis, they are easily killed by drying.

This description applies to most ferns of both temperate and tropical regions but does not include all species. In some ferns the gametophyte is thicker and more massive and may be strap- rather than heart-shaped. Tuberous subterranean gametophytes much like those of *Psilotum* are found in some ferns that are believed to be primitive. At the other extreme are gametophytes consisting entirely of branching filaments. Some prothalli reproduce vegetatively by gemmae that readily establish new prothalli, sometimes leading to large clonal patches in which sporophytes are rare or even absent. One such species (*Vittaria lineata*) is not known to produce sporophytes at all, over a large part of its range in the Appalachian region of the United States. Such a fern could easily be mistaken for a bryophyte.

The Sex Organs

The sex organs, archegonia and antheridia, are multicellular (Figure 29-16A, C) and resemble those of other lower vascular plants and of bryophytes. Each archegonium contains a single egg, whereas a number of motile sperms are formed in each antheridium. Antheridia are most numerous on the basal region of the prothallus scattered among the rhizoids, and archegonia are located chiefly on the central pad near the notch. In most ferns both kinds of sex organs can occur on the same prothallus, but because they often develop at different times they do not mature simultaneously, and cross fertilization is therefore favored. More than one egg on the same prothallus can be fertilized, but normally only one of the zygotes develops into a new independent sporophyte.

The mature antheridium (Figure 29-16A) is dome-shaped and projects from the lower surface of the gametophyte. It consists of a jacket of cells surrounding a cavity that contains the sperms. These male gametes are spirally coiled and bear numerous flagella (Figure 29-16B). The antheridial jacket is only one cell thick and is composed of three cells, two of which are donut-shaped and extend entirely around the antheridium while the third, the cap cell, forms a lid. The archegonium is flask-shaped and consists of an enlarged basal portion buried in the tissue of the gametophyte, with a neck projecting from the surface (Figure 29-16C). The basal portion contains a large egg and another cell, the ventral canal cell, just above the egg at the base of the neck. The neck itself is composed of a cellular jacket that surrounds a single neck canal cell that is binucleate (rarely, three of four nuclei may be present).

If water is present between the lower surface of the gametophyte and the soil, the mature antheridium absorbs water and swells. The cap cell either tilts upward or is detached, thus permitting the sperms to escape. The mature archegonium likewise absorbs water. The ventral and neck canal cells separate from the neck of the archegonium and are extruded to the outside through the ruptured terminal cells of the neck. The sperms, attracted by a diffusible secretion, swim to the mouth of the archegonium and move through the canal toward the egg (Figure 29-16D). Although several

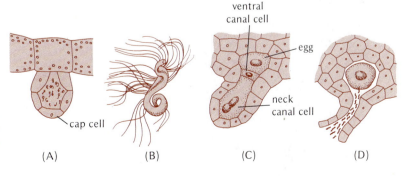

Figure 29-16.   Sex organs and fertilization. (A) Antheridium. (B) Sperm. (C) Mature but unopened archegonium. (D) Archegonium at the time of fertilization.

(A)   (B)   (C)   (D)

sperms may enter the archegonium, only one penetrates the egg, fuses with it, and forms a zygote.

The production of motile sperms and the necessity for water in the fertilization of ferns are ancient features that the ferns have retained from their evolutionary past. As a group they are relatively advanced, with a highly organized plant body and a long history of survival in a terrestrial environment. Nevertheless, they are in a sense amphibious in their mode of life. On the one hand, they depend upon a terrestrial environment for vegetative existence and for the release and dispersal of spores. On the other hand, like the algal progenitors of the land plants, they depend upon water as a medium through which the sperm can reach the egg. Ferns are not alone among living vascular plants in the retention of this primitive feature, for swimming sperms are found in other spore-producing plants (Chapters 27, 28) and even in some archaic seed plants. A somewhat analogous situation is found in the reproduction of some amphibians such as the frog, toad, and salamander. Such animals, although terrestrial or even arboreal as adults, deposit their eggs in the water where the progeny pass through larval stages. Like

the ferns, these animals have preserved an ancestral manner of life in certain phases of their life cycles.

The Embryo and the Young Sporophyte

The fertilized egg remains in the base of the archegonium and develops into a spherical mass of diploid cells. As growth proceeds, the organs of the sporophyte are developed from this embryo—the foot, first leaf, stem, and primary root (Figure 29-17). The foot is formed in the upper part of the embryo, and it grows between the cells of the gametophyte, from which it obtains food. The first leaf and the primary root are developed in rapid succession from other regions of the embryo, followed by the shoot apex that begins to form the shoot system. The first leaf and root soon emerge, the first leaf growing upward through the apical notch of the gametophyte and the root penetrating the soil. At this stage the sporophyte is an independent plant (Figure 29-18). The primary leaf, which is usually short-lived, is followed by a series of leaves of increasing size and complex-

Figure 29-17. Embryo development. (A) Early development of a spherical mass of cells. (B) Initiation of sporophyte organs. (C) Section through gametophyte and attached sporophyte. (D) Young sporophyte attached to gametophyte, seen from lower side (Figure 29-18).

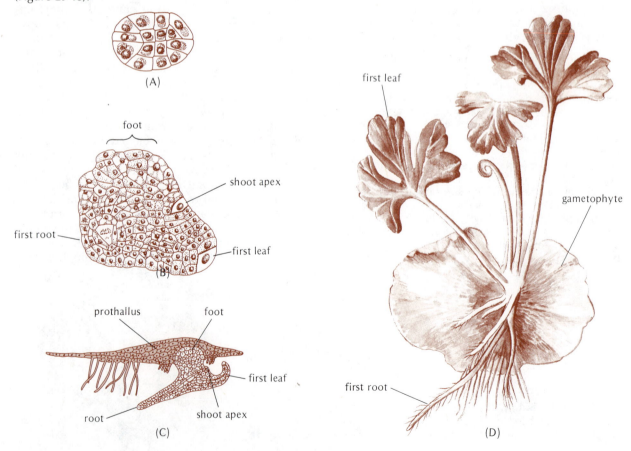

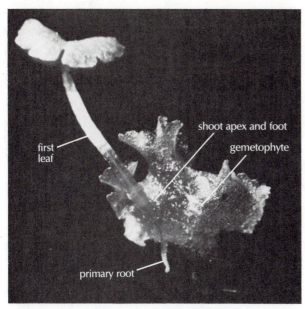

first leaf

shoot apex and foot

gemetophyte

primary root

Figure 29-18. Young sporophyte seen from the lower side of the prothallus. The sporophyte, although still attached, is essentially an independent plant at this stage.

ity until the adult form is attained. The primary root is soon replaced by adventitious roots arising from the stem. As the sporophyte develops, the prothallus shrivels and disappears.

Most fern sporophytes must grow for several years before they are able to produce sporangia and spores. However, one species of fern (*Anogramma leptophylla*) forms perennial subterranean gametophytes from which delicate, annual sporophytes arise.

Developmental Physiology of Fern Gametophytes

The typical heart-shaped fern prothallus has a simple structure consisting of relatively few cell types so arranged that they can be easily observed. There is evidence that their development is controlled by mechanisms similar to those that function in the more complex sporophytes of vascular plants. This, together with the ease with which they can be cultivated in large numbers, in sterile conditions if desired, has made them ideal for experimental studies on the physiology of development.

Although spores of a few fern species will germinate in the dark, most require light. This is a reasonable adaptation because gametophytes would be unlikely to survive without light for photosynthesis. In some cases red light promotes germination, and far-red light reverses this promotion. Thus the phytochrome system (see Chapter 14) apparently controls spore germination. Further, gibberellin can replace light in the germination process, as it does in some cases with seeds. Spores of at least one species of *Anemia* can be used as a sensitive and convenient bioassay for gibberellins.

The transition from filamentous to two-dimensional growth of gametophytes requires blue light; continued exposure to red light holds the gametophytes in the filamentous condition, blocking the nor-

mal transition (Figure 29-19). This represents a control of the orientation of cell division. Moreover, the direction of light influences development: light from below induces rhizoids and sex organs to form on the upper surface instead of in the usual position on the lower surface.

In old or injured gametophytes, cells far from the apical meristem can begin development anew, forming daughter prothalli. If separated from the meristem by cutting, even young cells undergo regeneration, indicating that the meristem exerts an apical dominance (inhibition) over its derivatives. In one species (*Pteris longifolia*), application of IAA can replace the action of the meristem as it does in the shoots of vascular plant sporophytes.

Sex Hormones

Hormones (**antheridiogens**) that induce the formation of antheridia were first discovered in bracken (*Pteridium aquilinum*) when it was noted that culture media from gametophyte cultures or water in which gametophytes had been floated for one or two days caused antheridia to develop prematurely in younger cultures. The active substance is effective in very dilute concentrations and in a wide range of species in 20 different genera, many or all of which produce the same substance. However, many other ferns do not respond to this substance, and several other groups of ferns apparently have their own antheridiogens. At least one of these groups shows a comparable response to gibberellic acid, but chemical data indicate that the natural hormone, although similar to gibberellic acid, does not correspond to any known gibberellin. Antheridiogens presumably serve under natural conditions to correlate in time the development of sexual structures on nearby gametophytes, improving the chances of cross fertilization.

Apospory and Apogamy

The leaves of sporophytes of some ferns can give rise to diploid gametophytes by the growth of individual cells into filaments and ultimately into two-dimensional structures (Figure 29-20A). This ordinarily happens when the leaf is in contact with moist soil or with the culture medium when sporophytes are grown in culture. This phenomenon is called **apospory**. Such diploid gametophytes may even form sex organs and give rise to tetraploid sporophytes after fertilization. Conversely, a gametophyte may give rise directly to a sporophyte, without the intervention of sex organs or fertilization, a phenomenon known as **apogamy** (Figure 29-20B). One or more cells of the prothallus grow into a kind of embryo that organizes shoot and root apices and develops into a vascular sporophyte having the same chromosome number as the gametophyte. Apogamy occurs naturally in the reproduction of several ferns, in which it is coupled with a lack of meiosis in the sporangium of the sporophyte, so that the spores and resultant gametophytes are diploid and the entire life cycle occurs at the diploid level. It may also be coupled with apospory, which yields the same result.

Gametophytes of many ferns that do not normally undergo apogamy may do so under experimental conditions. Apogamy and apospory, like the formation of

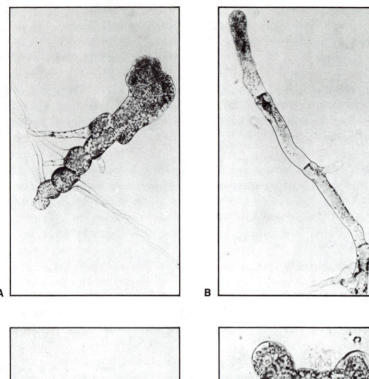

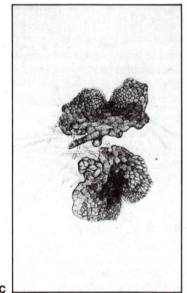

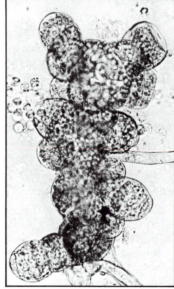

Figure 29-19. Young gametophytes (of the same age) of bracken (*Pteridium*) which have been grown in white light (A) and in red light (B). In white light, which includes wavelengths in the blue range, the growth has become two-dimensional, but in red light the filamentous condition persists. The lower pictures show gametophytes of the sensitive fern (*Onoclea*) cultured with (D) and without (C) antheridiogen from *Pteridium* in the culture medium. Note the antheridia and the release of sperms in (D). (C) and (D) are not at the same magnification.

embryos and haploid plants from pollen cells, show that the striking developmental differences between sporophyte and gametophyte generations are not really due to their difference in chromosome number, but result from different developmental programs whose expression is controlled in some other way, normally coupled to the haploid and diploid phases of the life cycle. One possible basis for the developmental divergence between the two generations is the different conditions under which the initiating cells develop— the spore released independently as a single cell, the zygote contained and nourished within the archegonium during its early development. Removal of the zygote from an archegonium before any divisions have occurred in it has been achieved experimentally in the fern *Todea*. Zygotes that were removed and cultured 4 to 5 days after fertilization developed slowly over a period of several months on a nutrient medium, giving rise to small structures that resembled gametophytes, not sporophytes. Embryos removed at a later stage, after they had become multicellular, gave rise to sporophytes, although their development was often abnormal. This experiment supports the idea that the environment in which development begins may determine which developmental program is expressed.

EVOLUTION AND DIVERSITY OF FERNS

The ferns, or **Filicopsida,** arose in the Devonian period, probably from *Psilophyton*-like predecessors (Trimerophytopsida; Chapter 28). This evolutionary line, in contrast to that leading to the seed plants, retained the free shedding of spores

Figure 29-20. Apospory and apogamy in *Pteridium*. (A) Apospory. Gametophytes have arisen from a sporeling leaf which was cut off and placed on an agar medium in a culture tube. These gametophytes are diploid. (B) Apogamy. A sporophyte has arisen directly from a gametophyte without sexual fusion. The sporophyte is haploid.

and failed to develop secondary tissues. The ferns flourished in the Carboniferous period, although they were not the dominant group, and many of the Coal Age ferns became extinct before the end of the Paleozoic era. Others have persisted to the present time as part of the modern fern flora. New families of ferns appear in the fossil record of the Mesozoic and Cenozoic eras, and a wide diversity of modern ferns is the result. Taxonomists have tried to group them naturally, making use of such structural features as the grouping of sporangia in sori, the position of sori on the leaf, the presence or absence of indusia and their form if present, and the type of annulus or opening and discharge mechanism of the sporangia. For some of the per-

sisting ancient families the story is relatively clear, but for the more modern ferns the relationships are obscured by the fact that the same kind of specialization seems to have occurred independently in parallel lines. It is not surprising, therefore, that there are a number of different classification schemes, no one of which is truly satisfactory.

The evolution of the sporangial discharge mechanism illustrates this difficulty (Figure 29-21). The most advanced ferns have a vertical annulus that, as explained earlier, lifts the top off the sporangium and hurls the spores outward from the sporangial cluster. Other ferns have an oblique annulus that approximates this situation but is probably less effective. Still others

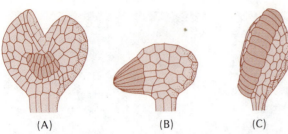

(A)   (B)   (C)

Figure 29-21. Sporangia of three ferns showing diverse opening mechanisms. (A) Cinnamon fern (*Osmunda cinnamomea*) with a lateral patch of cells of the annulus type. (B) Climbing fern (*Lygodium*) with cells of this type at the apex. (C) Filmy fern (*Hymenophyllum*) with an oblique annulus.

Figure 29-22. *Botrychium,* one of the grape ferns.

have cells of the annulus type at the apex of the sporangium or in a patch on the side wall; these cells open the sporangium laterally but do not discharge the spores. This mechanism is much less effective than the vertical annulus because it does not free the spores from the clustered sporangia. The most primitive ferns have no sporangial opening mechanism at all. Traditionally, all of the ferns with a vertical annulus were classified in one large family, the Polypodiaceae, but it is now recognized that this advanced condition has been achieved independently several or many times. Modern schemes of classification recognize from 5 or 6 to more than 30 families in this assemblage, depending on the interpretation of evolutionary lines.

Some Representative Fern Groups

The following incomplete survey of some of the orders and families of living ferns will give an idea of their diversity.

Adder's Tongue and Grape Ferns (*Ophioglossales*)

This order, with only one family and three genera, is morphologically primitive and apparently very ancient, but fossils are unknown. Only the grape ferns (*Botrychium*), with some 20 to 30 species, are considered here. They are mostly small plants, a few centimeters high, although one North American species may attain a height of more than 60 cm. The aerial portion of the shoot is usually a single frond composed of a sterile, expanded, leaf-like portion and an erect fertile spike (Figure 29-22). The sporangia are borne in two rows, one on each side of the divisions of the fertile spike, and are very large, containing from 1500 to 2000 spores. When mature they split open transversely, and the spores fall away—there is no annulus or mechanism of spore discharge. The gametophyte is a flattened tuber-like body up to 2 cm long, which is buried in the soil and obtains nutrients through the activities of an associated fungus.

Marattiales

This order of tropical ferns includes six genera and about 200 species. The leaves are large and coarse and arise either from stout upright trunks 60 cm or more in height or from creeping horizontal stems. The sporangia are large and contain many spores, but have little or no annulus; they occur in sori and are often fused into compound structures. The gametophytes are large, relatively thick, dark green thalloid structures. The history of this group can be traced back to the Carboniferous, and they are well represented in fossils of that period.

Filicales

Most living ferns are placed in this order, grouped into a number of families, only a few of which will be discussed here.

Osmunda Family (Osmundaceae). The osmundas, with three living genera and 20 species, are large, coarse ferns with a geological record extending back to the Permian. The common cinnamon fern (*Osmunda cinnamomea*) is strongly dimorphic (Figure 29-5). The sporangia, borne on the margins of the much-reduced segments of the fertile frond, are not arranged in sori. They are large, short-stalked, globose structures, each producing up to 500 spores. The annulus is rudimentary, and there is no mechanism of spore discharge. The gametophytes are heart-shaped, large, and long-lived.

Climbing Fern or Curly Grass Family (Schizaeaceae). This is another ancient family that can be traced back to the Carboniferous. The family includes five genera and more than 150 species, chiefly in the tropics, but two species, a climbing fern (*Lygodium*) (Figure 29-23) and a curly grass (*Schizaea*), occur in eastern North America. In a tropical climbing fern the

Figure 29-23. Climbing fern, *Lygodium.*

Figure 29-24. (A) An epiphytic filmy fern (*Hymeno-phyllum polyanthos*), Central America; about 7 inches high. (B) A different species of *Hymenophyllum*, leaf detail.

twining leaves, which have apparently unlimited growth, may attain a length of 30 m or more, the longest leaves of any plant. The large sporangia, characterized by a cap-like annulus (Figure 29-21B), are not clustered in sori. The sporangium dehisces longitudinally, and there is no mechanism of spore discharge.

Filmy Ferns (Hymenophyllaceae). The filmy ferns are so named because the leaves are extremely thin—usually only one cell layer thick except for the veins. The 600–650 species are mostly restricted to the moist tropics or subtropics, where they grow on the soil or as epiphytes, but a few species extend northward into eastern United States. They may be upright (Figure 29-24A) or creeping, with fronds ranging from a few centimeters to nearly 60 cm in length. The sporangia are short-stalked and have an oblique annulus (Figure 29-21C). They are attached to an elongated axis and enclosed at the base by a cup-like or two-lobed indusium. The filmy character of the leaf is due to evolutionary reduction and is an adaptation to conditions of high humidity. These ferns are, ecologically and in other ways, highly specialized. They probably have a long history, but the fossil record is scanty.

Tree Ferns. The tree-like habit appears in several families of ferns but is most characteristically developed in the two families Cyatheaceae and Dicksoniaceae. Some investigators place the two in a single large family. There are also members of these families that are not tree-like but are associated by other characteristics. The sporangia are grouped in sori with indusia, and the sporangium has an oblique annulus like that of the filmy ferns. The tree ferns are typically found in tropical mountain forests (Figure 29-25), and several species are cultivated in conservatories. Their fossil record goes back to the Jurassic.

Advanced Ferns. By far the largest number of fern species, both tropical and temperate, belong in this category, which formerly consisted of a single large family called the Polypodiaceae. It is now divided into several families, including a more restricted Polypodiaceae. Much of the general discussion of the ferns in the earlier part of this chapter, including the sporan-

Figure 29-25.   Tree ferns in Malasia.

Figure 29-26.   Cloverleaf fern, *Marsilea drummondii*.

Figure 29-27.   Two members of the Salvineaceae: *Azolla caroliniana* and *Salvinia auriculata* (larger, with leaf hairs).

gium with an incomplete vertical annulus and the gametophyte consisting of a green heart-shaped prothallus, pertains particularly to these advanced ferns. As previously explained, they represent the endpoints of a number of parallel evolutionary lines. Much of this evolution took place in the Cenozoic era.

Water Ferns (*Marsileales* and *Salviniales*)

Two divergent fern groups that have assumed an aquatic or semiaquatic habitat and have been distinct since at least the Cretaceous period must be mentioned. Each of them is placed in its own separate order, an indication of the gap dividing them from the other ferns and from each other. Unlike any other known ferns, both of these groups are heterosporous, and the gametophytes are reduced and retained within their respective spores, much as in *Selaginella* (Chapter 28). The Marsileaceae—represented by *Marsilea*, the cloverleaf fern—are rhizomatous and grow in shallow water or in extremely moist sites subject to periodic flooding (Figure 29-26). Their megaspores and microspores are produced within extremely hard and resistant sporocarps that resist drying and have been found to be viable after many years in an herbarium. The Salvineaceae are extremely reduced, free-floating aquatics that are scarcely recognizable as ferns to the casual observer (Figure 29-27).

SUMMARY

(1) Ferns have a long history extending back to the Devonian period. They are today widely distributed in many different environments and, although not a major component of the earth's vegetation, show evidence of evolutionary vitality.

(2) The fern stem is commonly a rhizome, but upright, tree-like forms occur. The leaves (fronds) develop by extensive apical growth and are coiled when young. The root system is adventitious. Fern vascular systems develop without a cambium, but they may become highly complex. Vegetative reproduction is common.

(3) The fern life cycle has two independent generations contrasting in size and complexity. The sporophyte is dominant, possessing stem, leaves, roots, and vascular system. The gametophyte is a small, liverwort-like organism with none of these structures.

(4) Sporangia are commonly grouped in clusters called sori, which are borne on the lower surfaces or margins of ordinary fronds, or on specialized fronds or parts of fronds. A sorus may be either naked or associated with an indusium. The most advanced ferns have a vertical annulus that opens the sporangium and ejects the haploid spores.

(5) The spore develops into a heart-shaped, haploid prothallus bearing rhizoids and multicellular sex organs on its lower side. The sperms released from the antheridium are motile and swim to the archegonium where the egg is fertilized. This aquatic method of fertilization is regarded as a retention of an ancestral feature.

(6) The fertilized egg develops into an embryo sporophyte composed of a foot, a primary leaf, a primary root, and a shoot apex. The young sporophyte is for a time dependent upon the gametophyte from which it absorbs water and nutrients through the foot. It soon becomes independent, and the gametophyte shrivels and disappears.

(7) The ferns appear to have evolved from the Devonian Trimerophytopsida and flourished in the subsequent Carboniferous period. Many types became extinct by the end of the Paleozoic era, but new families appeared during the Mesozoic and Cenozoic. The classification of ferns is difficult because similar evolutionary specializations occurred in parallel lines leading to the modern ferns.

Chapter 30

THE SEED PLANTS: GYMNOSPERMS

The gymnosperms are a diverse group of vascular plants, both living and extinct, whose common feature is the production of seeds borne on the surface of an appendage and not enclosed, as in the angiosperms, within a pistil. The name gymnosperm means "naked seed." The gymnosperms are believed to include several different evolutionary lines, and their classification reflects this. In this text three classes are recognized—Cycadopsida, Coniferopsida, and Gnetopsida; some botanists subdivide them even further.

Besides the production of naked seeds, the gymnosperms share another common feature—the development of a cambium and the formation of secondary tissues. They are all woody plants, and this feature has set them apart from the ferns since the Devonian origin of both groups. This does not mean that their body is stereotyped, however. In fact, gymnosperms range from the stately conifer trees, which produce some of the finest timber, to small prairie, desert, and mountaintop shrubs; they include the soft-wooded and often squat cycads with fern-like leaves and the strange desert plant *Welwitschia*, (Figure 30-31), a plant many botanists consider the most bizarre that has ever existed. With just over 700 modern species, the gymnosperms actually contribute less to the world's flora than either the ferns or the club mosses, but both economically and ecologically the gymnosperms are much more important than either of the latter groups. In some regions, as in mountains of the temperate zones and the boreal forest of the north, they are dominant. Moreover, the gymnosperms have been much more prominent in the past; for two-thirds of the Mesozoic era they dominated the entire earth's land vegetation.

NATURE OF THE SEED

Although the seed is an important departure from the lower vascular plants' type of reproduction by freely shed spores and independent gametophyte and sporophyte generations, seed development morphologically is simply a modification of the ancient pattern of alternation of generations. A seed plant is a sporophyte that is heterosporous, like *Selaginella*, producing both megaspores and microspores. The structure that produces megaspores is the **ovule,** which consists of a megasporangium (often called the **nucellus** in seed plants) surrounded by an **integument** (Figure 30-1). Within the megasporangium a megaspore mother cell undergoes meiosis, producing four megaspores, of which only one usually functions, the others soon degenerating. However, rather than developing a resistant wall and being released (as in *Selaginella*), the megaspore remains instead within the ovule attached to the parent sporophyte and there develops into a female gametophyte.

Microsporangia are also formed, producing numerous microspores. The microspores, as in *Selaginella*, form protective walls and are released, but before being shed each begins to develop into a male gametophyte enclosed within the spore wall, becoming a **pollen grain** (collectively, pollen). By some agent such as wind or insects, pollen is transferred to the ovules. In gymnosperms the transfer occurs directly to the ex-

Figure 30-A. Gymnosperms include some of the most impressive of all plants, the giant sequoias or Sierra Redwoods (*Sequoiadendron giganteum*), which occur as isolated groves in the Sierra Nevada mountains of California at elevations where a great deal of snow (often 3–6 m or 10–20 ft) falls each winter. Seen here, in the "Giant Forest" of Sequoia National Park, is the huge snow-decorated crown of a relatively young giant sequoia, a few hundred years old; to the right, part of the massive trunk of a mature tree, probably more than 1000 years old; these trunks can reach 12 m (40 ft) in diameter at the base. Some individual trees still alive and vigorous today were in their youth or early maturity at the time of Christ. These old trees are often scarred from forest fires, lightning strikes, and pest attacks they have sustained and survived over thousands of years. Part of their secret for longevity is their thick, fibrous, fire-retarding and pest-repelling red-brown bark.

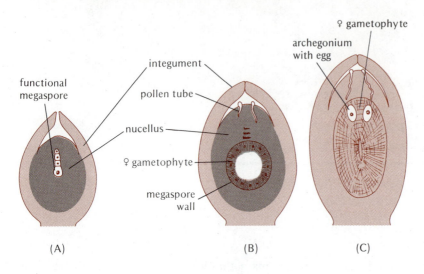

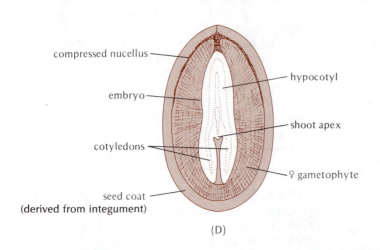

Figure 30-1. Development of the seed in gymnosperms. (A) Longitudinal section of an ovule showing a tetrad of haploid megaspores within the megasporangium or nucellus; only one megaspore functions; the nucellus is surrounded by an integument. (B) The functional megaspore is developing into a female gametophyte; pollen has been transferred to the ovule and pollen tubes are growing toward the female gametophyte. (C) A pollen tube grows to an egg in an archegonium and releases sperms, and fertilization occurs. (D) A diploid embryo sporophyte has developed, surrounded by nutritive tissue of the female gametophyte; the seed is essentially mature.

posed ovules, rather than to an enclosing pistil as in angiosperms.

Germination of a pollen grain results in the formation of a **pollen tube** that contains two sperms when mature (Figure 30-1). In a few apparently archaic gymnosperms the sperms are motile and, when released within the ovule, swim to the eggs. In most gymnosperms and all angiosperms, however, the sperms are nonmotile and are delivered to the egg by the growth of the pollen tube. Fertilization is followed by the development of an embryo sporophyte. This is surrounded by a tissue containing reserves of nutrients. In gymnosperms this tissue is simply the female gametophyte rather than the specifically developed endosperm that occurs in angiosperm seeds (Chapter 16). In either case the reserve nutrients that will feed the embryo during seed germination are manufactured not by the gametophyte, as in lower vascular plants, but are drawn from the parent sporophyte. The ovule has now matured into a seed.

The seed habit confers considerable advantage to those plants possessing it, a fact demonstrated by the seed plants' widespread success. Unlike lower vascular plants, development of the gametophyte does not depend on continuously moist conditions. The sexual cycle can be completed wherever the sporophyte generation can thrive. The seed habit also eliminates the necessity for external water for fertilization, an important restriction upon lower vascular plants. Even in those gymnosperms that have motile sperms the necessary fluid medium is provided internally within the ovule. The nonseed plants, even when otherwise well adapted to life on land, must, at least figuratively, "return to the water" to complete sexual reproduction. The seed plants, like the reptiles, birds, and mammals, do not require a watery environment for fertilization.

In Chapter 28 the evolution of the land plant spore with its cutinized wall was described as an essential aspect of colonizing the land. The spore evolved as an agent of dispersal. In seed plants the spore's function has been greatly modified. The megaspore no longer serves for dispersal and is very unspore-like in morphology. The microspore's dispersal function, however, is retained;

Figure 30-2. A cycad (*Encephalartos transvenosus*) growing in South Africa. The seed cones are clearly visible.

when it becomes the pollen grain it functions to bring the male and female gametes together. In essence, it disperses the male gametophyte to the site at which it can produce sperms with reasonable prospects of effecting fertilization.

THE CYCADOPSIDA

The Cycadopsida or cycadophytes are characterized by a palm-like growth habit with relatively large, often pinnately compound leaves, a weakly branched or unbranched and either columnar or short and conical axis, and a rather poor development of secondary xylem containing a great deal of parenchyma mixed with the conducting cells. Three orders are grouped in this class of gymnosperms, one still living but clearly a relict much more important in the past, and two already extinct. The living cycads are considered first.

The Cycads (Cycadales)

The cycads are a small group of 10 living genera and about 100 species, found in the tropics and subtropics. The fossil record shows that the modern cycads are the survivors of a group that was widely distributed and very abundant during the Mesozoic era, particularly from late Triassic to early Cretaceous times. Since the Cretaceous, they have been restricted in numbers and distribution. Four genera are found in the western hemisphere, the largest, *Zamia*, occurring in Florida and from Mexico to Chile. The other six genera are native to Australasia or South Africa.

The stem is typically unbranched, with a terminal crown of long, leathery, compound leaves (Figure 30-2). In some species the stem is very short and tuber-like; in others it is columnar. Although certain cycads attain a height of 18 m, most are usually not more than 1 m high. They also tend to be long-lived, with ages of as much as 1000 years estimated for individual specimens.

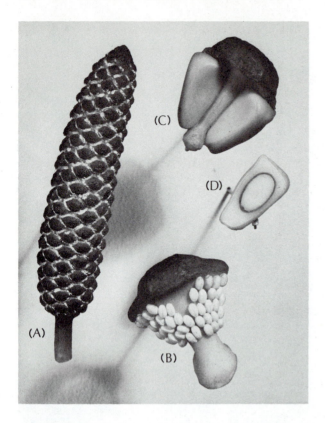

Figure 30-3. Reproductive structures of the cycad *Zamia floridana*. (A) An entire microsporangiate (pollen-producing) cone; (B) a single scale-like microsporophyll removed from a cone and shown with many microsporangia located on its lower (abaxial) surface; (C) a single megasporophyll from a megasporangiate cone with two ovules; (D) an ovule cut longitudinally to show the female gametophyte developing in its interior.

Several species are cultivated as ornamentals in conservatories, where nonbotanists often confuse them with palms. The starchy stems and seeds of some cycads are used locally as food, but they must be used with caution and are often specially prepared because of the presence of toxins that can cause severe gastrointestinal disorders and even paralysis. Similar effects have been reported in domestic animals that browse on cycad foliage.

REPRODUCTION

Cycad sporophytes are dioecious, separate individuals producing the microsporangia and megasporangia. Microsporangia are borne on scale-like microsporophylls in compact cones (Figure 30-3). The megasporophylls bearing ovules are also usually grouped in cones, but in the genus *Cycas* the megasporophylls resemble small foliage leaves and form a loose cluster around the stem apex (Figure 30-4). Megasporangiate cones are often spectacularly large. One Australian species produces cones nearly a meter long and weighing as much as 85 to 90 pounds.

Within each ovule a female gametophyte develops from one of the megaspores produced by meiosis, becomes filled with nutrients derived from the parent sporophyte, and initiates several archegonia, each containing a large egg. These are located on the surface of the gametophyte under the **micropyle,** an opening in the integument at the tip of the ovule. Pollen released from the microsporangia is transferred to the ovules, probably by the wind, although some observers believe that insects may be involved. Pollen grains are caught in a droplet of fluid exuded from the micropyle and are drawn into the ovule as the droplet dries, coming to rest on the megasporangium or nucellus. The pollen grain germinates, producing a pol-

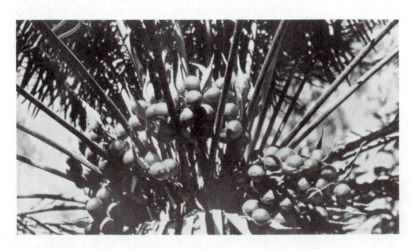

Figure 30-4. A cycad (*Cycas media*), with seeds borne on leaf-like structures arranged in loose clusters. The plants are commonly about 3 m tall with leaves that can be over 1 m long. This genus occurs from Japan to Australia, India, and Madagascar.

len tube that penetrates the megasporangium and absorbs nutrients from it. Two large motile sperms are produced which at first swim about in the base of the pollen tube and are ultimately released near the archegonia. These sperms may be as much as 180 μm in diameter with 20,000 spirally arranged flagella and are visible to the naked eye. Fertiliza- tion then occurs, often in more than one arche- gonium, but ordinarily only one embryo develops. This completes the development of the seed, which on germination yields a new cycad sporophyte (Figure 30-5). From the time of pollination, 4 to 6 months may elapse before fertilization and as much as a year before the seed is mature.

Figure 30-5. Life cycle of a cycad. Microsporangia and megasporangia are borne on sporophylls in cones on separate plants. Haploid microspores begin the development of male gametophytes and are shed as pollen grains containing several cells. In the megasporangium or nucellus within the ovule, one functional megaspore gives rise to a haploid female gametophyte with archegonia, each containing an egg. Pollen grains come to rest in a cavity of the nucellus called a pollen chamber. A pollen grain ger- minates, producing a pollen tube, and ultimately two large, motile sperms are formed. This is the fully developed male gametophyte. The sperms are released and swim to the archegonia. Fertilization is followed by the development of an embryo sporophyte surrounded by tissue of the female gametophyte. A new cycad sporophyte is estab- lished when the seed germinates.

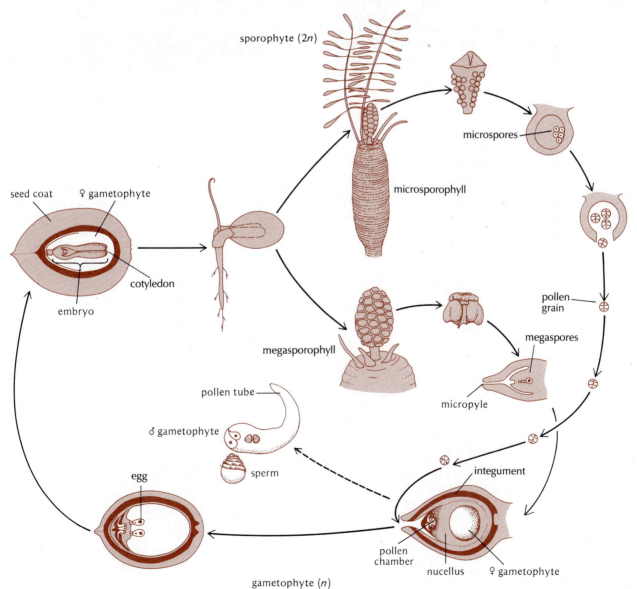

(A)

(B)

Figure 30-6. *Cycadeoidea,* an extinct cycadophyte of the Mesozoic. (A) Reconstruction of a plant as it probably looked when living. (B) A petrified trunk of *Cycadeoidea.* This speciment was discovered in Maryland in 1860. The trunk is covered by closely packed leaf bases with cones (the small, nearly circular clusters) scattered among them.

Symbiotic Nitrogen Fixation. Cycads are noteworthy for a number of species that harbor symbiotic, nitrogen-fixing cyanobacteria in their roots. The association apparently contributes significantly to the nitrogen supply of the plant communities to which these cycads belong. If, as seems likely, the Mesozoic cycads possessed this feature, they may have been the first vascular plants to conduct symbiotic N_2 fixation.

Extinct Cycadophytes

The Cycadopsida also include two groups of entirely extinct seed plants. The **Cycadeoidales** were a striking and prominent feature of the vegetation in many parts of the world during Upper Triassic, Jurassic, and early Cretaceous times. Their history is much like that of the cycads, which they resemble morphologically and to which they show considerable relationship. The best-known member of the group, *Cycadeoidea,* had a barrel-shaped trunk, usually a meter or less high, covered by an armor of persistent leaf bases (Figure 30-6A). It bore a terminal crown of palm-like leaves. Petrified trunks of these plants make beautiful specimens that have long been prized by collectors, including the Etruscans who included one among the objects of reverence in a burial chamber more than 4000 years ago (Figure 30-6B). The reproductive structures of cycadeoids were cones which differed from those of cycads in having both microsporangia and megasporangia in the same cone. These cones were complex structures that superficially had a rather flower-like appearance, leading to speculation that flowering plants might have evolved from cycadeoids. However, the organization of these cones actually is quite different from that of an angiosperm flower. The cycadeoids became extinct before the end of the Cretaceous.

The seed ferns (**Pteridospermales**) were a much older group that were most abundant and widely distributed in the Carboniferous period but persisted into the Jurassic. They were very fern-like in appearance, particularly their leaves, which, unless reproductive structures are present, are indistinguishable as fossils from those of true ferns. Some were undoubtedly large plants resembling modern tree ferns (Figure 30-7A); others were smaller with slender stems that were probably supported by surrounding vegetation. The seed ferns differed from true ferns, however, in having a cambium that often formed a considerable amount of secondary xylem. Seed ferns bore pollen sacs and seeds directly on their fronds in various positions (Figure 30-7B) but not in cones. The presence of naked seeds places these plants among the gymnosperms, distinct from the ferns that they resemble vegetatively. The seed ferns are generally believed to have given rise to both the Cycadales and the Cycadeoidales toward the end of the Paleozoic era, but fossil evidence documenting this transition has not yet been found.

THE CONIFEROPSIDA

The Coniferopsida, which differ in important characteristics from the cycadophytes, represent a second major line of gymnosperm evolution. The modern conifers, with their often tall trunks and abundant branches (Figure 30-22), perhaps best illustrate the habit of this group, although some

Figure 30-7.   Seed ferns (*Pteridospermales*). (A) *Medullosa,* a seed fern of the Carboniferous period. (B) Seeds of the seed ferns were borne in various positions on the fernlike leaves, including (top) the tips of pinnae or (below) directly on subdivisions of a pinna.

species do have a shrubby habit. The leaves are simple and in general relatively small, and are usually but not necessarily needle-like or scalelike. Moreover, the coniferophytes are characterized by vigorous cambial activity and the formation of tough secondary xylem containing only a small amount of parenchyma. There are three orders: the Coniferales, which, although more prominent in the past than at present, are still a flourishing group; the Ginkgoales with but one surviving species; and the extinct Cordaitales.

The Conifers (Coniferales)

Extensive conifer forests occur in both the northern and southern hemispheres but are found chiefly in temperate regions or, in the tropics, at high elevations. With some 550 species, the conifers show considerable diversity and include the tallest, the largest, and the longest-lived plants known. Although still a vigorous group, they are as ancient as the cycads, extending back into the Permian. The fossil record shows that the conifers reached the peak of their evolutionary development in mid-Mesozoic times, when, with the Cycadales and the Cycadeoidales, they dominated the earth's vegetation. Many of the petrified logs found in the Petrified Forest National Park in Arizona and elsewhere are remains of an extinct conifer (*Araucarioxylon*) of Upper Triassic age (Figure 21-1C). These extinct conifers have several living relatives, none of which occur in the northern hemisphere.

The conifers are mostly evergreen and include such well-known trees as pine, spruce, fir, cedar, Douglas fir, hemlock, and the sequoias. Two genera found in North America, larch and bald cypress, are deciduous. Some species of juniper and yew are shrubs, but neither the Coniferales

Figure 30-8.   Pollen cones of eastern white pine (*Pinus strobus*). About twice natural size. Note the dark, pointed bract at the base of each cone.

Figure 30-9.   A hard pine, western yellow pine (*Pinus ponderosa*). Two one-year (immature) cones and a cone that has opened and shed its seeds.

nor any other order of gymnosperms includes herbaceous plants. Conifer leaves are often needle-like or even scale-like, but the needles may be very long; some conifers of the southern hemisphere have relatively broad leaves.

The wood or secondary xylem of conifers was described in detail in Chapter 13. The conifers provide a very large proportion of our lumber and raw material for the pulp and paper industry. Their wood is adaptable to a great variety of uses and is easily seasoned and manufactured. Conifers include many species famous for their beauty as well as for their valuable timber.

The Coniferales ("cone-bearers") are characterized by the production of cones. As in the cycads, these are of two types, **microsporangiate** or **pollen cones** and **megasporangiate** or **seed cones**, also called **ovulate cones** when immature. The pollen cones are relatively small and are borne singly or in clusters (Figure 30-8). They are frequently a bright shade of red or yellow. They are ephemeral, lasting but a few days, and often go unnoticed. They shrivel and drop from the tree soon after the pollen is shed. The seed cones are usually woody (Figure 30-9). They differ greatly in size, depending upon the species, ranging in length from about 10 mm to nearly 90 cm (2 ft)—a length attained by the cones of the sugar pine (*Pinus lambertiana*) of the western United States. In certain conifers such as junipers and yews, the ovulate cone is highly modified and resembles a berry. Although some species of conifers are dioecious, most are monoecious. The two kinds of cones are usually borne on separate branches of the same tree.

LIFE HISTORY OF THE PINE

Pines (genus *Pinus*) are the best known and in many regions the most common of the conifers. They are plants of the northern hemisphere: only one species extends below the Equator as a wild plant (northern hemisphere pines have been planted extensively as a forest crop in New Zealand and Austrialia, however). The life cycle of the pine (Figure 30-10) provides another example of seed formation and illustrates the characteristic reproductive processes of all conifers.

Pollen Cones and Pollen.   The pollen cones of pines appear in the spring (Figure 30-8). Each cone is composed of a central axis to which are attached numerous spirally arranged scale-like microsporophylls, each bearing two elongated microsporangia on its lower surface (Figure 30-11). Within each microsporangium many microspore mother cells undergo meiosis, forming tetrads of microspores. Each microspore develops, over most of its surface, a thick wall composed of an inner and an outer layer. The outer layer expands on

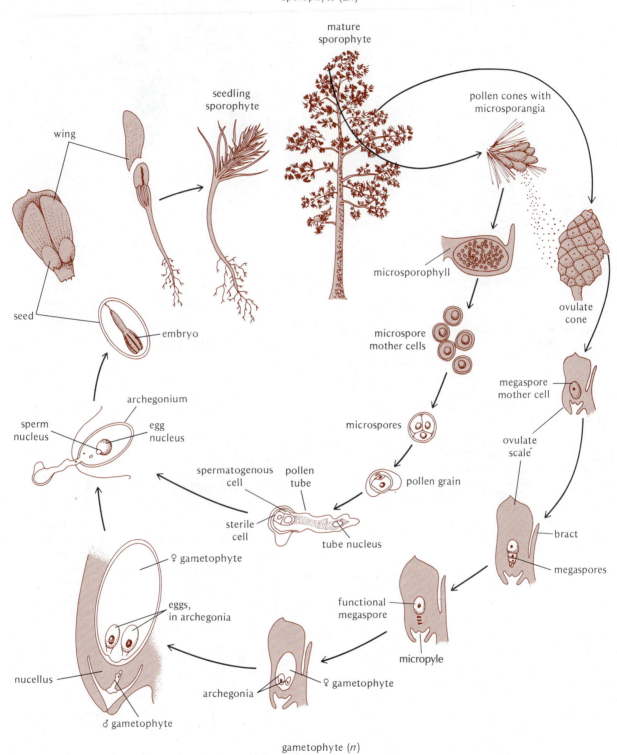

Figure 30-10. Life cycle of pine. Pollen cones and ovulate cones are produced on the same tree (sporophyte). Pollen is transferred by the wind to the ovules, where development of the male gametophyte is completed in the tissues of the nucellus. Unlike those of the cycads, the sperms are nonmotile and are carried to the archegonium by the pollen tube. There is a long interval between pollination and fertilization and the seeds mature near the end of the season after that in which the cones appeared.

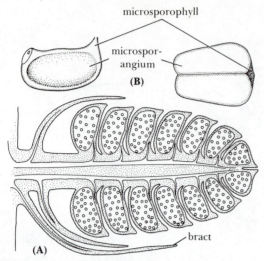

Figure 30-11. Male reproductive structures in pine. (A) Longitudinal section of pollen cone. (B) Microsporophyll and microsporangium, view from side and below.

either side, forming a wing or bladder filled with air (Figure 30-12).

While the microspore is still within the sporangium, its nucleus undergoes several divisions, forming a mature pollen grain (Figure 30-12B). This consists of a much reduced male gametophyte enclosed within the microspore wall.

Two of the four cells that compose the pollen grain are termed **prothallial cells.** These cells are vestigial and soon degenerate. The remaining two cells are a small **generative cell** and a larger **tube cell.** Winged pollen grains similar to those of pines are also found in the spruces, firs, and other conifers. On the other hand, in larch, hemlock, Douglas fir, and many others the pollen grains are wingless. The wings probably aid in dissemination by wind.

Each pollen cone is a short lateral branch, as shown by its position in the axil of a modified leaf (bract), which is conspicuous before the pollen is shed (Figures 30-8 and 30-11).

When the pollen grains are mature, the axis of the pollen cone elongates somewhat, separating the microsporophylls. The microsporangia (now called **pollen sacs**) open by a longitudinal slit, and the pollen grains are liberated and carried away by the wind. The pollen is produced so abundantly that it may form a cloud of sulfur-colored dust in the vicinity of the tree.

Ovulate Cone and the Ovule. The ovulate cones also appear in the spring, growing erect at the ends of branches (Figure 30-13). A cone consists of a central axis bearing scales (Figure 30-14), each carrying two ovules on the upper side. Their micropyles point inward toward the axis of the cone. Each ovule (Figure 30-15) consists of an integument surrounding and united with a megasporangium or nucellus. The opening of this integument forms the micropyle. Further prolonga-

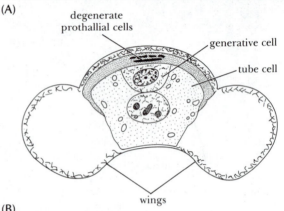

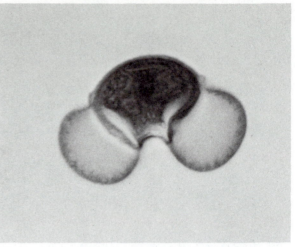

Figure 30-12. (A) Mature pollen grain of pine. Note the clear air bladders (wings) on either side. (B) Male gametophyte within a pollen grain.

tions of the integument result in the formation of an arm on either side of the micropyle. The megasporangium contains a single large megaspore mother cell; this cell, following meiosis, gives rise to four megaspores arranged in a row. Three of the megaspores disintegrate, but the fourth, that farthest from the micropyle, becomes the functional megaspore (Figures 30-10 and 30-16).

Each ovulate cone scale develops in the axil of a scale-like bract. The position of the scale in the axil of a bract indicates that it is a much modified, short lateral branch, not a sporophyll.

Female Gametophyte. Within the nucellus the megaspore grows and its nucleus divides, yielding a large number of free haploid nuclei. Cell walls then form between the nuclei. The resulting multicellular haploid tissue is the female gametophyte (Figure 30-15). During the later stages of its development, two to five archegonia, much reduced compared with the archegonium of a fern, differentiate at the micropylar end of the gametophyte. The archegonium consists chiefly of a small and variable number of neck cells located just above a large egg. A sheath of specialized cells

(A)

(B)   (C)

Figure 30-13.   (A) Ovulate cone of white pine at the time of pollination. (B) Ovulate cone at the end of the first season of growth. (C) Seed cone, end of the second season of growth; the cone has opened, allowing the seeds to fall away.

surrounds the egg. As the female gametophyte develops, its cells become filled with food materials derived from the parent sporophyte. These reserves support the development of the embryo after

fertilization and, remaining in the mature seed, nourish the seedling during germination.

Pollination and Fertilization.   Pollen is discharged for only a few days during the spring or early summer. At the time of pollen discharge, the scales of the small ovulate cones spread apart (Figure 30-13A). Pollen grains drift between the scales, lodging upon the rim of the micropyle and upon the micropylar arms on either side (Figure 30-16, left) in a fluid secretion ("pollination droplet") which exudes from the micropyle. The pollen grains apparently are carried to the nucellus by this fluid, as it evaporates or is reabsorbed.

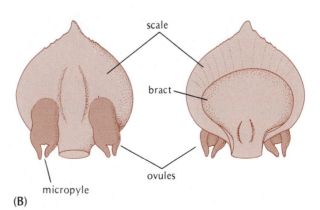

(B)

Figure 30-14.   (A) Ovulate cone of pine, longitudinal section. A, cone axis; B, bract; O, ovule; S, cone scales. (B) Scales of an ovulate cone of *Pinus* showing the upper (left) and lower (right) surfaces. The ovules are borne on the upper surface. From below the scale is seen to be located in the axil of a bract.

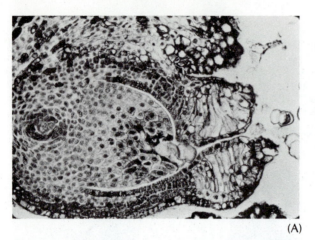

(A)

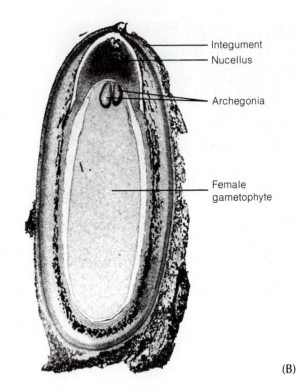

- Integument
- Nucellus

- Archegonia

Female gametophyte

(B)

Figure 30-15. (A) Pine ovule in longitudinal section. Note swollen micropylar arms, indicating that pollination has occurred. (B) An ovule of *Pinus* cut longitudinally to show the female gametophyte (with archegonia) inside the nucellus or megasporangium. This in turn is surrounded by the integument. As the female gametophyte enlarges, the nucellus is compressed except in the region above the archegonia. Here it forms a thick cap on which the pollen grains germinate and through which the pollen tubes grow.

After the pollen enters the ovule, the ovulate scales grow until they are closely pressed together (Figure 30-13B). In many pines, the cone gradually bends downward by curvature of its stalk (Figure 30-13C). Within the ovule the pollen grain germinates, producing a pollen tube that penetrates the tissue of the megasporangium (Figure 30-10). During this process, further mitotic divisions within the male gametophyte form two nonmotile male gametes or sperms, one commonly larger than the other.

The generative cell of the pollen grain divides into a **sterile cell** and a **spermatogenous cell,** the spermatogenous cell again dividing to yield the two sperms. A fully developed male gametophyte in pine thus contains two degenerated prothallial cells, a tube nucleus, a sterile cell, and two sperms.

The pollen tube eventually penetrates between the neck cells of an archegonium and discharges its sperms into the egg. The larger of the sperms fuses with the egg nucleus and the remaining male gametophyte nuclei disintegrate.

The method of achieving fertilization in conifers is considered to represent an evolutionary advancement over that in cycads. Conifer sperms lack the power of independent motion and are delivered to the archegonium by the pollen tube, which in the cycads serves only a nutritive function. In conifer reproduction the final vestige of the vascular plants' aquatic ancestry has been lost.

Embryo and Seed.   The fertilized egg gives rise, after nuclear divisions and cell wall formation, to a **proembryo** (Figure 30-17A) composed of 16 cells arranged in four tiers of four cells each. The proembryo is located at the end of the egg away from the neck of the archegonium. The subapical cells of the proembryo greatly elongate to form a **suspensor** (Figure 30-17B), pushing the tip cells of the proembryo deep into the female gametophyte. The four apical cells of the proembryo divide, each giving rise to a separate embryo plant

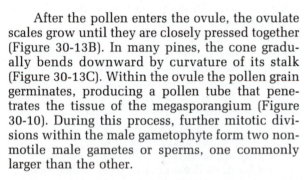

- nucellus
- megaspore mother cell
- integument
- micropyle
- pollen grain
- micropylar arm

- functional megaspore
- linear tetrad of megaspores
- nucellus
- pollen grain
- swollen portion of micropylar arms

Figure 30-16.   The pine ovule at the time of pollination. (Left) Pollen grains in the pollination droplet. (Right) Ovule shown in cross section after pollination has taken place.

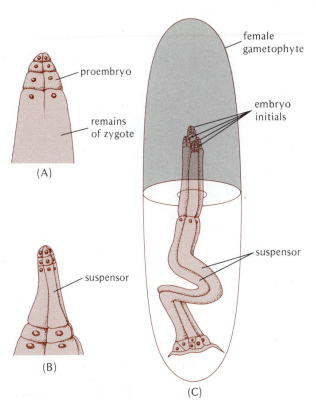

Figure 30-17. Embryo development in pine. (A) Proembryo with four tiers of cells. (B) Cells below the tip elongate to form the suspensor. (C) The tip cells, as they are forced into the nutritive tissue of the female gametophyte, give rise to four separate embryo plants, only one of which survives.

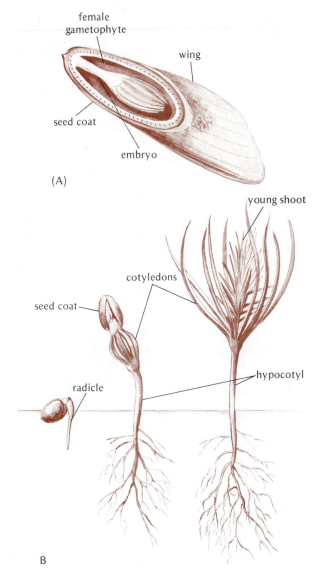

Figure 30-18. Seed and seedlings of pine. (A) Mature seed cut to show embryo embedded in the female gametophyte. (B) Stages in germination.

(Figure 30-17C). Only one of the embryos ordinarily survives, the remainder being eliminated by competition within the female gametophyte.

Pine embryo development begins with a phase of free nuclear division preceding any wall formation, which is a general characteristic of gymnosperm embryos. In conifers this phase is of limited extent (four nuclei in pine), but in cycads several hundred free nuclei are usually produced before wall formation begins. This remarkable process, unique among plants, is also found in insects.

The mature embryo of pine (Figure 30-18) is composed of a plumule, hypocotyl, radicle, and a variable number of cotyledons (about eight). It is embedded in the female gametophyte, which in turn is surrounded by a hard seed coat derived from the integument of the ovule. In most pines part of the upper surface of the ovulate cone scale splits off as a wing attached to the mature seed. These seeds, dispersed by wind, may be carried many meters from the parent tree.

The seeds of pine, and to some extent other conifers, form an important source of food for seed-eating birds, rodents, and other animals. The seeds of the piñon pine (*Pinus edulis*) of the southwestern United

States were long used as a staple in the diet of the Indians of that region and are still collected to some extent for sale.

In pines the seeds do not mature in the same growing season in which pollination occurs. The interval between pollination and fertilization is a very long one, in contrast to the condition in angiosperms, in which fertilization usually occurs within a few hours or days after pollination. In the eastern white pine (*Pinus strobus*), for example, fertilization takes place approximately a year after pollination, and the cones mature by late summer of the second season. However, most of the north temperate conifers other than pines, such as spruce, fir, larch, and hemlock, have an interval of only a few weeks between pollination and fertilization. The seeds of these conifers ripen in the same season in which pollination takes place.

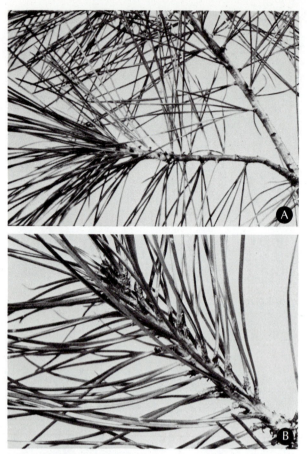

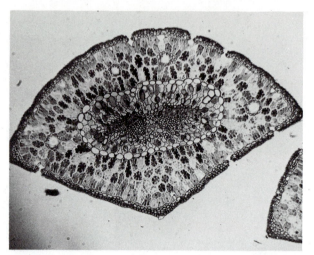

Figure 30-20. Cross section of a needle from a hard pine (*Pinus palustris*), showing the two vascular bundles characteristic of the hard pine leaf.

Figure 30-19. Branchlets of soft (A, white pine) and hard (B, Scotch pine) pines showing needle clusters.

Germination. Pine seeds germinate (Figure 30-18) similarly to those of many angiosperms. The radicle emerges from the seed coat and penetrates the soil. The hypocotyl becomes arched as it grows and appears above the soil. It then straightens and carries the cotyledons and seed coat upward. The cotyledons absorb nutrients from the female gametophyte until the supply is exhausted; the seed coat then drops away. The cotyledons become green and carry on photosynthesis, persisting for one or two growing seasons. The plumule, surrounded by the whorl of cotyledons, develops into the stem and leaves of the mature sporophyte.

Some Modern Conifers

Pine Family (Pinaceae). The pines (genus *Pinus*) have been a conspicuous feature of the earth's forests since the end of the Mesozoic. They are easily recognized by their long, needle-like leaves in clusters of usually two, three, or five. These clusters are actually greatly reduced lateral shoots. Pines fall into two groups—the soft, or white, pines with five needles in a cluster, and the hard pines with two or three needles

in a cluster (Figure 30-19). Another distinguishing feature is that the soft pines contain only one vascular bundle in each needle whereas the hard pines have two (Figure 30-20). The summer wood of hard pines is usually heavier and harder than that of soft pines, making the growth rings appear more pronounced in hard-pine wood.

One soft pine, the eastern white pine (*Pinus strobus*), was formerly one of the most important timber trees in North America. Accessible stands were mostly used up more than a century ago, and only remnants now remain of the great forests of this species that once covered much of southern Canada and the northeastern and Great Lakes states of the United States. A number of other white pines occur in the Rocky Mountain and Pacific Coast states, the sugar pine and the western white pine being currently important timber trees.

Among the more important eastern two- and three-needled pines are the red pine of northeastern and lake states and species of southern yellow pines. The latter grow very rapidly and flourish on cut-over and burned-over areas in poor, sandy soils. The yellow pines constitute one of the most important tree crops of the South and are used extensively in reforestation. Another southern hard pine is the long-leaf pine, *Pinus australis*, with needles up to 45 cm (18 in) long. This species and the slash pine (*Pinus elliottii*) are the source of most of our rosin and turpentine. Yellow pines, notably the ponderosa pine, are also important timber trees in western North America. The two-needled Austrian and Scotch pines, native to Europe, are commonly used for ornamental planting in the East and Midwest.

In 1957 it was discovered (by counting growth rings) that certain bristlecone pine trees (*Pinus longaeva*; Figure 19-A), growing in an arid mountain region in east-central California, surpass the longevity record of the giant sequoias, previously believed to include the oldest living trees. Several bristlecone trees are 4000 years old or more, and one more than 4600 years. In 1964 an even older tree—4844 years old—was found in eastern Nevada at an altitude of 3300 m (10,700 ft). The tree was

Figure 30-21. (A) Branchlet of American larch (*Larix laricina*). (B) Leaves and cones of hemlock (*Tsuga canadensis*). (C) White spruce (*Picea glauca*). (D) Douglas fir (*Pseudotsuga menziesii*).

6.5 m (21 ft) in circumference 45 cm above ground. Most of the tree was dead, but it bore a living shoot 3.5 m (11 ft) high. Tragically, the U.S. Forest Service allowed this oldest and most venerable of all living organisms on earth to be cut down and killed.

The Pinaceae include a number of important genera in addition to the pines. The larches (*Larix*), also known as tamaracks, are deciduous conifers of nothern forests. Most of the short needles of larch are borne in close clusters on short side branches (Figure 30-21A).

Figure 30-22. A redwood (*Sequoia sempervirens*) stand in California.

The hemlocks (*Tsuga*) are widely distributed in cool, moist forest soils. The needles are short, flat, and spirally arranged on the branchlets, but in most of the species the leaves, by a twist of their bases, seem to grow from opposite sides in two-ranked, flat sprays (Figure 30-21B). Firs (*Abies*) occur mainly in subarctic and cold-temperate regions of the northern hemisphere. The firs are popularly known as balsam or fir-balsam. The leaves are short, usually blunt and flat, resembling those of hemlock, but leave a round, flat scar when they fall. The cones, unlike those of other North American conifers, grow erect and do not fall complete; instead, the scales fall away singly, leaving intact a pointed central axis that usually persists for several years. Spruces (*Picea*), which also occur mainly at colder latitudes and altitudes, have short, flattened or four-sided, often sharply pointed needles. When the needles fall, they leave short, peg-like projections on the branchlets (Figure 30-21C). One of the most important timber trees in western North America is the Douglas fir (*Pseudotsuga*). The leaves are flat, resembling those of fir. The cones are readily recognized because the bracts protrude prominently from beneath

each scale (Figure 30-21D). Fir, spruce, and Douglas fir are widely used as Christmas trees.

Redwood Family (Taxodiaceae). This interesting family is today limited in its range and number of species. The most prominent North American species are the big tree, or giant sequoia (*Sequoiadendron giganteum*), the closely related redwood (*Sequoia sempervirens*) (Figure 30-22) of the West Coast, and the bald cypress (*Taxodium distichum*) found in the southeastern United States and extending up the Mississippi Valley to Indiana and Illinois. In ancient times, however, the family was widely distributed over the northern hemisphere. The ancestors of the present-day sequoia and redwood go back to the Jurassic. Both trees have reddish-brown fibrous bark, 30 cm or more thick on large trees and highly resistant to fire.

One of the most exciting scientific discoveries of modern decades was made in 1946 when the "dawn redwood" (*Metasequoia*), previously known only from fossils, was found growing in central China (Figure 30-23). This tree, a close relative of redwood and sequoia, was a prominent forest tree during the Tertiary. It dis-

Figure 30-23. Dawn redwood (*Metasequoia*).

Figure 30-24. A branch of western juniper, with "berries."

appeared from North America by the end of the Miocene and was believed extinct, but it persisted in a limited area in Asia where it escaped scientific notice. It is now cultivated widely for both its attractive form and its scientific interest.

Included in the family are the tallest and greatest in bulk of living things. The world's tallest tree is a redwood, 113 m (372 ft) high. The greatest in bulk is the General Sherman tree, a giant sequoia growing in Sequoia National Park. This tree is 83 m (273 ft) high, with a circumference of 25.25 m (84 ft) at the base. It is difficult to determine the age of a living sequoia, but the General Sherman is conservatively estimated to be at least 3500 years and perhaps as much as 4000 years old.

The bald cypress also attains considerable size and age. The leaves appear in two ranks and, still attached to lateral branchlets, fall away at the end of each growing season.

Cypress Family (Cupressaceae). In this family the leaves are usually small and scale-like, occurring in pairs or threes closely pressed against the stem. Among important western trees are western red cedar (*Thuja plicata*) and Port Orford cedar (*Chamaecyparis nootkatensis*). In the East, red cedar (*Juniperus virginiana*) and white cedar or arborvitae (*Thuja occidentalis*) are widely distributed.

Many species of cypress (*Cupressus*) and other members of the family are cultivated as ornamentals, and many horticultural varieties have been developed. All members of the family except the genus *Juniperus* bear small, dry cones. In *Juniperus* the cone scales are fleshy and grow together, forming a berry-like structure about the size of a small pea, blue at maturity (Figure 30-24). A number of junipers are low, hardy shrubs, extensively used as ornamentals in landscaping.

The use of the word cedar in the popular names of certain members of this family is misleading. The true cedars are members of the genus *Cedrus,* of the Pinaceae. They are native in the Himalayas, Asia Minor, and Lebanon. The Cedar of Lebanon, the famed cedar of the Bible, belongs to *Cedrus.* Cedar is used in North America as a common name, applied to various kinds of trees in different parts of the region.

Araucaria Family (Araucariaceae). The living members of this family occur naturally only in the southern hemisphere. The group is very ancient, however, and was abundant in North America during the Mesozoic. A species of *Araucaria,* called Parana pine, is one of the most valuable timber trees in South America, in certain areas taking the place of the pines of the north. Several araucarias, such as the monkey puzzle from Chile, the bunya-bunya from Australia, and the Norfolk Island pine (Figure 30-25), are cultivated in greenhouses and grown out-of-doors as ornamentals in the warmer parts of the United States, especially along the West Coast.

Yew Family (Taxaceae). The small family of yews includes both trees and shrubs. The female reproductive structure is not cone-like but consists of a single seed surrounded by a fleshy covering. In the yews (*Taxus*) this covering is cup-like, about the size of a large pea (Figure 30-26), and scarlet at maturity. Yews are widely employed in ornamental plantings. They are shapely and slow growing, and have attractive, glossy, short, flat leaves. Various species of the family are native to North America, Europe, and Asia. The English yew, a tree native to Europe, western Asia, and North Africa is long-lived, and numerous specimens of great girth are recorded. The very strong wood of this tree was once prized for making the bows that brought fame to English archers.

Figure 30-25. Norfolk Island pine (*Araucaria heterophylla*).

The Ginkgo (Ginkgoales)

The ginkgo or maidenhair tree, *Ginkgo biloba* (Figure 30-27) is the only living species of the order Ginkgoales. This unusual tree is native to China and apparently still exists in the wild state in a

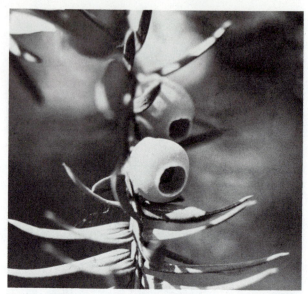

Figure 30-26.   A yew (*Taxus*) twig, with a red "berry."

mountainous area in the southeastern part of that country. Long cultivated in China, it was introduced to Japan and thence to Europe in the early part of the eighteenth century. As a group, however, the Ginkgoales are of very ancient lineage, and the order contains numerous extinct genera and species. The group originated in the Permian, was distributed worldwide by Jurassic times, and began to decline in the Cretaceous. The single remaining species is the sole survivor of a group that for millions of years occupied a prominent place in the world's vegetation.

The ginkgo is a large tree, known to reach a height of more than 30 m (100 ft). It is strongly excurrent in growth, at least for many years. The leaves are broadly fan-shaped, 5 to 10 cm across, and often divided by a deep depression into two

Figure 30-27. (A) Maidenhair tree (*Ginkgo biloba*). (B) Ginkgo branch, showing leaf shape and fleshy seeds.

Figure 30-29. *Ephedra* ("Mormon tea"), commonly found in the Sonoran Desert region of the United States.

Figure 30-28. *Dorycordaites*, a member of the extinct Paleozoic Cordaitales which are believed to have given rise to the modern conifers.

symmetrical halves (Figure 30-27). The dichotomous (division into two more or less equal parts) venation and general shape of the leaves resemble those of the leaflets of the maidenhair fern (*Adiantum*), accounting for the common name of the tree. Unlike most living gymnosperms, the ginkgo is deciduous. It is widely cultivated for shade and ornamental purposes in North America and other parts of the world. It is remarkably resistant to insect and fungus pests and seems to thrive in the polluted environment of modern cities. Although the ginkgo is not particularly conifer-like in general aspect, examination of its anatomical details, particularly those of its extensive secondary xylem, shows that it is properly placed among the Coniferopsida.

The ginkgo is dioecious, that is, the pollen and seeds are borne on different trees. The microsporangiate strobilus is a loose, catkin-like structure consisting of an axis with appendages, each of which bears a pair of microsporangia. The ovules are borne, not in strobili, but singly or in pairs on the end of a long stalk. After pollination by wind, pollen tubes with large, motile sperms are produced, and fertilization takes place. The ovules develop into yellowish, fleshy seeds (Figure 30-27), about the size of a large cherry but with a rancid

odor that people will cross the street to avoid. For this reason, microsporangiate or pollen-producing trees are favored for urban plantings. The seeds, however, are eaten like nuts in China, usually after roasting.

From the evolutionary point of view, the reproductive system of the ginkgo is highly significant because it is remarkably similar to that of the living cycads. The pollen tube functions only as an absorbing structure, and motile, flagellated sperms resembling those of cycads swim to the archegonia. This method of fertilization was probably common to all ancestral gymnosperms.

The Cordaitales

These entirely extinct Paleozoic trees resembled modern conifers in size and general form (Figure 30-28). Their tall trunks were as much as 30 m high. The high branches bore numerous strap-shaped leaves whose length varied in different species from 15 cm to as much as a meter. The leaves were tough and leathery with internal xeromorphic modifications such as abundant sclerenchyma tissue. Compressions of cordaitalean leaves are common fossils in the shales associated with bituminous coal. The wood structure was like that of some living conifers. This group attained its greatest development in the Upper Carboniferous, but lived from the Lower Carboniferous into the Permian, when it died out completely. It is generally believed that the Cordaitales constituted the ancestral stock from which the conifers arose. This interpretation is supported by a remarkably complete fossil record of Permian and Triassic age that documents the transition from cordaite to conifer characteristics.

THE GNETOPSIDA

This small group of only three genera, very different from one another as well as from all other gymnosperms, is apparently an evolutionary sideline. *Ephedra* (Figure 30-29) is an extremely xeromorphic shrub of temperate, mostly arid regions, with scale-like leaves. *Gnetum* (Figure 30-30) is a woody vine, rarely a tree, of tropical rain forests, with leaves resembling those of many dicotyledons. *Welwitschia mirabilis* (Figure 30-31),

Figure 30-30. A branch of *Gnetum gnemon*.

line, but there is essentially no fossil record to verify this hypothesis. The presence of vessels in the xylem, an angiosperm characteristic not encountered elsewhere among the gymnosperms, has led some botanists to postulate a relationship to the angiosperms. This similarity, however, is more commonly regarded as an example of convergent evolution, not as an indication of any real affinity.

ORIGIN OF THE GYMNOSPERMS

The Progymnospermopsida (Chapter 28) are a distinct class of Middle and Upper Devonian plants apparently derived from ancestors of the *Psilophyton* type, the Trimerophytopsida. This group is regarded as the ancestral stock from which the gymnosperms evolved. The progymnosperms, like the early ferns that arose as a parallel group, developed megaphylls by the subordination of lateral branch systems. Unlike the ferns, however, they developed a cambium and secondary vascular tissues. They also tended to become heterosporous, although some were homosporous, and before the end of the Devonian period the seed habit had been achieved. It is believed that within the progymnosperm group portions of the frond-like lateral branches ultimately evolved into leaves. In one group, these large, lateral branches themselves became fernlike leaves; this group probably became the seed ferns (Pteridospermales) of the Carboniferous period. From these, both the Cycadales and the Cycadeoidales are thought to have descended in later times. In another major group, smaller portions of the lateral branches became leaves of much more extensively branched plants, the Coniferopsida. The Cordaitales are the oldest representatives of this group, and the descent of the Coniferales from them seems reasonably clear. Many botanists are convinced that the Ginkgoales are also derived from the cordaitalean stock, but an independent origin from the progymnosperms cannot be ruled out. Thus the two main lines of gymnosperm evolution, the cycadophyte and the coniferophyte, despite the wide divergence of later representatives, can at least tentatively be traced to a common origin in the Devonian progymnosperms (Figure 30-32). However, many gaps remain in the story of their early evolution.

confined to extreme desert areas of southwest Africa, consists of a low woody trunk bearing only two enormous, strap-like leaves that continue to grow at the base throughout the the life of the plant, which is estimated to be as much as 1000 years in some cases. The short, ground-level trunk springs from a very long taproot, no doubt an advantage in an almost rainless climate.

The anatomy of the Gnetopsida, particularly of their secondary xylem, indicates that they are more closely related to the Coniferopsida than to the Cycadopsida. They are considered an early offshoot of the conifer

Figure 30-31. *Welwitschia mirabilis*, southwest Africa.

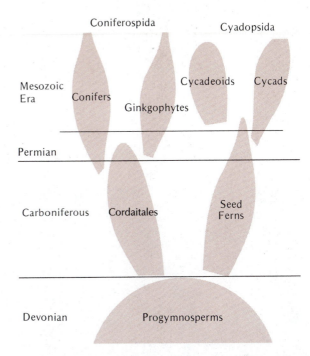

Coniferospida

Cyadopsida

Mesozoic
Era

Conifers

Cycadeoids   Cycads

Ginkgophytes

Permian

Carboniferous   Cordaitales   Seed
Ferns

Devonian   Progymnosperms

Figure 30-32. A diagrammatic representation of the origin of cycadophytes and coniferophytes from the progymnosperms, and their subsequent evolution.

SUMMARY

(1) The gymnosperms comprise a diverse group of woody plants that produce naked seeds. They are classified in three classes: the Cycadopsida, the Coniferopsida, and a small bizarre group, the Gnetopsida.

(2) The seed habit is an evolutionary specialization of the pattern of alternation of generations common to all vascular plants. Seed plants are heterosporous. The ovule consists of an integument-enclosed megasporangium within which one functional megaspore produces a female gametophyte without being shed. The microspores produce male gametophytes that are transferred as pollen to the ovules. Fertilization occurs within the ovule, which then develops into a seed containing an embryo sporophyte. In gymnosperm seeds the embryo is embedded within the female gametophyte, which stores food reserves as the endosperm does in an angiosperm seed.

(3) The Cycadopsida include the living cycads as well as two extinct orders, the seed ferns and the Cycadeoidales. They have large, often pinnately compound leaves, a weakly branched or unbranched axis, and soft secondary xylem containing much parenchyma. The Cycadales are a small group of tropical or subtropical plants, much more prominent in the Mesozoic era than at present. The pollen sacs and ovules are usually borne in cones. Fertilization is accomplished by motile sperms formed in the pollen tube.

(4) The Coniferopsida include the Coniferales, the most important living order of gymnosperms. All are woody plants, and most are evergreen. Many are of economic importance as a source of lumber, paper pulp, and other products. Conifers produce short-lived pollen cones and ovulate cones that develop into woody seed cones. Most conifers are monoecious. The life cycle of pine illustrates the reproductive process in conifers. The sperms are nonmotile and are carried to the egg in the archegonium by the growth of the pollen tube.

(5) Also included in the Coniferopsida are the ginkgo and the extinct Cordaitales. The ancient Ginkgoales now survive as a single species that is broad-leaved and deciduous. Like the cycads, it produces motile sperms. (The Cordaitales are believed to be the stock from which the conifers descended. The diverse types of gymnosperms are thought to have had a common origin in the Devonian progymnosperms.)

Chapter 31

THE SEED PLANTS: ANGIOSPERMS

Through the long span of geological time since the Devonian period, major groups of vascular plants have risen to ascendancy only to decline to a minor persisting remnant in the earth's flora or even to become extinct. The angiosperms, or flowering plants, are a group that is now in the full flush of its evolutionary development. This is the "age of the angiosperms," reflected in the number and diversity of flowering plants—more than 250,000 species, grouped into 12,500 genera and some 300 families. They include trees, shrubs, vines and herbs, annuals and perennials, succulents, and forms with underground storage organs. In some, such as the desert cacti, the body form has been modified so much that recognizing the basic organs is difficult. Some angiosperms, such as the Australian gum trees (Figure 31-1A), rival the largest conifers in size; at the other extreme, the tiny, floating water-meal (*Wolffia*) (Figure 31-1B) is barely visible to the naked eye. The angiosperms have occupied almost every terrestrial habitat and now predominate in most of them, the tundra and the northern coniferous forests being among the few exceptions. They occur in vast stands of relatively few species, as in the great prairies of North America, and form complex associations of innumerable species, as in the tropical rain forests. They inhabit arid deserts, lakes and streams, and even the oceans. Although the overwhelming majority of angiosperms are autotrophic, some species are mycorrhizally saprophytic (Figure 31-1C), and others obtain all or part of their nutrition through parasitic associations with other plants (Figure 31-1D).

Because of their abundance and widespread distribution, angiosperms constitute a major part of the environment of other organisms. As such, they have exerted a profound effect on other organisms, particularly the animals that feed on them. Since the flowering plants rose to prominence in the Cretaceous period more than 100 million years ago, their influence has been particularly evident in the evolution of mammals, birds, and insects. For example, the evolution of herbivorous mammals in mid-Tertiary times correlated with the worldwide spread of grasses, legumes, and other herbs. Most modern birds have evolved as organisms that feed on either the seeds, fruits, or buds of flowering plants or on insects, which in turn feed on angiosperms. The rapid development of the angiosperms also favored the evolution of several insect orders; the interdependence of flowers and insects is one of the most remarkable examples of co-evolution (Chapter 20) in the biological world.

DISTINGUISHING FEATURES OF THE ANGIOSPERMS

Because flowering plants are the most common and familiar plants, their structure, reproduction, and physiology have been considered in some detail in earlier chapters. In this discussion of the angiosperms in an evolutionary context, it is necessary to summarize only the more important features and to make comparisons with other groups of vascular plants, particularly with the other great group of living seed plants, the Coniferales. Despite their large numbers and diversity, the flowering plants are believed to be a unified group with a common origin. Thus they are usually treated as a single class, the Angiospermopsida. However, since the time of the English botanist John Ray in the late 1600s, it has been customary to recognize two groups or subclasses, the dicotyledons, with some 200,000 species, and the monocotyledons, with about 50,000 species. Although united by many common fundamental features, these subclasses also have distinctive differences, as noted in Table 31-1.

Vegetative Organs

The conifers are all woody perennials, mostly large trees. So also are thousands of angiosperm

---

Figure 31-A. A flowering apple tree.

615

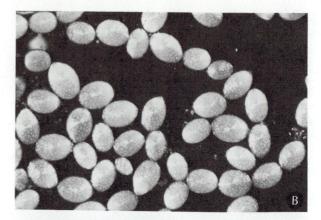

Figure 31-1.   Diversity of body forms and types of nutrition among angiosperms. (A) A giant forest tree, *Eucalyptus viminalis,* in Australia. Eucalyptus trees number among the tallest of recorded trees, over 100 m (300 ft) in height. (B) "Water meal" (*Wolffia*), a rootless and leafless aquatic angiosperm with a globose photosynthetic stem less than 1 mm across, which floats on the surface of ponds. (C) A "saprophytic" nongreen angiosperm, *Sarcodes sanguinea* ("snowplant," so named because it usually appears in early spring as the snow is melting). The plant actually obtains its nutrients by associating mycorrhizally with soil fungi that either decompose forest litter or parasitize the roots of other plants. (D) A parasitic nongreen angiosperm with minute white flowers, dodder (*Cuscuta*), twining around and parasitizing the stems of a host plant.

TABLE 31–1 Summary of Features Distinguishing Monocotyledons from Dicotyledons

Monocotyledons	Dicotyledons
One cotyledon in embryo	Two cotyledons in embryo
Flower parts in threes	Flower parts usually in fours or fives
Vascular bundles scattered throughout stem	Vascular bundles in a ring around pith of stem
Principal veins of leaf parallel to one another	Principal veins of leaf forming a network

Figure 31-2. Wildflower display on the Alaskan tundra.

species, but many thousands of others are perennial or annual herbs with limited cambial activity or none. Monocots generally are herbs, but some, such as palms, attain considerable size, and in a few there is a kind of cambial activity. Stems of angiosperms display many departures from the typical form, including climbing stems; horizontal stems that, as in the case of rhizomes, may be subterranean; storage organs such as tubers, which also function in vegetative propagation; and swollen, succulent stems such as those in the cacti. Roots also show many modifications useful in accumulating food and in vegetative reproduction. These and other modifications of the plant body have enabled the angiosperms to occupy many habitats to which conifers and other vascular plants are not adapted.

The leaves of the flowering plants, like the stems and roots, exhibit a high degree of variation and adaptation to various environmental conditions. They are mostly thin and expanded, although in general form they vary widely from the long, narrow leaves of the grasses to oval or round leaves.

Great variations in the size of leaves are also found. In some plants of specialized habitats, the leaves are so small that they can be recognized only with difficulty. At the other end of the scale are the leaves of the cultivated banana, the blade of which alone may be more than 4 m (12 ft) long and 1 m (3 ft) wide. The traveler's-tree, a relative of the banana, has leaf blades 6 m or more in length. The circular leaves of the royal water lily of Guyana and the Amazon may attain a diameter of more than 2 m. Even these dimensions may be exceeded by certain perennial herbs of the tropics.

Vascular System

The xylem of angiosperms is more specialized and complex than that of conifers. Conifer wood is composed mainly of thick-walled tracheids and ray cells. The wood of angiosperms contains a greater variety of cells, including the important vessel members, which are not found in conifers. Vessel members form long conducting tubes, the vessels, which provide efficient chan-

nels for water conduction. The phloem is also more complex, containing sieve tubes composed of sieve-tube members in contrast to the simpler sieve elements of conifers and lower vascular plants. In dicots the vascular bundles of primary xylem and phloem are usually arranged in a ring around the central pith, whereas in monocots the bundles are scattered throughout the stem so that the pith and the cortex are indistinguishable.

The Flower

The most obvious and popularly recognized characteristic of the angiosperms is the formation of reproductive structures called flowers (Figures 31-A and 31-2). The flower is a determinate shoot (i.e., a shoot of limited or restricted growth) that produces pollen and ovules and is constructed to facilitate pollination. The basic organization of the flower was described in Chapter 16, so only particularly characteristic features are summarized here.

The flower is formed by the activity of a modified shoot apex. The axis, or **receptacle,** of the flower is thus a stem, and the floral appendages are regarded as foliar. This does not mean that they are necessarily modified foliage leaves but rather that they are homologous with leaves. The petals and sepals (collectively termed the **perianth**) may simply be modified leaves, but the reproductive organs have probably evolved parallel to leaves from a common ancestral organ, undergoing transformation related to their reproductive function. The pistil is composed of one or more fundamental foliar units called **carpels.** A **simple pistil** is a single carpel, and a flower may have only one of these, as in the pea, or several, as in the buttercup, peony, or magnolia. In many other flowers, carpels are fused forming a **compound pistil** in which the individual carpels are sometimes difficult to recognize. The sepals and petals (or **tepals,** if petals and sepals are not differentiated from one another) protect the inner fertile parts in the unopened bud and later attract insects or other pollinating agents by their coloration and often direct them toward the fertile regions.

The stamens and carpels are regarded as modified sporophylls, that is, foliar organs that produce spores. The ovules are enclosed within the carpel that bears them so that, unlike the situation in conifers, the pollen is transferred not to them directly but to the stigmatic surface of the carpel. There the pollen germinates, and the pollen tubes grow down the style to the ovules. The enclosure of the ovules within a protective covering is a unique characteristic of the angiosperms and is responsible for this name (from Greek *angeion*, vessel; *sperma*, seed). The pistil, or at least the ovary, sometimes with other flower parts associated, matures into a fruit, a structure found only in the flowering plants. At the microscopic level extreme structural reduction of the gametophyte generation and the processes of double fertilization and endosperm development (Chapter 16) are distinguishing features of the angiosperms. These features are discussed in the next section.

More efficient conduction in both xylem and phloem, a highly effective reproductive system in the flower, and diverse mechanisms of vegetative propagation contribute to the angiosperms' overwhelming success. Angiosperms have been able to modify and adapt these and other features evolutionarily to a wide range of habitats and environmental conditions, giving them an advantage over all other terrestrial plant groups.

ALTERNATION OF GENERATIONS IN THE ANGIOSPERMS

The pattern of alternating sporophytic and gametophytic generations that we have traced through successive vascular plant groups is also present in the angiosperms. It is important to compare the angiosperm life cycle (outlined in Chapter 16) with those of other groups to understand the real nature of the reproductive processes. Historically, this type of comparison enabled the German botanist, Wilhelm Hofmeister, to elucidate these processes with great clarity in the middle of the last century.

The angiosperms, like *Selaginella* and all gymnosperms, are heterosporous. Microspores produced following meiosis in the anther of the stamen are released as pollen grains only after the male gametophyte begins developing. As in the conifers, in each ovule only one megaspore mother cell undergoes meiosis, and of the resulting tetrad only a single haploid megaspore functions. This spore is not shed but undergoes development into the female gametophyte within the megasporangium or nucellus. This, as noted earlier, is an essential feature of the seed habit (see Figure 16-8).

In the angiosperms the male and female gametophytes have reached an extreme stage of structural reduction. The male gametophyte at its fullest development consists of only three cells, two sperms and a tube cell. Yet as it grows to form the pollen tube the male gametophyte may attain considerable size, or at least length, when the pathway to the ovule is long, as in corn. The rapidly growing pollen tube, drawing nutrients from the pistil tissues through which it grows, is a very active organism. The female gametophyte, or embryo sac, consists of seven cells, but there are eight nuclei because two polar nuclei occupy the common cytoplasm of the central cell. All traces of archegonial structure, such as are found in the gymnosperms, are lacking. The very limited development of the female gametophyte greatly accelerates the reproductive process because the egg is ready for fertilization much more quickly than in the gymnosperms, in which extensive gametophyte development precedes fertilization. This may offer a real advantage to the flowering plants, particularly in habitats where the season favorable for growth is brief.

The process of fertilization after the pollen tube enters the embryo sac restores the diploid condition and initiates development of the sporophytic generation. In contrast to gymnosperms no initial stage of free nuclear division precedes the formation of an organized proembryo. Instead, the early divisions follow a precise pattern, and cell wall formation always accompanies mitosis. This again accelerates the reproductive process because the embryo is normally ready for germination much more quickly than in the gymnosperms.

The unique phenomenon of double fertilization, which results in a triploid nutritive tissue, the endosperm, has intrigued botanists ever since its discovery in 1898. Functionally, it provides a nutritive base for the developing embryo, replacing the haploid female gametophyte, which plays this role in gymnosperms. It is often regarded as a sort of delayed gametophyte development in which the introduction of genes from the male parent may facilitate the interaction between the embryo and its supporting tissue. However, this remains a matter of speculation because the evolutionary origins are unknown. The occurrence of double fertilization throughout the angiosperms and nowhere else, as far as is known, is often used to support the view that the flowering plants, monocots and dicots, are a unified group with a common origin. The existence of double fertilization disturbs to some extent the simple pattern of alternation of generations because it is difficult to assign the triploid endosperm to either the haploid gametophytic or the diploid sporophytic generation. Except for the endosperm, however, an-

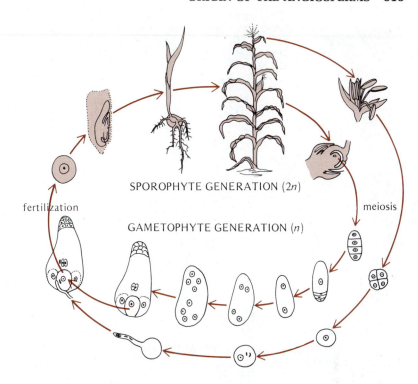

SPOROPHYTE GENERATION (2n)

fertilization

meiosis

GAMETOPHYTE GENERATION (n)

Figure 31-3. Alternation of generations in an angiosperm (corn). The sporophyte generation is shown in color.

giosperm reproduction conforms to the basic plan of the vascular plants (Figure 31-3).

The pattern of embryo sac development described in Chapter 16 is widespread in both dicotyledons and monocotyledons, but it is not the only pattern. It is found in about 80% of the species thus far examined and is often referred to as the "normal" or "common" type. However, at least 10 other patterns have been described, and it is certainly possible that more will be found. These differ from one another in the number of cells and nuclei in the fully developed embryo sac and in their origin and distribution. One example of a type deviating rather sharply from the common pattern occurs in the lily (Figure 31-4). In this plant, following meiosis in the megaspore mother cell, the four haploid megaspore nuclei remain in a common cytoplasm without cell wall formation, and *all* participate in the formation of the female gametophyte. This is a most curious phenomenon, departing from the usual plan of alternation of generations in which a gametophyte arises from a single spore. In some other cases the embryo sac develops from two megaspores, and the remainder degenerate. In lily, following meiosis one nucleus migrates to the micropylar end of the embryo sac, the other three move to the opposite end, and all divide. However, the three dividing nuclei at the opposite end merge, and thus from their division two *triploid* nuclei are formed. The embryo sac then contains two haploid and two triploid nuclei. Following another division there are four of each. Three haploid nuclei then form the egg and two synergids, while three triploid nuclei at the opposite end form antipodals. One haploid and one triploid nucleus become the polar nuclei, which, after fusion with one of the sperms, give rise to a *pentaploid* (5n) endosperm. The fully developed female gametophyte of lily thus resembles the normal type, but the development is very different, as are the chromosome numbers of some of the cells. The egg, how-

ever, is haploid, and no abnormalities in embryo development occur after fertilization. Other deviations from the normal type are equally striking; in one case the endosperm is 15-ploid. There has been much speculation but little agreement about the origin and evolutionary significance of these deviant forms. They are not abnormal because they are entirely typical of the species in which they occur.

ORIGIN OF THE ANGIOSPERMS

A century ago Charles Darwin referred to the angiosperms' origin as an "abominable mystery." Unfortunately, despite much additional evidence about plant evolution gained since Darwin's time, the evolution of this most modern vascular plant group, so overwhelmingly significant in the earth's present vegetation, still remains basically a mystery. As explained in the following sections, the reason for this lack of understanding is the gap between living or fossil angiosperms and any other plant group. Although the angiosperms probably evolved from some group of gymnosperms, the fossil record has not disclosed the missing links.

Evidence from Fossils

The first fossil remains generally accepted as genuine angiosperm appear rather abruptly in the early Cretaceous period in such widely separated localities as Europe and eastern North America. These earliest fossil angiosperms represent only a

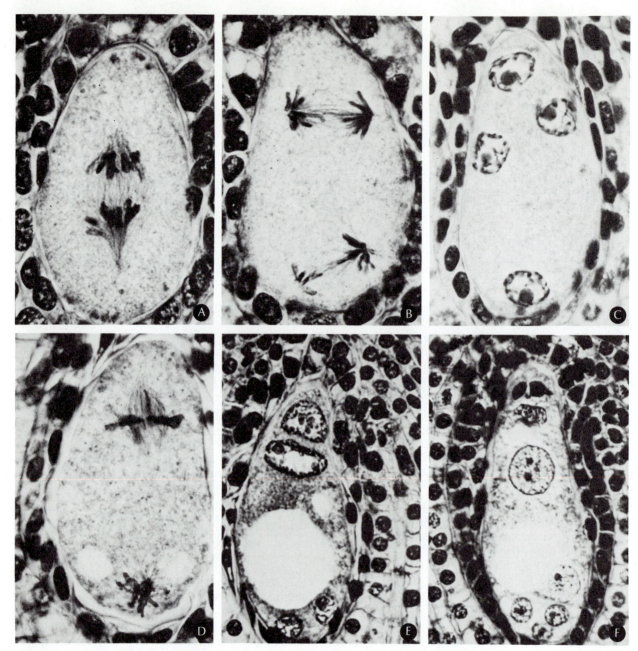

Figure 31-4.   Development of the lily embryo sac within ovule (micropyle is beyond bottom of each picture). (A) First meiotic division of megaspore mother cell. (B) Second meiotic division. (C) Tetrad of haploid nuclei, of which the three above are associating to yield a triploid nucleus. (D) Division of the merged nuclei (above) and of the remaining haploid nucleus (below). (E) After this division, two triploid nuclei above, two haploid nuclei below. (F) After a further division and some differentiation, haploid egg and two synergids below, plus a haploid polar nucleus that will migrate and fuse with the triploid polar nucleus located above center; three degenerating triploid antipodal cells at top.

small component of floras dominated by ferns and gymnosperms. Well before the end of the Cretaceous period angiosperms became the dominant vegetation nearly everywhere that plant fossils are found. There are scattered reports of older angiosperm fossils as far back as the Triassic period,

but none of these can be accepted as undisputed flowering plants, and they probably are not. The early angiosperm fossils consist mostly of leaves (Figure 31-5A) and pollen grains, with some wood specimens also appearing here and there. Flowers, which would answer so many important ev-

Figure 31-5. Angiosperm fossils. (A) Fossil leaves in a piece of Dakota sandstone of mid-Cretaceous age. (B) Mid-Cretaceous fossil flowers, also from the Dakota formation.

olutionary questions, are rare throughout the angiosperm fossil record. Fortunately, however, they are being discovered with increasing frequency, extending as far back as the middle of the Cretaceous period (Figure 31-5B).

On the basis of pollen and leaf characteristics, there is no doubt that the early Cretaceous fossils were angiosperms. In fact, the leaves were originally assigned to modern families and even to genera in some cases. In the absence of more knowledge of the plants as a whole, particularly flowers, this practice is probably unwise, and the earlier identifications are now being reevaluated. Clearly, however, these fossils are angiosperms and not intermediate between them and other, more ancient groups of vascular plants. Faced with this incomplete record, botanists have speculated about what they cannot or at least do not know directly.

Current Hypotheses

For several decades the more widely held interpretation has been that the angiosperms had a long evolutionary history before their relatively late appearance in the fossil record. It is postulated that they may even have been present before the end of the Paleozoic era but were restricted to upland areas away from locations where fossil deposits were being formed. The sudden appearance of well-developed angiosperms is thus viewed as an artifact of the fossil record.

Another viewpoint, now rapidly gaining acceptance, is that the angiosperms evolved as a distinct group only shortly before their appearance in the fossil record. The explosive development following their appearance indicates that they were a rapidly evolving group that did not require 50 million years or more to become distinct. Changes separating the new group from ancestral forms may have occurred relatively quickly. Fossils of the actual transitional forms may never be found if these early angiosperms were, as is probable, relatively unimportant in the flora of that time and if they occurred in areas not suited to the preservation of fossils.

Early Evolution of Angiosperm Features

G. L. Stebbins, a prominent evolutionist, has postulated that the angiosperms probably arose in a mountainous region subjected to seasonal drought (the kind of locality in which fossil preservation would be unlikely). Under these conditions the season for pollen and ovule development, pollination, and seed maturation may have been relatively brief, giving selective advantage to any changes accelerating these processes or increasing the efficiency of pollination. The marked reduction of the female gametophyte and the correlated development of endosperm may have resulted from this environmental pressure. Certain kinds of insects, possibly beetles at the outset, may have begun feeding on pollen. If the pollen-producing and ovule-producing organs were brought together in groups in the same structure, essentially a bisporangiate cone or flower (in contrast to the separate pollen- and ovule-bearing cones of most gymnosperms), pollen transfer by the insects would be favored, with a probable increase in efficiency. Subsequent modifications of the reproductive structure favoring the attraction of insects would have initiated the long and spectacular history of floral evolution. The production of nectar and its collection by insects such as bees, moths, butterflies, and flies and also by birds and bats led to a diversity of pollination mechanisms, as described in Chapter 16. Not only did the flowering plants become dependent on these biological pollinating agents, but, by co-evolution, the pollinators came to rely on flowers as their primary food source and evolved corresponding modifications. In several angiosperm groups wind pollination evolved secondarily at a later time, but this does not diminish the significance of insect pollination in the early evolution of the group as a whole.

Because the original insect pollinators probably ate pollen, they may have eaten ovules as well. Enclosure of the ovules within a carpel (i.e., angiospermy) could have been favored because it protected these delicate structures from being fed upon. Enclosure would also have protected them from desiccation and other adverse environmental factors. With an effective mechanism of cross-pollination by insects and a relatively rapid life cycle, early angiosperms would have tended to evolve and diversify rapidly if they occurred in a mountainous region with diverse habitats. Subsequent migration of this dynamic and efficient group into regions where fossilization was possible would have resulted in the kind of fossil record that in fact exists. This interpretation is of course speculative and will remain so unless actual transitional forms are discovered.

Possible Progenitors

Equally speculative is the question of which vascular plant group gave rise to the angiosperms. Bota-

nists generally agree that their origins should be sought among the more ancient seed plants, the gymnosperms. Angiosperms and gymnosperms share a number of common features, including the seed habit and the occurrence of vascular cambium and secondary growth. As discussed in Chapter 30, gymnosperms were abundant and richly diversified in the Mesozoic era when the angiosperms arose. The Gnetopsida, which like most angiosperms contain vessels in their xylem, and the extinct Cycadeoidales, which produce superficially flower-like cones containing both microsporophylls and megasporophylls, have been proposed as possible angiosperm ancestors. Closer study of these specializations, however, has convinced most specialists that these features have resulted from convergent evolution and do not indicate a close relationship. Increasingly, attention has focused on the seed ferns, an ancient and relatively unspecialized gymnosperm line that flourished in the Carboniferous period but persisted into the middle of the Mesozoic era, as the probable progenitors of the flowering plants. No fossils yet document the transition from seed ferns to flowering plants, and the case for such a transition rests not so much on positive evidence as on the lack of evidence against it.

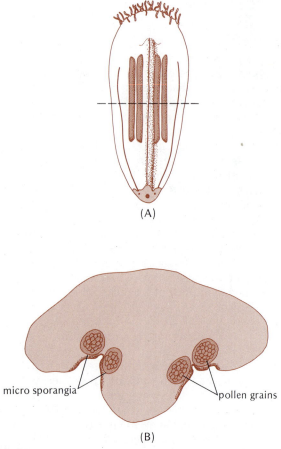
(A)

(B)

Figure 31-7. The leaf-like stamen of *Degenaria*. (A) Viewed from the lower surface showing four embedded microsporangia or pollen sacs. (B) Cross section cut at the broken line shown in (A).

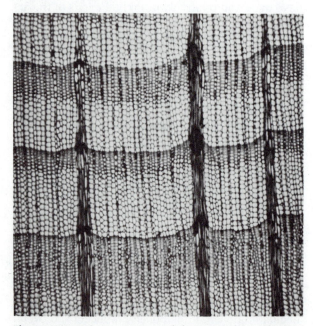

Figure 31-6. Cross section of the secondary xylem of *Trochodendron*, a vessel-less dicotyledon.

Characteristics of Primitive Angiosperms: The Ranalian Complex

One approach to deciphering the angiosperms' origin is attempting to define the nature of the early angiosperms. This is also essential for tracing lines of evolutionary advancement in the angiosperms and for understanding relationships among the diverse families of this large and complex group. Since the fossil record has not revealed the original angiosperms, their characteristics must be inferred from less direct evidence. Comparative study of living angiosperms from all over the world, but particularly from tropical regions, has revealed a number of families that have apparently retained unspecialized and therefore presumably primitive features. Since evolution may proceed at different rates in different plant parts, these primitive features do not necessarily all occur together in the same family. Plants with an advanced body form, for example, may bear flowers of a primitive type. However, primitive features do tend to occur together. This has led to the recognition of an assemblage of families, sometimes called the *ranalian complex*, that are believed to embody the essential features of primitive angiosperms. The ranalian complex (formerly called the Order Ranales, but now considered too diverse to be grouped in one order) includes a number of families distributed chiefly in the tropics, but with some temperate zone representatives. Examples of woody ranalian families are the magnolia (Magnoliaceae) and custard apple (Annonaceae) families, and herbaceous families include the water lily (Nymphaeaceae) and crowfoot (Ranunculaceae) families, the latter name being the source of the term ranalian complex. The ranalian families are not necessarily to be regarded as ancestral to other, more specialized angiosperms, but merely seem to have retained features of the now extinct ancestors—that is, they resemble the ancestral forms more closely than do other, more specialized families. Some evolutionists call them archaic rather than primitive. Study of this assemblage of families, particularly some of the less familiar tropical representatives, has revealed much interesting information about the probable nature of the original flowering plants.

LEAVES

Probably the primitive leaves were simple, pinnately veined, and occurred in an alternate or helical arrangement. This type of leaf and leaf arrangement predominates in woody ranalian families as well as in the earliest angiosperm fossils.

XYLEM

The primitive angiosperms were almost certainly woody—that is, plants with cambial activity and a significant development of secondary tissues, and this body form predominates in the ranalian complex. In fact, approximately half of the modern angiosperm families have no non-woody or herbaceous members. Also, the gymnosperms, from which the flowering plants probably evolved, are entirely woody plants.

A rather surprising conclusion from comparative studies is that the original angiosperms lacked vessels and, like all gymnosperms apart from the Gnetopsida, had only tracheids as water-conducting elements in their xylem. This conclusion is based on the discovery that, although vessels are a general characteristic of the flowering plants, some 100 species in five ranalian families are without them (Figure 31-6). Because these species also show other primitive features, they are believed to have retained the original, vessel-less condition of the ancestral angiosperms.

THE FLOWER

As already noted, the primitive flower seems to have been insect-pollinated, and adaptation to other agents of pollen transfer are considered derived. Flowers such as those of the tulip tree, buttercup, and marsh marigold (Figures 31-14, 31-15, 31-16) have retained many of the original features. The flowers have simple radial symmetry and possess a perianth either of sepals and petals or of undifferentiated tepals. They have numerous stamens and carpels, and the flower parts are not fused as they are in many more advanced flowers. Further, the carpels and stamens, and sometimes the perianth parts as well, are often helically ("spirally") arranged along a somewhat elongated receptacle (floral axis). This last characteristic reflects the relationship of the primitive flower to a gymnosperm strobilus or ultimately a vegetative shoot.

In line with this presumed derivation, the floral parts are regarded as foliar in nature. Recognizing the sepals, or tepals, as modified leaves is not difficult because of their leaf-like form. Petals are also modified leaves in some cases; in others they evidently arose from stamens that became sterile, broadened, and leaf-like. The foliar origin of stamens and carpels is supported by the remarkably leaf-like nature of these organs in some plants of the ranalian complex. In the Fijian tree *Degeneria*, for example, the stamens are not differentiated into filament and anther but are shaped much like miniature leaves with two pairs of microsporangia embedded on the lower side (Figure 31-7). In another ranalian genus, *Drimys*, which is also vessel-less, the carpels are little more than leaves folded inward along the midrib, thereby enclosing the ovules, which are borne on the upper surface (Figure 31-8). In fact, at pollination the ovules are not truly enclosed because the edges of the folded leaf are not fused. Rather, a profuse growth of interlocking hairs, which constitute a **stigmatic crest,** blocks the cleft. The pollen germinates on this crest, and the pollen tubes grow

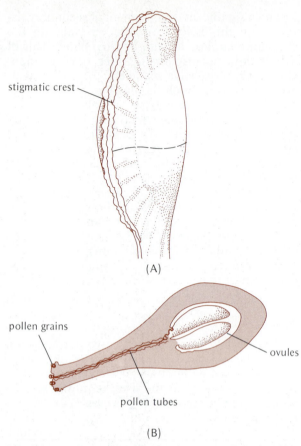

(A)

(B)

Figure 31-8. The carpel of *Drimys* represents a leaf folded inward so that ovules borne on its upper surface are enclosed. (A) Whole carpel; (B) a section cut at the broken line shown in (A).

through a mat of hairs to the ovules within. As the fruit matures, the marginal cleft becomes sealed. In other related plants the stigmatic crest is more restricted toward the apex and transitions to the typical closed carpel with a terminal stigma occur (Figure 31-9).

Evolutionary Trends in the Angiosperms

The existence of so many angiosperm species in so many different habitats clearly reveals the evolutionary plasticity of this group. Earlier chapters discussed some of their diversity in vegetative and reproductive features. It was mentioned that the earliest angiosperms were almost certainly woody. If they first appeared in a mountainous region with a seasonally arid climate, they may well have been shrubs rather than large trees. The herbaceous habit, in which the aerial portion of the plant is short-lived and the tissues are relatively soft, is thought to have evolved through a reduction of cambial activity. This change seems to have begun early in angiosperm history in certain groups that became adapted to moist or even aquatic habitats. The major development of her-

baceous forms, however, probably occurred in response to the cooling and drying climate in Tertiary times, as an adaptation for seasonal growth, the aerial, exposed parts of the plant simply dying down when conditions are unfavorable. The most extreme specialization is the annual habit, in which the entire plant dies and seeds or other propagules reestablish the new generation.

Vascular Evolution

The dicot type of vascular pattern, which is similar to that of gymnosperms, is evidently primitive in angiosperms, the monocot pattern representing an evolutionary specialization. Within monocots and dicots several other evolutionary trends of vascular tissue modification can be recognized, of which we shall note here only the progressive evolutionary improvement of vessels, explained previously in Chapter 10 (see Figure 10-10). Primitive angiosperm vessel members, found for example in those ranalian groups that possess vessels, are narrow, elongated, tracheid-like cells, with several to many slit-like perforations at each end, evidently derived from pits. More advanced angiosperms develop wider and shorter vessel members with fewer, larger perforations, culminating in very wide vessel members with a single large (simple) perforation at each end. These features minimize resistance to water flow and hence maximize the water conductivity of the xylem.

Floral Evolution

The flowers of various families provide abundant evidence of evolutionary modification from the basic or primitive type. Starting with the primitive flower outlined in the previous section, these variations may be seen as evolutionary trends leading in some cases from relative simplicity to greater complexity, as in many of the modifications associated with pollination by insects; in other cases they may lead to greater simplicity through reduction. These trends are important because floral structures are the major characters used in angiosperm classification and provide the most useful clues to relationships among diverse families. Some of the more outstanding directions in which floral evolution has proceeded are summarized below.

NUMBER AND ARRANGEMENT OF PARTS

A reduction in the number of floral parts to a small and definite number seems to have been a general trend. Associated with this reduction was a change from a helical to a whorled arrangement of floral organs accompanied by a shortening of the floral axis, thus bringing the parts closer to-

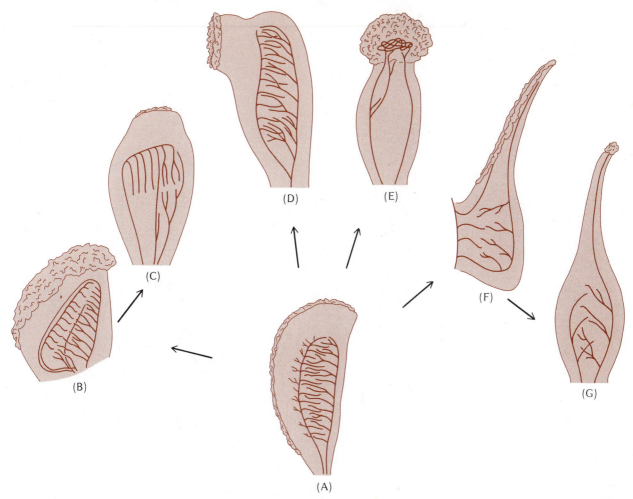

Figure 31-9. Modifications of the primitive carpel found in currently living forms. Beginning with the original type represented by *Drimys* (A), the stigmatic crest is restricted in various ways and the typical closed carpel results.

gether. In the dicotyledons the whorls usually contain four or five members or multiples of these numbers. A flower may have, for example, five sepals, five petals, ten stamens, and five carpels. However, there are often only two or three carpels in a compound pistil. In the monocotyledons the number of organs in a whorl is usually three or a multiple of three. In both subclasses, however, certain families have numerous stamens and carpels.

FUSION OF PARTS

Among the important modifications found in the flower are fusions of the organs of the same whorl and of different whorls. In many flowers the petals are fused along their margins into a **corolla tube** (Figure 31-10); the term **corolla** refers to all the petals, collectively. This tube is commonly lobed, the number of lobes corresponding to the number of petals. The sepals, collectively

termed the **calyx**, are usually separate but are sometimes fused at the base into a **calyx tube.** The stamens may become fused by their filaments or by their anthers into one or several groups. Sometimes the filaments of the stamens are partially or completely fused to the corolla tube (Figure 31-10). As noted previously the carpels commonly are united into a compound pistil.

In a flower with distinct parts, the sepals, petals, and stamens are clearly attached to the receptacle beneath the ovary. Because the ovary is situated above the zone of attachment, it is said to be **superior** (Figure 31-11A); this is the condition in a large number of flowering plants. In some species such as plum, cherry, and blackberry, the basal portions of the sepals, petals, and stamens are united, forming a cup-shaped **floral tube.** The lobes of the sepals and petals and the unfused portions of the stamens extend from the margins of this tube (Figure 31-11B). The ovary is still superior.

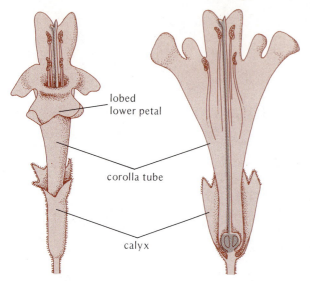

Figure 31-10. Flower of ground ivy (*Glechoma hederacea*) of the mint family (Lamiaceae). The petals are joined into a corolla tube and the sepals form a calyx tube. The filaments of the stamens are partially fused to the corolla tube and the flower is irregular.

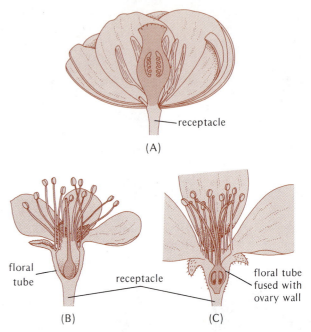

Figure 31-11. Flowers with superior and inferior ovaries. (A) Flower of mayapple (*Podophyllum peltatum*) of the barberry family (Berberidaceae); ovary superior. (B) Flower of sour cherry (*Prunus cerasus*); ovary superior, floral parts fused forming a floral tube. (C) Flower of apple (*Pyrus malus*); ovary inferior, the result of fusion of the floral tube to the ovary wall. *Prunus* and *Pyrus* are members of the rose family (Rosaceae).

Another type of fusion is that of the floral tube with the ovary. In this type, the sepals, petals, and stamens appear to arise from the top of the ovary, which is then said to be **inferior** (Figure 31-11C). In fact, the pistil is still the uppermost structure of the flower, but it appears to be inferior because of its union with the floral tube. The position of the ovary, whether inferior or superior, is significant in plant classification and also helps in interpreting the nature of many kinds of fruits.

The union of floral parts occurs in different ways in the development of the individual flower. In some cases, organs such as carpels are initiated as separate primordia and grow together in the course of their development. In many other flowers, however, they are united from the very beginning of growth in the meristematic cells of the floral axis. In such cases the term fusion has an evolutionary meaning only and does not refer to a developmental process.

SYMMETRY

In many flowers, all the members of each whorl are alike in size and shape, so the flower appears circular when viewed from above. Such radially symmetrical flowers are termed **regular.** The flowers of other species are bilaterally symmetrical, capable of being divided into two identical parts along only one longitudinal plane. Such flowers are termed **irregular.** Commonly, the corolla is more modified than other floral parts. Familiar examples of irregular flowers are those of the legume family (Chapter 32), snapdragons, mints, violets, and orchids (Figures 31-10, 31-19, 31-25).

REDUCTION

A flower in which all four kinds of floral organs occur—sepals, petals, stamens, and pistil—is called a **complete** flower. The flower, however, may be reduced so that some organs are no longer evident, although they may be present in vestigial form. In many plants, for example, the petals have disappeared and the sepals may be represented only by scales, bristles, teeth, or ridges (Figure 31-12). A flower lacking a calyx, corolla, or both is said to be **incomplete.** This evolutionary reduction is often associated with wind pollination (Chapter 16) and is found in the grasses and many trees such as the willows, poplars, oaks, hickories, and elms.

The evolutionary loss of floral parts may also extend to the stamens and pistils. A flower in which both kinds of organs are present and functional is said to be **perfect.** If either stamens or pistils are lacking or nonfunctional, the flower is termed **imperfect** and is either **staminate** if it lacks a pistil or **pistillate** if it lacks stamens (Figure 31-12). A perianth may be present or absent in imperfect flowers. The importance of imperfect flowers as a mechanism for cross-pollination has been stressed in Chapter 16.

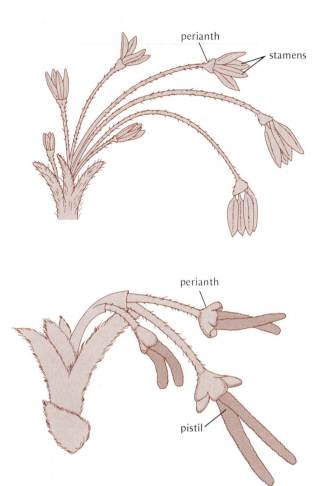

Figure 31-12. Flowers of box elder (*Acer negundo*) of the maple family (Aceraceae). Flowers are imperfect and the plant is dioecious. Staminate flowers above, pistillate below.

CLASSIFICATION OF THE ANGIOSPERMS

In a group as large and diverse as the angiosperms the problem of devising a reasonably natural phylogenetic scheme of classification is challenging, and major differences of opinion are not surprising. Recognizing the approximately 300 families is a problem in itself, but even greater difficulties arise at the next level of grouping, the order, each composed of one to several families. Floral characteristics provide the most useful and widely used criteria of relationship among the families, but other features such as pollen structure, chromosome number and form, internal anatomy, and the presence or absence of particular organic compounds may be important in revealing genetic relationships. Those families with the greatest number of features in common are presumed to be the most closely related, and those with the fewest features in common are the most distantly related. Some taxonomists recognize a large number of orders, whereas others prefer to deal with fewer, more inclusive orders.

Although the angiosperms are generally agreed to have had a common origin, for more than 100 million years they have been diverging along many evolutionary lines. The characteristics of the ancestral forms have been deduced from a study of primitive or archaic features of existing plants, but these retained attributes occur in species or families that form part of several different evolutionary lines within the angiosperms. The American botanist, Arthur Cronquist, and the Soviet botanist, Armen Takhtajan, independently proposed a scheme recognizing six major groupings of dicotyledons and four of monocotyledons, each identified as an important evolutionary line. Each has relatively primitive and more advanced members. Such evolutionary trends as the development of the herbaceous habit, the fusion of floral parts, or the appearance of the inferior ovary have occurred independently or in parallel fashion in different lines. For example, the corolla of the orchids, the snapdragons, and the legumes (pea family) is irregular or zygomorphic, but this does not indicate a close relationship among these families.

Although it is beyond the scope of this text to present a complete and detailed classification of the angiosperms, it may be helpful to look briefly at a proposed scheme of classification. Figure 31-13 is adapted from a recent book on angiosperm evolution by G. L. Stebbins. It is a scheme based on the system of Cronquist and Takhtajan but presents the numerous orders on a horizontal plan grouped according to presumed relationships. If the necessary information were available for constructing past history, these groups would be tied together in the third dimension in a traditional, much branched phylogenetic tree. Because this information is not available, the plan reveals only present relationships. Relative primitiveness is shown by proximity to the center and the ancestral complex, and distance from the center indicates evolutionary advancement.

Within the Class Angiospermopsida there are two major subclasses, the dicotyledons and the monocotyledons. Probably the monocotyledons arose from a primitive dicot group at an early stage and have been distinct for many millions of years. Fortunately, primitive members of both groups suggest how the divergence may have taken place. Among the monocots are two families, the Alismataceae, or water plantains, and the Butomaceae, or flowering rushes, with a floral structure not far removed from that of the "ranalian" type as illustrated by the buttercups (Figure 31-15). These monocot families are herbaceous and are typically marsh plants or even aquatics, as are many of the buttercups. The monocots probably arose as a group adapted to moist habitats from the same dicot line that gave rise to the buttercup, marsh

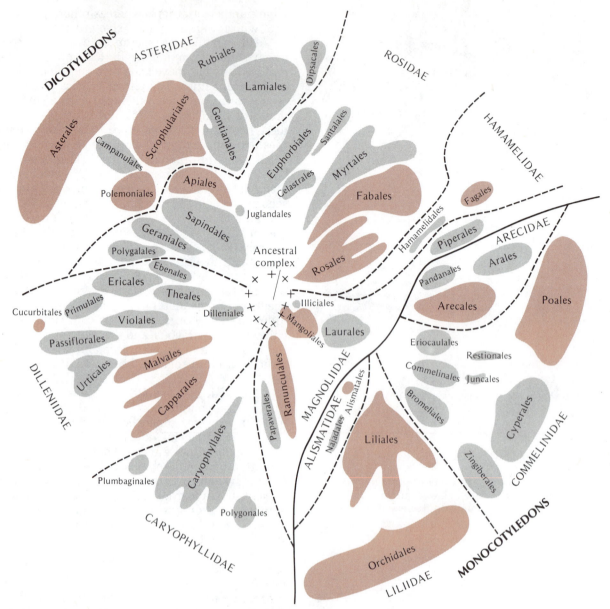

Figure 31-13. A diagrammatic scheme which attempts to place the orders of angiosperms in their appropriate relationship on a horizontal plane. Such a diagram is the equivalent of a cross section of a conventional phylogenetic tree and is a more realistic approach to present-day relationships. The size of each group reflects the number of species; proximity to the center (ancestral complex) indicates primitiveness; and the radial extent represents the spectrum of primitive to advanced members of the order. Six major groupings of dicotyledons (to the left of the solid line) and four of monocotyledons are shown. Each of these groupings represents a major evolutionary line with relatively primitive and more advanced members. The orders which are mentioned in Chapters 31 and 32 are shaded in gray. A number of smaller orders have been omitted for the sake of simplicity.

marigold, and water lily. Subsequent evolution in different habitats has resulted in the diversity of present-day monocotyledons, including some such as palms and tree lilies that secondarily adopted an arborescent growth form. Monocot trees are organized much differently from typical dicotyledonous trees, as would be expected (see Chapter

13). The trends in floral evolution in monocots parallel those in dicots very closely.

To permit some appreciation of angiosperm diversity and to illustrate how evolutionary trends have led to this diversity, a few selected families of dicotyledons and monocotyledons are surveyed in the following section. This very limited survey focuses attention on cer-

Figure 31-14. Flower of the tulip tree (*Liriodendron tulipifera*). The three sepals beneath the petals cannot be seen.

Figure 31-15. The common buttercup (*Ranunculus acris*) of the crowfoot family (Ranunculaceae). The flower parts are numerous and spirally arranged, characteristics believed to be primitive.

tain groups of evolutionary importance. In the next chapter some other families of economic and botanical significance are discussed in the context of human dependence on plants. The order to which each family is assigned is noted so that it can be located in the plan of Figure 31-13.

Selected Families of Dicotyledons

Magnolia Family (*Magnoliaceae*)

Order Magnoliales. As mentioned before, the magnolia family is relatively primitive. All the hundred or so species in ten genera are either trees or shrubs, often with large, attractive flowers. A number are widely planted as ornamentals. The stamens and carpels (simple pistils) are usually numerous and spirally arranged on an elongated axis (receptacle). Several species of *Magnolia,* as well as the tulip tree (*Liriodendron*) (Figure 31-14), one of the largest and most valuable of our timber trees, are native to the eastern and southeastern United States. The family as a whole is largely restricted to eastern North America and eastern Asia. The only existing relative of the tulip tree is a Chinese species.

Fossil remains of *Magnolia* and *Liriodendron* indicate, however, that these genera were widely distributed throughout the northern hemisphere in late Cretaceous times, and that the family is probably among the most ancient of the living dicotyledons. Like the sequoia and ginkgo, they are survivors of the past and have become extinct over great areas of their former range. This family, and the next, belong to the ranalian complex as noted earlier.

Crowfoot Family (*Ranunculaceae*)

Order Ranunculales. Among the better-known members of this family are the buttercup (Figure 31-15) and crowfoot (both are *Ranunculus* species), marsh marigold (Figure 31-16), hepatica, anemone, columbine, larkspur, meadowrue, and clematis. A number of species contain substances causing forage poisoning in sheep, cattle, or horses. The root of the cultivated aconite or wolfsbane is extremely poisonous and is the source of the drug aconitine. The leaves of this plant are also poisonous when eaten. Most of the 1500 species of the family are perennial herbs, widely distributed in the temperate and arctic regions of the northern hemisphere. A number of crowfoots are aquatic or

Figure 31-16. Flowers of marsh marigold (*Caltha palustris*), of the crowfoot family (Ranunculaceae). Numerous stamens surround a cluster of simple pistils in the center. The tepals are petal-like.

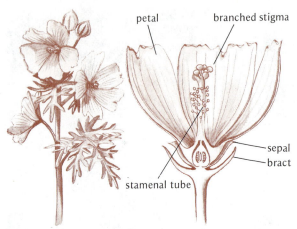

Figure 31-17. Flowers of musk mallow (*Malva moschata*), of the mallow family (Malvaceae). The enlarged flower on the right has been partially cut away to show how the numerous stamens are fused into a tube.

The flowers of the Malvaceae are commonly large and showy, with five sepals and five petals. The filaments of the numerous stamens are fused into a tube that surrounds the superior ovary, which is composed of three to many fused carpels (Figure 31-17).

Parsley Family (*Apiaceae* or *Umbelliferae*)[1]

Order Apiales. The members of the parsley family, comprising nearly 3000 species, are almost all herbaceous. The family includes a number of aromatic herbs used for seasoning—anise, caraway, dill, fennel, parsley, and coriander. Other members of the family are used as vegetables, such as carrot, parsnip, and celery. A final group are weeds, some of which are poisonous. Among the latter is the poison hemlock (*Conium maculatum*), introduced from Europe and common in many parts of the United States. The toxic character of the hemlock was known to the ancients, and this plant is believed to have been the source of the poison administered to Socrates by the Athenians in 402 B.C. The several species of the genus *Cicuta* (water hemlock), native to North America, are also very poisonous.

The traditional scientific name of the parsley family, Umbelliferae, is derived from the arrangement of the flowers, which are grouped in umbrella-like clusters (umbels). The individual flowers (Figure 31-18) are usually small, but this is compensated by the showiness of the assemblage. The sepals are minute or absent, and the five petals are commonly white in color. The five stamens are attached to the top of the inferior ovary,

semiaquatic, growing in ponds or wet places. Many species are cultivated as ornamentals.

The flowers of the crowfoot family are typically regular and perfect; the stamens and carpels, and in some species the perianth members also, are spirally arranged. The stamens are commonly numerous, and the carpels are usually free and numerous. These, as previously noted, are primitive floral characteristics. Some members exhibit advanced characters, such as irregular flowers in *Aconitum* and *Delphinium,* but the family as a whole is generally considered the most primitive of herbaceous dicotyledons.

Mallow Family (*Malvaceae*)

Order Malvales. The mallows include about 1000 species, mostly herbs but with a few tropical trees or shrubs, and are widely distributed in both tropical and subtropical climates. The family contains a number of attractive wild flowers, together with some troublesome weeds. Among the forms cultivated as ornamentals are hollyhock and rose of Sharon. The capsules of okra, or gumbo, are used in soups and stews. Several members of the family yield fibers of commercial importance. The cottons, of which there are many species, are the most valuable plants of the family. The hairs on the seed form the cotton of commerce (Chapter 32).

---

[1]Here and elsewhere in this chapter and in Chapter 32 an alternate family name is sometimes given (i.e., Umbelliferae). The second names are traditional family names, mostly dating back to or before Linnaeus, which do not conform to the International Code of Botanical Nomenclature (1972) rule that family names are to be formed from a genus name plus -aceae. The traditional names are still widely used and are often more familiar than the strictly correct "-aceae" names given first. Some of the families mentioned, like most of the other families now recognized, do not have a traditional, nonconforming botanical name.

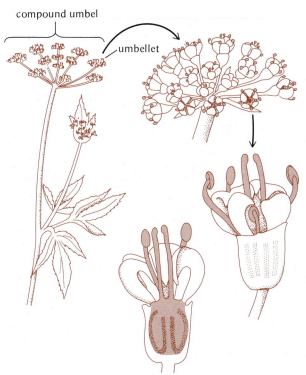

Figure 31-18. Flowers of golden alexander (*Zizea aurea*) of the parsley family (Apiaceae). The small individual flowers are grouped into umbellets which in turn are grouped into a compound umbel. The flower at the bottom is split to show inferior ovary composed of two fused carpels.

Figure 31-19. The monkey flower (*Mimulus ringens*) of the foxglove family (Scrophulariaceae). The five petals form a corolla tube which is irregular. The flower at the lower right is split to show two long and two short stamens and the compound pistil. The ovary is superior.

which is composed of two fused carpels. The fruits are dry and split into two single-seeded parts at maturity.

Birch Family (*Betulaceae*)

Order Fagales. This family of slightly more than 100 species of northern hemisphere trees and shrubs illustrates the reduction of floral structure associated with wind pollination. It includes the birches, alders, and hazelnuts that are widespread on the North American continent. In the same order is the beech family (Fagaceae) including the oaks, beeches, and chestnuts, which like the birches are very important in our deciduous broadleaved forests.

Birch flowers are monoecious (staminate and pistillate flowers separate but borne on the same plant) and for the most part are assembled in separate, many-flowered catkins (Figure 16-19). The perianth is either lacking or extremely reduced. The staminate flowers have two to many stamens, and the pistillate flowers contain an inferior ovary of two fused carpels. Large quantities of pollen are released in the spring and are transported by wind.

Figwort or Foxglove Family (*Scrophulariaceae*)

Order Scrophulariales. The members of this large family, about 4000 species, are widely distributed over the world. Most species are herbs, but some are small shrubs, trees, or vines. The family is chiefly important

for the large number of houseplants and garden ornamentals, including *Calceolaria*, snapdragon, the speedwells, pentstemons or beard tongues, Kenilworth ivy, and many others. The common foxglove (*Digitalis*) is a frequently cultivated biennial, and preparations made from the leaves have important uses in treatment of heart diseases. The common mullein (*Verbascum thapsus*), a conspicuous biennial weed of this family introduced from Europe, is found throughout temperate North America in pastures and waste places. Its flowers, borne on tall unbranched stems, are pale yellow, and the stem and leaves are covered with a heavy mat of hairs, giving them a woolly appearance.

The family has perfect flowers, and the perianth is usually irregular, the five fused petals forming a corolla tube (Figure 31-19). The compound ovary, containing numerous ovules, is superior, and the carpels have been reduced to two and are fused. There are typically four stamens, in some species two, rarely five; a fifth stamen is occasionally represented by the filament only (e.g., penstemon). The stamens are usually in two pairs, one with longer and one with shorter filaments. The flowers are usually pollinated by insects, commonly bees.

The Composites (*Asteraceae* or *Compositae*)

Order Asterales. The composites are the second largest family of flowering plants, with about 20,000

Figure 31-20.   Sunflower plant with flowering "head."

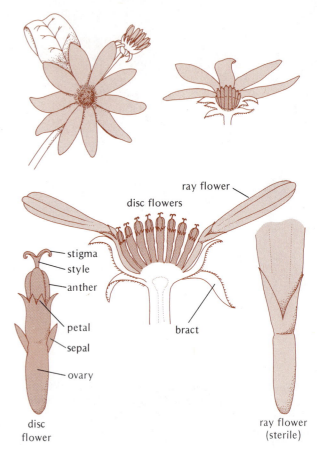

Figure 31-21.   Inflorescence and flowers of the thin-leaf sunflower (*Helianthus decapetalus*) of the composite family (Asteraceae). The inflorescence in the center is split to show the arrangement of disk and ray flowers.

species, mostly annual or perennial herbs. To list the more common composites is a task in itself. Many familiar food plants are found in this family (Chapter 32). A large number of composites, such as chrysanthemum, cosmos, zinnia, dahlia, calendula, and marigold, are garden ornamentals. Burdock, boneset, grindelia, camomile, and arnica have been used medicinally. The white snakeroot (*Eupatorium rugosum*), when eaten by cattle, horses, and sheep, causes trembles, or milk sickness. The disease is transmitted to people who drink milk from affected cows or use butter made from such milk. The flowers of pyrethrum, a species of *Chrysanthemum,* are the source of a well-known insecticide. The sagebrushes, ragweeds, and their relatives are important causes of hay fever. Finally, a number of composites are weeds, troublesome to farmers and gardeners—thistles, dandelions, hawkweeds, cockleburs, and many others.

The flowers of the composites are very small and are closely grouped into compact heads that are commonly mistaken for a single flower (Figure 31-20). The individual flowers are of two kinds. In one, the **disk** flowers, the corolla is tubular. In the other, the **ray** flowers, the corolla is strap-shaped and petal-like on one side. The head of composite flowers is commonly composed of disk flowers in the main body of the head and ray flowers along the margins (Figure 31-21). The ray flowers in such heads are frequently sterile and

produce no seeds. In some genera only disk flowers or only ray flowers are present.

The stamens are fused by their anthers and form a cylinder around the style; the ovary is inferior and is composed of two united carpels. The fruit, an achene, is popularly called a seed. A calyx is either absent or modified into plumes, hairs, spines, or hooks, which are often useful in dissemination of the fruit. An example is the downy parachute of the dandelion (Figure 17-A).

The composites are a highly evolved group of plants. Few groups are more familiar and widespread, from the asters and goldenrods of eastern autumn roadsides to the sagebrush of western plains and the golden yellow composite wildflowers of the southwestern deserts.

Selected Families of Monocotyledons

Water Plantain Family (*Alismataceae*)

Order Alismatales. As noted earlier, this family of about 50 species of marsh or aquatic herbs has played an important role in interpreting the origin of the monocotyledons. In addition to the water plantain (*Alisma*),

Figure 31-23. Flower of the Canada lily (*Lilium canadense*).

Figure 31-22. Arrowhead flowers (*Sagittaria latifolia*) of the water plantain family (*Alismataceae*).

it includes the common arrowhead (*Sagittaria;* Figure 31-22) of pond margins. The flowers (with three green sepals, three whitish petals, six or more free stamens, and numerous separate carpels) differ little from those of the Ranunculaceae and are believed to show a very primitive condition of the monocot flower.

Lily Family (*Liliaceae*)

Order Liliales. Although named for a common conspicuous genus (*Lilium*), most of the 4000 species of the lily family are not lilies at all. Asparagus is an important food plant. Onions, leeks, chives, and garlic are usually classed with the lilies, but some botanists place them in the closely related amaryllis family. Many lilies are prized as garden or houseplants. These include the easter lily, regal lily, tiger lily, and other colorful species. The day lily, lily-of-the-valley, and the Mariposa lily are not included in the genus *Lilium* but belong in the family. Other ornamentals are the tulip, hyacinth, and autumn crocus (*Colchicum*). The alkaloid colchicine from this plant is widely employed in the production of polyploids. Plants poisonous to humans or livestock include red squill (*Urginea*), used as a rat poison; false hellebore (*Veratrum*); and species of death camas (*Zigadenus*). Conspicuous among the attractive

wild plants in this family are trilliums, bellworts, adder's-tongues, and clintonia.

The Liliaceae are mostly perennial herbs with a rhizome, bulb, or corm. The flower is regular, with six petal-like perianth parts in two whorls. There are usually six stamens and a superior ovary composed of three united carpels (Figure 31-23).

Palm Family (*Arecaceae* or *Palmae*)

Order Arecales. Several thousand species of palms are known. Many are small, but some grow more than 30 m high, rivaling many dicot trees in size. Unlike dicot trees, however, they have no cambium and achieve their trunk diameter through the extensive development of primary tissue. Most palms are tropical, but a dozen or so are native to the subtropical parts of North America. In the late Cretaceous and early Tertiary, palms were widely distributed in the northern hemisphere, extending into Canada. Modern palms are an important source of food (coconuts and dates), lumber, fibers for clothing, thatch for native houses, vegetable ivory, oils and waxes, sago starch, and many other products in the tropics (Chapter 32).

The palms as a whole are wind pollinated, although there are well-documented cases of insect pollination. The individual flowers are small, clustered in large inflorescences, and either imperfect or perfect (Figure 31-24). There is a calyx of three sepals and a corolla of three petals and usually at least six stamens. The three carpels may be fused or separate. The ovary is superior. In the imperfect flowers vestiges of the nonfunctional organs often occur. The flower is relatively primitive but shows the tendency to reduction associated with wind pollination.

Orchid Family (*Orchidaceae*)

Order Orchidales. The orchids comprise the largest family of flowering plants, with about 25,000 species. Like the composites, the orchids are a very diverse group. All are herbaceous perennials. Contrary to popular opinion, they are not all tropical nor do they all grow upon other plants. Many are subtropical, and many

Figure 31-24. Flowers of the coconut palm (*Cocos nucifera*).

others thrive in temperate regions. About 140 species are native to North America. Many orchids grow as epiphytes, but many others, including most temperate zone species, are terrestrial, growing in the soil. A few are colorless mycorrhizal saprophytes or epiparasites (Chapter 25).

The orchids have little economic value except as ornamentals. Vanilla flavoring is extracted from the pods of a tropical orchid, but a synthetic vanilla flavoring is now commonly used. A very large number of tropical species, especially of the genus *Cattleya*, are cultivated for their spectacular flowers, and many hundreds of horticultural varieties have been produced by hybridization. Among our native orchids, some of which are cultivated out-of-doors, are the lady's slipper, showy orchid, pogonia, calypso, and lady's tresses. The picking of native orchids is forbidden by conservation laws in many states.

The orchid flower is highly modified and adapted to cross-pollination by insects. It is constructed upon the usual monocotyledonous plan, but this is obscured

Figure 31-26. A lady's slipper orchid (*Cypripedium calceolus*) growing in a bog in Cheboygan County, Michigan.

by the fusion of various parts. The ovary is inferior. The three sepals are usually distinct, but in some species two sepals are fused into one organ. The three petals are modified so that two form "wings" and the central

Figure 31-25. *Cattleya*, a spectacular, widely cultivated tropical epiphytic orchid.

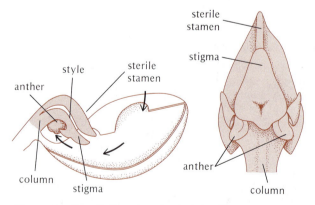

Figure 31-27. Pollination in lady's slipper (*Cypripedium*). (Left) Flower from the side, with portion of the slipper removed; the arrows show the route of a small bee as it enters and leaves the flower. (Right) The column, which terminates in the stigma and bears the stamens.

one forms a conspicuous cup-like **lip** on one side of the flower (Figure 31-25). Usually only one of the stamens is functional, but in some species two are.

An important feature of the orchid flower is the **column** formed by the fusion of the stamens with the style. In the lady's slipper the central lip, much enlarged and slipper-shaped, encloses a large cavity open at the top (Figure 31-26). The column lies at the back of the flower. One of the three stamens is sterile and arches over the stigma. The two fertile stamens are located on either side of the stigma (Figure 31-27). Small bees, entering the oval opening at the top of the lady's slipper, feed upon juicy hairs at the bottom. Because

escape by the path by which they entered is prevented by the overarching wall of the slipper, the insects can leave only through two small openings at the back of the flower, one on either side of the column. They creep under the stigma and squeeze through one of these openings, but in doing so, they brush against an anther, picking up some of the sticky pollen. When they visit another flower, the bees escape in the same way, this time depositing pollen on the stigma as they creep under it. In the absence of insect visits, pollination fails and no seeds form. The orchids and composites each illustrate different culminations in the evolution of the modern flower.

SUMMARY

(1) The angiosperms are the dominant plants of the earth's land vegetation. They occupy almost every habitat and are extremely important to other organisms, including humans.

(2) In spite of their number and diversity, the angiosperms are believed to be a unified group (class) with a common origin. They have a wide range of body forms, diversity of leaf size and morphology, vessels and sieve tubes in their xylem and phloem respectively, and distinctive reproductive structures known as flowers.

(3) The angiosperms are separated into two subclasses, the monocotyledons and the dicotyledons. The two groups are distinguished by number of cotyledons, leaf venation, arrangement of vascular tissue in the stem, and number of parts in each whorl of the flower.

(4) The flower is a reproductive shoot, the axis or receptacle being a stem and the appendages foliar in nature. The stamens and carpels are sporophylls. Ovules in the angiosperms are enclosed, and, unlike the condition in gymnosperms, pollen is transferred to the stigmatic region of the pistil rather than to the ovules. The nonmotile sperms are carried to the ovule by growth of the pollen tube.

(5) Like all other vascular plants, the angiosperms have an alternation of generations in their reproductive cycle. They share the seed habit with the gymnosperms, but the gametophyte stages are even further reduced. Double fertilization leading to endosperm development, uniquely found in the angiosperms, is superimposed upon the pattern of alternation.

(6) The origin of the angiosperms is not known with certainty because of the failure to find transitional forms from presumed gymnosperm ancestors in the fossil record and because of the absence of a close relationship to any known gymnosperm group. They are believed to have arisen shortly before their abrupt appearance in the fossil record of the Cretaceous period.

(7) Primitive angiosperm characteristics have been deduced by comparative study of living plants that are regarded as archaic (ranalian complex). These features include a woody habit, absence of vessels, and insect-pollinated, perfect flowers with regular symmetry and many free parts helically arranged on an elongated receptacle.

(8) The herbaceous habit is believed to have been derived from the woody habit by reduction of cambial activity and secondary tissues. Its widespread occurrence is an adaptation to deteriorating climatic conditions beginning in the Tertiary period.

(9) Trends in floral evolution include (a) reduction in the number of floral parts; (b) whorled arrangement of parts on a shortened axis; (c) fusion of parts both within whorls and with other whorls; (d) loss of floral parts, either perianth parts or reproductive parts, the latter resulting in imperfect flowers; and (e) development of irregular symmetry.

(10) Classification of so large and diverse a group according to natural relationships is a difficult problem. Floral characters provide the best criteria. In one modern system six major groupings of dicotyledons and four of monocotyledons are recognized, each including both primitive and advanced members and representing a major line of evolutionary development. The monocotyledons are believed to have arisen from dicotyledonous ancestors.

Chapter 32

PLANTS AND PEOPLE

Around the world people appear to have used more than 10,000 species of plants, at least historically, but the number that have attained real economic importance is closer to 1000. Large though that seems, it is a small proportion indeed of the more than 350,000 species of plants that are available in nature. All but about 50 of the economically significant plants are seed plants, and the great majority of these are angiosperms. Thus it is appropriate to consider plants that are important to people in connection with our study of the angiosperms, begun in the preceding chapter.

Of the 1000 or so economically significant plant species, not more than about 150 are extensively cultivated in modern agriculture. If we consider not just the staple crops that provide the bulk of our calories and protein but others that provide nutritionally and gastronomically valuable vegetables and fruits, we find that most of the common food plants fall into about a dozen principal botanical families. The features of several of these families were pointed out in the preceding chapter. We shall consider here a number of the others. It should be kept in mind that this will still be a very small selection from among the more than 300 families of flowering plants that have been recognized.

Useful or economic plants may be grouped in categories based upon the use to which they are put. This utilitarian classification is totally different from the natural or evolutionary approach that we have followed up to this point. It is, in fact, anthropocentric or human-centered, but it is justified by the purpose of this aspect of the study of botany.

PLANTS AS FOOD

Food is one plant product for which there is no substitute. Furthermore, an adequate food supply is not just a matter of quantity. The essential components of the diet are energy sources, pro-

tein, vitamins, and minerals. The main energy foods are carbohydrates (starch and sugars) and fats. Proteins can also be metabolized for energy, but they have the essential dietary role of supplying the amino acids that are needed to build the proteins of the human body. Inadequate protein is one of the major factors in malnutrition in the poorer regions of the world, which depend almost entirely upon plant foods. Many of the staple plant foodstuffs such as cereals and potatoes are high in carbohydrate content and low in protein compared with the needs of a nutritionally balanced diet.

Animal proteins such as those in milk, eggs, and meat are much like those of the human body in amino acid composition, and are ideal sources of essential amino acids for humans. Unfortunately, for a majority of the world's people, animal products are a seldom enjoyed luxury. No single plant food contains all of the essential amino acids in desirable proportions, although the soybean comes very close. However, different classes of plant foods differ in their amino acid deficiencies, so one can obtain adequate protein nutrition, without eating animal products, by using the right combination of food plants. To an impressive extent, primitive peoples learned to do this. The cultivation by American Indians of maize, beans, and squash, often together in the same plots, provided a satisfactory diet, beans being especially rich in those essential amino acids that are deficient in maize. Similarly, in Asia, a diet based on rice as the energy source also included soybeans. This principle is being used in current efforts to develop dietary supplements for improving the protein nutrition of people in areas where cereal grains are the principal protein source.

The Cereals

The cereal grains are, and have been since the dawn of agriculture, man's most important food plants (Table 32-1). Wheat, rice, and maize (corn) are the mainstays of world agriculture; rye, barley, oats, sorghum, and millet are also important grain crops globally. Some 70% of the world's crop land

Figure 32-A. Wheat being harvested in the American Great Plains.

TABLE 32-1   World Production of Plant Products

Data for 1977, in millions of metric tons,[1] from FAO Production Yearbook 31 and 1977 Yearbook of Forest Products, United Nations Food and Agriculture Organization. Data in roman type are for air-dry products. Data in italics are for produce, the dry weight or biomass of which would be only about one fifth to one tenth of the fresh weight listed.

Product Class	Total Production	Major Crops or Crop Groups	Production
Cereal grains	1459	Wheat	387
		Rice	366
		Maize (corn)	350
		Barley	173
Starchy roots and tubers	*570*	White potato	*293*
		Sweet potato	*138*
		Manioc (cassava)	*110*
Sugar (raw crystallized)	106	Sugar cane (unprocessed)	*737*
		Sugar beets (whole)	*290*
Legumes (dry seeds)	143	Soybean	77
		Beans and peas (several species)	38
		Peanut	17
Oil crops	66	Sunflower	12
		Rape, mustard	8
		Oil palm seed	5
		Cotton seed	27
		Olive	8
Fruits	*257*	Grape (mostly for wine production)	*57*
		Banana	*37*
		Plantain	*20*
		Orange	*33*
		Apple	*21*
		Soft fruits (pears, peaches, apricots, etc.)	*20*
		Melons	*28*
		Mango	*13*
		Pineapple	*6*
		Date	*2*
Tree nuts	36	Coconut	32
		Deciduous trees (walnut, almond, etc.)	2.6
Vegetables	*291*	Tomato	*45*
		Cabbage group	*26*
		Onions and garlic	*19*
		Cucurbits (squash, cucumber, etc.)	*14*
		Peppers (chili, green)	*6*
		Green peas and beans	*7*
		Carrot	*7*
Beverage crops	7.5	Coffee	4.3
		Tea	1.8
		Cocoa bean	1.4
Tobacco	5.6		
Fiber crops	21	Cotton	14
		Jute	4
		Flax	0.7
		Hard fibers (abacá, sisal, etc.)	0.5
Rubber (natural)	3.6	(Almost all para rubber)	

		Softwoods	Hardwoods
Wood and related products:	1844		
Fuelwood		102	753
Saw logs for lumber, plywood, veneer, etc.		431	189
Pulpwood[13]		152	64
Other (posts, pilings, match wood, etc.		86	67
Cork	0.3		

[1] 1 metric ton = 1000 kg = 1.1 short ton (s.t.) of 2000 lb/s.t. Because of many minor crops that cannot be listed, the total production given is usually greater than the sum of the individual crops or crop groups that are listed, especially in the case of vegetables.

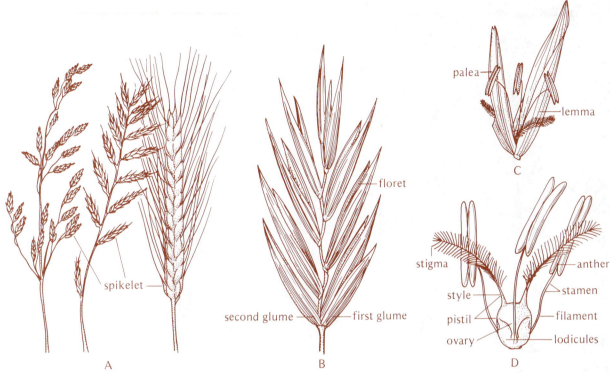

Figure 32-1. Flower structure in grasses. (A) Typical inflorescences, showing spikelets (condensed flower-bearing branch tips). (B) One spikelet showing basal bracts (glumes) and individual bract-enclosed flowers (florets). (C) One floret open for pollination, with its bracts spread apart. (D) Flower parts as seen after removal of bracts.

is devoted to the cultivation of cereal grains, and more than 50% of the calories consumed by man are provided by them. The cultivation of cereal crops is, in the popular mind, almost synonymous with agriculture; the word cereal is derived from Ceres, the Roman goddess of agriculture. All cereals belong to the monocot family Poaceae (Gramineae), the grass family. The grain they produce is nutritious, palatable, and well suited to both storage and transport. Moreover, the cereals are highly productive and can be adapted easily to large-scale cultivation with mechanization.

Grass Family (Poaceae or Gramineae)

Order Poales.[1] The grasses are a large family of about 8000 species, found under all climatic conditions from the warm tropics to the Arctic, and from swamps to deserts. Grass flowers (Figures 16-18B and 32-1) are wind-pollinated and therefore lack showy petals. They are small and borne either in dense spikes or open branching clusters. The spikes or clusters are composed of units called **spikelets.** Each spikelet consists of an axis bearing one to several flowers together with a number of specialized bracts (reduced, modified leaves).

---

[1]As in Chapter 31, the order to which each family is assigned is given so that it may be located on the plan of Figure 31-13.

Each individual grass flower is tightly enclosed within two bracts, comprising the "hulls" which must be removed from grain by threshing. The flower is composed of a pistil, three stamens, and usually two scales or **lodicules,** below the ovary, representing the highly reduced perianth. The stamens emerge from the bracts at flowering and shed their pollen into the air. Two feathery stigmas protrude and catch the windborne pollen. The grain, or one-seeded fruit, is a matured ovary in which the ovary wall is fused to the seed coat surrounding the embryo or "germ" and the extensive endosperm.

Wheat (Figure 32-2) is the main cereal crop of the temperate regions of the world, but it is also the most widely cultivated crop plant, grown on every continent. With a history spanning some 9000 years, it is one of our oldest cultivated plants. Worldwide production today exceeds that of any other food plant and constitutes the most important agricultural product in world trade, although only a few nations still have a surplus for export. Wheat is adapted to cool, relatively dry climates and is grown on a large scale in the temperate grassland regions of the world such as the Great Plains of North America (Figure 32-A). It is, however, successful in warmer regions, often as a winter crop, but it does not thrive in the moist tropics. The numerous cultivars of wheat are considered to constitute about a dozen species of the genus *Triticum*, falling into three classes, the diploid,

Figure 32-2. Major species of wheat and wild grasses figuring in their ancestry. (A) Primitive diploid wheat ("Einkorn," *Triticum monococcum*), which carries genome *A*. (B) The weedy diploid "goat grass," *Aegilops speltoides*, carrying genome *B*. (C) Wild tetraploid wheat ("wild emmer," *T. dicoccoides*), combining genomes *A* and *B*; note the fragile axis of its head, like that of many weedy grasses, making their grain difficult to harvest without loss. (D) A cultivated tetraploid wheat ("emmer," *T. dicoccum*), with a much firmer head axis. (E) Another weedy diploid goat grass, *Aegilops squarrosa*, carrying genome *D*. (F) Common hexaploid "bread wheat" (*T. aestivum*), combining genomes *A, B,* and *D*. The grasses commonly called *Aegilops* (B and E) are sometimes classified instead as species of *Triticum* (*T. speltoides* and *T. squarrosa*, respectively).

tetraploid, and hexaploid wheats with 2n chromosome numbers of 14, 28, and 42, respectively (Figure 32-2). Today the hexaploid wheats, especially the bread wheats (*T. aestivum*), are by far the most widely grown, although the tetraploid durum wheat (*T. durum*) is still produced on a large scale for the manufacture of macaroni and similar products.

The wheat genus includes a few wild species occurring in the Middle East, the general region where wheat was apparently first domesticated. Some of these species, as well as wild grasses from genera closely related to *Triticum,* appear to have participated, by hybridization followed by chromosome doubling, in the evolution of domesticated wheats as a complex allopolyploid series (Figure 32-3). Some of this hybridization and chromosome doubling has been duplicated experimentally in modern times to yield an artificial hexaploid wheat closely resembling certain of the historically evolved hexaploid cultivars.

The wheat grain or kernel is high in nutritional value, containing about 70% carbohydrate, 8–16% protein, and about 2% fat as well as several of the important vitamins. If the kernel is eaten whole or converted into flour by traditional stone grinding, all of the nutritional value is obtained. However, modern milling methods eliminate the "germ" or embryo, which contains nearly all the fat and is high in protein, and the "bran," consisting of the ovary wall, seed coat, and outermost layer of endosperm (aleurone layer), which contains more protein and nearly all the vitamins. The resulting white flour is mostly carbohydrate and, although pleasing to the taste, is nutritionally impoverished. It is often artificially "enriched" with vitamins to improve its food value. Health food enthusiasts today advocate the use of whole grain flour for baking, but their campaign is not a new one. In the early nineteenth century the nutritional deficiencies of refined white flour were vigorously publicized by a Massachusetts preacher named Sylvester Graham, for whom the graham cracker is named.

The starchy endosperm, however, does contain some protein, and it is this that, in the case of bread wheats, gives to the flour its desirable baking qualities. The two proteins glutenin and gliadin cause the dough made when flour is moistened to become sticky and elastic and thus to trap bubbles of carbon dioxide released by the yeast used as leavening. This causes the dough to "rise" and to form, when baked, light, highquality bread. Other grains do not contain these pro-

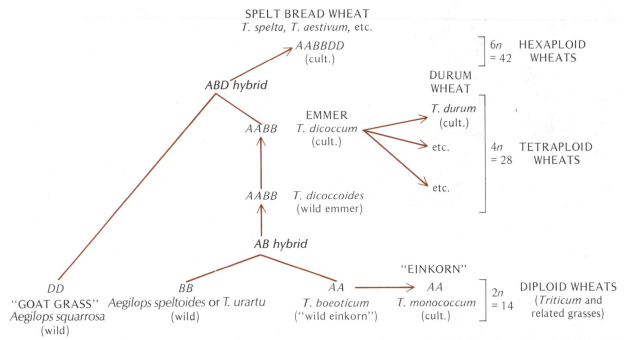

Figure 32-3. Simplified diagram of the history of diploid, tetraploid, and hexaploid cultivated wheats. A, B, and D denote the haploid genomes (7 chromosomes each) of (A) wild einkorn, (B) *Aegilops speltoides* (or *Triticum urartu*), and (D) goat grass.

teins and thus are not good sources of bread flour, except for rye, which is inferior in other respects.

Rice (Figure 32-4), in contrast to wheat, is a cereal crop of the humid tropics, requiring warm temperatures and abundant moisture. In quantity produced, rice approximately equals wheat, but 90% of it is grown in Asia and also consumed there, usually not far from where it is harvested. As a crop of one of the world's most populous regions, it is the dietary mainstay of more than 50% of the world's people. It is also grown in Africa, Latin America, and the United States. Rice is believed to be of Asiatic origin. It has been found in archaeological sites in Thailand as old as 7000

years. There are thousands of cultivars, which are usually placed in the single species *Oryza sativa*, but this is only a convenient catch-all for a complex that has not been fully worked out. The wild rice of North America is unrelated, belonging to a different genus (*Zizania*).

Rice cultivation requires a great deal of water. Most kinds of rice are grown in flooded fields, at least for much of their development. (There are, however, upland rice cultivars that are grown without flooding where rainfall is adequate, mostly outside Asia.) Much of the rice in Asia is grown in paddies—small, diked fields that can be flooded, often on terraced hillsides (Figure 32-5). Enormous amounts of hand labor are required, but the yield is considerable. Sowing is first done in seed beds, and the seedlings are later transplanted to the flooded paddies by hand, the only cereal crop that is accorded this kind of attention. When the grain is nearly mature, the paddies are drained and allowed to dry out before harvesting, usually by cutting the heads individually. On the other hand, rice cultivation can be mechanized in large-scale operations, and this has occurred in the United States, particularly in California.

Like all cereals, rice has a low protein content, but with only 7.5% it is one of the lowest. Most rice is consumed as the whole grain and, when only the hulls are removed by threshing, the resulting brown rice contains all the nutritional value. However, much rice is further processed by polishing, which removes the outer layers (the equivalent of bran) and the embryo, leaving little more than carbohydrate. White rice, like white flour, is produced with the sacrifice of much potential food value. Rice can be ground into flour, but this is not suitable for breadmaking.

Figure 32-4. Rice plant, Japan.

Figure 32-5. Terraced, flooded rice paddies into which young rice shoots have recently been transplanted. Bali, Indonesia.

and also in the separation of its pollen-producing flowers (in a terminal tassel) from its pistillate or seed-producing flowers (on swollen lateral axes called cobs; Figure 16-21).

More than half of the world maize crop is used as animal feed. For human consumption maize kernels can be ground to produce corn meal, but this is not a good bread flour, affording only a crumbly (but tasty) cake-like product known as corn bread or johnny cake. The traditional Mexican method of preparation consists of soaking the kernels in lime water, squeezing off the hulls, and grinding the remainder into a paste from which the flat, pancake-like *tortilla* is made. The corn embryo is rich in oil, and from it corn oil, one of the most important vegetable oils, is produced commercially. Unripe ears are consumed fresh as a vegetable, especially those of certain cultivars with a high sugar content, known as sweet corn. Many food and industrial products are manufactured from maize.

Maize (*Zea mays*), the third major cereal plant, is a native of the western hemisphere and was not known elsewhere until after the voyages of Columbus. Maize cultivation is now widespread but still is most extensive in the Americas, where in pre-Columbian times it was grown from Chile well into Canada. Originally known to English colonists as "Indian corn," it is usually called simply corn by North Americans, but this common name is confusing because in Europe corn means any cereal. Numerous cultivars of maize exist. Some believe that maize arose in Mexico from a now extinct wild progenitor, probably about 7000 years ago. Others consider that teosinte (*Zea mexicana*), a Mexican weed with certain resemblances to maize, is the ancestor. Among the important cereal grasses, maize is distinctive for its large size

The Legumes

As food plants the legumes are second in importance only to the cereals (Table 32-1). They are rich sources of energy and are high in protein content and often in oil as well. For people whose diet can contain little or no animal material they provide an excellent nutritional counterbalance to the overly carbohydrate-rich cereals. Legumes are unique among crop plants in their ability to conduct symbiotic nitrogen fixation (Chapter 25), enabling them to be grown successfully in nitrogen-poor soils. Their nitrogen fixation also benefits other crops grown in rotation with legumes, an important consideration in many parts of the world where people cannot obtain or afford expensive nitrogen fertilizers.

Figure 32-6. Butterfly-shaped (papilionaceous) legume family flower, and fruit. (A) External view of flower, showing its five petals ("banner," two "wings," two petals fused into a "keel"). (B) Side view of interior, with keel bent down to expose carpel and partially fused stamens. (C) Fruit (legume pod). (D) Fruit at time of opening to release seeds.

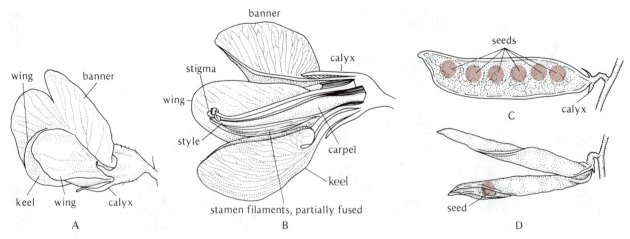

Legume Family (*Fabaceae* or *Leguminosae*)

Order Fabales. The legume family, with more than 13,000 species, is the third largest family of flowering plants (after the orchids and composites). Legumes are enormously diverse, including herbs, shrubs, vines, and trees growing under a great variety of conditions from deserts and grasslands to forests of the temperate and tropical regions. In addition to the many legumes used directly as human food, there are several other useful species. Alfalfa, vetch, clover, and lespedeza are among the most important forage crops for domestic animals. The important ornamentals include wisteria, lupines, redbuds, locust trees, and acacias.

The leaves of legume plants are usually alternate and compound. The pod-like fruit (e.g., pea or bean pod), termed a **legume** (Figure 32-6C, D), is derived from a single carpel, and when ripe typically splits along two sides to shed its seeds. The structure of the flower varies to such an extent that three subfamilies of legumes are recognized. The largest subfamily has a distinctive ''butterfly-shaped'' flower such as that of the garden pea or the sweet pea (Figure 32-6A). Legume flowers are adapted for bee pollination (Figure 32-6B).

Long the major legume crop of Asia, especially in its native China, where it has been cultivated for nearly 5000 years, the **soybean** (*Glycine max*; Figure 32-7) in the last 50 years has

Figure 32-8. Peanut plant, Georgia. After pollination the developing fruits (legumes) are pushed into the soil by intercalary growth.

risen to prominence in the western hemisphere. The United States now produces half or more of the world crop, mostly in the Central Plains area, the corn belt, where mechanized agriculture is well developed. Traditionally, soybeans are eaten fresh, boiled, roasted, or sprouted, and they constitute an excellent complement to the rice-based diet of the Orient. They also form the basis of many fermented foods and are an important source of vegetable oil. In fact, the oil content (13–25%) is the major reason for their cultivation in North America. Soya oil is used in salad oils, margarine, and shortening, and has diverse industrial uses including the manufacture of plastics. After oil extraction the soybean meal that remains is a valuable livestock food. It is being used increasingly as a high-protein supplement for human consumption, often mixed with wheat flour or used in the preparation of meat substitutes that are processed and flavored to resemble fresh meat.

Other important legume food plants include the **common bean** (*Phaseolus vulgaris*), cultivated since at least 7000 years ago in both Mexico and Peru; **peas** (*Pisum sativum*), which have been found in the oldest agricultural sites in the Near East (9000 years old); and the **chick pea** or **garbanzo** (*Cicer arietinum*), which was known to the ancient Egyptians, Greeks, and Hebrews and is today a major food crop in India. The **peanut** (*Arachis hypogea*; Figure 32-8) was originally domesticated in tropical South America, where it has been found in Peruvian tombs as old as 4000 years. Soon after Europeans arrived in America, they took the plant to Africa and the Far East, and it was later introduced into North America from Africa. It is now grown extensively in the tropics and subtropics, not just for snacks and confections (its familiar use in North America) but as an oil source and in many parts of the world as a staple food crop.

Figure 32-7. Soybean plant (*Glycine max*), Korea.

Starchy Roots and Stems

Most plants contain some starch, but certain species store up particularly large amounts of it in swollen roots or stems, especially underground stems called tubers. Many common vegetables fall into this category, and a few, such as the potato, have become staple foods for large numbers of people.

The **potato,** or white potato (*Solanum tuberosum*; Figure 15-8), is one of the most important food plants (see Table 32-1). It is cultivated in temperate regions around the world and at higher elevations in the tropics. The potato thrives both in small garden plots and in large acreages under highly mechanized agricultural practices. Although often called the Irish potato because of its close association with Ireland and its people, the potato originated in the high Andes of Peru and Bolivia, where it was cultivated for centuries before the arrival of Europeans. Potato tubers were a staple food that could be grown at elevations too high and cold for the cultivation of maize. Unlike maize, it had spread little or not at all from the site of its domestication when the Spaniards first discovered it and carried it back to Europe around 1570. Today nearly 90% of the annual crop is produced in northern Europe. Recognition of its value developed slowly, however, perhaps because of the poisonous properties of many related species in the family Solanaceae to which it belongs.

The potato tuber is a modified stem which develops as the enlarged tip of an underground rhizome. Potatoes are propagated by planting pieces of tuber, each of which must bear at least one "eye"

(Figure 15-8). Almost all of the more than 400 potato cultivars are tetraploids which, although they flower profusely, rarely set seed.

Nightshade Family (*Solanaceae*)

Order Polemoniales. The nightshade family includes more than 2000 species of herbs, shrubs, and small trees in the tropics and temperate regions. Other important vegetables from this family are tomato, eggplant, green or bell pepper, and red or chili pepper. Tobacco also belongs to this family, as do a number of drug plants (discussed below).

The flowers of nightshades possess five petals fused together into a wheel-shaped or tubular corolla (Figure 32-9). The (usually five) stamens are attached to the corolla, and the ovary is superior.

Most North Americans use the **sweet potato** only as an occasional vegetable, but in many parts of Asia, Africa, and Polynesia this root crop is a staple food of major importance, taking the place of the white potato. The sweet potato plant (*Ipomea batatas*; Figure 32-10) is a vine, a member of the morning glory family (Convolvulaceae), not closely related to the white potato. The sweet potato is an enlarged, carbohydrate-filled storage root, in contrast with the tuber of the white

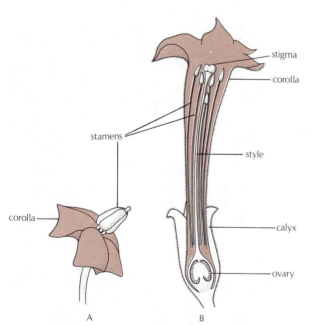

Figure 32-9. Flowers of Solanaceae. (A) Potato (*Solanum tuberosum*); (B) tobacco (*Nicotiana tabacum*).

Figure 32-10. Sweet potato vine and roots.

Figure 32-11. A yam (*Dioscorea*).

Figure 32-12. (A) Cassava plant, growing in Surinam (formerly Dutch Guyana), and (B) tubers.

potato. It is propagated vegetatively by planting slips or adventitious shoots arising from the storage roots.

The sweet potato had its origin in the lowland tropics of Central or northern South America and, unlike the white potato, it was widely distributed in the warmer parts of the western hemisphere before the arrival of Europeans. The word "potato" appears to have been derived from *batata*, the name of the sweet potato in the Caribbean islands and mistakenly applied to the white potato upon its introduction into Europe.

Often confused with the sweet potato by name is the **yam** (*Dioscorea*, family Dioscoreaceae; Figure 32-11). Botanically the yam is entirely different, being a monocot whereas the sweet potato is a dicot. *Dioscorea* is a tropical vine, many species of which are cultivated for their starchy roots in Africa and China.

Cassava, manioc, or mandioca (*Manihot esculenta*; Figure 32-12) of the spurge family (Euphorbiaceae) is another native of the American tropics which has become one of the tropical world's important starchy root crops. It is grown extensively in South America and the West Indies and is also a vital food plant in Africa and Indonesia. Cassava is a shrubby plant up to 3 m (9 ft) tall which produces as much as 20 kg (45 lb) of large, starch-filled roots per plant. The roots are peeled and shredded, the juice is pressed out, and the residue is dried with heat and pulverized to produce a meal called farinha. This may be eaten with beans or gravies or baked in thin cakes known as cassava bread. Cassava is also the source of tapioca, used in puddings.

The roots of many cassava cultivars contain a poisonous, cyanide-releasing glycoside which must be removed or destroyed. This is done by carefully express-

ing the juice or, where the concentration of the poison is not too high, by heating. One wonders how primitive people discovered how to prepare cassava so that it could be eaten without poisoning its consumers.

Of similar importance in some parts of the tropics, notably the Pacific islands where it is traditionally the major staple crop, is **taro** (*Colocasia esculenta*; Figure 32-13). It belongs to the aroid family (Araceae) of monocots, familiar temperate-zone members of which are jack-in-the-pulpit, skunk cabbage, and (as an ornamental) the calla "lily."

Figure 32-13.   Taro harvesting on Kauai Island, Hawaii.

Figure 32-14.   Sugar beet (*Beta vulgaris*), upstate New York.

Sugar Plants

The human diet has no absolute need for sugar if starchy foods are in adequate supply, but people appear to have a universal sweet tooth. The annual estimated global production of about 100 million tons of sugar works out to about 50 pounds per person (Table 32-1). (Since annual consumption in North America is in excess of 100 pounds per person, some people in the world are not getting their fair share.) The original source of sugar, at least back to Mesolithic times, was honey, and this remained the most important sweetening substance for food and drink in Europe until nearly the beginning of the eighteenth century. Honey is, of course, a plant product, being processed by bees from the sugar-containing nectar of flowers. Today two plant species directly provide all but a minor part of the world's sugar supply, one a member of the grass family and the other a close relative of the common garden beet.

Sugarcane (*Saccharum officinarum*, Figure 1-A) is a robust perennial grass with broad leaves and a solid stem 3–7 m (9–21 ft) high. The content of sucrose in the stem can become remarkably high, reaching as much as 20% in the best cultivars. Extraction involves crushing the cane, pressing out the juice, and crystallizing the sugar by evaporation. This may be done with crude implements, which produce a highly impure but relatively nutritious product, or with modern refining processes that yield almost pure sucrose.

Sugarcane originated in Southeast Asia and was apparently used in India as early as 3000 B.C. It was probably first chewed as a confection. Sugar became known in Europe only during the Middle Ages, brought from India by Arab traders, and was for a long time an exorbitantly expensive luxury. Today sugarcane is grown extensively in the moist tropics around the world.

The other important sugar plant, accounting for approximately 40% of the world's supply, is a root crop grown widely in temperate regions, especially in northern Europe and North America. The sugar beet (*Beta vulgaris*, Figure 32-14) belongs to the same species as the red root vegetable we call a beet. The sugar beet was deliberately selected from white cultivars of beet early in the nineteenth century to provide a domestically produced sugar source for Europe. Sugar beet is well adapted to mechanized agriculture and yields, after refining, a product that is virtually identical to cane sugar.

In the northeastern United States and adjacent Canada the **sugar maple** (*Acer saccharum*) is a minor sugar source but nevertheless an important local specialty item. In colonial times, and even earlier among the native people, sugar from this source was an important food product. In the spring of the year when daytime temperatures rise above freezing and the nights are cold, starch in the living (parenchyma) cells of the wood is converted to sugar. As a result, pressure builds up osmotically in the xylem, and sap will flow from holes drilled into the trunk, the technique called tapping (Figure 32-15). The sap contains a fair concentration of sugar, which may be crystallized by concentrating it by evaporation (boiling down the sap), although much of the concentrated product is now simply sold as maple syrup. No attempt is made to refine maple sugar because the distinctive flavor that gives it its value results from the nonsugar components.

Vegetables

A great diversity of species supply us with what we call vegetables. Although except for starchy roots and tubers they rarely constitute staple foods, they are important sources of vitamins and minerals that frequently are deficient in staple foods. Among the common vegetables are represented virtually every part of the plant body—stem, root, leaf, bud, inflorescence, fruit, and seed. The edible portion is an enlarged tap root in the carrot (*Daucus carota*) and parsnip (*Pastinaca sa-*

Figure 32-15.   Sugar maple being tapped, Maryland.

Figure 32-16.   The globe artichoke is the immature flower head of the giant thistle plant (*Cynara scolymus*) belonging to the composite family, in which many small flowers are borne in a compact flower-like head (Chapter 31). The edible portion is the expanded tip of the stem (the artichoke "heart") to which the flowers are attached, as well as the bases of the modified leaves (bracts) enclosing the flowers in the head.

tiva) of the family Apiaceae (Chapter 31); the radish (*Raphanus sativus*), turnip (*Brassica rapa*), and rutabaga (*Brassica napobrassica*) of the family Brassicaceae; and the garden beet, which belongs to the same species as the sugar beet.

Few plants are valued for the aerial or aboveground stem, a notable exception being asparagus (*Asparagus officinale*) of the lily family (Liliaceae) (Chapter 31). The young aerial stems of asparagus, which arise from a perennial rhizome system, constitute the highly esteemed vegetable.

Leaf vegetables provide both raw salad and cooked greens, many being consumed both ways. The onion (*Allium cepa*), which belongs, like asparagus, to the family Liliaceae, produces not only the leaves ("green onions") that are often used to spice up salads but also the well known edible bulb. This is, in fact, a shortened stem bearing many ensheathing storage leaves, the scales. From the dicot family Chenopodiaceae come the leafy vegetables chard, which belongs to the same species (*Beta vulgaris*) as garden and sugar beet, and spinach (*Spinacia oleracea*), in which the whole leaf is ordinarily eaten after boiling. In the case of celery (*Apium graveolens*) (Apiaceae, Chapter 31), it is the leaf stalks or petioles that are used. An-

other leafy vegetable from the Apiaceae is parsley (*Petroselinum sativum*), which is very useful as a flavorful garnish.

One of the most widely used salad vegetables is lettuce (*Lactuca sativa*) of the family Asteraceae, described in Chapter 31. Many cultivars of lettuce form a "head," which is a greatly enlarged terminal bud. Other leafy vegetables from the Asteraceae include two species of the genus *Cichorium*, the lettuce-like common endive or escarole (*C. endivia*), and French endive (*C. intybus*), which is an expensive delicacy. The artichoke (*Cynaria scolymus*; Figure 32-16) is another vegetable delicacy from the Asteraceae, in this case the immature flower head. After steaming it, the bracts or modified leaves around the head, and the central stem structure or receptacle ("heart") provide a very delicious food.

A particularly important source of vegetables is the Brassicaceae, or mustard family. The root crops radish, turnip, and rutabaga were mentioned above. The single species *Brassica oleracea* provides a whole series of vegetables including heading and nonheading cabbage, Brussels sprouts, cauliflower, broccoli, and kohlrabi, which are classified as separate botanical varieties (Table 32-2), each of which may include several to many cultivars. Among the different varieties, all the parts of the plant except its roots and fruits are used as a vegetable.

Crucifer or Mustard Family (*Brassicaceae or Cruciferae*)

Order Capparales. The mustard family is worldwide in distribution with about 3000 species of annual,

TABLE 32-2   Cultivated Species and Varieties of *Brassica* (Brassicaceae)[1]

Species	Variety (var.-)	Common name	Part(s) used	Origin[2]
B. nigra	—	Black mustard	Seeds (condiment)	M, I
B. carinata[3]	—	Ethiopian mustard	Seeds (oil)	A
B. juncea[4]	—	Indian mustard	Seeds (oil)	C or I
B. hirta[5]	—	White mustard	Seed (condiment)	M
B. oleracea	*fruticosa*	Wild or bush kale[6]	Leaves	M
B. oleracea	*acephala*	Kale, collards	Leaves	M
B. oleracea	*capitata*	Head cabbage	Terminal bud (head)	E
B. oleracea	*gemmifera*	Brussels sprouts	Lateral buds	E
B. oleracea	*italica*	Broccoli	Inflorescence	M
B. oleracea	*botrytis*	Cauliflower	Inflorescence	M
B. oleracea	*caulo-rapa*	Kohlrabi	Tuberous stem	E
B. campestris	*rapa*	Turnip	Root	M[7]
B. campestris	*oleifera*	Turnip rape	Seed (oil)	M, I[7]
B. campestris	*pekinensis*	Chinese "pak-choi" cabbage	Leaves	C
B. campestris	*chinensis*	Chinese "pe-tsai" cabbage	Terminal bud (head)	C
B. campestris[8]	*napobrassica*	Rutabaga (Swede)	Root	E
B. napus	*napus*	Rape	Seed (oil)	M

[1]Data mainly from J.G. Vaughan, *Bioscience* 27, 35–40 (1977); B. Brouck, *Plants Consumed by Man*, Academic Press, 1975; and J.R. Harlan, *Crops and Man*, American Society of Agronomy, 1975.

[2]A = Africa, C = China, E = Central or Northern Europe, I = India/Pakistan area, M = Mediterranean and Near East.

[3]*B. nigra* × *B. oleracea*, amphidiploid.

[4]*B. nigra* × *B. campestris*, amphidiploid. Besides use as an oil seed, cultivars of *B. juncea* have evolved in China as leaf vegetables and one root vegetable.

[5]Known more widely as *Sinapis alba*. Recent work supports the maintenance of a generic distinction between white mustard and the *Brassicas*. White mustard, which is the most important commercial source of the condiment called mustard, is included here so as not to give the impression that "mustard" comes only from the species of *Brassica* that are called mustards and are listed above. White mustard also yields the principal "mustard green" used in salads.

[6]Resembles wild cabbage, which is technically *B. oleracea* var. *oleracea*, and which occurs along coasts of Spain, France, and England.

[7]May have been independently domesticated from widespread weedy *B. campestris* var. *campestris* in more than one area as indicated.

[8]*B. oleracea* × *B. campestris*, amphidiploid.

biennial, or perennial herbs. The traditional botanical name Cruciferae (Latin *crucis*, of a cross; *ferere*, to bear) refers to the characteristic cross-like form of the flowers (Figure 32-17), which have four petals, four sepals, and six stamens, four of which are longer than the other two. In addition to the common vegetables already mentioned, the family includes the important vegetable oil plant rape (*Brassica napus*), the Chinese cabbages (*B. campestris*, varieties *chinensis* and *pekinensis*), and several species whose seeds are ground to produce the condiment mustard.

The forms of *Brassica oleracea* illustrate strikingly how much morphological evolution human selection can bring about under domestication (Figure 20-11). Wild cabbage (*Brassica oleracea*) is a native of the eastern Mediterranean region and the shores of the North Sea in Europe. Seeds of *B. oleracea* have been found in Neolithic European lake dwellings, and more than 4000 years ago derivatives of this plant with more succulent leaves were being cultivated in the Middle East as a green vegetable. With the cabbage group we can piece together from historical records an unusually detailed timetable for evolutionary change and diversification of a species under domestication.

Vegetables that are botanically fruits, that is, seed-containing structures of flowering plants, may be called **fruit vegetables.** From the family Solanaceae come tomato (*Lycopersicon esculentum*), eggplant (*Solanum melongena*), and the green (bell) and chili peppers (*Capsicum annuum* and related species). Eggplant is an Old World vegetable, apparently from India or China, whereas the other two are New World plants. They all come in a wide range of forms or cultivars. According to tradition, Columbus found, in the pungent fruits of the chili or cayenne type of *Capsicum* that were being grown in the New World, something equivalent to the much-valued pepper he had been hoping to get in the Orient. As a result, the name pepper was applied to *Capsicum*, a plant totally different from the oriental or black pepper (*Piper nigrum*).

Another notable group of fruit vegetables are the "vine crops" or cucurbits, members of the family Cucurbitaceae. These include cucumber (*Cucumis sativus*) and the diverse squashes (*Cu-*

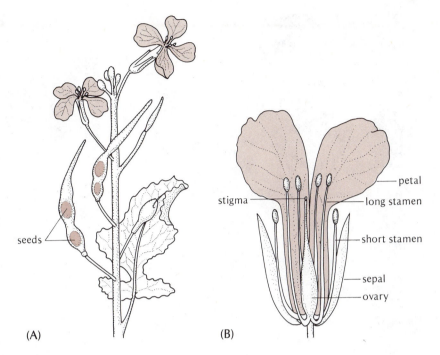

Figure 32-17. Typical crucifer flower structure (*Raphanus*, radish). (A) Inflorescence showing four-petaled flowers, and seed capsules. (B) Flower with one sepal and two petals removed, showing its single pistil and its six stamens, four of which are long and two short.

(A)

(B)

curbita). The latter were important domesticated plants of the native peoples of the New World and reached Europe only after the Spanish conquest. They thrive under very warm conditions and are thus eminently suited to summer cultivation in eastern North America. They afford a valuable range of vegetable styles, from the delicate "summer squashes," which should be used soon after harvest, to the tough-skinned, long-lived "winter squashes" that may be kept for months.

Cucumber Family (*Cucurbitaceae*)

Order Cucurbitales. This family of more than 400 species is distinctive in having usually monoecious flowers, with separate "male" (staminate) and "female" (pistillate) flowers borne on the same plant (Figure 16-22). The flowers have a showy corolla of fused petals, the ovary is inferior, and in the staminate flowers the stamens are fused together. Thus the flowers show many evolutionarily advanced features (Chapter 31). Besides fruit vegetables, this family provides some important sweet fruits, the melons, which appear to be of Old World origin (mainly Africa). Most members of the family are sprawling or climbing herbaceous vines, hence the name "vine crops" applied to them by horticulturists.

Fruits and Nuts

In everyday language the word fruit refers to sweet edible fruits. In the temperate regions of the world the single most important sweet fruit is the **apple** (*Pyrus malus*), and indeed even in tropical and subtropical zones it is grown at higher ele-

vations. A native of the Caucasus Mountains, the apple tree has been cultivated for thousands of years, and more than 2000 cultivars are now recognized. It is a member of the rose family (Rosaceae). This includes a wide variety of temperate zone fruits, including the pear, blackberry, raspberry, and strawberry, as well as the stone fruits plum, cherry, peach, and apricot, as well as wild fruits valued for direct consumption or for the making of jams and jellies.

Rose Family (*Rosaceae*)

Order Rosales. The rose family includes some 3000 species of herbs, shrubs, and trees. The five sepals, five petals, and usually numerous stamens of the flowers are attached to the rim of a floral tube (Figure 31-11). The number of carpels varies from one to many, and these may be free, comprising one or more simple pistils, or fused to form a compound pistil. The ovary may be superior as in the plum and peach or, if the floral tube is fused to it as in the apple, inferior. In addition to fruits, the rose family includes the almond and many species of highly valued ornamentals.

Tropical Fruits

For those who live in temperate or colder regions of the earth, it is difficult to imagine the bewildering array of fruits that exist in tropical and subtropical climates. This diversity is a reflection of the richness of the angiosperm floras of such areas. Probably the most important of these warm climate fruits are from the genus *Citrus* of the family Rutaceae, which includes oranges, lemons, and grapefruits. All are Asiatic in origin and produce a distinctive fleshy fruit in which the outer part of the fruit wall is leathery. The juicy pulp is the result of outgrowth, from the inner surface of the

Figure 32-18.   Harvesting bananas.

Figure 32-19.   Date palms, California.

carpel walls, of multicellular, hair-like structures that become enlarged and filled with fluid.

One of the most valued tropical fruits is the **banana,** *Musa* (Figure 32-18), of the banana family (Musaceae). The more than 300 varieties fall into two main groups, the sweet or dessert bananas and the large, starchy **plantains** that require cooking before being eaten. Both are high in nutritional value, with approximately 20% carbohydrate in the form of starch or sugar. The fruit ordinarily develops parthenocarpically and hence contains no seeds. Originally from southeast Asia, cultivated bananas are mostly sterile triploids and were domesticated thousands of years ago. They first became known to Europeans when seen by the army of Alexander the Great in 327 B.C. during its incursion into India.

In the drier parts of the Old World tropics and subtropics, the **date** (*Phoenix dactylifera;* Figure 32-19) is a major food source. It is a member of the palm family, Arecaceae (Chapter 31). With approximately 70% carbohydrate but only 2% protein the date, like the potato, is energy-rich but nutritionally unbalanced. The date palm can tolerate heat and salinity better than almost any other crop plant. A native of southwestern Asia, it has been cultivated for 5000 to 8000 years. Date palms are dioecious, that is, with staminate and pistillate flowers on separate plants. The ancient Babylonians, Hebrews, and Egyptians are known to have practiced artificial pollination to assure a good set of fruit.

Nuts

Botanically speaking, nuts are fruits with a stony wall (Chapter 17), and some of the foods popularly called nuts, such as the chestnut and the walnut, fit this definition. The peanut, however, is the seed of a dry but not stony-shelled legume, and the Brazil nut is a seed with a stony seed coat. Coconuts and almonds consist of a seed enclosed by the stony inner wall of a fruit (drupe) whose outer fleshy or fibrous layers are re-

moved in processing. The almond is equivalent to the "pit" of a peach or apricot. Other nut crops come from diverse botanical families and are mostly tree species.

By far the most important nut crop is the **coconut** (*Cocos nucifera;* Figure 3-4C), the most valuable member of the palm family. Widely distributed along tropical shorelines, it is indispensable to millions of tropical coast and island people. The outer fibrous part of the coconut fruit is the source of a fiber (coir), used for stuffing, caulking, and producing mats and rope. The endosperm of the single large seed is extremely nutritious, rich in all major dietary requirements including vitamins. Liquid at first, it progressively becomes cellular, forming the coconut "meat." This can be dried to yield **copra,** from which the valuable vegetable oil, coconut oil, is pressed for use in foods as well as in soaps and cosmetics.

CONDIMENTS: SPICES AND HERBS

People from very early times have used a variety of accessory materials that themselves have little or no food value to enhance the taste of foods. Many of these are derived from plants and are referred to as **spices** (Figure 32-20). They are an important component of the diet in most of the world, a significant item of trade and commerce, and a source of badly needed foreign exchange for several developing nations. Spices are valued for their pungent, aromatic flavors, which in most cases are caused by the presence of essential oils (terpenoids, Chapter 6), the same substances that give the aromas of perfumes.

Some 50 different spices are considered important, and this is only a fraction of the species that are used at least locally. Spices and herbs are derived from a wide range of botanical families. The term savory or culinary herb, or often just **herb,** is applied to flavoring materials such as thyme, sweet basil, and mint. The so-called herbs are in reality no different from spices, and the conventional distinction is merely that spices are

Figure 32-20.   Spice plants. (A) Black pepper; (B) cloves; (C) nutmeg; (D) thyme.

tropical in origin whereas herbs are grown in temperate regions. Almost any part of a plant—root, stem, leaf, bark, flower, fruit, or seeds—may be used as a source of spice (Table 32-3). One of the most extravagant of the spices is the brilliantly yellow and delicately flavored **saffron,** used in baking, which consists of the stigmas of flowers of a species of *Crocus.* Some 70,000 flowers are required to produce one pound of spice. In Table 32-3 the diversity of botanical origin of spices and herbs is illustrated by presenting pertinent information about 12 of the more important.

The use of spices has been documented in ancient Egypt as early as 2600 B.C., when they were used for embalming the dead and as offerings in religious ceremonies, and there is evidence of their importance in the early civilizations of Babylon, China, and India. Oriental spices such as black pepper, cinnamon, and gin-

ger were highly valued in classical Greece and later in Rome, in both cases obtained from Arab traders who preserved a virtual monopoly by guarding the secret of their eastern sources. Driven by exorbitant prices, the Romans established direct sea trade with India from ports in Egypt in the first century A.D. After the decline of the Roman Empire in the 5th century, the Arabs again regained control, trading primarily with the Venetian and Genoese city states, but always obscuring their sources with fantastic tales. The return of the Crusaders to Europe stimulated medieval interest in the exotic luxuries of the east. The 26-year journey of Marco Polo acquainted Europeans with the true source of Oriental spices and stimulated the drive to find a direct sea route. The challenge was taken up first by the Portuguese and then by the Spaniards with well known results, including the discovery of the New World. The search for spices can certainly be credited with an important share of the motivation for two centuries of exploration which ushered in the modern era.

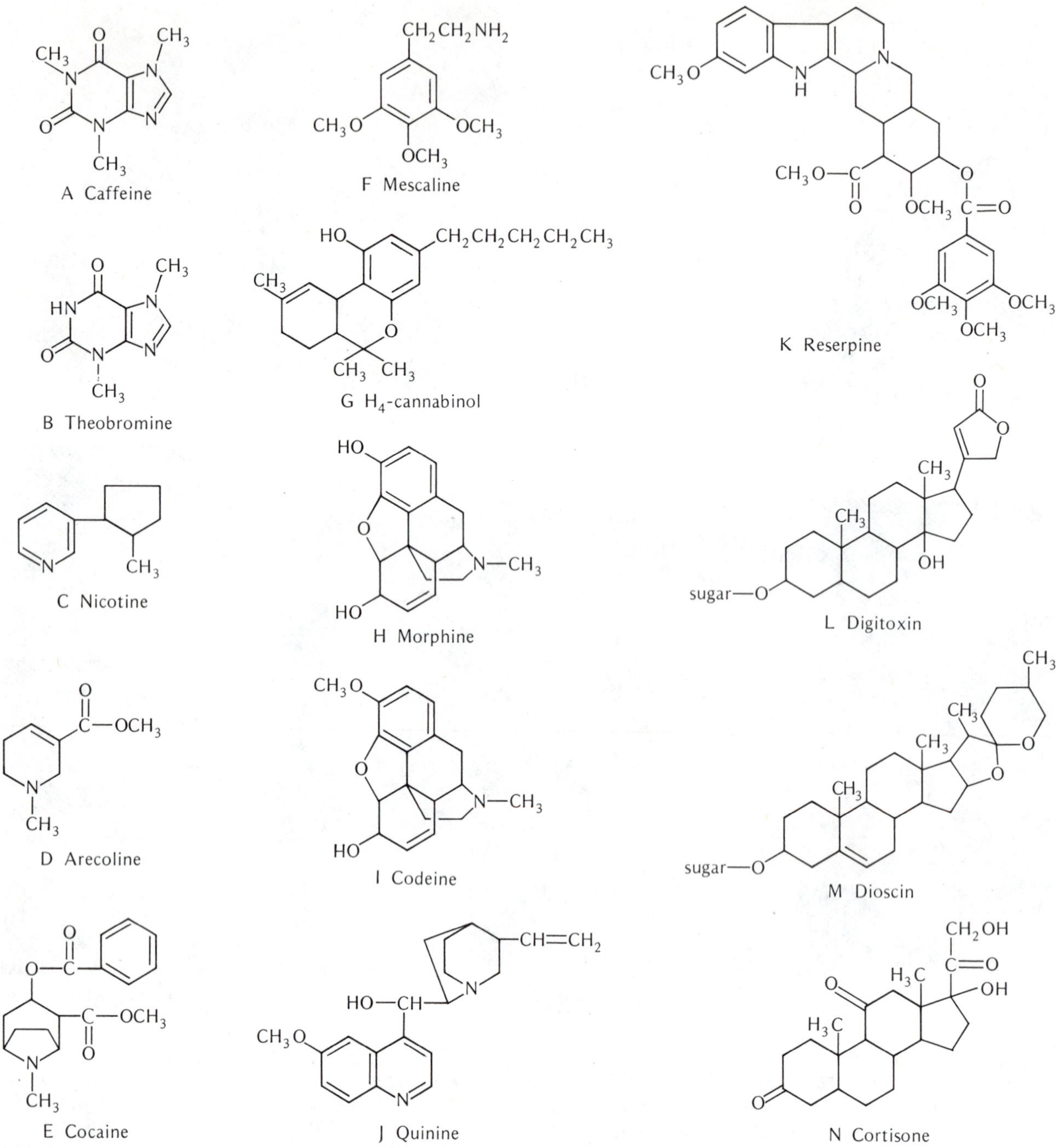

Figure 32-21. Structural formulas of various secondary plant products active as drugs.

STIMULANTS, NARCOTICS, AND MEDICINES

From very early times people have made use of plants and plant derivatives that have a physiological effect upon the body other than in nutrition. Many of these act upon the nervous system, producing varying degrees of stimulation, depression, anesthesia, or hallucination, depending upon the substance contained and its concentration. They can in some cases become habit-forming or addictive, that is, the body develops a dependence upon them so that their use can be discontinued only to the accompaniment of unpleasant or even dangerous withdrawal symptoms. The most common active constituents of such physiologically active plants are alkaloids, nitrogen-containing ring-structured molecules that are usually alkaline and have a bitter taste (Figure 32-21). Other active principles, such as glycosides, have a variety of structures.

TABLE 32-3   Some Common Spices and Herbs

Spice	Species	Family	Type of Plant	Part Used	Major Source
Pepper	*Piper nigrum*	Piperaceae	Woody vine	Dried unripe fruit	Far East
Clove	*Eugenia caryophyllata*	Myrtaceae	Tree	Dried unopened flower bud	Malagasy Republic, Indonesia
Ginger	*Zingiber officinale*	Zingiberaceae	Perennial herb	Rhizome	World tropics
Saffron	*Crocus sativus*	Iridaceae	Cormous herb	Stigma	Spain, Turkey, India
Vanilla	*Vanilla planifolia*	Orchidaceae	Nonwoody vine	Fermented pod (fruit)	Mexico, Africa
Cinnamon	*Cinnamomum zeylanicum*	Lauraceae	Tree	Inner bark	Ceylon, Sechyelles
Turmeric	*Curcuma longa*	Zingiberaceae	Perennial herb	Rhizome	India, China
Nutmeg, mace	*Myristica fragrans*	Myristicaceae	Tree	Seed (endosperm, nutmeg; aril, mace)	Far East, West Indies
Allspice	*Pimenta officinalis*	Myrtaceae	Tree	Dried fruit	Jamaica, Mexico
Mustard	*Brassica* sp.	Brassicaceae	Annual herbs	Seed	Europe, N. America, China
Dill	*Anethum graveolens*	Apiaceae	Annual herb	Stem, leaf, "seed"	Europe, Asia
Thyme	*Thymus* sp.	Lamiaceae	Small shrub	Leaves	Europe

Caffeine Beverages

Among the most widely used of the physiologically active substances is the alkaloid caffeine (Figure 32-21A), a mild and nonaddictive (except in a cultural sense) stimulant. Caffeine is derived from the leaves, buds, seeds, and even bark of a variety of plants from many parts of the world. All of them, used primarily in beverage form, relieve fatigue and promote a sense of well-being, harmlessly unless taken in excess. Coffee, tea, and cacao (chocolate) are the most widely used, but there are many other important caffeine beverages such as cola, from the seeds of a West African tree (*Cola nitida*), and maté, from leaves of a South American holly (*Ilex paraguariensis*). The desirable flavors of these beverages are not due to caffeine itself but to essential oils, tannins, and other secondary products.

Tea, used by about half of the world's people, consists of dried leaves or buds of a small tree or shrub [*Camellia (Thea) sinensis*, Figure 32-22A] native to southern China and possibly northern India. Its shoots are carefully picked by hand, the quality decreasing with the age of the leaf, and either dried immediately to produce green tea or wilted, crushed, and allowed to stand moist for several hours before drying to produce black tea. Dry tea contains 2–5% caffeine, more than coffee, but is ordinarily used in a weaker infusion. First used as a medicine for a variety of ailments at least as early as 200 B.C., it was introduced to Europe in the late 1600s by Dutch and Portuguese traders. Growing European demand led to widespread cultivation where the climate is warm with abundant rainfall, and today over 1 million tons are produced annually, half of it in China.

Coffee is similarly a product of warm climates and is only slightly less popular than tea. The fa-

Figure 32-22. (A) Tea being harvested in Japan. (B) Coffee bush growing near Bogota, Colombia.

Figure. 32-23.   Pods on cocoa tree, Trinidad Island.

Figure 32-24.   Tobacco plants, Kentucky.

miliar coffee "bean" is the seed of a small tree (*Coffea arabica*; Figure 32-22B) that is native to the highlands of Ethiopia. Carried to Arabia in the sixth century, it became a popular beverage frequently associated with the Arab world. Today nearly half of the world production is in Brazil, but the crop is grown in the tropics around the world, mainly in the cooler highlands. The tree produces bright red, berry-like fruits, each containing usually two seeds. These must be removed from the pulp and then roasted at a high temperature in order to develop the characteristic flavor. Since the flavor of the unroasted beans is not attractive, one must wonder how this method of processing was initially discovered.

A small tree native to the tropical lowlands of America is the source of chocolate, used both as a beverage and as a confection. **Cacao** (*Theobroma cacao*) is today cultivated mainly in West Africa, where it was introduced after the Spanish conquerors found it in use in Mexico early in the sixteenth century. The fruit, borne directly on the main trunk and branches (Figure 32-23) is a capsule containing 40–60 seeds embedded in a pulp. The seeds are removed from the pulp after a fermentation process is allowed to break it down. They are then roasted before use, shelled, and ground to produce chocolate. The seeds contain theobromine (Figure 32-21B), an alkaloid closely related to caffeine, and some caffeine as well. They also contain a high percentage of fat, known as cocoa butter, which, when pressed out, leaves the product that we use as cocoa.

Some More Potent Products

Of the numerous plant products other than caffeine that are used because of their agreeable effect upon the body, some are mild and seemingly harmless, others are extremely dangerous, and the status of many is in dispute. For good or ill these products have played an important role in human affairs and continue to do so.

Tobacco (*Nicotiana tabacum*, Figure 32-24) is one of the most widely used plants in the world, estimates indicating that as many as half the world's people smoke, chew, or snuff (in powdered form) its leaves for the effects of the nicotine that they contain. *Nicotiana tabacum* is another member of the family Solanaceae, which includes a number of alkaloid-bearing species. Tobacco is grown in many countries, and its production and processing constitute a major industry which appears to be thriving in spite of energetic campaigns against smoking for health reasons. The alkaloid nicotine (Figure 32-21C) is rated as a mild and soothing stimulant, but it appears to be habit-forming for some people. Many of the harmful consequences of smoking, however, result not from the nicotine but from other volatile products of burning, which are proven causes of cancer and suspected contributors to heart disease and other illnesses.

Tobacco is native to the American tropics and was widely cultivated when encountered by Columbus' men in the West Indies. All the major methods of tobacco use had apparently been invented by that time. The Spaniards and Portuguese introduced tobacco to Europe, from where it was spread around the world. The tobacco of commerce is prepared from the leaves of the tobacco plant by "curing" or slow air-drying with or without artificial heat, during which the leaves turn brown and lose much of their protein and starch. Various aromatic materials are often added to pipe and chewing tobaccos.

Throughout the Far East the use of tobacco is rivaled by the chewing of **betel,** the most extensively used masticatory. The main constituent of a "betel chew" is derived from the seeds of the betel palm (*Areca catechu*), a widely cultivated native of Malaya. Slices of the seed are placed on a fresh leaf of betel pepper (*Piper betel*) that has been smeared with lime paste, often together with resin from the palm and assorted spices for flavor, and the leaf is folded to form a wad that is held in the mouth and gently chewed or sucked. The alkaloid arecoline (Figure 32-21D), from the betel nut, produces a mild stimulation and a general sense of well-being that is considered to be without ill effects.

Figure 32-25.   Peyote cactus.

Figure 32-26.   Female (left) and male (right) marijuana plants.

For many centuries native people in the Andes Mountains of South America have chewed the leaves of a small native tree or shrub called **coca** (*Erythroxylon coca*) to alleviate pangs of fatigue and hunger and to experience a general stimulation. Its effect is produced by the alkaloid cocaine (Figure 32-21E). The extracted alkaloid was formerly used as a local anesthetic, and it is generally regarded as a dangerous narcotic. Whether, however, its harmful effects, when taken in the form of leaves, justify the attempts that have been made to stamp out its use is a matter of debate. Besides wild sources, the plant is now cultivated in South America and the East Indies. Although it is still used extensively in the traditional manner in its native region, the medicinal uses of cocaine have declined as synthetic substitutes have been developed. A certain amount of coca is also used, after extraction of the alkaloid, as a flavoring in cola beverages (whence the name Coca-Cola).

Peyote, a low spineless cactus (*Lophophora williamsii*; Figure 32-25) native to arid regions of Mexico and the southwestern United States, is used ceremonially by native people for its hallucinogenic effect. The plant is small, somewhat carrot-shaped, and largely underground. The traditional preparation method is to cut off and dry the exposed top of the plant to form a peyote or mescal "button," which is held in the mouth until soft and then rolled and swallowed. Peyote contains some nine different alkaloids, the most important of which is mescaline (Figure 32-21F). The use of peyote

as a focal point of religious ritual has been increasing, but excessive harvesting is now threatening the plant with extinction.

One of the most controversial drug plants today is the hemp plant, *Cannabis sativa* (Figure 32-26), a dioecious annual, the source of a valuable fiber and, from the seeds, an edible and industrial oil. But it also yields **marijuana** and **hashish.** Of Asiatic origin, hemp was cultivated and used in China possibly as far back as 3000 B.C. and today has an estimated 200 million users around the world. The active hallucinogenic ingredient (tetrahydrocannabinol or THC, Figure 32-21G) is unusual in that it is not an alkaloid. It is contained in a resin synthesized in special gland cells on the leaves and especially on bracts of the female inflorescence. The resin itself constitutes hashish and is relatively potent. Marijuana consists of dried fragments of the plant and contains much less of the active ingredient, the content depending upon the preparation. The most potent consists of the tops of female plants. Lower grades include the entire female plant or even whole male and female plants.

Many popular and sometimes exaggerated ideas about the evils of plant narcotics in general are derived from the very real dangers associated with **opium,** the latex of the opium poppy (*Papaver somniferum,* Figure 32-27). Its use in Asia Minor goes back to at least 2500 B.C. It was widely distributed in ancient times and was well known to Egyptians, Greeks, and Romans. It is one of the most frequently used narcotics today, having an estimated 900 million users, largely in Asia, although this number may have been decreased by the concerted efforts of modern China to prevent opium addiction. Opium is obtained by slashing the capsules of the poppy just after the petals fall and scraping off the exuded latex as it hardens. The latex contains about 30 different alkaloids, of which morphine (Figure 32-21H) and codeine (Figure 32-21I) are the most important. Both are extremely addictive and hence dangerous. However, morphine and codeine are among our most im-

Figure 32-27. (A) Opium capsules and poppy. (B) Latex exuding from slashed capsules.

portant medicinal products, and their legitimate use as pain-killers continues to reduce suffering around the world.

Alcoholic Beverages

Alcohol is a product of yeast fermentation of sugar derived from plant sources. The same alcohol (ethanol) is found in all alcoholic beverages; the wide variety of these beverages results from the plant materials used as substrates and the type of processing. These fall into two groups, the simple fermented beverages, such as beer and wines, and the distilled spirits such as brandy and whiskey.

When the carbohydrate substrate is provided by cereals the simple fermented beverage is **beer.** Barley is the preferred grain, at least in Europe and North America. It is first germinated, or "malted," so that the enzyme amylase will convert starch into fermentable sugar (Chapter 6). Ungerminated grain is usually added to the malt, to increase the amount of carbohydrate material. During brewing, beer is ordinarily flavored with an infusion of hops (dried inflorescences of the plant *Humulus lupulus*), which adds the characteristic slightly bitter taste. After the brew has been fermented, its alcoholic content generally reaches 5–8%.

Wine refers, in its more inclusive sense, to all simple fermented beverages made from fruits. The art of wine-making, like that of brewing, dates back nearly 6000 years. Since wild yeasts are normally present on fruit, it is easy to see how the art was discovered. The preeminent type of wine is that made from the European wine grape, *Vitis vinifera* (Figure 32-28). Wines made from other fruits are called fruit wines, and the kind of fruit is specified (e.g., peach wine). The unqualified term

wine normally means the product of the grape. Since grapes contain sugars that yeasts can ferment directly, no preliminary hydrolysis of carbohydrate is required as it is in beer production. The relatively high sugar content of ripe grapes normally yields alcohol contents in natural wines of 9–14%.

Vitis vinifera has given rise to numerous cultivars, which are employed regionally to yield the wines of characteristically different flavors and aromas that are traditionally associated with different wine-growing regions. White wines are made by pressing out the juice of white grapes (i.e., nonpigmented cultivars) and allowing it to ferment. Red wines are made by crushing red-colored grapes ("black" cultivars) and fermenting the mixture of juice, grape skins, and seeds, which is called **must.** This procedure extracts from the skins the red anthocyanin pigment and tannins they contain, which are important for the character and body of red wines. After fermentation the must is pressed to obtain

Figure 32-28. Part of a mountain vineyard in the coast range of California. The vines are *Vitis vinifera* cv. Pinot noir, one of the finest cultivars for winemaking, the traditional cultivar of the Burgundy region of France.

the wine, and a process of clarification and aging is begun, to develop a product with the desired characteristics of flavor and aroma.

In the production of "dry" wines or table wines, which are used with meals, yeasts convert the grape sugar completely into alcohol during fermentation. Sweet wines, used with fruit or desserts, or as an apéritif, are made in several ways. For sauternes and extra-fine Rhine wines, the grapes are brought before harvest into an over-ripe condition by drying out as a result of infection by the "noble rot" mold, *Botrytis cinerea*. Their juice then contains so much sugar that the maximum alcohol content that can be produced by yeast fermentation, about 14%, is attained while substantial sugar still remains in the wine. Other sweet wines of lower alcohol content, especially in Germany, are produced by adding to a dry wine a reserve of unfermented grape juice, then filtering the wine through a sterile (yeast-retaining) filter and bottling it in sterilized bottles. Such wine would ferment further if yeast cells got access to it. Sherry and port are produced by stopping the yeast fermentation, before it uses up all the sugar in the grape juice, by adding distilled spirits (brandy) to raise the alcohol content above 14%. These fortified wines are stable, since yeasts cannot grow in them.

Most of the wine produced today is made in large quantities by a highly technical industry that puts out more than 7 billion gallons annually. Production, by small-scale traditional methods, of wines with extraordinary flavor and bouquet still actively continues, however, in the fine-wine districts of Europe such as Burgundy, the Bordeaux region, and the Rhine and Rhone Valleys, and also to a growing extent in the coastal valleys of California.

Some 2000–3000 years ago in China, Egypt, and Arabia the art of **distillation** was discovered, leading to the production of beverages with a much higher alcohol content, usually 40–50% by volume. If the equivalent of beer, brewed especially for the purpose, is distilled, the product is whiskey or vodka. Whiskey differs from vodka by being aged in oak barrels, much of the characteristic flavor and color of whiskies being due in fact to tannins from the oak wood. The type of whiskey depends in large part upon the grain used for fermentation. Scotch and Irish whiskey are made mainly from barley. Rye, wheat, and maize may also be used, the latter being the main component of bourbon, and ordinarily malt is used to prepare the mash prior to the fermentation. The distillation of wine or of other fruit mashes yields brandy, the character of the distillate depending upon the fruit because other constituents in addition to alcohol are distilled. Rum is produced by distilling fermented molasses (from sugarcane). Gin is made like whiskey except that aromatic materials from juniper "berries" and other flavoring substances are added either before or after distillation. A variety of highly flavored liqueurs or cordials are prepared by adding spices, herbs, or fruit extracts and sugar to brandy or to the equivalent of whiskey.

Plant Medicines

In this day of advanced medical science, plants play a decreasing role in the treatment of human ailments. Throughout history, however, from ear-liest times to the present century, the situation has been very different. For centuries botany and medicine were almost the same discipline, and many great historical figures of botany, including Linnaeus, were professors of medicine.

Many drug plants are known to have been used in China as far back as 4000–5000 B.C. In fifth century B.C. Greece, Hippocrates, often cited as the father of medicine, is reported to have recommended 300 to 400 medicinal plants. The *De Materia Medica* of Dioscorides in the first century A.D. listed about 500 botanicals. This famous volume was the standard medical reference throughout the Middle Ages. However, it gradually became clear that many of the effects traditionally claimed for botanicals were not reliable, especially those that had been inferred from the once widely believed "doctrine of signatures" that plants are "signed" to indicate their usefulness. (Heart-shaped leaves suggested a cure for heart diseases, yellow juice efficacy in treating jaundice, red juice an aid for blood disorders, and so on.) Today only about a hundred plant species are officially recognized for medical use, and the roles of many of these have been taken over by microbial products or synthetic drugs. Interest in plant medicines has nevertheless been rekindled in recent years because identification of their active principles has often pointed the way to new synthetic drugs. Moreover, certain plant drugs have yet to be supplanted by synthetic products; indeed, two of the ten most frequently prescribed drugs (codeine, mentioned above, and digitalis) are derived from plant sources.

Digitalis is obtained from the foxglove, *Digitalis purpurea* (family Scrophulariaceae, Chapter 31; Figure 32-29), a European wild flower considered in the Mid-

Figure 32-29. Foxglove plant in flower.

dle Ages to be a cure for many ailments. In the latter part of the eighteenth century it became a specific remedy for dropsy, a condition in which fluids accumulate in the tissues. The recognition of its specific effects was the result of the studies of an English physician, William Withering (1741–1799), who carefully analyzed the constituents of an herb concoction found to be effective against dropsy. He concluded that, among the many components, foxglove was the important one and that the leaves were the most potent part of the plant. He also worked out methods of preparation and determined safe and effective dosages, thus perfecting a specific remedy for a debilitating disease. Later investigators showed that the primary effect of digitalis is as a heart stimulant, which improves circulation. Digitalis contains a number of glycosides, the most important being digitoxin (Figure 32-21L). Today digitalis is widely used to alleviate cardiac insufficiency and related problems. A few hundred hectares of the plant supply the world's total requirement.

Another important plant drug is **quinine** (Figure 32-21J), specific for malaria. The drug is derived from the bark of several species of *Cinchona,* trees from the same family as coffee (Rubiaceae) and native to the Andes of South America but also cultivated in Java. The use of quinine was discovered by the Spanish conquerors of Peru, who found it in use as a native cure for malaria and introduced it to Europe as Jesuit's Bark. It quickly became a highly valued medicine. Quinine has been synthesized, but the process is too expensive for commercial production. A synthetic substitute, Atabrine, is widely used although it is less effective than the natural product. Quinine is also the main flavoring constituent of tonic water.

An example of how traditional plant medicines can lead to new pharmaceutical developments is the snakeroot plant (*Rauwolfia serpentina*), used in India for more than 3000 years for many ailments and as a sedative. In 1952 the alkaloid **reserpine** (Figure 32-21K) was isolated from its roots and, still derived from natural sources, has become an important drug for the treatment of high blood pressure (hypertension) and a variety of nervous and mental disorders, including schizophrenia. The discovery of reserpine stimulated the development of many synthetic behavior-modifying drugs, ushering in a new era in chemical treatment of nervous and mental illnesses.

Another example was the finding that a terpenoid glycoside called **dioscin** (Figure 32-21M), obtainable from several species of yam (*Dioscorea*), can be converted easily into various animal steroid hormones previously obtainable only in limited amounts, and at high cost, from the adrenal glands of animals. This made economical the large-scale production of valuable steroids such as the anti-inflammatory drug **cortisone** (Figure 32-21N) and the male and female sex hormones used in oral contraceptives. Approximately 60,000 tons of fresh yams are now used each year in the manufacture of medicinal drugs.

PLANT EXTRACTS AND ESSENCES

The metabolism of plants results in the synthesis of a broad spectrum of substances that in particular species may be accumulated in sub-

stantial concentrations. Some of these have food or food-related uses, and others serve as raw materials for industry. People long ago began to extract or otherwise separate these valuable products from their natural sources. A number of them play an important role in manufacturing today, while others were of great significance in the past but have now been replaced by synthetic substitutes.

Vegetable Oils

Oils and fats (the chemical nature of which is discussed in Chapter 5) are an important component of the human diet. They also find use as vehicles for cooking. Some food plants such as corn, soybean, peanut, and coconut are important sources of useful oils. Corn oil is obtained from the maize embryo or "germ," removed when the grain is milled in producing corn starch. Cotton seeds, left over from the ginning of cotton fiber (see below), are also an important oil source, both cottonseed oil and corn oil being used extensively for manufacturing margarine. Among a diversity of other crops grown mainly for their seed oils are the sunflower and the safflower of the composite family (Chapter 31) and the rape seed from the mustard family. Two tree crops grown largely or entirely for oils yielded by their fruits are the African oil palm and the olive (*Olea europea*, Figure 32-30), a tree that has been cultivated in the Mediterranean region for more than 4000 years.

Many oils are also widely used industrially in the manufacture of such products as soap, lubricants, printers' ink, linoleum, oilcloth, and many other items. Drying oils, which absorb oxygen and form a tough, resistant film upon exposure to air, are the basis of oil paints and varnishes. Linseed oil, derived from the seeds of flax (*Linum usitatissimum*), is primarily an industrial oil, as are castor and tung oil. The castor bean and tung oil

Figure 32-30. Olive tree bearing fruit, northern Syria.

Figure 32-31. Chipping southern pine for oleoresin, Georgia.

tree are both tropical members of the spurge family (Euphorbiaceae).

Essential Oils

These aromatic compounds have already been mentioned as the source of the pungency of spices and herbs. They are not chemically related to vegetable oils, and the name essential is in reference to an "essence" owing to their volatility and resultant aroma. They are the foundation of perfumes, which appeal to the sense of smell. Widely distributed in plants, they are extracted from flowers (rose, violet, jasmine, lavender), leaves and stems (mint, geranium, citronella), fruits (bitter orange), and wood (sandalwood and camphor). Some of the flower oils used in fine perfumes are extremely costly, but organic chemists have acquired great skill in converting cheaper oils, such as the antiseptic-smelling oil of citronella (from an Asiatic grass, *Cymbopogon nardus*), into fragrant essences resembling flower perfumes.

One volatile oil with considerable industrial significance is **oil of turpentine,** which is important in the chemical industry, especially in the manufacture of paints and paint thinners. It is produced as part of the resin formed by the resin ducts of coniferous trees and is harvested from several species of pine in North America and in Europe. Standing trees may be incised or tapped (Figure 32-31) and the exudate, a mixture of essential oils and resins, collected. This oleoresin is distilled, separating the oil of turpentine from the resin, called rosin, which has a variety of uses. Turpentine may also be distilled from wood, and considerable

quantities are recovered as by-products of the wood pulp industry (see below) when pine species are used.

Latex Products

One of the most intriguing products of plants is a material known as latex, a juice that is often milky and is produced in specialized cells or multicellular tubes (called latex ducts) in about 20 families of dicotyledons. Latex contains a wide range of organic compounds in solution, suspension, or colloidal dispersion. The latex of some species (about 1800 of them) contains suspended particles of a polymerized hydrocarbon consisting of many units of isoprene (C_5H_8). This polyisoprene can be coagulated by heat or acid to yield **rubber,** one of the mainstays of modern technology. At least 50,000 different articles are made from rubber, but approximately three quarters of world production is used in automotive tires. Several types of synthetic rubber have been developed from coal and petroleum for use alone or, more commonly, mixed with natural rubber, and at present about half of the rubber used is of synthetic origin. However, the demand for natural rubber has not declined.

About a half dozen different plant species have been used as commercial sources of rubber. One species, however, has predominated over all others, para rubber, *Hevea brasiliensis* (Figure 32-32) of the family Euphorbiaceae. Para rubber, named for the port of Para, now Belem, in Brazil, is derived from a large tree 20–50 m (60–150 ft) tall native to the rain forests of the Amazon and Orinoco basins. Originally collected from scattered wild trees over a vast area, today 98% of the world's rubber is produced on plantations in Malaysia and Indonesia, and to a limited extent in West Africa and South America.

The latex is obtained by incising the bark down to the cambium, which intercepts latex ducts in the phloem (Figure 13-22). The cuts are opened daily during the tapping periods; the latex flows into collecting pots and is subsequently coagulated.

CELLULOSE PRODUCTS

From the point of view of human utility the polysaccharide cellulose, which occurs in cell walls, is one of the most important plant products. Fortunately, cellulose is the most abundant organic material in nature (Chapter 5). Secondary cell walls, which are generally thicker and contain a higher percentage of cellulose than primary walls, provide the principal useful cellulose products. In some instances the wall may consist largely of cellulose, but usually there are other substances present. Chief among these is lignin, which for many uses (such as the manufacture of paper) is undesirable and must be removed.

Figure 32-32. Rubber tree being tapped for latex, Malaysia.

Wood

Among human resources, wood is probably second only to food in global significance. Wood is the secondary xylem of conifer and dicot (hardwood) trees (Chapter 13), and is used both directly as lumber, and indirectly as wood pulp and chips for many products such as paper, cardboard, fiberboard, and cellulose plastics. Although many substitutes and replacements have been developed, world wood consumption, paced by a growing world population, still exceeds the total consumption of foods (Table 32-1). At current rates of world consumption forest stands are rapidly shrinking, pointing to a future wood crisis comparable to the current petroleum crisis.

Fuel

In the technologically advanced regions of the world where so much emphasis is placed upon fossil fuels it is difficult to appreciate that almost half of the wood consumed is still used as fuel (Table 32-1). In underdeveloped regions it is the major source of energy for cooking, heating, and even for the generation of electricity. Use of wood for fuel is increasing even in developed nations. Fuel wood has few requirements, and much of it consists of sticks unsuitable for other purposes and often of cull trees or remnants from lumbering operations. However, fuel gathering in many

areas of the tropics contributes to the progressive destruction of vegetation and expansion of desert-like conditions called desertification.

Lumber

Wood is still the preferred structural material for private dwellings and even many public buildings. Even when the basic construction is of some other material, wood is almost always employed in interior and exterior finishing because of its beauty and the ease with which it can be worked. Wood also continues to be preferred for furniture.

One of the most rapidly developing aspects of wood usage is the production of laminated wood or plywood because of its efficiency and strength. The sheets shaved from a log by rotating it against a knife are cut into slabs and glued together with resistant adhesives. There is also increasing use of wastes from lumber mills, such as sawdust, chips, or flakes, which are pressed with a binding agent into panels known as particle board. Such processes lead to greater efficiency in the utilization of dwindling forest resources.

PAPER

It is difficult to imagine the functioning of our society without paper, the substrate of almost all written communication, besides many other uses ranging from wrapping to construction.

Before 2500 B.C. the Egyptians had developed a writing surface from **papyrus** (from which the word paper is derived). It was prepared by cutting strips of the pith of a large reed, *Cyperus papyrus* (Figure 32-33), of the monocot family Cyperaceae, and pounding them together into sheets.

True paper was invented in China around 100 A.D. It consists of matted or felted plant fibers that have been compressed into sheets. It is made by reducing fibrous material to a pulp, suspending the pulp in water, then sieving off the water and pressing and drying the resulting thin sheet of fibrous material. Pulp was first made by pounding bark. As the technique spread, ultimately being carried to Europe by the Arabs, a variety of fiber sources were introduced including linen rags, themselves of plant origin. In fact, linen rag paper is still considered to be of the highest quality. At the present time, however, by far the most important source of fiber for paper is the wood of various conifers and hardwood trees. Wood pulp from conifers is preferred because of the length of the tracheids that constitute the fibers, but many dicots, including aspen (*Populus tremuloides*) in North America and increasing numbers of tropical hardwoods, find widespread usage.

The use of wood for paper making was first suggested early in the 1700s by a Frenchman, René de Reaumer, as a result of his observing that wasps build their "paper" nests by rasping fibers from fence posts or dead tree limbs and gluing them together. It was not, however, until 1840 that the first all-wood paper was made in Germany from pulp produced by a mechanical grinding process. Today the wood is usually pulped by chemical treatment after first being reduced

Figure 32-33.   Papyrus (*Cyperus papyrus*) growing in (A) Africa and (B) California.

to chips. The pulping process yields raw material not only for paper and cardboard but also for various cellulose derivatives including nitrocellulose and cellulose acetate plastics.

Plant Fibers

People first began to use fibers of plant origin for textiles, rope, and twine long before recorded history and almost certainly before the beginning of cultivation. Fiber plants were early domesticated, our major fiber sources such as cotton and flax being among the most ancient cultivated plants. In the twentieth century the development of syn-

Figure 32-34.   Cotton bolls ready for harvest in Tennessee.

thetic fibers such as nylon and polyester has brought about a marked change in plant fiber utilization, but it has by no means reduced them to insignificance. The synthetic fibers are very often blended with 30–40% cotton to make clothes that are more comfortable in warm weather, and some two thirds of all rope and twine is still made from plant fibers, which are less expensive than the synthetic ones. At least 200 different plant species have been used as fiber souces at some time, and some 25–30 may be considered to be of commercial importance today.

Fibers may be classified according to the part of the plant from which the fiber comes. **Surface** fibers are usually single-celled, hair-like outgrowths from the surfaces of plant parts, especially seeds and the inner chambers of fruits. Cotton is the most important of the surface fibers. **Soft** or **bast** fibers are multicellular strands composed of elongated, thick-walled sclerenchyma cells from the phloem and cortical regions of the stems of dicots. Flax, jute, hemp, and ramie are important examples of this type. By contrast **hard** or **structural** fibers are vascular bundles from the leaves or leaf sheaths of monocot species. Manila hemp, or abacá, and sisal are hard fibers. Finally, a miscellaneous category encompasses leaves or leaf strands, roots, split stems, and even inflorescences, which are used as fibers for special purposes such as basket weaving.

Cotton is our chief fiber plant, and literally clothes most of the world's people. Cotton fibers are surface hairs on the seeds of several species of the genus *Gossypium*, small shrubs (Figure 32-34) of the family Malvaceae (mallow family). The seeds with their hairs develop inside a seed pod or capsule, the cotton boll. After growing to a length of several centimeters, the single-celled hairs lay down a thick secondary wall that is unusual in consisting of over 90% cellulose and containing no lignin. The fiber is therefore strong yet pliable, and can be used directly for spinning and weaving after removal from the seeds (ginning). Different species of cotton were domesticated in the Old

Figure 32-35. Flax.

Figure 32-36. Sisal plantation, eastern Africa.

and New Worlds in ancient times, originally in Peru and Mexico and in India. When Europeans arrived in the western hemisphere they found the species cultivated there (*G. hirsutum* and *G. barbadense*) to be superior in fiber length to Old World cottons, and therefore spread the New World species to other continents.

The stems of **flax** (*Linum usitatissimum*; Figure 32-35) were once the world's leading soft fiber source for textiles (linen), but flax has long since been displaced from this position by cotton. Much of the remaining linen industry is centered in northern Europe, but large acreages of flax are grown elsewhere for the production of linseed oil.

Of greater economic importance today is **jute**, also a soft fiber, obtained from a 2–3 m tall plant (*Corchorus capsularis*) native to southeast Asia. Because of their lignin content, the fibers become discolored and also are not very strong, but they are extensively used for burlap and gunny sacks, and are a major export of India and Bangladesh.

The important hard fibers today are **Manila hemp** or **abacá,** and **sisal.** Manila hemp is obtained from a species of inedible banana (*Musa textilis*) native to the Philippines, from which almost all of the world supply still comes. The fibers are used chiefly in the manufacture of strong and elastic rope for ships and boats because it is resistant to salt water. Sisal is derived from the leaves of another monocot, *Agave sisalana* (Figure 32-36), one of the "century plants," in Central America.

It is employed mostly in heavy twine such as that used in hay and straw balers, and its cultivation is spreading to other parts of the world, especially Africa.

Cork

Commercial cork is the outer bark of the cork oak, *Quercus suber*, an evergreen tree native to the western Mediterranean region in both southern Europe and North Africa. The cork cambium of this species produces a thick layer of cork, an adaptation to the low rainfall and drying winds characteristic of its habitat. The cork can be harvested by stripping it away in sheets (Figure 32-37). The stripping destroys the cork cambium, but a new meristem forms within the inner bark and produces a new layer of cork that can in turn be removed 8–10 years later. After boiling to soften it and remove impurities, the rough outer surface is scraped away and the cork is ready for use.

Cork is valued as a closure for bottles because it is elastic, impervious to liquids, and imparts no flavor to the contents. It is also used to make buoyant floats, insulators for both heat and electricity, and attractively elastic and durable floor and wall coverings. Although the walls of cork cells, like other cell walls, contain cellulose, the unique properties of cork are due to impregnation of its walls with **suberin,** a waterproofing biopolymer which confers on cork tissue its usefulness in protecting tree trunks and roots against dessication, fire, and mechanical damage. A satisfactory artificial substitute for cork has yet to be perfected. Today the demand for cork chronically exceeds the world supply, and its price is steadily and steeply on the rise.

CONCLUSIONS

This chapter has attempted to show how dependent the human species is upon plants for fundamental necessities like food, for many products that add to the enjoyment of life, and for raw materials of modern industry. Since long before the dawn of recorded history this dependence has

Figure 32-37. Cork oak trees, in the foreground and on the hill beyond, in a cork plantation in southern France. The cork oak is an evergreen species ("live oak") that forms an unusually thick outer bark, helping the tree survive the fires prevalent in summer-dry Mediterranean vegetation. The pale lower parts of the trunks of the two trees on the left have recently been harvested of cork by removing their outer bark, a pile of which can be seen beside the road. After stripping, the trees produce new layers of outer bark which can be harvested again after some years.

played a major role in shaping human cultural development, and there is no evidence that it will diminish in importance in the future. As older uses are abandoned, new ones appear. Because they are a renewable resource, plants offer real hope of replacing rapidly declining stocks of non-renewable resources. These plant resources, however, must be managed with great care, and the science of botany in all its aspects provides the knowledge required for this management. It is also important to recognize that many of our most essential plants have been so drastically changed by human intervention that they are now as dependent upon us as we are upon them. This interdependence of plants and people is all too easily overlooked in a technological and largely urban society.

SUMMARY

(1) Food is basically derived from plants even when secondarily obtained from animal products. An adequate diet can be obtained from plants alone, but animal protein is an ideal source of essential amino acids.

(2) The cereal grains, members of the grass family (Poaceae), are the most important food plants. They provide more than half of all the calories consumed by man. Wheat, rice, and maize are the major cereal crops.

(3) Legumes (family Fabaceae) are major sources of protein in the human diet and are among the world's most important food plants.

(4) Starchy roots and underground stems (tubers) such as potato provide carbohydrates for the diet of millions of people in both temperate and tropical regions.

(5) Most of the world's sugar is obtained from two plants, sugarcane, a grass, and sugar beet, which is a close relative of the common garden beet.

(6) Vegetables are an important source of vitamins and minerals in the diet. Common vegetables represent a great diversity of plant parts: stem, root, leaf, bud, inflorescence, fruit, and seed.

(7) Many edible fruits such as tomato, pepper, squash, and cucumber are considered vegetables. Among the sweet fruits the apple is the most important in temperate regions.

(8) Spices and herbs, which owe their pungent, aromatic quality mainly to essential oils, have long been used to flavor foods.

(9) Many plants contain widely used physiologically active substances, mainly alkaloids and glycosides. These include the caffeine beverages such as tea and coffee and more potent products such as tobacco and marijuana. Alcohol is derived from a variety of plant sources by yeast fermentation of carbohydrates. Although less important than in earlier times, plant medicines are still used and provide valuable clues in the search for new curative agents.

(10) The metabolism of plants produces many products that are useful in industry. Among these are vegetable oils, essential oils, and rubber.

(11) Cellulose from cell walls is one of the most useful plant products and is used in a variety of forms. Wood, the secondary xylem of conifer and dicot trees, provides fuel, lumber, and the raw material for the manufacture of paper and plastics. Plant fibers such as cotton are still important in spite of the development of synthetic substitutes. Cork, from the bark of the cork oak, is in great demand for many uses.

Chapter 33

DYNAMICS OF ECOSYSTEMS

An ecosystem comprises a set of organisms, or **community,** plus the physical environment (**habitat**) in which the organisms live. The members of a community share the resources of their habitat. Although individuals compete for these resources, in the course of their life and death they pass some of their resources from one to another, resources being reused again and again, or "cycled." Cycling enables ecosystems to be more productive than they could be if their resources were thrown away after use.

The organisms in a community are themselves in a state of flux, new individuals continually replacing older ones. Often, the composition of the community gradually changes by replacement of some or all of its species by others, a process called ecological **succession** (Figure 33-A). This chapter considers these dynamic processes in ecosystems, building on the principles of ecological adaptation covered in Chapter 19 and making use of information on plant diversity contained in Chapter 22 and succeeding chapters.

FLOW OF CARBON AND ENERGY THROUGH ECOSYSTEMS

Food Chains and the Carbon Cycle

Virtually all the organic matter of most ecosystems derives from photosynthetic products produced by plants. The carbon fixed in photosynthesis flows through a community, from organism to organism, in a set of **food chains,** in which each organism in the sequence is food for the next member of the chain (Figure 33-1). **Producer** organisms, plants ranging in size from microscopic algae to giant forest trees, are the base of the food chains. **Consumer** organisms such as animals and heterotrophic plants utilize organic matter made by the producer organisms. Herbivorous animals, such as cattle, rabbits, and squirrels, are important **primary consumers** of plants and plant products. There are thousands of plant-eating insects and many seed- and fruit-eating birds. More than half of all the kinds of animals, vertebrate and invertebrate, obtain their food directly from plants. Many of these **herbivores,** in turn, are fed on by **carnivores** (predators), consumers of animal food only, and by **omnivores** (e.g., humans), which feed on a mixed diet. Carnivores may be secondary or tertiary consumers. Thus the herbivorous aphids are eaten by beetles, which are consumed by small birds, which are in turn prey for larger predators such as hawks. The producer organisms and the primary, secondary, tertiary, and subsequent consumers comprise the **trophic levels** (Greek trophē, nourishment) of an ecosystem.

The organisms at each trophic level may be fed on by parasites such as disease organisms, and each of these can in turn be preyed on and parasitized, adding further dimensions to the food chain. Moreover, at each trophic level some of the carbon received from the next lower level is discarded in the form of detritus (e.g., falling leaves, feathers, hair, feces), and some of the organisms at each level die without being used as food for the level above. This dead organic matter provides food for (1) **saprophages,** or detritus-feeding animals, such as earthworms, millipedes, sow bugs, and carrion feeders like vultures, and (2) **decomposers,** saprotrophic bacteria and fungi that break down detritus and body remains to obtain organic matter and energy. Saprophages and decomposers constitute prey for other organisms (e.g., birds eat earthworms, and various soil animals consume soil fungi and bacteria), so their activities recycle some carbon back into the food chain. The interlocking food chains within an ecosystem comprise the **food web** (Figure 33-1).

Figure 33-A. Mountain scene illustrating the process of vegetational change, or plant succession. In the middle foreground a rock slope has been colonized by grasses and shrubs, including willows (foreground). Beyond, seedlings of pioneer pine species are invading, leading to a pine forest as seen behind them. Within the forest, shade-tolerant hemlock (*Tsuga mertensiana*) and red fir (*Abies magnifica*) seedlings have grown up (trees with feathery tops) and are beginning to replace the pines. The final or climax vegetation in this site will be a hemlock/fir forest like that in the further distance. Sierra Nevada, California.

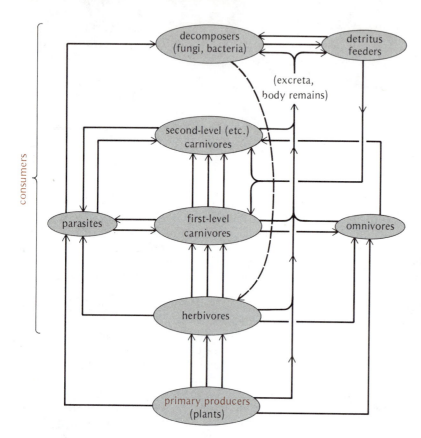

consumers

decomposers (fungi, bacteria)

detritus feeders

(excreta, body remains)

second-level (etc.) carnivores

parasites

first-level carnivores

omnivores

herbivores

primary producers (plants)

Figure 33-1. Interlocking complex of food chains, or food web, of an ecosystem. Hervibores here include not only plant-eating animals like sheep and cattle but seed-eating birds, plant-eating insects, and organisms such as slime molds that feed on saprotrophic bacteria and insects that feed on fungi. Detritus feeders are animals that feed on excrements or body remains of other animals. Although most matter and energy flow is upward through successive trophic levels, with the loss of much carbon and energy at each step, some is recycled to lower levels by herbivores that eat decomposers, and carnivores that eat parasites or detritus feeders. The arrows mutually interconnecting decomposers and detritus feeders show that the latter often eat debris containing the former, while the former break down the remains of the latter when they die.

Ultimately, almost all the carbon fixed in a given day's photosynthesis is reconverted into carbon dioxide by decomposers (a small fraction of it may instead eventually become fossil carbon—coal, oil, and so on). The flow of carbon from CO_2 via photosynthesis, the food web, and eventual decomposition back to CO_2 constitutes the **carbon cycle** (Figure 33-2).

Except for migration of some animals and long-distance food transport by humans, the flow of organic carbon in the food web is usually local. But the carbon of an ecosystem derives from, and is returned to, the carbon dioxide pool in the atmosphere or (for marine ecosystems) the ocean, which all ecosystems share [atmospheric CO_2 exchanges slowly with CO_2 dissolved, mainly as bicarbonate (HCO_3^-), in the oceans]. Thus the complete carbon cycle (Figure 33-2) is a global one.

Energy Flow

The food web represents a flow not only of organic matter but also of energy, derived from sunlight by photosynthesis. At each trophic level part of this energy is released, by respiration, for the growth, physical and chemical activity, and

maintenance of the organisms at that level. This release begins at the producer level: land plants respire one third to two thirds of the photosynthate they produce (Chapter 6). Most animals respire to carbon dioxide at least this large a proportion of what they eat. Because of the carbon lost by respiration at each trophic level, the amount of biomass that can pass to successive trophic levels up a food chain becomes a smaller and smaller fraction of the biomass at the producer level. The distribution of biomass with height in the food chain is thus usually like a rapidly tapering pyramid (Figure 33-3).

The energy organisms obtain by respiring organic matter derived from photosynthesis ends up as heat, eventually dissipated into their surroundings and not usable as an energy source. Thus, unlike carbon, energy is not cycled in the ecosystem. It comes from sunlight and must be replaced from this source with every turn of the carbon cycle.

An economically important consequence of the food chain pyramid principle is that far more agricultural resources are needed to produce a gram of animal biomass for human food than a gram of plant biomass. The practice of raising animals for food has been widely criticized as an example of profligate use of resources

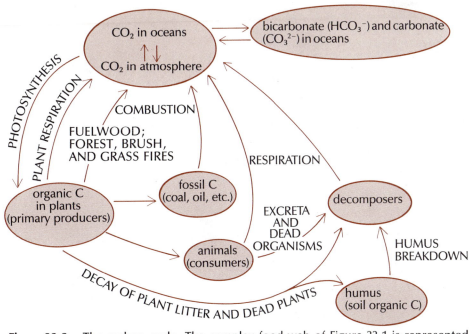

Figure 33-2. The carbon cycle. The complex food web of Figure 33-1 is represented here simply by the arrows from primary producers to consumers and decomposers.

by western societies in which meat is an important component of the human diet. However, it should be noted that animals raised on range land and pasture convert feed that is indigestible by humans into a form that we can eat, and domestic animals similarly use feed grains that, because of their nutritional imbalance, are not a very efficient source of human nutrition.

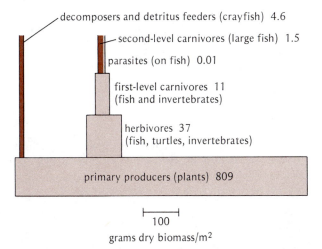

Figure 33-3. Biomass pyramid estimated by Howard T. Odum for aquatic community in Silver Springs, Florida. Biomass of omnivores was divided between the herbivore and first level carnivore categories in calculating the total biomass at each trophic level (numbers). From *Ecological Monographs*, Vol. 27, p. 84, 1957.

MINERAL NUTRIENT CYCLING

The mineral nutrients such as nitrogen, phosphorus, and sulfur that plants need and obtain from soil (or from water in an aquatic ecosystem) are passed, along with carbon, up each food chain and are utilized in turn by the organisms at each trophic level. Animals **excrete** the mineral nutrients left over from breakdown (turnover) of body constituents and use of food materials for respiration. At successive trophic levels, therefore, the mineral nutrients derived from plants are progressively returned to the soil or the waters from which they came. Decomposers release in inorganic form the nutrient elements contained in organic excretion products and in dead plant and animal material. This **mineralization** of nutrients makes them available for reuse by plants, which send the nutrients up the food chain once again. Thus the nutrients experience a regularly repeating cycle of utilization and reutilization (Figure 33-4).

Under natural conditions, mineral nutrients on land experience mainly local cycling between the soil and the biomass of the local ecosystem. An ecosystem builds up, as it develops, a "capital" of available and potentially available nutrients free in the soil as well as those temporarily locked up in the humus of the soil and in the biomass of plants, animals, and microorganisms in and above the soil. Each year nutrients are added to an ecosystem's nutrient capital by weathering

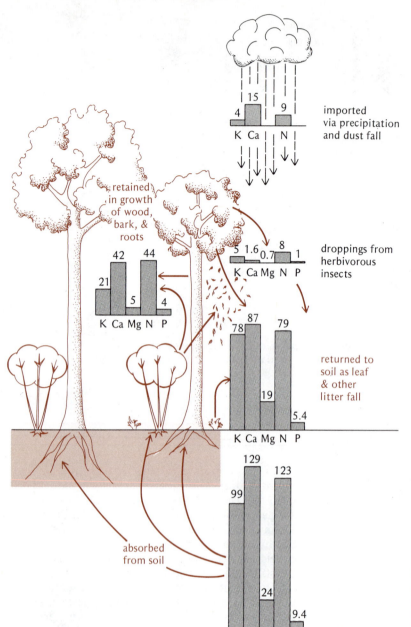

imported
via precipitation
and dust fall

retained
in growth
of wood,
bark, &
roots

droppings from
herbivorous
insects

returned to
soil as leaf
& other
litter fall

absorbed
from soil

Figure 33-4. Annual cycling of mineral nutrients (in kilograms per hectare) as estimated for an oak-ash woods in Belgium. Data of P. Duvigneaud and S. Denaeyer-de Smet, *Ecological Studies,* Vol. 1, pp. 199-225 (1970).

of rock minerals in the subsoil, nitrogen fixation, and fall of dust and precipitation from the air. But in a fully developed ecosystem the input of these nutrients normally serves just to offset losses due to such processes as leaching and erosion. Most of the nutrients the community consumes each year derive from cycling of its nutrient capital (see Figure 33-4).

In river-bottom lands, deposition of nutrient-laden clay and silt particles from streams during annual flooding brings in substantial amounts of nutrients from outside the area. Bottom lands consequently are often richer and more productive agriculturally than upland soils.

This fact was the basis of the Nile Valley agriculture that made possible the civilization of ancient Egypt.

The Nitrogen Cycle

The nitrogen cycle (Figure 33-5) is more complicated than other nutrient cycles primarily because of the transformation of inorganic nitrogen in the soil, shown in Figure 33-5 and explained previously in Chapters 11 and 22. Decay of plant and animal remains mineralizes, as ammonium (NH_4^+), most of the nitrogen they contain; the rest

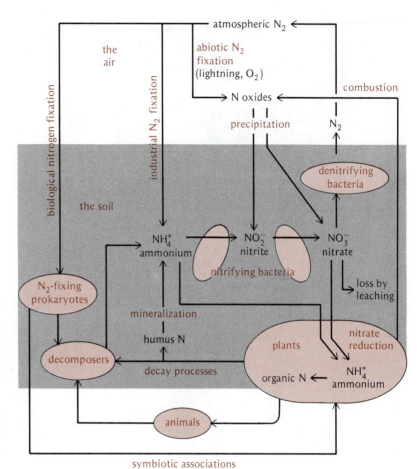

Figure 33-5. The global nitrogen cycle is actually two interlocking cycles, one the cycling of combined forms of nitrogen by organisms in and above the soil (center part of diagram), the other the grand cycling of elemental nitrogen from the atmosphere to organisms by nitrogen fixation and from the soil back to the air by denitrification and combustion (perimeter of diagram). Much more nitrogen per year flows around the central cycle than the peripheral one, but by far the largest pool of nitrogen is the atmospheric pool of N_2.

becomes locked up in soil organic matter or humus, which roots cannot absorb or use directly. Humus nitrogen is slowly mineralized to $NH_4{}^+$. However, because of nitrification (Chapter 22), in many soils the principal form of nitrogen available to plant roots is nitrate, which plants use by reducing it back to $NH_4{}^+$, from which they can make organic nitrogen compounds. This process provides the nitrogen needed by the consumer levels of the food chain.

As described in Chapter 22, nitrogen is more subject than other mineral nutrients to losses from terrestrial ecosystems. Because of these losses (Figure 33-5) the nitrogen capital of any ecosystem would gradually run down, and with it the system's productivity, if lost nitrogen were not replaced. Unlike other mineral elements, weathering of parent rock does not furnish nitrogen to the soil because most rocks lack any substantial amount of nitrogen. Replacement nitrogen for the soil comes in part from precipitation, which carries oxides of nitrogen formed by oxidation of atmospheric nitrogen gas (N_2) during lightning discharges, and nowadays also from automobile exhaust and other kinds of air pollution. The most important addi-

tion of nitrogen to natural ecosystems, however, comes from biological nitrogen fixation.

Role of Nitrogen Fixers

In many natural communities wild leguminous plants play a role similar to that of legume crops in agriculture in maintaining the nitrogen capital of the ecosystem (Figure 33-6). In certain communities a comparable role is played by various nonleguminous woody plants that bear root nodules containing nitrogen-fixing actinomycetes (*Frankia*). Nonlegume symbiotic nitrogen fixers include alder trees (*Alnus*), sweet gale (*Myrica*), wild lilac (*Ceanothus*), and several genera of shrubs belonging to the rose family (Figure 34-1).

In areas where productivity (quantity of biomass produced per year) is low, such as arctic tundra, rocky terrain, and the open ocean, the major nitrogen input apparently comes from free-living nitrogen fixers, frequently cyanobacteria. (However, nitrogen-fixing blue-greens can also be important in more productive environments, as noted in Chapter 22.)

Figure 33-6. A montane conifer and broad-leaved forest with a ground cover of lupines, nitrogen-fixing legumes which maintain the nitrogen supply of the community, encouraging the growth of new trees such as those at left. In middle foreground a fallen tree has been decayed by wood-rotting fungi, releasing its mineral nutrients for use by other plants in the community, especially the young trees. The visible decay-resistant remains of the trunk (mainly lignin residues) will become part of the soil's humus content, aiding in retention of water and available mineral nutrients.

Figure 33-7. Beginnings of primary succession on bare rock in (A) a dry site and (B) a moist site. In (A), lichens are the principal primary colonizers, followed later in this locality by a xerophytic species of *Selaginella*, a lower vascular plant visible as brownish-green colonies where soil first starts to accumulate. Herbaceous flowering plants, including grasses, follow, and eventually shrubs and tree seedlings (not shown in picture). In (B) some lichens occur on the bare rock but mosses are advancing across the surface, creating a continuous cover that permits herbaceous vascular plants to succeed, followed by shrubs (willows) and pine seedlings (one visible in the dense shrubbery).

SUCCESSION

The community of organisms inhabiting a given area is usually the product of a gradual development. Originally, colonizer or **pioneer** species (Chapter 19) occupy the site, then are replaced successively by other species, leading eventually to the community that now exists. Development of vegetation on unoccupied areas, such as bare rock surfaces, landslides, new volcanic material, sand dunes, or lakes being filled in by sediment is called **primary succession.** Redevelopment of vegetation after partial destruction such as by fire, flood, or windstorm, or when farm or pasture land is abandoned, is called **secondary succession.**

Primary Succession

On bare rock the primary colonizers are usually lichens and, at least in moist areas, mosses and microscopic algae (Figure 33-7). These small plants can satisfy their limited mineral nutrient needs from dust fall and precipitation, and by encouraging weathering of the rocks to which they attach. Some lichens containing cyanobacteria get their nitrogen by nitrogen fixation (Chapter 25), adapting them especially well to primary colonization. The surface layer of lichens and/or mosses

Figure 33-8. In this grove of aspens (*Populus tremuloides*), which are sun-requiring pioneer trees, shade-tolerant spruce seedlings have become established. As they grow they will eventually replace the aspens in this site.

on rocks provides a beginning for the development of soil. When small vascular plants such as grasses and succulents that can tolerate desiccation colonize cracks or crannies in rocks or become rooted in the superficial crust of moss or lichens, their roots and leaf debris help an actual soil start to accumulate. As the soil increases, larger plants can become established. Eventually shrub and tree seedlings take hold (Figure 33-A) and, if the climate permits, a forest cover starts to develop.

The initially colonizing herbs and tree seedlings are sun plants, being photosynthetically adapted to (and requiring) full sunlight. Because of the limited amount and quality of the soil, they usually must be somewhat xerophytic, that is, adapted to relatively dry situations. As trees develop, their shade weakens the previously developed ground cover of herbaceous sun plants and permits shade-tolerant forest floor species (shade plants) to become established. Shade also prevents seedlings of the pioneer trees themselves, which are typically shade-intolerant, from succeeding beneath the tree canopy. Other tree species whose seedlings tolerate shade eventually replace the pioneer trees (Figures 33-A and 33-8).

Succession varies greatly with the nature of the starting site. On a loose substrate such as raw volcanic ash or glacial debris, which will retain water and can be penetrated by vascular plant roots, primary succession can start at a more advanced stage and progress much more rapidly than when it starts with bare rock or open water. It apparently takes some 400 years for rain forest to develop fully on fresh lava (volcanic rock) extruded by the volcanoes in Hawaii, and about 700 years for bare granite outcrops in Georgia to become covered with pine woods, whereas a spruce-fir forest can develop on debris left by retreating glaciers in Alaska in little more than 100 years (Figure 33-9).

On land formed by the filling of ponds or lake basins, succession involves first hydrophytes and then mesophytes (Figure 33-10), rather than the xerophytes prominent in early stages of succession starting from bare rock. However, the later progression from sun-requiring to shade-tolerant species is similar in both cases.

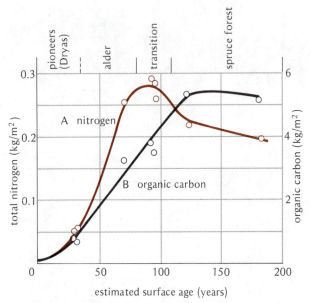

Figure 33-9. Soil development, shown by soil nitrogen and organic carbon accumulation per square meter of land area, during primary succession on land exposed by glacial retreat at Glacier Bay, Alaska. The ages of different sites sampled to obtain these data could be determined from historical records and other evidence. As indicated above the graph, in the early part of the succession symbiotically nitrogen-fixing non-legumes (*Dryas* and alder) are major colonizers. The decline in soil nitrogen later in the succession is probably due to nitrogen accumulation in the biomass of the developing spruce forest. (Data of R. L. Crocker and J. Major, *Journal of Ecology*, Vol. 43, pp. 427–448, 1955.)

Figure 33-10. Primary succession on the margins of a shallow pond: floating and submerged hydrophytes out in the pond, amphibious sedges and grasses making the transition to emergent ground, behind them a shrub zone subsequently invaded by pioneer trees (here, pines), saplings of which can be seen at the forest's edge. (Sierra Nevada, California). In more humid, temperate climates pioneer conifers would usually become replaced later by broad-leaved trees.

If the physical conditions of the habitat are hostile, such as those of deserts, very high mountains, or subpolar latitudes, succession is usually less marked. In extreme cases the ultimate vegetation consists just of species that are simply able to colonize and survive in the habitat.

NUTRIENT CAPITAL AND SOIL DEVELOPMENT

Pioneer species must often tolerate a poor nutrient supply. Many of them are mycorrhizal (Chapter 25). Species with nitrogen-fixing symbionts, either legumes or nonleguminous nodulated plants such as alder and *Dryas*, tend to be prominent. The activities of these plants build up the system's nutrient capital (Figure 33-9), permitting more nutrient-demanding species to succeed. Similarly, finer-textured, more organic soil develops in the course of succession, as a result of the accumulation of plant debris and its transformation into humus by soil microorganisms. This improves soil water retention and permits more lush growth, for the given precipitation pattern, than would otherwise be possible. Biological ac-

tivities during succession thus contribute positive feedback, helping the ecosystem progress toward an optimum vegetation and accompanying animal life.

Soil development is at best a slow process. Under the most ideal conditions for humus accumulation and clay development by mineral weathering it appears that raw mineral debris can be transformed into a recognizable soil in as little as 200 years, but more generally thousands of years seem to be required.

Succession in Lakes and Ponds

In freshwater habitats nutrient changes also tend to occur with time, leading to a characteristic succession. Lakes that form in barren areas such as those uncovered by glacier retreat, initially fed by waters very low in mineral nutrients, are called **oligotrophic** (Greek *oligos*, small; *trophē*, nourishment). Because very little algal growth can occur in them, the water in these lakes is very clear. In a high mountain lake fed by a limited watershed developing only modest vegetation, the oligotrophic condition can persist for the life of the lake if artificial introduction of nutrients is avoided. In a lowland lake, as succession occurs in its watershed and the ecosystem's nutrient capital increases, more nutrients flow into the lake, so algal growth and productivity increase, making the water somewhat cloudy. A modest nutrient supply and algal production, sometimes called a mesotrophic state (Greek *mesos*, intermediate), can persist in lakes for thousands of years. Eventually, as the lake or pond becomes shallow through sediment accumulation on its bottom and nutrient input increases, for example, from encroaching shoreline vegetation, the lake waters attain a relatively high nutrient level, or **eutrophic** condition (Greek *eu-*, good;

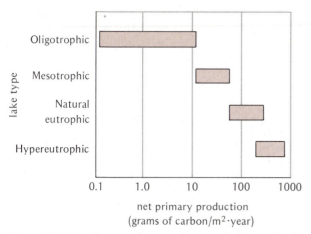

Figure 33-11. Range of net primary photosynthetic production (plotted on a logarithmic scale) for different successional types of lakes, with increasing levels of mineral nutrients from the top of the diagram downward. Hypereutrophic lakes are those polluted with nutrients by human activities (called eutrophication in much popular writing).

Figure 33-12. Aerial view of hills toward the periphery of the "blast zone" around Mt. St. Helens, Washington, where the conifer forest was destroyed by the eruption of May 18, 1980. Patches of green in the picture, taken in the summer of 1981, show where colonizing plants have invaded and perennials surviving by means of underground organs have regenerated, beginning the succession back to a conifer forest.

trophē, nourishment), permitting an extensive algal population to develop, which in turn causes the lake to turn green and heavily cloudy. Primary production in such a lake can be hundreds of times that in an oligotrophic lake (Figure 33-11).

"Eutrophication." A shallow eutrophic lake is usually on its way to the land plant succession sequence described above. However, even in a deep lake any substantial increase in the availability of nutrients, such as from a fire or a logging operation that destroys and releases nutrients from adjacent forest cover, can bring the surface waters of the lake into a eutrophic condition, permitting algal proliferation. Release of waste or surplus nutrients from human activities (sewage, phosphate detergents, excess agricultural fertilizer nutrients, and so on) have created an exaggerated eutrophic condition in many lakes (e.g., Lake Erie), causing development of heavy, unpleasant algal "blooms" with serious consequences for aquatic life and water quality owing to the ensuing decay of the masses of accumulated algae. This is commonly called eutrophication but is better called cultural eutrophication or hypereutrophication because, as noted above, a eutrophic state develops naturally in many older lakes without an artificial excess of nutrients.

Secondary Succession

When vegetation is disturbed or destroyed by human activities such as logging, or by accidents such as fires, hurricanes, or disease epidemics, colonizer species invade and occupy the area opened up (Figure 33-12). This starts a secondary succession that in time can regenerate the previously existing type of community. The level at which the succession begins and the speed with which it progresses depend on the nature and extent of the disturbance and how much it affects the vegetation, the condition of the soil, and the available nutrient capital. Normally, only after an extreme disturbance must succession start with primary pioneers like mosses and lichens. Often secondary succession begins with a meadow or a shrubby plant community, or involves colonizer tree seedlings from the outset.

In the mid-Atlantic and northeastern United States, for example, where much formerly cultivated or pastured farm land has been abandoned in the course of the last century, a characteristic "old field" succession occurs (Table 33-1). After being initially occupied primarily by herbaceous

TABLE 33-1   Secondary Succession in Old Fields

Years after Abandonment (approx.)	Dominant Plant Types
1–2	annual weeds (crabgrass, ragweed, etc.), some perennials
2–5	perennial herbs, tall grasses especially
5–10	pine seedlings in tall grass
10–60	young to mature pine woods
60–150	pine forest with understorey of hardwoods
150+	hardwood forest (oak, hickory, dogwood, etc.)

This specific sequence refers to observations on old field succession in North Carolina, but similar changes occur elsewhere in the eastern United States.

Figure 33-13. Example of "old field" succession in progress. In left foreground, grasses and other herbaceous plants have occupied the land, while in the right foreground shrub species are beginning to invade. In the middle distance at right pioneer conifers are growing up and some young hardwood trees have also gotten a start. In far distance at left, a fully developed deciduous woods, possibly approaching climax vegetation.

and then shrubby colonizing species, seedlings of sun-loving pioneer tree species such as pines become established (Figure 33-13) and start to build a forest canopy. The pioneer trees are eventually replaced by hardwoods whose seedlings are shade-tolerant. This kind of succession has created the principal nonagricultural landscapes in rural parts of the eastern United States.

Neither primary nor secondary succession follow a rigid or fully predictable course; a significant element of chance enters into these processes. For example, which colonizer species initially occupy an open area, influencing the speed and nature of later successional steps, depends not only on the nature of the site but also upon the chance immigration of dispersal units (seeds, spores, etc.) of different species and chance variations in weather affecting which of these propagules become successfully established in the site. Thus, although a general pattern of succession like that in Table 33-1 can be described for a given region and type of site, the duration and other details of the steps including the principal species involved can differ even between closely comparable sites, due to chance factors.

FRAGILITY

As with primary succession, environments that permit only limited photosynthetic production allow only a slow secondary succession after a disturbance, compared with the vigorous regrowth of vegetation in moist, warm climates. Ecosystems with limited capability for secondary succession and self-regeneration are termed **fragile;** the scars

of damage done to them are long-lasting and in some cases seem to persist almost indefinitely. Besides climatic constraints, edaphic consequences of disturbance can also make an ecosystem fragile. For example, on steep hillsides soil denuded of vegetation rapidly erodes away under heavy rains, impoverishing the plant growth substratum so much that the rapid secondary succession that would otherwise occur under moist conditions can be slowed or almost completely arrested. A heavy forest can become replaced by a useless tangle of brush or even a virtual badlands. Fragility and the reasons for it are among the important ecological issues that should be taken into account during human disturbance of the environment.

End Point of Succession: Climax

Through succession, given enough time free from destructive interference, the vegetation of any area should eventually become composed of species that can reproduce and maintain themselves in the habitat they occupy rather than being replaced by others. This permanent vegetation is called a **climax.** Its general character and species composition depend primarily upon climate and the nature of the ultimately developed soil and root environment. The standing crop of plant biomass usually reaches a maximum in a climax community. However, measurements show that photosynthetic primary productivity is actually greater at intermediate stages of succession, when there is a complete cover of fast-growing pioneer species, rather than the mature or old-aged, slowly growing individuals dominating a climax community.

There are several reasons for the reduced productivity of a climax community such as a mature forest. One is the large maintenance respiration cost (see Chapter 6) of the community's large nonphotosynthetic biomass (e.g., tree trunks, branches, massive root systems). Another is that during succession an increasing proportion of the ecosystem's nutrient capital tends to become locked up in its accumulating biomass, compared with intermediate successional stages when more is often available in the soil. Productivity in a climax often is limited by the rate of nutrient cycling, depending mainly on the death and decay of long-lived dominant plants, especially forest trees.

PRECLIMAXES

In areas chronically subject to a disturbing or destructive influence such as fire, severe wind, flooding, salt spray, heavy grazing, or cutting and fuel gathering by humans, succession usually cannot progress to the climax vegetation that the area's climate could sustain. Such a state of arrested succession is called a **preclimax.** Many grasslands, for example, represent a fire and grazing

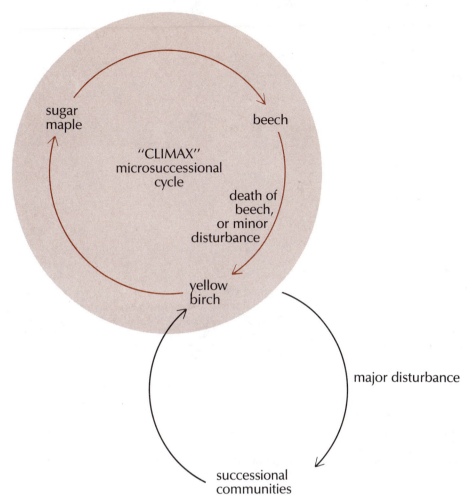

Figure 33-14. Successional reproductive cycle of beech, maple, and birch in a climax beech-maple forest in New Hampshire. Colored arrows show the cyclical replacement of the dominants in forest during minimum natural disturbance, as interpreted by L. K. Forcier (*Science,* Vol. 189, pp. 808-810, 1975). Black arrows show secondary succession following forest destruction by fire, lumbering, severe windstorm, and so forth.

preclimax, from which trees are excluded by recurrent grass fires and by their seedlings being eaten by herbivorous animals (see Chapter 34). Agricultural ecosystems represent a special class of preclimax in which succession is prevented by human management practices.

"CLIMAX" INCLUDES OPPORTUNIST SPECIES

Every ecosystem is subject to some degree of disturbance, if not by humans then at least by other animals, especially herbivores and insects, by plant diseases, and by meteorological accidents such as windstorms or unseasonable frosts or droughts. By weakening or killing individual plants these disturbances open up local sites that can be, and often are, occupied by colonizer species. Thus virtually all mature vegetation includes elements of succession, and the concept of an undisturbed climax vegetation is really almost a theoretical abstraction.

Cyclic Succession. In some types of climax forest, seedlings of dominant trees such as beech normally do not succeed beneath their parents, so when a mature tree dies, its place is often taken by an opportunist such as birch, beginning a local succession through which beech eventually recovers this place in the canopy (Figure 33-14). In other forests such as the northern coniferous belt (taiga, Chapter 34) most of the climax trees in any local area pass together through a rather synchronized cycle of rapid juvenile growth, maturity, senescence, and death, allowing colonizers and a new population of conifer seedlings to begin growth. Thus the mature vegetation is a patchwork of groves of different ages plus areas of dying trees rather than a uniform forest.

In some kinds of habitats and regions a more extreme cycle of succession occurs, producing no one type of permanent vegetation that could be regarded as climax. In arctic lowlands, for example, local areas of land subside and become ponds as a result of ground motion caused by frost-heaving and melting. Land plants invade the ponds, which are gradually filled and revegetated, only to subside again later and repeat the cycle. Species replacement and erosion can similarly

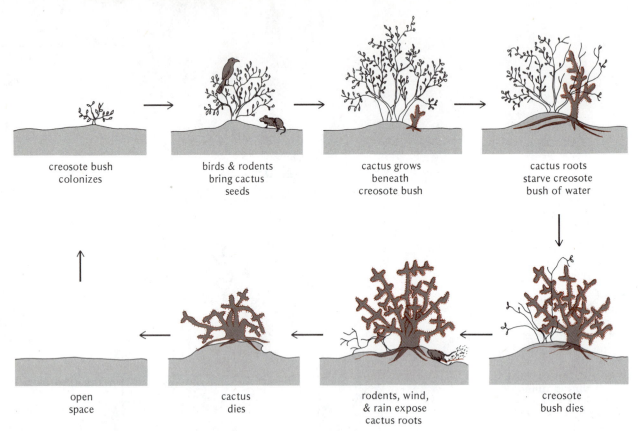

Figure 33-15. Successional cycle in hot desert creosote bush (*Larrea tridentata*)-cholla cactus (*Opuntia leptocaulis*) community in Big Bend National Park, Texas. The cactus produces roots near the soil surface which trap the scant rainfall and keep water from reaching the deeper-rooted creosote bush, allowing the cactus to replace the creosote bush. However, the cactus' superficial roots easily become exposed by erosion, leading to a cyclical succession as shown. After R. I. Yeaton, *Journal of Ecology*, Vol. 66, 651–656, 1978.

cause a repeating cycle of succession in semidesert communities (Figure 33-15).

Succession Induced by Climatic Changes

Geological and fossil plant evidence shows that the earth's climate is subject to change. Very extensive changes occurred during the Cenozoic era and up to the present (Chapter 21). Both human and biological records (e.g., tree growth ring data) show substantial temperature fluctuations even within the last several hundred years. Such changes and attendant changes in precipitation affect the success of different plant species, leading to changes in relative abundance and to species replacements that are somewhat analogous to the successional changes described above, but occur normally over a longer time scale (depending upon how gradually or abruptly the climate changes). Owing to climatic fluctuations, undisturbed vegetation may rarely have enough time under any one specific climatic regime to reach a true climax condition, another reason why the climax concept is probably more of a theoretical abstraction than a reality.

From the pollen deposited in lake sediments by wind-pollinated plants and trees growing in the vicinity of the lake one can estimate the species composition of the surrounding vegetation, a procedure called **pollen analysis.** These records demonstrate dramatic post-Pleistocene changes in vegetation over time in northern Europe and North America (Figure 33-16), presumably due to climatic changes.

Climatically induced vegetation changes are often excluded from the definition of succession, which according to this restricted definition constitutes only vegetational change due to colonization and modification of an unoccupied or incompletely occupied site. In an actual situation it is not always easy to tell whether a change in vegetation is part of a successional sequence, is due to climatic change, or is the result of disturbance. This question becomes crucial in evaluating human impact on ecosystems.

HUMAN MODIFICATION OF ECOSYSTEM PROCESSES

Even primitive humans affected their environment significantly in ways that either initiated

Figure 33-16. Fossil pollen diagram for lake sediments at successive depths in a bog (naturally filled lake basin) in Nova Scotia, Canada. Abundance of each pollen type is expressed as percent of total pollen at that depth. Age of sediments was estimated by ^{14}C dating. Note the indications of climatic changes about 8000 years ago increasing the importance of oak, hemlock, and pine with a decline of spruce, fir, and birch, then a resurgence of these latter species about 5000 years ago, followed by an increase in beech. (Data of D. A. Livingstone, *Ecological Monographs*, Vol. 38, pp. 87-125, 1968).

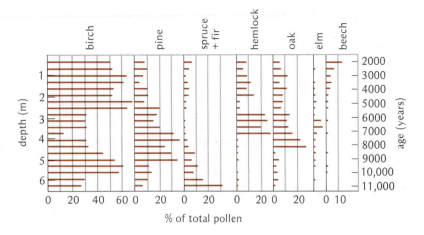

or retarded ecological succession. As agriculture, technology, and human populations have grown, human impact on ecological processes has become predominant worldwide. Some important aspects of this impact are discussed briefly here in relation to the principles of community and ecosystem dynamics.

Interference with Food Chains

Agriculture substitutes, for the complicated food web connecting the numerous organisms in a natural community, a simplified food chain from crop plant to humans, or from crop plant to domestic animals to humans. The "monoculture" of a single domesticated plant or animal species gives disease organisms and pests capable of attacking the crop or the domesticated animal a much greater opportunity for proliferation, permitting their populations to explode with destructive consequences for the food supply. In a natural community comprised of many species such explosions tend to be held in check both by the relative scarcity of any one species as a food source, and by predator organisms at trophic levels above those of the diseases and pests using the latter as food sources and thereby controlling them. Modern humans, in trying to control diseases and pests of domesticated plants and animals, have turned to the use of chemical pesticides, which in most cases also kill off the predators that could control the disease or pest naturally. Thus, by breaking the natural food chain, pesticides steadily increase the pest problem rather than solving it.

Altering the Carbon Cycle

The combustion of fossil fuels (Figure 33-2), which represent carbon fixed in previous geological ages, shifts the balance of the natural carbon cycle, steadily raising the carbon dioxide content of the atmosphere (Figure 33-17). Because photosynthesis depends on atmospheric carbon dioxide concentration, the rate of global photosynthesis is probably increasing, which of itself might be beneficial. However, because carbon dioxide absorbs infrared energy, it tends to trap incoming solar energy near the earth's surface by keeping heat from being radiated back into space, a "greenhouse effect" that tends to increase the earth's surface temperature.

How great a temperature rise should be anticipated because of the substantial rise in atmospheric carbon dioxide that will occur by the year 2000 is still somewhat in doubt, but any significant increase (1°C or more) will have important ecological consequences. These could be favorable in some areas but serious in many others. If the warming melts the Antarctic and Greenland ice caps, the earth's sea level would rise some 60 m (about 200 ft), flooding all lowland areas including most major cities and agricultural areas. This can be expected to happen if the earth's mean temperature increases by as little as 3°C.

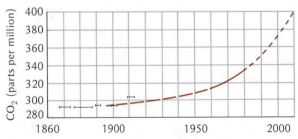

Figure 33-17. The earth's changing atmospheric CO_2 concentration caused by fossil fuel combustion since the industrial revolution. Bars at left indicate early observations, the solid line between 1960 and 1980 shows precise yearly mean values obtained by regular sampling since 1960, and the dashed line to the right of 1980 shows projected future changes based on anticipated further consumption of fossil fuels.

Interference with Nutrient Cycles

Agricultural practices break the basic ecological mineral nutrient cycles, first by removing harvested crops, with the nutrients they contain. Rather than returning to the land from which they came, the nutrient elements in the wastes from food consumption end up mostly in rivers and streams, where excess of mineral nutrients creates problems (eutrophication, see above) while the agricultural land is correspondingly depleted. Fertilization (including legume crop rotation for nitrogen fixation) must be practiced to keep the land productive.

Animals such as cattle and hogs raised for human food consume nearly 10 grams of feed for every gram of animal protein that can be harvested. The excreted nitrogen and other nutrient elements in the unconserved feed are returned to the land and recycled for further plant growth and feed production if the animals are being raised on pasture or range land. But in present methods of economically efficient animal production the animals are kept in local areas such as feedlots and hog or chicken houses; the nutrient elements they excrete are often not returned to the land but mostly find their way, as nutrient pollution, into streams and rivers. The loss of nitrogen and other nutrients from agricul-

tural ecosystems by this practice is very large, roughly 10 times the loss that would result from production of an equivalent amount of vegetable food.

Soil Erosion and Deterioration

Agricultural cultivation accelerates erosion, the washing or blowing away of the soil (Figure 33-18). Cultivation loosens the soil and exposes it directly to the action of wind, falling rain, and runoff. By accelerating humus breakdown, cultivation also promotes erosion because, as explained in Chapter 11, humus binds soil particles together, which tends to keep them from washing or blowing away. Overgrazing and destruction of vegetation by forest fires, fuel cutting, logging, and similar practices also promote erosion by exposing the soil to the direct action of wind and water. Erosion not only removes nutrients but also reduces the capacity of the ecosystem to store and deliver both nutrients and water to plants, reversing the developmental processes by which a productive ecosystem arises. Erosion, humus depletion, and other retrogressive soil changes caused by continuous agriculture degrade the productive

Figure 33-18. Uncontrolled erosion on formerly productive farm land. Secondary succession is very poor in such a badly deteriorated site and the few colonizing plants that succeed can do little to prevent further erosion.

(A)

(B)

Figure 33-19. Progress of desertization in sub-Saharan Africa (Sahel Region). (A) Healthy, moderately grazed savanna vegetation during the dry season. (B) Similar area after overgrazing during drought cycle: hardly any vegetation or soil remains; human and animal populations that have caused the deterioration are collapsing.

potential of an ecosystem and can eventually lead to a sterile environment incapable of supporting significant human or other animal life. Many earlier human civilizations appear to have declined and died because they degraded their soil and forest resources until they became unproductive. This is happening today in many parts of the world, most dramatically with the human-caused, year-by-year expansion of deserts such as the Sahara in Africa and the Thar Desert in India; such desert expansion is due to overgrazing, fuel gathering, and attempts to cultivate erosion-prone terrain, a combination of phenomena known collectively as **desertization** (Figure 33-19). However, ecosystem deterioration is not confined to underdeveloped regions but is also occurring at an alarming rate in agriculturally advanced countries. This is due in part to technology, including the fact that mechanized agriculture normally does not return enough crop residues to the soil to balance its yearly loss of humus. This is especially true under current high rates of chemical fertilization, which stimulate humus breakdown by promoting growth of soil microorganisms that use humus as a carbon source.

The "dust bowl" era of the 1930s, when large areas of prairie land plowed up for farming in the southern Great Plains lost their soil and became useless as a result of wind erosion in dry years, spurred national recognition of the seriousness of the soil erosion problem. Surveys made at that time indicated that already 775 million acres or 41% of the land area of the United States had lost between a quarter and three quarters of its original surface soil, a degree of erosion labeled "moderate." "Severe" erosion—loss of more than three quarters of the soil—had occurred on 225 million acres or 12% of the land, and 3%, about 57 million acres, had been totally lost to cultivation. By current estimates, southern Iowa has now lost half of its original topsoil after just 100 years of cultivation, and the food-producing potential of the United States as a whole is now estimated to have fallen by at least 10–15%, and possibly as much as 35%, owing to erosion. Water-borne soil derived from erosion dwarfs in magnitude all other kinds of river and lake pollution in the United States, amounting to about 3.5 billion tons per year. Globally, at known present rates of erosion, it appears that one third of the entire world's crop land will disappear within the next 20 years.

CONTROL OF EROSION

Traditional methods of trying to control erosion include terracing (Figure 33-20), contour farming (Figure 33-21), and alternation (rotation) of row crops (e.g., corn) with sod or pasture, which both hold back erosion while they are growing and raise the organic content of the soil, helping reduce erosion for a year or two after plowing them under. However, these practices generally just retard the rate of erosion and do not bring it even close to a balance with the very slow rate of new soil formation.

A new approach to erosion control is **minimum-tillage** or **zero-tillage** agriculture, noted in Chapter 1. Specific herbicides or hormonal growth retardants are used to kill or to hold back the growth of a winter or spring weed cover or an intentionally planted cover crop, and the commercial crop is then planted in the unplowed ground using a seed drill. This avoids the soil-loosening and hu-

Figure 33-20. Terracing, ancient and modern, for erosion control on steep agricultural land. (A) Pre-Columbian terraced farm benches built by the Incas at Machu Pichu in the Andes Mountains of Peru. (B) Modern terraced vineyards along the Rhine River in Germany (Kaiserstuhl winegrowing district, Baden).

mus-depleting actions of cultivation that promote erosion, and retains a cover of dead or living undergrowth below the crop (Figure 33-22), which not only protects the soil from the erosive action of falling rain but also acts as a mulch, keeping the soil cooler and slowing its drying out, thus saving water. Experiments show that this technique can virtually prevent erosion on sloping farmland subject to a substantial rate of soil loss under conventional cultivation and can give a crop yield as high as or even slightly above that of conventional agriculture, with significantly less expense and tractor fuel consumption. Use of minimum-tillage practices is expected to increase greatly in the United States in coming decades, even though one might question whether a system completely based on large-scale application of phytotoxic chemicals is entirely wise ecologically.

Salinization

Salinization, or buildup of a high salt concentration in the soil (Figure 33-23), is another consequence of agriculture that has had serious consequences for past civilizations, leading to desertization, and is adversely affecting much agricultural land today in arid and semi-

arid irrigated regions. Salt buildup occurs because salts contained in irrigation water remain behind after the water has evaporated or been transpired by the crop, and too little rain falls to wash them away. Salt accumulation creates conditions that are suitable for halophytes (Chapter 19) but reduce the growth and productivity of most agricultural plants or are completely toxic to them. When salinization becomes severe enough the land must be virtually abandoned for conventional agriculture. Programs to breed salt-tolerant strains of crop plants are attempting to cope with this problem, but probably no solution will permit land to be farmed under irrigation indefinitely in arid regions, except when unusually copious irrigation water is available for leaching accumulated salt below the root zone.

Other Adverse Effects and the Future of *Homo sapiens*

Besides deterioration of ecosystems due to agriculture itself, prime agricultural, forest, and range land that is needed to support present and future human populations is being rapidly lost today by urban expansion, road and reservoir construction, and wastage by strip mining. Agricultural and forest productivity around expanding urban centers is also being affected adversely or eliminated entirely by air pollution, because the photosynthetic systems of leaves are extraordinarily sensitive to damage by sulfur dioxide and by the oxidants in photochemical smog, two of the most widespread types of air pollution. In other words, as the human population, with its activities and its needs for agricultural and forest products, steadily grows, this very growth leaves less and less of an ecological base for meeting these needs.

In considering the future of the human species, the frightening aspects of current human ecological impact are its speed and universality. Agricultural soils erode and forests are destroyed on a time scale enormously shorter than that of the natural processes of ecosystem regeneration, processes for which there is still no physically or economically feasible artificial sub-

Figure 33-21.   Contour farming of hilly land in North Carolina. The contour plowing and alternation of plowed strips with strips planted to grain or sod slows runoff water flow and improves its infiltration into the soil.

stitute. The effects are therefore essentially irreversible within the time scale of human life spans and yearly needs. Moreover, because of our current human population of billions, these effects are no longer occurring just locally, as they did in the historical past, but virtually everywhere on earth.

It seems ecologically inevitable that if we do not voluntarily limit our population size and cut back our ecological impact to reach an equilibrium with regenerative processes in the ecosystem, nature will eventually and very heavy-handedly do it for us. With a growing population and a simultaneously declining basis for primary production at the base of the food chain, production cannot avoid eventually—indeed, probably within the next 100–200 years at most—falling seriously short of demand, leading to serious human population crashes likely to be accompanied and abetted by political and social upheavals as the technological system

Figure 33-22.   Corn being grown by a no-tillage technique on a farm in central Kentucky. Between the rows is an earlier seeded cover crop of grass, subsequently killed by weed-killer treatment. Its remains act as an erosion- and evaporation-controlling mulch at the soil surface, greatly reducing soil loss by erosion on this hilly farmland. Remains of corn stalks visible on the ground show that this land has not been plowed since the previous growing season. [For details see *Science*, Vol. 208, pp. 1108–1113 (1980).]

Figure 33-23. Salinized pasture land in the Great Basin. Evaporation of irrigation water has concentrated the salts contained in it until they form a salty surface crust, resembling snow, inhibitory to grass growth. Saltbush (*Atriplex*) shrubs, which are salt-tolerant halophytes, are invading at the left.

that has supported excess population growth collapses. Indeed, advanced technology, by increasing the productivity of land that is still usable, can only delay the balancing of the equation, and, by meanwhile encouraging further population growth and further deterio-ration of the earth's ecological base, technology itself is contributing to the potential disaster. We must act promptly and effectively to achieve a stable population and a sustainable rate of resource utilization if our species is to have a long-term future.

SUMMARY

(1) Carbon and mineral nutrient elements flow through communities from primary producers (plants) via food chains. Energy flows through the same channels, with smaller and smaller amounts of energy and carbon reaching higher trophic levels. Chemical elements are cycled back to primary producers via the the atmosphere and oceans (carbon), the soil and surface waters (mineral nutrients other than nitrogen), or both (nitrogen). Energy must be replaced from sunlight with each turn of this cycle.

(2) Farming interrupts the carbon and mineral nutrient cycles, diverting carbon, energy, and nutrients away from ecologically normal food chains and tending to deplete soil systems of nutrients and overload fresh waters with nutrient-containing wastes. Agricultural fertilization at current levels exacerbates this problem by supplying crops with much higher levels of nutrients than they can use, causing eutrophication of streams and lakes.

(3) By primary succession land vegetation develops on raw unoccupied land and on lake sed-iments, with gradual modification and improvement of the habitat by successive vegetational stages. The sequence varies with the type of site and the climate, but common to almost all successional sequences are slow development of soil, accumulation of soil nutrients and organic matter, and an increasing standing crop of plant biomass. The eventual product of succession under undisturbed conditions is a climatically and edaphically determined climax vegetation, or, under chronic disturbance, a preclimax.

(4) Secondary succession is the recolonization of land by vegetation after its partial destruction or after the abandonment of farm land, that is, where a soil already exists and can be colonized. Secondary succession should eventually regenerate a climax or preclimax if the terrain, especially the soil, has not been too severely degraded by the experience.

(5) In both primary and secondary succession, opportunists or colonizer species that grow and reproduce rapidly in full sun dominate the

earlier stages, whereas once a complete tree canopy has developed, plants and trees with slower growing but shade-tolerant seedlings eventually take over.

(6) Agricultural activity depletes soil humus and causes loss of soil by erosion, these processes together progressively impoverishing agricultural soils much faster than natural processes can regenerate them. Various methods of soil conservation are known, but the global loss of agricultural soil continues at a rapid pace. Erosion and salinization are leading to desert expansion or desertization in many areas; deforestation is making many mountain areas unproductive; and air pollution is impairing productivity around urban centers. Urban and highway development are destroying huge areas of useful agricultural land. In the longer term the human species cannot survive on earth even at its present population level if it continues to destroy earth's ecological life-support systems.

Chapter 34

NATURE AND DIVERSITY OF PLANT COMMUNITIES

Even a casual acquaintance with nature reveals not only that vegetation differs greatly from one environment to another but also that in any given environment certain species or species groups often occur together—fir and spruce in the north woods, pinyon and juniper in the Great Basin, oak and hickory in midwestern woodlands, and so forth. The plants in an **association** or plant community that populates a given habitat are obviously united by the fact that all of them not only tolerate but can grow and reproduce under the conditions of that habitat, evidently outcompeting others that might also be able to grow there. Although the members of a community compete among themselves for its resources, the utilization of different adaptive strategies by different life forms such as annuals, herbaceous perennials, shrubs, trees, and vines tends to reduce this competition and permit them to coexist. Some community members may actually depend upon others that occur there for their ability to exist in the habitat. Some members may require a modified environment created by others—for example, obligate shade plants depend on the shade of a forest canopy. Others may depend more specifically on associated species, such as forest trees that need mycorrhizal fungi to assist in their mineral nutrition, or parasites that of course depend on their hosts.

These multitudinous interactions, along with the nutritional cycling processes considered in Chapter 33, make for a certain degree of structure and of functional integration within a plant community. This chapter considers some of the general organizational and functional features of plant communities and then surveys the most important major community types (biomes) making up the earth's vegetation, considering how these communities are adapted to the types of environments they inhabit. Finally, we consider some ecological features of agricultural plant communities managed by humans, which comprise an ever-growing proportion of the earth's plant cover.

STRUCTURE OF PLANT COMMUNITIES

Species Composition

Those species that by virtue of their size and abundance occupy the majority of a habitat's area and capture the majority of its light, water, and mineral nutrient resources are regarded as its **dominants.** The dominants in a forest are the most abundant tree species; in brushy vegetation, the most abundant shrubs; and in a herbaceous vegetation such as grassland, the most abundant grass (or other herb) species. In some types of vegetation it may be hard to decide which species are dominants—for example, a savanna (see below) consisting of grassland with scattered large trees. In an agricultural community, the crop should be dominant; if it is not, weeds have gotten out of control.

In a nonmanaged community, the dominants are presumably species that are better adapted, on the average, to the conditions of the habitat than other, rarer species are. Since it is very unlikely that even two species will be exactly equivalent in adaptation or fitness for a particular habitat, in terms of natural selection it might be expected that the one species with the greatest fitness would eventually win out and completely preempt the habitat, much as humans now do in many areas they occupy. Although examples of this can be found—for example, the pure stands of Douglas fir, red fir, western white pine, or lodgepole pine in mountain areas in the western United States (Figure 34-1), most forests and other vegetation types include several dominants (Figure 34-A) and several to many less abundant or secondary spe-

Figure 34-A. Mixed deciduous forest in the Hudson Highlands, New York State. The patchwork pattern caused by the autumn colors in this photograph reveals a considerable variety of tree species occurring side by side in this community.

Figure 34-1. A pure stand forest, consisting of a single tree species, western yellow pine or ponderosa pine, *Pinus ponderosa*. Undergrowth is antelope brush (*Purshia tridentata*), a nonleguminous symbiotically nitrogen-fixing shrub belonging to the rose family; this shrub is important in maintaining the nitrogen supply in this forest community. Eastern slope of the Cascade Mountains, Oregon.

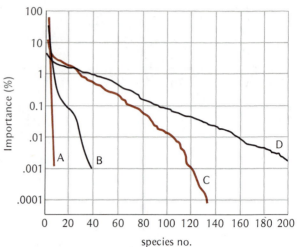

Figure 34-2. Species diversity and relative dominance in (A) a subalpine boreal conifer forest, (B) a moist temperate deciduous forest, (C) a tropical seasonal (dry-season deciduous) forest, and (D) a tropical rain forest. The species in each community were arranged and numbered in order of decreasing relative importance (percent of total primary production or of total biomass), species no. 1 being the most important, and the importance values are shown plotted (on a logarithmic scale) against species number. In (B) and especially (A) there are a few dominant species, responsible for the majority of the community's primary production, whereas in (C) no species accounts for more than about 10%, or in (D) more than 5% of the community, and many more species in all, running into the hundreds, occur in these communities. The contrast is even greater than appears in the graph because for (C) and (D) only trees and shrubs are included, whereas the plots for (A) and (B) also include the herbaceous plants of these communities. After S. P. Hubbell, *Science*, Vol. 203, p. 1299, 1978.

cies (Figure 34-2). Tropical forests indeed often contain hundreds of different tree species (Figure 34-2C, D). Species diversity is one of the important features making natural communities attractive; although several reasons can be given for it, a full explanation of species diversity in natural communities is not yet available.

Ecological Niches

One reason for species diversity in communities is evolutionary specialization or preference of different species for different microenvironments (local variations in conditions) within the area. Hilltops, slopes, swales, valleys, stream sides, and lakeshores, with different amounts of soil drainage, available water, and amount of exposure to the sun are examples illustrating microenvironmental variations that different species can exploit, as are local differences in bedrock or soil development, which make for different soil types and nutrient supplies. These different microhabitats can be called different **ecological niches** within the ecosystem.

Another important kind of niche differentiation has already been mentioned. Different life forms, growth and reproductive patterns, and physiological adjustments enable different plants to exploit different physical or temporal aspects of a habitat, for example, forest floor shade plants that grow successfully beneath a tree canopy. Opportunists are usually present, ready to exploit the results of any human or natural disturbance of the community, any opening created by death or re-

moval of one or more existing individuals in the stand. Thus opportunistic species form part of the vegetation, because no area is entirely free of natural disturbances or accidents, and certainly every member of the community eventually dies, creating a temporary opening.

Stratification

The more luxuriant the vegetation, the more the community tends to become stratified into successive layers composed of different species differing in shade adaptation, tolerance to water stress, and other features. Although some dense conifer woods exclude all undergrowth (Figure 34-10), most coniferous and especially broad-leaved forests have an **understorey** of shade-tolerant shrubs and herbs (Figure 34-3A). Moist tropical

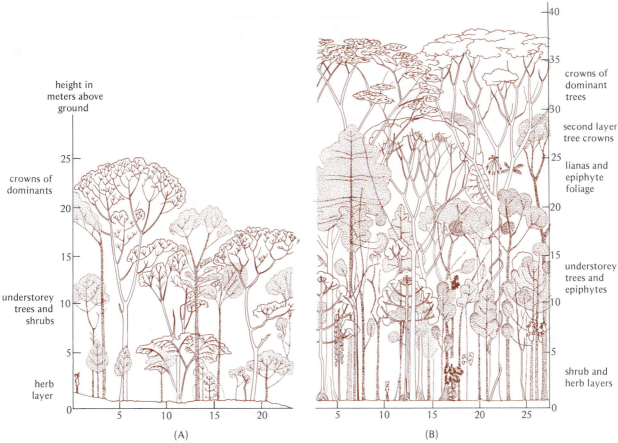

Figure 34-3. Vertical structure of two forest communities. (A) Eastern broadleafed deciduous forest, Massachusetts. (B) Tropical rain forest, French Guiana. Different species are depicted by the different stem and crown forms drawn. Herbs and low shrubs of the ground layer are not shown. After F. Halle, R. Oldeman and P. B. Tomlinson, Tropical Trees and Forests, an Architectural Analysis, Springer Verlag, 1978.

forests develop several layers, from the canopy of tallest trees in full sun through one or more layers of understorey trees in partial shade, down to the forest floor layer of shrubs, herbs, and bryophytes in heavy shade (Figure 34-3B). This layered structure captures virtually all the light energy falling on the area and creates a diversity of ecological niches for different animals within the vegetation.

Shrubby and herbaceous vegetation types often show some, but less pronounced, aboveground stratification. Even these vegetation types, however, are often extensively stratified below ground, different species possessing root systems of different depths and exploiting different soil or subsoil layers (Figure 19-29).

Yet another reason for species diversity in plant communities is that increased abundance of a given species exposes it increasingly to biological hazards, because substantial predator and parasite populations feeding specifically on it can build up. The same type of hazard is involved in crop monocultures (Chapters 18, 33). Especially where parasites and predators are relatively free from climatic restraints, as in the tropics, a high population density of any one species may be selected against, and a wide spacing of the individuals of a given species, which makes it more difficult for predators and parasites to exploit that species, may be favored.

Species Distribution

In the early decades of this century the pioneering American plant ecologist, Frederick W. Clements, taught that plant communities are organic entities, almost superorganisms, with an organization and functional integrity of their own. This view, although influential among some ecologists even today, has been tempered by results

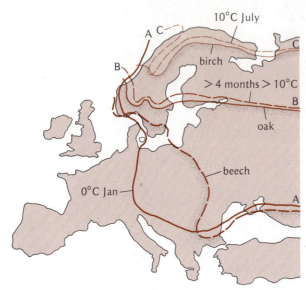

Figure 34-4. Correlation of the limits of different European tree species with temperature features of the environment. The northern limits of beech (*Fagus silvatica*), oak (*Quercus robur*), and birch (*Betula odorata*) are compared with (A) January mean temperature of 0°C, (B) northern limit of area with more than 4 months per year with mean temperatures above 10°C, and (C) July mean temperature of 10°C. The imperfect correlation of the beech limit with (A) implies that other factors besides midwinter temperature contribute to limiting this species' range. After A. S. Collinson, *Introduction to World Vegetation*, Allen & Urwin, 1977.

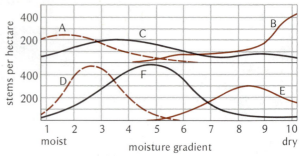

Figure 34-5. Distribution patterns for abundance of different tree or shrub species (designated by different letters) along moisture gradients in a Pacific Northwest mountain forest (upper graph) and an Arizona mountain woodland (lower graph). While certain species may occur under all conditions, others are restricted to wetter or drier locations and each has its own optimum for competition with the other species that are present. Many species show different but overlapping, not mutually exclusive, patterns of distribution. (After R. H. Whittaker, *Communities and Ecosystems*, 2/e, Macmillan, 1975)

of careful research showing that the different species making up a recognized plant association usually have quantitatively different geographical and climatic distributions (Figures 34-4 and 34-5). The distribution of each species is apparently set primarily by its own tolerance range and competitive ability rather than by its role in the plant association. Thus the species makeup of a given type of plant community gradually changes over broad geographical distances, as some species drop out (Figure 34-4) and other new ones appear. The reason different species may occur together in a characteristic association over a considerable area is that their respective climatic adaptations and tolerance ranges happen to be relatively similar, and match a widespread climatic regime. This is the **individualistic** view of the nature of plant associations. Although interactions and interdependencies between community members exist and are important, they are apparently not the principal basis for the characteristic association of plant species in a given environment.

In actuality, many of the interactions between community members are probably negative and would tend to break up the community if it were not for the competitive efforts of its different members to succeed in the habitat. Parasitism and competition for resources are obviously negative interactions. Another is **allelopathy** (Greek *allelon*, of one other; *pathos*, disease), the negative chemical interaction between different plant species mentioned in Chapter 1. Allelopathic effects are due to production and release, usually into the soil (e.g., from leaves when shed), of specialized secondary products that are inhibitory or toxic to germination and growth of seedlings of other species. The great majority of known plant secondary product effects, however, are directed against insects and other animals as defenses against herbivory. Whether allelopathy between plant species is of wide ecological importance cannot yet be said.

The breadth of its tolerance range—in fact, the sum of its genetic variation in tolerance and the plasticity of individual genotypes—determines how broad a geographical range a species could occupy, whereas negative biological interactions such as competition, parasitism, and herbivory reduce this potential range in practice to a narrower one. Species with a narrow tolerance range or limited competitive ability may occur over a considerable area in a specialized ecological niche, or they may be restricted to an unusually limited geographical range, species of the latter type being called **endemics** (Figure 34-6).

Historical factors also enter importantly into endemism. Species or even genera and families that have evolved in an isolated habitat such as a remote island, a landlocked lake or stream system, or a high mountain surrounded by extensive lowlands may be endemics merely because their habitat is of restricted

Figure 34-6. The Monterey cypress (*Cupressus macrocarpa*), an example of a narrow endemic species, at Cypress Point, Monterey Peninsula, California. This photo takes in virtually all of this species' natural range, limited to the seacoast bluffs at Cypress Point and at Point Lobos, seen in the distance on the other side of Carmel Bay. The cypress is restricted to a narrow zone where salt spray from the ocean drenches the land during storms.

TABLE 34-1. Biome-type Classification

Biome Type	Growth Form of Dominants
Closed forest	complete tree canopy, evergreen or deciduous
Woodland (open forest)	tree canopy covers >40% of area, but crowns mostly not touching one another
Savanna	grass or other herbs plus <40% tree cover
Scrub	woody dominants 0.05–5 m tall
Dwarf scrub (heath)	woody dominants <0.5 m tall
Grassland	grass + other herbs, few if any shrubs or trees
Swamp (marsh, bog)	submerged and/or emergent herbaceous hydrophytes, sometimes + shrubs or trees

Simplified from a world biome classification scheme adopted by UNESCO (1973).

extent. Endemism is particularly common on remote islands, on which evolution has proceeded independently of any contact with a mainland for a considerable period, new species arising from a few immigrant species to fill ecological niches already occupied by other species on the mainland. For example, over half of the woody plant species on Lord Howe Island, a small oceanic island about halfway between Australia and New Zealand (over 500 km from either), are endemic species that occur nowhere else, and the island boasts no less than three endemic *genera* of palms. Such situations helped Darwin deduce his theory of organic evolution (Chapter 21).

Biomes

Although in different areas different plant species may dominate a particular vegetation type such as deciduous forest, coniferous forest, or grassland, each of these kinds of vegetation represents a particular form of plant growth adapted to a certain broad environmental regime, and each tends to be inhabited by certain kinds of animals adapted to that kind of environment and habitat. Each of these great groups of communities with a

more or less common form and structure is called a **biome.** A biome includes both the plants and the animals of the region but is named according to the type of climax (or preclimax) vegetation characterizing it, such as the deciduous forest biome or the grassland biome.

Biomes can be classified (Table 34-1) according to their extent of tree or shrub development (related mainly to moisture supply but affected also by temperature) and the vegetative form of their dominant woody plants (or herbaceous plants, if dominant). This is a very oversimplified classification; a number of biome types are recognized within or between some of the categories in Table 34-1, which provides only the most general overview of biome diversity.

PLANT COMMUNITIES OF MAJOR TERRESTRIAL BIOMES

Figure 34-7 shows the world distribution of major biome types. The general zonation of biomes from north to south is due to variation with latitude in temperature and length of growing season. Toward the equator this latitudinal zonation is greatly modified, however, by varying amounts of precipitation.

Tundra

Tundra is the low-growing vegetation of arctic and alpine areas beyond the tree limit (Figure

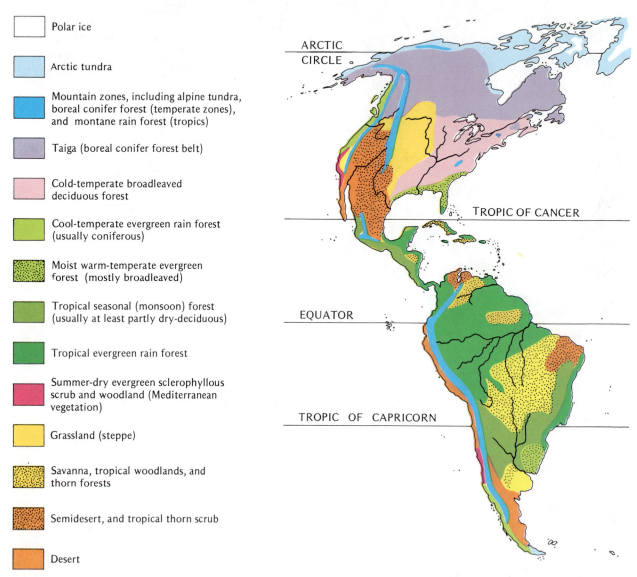

Polar ice

Arctic tundra

Mountain zones, including alpine tundra, boreal conifer forest (temperate zones), and montane rain forest (tropics)

Taiga (boreal conifer forest belt)

Cold-temperate broadleaved deciduous forest

Cool-temperate evergreen rain forest (usually coniferous)

Moist warm-temperate evergreen forest (mostly broadleaved)

Tropical seasonal (monsoon) forest (usually at least partly dry-deciduous)

Tropical evergreen rain forest

Summer-dry evergreen sclerophyllous scrub and woodland (Mediterranean vegetation)

Grassland (steppe)

Savanna, tropical woodlands, and thorn forests

Semidesert, and tropical thorn scrub

Desert

Figure 34-7. Distribution of major biomes on the larger land masses. Note that on such a small scale map many significant variations in vegetation cannot be shown.

34-8). The growing season is very short, often less than 8 weeks; frost and snow can occur at any time, even in the middle of the growing season. The soil temperature usually remains cold throughout the growing season, and in many areas permanent ice (permafrost) occurs a short distance below the soil surface, only the top few centimeters of the soil thawing out during the brief summer and being available to roots. Severe winds are frequent. Although much precipitation usually falls in the alpine zone of high mountains, relatively little falls on lowland areas in much of

the arctic, often as little as 10 cm, which is as little as in many deserts. This vast region, in which not only cold but also water stress is often severe, is called the "polar desert." Because of the combination of short growing season, frequently unfavorable conditions even during that season, and the sparseness of the tundra vegetation, its productivity is rather low (Table 34-2).

Compared with other kinds of vegetation, the tundra is poor in numbers of plant species, no doubt because of the severity of the conditions that would-be inhabitants must face. Tundra vege-

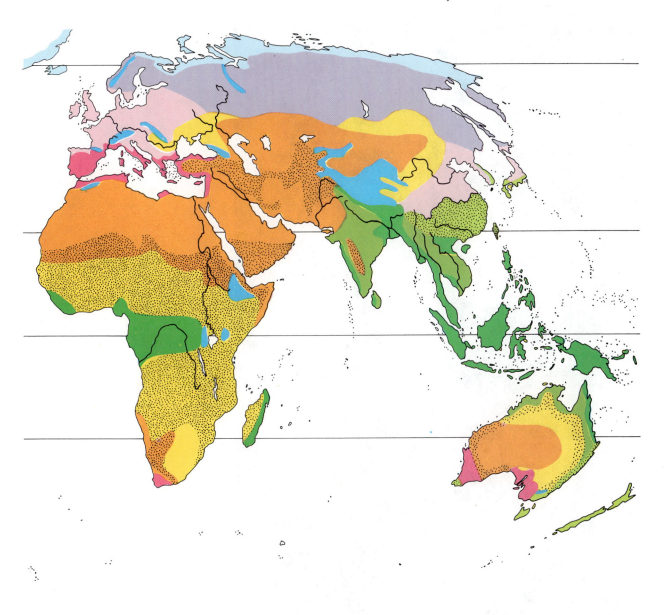

tation consists chiefly of grasses, sedges, and dwarf shrubs such as the arctic dwarf birch (*Betula nana*, Figure 34-9A). In the arctic, lichens (Figure 34-9B) and mosses comprise an important part of the ground cover and are a major food source for herds of caribou and reindeer. Another distinctive growth form is "cushion plants," densely tufted and low to the ground with an attractive pincushion appearance that is evidently adaptive against the harsh winds. Many cushion plants are dicots, whose bright flowers often open right at the surface of the cushion (Figure 34-9C).

A number of arctic species are circumpolar in distribution, occurring in North America, northern Scandinavia, and northern Siberia. This is notable because very few other land plant species occur on different continents, except for crop plants, weeds, and ornamentals that have been transported by humans. Some of the circumpolar arctic species such as alpine sorrel (*Oxyria digyna*, Figure 34-9D) extend southward into the alpine zones of major mountain ranges like the White Mountains of New England, Rocky Mountains and Sierra Nevada of western North America, and the Alps and Pyrenees of Europe. The alpine populations of these species are ecotypes that are physiologically

Figure 34-8.   Tundra vegetation, Alaska.

and genetically different from the arctic ecotypes of the same species, with differences adapting them to these respective kinds of tundra environments.

In the tropics a tundra-like biome occurs only on the highest mountains. It is very different from the arctic–alpine tundra, however, because near the equator seasonal changes are minimal. The high mountain plants experience severe winds, and night-time frosts throughout the year that are sufficient to prevent tree growth, but they are never subject to the extremely low temperatures that occur in winter in temperate zone alpine areas or in the arctic. The equatorial high-mountain, tundra-like vegetation is made up of genera and species almost entirely different from those of the arctic–alpine tundra.

Very little tundra vegetation occurs in the southern hemisphere, except in high mountains such as the Andes and the highest mountaintops in New Zealand and Australia, because most of the subpolar zone that in the northern hemisphere is home for the tundra is occupied by ocean in the southern hemisphere. Antarctica has such a forbidding climate and nearly complete

TABLE 34-2 Primary Productivity of Major Biomes

Biome Type	Mean Net Primary Productivity g/m²·yr	Global Area 10⁶ km²	Global Annual Production 10⁹ tons/yr
Tundra (arctic + alpine)	140	8	1.1
Boreal forest (taiga)	800	12	9.6
Temperate deciduous forest	1200	7	8.4
Temperate evergreen forest	1300	5	6.5
Tropical rain forest	2200	17	37.4
Tropical seasonal forest	1600	7.5	12.0
Tropical/subtropical woodlands	600	7	4.2
Evergreen mediterranean scrub[a]	800	1.5	1.2
Savanna & tropical grassland	900	15	13.5
Temperate grassland	600	9	5.4
Semidesert & desert scrub	90	18	1.6
Extreme dry desert	3	8.5	0.1
Agricultural land (cultivated)	650	14	9.1
Lakes and streams	400	2	0.8
Swamps & marshes	3000	2	6.0
Total continental	*875*	*133.5*	*116.9*
Oceans (overall average)	*152*	*361*	*55.0*

[a]Includes chaparral

Data from H. Lieth and R. H. Whittaker, *Primary Productivity of the Biosphere*, Springer, 1975, Tables 10-1 and 15-1. Available data for many of the biome types are limited and/or include a wide range of productivity values, so the means and the global annual production figures calculated therefrom are at best only approximate.

blanket of perpetual snow and ice that no proper vegetation exists there, no more than two species of flowering plants being known to occur on the continent, plus a number of mosses and lichens.

Coniferous Forest

Extending across North America, northern Europe, and Siberia south of the tundra is a vast belt of coniferous forest called the **boreal forest** or, from its Russian name, the **taiga** (Figure 34-10). Spruces (*Picea*) and firs (*Abies*) usually dominate the climax, and birches and alders are important pioneer species.

This zone generally has a climate only a little less harsh than that of the tundra, with short, mostly cool, moist summers and long, extremely cold winters. Although the dominant trees are evergreen, their productivity is relatively low because they cannot photosynthesize to any significant extent during much of the year. Not only does the temperature in winter usually not permit it, but these conifers' capacity for photosynthesis is actually shut down in winter as part of their adaptation to winter conditions.

From the boreal belt, zones of coniferous forest extend southward along mountain ranges at increasing elevations. In the drier parts of the coniferous forest zone, pines, which can tolerate less moist conditions than those required by firs and spruces, often replace the latter and form extensive forests, for example at middle elevations and on the eastern sides of high mountains in the western United States (Figure 34-1).

Another arm of coniferous forest extends south from the boreal region along the cool moist Pacific Coast in British Columbia, Washington, Oregon, and northern California. This region has a climate much milder than that of the boreal belt. The spruces and first of the latter are replaced farther south by other conifers, especially Douglas fir (*Pseudotsuga*, not a true fir) and, in the southern part of this region, the California Coast Redwood (*Sequoia sempervirens*). This type of giant coniferous rain forest is sometimes called a **temperate evergreen forest,** a term that encompasses the middle-elevation mountain conifer forests mentioned above. It also includes both broad-leaved and coniferous cool-climate evergreen rain forests of the southern hemisphere, including very tall Eucalyptus forests in southeastern Australia and forests of exotic southern-hemisphere conifers such as *Araucaria* (the "monkey-puzzle tree" and its relatives, Chapter 30) in the Andes and elsewhere. A typical taiga biome of significant size does

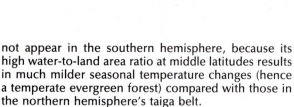

Figure 34-9. Tundra plants. (A) Dwarf shrubs, including arctic birch (*Betula nana,* with yellow catkins) growing almost flat along the ground, and crow-berry (*Empetrum nigrum*), the compact reddish heather-like subshrub bearing occasional black berries. (B) Fruticose lichens, including reindeer moss (*Cladonia alpestris*). Crustose lichens occur on the rocks in (A). (C) *Diapensia lapponica,* a "cushion plant" found in both the European (Lappland) and North American arctic, and in alpine habitats as far south as Mt. Washington in New Hampshire. (D) Alpine sorrel (*Oxyria digyna*), also circumpolar and in the alpine zone of the Rocky Mountains and ranges of the American west coast as far south as southern California.

is little utilized for agriculture compared with the zones of broad-leaved forests and grasslands.

Deciduous Broad-leaved Forest

South of the boreal forest in the more humid parts of the northern hemisphere, including eastern North America, Europe, and parts of China and Japan, is a broad zone originally occupied by deciduous, broad-leaved forest adapted to a cold-temperate climate (Figure 34-11). Because of extensive exploitation of this zone for agriculture, today this forest in most areas consists only of isolated remnants and successional stages on abandoned farm land. This comprises the arbo-

not appear in the southern hemisphere, because its high water-to-land area ratio at middle latitudes results in much milder seasonal temperature changes (hence a temperate evergreen forest) compared with those in the northern hemisphere's taiga belt.

Microbial decay of coniferous litter leads to a very acidic soil (pH 4.5–5.0) compared with the soils under broad-leaved forests or grasslands (mostly pH 6–7). This is another reason, beside the climate, why the taiga belt

Figure 34-10. Boreal conifer forest, or taiga vegetation, interior Alaska. The forest is a mixed stand of spruces and firs.

Figure 34-11. A moist deciduous hardwood forest in the Palisades, New Jersey. Trees present include tulip tree, maple, birch, and hemlock (right), with a shrub understorey.

rescent vegetation of much of the United States east of the Mississippi River.

The deciduous forest is characterized by a seasonality that looks, because of its winter leaflessness, more extreme than that of the evergreen boreal forest. Both kinds of vegetation, as noted already, are actually dormant during the winter. The growing season is considerably longer and warmer in most parts of the deciduous forest belt than in the boreal forest zone. Compared with most conifer forests, deciduous broad-leaved forests generally have a well-developed understorey of forest floor shade plants (Figure 19-11). Because of the large amounts of water required to support the transpiration of broad-leaved deciduous trees, this kind of vegetation depends upon summer rainfall and cannot develop extensively in areas such as central Asia and western North America where dry summer conditions prevail on the warm side of the boreal coniferous belt.

Although the Asian, European, and American deciduous forests consist almost entirely of different species, a number of genera occur throughout much or all of the area, including the oaks (*Quercus*), chestnuts (*Castanea*), beeches (*Fagus*), birches (*Betula*), ashes (*Fraxinus*), and elms (*Ulmus*). A common origin for these forests can be perceived, therefore, even though they have now substantially diverged evolutionarily.

The hardwood species composition varies within the American deciduous forest belt. Maple, beech, and birches are prominent in the northeast, while oak and hickory are more prominent in the west and south, where other hardwoods such as tulip tree (*Liriodendron*) and sweetgum (*Liquidambar*) are also important.

Conifers occur to some extent in many deciduous forests. In the northeastern hardwood forests of the United States and Canada, white pine (*Pinus strobus*) occurs widely, mainly as a successional or pioneer species, and hemlock (*Tsuga canadensis*), a shade-tolerant species, is an important and characteristic member of the climax community (Figure 34-11). In the southeast-

ern states, various other pines occur either as successional stages (Table 33-1) or on poorer soils to which hardwoods are not well adapted.

SUBTROPICAL AND TROPICAL SEASONAL FORESTS

At lower latitudes where winter temperatures stay mostly above freezing, the deciduous forest becomes replaced, in humid regions, by a subtropical, evergreen broad-leaved forest (see below). However, even in the tropics, drought-deciduous "seasonal" forests are encountered in areas that have a dry season alternating with a season of copious precipitation. Their trees generally belong to tropical groups and are unrelated to the dominant trees of the cold-temperate deciduous forest. Southeast Asia and India carry extensive areas of tropical seasonal forest, called "monsoon forest" because the rain-bearing monsoon winds set its seasonal growth cycle.

Although tropical drought-deciduous forests are found widely in the southern hemisphere, cold-temperate deciduous forests do not occur there to any substantial extent for the same reason that a taiga biome is lacking. As noted above, the moist forests of cooler latitudes in the southern hemisphere tend to be predominantly evergreen.

Tropical Broad-leaved Evergreen Forest (Tropical Rain Forest)

Where humid conditions and temperatures favorable for photosynthesis occur year round, as in subequatorial and equatorial regions such as Central America, the Amazon basin in South America, the Congo basin in Africa, and southeast Asia and Indonesia, plant productivity (Table

Figure 34-13. Tropical epiphytes. (A) Epiphytic orchid, showing thick, spongy roots which soak up rain when it falls on them. (B) A bromeliad (family Bromeliaceae) with cup-shaped leaf bases that catch and retain rainwater.

Figure 34-12. Interior of a tropical rain forest in the upper Amazon basin. Trees carry lianas and epiphytes, including *Monstera* (familiar as a houseplant); ferns are prominent in the herbaceous understorey.

34-2) and the complexity of vegetation reach a zenith. Dense, tall tropical rain forests or jungles (Figure 34-12) can consist of hundreds of tree species forming a multilayered canopy (Figure 34-3). Ferns, including tree ferns, are often prominent. Woody vines (*lianas*) clamber up into the treetops. **Epiphytes,** plants specialized for growing attached to the branches of trees (Figure 34-13), often festoon the forest.

Epiphytes (Greek *epi*, upon; *phyton*, plant) are herbaceous plants that grow with their roots attached to the bark of trees or embedded in mosses and liverworts growing on the trees. Mosses and lichens, which of course lack true roots, occur as epiphytes in most forest biomes, especially the more moist ones. Epiphytic vascular plants are important mainly in tropical forests, where frequent precipitation occurs. Because of the extreme absorption of light by the multilayered canopy of a tropical rain forest, in competing for light it is advantageous for a herbaceous plant to locate itself far above the ground, closer to the treetops. An ephiphyte depends on concurrent precipitation for its water supply. Many epiphytes possess stocky roots, often as thick as a lead pencil, with an extensive

spongy cortex that soaks up rain as it falls (Figure 34-13A). Others possess cup-shaped leaves or leaf bases that catch pools of rain water (Figure 34-13B). However, because they lack access to soil water storage, epiphytes soon run short of water during rainless periods. Presumably for this reason, almost all epiphytes have xeromorphic features. Many of them have CAM metabolism, which improves their water conservation (Chapters 8, 19). Certain flowering plant families have specialized as epiphytes or at least have an unusually high proportion of members that are epiphytes, notably the orchid family and the bromeliads (Bromeliaceae) among the monocots and the cactus family among dicots. Many ferns also occur as epiphytes.

A tropical rain forest requires at least 150 cm (60 in) of rainfall, well distributed over the year, to flourish. The humidity is often above 80% and the temperature above 25°C. Although seasonality is much less marked than in the temperate or boreal zones, the rate of precipitation commonly varies considerably over the course of the year.

Even in continuously moist tropical forests some trees may shed their leaves and become bare for a more or less brief period before beginning a new growth cycle. Different trees do not shed at the same time, however, so the forest presents an evergreen ap-

Figure 34-15. Savanna vegetation, with native village, Upper Volta, Africa. Picture was taken during the dry season; some trees are shedding their leaves while others remain green.

Figure 34-14. A *Eucalyptus* woodland in Australia, with grass understorey, dried up because the photo was taken during the dry season. The broad-leaved trees are evergreen, and have survived fires as shown by the blackening of their trunks. Many eucalypts can survive a crown fire that burns their entire tops, sprouting new buds from the bark of the unkilled trunk and larger branches.

pearance. Annual flowering cycles may be tied in with this seasonal leaf fall and new growth; some trees flower only when bare of leaves. Certain trees shed the leaves on different branches at different times, so one branch may be bare while another on the same tree is clothed with leaves and a third is leafless but in full bloom. What controls these irregular cycles of dormancy and growth of tropical trees is not well understood.

Unlike temperate deciduous forests, tropical forests on different continents have a widely different composition in terms of the genera and botanical families represented. Most tropical forest trees belong to botanical families that are scarcely represented in temperate latitudes. The exotic character of the plant growth impresses visitors from the north.

Although the lushness of a tropical rain forest suggests a very fertile soil, this appearance is deceiving. The forest indeed typically contains a large amount of nutrients such as nitrogen and phosphorus per hec-

tare, but intense competition of the roots for these nutrients in the soil, plus heavy leaching by precipitation that washes away any nutrients that are not absorbed, keeps the soil itself in a nutritionally depleted state. Most of the ecosystem's nutrient capital is locked up in its vegetation. Growth of the vegetation depends on immediate cycling of mineral nutrients from litter (falling leaves, branches, fruit, dead trees, and so on) by decomposers, which are very active at the prevailing warm temperatures. Consequently, very little humus accumulates in the soil.

Woodlands and Savanna

Vast areas in the tropics and subtropics lie outside the belts of high year-round precipitation that permit a tropical evergreen forest to develop; they receive a highly seasonal and often erratic precipitation of about 50–150 cm, with a lengthy dry season. This climatic regime supports a vegetation of scattered, usually relatively short, trees standing in a ground cover usually of grasses (sometimes shrubs). This vegetation is called a **woodland** (Figure 34-14) if the tree canopy covers more than about 40% of the area, and a **savanna** (Figure 34-15; from Spanish, *sabana*) if the coverage is less than 40%. The grasses and other herbs grow during the wet season and dry up during the dry season, whereas the trees are deep-rooted and obtain enough water to remain photosynthetically active during at least a part of the dry season. The famous African big-game regions south and east of the Congo are part of this biome complex, as is the Sahel belt between humid central Africa and the Sahara Desert to the north. Woodlands and savannas also occur south of the Amazon rain forest in Brazil and north of it in Venezuela (the Llanos). In North America, wood-

Figure 34-16. Oak savanna or sparse woodland, in lowland California. Vegetation on hill to left has recently experienced a dry-season grass fire. Note that at least part of the top of most of the trees remains green, showing that they survived the fire.

Figure 34-17. Chaparral scrub vegetation, in the mountains of southern California. A few trees are growing on the shaded slope to the left, which is more protected from the sun as evidenced by the persistence of recent snow from an unusual January snowfall. The chaparral is a mixed population including several shrub species that do most of their annual photosynthesis during the winter rainy season. Many of them can survive and regenerate after fires by sprouting from thickened root structures located just below ground.

lands and savannas extend from the semiarid parts of Texas south into Mexico and on into relatively dry interior locations in Central America; they cover extensive areas in lowland California (Figure 34-16), upland Arizona, and southern Oregon away from the coast.

The trees and herbaceous plants making up savanna and woodland vegetation vary greatly in different areas. The trees may be winter-deciduous (in cool- or cold-winter woodlands), evergreen [either broad-leaved (Figures 34-14 and 34-16) or conifers (Figure 34-22)], or drought-deciduous. During the dry season, when the ground cover is dry, savannas frequently (annually, in many regions) experience fire, which is often started intentionally by humans as a means of favoring growth of grasses that can be used for pasture. Savanna tree species must be able to tolerate the occurrence of fire around their trunks. Because the trees are mostly isolated from one another, grass fires usually do not burn their tops (Figure 34-16).

The density of tree cover in a savanna tends to increase with the amount of precipitation, and on the high precipitation side it usually grades into a woodland. However, fire also influences tree density, since the closer together the trees are, the more likely a grass fire is to become a crown fire, moving through the woodland and killing the trees. Repeated human-started fires in many savanna regions kill most tree seedlings each year and keep the tree density from increasing. The savanna is then a fire preclimax. Much of the very extensive area of savanna in Africa (Figure 34-7) would become a seasonal woodland or tropical seasonally deciduous forest were it not for heavy grazing and frequent fires.

Many woodland and savanna trees possess some degree of xeromorphic specialization. An extreme is

reached in certain very thorny tropical woodlands called thornwoods or thorn forests (if denser), vegetation that is difficult or impossible for humans to travel through. The woodlands of less extremely xeromorphic, broad-leaved but sclerophyllous (see Chapter 19) evergreen oaks ("live oaks") and other small trees of summer-dry mountains back of the Pacific Coast in California and southern Oregon yield at lower elevations to a dense broad-leaved evergreen scrub (shrubby) community called the **chaparral** (Figure 34-17); it occurs extensively in the Pacific Coast ranges, especially in central and southern California. Communities of very similar form but entirely different species occur in the similar cool moist winter/warm dry summer climates of the Mediterranean region, South Africa, and coastal Chile. Because it becomes very dry in summer, this is essentially a "fire vegetation," adapted to suffering fire at irregular intervals, as the economically destructive summer and autumn brush fires of recent years in California have brought to national attention. Many of the shrubs are specialized for surviving fires by subsequently sprouting from unkilled, much-thickened roots beneath the ground (Figure 19-3).

The chaparral and similar evergreen sclerophyll scrub vegetation types, plus broad-leaved evergreen sclerophyllous woodlands such as those found widely in the southwestern United States and similar extensive woodlands in Australia (Figure 34-14), are collectively called **broad sclerophyll vegetation.**

Grassland

In the temperate and cold-temperate latitudes, precipitation that is marginal (30–100 cm),

Figure 34-18. Short-grass prairie vegetation near Browning, Montana; one of the Rocky Mountain ranges in the distance.

Figure 34-19. Severe desert landscape on the perimeter of the Sahara in Algeria. A few small trees survive in the Wadi (dry streambed), probably as phreatophytes (plants that obtain water from a water table deep underground).

seasonal, and erratic engenders a treeless vegetation comprising a dense growth of perennial grasses plus other herbaceous plants (Figure 34-18). The prairies of the Great Plains of central North America, the steppes of southern Russia and central Asia south of the taiga forest, and the pampas of Argentina are important examples. Modern humans have destroyed much of the grassland biome for use in agriculture because it occurs where conditions are favorable for growing wheat and other small-grain crops.

In its primordial condition the grassland was, like savannas, extensively grazed by herbivorous mammals and was regularly subject to dry-season fires. Drought-dormant grasses tolerate these fires, which limit the invasion of trees along the grassland's more humid margins and of desert shrubs along its drier borders, so the grasslands are in part a fire preclimax. The grass species involved vary from those standing more than a meter tall in the "tall-grass prairies" of the moistest grassland regions to short bunchgrasses (grasses occurring as clumps or tufts) in the "short-grass prairies" of the driest areas, with species of intermediate stature occurring between these extremes. Although the general impression of this biome is one of treelessness, trees occur locally through most of the grassland biome, along stream banks and at breaks in the level of the land, where steep slopes partly sheltered from the sun afford a more moist habitat that offers some protection from fires.

Grasses tolerate grazing very well, but only as long as the grazing animals are so few that grass growth can keep pace with consumption by the animals. The lower the water supply, the lower this cutoff point, so the problems of overgrazing are worst at the dry end of a rangeland belt, where it borders on a desert. Since the tough and often thorny or aromatic (from secondary products) shoots of desert shrubs are unpalatable to cattle, excess grazing pressure selects in favor of the

shrubs. Overgrazing has thus converted large areas of range land in the western United States and elsewhere essentially into desert scrub vegetation.

Desert

An annual precipitation of less than about 25 cm in areas of warm climate can support only a sparse growth of usually shrubby, often thorny xerophytic plants, called a desert scrub. The most extreme, almost lifeless deserts (Figure 34-19) occur in a belt at about 20° N latitude extending from North Africa (the Sahara Desert) through the Red Sea and Persian Gulf area to the Gobi Desert of Mongolia, and in a band about 100 km wide along the west coast of South America in Peru and Chile. These areas usually receive less than 10 cm of rain per year, many places getting no rain at all for years at a time. The North American "semideserts" of northwestern Mexico and the southwestern United States are somewhat less extreme because many areas receive an average of as much as 25 cm of rain, which is, however, very erratic and unreliable from year to year. Semideserts are inhabited by a wide variety of specialized plants spread rather thinly over the surface (Figure 34-20). Large semideserts also occur in southwest Africa and in central to western Australia. Because of the sparse cover of vegetation and factors inhibiting its photosynthesis (noted below), the productivity of desert vegetation is very low (Table 34-2). Recovery from disturbance is correspondingly slow, and the ecosystem is consequently very fragile.

Chapter 19 considered various adaptations of desert plants, or xerophytes, to water shortage. Strikingly, most of these adaptations occur in all desert areas, even though the species and usually

Figure 34-20. Southwest American warm semidesert ("low desert") vegetation. In the foreground is a creosote bush (*Larrea tridentata*), an evergreen xeromorphic shrub with yellow flowers; the darker shrubs in the background are the same species, which predominates over much of the warm semidesert area. The tall straight-stemmed shrub is the fiercely thorny, drought-deciduous "ocotillo" (*Fouquieria splendens*; see Figure 19-25). Cacti, although not seen in this view, are also commonly encountered in the low desert.

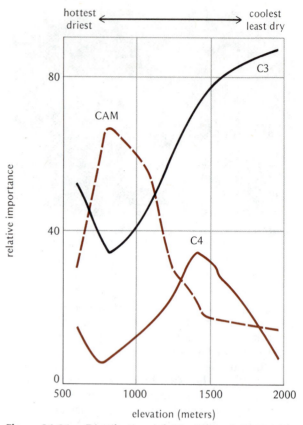

Figure 34-21. Distribution of C3, C4, and CAM (succulent) plants in the desert and semidesert vegetation of Big Bend National Park, Texas, along a moisture and temperature gradient due to differences in elevation. CAM plants with nocturnal gas exchange for water conservation predominate in the hottest, driest environment, C4 plants are maximally important under intermediate temperature and moisture conditions, and C3 plants predominate at the cooler, least dry end of the gradient. After data of W. B. Eickmeier, *Photosynthetica*, Vol. 12, pp. 290–297, 1978.

even the botanical families involved are different. Desert plants thus provide numerous examples of convergent evolution.

The typically high summer temperatures in most deserts compound the plants' water and photosynthetic problems. C4 photosynthesis, with its relatively high temperature optimum, helps many desert plants adapt to this. Surveys indicate that the proportion of C4 species in desert vegetation increase at lower, hotter elevations (Figure 34-21). However, C4 photosynthesis is no more useful than the C3 type when no water is available at all. In the warmest areas, water-storing succulents with CAM metabolism, which permits nocturnal instead of diurnal gas exchange resulting in substantial water conservation, are favored (Figure 34-21). But even in the warmest desert areas many of the plants still have C3 photosynthesis, remaining dormant photosynthetically during the long season of high temperature and water shortage.

In North America, desert or semidesert conditions extend from the low southwestern hot desert area (southern Arizona and New Mexico, and adjacent Mexico), where the dominant plant is the creosote bush (*Larrea tridentata*, Figure 34-20), to the high or "cool" desert of the upper Great Basin, where sagebrush (*Artemisia tridentata*) dominates vast open spaces. Elevated areas with somewhat more precipitation support an open "pinyon-juniper woodland" of small conifers (pinyon pine, *Pinus edulis*, and juniper, *Juniperus* sp.) with sagebrush between them (Figure 34-22).

Semidesert vegetation can include small, sometimes phreatophytic, xeromorphic trees ("subtrees") such as Joshua trees (Figure 13-28), mesquite, and palo verde. Large trees, however, usually occur only at oases (springs) or along water courses, which are usually dry at the surface during the dry season but offer subsur-

face water that can be tapped by deep-rooted phreatophytes (Chapter 19). The desert cottonwood (*Populus fremontii*) commonly occupies this ecological niche in the American deserts.

PHYSICAL DETERMINANTS OF BIOMES AND THEIR BOUNDARIES

Figure 34-23 shows roughly how different biomes relate to annual precipitation and mean temperature. Regions of lower mean temperature receive a lower range of possible precipitation values because cool air holds less moisture; this explains why the pattern of biomes rises higher on the right side of Figure 34-23. However, at cooler temperatures evaporation is less, so the same total precipitation supports a much more mesic biome

Figure 34-22. Great Basin high semi-desert vegetation near Scipio, Utah. The low grey shrubs are sagebrush (*Artemisia tridentata*), dominant over much of the "cool desert." In the foreground are low-growing juniper (center) and pinyon pine (*Pinus edulis*, right) trees, forming pinyon-juniper woodland on lower hills in distance.

in a cool climate than in a warm one, as can be seen by comparing the biomes intersected by any horizontal line drawn across the lower part of the diagram. An important factor that Figure 34-23 does not take into account is seasonality. Differences in seasonal distribution of precipitation and heat can lead to different biomes in different regions of equal mean temperature and total precipitation.

Boundaries Between Biomes

Contrary to the impression given by Figures 34-7 and 34-23, different biomes are not always separated by sharp boundaries but often intergrade. For example, forest often grades into woodland and woodland into savanna simply by a decrease in the density and height of the tree cover; savanna grades into grassland as trees become scarce and finally disappear altogether. Such gradual changes in vegetation over distance are

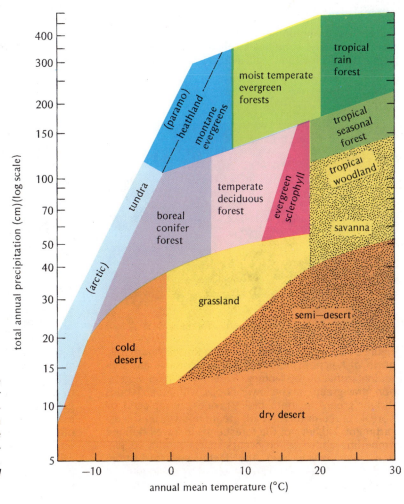

Figure 34-23. Approximate temperature and precipitation relationships of major biome types. The diagram is blank at upper left because this much precipitation is never received in colder regions, due to the smaller water vapor holding capacity of the air at lower temperatures. (Adapted from O. Stocker, *Israel Journal of Botany*, Vol. 13, pp. 154–165, 1964.)

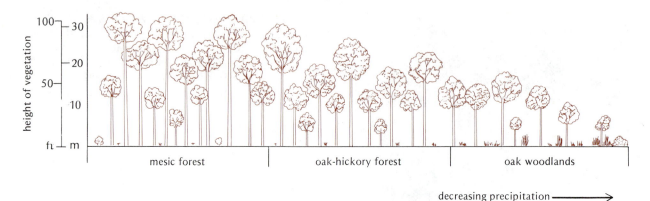

Figure 34-24. Diagram of an ecocline, representing the transition from moist cold-temperate deciduous forest through grassland to desert scrub along a gradient of precipitation. The transition pictured might take place over a distance varying from dozens to hundreds of miles depending upon how rapidly precipitation changes with location in the region in question. (After R. H. Whittaker, *Communities and Ecosystems*, 2/e, 1975.)

called **ecoclines** (Figure 34-24). They generally occur along a climatic gradient, such as a line placed in any direction on Figure 34-23. Ecoclines presumably reflect how climatic and biotic selection pressure change gradually over geographical distances.

Certain communities or biomes interface across a sharp boundary, sometimes called an **ecotone**, rather than gradually intergrading (Figure 34-25). In some cases

the transition is due to a locally rapid change in one or more determining conditions. For example, a coastal salt marsh biome (a specialized biome type not covered in the preceding survey) normally has sharp boundaries because it extends just up to the high tide line. Boundaries between soil types can engender sharp changes in vegetation (Figure 19-8). Sharp boundaries between grassland (or savanna) and woodland (or forest) apparently often represent the limit of regular dry-season grass fires.

Figure 34-25. An escarpment in the Rocky Mountains, north of Ketchum, Idaho, showing ecotones and altitudinal zonation of biomes. Sagebrush scrub vegetation (gray-green) on the lower slopes, with aspen woods (bright green) in more moist ravines, is replaced by boreal spruce-fir forest (dark green) midway up the escarpment. Higher up, the forest ends in a well-defined timberline above which occurs sparse tundra vegetation (seemingly barren). The deeper soils of bottom lands in the foreground are occupied by meadows and a lodgepole pine forest community (olive green).

Figure 34-26. At timberline on a high mountain peak at an elevation of about 3000 m (10,000 ft). The trees (white-bark pines, *Pinus albicaulis*) here are reduced by severe winds and winter cold to almost a carpet, a growth form called krummholz ("bent wood," in German), any shoots that grow taller being periodically killed off (dead snags visible in picture). Tundra vegetation on peak above consists of sparse clumps of grass and other herbaceous plants. Timberline presumably represents the tolerance limit for winter survival of tree species. San Joaquin Mt., Sierra Nevada, California.

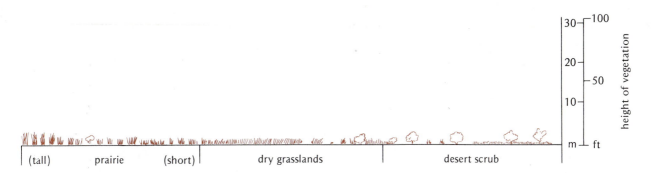

| (tall) | prairie | (short) | dry grasslands | desert scrub |

Effects of Topography

Mountains, valleys, escarpments, and other kinds of relief are responsible for many of the noticeable and attractive variations in natural vegetation; they affect biome boundaries, making them actually much more complex than can be shown in a small-scale map like Figure 34-7. One factor, drainage of water downhill, which tends to make low areas more moist than the surrounding higher ground, has been mentioned above and in Chapter 19. A second and dominant influence is that of elevation, or altitude. The average temperature decreases by about 6°C per 1000 m of rise in altitude (about 3.3°F per 1000 ft). Total precipitation, furthermore, normally increases markedly with altitude. The combination of these changes causes a zonation of different communities or biomes with altitude in mountain ranges (Figure 34-25), which is equivalent to moving upward and to the left in Figure 34-23. This zonation at least superficially resembles that of biome types with latitude, 300 m (1000 ft) of altitude being equivalent to about 400 km (250 miles) of north-south difference in location. Thus, similar vegetation zones, and the tree limit or **timberline** (Figure 34-26), occur at lower elevations farther north (in the northern hemisphere) (see Figure 34-27).

Another biologically important influence of mountains is the **rain shadow effect.** Air cools as it rises over a mountain, causing it, if it is moist, to lose moisture as precipitation. As the air descends on the leeward side of the mountain it warms up, increasing its moisture-holding capac-

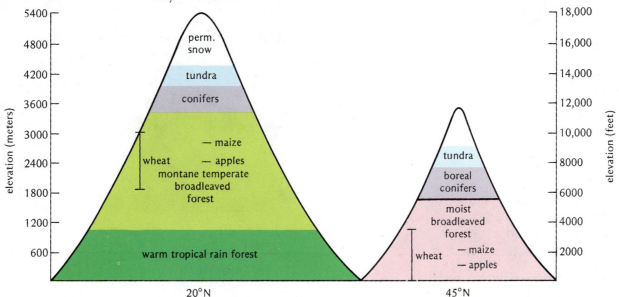

Figure 34-27. Montane zonation of biomes and crops, (left) in a moist tropical region (latitude 20°, e.g., central Mexico) and (right) in a moist temperate region (latitude 45°, e.g., northern United States or central Europe). Marks show the upper limits of cultivation of maize and apples, and the bar shows the altitudinal range over which wheat may be cultivated.

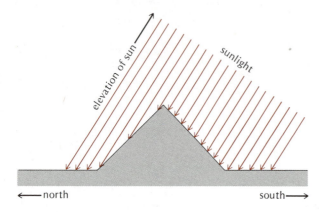

Figure 34-28. Effect of direction of slope on amount of solar radiation received in the north temperate zone. Because of the southerly inclination of the sun in the sky, a uniform flux of solar energy is spread much more thinly on a north-facing than on a south-facing slope, which is consequently warmer and drier.

ity, which causes precipitation to stop. Thus the front side of a mountain range (toward prevailing winds) always receives more precipitation than the lee side, shifting the biome zones relative to altitude on the two sides of the mountain or resulting in entirely different biomes on the two sides of a mountain range. Beyond a mountain range there is usually a more or less extensive, relatively dry or even arid region, the "rain shadow." This effect is responsible, for example, for the semi-desert conditions prevailing over most of the Great Basin.

Still another important effect of topography is that of slope exposure to the sun. At latitudes well away from the equator (above about 25° latitude, which includes all of the United States), the sun always stands south of the zenith at midday (in the northern hemisphere; vice versa in the southern hemisphere). In northern latitudes, therefore, south-facing slopes get more intense sunlight and more warmth than north-facing ones (Figure 34-28) and, because of increased evaporation, are markedly drier, as if they were at a lower elevation. Consequently, different communities or even different biomes commonly occur next to one another on slopes facing in different directions (Figure 34-29). This effect is most dramatic in semiarid climates and at higher latitudes, where differently facing slopes receive very different amounts of sunlight. It is also sometimes important in agriculture. For example, in the more northerly wine-growing regions of Europe and America the vineyards are always planted on south-facing slopes, not on north-facing ones, to assure the vines of enough warmth to ripen their grapes.

AGRICULTURAL ECOSYSTEMS

With the advent of agriculture, humans began to alter the earth's vegetation by clearing away wild species and planting crops in their place. Most of northern Europe, still largely forested at the time of the Romans, was converted into open agricultural fields during the Middle Ages, a transformation that much earlier affected the landscapes of ancient civilized areas such as Italy, Greece, India, and China. In the last century set-

tlers cleared, with enormous effort, the vast temperate broad-leaved deciduous forests of central North America and turned them into farm land. The once seemingly endless prairies of the midwestern plains were gradually put to the plow until now almost nothing remains of their original vegetation. Today, about 10% of the earth's ice-free land area (almost 50%, however, of all the land regarded as "potentialy arable") is under cultivation. Much of the rest of the "potentially arable" land is actually too dry or has too short a growing season for profitable agricultural use. At

Figure 34-29. Example of slope direction effect on vegetation. Coolest, north-facing side of hill, to left, bears pine and fir forest; west-facing side with intermediate exposure to the sun, to right, carries deciduous oak woods (lighter green, hummock-like tree crowns); warmest, south-facing slope along right skyline is covered with scrub vegetation (chaparral). San Jacinto Mts., southern California.

present the last stronghold of extensive natural warm-climate lowland vegetation on earth is being destroyed as Brazil and Peru clear the Amazon Basin for cattle raising, agriculture, and other uses (Figure 34-30); it is expected that within a generation practically none of its presently still vast rain forest will remain.

Structure and Management of Agricultural Communities

Since most agricultural fields are monocultures, the plant community usually has a simple structure compared with that of most natural communities; it is normally dominated by a single crop canopy. Some cropping arrangements are more complex, for example, coffee, a small sun-intolerant tree, is often grown in the shade of larger trees (Figure 32-22B). **Mixed cropping** is interplanting of different crop species. Beans and corn were planted together by American Indians and still are in traditional central and south American subsistence agriculture; it is a happy combination because nitrogen fixation associated with the beans helps fertilize the corn, and the amino acid composition of bean seed protein tends to make up for the nutritional inadequacies, for humans, of corn protein. Mixed cropping or "companion planting" is now being advocated as a way of reducing the hazards of monocultures, increasing and diversifying production on small hand-managed farms and gardens, and reducing the impact of certain pests. It is still impractical, however, for large-scale mechanized farming. **Multiple cropping,** the planting in succession of several crops, suitably adapted to the different seasons through the year, is an important goal for increasing food production in tropical and subtropical climates.

Although agricultural communities are usually simple, they are not usually as simple as the farmer might want. As noted in Chapters 19 and 20, opportunists or colonizer plant species evolved as weeds hand in hand with the evolution of agriculture itself, competing with the crop for the unoccupied area opened up by human cultivation. To combat the depressed crop productivity resulting from this competition, the farmer has had to combat weeds from time immemorial, and weeds have thereby been selected evolutionarily for physiological features that help them circumvent the effects of the hoe and the plow, notably their seed germination responses (Chapter 17). Modern use of herbicides for weed control is just the latest round in this battle, and weeds have already begun to evolve herbicide resistance.

Besides vegetation removal and weed control, perhaps the most basic modification humans effect in agricultural ecosystems is the alteration of the water status of habitats. Ancient peoples discovered how to irrigate dry lands and did so on an extensive scale, for example, in Mesopotamia

Figure 34-30. Destruction of tropical rain forest in progress in the lower Amazon basin (Brazil). Note lianas attached to the trees. Many hundreds of tree and other plant species will become extinct as this land exploitation is pressed to completion in the years ahead.

and Egypt. Irrigation permits a lush crop cover to achieve much greater primary productivity than that of the xerophytic vegetation of an arid or semiarid region. Humans also learned to drain swampy land, thereby making it usable for farming, an enterprise that over the centuries has added a great deal to the useful acreage of arable lands in humid climates. Fertilization and liming are also important human activities that help to achieve and maintain a higher productivity by the land than would otherwise be possible under sustained agriculture.

Types of Agricultural Ecosystems

The principal kinds of agricultural ecosystems humans have developed correspond generally with the climatically determined major biome areas. In the humid tropics, **shifting cultivation** or slash-and-burn agriculture, an agricultural system developed in prehistoric times, continues as a method of subsistence farming in many poorly populated areas. Sustained farming, however, is becoming more widespread in the humid tropics with the improvement of fertilization practices and better knowledge of tropical soil management.

Shifting cultivation involves a repeating cycle of secondary succession. Farmers cut down and burn the forest or jungle in a small patch of a few acres, then plant a crop on the land (Figure 34-31). The nutrients contained in the ashes of the burned vegetation confer a relatively high fertility for crops at first. But because of the heavy leaching of nutrients by rainfall, fertility rapidly declines, and after 3 or 4 years it is usually necessary to abandon the land and clear another forest patch for planting. The former crop patch is rapidly invaded by pioneer species and goes through secondary succession leading again to a dense tropical forest. The development of the forest conserves mineral nutrients that become available in the ecosystem and builds up a new nutrient capital that humans eventually exploit several decades later by cutting and burning the second-growth forest for another crop patch.

By making use of the phenomena of succession, the slash-and-burn system makes sustained human food production possible without the steady and permanent deterioration of the soil and the ecosystem's productivity that more intensive agricultural systems cause. It is well regarded, therefore, by many ecologists, even though it cannot support a large human population because so little of the land area can be in cultivation at any one time.

Tropical agriculture makes use of a wide variety of fruit, vegetable, and staple plants that are consumed locally and are almost unknown in temperate latitudes—for example, the root crops manioc (*Manihot esculenta*) and taro (*Colocasia esculenta*) (Figure 32-13). "Plantations" of tree crops (coffee, tea, cocoa, rubber) and of sugarcane are an important source of cash income in tropical agriculture. Grain production, especially rice and wheat, on floodplain or terraced land (Figure 32-5) is also a very important aspect of tropical agriculture and feeds many more people than the more exotic tropical crops do. Much grain is raised in tropical wet/dry seasonal savanna climates, like the monsoon area of India and southeast Asia, but with this type of agriculture disastrous famine cycles occur in years when the rains come too little or too late. Development of deep-well irrigation is being promoted to reduce this hazard and permit expanded use of high-yielding "miracle" wheat and rice strains in tropical agriculture. In addition to grains, tropical agriculture tends to emphasize legume seed crops as staples (beans, chick peas, and a considerable variety of other relatively unfamiliar legumes, especially in India).

In temperate latitudes where sufficient precipitation falls, the traditional "**general farming**" methods of western culture raise a variety of grains—wheat for human food, other small grains mainly for livestock feed—plus root crops (potatoes, turnips, and so on) and some green vegetables primarily for local consumption. Corn

Figure 34-32. Aerial view of irrigated wheat fields in semi-arid eastern Oregon. Each green circle is irrigated by a boom that pivots around the center point, carrying sprinkler heads fed from a deep well.

(maize, *Zea mays*), originally a crop of tropical Central and South America, has proved very successful in temperate latitudes with sufficiently warm summers and is now raised on a vast scale (exceeded worldwide only by wheat and rice), primarily for livestock feed. In recent years the soybean (*Glycine max*), originally from the Orient, has enjoyed a similar popularity as a feed and oil crop.

Whereas fresh produce used to be available in cold-winter climates only during the summer and to a limited extent from greenhouse cultivation, modern transportation permits vegetables to be raised for the winter market in specialized intensive agriculture in warm-winter areas such as Florida, Mexico, and California's Imperial Valley. Locally, where the climate is suitable, orchard fruits and nuts make important seasonal cash crops in temperate latitudes. In cold-temperate regions a widespread cash crop is sugar beets, grown for commercial extraction of table sugar (sucrose, which is identical with cane sugar). In warm-summer regions a very important crop is cotton, which

Figure 34-31. Shifting cultivation agriculture. Tropical rain forest has been cleared for planting food crops by subsistence farmer residing in the hut. After 2–3 years this land will be abandoned and another clearing made elsewhere, allowing the area in view to revert to jungle by the process of secondary succession (Chapter 33).

Figure 34-33. Vegetables being grown in a geothermally heated greenhouse in volcanically active but climatically cold Iceland. The heat is free.

is ecologically notorious for the large quantities of insecticides usually used on it.

In the semiarid prairie and steppe areas of temperate latitudes much land is used for raising grains, primarily wheat, by "dry farming" without irrigation. This employs practices such as clean-cultivated fallowing of the land every other year to store moisture for the crop to be raised the following year. The success of this kind of farming depends so heavily on yearly accidents of the weather that farmers are turning increasingly to irrigation from deep wells (Figure 34-32). Irrigation has enabled agriculture to expand greatly in arid temperate and tropical regions such as Arizona, interior California, and Pakistan, in areas where almost no significant agriculture could otherwise exist.

Wheat raising extends into relatively high latitudes in some areas such as central Canada and Scandinavia, but in the boreal belt agriculture is generally secondary to forest exploitation and other livelihoods. In the tundra climate there is little or no significant agriculture, an interesting exception being volcanic Iceland where crops are extensively raised in geothermally heated greenhouses (Figure 34-33).

SUMMARY

(1) A plant community or assocation is a set of species inhabiting a particular area together, adapted to its climatic and edaphic conditions. In some cases one or two species completely dominate or occupy almost all the space, but more often there are several dominants and many secondary species.

(2) Species diversity in communities results in part from the fact that different species are specialized for different ecological niches in the habitat, including occupying different microhabitats as well as performing different biological roles such as those of understorey shade plants, vines, and opportunist species. The action of species-specific predators and parasites also encourages species diversity.

(3) Communities tend to be stratified into layers representing different roles or adaptations (different niches) of different subsets of member species—for example, the tallest or dominant trees, one or more layers of understorey trees, a shrub layer, and a herbaceous ground cover. Stratification is less pronounced but still exists in scrub or entirely herbaceous communities.

(4) A biome is a set of communities in different areas of similar climate, having vegetation with similar growth forms and morphology but not necessarily similar species, and inhabited by similar kinds of animals. Biomes are named according to their characteristic vegetation type.

(5) Major biomes are differentiated on land primarily by climatic factors, especially average temperature and precipitation, and the seasonality of these factors. In seasonally dry climates fire influences the makeup of the vegetation, and herbivorous animals significantly affect the character of some biomes.

(6) Cold-climate biomes include the treeless arctic and alpine tundras, and the taiga or boreal conifer forest. Humid temperate-climate biomes include cold-temperate deciduous broad-leaved forests and, in less seasonal climates, temperate evergreen coniferous and broad-leaved forests. Humid tropical climates with year-round precipitation support evergreen rain forests and, in seasonally dry areas, drought-deciduous (monsoon) forests. Semiarid or more severely seasonally dry areas in both tropical and temperate zones carry woodland or savanna biomes, or grasslands. Arid regions support only a sparse desert scrub.

(7) The boundaries between communities or biomes are affected by local soil differences and by variations in water supply between low-lying and elevated ground, between slopes facing toward and away from the sun (exposure effect), and between windward and leeward sides of mountains (rain shadow effect). Vegetation zones are created by variation in altitude owing to its effect on temperature and precipitation.

(8) Agricultural communities include a dominant crop (or crops, in intercropping, companion planting, or multiple cropping systems), plus weeds. The extent and character of agricultural systems are affected by the same geographical variations in climate and local variations in physical features that determine natural biomes. Agriculture is most extensively developed in cold-temperate deciduous forest and grassland biomes, in tropical seasonal forest regions, and in semiarid areas where extensive irrigation is possible. The traditional crops and suitable cropping systems vary considerably among different biome areas, but in recent years corn, wheat, and soybeans have been grown more widely in many areas (especially in the tropics and subtropics) that formerly grew other crops or none.

APPENDIX:

BASIC CHEMISTRY NEEDED FOR BOTANY

Although gases, liquids, and many solids seem macroscopically to be continuous, homogeneous substances, chemical and physical data show that matter is actually made up of minute particles called **atoms,** joined together in highly specific combinations called **molecules,** which cannot be subdivided without altering the chemical and physical nature of the material.

Atoms and Chemical Elements

About 80 chemically distinguishable kinds of atoms, called the **chemical elements,** occur in nature, familiar examples being oxygen, hydrogen, carbon, iron, copper, and sulfur. Atoms are composed of **elementary particles** called electrons, protons, and neutrons, and the atoms of different elements differ in the numbers of these particles

that they contain (Figure A-1). Electrons are very light (almost weightless) particles carrying a negative (−1) electrical charge (electrons are what constitute ordinary electric current). Protons and neutrons are relatively heavy particles, both with a weight, or **mass,** of about 1 in the conventional units (**daltons**) used for the weights of elementary particles, atoms, and molecules. A proton carries a positive (+1) electrical charge, while a neutron is electrically neutral (no charge).

Every atom consists of a positively charged **nucleus,** composed of protons and (usually) neutrons, plus one or more electrons that circulate around the nucleus, somewhat like a miniature planetary system. The simplest atom is hydrogen, consisting of one proton (its nucleus) plus one electron (Figure A-1 [1]). Other elements are formed by adding neutrons and protons to the nucleus plus additional electrons around it (Figure A-1 [2],

1: Hydrogen (**H**) 2: Helium (**He**) 3: Lithium (**Li**) 4: Beryllium (**Be**) 5: Boron (**B**) 6: Carbon (**C**)

    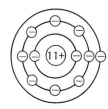 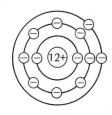

7: Nitrogen (**N**) 8: Oxygen (**O**) 9: Fluorine (**F**) 10: Argon (**A**) 11: Sodium (**Na**) 12: Magnesium (**Mg**)

Figure A-1. Elementary structure of atoms of the 12 lightest (smallest) chemical elements. Each element's "atomic number," shown to the left of its name, equals the number of electrons (⊖) or protons (⊕) in its atoms. Neutrons (○) are shown just for the first four elements. The number of protons in the larger atoms is indicated just by the figure followed by + in the nucleus; the number of neutrons (not shown) usually approximately equals the number of protons.

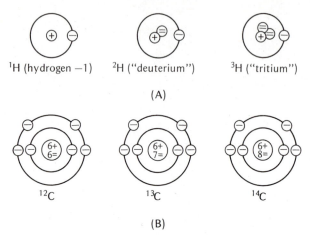

^1H (hydrogen —1) ^2H ("deuterium") ^3H ("tritium")

(A)

^{12}C ^{13}C ^{14}C

(B)

Figure A-2.   Isotopes of (A) hydrogen and (B) carbon; =, neutrons; +, protons; ⊖, electrons.

[3], etc.). Each type of atom has a particular weight or mass, called its **atomic mass,** which is given in daltons and is essentially equal to the total number of protons plus neutrons in its nucleus. By their charge the protons attract an equal number of electrons (since opposite electrical charges attract) to yield an electrically neutral atom (no net charge, because the opposite charges cancel). Since the chemical properties of an atom, as we shall see, are due to the number of electrons it has, the number of protons in the nucleus, by determining the number of electrons held by the atom, determines the chemical identity of the atom—that is, what element it represents. This number is the **atomic number** of the element.

Each element is designated by a conventional chemical **symbol** of one or two letters (Figure A-1); when the symbol is two letters the first is always capitalized, and the second is not.

Isotopes. Atoms of the same chemical element can differ in the number of neutrons in their nuclei, and therefore in their atomic mass. These atoms of different mass are called **isotopes** of the element (Figure A-2). Isotopes are designated by a superscript number indicating their atomic mass. For example, the most common isotope of carbon is ^{12}C (mass 12 daltons), but isotopes ^{13}C and ^{14}C also occur (Figure A-2). The atomic masses ("atomic weights") given in tables of the elements represent in effect an average of the masses of the different naturally occurring isotopes of each element.

Because different isotopes of an element have essentially the same chemical properties, we can use uncommon or artificially produced isotopes as biological or biochemical "tracers" for their elements. "Heavy oxygen," ^{18}O (ordinary oxygen is mostly ^{16}O), for example, allows an investigator to determine where the oxygen produced by photosynthesis comes from (Chapter 7), and heavy nitrogen, ^{15}N (normal nitrogen is mostly ^{14}N), has been used to follow the capture of nitrogen from the air by nitrogen-fixing organisms

(Chapters 22 and 25). However, because a complex, expensive instrument called a mass spectrometer is needed to detect these heavy but stable isotopes, they find only limited use in biology. Much more commonly used as tracers are **radioactive** isotopes (usually produced by nuclear reactors), which have spontaneously unstable nuclei that disintegrate at known rates, giving off atomic radiation that allows even minute amounts of the isotope to be detected. The radioactive carbon isotope ^{14}C, for example, when combined with oxygen as carbon-14 dioxide, allowed the pathway of photosynthetic carbon metabolism to be traced (Chapter 7). Similarly, the radioactive isotope of hydrogen, ^3H (called **tritium**), in the form of tritium oxide (tritiated water), allows movement of water into and out of cells to be measured.

Electronic Structure of Atoms

The electrons of an atom are restricted to a series of specific energy levels, traditionally called "shells" and represented in Figure A-1 as circular (spherical in three dimensions) zones at specific distances from the nucleus. Successive shells of increasing diameter have successively greater energies associated with them. Each electron added to an atom falls into the lowest-energy shell that is not already fully occupied. Thus, in the sequence of elements shown in Figure A-1, electrons fill first the smallest shell (called shell 1), which accepts only two electrons, then the second shell, which accepts eight electrons, and then begin to fill the third shell. Although occupancy of the shells becomes more complicated as atoms grow larger than those shown in Figure A-1, one general rule is that the *outermost* shell or energy level can hold only eight electrons (except shell 1 which holds only two electrons). This is important because chemical interactions between any atom and others are due mainly to the electrons in the atom's outermost shell, called its **valence** electrons (valence refers to an element's capacity to form chemical bonds). For most chemical purposes, therefore, instead of the relatively complicated atomic models given in Figure A-1 we may substitute simple ones showing just the atom's valence electrons. This is conventionally done by giving the chemical symbol for the element in question surrounded by as many dots as it has valence electrons (examples are shown in Figure A-3).

The electrons within one electron shell do not all move in the same path, as they appear to do in Figure A-1, but occupy separate domains called **orbitals**, of different shapes and positions, around the nucleus. For the sake of simplicity we shall not go into the geometry of electron orbitals but will merely state that this geometry becomes important in the structure of molecules because it determines the angles between the bonds that an atom can make with other atoms, and these bond angles determine the three-dimensional shapes of molecules.

Chemical Compounds

Different elements react with one another via the valence electrons of their atoms to form chemical **compounds.** The atoms in a compound are held together by chemical **bonds** in specifically organized atomic aggregates, or **molecules.** An inorganic (noncarbon) compound is designated by the chemical symbols of the elements its molecules contain, with subscripts to show how many atoms of each (if other than 1) are present; for example, MgO (magnesium oxide), H_2O (hydrogen oxide or water), NaCl (sodium chloride or table salt), Al_2O_3 (aluminum oxide), and H_2S (hydrogen sulfide). Many inorganic compounds are formed either from a metal or metal-like element or from hydrogen, plus a nonmetallic element such as oxygen, chlorine, or sulfur, the name of the nonmetal in the compound being given the ending *ide*, as in the previous examples. Many other compounds contain a metal plus a complex (**radical**) composed of oxygen and some other element. Such radicals are usually named with the ending *ate* or *ite* after the root of the other element's name—for example, $-NO_2$, nitrite and $-NO_3$, nitrate (from nitrogen), $-SO_4$, sulfate (from sulfur,—CO_3, carbonate (from carbon), and $-PO_4$, phosphate (from phosphorus). Examples of their compounds are sodium phosphate (Na_3PO_4), calcium nitrate ($Ca[NO_3]_2$), and potassium sulfate (K_2SO_4).

Formation of compounds is dictated by a tendency of atoms to either gain, lose, or share electrons until the outer electron shell of each atom is completely filled, that is, holds eight electrons (except for the first shell). A strong **nonmetal** such as oxygen or chlorine has a nearly filled valence shell (Figure 4-3) and gains electrons from other atoms (e.g., metals or hydrogen) until its valence shell contains eight electrons (Figure A-4A,D). A **metal,** by contrast, has only one to at most a few relatively weakly held electrons in its valence shell (Figure A-3) and tends to *lose* electrons until its valence shell becomes empty. The number of atoms of each element in a compound is dictated by how many electrons the metal-like element donates and how many electrons the nonmetal receives. With NaCl, one atom of each element forms the compound because Na· can donate just one electron and :C̈l· needs just one (Figure A-4A). With water (H_2O), two atoms of hydrogen are involved because :Ö· needs two electrons to fill its valence shell, but a hydrogen atom has only one electron to donate (Figure A-4D).

Ions and Ionic Bonding

When metals and strong nonmetals such as sodium and chlorine, respectively, react, the nonmetal takes electron(s) completely away from the metal atom, leaving the metal with a positive charge (because it has lost one or more negative charges), and the nonmetal with a negative charge (because it has gained electrons; see Figure A-4B). These charged atoms are called **ions** and are designated by the symbol for the element followed by a superscript number indicating how much positive or negative charge the ion carries (e.g.,

number of valence electrons

1	2	3	4	5	6	7
H· hydrogen	Be· beryllium	·B· boron	·C· carbon	:N· nitrogen	:O· oxygen	:F· fluorine
Na· sodium	Mg· magnesium	·Al· aluminum	·Si· silicon	:P· phosphorus	:S· sulfur	:Cl· chlorine
K· potassium	Ca· calcium					

Figure A-3. Simplified representation and classification, by number of valence electrons, of atoms of some biologically important elements. The members of each vertical row have relatively similar chemical properties. Metals are shown at left (except hydrogen), nonmetals at right, and elements of intermediate properties in between. The "transition metals" such as manganese, iron, copper, and zinc, which occur above calcium in the sequence of elements, are omitted because, although biologically important, they have a more complicated electronic structure than that of the lighter elements displayed here. Also omitted are the "noble gases" (helium, argon, etc.), which have a filled valence shell and are therefore chemically unreactive and biologically unimportant; lithium (no. 3 in Figure A-1), also biologically unimportant, is omitted merely to reduce the number of horizontal rows. (This chart is a very abbreviated and simplified version of the "Periodic Table of Elements" to be found in any chemistry textbook.)

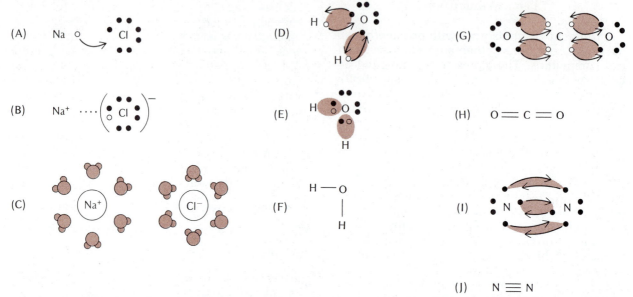

Figure A-4. Formation of, and conventional notation for, chemical bonds between atoms. Valence electrons of different elements are indicated by black or open dots, respectively. *A,* Electron transfer from one atom (sodium, Na) to another (chlorine, Cl), leading to formation of ions (*B*) held by ionic bond (common salt, NaCl). *C,* Salt in solution, its ions separated and nearly independent of one another due to associated water molecules. *D,* Formation of covalent bonds by electron sharing between atoms, a shared pair of electrons being indicated by two dots between the sharing atoms (*E*). *F,* Conventional representation of single covalent bonds (water, H_2O). *G,* Formation of double bonds (carbon dioxide, CO_2) and (*I*) a triple bond (molecular nitrogen gas, N_2) by multiple electron pair sharing; the conventional representation of such bonds is shown in *H* and *J*.

sodium ion Na^+, magnesium ion Mg^{2+}, aluminum ion Al^{3+}, chloride ion Cl^-, and sulfide ion S^{2-}). During their formation the common inorganic oxygen-containing radicals take one or more electrons from other atoms and therefore occur as specifically charged ions—for example, NO_3^- (nitrate ion), SO_4^{2-}, (sulfate ion), or PO_4^{3-} (phosphate ion).

Because unlike charges attract, + and − ions tend to stick together by what is called an **ionic bond** (Figure A-4B). Compounds made up of ions held in ionic bonding are called **salts,** by analogy with common table salt (NaCl). Many inorganic compounds are salts, common examples being epsom salts (magnesium sulfate, $MgSO_4$), saltpeter (sodium nitrate, $NaNO_3$), and washing soda (sodium carbonate, Na_2CO_3).

Many salts, such as those just mentioned, are soluble in water. When they dissolve, they **dissociate** into their component ions because water molecules associate with each ion, shielding it from the electrical attraction of other ions. Thus in solution, sodium chloride consists of Na^+ and Cl^- ions (each associated with H_2O), not NaCl molecules (Figure A-4C); magnesium sulfate dissociates into magnesium (Mg^{2+}) and sulfate (SO_4^{2-}) ions, etc. Therefore, in cell fluids such as cytoplasm we are not concerned with salts as such

but with the concentrations of essentially independent inorganic ions such as Na^+, K^+, Mg^{2+}, Cl^-, and SO_4^{2-}.

Some elements can lose or gain a variable number of electrons in the chemical reactions they undergo, leading to ions of different charges. For example, iron (Fe) can occur either as **ferrous** ion (Fe^{2+}) or as **ferric** ion (Fe^{3+}).

Covalent Bonding

Nonmetal atoms can exchange and *share* electrons from their valence shells (Figure A-4D) until the shared electrons effectively fill up the valence shells of the participating atoms (Figure A-4E), thereby satisfying the filled-shell rule. Each pair of shared electrons acts as a **covalent bond** holding the two atoms together, conventionally denoted by a line drawn between the chemical symbols for the atoms (Figure A-4F). Two pairs of electrons can be shared between two atoms (Figure A-4G), yielding a **double bond** denoted by a double line (Figure A-4H). In some compounds three pairs of electrons are shared (Figure A-4I), yielding a triple bond (Figure A-4J).

The number of covalent bonds an atom makes with others normally equals the number of *un-*

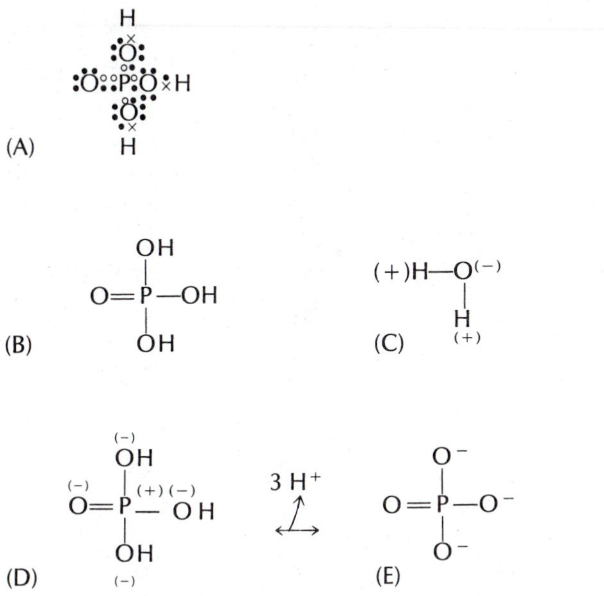

(A)

(B)

(C)

(D)

(E)

phosphoric acid
(H₃PO₄)

phosphate ion (PO₄³⁻)

Figure A-5. *A*, Covalent bonds in phosphoric acid (H_3PO_4) formed by sharing valence electrons of phosphorus ($\bigcirc$), oxygen ($\bullet$), and hydrogen (x); *B*, conventional representation of bonds. *C*, Polar character of water molecule and (*D*) of phosphoric acid molecule, signs in ( ) indicating partial positive and negative charges. *E*, Phosphate ion formed by dissociation of H_3PO_4, very strongly polar because of the full negative charges (minus signs) on the oxygen atoms.

filled places in its valence shell, because it must receive this number of (shared) electrons in order to fill its valence shell fully and thus yield its most stable electronic state. Thus hydrogen always forms but one covalent bond, oxygen forms two, nitrogen forms three, and carbon forms four (Figure A-4D–J).

Phosphorus (P), another biochemically important element, is an exception to this rule. Like nitrogen, it has three unfilled places in its valence shell (Figure A-3), but it normally forms five covalent bonds (Figure A-5B). All five electrons in its valence shell actually can participate in covalent bonding (Figure A-5A).

Polarity in Covalent Bonds. When elements that differ in their attraction for electrons bond together covalently, the more electron-attracting atom draws the shared electrons unequally toward it and acquires a partial negative charge while the other atom acquires a fractional positive charge. Such electrically polarized, or **polar,** covalent bonds are illustrated by those in the water molecule, in which the electron-attracting O atom bears a partial − charge, and the two H atoms bear fractional + charges (Figure A-5C). Since the two H atoms are bonded almost at right angles to one an-

other, this makes the entire H_2O molecule electrically polar, the side bearing the H atoms being somewhat positive and the side on which the O atom is exposed being somewhat negative. This is an extremely important property, because it allows water to associate with ions and thus to dissolve salts, as mentioned above, and also to interact with and dissolve many organic molecules vital to life. The predominance of water, with its polar solvent properties, was certainly one of the most important factors that permitted life to arise on earth.

Organic Compounds

Covalent bonding is the basis of the complex chemistry of carbon, allowing formation of the complicated molecules on which life is based. As noted above, carbon has four valence electrons, which permit it to form four covalent bonds with other atoms, including other carbon atoms. This allows chains and rings of carbon atoms to form, with other atoms such as hydrogen, oxygen, and/or nitrogen attached to them. These multicarbon compounds are called **organic compounds** because they were first known to chemists only as components or products of living organisms. Chemists later discovered how to synthesize certain organic compounds artificially and eventually developed elaborate procedures for determining their atomic arrangements or structure and for achieving the synthesis of very complex organic molecules, including many that differ from any known to exist in nature. Much of the modern chemical industry is based on this knowledge of organic chemistry.

The names of organic compounds are often derived from the name of a plant from which the compound was first isolated—for example, citric acid from *Citrus* (lemon, orange), caffeine from coffee, nicotine from *Nicotiana* (botanical name of tobacco), cannabinol from *Cannabis* (botanical name of marijuana), and so on.

Organic compounds can be represented either by an **empirical formula** (such as $C_6H_{12}O_6$ for dextrose sugar, or glucose), which shows how many atoms of each kind form the molecule but not how they are bonded, or by a **structural formula** in which the bonding arrangements, deduced from the results of experimental procedures just alluded to, are represented by the conventions described above for covalent bonds. Figure A-6A shows some examples of structural formulas of organic compounds. Essentially, every different organic compound has a different structural formula even though different compounds may have the same empirical formula. For example, several different sugars have the empirical formula $C_6H_{12}O_6$, including glucose (dextrose), fructose, galactose, and mannose, all of which are common plant constituents.

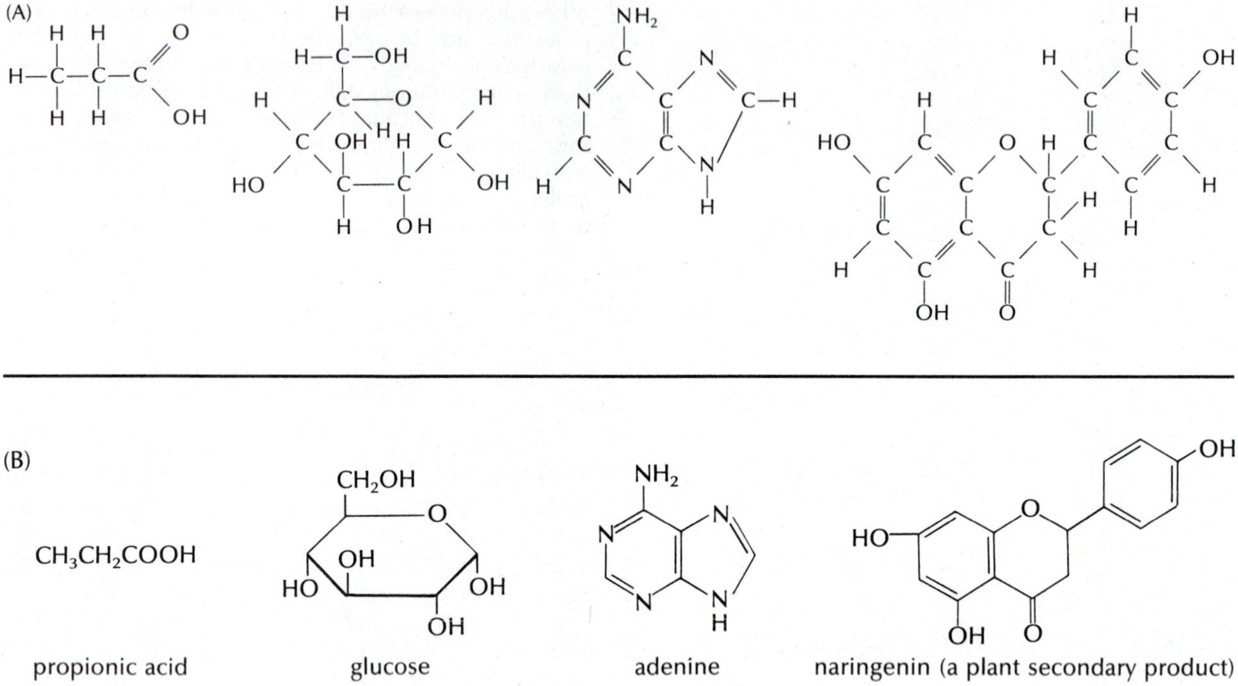

(A)

(B)

CH₃CH₂COOH

propionic acid   glucose   adenine   naringenin (a plant secondary product)

Figure A-6. Examples of structural formulas of some organic compounds. Upper row (A), complete structure; lower row (B), conventional, abbreviated, or simplified representation of structure. The heavy line drawn along the lower edge of the glucose ring (and other sugars) indicates that the ring is viewed tipped away from you, so that the positions of groups projecting above and below the plane of the ring can be visualized.

RADICALS

Certain small groupings of atoms, called radicals or functional groups, appear repeatedly in the structures of organic compounds. The most commonly encountered of these are named and illustrated in Figure A-7; several of these are seen in the structural formulas in Figure A-6. These radicals are often much more important to a compound's properties than are other details of its structural formula. For example, sugars such as dextrose bear several hydroxyl (—OH) groups, which make sugars water-soluble and allow them to build up larger chain-like molecules such as cellulose.

Groups such as —OH, —NH₂, —C=O, and —COOH are polar, for reasons explained above for H₂O. They have a profound effect on the properties of organic compounds containing them, first because they tend to associate with water (because of its polarity), increasing the compound's solubility in water, and second because some of them tend to associate with one another owing to an electrical attraction called **hydrogen bonding,** the same kind of attraction that is involved in their tendency to associate with water. Although individually weak, the numerous hydrogen bonds that can form between the many polar groups in a large organic molecule such as a protein are extremely important in conferring a specific three-dimensional shape to the molecule, essential to its biological function (Chapter 5).

SIMPLIFIED CONVENTIONAL STRUCTURAL DIAGRAMS

Ring compounds are usually depicted not by complete structural formulas such as those in Figure A-6A but by diagrams (Figure A-6B) containing certain simplifying conventions: (1) the sides of polygons represent single, double, or triple bonds according to the number of parallel lines drawn, (2) the angles between these bonds represent carbon atoms unless a chemical symbol for some other element is shown at that point, (3) radicals attached to the rings are represented by conventional notation (shown on the right side of Figure A-7), joined to the angles (atoms) of the rings, and (4) hydrogen atoms attached to ring carbon atoms are usually not shown.

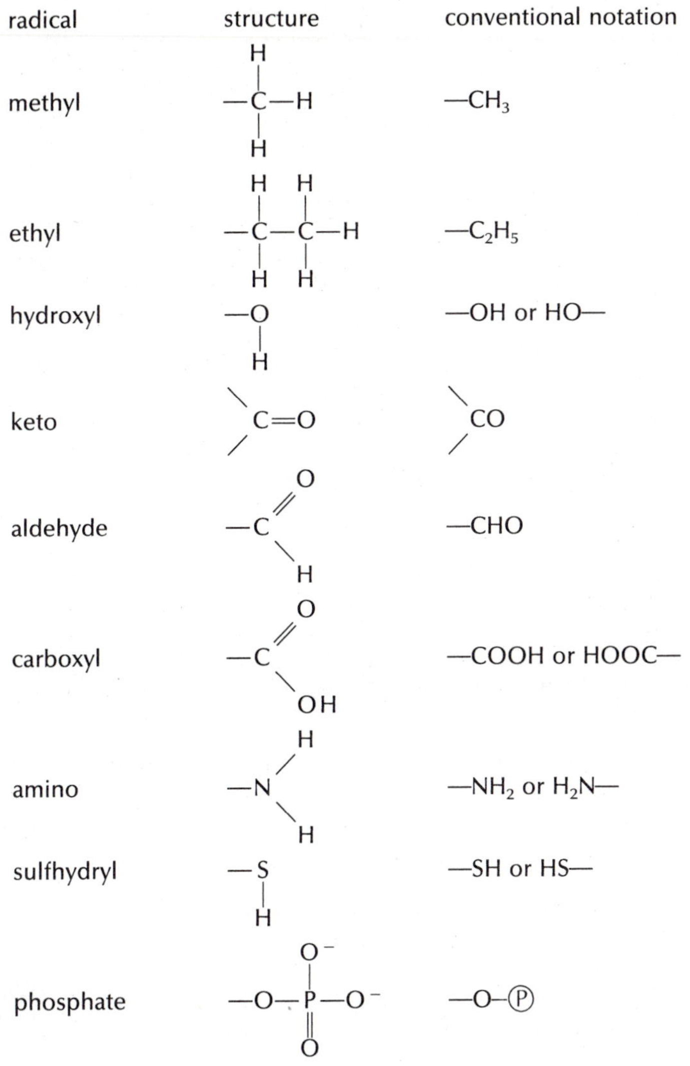

radical	structure	conventional notation
methyl		$-CH_3$
ethyl		$-C_2H_5$
hydroxyl		$-OH$ or $HO-$
keto		
aldehyde		$-CHO$
carboxyl		$-COOH$ or $HOOC-$
amino		$-NH_2$ or H_2N-
sulfhydryl		$-SH$ or $HS-$
phosphate		

Figure A-7. Examples of functional groups (radicals) commonly occurring in biologically important organic compounds (not a complete list).

Molecular Weights; Molar Units

Atoms and molecules, as noted earlier, are extremely minute, and any macroscopic sample of material contains enormous numbers of them. In practice, macroscopic quantities of molecules are given in units of **moles,** 1 mole consisting of 6.02×10^{23} molecules. A mole is actually defined relative to the **molecular weight** of any compound: the molecular weight, in daltons, is simply the sum of the atomic masses ("atomic weights") of all the atoms in the molecule. A mole is the number of molecules in one "gram molecular weight" of a substance, that is, its molecular weight expressed in grams. For example, water, which contains per molecule two atoms of hydrogen of atomic mass about 1 and one atom of oxygen of atomic mass 16, has a molecular weight of 18 daltons. Thus 18 grams of water (which equals 18 cubic centimeters or about 0.6 fluid ounce) is 1 mole, containing 6.02×10^{23} molecules. One mole, or 6.02×10^{23} molecules, of dextrose (glucose), on the other hand, weighs 180 grams because glucose has a molecular weight of 180 daltons.

The reason for the strange number, 6.02×10^{23} molecules per mole ("Avogadro's number"), is merely that the gram scale of weights is not based on the masses of atomic particles, and 1 dalton of atomic mass happens to equal $1/(6.02 \times 10^{23})$ of a gram. Thus when a compound's molecular weight is weighed in grams, it contains 6.02×10^{23} molecules.

Millimoles (1 thousandth of a mole) and micromoles (1 millionth of a mole) are used for smaller amounts of substances. The concentration of a substance in a solution is normally given in **molar** (abbreviated M) units: a 1.0 M solution contains 1 mole of the dissolved substance per liter of solution. For dilute solutions we commonly use millimolar (mM, 10^{-3} mole per liter) or micromolar (μM, 10^{-6} mole per liter) units. Amounts and concentrations of other types of particles, such as ions, are also given in molar units.

SMALL MOLECULES AND MACROMOLECULES

Inorganic and organic compounds with a limited number of atoms, and consequently a relatively limited molecular weight (up to about 1000 daltons) are called "small molecules" to distinguish them from a class of very much larger organic molecules called polymers (from Greek *polys*, many, and *meros*, part) or macromolecules (from Greek *macro*, large) produced by living organisms, and now also artificially, by combining many small-molecule building blocks together into a long chain. Macromolecules generally have molecular weights ranging from 10,000 up to many million daltons, depending on the polymer, and their properties are fundamentally different from those of most small molecules. The functional and structural molecules most essential to life such as nucleic acids, proteins, and cellulose, are macromolecules; they are considered in Chapter 5.

CHEMICAL REACTIONS

Chemical reactions are processes that form or change chemical bonds between atoms, converting elements into compounds (or vice versa) and existing compounds into different compounds. In general, for bonds to form or change, the atoms or molecules involved must collide with one another. At ordinary temperatures atoms and molecules in fact frequently collide because they are constantly in motion owing to their thermal (heat) energy, which is simply kinetic energy (energy of motion) of the atoms and molecules of a substance. Because they collide not only more frequently but also harder (with more energy) as the temperature is raised, chemical (and biochemical) reactions are speeded up by higher temperatures and slowed by cool or cold temperatures. Because of this the activity of organisms varies greatly with their body temperature; this principle is also involved in preserving food by refrigeration or freezing.

We may write a generalized chemical reaction as

$$A + B \longrightarrow C + D \qquad \text{(A-1)}$$

where A, B, C, and D are different elements or compounds. The rate of this reaction, that is, the rate of formation of reaction **products** C and D, will increase if we raise the concentration of either of the **reactants,** A or B, in the system, because the greater the concentration of either A or B, the more often A and B will collide with one another and react. This is called the **law of mass action.**

Reversible and Irreversible Reactions

Many reactions proceed in either direction, that is,

$$A + B \Longleftrightarrow C + D \qquad \text{(A-2)}$$

depending upon the relative concentrations of the substances on the left and right sides of the equation. A high concentration of A and B tends by mass action to drive the reaction from left to right, forming C and D, and a high concentration of C and D drives the reaction from right to left, producing A and B. Such a reaction is said to be **reversible.** Whatever the initial concentrations of A, B, C, and D happen to be, the overall reaction will continue until the rate of the forward (left to right) reaction becomes equal to the rate of the reverse (right to left) reaction. Thereafter the concentrations of A, B, C, and D will not change further, and a state of **equilibrium** will exist in the system. The concentrations of A, B, C, and D are not necessarily *equal* at equilibrium; the equilibrium ratio of concentrations depends on the nature of the reaction and on environmental conditions such as temperature.

Many other reactions, however, are **irreversible,** effectively proceeding in only one direction as represented by the unidirectional arrow in Eq. (A-1). An example is the combustion of carbon (e.g., burning of charcoal),

$$\underset{\text{carbon}}{C} + \underset{\text{oxygen}}{O_2} \longrightarrow \underset{\text{carbon dioxide}}{CO_2} \qquad \text{(A-3)}$$

We do not expect, and do not observe, that by increasing the concentration of carbon dioxide in the system we can reverse the reaction and make carbon dioxide turn into carbon and oxygen.

Biological processes include both reversible and irreversible reactions. The most familiar example of an irreversible reaction is respiration, in which oxygen reacts in effect with food materials, releasing carbon dioxide somewhat as in Eq. (A-3).

Chemical Energetics

The formation of chemical bonds between elements normally releases energy, as the combustion of carbon [Eq. (A-3)] illustrates. Different bonds differ in their energy status, however, so a chemical reaction such as that in Eq. (A-2) generally involves either a release or an uptake of energy (mainly heat energy), depending upon whether the bonds in C and D have, respectively, less or more energy associated with them than the bonds in A and B. If the reaction releases energy it is called **exergonic** (exo, out of) or energetically downhill. If it consumes energy (and thus *stores* energy in its products) it is **endergonic** (endo,

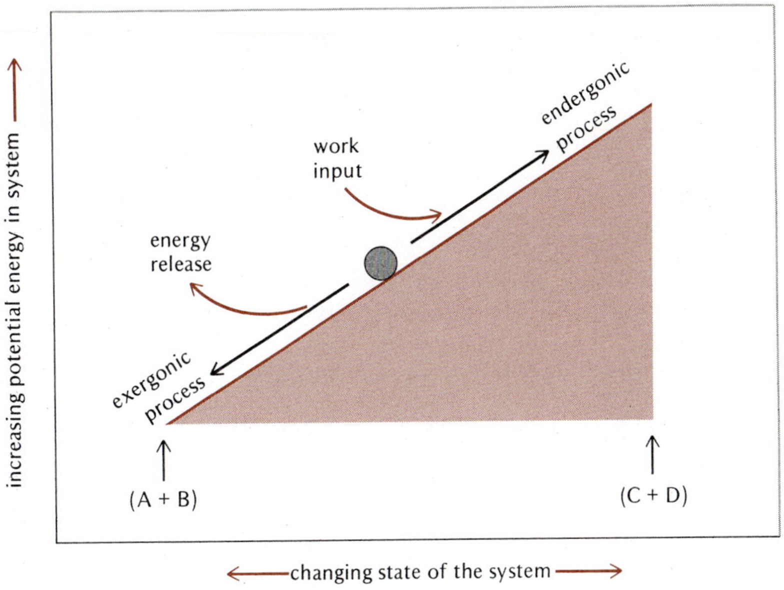

Figure A-8. Energetics of chemical changes, analogy with a ball rolling downhill (exergonic) or being pushed uphill (endergonic). If the energy of products C + D in Eq. (A-4) were greater than that of reactants A and B, as shown in this diagram, the reaction from left to right would be endergonic and could occur only with an input of energy (work).

within or into) or energetically uphill, as in pushing a ball uphill, storing energy as potential energy in the ball (Figure A-8).

If a reaction is energetically downhill (releases energy) in one direction, it must be uphill in the reverse direction (Figure A-8). Otherwise, energy could be created out of nothing. For a reaction to be reversible, its energy change must be relatively small. Reactions with a large energy change are irreversible and go spontaneously only in the downhill (energy-releasing) direction, as the combustion of carbon [Eq. (A-3)] well illustrates. Such reactions can be driven in the opposite direction if a special mechanism can be devised to put energy into the molecules involved—that is, work must be done on them just as it must be done on a ball to move it uphill (Figure A-8). If a reaction is exergonic, on the other hand, it has at least the potential for working on other chemical (or physical) systems. These principles are of central importance in the function of cells and organisms, because cells must build many molecules and structures that contain more energy than their component building blocks. Therefore, cells have had to evolve mechanisms to specifically couple exergonic processes (energy sources) to endergonic biosynthetic reactions. The mechanisms are described in Chapters 5, 6, and 7.

Although chemical energy changes in most cases comprise mainly heat energy, an additional factor called **entropy,** which is essentially a measure of the randomness of the system, also enters decisively into the overall energetics (called **free energy** changes) that determine the direction of a chemical reaction, and the work that it can do or that must be done on it to make it take place. The quantitative analysis of these relations is called **thermodynamics.** Throughout this book, wherever chemical energy changes are mentioned, change in free energy (which includes both entropy and ordinary energy changes) is meant. In the most general, thermodynamic, sense living cells and organisms comprise systems that expend energy (ultimately from light, via photosynthesis) to reduce their own entropy—that is, they reduce randomness or disorganization in their components and thus build up and maintain their highly organized state. This behavior is almost unique to life, nonliving systems generally progressing toward a state of increasing randomness or disorganization (increasing entropy).

Catalysis

Reactions can often be speeded up by agents called **catalysts,** which are not used up in the reaction, as the reactants are. Some catalysts such as acids are very simple and nonspecific in their action while others are more specific or, in the case of enzymes (the catalysts upon which cell functions depend [Chapter 5]), highly specific to particular molecules and one type of reaction. Catalysts combine temporarily with one or more of the reactants, providing a means by which the participating molecules can react more easily (i.e., with a less severe collision) than they can react in the absence of the catalyst.

Types of Chemical Reactions

There are countless possible chemical reactions between the virtually innumerable chemical compounds existing on earth, but most of them fall into one of three classes: (1) **addition** and (its reverse) **elimination** reactions,

$$A + B \rightleftharpoons AB \qquad (A\text{-}4)$$

in which A and B represent atoms, ions, or molecules;
(2) **substitution** reactions

$$AB + CD \rightleftharpoons AD + BC \qquad (A\text{-}5)$$

in which A, B, C, and D represent atoms, functional groups (radicals), or other parts of molecules; and
(3) **oxidation-reduction** reactions, in which electrons are transferred from one atom, ion, or molecule to another, for example,

$$A + B \rightleftharpoons A^+ + B^- \qquad (A\text{-}6)$$

where from left to right A loses an electron and B gains one, thus acquiring a negative charge.

Acid-base reactions are important examples of addition and elimination reactions. An acid is a compound that can **dissociate** (an elimination reaction) to release a hydrogen ion (H^+)

$$AH \rightleftharpoons A^- + H^+ \qquad (A\text{-}7)$$

(A more general and abstract definition of acids is now used in chemistry but is not needed for our purpose.) Hydrochloric acid (HCl), for example, dissociates into H^+ and Cl^- (chloride ion); dissociation of phosphoric acid is illustrated in Figure A-5D,E. Organic compounds containing a carboxyl group (Figure A-6) are acids because the —COOH group can dissociate,

$$R-C{\overset{O}{\underset{OH}{}}} \rightleftharpoons R-C{\overset{O}{\underset{O^-}{}}} + H^+ \qquad (A\text{-}8)$$

where R stands for the rest of the molecule. Acetic acid (CH_3—COOH), the acid of vinegar, for example, dissociates into H^+ plus *acetate* ion (CH_3—COO^-), the ending *ate* denoting the organic ion resulting from dissociation of an organic acid (malic acid yields malate, citric acid, citrate, etc.). These organic acids are called **weak acids** because the carboxyl group does not dissociate or ionize very strongly, in contrast to HCl (a strong acid), which in water dissociates essentially completely into H^+ and Cl^-.

A **base** is a compound or ion that tends to associate with (add) one or more H^+ ions, for example, ammonia (NH_3) is a base,

$$NH_3 + H^+ \rightleftharpoons NH_4^+ \qquad (A\text{-}9)$$

The amino group (—NH_2, Figure A-6), acts similarly as a base. The most common base is hydroxyl ion (OH^-) or compounds such as sodium hydroxide (NaOH) that dissociate to release OH^- ion, which combines avidly with H^+ ion, yielding water:

$$OH^- + H^+ \rightleftharpoons H_2O \qquad (A\text{-}10)$$

This is a reversible reaction but proceeds only very weakly from right to left, plain water dissociating to release only about 10^{-7} M (one-tenth of a millionth of a mole per liter) of H^+ and OH^- ions (a liter contains 55 moles of water, so at equilibrium only 1 water molecule in 550 million actually dissociates). This H^+ concentration is called **neutrality.**

If H^+ ions are added to a solution, they shift the dissociation/association equilibria of weak acids and bases by mass action. H^+ ions drive Eq. (A-8) from right to left, reducing the dissociation of a weak acid, and they drive reactions like Eq. (A-9) from left to right, increasing the association of a base with H^+ ions. Addition of OH^- ions, on the other hand, by driving Eq. (A-10) from left to right, decreases the H^+ concentration and as a result shifts acid and base reactions in the direction opposite to that cited respectively for Eq. (A-8) and (A-9) in the previous sentence. Changes in H^+ or OH^- concentration, by altering in this way the extent to which the acidic and basic groups of organic molecules are ionized, profoundly affect the function of cell components and cells. Special attention must be given, therefore, to H^+ concentration or acidity (**pH**) as a factor affecting biological performance (Chapter 5).

SUBSTITUTION AND OXIDATION-REDUCTION REACTIONS

Most reactions of organic compounds are substitution reactions (Eq. (A-5)], in which functional groups (Figure A-6) or other parts of the reacting molecules are exchanged or altered by transferring or interchanging certain atoms. Many examples appear in Chapters 5–7.

Oxidation-reduction reactions are of special importance to cells because they often involve a relatively large energy change; photosynthesis is an example. Oxidation originally meant reaction with oxygen [e.g., Eq. (A-3)], but since such reactions proceed because of the tendency of oxygen to remove electrons from other atoms, the term oxidation is now used for all reactions involving electron transfer between atoms, such as the formation of NaCl from sodium and chlorine (Figure A-4a). An atom or molecule that donates electrons is said to become **oxidized,** and one that receives electrons from an electron donor is said to become **reduced.** The electron donor is thus a **reductant** (causes reduction), and the electron acceptor (such as oxygen or chlorine) is an **oxidant.** In every oxidation reaction a reductant becomes oxidized and an oxidant (e.g., oxygen) becomes reduced:

$$\begin{array}{ccc} \underset{\text{reductant}}{A} & + & \underset{\text{oxidant}}{B} & \xrightarrow{\substack{\text{transfer of}\\\text{electron}\\\text{from A to B}}} \\[2em] \underset{\substack{\text{oxidized}\\\text{reductant}}}{A^+} & + & \underset{\substack{\text{reduced}\\\text{oxidant}}}{B^-} & (A\text{-}11) \end{array}$$

thus the name oxidation-reduction reaction ("redox" reaction, for short).

Simple electron transfer [Eq. (A-11)] is an important type of redox reaction in respiration and photosynthesis (Chapters 6 and 7). In these processes cells use metals such as iron, copper, and manganese that can donate a variable number of electrons (i.e., can exist in more than one oxidation state). For example, the previously mentioned ferrous (Fe^{2+}) and ferric (Fe^{3+}) forms of iron differ in oxidation state by one electron (Fe^{2+} has one more electron than Fe^{3+}). Fe^{3+} thus can act as an electron carrier, receiving an electron from some other compound and being thereby reduced to Fe^{2+}, then giving up the electron to another, stronger electron acceptor such as oxygen and being thereby reoxidized to Fe^{3+}, each of these redox reactions releasing some energy.

Most biological oxidation-reduction reactions take place by removing two hydrogen atoms (two protons *plus their associated electrons*) from an organic compound, which thereby becomes oxidized. The oxidant can be either another organic compound that becomes reduced by accepting the hydrogen atoms with their electrons,

$$\underset{\text{reductant}}{AH_2} + \underset{\text{oxidant}}{B} \xrightarrow[\text{from } AH_2 \text{ to } B]{\text{transfer of } 2H}$$

$$\underset{\substack{\text{oxidized} \\ \text{reductant}}}{A} + \underset{\substack{\text{reduced} \\ \text{oxidant}}}{BH_2} \qquad \text{(A-12)}$$

or the oxidant can be an oxidized metal such as Fe^{3+} that becomes reduced by accepting an electron but not the associated proton,

$$\underset{\text{reductant}}{AH_2} + \underset{\substack{\text{ferric} \\ \text{ions}}}{2\,Fe^{3+}} \longrightarrow$$

$$\underset{\substack{\text{oxidized} \\ \text{reductant}}}{A} + \underset{\substack{\text{ferrous} \\ \text{ions}}}{2\,Fe^{2+}} + \underset{\substack{\text{hydrogen} \\ \text{ions}}}{2\,H^+} \qquad \text{(A-13)}$$

Reactions of this type, occurring in respiration and photosynthesis, produce or (when running in reverse) consume H^+ ions, a feature important in the extraction of useful energy from oxidation-reduction processes by cells as explained in Chapters 6 and 7.

FURTHER READING

CHAPTER 1.   Basic Plant Functions: Discovery and Utilization

Asimov, I. 1968. *Photosynthesis*. Basic Books, New York. Chapter 1 covers very readably the steps in discovery of photosynthesis, more extensively than this chapter

Bienz, D.R. 1980. *The Why and How of Home Horticulture*. W.H. Freeman & Co., San Francisco. Introduction to gardening practices and procedures with some explanation of their physiological and environmental basis

Butcher, D.N. and D.S. Ingram. 1976. *Plant Tissue Culture*. Studies in Biology, No. 65. University Park Press, Baltimore. Brief introduction to methods and applications of plant tissue and cell culture

Calvin, M. 1976. "Photosynthesis as a Resource for Energy and Materials." *Photochemistry and Photobiology*, Vol. 23. pp. 425–444. Partly nontechnical, partly uses background in Chapter 7

Carlson, P.S. 1980. *The Biology of Crop Productivity*. Academic Press, New York. Chapter 11 gives a good overview of major agricultural problems and future botanical applications touched on in this chapter; other chapters cover specific aspects in more detail, often depending on background contained in later chapters of this textbook

Chrispeels, M.J. and D. Sadava. 1977. *Plants, Food, and People*. W.H. Freeman & Co., San Francisco. Introductory text covering basics of botany and their role in the green revolution, its consequences, and other aspects of technological improvement of agriculture, as well as the world food problem

Mitsui, A., et al., (eds.). 1977. *Biological Solar Energy Conversion*. Academic Press, New York. A collection of articles by specialists on aspects of photosynthesis and its application to energy and nitrogen supply

Pimentel, D., et al. 1978. "Biological Solar Energy Conversion and U.S. Energy Policy." *BioScience*, Vol. 28. pp. 376–381. Evaluates the potential of photosynthesis for alleviating energy shortages

Rabinowitch, E., and Govindjee. 1969. *Photosynthesis*. John Wiley & Sons, New York. Chapter 1 gives a brief but cogent history of discovery of photosynthesis. Other chapters cover material relevant to Chapters 2, 6, 7, and 33 of this book

Sanderson, F.H., et al. 1975. "Food." *Science*, Vol. 188. pp. 503–649. An entire issue of articles on the food problem, and on botanical and agricultural research relevant thereto. Covers many of the topics touched on in last section of Chapter 1

San Pietro, A. (ed.). 1980. *Biochemical and Photosynthetic Aspects of Energy Production*. Academic Press, New York. Short articles on potential uses of plant photosynthesis and biomass for fuels, and related topics. Chapter 1 provides an excellent overview

Scott, T.K. (ed.). 1979. *Plant Regulation and World Agriculture*. Plenum Press, New York. Articles on general and specific aspects of many of the agricultural applications of botanical discoveries noted in this chapter

Smith, J.E. 1981. *Biotechnology*. Studies in Biology, No. 136. University Park Press, Baltimore. Short book covering many of the new areas of innovative applied biology noted in this and later chapters of this textbook, as well as other areas further removed from botany

Thomas, E., and M.R. Davey. 1975. *From Single Cells to Plants*. Springer-Verlag New York, New York. A short but information-packed book on plant tissue and cell culture and its genetic applications

Wittwer, S.H. 1979. "Future Technological Advances in Agriculture and Their Impact on the Regulatory Environment." *BioScience*, Vol. 29. pp. 603–610. Brief coverage of potential major developments in agriculture which are noted in this chapter

CHAPTER 2.   Nature, Kinds, and Names of Plants

Anderson, J.W. 1980. *Bioenergetics of Autotrophs and Heterotrophs.* Studies in Biology, No. 126. University Park Press, Baltimore. Compact but systematic coverage of the basic kinds of energy and carbon nutrition among organisms

Bailey, L.H. 1933. *How Plants Get Their Names.* Yale Research, Detroit; also Dover reprint. Explains the origin and nature of the plant naming system, with accounts of Linnaeus and his predecessors

Blunt, W. 1971. *The Compleat Naturalist, A Life of Linnaeus.* Viking Press, New York. Fascinating, beautifully illustrated account of Linnaeus and his times, showing how his profound influence on botany and indeed all biology developed

Corner, E.J.H. 1964. *The Life of Plants.* World Publishing, New York. A philosophical view of the plant kingdom and its evolution

Healey, B.J. 1972. *A Gardener's Guide to Plant Names.* Charles Scribner's Sons, New York. Introduction covers the rationale of naming plant groups and classifying cultivars

Harrington, H.D., and L.W. Durrell. 1957. *How to Identify Plants.* Swallow Press, Chicago. Gives concise explanation of procedures, terminology and information needed to identify plants using standard botanical manuals, and a summary of the naming system

Harris, R.M. 1969. *Plant Diversity.* Wm. C. Brown, Dubuque, IA. Brief illustrated introduction to plant groups

Hylander, C.J. 1956. *The World of Plant Life,* 2nd ed. Macmillan, New York. Chapters 1–9 give attractive non-technical introduction to thallophytes and lower vascular plants; rest of book extensively covers seed plant groups

Milne, L., and M. Milne. 1971. *The Nature of Plants.* J.B. Lippincott, Philadelphia. Chapters 1–3 give appealing, discursive coverage of major plant groups and their significance in nature, evolution, and human affairs

Tippo, O., and W.L. Stern. 1977. *Humanistic Botany.* W. W. Norton & Co., New York. Chapter 3 gives a more extensive introductory account of the principles of botanical naming and classification. Chapter 4 is devoted entirely to Linnaeus' life and accomplishments

CHAPTER 3.   The Seed Plant Plan and its Development

The subject of seed plant external form and development is mostly classical botanical understanding, now rarely if ever presented in one place in more depth and detail than the elementary textbook coverage. The following references cover at least some additional aspects of this basic subject.

Eames, A.J. 1961. *Morphology of the Angiosperms.* McGraw-Hill, New York. Chapter 1

Fritsch, F.E., and E.J. Salisbury. 1938. *Plant Form and Function.* G. Bell and Sons, London. Chapters 1–4 and 11–12

Gray, A. 1858. *How Plants Grow, A Simple Introduction to Structural Botany.* Ivison, Blakeman and Co., New York. Chapter 1 of this once popular introduction to botany by the leading American botanist of the last century covers systematically the material of our Chapter 3. Although long out of print, this book was very widely used and may still be found in many libraries

Jones, S.B., and A.E. Luchsinger. 1979. *Plant Systematics.* McGraw-Hill, New York. Chapter 10 gives systematically organized brief descriptions and illustrations of many points of vegetative and reproductive morphology and organ modification that are important for understanding and identifying plants

CHAPTER 4.   Organization and Function of the Cell

Buvat, R. 1969. *Plant Cells, An Introduction to Plant Protoplasm.* McGraw-Hill, New York. An information-packed, attractive paperback by an eminent French cytologist

Gunning, B.E.S. and M.W. Steer. 1975. *Ultrastructure and the Biology of Plant Cells.* Edward Arnold, London. Beautifully illustrated account of plant cells, cell types, and organelles

Gunning, B.E.S. and M.W. Steer. 1975. *Plant Cell Biology, An Ultrastructural Approach.* Crane, Russak and Co., New

York. Beautiful electron micrographs taken from the preceding work but with a much condensed text

Hall, J.L., T.J. Flowers, and R.M. Roberts. 1974. *Plant Cell Structure and Metabolism.* Longman Group Ltd., London. Thorough and well-illustrated coverage of material in our Chapters 4–7

Jensen, W.A. and R.B. Park. 1967. *Cell Ultrastructure.* Wadsworth, Belmont, CA. Collection of electron micrographs in large size format illustrating many features considered in this chapter

Ledbetter, M.C., and K.R. Porter. 1970. *Introduction to the Fine Structure of Plant Cells.* Springer-Verlag New York, New York. Handsome electron micrograph selection covering general cell organelles and features of differentiated cells

Northcote, D.H. 1974. *Differentiation in Higher Plants.* Oxford Biology Readers, No. 44. Carolina Biological Supply, Burlington, NC. Short paperback covering cell organelles and structure of differentiated cell types

Reinert, J. 1980. *Chloroplasts.* Springer-Verlag New York, New York. Chapters 1–2 give short but detailed coverage by

specialists, of structure, development, and interconversions of the various types of plastids

Robards, A.W. 1970. *Electron Microscopy and Plant Ultrastructure.* McGraw-Hill, New York. Introduction to the electron microscope and to specimen preparation, as well as excellent chapters on organelles, membrane systems, and the cell wall, with many helpful diagrams and micrographs

Robards, A.W. (ed.). 1974. *Dynamic Aspects of Plant Ultrastructure.* McGraw-Hill, New York. Valuable upper division level articles by specialists on all aspects of plant organelles and on differentiated cell types

Roland, J.C., A. Szöllösi, and D. Szöllösi. 1977. *Atlas of Cell Biology.* Little, Brown & Co., Boston. Attractive illustrations of cell structures, with considerable emphasis on plant cells. Useful introduction to ultrastructure methods at end of book

Swanson, C.P., and P.L. Webster. 1977. *The Cell.* Prentice-Hall, Englewood Cliffs, NJ. Relatively short but detailed coverage of material in this chapter as well as our Chapters 5, 6 and part of 21

CHAPTER 5. Cell Matter and Information

Albersheim, P. 1975. "The Walls of Growing Plant Cells." *Scientific American,* Vol. 232 (April). pp. 80–95. Unique, functionally important molecular features of the plant cell wall, by a principal investigator thereof

Chambon, P. 1981. "Split genes." *Scientific American,* Vol. 244 (May). pp. 60–71. Introduction to introns and processing by splicing in the expression of eukaryotic genes

Clark, B.F.C. 1977. *The Genetic Code.* Studies in Biology, No. 83. University Park Press, Baltimore. This small book thoroughly covers protein and nucleic acid structure, information transfer, and protein synthesis mechanism

Clarkson, D.T. 1974. *Ion Transport and Cell Structure in Plants.* John Wiley & Sons, New York. Chapter 2 covers structure of cellular membranes. Further chapters deal with energetics and mechanisms of ion transport, spelling out the complexities only alluded to in our Chapter 6

Hess, D. 1975. *Plant Physiology: Molecular, Biochemical, and Physiological Fundamentals of Metabolism and Development.* Springer-Verlag New York, New York. Pages 1–34 and 163–169 give a fairly detailed introduction to informational molecules. Pages 57–162 extensively cover material of our Chapter 6

Hollaway, M.R. 1976. *The Mechanism of Enzyme Action.* Oxford Biology Readers, No. 45. Carolina Biological Supply, Burlington, NC. Brief, cogent presentation of principles and examples of enzyme structure, active site, and energetics

Jordan, E.G. 1978. *The Nucleolus,* 2nd ed. Oxford Biology Readers, No. 16. Carolina Biological Supply, Burlington, NC. Fifteen pages of current information on the role of the nucleolus in ribosome formation, with illustrations mainly taken from plant cells

Kornberg, R.D., and A. Klug. 1981. "The Nucleosome." *Scientific American,* Vol. 244 (February). pp. 52–64. An account of the discovery and structure of the repeating organizational unit of chromatin

Leshem, Y. 1973. *The Molecular and Hormonal Basis of Plant-Growth Regulation.* Pergamon Press, New York. Chapters 1–6 give a brief introduction to informational macromolecules, their processing and regulation; more detailed than present chapter

Phillips, D.C., and A.C.T. North. 1973. *Protein Structure.* Oxford Biology Readers, No. 34. Carolina Biological Supply, Burlington, NC. Principles and examples of protein shape and its functional significance are revealingly illustrated with 3-D stereoscopic diagrams

Smith, H. (ed.). 1977. *The Molecular Biology of Plant Cells.* University of California Press, Berkeley, CA. Detailed and valuable advanced articles on cell components, organelles, and information processing, plus material on regulation pertinent to Chapters 12 and 14; requires biochemistry background

Stace, C.A., and A.C. Shaw. 1971. *A Guide to Subcellular Botany.* Longman Group, London. Chapters 1–6 give a brief, useful introductory survey of plant biochemistry; more detailed than this textbook

Travers, A.A. 1978. *Transcription of DNA,* 2nd ed. Oxford Biology Readers, No. 75. Carolina Biological Supply, Burlington, NC. Fifteen pages of condensed current information on the mechanism of transcription

Some of the references cited for Chapter 6 also cover material pertinent to this chapter.

CHAPTER 6. Cell Metabolism, Energy, and Transport

Becker, W.M. 1977. *Energy and the Living Cell, An Introduction to Bioenergetics.* J.B. Lippincott, Philadelphia. Chapters 1–2 give general background and introduction to the theory of energetics (thermodynamics). Chapters 4–6 and 8–9 cover excellently the areas of this chapter, with considerably more sophistication. Chapter 3 covers principles of enzyme action

Bell, E.A. 1980. "The Non-Protein Amino Acids of Higher Plants." *Endeavor.* New Series, Vol. 3. pp. 102–107. Nature and significance of a bizarre but widespread class of plant secondary products

Hinkle, P.C., and R.E. McCarty. 1978. "How Cells Make ATP." *Scientific American,* Vol. 238 (March). pp. 104–123. Explains evidence for chemiosmotic mechanism of phosphorylation in readily understandable terms

Kumar, H.D. and H.N. Singh. 1976. *Plant Metabolism.* Affiliated East-West Press Ltd, New Delhi. Brief and readable introductory account of most aspects of plant metabolism including secondary products

Nicholls, P. 1975. *Cytochromes and Biological Oxidation.* Oxford Biology Readers, No. 66. Carolina Biological Supply, Burlington, NC. Very interesting short account of the discovery of cytochromes and the evidence for their role in mitochondrial respiration

Öpik, H. 1982. *The respiration of higher plants.* Studies in Biology, No. 120. University Park Press, Baltimore. Short book covering plant respiration and fermentation, more detailed than this chapter

Richter, G. 1978. *Plant Metabolism.* G. Thieme, Stuttgart. Very detailed but compact coverage of plant respiration and other aspects of metabolism; requires chemistry background

Robinson, T. 1980. *The Organic Constituents of Higher Plants,* 4th ed. Cordus Press, N. Amherst, MA. Surveys the diversity of plant metabolites and secondary products of all kinds, plus information on their biosynthetic pathways

Street, H.E. and W. Cockburn. 1972. *Plant Metabolism,* 2nd ed. Pergamon, New York. Fairly discursive introductory coverage of most aspects of plant metabolism as of its date

Vickery, M.L., and B. Vickery. 1981. *Secondary Plant Metabolism.* University Park Press, Baltimore. Thorough but condensed coverage of secondary product classes, examples, their biosynthesis, ecological roles and chemotaxonomic uses, for students with organic chemistry background

Walker, J.R.L. 1975. *The Biology of Plant Phenolics.* Studies in Biology, No. 54. University Park Press, Baltimore. Short book covering these important secondary products and their ecological and functional significance

See also references cited for Chapter 5.

CHAPTER 7. Photosynthesis: The Chloroplast

Buvet, R., M.J. Allen, and J.P. Massué (eds). 1977. *Living Systems as Energy Converters.* North-Holland Publishing, Amsterdam, New York. Articles, by specialists, on mechanisms as well as applications of photosynthetic energy conversion

Clayton, R.D. 1974. *Photosynthesis: How Light is Converted to Chemical Energy.* Modules in Biology, No. 13. Addison-Wesley, Reading, MA. Forty-page introduction to the photoprocesses of photosynthesis and their experimental evidence, with short appendix on relevant physical principles

Clayton, R.D. 1980. *Photosynthesis: Physical Mechanisms and Chemical Patterns.* Cambridge University Press, New York. This small book goes deeply into current views of the biophysics of photosynthesis and their historical development, beginning with basic principles. For advanced students

Govindjee, and R. Govindjee. 1974. "The Primary Events of Photosynthesis." *Scientific American,* Vol. 231, (December). pp. 68–82. Basics of photosynthetic light capture, energy transfer, and conversion, by outstanding investigators of this field

Gregory, R.P.F. 1977. *Biochemistry of Photosynthesis,* 2nd ed. John Wiley & Sons, Wiley-Interscience, New York. Chapters 2–5 give a general introduction to physical and biochemical aspects of photosynthesis; requires chemistry background

Hall, D.O., and K.K. Rao. 1981. *Photosynthesis,* 3rd ed. Studies in Biology, No. 37. University Park Press, Baltimore. Good short account of principles as well as techniques of studying photosynthesis

Krogmann, D. 1973. *The Biochemistry of Green Plants.* Prentice-Hall, Englewood Cliffs, NJ. Readable, lucid explanation of the physical and biochemical aspects of photosynthesis, plus coverage of some of the material of Chapters 6 and 11

Miller, K.R. 1979. "The Photosynthetic Membrane." *Scientific American,* Vol. 241 (October). pp. 102–113. Recent information on particle structure of thylakoid membranes and how they function

Whittingham, C.P. 1974. *The Mechanism of Photosynthesis.* Edward Arnold, London. Although no longer fully up to date, this is still one of the best short, readable coverages of the basics of the subject. Chapter 1 covers briefly but quantitatively the gas exchange aspects given in Chapters 8–9 of this textbook

Several valuable recent books on photosynthesis and its applications are cited in the reading list for Chapter 1.

CHAPTER 8. The Leaf: Structure and Photosynthetic Activity

Björkman, O., and J. Berry. 1973. "High-efficiency Photosynthesis." *Scientific American*, Vol. 10 (October). pp. 80–93. Introductory discussion of C4 photosynthesis and its ecological significance

Burris, R.H. and C.C. Black. 1976. *CO₂ Metabolism and Plant Productivity.* University Park Press, Baltimore. Much material on C4 photosynthesis and photorespiration, mostly for advanced students

Goldsworthy, A. 1976. *Photorespiration.* Oxford Biology Readers, No. 80. Carolina Biological Supply, Burlington, NC. Fifteen-page discussion of the causes and mechanism of photorespiration and the C4 photosynthetic adaptation to reduce it

Good, N.E., and D.H. Bell. 1980. "Photosynthesis, Plant Productivity, And Crop Yield." *In The Biology of Crop Productivity.* ed. P.S. Carlson. Academic Press, New York, Chapter 1. Applications of principles covered in this chapter and Chapter 9

Heath, O.V.S. 1969. *Physiological Aspects of Photosynthesis.* Stanford University Press, Stanford, CA. Small book on pigment systems and gas exchange mechanism of leaves, by a long time investigator of this subject

Woolhouse, H.W. 1978. "Light-Gathering and Carbon Assimilation Processes in Photosynthesis; Their Adaptive Modifications and Significance for Agriculture." *Endeavor.* New Series, Vol. 2. pp. 35–46. Nicely illustrated account of C4 photosynthesis and its ecological and agricultural implications

CHAPTER 9. Leaf Transpiration and Water Balance

Gates, D.M. 1965. "Heat Transfer in Plants." *Scientific American*, Vol. 213 (December). pp. 76–84. Shows the importance of transpiration and other heat exchange mechanisms in fixing the temperature of a leaf, an important determinant of its photosynthetic performance

Gates, D.M. 1980 *Biophysical Ecology.* Springer-Verlag New York, New York. Chapters 2, 3, 11, and 14 cover leaf gas exchange, transpiration, and energy balance to a considerable level of complexity, for advanced students

Heath, O.V.S. 1975. *Stomata.* Oxford Biology Readers, No. 37. Carolina Biological Supply, Burlington, NC. Thorough 30-page introductory exposition of methods, phenomena, and mechanisms of stomatal responses

Kluge, M., and I.P. Ting. 1978. *Crassulacean Acid Metabolism.* Ecological Studies, Vol. 30. Springer-Verlag New York, New York. Small book on succulent plant CAM and its ecological significance

Martin, J.T. and B.E. Juniper. 1970. *The Cuticles of Plants.* St. Martin's Press, New York. Monograph on all aspects of this essential external barrier

Meidner, H., and D.W. Sheriff. 1976. *Water and Plants.* John Wiley & Sons, New York. Chapter 2 is a good introductory quantitative treatment of leaf gas exchange and transpiration

Nobel, P.S. 1974. *Introduction to Biophysical Plant Physiology.* W. H. Freeman & Co., San Francisco. Sophisticated, valuable quantitative treatment of subjects of this chapter and Chapters 10–11, for students with physics and chemistry background

Rutter, A.J. 1972. *Transpiration.* Oxford Biology Readers, No. 24. Carolina Biological Supply, Burlington, NC. Short presentation of the quantitative principles of stomatal diffusion and resultant transpiration performance

Sheriff, D.W. 1979. "Stomatal Aperture and the Sensing of the Environment by Guard Cells." *Plant, Cell, and Environment*, Vol. 2. pp. 15–22. Short, critical review of stomatal control mechanisms

CHAPTER 10. The Stem and Vascular System

Baker, D.A. 1978. *Transport Phenomena in Plants.* Chapman and Hall, London. Chapter 5 and pp. 66–72 give an excellent brief evaluation of competing theories of phloem translocation

Meidner, H., and D.W. Sheriff. 1976. *Water and Plants.* John Wiley & Sons, New York. Chapter 1 and 3 are an excellent quantitative introduction to water transport. Chapter 4 enlarges on the subject of soil water covered in our Chapter 11

Milburn, J.A. 1979. *Water Flow in Plants.* Longman, New York. Compact, somewhat mathematical coverage of water transport and its physical principles, including water potential. Chapters 6 and 7 cover material of our Chapter 9

Richardson, M. 1968. *Translocation in Plants.* Studies in Biology, No. 10. University Park Press, Baltimore. Short book covering xylem and phloem transport in more depth than this chapter

Wardlaw, I.F. 1980. "Translocation and source-sink relationships." *In The Biology of Crop Productivity.* ed. P.S. Carlson. Academic Press, New York, Chapter 8. Nice presentation of the phenomena of phloem translocation and their relation to plant performance

Wooding, F.B.P. 1977. *Phloem.* Oxford Biology Readers, No. 15. Carolina Biological Supply, Burlington, NC. Strikingly illustrated account of sieve tube structure and the disputed translocation mechanism

Zimmermann, M.H. 1967. "How Sap Moves in Trees." *In From Cell to Organism.* ed. D. Kennedy. Readings from *Scientific American*, W.H. Freeman & Co., San Francisco, pp. 71–79. Short, clear and well-illustrated presentation of the problems about xylem and phloem transport

Zimmermann, M.H., and C.L. Brown. 1971. *Trees: Structure and Function.* Springer-Verlag New York, New York. Chapters 4–7 give one of the best general presentations of long distance transport phenomena in plants

CHAPTER 11. The Root, The Soil, and Plant Nutrients

Baker, D.A. 1978. *Transport Phenomena in Plants.* Chapman and Hall, London. Chapters 2 and 3 give an excellent short account of principles of ion uptake by cells and tissues.

Beevers, L. 1976. *Nitrogen Metabolism in Plants.* E. Arnold, London. Compactly but thoroughly covers all aspects of nitrogen utilization

Berger, K.C. 1972. *Sun, Soil, and Survival, An Introduction to Soils.* University of Oklahoma Press, Norman, OK. This and other soil science textbooks can be consulted to extend the limited information on soils for which there was space in this chapter

Epstein E. 1973. "Roots." *Scientific American*, Vol. 228 (May). pp. 48–55. Cogent exposition of water and nutrient uptake and their relation to the solution of practical problems

Harris, D. 1974. *Hydroponics, Growing Without Soil,* Revised ed. David and Charles (Holdings) Ltd, Vancouver and London. Practical manual explaining a variety of hydroponic techniques easily used by home- or apartment-dwellers for growing their own crops and flowers

Hewitt, E.J., and T.A. Smith. 1975. *Plant Mineral Nutrition.* John Wiley & Sons, New York. Good coverage of requirements, functions, and deficiency diseases

Lüttge, U., and N. Higinbotham. 1979. *Transport in Plants.* Springer-Verlag New York, New York. Extensive discussion of principles and experimental results on ion transport

Russell, R.S. 1977. *Plant Root Systems: Their Function and Interaction with the Soil.* McGraw-Hill, New York. Much compactly presented, interesting information, mainly from the whole-plant and soil science points of view

Sutcliffe, J.F., and D.A. Baker. 1974. *Plants and Mineral Salts.* Studies in Biology, No. 48. University Park Press, Baltimore. Chapters 1 and 2 are good general reading on mineral nutrition

Winter, E.J. 1974. *Water, Soil, and the Plant.* Macmillan, New York. Short introductory book on soil water relations and their practical significance

CHAPTER 12. Primary Growth and Its Hormonal Control

Clowes, F.A.L. 1972. *Morphogenesis of the Shoot Apex.* Oxford Biology Readers, No. 23. Carolina Biological Supply, Burlington, NC. Nicely illustrated brief exposition of apical meristems and their actions

Dale, J.E. 1982. *The Growth of Leaves.* Studies in Biology, No. 137. University Park Press, Baltimore. Condensed but revealing exposition of leaf initiation, growth, morphogenesis and differentiation processes, and their physiological regulation

Gunning, B.E.S., and A.R. Hardham. 1979. "Microtubules and Morphogenesis in Plants." *Endeavor.* New Series, Vol. 3. pp. 112–117. Short, well-illustrated account of the important developmental roles of these organelles

Hill, T.A. 1973. *Endogenous Plant Growth Substances.* Studies in Biology, No. 40. University Park Press, Baltimore. This short book covers discovery and diverse actions of plant hormones, including some information on experimental methods

Kiermeyer, O. (ed.). 1981. *Cytomorphogenesis in Plants.* Springer-Verlag New York, New York. Part II comprises several advanced review articles covering ultrastructural mechanisms of morphogenesis and cell differentiation in higher plants

Langer, R.H.M. 1972. *How Grasses Grow.* Studies in Biology, No. 34. University Park Press, Baltimore. Explains intercalary meristem growth and many other aspects of grass vegetative and reproductive development

Luckwill, L.C. 1981. *Growth Regulators and Crop Production.* Studies in Biology, No. 129. University Park Press, Baltimore. Short survey of agricultural and horticultural uses of plant hormones, including material pertinent also to our Chapters 14, 15, and 17

Moore, T.C. 1979. *Biochemistry and Physiology of Plant Hormones.* Springer-Verlag New York, New York. This small book compactly presents current aspects of action of plant hormones

Nickell, L.G. 1982. *Plant Growth Regulators—Agricultural Uses.* Springer-Verlag New York, New York. Extensive survey of practical applications of plant hormones, including aspects relevant to our Chapters 13–17

O'Brien, T.P. and M.E. McCully. 1969. *Plant Structure and Development. A Pictorial and Physiological Approach.* Macmillan, New York. Chapters 2–7 give beautiful light and electron micrographs illustrating material of this chapter; their Chapters 8–9 similarly illustrate aspects of our Chapters 16 and 17

Roland, J.-C., and F. Roland. 1980. *Atlas of Flowering Plant Structure.* Longman, New York. Valuable transmission and scanning electron micrographs show many features of plant development as well as reproduction

Thimann, K.V. 1977. *Hormone Action in the Whole Life of Plants.* University of Massachusetts Press, Amherst, MA. Broad but also personal coverage of many aspects of plant hormone phenomena by a pioneer of this field

Weaver, R.J. 1972. *Plant Growth Substances in Agriculture.* W. H. Freeman & Co., San Francisco. Chapters 1–4 give useful history of hormone discoveries and illustrated descriptions of simple laboratory experiments for testing their effects. The rest of the book is an extensive survey of many agricultural and horticultural uses

CHAPTER 13. Secondary Growth: Structure and Formation of Wood and Bark

Esau, K. 1977. *Anatomy of Seed Plants,* 2nd ed., John Wiley & Sons, New York. Chapters 8–12 and 17 give a much more complete account of secondary growth and secondary tissues than this chapter. Comparable material is covered in the same author's *Plant Anatomy,* 2nd ed. (Wiley 1965), Chapters 6 and 11–15

Fritts, H.C. 1976. *Tree Rings and Climate.* Academic Press, New York. Extensive book on the ecological and physiological basis of tree ring chronologies

Kramer, P.J., and T.T. Kozlowski. 1979. *Physiology of Woody Plants.* Academic Press, New York. Different aspects of the physiology of cambial growth are covered in several places in the book, consult index

Loewus, F.A., and V.C. Runeckles (eds.). 1977. *The Structure, Biosynthesis, and Degradation of Wood.* Recent Advances in Phytochemistry, Vol. 11. Plenum Press, New York. Articles on wood ultrastructure, wall polymer biosynthesis, and the utilization and post-harvest or pathological changes in wood components, mostly using considerable chemistry background

Meylan, B.A., and B.G. Butterfield. 1972. *Three-Dimensional Structure of Wood.* Chapman and Hall, London. Collection of dramatic and instructive scanning electron micrographs, with accompanying explanations, showing many aspects of diverse kinds of wood structure

Morey, P.R. 1973. *How Trees Grow.* Studies in Biology, No. 39. University Park Press, Baltimore. Small book devoted entirely to cambial growth, wood and bark formation and structure

Zimmermann, M.H., and C.L. Brown. 1971. *Trees: Structure and Function.* Springer-Verlag New York, New York. Chapter 2 gives extensive exposition of secondary growth; Chapters 1 and 3, on form and primary growth, are relevant to our Chapters 12 and 14

CHAPTER 14. Environmental and Genetic Control of Development

Audus, L.J. 1979. "Plant Geosensors." *Journal of Experimental Botany,* Vol. 30. pp. 1051–1073. Good readable review of gravity perception by plants in geotropism

Black, M., and J. Edelman. 1970. *Plant Growth.* Harvard University Press, Cambridge, MA. Readable and fairly complete short coverage of material of this chapter and Chapter 12; some points presented as facts here have been superseded by more recent research.

Galston, A.W., P.J. Davies, and R.L. Satter. 1980. *The Life of the Green Plant.* Prentice-Hall, Englewood Cliffs, NJ. Chapters 9–14 give a more extensive presentation of material of this chapter and physiological aspects of our Chapters 12 and 16

Gibor, A. 1966. "Acetabularia, A Useful Giant Cell." *Scientific American,* Vol. 215 (November). pp. 118–124. Reviews classical grafting and nuclear transplantation experiments on the nucleus' role in development

Kendrick, R.E., and B. Frankland. 1976. *Phytochrome and Plant Growth.* Studies in Biology, No. 68. University Park Press, Baltimore. Brief, readable survey

Leopold, A.D., and P.E. Kriedemann. 1975. *Plant Growth and Development,* 2nd ed. McGraw-Hill, New York. Good general presentation, as of its date, of all classes of hormone and environmental responses; well illustrated by experimental data

Mohr, H. 1977. "Phytochrome and Chloroplast Development." *Endeavor*. New Series, Vol. 1. pp. 107–114. Short but informative account of phytochrome action by a prominent German researcher thereof

Steeves, T.A., and I.M. Sussex. 1972. *Patterns in Plant Development*. Prentice-Hall, Englewood Cliffs, NJ. Compact review of experiments on genetic and hormonal basis of morphogenesis and differentiation in meristems and cultures, mostly at the non-molecular level

Torrey, J.G. 1967. *Development in Flowering Plants*. Macmillan, New York. Valuable short presentation as of its date, of genetic, cellular, and physiological aspects of developmental control, including cell culture appproaches

Wareing, P.F., and I.D.J. Phillips. 1981. *Growth and Differentiation in Plants*, 3rd ed. Pergamon Press, Elmsford, NY. Good introduction and short general coverage of hormonal and environmental aspects of this chapter and Chapter 12

Whatley, J.M., and F.R. Whatley. 1980. *Light and Plant Life*. Studies in Biology, No. 124. University Park Press, Baltimore. Short account of the diversity of developmental and tropic light responses, their mechanisms and ecological roles

References on plant tissue culture cited in the list for Chapter 1 contain material relevant to this chapter. Books on plant hormones cited for Chapter 12 contain much material enlarging on the environmental responses and growth regulator action topics covered in this chapter.

CHAPTER 15. Vegetative Reproduction

Abrahamson, W.G. 1980. "Demography and Vegetative Reproduction." *In Demography and Evolution in Plant Populations.* ed. O.T. Solbring. University of California Press, Berkeley. pp. 89–106. Excellent short overview of ecological and evolutionary significance of vegetative reproduction

Galil, J. 1980. "Kinetics of bulbous plants." *Endeavor*. New Series, Vol 5. pp. 15–20. Describes and illustrates multiplication of bulbs and corms and their positioning and migration by means of contractile roots

Hartmann, H.T. and D.E. Kester. 1975. *Plant Propagation*, 3rd ed. Prentice-Hall, Englewood Cliffs, NJ. Chapters 8–16 give a very informative survey of principles and practices of vegetative propagation, especially the many kinds of grafting

Murashige, T. 1974. "Plant Propagation Through Tissue Cultures." *Annual Review of Plant Physiology*, Vol. 25. pp. 135–166. Good introduction to this rapidly developing subject

Reinert, J., and Y.P.S. Bajaj. 1977. *Applied and Fundamental Aspects of Plant Cell, Tissue, and Organ Culture.* Springer-Verlag New York, New York. Chapter 1 treats propagation by tissue cultures much more extensively than the preceding reference

Rice, L.W., and R.P. Rice, Jr. 1980. *Practical Horticulture, A Guide to Growing Indoor and Outdoor Plants.* Saunders College Publishing/Holt, Rinehart and Winston, Philadelphia. Much well-presented practical information on vegetative propagation, grafting, seed propagation, etc.

CHAPTER 16. Sexual Reproduction: The Flower

Brady, J. 1979. *Biological Clocks*. Studies in Biology, No. 104. University Park Press, Baltimore. Covers role of circadian rhythms in photoperiodic time measurement, plus other clock phenomena

Bristow, A. 1978. *The Sex Life of Plants*. Holt, Rinehart and Winston, New York. Fascinating popular level excursion into plant sexual reproduction and important practical aspects

Clarke, A.E., and P.A. Gleeson. 1981. "Molecular Aspects of Recognition and Response in the Pollen-Stigma Interaction." *Recent Advances in Phytochemistry*, Vol. 15. pp. 161–211. Brief but cogent presentation of self-incompatibility mechanisms, plus biochemical work on nature of cell-cell recognition factors involved

Duddington, C.L. 1969. *Evolution in Plant Design*. Faber and Faber, London. Chapters 7–8 are a valuable discussion of pollination mechanisms and adaptations

Esau, K. 1977. *Anatomy of Seed Plants*, 2nd ed. John Wiley & Sons, New York. Chapters 20–21 cover flower development, structure and ultrastructure, and the reproductive process much more extensively than this chapter

Evans, L.T. 1975. *Daylength and the Flowering of Plants*. Benjamin/Cummings, Menlo Park, CA. Short but penetrating account of photoperiodic phenomenon, mechanism, and the flowering hormone issue

Galil, J. 1977. "Fig Biology." *Endeavor*. New Series, Vol. 1. pp. 52–56. Short fascinating account of a geologically ancient obligatory pollination symbiosis

Hillman, W.S. 1979. *Photoperiodism in Plants and Animals*. Oxford Biology Readers, No. 107. Carolina Biological Supply, Burlington, NC. Short account of phenomena and mechanisms in flowering and various other responses

Holm, E. 1979. *The Biology of Flowers*. Penguin Books, New York. Short, colorfully illustrated explanations of many kinds of pollination and other interesting aspects of flower reproduction

Jaeger, P. 1961. *The Wonderful Life of Flowers.* E.P. Dutton, New York. Marvellously illustrated account of many aspects of flower function, form, and structure

Knox, R.B. 1979. *Pollen and Allergy.* Studies in Biology, No. 107. University Park Press, Baltimore. This short book attractively covers pollen development, pollination mechanisms, recognition and fertilization, as well as hay fever

Lewis, D. 1979. *Sexual Incompatibility in Plants.* Studies in Biology, No. 110. University Park Press, Baltimore. Short but thorough book on genetics, physiology, and molecular nature of self-incompatibility systems in flowers, with a chapter on sexual incompatibility systems of lower plants

Meeuse, B.J.D. 1961. *The Story of Pollination.* Ronald Press, New York. Readable, short but informative account of the amazing diversity of adaptations for pollination

Pettitt, J., S. Ducker and B. Knox. 1981. "Submarine Pollination." *Scientific American,* Vol. 244 (March). pp. 135–143. Bizarre specializations for aquatic pollination, not covered in this chapter

Proctor, M., and P. Yeo. 1972. *The Pollination of Flowers.* Taplinger, New York. Extensive, marvellously illustrated book on diverse pollination behavior of insects, and floral adaptations thereto

CHAPTER 17. Embryo, Seed, and Fruit

Bewley, J.D., and M. Black. 1978. *Physiology and Biochemistry of Seeds.* Vol. 1. Development, germination, and growth. Springer-Verlag New York, New York. Detailed account of seed structure, reserves, germination, metabolism, and their control. A second volume on seed dormancy and longevity is forthcoming

Black, M. 1972. *Control Processes in Germination and Dormancy.* Oxford Biology Readers, No. 20. Carolina Biological Supply, Burlington, NC. Short account of experimental evidence for hormonal and phytochrome control of seed germination and food reserve mobilization

Cook, R. 1980. "The Biology of Seeds in the Soil." *In Demography and Evolution in Plant Populations.* ed. O.T. Solbrig. University of California Press, Berkeley. pp. 107–129. Compactly presented, interesting and ecologically relevant information on dynamics of seed populations in soils

Duddington, C.L. 1969. *Evolution in Plant Design.* Faber and Faber, London. Chapter 9 is useful short coverage of seed dispersal mechanisms

Esau, K. 1977. *Anatomy of Seed Plants,* 2nd ed. John Wiley & Sons, New York. Chapters 22–24 give fairly extensive in-

formation on characteristics and structure of fruit and seed types, and embryo development

Maheshwari, P. 1950. *Introduction to the Embryology of Angiosperms.* McGraw-Hill, New York. Synthesis of the classical research on the diverse features of embryo sac and embryo development in flowering plants

Mayer, A.M., and A. Poljakoff-Mayber. 1982. *Germination of Seeds,* 3rd ed. Pergamon Press, New York. General, relatively short account of seed structure, biology, and current research problems about seeds

Steeves, T.A., and I.M. Sussex. 1972. *Patterns in Plant Development.* Prentice-Hall, Englewood Cliffs, NJ. Chapters 2, 3, and 17 present experimental and culture work on embryo and embryoid formation and development

U.S. Department of Agriculture, 1961. "Seeds." *The Yearbook of Agriculture (1961).* U.S. Government Printing Office, Washington, D.C. A collection of short articles on many aspects of seed biology and technology

References on plant growth regulators in lists for Chapters 12 and 14 mostly include chapters on physiology of fruit and seed development and seed germination.

CHAPTER 18. Inheritance

Beale, G., and J. Knowles. 1979. *Extranuclear Genetics.* University Park Press. Baltimore. Short introductory account of organelle inheritance

Brettell, R.I.S., and W.S. Ingram. 1979. "Tissue Culture in the Production of Novel Disease-Resistant Crop Plants." *Biological Reviews,* Vol. 54. pp. 329–345. Thorough, critical review of principles of tissue culture applications to crop improvement

Brock, R.D. 1980. "Mutagenesis and Crop Improvement." *In The Biology of Crop Productivity.* ed. P.S. Carlson Academic Press, New York, Chapter 10. Good general coverage of theory and practice of current plant breeding methods for crop improvement. Accompanying Chapter 9 by V. Walbot presents prospects for molecular genetic-engineering approaches to crop improvement

Chaleff, R.S. 1981. *Genetics of Higher Plants—Applications of Cell Culture.* Cambridge University Press. Cambridge and New York. Chapters 1 and 2 give lucid, brief exposition of plant genetics, gene regulation, and their relationship to the potential for plant improvement by genetic engineering. Rest of this excellent small book covers current genetic work with plant cell and tissue cultures in some depth

Grant, V. 1975. *Genetics of Flowering Plants.* Columbia University Press, New York. Extensive information on plant genetics and its ecological and evolutionary implications

Harpstead, D.D. 1971. "High-Lysine Corn." *Scientific American,* Vol. 225 (August). pp. 34–42. Example of efforts to improve the protein nutritional value of cereal grains

Jinks, J.L. 1976. *Cytoplasmic Inheritance.* Oxford Biology Readers, No. 72. Carolina Biological Supply, Burlington, NC. Brief but informative discussion of extranuclear genomes, with emphasis on plants and fungi

John, B., and K.R. Lewis. 1973. *The Meiotic Mechanism.* Oxford Biology Readers, No. 65. Carolina Biological Supply, Burlington, NC. Valuable 30-page detailed account of meiosis and its mechanism and consequences, strikingly illustrated

Lawrence, W.J.C. 1968. *Plant Breeding.* Studies in Biology, No. 12. University Park Press, Baltimore. Short, useful systematic survey of the objectives and variety of methods of conventional plant breeding, i.e. using sexual hybridization

Markham, R., (ed.). 1975. *Modification of the Information Content of Plant Cells.* American Elsevier, New York. Symposium articles on most kinds of unconventional gene transfer contemplated for plant cells as of its date

Office of Technology Assessment. 1982. *Genetic Technology, A New Frontier.* Westview Press, Boulder, CO. Chapter 8, "The application of genetics to plants," gives terse current assessment of prospects for unconventional methods of genetic improvement

Reinert, J. (ed.). 1980. *Chloroplasts.* Springer-Verlag New York, New York. Advanced detailed articles, by specialists, on plastid genome and partial biosynthetic autonomy, etc.

Shepard, J.F. 1982. "The Regeneration of Potato Plants from Leaf-Cell Protoplasts." *Scientific American,* Vol. 246 (May). pp. 154–166. Single-cell cloning and its potential for isolating variants useful for crop improvement

Tippo, O., and W.L. Stern. 1977. *Humanistic Botany.* W.W. Norton & Co., New York. Chapter 18 gives an account of Mendel's life. The preceding chapter covers basic genetics and its origins

Vasil, I.K. 1980. *Perspectives in Plant Cell and Tissue Culture.* International Review of Cytology, Supplement 11A. Academic Press, New York. Advanced but relatively condensed articles covering phenomena and special applications including regeneration, clonal propagation, haploid plants from anthers, and isolation of mutants

CHAPTER 19.   Ecological Adaptation

Duddington, C.L. 1969. *Evolution in Plant Design.* Faber and Faber, London. Chapters 10–11 discuss the variety of morphological adaptations of plants to their environment

Edwards, P.J., and S.D. Wratten. 1980. *Ecology of Insect-Plant Interactions.* Studies in Biology, No. 121. University Park Press, Baltimore. Short book covering plant secondary products and their role in defense against herbivorous insects

Etherington, J.R. 1974. *Environment and Plant Ecology.* John Wiley & Sons, New York. Appealing, informative presentation of this chapter's subject and material of our Chapter 33

Fitter, A.H., and R.K.M. Hay. 1981. *Environmental Physiology of Plants.* Academic Press, New York. Excellent advanced coverage of mechanisms of plant adaptation to climatic, edaphic, and biotic factors

Gemmell, R.P. 1977. *Colonization of Industrial Wasteland.* Studies in Biology, No. 80. University Park Press, Baltimore. Short but chemically detailed account of industrial soil contamination and modification problems, and plant adaptations thereto

Grime, J.P. 1979. *Plant Strategies and Vegetation Processes.* John Wiley & Sons, New York. Part 1 concisely but very informatively covers stress tolerance and adaptive strategies of plants, going much beyond the introduction in this chapter

Hill, T.A. 1977. *The Biology of Weeds.* Studies in Biology, No. 79. University Park Press, Baltimore. Adaptive strategies of colonizing species, and plant competition experiments

Larcher, W. 1980. *Physiological Plant Ecology,* 2nd ed. Springer-Verlag New York, New York. Compact but information-packed coverage of background of our Chapters 19 and 33

Mussell, H., and R.C. Staples (eds.). 1979. *Stress Physiology in Crop Plants.* John Wiley & Sons, New York. Articles covering many aspects of stress tolerance and their importance and improvement in agricultural situations

Paleg. L.G., and D. Aspinall (eds.). 1981. *The Physiology and Biochemistry of Drought Resistance in Plants.* Academic Press, New York. General chapter on drought significance; detailed articles for advanced students on many aspects of water stress adaptation not covered in this chapter

Slack, A. 1979. *Carnivorous Plants.* MIT Press, Cambridge, MA. Beautifully illustrated short accounts of these curious, diverse specialized plants and how to grow them

Tivy, J. 1971. *Biogeography, A Study of Plants in the Ecosphere.* Oliver and Boyd (Longman Group), Edinburgh. Excellent, very perceptive enlargement on all aspects of this chapter as well as our Chapters 33–34; highly recommended

Villiers, T.A. 1975. *Dormancy and the Survival of Plants.* Studies in Biology, No. 57. University Park Press, Baltimore. Brief, readable coverage of the various dormancy phenomena and their ecological importance

Woodell, S.R.J. 1973. *Xerophytes.* Oxford Biology Readers, No. 39. Carolina Biological Supply, Burlington, NC. Attractive, brief presentation of morphological and physiological adaptations of desert plants

CHAPTER 20.   Evolution

Ayala, F.J. 1978. "The Mechanisms of Evolution." *Scientific American,* Vol. 239 (September). pp. 56–69. Molecular nature of gene mutations and genetic variability in populations, the raw material of evolution

Boulter, D. 1974. "The Evolution of Plant Proteins, With Special Reference to Higher Plant Cytochromes C." *Current Advances in Plant Science*, Vol. 4, Commentaries in Plant Science No. 8. pp. 1–16. Brief but useful discussion of methods and results of studying protein molecular evolution in plants

Darwin, C. 1962 (reprint). *The Voyage of the Beagle.* Doubleday & Co., Garden City, NY. This reissue of Darwin's classic chronicle gives a good introduction to evolution of his own concepts

Dickerson, R.E. 1980. "Cytochrome C and the Evolution of Energy Metabolism." *Scientific American*, Vol. 242 (March). pp. 136–153. Outstanding example of molecular evolution, enlarging greatly on material touched on in this chapter. Author's earlier article on same subject in April 1972 *Scientific American* is also valuable

Dobzhansky, T., F. Ayala, L. Stebbins and J. Valentine. 1977. *Evolution.* W.H. Freeman & Co., San Francisco. General text by leaders in the field

Ferguson, M. 1980. *Biochemical Systematics and Evolution.* John Wiley & Sons, New York. A wide range of material on molecular evolution

Gilbert, L.E. and P.H. Raven. 1980. *Coevolution of Animals and Plants,* Revised ed. University of Texas Press, Austin, TX. Valuable short articles on many facets of the evolution of plant/animal interactions

Grant, V. 1981. *Plant Speciation,* 2nd ed. Columbia University Press, New York. Discursive, readable account of mechanisms of plant species formation, illustrated by good examples drawn especially from the author's own extensive research; for more advanced students

Parkin, D.T. 1979. *An Introduction to Evolutionary Genetics.* University Park Press, Baltimore. Short but informative exposition of genetic mechanisms of evolution, extending well beyond material in this chapter

Stebbins, G.L. 1971. *Processes of Organic Evolution,* 2nd ed. Prentice-Hall, Englewood Cliffs, NJ. Excellent introduction to mechanisms of plant and animal evolution by a distinguished investigator of plant evolution. His 1950 treatise, *Variation and Evolution in Plants,* Columbia University Press, is still valuable reading, with many good ideas and points of information

Tippo, O., and W.L. Stern. 1977. *Humanistic Botany.* W.W. Norton & Co., New York. Chapter 20, devoted entirely to Darwin, is nicely illustrated. The preceding chapter on evolution recounts some of the contemporary social and political repercussions of the appearance of the theory

CHAPTER 21.   Plants Through Time: Origin, Evolutionary History, and Classification

Arber, A. 1953. *Herbals, Their Origin and Evolution,* 2nd ed. Cambridge University Press, Cambridge, MA. Fascinating account of herbals, herbalists, and early efforts at plant classification, with many reproductions of woodcuts illustrating the evolution of botanical art

Barghoorn, E.S. 1971. "The Oldest Fossils." *Scientific American,* Vol. 224 (May). pp. 30–42. First evidence of bacterial cell fossils older than 3 billion years; older finds have been made more recently

Bernal, J.D., and A. Synge. 1972. *The Origin of Life.* Oxford Biology Readers, No. 13. Carolina Biological Supply, Burlington, NC. Fifteen-page illustrated introduction to the evidence and theories of life's origin and early evolution

Day, W. 1979. *Genesis on Planet Earth.* House of Talos Publishers, East Lansing, MI. Sprightly, entertaining, yet serious and informative interpretation of the evidence regarding the origin of life and its early evolution and fossil history

Delevoryas, T. 1977. *Plant Diversification.* Holt, Rinehart and Winston, New York. Short coverage of early evolution and of major past and present groups of plants noted in this chapter and covered in more detail in subsequent ones

Eigen, M., et al. 1981. "The Origin of Genetic Information." *Scientific American,* Vol. 244 (April). pp. 88–118. Provocative results and speculations as to how RNA sequences and genetic coding may have arisen prebiotically

Folsome, C.E. 1979. *The Origin of Life: A Warm Little Pond.* W.H. Freeman & Co., San Francisco. Short book covering current concepts of prebiotic and early biotic evolution

Groves, D.L., et al. 1981. "An Early Habitat of Life." *Scientific American,* Vol. 245 (October). pp. 64–73. Presents evidence for life 3.5 billion years ago in the "North Pole" area of Australia

Heywood, V.H. 1967. *Plant Taxonomy.* Studies in Biology, No. 5. University Park Press, Baltimore. A short, well-organized exposition of modern methods of evolutionary analysis including population genetics, chromosome evolution, and biochemical and computer systematics

Jones, S.B., and A.E. Luchsinger. 1979. *Plant Systematics.* McGraw-Hill, New York. Brief history and detailed account of principles of phylogenetic classification, oriented largely toward vascular plants

Lewin, R.A. 1981. "The Prochlorophytes." In *The Prokaryotes,* Vol. I. eds. M.P. Starr, *et al.* Springer-Verlag New York, New York, pp. 257–266. Detailed, well illustrated review of information about *Prochloron*

Margulis, L. 1981. *Symbiosis in Cell Evolution: Life and Its Environment on the Early Earth.* W.H. Freeman & Co., San Francisco. Detailed presentation of theories of early cell evolution and the endosymbiotic origin of organelles, by a principal proponent of the concept

Mayr, E. *et al.* 1978. *Evolution.* Scientific American. W.H. Freeman & Co., San Francisco. Collection of attractive articles on origin and early evolution of life, and the broad sweep of later organismal and ecological evolution on earth as sketched in this chapter

Sheehan, A. (ed.). 1975. *The Prehistoric World.* Visual World Library, Warwick Press, London. Beautifully color-illustrated introductory accounts of origin of life; the formation, nature and study of fossils; the geological time scale; and the life of successive geological periods, by professional paleontologists

Spinar, Z.V. 1972. *Life Before Man.* American Heritage Press, McGraw-Hill, New York. Vividly and imaginatively illus-

trated landscapes of the geological periods and the plants and animals that inhabited them

Tribe, M., P. Whittaker and A. Morgan. 1982. *The Evolution of Eukaryotic Cells.* Studies in Biology, No. 131. University Park Press, Baltimore. Short account and evidence of the endosymbiotic theory

---

General References for Chapters 22–31 (Plant Diversity)

Bold, H.C., C.J. Alexopoulos and T. Delevoryas. 1980. *Morphology of Plants and Fungi,* 4th ed. Harper & Row, New York. Detailed survey including cyanobacteria as well as plants and fungi

Scagel, R.F., R.J. Bandoni, G.E. Rouse, W.B. Schofield, J.R. Stein and T.M.C. Taylor. 1965. *An Evolutionary Survey of the Plant Kingdom.* Wadsworth, Belmont, CA. A well-illustrated general text, approximately half of which is devoted to algae and fungi, the other half to bryophytes and vascular plants

---

CHAPTER 22.   Prokaryotes

Brill, W.J. 1977. "Biological Nitrogen Fixation." *Scientific American,* Vol. 236 (March). pp. 68–81. Basic information, including fixation by both symbiotic and free-living prokaryotes. Another article by author on symbiotic nitrogen fixation listed for Chapter 25

Brock, T.D. 1979. *Biology of Microorganisms,* 3rd ed. Prentice-Hall, Englewood Cliffs, NJ. A well-written textbook including diverse topics relating to microbial activities

Campbell, A.M. 1976. "How Viruses Insert Their DNA into the DNA of the Host Cell." *Scientific American,* Vol. 235 (December). pp. 103–113. Basic information concerning bacteriophages and the expression of their genes via insertion in the host bacterial chromosome

Fogg, G.E., *et al.* **1973.** *The Blue-Green Algae.* Academic Press, New York. Basic information on all aspects of these prokaryotes

Goodenough, U. 1978. *Genetics,* 2nd ed. Holt, Rinehart and Winston, New York. Chapter 5, "DNA Replication and the Transmission of Prokaryotic and Viral Chromosomes," and the initial part on plasmids and cloning of Chapter 14, "Extranuclear Genetic Systems," elaborate on second level material covering recombination in prokaryotes given in our Chapter 22

Gutschick, V.P. 1978. "Energy and Nitrogen Fixation." *BioScience,* Vol. 28. pp. 571–575. Consideration of the nitrogen cycle and of possible manipulation of various parts to increase crop productivity

Helinski, D.R. 1978. "Plasmids as Vehicles for Gene Cloning: Impact on Basic and Applied Research." *Trends in Biochemical Sciences,* Vol. 3. pp. 10–14. A concise article outlining plasmid use for cloning genes

Hopwood, D.A. 1981. "The Genetic Programming of Industrial Microorganisms." *Scientific American,* Vol. 245 (September). pp. 90–102. A discussion of genetic modifications used in designing microorganisms with desirable metabolic pathways for human utilization. Other interesting

articles on different aspects of industrial microbiology can be found in this issue of *Scientific American*

Huber, D.M. *et al.* **1977.** "Nitrification Inhibitors—New Tools for Food Production." *BioScience,* Vol. 27. pp. 523–529. Utilization of these compounds for enhancing nitrogen fertilizer efficiency

Lynch, J.M. and N.J. Poole (eds.). 1979. *Microbial Ecology: A Conceptual Approach.* John Wiley & Sons, New York. Information-packed, with chapters in latter part on microorganisms in their natural environments (soil, water, air, and animals) and on economic microbial ecology, e.g. food spoilage, nitrogen fixation, water pollution, etc.

Postgate, J. 1978. *Nitrogen Fixation.* Studies in Biology, No. 92. University Park Press, Baltimore. A brief overview, including a consideration of the nitrogen cycle, the reactions of fixation, and both free-living and symbiotic organisms

Reynolds, C.S. and A.E. Walsby. 1975. "Water Blooms." *Biological Reviews,* Vol. 50. pp. 437–481. Detailed consideration of blooms caused by blue-green algae

Stanier, R.Y., *et al.* **1976.** *The Microbial World,* 4th ed. Prentice-Hall, Englewood Cliffs, NJ. Coverage of all groups of microorganisms, with emphasis on the physiology and metabolism of prokaryotes

Stanier, R.Y. and G. Cohen-Bazire. 1977. "Phototrophic Prokaryotes: The Cyanobacteria." *Annual Review of Microbiology,* Vol. 31. pp. 225–274. A detailed, concise resume of these photosynthetic prokaryotes for the advanced student

Starr, M.P. *et al.* **(eds.). 1981.** *The Prokaryotes,* 2 Vols. Springer-Verlag New York, New York. Everything you might want to know about specific prokaryotes, written by researchers on the individual genera, plus general introductory chapters on their diversity, occurrence, isolation and identification. Section B, Vol. 1 is on phototrophic prokaryotes and includes several chapters on cyanobacteria;

Section I, Vol. 1, is on nitrogen-fixing bacteria and relatives, such as *Agrobacterium*; Section U, Vol. 2, covers the actinomycetes, with a chapter on the genus *Frankia*

Wilson, C.L. 1980. "Crown Gall: Plant Cancer," *Garden*, Vol. 4 (Mar/Apr). pp. 26–30. A brief descriptive article on this disease and its causative agent. A more recent article on crown gall control by competition with an antibiotic-pro-

ducing strain of *Agrobacterium tumefaciens* can be found in the April 1982 issue of the same journal: *Garden*, Vol. 6. pp. 18–21

Woese, C.R. 1981. "Archaebacteria." *Scientific American*, Vol. 244 (June). pp. 98–125. Presents evidence that this long-separate evolutionary line may have to be regarded as a third kingdom distinct from both prokaryotes and eukaryotes

CHAPTER 23.   Algae

Bold, H.C. and M.J. Wynne. 1978. *Introduction to the Algae.* Prentice-Hall, Englewood Cliffs, NJ. Detailed coverage of all algal divisions, emphasizing structure and reproduction. Representative genera are described for many groups

Boney, A.D. 1976. *Phytoplankton.* Studies in Biology, No. 52. University Park Press, Baltimore. Short, abundantly informative account of the kinds of phytoplankton and their roles and dynamics in the biology of oceans and lakes

Chapman, V.J. and D.J. Chapman. 1980. *Seaweeds and Their Uses*, 3rd ed. Chapman & Hall, London. A thorough consideration of the many uses of seaweeds and their products

Dawes, C.J. 1981. *Marine Botany.* John Wiley & Sons, New York. General coverage of marine algae, with ecological emphasis

Dodge, J.D. 1973. *The Fine Structure of Algal Cells.* Academic Press, New York. Discussion, with many illustrations, of structural features of algal cells

Kiermeyer, O. (ed.). 1981. *Cytomorphogenesis in Plants.* Springer-Verlag New York, New York. Part I is comprised of advanced articles on cell morphogenesis in unicellular

algae including flagellates, diatoms, desmids, and *Acetabularia*

Lee, R.L. 1980. *Phycology.* Cambridge University Press, Cambridge and New York. Compact, biologically oriented textbook presenting basic information on all algal groups

Pickett-Heaps, J. 1975. *Green Algae.* Sinauer Associates, Sunderland, MA. Photographic documentation of algal structure and of cell division and reproductive features in selected green algae, with transmission and scanning electron micrographs and many excellent light photomicrographs

Stewart, W.D.P., (ed.). 1974. *Algal Physiology and Biochemistry.* Blackwell, Oxford. Chapters by various researchers reviewing topics such as plastids, extracellular products, movement, vitamins, and growth regulators

Taylor, F.J.R. 1980. "Basic Biological Features of Phytoplankton Cells." *In The Physiological Ecology of Phytoplankton.* I. Morris ed., pp. 3–55. University of California Press, Berkeley, CA. Concise, informative introduction to the microorganisms comprising phytoplankton

Trainor, F.R. 1978. *Introductory Phycology.* John Wiley & Sons, New York. General information on major algal groups; for beginning students

CHAPTER 24.   Fungi

Alexopoulos, C.J. and E.W. Mims. 1979. *Introductory Mycology*, 3rd ed. John Wiley & Sons, New York. Comprehensive textbook with detailed information on the different classes of fungi

Ashworth, J.M. and J. Dee. 1975. *The Biology of Slime Molds.* Studies in Biology, No. 56. University Park Press, Baltimore. Brief introduction to both plasmodial and cellular slime molds, and to features that have made them popular for studying cell cycle events and differentiation

Christensen, C.M. 1975. *Molds, Mushrooms, and Mycotoxins.* University of Minnesota Press, Minneapolis. An entertaining account of some of the activities of fungi; includes chapters on wood-rotting fungi, ergotism, and aflatoxin

Ciegler, A. and J.W. Bennet. 1980. "Mycotoxins and Mycotoxicoses." *BioScience*, Vol. 30. pp. 512–515. A brief account of some of the more important mycotoxins by two authorities in the field

Cooke, R.C. 1977. *Fungi, Man and His Environment.* Longman Group, London. After a descriptive chapter on what fungi are, this small book considers some of the activities of

specialized groups, such as mushrooms containing hallucinogens, fungi gardened by ants, etc.

Demain, A.L., *et al.* 1981. "Industrial Microbiology." *Scientific American*, Vol. 245 (September). An entire issue of articles devoted to diverse applications of microorganisms for biosynthesis of useful products

Griffin, D.H. 1981. *Fungal Physiology.* John Wiley & Sons, New York. Includes much material relevant to this chapter and Chapter 26

Miller, O.K., Jr. 1977. *Mushrooms of North America.* E.P. Dutton, New York. One of a number of available mushroom field guides, this guide has many excellent color photographs alongside the species descriptions

Noble, W.C. and J. Naidoo. 1979. *Microorganisms and Man.* Studies in Biology, No. 111. University Park Press, Baltimore. Interesting presentation of the interfaces between microorganisms and humans, from foods to sewage sludge. Fungi figure prominently in the discussion

O'Day, D.H. and P.A. Horgen (eds.). 1977. *Eukaryotic Microbes as Model Developmental Systems.* Marcel Dekker,

New York. Chapters on development in different microorganisms, mostly fungi, by investigators of these systems

Ross, I.K. 1979. *Biology of the Fungi.* McGraw-Hill, New York. Introductory textbook with a developmental slant

CHAPTER 25. Symbiotic Associations

Ahmadjian, V., and M.E. Hale (eds.). 1973. *The Lichens.* Academic Press, New York. Chapters on lichen structure, physiology, reproduction, etc., are each written by an investigator in that area

Ashton, P.J. and R.D. Walmsley. 1976. "The Aquatic Fern *Azolla* and its *Anabaena* Symbiont." *Endeavor,* New Series, Vol. 35. pp. 39–43. A discussion of each of the symbionts and of their interdependence

Batra, L.R. (ed.). 1979. *Insect-Fungus Symbiosis.* Allanheld, Osmun & Co., Montclair, NJ. Small book describing a number of these associations, including several not discussed in Chapter 25. The entertaining introductory chapter is especially recommended

BioScience **Vol. 28 (Sept), 1978.** *Future of biological nitrogen fixation.* Especially recommended for Chapter 25 are: "Nitrogen fixation by actinomycete-nodulated angiosperms," pp. 586–592, "Legumes—Past, Present, and Future," pp. 565–570, and "Blue-green Algae and Algal Associations," pp. 580–585

Brill, W.J. 1981. "Agricultural Microbiology." *Scientific American,* Vol. 245 (September). pp. 199–215. Phenomena and importance of rhizobial nitrogen fixation and the prospects of improving and extending it by genetic engineering

Cooke, R.C. 1977. *Biology of Symbiotic Fungi.* John Wiley & Sons, New York. Informative coverage of beneficial and harmful associations of fungi with other organisms, including those mentioned in this chapter

Dommergues, Y.R. and S.V. Krupa (eds.). 1978. *Interactions Between Non-Pathogenic Soil Microorganisms and Plants.* Elsevier, Amsterdam. This book contains a number of well-written, in-depth resumes of symbiotic associations covered in Chapter 25 (mycorrhizae, nitrogen fixation), as well as other topics relating to roots and the rhizosphere microorganisms

Giles, K.L. and A.G. Atherly (eds.). 1981. "Biology of the Rhizobiaceae." *International Review of Cytology,* Suppl. 13. Academic Press, New York. Articles covering both *Rhizobium* and *Agrobacterium* genera and aspects of their interactions with plants

Emerich, D.W. and H.J. Evans. 1980. "Biological Nitrogen Fixation with Emphasis on Legumes." In *Biochemical and Photosynthetic Aspects of Energy Production.* A. San Pietro ed. Academic Press, New York. pp. 117–145. Excellent short coverage of symbiotic nitrogen fixation and possibilities for its improvement

Hale, M.E. 1969. *How to Know the Lichens.* Wm. C. Brown, Dubuque, IA. A pictured key to North American lichens, with a 25-page introduction describing a number of their features

Hardy, R.W.F. and U.D. Havelka. 1975. "Nitrogen Fixation Research: A Key to World Food?" *Science,* Vol. 188. pp. 633–643. Evaluation of the need for fixed nitrogen for crop production, and of possibilities for increasing the amount of fixation

Harley, J.L. 1971. *Mycorrhiza.* Oxford Biology Readers, No. 12. Carolina Biological Supply, Burlington, NC. Brief, well-

written introduction to the field by one of its best-known investigators

Heslop-Harrison, J. 1978. *Cellular Recognition Systems in Plants.* Studies in Biology, No. 100. University Park Press, Baltimore. Brief coverage of cell recognition phenomena in both mutualistic and parasitic symbioses

Marks, G.C. and T.T. Kozlowski (eds.). 1973. *Ectomycorrhizae.* Academic Press, New York. Ten chapters by specialists describe these mycorrhizae, their ecology and physiology

Richardson, D.H.S. 1981. "Lichens and Pollution Monitoring." *Endeavor,* New Series, Vol. 5. pp. 127–133. Brief, current review of the unusual sensitivity of lichens to air pollutants

Richardson, D.H.S. 1975. *The Vanishing Lichens.* David and Charles, Newton Abbot, UK. Interesting small book on some of the more unusual aspects of lichens, including past and present uses

Richmond, M.H., et al. 1979. "The Cell as Habitat." *Proceedings of the Royal Society of London,* B, Vol. 204. pp. 115–286. Articles devoted to various intracellular symbioses, including: "The *Rhizobium*-Legume Symbiosis," pp. 219–234, "Regulation of Numbers of Intracellular Algae," pp. 131–140, and "From Extracellular to Intracellular: The Establishment of a Symbiosis," pp. 115–130

Sanders, F.E., et al. (eds.). 1975. *Endomycorrhizas.* Academic Press, London. Reports on the ecology, physiology, fine structure, and other aspects of these mycorrhizae, with some of the articles providing background for other more specialized reports

Smith, D.C. 1973. *Symbiosis of Algae with Invertebrates.* Oxford Biology Readers, No. 43. Carolina Biological Supply, Burlington, NC. A few pages illustrating these symbioses, with facts about several of the most-investigated examples

Sprent, J.I. 1979. *The Biology of Nitrogen-Fixing Organisms.* McGraw-Hill, New York. Illustrated overview of the field with coverage of many of the symbiotic associations besides those forming root nodules

Stewart, W.D.P. 1978. "Nitrogen-Fixing Cyanobacteria and Their Associations with Eukaryotic Plants." *Endeavor,* New Series, Vol. 2. pp. 170–179. Nicely illustrated account of various symbioses involving blue-greens and their importance

Trappe, J.M. and R.D. Fogel. 1977. "Ecosystematic Functions of Mycorrhizae." In *The Belowground Ecosystem.* ed. J.K. Marshall. Colorado State University, Ft. Collins, CO. Excellent brief summary of facts concerning mycorrhizal associations

Weber, N.A. 1966. "The Fungus-Growing Ants." *Science,* Vol. 153. pp. 587–604. A thorough review article by a major contributor to information about this interesting group of insects

Additional articles on nitrogen fixation are listed under Chapter 22.

CHAPTER 26. Pathogens and Plant Diseases

Agrios, G.N. 1978. *Plant Pathology*, 2nd ed. Academic Press, New York. Comprehensive coverage, with descriptions of many common diseases and their causative agents

Bent, K.J. 1979. "Fungicides in Perspective: 1979." *Endeavor*, New Series, Vol. 3. pp. 7–14. A brief but detailed review of control problems in fungal diseases

Carefoot, G.L., and E.R. Sprott. 1967. *Famine on the Wind: Man's Battle Against Plant Disease.* Rand McNally & Co., Chicago. Discursive account of major plant diseases and their consequences

Dickinson, C.H., and J.A. Lucas. 1977. *Plant Pathology and Plant Pathogens.* John Wiley & Sons, New York. Small book outlining in readable style the basics of infectious plant diseases and plant-pathogen interactions

Diener, T.O. 1982. "Viroids: Minimal Biological Systems." *BioScience*, Vol. 32. pp. 38–44. A short article describing these smallest infectious disease agents

Ellingboe, A.H. 1980. "Pathogenic Fungi and Crop Productivity." *In The Biology of Crop Productivity.* ed. P.S. Carlson. Academic Press, New York, Chapter 5. Excellent general presentation of disease problem and genetics of plant resistance and parasite pathogenicity. Chapter 4 similarly covers plant virus diseases

Gibbs, A.J., and B.D. Harrison. 1976. *Plant Virology, The Principles.* John Wiley & Sons, New York. An informative basic textbook

Maramorosch, K. 1981. "Spiroplasmas: Agents of Animal and Plant Diseases." *BioScience*, Vol. 31. pp. 374–380. Current information on these wall-less prokaryotes

Matthews, R.E.F. 1981. *Plant Virology*, 2nd ed. Academic Press, New York. Thorough introduction to all aspects of the subject

McCoy, R.E. 1981. "Wall-Free Prokaryotes of Plants and Invertebrates." *In The Prokaryotes*, Vol. 2. eds. M.P. Starr *et al.* pp. 2238–2246. Springer-Verlag New York, New York. Consideration of MLO's as plant yellows agents. *Spiroplasma* is covered in pp. 2271–2284 of the same volume

Strobel, G.A. and G.N. Lanier. 1981. "Dutch Elm Disease." *Scientific American*, Vol. 245 (Aug). pp. 56–83. Informative article on this prevalent disease of our native elms

Wheeler, B.E.J. 1976. *Diseases in Crops.* Studies in Biology, No. 64. University Park Press, Baltimore. Small book presenting an overview of major factors affecting infectious plant diseases

Wood, R.K.S. 1974. *Disease in Higher Plants.* Oxford Biology Readers, No. 57. Carolina Biological Supply, Burlington, NC. Thirty-two pages, with many illustrations, introducing both fungal and bacterial plant-pathogen interactions

CHAPTER 27. Bryophytes

Doyle, W.T. 1970. *The Biology of Higher Cryptogams.* Macmillan, New York. Concise account of bryophytes and lower vascular plants, including developmental and experimental aspects as well as evolution; pertinent to our Chapters 28 and 29 as well as 27

Richardson, D.H.S. 1981. *The Biology of Mosses.* John Wiley & Sons, New York. Recent synthesis of research information about physiology, genetics, and ecology of mosses

Watson, E.V. 1971. *The Structure and Life of Bryophytes*, 3rd ed. Hutchinson University Library, London. A brief but comprehensive treatment of all aspects of bryophyte biology

Watson, E.V. 1972. *Mosses.* Oxford Biology Readers, No. 29. Carolina Biological Supply, Burlington, NC. Sixteen-page, well-illustrated account of moss structure, reproduction, and ecology

CHAPTER 28. Early Vascular Plants: Evolution and Modern Survivors

Andrews, H.N., Jr. 1961. *Studies in Paleobotany.* John Wiley & Sons, New York. Although somewhat out-of-date, this is a readable and still useful account of the fossil record of plants

Banks, H.P. 1970. *Evolution and Plants of the Past.* Wadsworth, Belmont, CA. Brief survey of plant fossils, stressing evolutionary relationships

Banks, H.P. 1975. "Early Land Plants: Proof and Conjecture." *Bioscience*, Vol. 25. pp. 730–737. Short review and critique of the enigmatic problem of land plant origins and the fossil evidence thereon

Delevoryas, T. 1962. *Morphology and Evolution of Fossil Plants.* Holt, Rinehart and Winston, New York. Brief, readable survey of the evolution of plants as revealed by the fossil record

Foster, A.S. and E.M. Gifford, Jr. 1974. *Comparative Morphology of Vascular Plants*, 2nd ed. W.H. Freeman & Co., San Francisco. This scholarly and authoritative text, widely regarded as the standard in the field, covers material of our Chapters 28–31

Taylor, T.N. 1981. *Paleobotany. An Introduction to Fossil Plant Biology.* McGraw-Hill, New York. Complete and up-to-date account of fossil plants and the evolutionary story they reveal

CHAPTER 29. Ferns

Bierhorst, D.W. 1971. *Morphology of Vascular Plants.* Macmillan, New York. Detailed, copiously illustrated general text on vascular plants, with very extensive accounts of ferns and fossil lower vascular plants

Dyer, A.F. 1979. *The Experimental Biology of Ferns.* Academic Press, New York. Chapters on fern ecology, reproduction, genetics, and developmental physiology, by specialists

Hoshizaki, B.J. 1975. *Fern Growers Manual.* Alfred A. Knopf, New York. Handsome large format book on cultivation of ferns, with extensive illustrations and identification of those ferns which are attractive as house plants

Mickel, J.T. 1979. *How to Know the Ferns and Fern Allies.* Wm. C. Brown, Dubuque, IA. An introduction to fern forms, reproduction, and cultivation, plus illustrated key to North American ferns, by a fern specialist

Perl, P. 1977. "Ferns." *Time-Life Encyclopedia of Gardening.* Time-Life Books, Alexandria, VA. Attractively illustrated coverage of ferns as ornamental plants

CHAPTER 30. Gymnosperms

Chamberlain, C.J. 1965 (reprint). *The Living Cycads.* Hafner Publishing, New York. A pioneer worker on cycads describes these remarkable plants and his experiences studying them in nature

Chamberlain, C.J. 1966 (reprint). *Gymnosperms. Structure and Evolution.* Dover Publishing, New York. This readable volume, first published in 1935, is a storehouse of information including many first-hand observations by the author

Engbeck, J.H. 1973. *The Enduring Giants.* University Extension, University of California, Berkeley, CA. Well-illustrated, popular exposition of the development, ecology, geological and political history of these most remarkable conifers, the giant sequoias

Hartesveldt, R.J., *et al.* 1975. *The Giant Sequoia of the Sierra Nevada.* U.S. Dept. of the Interior, National Park Service, Washington, D.C. Compact but detailed small book covering topics similar to the preceding reference

Johnson, H. 1973. *The International Book of Trees: A Guide and Tribute to the Trees of our Forests and Gardens.* Simon & Schuster, New York. The first half of this beautifully illustrated, popular level large-format book covers conifers and other gymnosperms, after a general introduction to tree biology. The second half covers broad-leaved trees

Mirov, N.T., and J. Hasbrouck. 1976. *The Story of Pines.* Indiana University Press, Bloomington, IN. Readable, personal, popular account of the biology, natural history, and esthetic wonder of the most important of the gymnosperms, co-authored by a long-time forestry researcher who devoted much of his life to studying them. Highly recommended

Sporne, K.R. 1965. *The Morphology of Gymosperms. The Structure and Evolution of Primitive Seed Plants.* Hutchinson University Library, London. A concise and well-written text

CHAPTER 31. Angiosperms

Carlquist, S. 1975. *Ecological Strategies of Xylem Evolution.* University of California Press, Berkeley, CA. Presents a breadth of information and illustrations of xylem evolution in seed plants as well as in lower vascular plants

Davis, P.H., and J. Cullen. 1979. *The Identification of Flowering Plant Families.* Cambridge University Press, New York. Explains necessary observations and terminology needed for identification and gives keys to, and compact descriptions of, major angiosperm families

Eames, A.J. 1961. *Morphology of the Angiosperms.* McGraw-Hill, New York. Invaluable source of information on structure, reproduction and evolution of the flowering plants, by one of the great plant morphologists of this century

Heywood, V.H. 1978 *Flowering Plants of the World.* Mayflower Books, New York. Valuable encyclopedia of numerous plant families lavishly illustrated in color and noting their economic uses

Milne, L., and M. Milne. 1975. *Living Plants of the World.* Random House, New York. Fabulously illustrated, discursive popular account of angiosperms of many families, emphasizing economically, ecologically, and esthetically noteworthy genera and species, also including gymnosperms

Porter, C.L. 1967. *Taxonomy of Flowering Plants,* 2nd ed. W.H. Freeman & Co., San Francisco. Relatively non-technical, well-illustrated coverage of the major angiosperm families, plus introductory chapters on history, methods, and terminology

Sporne, K.R. 1971. *The Mysterious Origin of Flowering Plants.* Oxford Biology Readers, No. 3. Carolina Biological Supply, Burlington, NC. Attractively illustrated 15-page analysis of this persistent botanical enigma

Stebbins, G.L. 1974. *Flowering Plants. Evolution Above the Species Level.* Belknap Press, Cambridge, MA. Fascinating account of the origin and diversification of flowering plants by one of the leading students of plant evolution

CHAPTER 32.   Plants and People

Altschul, S.v.R. 1977. "Exploring the Herbarium." *Scientific American*, Vol. 236 (May). pp. 96–104. Explains and illustrates how systematic plant collections are made and kept, and their use in prospecting for novel food and drug species

Baker, H.G. 1978. *Plants and Civilization*, 3rd ed. Wadsworth, Belmont, CA. A brief and stimulating account of useful plants, their history and their impact upon human society

Beadle, G.W. 1980. "The Ancestry of Corn." *Scientific American*, Vol. 242 (January). pp. 112–119. The still-disputed problem of where one of our most important crop plants came from, by a famous senior geneticist

Beard, B.J. 1981. "The Sunflower Crop." *Scientific American*, Vol. 244 (May). pp. 150–161. Interesting information on the crop that globally is second only to soybeans as a vegetable oil source

Brouk, B. 1975. *Plants Consumed by Man*. Academic Press, New York. Systematic coverage, by utility groups, of economic higher and lower plants and fungi

Carpenter, P.L., T.D. Walker and F.O. Lanphear. 1975. *Plants in the Landscape*. W.H. Freeman & Co., San Francisco. Good text on ornamental plants and their uses, not covered in this chapter

Emboden, W. 1979. *Narcotic Plants*. Macmillan, New York. Careful account of history, uses and actions of alkaloid-containing plant groups

Harlan, J.R. 1975. *Crops and Man*. American Society of Agronomy, Crop Science Society of America, Madison, WI. A scholarly treatment of the beginnings of cultivation and the origin of our major crop plants

Heiser, C.B., Jr. 1981. *Seed to Civilization: The Story of Food*, 2nd ed. W.H. Freeman & Co., San Francisco. Exciting treatment of historical background and modern improvement of crop plants

Hill, A.F. 1952. *Economic Botany: A Textbook of Useful Plants and Plant Products*, 2nd ed. McGraw-Hill, New York. Although old, this text remains a standard source of information about economic plants.

Janick, J., et al. (eds.). 1970. *Plant Agriculture*. W.H. Freeman & Co., San Francisco. A selection of *Scientific American* articles dealing with many aspects of plant cultivation

Janick, J., et al. 1981. *Plant Science: An Introduction to World Crops*, 3rd ed. W.H. Freeman & Co., San Francisco. Part V gives short, readable accounts of agricultural plants, including many crops that are not or are only barely mentioned in this chapter.

Klein, R.M. 1979. *The Green World: An Introduction to Plants and People*. Harper & Row, New York. A light-hearted, highly informative account of the human relationship with plants, replete with historical anecdotes

Langenheim, J.H., and K.V. Thimann. 1982. *Plant Biology and its Relation to Human Affairs*, John Wiley & Sons, New York. Chapters 3 and 16–31 give an informative, delightfully illustrated introduction to many crops and useful wild plants, and the history and ecology of agriculture. The references cited for these chapters at the end of the book include much valuable reading that cannot be listed here

Lewis, W.H., and M.P.F. Elvin-Lewis. 1977. *Medical Botany: Plants Affecting Man's Health*. John Wiley & Sons, New York. Extensive coverage of medicinal and poisonous plants by types of action

Masefield, G.B., et al. 1969. *The Oxford Book of Food Plants*. Oxford University Press, London. Description of the world's major food, beverage and flavoring plants, superbly illustrated by colored drawings.

Richardson, W.N., and T. Stubbs, 1978. *Plants, Agriculture, and Human Society*. Benjamin Cummings, Menlo Park, CA. Short synoptic account of cultivated plants, world agricultural systems, their prehistoric and historic origins, their consequences and their future

Rosengarten, F., Jr. 1969. *The Book of Spices*. Livingston Publishing, Wynnewood, PA. Beautifully illustrated account of history, production, and uses of spices, including recipes

Schery, W. 1972. *Plants for Man*, 2nd ed. Prentice-Hall, Englewood Cliffs, NJ. Comprehensive survey of economic plants including a major section dealing with forest products

Schultes, R.E. 1977. "The Odyssey of the Cultivated Rubber Tree." *Endeavor*, New Series, Vol. 1. pp. 133–138. The domestication of the rubber tree, which led to a major technological revolution

Schultes, R.E., and A. Hofmann. 1979. *Plants of the Gods: Origins of Hallucinogenic Use*. McGraw-Hill, New York. Beautifully illustrated, extensive historical and current account of psychedelic plants

Tannahill, R. 1973. *Food in History*. Stein and Day, Briarcliff Manor, NY. Fascinating description of food uses and preferences around the world from prehistoric times to the present

Tippo, O., and W.L. Stern. 1977. *Humanistic Botany*. W.W. Norton & Co., New York. Chapters 6–12 attractively cover many of the important useful plants and plant products that are only briefly touched on in this chapter.

CHAPTER 33. Dynamics of Ecosystems

Allen, R. 1980. *How to Save the World: Strategy for World Conservation.* Kogan Page, London. This small book lays on the line the major human ecological mistakes threatening our environment and the future of our species, and proposes solutions to these problems, based on a cooperative international inquiry sponsored by the UN

Aykroyd, W.R. 1975. *The Conquest of Famine.* E.P. Dutton, New York. Historical accounts of some of the human disasters stemming from ecological mistakes and overpopulation

Bilsky, L.J. 1980. *Historical Ecology: Essays on Environment and Social Change.* Kennikat Press, Port Washington, NY. Set of information-packed articles on human ecological effects and crises from prehistoric, ancient, and medieval times through the 19th century, with many significant lessons for present and future

Blaxter, K. (ed.). 1980. *Food Chains and Human Nutrition.* Applied Science Publishers, London. Collection of brief but systematic articles on many aspects of direct and indirect human utilization of photosynthetic production and plant resources available thereby, with emphasis on present ecological problems and future prospects

Carter, V.G., and T. Dale. 1974. *Topsoil and Civilization.* University of Oklahoma Press, Norman, OK. Vividly tells the history of some of humanity's important ecological mistakes, which teach such important lessons about our future

Cloudsley-Thompson, J.L. 1977. *Man and the Biology of Arid Zones.* University Park Press, Baltimore. Well-illustrated and documented information on desertification

Eckholm, E. 1976. *Losing Ground: Environmental Stress and World Food Prospects.* W.W. Norton & Co., New York. How intensive agriculture and forestry is degrading the ecosystem and mortgaging our future

Etherington, J.R. 1978. *Plant Physiological Ecology.* Studies in Biology, No. 98. University Park Press, Baltimore. Short but more detailed coverage of material from this chapter as well as some aspects of our Chapters 19 and 34. More extensive book by same author cited under Chapter 19

Garrels, R.M., F.T. Mackenzie, and C. Hunt. 1975. *Chemical Cycles and the Global Environment.* William Kaufmann, Los Altos, CA. Covers biogeochemical nutrient cycles and the various kinds of human interference including pollution

Grime, J.P. 1979. *Plant Strategies and Vegetation Processes.* John Wiley & Son, New York. Part 2 valuably covers competition, dominance, succession, and coexistence of plant species in communities.

Harper, J.L. 1977. *Population Biology of Plants.* Academic Press, New York. Extensive account of seed reproduction ecology, plant competition, effects of herbivores, and vegetation dynamics; for more advanced students

Hutchinson, G.E., *et al.* 1970. *The Biosphere.* W.H. Freeman & Co., San Francisco. Well-illustrated articles from *Scientific American*, Vol. 223, No. 3 (September), covering many aspects of water, nutrient and energy cycles and their impact on human problems

Jackson, R.M., and F. Raw. 1966. *Life in the Soil.* St. Martin's Press, New York. Short but useful coverage of soil properties and types, and major nutrient cycles, as well as soil organisms

Miles, J. 1979. *Vegetation Dynamics.* Chapman and Hall, London. Good small book analyzing succession phenomena, with valuable, diverse examples

Miller, G.T. 1975. *Energy and Environment: The Four Energy Crises.* Wadsworth, Belmont, CA. Nontechnical introduction to human energy use problems as related to ecological matter and energy cycles, and human interference therewith

Miller, G.T. 1979. *Living in the Environment,* 2nd ed. Wadsworth, Belmont, CA. Much detailed material on ecosystem function, pollution, etc., and their connection with socioeconomic issues

Phillips, R.E., *et al.* 1980. "No-Tillage Agriculture." *Science,* Vol. 208. pp. 1108–1113. Brief but informative exposition of erosion control, water and fertilizer use, and energy cost benefits

Ricklefs, R.E. 1976. *The Economy of Nature.* Chiron Press, Portland, OR. Penetrating but readable explanation of many aspects of the environment and of ecosystem dynamics, as well as principles of community ecology pertinent to Chapter 34

Skinner, B.J. (ed.). 1981. *Use and Misuse of the Earth's Surface.* William Kaufmann, Los Altos, CA. The impact of human populations and activities on the ecosystem and the hazards for our future created by this impact

West, R.G. 1971. *Studying the Past by Pollen Analysis.* Oxford Biology Readers, No. 10. Carolina Biological Supply, Burlington, NC. Brief, informative discussion of pollen analysis with good examples of data and their interpretation.

CHAPTER 34. Nature and Diversity of Plant Communities

Billings, W.D. 1978. *Plants and the Ecosystem,* 3rd ed. Wadsworth, Belmont, CA. Short, readable account of plant distribution and community types and their ecological basis

Collinson, A.S. 1977. *Introduction to World Vegetation.* Allen and Unwin, Boston. Excellent short coverage of community ecology

Cox G.W., and M.D. Atkins. 1979. *Agricultural Ecology: An Analysis of World Food Production Systems.* W.H. Freeman & Co., San Francisco. Extensively treats the ecological and historical context of agriculture, the dynamics of agricultural ecosystems, and the future prospects of agriculture

Ebeling, W. 1979. *The Fruited Plain: The Story of American Agriculture.* University of California Press, Berkeley, CA. Gives history, modern developments, and problems of agriculture for each major region of the U.S.

Edlin, H. 1973. *Atlas of Plant Life.* John Day, New York. Attractively illustrated account of world vegetation, crop plants, and their distribution by man

Grigg, D.B. 1974. *The Agricultural Systems of the World: An Evolutionary Approach.* Cambridge University Press, London and New York. Fact-filled, absorbingly written historical account of the major kinds of agriculture on earth

Janzen, D.H. 1975. *Ecology of Plants in the Tropics.* Studies in Biology, No. 58. University Park Press, Baltimore. Short but fascinating account of the special ecological and biological features of tropical vegetation

Jones, G. 1979. *Vegetation Productivity.* Longman, New York. Small book with valuable information on natural, managed, and agricultural ecosystem productivity and its relation to human needs

Pruitt, W.O. 1978. *Boreal Ecology.* Studies in Biology, No. 91. University Park Press, Baltimore. Good short survey of physical and biotic aspects of arctic and subarctic biomes

Rutger, N.G., and D.M. Brandon. 1981. "California Rice Culture." *Scientific American,* Vol. 244 (February). pp. 42–51. Example of high-technology agriculture involving operations conducted by airplane

Snaydon, R.W. 1980. "Plant Demography in Agricultural Systems." *In Demography and Evolution in Plant Population.* ed. O.T. Solbrig. University of California Press, Berkeley, CA, pp. 131–160. Valuable treatment of ecology and agricultural plant communities, both crop and weeds, for more advanced students

Spedding, C.R.W. 1975. *The Biology of Agricultural Systems.* Academic Press, New York. Information- and theory-packed coverage of botanical and ecological basis for agricultural practices and strategies

Vankat, J.L. 1979. *The Natural Vegetation of North America, An Introduction.* John Wiley & Sons, New York. Relatively short but well-illustrated account of North American biomes

Walter, H. 1979. *Vegetation of the Earth, and Ecological Systems of the Geobiosphere.* Springer-Verlag New York, New York. Detailed information on world biomes and the conditions that determine them

Whittaker, R.H. 1975. *Communities and Ecosystems.* Macmillan, New York. Penetrating exposition of the basics of population ecology, with much valuable information about vegetation types

Wortman, S., *et al.* 1976. *Food and Agriculture.* W.H. Freeman & Co., San Francisco. Articles from the Sept. 1976 issue of *Scientific American* on the world's different agricultural systems and on the food problem, human nutrition, and green revolution

CHEMISTRY APPENDIX

Baker, J.J.W., and G.E. Allen. 1981. *Matter, Energy, and Life,* 4th ed. Addison-Wesley, Reading, MA. Provides chemistry introduction for biology students, more extensive than this appendix, as well as material on cell functions covered in our Chapters 5–7

Fessenden, R.J., and J.S. Fessenden. 1976. *Chemical Principles for the Life Sciences.* Allyn and Bacon, Boston. Full length introductory chemistry textbook oriented toward biological applications

Goodman, M., and F. Morehouse. 1973. *Organic Molecules in Action.* Gordon and Breach, Science Publishers, New York. Delightful, readable book for the general reader, takes you on excursions into various exciting biological aspects of chemistry such as the origin of life, the structure and function of macromolecules, vitamins and hormones, smell and taste, and drugs

Nass, G. 1970. *The Molecules of Life.* McGraw-Hill, World University Library, New York. Very non-technical introduction to cell molecules and their functions

Stanitski, C.L., and C.T. Sears. 1976. *Chemistry for Health-Related Sciences, Concepts and Correlations.* Prentice-Hall, Englewood Cliffs, NJ. Unusually readable and appealing, valuable introduction to biologically important chemical principles. Strongly recommended for students who want, but have never had a good feel for the basic chemistry that underlies cell and organismal function

White, E.H. 1970. *Chemical Background for the Biological Sciences.* Prentice-Hall, Englewood Cliffs, NJ. Condensed but thorough coverage of basics of inorganic and organic chemistry

PHOTO CREDITS

COVER: Susan Kuklin, Photo Researchers, Inc.

740

5-17: Richard Kolodner, from R. Kolodner and K. K. Tewari, *Biochem. Biophys. Acta* 402:372–390 (1975)

6-A: Biophoto Associates
6-3: F. and M. Shaefer Brewing Company
6-6: Rahr Malting Company, Minneapolis, Minnesota
6-7: Eugene L. Vigil, University of Maryland

7-A: D. K. Shumway, Photo Researchers, Inc.
7-1: E. R. Degginger
7-3: Kenneth R. Miller, Brown University
7-14: Dr. Olle Bjorkman, Carnegie Institution of Washington, Dept. of Plant Biology, Stanford University
7-15: Eldon H. Newcomb, University of Wisconsin

8-A: Long Ashton Research Station, University of Bristol, England
8-1: Hugh Spencer, Photo Researchers, Inc.
8-3: E. R. Degginger
8-4: Walker England, Photo Researchers, Inc.
8-7: Library of the New York Botanical Garden, Bronx, New York
8-8: U. S. Forest Service
8-10: C. L. Prior, Iowa State University
8-13: (Left) Hort-Pix, Playa del Ray, CA, Photo Researchers, Inc.
(Right) John H. Gerard, Photo Researchers, Inc.
8-14: Carolina Biological Supply Company
8-15: Dr. Mary L. Parker, Plant Breeding Institute, Cambridge, UK

9-A: Dr. David M. Gates, University of Michigan, Ann Arbor
9-6: Ray Simons, Photo Researchers, Inc.
9-7: (A) Barry A. Palevitz, University of Georgia
(B) J. Troughton and L. A. Donaldson, *Probing Plant Structure*, McGraw-Hill, 1972
9-10: J. K. Wilson

10-A: Allen Rokach, Library of the New York Botanical Garden, Bronx, New York
10-2: Prof. J. N. A. Lott, McMaster University
10-4: Sass, *Botanical Microtechnique*, 3/e, Iowa State University Press
10-17: R. Anderson and J. Cronshaw, *Planta*, 91:173 (1970). Berlin-Heidelberg-New York: Springer 1970
10-18: Barry A. Palevitz, University of Georgia
10-19: M. H. Zimmerman, *Science 133*: cover and p. 76. Copyright American Association for the Advancement of Science
10-22: Dr. Brian E. S. Gunning, in *Science Progress*, Oxford, 64:539–568 (1977)

11-A: James Hanley, Photo Researchers, Inc.
11-1: Dr. B. E. Juniper, Oxford University
11-2: Jeremy Pickett-Heaps, Photo Researchers, Inc.
11-5: Omikron, Photo Researchers, Inc.
11-6: Russ Kinne, Photo Researchers, Inc.
11-7: Carolina Biological Supply Company
11-11: Shive and Robbins, N. J. Agricultural Experiment Station
11-12: USDA Misc. Publ. 923
11-13: O. Biddulph et al, *Plant Physiology* 33:293 (1958)
11-19: Ralph Walter, Grand Ridge, Illinois
11-22: Hoagland and Arnon, in *Univ. Calif. Agr. Experiment Stat. Bull. 447*

12-A: J. Troughton and L. A. Donaldson, *Probing Plant Structure*, McGraw-Hill, 1972
12-2: Douglas DesBrisay and John Waddington, University of Saskatchewan
12-3: Photo by Peter M. Ray, from *The Living Plant*, 2/e
12-4: D. Stein, Mount Holyoke College, and O. Stein, University of Massachusetts
12-5: James E. Hanneford, Dartmouth College
12-12: James Seago, SUNY Oswego Campus
12-19: Prof. B. O. Phinney, University of California, Los Angeles
12-20: S. H. Wittwer and M. J. Bukovac, in *Economic Botany*
12-21: R. N. Stewart, M. Lieberman, and A. T. Kunishi, *Plant Physiology* 54:1–5 (1974), as supplied by M. L.
12-24: E. H. Newcomb, University of Wisconsin
12-27: W. P. Jacobs, Princeton University

13-A: W. A. Cote, SUNY College of Environmental Science and Forestry, Center for Ultrastructure Studies
13-2: Sass, *Botanical Microtechnique*, 3/e, Iowa State University Press
13-3: (B) U. S. Forest Service
13-9: R. F. Evert, in *Univ. California Publ. in Botany*
13-13: W. A. Cote, SUNY College of Environmental Science and Forestry, Center for Ultrastructure Studies
13-14: W. A. Cote, SUNY College of Environmental Science and Forestry, Center for Ultrastructure Studies
13-15: W. A. Cote, SUNY College of Environmental Science and Forestry, Center for Ultrastructure Studies
13-16: U. S. Forest Products Laboratory
13-25: L. H. MacDaniels, Cornell University
13-28: Gladys Diesing, Photo Researchers, Inc.

14-3: Dr. C. R. Hawes, Oxford University
14-7: Jack Dermid, Photo Researchers, Inc.
14-8: Tom McHugh, Photo Researchers, Inc.

14-9: R. Satter, D. Sabnis, and A. Galston, *Am. J. Botany* 57:374–381 (1970)
14-10: Dr. Olle Bjorkman, Dept. of Plant Biology, Carnegie Institution of Washington, Stanford, CA
14-11: E. R. Degginger
14-13: H. A. Borthwick and S. B. Hendricks, *Science* 132:1223 (1960) Copyright 1960 American Association for the Advancement of Science
14-14: Photos by Peter M. Ray and Margery Marsden, from *The Living Plant*, 2/e
14-15: Dr. R. J. Downs, North Carolina State University
14-16: Omikron, Photo Researchers, Inc.
14-19: (A), (C), (D) Steward, Mapes, and Holsten, *Science* 143:18–27. Copyright 1964 American Association for the Advancement of Science (B) Walter Holperin, in *Am. J. Botany*
14-26: (A) T. A. Steeves and P. M. Sussex, in *Am. J. Botany* (B) T. A. Steeves, in *Phytomorphology*

15-A: Russ Kinne, Photo Researchers, Inc.
15-2: *USDA Technical Bulletin* 1286
15-7: Photo by Peter M. Ray
15-9: Boyce Thompson Institute
15-13: Photo by Peter M. Ray

16-A: Mary M. Thatcher, Photo Researchers, Inc.
16-3: Manfred Kage — Peter Arnold, Inc.
16-4: Ward's Natural Science Establishment, Inc.
16-7: Carolina Biological Supply Company
16-9: R. Knauft, Photo Researchers, Inc.
16-10: H. J. Wilms, from *Acta Botanica Neerlandica* 30, 101–122 (1981)
16-12: (A) Dale Athanas, Photo Researchers, Inc. (B) Anita Sabarese (C) Library of the New York Botanical Garden, Bronx, New York (D) W. H. Hodge — Peter Arnold, Inc.
16-15: (A) Frank P. Bogel, Library of the New York Botanical Garden, Bronx, New York (B) Photo by Peter M. Ray (C) E. R. Degginger
16-16: M. W. F. Tweedie, Photo Researchers, Inc.
16-17: (A) N. H. Cheatham, Photo Researchers, Inc. (B) Russ Kinne, Photo Researchers, Inc. (C) Walter E. Harvey, Photo Researchers, Inc.
16-18: (A) A-2 Collection Ltd., Photo Researchers, Inc. (B) Noble Proctor, Photo Researchers, Inc.
16-21: Jesse Lunger, Photo Researchers, Inc.
16-26: H. A. Borthwick

17-A: Lynwood M. Chace, Photo Researchers, Inc.
17-5: From D. Fairchild, *National Geographic Magazine*, 1934
17-8: Frank C. Dennis, *Science* 156:71–73, 1967. Copyright 1967 American Association for the Advancement of Science

17-11: Pineapple Growers Association of Hawaii
17-12: (A) Thomas W. Martin, Photo Researchers, Inc. (B) Jerome Wexler, Photo Researchers, Inc. (C) Carolina Biological Supply Company
17-13: Tom McHugh, Photo Researchers, Inc.
17-15: H. A. Borthwick et al, *Proc. Natl. Acad. Sci.* 38:662–666 (1952)
17-17: Lewis Knudsen

18-A: Farrell Grehan, Photo Researchers, Inc.
18-1: The Bettman Archive
18-12: (A) From A. B. Novikoff and E. Holtzman, *Cells and Organelles*, 2/e, Holt, Rinehart and Winston, 1976. Photo courtesy of J. E. Trosko and S. Wolff
18-16: K. C. Vaughn and K. Wilson, *Cytobios* 28:71–83 (1980), as supplied by K.C.V.
18-17: USDA Photo
18-20: Connecticut Agricultural Experiment Station
18-21: Connecticut Agricultural Experiment Station
18-22: Asgrow Seed Company
18-24: F. Constabel et al, *C. R. Acad Sci.* 285:319–322 (1977) as supplied by F.C.

19-A: Sandra Grant, Photo Researchers, Inc.
19-3: Tom McHugh, Photo Researchers, Inc.
19-5: R. J. Downs, *Bot. Review* 46:447–489 (1980)
19-6: Dr. Joseph Berry, Dept. of Plant Biology, Carnegie Institution of Washington, Stanford, CA
19-8: Photo by Peter M. Ray
19-10: Leonard Lee Rue III, Photo Researchers, Inc.
19-11: E. R. Degginger
19-12: E. R. Degginger
19-13: From J. Clausen, D. Keck, and W. Hiesey, Carnegie Institution of Washington, Publication 520
19-20: E. R. Degginger
19-24: (C) Howard Miller, Photo Researchers, Inc. (D)– (G) Walter H. Hodge — Peter Arnold, Inc.
19-24: Richard Weymouth Brooks, Photo Researchers, Inc.
19-25: (A) Kenneth W. Fink, Photo Researchers, Inc. (B) Gilbert Grant, Photo Researchers, Inc. (C) Harry Rogers, Photo Researchers, Inc.
19-26: (A) Gilbert Grant, Photo Researchers, Inc. (B) Leonard Lee Rue III, Photo Researchers, Inc. (C) Russ Kinne, Photo Researchers, Inc. (D) E. R. Degginger
19-27: (A) Francois Gohier, Photo Researchers, Inc. (B) Bjorn Bolstad — Peter Arnold, Inc.
19-30: Photos by Peter M. Ray
19-31: Tom McHugh, Photo Researchers, Inc.
19-33: E. J. Kohl
19-35: (A) N. H. Cheatham, Photo Researchers, Inc. (B) Harry Rogers, Photo Researchers, Inc.

20-A: Photo by Peter M. Ray
20-1: The Granger Collection, New York

20-18: Dr. Malcolm Nobs, Dept. of Plant Biology, Carnegie Institution of Washington, Stanford, CA

20-19: Hugh Spencer, Photo Researchers, Inc.

21-A: Thomas R. Taylor, Photo Researchers, Inc.

21-1: (A) Field Museum of Natural History, Chicago
(B) Carolina Biological Supply Company
(C) F. Gohier, Photo Researchers, Inc.
(D) Charles R. Knight, Field Museum of Natural History, Chicago

21-2: Dr. J. William Schopf, University of California, Los Angeles

21-7: Oxford University Botany School

21-8: Charles R. Knight, Field Museum of Natural History, Chicago

21-9: Field Museum of Natural History, Chicago

21-10: Field Museum of Natural History, Chicago

21-13: Allen Rokach, Library of the New York Botanical Gardens, Bronx, New York

21-14: From Linnaeus, *C. Species Plantarum*, Facsimile of 1753 ed., Vol. I, p. 28, Ray Society, London, 1957

21-15: Hunt Institute for Botanical Documentation, Carnegie Mellon University

21-19: (B) E. R. Degginger
(C) Photo by Peter M. Ray

22-A: Biophoto Asociates

22-2: J. Staley

22-3: Douglas Inloes and Frances Thomas, Stanford University

22-7: Biophoto Associates

22-10: Biophoto Associates

22-11: E. S. Boatman

22-12: S. Fultz

22-13: (A) Herriott and Barlow, *J. Gen. Physiol.*

22-20: Dr. R. B. Simpson, ARCO Plant Cell Research Institute

22-23: Dr. S. W. Watson, from Watson & Mandel *J. Bacteriology* 107:563–569 (1971)

22-24: Dr. J. Hunter, Cetus Corporation, (C) A. Dietz, Upjohn Corporation

22-25: (A) S. Fultz, (B) Biophoto Associates, (C) Biology Media

22-26: (A) Walter Dawn, Photo Researchers, Inc.
(B) Biophoto Associates

22-27: Dr. E. Gantt, from Edwards and Gantt, *J. Cell Biol.* 50:896–900 (1971)

22-29: Photo Researchers, Inc.

22-30: Biophoto Associates

23-A: James Bell, Photo Researchers, Inc.

23-1: (A) S. Fultz
(B) Ray Simons, Photo Researchers, Inc.
(C) Marvin Winter, Photo Researchers, Inc.

23-3: Dr. J. Sears, Southeastern Massachusetts University

23-10: Dr. D. Markey, University of Massachusetts

23-11: (A) Prof. D. Myles, University of Connecticut School of Medicine

23-11: (B), (C) From O. Moestrup and P. L. Walne, *J. Cell Science* 36:437–459 (1979)

23-12: (A) S. Fultz
(B) Drs. M. Gowing and M. W. Silver, UCSC Center for Coastal Marine Studies
(C) J. Pickett-Heaps, Photo Researchers, Inc.
(D) J. Pickett-Heaps, Photo Researchers, Inc.

23-21: (A) W. Magruder, Hopkins Marine Station, Stanford University
(B) Chesher, Photo Researchers, Inc.
(C) W. Magruder, Hopkins Marine Station, Stanford University

23-24: (A) Carolina Biological Supply Company
(B) Biology Media, Photo Researchers, Inc.

23-25: (A) S. Fultz
(B) Winton Patnode, Photo Researchers, Inc.

23-27: (A) Walker England, Photo Researchers, Inc.
(B) Walker England, Photo Researchers, Inc.
(C) Harold R. Hungerford, Photo Researchers, Inc.

23-30: (A) Carolina Biological Supply Company
(B) L. Frado, University of Massachusetts

23-34: (A) Walker England, Photo Researchers, Inc.
(C) Prof. Arthur E. Linkins, Virginia Polytechnic Institute and State University

23-35: S. Fultz

23-36: Drs. M. Gowing and M. W. Silver, UCSC Center for Coastal Marine Studies

23-37: (B) Dr. H. B. Bigelow, in *Bull. U.S. Bureau of Fisheries*
(C) Prof. W. Evitt, Stanford University

23-38: (A) W. Magruder, Hopkins Marine Station, Stanford University
(B) Robert Perron, Photo Researchers, Inc.
(C) Leonard Lee Rue III, Photo Researchers, Inc.

23-39: (A), (B) Prof. L. M. Srivastava, from *Am. J. Botany* 63:679–693 (1976)

23-41: Dr. E. Gantt and S. F. Conti, *J. Cell Biol.*

23-42: W. Magruder, Hopkins Marine Station, Stanford University

23-43: Prof. Isabella Abbott, Stanford University

24-A: Manfred Kage — Peter Arnold, Inc.

24-1: Russ Kinne, Photo Researchers, Inc.

24-3: Dr. Wayne Wilcox, USDA Forest Service

24-4: S. Fultz and R. A. Woolf, *Mycologia* 64: 212–218 (1972)

24-5: (A) Dr. D. J. S. Barr, Research Branch, Agriculture Canada
(B) S. Fultz

24-6: (A) S. Fultz
(B) Robert Knauft, Photo Researchers, Inc.

24-7: S. Fultz

24-10: (A) S. Fultz
(B) P. W. Grace, Photo Researchers, Inc.
(C) S. Fultz

24-14: E. R. Degginger

24-15: (A) Ralph Willis, Photo Researchers, Inc.
(B) S. Fultz

24-19: Hitchcock Foundation, Hanover, New Hampshire

24-21: S. Fultz

28-33: (A) From Wilson et al: BOTANY, 5/e
(B) From H. Delentre, in *Mem. de l'Inst. Geol. de l'Univ. de Louvrain*
28-34: W. H. Hodge — Peter Arnold, Inc.
28-35: (A) Roche, Photo Researchers, Inc.
(B) Frank P. Bogel
28-39: From Wilson et al: BOTANY, 5/e
28-40: From Wilson et al: BOTANY, 5/e

29-A: Charles Marden Fitch
29-1: R. Van Nostrand, Photo Researchers, Inc.
29-2: E. R. Degginger
29-3: (A) Robert Dunne, Photo Researchers, Inc.
(B) Dr. Jack McLachlan, National Research Council of Canada, Halifax, Nova Scotia
29-5: Paul E. Taylor, Photo Researchers, Inc.
29-9: Russ Kinne, Photo Researchers, Inc.
29-11: (A) Hugh Spencer, Photo Researchers, Inc.
(B) Hugh Spencer, Photo Researchers, Inc.
29-15: From Wilson et al: BOTANY, 5/e
29-18: T. Steeves
29-19: (A)–(D) Dr. E. G. Cutter, University of Manchester
29-20: (A), (B) T. Steeves
29-22: Roger Wilder, Photo Researchers, Inc.
29-23: John R. MacGregor
29-24: (A) From Wilson et al: BOTANY, 5/e
(B) W. H. Hodge — Peter Arnold, Inc.
29-25: Richard L. Carlton, Photo Researchers, Inc.
29-26: Anita Sabarese
29-27: W. H. Hodge — Peter Arnold, Inc.

30-A: Photo by Peter M. Ray
30-2: W. H. Hodge — Peter Arnold, Inc.
30-3: Field Museum of Natural History, Chicago
30-4: From Wilson et al: BOTANY, 5/e
30-6: (B) From *American Fossil Cycads*, Vol. I, by G. R. Wieland, (1960), Carnegie Institution, Washington, D. C.
30-8: From Wilson et al: BOTANY, 5/e
30-9: U. S. Forest Service
30-12: Manfred Kage — Peter Arnold, Inc.
30-13: (A) Ward's Natural Science Establishment, Inc.
(B), (C) From Wilson et al: BOTANY, 5/e
30-14: (A) Ward's Natural Science Establishment, Inc.
30-15: (A), (B) Carolina Biological Supply Company
30-19: (A) E. R. Degginger
(B) E. R. Degginger
30-20: Cameron Thatcher, Photo Researchers, Inc.
30-21: (A)–(D) From Wilson et al: BOTANY, 5/e
30-22: Joe Munroe, Photo Researchers, Inc.
30-23: E. R. Degginger
30-24: Telford W. Cooper, Photo Researchers, Inc.
30-25: E. R. Degginger
30-26: Library of the New York Botanical Garden, Bronx, New York
30-27: (A) E. R. Degginger
(B) W. H. Hodge — Peter Arnold, Inc.
30-29: Tony Gauba, Photo Researchers, Inc.
30-30: W. H. Hodge — Peter Arnold, Inc.

30-31: Hubertos Kanus, Rapho/Photo Researchers, Inc.

31-A: E. R. Degginger
31-1: (A) K. B. Sandved, Photo Researchers, Inc.
(B) E. R. Degginger
(C) Photo by Peter M. Ray
(D) Russ Kinne, Photo Researchers, Inc.
31-2: Robert Hernandez, Photo Researchers, Inc.
31-4: (A)–(F) Carolina Biological Supply Company
31-5: (A) David Wong and Terry Bethune
(B) Dr. J. F. Basinger
31-6: David Wong and Terry Bethune
31-13: Adapted from G. L. Stebbins, 1974, *Flowering Plants: Evolution Above the Species Level.* Belknap Press, Cambridge, Mass
31-14: George Leavens, Photo Researchers, Inc.
31-16: S. J. Krasemann, Photo Researchers, Inc.
31-20: M. J. Manuel, Photo Researchers, Inc.
31-22: John Bova, Photo Researchers, Inc.
31-23: Pat Lynch, Photo Researchers, Inc.
31-24: W. H. Hodge — Peter Arnold, Inc.
31-25: Photo by Peter M. Ray
31-26: Photo by Sara Fultz

32-A: E. R. Degginger
32-2: H. G. Baker, University of California, Berkeley
32-4: H. W. Silvester, Photo Researchers, Inc.
32-5: Susan McCartney, Photo Researchers, Inc.
32-7: W. H. Hodge — Peter Arnold, Inc.
32-8: USDA Photo by Steve Wade
32-10: Jerome Wexler, Photo Researchers, Inc.
32-11: A. W. Ambler, Photo Researchers, Inc.
32-12: (A) Russ Kinne, Photo Researchers, Inc.
(B) Anita Sabarese
32-13: William Townsend, Jr., Photo Researchers, Inc.
32-14: W. H. Hodge — Peter Arnold, Inc.
32-15: M. E. Warren, Photo Researchers, Inc.
32-16: Russ Kinne, Photo Researchers, Inc.
32-18: United Fruit Company
32-19: Joseph Daniels, Photo Researchers, Inc.
32-20: (A) Lewis S. Maxwell, Photo Researchers, Inc.
(B), (C) W. H. Hodge — Peter Arnold, Inc.
(D) Winton Patnode, Photo Researchers, Inc.
32-22: (A) Bruno J. Zehnder — Peter Arnold, Inc.
(B) Kenneth W. Fink
32-23: Jacques Jangoux — Peter Arnold, Inc.
32-24: Peter Kaplan, Photo Researchers, Inc.
32-25: Allan Cruickshank, Photo Researchers, Inc.
32-26: Tom McHugh, Photo Researchers, Inc.
32-27: (A), (B) W. H. Hodge — Peter Arnold, Inc.
32-28: Photo by Peter M. Ray
32-29: Jane Latta
32-30: Paolo Koch, Photo Researchers, Inc.
32-31: U. S. Forest Service
32-32: Brian Brake, Rapho/Photo Researchers, Inc.
32-33: (A) Russ Kinne, Photo Researchers, Inc.
(B) Gilbert Grant, Photo Researchers, Inc.
32-34: Jane Latta
32-35: W. H. Hodge — Peter Arnold, Inc.

INDEX/GLOSSARY

Note: Page numbers in italics refer to illustrations; *t* refers to tables.

apart from the nucleus; also, the protoplasmic material in which the cell's organelles and internal membranes are suspended, 62

of prokaryotic cells, 408

Cytoplasmic inheritance. Inheritance of certain characters transmitted preferentially or exclusively through the female parent because certain genetic factors are carried in the cytoplasm, 322–323, *323*

Cytoplasmic male sterility. Genetic defect due to defective function of mitochondria in the pollen, an example of cytoplasmic inheritance, 322

Cytoplasmic streaming. Phenomenon in which the cytoplasm flows in regular streams, 62

Dark reactions. Reactions of carbon dioxide reduction in photosynthesis that do not directly involve light but are driven indirectly by the products of light reactions, 123

Darwin, Charles, 359–360

Date, 650

Day-neutral plants. Plants whose flowering is insensitive to day length, 287

de Candolle, A.P., 391

deJussieu, A.L., 391

de Saussure, Theodore, 10, *11*

Deciduous. Plants that lose their leaves at the end of each growing season, 35

Decomposers. Saprotrophic bacteria and fungi that break down detritus and body remains to obtain organic matter and energy.

Dehydrogenase. Enzymes that catalyze the hydrogen transfer reaction, 106

Denaturation. Disruption of the native shape of a protein.

Denitrification. Process of reducing nitrite and nitrate by adding electrons, carried out by denitrifiers in order to use nitrate or nitrite as a substitute for oxygen in anaerobic conditions, 417

Denitrifers. Bacteria that, under anaerobic conditions, are able to shift from using oxygen in respiration to using nitrite and nitrate as a substitute by

carrying on denitrification, 417

Density gradient centrifugation. Process of separating the different components of the microsomal fraction according to their inherent buoyant densities, 71, *72*

Deoxyribonucleic acid, base substitution in, 361, *362*
chloroplast, and cytoplasmic inheritance, 323
double helix of, structural model of, *83*
in organelles, and endosymbiotic origin of organelles, 384
in plasmids of prokaryotes, 411, *411*
mitochondrial, and cytoplasmic inheritance, 323
multiple copy. DNA base sequences repeated many times, 94, 365
noncoding sequences in, 94
nuclear, 68
polymerase. Enzyme that catalyzes DNA replication, 92
recombinant, in genetic engineering, 330
replication. Copying or duplication of chromosomal DNA during interphase, 78, 92, *93*
specification of nonprotein cell components of, 98
structure of, 91–92, *92, 93*

Deoxyribose. One of the constituents of the nucleotides in DNA, 92

Desert, 699–700, *700. See also* Xerophytes.

Desertization. Desert expansion due to overgrazing, fuel gathering, and attempts to cultivate erosion-prone terrain, 679, *679*

Desmids, characteristics of, 442, *443*

Determinate growth. Growth that is limited in extent, for instance, the growth of leaves as opposed to that of stems, 41

Determination. First step in cell differentiation in which a cell becomes committed to a particular developmental program, 255

Deuteromycetes, 472, *472*

Development, genetic bases of, 249–251

Developmental response, adaptation and, 341–345

competition and, 337
kinds of, 239

Devonian, evolution in, 552–554, *552–555*
land plants in, 387

Diageotropism. Growth curvature response that orients the growing tip perpendicular to the direction of gravity.

Diatomaceous earth, 450

Diatom(s), 449–450, *449*
sexual reproduction in, *438*

Dichotomous. Having two equal parts, 548

Dicots. Angiosperms that have two "seed leaves" within the seed; they include most broad-leaved plants, most vegetable plants, and broad-leaved trees, 26
embryo of, development of, 293, *294*
families of, 629–632
root of, cross section of, 180, *180*
vs. monocots, 617*t*

Dictyosomes. Stacks of cisternae that secrete protein and carbohydrate products from the cell to its exterior, 63, *65, 66*

Differential centrifugation. A process of centrifuging cells at successively greater speeds to sift out or "pellet" successively smaller organelles and membrane fragments, 70, *72*

Differentiation. Development of structural and functional specializations of cells and tissues, 197
and hormonal regulation, 216–217
in vascular cambium, 223, *226*
of primary tissues, 202, *201, 203*
of root cells, 206

Diffusion. Process by which plants passively take up substances from their surroundings owing to the random movements of molecules; dissolved particles pass from a region of high concentration to a region of lower concentration and tend to become evenly distributed.

Diffusion resistance. The impediment of the tissue structure to diffusion.

Diffusional gas exchange, and transpiration: photosynthesis ratio, 147

Digestion. Breakdown of energy reserves that cannot be respired directly for energy (sucrose, starch, fat) for use in

Electrophoresis. A protein separation technique that shows, in different individuals, the presence of different alleles specifying slightly different proteins of the same type, 361, *363*

Elongation zone, of root, 205–206

Embryo. Young sporophytic plant still retained in the seed or gametophyte.

culture of, 295
dormancy and, 296
development of, 293–297, *294*
dicot, development of, 293, *294*
dormancy of, 296
fern, 584, *584*
heart stage of, 293, *294*
monocot, development of, 293, *295*
of pine, 604–605, *605*
torpedo stage of, 293, *294*
See also Seed

Embryo sac. Female gametophyte of the angiosperms, in which the embryo plant later develops, 275

development of, in angiosperms, 619

Embryoid. Embryo-like structures developed under tissue culture conditions from cells derived from vegetative tissues, 294–295

Embryonta. Phylogenetic subkingdom consisting of land plants with green chloroplasts.

Emergent hydrophytes. Hydrophytes that grow into the air from roots or rhizomes anchored in mud or sand beneath the water, 347

Enations. Small leaf-like outgrowths from the stem in *Asteroxylon*, 554, *555*

End product inhibition. Process in which a given compound inhibits the action of an enzyme in its own biosynthetic pathway, preventing overproduction, 111, *111*

Endemic species. Species with a narrow tolerance range or limited competitive ability that is restricted to an unusually limited geographical range, 688

causes of, 688–689

Endodermis. In a root, a single cell layer separating the stele and the cortex.

Endomycorrhizae. Internal mycorrhizae in which the hyphae grow inside the root cells, *492*, *493*

Endoplasmic reticulum. System of paired membrane sheets ramifying through the cytoplasm and often bearing numerous ribosomes attached to the outer surfaces, 63, *64*

Endosperm. Triploid tissue in which the diploid zygote is embedded within the seed; it initially provides nourishment and hormones for the embryo and later may store up reserve products for use in seed germination, 277–278

embryo development and, 295
in later development of seed, 295–296
storage reserves in, utilization of by seedling, 308

Endospores. Dormant survival spores formed by some eubacteria, 402, *403*

Endosymbionts. Foreign cells residing symbiotically inside a host cell, 384

in algal-animal symbioses, 501–503
organelles as, 384–385

Energy, light, absorption and transfer of, 124–125, *125*. *See also* Light, absorption of
captured by leaves, 137–139
cellular, sources of, 103
efficient use of, in photosynthesis, 129–133
in ecosystems, 665–667, *667*
intensive fertilization and, 193
metabolism and, 105–108
on primordial earth, 382
plants and, 3, 15–18

Energy organelles, 64–65

Enrichment culture. Specific nutrient medium and specific, unusual conditions needed or tolerated by a desired bacterium to the exclusion of most others, 405–406

Environment, optimal, 337–338
selection pressure and, 367
species survival and, tolerance range and, 334–336
water in, ecological adaptation and, 345–355

Enzymes. Proteins that act as catalysts, speeding the chemical reactions of metabolism, 87

cofactors and, 90
effect of acidity or pH on, 91, *91*
gibberellin, induction of, in aleurone cells, 253, *253*
in cell metabolism, 89–91, *91*
temperature effects and, 90

Enzymatic reactions, mechanism of, 90, *91*

Ephedra, 611

Ephemerals. Xerophytic annuals that die as or before the dry season begins, 347; also, early spring annuals and bulb plants.

as desert annuals, 347, *349*

Epicormic regeneration, *358*

Epidermis. Cells of the surface layer of young plant tissues. In stem and leaf they secrete a cuticle onto the outer part of their exterior walls, in the root they do not.
of leaf, 76, 135

Epigeous germination. Germination in which the cotyledons are brought above ground, 54

Epiparasites, micorrhizal, 493

Epiphytes. Plants attached to other plants or to animals, 455, 696

Episomes. Plasmids with a short DNA base sequence complementary to sequences in the bacterial chromosome, 411

Equisetum, life cycle of, 569–570, *569*, *570*
structure of, 568–569, *568*, *569*

Erosion, soil, control of, 679–682

Essential elements. Chemical elements required by plants for healthy, disease-free growth, 185, *184t*
classes of, 186
roles of, 186
See also Mineral nutrients

Ethylene. A gaseous growth regulator that can inhibit elongation in growing tissues, 213

fruit ripening and, 298, *299*
in growth regulation, 212–213, *213*
produced by plant cells, 213

Etiolation. Syndrome of symptoms of plants grown in the dark, 246

Etioplasts. Partially differentiated organelles developed from chloroplasts in a plant grown in the dark; they contain no chlorophyll and no thylakoid membrane system, 247, *248*

Eubacteria. Group of prokaryotes that are unicellular, nonphotosynthetic organisms ("true bacteria").
vs. Archaebacteria, 409

Euglenoids, 445–446, *446*

Gill(s). Sheet-like structures radiating from the stipe on the underside of the cap in gilled fungi, 473

Gilled fungi, characteristics of, 473–474, *473*, *473*

Gingko, 610–611, *610*

Glaciation, Pleistocene, 390, *390*

Glauber, Johann Rudolf, 7

Glucose. Basic foodstuff or metabolite, the starting material in most cells for both energy production and biosynthesis, 84

as substrate of energy metabolism, 105

Glycolysis. The early phase of respiration in which sugar is converted to pyruvic acid, producing a small amount of useful energy, 106, *107*

Glycoprotein. Protein molecules with sugar units added, 98

Glyoxalate cycle. Metabolic cycle that allows fatty acids to be converted into sugar.

Glyoxysomes. Organelles containing the enzymes of the glyoxalate cycle that appear in fat-storing tissues during fat mobilization, *111*

Gnetum, *612*

Gnetopsida, 611–612, *611*, *612*

Golden-brown algae, 449–450, *449–450*

Golgi bodies, 63, *65*, *66*. *See also* Dictyosomes
products secreted by, 64

Gonorrhea, and resistance plasmids, 413

Grafting. Technique of vegetative reproduction in which a cutting from one plant is attached to the root or stem of another plant, 266, 268, *266*

Gram-negative bacteria. Bacteria that do not retain the dye used in the Gram stain and must be stained subsequently with a dye of a different color, 406

Gram-positive bacteria. Bacteria that retain the purple color of a dye used in the Gram stain, 406

Gramineae, food and, 639–642, *639*

Granal stacks. Aggregations of thylakoids in the chloroplasts of higher plants, 120

Grape ferns, 588, *588*

Grass family, for food, 639–642, *639*

Grassland, 698–699, *699*

Gravitational water. Water in the largest pores of the soil that drains freely under the force of gravity, 183

Gravitropism. *See* Geotropism

Green algae, as land plant progenitors, 550, *550*, *551*
coenocytic, 445, *440*
euglenoids of, 445–446
morphology and biology of, 439–447
multicellular, 442–445, *437*, *439*, *444–445*
stoneworts of, 446–447, *447*
unicellular and colonial, 441–442, *441–443*

Green bacteria. Photoautotrophic eubacteria possessing bacteriochlorophyll *c*, *d*, or *e*.

Green manuring. Turning under a seasonal cover crop such as winter rye or barley to increase the humus content of soil.

Green revolution. The development of high-yielding varieties of crop plants, 11, 327–328

Greenhouse effect, 677

Greenhouses, enclosed, in arid regions, 149, *151*

Grew, Nehemiah, 8

Ground meristem. Primary meristematic tissue that gives rise to the primary ground tissue (pith and cortex of stem, and mesophyll of leaf), 202

Ground tissue. Soft stem tissue in herbaceous plants composed mainly of parenchyma and comprising most of the stem's volume, 161

Growth, initiation of, timing and, 341
in thickness, in monocots, 234–235, *236*
metabolic costs of, 114–115
of seedling, early, 306–307. *See also* germination
primary. Plant growth involving an increase in length of root or shoot, 197
of root, 204–206, *204–207*
of shoot, 197–204
regulation of component processes in, 213–217
seasonality of, 203–204
regulators of, 206–213
chemical structures of, *209*
gene expression and, 251–254
induction of mRNA synthesis and, 253, *252*
membrane-level, 253–254
secondary. Plant growth involving an increase in diameter or thickness without elongation, occurring in the older part of stems and roots.

healing of wounds and, 233–234, *234*
of roots, 221–222, *223*, *224*
seasonality of, *220–221*

Growth correlation. Process brought about by hormone effects, by which different organs or growth centers regulate one another's growth, 206

Growth flush. Concentrated shoot growth by trees in a brief spring and early summer period that uses previously stored photosynthate, 341

Growth ring(s), of hardwood, 226
of trees, 221, *221*
dating by, 221, *222*

GTP, for energizing biosynthetic processes, 105

Guanosine triphosphate, for energizing biosynthetic processes, 105

Guard cells. Specialized epidermal cells on either side of a stomate that can open and close the pore, 155–156, *156*

of leaf, 135

Gunnera, symbiotic blue-greens in, 501

Guttation. Emergence of water from vein endings in leaves at night when transpiration is not occurring, 158

Gymnosperms. Vascular seed plants that bear their seed "naked" — not enclosed within a fruit; they include cone-bearing, needle-leaved trees, 26, *26*
age of, 388, *389*
characteristics of, 593
evolution of, 612, *613*
seed of, development of, 593–595, *594*
vs. flowering plant seeds, 296

Haber process. Process used to produce liquid ammonia (NH_3) by reacting nitrogen gas from the air with hydrogen at high temperatures, 192, *193*

Habit. General appearance and form of branching, 35

Habitat. The particular environment in which an organism or a community lives, 333, 665
of algae, 426–427, *426–427*
unoccupied, opportunists and, 339–340

Hales, Stephen, 8

Internal surface (of leaf). Mesophyll cell surface of leaf for absorption of CO$_2$ from the intercellular gas spaces.

Internodes. Intervals on a stem between nodes, 33

growth of, 199–200, *201*

Interphase. Nondividing state of the nucleus between one mitosis and the next.

Introns. Stretches of DNA in eukaryotes that do not code for any part of the protein, 94

Invertebrates, symbiotic with algae, 501-504

Ion transport, by roots, into xylem and shoot, 189, *190*

Irregular flowers. Flower that is bilaterally symmetrical, capable of being divided into two identical parts along only one longitudinal plane (e.g., snapdragons, violets, orchids).

Irrigation, 705
Isoetes, 567, *568*

Isogamous reproduction. Sexual reproduction in algae and fungi in which two gametes of similar size and form fuse together into a zygote, 434

Isomorphic alteration of generations. Life cycle in which both haploid and diploid generations are independent and free-living, 436

Isoprenoids, structure and biosynthesis of, 112, *113*
Isotopes, radioactive, lichens and, 487

June drop, 297
Jute, 662

Karyotype. Distinctive features of a haploid chromosome set in a species; it is the physical form of the species' linkage map or nuclear genome, 319

with evolution of diversity of species, 371, *371*
Kelp, 451, *452*. See also Brown algae

Knot. The base of a branch that has died and has been covered by secondary growth of the trunk, 227, *230, 231*

Kranz anatomy. Leaf anatomy in which large bundle sheath cells possessing conspicuous chloroplasts are surrounded by a sheath of chloroplast-containing mesophyll cells, forming a wreath in cross section, 141, *142, 143*

Krebs cycle. See *Citric acid cycle*

K-selection. Selection for maximum survival ability

vs. r-selection, 368–369

Lakes, succession in, 672
Lamarck, Jean Baptiste de, concept of evolution of, 359

Laminarin. Family of glucose polysaccharides that are structurally different from starch; stored by brown algae, 448, 454

Lammas shoots, 37, *39*
Land plants, early, fossil record of, 547

relationships among, *573*
evolution of, 387–390
nitrogen-fixing symbiosis with cyanobacteria and, 500–501, *500–501*
origin of, 550–551, *551*
See also Vascular plants; Seed plants; Angiosperms; Gymnosperms
Larch, 607

Latex ducts. Tube-like cells or cell systems that produce a milky juice (latex) that contains secondary products, 231, *232*

Latex products, 659, *660*
Laticifers. See Latex ducts
Lawes, J. B., fertilizers and, 191

Layering. Method of vegetative reproduction in which stems produce adventitious roots when they contact the soil, 264, *264*

Leader. A main vertical shoot, 35

Leaf (leaves), 33
angiosperm, 617
primitive, 623
arrangement of, *34*
C3, carbon dioxide relations of, vs. C4 leaves, 142
C4, carbon dioxide relations of, vs. C3 leaves, 142
canopy structure of, 138–139
carbon dioxide uptake from air and, 143–145
compound, 41, *41*
energy capture by, 137–139
energy exchange mechanisms in, 149, *150*
evolutionary origin of, 552–553
fern, growth of, 577
function of, 135
modifications of, 45–46, *45–47*
of *Selaginella*, 562, *562*
origin and growth of, 200–202, *198, 200*

phototropism in, 242, *242*
propagation by, 264–265, *265*
scale, 45, *45, 46*
shade, 139–141, *141, 142*
light compensation point of, 140
light saturation curves of, 140, *141*
simple, 41, *40*
sleep movement of, 243–244, *243, 244*
solar tracking of, 244–245, *245*
stomates of, 150, *151, 152*
diffusion of air through, 144–145
structure of, 41, 135–137, *40, 134, 135*
of xeromorphs, 353–355, *352–355*
sun, light saturation curves of, 140, *141*
temperature of, and transpiration, 149, *150*
touch-sensitive turgor movements in, 245, *245*
transpiration by, water and carbon dioxide relations in, 147–149
water stress in, 150
xeromorphic specializations of, 352t, *352*
Leaf abscission, 247, 249
and senescence, 249

Leaf axil. Angle between the leaf and the stem, in which a bud usually occurs, 33

Leaf buds. Buds that develop into strictly vegetative shoots, 39

Leaf chamber, for measuring photosynthesis, *338*

Leaf cushions. Spirally arranged structures on the stem and branches of some lycopods composed of the enlarged basal part of the leaf, 567, *567*

Leaf gap. Place where the leaf traces depart from the stem's vascular cylinder, leaving a parenchyma-filled gap.

in ferns, 578, *578, 579*

Leaf mosaic. Ability of some plants to position their leaves for minimal mutual shading, 138, *139*

Leaf primordium. An outgrowth of the apical meristem which will become a new leaf, consisting of embryonic petiole and the embryonic midrib or axis of the leaf, 197, 201

growth regulation and, 216, *216*

Leaf scar. Scar left on the stem after the detachment of a leaf, 36

tubulin that are active in the spindle structure during mitosis and are permanent components of flagella, 66

and direction of cell enlargement, 215
in formation of cell wall, 75

Middle lamella. Zone of contact between the cell walls of adjacent cells, 73

Mineral deficiency, 186
Mineral nutrients, essential, 185–186
classes of, 186
roles of, 186
for plants, 10, 14
in food chain, 667–668, *668*
in soil, 185–187, 190–191, *191*
internal recycling of, 186
newly absorbed, utilization of, 187
required by algae, 430
required by green plants, 184*t*
uptake of by roots, 187, 189
See also Fertilization

Mineral nutrition. The need by plants for nitrogen and other chemical elements.

Mineral soils. Soils containing only a minor proportion (1–10%) of humus.

Mineralization. A process carried on by soil microorganisms of breaking down humus, releasing ammonium ions from it, and gradually making the soil's insoluble nitrogen available to plant roots, 667

of nitrogen, 416–417, *416*

Minimum-tillage agriculture. Planting of a crop in unplowed land, 679, *681*

Mitochondria. Membrane-bounded organelles that function in cellular aerobic respiration and the oxidation of foodstuffs for the production of energy for cellular activities, 64, 67

and cytoplasmic inheritance, 322
DNA function in, 96
endosymbiotic origin of, 384–385

Mitosis. Process, occurring during the division of the nucleus of a eukaryotic cell, in which each daughter cell is provided with a complete set of all the genes possessed by the parent cell, 76, 78–79, *77*

stages in, 78, *78*
vs. meiosis, 315

Mixed cropping. Interplanting of different crop species, 705

Mobilization. Also called Digestion; breakdown of

energy reserves that cannot be respired directly (sucrose, starch, fat) because they are insoluble, into soluble products that can be used in glycolysis and the citric acid cycle, 667

in seeds, 307
Modifier genes, 321, *322*
Molds, slime. See Slime molds.
Molecular evolution, 361–365
Molecular specificity, in symbiotic interaction, 484
Molecule(s), in primordial soup, 382

Monocots. Angiosperms that have one cotyledon within the seed; they include the grasses, rushes, lilies, palms, bananas, and orchids, 26

embryo of, development of, 293, *295*
families of, 632–635
growth in thickness in, 234–235, *236*
root, cross section of, 180, *181*
vs. dicots, 617*t*

Monoculture(s). Genetically uniform populations of crops, 329

and susceptibility to disease, 521

Monoecious. Condition in which both male and female flowers are borne on the same plant, 284, *285*

Monohybrid cross. Cross involving a single genetic character, 317

Morels, 469, *469*

Morphogenesis. Development of the specific form and shape of cells and organs, 197

growth regulation and, 214–215

Morphogenetic response. Developmental response involving a change in the rate or the nature of a developmental process not related to the direction of an inducing stimulus, 239

to light, phytochrome and, 246–247, *246–247*

Morphology. Botany subfield that deals with the external form of plants and its development.

Moss(es), 536–537
bog, 542–543, *542, 543*
characteristics of, 25, *25*
club, 537. *See also* Lycopsida.
Iceland, 537
peat, 542–543, *542, 543*
reindeer, 537
sea, 537
Moss, Spanish, 537
sphagnum, 542–543, *542, 543*

true, 537–542, *536–541*
capsule of, 540
conducting tissues in, 542
life cycle of, *541*
reproduction in, 538–539, *537–538*
sporophyte generation in, 539–541, *539–541*
vegetative reproduction of, 543

Motor cells. Specialized cells in leaves that gain or lose turgor pressure in response to external stimuli, 243

Mountains, and biomes, 703, *703*
slope exposure to sun and, 704, *704*

Mucilage, algal, 433
Multiple copy DNA, 365
Multiple cropping. Planting of several crops in succession, 705

Musci, 536-537. *See also* Mosses.
Mushrooms, edible vs. poisonous, 475–476
life cycle of, 474–475, *475*
Mustard family, 647–648
Mutagens, 323

Mutant. An individual possessing an altered form of a particular gene, 323

as analytical tools, 323–324, *324*

Mutation(s). Alteration of genetic material; one of the sources of genetic variation within a species, 321–325, 360, 361

chromosomal. Alterations in the location and arrangement of genes on chromosomes, 361, 365

evolution and, 361–365
gene. Alterations of or within indiviudal genes, 361, *362*
incidence and types of, 361, *362*

plant genetics and, 15
useful, 324–325

Mutualism. A symbiotic relationship in which both partners benefit, 22

possible evolutionary stages of, *484*

Mycelium. Delicate filamentous body of fungi, 24, 459

Mycetangia. "Fungus enclosure"; a body chamber within some wood-boring insects for housing and feeding symbiotic fungi, 505, *511*

Mycetocytes. Specialized cells in various insects for carrying yeasts or yeast-like fungi in a symbiotic association, 505–506, *506*

Mycobiont. The fungal associate in a lichen, 485, 487

Myconta. Phylogenetic subkingdom consisting of fungi (molds and yeasts), 397, *397*

Mycoplasmas, and plant disease, 520, *520*

Mycorrhiza(ae). Association between plant roots and fungi that grow on the surface or penetrate the interior without killing the root. This association improves the plant's ability to take up nutrient ions, 14, 489–493

and fertilization, 14
function of, 490–493, *490–493*
practical value of, 493

Mycotoxins. Secondary products produced by some fungi that are poisonous to humans or other animals, 470

Myxomycetes, 478–479, *478, 479*

Myxonta. Phylogenetic subkingdom consisting of slime molds, 478–480

Narcotics, from plants, 652

Natural classification. *See* Phylogenetic classification.

Natural selection. The automatic selection of genes that, through competition, help individuals to survive and reproduce in their environment, 360

and evolution, 367–369
in action, *358*

Neck canal cells. Cells found in the neck of the archegonium in thallose liverworts, 531, *532*

Necrotrophic pathogens. Pathogens that absorb their nutrients from dead tissue, 509–510

Nectar. Sugar secretion produced by many flowers that aids in pollination.

Net venation. Vein arrangement in which the veins form a net-like pattern, as in dicots, 41

Nicotine, 654

Nightshade family, food and, 644–645, *644, 645*

Nitella, 446

Nitrate, in soils, 191, 417

Nitrate reductase system, 187, *189*

Nitrification. Process of oxidizing soil ammonium to nitrite and then to nitrate, carried on by nitrifiers, 191, 417

inhibition of, in fertilizer use, 418, *418t*

Nitrifiers. Bacteria in soil that oxidize ammonium to nitrite, then to nitrate, 417

Nitrogen, in fertilizer, 192
in soil, 191
bacteria transforming, 416–418, *416, 417*
replacement of, 669
mineralization of, 416–417, *416*

Nitrogen base. Nitrogen-containing molecule, one of the constituents of a nucleotide, 91, *92*

Nitrogen cycle, 688–669, *669*

Nitrogen fertilization, in urea foliar sprays, 194

Nitrogen fixation. Biological process carried out by bacteria and other prokaryotic organisms or by association of microorganisms with roots of some higher plants to bring nitrogen into combination with other elements, making it available to plants. Nitrogen-fixing organisms have enzymes that can reduce gaseous nitrogen to ammonium, 14, 187, *15*

by blue-greens, 422
improvement of, 416
in legumes, 495
significance of, 495–496
in nonlegume bacterial root nodules, 497–498, *498, 498t*
significance of, 498–499
origin of, 384
process of, 415–416
symbiotic, in cycads, 598

Nitrogen-fixing bacteria, 415–416, *415*
in primary succession, 672
role of, 669, *670*

Nitrogen-fixing symbiosis, between cyanobacteria and land plants, 500–501, *500–501*
types of, 494
with green plants, 493–501

Nitrogenase. Enzyme used by nitrogen-fixing bacteria to reduce nitrogen to ammonia, 415

in heterocysts of cyanobacteria, 422

Nodes. The points on a stem where a leaf or leaves are attached, 33

Nonseptate hyphae. Hyphae with few or no cross walls, 459

Nostoc, symbiosis with liverworts, 500

Nucellus. Megasporangium in seed plants.

Nuclear envelope. Double membrane enclosing the cell nucleus, 58, 68

Nuclear pores, 68

Nucleic acids, 91–99
See also DNA; RNA

Nucleoid. In prokaryotic cells, the location of the DNA in a central part of the cell, 408

Nucleolus(i). Spherical bodies in the nucleus consisting of RNA destined to become part of cytoplasmic ribosomes, 69

Nucleoproteins, in DNA replication in eukaryotic chromosomes, 92

Nucleoside triphosphates, for energy, 105

Nucleosomes. Units of globular aggregates of histone proteins in DNA replication in eukaryotes, 92, *95*

Nucleotides. Subunits of nucleic acids, 91

Nucleus. Organelle within the eukaryotic cellular protoplasm containing DNA and bounded by a double membrane, 57, 68–69

Nuts, 650

Nutrient(s), exchangeable. Soil nutrient ions bound to soil particles.

for pioneer species and, 672

Nutrient cycles, human interference with, 678

Nutrient stress, mycorrhizae and, 490

Nutrition, biological, kinds of, 21–22
holozoic, 22
of algae, 429–430
in sexual reproduction, 435
of pathogens, 509–511, *510, 511*
plant. *See* Plant nutrition; Mineral nutrients.

Nyctinasty. Sleep movements, or nocturnal changes in the orientation of leaves caused by expansion and contraction of motor cells, which gain or lose turgor pressure in response to an external stimulus, 243, *243, 244*

Obligate. Pertaining to the necessity for one way of growth over another, 404

Obligate anaerobes. Bacteria that are metabolically active only in the absence of oxygen, 404

Obligate photoperiodism. The condition of plants which

as house plants, 140–141
facultative, 343, *343*

Shoot. Above-ground portion of a
vascular plant, especially the
stem and leaves, 33, *34*

adventitious, 43, *44*
development of, 36–40
growth of, significance of, 40
herbaceous, growth of, 39
hook of, straightening of, 308
lammas, 37, *39*
positioning responses of,
308–309
primary growth of, 197–204
seasonality of, 203–204
rosette, 34, *35*
secondary growth of, 40
short and long, of woody plants,
37
young, supporting tissues of,
202–203
woody, growth of, 36–37, *38, 39*

Shoot apex. Mass of tissue in the
terminal bud composed of
cells capable of active division,
36

Short-day plants. Plants with a
photoperiodic response that
initiates flowers when the day
length becomes less than a
characteristic critical day
length.

Shrub. Small woody plants with
several stems, 35

Sieve plate. Sieve-like wall
between adjacent sieve tube
members.

Sieve tube. Longitudinal rows of
cells in angiosperm phloem,
168–169, *168–171*

plugging of, 172, 174
vs. xylem vessels, 169–170

Sieve tube member, development
of, *232*
life span of, 229–230

Sieve tube sap, composition of, 170
Sigillaria, 567, *566*
Silica, in diatom cell wall, 449, *449*
Silicoflagellates, 450

Silt. Soil containing fine-grained
particles.

Simple leaf. Leaf with blade that is
undivided, 41

Sisal, 662, *662*
Sleep movements, of leaves,
243–244, *243*

Slime molds. Organisms with a
semifluid body that moves by
a flowing process; they feed
phagotrophically on bacteria,
24, 478–480, *25*

Snakeroot plant, reserpine and,
658

Softwood. Wood of conifers, which
has a simpler structure than
that of dicot trees, 224

characteristics of, 225, *228*
Soil, composition of, 182
conductivity in 184–185
in tropical rain forest, 697
mineral nutrients in, 182, 185–
187, 190–191, *191*
nitrogen in, 191
bacteria transforming,
416–418, *416, 417*
organic, 182
seed pool in, 306
structure of, 182–185
types of, moisture in, 183t
water in, 183–184, *183*
Soil development, in primary
succession, 672
Soil erosion, control of, 679–682
humans and, 678–679
Soil pH, edaphic adaptation and,
338

Soil solution. Soil water and the
ions dissolved in it directly
available to plants.

Soil sterilization, 523
Solanaceae, 644–645, *644, 645*
Solar tracking, of leaves, 244–245,
245
Solute concentration, osmosis and,
154
Somatic cell genetics, 330
Somatic hybridization, 15, *17*

Soredia. Small, readily dispersed
fragments produced by lichens
in reproduction.

Sorocarp. Sporangium formed
vegetatively by cellular slime
molds, 480

Sorus(i). One part of a cluster of
sporangia, 579,581

Source-to-sink directionality.
Translocation of sugar up or
down in phloem, depending
on where it is needed, 170, *172*

Soybean, for food, 642
Spanish moss, 537

Spatial isolation. Geographical
isolation of races in a species
such that they do not
exchange genes at all, 369

Species. A group of actually or
potentially interbreeding
populations that are
reproductively isolated from
other such groups, 28, 369

arctic, 689–692. See also Tundra
diversity of, 686, 687
dominant, 685–686
ecological fitness range of, 338
endemic, 688–689
extinction of, 375–376
formation of, 369–371

relict, 376
survival of, in environmental
conditions, tolerance range
and, 334–336
Species distribution, 687–689, *688*
limitations to, 336
Species Plantarum, 29

Spermatia. Small "male" gametes
released by gametophytes of
red algae, 455

Spermatogenous cell. In pine, a
cell produced by a generative
cell of the pollen grain and
dividing again into two
sperms, 604

Sphagnum moss, 542–543, *542, 543*
Sphenophyllum, evolution of, 571
Sphenopsida, evolution of, 570–571
general characteristics of, 568
life cycle of, 569–570, *569, 570*
structure of, 568–569, *568, 569*
Spices, 650–651, *651, 653t*

Splicing. The deletion of introns
from RNA after transcription;
an example of
posttranscriptional
modification, 94

Spines. Modified leaves, 46, *47*
Spirogyra, characteristics of, 444,
444
Spiroplasma, diseases due to, 520

Spongy parenchyma cells. In
leaves, lower layer of
irregularly shaped cells of
mesophyll, 136

Sporangiophores. Stalked,
umbrella-shaped structures
containing sporangia, 569

Sporangium(a). Specialized
unicellular or multicellular
structure that produces
asexual spores, 462, 579

of ferns, 580–582, *582*
evolution of, 587–588, *588*

Spore(s). Single cells that can
germinate to produce new
individuals, 23, 434
formation of, in prokaryotes,
402–403, *403*
of ferns, discharge of, 581–582,
582
evolution of, 587–588

Spore mother cell. Diploid cell in
the sporophyte generation that
produces haploid cells (spores)
or haploid nuclei after
meiosis, 532

Sporocarps. Spore-producing
bodies composed of large
aggregates of hyphae, 462

Sporophylls. Leaves containing
sporangia, 557

Sporophyte. Diploid phase of a
plant resulting from the fusion

plant is able to survive, 334–336

and species distribution, 688
Topography, biomes and, 703–704, *704*

Totipotency. The potential in a fertilized egg for all the kinds of development of which the species is capable, 249–250, 294

Touch-sensitive turgor movements, 245, *245*

Toxic compounds, in plant cells, for disease resistance, 522

Tracheids. Elongated single cells overlapping others and possessing pits in their secondary walls; these pits provide a means of water transport through the xylem of a stem, 76, *76*

vs. vessels, 164, 165

Transcription. The production (by RNA polymerase) of an RNA base sequence complementary to the copied DNA strand, 93–94, *95*, *96*

and control of gene activity, 99, *99*

and gene expression, 250–251, *252*

by a regulatory molecule, *252*, *253*

Transduction. In bacteria, the transfer by a bacterial virus of a fragment of a bacterial chromosome to another bacterium, 412–413, *413*

Transfer cells. Cells adjacent to sieve tubes with elaborately infolded plasma membranes, providing much surface area for membrane transport into the sieve tubes, 172, *173*

Transfer RNA (tRNA). RNA designed to translate mRNA codons for a particular amino acid, which is carried at one end of the tRNA molecule, 95, *97*

Transformation. The transfer of genetic material by means of "taking up" foreign DNA from the external environment, 411, *411*

in prokaryotes, 411–412, *411*

Translation. Process, accomplished by ribosomes in combination with mRNAs, of "reading" the code on the mRNA and then inserting the correct amino acids in the correct sequence into the polypeptide chain being synthesized, 95, *97*

control of gene activity and, 99–100
of mRNA, 94–100, *97–99*

Translocation. (1) Transportation by vascular bundles of water into and photosynthetic products out of the leaf; (2) the interchange of chromosome segments between non-homologous chromosomes.

mechanism of, 170–172, *173–174*

Transmission electron microscopy. Electron microscopy in which the electron beam is transmitted through the specimen, 59

Transpiration. Evaporation of water from a leaf into air, 13

control of, 13–14
cuticular, 150, *152*
environmental influences on, 148, *150*
in leaf energy balance and temperature, 149, *150*
in xeromorphs, 353–354
magnitude of, 148
osmosis and, 153–155, *154*
stomatal control of, 150–153
water and carbon dioxide relations, 147–149
water uptake in response to, 156–158

Transpiration: photosynthesis ratio. Amount of water transpired by land plants for a given amount of biomass gained by photosynthesis, 147, 149*t*

Transpiration pull. Driving force, resulting from transpiration, for water transport through the xylem of vascular plants, *167*

and water uptake by roots, 182
Transplantation, technique of, 178–179
Transport, active, 114
electron, photosynthetic, 127
passive, 113
Trebouxia, in lichens, 486
Trees. Large woody plants with a single erect stem, 35

ectomycorrhizae and, 491
growth habit in, 35, *36*
growth rings of, 221, *221*
dating by, 221, *222*
Tree ferns, 589, *590*
Trichogyne. Receptive projection on egg cells of red algae and some ascomycetes and basidiomycetes, 456
Trichome. Epidermal outgrowths such as hairs or scales, 420, *420*
Trimerophytes, 552–553

Triploids. Plants with three chromosome sets formed by a combination of an ordinary haploid gamete with a diploid gamete, 373
Triploid nucleus. Union of three haploid nuclei (two polar nuclei and one haploid nucleus from the pollen tube), 277
Trophic level (of ecosystem). Producer organisms and primary, secondary, and subsequent consumers in an ecosystem.
Tropical fruits, 649–650
Tropical tundra, vs. arctic-alpine tundra, 692
Tropism. Developmental response involving a bending or curving toward or away from a directional stimulus such as light or gravity, 239–243, *240–243*
Truffles, 469
Tube cell, of pine pollen grain, 602, *602*, 604
Tuber. A short, much-enlarged portion of stem containing stored food reserves, 49, *48, 49*
Tubulin. Protein units forming the substance of microtubules, 66
Tundra. Low-growing vegetation of arctic and alpine areas beyond the tree limit, 689–693, *690–692*
tropical, vs. arctic-alpine tundra, 692
Tunica. Cell layers covering the surface of the apical meristem in flowering plants, 198
Turgid. A cell or tissue with substantial turgor pressure.
Turgor movements. Bending responses resulting from reversible changes in the size of certain cells due to an increase or decrease in turgor pressure, 239, 243–245, *243–245*
touch-sensitive, 245, *245*
Turgor pressure. Outwardly-directed pressure within a cell resulting from the absorption of water, 154
and cell enlargement, 213
stomatal action and, 155–156, *155*
Turnover. Continuous breakdown of cell components to maintain the cell's enzymes and membranes in a fully functional state.

Summary diagram of plant cell functions and substructure (following pages)

- Major groups of processes (all capital letters) and classes of compounds (all lower case letters) are included in the tinted color rectangles. These rectangles do not represent separate structures within the cell.

- For purposes of this summary different organelles are not drawn to the same scale. The ribosome, for example, is enlarged relative to other structures.

- Most of the processes and organelles occur in all cells but some, such as reserve protein or secondary product biosynthesis, may occur only in specialized cells.

- Except for the nucleus and vacuole, the organelles represented in this summary occur more than once in most cells.

- Cells usually possess either chloroplasts or amyloplasts, not both as shown here.

- The processes and structures shown here are explained mainly in Chapters 4 through 7.

SUMMARY OF PLANT CELL FUNCTIONS AND SUBSTRUCTURE

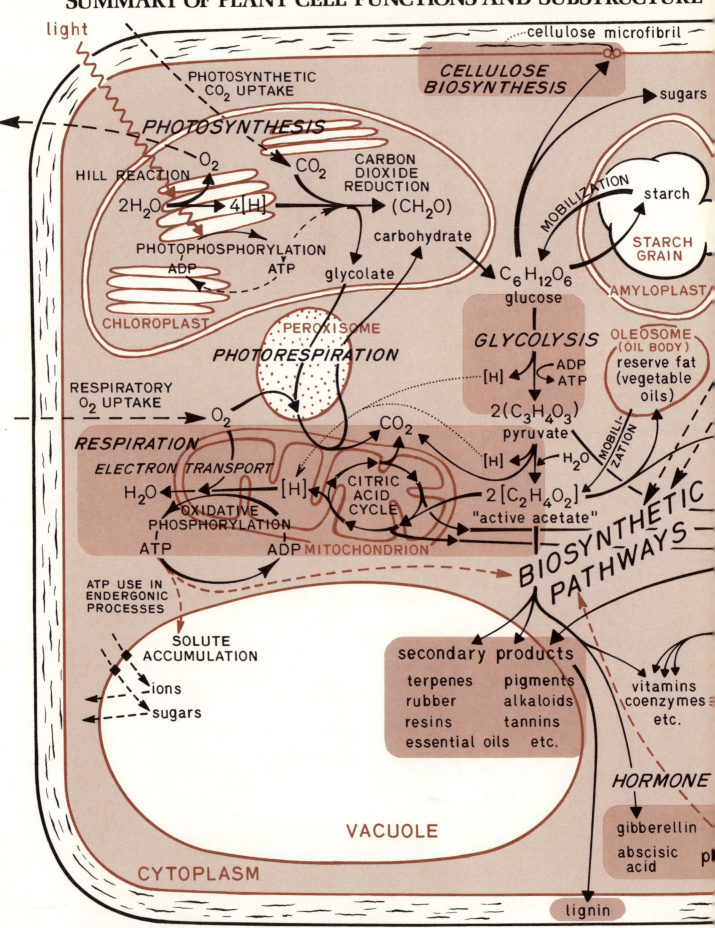

light

PHOTOSYNTHETIC CO_2 UPTAKE

cellulose microfibril

CELLULOSE BIOSYNTHESIS

sugars

PHOTOSYNTHESIS

HILL REACTION

O_2

CO_2

CARBON DIOXIDE REDUCTION

(CH_2O)

carbohydrate

$2H_2O$ ⟶ $4[H]$

MOBILIZATION

starch

STARCH GRAIN

PHOTOPHOSPHORYLATION

ADP   ATP

glycolate

$C_6H_{12}O_6$

glucose

AMYLOPLAST

CHLOROPLAST

OLEOSOME (OIL BODY)

reserve fat (vegetable oils)

PEROXISOME

PHOTORESPIRATION

GLYCOLYSIS

$[H]$ ADP   ATP

$2(C_3H_4O_3)$

pyruvate

RESPIRATORY O_2 UPTAKE

O_2

CO_2

$[H]$ H_2O

MOBILIZATION

RESPIRATION

ELECTRON TRANSPORT

CITRIC ACID CYCLE

$2[C_2H_4O_2]$

"active acetate"

H_2O $[H]$

OXIDATIVE PHOSPHORYLATION

ATP   ADP   MITOCHONDRION

BIOSYNTHETIC PATHWAYS

ATP USE IN ENDERGONIC PROCESSES

SOLUTE ACCUMULATION

ions

sugars

secondary products

terpenes   pigments
rubber   alkaloids
resins   tannins
essential oils   etc.

vitamins
coenzymes
etc.

HORMONE

gibberellin

abscisic acid

VACUOLE

CYTOPLASM

lignin

ALL *BLACK* CAPITALS = process names

lower case = compounds and

⟶ = metabolic pathways

---⟶ = transport and transfer